Encyclopedia of Plant Physiology

New Series Volume 7

Editors

A. Pirson, Göttingen

M. H. Zimmermann, Harvard

Physiology of Movements

Edited by

W. Haupt and M.E. Feinleib

Contributors

B. Bean F.W. Bentrup S.J. Britz D.S. Dennison
M.E. Feinleib Ph. Filner D.-P. Häder L.N. Halfen W. Haupt
A. Johnsson B.G. Kang R.M. Macnab G.K. Ojakian
K.L. Poff K. Raschke P. Satir R.L. Satter
K. Seitz W. Shropshire, Jr. A. Sievers B.M. Sweeney
D. Volkmann G. Wagner M.H. Weisenseel
B.D. Whitaker M.B. Wilkins N.S. Yadav

Springer-Verlag Berlin Heidelberg New York 1979

Professor Dr. WOLFGANG HAUPT
Institut für Botanik und Pharmazeutische Biologie
der Universität Erlangen-Nürnberg
Schloßgarten 4
8520 Erlangen/FRG

Professor Dr. MARY ELLA FEINLEIB
Department of Biology, Tufts University
Medford, MA 02155/USA

With 185 Figures

ISBN 3-540-08776-1 Springer-Verlag Berlin Heidelberg New York
ISBN 0-387-08776-1 Springer-Verlag New York Heidelberg Berlin

Library of Congress Cataloging in Publication Data. Main entry under title: Physiology of movements. (Encyclopedia of plant physiology; new ser., v. 7) Bibliography: p. Includes index. 1. Plants-Irritability and movements. I. Haupt, Wolfgang, 1921–. II. Feinleib, M.E., 1938–. III. Series. QK711.2.E5 vol. 7 [QK771].581.1′08s [581.1′8] 78-14724.

Typesetting, printing and bookbinding: Universitätsdruckerei H. Stürtz AG, Würzburg.
2131/3130–543210

Preface

In 1959 and 1962, the first edition of the Encyclopedia reviewed the physiology of plant movements in two volumes. These volumes tried to comprise all information on these topics then available.

At that time, against the background of the rapidly evolving field of molecular biology, plant movements were considered, by some scientists, as "classical" (=old-fashioned) topics which might not contribute much to "modern" biology. However, it soon turned out that some reaction chains which lead to plant movement can contribute importantly to our understanding of fundamental biological processes.

Consequently, since the publication of the first edition, ample effort and progress have been made, on different levels, in nearly all fields of plant movement, viz. mechanisms of movement, internal and external control of movement, and molecular processes of perception of environmental stimuli. Although movements are certainly not the only responses resulting from stimulus perception and transduction, they are frequently the most obvious indicators of those processes and are well suited for analysis.

This volume emphasizes those fields of plant movement where substantial progress in understanding has been made, or where major new aspects are evolving. It does not attempt to review all papers published since the first edition, nor does it have quite as "encyclopedic" a coverage of topics. A few important but isolated special topics may inadvertently have been omitted.

An intentional omission is the broad topic of *ballistic movements* (i.e., dispersal of spores, seeds, fruit), treated in detail in the first edition of the Encyclopedia. Although quite a few interesting investigations on the physiology of ballistic movements have been published since that time, they encompass such an inhomogeneous array of phenomena that a cohesive chapter could not easily have been written. The mechanisms of these movements should be adequately covered within this series in one of its forthcoming volumes on physiological ecology.

Despite impressive progress in the field, many questions are, of course, still unanswered. (A few major questions are noted in the Introduction.) Nonetheless, we hope it will become evident to the reader that the problems can be seen much more clearly because of the recent progress and that new lines of experimental approach are emerging. The authors and the editors will feel gratified if the present volume stimulates further research.

Finally the editors would like to thank all the authors for their cooperation and their gracious acceptance of editorial commentary. Special thanks are due to those authors who consented to write chapters in response to belated requests resulting from reorganization of the outline. The valuable assistance of Mr. R. Scheuerlein in preparing the subject index is acknowledged. The editors are grateful also to the series editors for their helpful advice, and last but not least to Springer-Verlag for patience, excellent organization, efficient production and proper layout. It was a pleasure for the editors to cooperate with Springer-Verlag.

Erlangen and Boston
February 1979

W. HAUPT
M.E. FEINLEIB

Contents

1. General Aspects of Plant Movement

1.1 Introduction

W. HAUPT and M.E. FEINLEIB

1.2 Stimulus Perception

W. SHROPSHIRE, Jr. (With 5 Figures)

2. Intracellular Movements

2.1 Role of Microtubules in Intracellular Movements

PH. FILNER and N.S. YADAV (With 5 Figures)

2.2 Actomyosin as a Basic Mechanism of Movement in Animals and Plants

G. WAGNER (With 3 Figures)

2.3 Cytoplasmic Streaming in *Physarum*

S.J. BRITZ

2.4 Cytoplasmic Streaming and Cyclosis of Chloroplasts

K. SEITZ (With 5 Figures)

3.1.2 Plant Cilia

P. SATIR and G.K. OJAKIAN (With 10 Figures)

3.1.3 Gliding Movements

L.N. HALFEN (With 4 Figures)

3.2 Control of Locomotion

3.2.1 Photomovement

D.-P. HÄDER (With 19 Figures)

5.2 Growth Movements Directed by Gravity

5.2.1 Gravitropism in Single Cells

A. SIEVERS and D. VOLKMANN (With 5 Figures)

5.2.2 Graviperception in Multicellular Organs

D. VOLKMANN and A. SIEVERS (With 8 Figures)

5.2.3 Growth-Control Mechanisms in Gravitropism
M.B. WILKINS (With 7 Figures)

5.3. Growth Movements Not Directed Primarily by External Stimuli

5.3.1 Circumnutation

A. JOHNSSON (With 9 Figures)

5.3.2 Epinasty
B.G. KANG (With 2 Figures)

List of Contributors

B. Bean
Department of Biology
Williams Hall Bldg. 31
Lehigh University
Bethlehem, PA 18015/USA

F.W. Bentrup
Abteilung für Biophysik der Pflanzen
Institut für Biologie I der Universität
Auf der Morgenstelle 1
7400 Tübingen/FRG

S.J. Britz
Department of Biology
Kline Biology Tower
Yale University
New Haven, CT 06520/USA

D.S. Dennison
Department of Biological Sciences
Dartmouth College
Hanover, NH 03755/USA

M.E. Feinleib
Department of Biology
Tufts University
Medford, MA 02155/USA

Ph. Filner
MSU-DOE Plant Research Laboratory
Michigan State University
East Lansing, MI 48824/USA

D.-P. Häder
Fachbereich Biologie-Botanik
der Phillips-Universität
Lahnberge, 3550 Marburg/FRG

L.N. Halfen
Manager, Analytical Services
Williams & Works,
611 Cascade West Parkway, S.E.
Grand Rapids, MI 49506/USA

W. Haupt
Institut für Botanik und
Pharmazeutische Biologie
der Universität Erlangen-Nürnberg
Schloßgarten 4
8520 Erlangen/FRG

A. Johnsson
Institute of Physics, N.L.H.T.
University of Trondheim
7000 Trondheim/Norway

B.G. Kang
Department of Biology
Yonsei University
Seoul 120/Korea

R.M. Macnab
Department of Molecular Biophysics
and Biochemistry
Box 1937 Yale Station
New Haven, CT 06520/USA

G.K. Ojakian
Downstate Medical Center
SUNY, 450 Clarkson Avenue
Brooklyn, NY 11203/USA

K.L. Poff
MSU-DOE Plant Research Laboratory
Michigan State University
East Lansing, MI 48824/USA

K. Raschke
MSU-DOE Plant Research Laboratory
Michigan State University
East Lansing, MI 48824/USA

Present address:
Pflanzenphysiologisches Institut
der Universität Göttingen
Untere Karspüle 2
3400 Göttingen/FRG

P. Satir
Department of Anatomy
Albert Einstein College of Medicine
1300 Morris Park Avenue
Bronx, NY 10461/USA

R.L. Satter
Department of Biology
Kline Biology Tower
Yale University
New Haven, CT 06520/USA

K. Seitz
Institut für Botanik und
Pharmazeutische Biologie
der Universität Erlangen-Nürnberg
Schloßgarten 4
8520 Erlangen/FRG

W. Shropshire, Jr.
Smithsonian Radiation Biology Laboratory
12441 Parklawn Drive
Rockville, MD 20852/USA

A. Sievers
Botanisches Institut
der Universität Bonn
Venusbergweg 22
5300 Bonn/FRG

B.M. Sweeney
Department of Biological Sciences
University of California
Santa Barbara, CA 93106/USA

D. Volkmann
Botanisches Institut
der Universität Bonn
Venusbergweg 22
5300 Bonn/FRG

G. Wagner
Institut für Botanik und
Pharmazeutische Biologie
der Universität Erlangen-Nürnberg
Schloßgarten 4
8520 Erlangen/FRG

M.H. Weisenseel
Institut für Botanik und
Pharmazeutische Biologie
der Universität Erlangen-Nürnberg
Schloßgarten 4
8520 Erlangen/FRG

B.D. Whitaker
MSU-DOE Plant Research Laboratory
Michigan State University
East Lansing, MI 48824/USA

M.B. Wilkins
Botany Department
Glasgow University
Glasgow, G 12 8QQ/United Kingdom

N.S. Yadav
MSU-DOE Plant Research Laboratory
Michigan State University
East Lansing, MI 48824/USA

1. General Aspects of Plant Movement

1.1 Introduction[1]

W. HAUPT and M.E. FEINLEIB

1. Recent Lines of Progress

Main lines of progress in the physiology of plant movements during the last fifteen years include:

1. better understanding of single reaction chains, starting with perception of a stimulus, continuing with signal transduction (or stimulus transduction; see 2.1) and resulting in the observed response, i.e., movement or change of movement.

2. recognition that similar or identical fundamental processes are shared by completely different types of movement or by completely different reaction chains (see 1.1).

3. realization that, conversely, the same type of response may be controlled by completely different mechanisms in different organisms or in different organs. This diversity is found at every level from perception through final response (see 1.2).

1.1 Common Molecular Mechanisms Shared by Different Types of Movements and Their Regulation

Various bending movements in multicellular systems are governed by unequal distribution of phytohormones. In coleoptiles, it is well established that phototropic and geotropic (gravitropic)[2] curvatures are both controlled by unequal distribution of auxin, which, in turn, depends on modifications in auxin transport due to the asymmetric action of light or gravity (DENNISON, WILKINS). Recent evidence shows that, in addition, epinastic movements are most probably brought about by similar mechanisms involving phytohormones (KANG).

In unicellular systems, different orientation responses obviously make use of at least part of a common reaction chain. Phototropism and polarotropism are both responses where differential light absorption results in differential growth, and, not surprisingly, whenever both responses are found in the same cell, they

[1] Unless stated otherwise, the references cited in the Introduction denote the chapters in this volume.

[2] Recently, it has been suggested that the prefix "gravi" be used instead of "geo" to denote the general dependence on mass acceleration rather than only on the gravity field of the earth (SIEVERS and VOLKMANN).

are controlled by light absorption in the same photoreceptor pigment (SHROPSHIRE, DENNISON).

Common photoreceptor pigments are not restricted to growth reactions. Judging from action spectra, a widespread phototropic blue-light receptor most probably also controls cytoplasmic streaming (photodinesis) and chloroplast orientation, and seems to be involved in many developmental (and even metabolic) processes as well (SHROPSHIRE, DENNISON, BRITZ, SEITZ). Obviously, once this pigment had been "invented" in evolution, it proved to be extremely useful and effective. A similar statement can be made about phytochrome; although primarily involved in photomorphogenesis, this pigment has also been put to use, by some plant systems, in the control of movement (DENNISON, BRITZ).

Recently, increasing evidence has been found that if pigments are to be photobiologically effective, they have to be bound to or associated with membranes. This has long been known to be true for the visual pigments of animals and for the photosynthetic pigments of plants. Now this principle has also been demonstrated for photoreceptor pigments in phototropism, chloroplast orientation, and photodinesis (SHROPSHIRE, DENNISON, BRITZ, SEITZ). In phototaxis, we at least know that the photoreceptor pigment is associated with cytoplasmic structures rather than being randomly distributed in the cytoplasm (HÄDER). One line of evidence for photoreceptor-membrane association is the dichroic orientation of pigment molecules, found in a wide variety of photoresponsive systems.

Membranes are the site not only of photoperception, but also of perception of other stimuli. In addition to intracellular statoliths (which shift their position in response to gravistimulation) *gravi*perception needs a "sensory" structure which perceives the pressure of the statoliths: deformations of membranes are almost certainly an important first step in the perception and transduction of gravitational as well as of mechanical stimuli (BENTRUP, VOLKMANN and SIEVERS, SATTER). In *chemo*perception, specific receptor proteins are associated with membranes, at least in some systems (MACNAB, BEAN).

Transduction processes are generally assumed to involve activities of cytoplasmic membranes, especially of the cell membrane (SHROPSHIRE, BENTRUP). One of the most important and widely used steps in transduction seems to be a change in ion transport across the cell membrane, resulting from the activation or inactivation of ion pumps or from alterations in membrane permeability. The change in influx and efflux rates of different ions can be observed as a change in the electrical properties of the membrane (BENTRUP). Amazingly, very similar ion transport reactions are assumed to be involved in movement responses as different as stomatal opening and closure, tendril curling, pulvinar response, chloroplast reorientation, flagellar or ciliary activity in plant and animal cells, and even induction of cell polarity; at least in responses based on turgor phenomena, the potassium ion seems to play the key role (RASCHKE, SATTER, SWEENEY, BRITZ, SATIR and OJAKIAN, BEAN, WEISENSEEL).

If we take changes in membrane transport of ions as the most important and general step in *transduction*, a similar significance can be attributed to protein microfilaments of the actomyosin type as the main molecular mechanism underlying *movement*. Actomyosin (or actomyosin-like protein) is not restricted to muscle contraction in animals, but has been found to be responsible for a wide range of plant movements too: it is involved in migration of slime molds, in gliding of blue-green algae, in cytoplasmic streaming of lower and higher plants, in orientation movements of chloroplasts and probably even in turgor mechanisms (WAGNER, POFF and WHITAKER, BRITZ, HALFEN, SEITZ, SATTER, BENTRUP).

1.2 Diversity of Mechanisms Underlying the Same Type of Movement in Different Systems

It is always tempting to generalize from one well-analyzed example to all other apparently similar responses. However, several chapters in this volume will demonstrate that the validity of such generalizations may be the exception rather than the rule. A few examples will be pointed out here.

At the *perception* level, it has long been assumed that the same blue-light photoreceptor pigment acts in all cases of phototropism. However, a few examples have recently been discovered which make use of phytochrome as the photoreceptor pigment (DENNISON). Several different photoreceptor pigments are also reported for the light effects on chloroplast rearrangement and on cytoplasmic streaming. In some of these cases, two independent pigment systems governing the same process coexist in one organism or even in one cell (BRITZ, SEITZ). The diversity of photoreceptor pigments seems to be greater still in photobehavior of motile microorganisms, where photosynthetic and/or nonphotosynthetic pigment systems can be involved (HÄDER).

Graviperception, according to the classical view, always involves amyloplasts (starch grains) as statoliths. But in the course of evolution, plants have evidently invented at least one completely different solution; this will be shown for *Chara* rhizoids, which use barium sulfate crystals as statoliths (SIEVERS and VOLKMANN).

At the *transduction* level, the importance of auxin in geotropism is presented by most textbooks as a general rule. Recent investigations have shown, however, that this concept is of rather limited value. Regulation of differential growth in the gravity field may be caused by any of a number of phytohormones depending on the organism and on the organ (WILKINS). The involvement of different hormones in different systems is also reported for epinastic and hyponastic movements (KANG).

Shortly after the discovery that potassium pumps are responsible for stomatal movement, it became evident that the accompanying anion is malate in some cases, and chloride in others (RASCHKE).

In the photokinesis of some microorganisms (e.g., the blue-green algae) the energy supply for increased rate of locomotion is generated by photosynthetic phosphorylation; other groups depend on a different energy source. Even within the blue-greens, the relative contribution of cyclic and noncyclic photophosphorylation differs greatly among species (HÄDER).

In circumnutation, the old alternative—"autonomous" vs. "induced" movement—has been answered, recently, in a quite unexpected way: either concept can be correct, depending on the organism, and depending even on the viewpoint of the investigator (JOHNSSON).

The *responses* proper can be brought about by different mechanisms. This can be seen, e.g., in the alternative "bulging or bending" in phototropism of single cells, whereby the faster growing flank will become either the concave or the convex side of the orienting cell (DENNISON).

Phototactic orientation may involve "steering" in the direction of a light source or may simply entail the favoring of movements toward the light source over those away from it (HÄDER). Similar diversity exists among chemotactic responses (MACNAB, BEAN, POFF and WHITAKER).

Apparently similar locomotor organelles can make use of fundamentally different mechanisms. The eukaryotic flagellum (cilium) bends in waves, whereas the bacterial flagellum rotates rigidly. Nevertheless, the resulting hydrodynamic effects are similar in the two cases (SATIR and OJAKIAN, MACNAB). Major differences in movement mechanisms evidently exist among gliding microorganisms as well, but in most of these cases, the mechanisms remain to be analyzed (HALFEN).

The diversity in movement mechanisms introduces fascinating new problems, and should stimulate further research. Phenomena that have been "explained" by extrapolation from one well-studied system will have to be reexamined in a variety of organisms before further generalizations can be made.

2. Terminological Problems

Discoveries of physiological diversity may also bring with them problems of a more mundane nature: terminological ambiguities. Although terminology is not science, scientific progress depends on unambiguous communication among investigators. Ideally, all discussion in this volume should be consistent in its terminology. However, some inconsistencies are unavoidable simply because of the constantly changing state of the art, and others because of valid differences in interpretation among investigators. The editors did not feel justified in demanding that the authors adhere to a rigid terminology.

At its best, consideration of terminology goes beyond semantics and can be useful in clarifying basic concepts of movement mechanisms. Several terminological issues that arise in this volume are introduced below.

2.1 Reaction Chains

Whenever an external stimulus (e.g., light, gravity) controls a movement, the entire phenomenon can be described in terms of a reaction chain containing three major "links":

a) *Perception:* the physical interaction between the stimulus and a receptor in the organism (*reception*), resulting in a change in the receptor. Some authors use "perception" and "reception" synonymously, while others treat "reception" and "susception" as interchangeable. If we compare perception processes for different stimuli, it becomes obvious that a truly homologous terminology cannot be developed at present. cf. BENTRUP, SHROPSHIRE.

b) *Transduction* (known also as "the black box"): used here as a collective term for the sequence of processes leading from stimulus perception to final response.

Sensu strictu, "transduction" refers to the conversion of one form of energy into another. For example, light energy may be "transduced" by a photoreceptor into chemical energy in the form of an alteration in molecular structure.

c) *Response*: the output of the reaction chain; i.e., the initiation, alteration, or cessation of movement. The term "response" may also be variously understood. For example if phytohormones are the objects of study, the investigated *response* may be alteration of auxin transport by gravity. But that same physiological event is interpreted as one of the last steps in *transduction*, if geotropic curvature is being analyzed. Where one draws the lines in a reaction sequence is partly a matter of convenience.

2.2 Nature of a Stimulus

Even the apparently straightforward concept of a *stimulus* can be the source of confusion. What, for example, is the stimulus leading to the photophobic response of a microorganism or to the light-growth response of *Phycomyces*? Is it simply light? But light will only elicit such responses if its intensity is changed quite suddenly. The stimulus would thus seem to be light-intensity change as a function of time. In the photokinesis of microorganisms, by contrast, response depends on the steady-state light intensity. Should one therefore define a stimulus by the kind of response it elicits? If so, one would speak of a "photophobic stimulus" (i.e., a photophobically effective stimulus) or a "photokinetic stimulus".

It has been suggested that a stimulus is better defined only in terms of its physical parameters (DIEHN et al., Photochem. Photobiol. *26*, 559–560, 1977), but final agreement on this issue has yet to be reached.

2.3 Behavior of Motile Microorganisms

Some agreement among specialists has been reached concerning the terminology for locomotion (DIEHN et al., Photochem. Photobiol. *26*, 559–560, 1977). Although photophobic responses can result in light-dependent patterns of distribution, neither the response nor the accumulation (e.g., in a light trap) need bear any relation to the light direction. The term "taxis" should thus not be applied in such cases; the classical term "photophobotaxis" has been abolished accordingly, and is no longer a problem (cf. HÄDER).

Nevertheless, the term *taxis* can also be problematic. On the one hand "taxis" can be defined in a broad phenomenological sense, i.e. for the accumulation of a population of microorganisms with respect to the direction of a stimulus. By this definition, e.g., any *accumulation* in a chemical gradient is a *chemotaxis*, regardless of how it is brought about. On the other hand, "taxis" can be defined more strictly, i.e. can be taken to imply *orientation* by the organism. According to this definition, the term chemotaxis would not be applied to an accumulation brought about by nonoriented behavior.

In bacterial chemotaxis, for example, the organism undergoes a randomly oriented direction change whenever it happens to move down-gradient in a chemical attractant (MACNAB). According to the stricter definition of taxis, the resulting redistribution of the population would be given the neutral name "chemoaccumulation".

Some terminological difficulties may even have a physical basis. There is, for example, a fundamental difference between chemical stimuli and light stimuli: the direction of a chemical stimulus is always defined by the direction of its concentration gradient. In the case of light, however, stimulus direction is separable from stimulus intensity (which is analogous to concentration). For example, movement from a dark region into a light trap represents movement up-gradient but does not occur in the direction of the stimulus. Conversely, in a parallel or convergent light beam, movement toward the light source is not equivalent to movement toward the region of highest intensity. Hence light, unlike a chemical substance, can act as a purely vectorial stimulus, without changing its intensity (cf. concentration) within the spatial region in which the organism moves.

2.4 Endogenous Versus Exogenous Control of Movement

In textbooks on botany, one usually finds a distinction drawn between "autonomous" and "induced" movements. Autonomous movements are thought to be regulated only by endogenous factors and do not require specific action of external factors (The nonspecific action of external factors is taken for granted, i.e., movement is only possible within the range of environmental conditions that allow life processes to go on). In contrast, an induced movement is elicited or controlled by a specific external factor and can be changed or abolished by changing that factor without drastically changing the overall life processes. However, it has recently become obvious that some of the so-called autonomous movements are strongly influenced by external factors.

Although there are still examples of circumnutation and epinasty that seem to be completely under endogenous control, exogenous factors are implicated in many cases. For instance, circumnutation of sunflower seedlings can be interpreted as a series of "overshoots" to gravitational stimulation (JOHNSSON), and, for some epinasties, the action of gravity or light is indispensable (KANG).

If we had wished to retain the old headings: autonomous *versus* induced movements, we would have had to bisect the chapters on circumnutation and epinasty accordingly. Worse still, we would not have known where to assign cases that remain to be analyzed or those that may be under dual control by endogenous and exogenous factors. Recent results thus reveal this classical distinction to be of limited value.

A similar problem arises in considering the "sleep movements" of leaves, leaflets, and petioles. In some cases, it is obvious that a movement is controlled directly by light (a "photonasty"[3]) or by temperature (a "thermonasty"). There are other, well-known cases in which the physiological clock (an "endogenous factor") is responsible for such movements, and external factors are effective only in resetting the clock. These movements are therefore considered to be mainly "endogenous". However, recent work on these phenomena indicates a complicated

[3] A nastic movement or "nasty" is induced by an external stimulus, but (unlike a tropism) its direction depends on the morphology of the organ (organism) and not on the direction of the stimulus. Nastic movement is analogous to phobic response in a motile organism, which (in contrast to taxis) is nonoriented with respect to stimulus direction.

interplay between the physiological clock and external factors (Satter, Sweeney). Again, "correct" classification of each phenomenon depends on the state of knowledge in that particular case.

Discussion of sleep movements can be further complicated by variations in terminology. Some investigators restrict "nyctinasty" to the folding movement induced by "light off". (This term is sometimes complemented by "photonasty" for the unfolding movement induced by "light on".) These authors distinguish nyctinastic/photonastic sleep movements from movements that continue under constant conditions by referring to the latter as "circadian movements". Again, these distinctions can only be applied in very well-analyzed cases. Other authors use "nyctinasty" as a general term to denote rhythmic opening and closure, irrespective of the controlling factors.

2.5 Miscellaneous Terminological Problems

Until recently, the term "flagellum" denoted a locomotor organelle of bacteria and of eukaryotic cells as well. The discovery that flagella in prokaryotes and in eukaryotes differ fundamentally, in ultrastructure and in movement mechanism, casts doubt on the validity of a common term for both organelles. Consequently, eukaryotic flagella can be called *cilia*, to denote their similarity to those organelles of animal cells. However, a consequence of this revision would be that the *flagellates* would no longer be characterized by *flagella*! The reader will find both terms, flagella and cilia, used for eukaryotes in this volume (Satir and Ojakian; Häder; Bean).

Another difficulty arises in connection with the term "photoreceptor". This word is used, on the one hand, to denote an *organelle* or a *structure* within the cell where a light stimulus is absorbed (e.g., the paraflagellar body in *Euglena*), and on the other hand, to denote the *pigment* in question. To avoid misunderstanding, it is recommended that the more precise terms "photoreceptor organelle" ("photoreceptor structure") and "photoreceptor pigment" be used (cf. Häder).

Taxonomists frequently decree changes in taxonomy, making well-known names of taxa "illegal". Sometimes this creates difficulties for the physiologist who is less-well trained in taxonomy. The editors felt justified, therefore, in retaining for example the name "dicotyledoneae" instead of the official "magnoliatae" and "cyanophyceae" instead of "cyanobacteria".

Physical units have also changed recently. Again, there may be many readers who are not yet completely familiar with the new S.I. units. Accordingly, the editors have not banned the use of "old" units such as *dalton* instead of the "legal" *g/mol*.

3. Problems for Future Investigation

The most important aim of this encyclopedia is to stimulate research. The authors have therefore made a special effort to point out unsolved problems. A few of those will be mentioned here.

3.1 Perception Mechanisms

In spite of substantial progress, our knowledge about organelles, structures, or substances underlying perception of stimuli is still superficial in most cases, and little is known about the primary action of stimuli on the receptor, but promising lines of research are being opened (SHROPSHIRE).

Just as puzzling are the mechanisms of detecting the *direction* of a stimulus whenever a response becomes oriented with respect to the stimulus. Even the simplest alternative is not yet solved in every case: whether a microorganism detects the direction of the stimulus by a one-instant mechanism (spatial gradient) or by a two-instant mechanism (temporal gradient); (cf. HÄDER, MACNAB, BEAN). The problem of perception of a vectorial stimulus is even more complicated in multicellular organs. In geotropism and phototropism, for example, we do not yet know which of two possible scenarios is correct:

1) Each sensor cell perceives the direction of the stimulus; the resulting *intracellular gradients* are integrated in the organ during transduction.

2. Due to the directional nature of the stimulus, different sensor cells are stimulated differently; the resulting *intercellular gradient* is transduced to the orientation response. cf. DENNISON, WILKINS.

3.2 Transduction Mechanisms

Increasing evidence points to biomembranes as general sites of transduction. This is true for responses as varied as bacterial chemotaxis, tendril curling, and geotropism (MACNAB, SATTER, VOLKMANN and SIEVERS). However, in few cases do we know what really happens at the membrane during the transduction processes and how membrane events link perception with response (BENTRUP). In bending movements of organs of higher plants, phytohormones are important links in the transduction processes. After a long period during which auxin was believed to be the key hormone in most tropisms and nasties, it is fascinating to learn that nearly each of the known phytohormones can play the key role (WILKINS, KANG). However, yet to be discovered are the internal conditions that make a particular phytohormone suited to a given reaction chain.

There are two additional interesting open problems which might belong either under perception or under transduction:

1. Are threshold phenomena properties of the perception process (as has commonly been held) or do they stem from transduction steps?

2. Do the adaptation processes observed in many different systems share some common mechanisms? The reader is referred in particular to SHROPSHIRE.

3.3 Response Mechanisms

Although increasing evidence points to actomyosin-like microfilaments underlying cell motility in plants as well as in animals (WAGNER), we still do not know how widespread this mechanism is in plant movement. What is the contribution of microfilaments to the different types of gliding movement? What role do microfilaments play in the contraction of seismonastic motor cells? Equally interesting and important is the question as to the function of microtubules; beyond a few particular cases, much remains to be determined about the significance of these structures for plant movement (FILNER and YADAV).

3.4 Emerging Fields of Inquiry

Several chapters in this volume contain reports on intriguing topics that are just beginning to emerge in the literature. Thermotaxis, for instance, often dismissed in older work as experimental artifact, now appears to be of fundamental interest: an amazingly high sensitivity to spatial temperature gradients has been found in slime molds (see chapter on Movement in Slime Molds, POFF and WHITAKER). Similarly, the existence of geotaxis in microorganisms, until recently held in doubt (BRINKMANN, Z. Pflanzenphysiol. *59*, 12–16, 1968), has now been reported anew for *Chlamydomonas* (see chapter on Chemotaxis in Unicellular Eukaryotes, BEAN). It is hoped that these brief reports will stimulate further research into these areas.

1.2 Stimulus Perception

W. SHROPSHIRE, Jr.

"Any philosopher of knowledge will have some kind of problem about perception."

and

"The area of perception is expecially replete with perplexities."

K.T. GALLAGHER (1964)

1. Terminology of Perception

Plant movements may be elicited by external stimuli. It is obvious that for the movement to be initiated, the stimulus must be perceived. Such stimulus perception requires that specialized molecules or structures known as receptors be present in the plant to interact with the stimulus.

The term "perception" has been used frequently (FITTING, 1908) in the botanical literature for more than one hundred years (DARWIN, 1859). The idea is also credited in the ancient philosophical literature to Aristotle and Plutarch who spoke of plants having perception analagous to the senses of man (McKEON, 1941; WRIGHT, 1907).

The term is also frequently being redefined in modern publications as authors working in the field of plant sensory responses recognize the necessity to clarify its usage. For example, in the recent literature on geotropism: "... the opening phase, i.e., the receipt of the stimulus, is a physical phase. This is commonly termed the *stimulation*. An altered physical state is assumed to pass into the first physiological phase which is termed *perception*." (JUNIPER, 1976).

The term does serve the useful function of drawing the attention of physiologists to the similarities which exist between sensory systems in plants, animals, and microorganisms, particularly at the molecular level. However, the term perception can mislead. In the strictest sense of its psychophysical definition it is inappropriate to use for many zoological examples, and certainly for all botanical systems. WEBSTER (1960) defines perception as "the awareness or consciousness of objects or other data through the medium of the senses". Plants and many animals do not possess consciousness despite the claims of the popular literature (TOMPKINS and BIRD, 1973) and despite the experimentally uncontrolled observations of some workers which purport that plants have feelings, emotions, and even extra-sensory perception.

In reality the sensory systems of plants are remarkable in their own right. They are complex enough to challenge the best experimentalist, but offer apparently simpler sensory systems for study and characterization at the molecular level than do zoological systems with their added complexity of neural networks for intercellular communication.

The literature of plant movements is replete with discussions of terminology over the last one hundred years, and these discussions have frequently been fruitless or even worse, misleading (BÜNNING, 1977). Such discussions arise when the state of a field lacks precision, definition, and factual data. The field of stimulus perception for plant movements is certainly in this category. However, the term perception is thoroughly imbedded in the botanical literature, and there does not appear to be any other more acceptable term.

2. Boundary Conditions

In this chapter I seek to present some general remarks about stimulus perception in plant movements. Because more is known in plant physiology about the reception of light than about the reception of other types of stimuli, most of the examples used will be from photoreception. However, some examples will be given of analogous principles from other types of perception, as well as examples of similar processes in animals which may have a molecular basis in common with that of plants. I also wish to draw attention to data that indicate ubiquitous mechanisms.

The only way we have of knowing that an organism has perceived a stimulus is by a response which we observe and correlate with the particular stimulus given. The stimulus interacts with a receptor within the organism and produces a change in a receptor which initiates a sequence of events ultimately resulting in a response. The initial events of this process are termed perception. In short, a linear sequence of events is assumed to occur when a plant is exposed to a stimulus.

This linear sequence may be written:

Stimulus →

Perception

Receptor → Altered receptor →

Transduction

Transduction to first physiological events → Integration of physiological events →

Observed response.

Or more simply:

Stimulus → Perception → Transduction → Response.

This chapter will be limited to the stimulus and perception portions of this simplistic pathway. The transduction processes leading to responses are dealt with in subsequent chapters.

3. Perception

The perception of stimuli by organisms can be looked at in at least two different ways. First, the stimulus may produce a single product which then initiates a process which was not occurring previously in the organism; for example, the production of chlorophyll from its precursor protochlorophyll. The quantum of light energy is utilized to convert a protochlorophyll molecule to a chlorophyll molecule, a product which was not present previously.

Second, the perception of most stimuli occurs in cells in which many metabolic processes are occurring. The stimulus produces a change that alters the rate of flow of materials in these processes from one pathway to another. That is, the perception of the stimulus may be thought of as the opening and closing of valves on processes already taking place. In this way, the perception of a stimulus may result either in a number of observed responses or in only one observable change.

3.1 Nature of Stimuli

Stimuli impart energy which alters or modifies a receptor. Thus, any form of energy which can interact with matter may alter a receptor. Stimulation may be photo-, gravi-, mechano-, thermo-, chemo-, electro-, etc. The stimulus may be characterized by its qualitative nature, as well as its magnitude, duration, and time rate of incidence.

3.2 Receptors

It follows, therefore, that a receptor is a molecule or structure which interacts with the energy contained within the stimulus. Thus, receptors are named for the quality of the stimulus they are sensitive to: photoreceptors, gravireceptors, mechanoreceptors, thermoreceptors, etc.

It should be noted that by this definition plant receptors are not analogous to the senses for human physiology. In man, a distinction traditionally is made between the senses as *passive* receivers of energy (GIBSON, 1966) and the perceptual systems. The perceptual systems are said to be *active* perceivers of energy, defined not only by the quality of the stimulus that is being perceived, but also by the way they interact with the stimulus. For plants the distinction is not so finely drawn, and little emphasis is placed on the role of integrated receptors as compared to individual receptors.

Such a definition for a receptor may seem trivial, but implicit within this definition is a kind of Grotthus-Draper law that holds for all stimulus perception. The photobiological Grotthus-Draper law states that only light quanta which are absorbed may affect a system. Similarly, to be effective, every stimulus must be received by a receptor. Even though this statement is an obvious common sense one, it is not so easy to prove rigorously and experimentally.

3.3 Interaction of Stimuli with Receptors

In order to ascertain the nature of the receptor involved with a particular stimulus, information may be gained from the characteristics of the interaction.

3.3.1 Threshold Values

Characteristics. One of the earliest ways that this interaction has been described is in measurements of threshold values of stimuli.

The range of threshold values for a number of stimuli has been compared in Table 1.

These data allow several interesting conclusions to be drawn about stimulus perception in general. The first conclusion is that the sensory systems of man, plants, and microorganisms are not so very different in the threshold values which have been reported insofar as dimensions are comparable.

The second conclusion is that for several reasons the construction of such a table of threshold values is misleading. The values measured depend in large part upon the sensitivity of the measuring system of the observer. Hence it can always be argued that if the sensitivity of the measuring system can be increased, a lower threshold will be found. This is certainly the case for chemotactic responses where the threshold values vary not only for various substances tested, but also because the number of molecules absorbed cannot be accurately determined. An additional limitation is the methodology used. As the methodology improves, the apparent threshold decreases.

Similarly, in order for the observed end response to occur, a sufficient number of receptors must be excited to lead to an observable response. Again, if the external detector is sensitive enough to measure absorption shifts in a single pigment molecule it would be possible to detect a response. This is apparently the case for dark-adapted retinal rods (HECHT et al., 1942). *In theory, however, there is no threshold for any stimulus perception, since the smallest physically detectable unit of stimulus causes a change in the receptor.*

In practice, this conclusion is not so easy to demonstrate. For psychophysical measurements where the observer's response is being recorded the application of Signal Detection Theory (SWETS, 1961; GREEN and SWETS, 1966) results in the conclusion that there is no practical measurable threshold and "a threshold" is not a very useful concept in experimental practice.

Interaction with Noise. The other reason why a threshold concept is not very valuable is that in order for a stimulus to have any meaning for an organism, the stimulus must be larger than the noise of the system operating without any incoming stimuli. For example, in the case of audition the incoming stimulus must produce displacements larger than those produced by the Brownian movements in the fluid molecules of endolymph fluid bathing the inner ear.

A similar situation must occur also for gravireceptors in plant cells. The cyclosis of the cytoplasm within the cell must be stimulating the gravireceptors continuously in a "noisy" fashion. If the false alarms produced by the cyclosis noise are truly random, then no preferred direction of growth occurs. In order for the gravireceptors to be stimulated enough to lead to an observable directional growth response, the sum of all newly stimulated gravireceptors must be greater than the background noise level.

Table 1. Range of response thresholds

Sense	Man	Receptor	Plants and Microorganisms
Vision	100 quanta to the eye or 10 quanta absorbed by rods (MCBURNEY and COLLINGS, 1977) 1 quantum, dark-adapted retinal rod (HECHT et al., 1942)	Photo-receptor	0.19 erg cm^{-2} (458 nm) *Avena* coleoptiles 4.4×10^{10} quanta cm^{-2} (phototropism) (BLAAUW and BLAAUW-JANSEN, 1970) 500 quanta absorbed (440 nm) *Phycomyces* sporangiophore (phototropism) (BERGMAN et al., 1969) 7 quanta absorbed (660 nm) seed of *Lactuca sativa* (seed germination) (BLAAUW et al., 1976)
Audition	0.0002 dynes cm^{-2} (MCBURNEY and COLLINGS, 1977) an order of magnitude above Brownian movements in fluid molecules of endolymph bathing the inner car (DAVIS, 1959)	Audio-receptor	None demonstrated at present
Taste	0.0001 M NaCl in H_2O (MCBURNEY and COLLINGS, 1977)	Chemo-receptor	6×10^{-8} M L-aspartate (*E. coli*) 2.5×10^{-4} M valine-chemotaxis (ADLER and WUNG-WAI, 1974) 4.4×10^{-7} Methylene (*Phycomyces*) growth response (RUSSO, 1977) 2×10^{-12} M antheridiol (*Achlya*) initiator of sexual reproduction (BARKSDALE, 1969) 0.4% per cm (*Lupinus*) gradient for hydrotropism of roots (BALL, 1969)
Smell	1×10^{-13} M 2-furaldehyde (BEIDLER, 1953) 1×10^{-12} M ethyl mercaptan (MCBURNEY and COLLINGS, 1977)		
Touch	10 mg force (facial) (MCBURNEY and COLLINGS, 1977)	Mechano-receptor	0.5 mg load per sporangiophore (*Phycomyces*) growth response (DENNISON and ROTH, 1967) < 1 mg per tendril for a few seconds (pea tendrils). (MEYER and ANDERSON, 1952) 0.32 mg (*Drosera*) curving of tentacles (DARWIN, 1896)
Temperature	0.001° C s^{-1} cutaneous temperature change 1.5×10^{-4} g-cal s^{-1} cm^{-2} (HARDY and OPPEL, 1937)	Thermo-receptor	0.04° C cm^{-1} half-maximal thermotactic response to gradient (*Dictyostelium discoideum*) 0.0004° C across average pseudoplasmodium (POFF and SKOKUT, 1977) Δ 0.5° C (*Crocus*) flower opening and closing (BALL, 1969)

Table 1 (continued)

Sense	Man	Receptor	Plants and Microorganisms
Gravity and Acceleration (Motion)	$0.5° s^{-2}$ angular acceleration $0.04 cm s^{-2}$ tangential acceleration at the labyrinth (GUALTIEROTTI, 1971)	Gravireceptor	240 gravity s^{-1} (oat coleoptiles) (JOHNSON, 1965) 0.0014 gravity (68 h) $\equiv 340$ gravity s^{-1} (oat coleoptiles) (SHEN-MILLER, 1970) 2–10 dyne cm^{-2} force of amyloplast on a membrane in a cell (*Helianthus*) (AUDUS, 1975)

3.3.2 Weber-Fechner Relationships

We know very little about the chemical structure or the intracellular location of receptor molecules in plant movement. However, there are a few generalizations which may be stated about the way in which they function. They all seem to be analogous to sensory systems in animals, in that the Fechner law is obeyed. The Fechner law states that for increases in the response which are of equal size (that is for an arithmetical progression) the stimulus must increase by a geometrical progression. Coupled with this law is the Weber Law which states that the variation in stimulus which causes the smallest detectable change in response maintains an approximately fixed ratio to the whole stimulus. ($\Delta I/I = k$).

3.3.3 Power Law

Although most sensory physiologists accept the Weber-Fechner relationships as being applicable, there has been a recent challenge to their validity. The question of course relates to how intensity discrimination occurs over such a wide range of magnitude. The magnitude of the response is presumed to grow proportional to the logarithm of the stimulus intensity. From the work on audition in the early 1930s (STEVENS, 1971) it was found that such a logarithmic relationship does not hold. The formalized replacement proposed is the power law (STEVENS, 1971). The test of this law has been applied in animal physiology to more than three dozen sensory and perceptual systems and holds in every case examined. For each response-stimulus system a different exponent is obtained but the data fit a power law:

$$\text{Response} = k(\text{stimulus intensity})^{\beta}$$

where β is a constant exponent for each system and for $\beta > 0$. To date it is not known whether such power functions are due to the inherent molecular properties of the receptors, but it is clear that sensory systems are capable of power function transformations. One of the central problems of all sensory physiology is the question of the enormous dynamic ranges over which the systems function. They have evolved with the capacity of handling environmental stimuli with billion-fold differences in magnitude. To my knowledge, no attempt has been made to apply power law concepts to stimulus perception in plants.

3.3.4 Refractory Periods and Range-Adjustment

Another general characteristic is that all of these receptor systems have built into them a mechanism for producing a change of sensitivity at the receptor level with different time constants. For most of these plant movements, response to an initial stimulus is followed by a refractory period; i.e., a period of time during which a second stimulus of the same size will produce little or no response. After a sufficient period of time elapses, the system regenerates full sensitivity. Such changes in sensitivity are variously referred to as range-adjustment or adaptation. To date almost nothing is known about the mechanisms of adaptation in any of the plant movement systems, but without exception they all demonstrate this general characteristic.

The most difficult question is whether the range-adjustment mechanism and the refractory periods are an intrinsic property of the receptor itself, or are part of the responding system after transduction has occurred. For photoexcitation, for example, the initial photochemistry must be completed in the range of 10^{-12} s. Therefore, if pigment excitation and bleaching are involved, light adaptation must be very fast. Dark adaptation and sensitivity changes which require on the order of minutes or even hours for regeneration are slow chemical reactions for resynthesis of receptor, or are part of the transducing systems. In several cases elaborate models have been constructed in terms of substrate-flow and limiting materials to explain adaptation to successive stimuli, but in no case has an adequate molecular model been derived, tested, and demonstrated.

3.3.5 Reciprocity

Reciprocity (stimulus for constant response = time × intensity) also holds for most stimulus perception involved in plant movements, at least over narrow ranges of time of presentation and intensity magnitude.

Typical examples are for phototropism which has been shown to obey the reciprocity relationship (BLAAUW, 1909). The reciprocity relationship is also valid for geotropism in which an exposure of 1 g for 240 s is nearly equivalent to 0.0014 g given for 68 h (see Table 1). If extreme ranges are examined either for very short times and large stimuli, or for very long presentation times and small stimuli, the relationship fails. For complex systems such as phototropism in *Avena* seedlings, reciprocity is valid for first positive curvatures, but not for second positive curvatures (STEYER, 1967). The reason for this failure is not yet clearly understood.

3.3.6 Action Spectroscopy

For photoreceptors the powerful tool of action spectroscopy is available. The probability that a light quantum will be absorbed by a pigment molecule is related to its structure. By measuring carefully the spectral sensitivity of the response to light it is often possible to characterize the nature of the photoreceptor (SHROPSHIRE, 1972; HARTMANN and COHNEN-UNSER, 1974).

Similarly, we imagine that a gravitational stimulus interacts by a displacement of a receptor within the organism and we postulate a gravireceptor. Unfortunately, every molecule possessing mass within the organism which is subjected to a gravitational stimulus is affected, whereas the light stimulus is specifically absorbed by discrete molecules that possess a chromophore for absorbing quanta. This lack of specificity for gravitational stimuli has made it difficult to identify the nature of the gravireceptor which is responsible for geotropism.

3.3.7 Characterization of Receptors

In general, the nature of receptors for plant movements have been sought by looking for changes which occur rapidly after presentation of the stimulus. For example, after photostimulation, rapid absorbance changes may be observed in the receptor pigment, or evidence for the first physiological change may be sought, although these physiological changes may already be well along the pathway, and may involve numerous transduction steps.

For none of the plant movements do we know for certain at the molecular level the nature of the receptor. Partly, this is due to the fact that many of the stimuli are neither quantized nor specific. We may, in fact, be fooling ourselves to look for a single receptor of a gravitational or a thermal stimulus. "There may be no gravity receptor either" (POLLARD, 1971). It may well be that a thermo- or gravi-stimulus causes a generalized perturbation to the already occurring processes, which then results in a changed response and in there being no discrete identifiable receptor. The entire biosystem is the receptor.

3.4 Common Attributes of Perceptive Systems for Plant Movements

3.4.1 Differential Growth or Elongation

There is one attribute common to all plant bending movements. In order for a change in the spatial orientation of the plant to occur there must be a differential growth or elongation caused by the stimulus. This asymmetry in growth or elongation most often results from an asymmetry in the spatial distribution of the stimulated receptors within the plant.

A gradient may exist in the exernal stimulus, but it is obvious that the gradient produced within the plant is the one which determines the direction of movement.

The structural properties of the plant are also important for the movement produced. In the nastic movement of leaves, for example, the asymmetry in response does not result from a gradient in the light stimulus. Instead, the asymmetry is developed because of the differential flow of ions, initiated by the light, in and out of specialized cells (SATTER, Chap. 4.2, this vol.). The direction of nastic movement is determined by the anatomy of the plant.

3.4.2 Dioptrics and Absorption

In a more specialized way the direction of phototropism is determined frequently by dioptrics. If a cylindrical cell such as a *Phycomyces* sporangiophore is placed in a parallel beam of blue light, the rays of light are refracted by the cylindrical structure in such a way that an asymmetrical light distribution is produced within the cell. The photoreceptors are stimulated by this asymmetric distribution, and asymmetrical bending results with the plant bending toward the source of light.

If this internal ray distribution is altered by inserting a cylindrical lens into the light path, the cell responds by bending away from the direction of the light source (SHROPSHIRE, 1962). This same dioptric effect may be important for the light perception and absorption gradient produced in *Avena* coleoptiles (MEYER, 1969a, 1969b; SHROPSHIRE, 1974).

In the filamentous green alga *Hormidium*, chloroplast orientation is also determined for a limited range of light intensities by a lens effect. The evidence (SCHOLZ, 1976a) indicates that the photoreceptor pigment is localized in the plasmalemma. For moderate intensities the chloroplast moves to the rear cell wall with respect to the incident light, but for very low and very high light intensities the chloroplast approaches the front of the cell. Under these conditions the light direction is perceived by an absorption gradient established as the light traverses the cell. A second photoreceptor is postulated to be localized within the chloroplast itself which responds to the absorption gradient within the cell (Fig. 1). These two responses can further be separated by their action dichroism (SCHOLZ, 1976b).

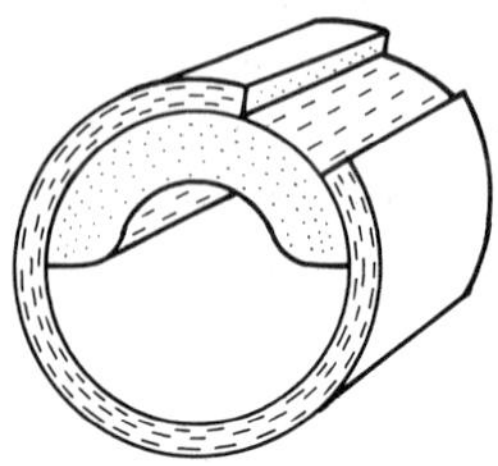

Fig. 1. Schematic representation of the dichroic orientation of the blue light photoreceptor molecules in a cell of *Hormidium*. Each *black dash* represents the relative absorption vectors of the pigment molecules. The outer cyclindrical layer is the plasmalemma, and the inner half-cylinder represents the chloroplast which can change its position. (Redrawn from SCHOLZ, 1976b)

The lens effect is observed if light is polarized with the electric vector perpendicular to the cell axis. The absorption system ("shadow system") operates mainly with polarized light vibrating parallel to the cell axis. From inhibitor studies the lens effect pigment appears to be a flavin. From measurements of the spectral sensitivity of and inhibitor effects on the shadow system, it is observed to function through two pigments. These are probably dichroically oriented chlorophyll molecules or a carotenoid molecule localized in the chloroplast.

The system is a complex one and serves very well to illustrate that plant movement responses may occur by the interaction of different photoreceptor systems which establish different absorption gradients, depending upon the particular wavelength of the light, the dioptrics which are in operation, the irradiance levels, or the polarization of the stimulating light. It remains to be described how the transduction steps occurring from each of these pigment receptors are integrated to give the final reorientation of the chloroplast.

3.4.3 Geometrical Structure

It is probably true that for every bending movement in plants the geometrical structure of the responding system is important for determining the direction of bending. In a trivial sense, the responding structure must have the capacity for differential elongation or growth. If these responding structures are distributed asymmetrically, the direction of plant movement will be determined by their distribution, and not primarily by the asymmetry in the distribution of receptors which are stimulated.

3.5 Localization of Receptors

The receptors may be concentrated in specific cells or organelles, or they may be distributed throughout the responding systems. For example, in grass seedlings the maximum phototropic sensitivity is usually localized in the tip of the coleoptile, although some sensitivity occurs in the rest of the coleoptile. In many dicotyledonous plants the region of phototropic sensitivity occurs in the leaves and cotyledons (STEYER, 1965). In Mung beans, however, the region of maximum sensitivity is in the apex of the stem (BRENNAN et al., 1976). With de-etiolated seedlings of *Cucumis* and *Helianthus* in which one cotyledon has been covered with aluminum foil, the cotyledon has been shown to induce curvatures. Under these conditions, curvature is mediated indirectly by the phytochrome system by asymmetric inhibition of hypocotyl elongation. In blue light, however, direct exposure of the hypocotyl is necessary to induce phototropism (SHUTTLEWORTH and BLACK, 1977).

It is often difficult to ascertain whether the photoreceptors are localized or distributed thoughout the organism. For example, in the fungus *Phycomyces*, the region of maximum sensitivity occurs just below the sporangium in the growing zone of the sporangiophore. For some time it was thought that the photoreceptors must be limited to this region. However, the mycelium was demonstrated also to contain the blue light photoreceptor, since initiation of new sporangiophores by the mycelium could be regulated by blue light exposure (BERGMAN, 1972). Therefore, the photoreceptor molecules may well be distributed throughout the fungus, and only when the photoreceptor is near a responsive region of the cell may the perceived stimulus be transduced into a response.

3.6 Orientation of Receptors

Information about intracellular localization can be obtained, in the case of photoreceptors, if they are oriented in a definite pattern, as discussed in the next two sections.

3.6.1 Dichroic Receptors

That the photoreceptor molecules have a fixed spatial orientation in an internal structure has been reasonably well established in a number of systems studied. A long series of experiments beginning in the 1930s and just recently summarized along with careful experiments for responses to polarized light with photosensitive albino mutants of *Phycomyces* demonstrate (JESAITIS, 1974), that the photoreceptors in this fungal system are dichroic and oriented within the cell.

However, is has been suggested (DENNISON, Chap. 5.1.2, this vol.) that the small magnitude of these responses could be explained with dichroic photoreceptors distributed in a purely random fashion.

Similarly, for corn coleoptiles, exposures to linearly polarized red and far red light were used to produce the photoconversion of the phytochrome photoreceptor. The data indicate that for both red and far red light the efficiency of photoconversion is 15% to 20% more effective when the light vibrates normal to the long axis of the coleoptile than when it vibrates parallel to this axis (MARMÉ and

SCHÄFER, 1972). It is concluded that the phytochrome photoreceptor molecules are located in the plasmalemma membrane or at least associated with this membrane. The phytochrome molecules are also concluded to be oriented along a screw line which has an angle of inclination of about 35°. This screw-line interpretation is a reasonable one for corn, but is not as firmly established as in the case of the alga *Mougeotia* (HAUPT et al., 1969). Detailed analysis of experimental data for chloroplast orientation leads strongly to the conclusion of an orientation of dichroic photoreceptors along a screw-line.

3.6.2 Motion of Receptors

For much stimulus perception a common property which has emerged in recent years is that motion of the receptor itself occurs as part of the perception system.

For geotropism this idea is, of course, the most easily accepted, because it is the simplest way to imagine that perception of the stimulus can occur. In the presence of a gravitational stimulus something within the organism must be displaced as a result of the gravitational force. Statoliths are imagined, have been searched for, and as some authors claim, have been observed and localized (JUNIPER, 1976; SIEVERS and VOLKMANN, Chap. 5.2.1, this vol.; VOLKMANN and SIEVERS, Chap. 5.2.2, this vol.). It is easy to imagine that as a result of the displacement of the gravireceptors some irritation must be conveyed to an adjacent membrane which then alters the rate of synthesis or rate of transport of some product or constituent. This change in rate results in the observable growth response.

However, it is also becoming apparent that for many photoreceptor systems, motion is also an underlying commonality. In gross macroscopic systems, for example, the eye motion of vertebrates in which rapid eye motion is required for the perception of stimuli has been reasonably well worked out in the past few years, and appears to be an integral requirement for the continual recognition of images (MCBURNEY and COLLINGS, 1977).

Similarly in plants, there seems to be a requirement for some photoresponses to require motion of the receptors. The organizational level at which motion is required varies from individual receptor molecules to whole integrative systems. For example, at the molecular level, it has been postulated that after red irradiation of the fern *Dryopteris* (ETZOLD, 1965) the dichroic photochromic pigment phytochrome undergoes a 90° rotation in orientation from a position tangential to the cell wall to a radial position. Observations for the movement of *Mougeotia* chloroplasts (HAUPT, 1968) in response to polarized and unpolarized red and far red light stimuli are similar. The dose–response characteristics for the control of these chloroplast orientation movements can be explained on the basis of phytochrome orientation after stimulation.

However, an analysis of photogrowth and phototropism of moss protonemata (*Physcomitrium*) concluded (NEBEL, 1969) that in this system rotation of the phytochrome photoreceptor molecule is not necessary to explain the results. A postulated disc-shaped photoreceptor takes into account all of the observed responses to polarized light stimuli. There is no requirement for a 90° rotation of the phytochrome molecule between the P_r and P_{fr} forms but, of course, these data are not conclusive proof against movement of the phytochrome molecule in *Physcomitrium*. The problem remains unresolved.

Similarly in the phototactic systems of *Euglena* and *Chlamydomonas* the rotation of the entire organisms appears to be necessary for the directional location of the stimulus to be resolved (FEINLEIB and CURRY, 1971; HÄDER, Chap. 3.2.1, this vol.). As the organism rotates, the carotenoid-rich eyespot is brought into the light path and shadows the photoreceptor, bringing about a tactic response. By a nulling method for continuous light stimuli, the organism can fix its direction of motion toward the light stimulus. This method is known for other algae (HAUPT, 1959) and has been carefully worked out for *Volvox* (SCHLETZ, 1975). *Chlamydomonas* responds phototactically to very brief single flashes, initially suggesting that rotation of the cell is not necessary for orientation. It is not known at present, however, if the only cells which become oriented are those in which the photoreceptor is not shaded at the time of the flash (FEINLEIB, 1975).

In *Volvox* the phototactic orientation is the result of periodic shading of the photoreceptors as the colony rotates. The colonies do not move precisely toward the light source, but consistently deviate to the left. The amount of deviation depends upon the wavelength of the source of light. If an action spectrum (SCHLETZ, 1975) is determined for the minimum deviation from the direction of the light source the action spectrum maximum occurs near 434 nm. If the absorption of the two major pigments which are responsible for the periodic shading are combined a close match to the action spectrum is obtained. These two pigments are found separately in the stigma (maximum at 465 nm) and in the chloroplast (maximum at 440 nm). Thus, for maximum combined absorption by these two pigments, the deviation caused by the periodic shading would be expected to be a minimum and occur at 434 nm. The colony apparently uses both shading pigments in conjunction with the photoreceptor pigment for perceiving the light direction.

It is also of interest that the colonies exhibit a typical stop response for large, sudden increases in irradiance. This response is mediated directly by the phototactic receptor itself, and an action spectrum for the stop response is largely unaffected by the periodic shading effects for determining the direction of the light source. That is, when a sudden increase in irradiance is given, many photoreceptors are unshaded by either stigma or chloroplast. The action spectrum for the stop response has a maximum near 480 to 490 nm and extends into the red to 626 nm. It is concluded that the pigment responsible for the stop response is a carotenoprotein (SCHLETZ, 1975).

When sporangiophores of *Phycomyces* are exposed to unilateral light stimuli in which the incident direction of the light is changed slowly, the sporangiophores continue to bend in the new direction. The fact that phototropism continues for many hours (Fig. 2, see also DENNISON, Chap. 5.1.2, this vol.) is surprising. On the basis of the range-adjustment process (Sect. 3.3.4, above) the responding system would be expected to adapt to the incident unilateral light intensity and cease to respond. There is evidence that the photoreceptors are somehow imbedded in a moving matrix which rotates around the longitudinal axis of the cell (DENNISON and FOSTER, 1977, and DENNISON, Chap. 5.1.2, this vol.). When the cell is exposed to unilateral light, it is this rotation which sweeps the receptors, successively, through the bright and the dimly lit regions of the cell, and allows for a continuous phototropic response. Experimentally, a rate of position change in the incident light of about 10° per minute can be found in which sporangiophores cease to bend. The conclusion from such physiological data is that the photoreceptors

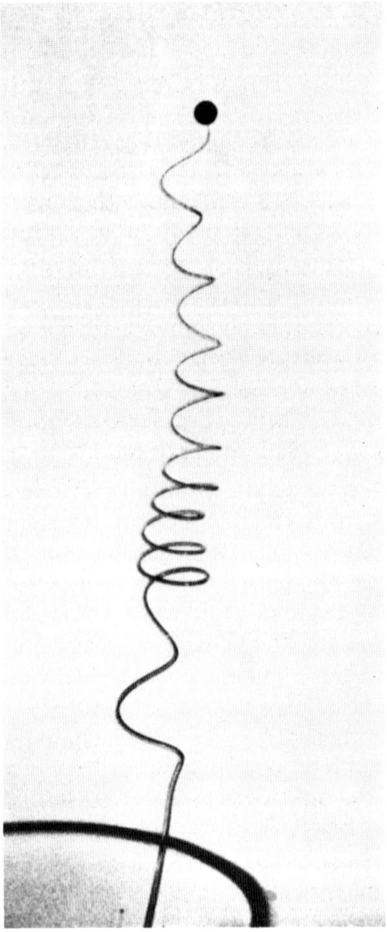

Fig. 2. *Phycomyces.* Frozen wave helix formed by phototropism at base of growing zone of sporangiophore during externally applied rotation (1 rev/6 h). The sporangiophore was exposed continuously to unilateral 10 μw cm^{-2} blue light during the rotation. The sporangiophore was growing on 5% potato dextrose agar in a shell vial. The diameter of the spherical sporangium at the top of the cell was 0.5 mm and the sporangiophore stalk had a diameter of 150 μm

are rotating endogenously within the cell at a rate of about 10° per minute around the longitudinal axis. A few attempts using electron microscopy and light microscopy have been unable to identify any structure within the cell which might be rotating at such a rate, but no one has carefully and systematically searched for one.

Although such data for the in vivo photoconversion of phytochrome indicate that both forms of the molecule are attached to a membrane, there is also considerable data for phytochrome which indicate that the red absorbing form (P_R) may be unbound. After formation by exposure to red light, the far red absorbing form may move to specific sites within the cell and become bound to membranes where it then exerts its physiological action. The principal in vivo evidence for such movement of phytochrome molecules after excitation is immunocytological evidence (PRATT et al., 1976) in which red light exposure causes accumulation of the phytochrome in discrete areas of the cell associated with membranes. Similarly, the displacement of the fluence-response curves for chlorophyll synthesis in pea seedlings by sequential exposure to light can best be explained on the basis of the migration of phytochrome molecules from a general, unlocalized pool to specific sites within the cell (RAVEN and SHROPSHIRE, 1975).

In vertebrate rods there is some evidence that the rhodopsin molecules themselves diffuse and rotate within the membrane matrix (CONE, 1972). The rhodopsin molecules are deeply imbedded in the disk membranes of frog rod outer segments. By means of high-speed flash photometry the rhodopsin is observed to undergo rapid rotational diffusion. This rotational diffusion occurs about an axis perpendicular to the membrane with a relaxation time of about 20 μs at 20° C (CONE, 1972). Since the rhodopsin molecules appear from X-ray evidence to be distributed in a liquid-like array in the plane of the membrane, they may also be able to undergo rapid lateral diffusion. Rapid microspectrophotometric measurements of rapid lateral diffusion in frog and mudpuppy rods confirm such lateral diffusion and give a value for the disk membrane viscosity of about 1 poise (POO and CONE, 1974). It is not known at present whether or not such rotation or diffusion in the membrane is involved with the generation of the first transduction step which generates receptor potentials needed for vision.

4. Integration of Perceiving Systems

4.1 Biophysical Transmission of Perceived Stimulus

Biophysical mechanisms for the transmission of a perceived stimulus from one part of a plant to another probably takes place by means of bioelectric potential changes which will be discussed in a subsequent chapter (BENTRUP, Chap. 1.3, this vol.). In some photomorphogenic responses such as the hook opening of bean seedlings, the size of the hook-opening response may be regulated by giving exposures to red and far red light either to the whole seedling or to isolated portions (DEGREEF et al., 1976). The hook opens if it is directly irradiated with red light. It may also open if the leaves alone are irradiated with red light, but the extent of opening is smaller than if the hook is directly exposed. If the entire plant system is exposed to far red light, the hook opening can be photoreversibly controlled. However, if the leaves alone are exposed to far red light, there is no effect on the subsequent opening of the hook.

In other words, there is a rapid transmission from the site of perception in the leaves to the site of response in the hook within a few seconds (Fig. 3; DEGREEF et al., 1976). However, there are ways in which integration of the sensory systems in plants may occur by other mechanisms.

4.2 Gaseous Interactions

Two or more responding systems may interact in such a way that the sensitivity to successive stimuli can be markedly altered. Sensitivity to stimuli can be modified by the production of gases by the organism or by interference with the response by the presence of exogenous gases.

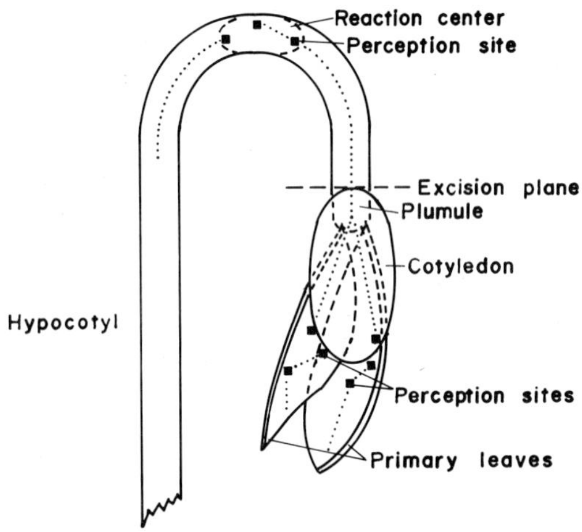

Fig. 3. Schematic diagram of bean hypocotyl hook. The postulated relay system explains the phytochrome-mediated inter-organ cooperativity for the light-induced hook opening. There is an explicit distinction between "reaction center" (=hook for hook opening) and "perception site" (where the signal is perceived in the organ). Red light signals are transmitted in both directions between different perception sites, but far red light is effective only when its signal can be integrated in the system. (Redrawn from DEGREEF et al., 1976)

The sporangiophore of *Phycomyces* is known to produce a gaseous material which can produce bending in the sporangiophore when a barrier is placed near the cell. This autochemotropic response is one in which the gaseous substance is thought to be emitted by the growing region of the cell. Introduction of a barrier near the cell modifies the distribution of the effective gas around the growing portion of the cell such that asymmetrical growth occurs, which results in a bending away from the barrier (BERGMAN et al., 1969).

The nature of this gaseous material has not been fully determined although water vapor, nonanal, and ethylene have been proposed as the effector molecule (COHEN et al., 1975; LAFAY et al., 1975). The important fact is that the cell itself perceives the chemotropically active material in the same region as it synthesizes it. This means that the cell must possess a mechanism for preventing a short circuit between the presence of molecules of the effective gas and the production of these same gas molecules. Examination of the growing zone where both receptors and producers are located does not reveal any specialized structures.

In phototropic responses of higher plants such as mung beans (BRENNAN et al., 1976) the magnitude of the phototropic response is reduced under hypobaric conditions. Ethylene synthesis is known to be induced by light and is probably mediated by phytochrome through the high irradiance system (CRAKER et al., 1973). Ethylene is known to inhibit auxin transport (MORGAN and GAUSMAN, 1966), and both lateral transport and basipetal transport of auxin are implicated in the formation of the asymmetric gradients of auxin which are involved in phototropism (GARDNER et al., 1974). Thus, each of these systems may be interacting in bending processes. The entire subject of the effects of growth regulators will be reviewed by PHARIS and REID in volume XI; see also KANG, Chap. 5.3.2, this vol.

Similarly, the dose–response relationships for phototropism can be altered markedly by exposures to red and far red light through the phytochrome system (BRIGGS, 1963). It may well be that these modifications in the sensitivity of phototropism are mediated indirectly through the production of ethylene (KANG and BURG, 1974). However, the ethylene probably interferes in the transduction steps and not in the perception mechanism. It is extremely difficult to disentangle the processes following the reception of a light stimulus, particularly when the light stimulus itself may induce several concurrent processes. Each of these processes may interact with one another during the transduction steps.

In roots that are ageotropic when grown in the dark, exposure to blue or red light induces the development of geotropic sensitivity (SCOTT and WILKINS, 1969; cf. also VOLKMANN and SIEVERS, Chap. 5.2.2, this vol.). It is not known if light induces the capacity for geotropism at the receptor level, or whether hormones such as ethylene that are induced by the light interact at either the receptor or transduction steps, or both. In any case, great caution must be exercised in interpreting the modification of sensitivity of any response at the perception level.

5. Changes in Sensitivity

5.1 Receptor Level

5.1.1 Bleaching

There are many ways in which the sensitivity of perception may be imagined to be modified. However, for plant movements, it is very difficult to find data that convincingly demonstrate the receptor itself has had a change in sensitivity (STILES and COCKING, 1969). A possible basis for change in photoreception sensitivity is bleaching of the pigments by a strong stimulus. Such is the explanation usually given for light–dark adaptation in human vision (FISHER et al., 1970). Examination of vision-sensitivity changes in frog rods suggest that these changes are directly linked to the amount of light activated rhodopsin phosphorylation. By inhibitor studies directly affecting such phosphorylation, the bleaching of 5×10^2 to 5×10^4 rhodopsin molecules per outer segment per second could be correlated with decreased sensitivity changes over approximately 4 log units of input light intensity (BRODIE and BOWNDS, 1976).

Clearly, if most of the photoreceptors are altered in such a way that they no longer absorb as they did prior to stimulation, there will be a reduction in the sensitivity to subsequent stimuli. That is, there must be a refractory period in which pigment must be regenerated before another response can be elicited (see Sect. 3.3.4). In vivo measurements of phototropically sensitive systems (POFF and BUTLER, 1974; POFF and WHITAKER, Chap. 3.2.4, this vol.) have indicated small absorption changes after large stimuli, but these changes are not known to be correlated with the phototropic response. In most instances changes in sensitivity are attributable to transduction steps occurring after perception or to responses not related to the receptor (LIPSON, 1977). However, the fact that the adaptation level overshoots the intensity level for a large irradiance increase suggests that both the rate of change of pigment concentration and the pigment concentration itself could be involved (FOSTER, 1977).

5.1.2 Hydration

A change in sensitivity at the receptor level is indicated in the control of seed germination by the photochromic pigment phytochrome as a function of the hydration level. In this case, the water content must reach a level of 15% or more in the seed for the phytochrome molecule to undergo photoreversible changes between the red and far red forms as measured by germination (Hsiao and Vidaver, 1971). The water microenvironment probably exerts its effect on phytochrome directly since these levels of water content will not result in germination even if the phytochrome is present in the active far red absorbing form. In addition, the form of the phytochrome molecule following a light exposure given when the seed is moist has been demonstrated to persist for at least a year after the seed is dried (Vidaver and Hsiao, 1972). If the seed is remoistened it will germinate in the dark to a percentage germination depending on the form of the phytochrome molecule prior to dehydration. The role of hydration in the phototransformation of phytochrome in vitro has also been demonstrated (Tobin et al., 1973).

5.1.3 Synthesis

The changes in germination sensitivity of seed to light exposure have also been attributed to phytochrome synthesis (Taylorson and Hendricks, 1971). Recent experiments (Duke et al., 1977) on the light sensitivity for seed germination indicate that the level of phytochrome does not change during dark incubation periods of varying duration. Since the level of phytochrome molecules did not change, it was concluded that any changes in sensitivity must occur on transduction steps and not at the level of perception. However, for many chemotactic responses in bacteria the synthesis of the receptor is known to be induced by the presence of the chemical stimulus (Koshland, 1977).

5.2 Screening Pigments

For photoreception the sensitivity may clearly be altered by screening pigments. The sensitivity of the response observed may be altered spatially or temporally by the presence of screening pigments which do not allow quanta to reach the active photoreceptor. For phototropism such an effect is on the asymmetric light intensity established in the organism and does not produce a sensitivity change in the receptor itself. An extreme example of this effect is clearly demonstrated by the reversal in the direction of phototropic response in *Phycomyces* sporangiophores to uv stimuli (Bergmann et al., 1969). The central vacuole contains large amounts of gallic acid which selectively absorb in the uv. For stimuli at 280 nm the two sides of the cell are effectively screened from each other, and only the side nearest the source is stimulated, resulting in strong negative curvatures. If the growth rate to an increase in either blue or uv light is measured it is seen to be strongly increased, indicating a positive growth output due to the stimuli (the transient change in growth rate induced by light exposure is defined as the light–growth response). The negative curvature occurs for unilateral uv light because the gradient established in light intensity across the cell by its dioptric properties is reversed compared to that for blue light by the internal screen (cf. Sect. 3.4.2 on dioptrics; Dennison, Chap. 5.1.2, this vol.).

5.3 Illusions

To my knowledge no one working in sensory physiology has used the concept of illusions to apply to stimulus perception in plants. However, the idea is an interesting one to keep in mind. Illusions for animal systems are described as responses observed and not expected for the specific input stimuli.

One interesting recent example is the "pincushion grid" illusion (SCHACHAR, 1975), Figure 4. The diagonal lines which are seen are not part of the printed page. They arise as part of the signal handling system beyond the retina and perhaps in the brain.

If the grid is rotated 45°, the illusion of diagonal lines is intensified. More spurious information appears in the grid! The reason for this intensification is probably due to the observation that the human visual system has greater visual resolving power for vertical and horizontal lines than for diagonal ones. Previously, it had been suggested that Fourier transform analysis of the pincushion grid gave diagonal components. However, when a two-dimensional optical Fourier transform of the pincushion grid was produced and photographed with coherent, collimated light no diagonal components were evident.

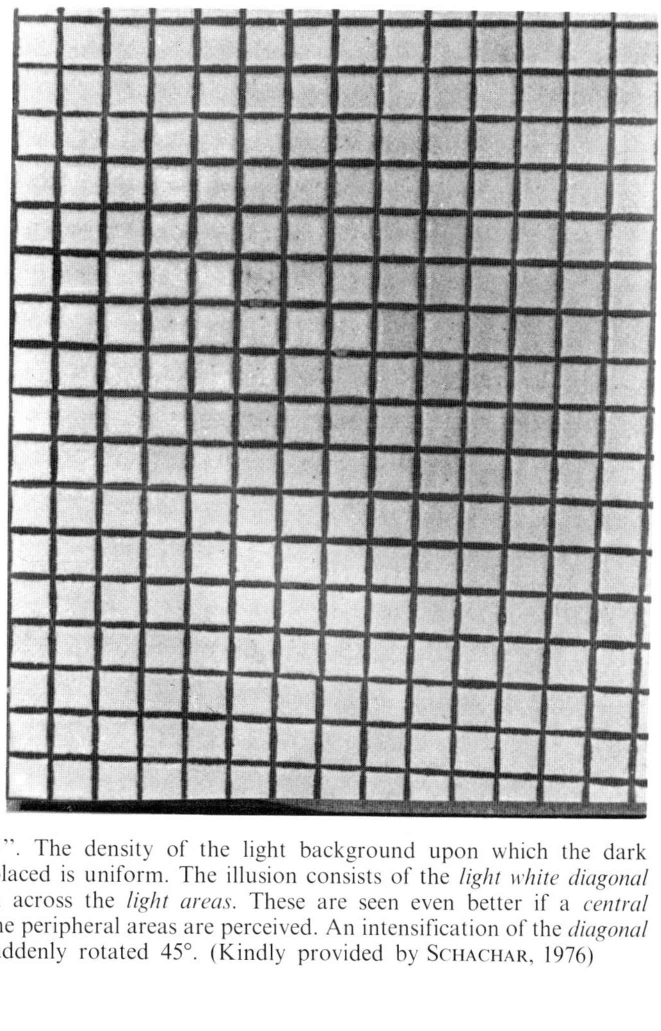

Fig. 4. A "pincushion grid". The density of the light background upon which the dark intersecting black lines are placed is uniform. The illusion consists of the *light white diagonal lines* which are seen to run across the *light areas*. These are seen even better if a *central square* is focused upon and the peripheral areas are perceived. An intensification of the *diagonal line* occurs if the grid is suddenly rotated 45°. (Kindly provided by SCHACHAR, 1976)

"The lack of diagonal components in the two-dimensional Fourier transform of the pincushion grid demonstrates that the relationship between visual perception and Fourier theory may be fortuitous and prediction on the basis of Fourier analysis may be unwarranted." (SCHACHAR, 1975).

I include this small digression into this illusion of sensory perception in man only to draw attention to the fact that observed anomalous behavior in stimulus perception in plants should be examined closely. Examination of such anomalous results coupled with suggestions for testing the origin of the illusion may lead to a separation of the receptor system from the first transduction steps. They may also give insights into the inherent properties of the receptor itself. I would suggest that the dioptric reversal of phototropism described for *Phycomyces* was in the class of illusion. Let me illustrate.

Some photobiologists postulated that two photoreceptors must be present in *Phycomyces*: one for negative curvatures, induced by uv stimuli, and a second one for positive curvatures induced by blue light (CURRY and GRUEN, 1957). After a careful examination was made of the optical properties of the image-forming cylindrical cell and action spectra for both phototropism and the light–growth response (the change in growth rate induced by light exposure) were determined and found to be the same, it became apparent that one photoreceptor pigment is sufficient to explain the cell's behavior (BERGMAN et al., 1969).

5.4 Endogenous and Exogenous Oscillations

Differences in the sensitivity to stimuli are also known to be caused by endogenous or externally entrained oscillations (SWEENEY, Chap. 1.4, this vol.). These differences have been suggested to be due to changes in the internal level of energy available to the transduction and responding portion of the system (SEITZ, 1974; WAGNER, 1976). If this explanation is correct, it is unlikely that the receptor itself is affected by such rhythmic changes, but rather that the first interaction with the transduction system is affected by the changes in energy level. It is possible that the number of receptor molecules fluctuates periodically, but no data exist to substantiate this hypothesis.

6. Connecting the Receptor to the Response

6.1 Absorbance Changes

One of the central problems in analyzing a response system beyond the level of stimulus perception is in connecting the suspected receptor with the observed response. In general, the approach is to look for responses which occur as rapidly as possible after the stimulus is given. For example, in photoreception a search is made for absorbance changes in vivo. Such changes have been observed after blue light stimulation (POFF and BUTLER, 1974; see also Sect. 5.1.1) and attempts have been made to correlate these changes with the photoreceptor for phototropism

(GALSTON et al., 1977). Extracts are made and absorbance changes are sought in vitro. The molecules producing the absorbance shifts can be identified (BRAIN et al., 1977). However, the problem still is not resolved in most cases because spurious reactions may occur after homogenization and extraction. Even though the stimulus for the response in question can elicit the absorbance changes in both in vivo and in vitro, it cannot be definitively proven that these changes are causally connected to the response.

Phototropically abnormal mutants of *Phycomyces* exist and have been examined for cytochrome absorbance changes (LIPSON and PRESTI, 1977). The phototropically abnormal strains gave blue light-induced absorbance changes which were identical to those observed in the phototropically normal strains.

In addition, quantum-yield measurements for the presumed flavin-mediated cytochrome absorbance changes in *Phycomyces* gave a value of only 0.015. Most biological photoresponses have evolved with high sensitivity and usually have quantum yields near one. The low value of the blue light-induced absorbance change would argue against it being relevant for phototropism (cf. also DENNISON, Chap. 5.1.2, this vol.).

From such results it can be concluded that either the absorbance changes are unrelated to the phototropic response, or that the defects in the photomutants are at a later step than the pigment receptor.

6.2 Gravitational Displacements

Another example of this type of approach has been to attempt to correlate intracellular displacements with observed geotropism. It is assumed that if a direct correlation could be demonstrated between such displacements and the observed responses, the gravireceptor could be identified.

The nature of this gravireceptor has long been pursued (SIEVERS and VOLKMANN, Chap. 5.2.1, this vol.; VOLKMANN and SIEVERS, Chap. 5.2.2, this vol.) and one specific identifiable gravireceptor may not exist. Most authors in this field are convinced that the mechanism, at least for higher plants, involves the sedimentation of amyloplastic starch grains (WILKINS, 1976). The sedimentation of these starch grains establishes an asymmetry within the cell which then is transduced into unequal growth rates through the altered transport of growth-regulating hormones. In Table 1, a value is presented for the calculated pressure exerted on a membrane in a *Helianthus* cell by such a starch grain (AUDUS, 1975). The concept that pressure on a cell membrane is the basis of gravireception has recently been challenged (PERBAL and PERBAL, 1976). By a careful biophysical study of the forces and motions generated after gravitational stimuli were applied to lentil root caps, it was concluded that, in *Lens*, geotropism cannot be explained by the simple action of sliding, pressure, or contact of amyloplasts on a sensitive surface near the longitudinal wall. These authors conclude that a number of parameters must be taken into account, and that the most significant is the sedimentation of the amyloplasts. A formal relationship is derived for the initial rate of curvature, $V = a(\log b) d \cdot c \cdot \sin \alpha$ where a and b are constants, d is the distance the amyloplasts move, c is the number of contacts with parietal cytoplasm and $\sin \alpha$ is the transverse component of the statolith weight. The results fit this formal description well. Although most authors agree that the site of gravireceptors in roots is in the core of the root cap (JUNIPER, 1976), they are not in agreement about how displacements within stimulated cells can effectively be transduced into the chemical gradients necessary for differential growth. The conclusion is that despite the concerted effort of many workers, the molecular nature of the gravireceptor remains an enigma.

7. Separation of Reception and Transduction

Another approach to analyzing a response system is to pull apart the sequence of events in the stimulus response chain by altering environmental conditions in such a way that the transduction events are blocked at various distances from the initial sensory event. That is, to stop somehow the flow of information from the stimulus to the response. If the block can be inserted rapidly enough and specifically enough, time is gained to examine the state of the pathway at that point or to examine the nature of the materials which pile up at the blocking point. Ideally, the block should be a reversible one so that after a period of time the block can be released and the flow of information continued with the expected response occurring.

Classically, metabolic inhibitors and temperature changes have been employed. A recent example is the use of phenylacetic acid to inhibit phototropism in corn seedlings. The phenylacetic acid is known to interact with flavins. Of additional interest is the fact that treated seedlings which give no phototropism for blue light exposures retain normal geotropism (SCHMIDT et al., 1977). The conclusion is drawn that the receptor for phototropism has been inhibited by the phenylacetic acid but that the gravireceptors and the growth readout for differential growth have been unaffected. However, it is rare that it can be demonstrated that the sensory input can be recovered after the block has been removed and it has not been demonstrated in this case.

Separation of the sensory and motor functions in the coiling movements of pea tendrils to mechanical stimulation has been demonstrated (JAFFE, 1977). If the pea tendrils of light-grown plants are rubbed on the abaxial side, they rapidly form a spiral coil. However, if the pea tendrils from similar plants are kept in the dark for three days and then mechanically rubbed they will not coil. If these dark-adapted tendrils which have been stimulated are exposed subsequently to light they will coil immediately. The ability to coil upon exposure to light is retained for at least 2 h after the mechanical stimulation. The conclusion is drawn that the sensory function is operating normally, but that the transduction to the motor response has been blocked in the dark, and that light releases this block.

Adenosine triphosphate (ATP) is known to substitute for light in tendrils which have been held in the dark, and is also known to be used by coiling tendrils (JAFFE and GALSTON, 1967). It could be that the ATP supply is depleted during the dark adaptation time, and that the coiling which occurs upon exposure to light is brought about by photosynthetic production of ATP. Alternately, it has been suggested (JAFFE, 1977) that an inhibitor accumulates in the dark period and is subsequently removed by brief exposure to light. In any case, the important point is that it is possible to dissect the pathway from stimulus to response by this strategem.

The tendril-coiling system can also be affected by temperature (JAFFE, 1977). If the tendrils are rubbed and held at 25° C they will coil, but if after rubbing they are kept at 5° C or at 10° C they will not coil. The information from the stimulus is retained for at least an hour at 5° C because rubbed tendrils will immediately coil when returned to 25° C. By the application of these two treatments (darkness and temperature) in various combinations, it should be possible to study the sensory portion of the mechanoreceptor independent from the motor response.

8. Characterization of Sensory Pathway

8.1 Dose–Response Curves and Action Spectra

Most plant movement systems can be affected by a number of stimuli, and the responses to these different stimuli can often be shown to interact. In the case of responses to light an approach to separate out these interactions and determine the underlying common causal mechanisms has been to measure carefully the dose–response curves and action spectra for several processes independently and then to correlate them. A recent example is the similarity reported for the action spectra and dose–response curves for phototropism, for photoinhibition of growth, and for auxin transport in *Avena* seedlings (ELLIOTT and SHEN-MILLER, 1976). The authors concluded that these processes are related phenomena and that, because the responses have action spectra which are similar in the blue portion of the spectrum, they are all mediated by the same photoreceptor. In addition, the maximum responses for each process observed occurred at a dose of 10^{13} photons cm^{-2}. As a further argument in favor of a common photoreceptor, these authors point out that both phototropism and growth inhibition are affected by prior exposure to red light, which changes the sensitivity of the responding system to subsequent blue light exposures (BRIGGS, 1963). Such correlations may be circumstantial, and may not indicate direct causal links (see discussion above on gas effects for an indirect linkage). For example, it is easy to imagine that in the transduction process metabolic energy is required for all three of these responses. Assume that there are three parallel pathways from photoreceptor to responses. Each pulls on the same pool of ATP for energy and obviously cannot be independent. A heavy drain on the ATP supply when one pathway is induced will necessarily compete against the second for the supply of ATP and will affect its subsequent sensitivity to an additional stimulus.

This relationship can be seen especially clearly by examining the responses in the red portion of the spectrum. No phototropic responses occur in seedlings for unilateral red exposures unless some artificial mechanism such as shielding a cotyledon is imposed to generate a differential in growth rate which can lead to curvature (SHROPSHIRE and MOHR, 1970). The fact that the action spectra both for light–growth response and for auxin transport have activity in red light argues that they cannot be causally related to phototropism at the receptor level. Since red light can only alter the sensitivity to subsequent blue light stimuli, and cannot directly induce phototropism, the responses again cannot have the same receptor. They are probably only connected late in the transduction and response pathways.

Such interactions are clearly demonstrated for the induction of curvatures in de-etiolated seedlings of *Helianthus* (SHUTTLEWORTH and BLACK, 1977), by red light exposures of the cotyledons (see Sect. 3.5) and blue light exposure of the hypocotyl. These observations clearly contradict the assumption that blue and red light act via the same pathway in this system. Therefore, it is concluded that great caution must be exercised in making correlations between dose–response curves and then extrapolating all the way back to the initial photoreceptor.

8.2 Time and Irradiance

Another caution which should be inserted is that the dose–response curves for photoresponses can be greatly distorted depending upon the time and irradiance variables. For phototropic curvatures of *Avena* coleoptiles, these distortion problems have been elegantly demonstrated (BLAAUW and BLAAUW-JANSEN, 1970). For many years there has been much confusion in the literature about first, second, and third positive curvatures of *Avena* coleoptiles when exposed to unilateral blue light stimuli. In many cases the controversy has been exacerbated by the techniques of handling the seedlings and by the effects of prior exposure to red light before phototropism was induced (BRIGGS, 1963). In a carefully detailed three-dimensional graph the phototropic curvatures were plotted as a function of irradiance and of exposure time (see graph and detailed discussion, DENNISON, Chap. 5.1.2, this vol.). If a cross-section is made for any particular exposure time or for a specified irradiance value, it is possible to generate almost any response curve which has been reported in the literature. The important point is that all of the variables involved in generating the dose–response curve must be carefully evaluated if spurious conclusions are to be avoided.

8.3 White Noise

Another approach for attacking the characterization of the pathway for the perception of stimuli is to use a white noise-stimulus technique (LIPSON, 1975a, b, c). Here, for a wide range of boundary conditions, random stimuli are given in which the response shape, observed as a continuous output may be derived as a function of a number of possible stimuli. By mathematical techniques and the assumption that the biological system must show time-invariance, boundedness and finite memory (SNEDDON, 1976), the relationship between all stimuli and all responses can be derived. Necessarily this technique requires that the response time constants of the organism and the output being observed are amenable to such continuous recording. It should also be pointed out that the model is a "best fit" one. It is unlikely that any developing biological system is truly time-invariant except over an arbitrarily selected narrow increment of time. For example, the technique would be very difficult to apply to a response in which an accumulated product is produced, such as the photoinduction of carotenoid synthesis. However, there has not been enough exploitation of this powerful technique for plant movements other than for phototropism and light–growth responses in *Phycomyces* and to some extent for chemo- and phototaxis in bacteria (BERG, 1975).

8.4 Genetics

Another powerful technique for pulling apart the pathway and obtaining early information about the receptor itself is genetic analysis. An outstanding example of this approach is the use of mutants of *Phycomyces* which have altered sensitivities to a number of stimuli which result in growth, movement, and differentiation (BERGMAN et al., 1973). The fungus *Phycomyces* demonstrates a number of such responses to stimuli including phototropism, light–growth rate changes, sporangiophore photoinduction, autochemotropism (avoidance), photocarotenogenesis, and geotropism. Most of the behavioral mutants with abnormal responses were derived by mutagenesis with nitrosoguanidine. A network model was proposed (BERGMAN et al., 1973) in which the defects were classified on the basis of whether the defect occurs on the input side (photoreceptor) or on the output side (response). "Input" mutants are normal on the output side, since they respond normally to other stimuli, although they fail to respond to light. "Output" mutants show normal mycelial responses to light, but no tropic responses to any stimuli. A third class of mutant is one in which a defect occurs somewhere between input and output. This scheme of the present classification of mutants tested is given in Figure 5.

Such genetic approaches have also been used for other behavioral systems (PAK and PINTO, 1976). Clear examples are found in the genetic dissection of the nervous system of insects and of mammals such as the mouse. The technique is also being used with bacteria, paramecia, and nematodes. To my knowledge

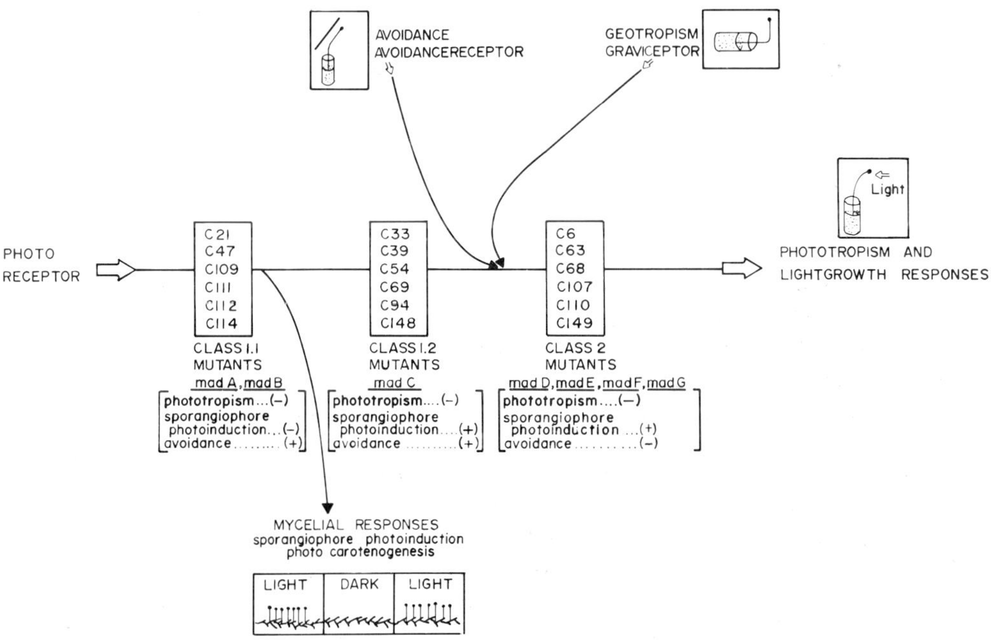

Fig. 5. Summary diagram for the network of stimulus-response systems in the fungus *Phycomyces*. The symbol (+) denotes a response to the stimulus and (−) denotes no response. The *C* numbers are strain designations. (Combined from BERGMAN et al., 1973 and OOTAKI et al., 1977)

little is being done to use such a genetic approach to studying higher plant stimulus perception, and it is an area that merits exploration.

9. Promising Technological Approaches

9.1 Immunocytochemical

Once the suspected receptor can be isolated and purified, a great deal of information can be obtained about the specific intracellular localization of receptors (PRATT et al., 1976). Also, if there is a modification in the chemical structure of the receptor after stimulation as in the phytochrome case described above (Sect. 4.1), then immunocytochemical techniques can be used to follow the rate of modification in the receptor as well as changes in location if they occur. They can also be used to map out the time course of regeneration of the initial molecular form of the receptor.

Another use is the detection of similar receptors between plant species. The suspected receptor, if successfully isolated from one tissue, may be used to prepare specific label for locating this receptor in other tissue. Also, because of their high specificity, immunological techniques can be used for selective inhibition of receptors as part of mapping the initial transduction steps.

9.2 Flash Photolysis and "Jiffy" Chemistry

Extremely high-speed, high-irradiance flashes of radiant energy followed by rapid time-scale measurement of the changes in the receptor molecules can give a great deal of information about the formation of intermediates. These labile intermediates are often formed very rapidly in the photochemistry of photoresponses and ultimately are converted to stable intermediates, which then lead into the transduction steps. Such flash photolysis has yet to be utilized extensively for studying plant photoresponses.

The time scale of resolution as indicated above in the section on absorbance-change measurements is continually being pushed to shorter and shorter times. Between 1941 and 1972 the time scale of following chemical reactions decreased from milliseconds (10^{-3}) to picoseconds (10^{-12}) (BIGELEISEN, 1977). The first biological process studied by this means was the mechanism of retinal isomerization (BUSCH, 1972). The rate of formation of prelumirhodopsin and its decay was determined. Prelumirhodopsin is formed within 6 picoseconds after the photoexcitation of bovine rhodopsin. As the time scale shortened, the physical chemist Gilbert N. Lewis coined the term "jiffy"—the time for a photon to travel 1 cm, or 33 picoseconds (BIGELEISEN, 1977).

Pulse radiolysis of water produces, within a picosecond almost exclusively OH radical, H^+ ion and e_{aq}^-, the hydrated electron. If two of these products are removed by scavengers, the reactions of the remaining product can be studied. The concentration of hydrated electrons, e_{aq}^- can be followed easily with picosecond spectroscopy because of their intense absorption at 720 nm in the far red region of the spectrum.

Such high speed measurements may be pushed even further, and femtosecond measurements (10^{-15}) have been suggested (BIGELEISEN, 1977). This time range is near the limit for such time measurements because of the Heisenberg uncertainty principle. The limits in energy measurements in this time range approach the spacing of the vibrational energies in molecules and the energy changes in chemical reactions. Thus, measurements in this time range would be limited by the uncertainties of measurement. However, it is postulated that new classes of reaction in this fast time range might be discovered. These could yield information pertinent to the deformation of receptors in stimulus perception in much the same manner that broadening of resonance signals has provided information on the conformational changes in molecules.

9.3 Acoustic Spectroscopy

Photoacoustic spectroscopy has been recently developed for examining solid biological materials (ROSENCWAIG, 1973a). The technique is particularly useful for determining spectra of materials in situ on solid surfaces or on membranes which are hard to examine by absorption spectroscopy. The photoacoustic spectra are very similar to absorption spectra of the same materials in solution. The acoustic signal arises from the cyclic heating and cooling of the solid sample by the absorbed pulsed light followed by the direct transfer of this heat energy to the surrounding gas. The technique is useful over the range of 200 nm to 900 nm. Using the technique it was possible (ROSENCWAIG, 1973b) to distinguish between reduced and oxidized cytochrome c in powdered form, and also to follow the state of hemoglobin in red blood cells. It is difficult to measure this redox state by absorption techniques because of interference from uv-absorbing materials and because of serious scattering problems. Such measurements of receptors located on membranes should be a powerful tool for following their properties before and after stimulation.

9.4 Tuneable Lasers

Tuneable lasers have the ability to vary wavelength of the emitted light over wide ranges. Such lasers are extremely useful not only for localization of receptors by localizing stimuli into very small volumes (focused into areas of 1 μm), but also for direct optical excitation of receptor molecules. Such excitation has been reported (DELBRÜCK et al., 1976) as evidence that the blue light receptor for phototropism is a flavin.

Similarly, such lasers, with their capacity for producing extremely intense monochromatic stimuli, have opened a number of phenomena for study (BERNS, 1976). For example, it has been suggested that two-photon effects result from the focusing of such pulsed laser beams onto chromosomes of living cells. An action spectrum of a two-photon effect was derived (BERNS, 1976) for photon densities which varied from 10^{21} to 10^{25} photons s^{-1} cm^{-2} for wavelengths between 473 and 562 nm. The failure of reciprocity for the response also indicated that an uncommon physical effect was occurring such as a two-photon effect. The energy for the receptor deformation is exactly twice the energy value for a single quantum of the incident light. Such lasers will become more and more useful for examining high irradiance responses for which reciprocity fails.

10. Conclusions

10.1 Paucity of Knowledge of Receptors and Stimulus Perception

For no plant movement is the molecular nature of the receptor known. For photoresponses there is an approach to knowing the photoreceptor. However, it will be some time before the sequence of events required for any stimulus perception will be known in molecular detail.

10.2 Need for Pushing Measuring Systems to Their Physical Limits

In almost every conceivable case of stimulus perception, for animals, as well as plants and microorganisms, the physical limits of the measuring system must be pushed to at least the noise level limits of the receptors. In addition, the time constants of the measuring technology must be pushed to the Heisenberg uncertainty limits, since the molecular events which are important for stimulus perception occur in the picosecond range at least for photoperception and probably for others. If there is to ever be a detailed mapping of the molecular events occurring in stimulus perception, followed by the events of transduction, it will be based upon solving such resolution problems.

10.3 Cell-Free Systems

Ultimately the test for an operational stimulus perception system is to be able to reconstitute such a system from known components. Coupling genetic techniques with in vitro biochemical techniques will be necessary. In almost every case of stimulus perception being studied one of the major barriers to present progress is the lack of a nonbiological assay which is certain to be coupled to the biophysical and biochemical events of interest. Where such assays have been developed that do not depend upon delayed physiological events, it has been possible to approach the isolation of the receptors.

10.4 Membrane Data as Carriers of Receptors

Lastly, it appears reasonably certain from what is presently known that, for stimulus perception, most of the initial events probably occur on or are closely associated with membranes. An increase in the specific knowledge of membrane structure and function would greatly aid in the understanding of stimulus perception.

10.5 Beginnings of Exploration of Stimulus Perception

Although the study of stimulus perception has a long history of significant progress in many areas, it has had to await technological advances. Stimulus perception in plants and microorganisms is beginning to be unravelled and the "terra incognita" of signal transducers (DELBRÜCK, 1972) is beginning to reveal its rich treasures. The current detailed maps are described in subsequent chapters of this volume.

References

Adler, J., Wung-Wai, T.: "Decision"-making in bacteria: chemotactic response of *Escherichia coli* to conflicting stimuli. Science *184*, 1292–1294 (1974)

Audus, L.J.: Geotropism in roots. In: The Development and Function of Roots. Torrey, J.G., Clarkson, D.T. (eds.). London: Academic Press, 1975, Ch. 16, pp. 327–363

Ball, N.G.: Tropic, nastic, and tactic responses, plant physiology, a treatise. In: Analysis of Growth: Behavior of Plants and Their Organs, Steward, F.C. (ed.). New York: Academic Press, 1969, Ch. 3, Vol. Va

Barksdale, A.W.: Sexual hormones of *Achlya* and other fungi. Science *166*, 831–837 (1969)

Beidler, L.M.: Physiological problems in odor research. Ann. N.Y. Acad. Sci. *58*, 52–57 (1954)

Bentrup, F.-W.: Lichtabhängige Membranpotentiale bei Pflanzen. Ber. Dtsch. Bot. Ges. *87*, 515–528 (1974)

Berg, H.C.: Chemotaxis in bacteria. Annu. Rev. Biophys. Bioeng. *4*, 119–136 (1975)

Bergman, K.: Blue-light control of sporangiophore initiation in *Phycomyces*. Planta *107*, 53–67 (1972)

Bergman, K., Burke, P.V., Cerdá-Olmedo, E., David, C.N., Delbrück, M., Foster, K.W., Goodell, E.W., Heisenberg, M., Meissner, G., Zalokar, M., Dennison, D.S., Shropshire, Jr.,W.: *Phycomyces*. Bacteriol. Rev. *33*, 99–157 (1969)

Bergman, K., Eslava, A.P., Cerdá-Olmedo, E.: Mutants of *Phycomyces* with abnormal phototropism. Mol. Gen. Genet. *123*, 1–16 (1973).

Berns, M.W.: A possible two-photon effect in vitro using a focused laser beam. Biophys. J. *16*, 973–977 (1976)

Bigeleisen, J.: Chemistry in a Jiffy. Natl. Res. Counc. 1977, Curr. Issues and Studies, 99 112 (1977)

Blaauw, O.H.: Die Perzeption des Lichtes. Recl. Trav. Bot. Neerl. *5*, 209–372 (1909)

Blaauw, O.H., Blaauw-Jansen, G.: The phototropic responses of *Avena* coleoptiles. Acta Bot. Neerl *19*, 755–763 (1970)

Blaauw, O.H., Blaauw-Jansen, G., Elgersma, O.: Determination of hit numbers from dose-response curves for phytochrome control of seed germination (*Lactuca sativa* cv. NORAN). Acta Bot. Neerl *25*, 341–348 (1976)

Brain, R.D., Freeberg, J.A., Weiss, C.V., Briggs, W.R.: Blue light-induced absorbance changes in membrane fractions from corn and *Neurospora*. Plant Physiol. *59*, 948–952 (1977)

Brennan, T., Gunckel, J.E., Frenkel, C.: Stem sensitivity and ethylene involvement in phototropism of Mung bean. Plant Physiol. *57*, 286–289 (1976)

Briggs, W.R.: The phototropic responses of higher plants. Annu. Rev. Plant Physiol. *14*, 311–352 (1963)

Brodie, A.E., Bownds, D.: Biochemical correlates of adaptation processes in isolated frog photoreceptor membranes. J. Gen. Physiol. *68*, 1–11 (1976)

Bünning, E.: Fifty years of research in the wake of Wilhelm Pfeffer. Annu. Rev. Plant Physiol. *28*, 1–22 (1977)

Busch, G.E., Applebury, M.L., Lamola, A.A., Rentzepis, P.M.: Formation and decay of prelumirhodopsin at room temperatures. Proc. Natl. Acad. Sci. USA. *69*, 2802–2806 (1972)
Carlile, M.J.: Primitive Sensory and Communication Systems. London: Academic Press, 1975, 258 pp
Cohen, R.J., Yan, Y.N., Matricon, J., Delbrück, M.: Avoidance response, house response, and wind responses of the sporangiophore Phycomyces. J. Gen. Physiol. *66*, 67–95 (1975)
Cone, R.A.: Rotational diffusion of rhodopsin in the visual receptor membrane. Nature (New Biol.) *236*, 39–43 (1972)
Craker, L.E., Abeles, F.B., Shropshire, Jr.,W.: Light-induced ethylene production in *Sorghum*. Plant Physiol. *51*, 1082–1083 (1973)
Curry, G.M., Gruen, H.E.: Negative Phototropism of *Phycomyces* in the Ultra-violet. Nature (London) *179*, 1028–1029 (1957)
Darwin, C.: On the Origin of Species. 1859: A Facsimilie of the first edition. Introduction by Ernst Mayr. Boston: Harvard Univ. Press, 1975
Darwin, C.: The Power of Movement in Plants. New York: D. Appleton and Company. 1896, p. 194
Davis, H.: Excitation of auditory receptors. In: Handbook of Physiology. Neurophysiology Field, J., Hall, J.E., Magoun, H.W. (eds.). Washington: American Physiological Society 1959, Sec. 1, Vol. I, pp. 565–584
DeGreef, J.A., Caubergs, R., Verbelen, J.P., Moereels, E.: Phytochrome-mediated inter-organ dependence and rapid transmission of the light stimulus. In: Light and Plant Development. Smith, H. (ed.). London: Butterworth, 1976. pp. 295–316
Delbrück, M.: Signal transducers: Terra incognita of molecular biology. Angew. Chem. Int. Ed. *11*, 1–6 (1972)
Delbrück, M., Katzir, A., Presti, D.: Responses of *Phycomyces* indicating optical excitation of the lowest triplet state of riboflavin. Proc. Natl. Acad. Sci. USA. *73*, 1969–1973 (1976)
Dennison, D.S., Foster, K.W.: Intracellular rotation and the phototropic response of *Phycomyces*. Biophys. J. *18*, 103–123 (1977)
Dennison, D.S., Roth, C.C.: *Phycomyces* Sporangiophores: fungal stretch receptors. Science *156*, 1386–1388 (1967)
Duke, S.O., Egley, G.H., Reger, B.J.: Model for variable light sensitivity in imbibed dark-dormant seeds. Plant Physiol. *59*, 244–249 (1977)
Elliott. W.M., Shen-Miller, J.: Similarity in dose responses, action spectra, and red light responses between phototropism and photoinhibition of growth. Photochem. Photobiol. *23*, 195–199 (1976)
Engelmann, T.W.: Über Licht- und Farbenperception niederster Organismen. Pflüger's Arch. Physiol. *29*, 387–400 (1882)
Etzold, H.: Der Polarotropismus und Phototropismus der Chloronemen von *Dryopteris Filix-mas* (L.) Schott. Planta *64*, 254–280 (1965)
Feinleib, M.E.: Phototactic response of *Chlamydomonas* to flashes of light. I. Response of cell populations. Photochem. Photobiol. *21*, 351–354 (1975)
Feinleib, M.E., Curry, G.M.: The Nature of the Photoreceptor in Phototaxis. In: Handbook of Sensory Physiology Vol. I: Principles of Receptor Physiology. Loewenstein, W.R. (ed.). Berlin-Heidelberg-New York: Springer 1971, Ch. 13, pp. 366–395
Fisher, K.D., Carr, C.J., Huff, J.E., Huber, T.E.: Dark adaptation and night vision. Fed. Proc. Am. Soc. Exp. Biol. *29*, 1605–1638 (1970)
Fitting, H.: Lichtperzeption und phototropische Empfindlichkeit, zugleich ein Beitrag zur Lehre vom Etiolement. Jahrb. Wiss. Bot. *45*, 83–136 (1908)
Foster, K.W.: Phototropism of coprophilous zygomycetes. Annu. Rev. Biophys. Bioeng. *6*, 419–443 (1977)
Gallagher, K.T.: The philosophy of knowledge, Chapter 4, The Problem of Perception. New York: Sheed and Ward, Inc., 1964. pp. 68–127
Galston, A.W., Britz, S.J., Briggs, W.R.: Photoinduced spectral changes in subcellular fractions from etiolated peas. Plant Physiol. *59*, Suppl. 48 (Abst. 266) (1977)
Gamow, R.I., Harris, J.F.: The infrared receptors of snakes. Sci. Am. *228*, 94–100 (1973)
Gardner, G., Shaw, S., Wilkins, M.B.: IAA transport during the phototropic reponses of intact *Zea* and *Avena* coleoptiles. Planta *121*, 237–251 (1974)
Gibson, J.J.: The Senses Considered as Perceptual Systems. Boston: Houghton Mifflin, 1966

Green, D.M., Swets, J.A.: Signal Detection Theory and Psychophysics. New York: Wiley, 1966
Gualtierotti, T.: The Gravity Sensing Mechanism of the Inner Ear. In: Gravity and the Organism. Chicago: The University of Chicago Press, 1971, pp. 263–281
Hardy, J.D., Oppel, T.W.: Studies in temperature sensation. III. The sensitivity of the body to heat and the spatial summation of the end organ responses. J. Clin. Invest. *16*, 533–540 (1937)
Hartmann, K.M., Cohnen-Unser, I.: Analytical action spectroscopy with living systems: photochemical aspects and attenuance. Ber. Dtsch. Bot. Ges. *85*, 481–551 (1974)
Haupt, W.: Die Phototaxis der Algen. In: Handbuch der Pflanzenphysiologie Band XVII/1. Berlin-Heidelberg-New York: Springer, 1959, pp. 318–370
Haupt, W.: Die Orientierung der Phytochrom-Moleküle in der Mougeotiazelle: Ein neues Modell zur Deutung der experimentellen Befunde. Z. Pflanzenphysiol. *58*, 331–346 (1968)
Haupt, W., Mörtel, G., Winkelnkemper, I.: Demonstration of different dichroic orientation of phytochrome P_R and P_{FR}. Planta *88*, 183–186 (1969)
Hecht, S., Shlaer, S., Pirenne, M.H.: Energy, quanta, and vision. J. Gen. Physiol. *25*, 819–840 (1942)
Hsiao, A.I., Vidaver, W.: Seed water content in relation to phytochrome-mediated germination of lettuce seeds (*Lactuca sativa* L. Var. Grand Rapids). Can. J. Bot. *49*, 111–115 (1971)
Iversen, T.-H.: Elimination of geotropic responsiveness in roots of Cress (*Lepidium sativum*) by removal of statolith starch. Physiol. Plant. *22*, 1251–1262 (1965)
Jaffe, M.J.: Experimental separation of sensory and motor functions in pea tendrils. Science *195*, 191–192 (1977)
Jaffe, M.J., Galston, A.W.: Physiological studies on pea tendrils. II. The role of light and ATP in contact coiling. Plant Physiol. *41*, 1152–1158 (1966)
Jaffe, M.J., Galston, A.W.: Physiological studies on pea tendrils. III. ATPase activity and contractility associated with coiling. Plant Physiol. *42*, 845–847 (1967)
Jesaitis, A.J.: Linear dichroism and orientation of the *Phycomyces* photopigment. J. Gen. Physiol. *63*, 1–21 (1974)
Johnson, A.: Investigations of the reciprocity rule by means of geotropic and geoelectric measurements. Physiol. Plant *18*, 945–967 (1965)
Juniper, B.E.: Geotropism. Annu. Rev. Plant Physiol. *27*, 385–406 (1976)
Kang, B.G., Burg, S.P.: Red light enhancement of the phototropic response of etiolated pea stems. Plant Physiol. *53*, 445–448 (1974)
Koshland, D.E., Jr.: A response regulator model in a simple sensory system. Science *196*, 1055–1063 (1977)
Lafay, J.F., Matricon, J., Bodéré, C.: The avoidance response of *Phycomyces*: distance dependance of the response. Physiol. Veg. *13*, 259–263 (1975)
Lipson, E.D.: White noise analysis of *Phycomyces* Light growth response system. I, II, and III. Biophys. J. *15*, 989–1045 (1975)
Lipson, E.D., Presti, D.: Light-induced absorbance changes in *Phycomyces* photomutants. Photochem. Photobiol. *25*, 203–208 (1977)
Makous, W.L.: Cutaneous color sensitivity: explanation and demonstration. Psych. Rev. *73*, 280–294 (1966)
Marmé, D., Schäfer, E.: On the localization and orientation of phytochrome molecules in corn coleoptiles (*Zea mays* L.) Z. Pflanzenphysiol. *67*, 192–194 (1972)
Mayer, W., Moser, I., Bünning, E.: Die Epidermis als Ort der Lichtperzeption für circadiane Laubblattbewegungen und photoperiodische Induktionen. Z. Pflanzenphysiol. *70*, 66–73 (1973)
McBurney, D.H., Collings, V.B.: Introduction to Sensation/Perception. New Jersey: Prentice Hall, 1977, 296 pp
McKeon, R.P.: Introduction to Aristotle. Chicago: University of Chicago Press, 1941
Meyer, A.M.: Versuche zur 1. positiven und zur negativen phototropischen Krümmung der *Avena*koleoptile. 1. Lichtperception und Absorptionsgradient. Z. Pflanzenphysiol. *60*, 418–433 (1969a)
Meyer, A.M.: Versuche zur 1. positiven und zur negativen phototropischen Krümmung der *Avena*koleoptile: II. Die Inversion durch Paraffinöl. Z. Pflanzenphysiol. *61*, 129–134 (1969b)
Meyer, B.S., Anderson, D.B.: Plant Physiology. Toronto: D. Van Nostrand Co., Inc., 1952, p. 747

Morgan, P.W., Gausman, H.W.: Effects of ethylene on auxin transport. Plant Physiol. *41*, 45–52 (1966)

Nebel, B.J.: Responses of moss protonemata to red and far-red polarized light: evidence for disc-shaped phytochrome photoreceptors. Planta *87*, 170–179 (1969)

Ootaki, T., Kinno, T., Yoshida, K.: Complementation between *Phycomyces* mutants of mating type (+) with abnormal phototropism. Mol. Gen. Genet. *152*, 245–251 (1977)

Pak, W.L., Pinto, L.H.: Genetic approach to the study of the nervous system. Annu. Rev. Biophys. Bioeng. *5*, 397–448 (1976)

Perbal, G., Perbal, P.: La perception géotropique dans la coiffe des racines de Lentille. Physiol. Plant. *37*, 42–48 (1976)

Poff, K.L., Butler, W.L.: Absorbance changes induced by blue light in *Phycomyces blakesleeanus* and *Dictyostelium discoideum*. Nature (London) *248*, 799–801 (1974)

Poff, K.L., Skokut, M.: Thermotaxis by pseudoplasmodia of *Dictyostelium discoideum*. Proc. Natl. Acad. Sci. USA. *74*, 2007–2010 (1977)

Pollard, E.C.: Physical determinants of receptor mechanisms. In: Gravity and the Organism. Gordon, S.A., Cohen, M.J. (eds.). Chicago: University of Chicago Press, 1971 pp. 25–34

Poo, M., Cone, R.A.: Lateral diffusion of rhodopsin in the photoreceptor membrane. Nature (London) *247*, 438–441 (1974)

Pratt, L.H., Coleman, R.A., Mackenzie, J.M., Jr.: Immunological visualization of phytochrome. In: Light and Plant Development, Ch. 7. Smith, H. (ed.). London: Butterworth 1976. pp. 75–94

Presti, D., Hsu, W.-J., Delbrück, M.: Phototropism in *Phycomyces* mutants lacking β-carotene. Photochem. Photobiol. *26*, 403–405 (1977)

Raven, C.W., Shropshire, W.; Jr.: Photoregulation of logarithmic fluence-response curves for phytochrome control of chlorophyll formation in *Pisum sativum* L. Photochem. Photobiol. *21*, 423–429 (1975)

Rosencwaig, A.: Photoacoustic spectroscopy of solids. Opt. Commun. *7*, 305–308 (1973a)

Rosencwaig, A.: Photoacoustic spectroscopy of biological materials. Science *181*, 657–658 (1973b)

Russo, V.E.A.: Ethylene-induced growth in *Phycomyces* mutants abnormal for autochemotropism. J. Bacteriol. *130*, 548–551 (1977)

Schachar, R.A.: The "pincushion grid" illusion. Science *192*, 389–390 (1976)

Schletz, K.: Phototaxis bei *Volvox*-Pigmentsysteme der Lichtrichtungsperzeption. Z. Pflanzenphysiol. *77*, 189–211 (1975)

Schmidt, W., Hart, J., Filner, P., Poff, K.L.: Specific inhibition of phototropism in corn seedlings. Plant Physiol. *60*, 736–738 (1977)

Scholz, A.: Lichtorientierte Chloroplastenbewegung bei *Hormidium flaccidum*: Perception der Lichtrichtung mittels Sammellinseneffekt. Z. Pflanzenphysiol. *77*, 406–421 (1976a)

Scholz, A.: Lichtorientierte Chloroplastenbewegung bei *Hormidium flaccidum*: Verschiedene Methoden der Lichtrichtungsperception und die wirksamen Pigmente. Z. Pflanzenphysiol. *77*, 422–436 (1976b)

Scott, T.K., Wilkins, M.B.: Auxin transport in roots. IV. Effects of light on IAA movement and geotropic responsiveness in *Zea* roots. Planta *87*, 249–258 (1969)

Seitz, K.: Lichtabhängige Orientierungsbewegungen und ihre Regelung. Ber. Dtsch. Bot. Ges. *87*, 195–206 (1974)

Shen-Miller, J.: Reciprocity in the activation of geotropism of oat coleoptiles grown on clinostats. Planta *92*, 152–163 (1970)

Shropshire, W., Jr.: The lens effect and phototropism of *Phycomyces*. J. Gen. Physiol. *45*, 949–958 (1962)

Shropshire, W., Jr.: Action Spectroscopy. Chapter 8 In: Phytochrome. Mitrakos, K., Shropshire, W., Jr. (eds.). London: Academic Press, 1972, pp. 161–181

Shropshire, W., Jr.: Phototropism. In: Progress in Photobiology, Proc. 6th. Int. Cong. Photobiol. Dtsch. Gesellschaft für Lichtforschung e.V., Frankfurt 1974, Abst. 024 Schenck, G.O. (ed.). Berlin-Heidelberg-New York: Springer, 1975, pp. 1–5

Shropshire, W., Jr., Mohr, H.: Gradient formation of anthocyanin in seedlings of *Fagopyrum* and *Sinapis* unilaterally exposed to red and far-red light. Photochem. Photobiol. *12*, 145–149 (1970)

Shuttleworth, J.E., Black, M.: Role of cotyledons in phototropism of de-etiolated seedlings. Planta *135*, 51–55 (1977)

Sneddon, I.N., Ed.: Encyclopaedic Dictionary of Mathematics for Engineers and Applied Scientists. Oxford: Pergamon Press, 1976, 800 pp

Stevens, S.S.: Sensory Power Functions and Neural Events, Chapter 7. In: Handbook of Sensory Physiology, Vol. I. Lowenstein, W.R. (ed.). Berlin-Heidelberg-New York: Springer 1971, Ch. 1

Steyer, B.: Der Phototropismus dicotyler Keimpflanzen (Sammelreferat). Wiss. Z. Univ. Rostock *14*, 493–502 (1965)

Steyer, B.: Die Dosis-Wirkungsrelationen bei geotroper und phototroper Reizung: Vergleich von Mono- mit Dicotyledonen. Planta *77*, 277–286 (1967)

Stiles, W., Cocking, E.C.: An Introduction to the Principles of Plant Physiology, Chap. 23. 3rd Ed. London: Methuen & Co. Ltd., 1969, pp. 498–524

Swets, J.A.: Is there a sensory threshold? Science *134*, 168–177 (1961)

Taylorson, R.B., Hendricks, S.B.: Changes in phytochrome expressed by germination of *Amaranthus retroflexus* L. seeds. Plant Physiol. *47*, 619–622 (1971)

Tobin, E.M., Briggs, W.R., Brown, P.K.: The role of hydration in the phototransformation of phytochrome. Photochem. Photobiol. *18*, 497–503 (1973)

Tompkins, P., Bird, C.: The Secret Life of Plants. New York: Harper-Row 1973

Vidaver, W., Hsiao, A.I.-H.: Persistence of phytochrome-mediated germination control in lettuce seeds for 1 year following a single monochromatic light flash. Can. J. Bot. *50*, 687–689 (1972)

Wagner, E.: Kinetics in metabolic control of time measurement in photoperiodism. J. Interdiscip. Cycle Res. *7*, 313–332 (1976)

Webster's New World Dictionary of the American Language. Cleveland, Ohio: The World Publishing Co., College Edition, 1960

Wilkins, M.B.: Gravity-sensing guidance systems in plants. Sci. Prog. (Oxford) *63*, 187–217 (1976)

Wright, W.C.: A Short History of Greek Literature: From Homer to Julian. New York: R. West Co., 1907

1.3 Reception and Transduction of Electrical and Mechanical Stimuli

F.W. BENTRUP

1. Introduction

Like animals, plants are able to perceive and respond to environmental stimuli. Unlike metazoan animals, however, plants have not developed a specialized sensory apparatus equipped with a standard type of receptor *cell* and a nervous network to process stimuli. Instead, they employ a diverse array of receptive structures–including, in a few cases, a receptor cell. In spite of this, the causal chain of a plant movement can be analyzed in terms of the three indispensible steps known from general sensory physiology:

Stimulus reception→stimulus transduction→physiological response

A. *Stimulus reception* (or *susception*) denotes the stimulus-specific physical or physicochemical interaction of the stimulus and the living matter: a *sensor* is activated by the physical or chemical energy of the *stimulus* (or *signal*). Reception may occur independently of metabolic activity and show a Q_{10} near 1. (Some authors refer to the activation or alteration of the receptor by the stimulus as *stimulus perception;* see SHROPSHIRE, Chap. 1.2, this vol.)

B. *Stimulus transduction* concerns strictly the control of a specific cellular (biophysical, biochemical) process by the activated sensor (or *sensory transducer, Reizwandler*). Frequently the term transduction is broadly used to cover the complete sequence of sensory processes connecting stimulus reception to the physiological response.

This article deals with transduction in this latter, broader sense; that is, besides sensory transducer mechanisms, it includes signal transmission and dose-response relationships.

C. The *physiological response* is the initiated or modified plant movement or developmental process. See HAUPT and FEINLEIB, Chapter 1.1, this volume. (The term response will be used also to refer to intermediate steps, e.g., to a receptor response.)

This article opens with a section (Sect. 2) devoted to current concepts of plant cell electrophysiology, because the sensory physiology of many plant movements includes bioelectric phenomena. In Section 3, electrical control of cell morphogenesis will be discussed briefly; for an extensive treatment of this topic see WEISENSEEL, Chapter 5.1.1, this volume. Section 4 treats plant movements elicited by mechanical stimuli. For this purpose, the rapid movements of carnivorous plants are reviewed in detail. In Section 5, a comparative sensory physiology for the major stimuli triggering plant movements is attempted.

In the 1959 Encyclopedia, Vol. XVII, the topic of this article was treated in several chapters by BÜNNING and by UMRATH. Both authors still are keen observers of the scene, and I wish to dedicate this review to them.

2. Current Concepts of Plant Cell Electrophysiology

2.1 Symbols, Units, and Abbreviations

E_m	transmembrane electrical potential difference (mV)
E_{diff}	electromotive force (e.m.f.) of the diffusion pathway of the membrane (mV)
E_H etc.	Nernst diffusion potential difference for H^+ etc. (mV)
E_p	e.m.f. of the electrogenic pump (mV)
F	Faraday constant (96490 C mol^{-1})
g_m	electrical membrane (slope) conductance (Ω^{-1} cm^{-2}=S cm^{-2})
g_{diff}	electrical conductance of the diffusion pathway (S cm^{-2})
g_H etc.	partial conductance due to the diffusion of H^+ etc. (S cm^{-2})
g_p	electrical conductance through the electrogenic pump pathway (S cm^{-2})
i_p	density of the electrical current through the pump pathway (A cm^{-2})
PD	electrical potential difference (mV)
P_H etc.	membrane permeability to H^+ etc. (cm s^{-1})
R	gas constant (8.314 J mol^{-1} K^{-1})
T	absolute temperature (K)
RT/F	is equal to 25.7 mV at 25° C
subscripts:	i, inside; o, outside; c, cytoplasmic; v, vacuolar

2.2 Membrane Potential and Conductance

In plant cells or tissues, electrical potential differences (PD's) are mostly generated by membranes, a notable exception being the Donnan potential (see below). A membrane potential (E_m) is assumed to arise from (1) passive ion diffusion in parallel with (2) an ATP-driven electrogenic ion pump, probably proton export at the plasmalemma (cf. BENTRUP, 1975).

1. Transmembrane diffusion of H^+, K^+, Na^+, and Cl^- governs the diffusion potential, E_{diff}; E_{diff} across the plasmalemma, E_{co}, is given by (GOLDMAN, 1943):

$$E_{diff}=E_{co}=-\frac{RT}{F}\ln\frac{P_H[H^+]_c+P_K[K^+]_c+P_{Na}[Na^+]_c+P_{Cl}[Cl^-]_o}{P_H[H^+]_o+P_K[K^+]_o+P_{Na}[Na^+]_o+P_{Cl}[Cl^-]_c} \tag{1}$$

R, T, and F denote gas constant, absolute temperature and Faraday constant, respectively, and the subscripts refer to the cytoplasmic (c) and the outside compartment (o). For the definition of the permeability coefficients, P_H etc., and the assumptions and limitations pertinent to this *constant-field* equation see, for instance, WALKER (1976) Vol. 2 in this Series. (Actually, the assumption of a constant electric field across the membrane is probably incorrect.) Equation (1) reduces to the Nernst equation, if the membrane is predominantly permeable to only one ion, say, the K^+ ion:

$$E_{co} = E_K = -(RT/F) \cdot \ln([K^+]_c/[K^+]_o). \quad (1a)$$

Note that E_{diff} [Eq. (1)] is governed by terms of permeability times concentration. Thus, even if $P_H > P_K$, the product $P_K[K^+]$ usually exceeds $P_H[H^+]$, because $[K^+]$ exceeds $[H^+]$ by several orders of magnitude in most cells, e.g., 100 mM K^+ vs. 0.001 mM H^+ in a vacuole. Therefore, under most physiological conditions, K^+ diffusion governs the passive electrical membrane properties.

In practice, relative permeabilities can be obtained from Eq. (1) by measuring the membrane potential upon changes of the external ion concentrations.

The ratio $P_{Na}:P_K$ ranges from 0.01 to 0.001 in marine algae, and from 0.2 to 0.02 in freshwater plants (cf. FINDLAY and HOPE, 1976). The plasmalemma of *Avena* coleoptile cells exhibits a $P_{Na}:P_{Cl}:P_K$ ratio of 0.68:0.037:1.00 (PIERCE and HIGINBOTHAM, 1970). Barley root cells have a $P_{Na}:P_K:P_H$ ratio of 0.21:1.00:20.6 (PITMAN et al., 1977). Generally, the plant cell membrane seems to follow the pattern $P_H > P_K > P_{Na} > P_{Cl}$.

2. In recent years it has been well established that in plant cell plasmamembranes an ATP-fueled electrogenic mechanism ("pump") operates in parallel with the diffusion pathway. This apparently ubiquitous electrogenic pump probably exports protons across the plasmalemma to the outside medium (a comparable mechanism is needed to transport protons across the tonoplast into the vacuole).

The equilibrium potential *(electromotive force)* of the proton pump, E_p, follows from the equation for the *protonmotive force*, p.m.f., across a membrane (in volts):

$$\text{p.m.f.} = \Delta\tilde{\mu}_H/F = E_p + E_H, \quad (2)$$

where E_H is defined in analogy to E_K by Eq. (1a).

The p.m.f. is the electrochemical equivalent of the free energy gradient available from ATP-hydrolysis when n protons are transported per ATP-molecule split, that is,

$$\Delta\tilde{\mu}_H/F = -\Delta G_{ATP}/n \cdot F.$$

The *protonmotive force* is known as the key parameter in the chemiosmotic hypothesis of energy conservation in organellar and bacterial membranes (see MITCHELL, 1966). It generates ATP by driving a proton current through a membrane-bound ATPase. This is, at least formally, equivalent to the plasmalemma proton pump running backwards. No net current flows through the pump sites when the membrane potential is equal to the pump equilibrium potential (E_p). At the plasmalemma, the ATPase pumps protons from the cell, because generally E_{co} is smaller (less negative) than $E_p + E_H$. The proton current which is driven back into the cell by the p.m.f., may do work. An example for this "secondary active transport" in plant cells is the proton-sugar cotransport where a proton and a glucose molecule bind to and are transported by a common carrier. For details of cotransport in plants see SLAYMAN (1974), and BENTRUP (1975, 1978).

Using $n=1$, E_p values of -300 to -350 mV have been obtained from *Neurospora* (SLAYMAN et al., 1973) and -360 mV from *Riccia* (FELLE, unpublished data).

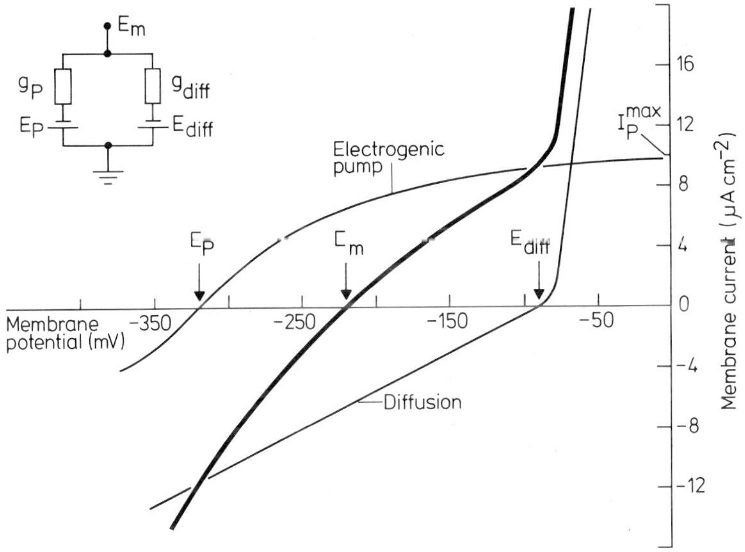

Fig. 1. Schematic current–voltage (*I–V*) relationship of a plant cell membrane (*bold line*). The *I–V* curve was obtained by adding, at each membrane potential, the component *I–V* curves for an electrogenic pump and a passive diffusion pathway. Note that the slopes of the curves represent conductances (g_p, g_{diff}, and g_m). The diffusion pathway mainly reflects K^+ diffusion, hence $E_{diff} \approx E_K$, and $g_{diff} \approx g_K$. *Ordinate*, Values are positive for outward membrane current

In fact, membrane potentials up to -300 mV have been found (see Sect. 4.1.3). Under certain conditions, the pump seems to operate like a constant current source (low conductance) with maximal membrane current densities i_p, of about 10 $\mu A/cm^2$. This range has been derived from current–voltage relationships of *Neurospora, Acetabularia* (Gradmann, 1975) and *Riccia* (Felle and Bentrup, 1976). From these plants an imaginary *I–V* relationship for E_m and its components has been reconstructed in Figure 1, using the equivalent circuit given there. The circuit yields (cf. Bentrup, 1975):

$$E_m = E_{diff} + i_p/g_{diff}, \tag{3}$$

where g_{diff} is the electrical conductance through the diffusion pathway of the membrane. Equation (3) implies (1) that in the absence of a pump current i_p, i.e., when the ATP synthesis is blocked, the membrane potential E_m should approach E_{diff}. Experiments of many authors confirm this (for references see Findlay and Hope, 1976). Furthermore, when $E_{diff} \approx E_K$, the diode-like *I–V* curve for the diffusion pathway (Fig. 1) in fact reflects g_K, the conductance through the K^+ channels. (2) The circuit in Figure 1 and Eq. (3) show, on the other hand, that E_m will shift toward E_p, if the pump conductance g_p becomes large compared with g_{diff} (note that $g_m = g_p + g_{diff}$).

Evidently, a reliable analysis of any E_m or changes thereof requires evaluation of Eqs. (1) to (3) and must include *I–V* measurements across a wide range of E_m, because the current–voltage relationships are nonlinear (Fig. 1). Ordinary resistance measurements cause a shift of only a few millivolts in the E_m, so that

$g(V) = I/V$ can be taken as constant. Wide-range I–V analysis seems at present not feasible for cells of higher plant tissue, because two electrodes (voltage, current) have to be inserted into a single cell, and because the voltage response of a cell to an injected current not only is nonlinear due to the nonlinear I–V characteristics of the membrane (Fig. 1), but varies also in a nonlinear fashion with the distance between the two electrodes. This is regrettable, because the identification or change of an E_m of higher plant cells could present valuable in vivo evidence for or against the hypothesis that a certain membrane pathway is controlled by, say, phytochrome or auxin. In the following, examples and possible sources of an observed change of E_m are briefly outlined.

Changes of Membrane Potential. In association with plant movements and morphogenesis there occur changes of membrane potential in time, ΔE_m, and membrane potential gradients (bioelectric fields). Several types of a ΔE_m may be discerned:

1. Propagated all-or-none *action potentials* (AP's) from excitable membranes (see Sect. 2.3).
2. Propagated *variation potentials* which are linked to the transpirational stream (Sect. 4.1).
3. Nonpropagated graded *receptor potentials* from nonexcitable membranes of sensory cells (p. 51).
4. A light-induced ΔE_m (quite common in photosynthetically active cells) generated at the plasmalemma (see review by BENTRUP, 1974a). A phytochrome-dependent ΔE_m has been reported by RACUSEN and SATTLER (1975) and by RACUSEN (1976). Its subcellular origin still is a matter of debate (see SATTER, Chap. 4.2, this vol.).
5. A hormone-induced ΔE_m presumably of the plasmalemma. Such changes have been found by several authors (see the book edited by MARRÉ and CIFERRI, 1977). A transient depolarization and a subsequent hyperpolarization upon addition of indoleacetic acid (IAA) has been well established, as well as a hyperpolarization by fusicoccin, the major phytotoxin of the fungus *Fusicoccum amygdali*.

   Except for the AP's, the ionic mechanism, or even the subcellular origin of these cases of a ΔE_m remains more or less obscure. In the range of a millisecond or longer, a ΔE_m would be influenced, in terms of Eq. (1), only by changes of the permeabilities, P_H etc., hence by g_{diff}, because changes of concentration cannot occur so rapidly. In the second-to-minute range, changes of $[H^+]$ in a bulk phase (cytoplasm etc.) may also occur. Changes of the other ion concentrations, and of the e.m.f. and i_p of the electrogenic pump may add to an observed ΔE_m in a still longer range of time.

For instance, the above-quoted action of IAA upon E_m may serve to illustrate the difficulty in assessing a ΔE_m. On the one hand, IAA is assumed to activate the electrogenic proton pump (HAGER et al., 1971); this would explain that addition of IAA hyperpolarizes a membrane, say, via i_p in Eq. (3). On the other hand, the transmembrane diffusion hypothesis of auxin movement recently revived by RUBERY and SHELDRAKE (1974) and RAVEN (1975) implies that a passive auxin flux equilibrium exists across the plasmalemma (see also the review by GOLDSMITH, 1977). Accordingly, transmembrane fluxes of the weak acid IAAH and its ion, IAA^-, are governed by E_m, the pH gradient (in fact, E_H), the auxin permeability ratio, $P_{IAA}:P_{IAAH}$, and the concentrations of IAA^- and IAAH on both sides of the membrane. The observed initial depolarization after application of auxin could be explained, for instance, by the assumption that the lipid-soluble IAAH operates as a protonophore, viz. by increasing P_H of the plasmalemma.

Similar results have been obtained with the uncoupler carbonyl cyanide m-chlorophenylhydrazone (CCCP), which is also a lipid-soluble weak acid (pK 5.95; IAAH:pK 4.7). This substance has been shown to depolarize the plasmalemma of *Riccia fluitans*, presumably by a 100-fold increase of P_H in Eq. (1) (FELLE and BENTRUP, 1977).

Furthermore, voltage effects outside the membrane must be considered, if a virtual ΔE_m is measured between the vacuole and the outside solution. The fixed-charge lattice of the cell wall may be source of a marked PD. The higher plant cell wall is a cation exchanger (pK ≈ 3) with fixed carboxylic groups of a concentration of 300–600 mM (WALKER and PITMAN, 1976). This Donnan free space of the wall may create a Donnan potential of -48 mV when the cell is bathed in a solution of 1 mM KCl and 1 mM $CaCl_2$. The Donnan potential critically depends upon the concentration of divalent cations in the solution; see Table 5.2 in WALKER and PITMANN (1976).

Another source of ambiguity is the existence of unstirred layers at the membrane–solution interfaces; if depletion or enhancement of particular ions has occurred between plasmalemma and cell wall, due to different transference numbers, these layers could contain ion concentrations very different from those of the bulk solution (cf. BARRY and HOPE, 1969).

2.3 Action Potentials

Although action potentials in the strict sense (AP's) have been observed throughout the plant kingdom, the underlying ionic mechanism has been identified to some extent only in algal cells, namely, the Characeae and *Acetabularia*. At least two types of AP have evolved in plants (Fig. 2).

1. In the internode cell of a characean, say, *Nitella*, the AP is caused by a transient conductance increase in the diffusion regime, probably by a transient surge of P_{Cl}. The vacuolar potential, E_{vo} temporarily swings toward E_{Cl} [see Eq. (1)]; specifically, in *Chara corallina* E_{co} (plasmalemma) may decrease from -160 to -40 mV and E_{vc} (tonoplast) may rise from $+15$ to $+60$ mV (see FINDLAY and HOPE, 1976; KIKUYAMA and TAZAWA, 1976). The depolarizing Cl^--efflux is balanced by a rapidly increasing K^+-efflux (φ_K) when E_m drops below E_K (i.e., attains values more positive than E_{diff} in Fig. 1); φ_K increases according to $\varphi_K = g_K(E_m - E_K)/F$, because both variables, g_K and the difference $E_m - E_K$, increase during depolarization. (Note that this voltage-sensitive change of a passive K^+-flux does not infer a change of the K^+-permeability!) *Chara corallina* loses about 1.8 pM KCl per cm^2 cell surface during an AP (ODA, 1976).
2. The "metabolic" AP of *Acetabularia* (GRADMANN, 1976) and *Neurospora* (SLAYMAN et al., 1976) is characterized by an overall decrease in g_m. As illustrated in the equivalent circuit for *Acetabularia* in Figure 2, no change in P_{Cl} or in P_K has been found. Instead, the electrogenic Cl^- import features two voltage-sensitive nonlinear conductances in series with E_p. One of these conductance elements is linked to the ATP pool of the cell and responds to changes thereof, i.e., occurring after light-off or in low temperature. This AP presumably regulates the intracellular K^+ concentration (MUMMERT and GRADMANN, 1976). It regularly appears during cell growth, i.e., stalk regeneration (NOVAK and BENTRUP,

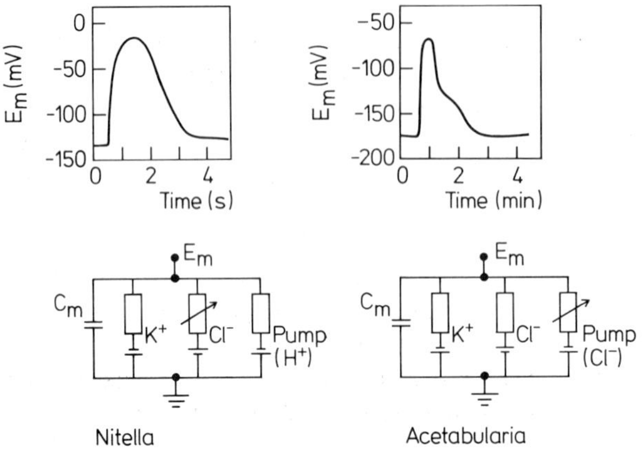

Fig. 2. Time course (*top*) of the action potential and equivalent circuit (*bottom*) for the membrane potential of the excitable membranes of *Nitella* and *Acetabularia*. *Arrows* denote the presumably excitable ion channel. See text

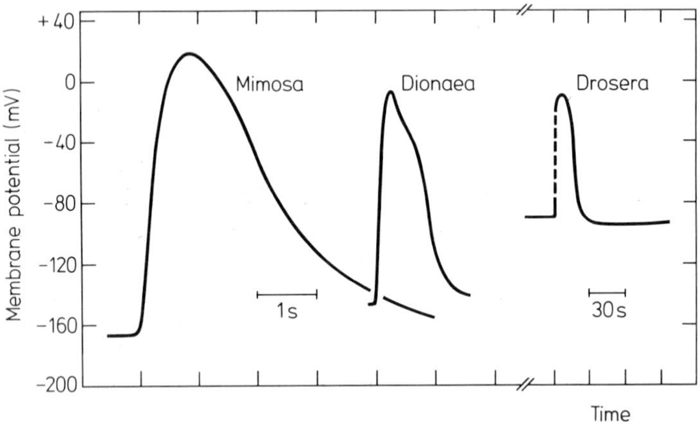

Fig. 3. Action potentials recorded intracellularly from a protoxylem cell in the petiole of *Mimosa pudica*, from a leaf cell of the trap of *Dionaea muscipula* (redrawn from SIBAOKA, 1966), and from an outer stalk cell of a tentacle of *Drosera rotundifolia*. (After WILLIAMS and SPANSWICK, 1976)

1972), and might serve to reduce the turgor by expulsion of KCl (see NUCCITELLI and JAFFE, 1976).

Two different AP's at the tonoplast, actually of inverted polarity with respect to the plasmalemma (vacuole is "outside") have been recorded from the marine dinoflagellate *Noctiluca miliaris*. One AP arises at the tentacle base and controls its movement (ECKERT and SIBAOKA, 1967; SIBAOKA and ECKERT, 1967); the other AP triggers the bioluminescence flash (ECKERT and SIBAOKA, 1968).

Higher plant AP's of major interest are shown in Figure 3. Obviously, large differences in amplitude and time course exist. Interestingly, in both *Mimosa* and *Dionaea*, excitable cells have a more negative resting potential than adjacent non-excitable cells. Average values are −154 to −161 mV in the protoxylem and phloem cells vs -50 mV in the other cells of the petiole of *Mimosa* (SIBAOKA, 1966), and -80 mV in the large sensory cells vs. -30 mV in the other cells of the sensory layer of the *Dionaea* trigger hair (BENOLKEN and JACOBSON, 1970; see Sect. 4.1). The excitable cells of the trap tissue, on the other hand, show values of about -150 mV (Fig. 3). The ionic mechanism of the higher plant AP is poorly understood. All resting potentials down to -185 mV in *Mimosa* pulvinar cells bathed in 0.05 mM K^+ (ABE and ODA, 1976) can be accounted for by K^+ diffusion (and thus require no electrogenic component); ALLEN (1969) found 120 to 160 mM K^+ in the pulvini of *Mimosa*. The membrane potential changes by 33 mV per tenfold change of K_0^+. A 48 mV change, that is, a relatively high K^+ permeability, was found by JACOBSON (1974) in *Dionaea* sensory hair cells. [Note that a K^+-selective membrane would yield a 59 mV change according to Eq. (1a).] The circuit given for *Nitella* in Figure 2 seems at present the best available guess for the AP in *Mimosa* and *Dionaea*.

3. Reception and Transduction of Electrical Stimuli

Naturally, a cell membrane may function as an electrosensor; that is, it may sense changes of its ionic milieu according to Eqs. (1) to (3). The plasmalemma of many plant cells has been shown to sense and transduce endogeneous and applied electrical signals into localized cell growth; cf. BENTRUP (1968), JAFFE (1968), JAFFE and NUCCITELLI (1977), and WEISENSEEL, Chapter 5.1.1, this volume. Effective electrical signals are direct (d.c.) or alternating (a.c.) electric fields (cf. NOVAK and BENTRUP, 1973; PENG and JAFFE, 1976), or ion concentration gradients established across developing cells. In the latter case, only ions which substantially shift E_m according to Eq. (1) may be expected to act electrically. This seems to be true for K^+ gradients which have been shown to control polar growth of *Fucus* zygotes (BENTRUP et al., 1967) and of *Acetabularia* stalk segments (CHRIST-ADLER and BENTRUP, 1976).

From a large array of electrophysiological and ion flux experiments, especially from his laboratory, JAFFE has proposed a mechanism of electrical control of localized cell growth of the fucacean zygote. Spontaneous or imposed depolarizations locally switch the plasmalemma from its general state of high K^+ permeability to a state facilitating an increased entry of Ca^{2+} ions. Thus a cellular gradient in membrane potential is transduced into a transcytoplasmic Ca^{2+} concentration gradient. This gradient is assumed to feed back and stabilize the initial local permeability shift of the membrane by pulling negatively charged particles electrophoretically to this region of Ca^{2+} entry. Finally, this region is determined to become the zygote's rhizoid pole.

A second mechanism by which a gradient of E_m could be transduced into a cellular growth pattern has been proposed by JAFFE (1977). This "perimembrane electrophoresis" hypothesis states that an E_m gradient electrophoretically segregates diffusible membrane components; these are assumed to float within the lipid layer of the membrane, and carry a net charge on one end, which protrudes from the membrane to the outside, and hence may sense the E_m gradient.

The transduction of action potentials by motor tissue cells of sensitive plants into the seismic response, i.e., sudden loss of turgor, will be discussed along with the mechanism of that movement by SATTER, Chapter 4.2, this volume.

4. Reception and Transduction of Mechanical Stimuli

4.1 The Rapid Movements of Carnivorous Plants

Since the 1959 edition of the Encyclopedia, relatively few studies have focused upon the sensory processes of the rapid movements some plants have evolved to catch animals, but these studies have been fruitful ones. Besides the broader reviews by SIBAOKA (1969) and PICKARD (1973), the reader should not miss the special treatise by WILLIAMS (1976).

4.1.1 The Venus Flytrap

The initial fast turgor-dependent component of *Dionaea* trap closure is triggered if one of the six sensory hairs on the trap surface is bent at least twice, or if two hairs are bent once (BURDON-SANDERSON and PAGE, 1876). The AP following each stimulation was also observed by these authors. DIPALMA et al. (1966) found that trap closure could also be elicited, independent of the trigger hairs, by mechanical stimulation of stellate trichomes protruding from the marginal hairs, from the outer surface, and from the petiole of *Dionaea* leaves.

An intracellular recording from a leaf lobe cell was given in Figure 3. The most important recent result is the discovery by JACOBSON (1965) and BENOLKEN and JACOBSON (1970) that the AP is always preceded by a graded nonpropagated transient depolarization, resembling the receptor potential of an animal sensory cell. Figure 4 shows intracellular recordings from sensory cells (S, Fig. 5) of an excised *Dionaea* hair, which was compressed by a known force: A saturating stimulus of 200 mg causes the response shown in Figure 4A, that is a fast depolarization followed by a two-phased slow repolarization. The initial spike appears to be an AP, masking the rise of the receptor potential that fired the AP. A stimulus of smaller amplitude (Fig. 4B) causes a smaller receptor potential and, for some unknown reason, an AP of smaller amplitude; a stimulus of lower rate of increase causes a small receptor potential but no AP (Fig. 4C). The stimulus-receptor relationship will be discussed below (Sect. 5.3 and Fig. 10). These electrical events arise only if the sensory layer of the hair is stimulated (Fig. 5, left). MUNK (1876) had successively sliced away the hair from its distal end, and noted that this surgery triggered the trap movement only when he cut the hair near to the podium. BENOLKEN and JACOBSON (1970) successively sliced away layers of tissue one or two cells thick from the podium end of the hair, and found that the receptor potential vanished when they had removed or destroyed the sensory layer.

Hence good evidence favors the notion of HABERLANDT (1906) that these indented cells (S in Fig. 5, left) are the mechanotransducer of the *Dionaea* trap. Electron micrographs by WILLIAMS and MOZINGO (1971) reveal a special feature of these cells: the endoplasmic reticulum is arranged in a whorled fashion around a vacuolar structure, and is located in the cell's basal portion (Fig. 6, left). Prima

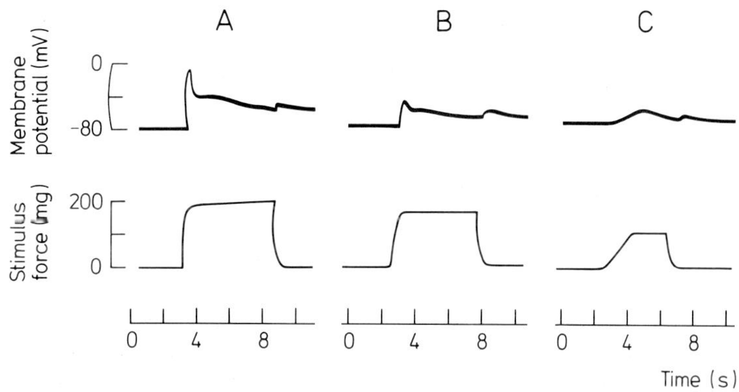

Fig. 4 A–C. Receptor potential and action potential recorded from a sensory cell of an excised trigger hair of *Dionaea* (*upper trace*) under three different stimulus conditions. A mechanical stimulus force of known rise time and amplitude was generated by, and linearly related to, the feed of a galvanometer driver, and was applied radially by a lever to the arrested sensory cell layer (see Fig. 5). A force of 100 mg exerts a pressure of about 0.1 bar upon an area of 10^{-3}cm^2 of the sensory layer. (After BENOLKEN and JACOBSON, 1970)

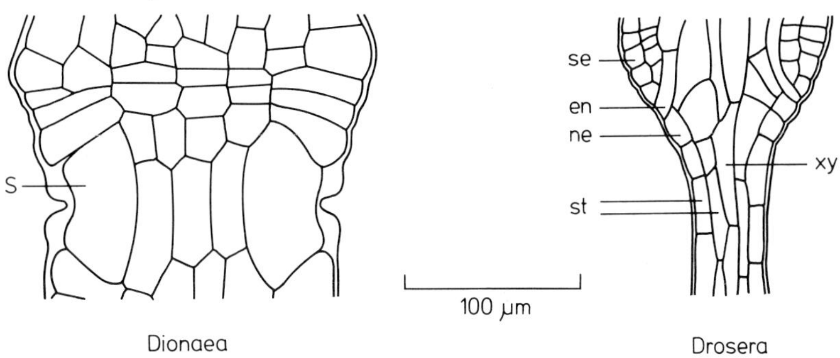

Fig. 5. Anatomy (longitudinal section) of the "joint" region of the trigger hair of *Dionaea* (*left*) and of a portion of a tentacle of *Drosera* (*right*); *S*, outer, sensory cell of the sensory layer; *se*, secretory cell; *en*, endodermoid cell; *ne*, neck cell; *st*, inner and outer stalk cells; *xy*, xylem vessel. Note that indentation of the sensory cells is due to the cell wall. (After HABERLANDT, 1906 and WILLIAMS, 1976)

facie, this points to a functional assymmetry known from animal receptor cells. In fact, BENOLKEN and JACOBSON have obtained evidence for a functional asymmetry, that is, a receptor potential gradient from the base of the cell to its apex, but they point out that the sensory cell probably becomes more deformed at the basal end near the soft podium than at the apical end adjacent to the woody hair. Therefore, it is open to question whether the cell membrane is differentiated into nonexcitable sensory and excitable AP-generating areas.

4.1.2 The Sundew Tentacle

In the genus *Drosera*, the initial rapid thigmonastic bending of a marginal secretory tentacle toward the center of the leaf also starts with a nonpropagated slow depolariza-

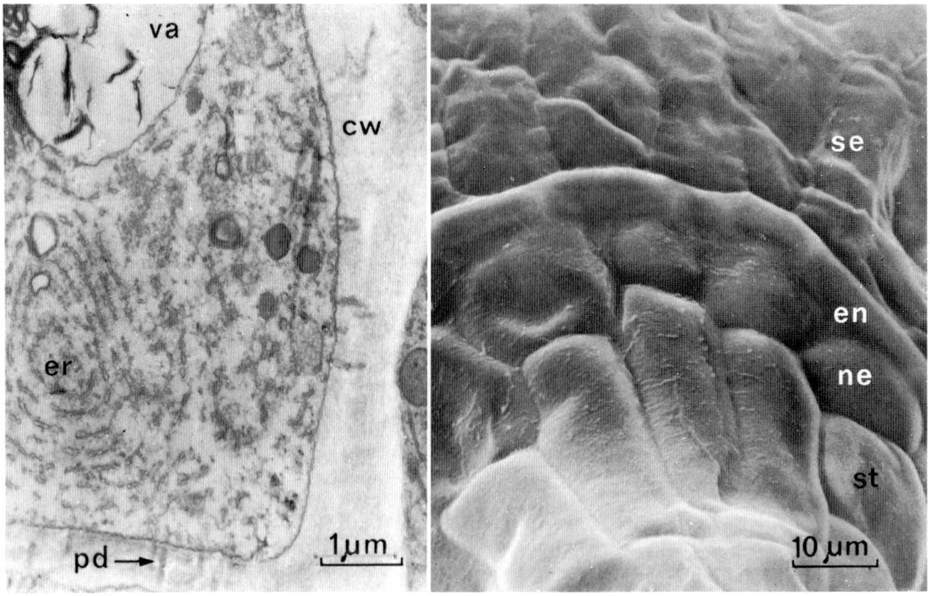

Fig. 6. *Left:* Electron micrograph of the lower part of a sensory cell (*S* in Fig. 5) of a *Dionaea* trigger hair; *cw*, cell wall; *er*, endoplasmic reticulum; *pd*, plasmodesmata, *va*, vacuole. After WILLIAMS and MOZINGO (1971). *Right:* Scanning electron microscope view of the neck and lower head area of a tentacle of *Drosera capensis*. Same symbols as in Figure 5. (After WILLIAMS and PICKARD, 1974)

tion of some cells in the tentacle head. This putative receptor potential, in turn, generates trains of AP's (WILLIAMS and PICKARD, 1972a, b). The electrical responses are illustrated for *D. intermedia* in Figure 7. In this recording, the receptor potential gains its full amplitude of 120 mV within 2 to 3 min and lasts for about 12 min. The AP's it generates appear, almost simultaneously, under the electrodes at the head and the lower stalk (Fig. 5, right). This lack of a significant time delay suggests that the sensory cells are located not far away from the stalk, and that their depolarization causes the externally measured potential of the head cells to become more negative. The neck cells (Fig. 5), which are morphologically homologous to the indented cells of the *Dionaea* hair, seem plausible candidates; in fact, a pronounced receptor potential can be recorded from the neck area (WILLIAMS, 1976), whereas the adjacent inner and outer stalk cells show AP's, but no receptor potential (WILLIAMS and SPANSWICK, 1976). If the neck cells were the only sensory cells, effective thigmic stimulation of the head would, in fact, be seismic as in *Dionaea;* namely, it would entail shearing or bending of the tentacle head relative to its stalk.

HABERLANDT (1906) ascribed the thigmic sensitivity of the head to papillate protoplasts isolated from lateral epidermal secretory cells of the head of tentacles of *D. rotundifolia* and *D. lingifolia*. However, no conspicuous sensory structures have been detected in recent studies by RAGETLI et al. (1972) and WILLIAMS and PICKARD (1974); particularly, the existence of papillate extensions of the secretory cells have not been confirmed in these studies (Fig. 6, right; see also Sect. 4.2.1).

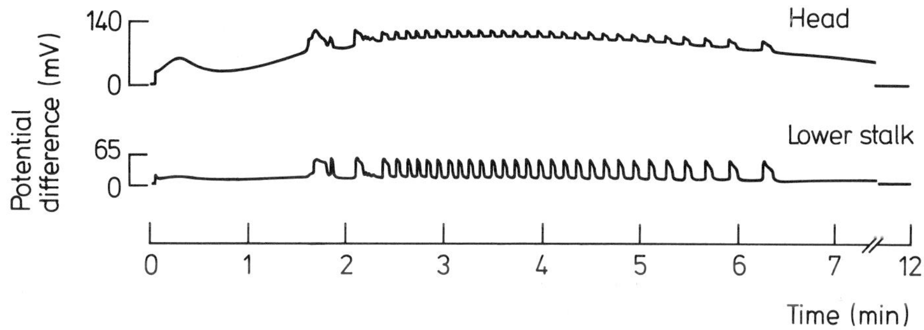

Fig. 7. Receptor potential and superimposed action potentials from the head and lower stalk of a *Drosera intermedia* tentacle after mechanical stimulation (at $t=0$) of the tentacle head surface with a wet cotton thread. Both voltages were picked up by <0.5 mm thick cotton threads, soaked with 0.1 M KCl compared against a reference electrode near the roots. The potentials were negative-going, as expected in a region of membrane depolarization, measured with external electrodes. (After WILLIAMS and PICKARD, 1972a)

4.1.3 The *Utricularia* Bladder

In the carnivorous aquatic plant *Utricularia*, triggering and resetting the bladder trap includes some amazing and ingenious details discovered by SYDENHAM and FINDLAY (1973, 1975).

Setting the Trap. SYDENHAM and FINDLAY (1975) found that the bladder is set by export of solutes, mainly Na^+, K^+, and Cl^-, and of water, from the lumen to the outside medium (Fig. 8A). During the setting process, the bladder loses 30% of its volume, and the luminal hydrostatic pressure drops by 0.12 bar, namely, from 0.05 to 0.17 bar below the atmospheric pressure of 1 bar. The salt concentration in the lumen of the bladder is always higher than in the outside medium; mean values (lumen/outside) from fired bladders are (mM): 7.8/2.0 Na^+, 3.5/0.2 K^+, and 4.7/2.7 Cl^-. Consequently, both gradients, in hydrostatic and in osmotic pressure, would drive water from the outside to the lumen. The observed outward movement of water thus occurs against the gradient of chemical potential of water.

Active water transport out of the lumen is possible either by an asymmetric "sandwich" membrane system or by coupling of the water flow to a metabolically driven ion pump in one of the membranes separating the lumen from the outside medium. The authors infer from their voltage and ion flux measurements that a respiration-dependent electrogenic Cl^- pump is located in the wall cell plasmamembranes facing the lumen. This pump transports Cl^- into the vacuole of the wall cells, hyperpolarizing the vacuole up to -300 mV. This voltage is visible as E_{vi} in the voltage profile of Figure 8B. The import of Cl^- into the wall cells is probably electrically balanced by a passive influx of Na^+. As the NaCl concentration in the wall cells rises, the chemical potential of water will decrease, and cause a water influx. This influx creates a turgor increase in the wall cells, and causes the observed drop of the hydrostatic pressure in the lumen. The involved membranes presumably have a high reflection coefficient, $\sigma \cong 1$. (Note that $\sigma = 1$ for a strictly semi-permeable membrane.) The wall cell membranes facing the outside solution, on the other hand, are assumed to have a lower reflection coeffi-

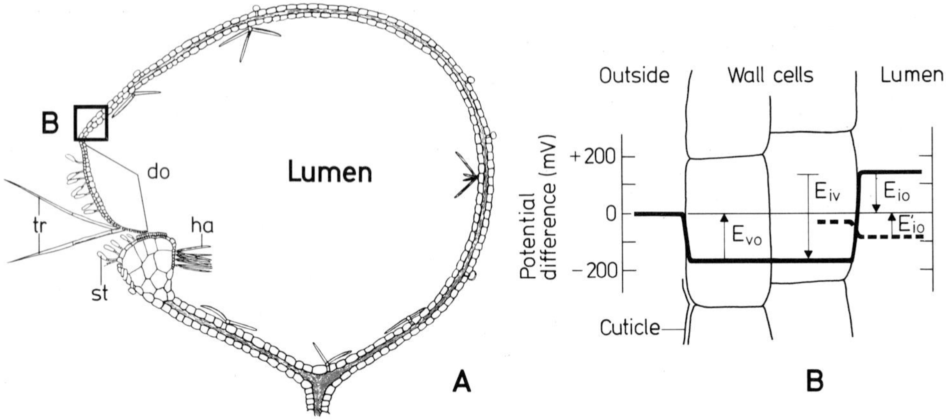

Fig. 8. A Anatomy of a bladder of *Utricularia flexuosa* after GOEBEL (1932); *do*, trapdoor; *ha*, two-armed hair cells; *st*, stalked cells; *tr*, trigger hairs. **B** Profile in electrical potential across the wall of an *Utricularia* bladder in the set state (*bold line*), and about 1 s after the bladder has been fired (*broken line*). E'_{io} is the peak of the transient change of E_{io} (see Fig. 9). Data on *Utricularia sp.* after SYDENHAM and FINDLAY (1973)

cient ($\sigma < 1$). Therefore, the outwardly directed gradient in hydrostatic pressure will overcome the inwardly directed gradient in osmotic pressure. As a result, there will be a bulk flow of NaCl and water from the cells to the outside solution via a subcuticular space. For details of this steady-state osmotic gradient hypothesis of water flow and active water transport see the review by ZIMMERMANN and STEUDLE (1977).

Bladders setting under paraffin oil expel fluid only at the mouth region. SYDENHAM and FINDLAY (1975) have suggested that the two-armed hair cells which are the only noncutinized cells inside the bladder near the mouth take up the fluid from the lumen; at the outside the stalked cells near the mouth might expel it into the outside medium (see Fig. 8A). Treatment with 1 mM sodium azide or exposure to 2° C which reduces the energy supply for the Cl^- pump, does not alter the width of a set bladder, indicating a high impermeability of the bladder wall to water. This amazing property points to a special role of the cuticle coating the bladder on the outside.

Firing the Trap. In a set bladder, the trapdoor opens inwardly if one of the sensory hairs on the door is touched (see Fig. 8A). The site of mechanoreception is not known. To record the trapdoor movement, SYDENHAM and FINDLAY (1973) clamped the luminal potential, E_{io}, to its resting value of +133 mV and fired the trap mechanically. The rise and fall of the clamp current to hold E_{io} indicated that the trapdoor opens within 10 to 15 ms and closes within 30 ms after firing. Values of 6 and 24 ms, respectively, had been obtained by photographic analysis (LLOYD, 1942). Similarly rapid (that is, completed within 10 to 30 ms) is the insect-triggered bending of the column of *Stylidium* (FINDLAY and FINDLAY, 1975).

Although SYDENHAM and FINDLAY (1973) failed to fire the trapdoor electrically, namely, by shifting E_{io} to +380 or to −120 mV, they have other evidence for an excitatory step in the firing process of *Utricularia*. The E_{io} and the membrane potential recorded from bladder wall cells undergo large transients which last

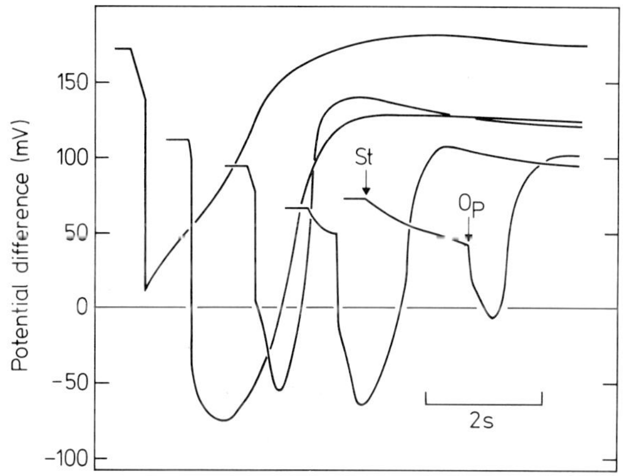

Fig. 9. Transients of the PD between lumen and outside medium, E_{io}, of a bladder of *Utricularia sp.* after five successive stimulations at intervals of 40 min. Each consecutive curve has been displaced in time. *St*, stimulation; *Op*, rapid opening of the trapdoor. (After SYDENHAM and FINDLAY, 1973)

much longer than the trap door opening. The voltage profile during the peak of the transients is also included in Figure 8 B: E_{vi} inverts from +300 to −40 mV and accounts for most of the shift of E_{io} to E'_{io}. Figure 9 shows the time-course of the transient change of E_{io} after five successive stimulations 40 min apart (usually traps may be fired a total of eight times). Clearly, E_{io} depolarizes in two phases: (1) during the increasing lag between stimulation and trap door opening, E_{io} depolarizes slowly, along with a slow inward movement of the trapdoor, and rapidly (2) when the trap door opens. The authors argue that phase (1) might indicate a loss of turgor by excitable trapdoor cells, which is not fully restored after each firing. After repetitive stimulation, cells more distant from the trigger hair now need to be fired in order to lower the rigidity of the trap door enough to cause opening. The graded depolarization between stimulation and movement (Fig. 9) also seems compatible with a receptor potential which becomes visible only after the threshold of excitation has been stepped up. The observation that the transient of E_{io} is essentially accounted for by a change of the membrane potential E_{vi} (see the voltage profiles in Fig. 8 B), raises the question whether a wave of excitation spreads from the door across the bladder wall, or whether the sudden pressure increase of 0.17 bar upon trapdoor opening triggers the electrical transient in all bladder cells facing the lumen. (See Sect. 4.2.2 for the discussion of turgor-dependent membrane properties.) Clearly, a further study of the sensory mechanism in *Utricularia* is of interest.

4.2 Mechanosensory Mechanisms

4.2.1 Mechanoreception

Compared to photo- and chemoreception, reception of seismic and thigmic stimuli by plants is a still poorly understood process, although amazingly detailed phenomena had already been reported by 19th century botanists, such as MOHL (1827), DARWIN (1875, 1876), PFEFFER (1885), BURDON-SANDERSON (1888), and HABERLANDT (1906), and some work on these topics had been published as early as in the 18th century; i.e., by GMELIN (1768). PFEFFER (1885) had worked out the difference in the phenomenology of seismic and thigmic stimulation.

Liquids, for instance, are thigmically ineffective: *Drosera* tentacles do not respond to application of a jet of water (DARWIN, 1875), or when rubbed with a drop of mercury (PFEFFER, 1885). PFEFFER also found that a piece of gelatine would not thigmically stimulate tendrils of *Sicyos* if it contained more than 86% water. He concluded that thigmic stimulation relies upon a localized shearing or differential deformation by the applied mechanical force.

Mechanoreception may reasonably be assumed to occur at the cell surface which is deformed by mechanical forces. Electrostatic interaction between a solid stimulus and the cell surface may also be involved. Specifically, the combination of cuticle, cell wall, and a closely associated plasmalemma might be a plausible candidate for a mechanosensor. This was already postulated by HABERLANDT (1906) when he found *Fühltüpfel* (sensory pits) on tendrils, which enlarge the membrane–cell wall interface. In sum, our understanding of mechanoreception has hardly improved since PFEFFER (1885). It may be characterized, for instance, by our failure to resolve whether the difference between thigmic and seismic stimulation is a matter of stimulus reception or subsequent steps.

4.2.2 Mechano-Electric Transduction

There are many observations of electrical activity of higher plant organs upon mechanical stimulation; see review by PICKARD (1973) and, for instance, her own account of AP's recorded from mechanically stimulated pea epicotyls (PICKARD, 1971). The droseracean genera *Drosera* and *Dionaea* obviously have invented mechano-electric sensory transduction. A causal relationship between the mechanical stimulation and the electrical events also follows from the observation that no AP could be detected in the genus *Drosophyllum,* which lacks the rapid tentacle movements (WILLIAMS, 1976).

Compression and deformation of the sensory layer of the *Dionaea* trigger hair elicits a receptor potential (Fig. 4). JACOBSON has carefully investigated the effect of both stimulus rise time (10 to 10^4 ms) and stimulus amplitude (10 to 10^3 µm) on the receptor response; that is, on the magnitude of the receptor potential and on the generation of an AP. He found that the magnitude of the receptor potential depends upon both stimulus parameters. If elicited by slow-rise stimuli ($\gtrsim$40 ms), the receptor potential is a function of the *velocity* (stimulus amplitude/rise time) of sensory hair displacement. Over a range of nearly two log units of both stimulus parameters (amplitude and rise time, respectively), a receptor potential of, say, 4 mV is produced by a stimulus velocity in the order of 0.05 to 0.5 $mm \cdot s^{-1}$. The receptor potential upon fast-rise stimuli ($\lesssim$40 ms), on the other hand, is a function of the *product* of amplitude and rise time. Figure 10 shows the relationship between the receptor response and the log of the stimulus amplitude for selected stimulus rise times. It illustrates (1) dose-response curves are of the quite common log type discussed below (Sect. 5.3.2). (2) generation of the AP obviously is not invariably fixed to a certain receptor potential level; rather it seems restricted to a limited range of stimulus amplitude (lower threshold 300 µm).

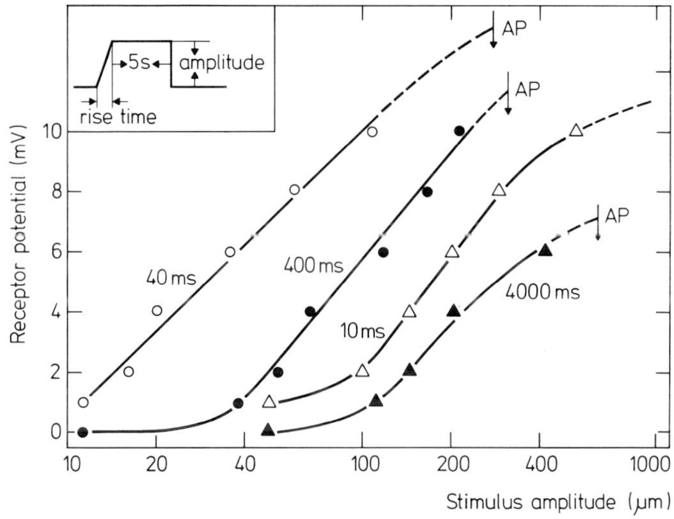

Fig. 10. Receptor potential of a sensory hair of *Dionaea* in situ as a function of the amplitude of the stimulus plotted on a logarithmic scale (*abscissa*) for different rise times given at each curve. A pipette was slipped over the cut tip of the hair and was used to record the extracellular voltage and to apply the stimulus. *Arrows*, stimulus amplitude sufficient to trigger an action potential (AP). *Inset*, stimulus characteristics: amplitude and rise time of the displacement of the sensory hair. (Data from Figure 6 in JACOBSON, 1965)

Stimuli of 10 ms rise time will not lead to an AP at all. (Remember that in excitable membranes the threshold for the AP increases as the duration of the electrical stimuli decreases.) Evidently mechano-electrical transduction of the *Dionaea* sensory cell is complex, because different characteristics of the input signal (i.e., amplitude and rise time) are processed. A further analysis including electrical recording from single cells (as in Fig. 4) upon changes of the ionic milieu, and turgor measurements during stimulation seems a challenging task.

Measuring of the cell turgor may well prove worthwhile, because an electromechanical model by ZIMMERMANN and coworkers considers the plasmalemma and tonoplast as pressure-transducers for turgor regulation in plant cells (see ZIMMERMANN, 1977). At least three forces act upon a plant cell membrane. (1) The turgor pressure, p_T; that is, the difference in hydrostatic pressure between the cell interior and the outside. The p_T, in fact exists as a pressure *gradient* across the plasmalemma; the *absolute* pressure also acts upon the tonoplast. (2) An electrical compressive force, p_E, created by the membrane potential. (3) A mechanical restoring force, p_m, arising from the molecular stress in the membrane and required for dimensional stability of the membrane. Thus in equilibrium:

$$p_T + p_E + p_m = 0.$$

p_m is a very complex parameter, because it infers several dimensionally different elastic compressive moduli of the anisotropic membrane. A joint elastic modulus of plasmalemma and cell wall has even been suggested to account for the joint stretching and shearing during turgor-induced cell volume changes or deformation, or (as we might add) after seismic and thigmic stimulation. A pressure of 0.1 bar can be roughly estimated for an effective seismic stimulus in *Dionaea* (see legend to Fig. 4). In *Valonia*, a change in p_T of 0.5 bar increases the membrane potential by 40 mV and the membrane conductance, g_m, by a factor of about 2.5 (ZIMMERMANN and STEUDLE, 1974). In fact, a sudden step-up or step-down of p_T in *Chara corallina* will cause a depolarization sufficient to elicit AP's (ZIMMERMANN and BECKERS, 1978). Therefore, it seems not pure speculation that similar electromechanical processes in the membrane might sense and transduce turgor and mechanical signals. TEORELL (1971) has worked out an electrokinetic model for animal mechanoreceptors. Here, turgor-dependent streaming potentials cause both receptor and action potential.

5. Comparative Sensory Physiology for Major Stimuli

5.1 Other Stimuli and Sensory Mechanisms

5.1.1 Gravity

Several variants of the statolith hypothesis have been proposed to account for the primary geosensory step ("graviperception") of plant cells and organs; see SIEVERS and VOLKMANN, Chapter 5.2.1; VOLKMANN and SIEVERS, Chapter 5.2.2, this volume. Gravity causes statoliths to exert a force upon their substratum, because statoliths have a higher density than cytoplasm ($\rho = 1$ g cm^{-3}). Higher plants make use of amyloplasts ($\rho = 1.5$ g cm^{-3}), whereas the statoliths (Glanzkörper) in the tip of the rhizoid cell of *Chara fragilis* are vacuoles containing barium sulfate crystals of $\rho = 4.5$ g cm^{-3} (SCHRÖTER et al., 1975). A priori, it seems reasonable to assume that the substratum of a statolith functions as the geosensor, in fact, as a pressure-transducer. However, other modes of interaction are also conceivable. (1) In the central cap cells of the *Lepidium* root, the amyloplasts seem to interact with an *ER complex* (VOLKMANN and SIEVERS, 1975). Apparently, the two organelles function jointly as the geosensor. The kind of interaction and thus the sensory output has not yet been established in this case. (2) HERTEL (1970) has suggested that the *plasmalemma* might be the geosensor in corn coleoptile cells, operating as a mechano-chemical transducer via auxin binding and transport. (A current version of this idea would be to assume that gravity causes amyloplasts to change the auxin permeability ratio, $P_{IAA}:P_{IAAH}$, of the plasmalemma by pressure or by some other way of interaction; see Sect. 2.2).

5.1.2 Light

Phototransduction is a particularly important but poorly understood process in plants. Although a close association of tropic photoreceptor molecules with the plasmalemma seems widely accepted, no cogent evidence exists that this membrane functions as a photo-electric transducer (see review by BENTRUP, 1974a); a receptor potential gradient would be an attractive mode of processing a light-absorption gradient (BENTRUP, 1974b). Experimental data compatible with a membrane-bound activity of phytochrome (P_{fr}) has been reviewed recently by MARRÉ and CIFERRI (1977); activity of ATPases and of enzymes such as choline acetyltransferase might be under control of P_{fr}. It appears that both photoreception and -transduction through phytochrome are intimately linked to the particular cellular site of the phytochrome-controlled physiological process. In any case, the currently most plausible mode of phototransduction seems to be a photochemical rather than a photo-electrical one. See SHROPSHIRE, Chapter 1.2, this volume.

5.1.3 Chemical Signals

Chemosensory transduction controlling chemotaxis and chemotropism in bacteria and eukaryotic lower plants is probably purely chemical; see MACNAB, Chapter 3.2.2, and BEAN, Chapter 3.2.3 this volume. In chemotaxis of *E. coli*, galactose binds to a specific receptor molecule. The allosterically changed receptor then initiates transmembrane galactose transport. MESIBOV et al. (1973) found that the effective concentration range in *E. coli* is specific for the stimulating substance: this range, from threshold to saturation, extends over five to six log units ($3–5 \cdot 10^{-7}$ M to 10^{-1} M) for α-methyl-aspartate, but over only one to two log units ($1–4 \cdot 10^{-8}$ M to 10^{-6} M) for galactose.

As for the possibility of a chemo-electric transducer mechanism, it should be added that SZMELCMAN and ADLER (1976) have detected a membrane hyperpolarization accompanying the chemotactic response of *E. coli* to various attractants and repellents, by measuring the distribution of triphenylmethyl-phosphonium ($TPMP^+$). The significance of this observation with regard to chemo-electric transduction seems questionable, however: a motile strain of *Streptococcus faecalis* responds chemotactically to leucine even if the membrane potential of this bacterium is effectively clamped to E_K by addition of the K^+ ionophore valinomycin (MANSON et al., 1977).

Firm evidence is also lacking for the occurrence of chemo-electric sensory transduction in higher plants; but some interesting observations have been made. Since the work of RICCA in 1916 and HOUWINK in 1935 (see review by PICKARD, 1973) it has been known that wounding or burning of *Mimosa* leaves releases substances which elicit a wave of electronegativity (*variation potential*) in adjacent intact leaves, travelling in the direction of the transpirational stream. The variation potential might trigger and send ahead AP's. VAN SAMBEEK and PICKARD (1976) also found this electrical activity in *Lycopersicon, Cucurbita, Gossypium* and *Xanthium*. At present it is not settled how the variation potential comes about in terms of Eqs. (1) to (3), and whether the unidentified released substances (*Ricca's factor*) act purely electrochemically or by means of chemoreception.

The slow movement of the *Dionaea* trap and the *Drosera* tentacle following capture of a prey by the initial rapid movement (Sect. 4.1) depends upon chemical stimulation, particularly by Na^+ and NH_4^+ in *Drosera* (DARWIN, 1875), and by Na^+, NH_4^+, but not by K^+, Ca^{2+} or glucose in *Dionaea* (LICHTNER and WILLIAMS, 1977). The sensory process is unknown. AP's are not involved in this slow movement. WILLIAMS and PICKARD (1972a) have suggested that the slow bending movement of the *Drosera* tentacle is mediated by growth hormones released upon chemical stimulation.

Table 1 provides an informal summary of current ideas and of available (mostly indirect) experimental evidence on sensory transducer mechanisms in plant move-

Table 1. Hypotheses and experimental results on sensory transduction in plant movements abstracted from Sections 3 to 5

Stimulus (=*input*)	Reception	Sensory transducer	Transducer activity (=*output*)	Transducer type (input/output)
Light	Absorption of photons	Molecular photoreceptor at membrane?	1. Modulation of an enzyme activity? 2. Change of a membrane function?	Photochemical
Chemical	Binding of signal to receptor molecule	Molecular chemoreceptor at plasmalemma		Chemical
Gravity	Interaction of amyloplast with ER-complex Pressure by amyloplast?	1. ER-complex (and amyloplast?) 2. Plasmalemma?	? Change of auxin flux equilibrium?	Chemical? Mechanochemical
Mechanical	Shift of the electromechanical equilibrium?	Plasmalemma? Tonoplast?	Generation of receptor potential	Mechanoelectrical
Electrical	Change of membrane potential	Plasmalemma	1. Change of voltage-sensitive ("gateable") membrane element 2. Electrophoresis by electric fields?	Electrochemical

ments. This table, which admittedly reflects the reviewer's choice of material, is meant to convey the ideas that (1) cellular membranes might be a prominent sensory site, but that satisfactory evidence for this statement is really scarce. (2) The transducer output is poorly established in most instances. Plants seem to have evolved a diversity of transducer types. The type might vary even for a given stimulus quality (gravity?). To date, receptor potentials, which are the regular output of animal sensory transducers (cf. THURM and KAISSLING, 1977), have been found only in mechanosensory plant cells. This transducer type facilitates signal transmission by action potentials, and hence presents itself when, as in the *Dionaea* trap, sensory and effector sites are at some distance from one another.

5.2 Signal Transmission

Most plant movements require one or several steps of signal transmission between stimulus reception and the physiological response. In fact, even preceding reception, the stimulus itself might be transmitted within the plant: light is frequently transmitted to the receptor site by suitable dioptrics (see SHROPSHIRE, Chap. 1.2, DENNISON, Chap. 5.1.2, this vol.), whereas gravity obviously needs no help.

Once a stimulus has been transduced by its sensor, the nature of the output signal and of the transmission line is of interest. Action potentials have been clearly shown to provide signal transmission in plants as well as in animals. Intercellular transmission of the AP's is assumed to occur by electrotonic coupling of adjacent cells through plasmodesmata (WILLIAMS and PICKARD, 1974; WILLIAMS and SPANSWICK, 1976). In *Mimosa* the AP moves from the site of generation exclusively along the excitable parenchyma cells of the protoxylem and phloem of the petioles to the excitable pulvinar cells at an average velocity of 2 to 3 cm s^{-1} (SIBAOKA, 1966). In *Drosera* the trains of AP's (Fig. 7) propagate from the head to the base of a stimulated tentacle at 0.4 cm s^{-1} (and at one cm s^{-1} upward if triggered electrically at the base); but they do not spread across the leaf to other tentacles (WILLIAMS and SPANSWICK, 1976).

In *Dionaea* the single AP evoked by stimulation of a sensory hair propagates across the trap tissue at 20 cm s^{-1} (BURDON-SANDERSON, 1888; SIBAOKA, 1966). A minimum of two AP's is needed to cause trap closure. However, closure is not complete after a second stimulus; BURDON-SANDERSON and PAGE (1876) already showed that each of a series of six stimuli, given at 1-min intervals, will elicit a successively increasing degree of closure, so that the trap is closed completely only after the sixth touch. (In fact, up to 19 stimuli are required, if the interval between successive stimuli is increased to 20 min; BROWN, 1916.) Therefore, as WILLIAMS (1973) has pointed out, the "memory" of the *Dionaea* trap actually resembles "facilitation" known from nerve physiology, because one stimulus increases the response to a subsequent stimulus. Evidence that "facilitation" occurs during transmission of the AP through the excitable trap cells was obtained by SIBAOKA (1966); he found that an AP which follows no later than 3 to 5 s upon a previous AP will be transmitted up to four times faster than the first AP. On the other hand, if the time between successive AP's is 30 to 60 s or more, the second AP will travel at the same velocity as the first (20 cm s^{-1}).

Movement of a phytohormone obviously is an ubiquitous mechanism of signal transmission in the control of growth and tropic movements, or of transport processes in a tissue or in the whole plant; see, for instance, VAN STEVENINCK (1976). It cannot be discussed adequately within the limits of this article. Note, however, that according to the transmembrane auxin diffusion hypothesis (Sect. 2.2) intercellular auxin movement and electrical signal conduction could be interdependent processes.

5.3 Dose–Response Phenomena

This section deals with some fundamental properties of stimulus-response chains, namely, signal amplification, adaptation, and signal storage. First of all, another well-established property, associated with stimulus reception, will be mentioned: in gravi- and photoreception the magnitude of a physiological response within certain limits is a function of the stimulus dose, $I_s \cdot t$, irrespective of the absolute magnitude of either the stimulus strength, I_s, or of the presentation time, t. Experimental evidence pertaining to this Bunsen-Roscoe reciprocity law will be given by SHROPSHIRE, Chapter 1.2, DENNISON, Chapter 5.1.2, and VOLKMANN and SIEVERS, Chapter 5.2.2, this volume. As for mechanoreception, no pertinent data seem to exist for the receptor potential of *Dionaea*. (In Fig. 10, t is kept at 5 s.) Another question is whether the observed reciprocity of stimulus amplitude and stimulus *rise time* is a phenomenon comparable to the Bunsen-Roscoe law and thus may also reflect a property of the unknown primary reception mechanism.

5.3.1 Signal Amplification

The energy delivered to the sensor by the stimulus may be lower than the energy required for generation of the output signal during transduction. Animal photoreceptor cells operate as photomultipliers by using the energy of one photon to move, say, 10^5 elementary charges needed for one receptor potential (see STIEVE, 1977). No comparable figures seem to exist for plant sensory cells. In *Dionaea*, a tentative estimate indicates that the input energy of an average mechanical stimulus (Fig. 4) is about 10^{-8}J. Let the sensory cells have a total surface area of 10^{-7}m^2, and a specific membrane capacitance of 10^{-2}F m^{-2}; then the cells represent a capacitor of 10^{-9}F. The energy needed for a receptor potential of 0.05 V, therefore, is $10^{-9} \cdot (0.05)^2\ \text{FV}^2 \approx 10^{-12}$J. Although the energy balance would indicate no amplification in that particular case, it may be assumed that examples of true amplification will be found in plants if they are searched for thoroughly.

A conceivable mechanism of signal amplification would be cooperativity, for instance, of membrane elements. Cooperativity of auxin transport sites in the plasmalemma of corn coleoptile cells during graviperception has been hypothesized by HERTEL (1970).

5.3.2 Adaptation

When the sensitivity of a sensory system decreases as the stimulus strength I_s increases, the system is said to show adaptation. In plants, adaptation processes generally occur rapidly relative to the latency and reaction time of the adapted response. In most cases, therefore, only the adapted *state* is observed rather than the adaptation *process*. A sufficiently slow adaptation process, observable via the physiological response, is known from the light–growth response of the *Phycomyces* sporangiophore (cf. BERGMANN et al., 1970; SHROPSHIRE, Chap. 1.2, and DENNISON, Chap. 5.1.2, this vol.). The rate of adaptation, dA/dt, after the light is turned off, stepped up or stepped down, can be described by the differential equation

$$dA/dt = (I - A)/b, \tag{4}$$

where I is the light intensity of the stimulus, A is the level of adaptation expressed as the light intensity to which the cell is adapted, and b is the time constant of the adaptive process. This constant gives the time during which A has changed by $1/e$ (by 37%). It is 4 min at 20 °C, and hence sufficiently large compared to the short latency of the light–growth response of 3 min.

Evidence for adaptation is commonly derived from a hyperbolic relationship between response (R) and stimulus strength (I_s),

$$R = k \cdot I_s/(mI_s + n), \tag{5}$$

where k, m, and n are constants.

Examples have been obtained for numerous chemo- and photoresponses of animals and plants, including bacterial chemotaxis (MESIBOV et al., 1973), and light–growth response of *Phycomyces* (cf. FOSTER and LIPSON, 1973). Figure 10 shows that the mechano-electrical receptor response of *Dionaea* could also be added to this list.

The mid-intensity range of such hyperbolic $R(I_s)$ curves generally shows a logarithmic relationship,

$$R = k' \cdot \log I_s, \; (k' = \text{const.}) \tag{6}$$

which has been known for more than 100 years as the *Weber-Fechner law*. It states that the increment in the response, ΔR, is proportional to the *relative* increment in stimulus strength, $\Delta I_s/I_s$, rather than to the absolute increment, ΔI_s. (WEBER had found that $\Delta I_s/I_s = \text{const.}$ for a given response magnitude, i.e., a threshold response; *Weber law*.)

The Weber-Fechner relationship has an impressive record for a broad variety of stimuli: chemotaxis of fern spermatozoids toward malate (PFEFFER, 1885) and of bacteria toward sugars and amino acids (PFEFFER, 1888; MESIBOV et al., 1973), phototropic responses (SHROPSHIRE, Chap. 1.2, DENNISON, Chap. 5.1.2, this vol.), rheotropism, tropic effects of ionic (H^+, K^+) gradients (see JAFFE, 1968; WEISENSEEL, Chap. 5.1.1, this vol.), and electrotropism (BENTRUP, 1968). However, cf. SHROPSHIRE's reference (Chap. 1.2, this vol.) to substitution of "power law" for the Weber-Fechner law.

What is the molecular basis of adaptation? Which stages between stimulus reception and the physiological response are subjected to adaptation? Equation (5) actually reflects the equilibrium of a bimolecular reaction according to the mass action law (cf. MESIBOV et al., 1973; THURM and KAISSLING, 1977). Therefore, adaptation has been interpreted frequently in terms of the interaction of a stimulus and its receptor according to the mass action law. It must be emphasized, however, that no mechanism can be predicted from a hyperbolic $R(I_s)$ curve (or a Weber-Fechner relationship, BÜNNING, 1953). In fact, sensory physiology of animal photoreceptor cells tells us that, just within that part of the chain from photoabsorption to generation of the receptor potential, several independent adaptation processes operate (see STIEVE, 1977). It seems desirable, therefore, to detect and analyze $R(I_s)$ relationships of steps that occur as early as possible in a physiological response, in order to minimize the number of processes involved. The receptor response of *Dionaea* is a relatively early event showing adaptation. Here, the adaptation process(es) is (are) obviously rapid; that is, likely to last less than 10 ms, the shortest stimulus rise time used (Fig. 10).

5.3.3 "Stimulus Storage"

The information derived from a stimulus may be processed at different rates by the consecutive steps of the stimulus–response chain. If at some stage the rate is very low, a product will be accumulated, that is, virtually "stored" (strictly speaking, the term "stimulus storage" is incorrect.)

Several authors employed a cold break to demonstrate "stimulus storage" (*Münchhausen experiment;* EBREY and CLAYTON, 1969; cf. also DENNISON, Chap. 5.1.2, this vol.). BRAUNER and HAGER (1958) found that *Helianthus* hypocotyls memorize a geotropic stimulus for about 6 h at 4° C. *Phycomyces* sporangiophores will "store" a phototropic stimulus for at least 2 h at 2° C (EBREY and CLAYTON, 1969; PETZUCH and DELBRÜCK, 1970). Whether "stimulus storage" occurs during transduction may be answered more specifically by experiments testing the reversibility of a stimulus effect by a subsequent stimulus. First of all, phytochrome P-730 appears as an unique example of photo-stimulus "storage" during reception, and maybe early transduction. No such storage can be expected with other photoreceptors for phototropism, i.e., flavoproteins or carotenoids, which have short-lived excited states (< 1 ms). However, photo-stimuli absorbed by these receptors may be "stored" transiently, as illustrated in the following example.

A single flash of light polarizes developing *Equisetum* spores (HAUPT, 1957), although the polarity is fixed only 1 to 2 h later (FEUCHT and BENTRUP, 1971; see also WEISENSEEL, Chap. 5.1.1, this vol.). When spores of *E. variegatum* are exposed to two sequential 10-s pulses of unilateral blue light, oriented at an angle of 90° to one another, the second pulse eliminates the effect of the first one if the two pulses are separated by an interval of >80 s. For intervals of <80 s, the spore polarity indicates that both pulses are integrated into a resultant absorption gradient; that is, the first pulse is transiently "memorized" for about 80 s at 25° C (BENTRUP, unpublished data).

Signal-specific "stimulus storage" at different stages of development is known from *Fucus* zygotes, which may be polarized by an unidentified rheotropically active chemical stimulus during a period from 2 to 6 h after fertilization. This response may be fully overridden by a subsequent (6–10 h) photostimulation (BENTRUP and JAFFE, 1968; see WEISENSEEL, Chap. 5.1.1, this vol.).

M.J. JAFFE (1970) found that tendril coiling in *Pisum* after a ventrally applied thigmic stimulus can be cancelled by a subsequent dorsal one only if the latter follows within 9 min (see SATTER, Chap. 4.2, this vol.). A hint that the thigmic stimulus might be reversibly stored before transduction is completed was already presented by UMRATH (1934) who recorded action currents (indicating successful transduction) from *Cucumis* tendrils only after they were stimulated ventrally. These action currents failed to occur, however, after dorsal or symmetrical stimulation, neither of which is physiologically effective.

5.4 The "Black Box" Approach

A stimulus–response system may be considered as a "black box" which receives a time-dependent input signal, $x(t)$, and delivers the output signal $y(t)$. A transfer function, that is, the relationship between $x(t)$ and $y(t)$, may then be worked out, representing the activity and cooperation of the elements in the "black box"—provided that the response, $y(t)$, can be modulated by the amplitude and frequency of a periodic stimulus, $x(t)$. The latter condition holds only for a few cases studied so far in plants.

Light of sinusoidally modulated intensity has been used by HANSEN (1971) to study the coupling of the membrane potential to photosynthesis in *Nitella,* and by HANSEN and GRADMANN (1971) in *Acetabularia*. The light–growth response of *Phycomyces* seems to be the only plant movement where an amplitude- and frequency-modulated output, viz. the growth response itself, has been found and analyzed (FOSTER and LIPSON, 1973; LIPSON, 1975).

FOSTER and LIPSON (1973) used light of 488 nm modulated sinusoidally over an intensity range, I, from 0.03 to 3 μW cm^{-2}. In the frequency range from 0.001 to 0.004 Hz (periods of 16.7 to 4.2 min) the growth velocity of the sporangiophore also varied sinusoidally, indicating a linear transfer system with respect to log I. At lower frequencies (<0.001 Hz), however, the nonlinear property of rectification occurred, that is, the growth velocity responded (viz. rose) only to the half-cycle of *increased* I. LIPSON (1975) used a random pattern of frequency and amplitude of the light intensity ("white gaussian noise") and suggested the following model to explain his experiments:

The light–growth system of *Phycomyces* is considered as a functional[1], **H**, with the input x and the output y:

$$x \rightarrow \boxed{\mathbf{H}} \rightarrow y.$$

The empirical functional **H** may be hypothetically split into two subsystems, **F** and **G**, operating in series. The "adaptation functional" **F** operates on the input x and produces the output z; the "output functional **G**", in turn, operates on z and produces y:

$$x \rightarrow \boxed{\mathbf{F}} \xrightarrow{z} \boxed{\mathbf{G}} \rightarrow y.$$

The input variable is $x = \log(I/I_0)$, where I_0 is a reference intensity; x accounts for the observed linearity of the system with log I and suggests that a "logarithmic transducer" is an early element in the light–growth pathway. It is another expression of the Weber-Fechner relationship [Eq. (6)]. A second, nonlinear adaptation element of **F** produces the output $z = (I - A)/A$; it controls the adaptation process given by Eq. (4). Note that $z = 0$ (no transfer) when $A = I$; that is, when the illuminated area of the sporangiophore has become adapted to the given light intensity I. This nonlinear adaptation element is assumed to account for the rectification mentioned above. In fact, LIPSON speculates that both elements of **F**, the "logarithmic transducer" and the nonlinear element, are related to the photochemical kinetics of the photoreceptor. The rectification of the growth velocity might arise from electrical rectification at a membrane (given by the passive element in Fig. 1) and hence might indicate that photoreception occurs at a membrane.

The output of the functional **G** is $y = V - V_0$, where $V(t)$ is the growth velocity and V_0 is the steady-state velocity. System **G** could be used in common by all growth stimuli of *Phycomyces*. This model is still highly speculative. Eventually, it might guide experiments to elucidate individual elements in the "black box" at the molecular level, for instance, by using mutants having deficiencies in the sensory apparatus (see DENNISON, Chap. 5.1.2, and SHROPSHIRE, Chap. 1.2, this vol.).

[1] A functional is a function of a function.

6. Conclusion

Plants have not developed a standard type of sensory cell with an amplitude-modulated receptor potential, generating frequency-modulated trains of action potentials—the known exception being the mechanosensory cells of the Droseraceae, notably of *Dionaea*. Indeed, in higher plant organs, some cells are equipped with auxiliary sensory devices to facilitate stimulus *reception;* namely, statoliths and Fühltüpfel (remember also the indented cell wall of the *Dionaea* sensory cells). Furthermore, elaborate dioptrics for photoreception is found throughout the plant kingdom. However, no cytoplasmic structure has been unambiguously identified to serve as a stimulus *transducer* for the plant's most relevant environmental signals: gravity, nonphotosynthetic light, and mechanical forces. There are suspected sensory structures, like the paraflagellar body of phototactic flagellated green algae, or the ER-complex in the receptor cell of *Dionaea* (Fig. 6) and in the statocyte of *Lepidium*. The apparent diversity of sensors is obviously matched by a diversity of transducer types (Table 1).

Progress in plant sensory physiology is closely linked to that of plant cell biology, particularly to the further development of concepts and techniques bridging the gap between in vivo and in vitro techniques. The use of protoplasts or membrane vesicles, for instance, seems promising. It should be kept in mind, however, that plant-cell membrane functions, especially sensory and regulatory functions, seem to rely upon the membrane's state of electromechanical equilibrium, which is lost in turgor-less naked protoplasts or vesicles.

The "black box" approach of systems theory could be extended to plant movements other than the light–growth response of the *Phycomyces* sporangiophore—if intermediate steps in the given stimulus–movement chain could be detected which display modulation by the stimulus amplitude and frequency. The receptor potential of *Dionaea* could possibly be subjected to this kind of analysis. After all, analysis of a specialized "one-sense" cell, such as the mechanosensory cell of *Dionaea,* might prove more feasible than analysis of "simple" multisensory lower plant cells, such as the *Phycomyces* sporangiophore. *Phycomyces* puzzles by its ability to sense and transduce virtually any kind of stimulus treated by this book.

References

Abe, T., Oda, K.: Resting and action potentials of excitable cells in the main pulvinus of *Mimosa pudica*. Plant Cell Physiol. *17*, 1343–1346 (1976)

Allen, R.D.: Mechanism of the seismonastic reaction in *Mimosa pudica*. Plant Physiol. *44*, 1101–1107 (1969)

Barry, P.H., Hope, A.B.: Electroosmosis in membranes: effects of unstirred layers and transport numbers. Biophys. J. *9*, 700–757 (1969)

Benolken, R.M., Jacobson, S.L.: Response properties of a sensory hair excised from Venus's flytrap. J. Gen. Physiol. *56*, 64–82 (1970)

Bentrup, F.W.: Die Morphogenese pflanzlicher Zellen im elektrischen Feld. Z. Pflanzenphysiol. *59*, 309–339 (1968)

Bentrup, F.W.: Lichtabhängige Membranpotentiale bei Pflanzen. Ber. Dtsch. Bot. Ges. *87*, 515–528 (1974a)

Bentrup, F.W.: Einfluß von K^+-Ionen auf die Polaritätsinduktion bei *Fucus serratus*. Ber. Dtsch. Bot. Ges. *86*, 365–369 (1974b)

Bentrup, F.W.: Cell electrophysiology and membrane transport. In: Progress in Botany. Ellenberg, H. et al. (eds.). Berlin-Heidelberg-New York: Springer, 1975, Vol. *37*, pp. 64–77

Bentrup, F.W.: Cell electrophysiology and membrane transport. In: Progress in Botany. Ellenberg, H. et al. (eds.). Vol. *40*. Berlin-Heidelberg-New York: Springer, 1978, pp. 84–98

Bentrup, F.W., Jaffe, L.F.: Analyzing the "group effect": Rheotropic responses of developing *Fucus* eggs. Protoplasma *65*, 25–35 (1968)

Bergmann, K., Burke, P.V., Cerdá-Olmedo, E., David, C.N., Delbrück, M., Foster, K.W., Goodell, E.W., Heisenberg, M., Meissner, G., Zalokar, M., Dennison, D.S., Shropshire, W., Jr.: *Phycomyces*. Bacteriol. Rev. *33*, 99–157 (1970)

Brauner, L., Hager, A.: Versuche zur Analyse der geotropischen Perzeption. I. Planta *51*, 115–147 (1958)

Brown, W.H.: The mechanism of movement and the duration of the effect of stimulation in the leaves of *Dionaea*. Am. J. Bot. *3*, 68–90 (1916)

Bünning, E.: Entwicklungs- und Bewegungsphysiologie der Pflanze. 3. ed. Berlin-Göttingen-Heidelberg: Springer, 1953

Burdon-Sanderson, J.S.: On the electromotive properties of the leaf of *Dionaea* in the excited and unexcited states. Phil. Trans. Roy. Soc. (London) *173*, 1–53 (1888)

Burdon-Sanderson, J.S., Page, J.F.M.: On the mechanical effects and on the electrical disturbance consequent on excitation of the leaf of *Dionaea muscipula*. Proc. Roy. Soc. (London) *25*, 411–434 (1876)

Christ-Adler, M., Bentrup, F.W.: Effect of K^+ and Cl^- ion gradients upon apex regeneration in *Acetabularia mediterranea*. Planta *129*, 91–93 (1976)

Darwin, Ch.: The Insectivorous Plants. London: Murray, 1875

Darwin, Ch.: The Movements and Habits of Climbing Plants. New York: Appleton & Co., 1876, 208 pp

DiPalma, J.R., McMichael, R., DiPalma, M.: Touch receptor of Venus flytrap, *Dionaea muscipula*. Science *152*, 539–540 (1966)

Ebrey, Th.G., Clayton, R.K.: *Phycomyces:* Stimulus storage in light-initiated reactions. Science *164*, 427–428 (1969)

Eckert, R., Sibaoka, T.: Bioelectric regulation of tentacle movement in a dinoflagellate. J. Exp. Biol. *47*, 433–446 (1967)

Eckert, R., Sibaoka, T.: The flash-triggering action potential of the luminescent dinoflagellate *Noctiluca*. J. Gen. Physiol. *52*, 258–282 (1968)

Felle, H., Bentrup, F.W.: Effect of light upon membrane potential, conductance, and ion fluxes in *Riccia fluitans*. J. Membr. Biol. *27*, 153–170 (1976)

Felle, H., Bentrup, F.W.: A study of the primary effect of the uncoupler carbonyl cyanide m-chlorophenyl-hydrazone on membrane potential and conductance in *Riccia fluitans*. Biochim. Biophys. Acta *464*, 179–187 (1977)

Feucht, U., Bentrup, F.W.: Über die photosensible Phase der Polaritätsinduktion bei *Equisetum*sporen und *Fucus*zygoten. Z. Pflanzenphysiol. *66*, 233–242 (1972)

Findlay, G.P., Findlay, N.: Anatomy and movement of the column in *Stylidium* Aust. J. Plant Physiol. *2*, 297–621 (1975)

Findlay, G.P., Hope, A.B.: Electrical properties of plant cells: Methods and findings. In: Encyclopedia of Plant Physiology, New Series, Vol. 2A. Lüttge, U. and Pitman, M.G. (eds.). Berlin-Heidelberg-New York: Springer, 1976, pp. 53–92

Foster, K.W., Lipson, E.D.: The light growth response of *Phycomyces*. J. Gen. Physiol. *62*, 590–617 (1973)

Gmelin, J.F.: Irritabilitatem vegetabilium, in singulis plantarum partibus exploratam, ulterioribusque experimentis confirmatam. Dissertation Tübingen 1768

Goebel, K.: Organographie der Pflanzen. 3rd. ed. Jena: Fischer 1932

Goldman, D.E.: Potential, impedance and rectification in membranes. J. Gen. Physiol. *27*, 37–60 (1943)

Goldsmith, M.H.M.: The polar transport of auxin. Ann. Rev. Plant Physiol. *28*, 439–478 (1977)

Gradmann, D.: Analog circuit of the *Acetabularia* membrane. J. Membr. Biol. *25*, 183–208 (1975)

Gradmann, D.: "Metabolic" action potentials in *Acetabularia*. J. Membr. Biol. *29*, 23–45 (1976)
Haberlandt, G.: Sinnesorgane im Pflanzenreich zur Perzeption mechanischer Reize. 2. ed. Leipzig: Engelmann, 1906, 207 pp.
Hager, A., Menzel, H., Krauss, A.: Versuche und Hypothese zur Primärwirkung des Auxins beim Streckungswachstum. Planta *100*, 47–75 (1971)
Haupt, W.: Die Induktion der Polarität bei der Spore von *Equisetum*. Planta *49*, 61–90 (1957)
Hansen, U.-P.: The frequency response of the action of light on the membrane potential of *Nitella*. Biophysik *7*, 223–227 (1971)
Hansen, U.-P., Gradmann, D.: The action of sinusoidally modulated light on the membrane potential of *Acetabularia*. Plant Cell Physiol. *12*, 335–348 (1971)
Hertel, R.: Aspects of the geotropic stimulus in plants. In: Gravity and the Organism. Gordon, S., Cohen, M. (eds.) Chicago: University Press, 1970, pp. 41–47
Jacobson, St.L.: Receptor response in Venus's Fly-trap. J. Gen. Physiol. *49*, 117–129 (1965)
Jacobson, St.L.: Effect of ionic environment on the response of the sensory hair of Venus's-flytrap. Can. J. Bot. *52*, 1293–1302 (1974)
Jaffe, L.F.: Localization in the developing *Fucus* egg and the general role of localizing currents. Adv. Morphogen. *7*, 295–328 (1968)
Jaffe, L.F.: Electrophoresis along cell membranes. Nature (London) *265*, 600–602 (1977)
Jaffe, L.F., Nuccitelli, R.: Electrical controls of development. Ann. Rev. Biophys. Bioeng. *6*, 445–476 (1977)
Jaffe, M.J.: Physiological studies on pea tendrils. VI. The characteristics of sensory perception and transduction. Plant Physiol. *45*, 756–760 (1970)
Kikuyama, M., Tazawa, T.: Tonoplast action potential in *Nitella* in relation to vacuolar chloride concentration. J. Membrane Biol. *29*, 95–110 (1976)
Lichtner, F.T., Williams, S.E.: Prey capture and factors controlling trap narrowing in *Dionaea* (Droseraceae). Am. J. Bot. *64*, 881–886 (1977)
Lipson, E.D.: White noise analysis of *Phycomyces* light-growth response system. Biophys. J. *15*, 989–1045 (1975)
Lloyd, F.E.: Carnivorous Plants. New York: Ronald Press, 1942
Manson, M.D., Tedesco, P., Berg, H.C., Harold, F.M.: A proton motive force drives bacterial flagella. Proc. Natl. Acad. Sci. USA *74*, 3060–3064 (1977)
Marrè, E., Ciferri, O. (eds.): Regulation of Cell Membrane Activities in Plants. Amsterdam-Oxford-New York: North-Holland, 1977, pp. 332
Mesibov, R., Ordal, G.W., Adler, J.: The range of attractant concentrations for bacterial chemotaxis and the threshold and size of response over this range. Weber law and related phenomena. J. Gen. Physiol. *62*, 203–223 (1973)
Mitchell, P.: Chemiosmotic coupling in oxidative and photosynthetic phosphorylation. Biol. Rev. Cambridge Phil. Soc. *41*, 445–502 (1966)
Mohl, H.: Über den Bau und das Winden der Ranken und Schlingpflanzen. Eine gekrönte Preisschrift. Tübingen: Laupp, 1827
Mummert, H., Gradmann, D.: Voltage-dependent potassium fluxes and the significance of action potentials in *Acetabularia*. Biochim. Biophys. Acta *443*, 443–450 (1976)
Munk, H.: Die elektrischen und Bewegungserscheinungen am Blatte der *Dionaea muscipula*. Leipzig, 1876
Novák, B., Bentrup, F.W.: An electrophysiological study of regeneration in *Acetabularia mediterranea*. Planta *108*, 227–244 (1972)
Novák, B., Bentrup, F.W.: Orientation of *Fucus* egg polarity by electric a.c. and d.c. fields. Biophysik *9*, 253–260 (1973)
Nuccitelli, R., Jaffe, L.F.: The ionic components of the current pulses generated by developing fucoid eggs. Dev. Biol. *49*, 518–531 (1976)
Oda, K.: Simultaneous recording of potassium and chloride effluxes during an action potential in *Chara corallina*. Plant Cell Physiol. *17*, 1085–1088 (1976)
Peng, H.B., Jaffe, L.F.: Polarization of fucoid eggs by steady electrical fields. Dev. Biol. *53*, 277–284 (1976)
Petzuch, M., Delbrück, M.: Effects of cold periods on the stimulus-response system of *Phycomyces*. J. Gen. Physiol. *56*, 297–308 (1970)

Pfeffer, W.: Locomotorische Richtungsbewegungen durch chemische Reize. Untersuchungen aus dem Botanischen Institut zu Tübingen. Leipzig: Engelmann, 1884, Vol. I, pp. 363–482
Pfeffer, W.: Zur Kenntnis der Kontaktreize. Untersuchungen aus dem Botanischen Institut zu Tübingen. Leipzig: Engelmann, 1884, Vol. I, pp. 483–535
Pfeffer, W.: Über chemotaktische Bewegungen von Bakterien, Flagellaten und Volvocineen. Untersuchungen aus dem Botanischen Institut zu Tübingen. Leipzig: Engelmann, 1888, Vol. II, pp. 582–662
Pickard, B.: Action potentials resulting from mechanical stimulation of pea epicotyls. Planta *97*, 106–115 (1971)
Pickard, B.: Action potentials in higher plants. Bot. Rev. *39*, 172–201 (1973)
Pierce, W.S., Higinbotham, N.: Compartments and fluxes of K^+, Na^+, and Cl^- in *Avena* coleoptile cells. Plant Physiol. *46*, 666–673 (1970)
Pitman, M.G., Anderson, W.P., Schaefer, N.: H^+ ion transport in plant roots. In: Regulation of Cell Membrane Activities in Plants. Marrè, E., Ciferri, O. (eds.). Amsterdam-Oxford-New York: North-Holland, 1977, pp. 147–160
Racusen, R.H.: Phytochrome control of electrical potentials and intercellular coupling in oat-coleoptile tissue. Planta *132*, 25–29 (1976)
Racusen, R.H., Satter, R.L.: Rhythmic and phytochrome-regulated changes in transmembrane potential in *Samanea* pulvini. Nature (London) *255*, 408–410 (1975)
Ragetli, H.W.J., Weintraub, M., Lo, E.: Characteristics of *Drosera* tentacles. I. Anatomical and cytological detail. Can. J. Bot. *50*, 159–168 (1971)
Raven, J.A.: Transport of indoleacetic acid in plant cells in relation to pH and electrical potential gradients, and its significance for polar IAA transport. New Phytol. *74*, 163–172 (1975)
Rubery, P.H., Sheldrake, A.R.: Carrier-mediated auxin transport. Planta *118*, 101–121 (1974)
Schröter, K., Läuchli, A., Sievers, A.: Mikroanalytische Identifikation von Bariumsulfat-Kristallen in den Statolithen der Rhizoide von *Chara fragilis* Desv. Planta *122*, 213–225 (1975)
Sibaoka, T.: Action potentials in plant organs. Symp. Soc. Exp. Biol. *20*, 49–74 (1966)
Sibaoka, T.: Physiology of rapid movements in higher plants. Ann. Rev. Plant Physiol. *20*, 165–184 (1969)
Sibaoka, T., Eckert, R.: An electrophysiological study of the tentacle-regulating potentials in *Noctiluca*. J. Exp. Biol. *47*, 447–459 (1967)
Slayman, C.L.: Proton pumping and generalized energetics of transport: a review. In: Zimmermann, U., Dainty, J. (eds.). Membrane Transport in Plants. Berlin-Heidelberg-New York: Springer, 1974, pp. 107–119
Slayman, C.L., Long, W.S., Gradmann, D.: "Action potentials" in *Neurospora crassa*, a mycelial fungus. Biochim. Biophys. Acta *426*, 732–744 (1976)
Slayman, C.L., Long, W.S., Lu, C.Y.H.: The relationship between ATP and an electrogenic pump in the plasma membrane of *Neurospora crassa*. J. Membrane Biol. *14*, 305–338 (1973)
Stieve, H.: Photorezeption und ihre molekularen Grundlagen. In: Biophysik. Hoppe, W. et al. (eds.). Berlin-Heidelberg-New York: Springer, 1977, pp. 482–501
Sydenham, P.H., Findlay, G.P.: The rapid movement of the bladder of *Utricularia spec.* Aust. J. Biol. Sci. *26*, 1115–1126 (1973)
Sydenham, P.H., Findlay, G.P.: Transport of solutes and water by resetting bladders of *Utricularia*. Aust. J. Plant Physiol. *2*, 335–351 (1975)
Szmelcman, S., Adler, J.: Change in membrane potential during bacterial chemotaxis. Proc. Natl. Acad. Sci. USA *73*, 4387–4391 (1976)
Teorell, T.: A biophysical analysis of mechano-electrical transduction. In: Handbook of Sensory Physiology. Beidler, L.M. (ed.). Berlin-Heidelberg-New York: Springer, 1971, Vol. I, pp. 291–339
Thurm, U., Kaissling, K.-E.: Sensorische Transduktionsprozesse. In: Biophysik. Hoppe, W. et al. (eds.). Berlin-Heidelberg-New York: Springer, 1977, pp. 391–414
Umrath, K.: Über die elektrischen Erscheinungen bei thigmischer Reizung. Planta *23*, 47–50 (1934)
Van Sambeek, J.W., Pickard, B.: Mediation of rapid electrical, metabolic, transpirational, and photosynthetic changes by factors released from wounds. Can. J. Bot. *54*, 2642–2671 (1976)
Van Steveninck, R.F.M.: Effect of hormones and related substances on ion transport. In:

Encyclopedia of Plant Physiology, New Series, Vol. 2B. Lüttge, U. and Pitman, M.G. (eds.). Berlin-Heidelberg-New York: Springer, 1976, pp. 307–342
Volkmann, D., Sievers, A.: Wirkung der Inversion auf die Anordnung des Endoplasmatischen Reticulums und die Polarität von Statocyten in Wurzeln von *Lepidium sativum*. Planta *127*, 11–19 (1975)
Walker, N.A.: Membrane transport: Theoretical background. In: Encyclopedia of Plant Physiology. New Series, Vol. 2A Lüttge, U. and Pitman, M.G. (eds.). Berlin-Heidelberg-New York: Springer 1976, pp. 36–52
Walker, N.A., Pitman, M.G.: Measurement of fluxes across membranes. In: Encyclopedia of Plant Physiology. New Series, Vol. 2A. Lüttge, U. and Pitman, M.G. (eds.). Berlin-Heidelberg-New York: Springer, 1976, pp. 93–128
Williams, M.E., Mozingo, H.N.: The fine structure of the trigger hair in Venus's flytrap. Am. J. Bot. *58*, 532–539 (1971)
Williams, S.E.: The "memory" of the Venus's flytrap. Carnivorous Plant Newslett. *2*, 23–25 (1973)
Williams, S.E.: Comparative physiology of the Droseraceae—the evolution of a plant sensory system. Proc. Am. Phil. Soc. *120*, 187–204 (1976)
Williams, S.E., Pickard, B.: Receptor potentials and action potentials in *Drosera* tentacles. Planta *103*, 193–221 (1972a)
Williams, S.E., Pickard, B.: Properties of action potentials in *Drosera* tentacles. Planta *103*, 222–240 (1972b)
Williams, S.E., Pickard, B.: Connections and barriers between cells of *Drosera* tentacles in relation to their electrophysiology. Planta *116*, 1–16 (1974)
Williams, S.E., Spanswick, R.M.: Propagation of the neuroid action potential of the carnivorous plant *Drosera*. J. Comp. Physiol. *108*, 211–223 (1976)
Zimmermann, U.: Cell turgor pressure regulation and turgor pressure-mediated transport processes. Symp. Soc. Exp. Biol. *31*, 117–154 (1977)
Zimmermann, U., Beckers, F.: Generation of action potentials in *Chara corallina* by turgor pressure changes. Planta *138*, 173–179 (1978)
Zimmermann, U., Steudle, E.: The pressure dependence of the hydraulic conductivity, the membrane resistance and membrane potential during turgor pressure regulation in *Valonia utricularis*. J. Membrane Biol. *16*, 331–352 (1974)
Zimmermann, U., Steudle, E.: Physical aspects of water transport in plants. Adv. Bot. Res. *46* (1978)

1.4 Endogenous Rhythms in the Movement of Plants

B.M. SWEENEY

1. Introduction

1.1 Advantages of Regulation of Orientation as a Function of Time

Since plants typically depend on light as an energy source, they are faced with the problem of arranging themselves in space in the most auspicious way for absorbing the sun's rays. These rays do not always approach a plant from a uniform direction, but impinge on it from different angles, depending on the location of the plant and the time of day. There is thus a selective advantage to plants in being able to make adjustments in the orientation of their light-absorbing structures. Motile plants such as phytoflagellates and the gametes of some plants accomplish this orientation by moving (HÄDER, Chap. 3.2.1, this vol.; POFF and WHITAKER, Chap. 3.2.4, this vol.), while larger plants which are permanently attached to the substrate orient to light by differential growth (WEISENSEEL, Chap. 5.1.1, this vol.; DENNISON, Chap. 5.1.2, this vol.). The difference in the angle of the sun's rays with time of day is highly regular and predictible. Plants have further refined their orientation to light to take advantage of this predictability. Rather than merely adjusting their position after perceiving the direction from which the light is coming, they have developed an endogenously controlled periodicity in movement that matches that in the environment. Such a mechanism allows them to adjust their position beforehand. Many examples of such an endogenous rhythmicity have been observed in plants. Most common are those in which the period matches that of the day–night cycle of the environment. These circadian rhythms of motility are the principal subject of this chapter.

In the intertidal zone, environmental fluctuations are further modified by the advancing and receding tides. Diatoms and flagellates inhabiting the intertidal part of beaches and mud flats might be expected to show time-dependent orientation in response to the tides. A few examples of such adaptations have been reported. These "tidal rhythms" will also be discussed.

In addition to endogenous rhythms with periods that match cycles in the environment, there are a few oscillations with periods of minutes or hours. The function of such rhythms is unclear.

1.2 Definition of an Endogenous Rhythm

An endogenous rhythm is a self-sustained oscillation, the source of which is within an organism.

1.2.1 Circadian Rhythms

Circadian rhythms are oscillations in the physiology, morphology or activity of an organism, with a 24 h period under natural illumination, which continue to be manifested under constant conditions of light and temperature (SWEENEY, 1976). Under constant conditions, the period is close to, but often not exactly 24 h (hence the name "circadian" meaning "about a day") and shows a slight dependence on ambient temperature, much less than required for a Q_{10} of 2. Circadian rhythms can be phase-shifted by exposure to a different light intensity or temperature for a time short relative to one 24 h cycle. This property allows the rhythmic function to be synchronized to the exactly 24 h periodicity of the environment.

The response of circadian rhythms to a single short light exposure in otherwise constant darkness is essentially the same in all organisms. A light exposure during the "day" part of the cycle has no effect, one in the early "night" phase delays the rhythm, while one late in the "night" causes the rhythm to jump ahead. This phenomenon can be seen most clearly when the change in phase is plotted as a function of the time in the circadian cycle. Such graphs are known as phase-response curves (Fig. 1).

While it is generally agreed that circadian oscillations arise endogenously, the question of greatest interest is: What is the physiology and biochemistry of the circadian oscillator which controls circadian rhythms? Two, and only two, characteristics are recognized as belonging irrefutably to the oscillator: the length of the period, and the phase relative to an external time reference. Without more indicators, it has been very difficult to make progress in identifying the oscillator. However, the study of motility rhythms in plants has proven useful in understanding this problem, as we shall see.

1.2.2 Tidal Rhythms

Tidal rhythms resemble circadian rhythms closely, with the exception that the phase is *not* changed by light pulses. As a consequence, tidal rhythms show periods of slightly more than 24 h under natural conditions, for example 24.8 h, so that they are synchronized with the tidal rather than with the solar environmental cycles. The factors responsible for the synchronization of tidal rhythms are not well understood at present, but they appear to differ from organism to organism.

1.2.3 Historical Note

The sleep movements of leaves are examples of typical circadian rhythms, the first to be discovered, and for many years the sole rhythms to be recognized as endogenously controlled. Only since the publication of Bünning's chapter in the *Encyclopedia of Plant Physiology* (BÜNNING, 1959) has the general importance of this phenomenon been fully appreciated. Now hundreds of circadian rhythms have been described in eukaryotes of every conceivable kind. So far however, none have been found in prokaryotes. This burgeoning field has been reviewed frequently during the last 20 years (ASCHOFF, 1965; BROWN et al., 1970; BÜNNING, 1973; CHOVNICK, 1960; BIERHUIZEN, 1972; CUMMING and WAGNER, 1968; DECOURSEY, 1976; HASTINGS and SCHWEIGER, 1976; HILLMAN, 1976; MENAKER, 1971;

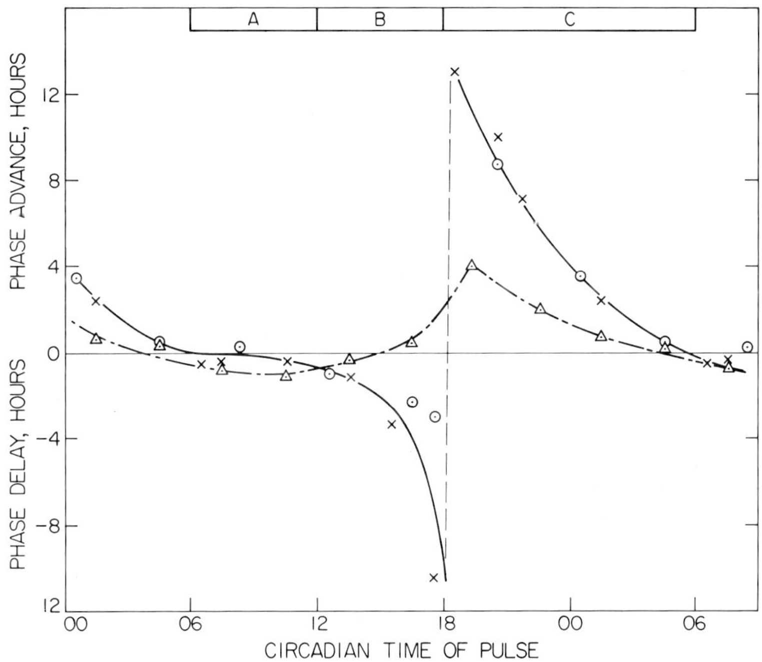

Fig. 1. Phase-response curve for shifting the circadian rhythm in bioluminescence in *Gonyaulax polyedra* with 3-h exposures to 10,000 lux white light when cells were otherwise in continuous darkness (○ and ×) or in light of relatively low intensity, 700 lux (△). Three regions can be distinguished in the phase response curve: region *A* where light pulses do not alter phase, region *B* (in cells in continuous darkness) where light pulses delay the next maximum in luminescence, and region *C* where the phase is advanced. Note the sudden transition between delays and advances at circadian time 18 in cells in continuous darkness. This is a typical phase-response curve for a circadian rhythm. (SWEENEY, 1969)

QUEIROZ, 1974; SWEENEY, 1963, 1969). It is not the purpose of this chapter to describe all the examples of endogenous rhythms in the movement of plants, but rather to point out new insights into the nature of this interesting phenomenon that have arisen from the study of plant movements during the last 20 years.

2. Motile Unicellular Organisms

2.1 Flagellates

2.1.1 Rhythms in Phototaxis

Bünning's article on the daily periodicity of plant movements in the *Encyclopedia of Plant Physiology* (BÜNNING, 1959) is concerned principally with the study of the sleep movements of the leaves and flowers of angiosperms, with only half a page describing the discovery by POHL (1948) of a rhythm in phototaxis in *Euglena gracilis*. Since 1959 however, more and more attention has been given to the circadian rhythms in flagellates and other algae. In part, this is the consequence of the ease with which these organisms can be cultured in large quantities. Many also have the advantage of being unicellular, hence their rhythms are not complicated by cell–cell interactions.

Both Bruce and Brinkmann have studied the rhythm in phototaxis in *Euglena* extensively. This flagellate moves rapidly into an illuminated region during the day part of a light–dark cycle, but hardly accumulates there at all at night. It is simple to record this behavior automatically by monitoring the signal from a photocell illuminated through the culture by a weak test light. Accumulation of cells in the light path is measured as a reduction in the signal from the photocell. When test lights are used at intervals during otherwise constant darkness, the rhythm in phototaxis continues for 4–5 days. Using this method, BRUCE and PITTENDRIGH (1956) were able to measure the period of the rhythm in phototaxis as a function of temperature, and to demonstrate the remarkable temperature independence of the period in this organism. They also measured the effect of single 4-h light exposures given at different times on the subsequent phase of the rhythm, and confirmed that the phase-response curve in *Euglena* is of the same shape as that in *Drosophila* (BRUCE and PITTENDRIGH, 1958), providing one item of evidence that circadian rhythms in all organisms share a common resetting behavior in response to light. The *Euglena* rhythm can follow a variety of light–dark cycles and cycles of high and low temperature, but light is the stronger entraining agent. This is another property common to circadian rhythms in all organisms (BRUCE, 1960).

The continuation of circadian oscillations under constant conditions, not only in higher plants, but also in unicellular flagellates, supports the contention that these oscillations arise endogenously within a single cell. That cellular biochemistry is responsible is confirmed by the observation that exposing *Euglena* to deuterated water lengthens the period of the rhythm (BRUCE and PITTENDRIGH, 1960), since the substitution of deuterium for hydrogen is well known to retard biochemical reactions. Attempts to identify the biochemistry of the oscillator early focused on inhibitors and other molecules of which the biological effects were known. Rhythmic organisms were exposed to these molecules, and changes in period or phase were sought. In many of these early experiments with inhibitors such as cyanide, azide, or the inhibitor of photosynthesis, dichlorophenyl dimethyl urea, no changes in the fundamental properties of the oscillator were detected.

The phototaxis rhythm of *Euglena* served as a test system for the assay of cycloheximide, an inhibitor of protein synthesis on 80 S ribosomes (FELDMAN, 1967). Exposure to cycloheximide caused lengthening of the period in direct proportion to the concentration added. This was the first evidence that oscillations depend in some way on the synthesis of a cytoplasmic protein.

One way to attack the problem of the nature of the circadian oscillators, a technique which has proven of great value in the solution of other biological problems, is the induction and analysis of mutations. Since *Euglena* has no sexual reproduction, Bruce turned his attention to the motile green flagellate, *Chlamydomonas reinhardi*, which does have a sexual cycle, and has been the subject of many genetic studies. He showed that *Chlamydomonas* shows a circadian rhythm in phototaxis essentially like that of *Euglena* (BRUCE, 1970). By mutagenesis and selection procedures, he isolated rhythmic mutants with periods significantly longer than that of the wild type (BRUCE, 1972). These mutants behaved as if a single

Table 1. The period in hours of the phototaxis rhythm in *Chlamydomonas reinhardi* in wild type (the two mating types, Lo^+ and Lo^-), four long-period mutants and one double mutant at three temperatures in constant darkness except for a test light. (BRUCE, 1974)

Temperature	Lo^+	Lo^-	per-1	per-2	per-3	per-4	per-3, per-4
16° C	21.5	22	24	24	24	25	26.5
22° C	24.5	24	27	26.5	26.5	28	30
25° C	24.5	24.5	27	29	28	28	30

gene were altered. In double or triple mutants (Table 1), the period changes were additive (BRUCE, 1974). Unfortunately, no altered gene product has been identified in the long-period mutants of *Chlamydomonas*.

Phototactic behavior has also been observed in marine dinoflagellates. A brackish water form, *Gyrodinium dorsum*, was studied by FORWARD and DAVENPORT (1970). In this cell, phototaxis is a response to blue light, which can be elicited during the daily light period. If, however, cells are taken from continuous darkness to the phototaxis test, they require a brief preirradiation with red light (620 nm), and this activation by red light is not equally effective at all times of day: it shows a circadian rhythm with a maximum in the early morning. Circadian rhythms in both phototaxis and the "stop response" to bright light were observed in *Gymnodinium splendens* maintained in continuous darkness (FORWARD, 1974).

Strong daily vertical migration patterns are observed in the ocean (EPPLEY et al., 1968). It is possible that circadian rhythms in response to light may play a part in these migrations. However, at least in one case, that of *Cachonina niei*, a rhythm other than in phototaxis seems to be responsible, since this organism has been shown to migrate up and down during at least 24 h of continuous darkness (EPPLEY et al., 1968).

2.1.2 Motility Rhythms in *Euglena*

Bruce studied rhythms in the "phototaxis" of *Euglena*, actually in its "photoaccumulation" (see HÄDER, Chap. 3.2.1, this vol.). The method used did not allow resolution of the response into its components (photokinesis, phototaxis or phobic response). BRINKMANN (1966) studied motility (photokinesis) independently of phototaxis or phobic response. He found both motility and other photoresponses to be rhythmic, but the motility rhythm was the more pronounced. For this reason, in subsequent studies he measured motility rather than a directional response to light. As in the phototactic rhythm, under autotrophic conditions the period of the motility rhythm is independent of ambient temperature between 15° and 35° C. He found, however, that under mixotrophic conditions with peptone, glucose and citric acid as carbon sources the period increases as the temperature is raised, from 24 h at 20° C to more than 27 h at 35° C. Such overcompensation to temperature changes has also been found in *Gonyaulax* (SWEENEY, 1969) and other algae. The interesting difference between autotrophic and mixotrophic culture conditions has not been explained. On the other hand, autotrophic and mixotrophic cultures of *Euglena* were found to behave similarly with respect to light (SCHNABEL, 1968), both in entrainment of the motility rhythm to different light–dark cycles and with respect to resetting following single irradiations. A colorless mutant of *Euglena gracilis* studied by KIRSCHSTEIN (1969) behaved essentially like a green *Euglena* growing under mixotrophic conditions.

BRINKMANN (1974) has recently used the motility rhythm of *Euglena* to study the effects of alcohols on the circadian oscillator. He finds that the period is lengthened when ethanol (50–200 mM) is present. Butanol has a similar effect, but, interestingly, neither methanol nor propanol causes changes in rhythmicity. Brinkmann concludes that the alcohol treatment acts via mitochondrial metabolism, which shows this characteristic sensitivity to alcohols. Hence mitochondria may play a role in the generation of circadian oscillations.

2.2 Gliding Organisms

2.2.1 Diatoms

Circadian Rhythms. Flagellar movement is not the only mechanism by which organisms are propelled. Pennate diatoms with a raphe are motile and move quite rapidly when in contact with a solid surface. Although the mechanism of motility is still in dispute, it is likely that it is a kind of slow-motion jet propulsion (see HALFEN, Chap. 3.1.3, this vol.). On mud flats in the estuary of the River Avon, both pennate diatoms and *Euglena limosa* have been observed on the mud surface exposed at low tide, from which they retreat before the incoming water reaches them (ROUND and PALMER, 1966). When samples of mud containing the diatoms, *Cylindrotheca signata*, *Navicula tryblionella* and *N. salinarum* (ROUND and PALMER, 1966) were brought into the laboratory and placed either under natural illumination or continuous light, the cells were observed to continue their upward and downward migrations, but now with a period of 24 h rather than 24.8 h. A circadian rhythm which was rephased each day by the coming and going of the muddy water of the river was apparently responsible for their movements in nature, rather than a true tidal rhythm, insensitive to light. A diurnal migration in constant darkness was observed in *Cylindrotheca signata* (ROUND, 1965). While in some species, movement in response to light is involved in these migrations, it is not the only rhythmic response in all cases.

Tidal Rhythm in Hantzschia. Of all the diatoms which appear at the surface at low tide in nature, only one, *Hantzschia virgata*, is known so far to possess a true tidal rhythm. This diatom is an intertidal species occupying the sand flats of Barnstable Harbor, Cape Cod, Massachusetts, U.S.A. (FAURÉ-FREMIET, 1951). Although it has not been possible as yet to culture this *Hantzschia*, its behavior in sand samples brought into the laboratory has been examined (PALMER and ROUND, 1967), both in 12-h-light–12-h-dark cycles and in constant illumination. Under both lighting regimes, *Hantzschia* moves up and down with a period of 24.8 h, rather than 24 h. The phase of the migration rhythm thus moves across the day, cells appearing at the sand surface later each day, as do the tides. However, upward migration does not take place in darkness when the cells are under a light–dark cycle. Instead, when the expected upward migration time reaches the beginning of the night phase, cells appear at the surface 12 h later than expected in the morning of the next day.

A similar behavior is observed when the diatoms are in continuous light. Palmer explains this odd behavior by postulating a tidal rhythm in phototaxis which is expressed only during the day phase of an underlying circadian oscillation. If this behavior of *Hantzschia* can be corroborated, it constitutes an example, unique among plants, of a true tidal rhythm which is not reset by light.

One other provocative example of tidal rhythmic behavior in a diatom has been reported by HOPKINS (1965) in the mudflat diatom from the estuary of the River Ouse, *Surirella gemma*. This organism can be observed at the surface at low tide. In the laboratory, in mud samples, migrations of *Surirella* continue, but with a circadian rhythm. However, this rhythm can be shifted, not only by light, but also by wetting the mud samples. Increase in the water content of the mud flats as the tide rose might then serve to synchronize the migration to the tide in nature. This diatom may represent a transitional stage between a purely circadian migration and a purely tidal rhythm like that seen in *Hantzschia*. It would be of great interest to study this organism further.

Although many blue-green algae are capable of gliding motion (see HALFEN, Chap. 3.1.3, this vol.), none is known to display either circadian or tidal migration rhythms. Many species of motile blue-green algae are maintained in culture and should be examined from this point of view.

3. Movement of Chloroplasts in Algae

3.1 Green Algae

While many motile algae adjust the amount of light they absorb by changing their position, some algae which are not motile have developed another mechanism to accomplish the same end. In these, the chloroplasts, rather than the whole plant, readjust their position in response to light (see BRITZ, Chap. 2.5, this vol.). The strap-shaped chloroplasts of *Mougeotia* present their full area to weak light, but rotate to an edge-on position at higher irradiance. The number of cells showing this response is somewhat different at different times of day (FLOHRS and HAUPT, 1971), and may represent a circadian rhythm (NEUSCHELER-WIRTH, 1971). Another example of chloroplast movement is found in the thin sheet–like thallus of *Ulva*, where the chloroplasts move from a face position during the day to a profile position at night. Because the light absorption of the whole thallus changes markedly with the position of the chloroplasts, it is possible to quantify the phenomenon by measuring the transmission of the thallus. BRITZ et al. (1976) have devised an automatic method for measuring changes in thallus transmission over time. These recordings show that the chloroplast movement in *Ulva* is circadian, the thallus transmission increases at night and decreases during the day in a light–dark cycle and continues this behavior under constant light of low intensity (3×10^{-4} W cm^{-2}) or darkness, for 3–6 cycles with a period of 25–26 h in constant darkness. The period shortens as the light intensity is increased (BRITZ et al., 1976). Unlike the chloroplasts in *Maugeotia* and other plants, *Ulva* chloroplasts in the face position could not be induced to reorient to the sides of the cell even in very bright light (80,000 lux). Thus, in *Ulva*, which lacks a direct response to illumination, the circadian rhythm seems to be the most important source of the regulation of light absorption.

The coenocytic green alga *Caulerpa prolifera* shows a different type of chloroplast reorientation. During the night in a light–dark cycle, the chloroplasts withdraw from the ends of the blades, leaving them white (DAWES and BARILOTTI, 1969). The chloroplasts return before the beginning of the following day. In continuous darkness, this behavior was observed for two additional cycles, and is presumably a manifestation of a circadian rhythm.

3.2 A Dinoflagellate, *Pyrocystis* (*Dissodinium*) *lunula*

Circadian rhythms in chloroplast movement are certainly not limited to the green algae, although not many other kinds of algae have been investigated in this respect. One interesting example is the movement of the chloroplasts in the large nonmotile cyst of the dinoflagellate *Pyrocystis* (*Dissodinium*) *lunula*. In this cell, the chloroplasts, and probably the entire cytoplasm gather in the center of the cell under high irradiance and spread out into the arms of the lunate cyst at lower light intensities (SWIFT and TAYLOR, 1967). At night, the chloroplasts are also concentrated near the center of the cell in four clusters, two in each arm of the cyst, just distal to the central nucleus. During the day, the chloroplasts are arranged more diffusely throughout the cell. In cultures in continuous darkness (Fig. 2), this change in the orientation of the chloroplasts continues for at least 4 days, although the number of cells which show the response is much smaller than in natural light. Another species of *Pyrocystis, P. fusiformis,* shows a similar rearrangement of the chloroplasts when grown in a light–dark cycle, but it has not been examined under constant conditions, so that the presence of a circadian rhythm has not been demonstrated in this species.

3.3 Diatoms

The chloroplasts of a number of diatoms change position with time of day. One of these is *Surirella gemma* (HOPKINS, 1965). In this estaurine mud flat diatom, the chloroplasts aggregate at night and disperse during the day. This change is observed for at least one day in the absence of light, and so may represent a circadian rhythm, although confirmation of this point is required. In a planktonic diatom, *Ditylum brightwellii*, the small chloroplasts are dispersed along the strands of cytoplasm during the day, and collected around the central nucleus at night. However, this movement is not observed in constant darkness (ANDERSON, 1974).

3.4 Changes in Chloroplast Shape and Rearrangement of the Thylakoids Within the Chloroplast

Not only can chloroplasts change their position in a cell, but they can also change their shape and even the arrangement of the thylakoids they contain. Some of

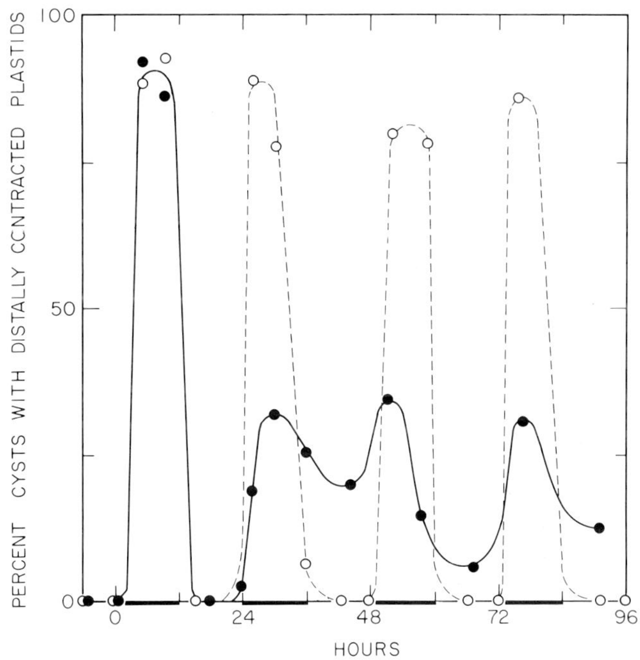

Fig. 2. The circadian rhythm in the position of the plastids in the lunulate cysts of *Pyrocystis (Dissodinium) lunula*. Cells which had been in light (340 μW cm^{-2}) for 85 h were placed in darkness at 0 time where they remained for 96 h *(solid circles)*. A control culture *(open circles)* received a 12-h light 12-h dark cycle starting with a dark period at 0 time. Four cycles of chloroplast movement can be distinguished in continuous darkness, chloroplasts occupy the center of the cyst during subjective night and are dispersed during the subjective day phase. (SWIFT and TAYLOR, 1967). *Bars on the abscissa* denote dark periods in the LD control

these movements are known to be rhythmic. If the chloroplasts of the large unicellular green alga, *Acetabularia,* are measured at different times of day in a light–dark cycle, the ratio of the length to the width varies from 4–5 at noon to only about 2 at midnight (VANDEN DRIESSCHE, 1966). These changes can still be observed when the cells remain in the light for one night. Accompanying these differences in shape were ultrastructural changes, most apparent in nucleated cells on a light–dark cycle. The number of appressed thylakoids increased at dawn as compared with other times (VANDEN DRIESSCHE and HARS, 1972a, 1972b). Whether or not these morphological variations could be observed in constant conditions was not reported. It would be interesting to know whether the rhythm in chloroplast shape contributes to the photosynthetic rhythm, which is well known in *Acetabularia*. Both show maxima at the same time of day.

In the dinoflagellate, *Gonyaulax polyedra*, the spacing of the thylakoids in the inner part of the chloroplasts can be observed with electron microscopy to change with time of day, being more widely spaced at noon and closer together at night (HERMAN and SWEENEY, 1975). This change also occurs in cells in continuous dim light, and thus constitutes a circadian rhythm in thylakoid position. However, the absorption of a cell suspension remains constant with time (PRÉZELIN et al., 1977). The rearrangement of the thylakoids in the central portion of the chloroplasts thus apparently serves another function. There is evidence that the chloroplast ribosomes are excluded from the space between the expanded thylakoids during the day phase, but are clearly present in this space at night. The nature of the material occupying the interthylakoid region during the day is unknown, but may represent a functional pyrenoid. Thus the change in the position of the thylakoids may be related to the effective use of the products of photosynthesis rather than to efficient light absorption.

Similar cyclic variation in thylakoid aggregation has been reported in synchronous cultures of *Scenedesmus*, but not under constant conditions (NILSHAMMAR-HOLMVALL, 1976).

4. Movement in the Leaves and Flowers of Angiosperms

4.1 Leaf-Movement Rhythms

4.1.1 *Phaseolus*

During the last 20 years, *Phaseolus* has not lost its popularity as an experimental tool in the study of circadian rhythms. The circadian properties of the sleep-movement rhythm have been defined with more precision. The species most commonly used have been *P. vulgaris* and *P. multiflorus*, now called *P. coccineus*. Using time-lapse photography, HOSHIZAKI and HAMNER (1964) recorded the persistence of the leaf-movement rhythm in *Phaseolus vulgaris* var. Pinto for four weeks in constant light of four different intensities at 27° C. Another study with time-lapse cinematography under carefully controlled constant conditions of light, temperature, CO_2, and relative humidity utilized *Phaseolus angularis* because the primary leaves are flatter, and thus better for photography (ALFORD and TIBBETTS, 1970). Very clear circadian rhythms were recorded, with a period of 27.4 $\pm$1.1 h, superimposed on which was a short-period oscillation, as is often seen with other species of *Phaseolus*. The seeds were germinated under constant conditions and the individual primary leaves even on the same plant were not in synchrony, since there was no entraining light–dark cycle. In *Phaseolus coccineus*, primary and secondary pulvini may not behave identically under low continuous irradiance (MAYER and SADLEDER, 1972). Between 6 and 6000 lux, the period of the movement rhythm is shorter in secondary pulvini, although the period is the same in both types of pulvini in constant darkness or in continuous bright light (30,000 lux).

A careful determination of the temperature dependence of the period of the *Phaseolus* rhythm has led to a revision of the Q_{10} from 1.2 to a value closer to 1. In one series of experiments (LEINWEBER, 1961), the period only changed from 28.0 to 28.3 h when the temperature was raised from 15° to 25 °C, and in another, the period remained essentially unchanged between 22° and 28 °C (BÜNNING et al., 1963). Changes in phase which follow short exposures to cold or warm temperatures or to light of different spectral composition were determined (MOSER, 1962; WAGNER, 1963; BÜNNING and MOSER, 1966; BÜNNING and MOSER,

1967). A rhythm can be initiated by a single stimulus of different kinds, such as darkening for at least 9 h or cooling to 5 °C for 18 h (WASSERMANN, 1959).

In connection with the reporting of experiments on changes in phase, it may be worthwhile to point out that the phase reference point considered zero circadian time in Bünning's laboratory is the time when the leaves are in their lowest position, that is, midnight, while it is customary in the United States to follow Pittendrigh's convention based on the time of eclosion of *Drosophila* or dawn as the phase reference or zero circadian time.

It has been difficult to understand the selective advantage of sleep movements in leaves. Various suggestions have been made, including water conservation at night, or the prevention of excessive heat loss. However, no difference in frost damage was found between leaves in the sleeping position and leaves prevented from closing (SCHWINTZER, 1971). The most original suggestion to date (BÜNNING and MOSER, 1969; BÜNNING, 1969, 1971) is that by folding together at night, leaves absorb less moonlight, and the light of a full moon is prevented from interfering with photoperiodic timing.

The most interesting recent work on the leaf movements of *Phaseolus* is directed toward gaining understanding of the circadian oscillator by examining the effects of dipping the cut end of the stem into solutions containing various chemicals and subsequently following the rhythm in constant conditions. While many of these substances were without effect, a lengthening of the period was observed in the presence of 2.5% ethanol (−4 h) and 3% methanol (−3 h), 0.1% theophyllin (−5 h) and 0.1% theobromine (−2 h) (KELLER, 1960). Two-hour exposure to 15–30% ethanol was followed by a phase shift, the amount and direction of which was dependent on when the treatment was given, in a manner reminiscent of a light phase-response curve (BÜNNING and BALTES, 1962; cf. also Fig. 1). Treatment for 4 h with theophyllin (10^{-2} M) shifted phase in *Phaseolus coccineus* in a manner dependent again on when it was administered (MAYER et al., 1975). Since theophyllin does not inhibit the phosphodiesterase of higher plants, the mechanism by which it exerted a phase effect was probably not the prevention of the destruction of cAMP, but may have been mediated via a change in membrane function caused by changes in the binding of Ca^{2+}. Caffeine also shifted phase of the circadian rhythm of leaf movement in this *Phaseolus coccineus* (MAYER and SCHERER, 1975). As in algae, deuteration leads to a lengthening of the period in the leaf-movement rhythm in *Phaseolus* (BÜNNING and BALTES, 1963). Fluctuations in magnetic field did not entrain this rhythm (TIBBITS et al., 1973). This experiment was undertaken because Brown has repeatedly suggested that diurnal fluctuations in the earth's magnetic field may be responsible for circadian rhythmicity (BROWN et al., 1970). In general, however, there is no evidence to support this view.

A most interesting series of phase changes was observed following exposure for 5 h to valinomycin via the transpiration stream at different times of day (Table 2a; BÜNNING and MOSER, 1972). This potassium ionophore caused large advance phase shifts in the early part of the day phase, changing to delays late in the day and during the night phase. Wilting for short times caused similar phase shifts. Advances in phase also followed 4-h exposure to 0.2–0.3 M KCl during the early day phase, but no significant phase delays were observed (Table 2c, BÜNNING and MOSER, 1973). These effects of valinomycin and potassium are not specific to the *Phaseolus* leaf-movement rhythm, since a very similar phase-response curve to valinomycin, displaced in time, was observed on treating *Gonyaulax* cul-

Table 2. Phase shifts in hours after treatment with valinomycin or KCl at different times in a circadian cycle. Times are given as the midpoint of the treatment in circadian time (zero is taken as the beginning of a subjective day, i.e., when behavior typical of dawn in a 12-h light–12-h dark cycle occurs)

a) *Phaseolus coccineus* (BÜNNING and MOSER, 1972) valinomycin 5 h 10^{-8} M in 1% EtOH		b) *Gonyaulax polyedra* (SWEENEY and HERZ, 1977) valinomycin 4 h 2×10^{-7} M in 0.1% acetone		c) *Phaseolus coccineus* (BÜNNING and MOSER, 1973) KCl 4 h 0.3 M		d) *Aplysia californica* (ESKIN, 1972) KCl 4 h 0.1 M	
Time	Phase shift (h)	Time	Phase shift (h)	Time	Phase shift (h)	Time	Phase shift (h)
0000						00	+5
0020				02	+1		
0040	+6.5	04	+1.3	04	+4	04	+2.5
0850	+0.5	08	+2.5	09	+2.8	12	−2
1650	−2	12	+1.6	16	+2	18	−5
2100	0	20	−1.5	21	+1.5	20	−9
2350	+4.5	23	0	23	−0.5	22	0

tures with valinomycin and measuring the circadian rhythm in bioluminescence (Table 2b; SWEENEY, 1974). In the rhythm in the frequency of spontaneous electrical signals from the optic nerve of *Aplysia*, adding KCl also shifts the phase depending on the time of application (Table 2d; ESKIN, 1972). The possible interpretation of these experiments with regard to the mechanism for the generation of circadian rhythms will be discussed later.

4.1.2 Legumes Other than *Phaseolus*

The sleep movements of leaves of many other legumes beside *Phaseolus* are known to be circadian. Kleinhoonte's beautiful records of *Canavalia ensiformis* leaf movements are reproduced in BÜNNING, 1959. Soybean leaf movements have also been studied (BREST et al., 1971) during light–dark cycles of 8 h light–64 h darkness to compare leaf behavior with flowering response rhythms in soybean, which have been extensively studied under this light–dark regime. The leaves showed three cycles of a sleep movement rhythm during the long dark period, like the rhythm in the inhibition of flowering by a light interruption. Interestingly, a light exposure (3 min at 10,000 lux from Growlux fluorescent lamps), effective in inhibiting flowering, did not change the phase of the leaf-movement rhythm at all. The leaf-movement rhythm was phase-shifted by longer light pulses of at least 2 h duration. This result clearly identifies two separate light responses, with different sensitivities, one concerned with flowering itself, and a second which acts on the circadian oscillator, as in other rhythmic systems. The leaf sleep movements in tropical trees such as *Albizzia julibrissin* and *Samanea saman* are particularly striking. These two species have been the subjects of a series of recent studies by SATTER and her colleagues (for a more detailed discussion, see SATTER, Chap. 4.2., this vol.). Pairs of leaflets cut from the compound leaves of *Albizzia* show sleep movements when floated on water and are convenient experimental material. In the larger *Samanea* leaves, sleep movements can be observed in a single terminal pulvinus from which the blade has been removed, if the remaining rachis is dipped in liquid in a small vial (SATTER et al., 1974). If sugar is supplied, movements will continue in darkness for at least seven cycles (SIMON et al., 1976). It is evident that the pulvinus alone is sufficient for the expression of rhythmic leaf movement.

When leaves of legumes open, they do so because the motor cells in the adaxial region of the pulvinus, the extensor cells, become more turgid, while those on the opposite side of the pulvinus, the flexor cells, lose water. Opening of the leaflets requires an intact respiration and protein synthesis and appears to be an active process (APPLEWHITE et al., 1973; SATTER et al., 1974). By microprobe analysis (SATTER and GALSTON, 1971a, 1971b), a redistribution of K^+ could be shown to accompany the turgor changes in the pulvini of *Albizzia* and in *Samanea* (SATTER et al., 1974). In both species, Cl^- accompanies K^+ (SCHREMPF et al., 1976; SATTER et al., 1977). It is also possible to detect changes in the transmembrane potential in *Samanea* following the ion flux (RACUSON and SATTER, 1975).

Work on the sleep movements of *Trifolium repens* has paralleled that on *Albizzia* and *Samanea*. The endogenous circadian nature of the oscillation was demonstrated in constant darkness and light (SCOTT and GULLINE, 1972), and both changes in K^+ and transmembrane potential accompany leaf movements (SCOTT and GULLINE, 1975; SCOTT et al., 1975). Changes in the distribution of K^+ associated with the circadian leaf movements of *Phaseolus vulgaris* have also been detected (KIYOSAWA and TANAKA, 1976). These ion and electrical changes in legumes occur rhythmically and may provide insight into the mechanism by which circadian oscillations are brought about in the pulvinus.

In *Samanea* pulvini excised and dipped into sucrose solution (50 mM sucrose), the rhythm persists long enough in continuous darkness to allow a study of the phase-shifting by light (SIMON et al., 1976). A 5-min red light pulse is sufficient, and a typical phase-response curve is obtained, phase delays suddenly changing to phase advances toward the end of the subjective night. Phase shifts by red light (5 min) can be negated by far-red light (1.5 min), so that phytochrome is undoubtedly the photoreceptor for phase-shifting this rhythm (Fig. 3).

4.1.3 Nonleguminous Plants

That a specialized pulvinus is not required for a circadian rhythm in leaf movement is clear. A leaf-movement rhythm which can be reset by 3-h pulses of white light has been studied in *Portulaca grandiflora* (KARVÉ and JIGAJINNI, 1965, 1966). A rhythmic movement of the leaves of cotton persisted in constant light for as long as 15 days with a circadian period (MILLER, 1970). Plants from different latitudes in Europe were screened for leaf-movement rhythms (MAYER, 1966) in constant light (1200 lux and 17 °C). Two *Taraxacum, T. arcticum* from Spitzbergen and *T. officinale* from middle Europe, and the tropical *Erythrina senegalensis* were rhythmic in continuous light with periods only slightly different from 24 h. No striking differences correlated with habitat latitude were observed in these species. Other tropical species in which leaf-movement rhythms could be detected in continuous light tended to show unusually long periods at 17 °C, up to 32 h in *Phaseolus*

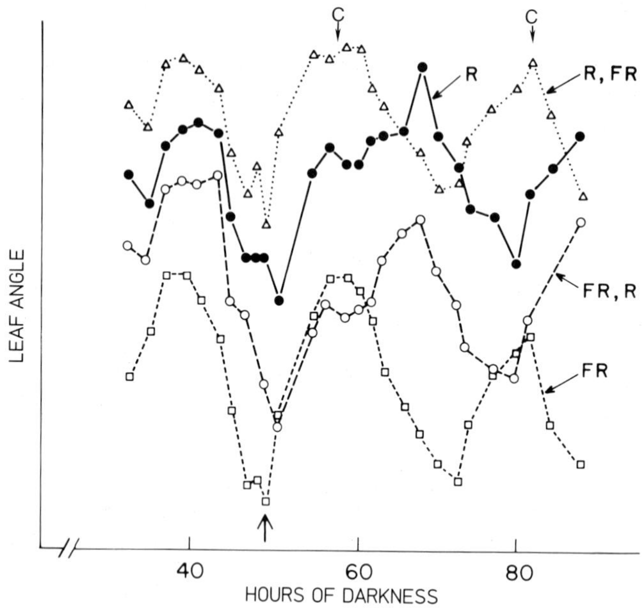

Fig. 3. Shifting the phase of rhythmic leaf movement in *Samanea saman* with red and reversal by far-red irradiation. Pulvini which had been in continuous darkness for 48 h were irradiated *(arrow)* with one of the following: red light for 5 min *(R)*, far-red light (1.5 min) *(FR)*, *R* followed by *FR* and *FR* followed by *R*. The phase was shifted by red light pulses either given alone or following *FR*, while *FR* light reversed the effect of *R*, showing that the phase shifting depends on phytochrome as photoreceptor. (SIMON et al., 1976). *Arrows C* (control) mark maximum leaf angle of control unirradiated pulvini

mungo. At 32 °C, however, a temperature approximately that of tropical habitats, the period of the leaf movements of these tropical plants was about 24 h.

The most thoroughly studied of leaf-movement rhythms in nonleguminous plants is that in *Coleus*, a hybrid of *C. blumei* × *C. frederici* which flowers as a short-day plant. The leaves of this *Coleus* continue to move up and down in constant darkness for about four cycles, and for a longer time in constant light of low intensity (HALABAN, 1968a). This rhythm is reset by light pulses in the same pattern as other typically circadian rhythms (HALABAN, 1968b). Especially interesting is the finding that blue light is effective during the time in the cycle when phase delays occur, while red light is active during the part of the cycle when phase advances are observed (HALABAN, 1969). The time in the cycle when light inhibits flowering in *Coleus* is always 5 h after the minimum position of the leaves, irrespective of the light–dark cycle in effect. The constant relationship between the phase of the leaf-movement rhythm and the maximum sensitivity of flowering to light interruptions suggests identity in the timing mechanisms for floral induction and leaf movement (HALABAN, 1968c). On the other hand, there is some evidence that the time relationship between the minimum leaf position and the maximum sensitivity to light interruption for flowering is not constant in at least some plants, for example *Xanthium strumarium* (SALISBURY and DENNY, 1971) and *Pharbitis nil* (BOLLIG, 1975).

4.2 Petal-Movement Rhythms

The petals as well as the leaves of flowering plants may open and close over the course of a day. In some of these plants, this behavior is known to be governed by a circadian timer. The rhythm in petal movement of the succulent plant, *Kalanchoë blossfeldiana,* first demonstrated by Bünsow in 1953, is the best known of these rhythms at present. It continues for four to five cycles in constant darkness, with a Q_{10} of 1.2–1.24 between 15° and 30°C (OLTMANNS, 1960). The phase-response curve for single light pulses has been determined and is of the typical shape for circadian systems generally (ZIMMER, 1962; BÜNNING and ZIMMER, 1962; BÜNNING and BLUME, 1963; ENGELMANN et al., 1973). Petal cycles in *Kalanchoë,* which is a short-day plant, are well correlated with floral induction during dark periods of different lengths (ENGELMANN, 1960). Flowers were open during the time when flowering was insensitive to light interruptions, and closed when light breaks inhibited flowering.

The flowers of *Kalanchoë* continue to open and close after they are cut from the plant. ENGELMANN et al. (1974a) have devised an automatic monitor for this rhythm, using detached flowers mounted on a rotating disc with their bases in sugar solution. As each flower passes under a photodetector the amount of light (from a green fluorescent lamp with green Cinemoid filter) which reaches the photodetector through the flower is recorded. When the flower is open, less light is recorded than when it is closed. With this device, readings of the position

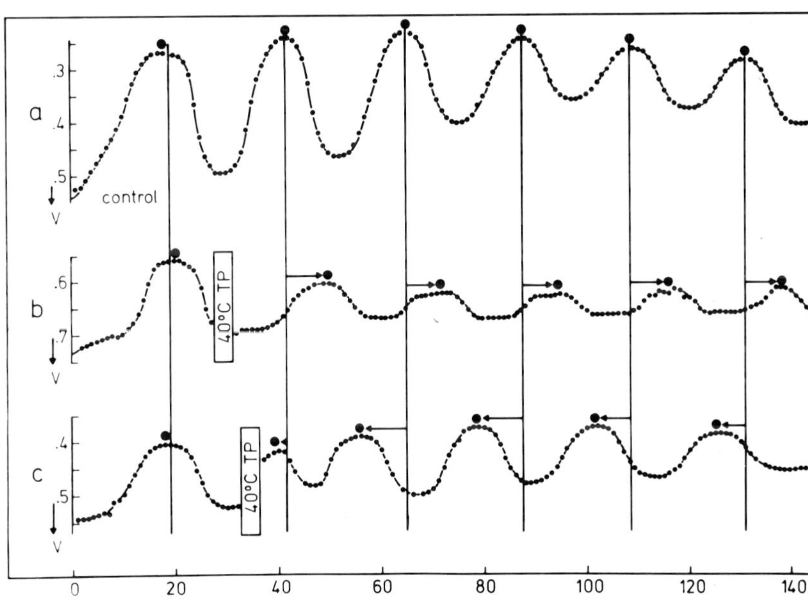

Fig. 4. Shifting the phase of the rhythm in petal movement in *Kalanchoë blossfeldiana* with a 40 °C pulse at different circadian times. *Abscissa,* hours after transfer of the flowers from 12-h light–12-h dark physiological darkness (weak green light). *Ordinate,* voltage of photoresister which decreases as flowers open. ●, denotes maximum flower opening; *a,* control; *b,* 40 °C pulse starts just before; *c,* just after maximum closing of flowers. (ENGELMANN et al., 1974b)

of each flower were made every hour. The presence of sucrose prolongs the rhythmicity as it does in excised pulvini.

With this equipment, the phase change in response to high temperature pulses (40 °C for 3 h) was measured (Fig. 4) and the phase-response curve was found to resemble a light phase-response curve (ENGELMANN et al., 1974b). These curves fitted rather closely calculated curves generated by a feedback mathematical model for the circadian oscillator (ENGELMANN et al., 1973). The presence of lithium ions and deuteration slow the clock in *Kalanchoë* (ENGELMANN, 1972; ENGELMANN et al., 1974a).

Less is known concerning petal-movement rhythms in other flowers, although the evening opening of the flowers of *Oenothera* continues for six cycles in constant light (5000 lux, 8–21° C) (ARNOLD, 1959). In *Cestrum* also, flower opening, as well as odor production, is circadian and the period in continuous light is only slightly temperature-dependent (OVERLAND, 1960).

5. The Opening and Closing of Stomata

The opening of stomata is mediated by an increase in the turgidity of the guard cells, brought about by an increase in their osmotic pressure[1]. While light and a low partial pressure of CO_2 favor stomatal opening, there is evidence for an endogenous component as well. Oscillations in the degree of opening with a short period, about 20 min, have been recognized in *Vicia faba* for a long time (STÅLFELT, 1929), and documented more recently (EHRLER, 1971) in *Citrus jambhiri* in constant darkness (period = 60–85 min). A circadian component in stomatal behavior has also been recognized (STÅLFELT, 1963, 1965), which persists for only a single cycle in either continuous darkness or constant light. The osmotic pressure of the guard cells in *Vicia faba* measured at incipient plasmolysis oscillates in synchrony with the aperture of the stomata (STÅLFELT, 1967) as would be expected, since an increase of turgor in the guard cells causes the stoma to open. Similar oscillations persisting for one cycle in constant conditions have also been detected in *Tradescantia virginiana* (MARTIN and MEIDNER, 1972), in *Xanthium pennsylvanicum* (MANSFIELD and HEATH, 1963), wheat (MEIDNER and MANSFIELD, 1965), and maize (KARVÉ and MISHAL, 1966) and may be present in *Antirrhinum majus* (PALLAS, 1969), although one wishes the data were more convincing. In constant light (3500 lux), two cycles of a circadian rhythm in the degree of opening of the stomata of *Stellaria media* and *Oxyria digyna* could be detected (HAAPALA, 1967).

The time required for the stomata of the leaves of the banana, *Musa acuminata*, to open when they are illuminated after a long dark period varies with the length of the dark period, the most rapid opening response being after 12, 36, and 60 h in darkness at 29 °C (BRUN, 1962). This constitutes evidence for a circadian rhythm in the rate rather than the degree of stomatal opening.

Active influx of K^+ is responsible for the increase in the osmotic pressure in the guard cells which, in turn, leads to water uptake and increase in turgidity and opening of the stomata (FISCHER, 1968). This is reminiscent of the potassium fluxes accompanying the turgor changes in the pulvinus of *Albizzia, Samanea* and *Trifolium* discussed above. In the case of guard cells, K^+ influx is across a membrane into a single cell, while in the pulvinus the path of K^+ is less clear.

[1] For a detailed discussion of stomatal movement see RASCHKE, Chap. 4.1, this vol.

6. The Significance of the Rhythms in Plant Movements for an Understanding of the Mechanism of Circadian Oscillations

The finding that leaves of *Albizzia, Samanea, Trifolium* and *Phaseolus* open in response to an active influx of potassium ions into the extensor motor region of the pulvinus suggests the likelihood that a similar mechanism is responsible for turgor changes in all pulvini. Leaves and petals of other species without pulvini that move up and down may also do so by turgor changes influenced by K^+ fluxes. Active K^+ fluxes are probably responsible for the opening of stomata. The existence of circadian rhythms in leaf sleep movements, petal movements and stomatal opening suggests that the membranes of the cells responsible for these movements change with time of day in their ability to transport K^+. It is possible, furthermore, that all the cells of the plant behave similarly, and the difference in membrane properties is not limited to the special cells responsible for these movements. Underlying the rhythms in movement in angiosperms may be a circadian rhythm in membrane function.

Two observations lead to the conclusion that this rhythm in membrane function may be fundamental to circadian timing rather than merely another example of a circadian rhythm in a physiological process. The first is the observation that, in *Phaseolus* (BÜNNING and MOSER, 1972) and *Gonyaulax* (SWEENEY, 1974a), the phase of the rhythm, a fundamental clock property, can be shifted by short exposures to the ionophore valinomycin which alters the permeability of membranes to K^+, and does so in a manner which is dependent on the time of treatment (Table 2). Thus valinomycin causes phase changes which reflect changes in clock time in two very different rhythmic systems. The second piece of evidence which suggests that membrane changes are a component of a circadian oscillator is that transforming the phytochrome from the P_r to the P_{fr} form changes the phase of the leaf-movement rhythm in *Samanea* (SIMON et al., 1976). There is evidence that phytochrome is located in membranes in vivo and that a change in electrical potential occurs during the P_r to P_{fr} transformation (JAFFE, 1968; NEWMAN and BRIGGS, 1972).

The membrane hypothesis for the generation of circadian rhythms was first advanced on empirical grounds in 1974 (NJUS et al., 1974; SWEENEY, 1974b). Mathematical models have shown that circadian rhythms with the properties observed including the typical phase-response curve can theoretically arise from a limit cycle (PAVLIDIS, 1973), in which two quantities vary, each depending on the other. The membrane hypothesis suggests that these quantities are the ion gradient across a membrane and the transport activity of the membrane with respect to this ion (NJUS et al., 1976). The importance of K^+ transport in leaf- and stomatal-movement rhythms and the phase shifts observed after short exposure to the K^+ ionophore, valinomycin, make K^+ an attractive candidate for the ion involved. The failure to find any prokaryote which displays a circadian rhythm suggests that membranes of internal compartments are the membranes in question, particularly since it would be difficult to generate a limit cycle in the plasma membrane because of the infinitely large external ion reservoir.

Membranes as the site of the oscillator are also attractive because of the diversity of the physiological processes which display circadian rhythms (photosynthesis, bioluminescence and cell division in *Gonyaulax,* for example). NJUS et al. (1974) pointed out that membranes are known to adjust the level of saturation of their fatty acids to compensate for changes in environmental temperature, maintaining fluidity constant, and this phenomenon might account for temperature compensation of the period in circadian rhythms. Phase shifts following a short irradiation may be understood as the discharge of the ion gradient by light. As yet there is little or no direct experimental evidence to verify or disprove this very general membrane hypothesis. For a more extensive discussion of this question, see SWEENEY (1976) and SWEENEY and HERZ (1977).

The rhythms in phototaxis, motility and chloroplast movement in algae cannot at the present time be interpreted in the light of the membrane hypothesis for the mechanism of circadian clocks. Not enough is known concerning how changes in K^+ or other ions and membrane properties might be expected to change the rate or the direction of movement in flagellates or chloroplasts. It is not unreasonable to predict, however, that both ion fluxes and membrane properties will be found to control these movements, as they do the leaf, petal and stomatal movements in angiosperms.

Acknowledgments. The author wishes to express her deep appreciation to the National Science Foundation for support through Grant No. BMS 72-02221A03 for her research on rhythms and during the writing of this review.

References

Alford, D.K., Tibbetts, T.W.: Circadian rhythm of leaves of *Phaseolus angularis* plants grown in a controlled carbon dioxide and humidity environment. Plant Physiol. *46,* 99–102 (1970)

Anderson, L.W.J.: Studies on the role of cellular lipids and inorganic ions in controlling sinking rates in *Ditylom broghtwelli,* a marine centric diatom. Ph. D. dissertation, Univ. California, Santa Barbara, 1974

Anderson, L.W.J., Sweeney, B.M.: Diel changes in sedimentation characteristics of *Ditylum brightwelli:* changes in cellular lipid and effects of respiratory inhibitors and ion transport modifiers. Limnol. Oceanogr. *22,* 540–552 (1977)

Applewhite, P.B., Satter, R.L., Galston, A.W.: Protein synthesis during endogenous rhythmic leaflet movement in *Albizzia.* J. Gen. Physiol. *62,* 707–713 (1973)

Arnold, G.G.: Die Blütenöffnung bei *Oenothera* in Abhängigkeit vom Licht-Dunkelrhythmus. Planta *53,* 198–211 (1959)

Aschoff, J. (ed.): Circadian Clocks. Amsterdam: North-Holland 1965

Bierhuizen, J.F. (ed.): Circadian Rhythms. Proc. Int. Symp. Circadian Rhythmicity, Wageningen the Netherlands, 26–29 April, 1971. Wageningen: Centre for Agricultural Publishing and Documentation, 1972

Bolling, I.: Photoperiodic time measurement and circadian leaf movement in *Pharbitis nil* controlled by the same clock? Z. Pflanzenphysiol. *77,* 54–69 (1975)

Brest, D.E., Hoshizaki, T., Hamner, K.C.: Rhythmic leaf movements in Biloxi Soybean and their relation to flowering. Plant Physiol. *47,* 676–681 (1971)

Brinkmann, K.: Temperatureinflüsse auf die circadiane Rhythmik von *Euglena gracilis* bei Mixotrophie und Autotrophie. Planta *70,* 344–389 (1966)

Brinkmann, K.: The effect of ethanol on the metabolism and circadian rhythm of *Euglena gracilis.* J. Interdiscip. Cycle Res. *5,* 186 (1974)

Brinkmann, K.: The influence of alcohols on the circadian rhythm and metabolism of *Euglena gracilis.* J. Interdiscip. Cycle Res. *7,* 149 (1976)

Britz, S.J., Briggs, W.R.: Circadian rhythm of chloroplast orientation and photosynthetic capacity in *Ulva*. Plant Physiol. *58*, 22–27 (1976)
Britz, S.J., Pfau, J., Nultsch, W., Briggs, W.R.: Automatic monitoring of a circadian rhythm of change in light transmittance in *Ulva*. Plant Physiol. *58*, 17–21 (1976)
Britz, S.J., Seliger, H.H.: Endogenous and photoperiodic diurnal rhythms of *in vivo* light absorption and scattering in the green alga *Ulva lactuca* L. Biol. Bull. *144*, 12–18 (1973)
Brown, F.A., Jr., Hasting, J.W., Palmer, J.D.: The Biological Clock, Two Views. New York-London: Academic Press, 1970
Bruce, V.G.: Environmental entrainment of a circadian rhythm. Cold Spring Harbor Symp. Quant. Biol. *25*, 29–48 (1960)
Bruce, V.G.: The biological clock in *Chlamydomonas reinhardi*. J. Protozool. *17*, 328–334 (1970)
Bruce, V.G.: Mutants of the biological clock in *Chlamydomonas reinhardi*. Genetics *70*, 537–548 (1972)
Bruce, V.G.: Recombinants between clock mutants of *Chlamydomonas reinhardi*. Genetics *77*, 221–230 (1974)
Bruce, V.G., Pittendrigh, C.S.: Temperature independence of a unicellular "clock". Proc. Natl. Acad. Sci. USA *42*, 676–682 (1956)
Bruce, V.G., Pittendrigh, C.S.: Resetting the *Euglena* clock with a single light stimulus. Am. Nat. *92*, 295–306 (1958)
Bruce, V.G., Pittendrigh, C.S.: An effect of heavy water on the phase and period of the circadian rhythm in *Euglena*. J. Cell. Comp. Physiol. *56*, 25–31 (1960)
Brun, W.A.: Rhythmic stomatal opening responses in banana leaves. Physiol. Plant. *15*, 623–630 (1962)
Bünning, E.: Tagesperiodische Bewegungen. In: Encyclopedia of Plant Physiology. Ruhland, W. (ed.). Vol. XVII/1 Berlin-Heidelberg-New York: Springer 1959, pp. 579–656
Bünning, E.: Die Bedeutung tagesperiodischer Blattbewegungen für die Präzision der Tageslängenmessung. Planta *86*, 209–217 (1969)
Bünning, E.: The adaptive value of circadian leaf movements. In: Biochronometry. Menaker M. (ed.). Washington, D.C.: Natl. Acad. Sci., 1971, pp. 203–211
Bünning, E.: The Physiological Clock. Revised 3rd ed. New York-Heidelberg-Berlin: Springer 1973
Bünning, E.: Critical remarks concerning the Q_{10}-values in circadian rhythms. Int. J. Chronometry *2*, 343–346 (1974)
Bünning, E., Baltes, J.: Wirkung von Äthylalkohol auf die physiologische Uhr. Naturwissenschaften *49*, 19–20 (1962)
Bünning, E., Baltes, J.: Zur Wirkung von schwerem Wasser auf die endogene Tagesrhythmik. Naturwissenschaften *50*, 622–623 (1963)
Bünning, E., Blume, J.: Zur Analyse der Wirkung von Störlicht und von abnormen Licht-Dunkel-Cyclen auf die endogene Tagesrhythmik. Z. Bot. *51*, 52–60 (1963)
Bünning, E., Kurras, S., Vielhaben, V.: Phasenverschiebungen der endogenen Tagesrhythmik durch Reduktion der Atmung. Planta *64*, 291–300 (1965)
Bünning, E., Moser, I.: Response-Kurven bei der circadianen Rhythmik von *Phaseolus*. Planta *69*, 101–110 (1966)
Bünning, E., Moser, I.: Weitere Versuche zur Lichtwirkung auf die circadiane Rhythmik von *Phaseolus multiflorus*. Planta *77*, 99–107 (1967)
Bünning, E., Moser, I.: Interference of moonlight with the photoperiodic measurement of time by plants, and their adaptive reaction. Proc. Natl. Acad. Sci USA *62*, 1018–1022 (1969)
Bünning, E., Moser, I.: Influence of valinomycin on circadian leaf movements of *Phaseolus*. Proc. Natl. Acad. Sci. USA *69*, 2732–2733 (1972)
Bünning, E., Moser, I.: Light-induced phase shifts of circadian leaf movements of *Phaseolus*: comparison with the effects of potassium and of ethyl alcohol. Proc. Natl. Acad. Sci USA *70*, 3387–3389 (1973)
Bünning, E., Moser, I., Wolf, G.: Über die Regelung der plagiotropen Blattlage. Z. Bot. *52*, 86–97 (1963)
Bünning, E., Zimmer, R.: Zur Deutung der Phasenverschiebungen und "Transients" nach exogener Störung endogener Rhythmen. Planta *59*, 1–14 (1962)

Chovnick, A. (ed.): Biological Clocks. Cold Spring Harbor Symp. Quant. Biol. *25*, 524 (1960)
Coursey, P. de (ed.): Biological Rhythms in the Marine Environment. Columbia: Univ. S. Carolina Press, 1976
Cumming, B.G., Wagner, E.: Rhythmic processes in plants. Annu. Rev. Plant Physiol. *19*, 381–416 (1968)
Dawes, C., Barilotti, D.C.: Cytoplasmic organization and rhythmic streaming in growing blades of *Caulerpa prolifera*. Am. J. Bot. *56*, 8–15 (1969)
Driessche, T. vanden: Circadian rhythms in *Acetabularia*: photosynthetic capacity and chloroplast shape. Exp. Cell Res. *42*, 18–30 (1966)
Driessche, T. vanden, Hars, R.: Variations circadiennes de l'ultrastructure des chloroplastes d'*Acetabularia*. I. Algues entières. J. Microsc. *15*, 85–90 (1972a)
Driessche, T. vanden, Hars, R.: Variations circadiennes de l'ultrastructure des chloroplastes d'*Acetabularia*. II. Algues anucléées. J. Microsc. *15*, 91–98 (1972b)
Ehrler, W.L.: Periodic nocturnal stomatal opening of *Citrus* in a steady environment. Physiol. Plant. *25*, 488–492 (1971)
Engelmann, W.: Endogene Rhythmik und photoperiodische Blühinduktion bei *Kalanchoe*. Planta *55*, 496–511 (1960)
Engelmann, W.: Lithium slows down the *Kalanchoe* clock. Z. Naturforsch. *27b*, 477 (1972)
Engelmann, W., Eger, I., Johnsson, A., Karlsson, H.G.: Effect of temperature pulses on the petal rhythm of *Kalanchoe*: an experimental and theoretical study. Int. J. Chronobiol. *2*, 347–358 (1974b)
Engelmann, W., Karlsson, H.G., Johnsson, A.: Phase shifts in the *Kalanchoe* petal rhythm caused by light pulses of different duration. Int. J. Chronobiol. *1*, 147–156 (1973).
Engelmann, W., Maurer, A., Muhlbach, M., Johnsson, J.: Action of lithium ions and heavy water in slowing circadian rhythm of petal movement in *Kalanchoe*. J. Interdiscip. Cycle Res. *5*, 199–205 (1974a)
Eppley, R.W., Holm-Hansen, O., Strickland, J.D.H.: Some observations on the vertical migration of dinoflagellates. J. Phycol. *4*, 333–340 (1968)
Eskin, A.: Phase shifting a circadian rhythm in the eye of *Aplysia* by high potassium pulses. J. Comp. Physiol. *80*, 353–376 (1972)
Fauré-Fremiet, E.: The tidal rhythm of the diatom *Hantzschia amphioxys*. Biol. Bull. *100*, 173–177 (1951)
Feldman, J.F.: Lengthening the period of a biological clock in *Euglena* by cycloheximide, an inhibitor of protein synthesis. Proc. Natl. Acad. Sci. USA *57*, 1080–1087 (1967)
Fischer, R.A.: Stomatal opening: role of potassium uptake of guard cells. Science *160*, 784–785
Flohrs, H., Haupt, W.: Daily fluctuations of sensitivity of the light-induced chloroplast movement in *Mougeotia*. Z. Pflanzenphysiol. *65*, 65–69 (1971)
Forward, R.B., Jr.: Phototaxis by the dinoflagellate *Gymnodinium splendens* Lebour. J. Protozool. *21*, 312–315 (1974)
Forward, R.B., Jr., Davenport, D.: The circadian rhythm of a behavioral photoresponse in the dinoflagellate *Gyrodinium dorsum*. Planta *92*, 259–266 (1970)
Haapala, H.: Studies on the endogenous rhythm of stomatal movements in *Oxyria digyna* (L.) Hill and *Stellaria media* (L.) Vill. Aquilo, Ser. Bot. *5*, 120–132 (1967)
Halaban, R.: The circadian rhythm of leaf movement of *Coleus blumei* × *C. frederici*, a short day plant. I. Under constant conditions. Plant Physiol. *43*, 1883–1886 (1968a)
Halaban, R.: The circadian leaf movement rhythm of *Coleus blumei* × *C. frederici*, a short day plant. II. The effects of light and temperature signals. Plant Physiol. *43*, 1887–1893 (1968b)
Halaban, R.: The flowering response of *Coleus* in relation to photoperiod and the circadian rhythm of leaf movement. Plant Physiol. *43*, 1894–1898 (1968c)
Halaban, R.: Effects of light quality on the circadian rhythm of leaf movement of a short-day-plant. Plant Physiol. *44*, 973–977 (1969)
Hastings, J.W., Schweiger, H.-G. (eds.): The Molecular Basis of Circadian Rhythms. Berlin: Life Sciences Research Report 1, Dahlem Konferenzen, 1976
Herman, E.M., Sweeney, B.M.: Circadian rhythm of chloroplast ultrastructure in *Gonyaulax polyedra*, concentric organization around a central cluster of ribosomes. J. Ultrastr. Res. *50*, 347–354 (1975)

Hillman, W.S.: Biological rhythms and physiological timing. Annu. Rev. Plant Physiol. *27,* 159–179 (1976)
Hopkins, J.T.: Some light-induced changes in behaviour and cytology of an estuarine mud-flat diatom. In: Light as an Ecological Factor. Bainbridge, R., Evans, G.C., Rockham, O. (eds.). John Wiley, New York: 1965, pp. 335–358
Hoshizaki, T., Hamner, K.C.: Circadian leaf movements: persistence in bean plants grown in continuous high-intensity light. Science *144,* 1240–1241 (1964)
Jaffe, M.J.: Phytochrome-mediated bioelectric potentials in mung bean seedlings. Science *162,* 1016 1017 (1968)
Karvé, A.D., Jigajinni, S.G.: Circadian rhythm in *Portulaca.* Z. Pflanzenphysiol. *53,* 169–172 (1965)
Karvé, A.D., Jigajinni, S.G.: Leaf movements of *Portulaca* under different photoperiods. Z. Pflanzenphysiol. *54,* 270–274 (1966)
Karvé, A.D., Mishal, B.D.: The phenomenon of overshoot in the case of stomatal opening. Naturwissenschaften *53,* 280 (1966)
Keller, S.: Über die Wirkung chemischer Faktoren auf die tagesperiodischen Blattbewegungen von *Phaseolus multiflorus.* Z. Bot. *48,* 32–57 (1960)
Kirschstein, M.: Das rhythmische Verhalten einer farblosen Mutante von *Euglena gracilis. Planta 85,* 126–134 (1969)
Kiyosawa, K., Tanaka, H.: Change in potassium distribution in a *Phaseolus pulvinus* during circadian leaf movement. Plant Cell Physiol. *17,* 289–298 (1976)
Leinweber, F.J.: Temperature coefficient of endodiurnal leaf movements in *Phaseolus.* Nature (London) *189,* 1028 (1961)
Mansfield, T.A., Heath, O.V.S.: Studies in stomatal behavior. IX. Photoperiodic effects on rhythmic phenomena in *Xanthium pennsylvanicu.* J. Exp. Bot. *14,* 334–352 (1963)
Martin, E.S., Meidner, H.: The phase-response of the dark stomatal rhythm in *Tradescantia virginiana* to light and dark treatments. New Phytol. *71,* 1045–1057 (1972)
Mayer, W.: Besonderheiten der circadianen Rhythmik bei Pflanzen verschiedener geographischer Breiten. Planta *70,* 237–256 (1966)
Mayer, W., Gruner, R., Strubel, H.: Period-lengthening and phase shifting of the circadian rhythm of *Phaseolus coccineus* by theophylline. Planta *125,* 141–148 (1975)
Mayer, W., Sadleder, D.: Unterschiedliche Lichtintensitätsabhängigkeit der Spontanperioden als Ursache interner Desynchronisation circadianer Rhythmen bei *Phaseolus coccineus.* Planta *108,* 173–178 (1972)
Mayer, W., Scherer, I.: Phase-shifting effects of caffeine on the circadian rhythm of *Phaseolus coccineus* L. Z. Naturforsch. *30c,* 855 (1975)
Meidner, H., Mansfield, T.A.: Stomatal responses to illumination. Biol. Rev. *40,* 483–509 (1965)
Menaker, M. (ed.): Biochronometry. Washington, D.C.: Natl. Acad. Sci. 1971
Miller, C.S.: Cotton leaf movements. Beltline Cotton Production Research Conference. 28 (1970)
Moser, I.: Phasenverschiebungen der endogenen Tagesrhythmik bei *Phaseolus* durch Temperatur- und Lichtintensitätsänderungen. Planta *58,* 199–219 (1962)
Neuscheler-Wirth, H.: Die lichtorientierte Chloroplastenbewegungen von *Mougeotia* in Abhängigkeit von der Vorkultur, vom Licht-Dunkel-Wechsel und vom Zellalter. Z. Pflanzenphysiol. *65,* 130–139 (1971)
Newman, I.A., Briggs, W.R.: Phytochrome-mediated electric potential changes in oat seedlings. Plant. Physiol. *50,* 687–693 (1972)
Nilshammar-Holmvall, M.: The cyclic variation in the thylakoid aggregation in normal and calcium deficient cells of *Scenedesmus.* Protoplasma *87,* 263–271 (1976)
Njus, D., Gooch, V.D., Mergenhafen, D., Sulzman, F., Hastings, J.W.: Membranes and molecules in circadian systems. Fed. Proc. *35,* 353–357 (1976)
Njus, D., Sulzman, F.M., Hastings, J.W.: Membrane model for the circadian clock. Nature (London) *248,* 116–120 (1974)
Oltmanns, O.: Über den Einfluß der Temperatur auf die endogene Tagesrhythmik und die Blühinduktion bei der Kurztagpflanze *Kalanchoë blossfeldiana.* Planta *34,* 233–264 (1960)
Overland, L.: Endogenous rhythm in opening and odor of flowers of *Cestrum nocturnum.* Am. J. Bot. *47,* 378–382 (1960)

Pallas, J.E., Jr.: Endogenous diurnal and nocturnal activity of guard cells as indicated by stomatal aperture and protoplasmic streaming. Physiol. Plant. *22,* 546–559 (1969)
Palmer, J.D., Round, F.E.: Persistent, vertical-migration rhythms in benthic microflora. VI. The tidal and diurnal nature of the rhythm in the diatom *Hantzschia virgata*. Biol. Bull. *132,* 44–55 (1967)
Pavlidis, T.: The free-run period of circadian rhythms and phase response curves. Am. Nat. *107,* 524–530 (1973)
Pohl, R.: Tagesrhythmus im phototaktischen Verhalten der *Euglena gracilis*. Z. Naturforsch. *3b,* 367–374 (1948)
Prézelin, B.B., Meeson, B.W., Sweeney, B.M., : Characterization of photosynthetic rhythms in marine dinoflagellates. I. Pigmentation, photosynthetic capacity and respiration. Plant. Physiol. *60,* 384–387 (1977)
Prézelin, B.B., Sweeney, B.M.: Characterization of photosynthetic rhythms in marine dinoflagellates. II. Photosynthesis-irradiance curves and *in vitro* chlorophyll *a* fluorescence. Plant. Physiol. *60,* 388–392 (1977)
Queiroz, O.: Circadian rhythms and metabolic processes. Annu. Rev. Plant. Physiol. *25,* 115–134 (1974)
Racusen, R., Satter, R.: Rhythmic and phytochrome-regulated changes in transmembrane potential in *Samanea* pulvini. Nature (London) *255,* 408–410 (1975)
Round, F.E.: Persistent, vertical-migration rhythms in benthic microflora. V. The effect of artificially Imposed light and dark cycles. Proc. 5th Seaweed Symp. Oxford: Pergamon Press 1966, pp. 197–203
Round, F.E., Palmer, J.D.: Persistent, vertical-migration rhythms in benthic microflora. II. Field and laboratory studies on diatoms from the banks of the River Avon. J. Mar. Biol. Assoc. U.K. *46,* 191–214 (1966)
Salisbury, F.B., Denny, A.: Separate clocks for leaf movement and photoperiodic flowering in *Xanthium strumarium* L. In: Biochronometry. Menaker, M. (ed.). Washington, D.C., Natl. Acad. Sci., 1971
Satter, R., Applewhite, P.B., Galston, A.W.: Rhythmic potassium flux in *Albizzia*. Plant Physiol. *54,* 280–285 (1974)
Satter, R., Galston, A.W.: Potassium flux: a common feature of *Albizzia* leaflet movement controlled by phytochrome or endogenous rhythms. Science *174,* 518–519 (1971a)
Satter, R., Galston, A.W.: Phytochrome-controlled nyctinasty in *Albizzia julibrissin*. III. Interactions between an endogenous rhythm and phytochrome in control of potassium flux and leaflet movement. Plant. Physiol. *48,* 740–746 (1971b)
Satter, R., Geballe, G.T., Applewhite, P.B., Galston, A.W.: Potassium flux and leaflet movement in *Samanea saman*. I. Rhythmic movement. J. Gen. Physiol. *64,* 413–430 (1974)
Satter, R., Geballe, G.T., Galston, A.W.: Potassium flux and leaf movement in *Samanea saman*. II. Phytochrome controlled movement. J. Gen. Physiol. *64,* 431–442 (1975)
Satter, R., Schrempf, M., Chaudhri, J., Galston, A.W.: Phytochrome and circadian clocks in *Samanea:* rhythmic redistribution of potassium and chloride within the pulvinus during long dark periods. Plant. Physiol. *59,* 231–235 (1977)
Schnabel, G.: Der Einfluß von Licht auf die circadiane Rhythmik von *Euglena gracilis* von Autotrophic und Mixotrophic. Planta *81,* 49–63 (1968)
Schrempf, M., Satter, R.L., Galston, A.W.: Potassium-linked chloride fluxes during rhythmic leaf movement of *Albizzia julibrissin*. Plant Physiol. *58,* 190–192 (1976)
Schwintzer, C.R.: Energy budgets and temperatures of nyctinastic leaves on freezing nights. Plant. Physiol. *48,* 203–207 (1971)
Scott, B.I.H., Gulline, H.P.: Natural and forced circadian oscillations in the leaf of *Trifolium repens*. Aust. J. Biol. Sci. *25,* 61–76 (1972)
Scott, B.I.H., Gulline, H.P.: Membrane changes in a circadian system. Nature (London) *254,* 69–70 (1975)
Scott, B.I.H., Gulline, H.P., Robinson, G.R.: Circadian electrical and ionic changes at the membrane of pulvinal cells of *Trifolium repens*. Plant Physiol. *56* Suppl 76 (1975)
Simon, E., Satter, R.L., Galston, A.W.: Circadian rhythmicity in excised *Samanea* pulvini. I. Sucrose-white light interactions. Plant Physiol. *58,* 417–420 (1976)
Simon, E., Satter, R.L., Galston, A.W.: Circadian rhythmicity in excised *Samanea* pulvini. II. Resetting the clock by phytochrome conversion. Plant Physiol. *58,* 421–425 (1976)

Stålfelt, M.G.: Pulsierende Blattgewebe. Planta *7*, 720–734 (1929)
Stålfelt, M.G.: Diurnal dark reactions in the stomatal movements. Physiol. Plant. *16*, 756–766 (1963)
Stålfelt, M.G.: The relation between the endogenous and induced elements of the stomatal movements. Physiol. Plant. *18*, 177–184 (1965)
Stålfelt, M.G.: Endogenous diurnal rhythm in the osmotic surplus of the guard cells. Physiol. Plant. *20*, 57–64 (1967)
Sweeney, B.M.: Biological clocks in plants. Annu. Rev. Plant Physiol. *14*, 411–440 (1963)
Sweeney, B.M.: Rhythmic Phenomena in Plants. London: Academic Press, 1969
Sweeney, B.M.: The potassium content of *Gonyaulax polyedra* and phase changes in the circadian rhythm of stimulated bioluminescence by short exposures to ethanol and valinomycin. Plant Physiol. *53*, 337–342 (1974a)
Sweeney, B.M.: A physiological model for circadian rhythms derived from the *Acetabularia* rhythm paradoxes. Int. J. Chronobiol. *2*, 25–33 (1974b)
Sweeney, B.M.: Circadian rhythms, definition and general characteristics. In: The Molecular Basis of Circadian Rhythms. Hastings, J.W., Schweiger, H.-G. (eds.). Berlin: Life Sci. Res. Rep. 1, Dahlem Konf., 1976, pp. 77–83
Sweeney, B.M.: Evidence that membranes are components of circadian oscillators. In: The Molecular Basis of Circadian Rhythms. Hastings, J.W., Schweiger, H.-G. (eds.) Berlin: Life Sci. Res. Rep. 1, Dahlem Konf., 1976, pp. 267–281
Sweeney, B.M., Herz, J.M.: Evidence that membranes play an important role in circadian rhythms. Proc. Int. Soc. Chronol. 12th Int. Conf. Milan: Il Ponto, 1977, pp. 751–761
Swift, E., Taylor, W.R.: Bioluminescence and chloroplast movement in the dinoflagellate *Pyrocystis lunula*. J. Phycol. *3*, 77–81 (1967)
Tibbitts, T.W., Hoshizaki, H., Alford, D.K.: Movement responses of *Phaseolus* leaves to diurnal and other magnetic fluctuations. Bio. Sci. *23*, 479–484 (1973)
Wagner, R.: Der Einfluß niedriger Temperatur und die Phasenlage der endogentagesperiodischen Blattbewegungen von *Phaseolus multiflorus*. Z. Bot. *51*, 179–204 (1963)
Wassermann, L.: Die Auslösung endogen-tagesperiodischer Vorgänge bei Pflanzen durch einmalige Reize. Planta *53*, 647–669 (1959)
Zimmer, R.: Phasenverschiebung und andere Störlichtwirkungen auf die endogen-tagesperiodischen Blütenblattbewegungen von *Kalanchoë blossfeldiana*. Planta *58*, 283–300 (1962)

2. Intracellular Movements

2.1 Role of Microtubules in Intracellular Movements

PH. FILNER and N.S. YADAV

1. Introduction

Eukaryotic cells are not bags of enzymes. On the contrary, they have the capacity to define an intracellular spatial distribution of molecules and organelles and to change that distribution. It is becoming increasingly clear that eukaryotes use at least two types of fibrous structures, microtubules and microfilaments, to perform and regulate movements of intracellular material, of the cell surface and of the whole cell. Knowledge concerning microtubules has been accumulating rapidly, and has been the subject of several excellent symposia (BORGERS and DE BRABANDER, 1975; SOIFER, 1975; GOLDMAN et al., 1976), general reviews (MOHRI, 1976; SNYDER and MCINTOSH, 1976; STEPHENS and EDDS, 1976), and reviews concerned with microtubules in the plant kingdom (PICKETT-HEAPS, 1974; HEPLER, 1976; HEPLER and PALEVITZ, 1976). The role of cytoplasmic microtubules in intracellular movements is discussed here, while elsewhere in this volume (Chap. 3.1.2), SATIR and OJAKIAN deal with the role of microtubules in locomotion of eukaryotic cells with flagella or cilia.

2. What is a Microtubule?

Microtubules are a class of subcellular structures of eukaryotes which have the appearance of rigid cylinders about 250 Å in diameter, with a central lumen about 150 Å in diameter (Fig. 1). The wall of the cylinder is usually composed of 13 protofilaments which are parallel or nearly parallel to the cylinder axis, each a chain of globular subunits. Center-to-center spacing between subunits is about 40 Å within a protofilament, and about 50 Å between adjacent protofilaments. Optical diffraction patterns obtained from electron micrographs of microtubules have provided evidence of helical periodicities on the microtubule surface (AMOS et al., 1976). When microtubules are negatively stained, the lumen fills with stain, indicating that the microtubule is probably a hollow cylinder (BRYAN, 1974). A clear annular zone or halo about 100 Å wide is sometimes observed around microtubules in cross-section (LEDBETTER and PORTER, 1963; BEHNKE, 1975), possibly indicating the existence of a nonstaining outer component.

Although microtubules have strikingly similar morphology regardless of the organism, microtubule subunits can assemble into structures with other geometries.

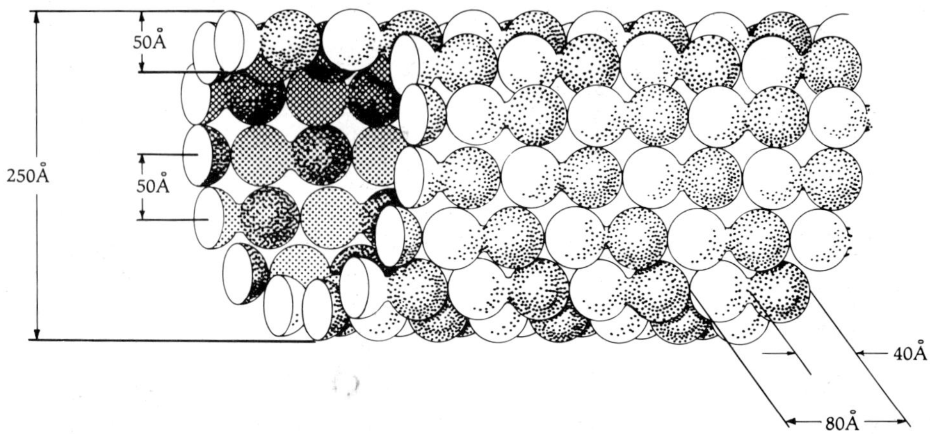

Fig. 1. Diagram of a portion of a microtubule. Transverse dimensions are from direct measurements of electron micrographs of thin sections. Longitudinal dimensions are from periodicities evident in optical diffraction patterns from electron micrographs of negatively stained microtubules. The existence of two kinds of subunits and of a staggered arrangement of protofilaments parallel to the microtubule axis were also deduced from optical diffraction patterns

Each outer doublet fiber in the axoneme of a cilium or a flagellum contains a complete microtubule called the A subfiber, and an incomplete microtubule, the B subfiber, which shares three protofilaments of the A subfiber. Thus at least one of the two subfibers cannot be circular in cross-section. In some organisms, microtubules composed of 12 to 15 protofilaments have been found (BURTON et al., 1975; NAGANO and SUZUKI, 1975). During disassembly of microtubules under certain conditions in vitro, e.g., in response to high salt (FILNER and BEHNKE, unpublished observations; Fig. 2) or to the antimitotic drug vinblastine (BROWN and BOUCK, 1973), the protofilaments separate from each other at the ends of the microtubules and curl back on themselves to form protofilament rings and helices. It can be inferred from the geometric variety that there are more subunit–subunit bonding orientations possible than those which occur in the basic microtubule structure.

In very orderly arrays of parallel microtubules, such as in axonemes, or in axopodia, axostyles, and cytopharyngeal baskets of protozoa, regularly spaced cross-bridges between microtubules are often observed (SOIFER, 1975). Less regular cross-bridges have been seen in less regular arrays of parallel microtubules; e.g., in mitotic spindles (MCINTOSH, 1974; HEPLER, 1976; HEPLER and PALEVITZ, 1976). Cross-bridges have also been observed between microtubules and membranes in animal and plant cells (FRANKE et al., 1972; SCHNEPF et al., 1975; HEPLER, 1976; HEPLER and PALEVITZ, 1976; HARDHAM and GUNNING, 1978). Since the cross-bridges are attached only to some subunits, they point to the existence of heterogeneities in subunit properties within a microtubule.

Microtubules in the cytoplasm are far more labile than those in axonemes (and in some less common structures, such as cytopharyngeal baskets). Cytoplasmic microtubules can be disrupted in vivo by exposure to low temperatures, high pressures, or antimitotic compounds, or by merely rupturing the cell in which they occur, whereas axonemal microtubules tolerate all of these conditions.

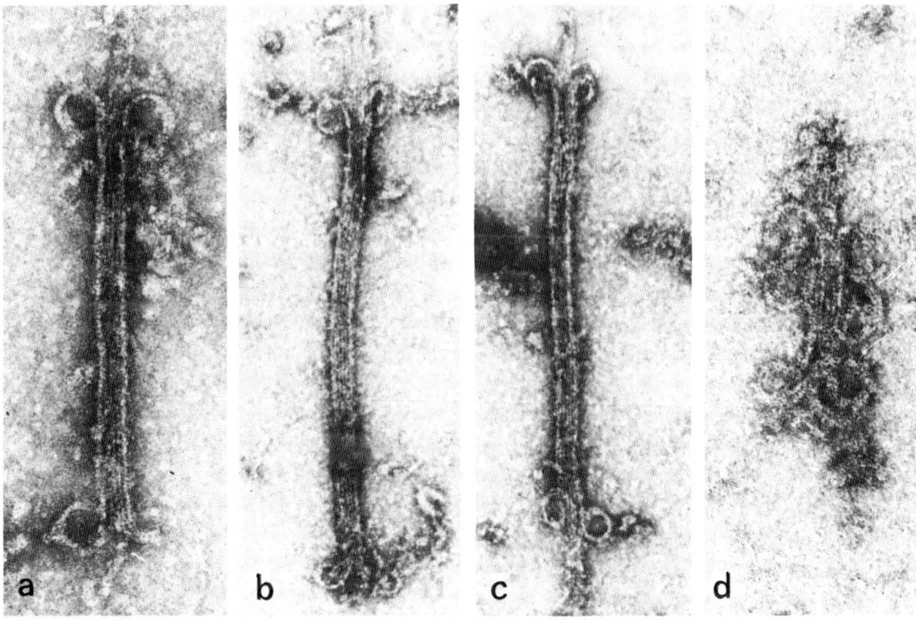

Fig. 2. Curling of protofilaments during depolymerization. Addition of 1 M NaCl (**a**) or KCl (**b–d**) to isolated cow brain microtubules in Microtubule Stabilizing Medium (BEHNKE, 1975) initiates slow depolymerization of the microtubules. After 1 hour at room temperature, separation of adjacent protofilaments and curling of the protofilaments can be seen at both ends of microtubules (FILNER and BEHNKE, unpublished observations). These images suggest that straight protofilaments within microtubules are under tension. Samples were negatively stained with 1% uranyl acetate. Electron micrographs were made by O. BEHNKE

The lability of cytoplasmic microtubules has been exceedingly useful for determining which cellular movements are dependent on microtubules. The most commonly used method for inducing the loss of microtubules in vivo has been exposure to colchicine, because this compound has a high (although not absolute; MIZEL and WILSON, 1972) specificity and a rather direct mechanism of action: binding to unpolymerized microtubule protein (WILSON and MEZA, 1973). Hence, blockage or alteration of a movement by colchicine is taken as strong evidence that the movement is dependent on microtubules. Lumicolchicine, a photolytic derivative of colchicine, is ineffective against microtubules. Consequently, an effect by colchicine coupled with a lack of effect by lumicolchicine at the same concentration provides even stronger evidence for involvement of microtubules. Colchicine-like effects with other kinds of microtubule poisons also add to the case.

3. The Microtubule Subunit

Microtubule subunits belong to a class of proteins called tubulins. Dissolution of axonemal microtubules yields proteins of molecular weight 110,000 daltons which are denaturable into two kinds of 55,000 dalton polypeptides, α and β. These can be separated by electrophoresis on polyacrylamide gels and occur in about a 1:1 ratio. The dimensions of the globular subunits of microtubules make it likely that they represent the 55,000 dalton polypeptides, i.e., tubulin is thought to be a dimer of the globular subunits. Although intact axonemes do not bind colchicine, solubilized axonemal tubulins from sea urchin sperm tails do have a binding site for this substance (WILSON and MEZA, 1973).

Soluble colchicine-binding tubulins have been purified from mammalian brain and from a variety of other animal tissues and cells. These proteins have properties that are strikingly similar to those of axonemal tubulin, including amino acid composition, sequence near the amino terminus, peptide fingerprints, and the presence of two binding sites for guanine nucleotides (STEPHENS and EDDS, 1976). Furthermore, the brain protein can undergo cell-free polymerization into structures which are virtually indistinguishable from naturally occurring labile microtubules (WEISENBERG, 1972). Nevertheless, microtubules present before cell rupture can, in general, account for only about one-third to one-fifth of the extractable tubulin (RUBIN and WEISS, 1975; PIPELEERS et al., 1977). The tubulin not in microtubules, as well as that obtained by dissolution of isolated microtubules, binds colchicine. In addition to tubulin in microtubules and unpolymerized, soluble tubulin, there also exists in cells membrane-bound tubulin (BHATTACHARYA and WOLFF, 1975; BRINKLEY and CARTWRIGHT, 1975). It remains to be determined if membrane-bound tubulin and soluble tubulin enter the tubulin pool of microtubules at some time in their existence. If they do not, i.e., if tubulins have other functions, blockage by colchicine may not be as specific an indicator of dependence on microtubules as is commonly assumed.

4. Microtubules in Plants

MANTON and co-workers (1952) did the pioneering work on microtubules in the plant kingdom, in their studies of the flagella of lower plants, and were the first to discern single microtubules in any system. Cytoplasmic microtubules do not appear to have been seen in plant cells prior to the advent of glutaraldehyde fixation, although they had been seen irregularly in animal cells, and the fundamental similarity of stable and labile microtubules had been recognized (ROTH, 1964).

Glutaraldehyde fixation of plant cells enabled LEDBETTER and PORTER (1963) to discover the labile cortical microtubules which often lie parallel to the newest layer of cellulose microfibrils in the nearby wall, and to observe microtubules in mitotic spindles of plant cells. While studying mitosis in wheat, PICKETT-HEAPS and NORTHCOTE (1966) discovered a band of microtubules which appears before or during prophase near the cell wall, usually in the plane in which the cell plate will form at the end of mitosis. HEPLER and JACKSON (1968) have provided evidence for another system of microtubules which develops at the time of cell plate formation, and is associated with the vesicles from which the cell plate appears to form. Up to the present, research on labile microtubules in plants has consisted largely of ultrastructural studies of the preprophase band, the mitotic spindle, the cell plate-associated microtubules, and the wall-associated microtubules.

Not all ultrastructures in plant cells with the morphology of a rigid hollow cylinder are microtubules. For example, some pure proteins (e.g. glutamic dehydrogenase) can polymerize into such structures (JOSEPHS and BORISY, 1972). Cylindrical structures have been observed in the extracellular matrix of plants (HEYN, 1972), in chloroplasts (PICKETT-HEAPS, 1968; SALEMA and BRANDÃO, 1978), and in prokaryotes (MARGULIS and CHASE, 1978) including blue-green algae (JENSEN and AYALA, 1976). The relationship, if any, of these structures to microtubules remains to be determined.

Although colchicine is the reagent of choice for detecting microtubules and tubulin in animal systems, it does not work so well in plant systems. Colchicine certainly blocks mitosis and other microtubule-associated phenomena in plants, but millimolar concentrations are usually required, compared to the micromolar concentrations effective in animals. Although this may, in part, reflect uptake differences, there have been repeated observations of poor colchicine binding activity (HART and SABNIS, 1976b) by proteins in extracts from fungi (HABER et al., 1972; DAVIDSE and FLACH, 1977) or higher plants (HOTTA and SHEPARD, 1973; LESCURE and FILNER, 1975; HART and SABNIS, 1976a, RUBIN and COUSIN, 1976). The native colchicine binding protein of cultured tobacco cells was at least twice as large as brain tubulin, but it was composed of the usual two species of 55,000 dalton subunits (LESCURE and FILNER, 1975). Only a small fraction of this protein had a detectable colchicine binding site.

Some progress has been made toward finding alternatives for work on the physiology and biochemistry of plant microtubules. For instance, yeast extracts contain a protein which binds colcemid, a derivative of colchicine, far better than it binds colchicine (HABER et al., 1972). Also, a structural relative of the fungicide benamyl was found to bind strongly to tubulin from *Aspergillus nidulans*, which has only a poor affinity for colchicine (DAVIDSE and FLACH, 1977).

An herbicide which can bind to the tubulin of higher plants would be useful. The herbicides isopropyl N-phenyl carbamate, trifluralin, oryzalin, and amiprophosmethyl have antimitotic activity in plants (HEPLER and JACKSON, 1969; BARTELS and HILTON, 1973; KIERMEYER and FEDTKE, 1977). However, isopropyl N-phenyl carbamate affects spindle orientation and number of spindle poles, rather than acting directly on tubulin (HEPLER and JACKSON, 1969, COSS et al., 1975). Amiprophosmethyl specifically inhibits tubulin synthesis in *Chlamydomonas reinhardtii* (COLLIS and WEEKS, in press). Radioactive oryzalin does not bind to soluble plant proteins of the size expected for tubulin (UPADHAYA and NOODÉN, 1978). Trifluralin has been reported to bind to flagellar tubulin from *C. reinhardtii*, (HESS and BAYER, 1977), but the reaction has not been characterized sufficiently to assess its specificity or its relevance to the antimitotic action of the herbicide. Trifluralin, oryzalin and amiprophosmethyl inhibit flagellar regeneration and induce flagellar shortening in *C. reinhardtii* (HESS and BAYER, 1977; QUADER et al., 1977). This assay may be useful in screening compounds for antimicrotubule activity in the plant kingdom.

5. Are Microtubules Involved in Movements?

The circumstantial evidence is quite striking. Plant microtubules are associated with movements of whole cells in the case of flagella, of chromosomes in the case of the mitotic spindle, of vesicles in the case of cell wall deposition and cell plate formation. The association between microtubules and movements is often temporal as well as spatial. For example, the mitotic spindle appears just before, and disappears just after chromosome movement; similarly, the cell plate-associated microtubules appear at the onset of the gathering of vesicles in cell-plate formation and disappear shortly afterward. Many more cases of association between microtubules and movement are known in animal cells than in plant cells (PORTER, 1966; TILNEY, 1971; STEPHENS and EDDS, 1976).

6. Microtubules and Cell Shape

Microtubules are also associated with cell shaping, a subtle kind of movement control (see also Volume 10 on Morphogenesis). The involvement appears to be rather direct in unwalled cells, where the rigidity of microtubules seems to deform the cell surface (BURGESS, 1970; MILLINGTON and GAWLIK, 1970; SANGER and JACKSON, 1971; SCHNEPF, 1972; BROWN and BOUCK, 1973; MARCHANT and PICKETT-HEAPS, 1976; VIAMONTES and KIRK, 1977). In such cases, microtubules appear to prevent movement (relaxation of the deformed surface) rather than to facilitate movement. Regardless of whether microtubules are associated with prevention or facilitation of movement in a particular instance, the common feature is that movement is controlled.

Cortical microtubules maintain the goblet shape of the wall-less alga, *Ochromonas danica*. Colchicine, isopropyl N-phenylcarbamate or high pressure reversibly induces a spherical shape, with concomitant loss of microtubules (BROWN and BOUCK, 1973). Similarly, the isodiametric generative cells in the pollen tubes of angiosperms become ellipsoidal, concomitant with the appearance of longitudinally arranged microtubules (BURGESS, 1970; SANGER and JACKSON, 1971). In the wall-less zooids of colonial algae in the Hydrodictyaceae, cortical microtubules underlie the sites of future adhesion, where a localized flattening occurs (MILLINGTON and GAWLIK, 1970; MARCHANT and PICKETT-HEAPS, 1976). It was concluded that microtubules are required for the normal species-specific pattern of linking of the zooids, since colchicine treatment results in disordered linking. MARCHANT and PICKETT-HEAPS (1976) postulated that the microtubules are involved in producing the required shape for adhesion. Sections through adjacent cells show a very striking one-to-one matching between the microtubules of the two neighboring cells, suggesting that the microtubules may also align surface receptors for the initial recognition between the zooids. Surface receptor movements within the cell membrane have been shown to be colchicine-sensitive in animal cells (YIN et al., 1972; YAHARA and EDELMAN, 1975).

Microtubules are also involved directly in organelle shaping in at least one situation: the elongation of sperm nuclei in some plants and animals during spermiogenesis. The elongated shape of the nucleus is apparently due to the spiral array of microtubules which ensheath it (MYLES and HEPLER, 1977).

The role of microtubules in shaping of walled cells is indirect. The nature of that role was predicted even before cytoplasmic microtubules were discovered. GREEN (1962) demonstrated in *Nitella* that colchicine inhibited cell elongation and caused lateral swelling, along with a loss of the usual transversely-oriented wall microfibrils. He postulated the existence of a transversely aligned, colchicine-sensitive, cytoplasmic filamentous system (analogous to the birefringent fibers of the mitotic apparatus) that would function to control microfibril orientation. His prediction was borne out by the discovery that, in cells that are forming primary or secondary walls, the orientation of newly deposited cellulose microfibrils is often (but not always) highly correlated with the orientation of the underlying cortical microtubules (NEWCOMB, 1969; O'BRIEN, 1972; HEATH, 1974; PICKETT-HEAPS, 1974; HART and SABNIS, 1976b; HEPLER, 1976; HEPLER and PALEVITZ, 1976). Such observations have been made in developing stomatal guard cells, xylem, phloem and cotton fiber as well as in the walls of algae (PICKET-HEAPS, 1967; PARTHASARATHY, 1974; PALEVITZ and HEPLER, 1976; WESTAFER and BROWN, 1976; ROBINSON and HERZOG, 1977; HOGETSU and SHIBAOKA, 1978).

The finding that colchicine almost always disrupts both cortical microtubules and the orderliness of the microfibrils in such systems has led to the hypothesis that microtubules control the orientation of microfibrils. The microtubules may influence the orientation in which wall precursors are deposited or, alternatively, may limit the deposition of precursors to discrete regions of the wall. In developing xylem cells, for instance, colchicine causes deposition to be smeared out rather than distinctly patterned, but the microfibrils remain largely parallel to each other (PICKETT-HEAPS, 1967; HEPLER and FOSKETT, 1971).

Granular complexes have been seen in the cell membranes of green and brown algae (BROWN and MONTEZINOS, 1976; PENG and JAFFE, 1976) and higher plants (MUELLER et al., 1976). Such complexes have also been seen at the ends of microfibrils in some of the same organisms (BROWN and MONTEZINOS, 1976; MUELLER et al., 1976). If these complexes are responsible for synthesis of the microfibrils, and if their movement in the cell membrane is controlled by microtubules or membrane-bound tubulin, it is easy to see how both the locale and orientation of microfibril deposition could be colchicine-sensitive (HEATH, 1974).

The conclusion that cortical microtubules control orientation and/or locale of deposition of cellulose microfibrils rests on the specificity of colchicine. ROBINSON and HERZOG (1977) have provided support for the concept by showing that, in addition to colchicine, also podophyllotoxin, trifluralin, oryzalin, and amiprophosmethyl disrupted microtubules and altered microfibril orientation in the alga *Oocystis solitaria*. Normally, the microfibrils in alternating wall layers are crossed; in treated cells, the layers were parallel. MONTEZINOS (1977) obtained a seemingly conflicting result with *Oocystis apiculata*. He found that a concentration of colchicine which disrupted the crossed layering of the microfibrils did not affect the cortical microtubules. Presumably either a small subpopulation of microtubules or tubulin (perhaps membrane-bound?) was affected by the colchicine.

Discrepancies have been reported between the orientation of microfibrils and underlying microtubules (PICKETT-HEAPS, 1974; HEPLER, 1976; HEPLER and PALEVITZ, 1976). Since the orientation of microfibrils is usually different in adjacent layers of cell wall, some of the observed discrepancies may simply reflect the dynamic nature of the process: at the beginning of the deposition of a layer, the microtubules will not be oriented parallel to the visible layer of microfibrils deposited previously. More difficult to explain is the reported absence of microtubules during deposition of transversely oriented microfibrils, e.g. in swarmers of *Cladophora* (ROBINSON et al., 1972). Perhaps the cortical microtubules are exceptionally difficult to preserve by fixation at certain stages.

In the alga *Glaucocystis*, the cortical microtubules are separated from the plasmalemma by a system of lacunae, except around the flagellar bases. If microtubules have a role in controlling the wall structure in this organism, it must be by a mechanism different from that operative in the cases discussed above. Recently, it has been reported that in this alga, the microfibrils are laid down as the cytoplasm rotates (WILLISON and BROWN, 1978).

7. How Are Microtubules Involved in Movements?

7.1 The Mitotic Spindle

Research on mitosis has produced many of the best tools for studying the relationship between microtubules and movements, as well as many of the dominant concepts. The anaphase movements offer a virtually unique opportunity to monitor in living cells the behavior of the entity being moved (the chromosomes) and the microtubule-containing apparatus responsible for the movement (the spindle) (Nicklas, 1971). This is possible because (a) chromosomes are large enough and their movements are slow enough to be monitored with a light interference microscope, and (b) in the polarized light microscope, the fibers of the mitotic spindle exhibit form birefringence, which provides a quantitative index of oriented material in the fibers (Inoué and Ritter, 1975). The intensity of form birefringence correlates fairly well with the number of microtubule bundles in the spindle fibers, and is widely accepted as an index of orientated microtubules (Inoué and Ritter, 1978). However, a decrease of birefringence without a proportionate decrease of microtubules and vice versa has been observed (Forer, 1976). Consequently, there may be another contributor to birefringence in the spindle, and one should be wary of applying to microtubules conclusions based solely on birefringence data. Independent confirming evidence, e.g., from electron microscopy, is needed.

Two kinds of fibers oriented parallel to the spindle axis can be seen with a polarized light microscope. They are (1) chromosomal fibers, which connect a specific differentiated region of the chromosome, the kinetochore (centromere) to a polar region and (2) interpolar fibers, which span all or part of the interpolar region. Evidence of actual fibers, as opposed to images which suggest fibers, has been obtained by micromanipulation (Nicklas and Staehly, 1967) and centrifugation (Shimamura, 1940); the chromosomes behaved as if tethered to a pole by a taut string or flexible rod.

Correspondingly, the principal oriented elements which can be seen in spindles with an electron microscope are loose bundles of microtubules, typically 20–100 microtubules μm^{-2}, which may also be classified into two major categories. They are (1) chromosomal (*k*inetochore) microtubules or kMTs, which are attached to a kinetochore and extend toward one pole and (2) interpolar (*p*olar) microtubules or pMTs, which have one end near a pole and the other end beyond the midplane of the spindle. They extend either to the vicinity of the far pole or end shortly beyond the midplane, where pMTs from opposite poles overlap and interdigitate. Some spindle microtubules have both ends free and may constitute a third category.

Although the kMTs and pMTs are morphologically indistinguishable, they differ in stability, the kMTs being the more stable when subjected to depolymerizing stresses, such as low temperature (Brinkley and Cartwright, 1975), or colchicine (Brinkley et al., 1967). It is not clear whether the stability difference reflects an intrinsic difference between the tubulins of kMTs and pMTs, or a difference in other spindle components. Microtubules are not alone in the spindle. Cross-bridges have been observed between microtubules within a bundle, between interdigitating pMTs at the midplane, and between pMTs and kMTs (McIntosh, 1974; Inoué and Ritter, 1975; Hepler and Palevitz, 1976). *Microfilaments* have also been observed in mitotic spindles. They occur in the spindle of *Haemanthus* with a distinct polarity (FORER and JACKSON, 1976; SANGER and SANGER, 1976), perhaps the strongest indication that they are an integral part of the spindle. Finally, spindles contain an assortment of uncharacterized granules, vesicles, etc. (HEPLER et al., 1970). One or more of these factors may affect microtubule stability.

7.2 Chromosome Movement

Chromosome movements occur during spindle formation (prometaphase) and during chromosome separation (anaphase), but the anaphase movements are the more orderly, hence the more susceptible to analysis. There are two kinds of anaphase movement: (RIS, 1949): (1) movement of a chromosome set towards each pole, resulting in decreased distance between chromosomes and poles, and (2) elongation of the spindle, causing an increase in the pole-to-pole distance. In most cases, both types of movements occur, though with varying degrees of temporal overlap; an extreme example is *Barbulanympha*, in which the two phenomena are entirely separated in time (INOUÉ and RITTER, 1975). In still rarer cases, only one of the two movements occurs (INOUÉ, 1976). Thus it is quite justifiable to treat the two as independent phenomena.

As a chromosome moves poleward, its associated spindle fiber appears to shorten by a length equal to the distance traversed by the chromosome. The shortened fibers do not thicken, nor do they increase in birefringence, in microtubule content, or microtubule dimensions. One is forced to conclude that the chromosomal spindle fibers and their constituent microtubules disassemble during anaphase at the same rate as the chromosome moves poleward. It has not been determined unequivocally at which end the fiber disassembles and whether chromosomes are pushed or pulled. The best evidence available was obtained by Forer (1965), who used an ultraviolet microbeam to bleach partially the birefringence from a portion of a chromosomal spindle fiber, thereby putting a reference mark on the fiber. He then observed that the bleached spot moved poleward. This result fits best with the idea that the fiber disassembles at the pole, and that the fiber pulls the chromosome poleward. It seems quite clear that the force which causes chromosomal movement is applied at the kinetochore. This is evident from the fact that the kinetochore is the leading part of the chromosome in the poleward movement, while the arms appear to be dragged along. It is noteworthy that the microtubules of the chromosomal fiber are attached to the chromosome at the kinetochore, making them the only obvious candidates for transmission of the force which moves the chromosomes poleward.

The nature of the spindle elongation process is less clear. Quantitative studies of changes in microtubule distribution and number during spindle elongation indicate that (1) there is little change outside the region of the midplane, other than elongation of the spindle, (2) the length of the overlap of pMTs at the midplane decreases, and (3) the increase in spindle length cannot be accounted for (completely) by the decrease in the overlap between pMTs. These observations necessitate addition of length to individual pMTs during spindle elongation. It is not known whether the spindle grows at the poles, or if the length is added to the end of pMTs near the midplane. In the latter case, the overlapping pMTs would presumably push past each other, in turn pushing the two halves of the spindle poleward.

Thus both chromosomal and polar movements are correlated with changes in microtubule length, the former being accompanied by microtubule shortening, hence depolymerization, and the latter by microtubule lengthening, hence polymerization. By observing the behavior of the chromosomes and microtubules in spindles which have been perturbed by a variety of antimitotic treatments, it has been possible to obtain evidence that active polymerization or depolymerization is required for the movements.

As noted in Section 7.1, spindle microtubules can be prevented from forming or can be induced to depolymerize by low temperature, high pressure, or any of several antimitotic compounds. The best known of these compounds are colchicine, vinblastine, and podophyllotoxin, all natural products of plants (Mohri, 1976; Snyder and McIntosh, 1976; Stephens and Edds, 1976). All of these agents stop chromosome movement. Obviously, it is rather difficult to claim that temperature or pressure act specifically on microtubules. However, there is excellent biochemical evidence that colchicine, vinblastine, and podophyllotoxin bind to specific sites on tubulins and prevent normal polymerization of partially purified tubulin in vitro (Wilson, 1975; Section 3 above).

The presence of microtubules is not a sufficient condition for movement to occur. Attempts to induce anaphase movements in what appear to be intact isolated spindles have not been successful. However, spindles in "lysed" cells may be functional, provided that a suitable chemical stimulus, such as tubulin, ATP, or polyethylene glycol, is introduced (Cande et al., 1974). Furthermore, deuterium oxide overstabilizes spindles so that both birefringence and microtubule content increase, but chromosome movements are arrested (Gross and Spindel, 1960). Presumably, what is needed in addition to microtubules is the expenditure of energy to drive the movements.

7.3 Possible Origins of the Force Responsible for Movement

There are at least three points of view on how the motive force may originate.
1. The microtubule could be a purely passive mechanical transducer of force applied to it, while some other system, e.g., microfilaments, would have to generate the force. Rigid rods are ideally suited for rapid and precise transduction of both rotational and translational motions over long distances.
2. The microtubule could generate the force by a mechanochemical transduction, but with the lengthening or shortening of the spindle microtubules being incidental to and not directly involved in the force generation mechanism. The sliding filament models (McIntosh et al., 1969; Nicklas, 1975; Margolis et al., 1978), which are patterned after the current model of muscle contraction, have these characteristics. In a sliding filament model, it is postulated that microtubules are moved past each other by a sequence of ATP-driven reactions of the cross-bridges: (a) change of angle between cross-bridges and the antiparallel microtubules to which they are attached, causing one microtubule to move relative to the other; (b) breakage of the linkage between each cross-bridge and one of the microtubules; (c) return of the cross-bridges to the original angle; (d) reformation of the cross-

bridges at the next attachment site. In this model, there is no force-generating function for shortening of chromosomal microtubules or lengthening of polar microtubules (MARGOLIS et al., 1978).

3. The microtubule could generate the force for movement by a mechanochemical transduction which employs polymerization-depolymerization in its mechanism. Inoué (INOUÉ and SATO, 1967; INOUÉ and RITTER, 1975) formulated a rather well-known model of this type on the basis of his studies of reversible changes in birefringence of mitotic spindles. He proposed that the oriented fibers of the spindle, which generate its birefringence, are polymers in equilibrium with their monomeric precursor. He further proposed that shifts in the equilibrium drive spindle movement, with depolymerization causing poleward movement of chromosomes and polymerization causing the increase in separation of poles. After the discovery of spindle microtubules, and of the seemingly reversible polymerizability of brain tubulin into microtubules, INOUÉ'S concept of a monomer-polymer equilibrium gained wide acceptance, although the concept of force generation by polymerization or depolymerization apparently did not.

Free energies of polymerization have been calculated for both in vivo polymerization based on birefringence data (INOUÉ and RITTER, 1975) and in vitro polymerization assayed turbidimetrically (LEE and TIMASHEFF, 1975). Such calculations are of course based on the assumption of a monomer-polymer equilibrium. However, there is now substantial evidence that cleavage of both alpha-beta and beta-gamma phosphate anhydride bonds of GTP accompanies polymerization-depolymerization of tubulin (SANDOVAL et al., 1978). Furthermore, tubulin polymerization seems to require binding of a guanine nucleotide, but not cleavage of a phosphate anhydride bond. Depolymerization, on the other hand, appears to require cleavage of both of the anhydride bonds in GTP.

Consequently, tubulin polymerization-depolymerization should be looked upon as a cyclic steady state driven by the free energy in GTP; it should not be interpreted as an equilibrium. In the absence of equilibrium, the meaning, if any, of calculations of free energies from ratios of reactants and products is not entirely clear.

7.4 How Can Polymerization-Depolymerization Generate Movement?

If chromosomal microtubules pull chromosomes poleward by depolymerization, as INOUÉ proposed, there are some obvious mechanical constraints. The microtubule must remain anchored at the pole continuously, and the microtubule must retain its integrity throughout the distance from chromosome to pole. Then where and how might the depolymerization occur? INOUÉ and RITTER (1975) have suggested depolymerization occurs subterminally, i.e. not at the anchored end, and as some subunits depart, the other subunits rearrange to fill the gaps. However, the available evidence, from studies of brain tubulin in vitro, indicates that microtubules depolymerize and polymerize at the ends, with depolymerization occurring preferentially at one end and polymerization at the other (DENTLER et al., 1974; MARGOLIS and WILSON, 1978).

The evidence that microtubules polymerize and depolymerize at the ends, coupled with the evidence that GTP breakdown is required for depolymerization, leads one to hypothesize a force-generating mechanism which operates at the depo-

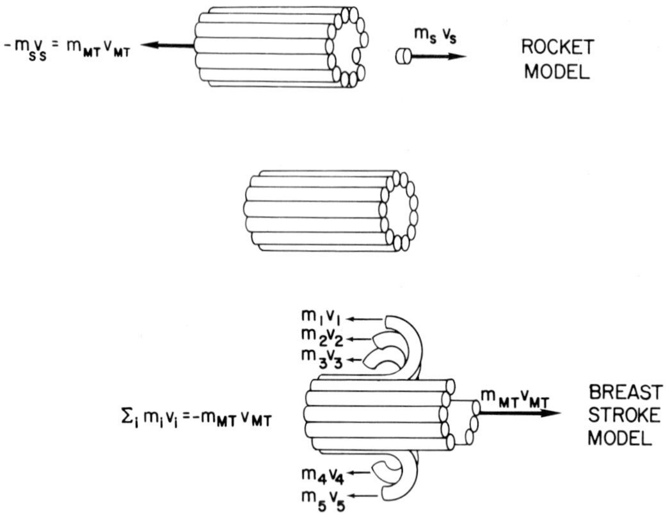

Fig. 3. Physical mechanisms for generation of microtubule movement through depolymerization at a microtubule end. The Rocket Model: the subunits of a depolymerizing microtubule leave with momentum, and the microtubule acquires equal and opposite momentum. The microtubule thereby moves away from the position at which depolymerization began. The Breast Stroke Model: the protofilaments separate, beginning at the microtubule end, and curl back on themselves. The microtubule acquires momentum equal and opposite to that of the backward curling protofilaments. The microtubule thereby moves toward or through the position at which depolymerization began

lymerizing end. One possibility is a "rocket" mechanism (Fig. 3). Microtubule subunits could function as "rocket fuel" if they departed from the ends of microtubules with force, rather than by merely dissociating and diffusing away. An equal and opposite force would be exerted on the microtubule. If the force vector had a projection on the microtubule axis, then movement of the microtubule would occur parallel to the axis, provided that the force was sufficient to overcome the viscous drag of the medium. However, if the chromosomal microtubules depolymerized into single subunits at the polar end, the recoil from departing subunits would move the microtubules away from the poles, not towards them. In order to generate poleward movement as a result of depolymerization at the pole, a force vector in the opposite direction is required.

Quite remarkably, microtubules seem to have precisely the property needed to achieve a 180° reversal of thrust (Fig. 3). As a microtubule depolymerizes at its end, the protofilaments separate from each other progressively from that end, and curve back on themselves, forming rings and helices (Fig. 2). That is, the subunits do not appear to detach from the protofilament until well after the bonds between adjacent protofilaments have been broken. The fact that the free protofilament end goes spontaneously into a tight curve (rather than becoming a limp string of beads or remaining a straight array of beads) indicates that the straight protofilaments in the microtubule are in a high-energy state, and that energy is released following the breakage of the bonds between adjacent protofilaments. As a subunit separates from adjacent protofilaments, it remains bound to a subunit in its own protofilament that is still integrated in the microtubule structure. In order

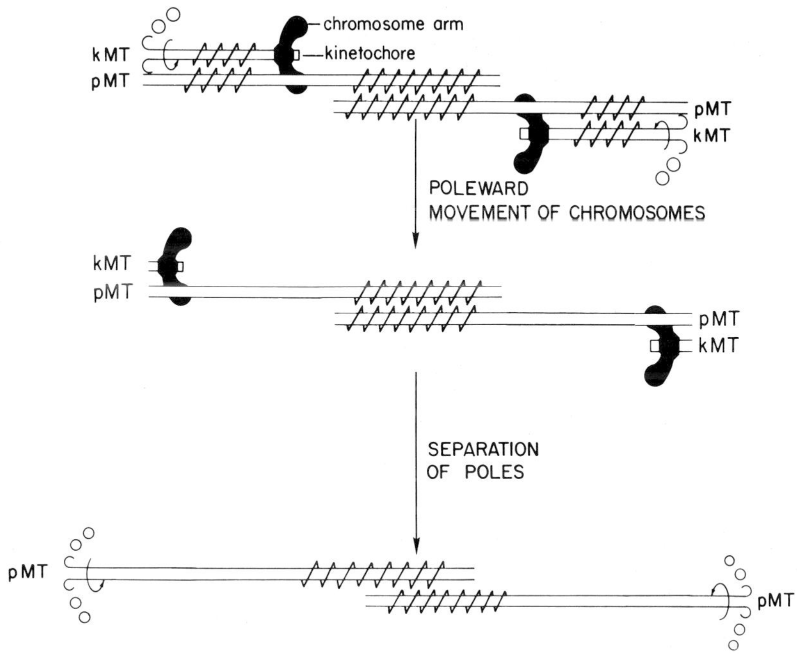

Fig. 4. The Worm Gear Model of Anaphase Movements: portions of microtubules have a helical outer component which functions as a worm gear. The outer components of pairs of microtubules mesh. Depolymerization occurs at one end of a microtubule, by curling of separating protofilaments, followed by subunit dissociation. Each protofilament curls out of the plane through the microtubule axis, thereby imparting angular momentum to the microtubule, so that it rotates about its axis. The meshed outer components translate the rotational motion into translational motion parallel to the microtubule axis. In the anaphase movements the kMTs depolymerize and rotate first, thus moving the chromosomes poleward; then the pMTs depolymerize and rotate against each other, thus moving the poles apart

for a subunit at the polar end to move out of the linear array while still attached to the integrated part of its protofilament, it must move away from the pole. Thus there must be a poleward component to the recoil force vector on the microtubule.

According to this model, the microtubule would resemble a 13-armed breast-stroke swimmer with a continuous power stroke. To account for the constant location of the ends of chromosomal microtubules at the pole, the mechanism must have the property that the combined force of the depolymerization of 13 subunits, one in each of the 13 protofilaments, generates just enough poleward movement to traverse one subunit diameter. As fanciful as the whole model may seem, the most difficult part to accept is the coincidental matching of the distance moved with the subunit diameter. One would like to have a model that accounts mechanically for this dimensional constraint. It is also very difficult to envision how hundreds of independent 13-armed breast strokers could manage to drag the chromosomes to the poles with the remarkable synchrony that they do.

If the curving protofilament moves out of the plane defined by its initial path and the axis of the microtubule, which must happen if it curves through more than 360°, the force vector should have a component perpendicular to the microtu-

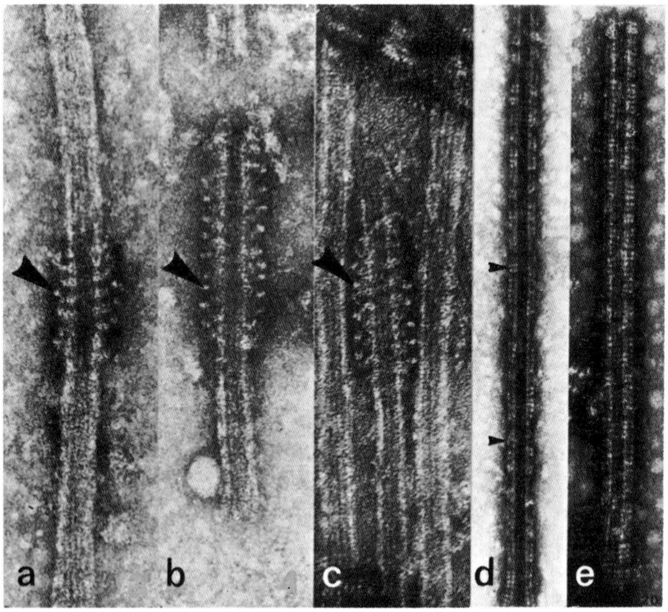

Fig. 5a–e. The helical outer component of microtubules. A helical wrapping can occasionally be seen surrounding short regions of isolated brain microtubules in Microtubule Stabilizing Medium (**a–c**). When microtubules formed in vitro from purified tubulin are treated with a polycation such as DEAE dextran, extensive ordered regions of helical wrapping are observed (**d, e**). The outer component has been identified tentatively as a second layer of tubulin. (From BEHNKE, 1975)

bule axis and tangential to its surface. If all the protofilaments developed an off-axis angle to their curvature with the same handedness around the microtubule, then all the tangential force vectors would reinforce one another, and cause the microtubule to rotate on its axis during depolymerization. Rotational motion suggests a mechanism for matching poleward movement with depolymerization at the polar end, as well as for coordination of the movements of many microtubules performing a complex orderly sequence of movements. That mechanism is the worm gear. Suppose that the poorly preserved "cross-bridges" which have been seen in mitotic spindles are, in reality, poorly preserved remnants of helical threads on the surface of the microtubules. As depolymerization at the pole of a chromosomal microtubule causes it to rotate on its axis, that thread need only be held against a similar helix on a nonrotating polar microtubule in order to transform the rotational motion into paraxial poleward motion (provided that the helix and the rotational motion have the same handedness); (Fig. 4).

The essential characteristics of the worm gear model are: (1) There exist two classes of microtubule in a movement-producing microtubule system. Each has a screw thread on portions of its surface and their screw threads are meshed. (2) One microtubule rotates against the second, which may be stationary or may rotate in the opposite direction. The rotational motion is generated by the off-axial backward curving of protofilaments as they separate from each other at the microtubule end. The energy which drives the protofilament curling is derived from cleavage of phosphate anhydride bonds in tubulin-bound GTP in microtubules. (3) The pitch of the helix and the rotation per tier of subunits depolymerized are matched in such a way that the end of the microtubule remains in the same place as depolymerization progresses.

Two clear predictions of this model which may be amenable to testing are: (1) there should exist regions of spindle microtubules which have a helical wrapping (outer component?); and (2) during poleward movement of chromosomes, a polar microtubule should rotate about its axis. BEHNKE (1975) has already shown that a helical outer component can be seen on isolated brain microtubules and on microtubules formed in vitro from purified tubulin (Fig. 5).

We have speculated on how microtubules might function in intracellular movements, but as K. PORTER wrote recently (1976), "We may be misinterpreting the observations, or overinterpreting them, but in either event, the fault would be greater were we not to attempt an interpretation".

Acknowledgments. This work was supported by United States Department of Energy Contract EY-76-C-02-1338. The concept of the backward curling of protofilaments as the driving mechanism for movement in microtubule systems originated in discussions between P.F. and O. BEHNKE, in the latter's laboratory in 1973, where the curling phenomenon was first observed.

References

Amos, L.A., Linck, R.W., Klug, A.: Molecular structure of flagellar microtubules. In: Cell Motility. Book C. Goldman, R., Pollard, T., Rosenbaum, J. (eds.). Cold Spring Harbor Laboratory, 1976, pp. 847–867

Bartels, P.G., Hilton, J.L.: Comparison of trifluralin, oryzalin, pronamide, propham and colchicine treatments on microtubules. Pest. Biochem. Physiol. *3*, 462–472 (1973)

Behnke, O.: Studies on isolated microtubules. Evidence for a clear space component. Cytobiol. *11*, 366–381 (1975)

Bhattacharya, B., Wolff, J.: Membrane-bound tubulin in brain and thyroid tissue. J. Biol. Chem. *250*, 7639–7649 (1975)

Borgers, M., De Brabander, M.: Microtubules and Microtubule Inhibitors. Amsterdam-Oxford: North-Holland, 1975

Brinkley, B.R., Cartwright, J.: Cold labile and cold stable microtubules in the mitotic spindle of mammalian cells. Ann. N.Y. Acad. Sci. *253*, 428–439 (1975)

Brinkley, B.R., Stubblefield, E., Hsu, T.C.: The effects of colcemid inhibition and reversal on the fine structure of the mitotic apparatus of chinese hamster cells in vitro. J. Ultrastruct. Res. *19*, 1–18 (1967)

Brown, D.L., Bouck, G.B.: Microtubule biogenesis and cell shape in *Ochromonas* II. The role of nucleating sites in shape development. J. Cell Biol. *56*, 360–378 (1973)

Brown, Jr., M.R., Montezinos, D.: Cellulose microfibrils: visualization of biosynthetic and orienting complexes in association with the plasma membrane. Proc. Natl. Acad. Sci. U.S.A. *73*, 143–147 (1976)

Bryan, J.: Microtubules. Bioscience *24*, 701–711 (1974)

Burgess, J.: Cell shape and mitotic spindle formation in the generative cell of *Endymion non-scriptus*. Planta *95*, 72–85 (1970)

Burton, P.R., Hinkley, R.E., Pierson, G.B.: Tannic acid-stained microtubules with 12, 13 and 15 protofilaments. J. Cell. Biol. *65*, 227–233 (1975)

Cande, W.Z., Snyder, J., Smith, D., Summers, K., McIntosh, J.R.: A functional mitotic spindle prepared from mammalian cells in culture. Proc. Natl. Acad. Sci. U.S.A. *71*, 1559–1563 (1974)

Collis, P.S., Weeks, D.P.: Rapid and selective inhibition of tubulin synthesis by APM during flagellar regeneration in *Chlamydomonas reinhardi*. Science, in press

Coss, R.A., Bloodgood, R.A., Brower, D.L., Pickett-Heaps, J.D., McIntosh, J.R.: Studies on the mechanism of action of isopropyl N-phenyl carbamate. Exp. Cell Res. *92*, 394–398 (1975)

Davidse, L.C., Flach, W.: Differential binding of methyl benzimidazol-2-yl carbamate to fungal tubulin as a mechanism of resistance to this antimitotic agent in mutant strains of *Aspergillus nidulans*. J. Cell Biol. *72*, 174–193 (1977)

Dentler, W.L., Granett, S., Witman, G.B., Rosenbaum, J.L.: Directionality of brain microtubule assembly in vitro. Proc. Natl. Acad. Sci. U.S.A. *71*, 1710–1714 (1974)

Forer, A.: Local reduction of spindle fiber birefringence in living *Nephrotoma saturalis* (Loew) spermatocytes induced by ultraviolet microbeam irradiation. J. Cell Biol. *25*, 95–117 (1965)

Forer, A.: Actin filaments and birefringent spindle fibers during chromosome movements. In: Cell Motility. Book C. Goldman, R., Pollard, T., Rosenbaum, J. (eds.). Cold Spring Harbor Laboratory, 1976, pp. 1273–1293

Forer, A., Jackson, W.T.: Actin filaments in the endosperm mitotic spindles in a higher plant, *Haemanthus katherinae* Baker. Cytobiol. *12*, 199–214 (1976)

Franke, W.W., Herth, W., Van Der Wonde, W.J., Morré, D.J.: Tubular and filamentous structures in pollen tubes: possible involvement as guide elements in protoplasmic streaming and vectorial migration of secretory vesicles. Planta *105*, 317–341 (1972)

Goldman, R., Pollard, T., Rosenbaum, J. (eds.): Cell Motility. Cold Spring Harbor Laboratory, 1976

Green, P.B.: Mechanism for plant cellular morphogenesis. Science *138*, 1405–1406 (1962)

Gross, P.R., Spindel, W.: Mitotic arrest by deuterium. Science *131*, 37–39 (1960)

Haber, J.E., Peloquin, J.G., Halvorson, H.O., Borisy, G.G., Colcemid inhibition of cell growth and the characterization of a colcemid-binding activity in *Saccharomyces cerevisiae*. J. Cell Biol. *55*, 355–367 (1972)

Hardham, A.R., Gunning, B.E.S.: Structure of cortical microtubule arrays in plant cells. J. Cell Biol. *77*, 14–34 (1978)

Hart, J.W., Sabnis, D.D.: Colchicine binding activity in extracts of higher plants. J. Exp. Bot. *27*, 1353–1360 (1976a)

Hart, J.W., Sabnis, D.D.: Colchicine and plant microtubules: a critical evaluation. Curr. Adv. Plant Sci. *26*, 1095–1104 (1976b)

Heath, I.B.: A unified hypothesis for the role of membrane bound enzyme complexes and microtubules in plant cell wall synthesis. J. Theor. Biol. *48*, 445–449 (1974)

Hepler, P.K.: Plant microtubules. In: Plant Biochemistry, 3rd ed. Bonner, J., Varner, J.E. (eds.). New York-London: Academic Press, 1976, pp. 147–187

Hepler, P.K., Foskett, D.E.: The role of microtubules in vessel member differentiation in *Coleus*. Protoplasma *72*, 213–236 (1971)

Hepler, P.K., Jackson, W.T.: Microtubules and early stages of cell-plate formation in the endosperm of *Haemanthus katherinae* Baker. J. Cell Biol. *38*, 437–446 (1968)

Hepler, P.K., Jackson, W.T.: Isopropyl N-phenylcarbamate affects spindle microtubule orientation in dividing endosperm cell of *Haemanthus katherinae* Baker. J. Cell Sci. *5*, 727–743 (1969)

Hepler, P.K., McIntosh, J.R., Cleland, S.: Intermicrotubule bridges in mitotic spindle apparatus. J. Cell Biol. *45*, 438–444 (1970)

Hepler, P.K., Palevitz, B.A.: Microtubules and microfilaments. Annu. Rev. Plant Physiol. *25*, 309–362 (1976)

Hess, F.D., Bayer, D.E.: Binding of the herbicide trifluralin to Chlamydomonas flagellar tubulin. J. Cell Sci. *24*, 351–360 (1977)

Heyn, A.N.J.: Intra- and extracytoplasmic microtubules in coleoptiles of *Avena*. J. Ultrastruct. Res. *40*, 433–457 (1972)

Hogetsu, T., Shibaoka, H.: Effects of colchicine on cell shape and on microfibril arrangement in the cell wall of *Closterium acerosum*. Planta *140*, 15–18 (1978)

Hotta, Y., Shepard, J.: Biochemical aspects of colchicine action on meiotic cells. Mol. Gen. Genet. *122*, 243–260 (1973)

Inoué, S.: Chromosome movement by reversible assembly of microtubules. In: Cell Motility. Book C. Goldman, R., Pollard, T., Rosenbaum, J. (eds.). Cold Spring Harbor Laboratory, 1976, pp. 1317–1328

Inoué, S., Ritter, H.: Dynamics of mitotic spindle organization and function. In: Molecules

and Cell Movement. Inoué, S., Stephens, R.E. (eds.). New York: Raven Press, 1975, pp. 3–30
Inoué, S., Ritter, H.: Mitosis in *Barbulanympha* II. Dynamics of a two-stage anaphase, nuclear morphogenesis, and cytokinesis. J. Cell Biol. *77*, 655–684 (1978)
Inoué, S., Sato, H.: Cell motility by labile association of molecules. The nature of the mitotic spindle fibers and their role in chromosome movement. J. Gen. Physiol. *50*, 259–292 (1967)
Jensen, T.E., Ayala, R.P.: The fine structure of striated microtubules and sleeve bodies in several species of *Anabena*. J. Ultrastruct. Res. *57*, 185–193 (1976)
Josephs, R., Borisy, G.G.: Self assembly of glutamic dehydrogenase into ordered superstructures: multichain tubes formed by association of single molecules. J. Mol. Biol. *65*, 127–155 (1972)
Kiermayer, O., Fedtke, C.: Strong anti-microtubule action of amiprophosmethyl (APM) in *Micrasterias*. Protoplasma *92*, 163–166 (1977)
Ledbetter, M.C., Porter, K.R.: A microtubule in plant cell fine structure. J. Cell Biol. *19*, 239–250 (1963)
Lee, J.C., Timasheff, S.N.: The reconstitution of microtubules from purified calf brain tubulin. Biochemistry *14*, 5183–5187 (1975)
Lescure, A.M., Filner, P.: Partially purified tubulin from cultured tobacco cells. In: Plant Research '74. East Lansing: MSU-AEC Plant Research Laboratory, 1975, p. 126
Manton, I., Clarke, B., Greenwood, A.D.: Further observations on the structure of plant cilia, by a combination of visual and electron microscopy. J. Exp. Bot. *3*, 204–215 (1952)
Marchant, H.J., Pickett-Heaps, J.D.: The effect of colchicine on colony formation in the algae, *Hydrodictyon, Pediastrum* and *Sorastrum*. Planta *116*, 291–300 (1976)
Margolis, R.L., Wilson, L.: Opposite end assembly and disassembly of microtubules at steady state in vitro. Cell *13*, 1–8 (1978)
Margolis, R.L., Wilson, L., Kiefer, B.I.: Mitotic mechanisms based on intrinsic microtubule behavior. Nature (London) *272*, 450–452 (1978)
Margulis, L., Chase, D.: Microtubules in prokaryotes. Science *200*, 1118–1124 (1978)
McIntosh, J.R.: Bridges between microtubules. J. Cell Sci. *61*, 166–187 (1974)
McIntosh, J.R., Hepler, P.K., Van Wie, D.G.: Models for mitosis. Nature (London) *224*, 649–663 (1969)
Millington, W.F., Gawlik, S.R.: Ultrastructure and initiation of wall pattern in *Pediastrum boryanum*. Am. J. Bot. *57*, 552–561 (1970)
Mizel, S.B., Wilson, L.: Nucleoside transport in mammalian cells. Inhibition by colchicine. Biochemistry *11*, 2573–2578 (1972)
Mohri, H.: The function of tubulin in motile systems. Biochim. Biophys. Acta *456*, 85–127 (1976)
Montezinos, D.L.: Cell Wall Biogenesis in *Oocystis apiculata*. Univ. North Carolina. Ph. D. thesis, 1977
Mueller, S.C., Brown, R.M., Jr., Scott, T.K.: Cellulosic microfibrils: nascent stages of synthesis in a higher plant cell. Science *194*, 949–951 (1976)
Myles, D.G., Hepler, P.K.: Spermiogenesis in the fern *Marsilea*: microtubules, nuclear shaping and cytomorphogenesis. J. Cell Sci. *23*, 57–83 (1977)
Nagano, T., Suzuki, F.: Microtubules with 15 subunits in cockroach epidermal cells. J. Cell Biol. *64*, 242–245 (1975)
Newcomb, E.H.: Plant microtubules. Annu. Rev. Plant Physiol. *20*, 253–258 (1969)
Nicklas, R.B.: Mitosis. Adv. Cell Biol. *2*, 225–297 (1971)
Nicklas, R.B.: Chromosome movement: current models and experiments on living cells. In: Molecules and Cell Movement. Inoué, S., Stephens, R.E. (eds.). New York: Raven Press, 1975, pp. 97–117
Nicklas, R.B., Staehly, C.A.: Chromosome micromanipulation I. The mechanics of chromosome attachment to the spindle. Chromosoma *21*, 1–16 (1967)
O'Brien, T.P.: The cytology of cell wall formation in some eukaryotic cells. Bot. Rev. *38*, 87–118 (1972)
Palevitz, B.A., Hepler, P.K.: Cellulose microfibrils orientation and cell shaping in developing guard cells of *Allium*: the role of microtubules and ion accumulation. Planta *132*, 71–93 (1976)

Parthasarathy, M.V.: Ultrastructure of phloem in palms. Protoplasma *79*, 59–91 (1974)
Peng, H.B., Jaffe, L.F.: Cell-wall formation in *Pelvetia* embryos. A freeze-fracture study. Planta *133*, 57–71 (1976)
Pickett-Heaps, J.D.: The effects of colchicine on the ultrastructure of dividing cells, xylem wall differentiation and distribution of cytoplasmic microtubules. Dev. Biol. *15*, 206–236 (1967)
Pickett-Heaps, J.D.: Microtubule-like structures in the growing plastids or chloroplasts of two algae. Planta *81*, 193–200 (1968)
Pickett-Heaps, J.D.: Plant microtubules. In: Dynamic Aspects of Plant Ultrastructure. Robards, A.W. (ed.). New York-London: McGraw-Hill, 1974, pp. 219–255
Pickett-Heaps, J.D., Northcote, D.H.: Organization of microtubules and endoplasmic reticulum during mitosis and cytokinesis in wheat meristems. J. Cell Sci. *1*, 109–120 (1966)
Pipeleers, D.G., Pipeleers-Marichal, M.A., Sherline, P., Kipnis, D.M.: A sensitive method for measuring polymerized and depolymerized forms of tubulin in tissues. J. Cell Biol. *74*, 341–350 (1977)
Porter, K.R.: Cytoplasmic microtubules and their function. In: Ciba Foundation Symposium, Principles of Biomolecular Organization. Wolstenholme, G.E.W., O'Connor, M. (eds.). London: Churchill, 1966, pp. 308–345
Porter, K.R.: Introduction-motility in cells. In: Cell Motility. Book A. Goldman, R., Pollard, T., Rosenbaum, J. (eds.). Cold Spring Harbor Laboratory, 1976, pp. 1–28
Quader, H., Cherniack, J., Filner, P.: Chemically induced shortening of flagella of *Chlamydomonas reinhardtii*. Plant Physiol. *59*, suppl. 19 (1977)
Ris, H.: The anaphase movement of chromosomes in the spermatocytes of the grasshopper. Biol. Bull. *96*, 90 (1949)
Robinson, D.G., Herzog, L.W.: Structure, synthesis and orientation of microfibrils. III. a survey of the action of microtubule inhibitors on microtubules and microfibril orientation in *Oocystis solitaria*. Cytobiol. *15*, 463–474 (1977)
Robinson, D.G., White, R.K., Preston, R.D.: Fine structure of swarmers of *Cladophora* and *Chaetomorpha* III. wall synthesis and development. Planta *107*, 131–144 (1972)
Roth, L.E.: Motile systems with continuous filaments. In: Primitive Motile Systems in Cell Biology. Allen, R.D., Kamiya, N. (eds.). New York-London: Academic Press, 1964, pp. 527–548
Rubin, R.W., Cousins, E.H.: Isolation of a tubulin-like protein from *Phaseolus*. Phytochemistry *15*, 1837–1839 (1976)
Rubin, R.W., Weiss, G.D.: Direct biochemical measurements of microtubule assembly and disassembly in Chinese hamster ovary cells: effect of cold D_2O and $N^6,O^{2'}$-dibutyryl cyclic adenosine monophosphate. J. Cell Biol. *64*, 42–53 (1975)
Salema, R., Brandão, I.: Development of microtubules in chloroplasts of two halophytes forced to follow crassulacean acid metabolism. J. Ultrastruct. Res. *62*, 132–136 (1978)
Sandoval, I., Jameson, J.L., Niedel, J., MacDonald, E., Cuatrecasas, P.: Role of nucleotides in tubulin polymerization: effect of guanosine 5′ methylene diphosphonate. Proc. Natl. Acad. Sci. U.S.A. *75*, 3178–3182 (1978)
Sanger, J.M., Jackson, W.T.: Fine structure study of pollen development in *Haemanthus katherinae* Baker II. microtubules and elongation of the generative cell. J. Cell Sci. *8*, 303–315 (1971)
Sanger, J.W., Sanger, J.M.: Actin localization during cell division. In: Cell Motility. Book C. Goldman, R., Pollard, T., Rosenbaum, J. (eds.). Cold Spring Harbor Laboratory, 1976, pp. 1295–1316
Schnepf, E.: The arrangement of microtubules after cell division in young *Sphagnum* leaves. Naturwissenschaften *59*, 471 (1972)
Schnepf, E., Röderer, G., Herth, W.: The formation of the fibrils in the lorica of *Poteriochromonas stipitata*: tip growth, kinetics and orientation. Planta *125*, 45–62 (1975)
Shimamura, T.: Studies on the effect of the centrifugal force upon nuclear division. Cytologia (Tokyo) *11*, 186–216 (1940)
Snyder, J.A., McIntosh, J.R.: Biochemistry and physiology of microtubules. Ann. Rev. Biochem. *45*, 699–720 (1976)
Soifer, D.: The Biology of Cytoplasmic Microtubules. Ann. N.Y. Acad. Sci. *253*, New York, 1975

Stephens, R.E., Edds, K.T.: Microtubules: structure, chemistry and function. Physiol. Rev. *56*, 709–777 (1976)
Tilney, L.G.: The origin and continuity of microtubules. In: Origin and Continuity of Cell Organelles. Reinert, J., Ursprung, H. (eds.). Berlin-Heidelberg-New York: Springer, 1971, pp. 222–260
Upadhaya, M.K., Noodén, L.: ^{14}C oryzalin binding in the roots of a resistant and sensitive species. Plant Physiol. *61*, suppl. 56 (1978)
Viamontes, G.I., Kirk, D.L.: Cell shape changes and the mechanism of inversion in *Volvox*. J. Cell Biol. *75*, 719–730 (1977)
Warfield, R.K.N., Bouck, G.B.: Microtubule-macrotubule transitions: intermediates after exposure to the mitotic inhibitor vinblastine. Science *186*, 1219–1220 (1974)
Weisenberg, R.C.: Microtubule formation in vitro in solutions containing low calcium concentrations. Science *177*, 1104–1105 (1972)
Westafer, J.M., Brown, Jr., M.R.: Electron microscopy of the cotton fibre: new observations on cell wall formation. Cytobios *151*, 111–138 (1976)
Willison, J.H.M., Brown, Jr., M.R.: A model for the pattern of deposition of microfibrils in the cell wall of *Glaucocystis*. Planta *141*, 51–58 (1978)
Wilson, L.: Action of drugs on microtubules. Life Sci. *17*, 303–310 (1975)
Wilson, L., Meza, I.: The mechanism of action of colchicine. J. Cell Biol. *58*, 709–719 (1973)
Yahara, I., Edelman, G.M.: Modulation of lymphocyte receptor mobility by concanavalin A and colchicine. Ann. N.Y. Acad. Sci. *253*, 455–469 (1975)
Yin, H.H., Ukena, T.E., Berlin, R.D.: Effect of colchicine, colcemid and vinblastine on the agglutination by concanavalin A of transformed cells. Science *178*, 867 868 (1972)

2.2 Actomyosin as a Basic Mechanism of Movement in Animals and Plants

G. WAGNER

1. Introduction

In the chapters on shuttle streaming in slime molds (BRITZ, Chap. 2.3, this vol.), on cyclosis (SEITZ, Chap. 2.4, this vol.) and on chloroplast movement (BRITZ, Chap. 2.5, this vol.), evidence will be given that protein filaments organized to a contractile complex [1] are responsible for these cytoplasmic movements. To characterize the structures responsible, research was undertaken at the molecular level, and there proved to be a close chemical similarity between these proteins and actomyosin in muscles of animals, including man. Hence, the term "actomyosin", first introduced in muscle research, is applied quite correctly to these plant proteins, as well. Accordingly, it seems reasonable in this introductory chapter to start with a summary of the chemical, structural, and dynamic properties of muscle actomyosin. This survey will be based mainly on the excellent review articles of GERGELEY (1973), SZENT-GYÖRGYI (1975), HUXLEY (1976), and MANNHERZ and GOODY (1976), which are also recommended for a complete list of references. Following this survey, plant actomyosin will be treated briefly in the present chapter, and in details in the contributions of BRITZ, POFF and WHITAKER, and SEITZ (Chaps. 2.3, 3.2.4, 2.4, this vol.).

2. Muscle Actomyosin

The striated muscle of vertebrates basically consists of two structural proteins, the fibrous polypeptide of actin and the paddle-shaped protein of myosin.

Actin forms the "thin filaments", which are anchored in the membraneous structure of the Z-line, whereas myosin builds, by side-to-side bonding, the "thick filaments" that lie parallel to the thin filaments. The complex of actin and of myosin is called "actomyosin".

2.1 Myosin and Localization of ATPase Activity

Biochemical and biophysical methods of structural identification have revealed that the myosin molecule (see Fig. 1) consists of two very large polypeptide chains

[1] The convenient, but misleading short term "contractile proteins", often found in the literature, is not used in this context: the proteins involved in contraction do not contract, but slide past each other; contraction of the entire protein complex is the result of this sliding mechanism.

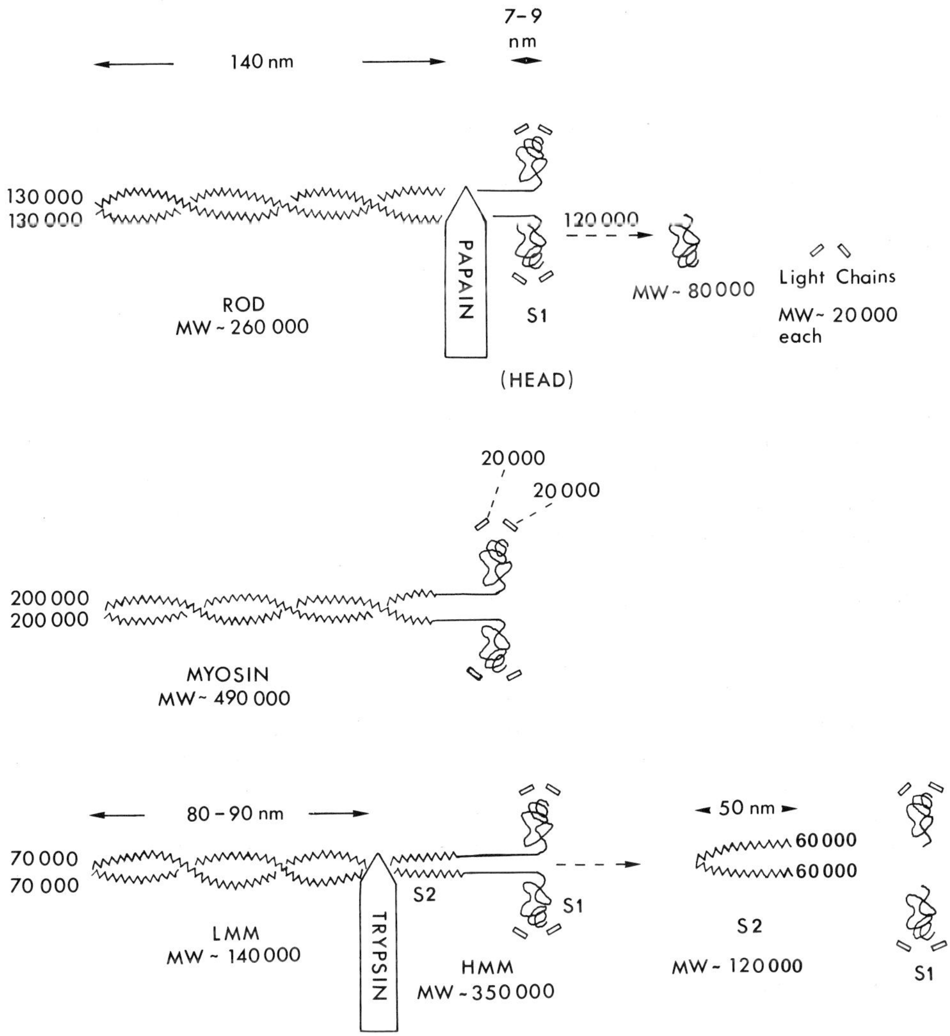

Fig. 1. *Center:* Scheme of the paddle-shaped structure of the myosin molecule (modified after LOWEY et al., 1969). It should be noted that the papain-cleaved subfragment S1 (*upper part*) contains one main polypeptide chain, while HMM and the remaining LMM subfragment obtained by trypsin digestion (*lower part*) consist of two main chains

of molecular weight of about 200,000 daltons each (the "heavy chains") and four smaller polypeptide chains having molecular weights in the 20,000-dalton range (the "light chains"). Along part of their length, the two heavy chains are coiled around each other to form a double-stranded structure about 140 nm in length and 2 nm in diameter. The rest of each heavy chain is folded up to form a globular "head", together with some or all of the light chains. By proteolytic attack, the large polypeptide chains of myosin can be split into well-characterized subfragments, the size of which is dependent on the splitting enzyme used; e.g. papain cleavage tears off the two globular myosin heads, each of which is commonly

Table 1. Magnesium effect on myosin ATPase: dependence on ionic strength of extraction buffer and presence of actin

Ionic strength	Magnesium effect on myosin ATPase	
	with actin	without actin
[KCl] ~ 0.1 M	Stimulation	Inhibition
[KCl] ≧ 0.5 M	Inhibition	Inhibition

defined as subfragment S1 (molecular weight 120,000); the remainder is the myosin rod (molecular weight 260,000). Trypsin digestion, by contrast, yields heavy meromyosin, which consists of the two S1 "heads" (S1 units) and part of the myosin rod, the S2 unit, which clamps the two S1 heads together (HMM; molecular weight 350,000). The remaining part of the myosin rod is called light meromyosin (LMM; molecular weight 140,000).

ATPase activity, and the ability to interact with actin, are present only in those fragments that contain globular heads of myosin, i.e., HMM or S1 itself (see Fig. 1). Stimulation or inhibition of myosin ATPase activity depends on the presence of actin as well as on the ionic strength of the extraction buffer and on the presence of magnesium ions. These relationships are compiled in Table 1.

2.2 The Actin Filament

In contrast to the remarkably structured myosin molecule, the actin *f*ilament (F-actin) is rather monotonously composed of identical *g*lobular actin monomers (G-actin) like pearls on a string. Two of these actin strands are twisted around each other with a helical repeat of 36–38 nm, providing space for seven actin monomers along each chain; the diameter of the actin monomer is about 5.5 nm, with a molecular weight of about 42,000 daltons. Pairs of globular troponin molecules are attached to the double-stranded actin filament, in a periodical repeat, seven actin monomers apart (see Sect. 2.2.3). The third component, finally, is a double-stranded structure of tropomyosin, interlaced with the actin filament and in close contact with the troponin molecules (see also Fig. 3).

2.2.1 Monomeric and Polymeric Properties of Actin

The actin monomer is a single-chain protein of 376 amino acid residues whose complete sequence has been reported recently for rabbit skeletal muscle actin (ELZINGA et al., 1973). At position 73, one molecule of the unusual amino acid N^{τ}-methylhistidine was found; its biological significance, however, is not yet known. Another remarkable feature is the large number of bulky proline residues (19), which decrease the α-helical content to about 30%. When the ionic strength of the extraction buffer for the monomeric G-actin is increased to the physiological range of 0.1 M KCl, G-actin polymerizes to filamentous F-actin, splitting off the terminal phosphate of one specifically bound ATP molecule. The scheme of this reaction is

$$n\ (\text{G-actin-ATP}) \rightleftharpoons \text{F-actin} \cdot \text{ADP} + nPO_4.$$

The role of ATP, however, and its interaction with actin is far from clear. One may speculate that the nucleotide serves as an energy source for conformational changes at the acetylated N-terminal end of the monomer. The existence of such conformational changes during the G- to F-actin transition has been clearly demonstrated by various spectroscopic techniques. (For a review see TAYLOR, 1972.) However, polymerization seems also to proceed if nonhydrolyzable ATP-analogs, especially ATP (β, γ-NH), occupy the respective binding site.

It may be mentioned at this stage that nonmuscle actins polymerize very readily when highly pure, but less so in crude extracts. In vivo, nonmuscle actins can change polymerization state very rapidly depending on physiological conditions (see Sect. 3.3).

2.2.2 Actin Conservation Throughout Evolution

Actin and actin-containing filaments are not limited to striated muscle, but have been identified in nonmuscle cells as well, and even in representatives of the plant kingdom (see Sect. 3). Numerous reports indicate gross sequence homologies in the primary structure of the polypeptide chain, not only of actins prepared from different muscle types such as bovine cardiac and trout muscle, but also of actins from slime mold, amoeba, and human platelets. From this overwhelming amount of evidence, one has to agree with the conclusion of MANNHERZ and GOODY (1976) that "undoubtedly, the physicochemical behavior of actin and its mode of interaction with myosin is little if at all changed ... (during evolution) ..., indicating that actin is one of the most conserved protein molecules in nature. This appears pausible, since a large part of the actin surface has to interact with other contractile proteins, and therefore has to fulfill active-site functions". It might be added that myosin has diverged far more than actin during evolution. This divergence can be shown, for example, by the incomplete antigen-antibody cross reactions among myosins from different sources, presumably due to differences in the size of the heavy chains.

2.2.3 Troponin and Tropomyosin

The troponin component of the actin strand is a globular protein complex (molecular weight 76,000) that regulates contractile activity in vertebrate skeletal muscle. The three subunits which make up troponin are named according to their function (see Fig. 2): Tn T (tropomyosin-binding subunit), Tn I (troponin-inhibitory subunit), and Tn C (calcium-binding subunit).

The calcium-binding subunit (Tn C; molecular weight 18,000) is the most acidic troponin component, due to its high content of aspartic and glutamic acid residues (pK about 4.1). This feature is apparently related to the calcium binding properties of Tn C: four repeating amino acid units have been found to contain mainly acidic residues, which presumably represent the high affinity sites for calcium. In addition, there is evidence for two Mg^{2+} binding sites. Ca^{2+} binding induces

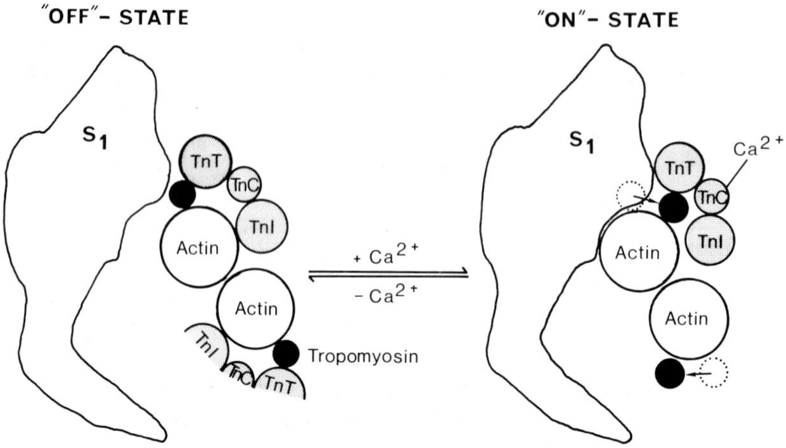

Fig. 2. Composite end-on view of actin-tropomyosin-troponin together with one myosin subfragment S1 (modified after HUXLEY, 1976). *Left*, in the "off" state, i.e., in the absence of calcium, the actin binding site to myosin subfragment S1 is sterically blocked by tropomyosin. *Right*, in the "on" state, i.e., in the presence of calcium, tropomyosin is shifted aside by troponin, to allow S1 binding to actin

conformational changes of Tn C that can be detected by a number of physicochemical techniques, such as changes in the intrinsic fluorescence of the subunit, in the fluorescence intensity of covalently linked fluorophores, and in the mobility of attached spin labels. These changes occur in the physiological calcium concentration range and are complete at 10^{-5} M. Most importantly, isolated Tn C changes its mobility on polyacrylamide gels in the presence of Ca^{2+}, an effect that is not explainable simply by a charge effect. The biological function of Tn C is to govern muscular contraction depending on the presence of calcium.

The troponin-inhibitory subunit (Tn I) is a slightly basic polypeptide chain of 179 residues (molecular weight about 21,000). Its basic region, from residues 102 to 135, probably represents the actin binding site, whereas the N-terminus represents the Tn C interacting site. Tn I acts in concert with tropomyosin and the tropomyosin-binding subunit to inhibit actin-myosin interaction.

The third component, the tropomyosin-binding subunit (Tn T, molecular weight about 37,000) is the most basic troponin subunit (pK about 8.8), due to its high arginine and lysine content. As its name implies, Tn T binds strongly to isolated tropomyosin and to F-actin-tropomyosin as well, probably by its N-terminal end and independently of calcium. Its biological function is to form a second connection between the Tn C·Tn I complex and the actin filament, via tropomyosin, thus conferring full calcium sensitivity to the myosin-actin interaction.

The subunits of troponin appear to be linked in a serial manner; their interaction and their affinity to actin depend on changes in calcium concentration in the physiological range (see Fig. 2).

2.3 Calcium Regulation of Muscle Contraction

At rest, i.e., in the absence of calcium, but in the presence of ATP, cross links are not made between actin and myosin. The muscle stretches readily, and the actin filaments may be pulled out with ease from the array of the myosin filaments. The resting state is maintained by preventing interaction of myosin and actin, and the role of calcium is to remove this inhibition. The binding of calcium to one Tn C subunit results in switching about seven actin monomers from the "off" state to the "on" state, thus the actin monomers function in a cooperative manner. The changes in the state of troponin are communicated to distant actin monomers via tropomyosin. X-ray diffraction and three-dimensional reconstitution of electron microscopic images of actin filaments suggest that, at rest, tropomyosin is situated at some distance from the center of the actin groove and positioned in such a way that it is likely to interfere with the binding of the S-1 portion of myosin (see Fig. 2).

Hence the simplest picture at present of calcium regulation at actin is the assumption that in the "off" state, the actin sites are blocked sterically by tropomyosin. The "on" state, by contrast, is characterized by a calcium-triggered and troponin-governed movement of tropomyosin in direction of the actin groove. Thus, the immediate effect of calcium is a conformational change of troponin which, in turn, shifts tropomyosin aside to allow myosin-binding to actin (SZENT-GYÖRGYI, 1975). This type of calcium regulation is commonly designated as "thin filament control", because it works on actin.

A different system of calcium regulation, associated with myosin, i.e., "thick filament control" acts in molluscan muscles (KENDRICK-JONES, 1976). The best-studied examples so far are the striated adductor muscles of scallop, the myosin of which consists of two heavy and four light chains, similar to myosin in vertebrate striated muscles. Unlike the vertebrate system, however, after detachment of one mol of light chain from myosin by 1–10 mM EDTA (Ethylenediaminetetraacetic acid), the residual myosin is desensitized: its ATPase activity is no longer regulated by calcium. The calcium binding of the desensitized myosin is reduced to about one half compared to the value measured before. The missing calcium-binding activity, however, is not found in the isolated light chain which also fails to show ATPase activity. Hence, one may speculate that the EDTA-removable light chain acts like a sensitizing cofactor to the remaining light chains. These, in turn, sterically block actin-binding sites on myosin in the "off" state, and move out of the way during the "on" state.

2.4 Interaction of Myosin with Nucleotides and Actin

Ample evidence from structural and kinetic studies support the so-called rowboat model of muscular contraction (HUXLEY, 1969) according to which the myosin molecule acts as the rowing component within a "fluid" of actin (see Fig. 3).

Charged by the energy of ATP hydrolysis, the myosin heads (S1 units) promptly bind to actin [see state (1) in Fig. 3A]. This binding simultaneously catalyzes a rapid release of the reaction products of ATP hydrolysis and a change in the angle between the S1 units and the actin filament from 90° to 45° [state (2)]. The flapping motion of the myosin heads may be regarded as a power stroke which leads to a shift of the actin filament past myosin by about 7–8 nm. At the bottom of the power stroke, binding of ATP to the myosin ATPase dissociates the myosin heads from actin, and the heads then return to the right-angled configuration, ready for another actin binding [state (3)]. If actin is absent, the reaction products of ATP hydrolysis stay myosin-bound, and are released only very slowly. Hence, the ($ADP + PO_4$) release is an actin-controlled rate-limiting step, keeping the myosin molecule in an "energized" state until the next actin molecule is ready

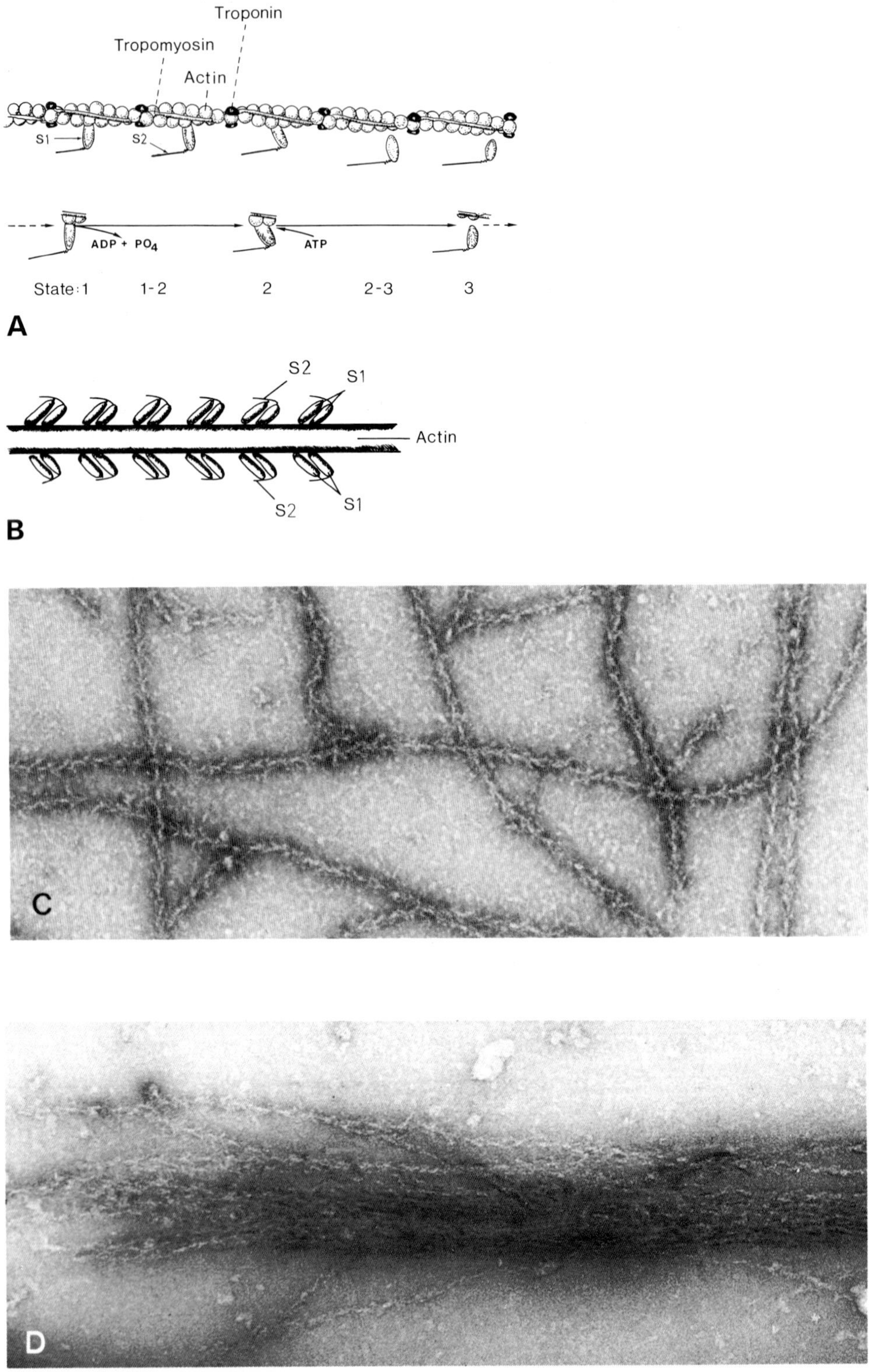
Troponin
Tropomyosin
Actin
S1
S2
ADP + PO4
ATP
State: 1
1-2
2
2-3
3
A
S2
S1
Actin
S2
S1
B
C
D

Table 2. Physicochemical states of actomyosin: dependence on ionic strength of extraction buffer and presence of ATP

Ionic strength	Actin	Myosin	Ionic strength	Actin	Myosin
[KCl] ~ 0.1 M	Insoluble	Insoluble	[KCl] ≧ 0.5 M	Slowly soluble	Rapidly soluble
↓	The two components together form functionally intact actomyosin complexes (synthetic actomyosin fibrils).		↓	Combining of the two components results in an anomously high viscosity, i.e. a viscosity higher than the sum of the viscosities of separate actin and myosin solutions.	
+ATP	Turbidity increases and settling is enhanced; volume of precipitate decreases by syneresis. The latter phenomenon is called "super-precipitation" and provides a simple type of contractile model. → *Main effect due to ATP:* Contraction of actomyosin complex		+ATP	Instantaneous drop in viscosity, interpreted as a result of dissociation of actin and myosin. The viscosity gradually increases over several minutes as ATP is hydrolysed by myosin. → *Main effect due to ATP:* Dissociation of actomyosin complex	

to be bound. If, on the other hand, ATP is absent, myosin heads remain stuck to actin, immobilizing the system in the contracted state. This state is known as rigor in muscles.

According to H.E. HUXLEY (1969), ATP explicitly plays a dual role: first, by catalyzing the dissociation of the actin-myosin complex, second, by delivering the energy to pivot the myosin head back into the energized 90° position after the power stroke. According to the view of A.F. HUXLEY and SIMMONS (1971), however, the essential role of ATP is only to

Fig. 3. A Diagram showing probable arrangement of actin, tropomyosin and troponin according to HANSON and LOWY (1963), and EBASHI and ENDO (1968). Three different states of a myosin head (S1 unit) rowing along the actin filament are shown, simultaneously. State *1*, myosin head has just attached to actin; products of splitting ATP (ADP+PO_4) are released. State *2*, the power stroke of the myosin head is completed; binding of ATP to the myosin ATPase dissociates the myosin head from actin. State *3*, myosin head is returned to the right-angled position, ready for another actin binding. **B** Schematic drawing of heavy meromyosin bound to an actin filament. Note that HMM consists of two S1 units clamped together by the S2 unit. **C** and **D** Micrographs of negatively stained actin filaments from rabbit skeletal muscle (**C**) and from *Nitella*, respectively (**D**), decorated with heavy meromyosin. The *arrowheads* formed by HMM point consistently in one direction due to a polarity in the myosin binding sites of F-actin. (**C** Magnification, 80,000×, photo courtesy of Dr. H. HINSSEN, Institut f. Cytol., Univ. Bonn, FRG; **D** Magnification, 120,000×, reprinted, with permission, from KERSEY et al., 1976)

dissociate the actin-myosin complex; in the detached form, the 45° and the 90° configurations of the myosin head are assumed to be energetically equal and hence reached with the same thermodynamic probability. Attachment to actin, however, favors the angled configuration, so that the driving force for the power stroke comes from this thermodynamic compulsion.

Simple "contractile models" provide an important tool for further analysis of the molecular interaction between actin and myosin (HINSSEN and D'HAESE, 1976). These reconstituted "muscle ghosts" basically consist of highly purified actomyosin, the physicochemical properties of which may depend on the presence of further components added at will. The dependence of some of these properties on the ionic strength of the extraction buffer and on the presence of ATP is summarized in Table 2.

3. Actomyosin in Plant Cells

3.1 Techniques of Identification

It has been proposed that microfilaments found in animal and plant cells are the less highly evolved ancestors of skeletal actin because of their fibrous appearance, their size (4–10 nm in diameter), and their presence in regions of cellular movement (GOLDMAN, 1975). Further substantiation of this hypothesis was derived from effects on movement by certain drugs like cytochalasin B (CARTER, 1967), N-ethylmaleimide, and p-chloromercuribenzoate (DREIZEN and GERSHMAN, 1970) or phalloidin (WIELAND, 1977) which are thought to interact with actomyosin-like microfilaments.

The most convincing and direct evidence for these microfilaments being actin, however, comes from the use of the heavy meromyosin (HMM) binding technique originally described by ISHIKAWA et al. (1969). This technique is based on H.E. HUXLEY's pioneering studies (1963) on the binding of rabbit skeletal muscle HMM to isolated rabbit F-actin to form "arrowhead" configurations or "decorated" actin strands. Arrowheads along individual strands point in one direction due to a polarity in F-actin (see Fig. 3). This distinctive pattern is most convincingly visualized in thin sections or after negative staining. Due to the amazingly accurate conservation of the actin molecule throughout evolution (see Sect. 2.2.2), actin decoration by HMM works using HMM even from a distantly divergent source no matter whether animal or plant. For convenience, HMM is usually derived from rabbit leg muscle or from chicken breast muscle.

The application of this technique to plant cells in situ was initiated by Wohlfarth-Bottermann and his coworkers, who identified in the plasmodium of *Physarum* 7-nm wide plasmafilaments as F-actin by in situ decoration with HMM (ALLÉRA et al., 1971, see also BRITZ, Chap. 2.3, this vol.). Similarly, a pioneering work with respect to green plants was done by PALEVITZ et al. (1974) and simultaneously by WILLIAMSON, who succeeded in 1974 in decorating actin filaments of the characean algae, *Nitella* and *Chara*, with rabbit HMM. Since then, successful actin decoration by HMM has only been reported from a few other species of green plants, namely from the green algae *Coleochaete* and *Mougeotia* (MARCHANT, 1976) and from the higher plants *Amaryllis belladonna* (CONDEELIS, 1974) and *Haemanthus katherinae* Baker (FORER and JACKSON, 1975, 1976; see Sect. 3.3).

3.2 Cyclosis in Characean Algae

The endless and somewhat mystifying protoplasmic streaming in the internodal cells of the giant characean algae has attracted scientists ever since the first report by Corti two hundred years ago (HOPE and WALKER, 1975). Modern light microscopy (phase contrast or Nomarsky optics) has revealed that protein fibers, possibly producing the cytoplasmic motive force, are positioned at the boundary between the stationary cortical cytoplasm (ectoplasm) and the flowing endoplasm. Electron microscopy shows that each of these fibers is composed of 50 to 100 microfilaments, which often fray out into individual microfilaments of 5 nm diameter. The fibers build straight, nonflexous structures parallel to the direction of streaming, and are anchored in the stationary ectoplasm. Although there is some association with the outer envelope of the chloroplasts, forming an immobile layer in the stationary ectoplasm, the real nature of anchorage is still unclear.

After treatment of the fibers with HMM, many of the microfilaments appear decorated with arrowheads, all of which point in the same direction. In suitable preparations conserving the in vivo direction of streaming, the arrowheads consistently point in the direction opposite to that of the cytoplasmic stream. This finding immediately raises the question of how these filaments are involved in motion. At present, a reasonable assumption is that the streaming endoplasm shears along the immobile ectoplasm driven by myosin molecules which move along the ectoplasmic actin filament bundles. These myosin molecules are either associated with structural proteins or free in the endoplasm (KAMIYA, 1959; WILLIAMSON, 1975; KERSEY et al., 1976; NAGAI and KAMIYA, 1977).

This shearing mechanism of movement seems clearly different from the sliding filament mechanism of contraction discussed with respect to striated muscle (see Sect. 2.4). However, careful consideration shows that the main difference is based on the structural organization of actin and myosin rather than on the mode of interaction. As in striated muscle, myosin "rows" along a membrane-anchored actin filament. In this case, however, the filament is endless, due to its circular organization within the cortical cytoplasm. From this shearing along the endless band of actin, a continuous cytoplasmic stream (cyclosis) automatically results, and is in fact observed as long as the cell is alive. Hence, cyclosis is independent of a trigger mechanism for starting or stopping, making it plausible that streaming is inhibited rather than stimulated by calcium (WILLIAMSON, 1975). Biochemical investigations, just started, may further elucidate these peculiarities of characean actomyosin (PALEVITZ, 1976; WILLIAMSON, 1976; KATO and TONOMURA 1977).

3.3 Outlook for Other Plant Systems and Modes of Actomyosin Regulation

Compared to the characean algae, the involvement of actin filaments in other plant systems is less well documented. However, some promising efforts have to be mentioned. In the green alga *Mougeotia*, HMM decoration of actin filaments in a cell homogenate was reported recently (MARCHANT, 1976). This in vitro evidence fits well with the in situ demonstration of 5–10 nm wide filaments in the cortical cytoplasm which may well be able to move the flat, ribbon-shaped chloroplast (WAGNER and KLEIN, 1978)—see also BRITZ, Chap. 2.5, this vol. Hence, the assumption that actin filaments are involved in rotating the single, large chloroplast seems reasonable (SCHÖNBOHM, 1969; WAGNER et al., 1972; WAGNER, 1974).

Still, the most intriguing aspects of this light-oriented and phytochrome-mediated chloroplast response remain to be clarified, namely the mechanisms of triggering and coordination of movement. An essential part of this reaction chain, as is generally true of nonmuscle contraction, is that the filaments must be built onto a prepared attachment site, comparable to an anchorage ground. It is also necessary that stimulation of filament formation be highly localized, occurring only in the immediate neighborhood of the attachment point; for instance, the *Mougeotia* chloroplast is able to rotate several times by 360° (HAUPT and HEYMANN, 1967). Taking this into consideration, it may well be that stimulation and localization of filament formation, rather than control of filament contraction, are the key processes governed by the external factors in question such as light in case of *Mougeotia*.

Better knowledge of the biochemistry of actin and myosin will increase our understanding of the different possibilities of actomyosin regulation realized in nonmuscle cells. A few examples may already be mentioned: for instance, polymerization of actin depends on the ionic balance among Ca^{2+}, H^+ and K^+; nucleotides bound to actin, and free sulfhydryl groups have a promoting effect on actin polymerization. Myosin enzymatic activity may be subject to post-transcriptional modification by the state of phosphorylation, and may depend on the presence of high molecular weight cofactor proteins (for a review see CLARKE and SPUDICH, 1977). Actin polymerization may be localized by membrane-bound proteins like plasmodium actinin in *Physarum polycephalum* (HATANO and OWARIBE, 1976), high molecular weight protein in *Acanthamoeba,* in *Dictyostelium,* in sea urchin eggs, in sperm of the echinoderm *Thyone,* in macrophages and in human platelets (POLLARD et al., 1976; COOKE et al., 1976; KANE, 1975; TILNEY, 1976; STOSSEL and HARTWIG, 1976; WEIHING, 1976), or by the low molecular weight protein profilin in calf spleen (CARLSSON et al., 1976). Comparable regulators should also be looked for in green plants. A technique like immunofluorescence microscopy (LAZARIDES and WEBER, 1974) may provide the essential tool. This antibody technique was first applied recently to plant cells in the amaryllidacea *Leucojum aestivum* to monitor the pattern of tubulin in the nucleus of the endosperm (FRANKE et al., 1977). Thanks to this experimental breakthrough, we surely will not have to wait long for a better insight into the dynamics of plant actomyosin as well.

Acknowledgements. Investigations performed in the authors' laboratory were supported by the Deutsche Forschungsgemeinschaft. Skillful drawing of the figures in this review article by Mrs. A. Mollwo is thankfully acknowledged.

References

Alléra, A., Beck, R., Wohlfarth-Bottermann, K.-E.: Weitreichende fibrilläre Protoplasmadifferenzierungen und ihre Bedeutung für die Protoplasmaströmung. VII. Identifizierung der Plasmafilamente von *Physarum polycephalum* als F-Actin durch Anlagerung von heavy meromyosin *in situ.* Cytobiologie *4*, 437–449 (1971)

Carlsson, L., Nyström, L.-E., Sundkist, I.: Profilin, a low-molecular weight protein controlling actin polymerisability. In: Contractile Systems in Non-Muscle Tissues. Perry, S.V., Margreth, A., Adelstein, R.S., (eds.). Amsterdam-Oxford-New York: Elsevier, 1976, pp. 39–49

Carter, S.B.: Effects of cytochalasins on mammalian cells. Nature (London) *213*, 261–264 (1967)

Clarke, M., Spudich, J.A.: Nonmuscle contractile proteins: The role of actin and myosin in cell motility and shape determination. Annu. Rev. Biochem. *46*, 797–822 (1977)

Condeelis, J.S.: The identification of F actin in the pollen tube and proteoplast of *Amaryllis belladonna*. Exp. Cell Res. *88*, 435–439 (1974)

Cooke, R., Clarke, M., von Wedel, R.J., Spudich, J.A.: Supramolecular forms of *Dictyostelium* actin. In: Cell Motility. Book B. Actin, Myosin and Associated Proteins. Goldman, R., Pollard, T., Rosenbaum, J. (eds.). Cold Spring Harbor, New York: Cold Spring Harbor, 1976, pp. 575–587

Dreizen, P., Gershman, L.C.: Molecular basis of muscular contraction. Myosin. Transact. New York Acad. Sci., Series II, Vol. *32*, 170–203 (1970)

Ebashi, S., Endo, M.: Calcium ion and muscle contraction. Prog. Biophys. Mol. Biol. *18*, 123–183 (1968)

Elzinga, M., Collins, J.H., Kuehl, W.M., Adelstein, R.S.: Complete aminoacid sequence of actin of rabbit skeletal muscle. Proc. Natl. Acad. Sci. USA *70*, 2687–2691 (1973)

Forer, A., Jackson, Wm.T.: Actin in the higher plant *Haemanthus katherinae* Baker. Cytobiologie *10*, 217–226 (1975)

Forer, A., Jackson, Wm.T.: Actin filaments in the endosperm mitotic spindles in a higher plant, *Haemanthus katherinae* Baker. Cytobiologie *12*, 199–214 (1976)

Franke, W.W., Seib, E., Osborn, M., Weber, K., Herth, W., Falk, H.: Tubulin-containing structures in the anastal mitotic apparatus of endosperm cells of the plant *Leucojum aestivum* as revealed by immunofluorescence microscopy. Cytobiologie *15*, 24–48 (1977)

Gergeley, J.: Muscle contraction, conformational changes and energy transduction. Ann. N.Y. Acad. Sci. *227*, 587–593 (1973)

Goldman, R.D.: The use of heavy meromyosin binding as an ultrastructural cytochemical method for localizing and determining the possible functions of actin-like microfilaments in nonmuscle cells. J. Histochem. Cytochem. *23*, 529–542 (1975)

Hanson, J., Lowy, J.: The structure of F-actin and of actin filaments isolated from muscle. J. Mol. Biol. *6*, 46–60 (1963)

Hatano, S., Owaribe, K.: Actin and actinin from myxomycete plasmodia. In: Cell Motility. Book B. Actin, Myosin and Associated Proteins. Goldman, R., Pollard, T., Rosenbaum, J. (eds.). Cold Spring Harbor, New York: Cold Spring Harbor, 1976, pp. 499–511

Haupt, W., Heymann, N.: Versuche zur Mechanik der Chloroplastenbewegung von *Mougeotia*. Z. Pflanzenphysiol. *57*, 68–71 (1967)

Hinssen, H., D'Haese, J.: Synthetic fibrils from *Physarum* actomyosin – self assembly, organization and contraction. Cytobiologie *13*, 132–157 (1976)

Hope, A.B., Walker, N.A.: The physiology of giant algal cells. Cambridge: University Press, 1975

Huxley, A.F., Simmons, R.M.: Proposed mechanism of force generation in striated musle. Nature (London) *233*, 533–538 (1971)

Huxley, H.E.: Electron microscope studies on the structure of natural and synthetic protein filaments from striated muscle. J. Mol. Biol. *7*, 281–308 (1963)

Huxley, H.E.: The mechanism of muscle contraction. Science *164*, 1356–1366 (1969)

Huxley, H.E.: The structural basis of contraction and regulation in skeletal muscle. In: Molecular Basis of Motility. Heilmeyer, L.M.G., Jr., Rüegg, J.C., Wieland, Th. (eds.). 26. Mosbach-Colloquium. Berlin-Heidelberg-New York: Springer, 1976, pp. 9–25

Ishikawa, H., Bishoff, R., Holtzer, H.: Formation of arrowhead complexes with heavy meromyosin in a variety of cell types. J. Cell Biol. *43*, 312–328 (1969)

Kamiya, N.: Protoplasmic streaming. Protoplasmatologia. Springer, 1959, Vol. VIII, pt. 3a, pp. 1–199

Kane, R.E.: Preparation and purification of polymerized actin from sea urchin egg extracts. J. Cell Biol. *66*, 305–315 (1975)

Kato, T., Tonomura, Y.: Identification of myosin in *Nitella flexilis*. J. Biochem. *82*, 777–782 (1977)

Kendrick-Jones, J.: Myosin-linked calcium regulation. In: Molecular Basis of Motility. Heilmeyer, L.M.G., Jr., Rüegg, J.C., Wieland, Th. (eds.). Mosbach-Colloquium. Berlin-Heidelberg-New York, Springer, 1976, pp. 122–136
Kersey, Y.M., Hepler, P.K., Palevitz, B.A., Wessells, N.K.: Polarity of actin filaments in characean algae. Proc. Natl. Acad. Sci. USA *73*, 165–167 (1976)
Lazarides, E., Weber, K.: Actin antibody: the specific visualization of actin filaments in non-muscle cells. Proc. Natl. Acad. Sci. USA *71*, 2268–2272 (1974)
Lowey, S., Slayter, H.S., Weeds, A.G., Baker, H.: Substructure of the myosin molecule. I. Subfragments of myosin by enzymatic degradation. J. Mol. Biol. *42*, 1–29 (1969)
Mannherz, H.G., Goody, R.S.: Proteins of contractile systems. Annu. Rev. Biochem. *45*, 427–465 (1976)
Marchant, H.J.: Actin in the green algae *Coleochaete* and *Mougeotia*. Planta *131*, 119–120 (1976)
Nagai, R., Kamiya, N.: Differential treatment of *Chara* cells with cytochalasin B with special reference to its effect on cytoplasmic streaming. Exp. Cell Res. *108*, 231–237 (1977)
Palevitz, B.A.: Actin cables and cytoplasmic streaming in green plants. In: Cell Motility. Book B. Actin, Myosin and Associated Proteins. Goldman, R., Pollard, T., Rosenbaum, J. (eds.). Cold Spring Harbor, New York: Cold Spring Harbor, 1976, pp. 601–611
Palevitz, B.A., Ash, J.F., Hepler, P.K.: Actin in the green alga, *Nitella*. Proc. Natl. Acad. Sci. USA *71*, 363–366 (1974)
Pollard, Th.D., Fujiwara, K., Niederman, R., Maupin-Szamier, P.: Evidence for the role of cytoplasmic actin and myosin in cellular structure and motility. In: Cell Motility. Book B. Actin, Myosin and Associated Proteins. Goldman, R., Pollard, T., Rosenbaum, J. (eds.). Cold Spring Harbor, New York: Cold Spring Harbor, 1976, pp. 689–724
Schönbohm, E.: Die Hemmung der lichtinduzierten Bewegung des *Mougeotia*-Chloroplasten durch p-Chloromercuribenzoat (Versuche zur Mechanik der Chloroplastenbewegung). Z. Pflanzenphysiol. *61*, 250–260 (1969)
Stossel, Th.P., Hartwig, J.H.: Phagocytosis and the contractile proteins of pulmonary macrophages. In: Cell Motility. Book B. Actin, Myosin and Associated Proteins. Goldman, R., Pollard, T., Rosenbaum, J. (eds.). Cold Spring Harbor, New York: Cold Spring Harbor, 1976, pp. 529–544
Szent-Györgyi, A.G.: Calcium regulation of muscle contraction. Biophys. J. *15*, 707–723 (1975)
Taylor, E.W.: Chemistry of muscle contraction, Annu. Rev. Biochem. *41*, 577–616 (1972)
Tilney, L.G.: Nonfilamentous aggregates of actin and their association with membranes. In: Cell Motility. Book B. Actin, Myosin and Associated Proteins. Goldman, R., Pollard, T., Rosenbaum, J. (eds.). Cold Spring Harbor, New York: Cold Spring Harbor, 1976, pp. 513–528
Wagner, G.: Some physiological properties of phytochrome in the alga *Mougeotia*, as studied by cytochalasin B and aminophylline. In: Proc. Annu. Eur. Symp. Plant Photomorphogenesis. De Greef, J.A. (ed.). University Press: Antwerpen, 1974, pp. 22–25
Wagner, G., Klein, K.: Differential effect of calcium on chloroplast movement in *Mougeotia*. Photochem. Photobiol. *27*, 137–140 (1978)
Wagner, G., Haupt, W., Laux, A.: Reversible inhibition of chloroplast movement by cytochalasin B in the green alga *Mougeotia*. Science *176*, 808–809 (1972)
Weihing, R.R.: Membrane association and polymerization. In: Cell Motility. Book B. Actin, Myosin and Associated Proteins. Goldman, R., Pollard, T., Rosenbaum, J. (eds.). Cold Spring Harbor, New York: Cold Spring Harbor, 1976, pp. 671–684
Wieland, Th.: Modifications of actins by phallotoxins. Naturwissenschaften *64*, 303–309 (1977)
Williamson, R.E.: Actin in the alga, *Chara corallina*. Nature (London) *248*, 801–802 (1974)
Williamson, R.E.: Cytoplasmic streaming in *Chara:* A cell model activated by ATP and inhibited by cytochalasin B. J. Cell Sci. *17*, 655–668 (1975)
Williamson, R.E.: Actin and motility in plant cells. In: Contractile Systems in Non-Muscle Tissues. Perry, S.V., Margreth, A., Adelstein, R.S. (eds.). Amsterdam-Oxford-New York: Elsevier, 1976, pp. 91–101

2.3 Cytoplasmic Streaming in *Physarum* [1]

S.J. BRITZ

1. Introduction

The unique streaming patterns of the acellular (i.e., true) slime molds have long fascinated biologists. In the plasmodium-stage of *Physarum,* the fluid, internal cytoplasm, or endoplasm, streams with speeds up to 1.5 mm s^{-1} through channels in the more gel-like peripheral cytoplasm, or ectoplasm (KAMIYA, 1959). These channels run both through the cytoplasmic mass at the advancing front of the plasmodium, as well as through the trailing anastomosing network of cytoplasmic veins. The direction of cytoplasmic streaming reverses periodically every few minutes in a pattern known as "shuttle" or "pendulum" streaming. Net migration of motile plasmodia results from more streaming on the average in one direction than in the other (see POFF and WHITAKER, Chap. 3.2.4, this vol.). Although the streaming pattern in adjacent veins or in different parts of a single vein may seem independent, shuttle streaming is obviously coordinated at the organismic level.

Contraction of the ectoplasm is thought to provide the motive force for endoplasmic streaming (see KAMIYA, 1959). The obvious analogy to muscle contraction made *Physarum* the first nonmuscle tissue to be examined for actomyosin-like proteins, a search which has proved successful. The first part of this review presents evidence for the presence of actin and myosin in *Physarum* and the ways in which these proteins differ from their muscle analogs. Possible actomyosin contractile mechanisms and their control are discussed, especially as they pertain to relatively nondifferentiated tissue such as *Physarum.* The second part of the review considers the location and organization of actin and myosin in the plasmodium, the evidence for the involvement of this actin and myosin in actual ectoplasmic contraction, and evidence for possible control mechanisms, including the coordination of shuttle streaming.

Easy culture of large quantities of *Physarum* plasmodia plus the ability to combine whole organism physiology with biochemical analyses have resulted in the emergence of *Physarum* as a favored organism for the investigation of cellular motility. The emphasis on *Physarum,* however, may have been at the expense of our understanding of other acellular slime molds such as *Reticulomyxa* in which

[1] CIW-DPB Publication No. 633.

Abbreviations: ATP, adenosine triphosphate; AMP, adenosine monophosphate; EDTA, ethylene diamine tetraacetic acid; EGTA, ethyleneglycol-*bis*-(β-aminoethyl ether)-N,N′-tetraacetic acid; PCMB, p-chloromecuribenzoic acid; PP_i, pyrophosphate; SDS, sodium dodecyl sulfate.

the mechanism for cytoplasmic streaming may not be the same (see KAMIYA, 1959).

Streaming in the acellular slime molds was discussed extensively by KAMIYA (1959) and was therefore treated only briefly in the last edition of the *Encyclopedia* (DREWS and NULTSCH, 1962). Since that time several reviews on *Physarum* have appeared (WOHLFARTH-BOTTERMANN, 1964; KAMIYA, 1968; KOMNICK et al., 1972, 1973; HATANO, 1973). Obviously it is impossible to dissociate the study of actomyosin and motility in *Physarum* from that in other organisms. Additional coverage of *Physarum* streaming mechanisms and of actomyosin has been included in reviews and symposia dedicated to the topic of nonmuscle motility (ALLEN and KAMIYA, 1964; JAHN and BOVEE, 1969; DRYL and ZURZYCKI, 1972; DURHAM, 1974; HEPLER and PALEVITZ, 1974; POLLARD and WEIHING, 1974; GOLDMAN et al., 1976; CLARKE and SPUDICH, 1977). Further information on actomyosin is provided by other sources (NEEDHAM, 1971; Ciba foundation symposium No. 14, 1973; Cold Spring Harbor Symposium on Quantitative Biology Vol. 37, 1973; INOUÉ and STEPHENS, 1975; and WAGNER, Chap. 2.2, this vol.).

2. *Physarum* Actin and Myosin

2.1 Actomyosin Complex

Incubation of muscle in high salt concentrations ($\geqq 500$ mM KCl) results in the progressive extraction first of myosin and then of actin; the combined proteins are termed actomyosin or Myosin B. The properties of actomyosin in solution can be related to the mechanism of muscle contraction (for comparison, see WAGNER, Chap. 2.2, this vol.). In the absence of ATP, myosin and actin bind to form an actomyosin solution of anomalously high viscosity (i.e., a viscosity higher than the sum of the viscosities of separate actin and myosin solutions). Added ATP or PP_i causes a rapid drop in viscosity which is interpreted as the result of dissociation of actin and myosin. With ATP, but not with PP_i, the viscosity then increases over several minutes as the nucleotide is hydrolyzed by myosin. At moderate ionic strengths (100 mM KCl, physiological for muscle) actomyosin is insoluble. Added ATP then causes increased turbidity, enhanced settling of the precipitate, and a decrease in precipitate volume by syneresis (this phenomenon is called superprecipitation and is a type of simple contractile model). Note the dual role of ATP: dissociation of the actomyosin complex and supply of energy for contraction (see WAGNER, Chap. 2.2, this vol.). The identification of nonmuscle proteins as actomyosin-like relies heavily on the above characteristic interactions.

Initial evidence for actomyosin in *Physarum* was accumulated over many years. LOEWY (1952) first found a reversible viscosity decrease upon addition of ATP to high salt extracts of *Physarum*. Later the existence in such extracts of ATPase activity and 7 nm wide filaments similar to the thin filaments of actin was demonstrated (TS'O et al., 1957). The latter authors termed their product "myxomyosin," but they were unable to separate it into component proteins. Further progress came when *Physarum* extracts were prepared in alkaline 600 mM KCl (NAKAJIMA, 1960). The actomyosin-like activity in these extracts could be precipitated at moderate ionic strengths and thereby partially purified by repeated cycles of precipitation and resolubilization. The resulting turbid product demonstrated a reversible decrease in viscosity upon addition of ATP. Consistent with the involvement of an actomyosin, AMP had no effect on viscosity, while PP_i caused an irreversible viscosity decrease. Furthermore, Ca^{2+} and Mg^{2+} enhanced and inhibited, respectively, the recovery of increased viscosity after addition of ATP. These results are in keeping with the effects of these ions on myosin ATPase (see also WAGNER, Chap. 2.2, this vol.). In the absence of ATP the ions had no effect on viscosity. At moderate ionic strength, ATP caused superprecipitation which was enhanced by both Ca^{2+} and Mg^{2+}. Unlike that of muscle, the ATPase of the *Physarum* preparations was inhibited by EDTA at all ionic strengths.

HATANO and TAZAWA (1968) provided the first physical evidence for the presence in *Physarum* extracts of two proteins with properties like actin and myosin. Through the addition of EDTA in the initial high salt extract, they obtained purer preparations with enhanced yields of ATPase. Ultracentrifugation then revealed a band sedimenting at 21 S in 500 mM KCl. In the presence of Mg-ATP the band separated into a 6 S component (presumptive myosin) and a 19 S component (presumptive actin filaments). In a separate experiment the precipitates from high-speed centrifugation were analyzed by electron microscopy. In the absence of ATP, 10 nm-wide filaments with attached material were pelleted. The attached material was presumably myosin. In the presence of ATP only 7.5 nm-wide helical filaments resembling thin filaments of actin were observed.

Further characterization, discussed below, reveals these actomyosin-like proteins to be almost identical with their muscle counterparts. Earlier terms such as "presumptive actin" or "myxomyosin" are rejected in favor of the suggestion (POLLARD and WEIHING, 1974) that nonmuscle actins and myosins be prefixed by the name of the appropriate genus (e.g., *Physarum* myosin).

2.2 Actin

2.2.1 Characterization of *Physarum* Actin

Column chromatography indicates that *Physarum* actin, like muscle actin, is a globular protein with a molecular weight of about 40,000 daltons (ADELMAN and TAYLOR, 1969b). In SDS polyacrylamide gel electrophoresis, *Physarum* actin comigrates with muscle actin yielding molecular weight estimates of 43,000 to 46,000 daltons (LESTOURGEON et al., 1975; HATANO and OWARIBE, 1976; JACOBSON et al., 1976).

The amino acid content of *Physarum* actin is similar to that of muscle actin (HATANO and OOSAWA, 1966b). It includes the presence of the unusual amino acid 3-methyl histidine and it shows a partial sequence homology with muscle actin (JOCKUSCH et al., 1974; LESTOURGEON et al., 1975).

In good agreement with the properties of muscle actin, *Physarum* actin binds ATP (HATANO and OOSAWA, 1966b; ADELMAN and TAYLOR, 1969b) and divalent cation, probably Ca^{2+}, in vivo (HATANO and TOTSUKA, 1972). The polymerization behavior of the two actins is also quite similar. Under low salt conditions, *Physarum* actin is present in a 3 S form, probably the G-actin monomer (HATANO and OOSAWA, 1966; ADELMAN and TAYLOR, 1969b). As measured by viscometry, 2 mM $MgCl_2$ or 100 mM KCl induces polymerization of column-purified *Physarum* actin (ADELMAN and TAYLOR, 1969b; HATANO and OWARIBE, 1976), with a resultant specific viscosity (10 dl/g) similar to that of muscle F-actin (HATANO and OWARIBE,

1976). ATP hydrolysis proceeds parallel to the viscosity increase in 100 mM KCl (HATANO et al., 1967).

Electron microscopy of *Physarum* F-actin filaments confirms their structural resemblance to muscle F-actin (HATANO et al., 1967; HATANO and TAZAWA, 1968; NACHMIAS, 1970). The *Physarum* F-actin filaments are double helices 1–2 μm long and 5–8 nm wide with a half-pitch of 35 nm. Each strand of the helix is composed of globular subunits with about 13 subunits per half-turn.

Nevertheless, a number of factors appear to cause the behavior of *Physarum* actin, especially in crude preparations, to differ from that of muscle actin. Among the first differences to be noted was that *Physarum* actomyosin, unlike that of muscle, is soluble at low ionic strengths in the presence of PP_i (ADELMAN and TAYLOR, 1969a). In fact, large amounts of *Physarum* actin are extractable and soluble under conditions where muscle actin would polymerize (JACOBSON et al., 1976). Crude extracts of *Physarum* in dilute buffer revealed only a small amount of pelletable (i.e., polymerized) actin at 100 mM KCl. However, the addition of millimolar amounts of ATP and Ca^{2+} resulted in the bulk of the actin becoming pelletable. Since bound nucleotide and divalent cation enhance the transition of G-actin to F-actin (see WAGNER, Chap. 2.2, this vol.), the authors speculated that *Physarum* actin bound these cofactors more poorly than muscle actin, and therefore polymerized less readily. Note that actin in this study was determined as material migrating like actin on SDS polyacrylamide gels.

Ultracentrifugation of *Physarum* cell sap revealed a component sedimenting like F-actin (ADELMAN and TAYLOR, 1969a). On the other hand, cell saps may also contain large amounts of G-actin (HINSSEN, 1972). Negative stain electron microscopy of the supernatant of a high-speed spin of crude cell sap revealed no filamentous material either before or after the addition of KCl. However, paracrystalline arrays of 5 nm-wide F-actin filaments were observed after incubation of the extract in 200 mM KCl plus millimolar ranges of ATP, $MgCl_2$ and EGTA. Again, the addition of cofactors appears important for polymerization.

The difference between the polymerization of crude and purified actin preparations may result from the use of ATP in the column chromatography steps (ADELMAN and TAYLOR, 1969b). On the other hand, the availability of some divalent cation is important too, since little F-actin (electron microscopy evidence) but large amounts of G-actin (gel evidence) were found in high salt extracts prepared with ATP and the chelator EGTA (HINSSEN and D'HAESE, 1974). Cofactor proteins may also be important (see Sect. 2.3).

ATP enhancement of polymerization may not be direct (NACHMIAS, 1976). Added millimolar 5′-AMP, but not 3′-AMP, caused F-actin formation in the supernatant after high speed centrifugation of a *Physarum* crude extract. NACHMIAS hypothesized that the ATP enhancement of F-actin polymerization may actually be mediated through 5′-AMP formed by a Ca-activated ATP-pyrophosphohydrolase present in *Physarum* extracts (KAWAMURA and NAGANO, 1975; NACHMIAS, 1976). The mechanism for a 5′-AMP effect is not known.

2.2.2 Magnesium-Polymer and *Physarum* Actinin

Physarum contains large amounts of an accessory protein, termed actinin, that causes actin to polymerize into an unusual form in the presence of Mg^{2+}. This so-called Mg-polymer is represented by a small viscosity increase over G-actin and continued hydrolysis of ATP after the viscosity has stabilized (HATANO et al., 1967). Electron microscopy of Mg-polymer revealed short filaments of 0.3 to 0.5 μm length (HATANO and TOTSUKA, 1972). Light scattering measurements indicate Mg-polymer is more flexible than F-actin (FUJIME and HATANO, 1972). TOTSUKA and HATANO (1970) suggested that the continuous hydrolysis of ATP is caused by breakage and reformation of intrapolymer bonds.

It is possible to convert between F-actin and Mg-polymer by long dialysis at room temperature (HATANO et al., 1967), but at 55° C in the presence of KCl and ATP, Mg-polymer will convert to F-actin in 30 min (TOTSUKA, 1971). The conversion is reversible if F-actin is first transformed to G-actin. Thus, the protein is not denatured by the heat treatment.

The high intracellular concentration of Mg^{2+} at first suggested to HATANO and TOTSUKA (1972) that Mg-polymer would be the in vivo form of *Physarum* actin. However, later investigations showed that a F-actin in modified form is in fact made at physiological concentrations of Mg^{2+}, KCl, and ATP (FUJIME and HATANO, 1972). Under the latter conditions, continuous ATP hydrolysis stops (HATANO, 1972). The results are, of course, consistent with the evidence of HATANO and TAZAWA (1968) and HINSSEN (1972) that F-actin forms in KCl and Mg-ATP.

Actinin was isolated from an acetone powder of *Physarum* by ammonium sulfate precipitation and column chromatography (HATANO and OWARIBE, 1976). The protein had a molecular weight of 95,000 daltons and was one-third as abundant as actin, on the basis of weight. On SDS polyacrylamide gels the protein comigrated with actin at 45,000 daltons, indicating the presence in the native protein of two subunits. Unlike actin, actinin did not polymerize or alter its viscosity in moderate salt solutions, it did not react with anti-actin antibody, and it activated the *Physarum* myosin ATPase only slightly. But the addition of actinin to column-purified actin resulted in the renewed ability of the product to form Mg-polymer.

Although actin and actinin seem to be different proteins, CLARKE and SPUDICH (1977) caution that actin dimers formed from sulfhydryl oxidation can also mediate the formation of Mg-polymer. The lack of reactivity of the presumptive cofactor protein with anti-actin may be the result of a masked antigenic site in the dimer. The subunit molecular weight data are also suggestive of the cofactor being an actin dimer. On the other hand, the cofactor protein can be isolated directly from fresh plasmodia in a high cysteine buffer in which sulfhydryl oxidation should be blocked (HATANO and OWARIBE, 1976).

In muscles, two quite different actinins are distinguished (MARUYAMA, 1971): α-actinin is a presumptive actin binding component of muscle Z-lines; it results in increased cross-linking and increased viscosity if added to F-actin (HOLMES et al., 1971). β-actinin is a presumptive muscle thin-filament chain-termination factor and can mediate the formation of Mg-polymer from muscle actin (KAMIYA et al., 1972). *Physarum* actinin (termed β-actinin by MARUYAMA et al., 1976) is physiologically comparable to β-actinin, but has a molecular weight more like that of α-actinin (HATANO and OWARIBE, 1976).

2.2.3 The Interaction of *Physarum* Actin With Myosin

The characteristic feature of any actin is its interaction with myosin, and *Physarum* actin behaves accordingly. Acetone dehydration of *Physarum* (or muscle) denatures the myosin but the actin survives and can be extracted in high salt buffer. Remarkably, when pure muscle myosin is added to such *Physarum* actin, a functional hybrid actomyosin results (HATANO and OOSAWA, 1966a). The viscosity of the hybrid actomyosin solution is anomalously high. At high ionic strengths ATP causes a viscosity decrease, while at low ionic strengths Mg-enhanced superprecipitation occurs with added ATP. The ATPase of the muscle myosin was activated by the *Physarum* actin.

Physarum F-actin can be decorated by the S-1 proteolytic fragment of muscle myosin (NACHMIAS et al., 1970), a highly specific cytochemical test for the interaction of actin and myosin (see WAGNER, Chap. 2.2, this vol.). The characteristic arrowhead pattern on the F-actin filaments is formed with all arrowheads pointing in the same direction, indicating uniform polarity of the *Physarum* F-actin. The spacing between the arrowheads is about 35 nm, the same as the half-pitch of the actin filaments. *Physarum* F-actin can also be decorated with *Physarum* myosin (HATANO and TAZAWA, 1968; NACHMIAS and INGRAM, 1970; NACHMIAS, 1972a).

2.2.4 Antibody to *Physarum* Actin

The ubiquity and highly conserved properties of actin (see WAGNER, Chap. 2.2, this vol.) have resulted in great difficulty in the preparation of antibodies to actin. Further evidence for subtle differences between muscle actin and *Physarum* actin is the fact that antibody to *Physarum* actin can be prepared relatively easily by conventional techniques (OWARIBE and HATANO, 1975). These authors were able to raise a monospecific antibody that inhibited G- to F-actin transformation and which decorated F-actin from *Physarum*. The antibody also reacted with sea urchin actin, but not with muscle actin.

2.2.5 The Actin Content of *Physarum*

Estimates of actin in crude extracts, determined colorimetrically from stained SDS polyacrylamide gels, vary between 2%-5% (JACOBSON et al., 1976) and 15%–22% (KESSLER et al., 1976) of total protein. Of course, these estimates will be exaggerated by contaminants (e.g., actinin) which comigrate with actin on SDS gels. Nonetheless, there are probably at least 100 mg of actin per 100 g of plasmodium (JACOBSON et al., 1976). For further discussion, see Section 2.3.4.

2.3 Myosin

2.3.1 General Characterization of *Physarum* Myosin

Muscle and nonmuscle myosins are generally less similar to one another than are the various actins (see WAGNER, Chap. 2.2, this vol.). Nonetheless, *Physarum* myosin has properties much like those of muscle myosin. Ultracentrifugation of *Physarum* myosin preparations in 500 mM KCl showed a band sedimenting at 6 to 6.4 S; the estimated molecular weight of 460,000 daltons is close to that for muscle myosin (HATANO and TAZAWA, 1968; ADELMAN and TAYLOR, 1969b). In column chromatography both types of myosins behave as asymmetric molecules. Confirming this assessment, negative stain electron microscopy revealed the *Physarum* myosin to be about 120 nm long, with a tail about 1.5 to 2 nm wide and with two 11 nm-diameter heads (HATANO and TAKAHASHI, 1971).

SDS polyacrylamide gel electrophoresis of *Physarum* myosin indicates the existence of a heavy chain of about 225,000 to 250,000 daltons which does not comigrate with muscle myosin heavy chain (NACHMIAS, 1974; LESTOURGEON et al.,1975; JA-

COBSON et al., 1976). It is not resolved whether *Physarum* myosin has functional light chains like muscle myosin.

Two heavy chains alone would be sufficient to account for the molecular weight of *Physarum* myosin (ADELMAN, 1975). Although HINSSEN and D'HAESE (1974) resolved no light molecular weight bands on gels of myosin, both NACHMIAS (1974) and JACOBSON et al. (1976) observed light molecular weight polypeptides, apparently copurifying with myosin. The spectrum of polypeptides was not altered by inclusion of protease inhibitors in the extraction medium.

The amino acid content of *Physarum* myosin was analyzed by HATANO and OHNUMA (1970) and found to be similar to that of muscle myosin with the notable exception of a lower cysteine content. A low cysteine content has also been reported for *Acanthamoeba* myosin (POLLARD and KORN, 1973a).

In moderate salt conditions *Physarum* myosin forms small aggregates of about 12.2 S (HATANO and TAZAWA, 1968), but does not, as is typical of muscle (HUXLEY, 1963) and many nonmuscle (POLLARD and WEIHING, 1974) myosins, form the "bone-shaped," bipolar tail-to-tail aggregates resembling thick filaments. Also unlike muscle myosin, *Physarum* myosin can be extracted with a low salt buffer containing PP_i (JACOBSON et al., 1976). Since the ability to form supramolecular aggregates may be crucial to the role of myosin in force transduction (see Sect. 2.5.2), a search was made to find conditions permissive or promotive for the formation of aggregates.

Initial studies reported that bone-shaped aggregates of *Physarum* were formed in the presence of Ca^{2+} (NACHMIAS, 1972b), known to promote aggregation in platelet myosin (COHEN and COHEN, 1972), or in the presence of ATP (HINSSEN, 1970). However, these ideas may have to be revised in light of more recent data. *Physarum* myosin when purified to a greater extent than in these earlier studies will form 2–10-fold longer thick filaments which are not bone-shaped (HINSSEN and D'HAESE, 1974; NACHMIAS, 1974). NACHMIAS suggested the formation of bone-shaped aggregates may be the result of impurities.

2.3.2 Enzymatic Properties of *Physarum* Myosin and Its Interaction with Actin

The highest specific activity reported to date for the Ca-activated ATPase of *Physarum* myosin is 2 μmol P_i-split min^{-1} mg^{-1} protein (JACOBSON et al., 1976). Highest substrate specificity was found for ATP, followed, in decreasing order of activity, by ITP and GTP. No activity was observed for ADP, AMP, or PP_i (ADELMAN and TAYLOR, 1969b).

Actin-binding and actin-activation of the ATPase are, of course, the characteristic features of a myosin. *Physarum* myosin hybridizes with muscle actin to yield a solution with anomalously high viscosity (HATANO and TAZAWA, 1968). *Physarum* myosin also decorates muscle F-actin (HINSSEN and D'HAESE, 1974). Moreover, actin will enhance the *Physarum* myosin ATPase 5–8 fold if the myosin is purified carefully and in the presence of ATP to remove possible residual, denaturated actin bound to the myosin (NACHMIAS, 1974; JACOBSON et al., 1976). This residual actin may otherwise interfere in the elicitation of the actin-activated ATPase.

Some of the reported enzymatic properties of *Physarum* myosin may be the result of sulfhydryl oxidation during isolation and purification. KATO and TONOMURA (1975a) found that the crude myosin preparations made with high concentrations of mercaptoethanol to protect free sulfhydryl groups showed substantial activation of myosin ATPase by actin and by EDTA, although earlier studies had shown the ATPase to be inhibited by EDTA (ADELMAN and TAYLOR, 1969b; HATANO and OHNUMA, 1970). EDTA also activates muscle myosin ATPase. In fact, modification of skeletal muscle myosin sulfhydryls can result in lowered actin-activation as well as EDTA inhibition (SEKINE and KIELLEY, 1964). The sulfhydryl reagent PCMB at low concentrations (about 0.03 μmol mg^{-1} myosin) activates muscle myosin ATPase, but inhibits *Physarum* myosin (HATANO and OHNUMA, 1970). Again, this difference may be related to the sulfhydryl oxidation state of the respective myosins. Since muscle myosin copurified with *Physarum* myosin remained activated by EDTA while the *Physarum* myosin did not (JACOBSON et al., 1976), *Physarum* myosin is presumed to be especially sensitive to sulfhydryl modification.

At high ionic strengths, pure muscle myosin is activated by Ca^{2+} and inhibited by Mg^{2+}, while with *Physarum* myosin, Ca^{2+} is relatively more stimulatory and Mg^{2+} is less inhibitory (ADELMAN and TAYLOR, 1969b; HATANO and OHNUMA, 1970).

2.3.3 Antibody to *Physarum* Myosin

An antibody to *Physarum* myosin has been prepared which inhibits enzymatic activity (NACHMIAS and KESSLER, 1976). Consistent with the greater diversity among nonmuscle myosins as compared to actins, the antibody reacted only with presumptive myosins from other species of the Physarales (KESSLER et al., 1976).

2.3.4 The Myosin Content of *Physarum* and Its Relation to Actin Content

As determined colorimetrically from stained SDS polyacrylamide gels, myosin constitutes about 0.5%–1.0% of total *Physarum* protein (JACOBSON et al., 1976). Based on specific activities of purified myosin and on enzyme activities in cell homogenates, estimates of about 25 mg of myosin per 100 g of plasmodium have been made (JACOBSON et al., 1976; HINSSEN and D'HAESE, 1974). This approach assumes that no activators or inhibitors are present in the crude homogenates. Alternatively, 60 mg of myosin per 100 g of plasmodium have been estimated with a complement-fixation technique utilizing anti-myosin antibody (KESSLER et al., 1976).

Like other nonmuscle motility systems (see, for example, POLLARD and WEIHING, 1974), *Physarum* contains much more actin than myosin (for comparison, see Sect. 2.2.5). Estimates range between 4 to 5 (JACOBSON et al., 1976) and 19 to 30 (KESSLER et al., 1976) times as much actin as myosin by weight. The lower amount of myosin has been held to be consistent with the lesser motive force required for streaming as compared to muscle contraction (HATANO and TAZAWA, 1968). Alternatively, the large amounts of actin may be related to the requirements of cellular motility systems where the contractile apparatus may undergo rapid cycles of assembly and disassembly. Excess actin may also have functions in the cell unrelated to motility (LESTOURGEON et al., 1975).

2.4 Contractile Models

It is reasonable to hypothesize that in *Physarum*, as in muscle, the large amounts of actin and myosin participate in the transduction of chemical to mechanical energy, for example, during ectoplasmic contraction. The actin and myosin of muscle are organized into filaments which slide with respect to each other, giving rise to contraction (see WAGNER, Chap. 2.2, this vol.). Yet ultrastructural studies of *Physarum* (see Sect. 3.2) show that actin and myosin form only transitory and disordered structures in situ. WOHLFARTH-BOTTERMANN and his coworkers have been concerned with the ability of unorganized actomyosin systems to support actual movement.

BECK et al. (1970) made actomyosin thread models by spraying high salt extracts of *Physarum* into water, where the actomyosin precipitated in the form of long threads. The threads decreased in length by 55% and in volume by 83.5% upon additions of 5 mM ATP. Electron microscopy of threads before contraction showed an unordered network of F-actin, whereas after contraction the network, although still unordered, had become much denser. Thus, even unoriented arrangements of filaments can produce movement. Thread models have been studied for many types of actomyosin and it has been presumed that a type of sliding filament mechanism is operative (D'HAESE and KOMNICK, 1972).

The nature of the myosin-myosin and actin-myosin interactions in these models is critical. D'HAESE and KOMNICK (1972) observed uncontracted threads to contain 7.5 to 10 nm filaments of F-actin, apparently decorated with myosin, as well as small aggregates, presumably of myosin. After contraction, the F-actin filaments were 5.5 nm in diameter and no longer decorated. Also seen were thick filaments, 12 nm in diameter, possibly representing aggregated forms of myosin. NACHMIAS (1973) speculated that F-actin filaments were cross-linked by tail-to-tail interactions of myosin oligomers. Both NACHMIAS and HINSSEN and D'HAESE speculated that oligomeric myosin was responsible for contraction. This idea is important in light of the reluctance of purified *Physarum* myosin to form thick filaments (see Sect. 2.3). Perhaps nonmuscle motility systems do not require the organized forms of myosin found in muscle.

OPLATKA and TIROSH (1973) and TIROSH et al. (1973) developed a different type of contractile model. They found that crude *Physarum* actomyosin in glass capillaries could stream upon addition of ATP. Particle movements with rates up to 1 mm s^{-1}, as well as oscillations reminiscent of shuttle streaming, were observed. The authors thus questioned whether contraction of the ectoplasm, or indeed, contraction at all, was required for streaming. MATSUTAKE et al. (1975) repeated these experiments and found that "streaming" could be induced with PP_i or even when the actin-myosin interaction was blocked. Streaming in capillaries may therefore be the result of viscosity changes unrelated to the active movement of myosin with respect to actin. Whether these findings have relevance to the situation in vivo remains to be assessed.

2.5 Regulation of Actin and Myosin

In muscle, Ca^{2+} regulates the interaction between actin thin filaments and myosin thick filaments through accessory proteins, either tropomyosin–troponin ("thin filament control") or an EDTA-dissociable myosin light chain ("thick filament control"), which, at submicromolar Ca^{2+} concentrations, cause steric inhibition of actin-myosin binding. ATP, also required for contraction, probably does not participate in regulation because the intracellular concentration of Mg^{2+} is high enough to inhibit the myosin ATPase in the absence of actin-activation. Thus, the concentration of ATP can remain at a high level in muscle. Among nonmuscle motility systems there is some evidence for thin filament control (e.g., platelets; COHEN et al., 1973). Thick filament control has not been found to date in any nonmuscle system (CLARKE and SPUDICH, 1977).

The emphasis on similarities to muscle in *Physarum* should not obscure the problem that control mechanisms for muscle, a highly specialized state of actin and myosin, may not be sufficient or solely responsible for control of nonmuscle systems with their greater organizational flexibility (JACOBSON et al., 1976). Thus, alternative modes of control are also discussed.

2.5.1 Tropomyosin-Troponin

Muscle tropomyosin-troponin added to *Physarum* actomyosin can confer a requirement for micromolar Ca^{2+} in the superprecipitation reaction, so-called Ca-control (TANAKA and HATANO, 1972). It also appears that *Physarum* actin can bind or otherwise interact with muscle tropomyosin. These studies further emphasize the structural similarities between the actins, but they do not prove that Ca-control operates in *Physarum* in vivo.

Physarum does appear to contain proteins similar to tropomyosin-troponin. The superprecipitation of pure muscle actomyosin acquires Ca-control after incubation in a *Physarum* extract (TANAKA and HATANO, 1972). Addition of a tropomyosin-troponin-like protein complex isolated from *Physarum* permits Ca-control in the superprecipitation of pure muscle actomyosin (KATO and TONOMURA, 1975b). Finally, a cofactor has been isolated from *Physarum* which, together with *Physarum* actin and muscle myosin, will form a hybrid actomyosin requiring Ca^{2+} for superprecipitation (NACHMIAS, 1975).

Although early studies were able to demonstrate Ca-control in *Physarum* actomyosin only in the presence of undefined additives from muscle, it now appears that reduced sulfhydryl groups are important for the expression of Ca-control. Superprecipitation of *Physarum* actomyosin did have an absolute requirement for micromolar Ca^{2+} if prepared in the presence of high concentrations of mercaptoethanol (KATO and TONOMURA, 1975a). A similar effect was observed for platelet actomyosin (COHEN and COHEN, 1972). In *Physarum*, the inclusion of EDTA in the extraction buffer argues against the involvement of thick filament control where an EDTA-dissociable myosin light chain is involved (KATO and TONOMURA, 1975a).

Although Ca^{2+} increased the Mg-ATPase activity of *Physarum* actomyosin at moderate ionic strengths (KATO and TONOMURA, 1975a), this test for thin filament control is a poor one for two reasons: Ca^{2+} also stimulates ATPase of myosin without bound actin, and Ca-activated nonmyosin ATPases may be present in *Physarum*, as evidenced by several observations (JACOBSON et al., 1976; NACHMIAS and ASCH, 1976). Since anti-myosin antibody inhibits myosin ATPase (NACHMIAS and KESSLER, 1976), it is theoretically possible to evaluate the extent to which the Ca-activated ATPase is non-myosin.

2.5.2 Alternative Modes of Regulation

Myosin ATPase, which is required for movement, can be modified by phosphorylation (e.g., in platelets; ADELSTEIN and CONTI, 1975), nonactin cofactor proteins (e.g., in *Acanthamoeba;* POLLARD and KORN, 1973b), or by a probable sulfhydryl oxidation which alters the actin cofactor requirement (see Sect. 2.3.2). Because of the high ATPase activity of pure *Physarum* myosin, only the latter possibility seems likely in *Physarum*.

Actin and myosin are structural proteins and their role in nonmuscle motility could be regulated through control of assembly and disassembly. This possibility stands in sharp contrast to muscle where the component proteins are more or less locked into a fixed structure.

Myosin aggregates are probably required to span the gap between F-actin filaments in order to generate sliding. In *Physarum,* ATP and divalent cations promote the formation of large myosin aggregates (see Sect. 2.3), and could therefore help regulate contractility. However, it is not clear whether such large aggregates are strictly required. In *Acanthamoeba* the myosin is a large globular protein which lacks a tail and the ability to form aggregates, but which can nonetheless cross-link F-actin filaments (POLLARD and KORN, 1973a).

Structural modes of control are discussed most often with respect to the polymerization of actin. Although nonmuscle actins polymerize readily under optimal conditions (BRAY and THOMAS, 1976), many nonmuscle motility systems are distinguished by rapid in vivo changes of the polymerization state of actin (TILNEY, 1976). Examples are the formation of the contractile ring in dividing cells (SCHROEDER, 1972) and the disappearance of microfilament bundles in cultured mammalian cells detached from their substrate (GOLDMAN, 1975). That soluble actin can be isolated so easily from *Physarum* suggests that rapid shifts in the G-to-F equilibrium can occur.

Note that actin polymerization is important not only for activation of the myosin ATPase and as a scaffold for myosin, but also for the continuous interconversion between ecto- and endoplasm which characterizes shuttle streaming. While it seems unlikely in the case of *Physarum,* in some organisms, changes in actin polymerization (TILNEY, 1976) or F-actin packing (TILNEY, 1975) can cause movement without myosin.

As discussed in Section 2.2, Ca^{2+} and ATP may promote actin polymerization in *Physarum*. Though not yet observed in *Physarum,* actin polymerization in various nonmuscle systems is affected by high molecular weight actin-binding proteins (e.g., STOSSEL and HARTWIG, 1976; TILNEY, 1976), the binding activity of which may be controlled by pH. In *Physarum,* actin does interact with an apparent β-actinin-like protein (see Sect. 2.2.2) to form Mg-polymer. The amount of Mg-polymer, which could influence cytoplasmic viscosity, could be regulated by Mg^{2+} and ATP levels.

The attachment of actin filaments to membranes, demonstrated in the cellular slime mold *Dictyostelium* (CLARKE et al., 1975) and in the tip of intestinal microvilli (MOOSEKER and TILNEY, 1975), could obviously be an important step in the regulation of ectoplasmic contraction in *Physarum*. A presumptive actin-binding protein of muscle Z-lines, α-actinin, has been localized at apparent attachment sites in some systems (SCHOLLMEYER et al., 1976) and could serve a regulatory function. α-actinin, which cross-links F-actin filaments in solution (HOLMES et al., 1971) , may be involved in changes in actin packing in *Limulus* sperm (TILNEY, 1975) and in the arrangement of actin microfilament bundles in mammalian cells (LAZARIDES, 1976). Note that neither α-actinin nor actin-membrane binding have been conclusively identified in *Physarum*.

3. The Physiology of Shuttle Streaming

3.1 Ectoplasmic Contraction

Differential and rhythmically changing ectoplasmic contraction and relaxation remains the most plausible theory to explain the generation of motive force for the shuttle streaming of endoplasm (JAHN, 1964; KOMNICK et al., 1973). In essence, the endoplasm is squeezed along. The extrusion of endoplasm after puncture of the plasmodium indicates that it is under pressure (JAHN et al., 1964).

Excised cytoplasmic veins are able to contract rhythmically (see KAMIYA, 1959) and, in analogy to muscle physiology, it has been possible to make tensiometric measurements of contraction (KAMIYA, 1970; FLEISCHER and WOHLFARTH-BOTTERMANN, 1975) under conditions of constant load (isotonic) or constant length (isometric). Such measurements have also been made on the lateral expansion and contraction of veins in situ (WOHLFARTH-BOTTERMANN, 1975). Cutting the vein on either side of the point of measurement stops streaming, but the contractions persist, thus demonstrating that the tensiometric changes are the result of actual localized contraction and not of passive swelling caused by the shuttle streaming itself.

Consistent with active processes in the ectoplasm forcing out endoplasm is the observation that presumptive contracting regions of ectoplasm are slightly warmer ($1–500 \times 10^{-4}$ °C) than relaxing regions (ALLEN et al., 1963).

3.2 The Identification of Actin and Myosin in situ

To assess the role of actin and myosin in ectoplasmic contraction and shuttle streaming it is first necessary to know the location and supramolecular state of these proteins in the cell. Starting at an ultrastructural level it is well established that microfilaments, 5–7 nm in diameter, similar to actin thin filaments, are found mainly in the ectoplasm of *Physarum* (WOHLFARTH-BOTTERMANN, 1962; CRAWLEY, 1966) and other slime molds (MCMANUS and ROTH, 1965). Bundles of filaments are also observed. Improved methods, notably the use of OsO_4 which allows very rapid fixation, have revealed several microfilament systems (RHEA, 1966; DANIEL and JÄRLFORS, 1972; WOHLFARTH-BOTTERMANN, 1974, 1975a). One system may run longitudinally, possibly underlying the plasma membrane. Other systems run circumferentially or radially in possible association with a system of plasma membrane invaginations that lace the ectoplasm. Some filaments may be bound to membranes and, of course, such attachments may aid in the transduction of actomyosin contraction into ectoplasmic contraction. WOHLFARTH-BOTTERMANN (1974) considers the invagination system to provide increased area for attachment and increased flexibility for cytoplasmic contraction. Freeze-etch electron microscopy, which aims for natural preservation of specimens, has yet to be successful in *Physarum* (ISENBERG et al., 1975).

The identification of the *Physarum* microfilaments as actin was confirmed by heavy meromyosin decoration (ALLÉRA et al., 1971; see also, WAGNER, Chap. 2.2, this vol.). The authors did not report whether ATP or PP_i reversed the decoration, a test which would have provided additional confirmation.

Light microscopy is useful to obtain a broader view of ectoplasmic contraction. Examination of fixed and sectioned *Physarum* by phase contrast microscopy revealed the presence of fibrils apparently inserted on invaginations of the plasma membrane (WOHLFARTH-BOTTERMANN, 1963; BAUER, 1967). These fibrils are presumably the same as the microfilament bundles visualized by electron microscopy (see above). Unfortunately, because *Physarum* plasmodia are thick and heavily pigmented, it has been difficult to apply direct in vivo light microscopy to best advantage. Recently, however, HÜLSMANN et al. (1974) used phase contrast and differential interference contrast microscopy on a less pigmented strain of *Physarum*. Filamentous material was seldom seen in strongly streaming channels, but rather more at the end of channels. Branched networks of filaments up to 100 µm in length and 1 µm in diameter were observed. To obtain these results, it was necessary to decrease the optical pathlength by pressing plasmodia, a procedure the authors felt could induce artifacts. In another in vivo study, positive axial birefringence along the veins, indicative of oriented structures, was observed with polarized light microscopy (NAKAJIMA and ALLEN, 1965). Again, pressed plasmodia were used.

Although biochemical studies confirm the presence of myosin in *Physarum*, direct localization of the protein, especially in relation to actin filaments, has been a problem. Since myosin does not readily polymerize, and since small aggregates might be involved in contraction, ultrastructural visualization could be difficult. ATPase activity in microfilament bundles has, however, been identified (WOHLFARTH-BOTTERMANN, 1964).

Slower fixation of *Physarum* plasmodia (as occurs with glutaraldehyde) or glycerination causes "convulsive" contraction. Under these conditions probable myosin thick filaments have been observed (KESSLER and NACHMIAS, 1970; ALLÉRA and WOHLFARTH-BOTTERMANN, 1972; NAGAI and KATO, 1975). The thick filaments were 13–26 nm wide, 700 nm long, and, although they formed within bundles of actin microfilaments, the thick filaments did not decorate with heavy meromyosin (ALLÉRA and WOHLFARTH-BOTTERMANN, 1972). The identification of these filaments as myosin is further supported by the appearance of cross-bridges between the thick filaments and the actin microfilaments, and the fact that the thick filaments could be extracted more easily than the microfilaments by high salt and ATP solutions (ALLÉRA and WOHLFARTH-BOTTERMANN, 1972). In material rapidly fixed with OsO_4, small filaments of 2–5 nm diameter were observed which may be monomers or dimers of myosin. Although the formation of thick filaments in microfilament bundles demonstrates the proximity of actin and myosin, the actual role of myosin aggregates in contraction remains unresolved.

The above studies have yielded information on the ultrastructural location of actin and myosin polymers only. Additional information on the monomeric forms should be obtainable by staining with fluorescent-labeled antibodies or heavy meromyosin. Such studies have not yet been reported.

3.3 Dynamic Behavior of Actin During Ectoplasmic Contraction

Considerable evidence exists that the regulation of the supramolecular state of actin, either as a filament composed of G-actin monomers or as a bundle of filaments, is an important step for the control of nonmuscle motility (see Sect. 2.5.2).

Fast fixation of excised cytoplasmic veins showed the most microfilament bundles in what were presumably contracting regions of ectoplasm, that is, portions of the vein from which the cytoplasm was leaving (WOHLFARTH-BOTTERMANN, 1965).

Actin filament bundles can form rapidly from what is presumed to be a G-actin pool in the endoplasm (KOMNICK et al., 1973; NAGAI and KATO, 1975). Bundles of filaments develop around the periphery of drops of endoplasm exuded from punctured cytoplasmic veins (WOHLFARTH-BOTTERMANN, 1962). The time of formation, about 10 min after puncture, correlates with resorption of the drop by the vein. WOHLFARTH-BOTTERMANN speculated that the filaments were part of a contractile network that forced the cytoplasm back into the vein. In isolated protoplasmic drops, filament formation and the resumption of streaming, observed by electron and light microscopy, respectively, were both inhibited by the sulfhydryl reagent Salyrgan (ACHTERRATH, 1969a). In both cases the inhibition was relieved by cysteine.

The environment of the endoplasm can apparently be shifted rapidly to a state favorable for microfilament formation. Abnormal stress on cytoplasmic veins can give rise to microfilament formation in the endoplasm, either by stretching of excised veins (FLEISCHER and WOHLFARTH-BOTTERMANN, 1975) or by prolonged isometric contraction of veins (WOHLFARTH-BOTTERMANN and FLEISCHER, 1976). KAMIYA (unpublished data cited in KAMIYA, 1970) saw increased birefringence after stretching excised veins, but this latter effect could be strain birefringence (KAMIYA et al., 1972). In microplasmodia from shake culture the appearance of filament bundles is concomitant with the initiation of surface contact (USUI, 1971; for comparison with mammalian cell systems, see GOLDMAN, 1975).

Changing the distribution of endoplasm (and hence the G-actin pool) can apparently influence streaming and microfilament formation (TAKATA et al., 1967). Excised cytoplasmic veins were plugged into a single opening between dual pressure chambers (see KAMIYA, 1959). Pressure applied to one chamber forced the endoplasm from the half of the vein in that chamber into the other half of the vein in the other chamber. Electron microscope analysis of the endoplasm-depleted half-vein showed that it lost its normal complement of ectoplasmic filaments, while the endoplasm-enriched half-vein contained more than the usual number of filaments. This latter side was gradually able to overcome the pressure gradient and to force cytoplasm back into the endoplasm-depleted half-vein. Thus, an increase in microfilament number was associated with an increase in motive force.

Further evidence relating filaments and contraction has been obtained with glycerinated cell models. These experiments are the in situ counterpart of contractile models using isolated actomyosin (see Sect. 2.4). The glycerinated cell systems are probably reasonable models since even the complex torsional movements of hanging, excised strands can be observed in glycerinated specimens (KAMIYA, 1968).

ATP causes filament formation and contraction in glycerinated plasmodial drops (KAMIYA and KURODA, 1965), as well as a condensation of filament bundles similar to superprecipitation (NAGAI and KAMIYA, 1966). Before addition of ATP, 6–12 nm filaments were present, probably F-actin decorated with myosin, whereas, after ATP treatment, only 6 nm filaments were seen (NAGAI and KAMIYA, 1968).

The authors speculated that ATP had caused the dissociation of actin and myosin, similar to observations with contractile models of isolated actomyosin (D'HAESE and KOMNICK, 1972).

Actin microfilament bundles isolated from mammalian cells by laser microbeam dissection are able to contract (ISENBERG et al., 1976). However, it is not clear that organized, large bundles are required for contraction, at least in *Physarum*. Although the natural contraction of exuded cytoplasmic drops approximately 10 min after puncture is correlated with filament bundle formation (see above), protoplasmic drops glycerinated immediately after puncture and lacking filament bundles were nonetheless able to contract later upon addition of ATP (ACHTERRATH, 1969b). Electron microscope analysis of these drops revealed only a fibrous network which condensed after ATP treatment. On the other hand, extraction of filament bundles from appropriate glycerinated drops did inhibit ATP-induced contraction (KOMNICK et al., 1970). Although this latter evidence argues for the importance of bundles in contraction, the simultaneous extraction of a residual diffuse contractile network cannot be excluded.

3.4 The Regulation of Shuttle Streaming

Regulation must be approached both temporally (the rhythmic change in streaming direction) and spatially (the generation of motive force in particular regions and the differentiation between endo- and ectoplasm). Ca^{2+} and ATP appear especially attractive as possible agents to control an actin-myosin system. The role of Ca^{2+} has received particular attention since *Physarum* seems to possess a tropomyosin-troponin complex similar to that of muscle. It should be clear, however, that many levels of control exist and that Ca^{2+} and/or ATP may be critical for more than one of them (see Sect. 2.5). Certainly, in demembranated *Chaos* cells both ATP and Ca^{2+} levels were important (TAYLOR et al., 1973).

3.4.1 The Role of ATP

In contrast to muscle contraction (Sect. 2.5), *Physarum* streaming may be controlled by ATP because ATP concentration in the plasmodium is apparently suboptimal. Thus, added ATP increases streaming rate (see KAMIYA, 1959) and local application of ATP to cytoplasmic veins causes contraction and flow of endoplasm away from the site of application (JAHN et al., 1964). ATP can also trigger contraction in *Physarum* cell models.

3.4.2 The Role of Ca^{2+}

Physarum protoplasmic drops treated with caffeine (which may make membranes more permeable to Ca^{2+}; WEBER and HERZ, 1968) exhibited streaming only when the external Ca^{2+} concentration was greater than 10^{-7} M (HATANO, 1970). Thus, a very low, minimal level of Ca^{2+} is necessary for movement. This requirement is almost the same as in muscle.

Critical to the hypothesis of Ca-control is the requirement for a Ca-sequestering system analogous to the sarcoplasmic reticulum of muscle. The activity of such a system should correlate with shuttle streaming patterns and be sufficient to lower free cytoplasmic Ca^{2+} concentrations below the threshold for activation (by whatever mechanism) of an actomyosin system. Ca-sequestering systems have been identified in *Amoeba* and *Chaos* (REINOLD and STOCKEM, 1972). Mitochondria sequester Ca^{2+} and, in smooth muscle, may help control contraction (BATRA, 1973).

In fact, a *Physarum* ectoplasmic vesicle system exists which accumulates Ca^{2+} (BRAATZ and KOMNICK, 1970, 1973; ETTIENNE, 1972). The accumulation is energy-dependent, inhibited by Salyrgan and capable of accumulating Ca^{2+} from micromolar external concentrations (BRAATZ and KOMNICK, 1970). The vesicles had relatively low Ca^{2+} contents in regions where extensive filament bundles were observed and where, presumably, Ca-activated ectoplasmic contraction was expelling the endoplasm (BRAATZ, 1975). Conversely, in relaxed regions of ectoplasm where endoplasm was entering, the vesicle contents of Ca^{2+} were high. In a related study (RIDGWAY and DURHAM, 1976), the amount of free Ca^{2+}, determined from the luminescence of the Ca-triggered photoprotein aequorin previously injected into the vein, was maximal at about the expected time of maximum motive force (based on electroplasmograms; KAMIYA, 1959). Thus, the *Physarum* Ca-sequestering system possesses the necessary attributes for it to account for Ca-control of shuttle streaming.

3.4.3 Changes in Potential Difference

Mechanisms which might trigger changes in Ca^{2+} or ATP are also being explored. It has long been known that the external potential difference between two ends of an excised plasmodial vein oscillates at an almost constant phase angle to the rhythm of motive force for shuttle streaming (see KAMIYA, 1959). One end of the vein is maximally negative with respect to the other end at about the time the cytoplasm changes direction and stops leaving the negatively charged side. The electrical changes may represent changes in membrane potential possibly involved in the control of shuttle streaming. For example, a reduction in membrane potential in sea urchin embryos results in Ca^{2+} influx and contraction (GINGELL, 1970). In *Physarum,* the potential change rhythm persists after the streaming rhythm is disrupted; thus, the electrical rhythm is not caused by the streaming. However, KAMIYA was concerned that the maximum of negative potential occurs $\frac{1}{4}\tau$ after the peak of motive force (τ = the time period of shuttle streaming). RIDGWAY and DURHAM (1976) point out that the electroplasmogram utilizes an external potential which does not necessarily represent the polarization of the cell membrane.

Little evidence exists to correlate the rhythm of potential change with shuttle streaming. Benzamide applied externally disrupted both the streaming and the potential change rhythm, whereas benzamide injected into the veins was without effect (KOROHODA et al., 1970). The authors felt the drug might be affecting membrane potential differences. An acetylcholinesterase activity has been found in *Physarum* (NAKAJIMA and HATANO, 1962). A nonspecific esterase activity has been localized near plasma membrane invaginations and possible sites of actin microfilament attachment (BAUER, 1967). Whether these two esterase activities are the same and whether the potential difference rhythm and shuttle streaming could ultimately be regulated by neurotransmitters remains unclear.

The ultimate cause of rhythmicity in shuttle streaming remains an open question. High-frequency metabolic rhythms (HESS, 1976) might be involved. Rhythmic fluorescence changes in phase with shuttle streaming and attributed to pyridine nucleotides have been described (DANIEL, 1970). Mitochondrial ATP production could affect myosin activity and the Ca^{2+} sequestering system. Energy-coupled sequestering of Ca^{2+} by mitochondria is also known and could affect contraction directly (BORLE, 1973).

However, changes in light transmission and fluorescence could be the result of changes in optical pathlength and light scattering during streaming (SACHSENMAIER et al., 1973). These authors measured transmission changes in plasmodia, but the changes lacked a pronounced wavelength dependency as would have been expected for absorption changes by specific chromophores. Further insight into the relationship between metabolism and shuttle streaming should benefit from application of more sophisticated microspectrophotometric methods, such as those possible in conjunction with on-line computers.

4. Conclusion

Physarum contains the proteins actin and myosin which, by a number of physicochemical criteria, are very similar to their muscle analogs. The similarity extends to the formation of functional actomyosin hybrids between *Physarum* and muscle proteins. Unlike those proteins in skeletal muscle, actin and myosin in *Physarum* do not form highly ordered structures. However, studies on actomyosin and glycerinated cell models from *Physarum* demonstrate that unordered arrays of actin and myosin can contract. This fact, the demonstration of ectoplasmic contraction in *Physarum* and the concentration of actin filaments in presumably contracting regions of the ectoplasm, support the theory that actomyosin provides the motive force for shuttle streaming. Oligomeric forms of myosin seem to be capable of generating the sliding of actin filaments that would lead to contraction. Although *Physarum* myosin can form thick filaments in vitro and in situ, it is not clear that such aggregates are actually used in contraction. Likewise, it is not clear whether actin filament bundles or even more disordered networks of filaments are active in ectoplasmic contraction.

Physarum contains proteins similar to the tropomyosin-troponin complex of skeletal muscle. In muscle, Ca^{2+} regulates the interaction of actin with myosin through the tropomyosin-troponin system. Under appropriate conditions, Ca^{2+} has been shown to be required both for shuttle streaming and for contraction of *Physarum* actomyosin models. *Physarum* also contains a vesicle system that sequesters Ca^{2+}. The activity of this system parallels the occurrence of shuttle streaming. Thus, regulation of streaming through a Ca-sensitive tropomyosin-troponin-like mechanism is quite possible.

Other possible regulatory mechanisms are not excluded. *Physarum* actin appears to transform readily between the monomeric and polymeric forms in vivo, a property consistent with the presumed greater flexibility inherent in nonmuscle contractile systems. Polymerized actin is necessary for the activity of myosin. The polymerization state of actin also influences cytoplasmic viscosity and the differentiation of cytoplasm into endo- and ectoplasm. Divalent cations, ATP, 5′-AMP and other factors, have all been shown to influence actin polymerization either directly or through actin-binding proteins such as a β-actinin-like protein which has been isolated in high yield from *Physarum*.

Acknowledgments. This review was supported in part by a Fellowship to the author from the Carnegie Institution of Washington and by a National Aeronautics and Space Administration grant to Dr. A.W. Galston.

References

Achterrath, M.: Hemmung der Entstehung des Fibrillenmusters von *Physarum polycephalum* durch Salyrgan. Cytobiologie *1*, 159–168 (1969a)

Achterrath, M.: Die Reaktion von glycerin-extrahierten Protoplasmatropfen von *Physarum polycephalum* auf Zugabe von ATP. Cytobiologie *1*, 169–183 (1969b)

Adelman, M.R.: Further studies on *Physarum* myosin. Biophys. J. *15*, 161a (1975)

Adelman, M.R., Taylor, E.W.: Isolation of an actomyosin-like protein complex from slime mold plasmodium and the separation of the complex into actin- and myosin-like fractions. Biochemistry *8*, 4964–4975 (1969a)

Adelman, M.R., Taylor, E.W.: Further purification and characterization of slime mold myosin and slime mold actin. Biochemistry *8*, 4976–4988 (1969b)

Adelstein, R.S., Conti, M.A.: Phosphorylation of platelet myosin increases actin-activated myosin ATPase activity. Nature (London) *256*, 597–598 (1975)

Allen, R.D., Kamiya, N. (eds.): Primitive Motile Systems in Cell Biology. New York: Academic Press, 1964

Allen, R.D., Pitts, W.R., Jr., Speir, D., Brault, J.: Shuttle-streaming: synchronization with heat production in slime mold. Science *142*, 1485–1487 (1963)

Alléra, A., Beck, R., Wohlfarth-Bottermann, K.-E.: Weitreichende, fibrilläre Protoplasmadifferenzierungen und ihre Bedeutung für die Protoplasmaströmung. VIII. Identifizierung der Plasmafilamente von *Physarum polycephalum* als F-actin durch Anlagerung von heavy meromyosin *in situ*. Cytobiologie *4*, 437–449 (1971)

Alléra, A., Wohlfarth-Bottermann, K.-E.: Weitreichende, fibrilläre Protoplasmadifferenzierungen und ihre Bedeutung für die Protoplasmaströmung. IX. Aggregationszustände des Myosins und Bedingungen zur Entstehung von Myosinfilamenten in den Plasmodien von *Physarum polycephalum*. Cytobiologie *6*, 261–286 (1972)

Batra, S.: The role of mitochondrial calcium uptake in contraction and relaxation of the human myometrium. Biochim. Biophys. Acta *305*, 428–432 (1973)

Bauer, L.G.: On the similar orientation of fibrillar structures and rows of esterolytic invaginations in plasmodia of the slime mold, *Physarum confertum* Macbr. J. Exp. Zool. *164*, 69–79 (1967)

Beck, R., Hinssen, H., Komnick, H., Stockem, W., Wohlfarth-Bottermann, K.-E.: Weitreichende, fibrilläre Protoplasmadifferenzierungen und ihre Bedeutung für die Protoplasmaströmung. V. Kontraktion, ATPase Aktivität und Feinstruktur isolierter Actomyosin-Fäden von *Physarum polycephalum*. Cytobiologie *2*, 259–274 (1970)

Borle, A.B.: Calcium metabolism at the cellular level. Fed. Proc. *32*, 1944–1950 (1973)

Braatz, R.: Differential histochemical localization of calcium and its relation to shuttle streaming in *Physarum*. Cytobiologie *12*, 74–78 (1975)

Braatz, R., Komnick, H.: Histochemischer Nachweis eines calcium-pumpenden Systems in Plasmodien von Schleimpilzen. Cytobiologie *2*, 457–463 (1970)

Braatz, R., Komnick, H.: Vacuolar calcium segregation in relaxed myxomycete protoplasm as revealed by combined electrolyte histochemistry and energy dispersive analysis of x-rays. Cytobiologie *8*, 158–163 (1973)

Bray, D., Thomas, C.: Unpolymerized actin in tissue cells. In: Cell Motility. Goldman, R., Pollard, T., Rosenbaum, J. (eds.). Cold Spring Harbor Laboratory, 1976, pp. 461–473

Ciba Foundation Symposium No. 14: Locomotion of Tissue Cells. Amsterdam: Elsevier, 1973

Clarke, M., Schatten, G., Mazia, D., Spudich, J.A.: Visualization of actin fibers associated with the cell membrane in amoebae of *Dictyostelium discoideum*. Proc. Natl. Acad. Sci. U.S.A. *72*, 1758–1762 (1975)

Clarke, M., Spudich, J.A.: Nonmuscle contractile proteins: the role of actin and myosin in cell motility and shape determination. Annu. Rev. Biochem. *46*, 797–822 (1977)

Cohen, I., Cohen, C.: A tropomyosin-like protein from human platelets. J. Mol. Biol. *68*, 383–387 (1972)

Cohen, I., Kaminski, E., DeVries, A.: Actin-linked regulation of the human platelet contractile system. FEBS Lett. *34*, 315–317 (1973)

Cold Spring Harbor Symposium on Quantitative Biology Vol. 37: The mechanism of muscle contraction. Cold Spring Harbor Laboratory 1973

Crawley, J.C.W.: Fine structure and cytoplasmic streaming in *Physarum polycephalum*. J. R. Microsc. Soc. *85*, 313–322 (1966)
Daniel, J.W.: Fluorescence oscillations accompanying cyclic protoplasmic flow. J. Cell Biol. *47*, 45a (1970)
Daniel, J.W., Järlfors, U.: Plasmodial ultrastructure of the myxomycete *Physarum polycephalum*. Tissue Cell *4*, 15–36 (1972)
D'Haese, J., Komnick, H.: Fine structure and contraction of isolated muscle actomyosin. I. Evidence for a sliding mechanism by means of oligomeric myosin. Z. Zellforsch. *134*, 411–426 (1972)
Drews, G., Nultsch, W.: Spezielle Bewegungsmechanismen von Einzellern (Bakterien, Algen). In: Encyclopedia of Plant Physiology. Ruhland, W. (ed.). Berlin-Göttingen-Heidelberg: Springer, 1962, Vol. XVII, pt. 2, pp. 876–919
Dryl, S., Zurzycki, J. (eds.): Symposium on motile systems of cells. Acta Protozool. (Warsaw) *11* (1972)
Durham, A.C.H.: A unified theory of the control of actin and myosin in nonmuscle movements. Cell *2*, 123–136 (1974)
Ettienne, E.: Subcellular localization of calcium repositories in plasmodia of the acellular slime mold *Physarum polycephalum*. J. Cell. Biol. *54*, 179–184 (1972)
Fleischer, M., Wohlfarth-Bottermann, K.-E.: Correlation between tension force generation, fibrillogenesis and ultrastructure of cytoplasmic actomyosin during isometric and isotonic contractions of protoplasmic strands. Cytobiologie *10*, 339–365 (1975)
Fujime, S., Hatano, S.: Plasmodium actin polymers studied by quasi-elastic scattering of laser light. J. Mechanochem. Cell Motil. *1*, 81–90 (1972)
Gingell, D.: Contractile responses at the surface of an amphibian egg. J. Embryol. Exp. Morphol. *23*, 583–609 (1970)
Goldman, R.D.: The use of heavy meromyosin binding as an ultrastructural cytochemical method for localizing and determining the possible functions of actin-like microfilaments in nonmuscle cells. J. Histochem. Cytochem. *23*, 529–542 (1975)
Goldman, R.D., Pollard, T., Rosenbaum, J. (eds.): Cell Motility. Cold Spring Harbor Laboratory 1976
Hatano, S.: Specific effect of Ca^{2+} on movement of plasmodial fragments obtained by caffeine treatment. Exp. Cell Res. *61*, 199–203 (1970)
Hatano, S.: Conformational changes of plasmodium actin polymers formed in the presence of Mg^{++}. J. Mechanochem. Cell Motil. *1*, 75–80 (1972)
Hatano, S.: Contractile proteins from the myxomycete plasmodium. Adv. Biophys. *5*, 143–176 (1973)
Hatano, S., Ohnuma, J.: Purification and characterization of myosin A from the myxomycete plasmodium. Biochim. Biophys. Acta *205*, 110–120 (1970)
Hatano, S., Oosawa, F.: Extraction of an actin-like protein from the plasmodium of a myxomycete and its interaction with myosin A from rabbit striated muscle. J. Cell. Physiol. *68*, 197–202 (1966a)
Hatano, S., Oosawa, F.: Isolation and characterization of plasmodium actin. Biochim. Biophys. Acta *127*, 488–498 (1966b)
Hatano, S., Owaribe, K.: Actin and actinin from myxomycete plasmodia. In: Cell Motility. Goldman, R., Pollard, T., Rosenbaum, J. (eds.). Cold Spring Harbor Laboratory, 1976, pp. 499–511
Hatano, S., Takahashi, K.: Structure of myosin A from the myxomycete plasmodium and its aggregation at low salt concentrations. J. Mechanochem. Cell Motil. *1*, 7–14 (1971)
Hatano, S., Tazawa, M.: Isolation, purification, and characterization of myosin B from myxomycete plasmodium. Biochim. Biophys. Acta *154*, 507–519 (1968)
Hatano, S., Totsuka, T.: The polymerization of plasmodium actin in the presence of divalent cations. J. Mechanochem. Cell Motil. *1*, 67–74 (1972)
Hatano, S., Totsuka, T., Oosawa, F.: Polymerization of plasmodium actin. Biochim. Biophys. Acta *140*, 109–122 (1967)
Hepler, P.K., Palevitz, B.A.: Microtubules and microfilaments. Annu. Rev. Plant Physiol. *25*, 309–362 (1974)
Hess, B.: Oscillations in biochemical systems. In: The Molecular Basis of Circadian Rhythms. Hastings, J.W., Schweiger, H.-G. (eds.). Berlin: Dahlem Konferenzen, 1976, pp. 175–191

Hinssen, H.: Synthetische Myosinfilamente von Schleimpilz-Plasmodien. Cytobiologie *2*, 326–331 (1970)
Hinssen, H.: Actin in isoliertem Grundplasma von *Physarum polycephalum*. Cytobiologie *5*, 146–164 (1972)
Hinssen, H., D'Haese, J.: Filament formation by slime mould myosin isolated at low ionic strength. J. Cell Sci. *15*, 113–129 (1974)
Holmes, G.R., Goll, D.E., Suzuki, A.: Effect of α-actinin on actin viscosity. Biochim. Biophys. Acta *253*, 240–253 (1971)
Hülsmann, N., Haberey, M., Wohlfarth-Bottermann, K.-E.: Vitalmikroskopische Darstellung cytoplasmatischer Actomyosin-Fibrillen. Microsc. Acta *76*, 38–47 (1974)
Huxley, H.E.: Electron microscope studies on the structure of natural and synthetic protein filaments from striated muscle. J. Mol. Biol. *7*, 281–308 (1963)
Inoué, S., Stephens, R.E. (eds.): Molecules and Cell Movement. New York: Raven, 1975
Isenberg, G., Giesbrecht, P., Wecke, J., Wohlfarth-Bottermann, K.-E.: Demonstration of cytoplasmic actomyosin fibrils by the freeze-etching technique. Microsc. Acta *77*, 30–36 (1975)
Isenberg, G., Rathke, P.C., Hülsmann, N., Franke, W.W., Wohlfarth-Bottermann, K.-E.: Cytoplasmic actomyosin fibrils in tissue culture cells. Direct proof of contractility by visualization of ATP-induced contraction in fibrils isolated by laser microbeam dissection. Cell Tissue Res. *166*, 427–443 (1976)
Jacobson, D.N., Johnke, R.M., Adelman, M.R.: Studies on motility in *Physarum polycephalum*. In: Cell Motility. Goldman, R., Pollard, T., Rosenbaum, J. (eds.). Cold Spring Harbor Laboratory, 1976, pp. 749–770
Jahn, T.L.: Protoplasmic flow in the mycetozoan, *Physarum* – II. The mechanism of flow: a reevaluation of the contraction-hydraulic theory and of the diffusion drag hypothesis. Biorheology *2*, 123–152 (1964)
Jahn, T.L., Bovee, E.C.: Protoplasmic movements within cells. Physiol. Rev. *49*, 793–862 (1969)
Jahn, T.L., Rinaldi, R.A., Brown, M.: Protoplasmic flow in the mycetozoan, *Physarum* – I. Geometry of the plasmodium and the observable facts of flow. Biorheology *2*, 123–131 (1964)
Jockusch, B.M., Becker, M., Hindennach, I., Jockusch, H.: Slime mould actin: Homology to vertebrate actin and presence in the nucleus. Exp. Cell Res. *89*, 241–246 (1974)
Kamiya, N.: Protoplasmic streaming. Protoplasmatologia VIII 3a, 1–199 (1959)
Kamiya, N.: The mechanism of cytoplasmic movement in a myxomycete plasmodium. Symp. Soc. Exp. Biol. *22*, 199–219 (1968)
Kamiya, N.: Contractile properties of the plasmodial strands. Proc. Jpn. Acad. Sci. *46*, 1026–1031 (1970)
Kamiya, N., Allen, R.D., Zeh, R.: Contractile properties of the slime mold strand. Acta Protozool. (Warsaw) *11*, 113–124 (1972)
Kamiya, N., Kuroda, N.: Movement of the myxomycete plasmodium. I. A study of glycerinated models. Proc. Jpn. Acad. Sci. *41*, 837–841 (1965)
Kamiya, R., Maruyama, K., Kuroda, M., Kawamura, M., Kikuchi, M.: Mg-polymer of actin formed under the influence of β-actinin. Biochim. Biophys. Acta *256*, 120–131 (1972)
Kato, T., Tonomura, Y.: Ca^{2+}-sensitivity of actomyosin ATPase purified from *Physarum polycephalum*. J. Biochem. (Tokyo) *77*, 1127–1134 (1975a)
Kato, T., Tonomura, Y.: *Physarum* tropomyosin-troponin complex. Isolation and properties. J. Biochem. (Tokyo) *78*, 583–588 (1975b)
Kawamura, M., Nagano, K.: A calcium ion-dependent ATP pyrophosphohydrolase in *Physarum polycephalum*. Biochim. Biophys. Acta *397*, 207–219 (1975)
Kessler, D., Nachmias, V.T.: Electron microscope studies of actomyosin and cytoplasmic filaments from the slime mold *Physarum polycephalum*. J. Cell Biol. *47*, 105a (1970)
Kessler, D., Nachmias, V.T., Loewy, A.G.: Actomyosin content of *Physarum* plasmodia and detection of immunological cross-reactions with myosins from related species. J. Cell Biol. *69*, 393–406 (1976)
Komnick, H., Stockem, W., Wohlfarth-Bottermann, K.-E.: Weitreichende, fibrilläre Protoplasmadifferenzierungen und ihre Bedeutung für die Protoplasmaströmung. VII. Experimentelle Induktion, Kontraktion und Extraktion der Plasmafibrillen von *Physarum polycephalum*. Z. Zellforsch. *109*, 420–430 (1970)

Komnick, H., Stockem, W., Wohlfarth-Bottermann, K.-E.: Ursachen, Begleitphänomene, und Steuerung zellulärer Bewegungserscheinungen. Fortschr. Zool. *21*, 1–74 (1972)
Komnick, H., Stockem, W., Wohlfarth-Bottermann, K.-E.: Cell motility: Mechanisms in protoplasmic streaming and ameboid movement. Int. Rev. Cytol. *34*, 169–249 (1973)
Korohoda, W., Rakoczy, L., Walczak, T.: On the control mechanism of protoplasmic streamings in the plasmodia of *Myxomycetes*. Acta Protozool. (Warsaw) *7*, 363–373 (1970)
Lazarides, E.: Actin, α-actinin, and tropomyosin interaction in structural organization of actin filaments in nonmuscle cells. J. Cell Biol. *68*, 202–219 (1976)
LeStourgeon, W.M., Forer, A., Yang, Y. Z., Bertram, J.S., Rusch, H.P.: Contractile proteins. Major components of nuclear and chromosome non-histone proteins. Biochim. Biophys. Acta *379*, 529–552 (1975)
Loewy, A.G.: An actomyosin-like substance from the plasmodium of a myxomycete. J. Cell. Comp. Physiol. *40*, 127–156 (1952)
Maruyama, K.: A study of β-actinin, myofibrillar protein from rabbit skeletal muscle. J. Biochem. (Tokyo) *69*, 369–386 (1971)
Maruyama, K., Kamiya, R., Kimura, S., Hatano, S.: β-actinin-like protein from plasmodium. J. Biochem. (Tokyo) *79*, 709–715 (1976)
Matsutake, Y., Sagara, J., Yamada, T., Shimizu, H.: On the active streaming experiment of actomyosin solutions in glass microcapillaries. Biochim. Biophys. Acta *405*, 347–352 (1975)
McManus, M.A., Roth, L.E.: Fibrillar differentiation in myxomycete plasmodia. J. Cell Biol. *25*, 305–318 (1965)
Mooseker, M.S., Tilney, L.G.: Organization of an actin filament-membrane complex. Filament polarity and membrane attachment in the microvilli of intestinal epithelial cells. J. Cell Biol. *67*, 725–743 (1975)
Nachmias, V.T.: Electron microscope observations on myosin from *Physarum polycephalum*. J. Cell Biol. *52*, 648–663 (1972a)
Nachmias, V.: Filament formation by purified *Physarum* myosin. Proc. Natl. Acad. Sci. U.S.A. *69*, 2011–2014 (1972b)
Nachmias, V.T.: *Physarum* myosin: Two new properties. Cold Spring Harbor Symp. Quant. Biol. *37*, 607–612 (1973)
Nachmias, V.T.: Properties of *Physarum* myosin purified by a potassium iodide procedure. J. Cell Biol. *62*, 54–65 (1974)
Nachmias, V.T.: Calcium sensitivity of hybrid complexes of muscle myosin and *Physarum* proteins. Biochim. Biophys. Acta *400*, 208–221 (1975)
Nachmias, V.T.: Adenosine 5′-phosphate and assembly of actin filaments *in vitro*. J. Mol. Biol. *107*, 623–629 (1976)
Nachmias, V., Asch, A.: Actin mediated calcium dependency of actomyosin in a myxomycete. Biochem. Biophys. Res. Commun. *60*, 656–664 (1974)
Nachmias, V.T., Asch, A.: Regulation and Polarity: Results with myxomycete plasmodium and with human platelets. In: Cell Motility. Goldman, R., Pollard, T., Rosenbaum, J. (eds.). Cold Spring Harbor Laboratory, 1976, pp. 771–783
Nachmias, V.T., Huxley, H.E., Kessler, D.: Electron microscope observations on actomyosin and actin preparations from *Physarum polycephalum*, and on their interaction with heavy meromyosin subfragment I from muscle myosin. J. Mol. Biol. *50*, 83–90 (1970)
Nachmias, V.T., Ingram, W.C.: Actomyosin from *Physarum polycephalum*: Electron microscopy of myosin-enriched preparations. Science *170*, 743–745 (1970)
Nachmias, V.T., Kessler, D.: Antibody of *Physarum* myosin. I. Preparation and functional effects. Immunology *30*, 419–424 (1976)
Nagai, R., Kamiya, N.: Movement of the mycomycete plasmodium. II. Electron microscope studies on fibrillar structures in the plasmodium. Proc. Jpn. Acad. Sci. *42*, 934–939 (1966)
Nagai, R., Kamiya, N.: Movement of the myxomycete plasmodium. VI. Fibrillar structures in the glycerinated plasmodium. Proc. Jpn. Acad. Sci. *44*, 1044–1047 (1968)
Nagai, R., Kato, T.: Cytoplasmic filaments and their assembly into bundles in *Physarum* plasmodium. Protoplasma *86*, 141–158 (1975)
Nakajima, H.: Some properties of a contractile protein in a myxomycete plasmodium. Protoplasma *52*, 413–436 (1960)
Nakajima, H., Allen, R.D.: The changing pattern of birefringence in plasmodia of the slime mold *Physarum polycephalum*. J. Cell. Biol. *25*, 361–374 (1965)

Nakajima, H., Hatano, S.: Acetylcholinesterase in the plasmodium of the myxomycete, *Physarum polycephalum*. J. Cell. Comp. Physiol. *59*, 259–263 (1962)
Needham, D.M.: *Machina carnis*. The biochemistry of muscular contraction in its historical development. Cambridge: University Press, 1971
Oplatka, A., Tirosh, R.: Active streaming in actomyosin solutions. Biochim. Biophys. Acta *305*, 684–688 (1973)
Owaribe, K., Hatano, S.: Induction of antibody against actin from myxomycete plasmodium and its properties. Biochemistry *14*, 3024–3029 (1975)
Pollard, T.D., Korn, E.D.: *Acanthamoeba* myosin. I. Isolation from *Acanthamoeba castellanii* of an enzyme similar to muscle myosin. J. Biol. Chem. *248*, 4682–4690 (1973a)
Pollard, T.D., Korn, E.D.: *Acanthamoeba* myosin. II. Interaction with actin and with a new cofactor protein required for actin activation of Mg^{2+} adenosine triphosphatase activity. J. Biol. Chem. *248*, 4691–4697 (1973b)
Pollard, T.D., Weihing, R.R.: Actin and myosin and cell movement. CRC Crit. Rev. Biochem. *2*, 1–65 (1974)
Reinold, M., Stockem, W.: Darstellung eines ATP-sensitiven Membransystems mit Ca^{2+}-transportierender Funktion bei Amöben. Cytobiologie *6*, 182–194 (1972)
Rhea, R.P.: Electron microscopic observations on the slime mold *Physarum polycephalum* with specific reference to fibrillar structures. J. Ultrastruct. Res. *15*, 349–379 (1966)
Ridgway, E.B., Durham, A.C.H.: Oscillations of calcium ion concentrations in *Physarum polycephalum*. J. Cell Biol. *69*, 223–226 (1976)
Sachsenmaier, W., Blessing, J., Brauser, B., Hansen, K.: Protoplasmic streaming in *Physarum polycephalum*. Observations of spontaneous and induced changes of the oscillatory pattern by photometric and fluorometric techniques. Protoplasma *77*, 381–396 (1973)
Schollmeyer, J.E., Furcht, L.T., Goll, D.E., Robson, R.M., Stromer, M.H.: Localization of contractile proteins in smooth muscle cells and in normal and transformed fibroblasts. In: Cell Motility. Goldman, R., Pollard, T., Rosenbaum, J. (eds.). Cold Spring Harbor Laboratory, 1976, pp. 361–388
Schroeder, T.E.: The contractile ring. II. Determining its brief existence, volumetric changes and vital role in cleaving *Arbacia* eggs. J. Cell Biol. *53*, 419–434 (1972)
Sekine, T., Kielley, W.W.: The enzymatic properties of N-ethylmaleimide modified myosin. Biochim. Biophys. Acta *81*, 336–345 (1964)
Stossel, T.P., Hartwig, J.H.: Interactions of actin, myosin, and a new actin-binding protein of rabbit pulmonary macrophages. II. Role in cytoplasmic movement and phagocytosis. J. Cell Biol. *68*, 602–619 (1976)
Takata, T., Nagai, R., Kamiya, N.: Movement of the myxomycete plasmodium. III. Artificial polarization in endoplasm distribution in a plasmodium and its bearing on protoplasmic streaming. Proc. Jpn. Acad. Sci. *43*, 45–50 (1967)
Tanaka, H., Hatano, S.: Extraction of native tropomyosin-like substances from myxomycete plasmodium and the cross reaction between plasmodium F-actin and muscle native tropomyosin. Biochim. Biophys. Acta *257*, 445–451 (1972)
Taylor, D.L., Condeelis, J.S., Moore, P.L., Allen, R.D.: The contractile basis of amoeboid movement. I. The chemical control of motility in isolated cytoplasm. J. Cell Biol. *59*, 378–394 (1973)
Tilney, L.G.: Actin filaments in the acrosomal reaction of *Limulus* sperm. Motion generated by alteration in the packing of the filaments. J. Cell Biol. *64*, 289–310 (1975)
Tilney, L.G.: Nonfilamentous aggregates of actin and their association with membranes. In: Cell Motility. Goldman, R., Pollard, T., Rosenbaum, J. (eds.). Cold Spring Harbor Laboratory, 1976, pp. 513–528
Tirosh, R., Oplatka, A., Chet, I.: Motility in a "cell sap" of the slime mold *Physarum polycephalum*. FEBS Lett. *34*, 40–42 (1973)
Totsuka, T.: Transformations of plasmodium actin polymers at high temperatures. Biochim. Biophys. Acta *234*, 162–169 (1971)
Totsuka, T., Hatano, S.: ATP-ase activity of plasmodium actin polymer formed in the presence of Mg^{2+}. Biochim. Biophys. Acta *223*, 189–197 (1970)
Ts'o, P.O.P., Eggman, L., Vinograd, J.: Physical and chemical studies of myxomyosin, an ATP-sensitive protein in cytoplasm. Biochim. Biophys. Acta *25*, 532–542 (1957)
Usui, N.: Fibrillar differentiation in a microplasmodium of the slime mold *Physarum polycephalum*. Dev., Growth Differ. *13*, 241–255 (1971)

Weber, A., Herz, R.: The relationship between caffeine on contracture of intact muscle and the effect of caffeine on reticulum. J. Gen. Physiol. *52*, 750–759 (1968)

Wohlfarth-Bottermann, K.-E.: Weitreichende, fibrilläre Protoplasmadifferenzierungen und ihre Bedeutung für die Protoplasmaströmung. I. Elektronenmikroskopischer Nachweis und Feinstruktur. Protoplasma *54*, 514–539 (1962)

Wohlfarth-Bottermann, K.-E.: Weitreichende, fibrilläre Protoplasmadifferenzierungen und ihre Bedeutung für die Protoplasmaströmung. II. Lichtmikroskopische Darstellung. Protoplasma *57*, 747–761 (1963)

Wohlfarth-Bottermann, K. E.: Cell structures and their significance for ameboid movement. Int. Rev. Cytol. *16*, 61–131 (1964)

Wohlfarth-Bottermann, K.-E.: Weitreichende, fibrilläre Protoplasmadifferenzierungen und ihre Bedeutung für die Protoplasmaströmung. III. Entstehung und experimentell induzierbare Musterbildungen. Wm. Roux' Arch. Entwicklungsmech. Org. *156*, 371–403 (1965)

Wohlfarth-Bottermann, K.-E.: Plasmalemma invaginations as characteristic constituents of plasmodia of *Physarum polycephalum*. J. Cell Sci. *16*, 23–37 (1974)

Wohlfarth-Bottermann, K.-E.: Weitreichende, fibrilläre Protoplasmadifferenzierungen und ihre Bedeutung für die Protoplasmaströmung. X. Die Anordnung der Actomyosin-Fibrillen in experimentell unbeeinflußten Protoplasmaadern von *Physarum polycephalum*. Protistologica *11*, 19–30 (1975a)

Wohlfarth-Bottermann, K.-E.: Tensiometric demonstration of endogenous, oscillating contractions in plasmodia of *Physarum polycephalum*. Z. Pflanzenphysiol. *76*, 14–27 (1975b)

Wohlfarth-Bottermann, K.-E., Fleischer, M.: Cycling aggregation patterns of cytoplasmic F-actin coordinated with oscillating tension force generation. Cell Tissue Res. *165*, 327–344 (1976)

2.4 Cytoplasmic Streaming and Cyclosis of Chloroplasts

K. SEITZ

1. Introduction

Cytoplasmic streaming is known in a great variety of cells of different organisms. Generally one can distinguish between streaming in cells or organisms having no rigid cell walls, such as slime molds (see BRITZ, Chap. 2.3, this vol.), amoebae and other protozoa, and streaming in plant cells that are surrounded by rigid cell walls. In plant cells, which will be considered here, different types of streaming (e.g., rotation, circulation, agitation, and others) are known. The characteristics of these types and the results of the many older investigations on cytoplasmic streaming were described by KAMIYA (1959, 1960, 1962). More recent reviews have been contributed by VIRGIN (1964), HOPE and WALKER (1975), and ALLEN and ALLEN (1978).

Most of the newer investigations deal with rotation, which is the typical streaming pattern of cells with a large central vacuole. Rotational streaming or cyclosis is a massive flow of cytoplasm around the periphery of the cell. The simplest example is the root hair, which does not contain chloroplasts: Here all the fluid endoplasm flows along the cell walls. In green cells, however, two types of streaming can be distinguished (see Fig. 1). In internodal cells of characean algae the chloroplasts remain stationary in the rigid ectoplasm, whereas the fluid endoplasm flows along the ectoplasmic layer. In leaf cells of *Elodea* or *Vallisneria,* in contrast, the chloroplasts participate in cytoplasmic streaming. Nevertheless the different patterns of cytoplasmic streaming share common features. There is a fluid endoplasm moving along a stationary ectoplasm; in those cases where the ectoplasm cannot be seen in the microscope, it may be in close contact to the outer cell membrane (plasmalemma). Furthermore, the molecular mechanism of movement probably is the same in all cases (KAMIYA, 1962). Thus the results from one organism may be generalized to others.

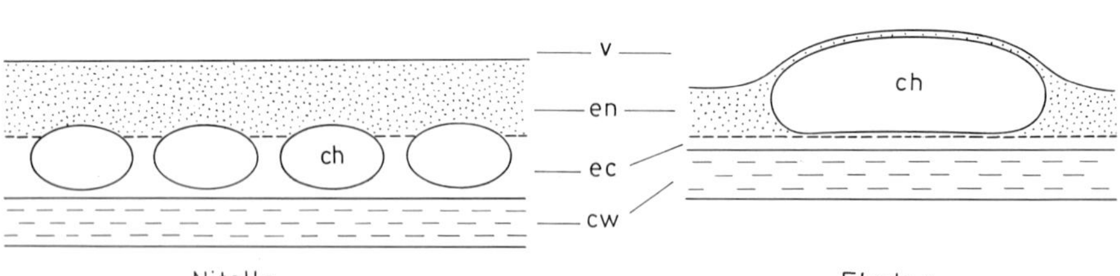

Fig. 1. Arrangement of streaming endoplasm (*en*), stationary ectoplasm (*ec*) and chloroplasts (*ch*) in cells of characean algae (e.g. *Nitella*) and leaf cells (e.g. *Elodea*). *v*=vacuole, *cw*=cell wall

2. Measurement of Cytoplasmic Streaming

Cytoplasmic streaming can be studied by the classical method of measuring the movement of organelles observed through the microscope. More detailed analysis is possible with cinematography or time-lapse photography, which allows accurate determination of rate and direction of movement of individual organelles (TAZAWA and KISHIMOTO, 1968; ALLEN, 1974).

Indirect evidence of cytoplasmic streaming can be obtained from measurements of the optical birefringence of the moving cytoplasm, which is caused by the alignment of structural elements, like membranes or filaments, parallel to the direction of streaming. Such changes in birefringence in relation to cytoplasmic streaming have been observed in amoeba (FRANCIS and ALLEN, 1971) and also in plant cells (PFEIFFER, 1953).

Another method that permits fast indirect registration of the speed of cytoplasmic streaming is laser light-scattering (MUSTACICH and WARE, 1974, 1976; SATELLE and BUCHAN, 1976). This technique measures the light scattered from a sample that is irradiated with a laser beam of about 50 µm diameter. If only random motion occurs in the sample, then the frequency spectrum of the scattered light is broadened around the incident frequency, and the extent of the broadening gives a measure of the diffusion coefficient of the scattering particles. However, if there is a directed motion in the sample, then the scattered light is Doppler-shifted and the magnitude of the shift in frequency is a direct measure of the velocity of the motion. The frequency spectrum of the scattered light shows a maximum at the frequency corresponding to the average speed, and it also gives a complete diagram of the velocity distribution of different particles in the cell. (Lower speeds correspond to slow-moving organelles in the vacuole, and higher speeds probably correspond to rapid motion of particles along actin filaments.) Yet, this velocity spectrum, which gives no information about the spatial distribution, is not equivalent to the intracellular velocity profile of the streaming endoplasm.

In vivo observation of streaming in characean cells, where chloroplasts in the stationary ectoplasm obstruct the optics, is greatly facilitated by the "window-technique," which locally removes chloroplasts by centrifugation after high-intensity microbeam irradiation (KAMITSUBO, 1972a).

As an index of cytoplasmic streaming the rate of movement of small cytoplasmic particles, granules, or organelles is measured. When the chloroplasts participate in streaming, their speed can be used as an index (TAGEEVA and KAZANTSEV, 1962). In such organisms streaming may also be evaluated by counting the percentage of cells in which cyclosis of chloroplasts can be observed (SEITZ, 1964).

3. Mechanism of Cytoplasmic Streaming in Characeae

3.1 Site and Molecular Basis of the Motive Force at the Ectoplasm

In internodal cells of *Chara* and *Nitella* the endoplasm flows on a constant course along the ectoplasm, in a direction parallel to the rows of chloroplasts. From careful analysis of the velocity distribution in the streaming endoplasm under different conditions, it is known that the motive mechanism lies at the boundary between the stationary ectoplasm and the moving endoplasm (KAMIYA, 1962). The endoplasm is driven by an active shearing or parallel-shifting force generated locally at the interface between ecto- and endoplasm. Indeed, streaming occurs only if the endoplasm is in contact with an undamaged ectoplasmic layer. Local damage of the ectoplasm induces a local cessation of streaming and accumulation of endoplasm in front of the damaged area (HAYASHI, 1964).

Actomyosin-like proteins are involved in the motive mechanism, as is suggested by the influence of inhibitors and by ultrastructural investigations. The inhibitor p-chloromercuribenzoic acid (PCMB), at concentrations of 10^{-4}–10^{-3} M, inhibits cytoplasmic streaming in *Nitella* and other organisms. The inhibition can be reversed by cysteine, indicating that PCMB specifically affects SH groups (ABE, 1964). In *Nitella*, PCMB inhibits streaming by affecting the motive force rather than cytoplasmic viscosity (TAZAWA, 1968). A recent investigation, differentiating experimentally between endo- and ectoplasm, showed, with N-ethylmaleimide (NEM) as an SH-specific reagent, that inhibition is more pronounced in the endoplasm than in the ectoplasm (CHEN and KAMIYA, 1975). The inhibitory effect of PCMB and NEM is thus considered to be caused by a specific blocking of the SH groups of an endoplasmic myosin-ATPase. However, one should perhaps keep in mind that the concentrations of PCMB used in these experiments also inhibit most other reactions that depend on SH groups and could thus block respiratory processes and synthesis of ATP as well.

Work with cytochalasin B has provided evidence for the involvement of filaments in the motive mechanism of streaming. The results of investigations with cytochalasin B have been reviewed and discussed by HEPLER and PALEVITZ (1974) and will thus not be treated in detail here. Cytochalasin B inhibits streaming in *Nitella*; it also inhibits streaming in *Chara* and *Zea* and cyclosis of chloroplasts in *Elodea* and *Vallisneria*, but it does not inhibit the centrifugability of endoplasm and chloroplasts in these organisms (GOLDSMITH and RAY, 1973; MAIER, 1975; NAGAI and KAMIYA, 1977; SEITZ, unpublished). These observations indicate that cytochalasin B inhibits only the motive mechanism of streaming and does not change the viscosity of cytoplasm. Cytochalasin B is effective only in the ectoplasm and not in the endoplasm (NAGAI and KAMIYA, 1977), indicating that the filaments essential for the generation of motive force must be located in the ectoplasm. In muscle, cytochalasin B inhibits actomyosin-ATPase by binding to actin in competition with myosin (SPUDICH and LIN, 1972); a similar mechanism may operate in cytoplasmic streaming.

Careful in vivo observations with the light microscope showed in *Nitella* and other plants that the streaming of endoplasm occurs along fibrils about 0.2 μm in diameter (O'BRIEN and MCCULLY, 1970; KAMITSUBO, 1972b; CHEN, 1973; ALLEN, 1974; PALEVITZ, 1976). These fibrils are located at the boundary between ecto- and endoplasm and are oriented parallel to the direction of streaming (KAMITSUBO, 1972b). The fibrils themselves do not move but organelle movement can be observed along the fibrils. The movement of organelles is unidirectional with respect to single fibrils, indicating a polarity of fibrils in relation to the direction of streaming (O'BRIEN and MCCULLY, 1970).

Electron-microscopic studies showed that these ectoplasmic fibrils are bundles or aggregates of microfilaments, each about 6 nm in diameter, a dimension like that of F-actin filaments in muscle (HEPLER and PALEVITZ, 1974; WAGNER, Chap. 2.2, this vol.). Similar filaments have also been found throughout the endoplasm (ALLEN, 1977; WAGNER and KLEIN, 1978). The filaments of *Nitella* combine with HMM from rabbit muscle to form arrowhead structures, which can be reversibly dissociated by addition of ATP. This indicates that the filaments consist of actin (PALEVITZ et al., 1974, 1975; WILLIAMSON, 1974, 1975; PALEVITZ, 1976). Moreover, these arrowhead structures show a marked polar orientation, pointing in the direction opposite from that of cytoplasmic streaming (KERSEY et al., 1976). A similar relation between polarity and motion is known from actin in muscle. In *Nitella*, the actin filaments are attached to the ectoplasm and to chloroplasts at the interface between ecto- and endoplasm (KERSEY and WESSELS, 1976; WILLIAMSON, 1976; HITCHCOCK, 1977). At this interface the filaments are involved in generating the motive force for cytoplasmic streaming.

The motive force in *Nitella* has been calculated to be about 1.6 dyne cm^{-2}, based on measurements of the balance acceleration necessary to bring the streaming endosplasm to a standstill in a centrifuge (KAMIYA, 1962). In agreement with this, the motive force was calculated to be 1.4–2 dyne cm^{-2}, based on measurements using the perfusion technique (TAZAWA, 1968). A motive force of the same order of magnitude (about 3.6 dyne cm^{-2}), was also found by DONALDSON (1972b). Based on a mathematical analysis, this author concluded that the motive force field must be restricted to a narrow layer of about 0.1 μm at the interface between endo- and ectoplasm, in order to explain the known velocity profile of streaming.

3.2 Transfer of Motive Force to the Endoplasm

The motive force of cytoplasmic streaming is generally considered to be caused by an actomyosin-like system. However, various models are under discussion for the mode of force transfer to the endoplasm, i.e., for the mechanism of the parallel shifting force at the interface between ecto- and endoplasm (HEPLER and PALEVITZ, 1974).

The actin filaments, which are polarly oriented and are anchored to the ectoplasm, could be involved in a direct interaction with a myosin-like protein in the endoplasm (KAMIYA, 1977). This myosin, which has been found in extracts from *Nitella* (ALLEN, 1976; KATO and TONOMURA, 1977), may be attached to organelles or to membranes of the endoplasmic reticulum (THAINE, 1965; REBHUN, 1967; DONALDSON, 1972a; BRADLEY, 1973). Such interaction could be effected by a cycle of reversible cross-bridge formation between the myosin and the actin filaments, as in the actomyosin system of muscle (HEPLER and PALEVITZ, 1974; WAGNER, Chap. 2.2, this vol.). Thus the activity of a myosin-like ATPase would cause the parallel shifting force and hence the motion of endoplasm and organelles along the filaments, when ATP is available (WILLIAMSON, 1975).

The force that drives cytoplasmic streaming in this model would be related to the number of myosin molecules active in cross-bridges. The velocity of streaming, like the speed of muscle contraction, would depend upon the rate of making and breaking cross-bridges, i.e., it would depend upon the turnover rate of the myosin-ATPase (HUXLEY and SIMMONS, 1971; TONOMURA and OOSAWA, 1972).

As the substrate of the myosin-ATPase, ATP would have a dual function at the interface between ecto- and endoplasm. ATP would control the motive force and the speed of cytoplasmic streaming. In addition ATP would dissociate the cross-bridges between actin and myosin and could thus lower the viscosity of the interface, i.e., the linkage between ecto- and endoplasm. Consequently, in the presence of ATP, the endoplasm could be moved more easily relative to the ectoplasm. Without ATP, many cross-bridges would be formed, as in the rigor state of muscle, so that the endoplasm could be moved less easily and organelles would become anchored to filaments (SZENT-GYÖRGYI, 1975; WILLIAMSON, 1975).

In contrast to the model of direct interaction between endoplasmic myosin and stationary actin filaments at the ectoplasm, other models of the motive mechanism assume that the filaments or fibrils move themselves and thus transfer motion to the endoplasm. The movement could be an undulatory beating of filaments as suggested by ALLEN (1974, 1976). In this model, filaments originating as branches from the filament bundles (fibrils) at the interface extend into the endoplasm, where they undulate like sperm tails in a sinusoidal fashion and thus drive cytoplasmic streaming. On the other hand, the movement could also be a rotation or torsional motion of the fibrils as suggested by the screw-mechanical theory of JAROSCH (1963, 1964).

Another hypothesis assumes that conformational changes caused by the ATPase activity could propagate like waves along the fibrils and could have an electric component. These propagating local electric fields could attract charged particles in the endoplasm and could thus transfer motion to the endoplasm (HEJNOWICZ, 1970; MOREL et al., 1976).

Finally an older hypothesis assumes that the filaments are distributed throughout the endoplasm and are associated with membranes of the endoplasmic reticulum (ER). Here the filaments could cause propagating undulations and pulsations of sheets and tubules of the ER resulting in a pumping action, like jet propulsion (KAVANAU, 1963).

Most evidence favors the first model, which involves a direct interaction between myosin in the endoplasm and actin filaments anchored to the ectoplasm. Accordingly, this model will be taken as a basis for the considerations in the following sections.

4. Mechanism of Cytoplasmic Streaming in Leaf Cells

4.1 Cyclosis of Chloroplasts

In leaf cells of *Elodea* and *Vallisneria* the chloroplasts are located in the endoplasm and participate in cytoplasmic streaming (Fig. 1). Nonetheless cyclosis of cytoplasm and chloroplasts in these organisms is similar to cyclosis in *Nitella*. The motive mechanism is again considered to be based upon an actomyosin-dependent process at the interface between ecto- and endoplasm as described above. However, in comparison to *Nitella, Elodea* shows greater variations in rate and pattern of

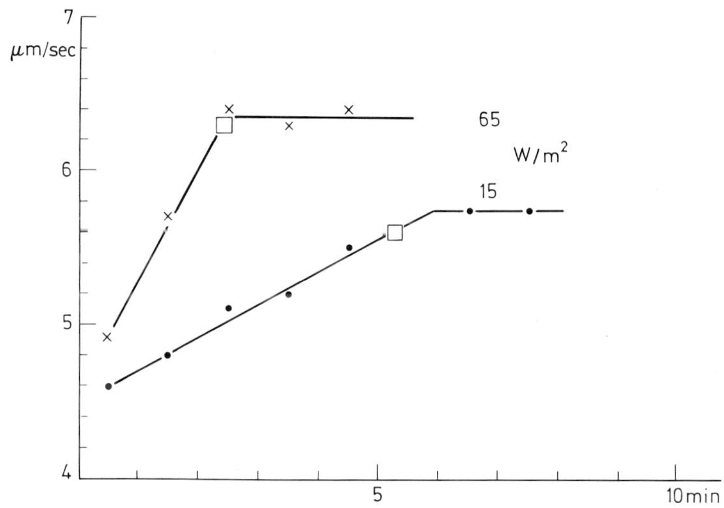

Fig. 2. Change in rate of cytoplasmic streaming and start of cyclosis of chloroplasts (□) in cells of *Elodea* under the influence of blue light. (Data from SEITZ, 1964)

streaming. Even changes in the type of streaming from rotation to circulation and agitation occur, especially when the rate of streaming is low (KAMIYA, 1959; BOLHAR-NORDENKAMPF, 1966).

In cyclosis the chloroplasts are moved passively by the streaming endoplasm, like stones in a river. This can be deduced from the slower and more irregular speed of chloroplasts in comparison to the surrounding cytoplasmic organelles (KAMIYA, 1962; ZURZYCKI, 1962b; SEITZ, 1964).

The relationship between the speed of chloroplasts and the speed of cytoplasmic organelles has been investigated for cyclosis in cells of *Elodea*. The results of this comparison showed that the speed of cytoplasm varies between about 5 and 11 µm sec^{-1}, and the speed of chloroplasts between about 2 and 8 µm sec^{-1}. The correlation between these speeds (in µm sec^{-1}) can be described by the formula $V_{chl}=0.8\,V_{cyt}-2$ (TAGEEVA and KAZANTSEV, 1962; TAGEEVA et al., 1965). This relation indicates that cyclosis of chloroplasts begins only when the speed of cytoplasmic streaming exceeds a certain minimum rate that is necessary to overcome the frictional resistances that oppose chloroplast movement. A closer analysis of the correlation between the speed of cytoplasm and the onset of chloroplast movement in *Elodea* has shown that light first induces an increase in the speed of cytoplasm. Cyclosis of chloroplasts starts in the cell when the speed of the cytoplasm has reached its maximum level, this level depending on the prevailing light conditions (Fig. 2; SEITZ, 1964).

If cyclosis is observed in a whole mount of a leaf, the chloroplasts are seen only at the anticlinal[1] walls of the cell. However, if the chloroplasts are not involved in cyclosis, they are located either almost randomly distributed at the different cell walls, e.g., in dark-adapted vein cells of *Elodea* (MAIER, 1975), or at the periclinal walls e.g., in epidermal cells of *Elodea* and *Vallisneria* after pretreatment

[1] Anticlinal walls are normal, periclinal walls parallel to the leaf surface.

with low intensity light (KAMIYA, 1959; TAGEEVA and KAZANTSEV, 1962). This means that before full cyclosis can be observed, at least some of the chloroplasts must undergo an orientation movement from the periclinal to the anticlinal walls of the cell. If cyclosis is induced by light, this movement is the so-called high-intensity movement of chloroplasts (SEITZ, 1967; MAIER, 1975; BRITZ, Chap. 2.5, this vol.). A consequence of this orientation movement preceding cyclosis is that cyclosis of chloroplasts can be observed only some time after an inducing stimulus. Another probable consequence is that in the state of full cyclosis more endoplasm is located at the anticlinal walls, i.e., in a thicker layer, than in a resting cell.

Theoretically stimulation of cytoplasmic streaming can be caused by an increase in the motive force or by a decrease in the viscosity of the cytoplasm, since the speed of cytoplasmic movement is given by the formula: speed = force/viscosity (KAMIYA, 1959; TAZAWA, 1968). In the case of induction of cyclosis of chloroplasts, their frictional resistance in the endoplasm and their adhesional contact to the ectoplasm are of importance for the rate of movement. This adhesion and the viscosity of cytoplasm must therefore be considered in relation to cytoplasmic streaming.

4.2 Centrifugability of Chloroplasts and Viscosity of Cytoplasm

There have been many investigations on the viscosity of cytoplasm (STÅLFELT, 1956; HEILBRUNN, 1958). In *Nitella*, cytoplasmic viscosity has been measured in vitro in isolated endoplasm. Here the velocity profile of the flowing endoplasm in a capillary was non-Newtonian in character and the apparent viscosity was determined to be about 2.5 to 10 poise, depending upon the shear stress (KAMIYA and KURODA, 1965). However, calculation of the in vivo viscosity of cytoplasm in *Nitella* gave a value of only about one-twelth that of the in vitro measurement (DONALDSON, 1972b).

In vivo estimates of the viscosity of cytoplasm in other organisms have been conducted mainly with the aid of centrifugation experiments. Here it is generally assumed that high centrifugability of cytoplasm and of chloroplasts indicates low viscosity and vice versa. The results of such experiments showed that in fact changes in the centrifugability can be observed in relation to changes in the activity of cytoplasmic streaming as expected (VIRGIN, 1949; ZURZYCKI, 1960, 1962a; SEITZ, 1967).

In *Nitella*, centrifugation displaces the entire endoplasm with respect to the ectoplasm; in *Elodea*, endoplasm and chloroplasts are displaced (VIRGIN, 1949; KAMIYA, 1959). The tonoplast remains intact and cyclosis resumes unchanged after centrifugation (LÄUCHLI and LÜTTGE, 1968; NAGAI and KAMIYA, 1977). The fact that the entire endoplasm moves along the ectoplasm without any apparent shear inside the endoplasm indicates that the viscosity of the shifting zone between endo- and ectoplasm is lower than the viscosity of the endoplasm (KAMIYA, 1959). This site of shifting during centrifugation is probably identical with the site where the parallel shifting force of the motive mechanism of cytoplasmic streaming is generated and where the frictional resistance opposing the motion of endoplasm and chloroplasts resides. Accordingly, centrifugation experiments do not record the viscosity of the endoplasm in general, but rather the viscosity or consistency of the interface between endo- and ectoplasm (KAMIYA, 1959; ZURZYCKI, 1962b). This consistency is determined mainly by the adhesion or linkage between endo- and ectoplasm and, according to the model of the motive mechanism of cytoplasmic streaming described above, it should depend upon the number of cross-bridges between myosin and actin.

5. Dependence of Streaming Upon Intracellular Conditions

5.1 Energetics and Control of Cytoplasmic Streaming

Cytoplasmic streaming depends upon the supply of oxygen, and streaming can be inhibited by uncouplers of oxidative phosphorylation such as dinitrophenol (DNP) and azide. In *Nitella* the motive force and the rate of cytoplasmic streaming are inhibited by the same concentrations of DNP that inhibit respiration (KAMIYA, 1962; TAIRBEKOV, 1970). The inhibition of streaming by DNP is reduced if the cells are irradiated with light of a sufficiently high intensity (BARR and BROYER, 1966). Such DNP-antagonizing effects of light can be observed only in cells that contain chloroplasts (LOMAGIN and ANTROPOVA, 1963). This suggests that, in the light, ATP from photophosphorylation replaces ATP from oxidative phosphorylation. A similar effect of light antagonizing the inhibition by DNP and other uncouplers is also known from the high intensity orientation movement of chloroplasts (SEITZ, 1970).

ATP appears to be a rate-limiting factor in cytoplasmic streaming: Adding ATP to the medium surrounding the cells induces an increase in the streaming rate (KAMIYA, 1962; POP et al., 1967; LAZAR-KEUL et al., 1974) and induces cyclosis as well as increased centrifugability of chloroplasts (SEITZ, 1971; MAIER, 1975).

In the motive mechanism of cytoplasmic streaming, as described above, ATP can control the rate and the effective motive force and it can control the adhesion of endoplasm and chloroplasts to the ectoplasm. In addition, ATP could be involved in reversible transformations of G-actin to F-actin, as in *Physarum* (BRITZ, Chap. 2.3, this vol.; WAGNER, Chap. 2.2, this vol.; HEPLER and PALEVITZ, 1974). Such changes in the degree of polymerization of actin could cause modifications of cytoplasmic viscosity. Finally ATP could play a role in the aggregation of filaments to fibrils (HEPLER and PALEVITZ, 1974; CLARKE and SPUDICH, 1977). The fibrils, which are known to be bundles of filaments, can best be seen under special conditions (CHEN, 1973); for example, after mechanical stimulation (ALLEN, 1976) or in isolated cytoplasm (KAMIYA, 1962). Under these conditions, cytoplasmic streaming stops. If the cessation of streaming is related to a lack of ATP, then in the presence of myosin, rigor-like filament complexes would be formed (WILLIAMSON, 1975). Such complexes of filaments can be formed easily if each myosin binds two actins (OFFER and ELLIOTT, 1978). Addition of ATP in fact causes all fibrils to disappear (PALEVITZ, 1976). Accordingly ATP could control the aggregation of filaments to fibrils. Such an ATP-dependent formation of filaments and subsequent aggregation to visible fibrils has been observed in cytoplasm of amoeba (POLLARD and ITO, 1970).

Besides ATP, Ca^{2+} could well be involved in the control of cytoplasmic streaming in plants as it is in other actomyosin-dependent nonmuscle systems (BRITZ, Chap. 2.3, this vol.; CLARKE and SPUDICH, 1977; HITCHCOCK, 1977; NUCCITELLI et al., 1977; TAYLOR, 1977; WAGNER, Chap. 2.2, this vol.). The influence of Ca^{2+} on cytoplasmic streaming will be described below together with the effect of other ions.

5.2 Ionic Conditions

The cytoplasm of plant cells is located in an electrochemical gradient between an external medium of low ionic concentration and an internal vacuolar medium of different ionic composition and higher concentration. Changes in the composition or concentration of ions in the external medium or in the cell sap could modify, at least temporarily, the ionic conditions in the cytoplasm and could affect cytoplasmic streaming.

Variations in the osmotic concentration of the cell sap have no influence on the rate of cytoplasmic streaming in *Nitella,* for concentrations between about 0.17 and 0.3 M. Below or above this range, however, the rate of cytoplasmic streaming decreases sharply (KAMIYA, 1962; TAZAWA, 1968; TAZAWA and KISHIMOTO, 1964). After some time osmoregulation begins to restore the normal osmotic value of 0.26 M, and the streaming rate increases to its normal level (KAMIYA, 1962).

Moderate changes in the ionic composition of the external medium do not cause changes in the rate of streaming (KAMIYA, 1959, 1962; BARR and BROYER, 1964). Addition of Ca^{2+} to the medium has no effect on streaming in *Nitella* (BARRY, 1968); it accelerates streaming in root hairs of *Hordeum* (VINTILA et al., 1973) and inhibits streaming in cells of *Elodea* (FORDE and STEER, 1976). These conflicting results may be caused by differences in the organisms or by the different concentrations used, since in *Acetabularia,* for instance, high concentrations of Ca^{2+} accelerate, whereas low concentrations inhibit streaming (KAMIYA, 1962). This observation probably suggests that Ca^{2+} is required for streaming. Recently, Ca^{2+} deficiency has been found to inhibit chloroplast movement in *Mougeotia* (WAGNER and KLEIN, 1978), but the reactions underlying this response might be completely different from those underlying cyclosis. In fact, addition to the medium of ethylenediaminetetraacetic acid (EDTA), which causes a decrease in the available Ca^{2+}, has no influence on the rate of cyclosis in *Nitella, Hordeum,* and *Coleochaete* (KISHIMOTO and AKABORI, 1959; LAZAR-KEUL et al., 1972; MARCHANT, 1977).

In *Nitella,* the vacuolar contents can be replaced by artificial cell saps, using the perfusion technique. Isotonic changes in the composition of the cell sap do not cause changes in the rate of streaming. Specifically, changes in the concentration of K^+ do not affect the rate of streaming, as long as the ratio of K^+ to Ca^{2+} (about 5 in the normal cell) remains between 2 and 50. K^+ can be replaced without effect by Li^+, Na^+, and Rb^+; Cs^+ is the only monovalent cation that induces a decrease in streaming rate (TAZAWA and KISHIMOTO, 1964). High concentrations of Cl^- inhibit streaming (WILLIAMSON, 1975). A change in the anion from Cl^- to NO_3^- does not modify streaming; only SO_4^{2-} causes a decrease in streaming rate (TAZAWA and KISHIMOTO, 1964). This effect of SO_4^{2-} is considered to be caused by a reduction in the level of Ca^{2+}, since the presence of more than 3 mM Ca^{2+} is a necessary condition for the maintenance of a normal streaming rate. Ca^{2+} concentrations higher than 50 mM, however, inhibit streaming (TAZAWA and KISHIMOTO, 1964). Such inhibition of streaming by a high Ca^{2+} concentration was also observed by WILLIAMSON (1975), but here no evidence for a Ca^{2+} requirement could be found, since streaming continued at full rate even in the presence of EGTA. Because of these conflicting reports regarding the influence of Ca^{2+} in the vacuole and in the external medium, a specific role for Ca^{2+} in the control of cytoplasmic streaming cannot be suggested at present. A Ca^{2+}-dependent control mechanism may not even be involved in the continuous streaming of cyclosis, in contrast to other responses.

The ionic gradient across the cytoplasm can be neutralized if the vacuole is filled with external medium of low ionic concentration made isotonic by the addition of mannitol. In this case there is a decrease in the rate of streaming, which has been attributed to a loss of K^+ from the cytoplasm. If the ionic gradient is reversed, however, by placing a cell with low ionic vacuolar medium into an external medium of artificial cell sap, a normal rate of streaming can be observed after some time, since the cell takes up ions (TAZAWA, 1975). Changes in the ionic gradient across the cytoplasm are also involved in the effect of action potentials on cytoplasmic streaming.

5.3 Action Potential and Cytoplasmic Streaming

Whenever the ionic gradient across the cytoplasm in *Nitella* is altered in the course of an action potential, a sudden cessation of streaming results (KAMIYA, 1959; KISHIMOTO and AKABORI, 1959; HOPE and WALKER, 1975). The action potential in *Nitella* can be triggered by an electric or by a mechanical stimulus or by other sudden changes in the external conditions (FINDLAY and HOPE, 1976). Subthreshold stimuli that do not elicit an action potential do not cause a change in the rate of cytoplasmic streaming.

An action potential is caused by transient permeability of the cell membrane and by resulting fluxes of ions. The ionic basis of action potentials in plant cells is mainly a large efflux of Cl^- and of K^+. An influx of Ca^{2+} has been considered to play a role also, but an increased influx of Ca^{2+} could not be measured during the action potential in *Nitella* (FINDLAY and HOPE, 1976). The efflux of Cl^- and K^+ is accompanied by a movement of water and by a decrease in turgor pressure and, as a consequence, a shortening of the cell can also be measured (HOPE and WALKER, 1975; ODA and LINDSTEAD, 1975).

A comparison of the time course of the action potential with the changes in streaming rate shows that the stoppage of streaming occurs somewhat later than the beginning of the action potential (KAMIYA, 1959). During the rising time of the action potential the rate remains unchanged; only when the peak of the action potential is reached, does the rate of streaming begin to slow down, and it comes to a standstill about 0.3 s later (Fig. 3; TAZAWA and KISHIMOTO, 1968). Streaming resumes slowly after a stop of about 30 s, and the normal rate is reached again after 4–10 min (KAMIYA, 1959; KISHIMOTO and AKABORI, 1959). TAZAWA

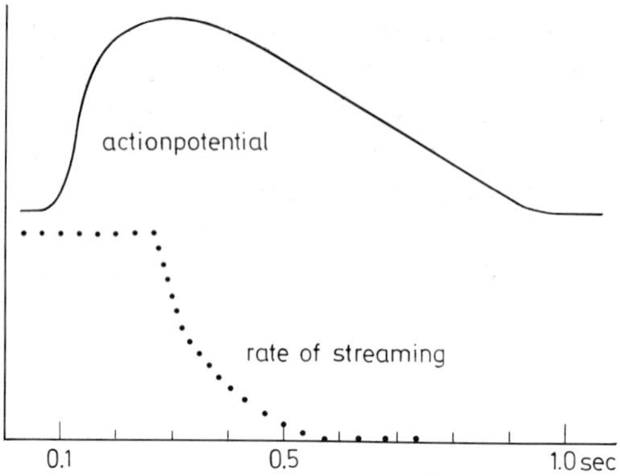

Fig. 3. Time course of action potential and cessation of cytoplasmic streaming in *Chara australis*. (Rate of streaming and membrane potential in arbitrary units on the ordinate, schematized from TAZAWA and KISHIMOTO, 1968)

and KISHIMOTO (1968) showed that the cessation of streaming is caused by a temporary disappearance of the motive force of streaming.

The disappearance of the motive force and the cessation of streaming must be caused by some change in the cytoplasm due to the action potential (KAMIYA, 1959). Factors that could be involved are changes in the electric properties of the cell membrane or changes in the ionic composition of the cytoplasm or changes in the total ion concentration of the cell. Changes in the electric parameters of the cell membrane apparently are not the cause for the stop in streaming, since electric currents below the threshold of the induction of action potentials do not modify the rate of streaming, and streaming is independent of the magnitude of the membrane potential of the cell (KISHIMOTO and AKABORI, 1959; PICKARD, 1969).

The large efflux of Cl^- and K^+ from the vacuole through the cytoplasm and into the external medium could lead to changes in the ionic composition of the cytoplasm and could cause the cessation of streaming, probably by modifying the charge and hydration of proteins (TAZAWA and KISHIMOTO, 1968). However, the molecular mechanism underlying the Cl^- and K^+ effect cannot yet be explained. Small, probably unmeasurable changes in Ca^{2+} have also been considered as a possible cause for cessation of streaming (TAZAWA and KISHIMOTO, 1968; PICKARD, 1972; HOPE and WALKER, 1975). Indeed the action potential and the cessation of streaming can be induced only if some Ca^{2+} is present in the medium (BARRY, 1968; FINDLAY and HOPE, 1976). Yet this probably can be explained by assuming that Ca^{2+} is necessary for triggering the rapid increase in membrane permeability to Cl^- (WEISENSEEL and RUPPERT, 1977).

As long as the effect of changes in single ions on cessation of streaming is not understood, one cannot rule out an alternative mechanism based on the bulk loss of ions during the action potential. The uptake of ions after an action potential and cytoplasmic streaming are both dependent upon the availability of ATP, so that the two processes compete for the limiting ATP supplied by the metabolism of the cell. Such close dependence of the ionic relations and of the membrane potential upon the metabolism of cells is well documented (RAVEN, 1971; BENTRUP et al., 1973; LÜTTGE and PITMAN, 1976; SHIMMEN and TAZAWA, 1977). Since much more ATP is consumed in ion uptake than in cytoplasmic streaming (HOPE and WALKER, 1975), a high rate of ion uptake will be accompanied by a great consumption of ATP. Thus less ATP will be available for the myosin ATPase, and cytoplasmic streaming will stop. This consideration is supported by experimental data showing that a decrease in the cellular ATP content can by measured concomitant with the action potential (GRADMANN, 1976). The time course of events is also in agreement with these considerations, since the recovery of streaming starts after the recovery of the membrane potential (KISHIMOTO and AKABORI, 1959) and after the recovery in turgor and cell length (ODA and LINDSTEAD, 1975). This indicates that streaming apparently begins again when the ionic relations of the cell have been restored.

6. Influence of External Factors on Streaming

6.1 Temperature

The rate of cytoplasmic streaming increases with increasing temperature between 5° and 30° C (KAMIYA, 1962; PICKARD, 1974). This increase in streaming rate may be caused by an activation of the motive force or by a decrease in the viscosity of the cytoplasm. This alternative has been analyzed in *Nitella* by a parallel investigation of the temperature dependence of streaming rate and of motive force (KAMIYA, 1962). The speed of cytoplasmic streaming increased with temperature with a Q_{10} of 2, but the motive force remained constant. It was concluded that a decrease in the viscosity must cause the increase in the speed of cytoplasmic streaming with increasing temperature, if the above-mentioned formula relating speed, viscosity, and motive force can be applied (TAZAWA, 1968). To explain the results, the viscosity of cytoplasm should decrease with a Q_{10} of 2, but in other investigations this temperature dependence of viscosity has a Q_{10} of only about 1.3–1.4 (WEBER and WEBER, 1916; WILSON and SCHWABE, 1964). Thus it is not certain that the change in speed is caused simply by a change in viscosity, and an alternative hypothesis may be considered.

If a myosin-actin interaction underlies cytoplasmic streaming, then the rate of making and breaking cross-bridges, i.e., the turnover rate of the myosin ATPase and thus the energy input per unit time would increase with temperature, as in any enzymatic reaction. The increase in the turnover rate of the actomyosin system would cause the increase in the speed of streaming without a change in viscosity, as long as the number of actomyosins that are involved remains constant.

6.2 Light

Light is known to stimulate cytoplasmic streaming and to cause induction of cyclosis of chloroplasts (HAUPT, 1959; KAMIYA, 1962).

In *Nitella* as well as in *Acetabularia* and *Elodea* a change from white to red irradiation causes the rate of cytoplasmic streaming to decrease to 35% of its initial value within about 1 h. Small quantities (ca. 27 w s m^{-2}) of additional blue then restore the streaming rate to its initial level in white light, after a lag period of 5–8 min (WILLIAMS and TERBORGH, 1970). These light-dependent changes in cytoplasmic streaming closely parallel the response of photosynthetic oxygen evolution to changes in irradiation (TERBORGH, 1966), and may be related to the known blue and red light-dependent changes in respiration and in carbohydrate and protein metabolism of plants (KOWALLIK, 1970).

The effect of light has been investigated in greater detail in *Elodea* and *Hordeum*. In root hairs of *Hordeum*, cytoplasmic streaming shows a constant rate in darkness. In dark-adapted material, light of high intensity induces an increase in the speed of streaming. In continuous light as well as after a short induction, this stimulation reaches its maximal level after about 20 min. The action spectrum of the response has maxima at 450 nm in the blue region and at 366 min in the near ultraviolet, with a minimum in between at 400 nm (Fig. 4, curve *a*). This suggests that a flavin is the photoreceptor for this response. A small, as yet unexplained, additional peak was found in the green region at 537–554 nm, but red light has no effect (KEUL, 1976). An analysis of the dose-response dependence of the effect of light shows a hyperbolic curve, which indicates that the primary action of light should be a photocatalytic effect (HARTMANN and COHNEN UNSER, 1972). Light is considered to induce an activation of metabolism, thus leading to an increased supply of ATP and stimulation of cytoplasmic streaming in *Hordeum* (KEUL, 1976).

Similarly, cytoplasmic streaming in leaf cells of *Elodea* continues at a moderate, constant speed in the dark. In dark-adapted cells, irradiation with light of high intensity induces an increase in the streaming rate, reaching saturation within a few min. When this level is reached, cyclosis of chloroplasts starts in the cell (SEITZ, 1964; see Fig. 2). The effect of light was further investigated in *Elodea* by taking the induction of cyclosis of chloroplasts as a measure for cytoplasmic streaming (SEITZ, 1964). The action spectrum of the induction of cyclosis of chloroplasts in *Elodea* has peaks at 450 and 370 nm, again indicating a flavin as photoreceptor (Fig. 4, curve *b*). In the long wavelength region, however, there is some difference in comparison to *Hordeum*. In *Elodea*, red light induces a small response while green light has no effect.

Cyclosis of chloroplasts can be induced, as is known, either by an activation of the motive mechanism of cytoplasmic streaming or by a decrease in the adhesion of chloroplasts to the ectoplasmic layer (see above). The question therefore arises which of these parameters is affected by light. Results from *Elodea* indicate that after high-intensity irradiation the centrifugability of chloroplasts, as well as the motive force of streaming, is increased (motive force measured as centrifugal force necessary to counterbalance rotational streaming of chloroplasts). After low-intensity irradiation, on the other hand, centrifugability and the motive force are decreased (VIRGIN, 1949).

The action of light in the control of these parameters was further investigated in *Vallisneria* (SEITZ, 1967). The same intensity range that causes the induction of cyclosis of chloroplasts induces an increase in their centrifugability. The action spectra for these two effects are nearly identical in *Vallisneria* (Fig. 4, curves *c* and *d*) and resemble the action spectrum for induction of increased centrifugability of chloroplasts in *Elodea* (Fig. 4, curve *e*; VIRGIN, 1952). Besides the main effect in the blue, there is, for both responses in *Vallisneria* and for the induction of cyclosis in *Elodea*, a small effect in the red. This suggests that two photoreceptor pigments are active: a flavin in the cytoplasm and chlorophylls in the chloroplasts. The location of the two pigments in different compartments of the cell is also

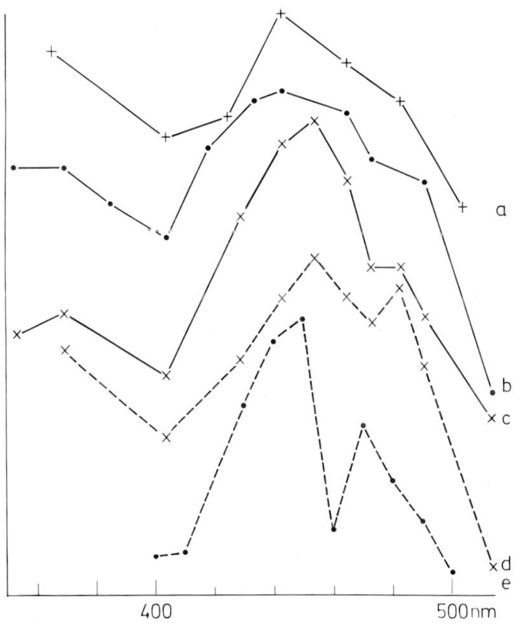

Fig. 4. Action spectra of light-induced stimulation of cytoplasmic streaming, induction of cyclosis of chloroplasts, and increase in centrifugability.
a Cytoplasmic streaming in root hairs of *Hordeum* (KEUL, 1976); *b* cyclosis of chloroplasts in leaf cells of *Elodea* (SEITZ, 1964); *c* cyclosis of chloroplasts in leaf cells of *Vallisneria* (SEITZ, 1967); *d* increase in centrifugability of chloroplasts in *Vallisneria* (SEITZ, 1967); *e* increase in centrifugability of chloroplasts in *Elodea* (VIRGIN, 1952). Spectra in relative units (peak: 100%) shifted along ordinate

demonstrated by the effect of linearly polarized light. Polarized blue light induces increased centrifugability of chloroplasts in *Vallisneria* only when the electrical vector is oriented parallel to the cell walls (Fig. 5; SEITZ, 1967). This indicates that the flavin photoreceptor must be localized in the resting ectoplasm or in the plasmalemma and must have a dichroic orientation parallel to the cell surface. In red light no such preferential effect of linearly polarized light could be found. The results with polarized blue light also show that light induces locally different effects inside a cell, as is apparent in Fig. 5.

Considering the similarity in the action spectra and in the effective intensity ranges in the results described above, it was concluded that light induces an increased rate of cytoplasmic streaming as well as cyclosis and increased centrifugability of chloroplasts via the same primary process (SEITZ, 1967).

The dose-response curve for the effect of light on chloroplast centrifugability shows three ranges: At very low intensities, centrifugability is increased (relative to the dark level), at moderately low intensities centrifugability is decreased, and at high intensities it is increased again (VIRGIN, 1951; ZURZYCKI, 1960, 1962a; SEITZ, 1971). This dose-response curve thus resembles that for phototropism in oat coleoptiles, with a first positive, a negative, and a second positive range of response.

In the low-intensity range (in contrast to the high-intensity range) blue and red light are about equally effective in decreasing centrifugability. Furthermore, the peaks in the blue are at 430 and 480 nm, i.e., this action spectrum is different from that for induction of increased centrifugability at high intensity (SEITZ, 1979). Experiments with metabolic inhibitors also show differences between the low and high intensity range. The increase in centrifugability in the high-intensity range is inhibited by uncouplers of oxidative phosphorylation and not by inhibitors

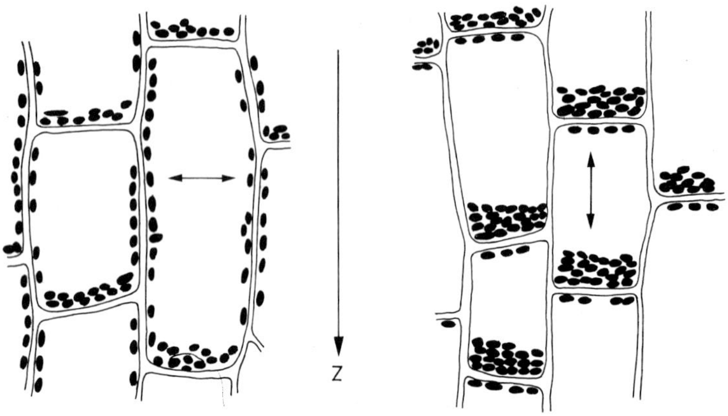

Fig. 5. Effect of linearly polarized light on the centrifugability of chloroplasts in cells of *Vallisneria* (Irradiation: 30 min with high-intensity blue. *Double arrow*=plane of vibration of electric vector. *Z*=direction of centrifugation. Schematized from SEITZ, 1967)

of photosynthetic electron flow, whereas the decrease in centrifugability in the low-intensity range is reduced only by inhibitors of photosynthetic electron flow (SEITZ, 1979). This indicates that different photoreaction systems are involved in the low and high intensity range.

The inhibition of the increase in centrifugability by uncouplers shows that ATP is rate limiting in this response, as was concluded above for cytoplasmic streaming. As a mechanism for the effect of light a hypothesis has been proposed for the control of intracellular availability of ATP via two photoreaction systems (SEITZ, 1971, 1975, 1979). One system induces an increase in the rate of respiration and in the availability of ATP via a flavin as photoreceptor. In the other system light is effective via photosynthetic processes, with chlorophylls as photoreceptors. Here two different reactions modify the availability of ATP as a function of irradiance. These are the different types of photophosphorylation and the noncyclic electron flow leading to CO_2 fixation, which consumes more ATP than is synthesized in the concurrent noncyclic photophosphorylation. At low intensity, under limiting light conditions, the electrons of the noncyclic flow are channeled preferentially into the reduction of NADP and not into cyclic or pseudocyclic photophosphorylation. A reduced availability of ATP will result. At high intensity, in the range of saturation of CO_2 fixation, the level of $NADPH_2$ rises, more electrons are diverted to cyclic or pseudocyclic photophosphorylation, and a surplus of ATP will be available. The blue light-induced increase in the rate of respiration together with the photosynthetic reactions could thus control the intracellular availability of ATP as a function of irradiance. The changes in ATP, in turn, could control the motive mechanism of cytoplasmic streaming and other blue light-dependent responses.

Based on its dual role in the actomyosin system (see above), ATP would control the rate of cytoplasmic streaming and the effective motive force as well as the adhesion of endoplasm and chloroplasts to the ectoplasmic layer. An increase in the availability of ATP would cause a higher rate of cytoplasmic streaming, cyclosis, and increased centrifugability of chloroplasts.

If light is effective by such a mechanism, then the rate of cytoplasmic streaming in cells with chloroplasts should depend upon the irradiance of the incident light, with a dose-response curve similar to that of the centrifugability of chloroplasts. In the low-intensity range, a decrease in the speed of cytoplasmic streaming should be expected, in addition to the known increase in the high-intensity range. Little is known about the irradiance dependence of cytoplasmic streaming. Only in cells of *Avena* coleoptiles, a decrease in streaming rate at low intensities and an increase at high intensities was observed (BOTTELIER, 1933); in *Vaucheria*, a decrease in cytoplasmic streaming rate at low intensity was reported (FISCHER-ARNOLD, 1963).

In conclusion, the effect of light on cytoplasmic streaming and centrifugability of chloroplasts is of great importance for the light-dependent orientation movement of chloroplasts (see BRITZ, Chap. 2.5, this vol.). In plant cells under unilateral light there is a gradient of light absorption between the different walls (HAUPT and SCHÖNBOHM, 1970). Consequently, a gradient of chloroplast mobility in cytoplasmic streaming is established, controlling chloroplast movement (ZURZYCKI, 1962a; SEITZ, 1971). The orientation movement, as is known, brings chloroplasts into a position in the cell that is assumed to allow optimal light absorption and optimal CO_2 fixation under changing external light conditions.

References

Abe, S.: The effect of p-Chloromercuribenzoate on rotational protoplasmic streaming in plant cells. Protoplasma *58*, 483–492 (1964)

Allen, N.S.: Endoplasmic filaments generate the motive force for rotational streaming in *Nitella*. J. Cell Biol. *63*, 270–287 (1974)

Allen, N.S.: Undulating filaments in *Nitella* endoplasm and motive force generation. In: Cell Motility. Goldman, R., Pollard, T., Rosenbaum, J. (eds.). New York: Cold Spring Harbor Lab. 1976, pp. 613–621

Allen, N.S., Allen, R.D.: Cytoplasmic streaming in green plants. Annu. Rev. Biophys. Bioeng. *7*, 497–526 (1978)

Allen, R.D.: Concluding remarks. In: International Cell Biology 1976–1977. Brinkley, B.R., Porter, K.R. (eds.). New York: The Rockefeller University Press 1977, pp. 403–406

Barr, C.E., Broyer, T.C.: Effect of light on sodium influx, membrane potential, and protoplasmic streaming in *Nitella*. Plant Physiol. *39*, 48–52 (1964)

Barr, C.E., Broyer, T.C.: Energy balancing in *Nitella* cells treated with dinitrophenol. Science *151*, 1245–1246 (1966)

Barry, W.H.: Coupling of excitation and cessation of cyclosis in *Nitella*: Role of divalent cations. J. Cell. Physiol. *72*, 153–160 (1968)

Bentrup, F.W., Gratz, H.J., Unbehauen, H.: The membrane potential of *Vallisneria* leaf cells: Evidence for light-dependent proton permeability changes. In: Ion Transport in Plants. Anderson, W.P. (ed.). London: Academic Press, 1973, pp. 171–182

Bolhar-Nordenkampf, H.: Cytomorphologische Studien am Plasma der Innenepidermiszellen von Allium cepa, unter besonderer Berücksichtigung der „Plasmaschläuche". Protoplasma *61*, 85–113 (1966)

Bottelier, H.P.: Über den Einfluß des Lichtes auf die Protoplasmaströmung von Avena. Proc. K. Ned. Akad. Wet. *36*, 790–794 (1933)

Bradley, M.O.: Microfilaments and cytoplasmic streaming: inhibition of streaming with cytochalasin. J. Cell Sci. *12*, 327–343 (1973)

Chen, J.C.W.: Observations of protoplasmic behavior and motile protoplasmic fibrils in cytochalasin B treated *Nitella* rhizoid. Protoplasma *77*, 427–435 (1973)

Chen, J.C.W., Kamiya, N.: Localization of myosin in the internodal cell of *Nitella* as suggested by differential treatment with N-Ethylmaleimide. Cell Structure Function *1*, 1–9 (1975)
Clarke, M., Spudich, J.A.: Nonmuscle contractile proteins: The role of actin and myosin in cell motility and shape determination. Annu. Rev. Biochem. *46*, 797–822 (1977)
Donaldson, J.G.: Cyclic longitudinal fibrillar motion as a basis for steady rotational protoplasmic streaming. J. Theor. Biol. *37*, 75–91 (1972a)
Donaldson, J.G.: The estimation of the motive force for protoplasmic streaming in *Nitella*. Protoplasma *74*, 329–344 (1972b)
Findlay, G.P., Hope, A.B.: Electrical properties of plant cells: methods and findings. In: Encycl. of Plant Physiol. N.S. Vol. 2A. Lüttge, U., Pitman, M.G. (eds.). Berlin-Heidelberg-New York: Springer, 1976, pp. 53–92
Fischer-Arnold, G.: Untersuchungen über die Chloroplastenbewegung bei *Vaucheria sessilis*. Protoplasma *56*, 495–520 (1963)
Forde, J., Steer, M.W.: Cytoplasmic streaming in *Elodea*. Can. J. Bot. *54*, 2688–2694 (1976)
Francis, D.W., Allen, R.D.: Induced birefringence as evidence of endoplasmic viscoelasticity in *Chaos carolinensis*. J. Mechanochem. Cell Motil. *1*, 1–6 (1971)
Goldsmith, M.H.M., Ray, P.M.: Intracellular localisation of the active process in polar transport of auxin. Planta *111*, 297–314 (1973)
Gradmann, D.: Metabolic action potentials in *Acetabularia*. J. Membr. Biol. *29*, 23–45 (1976)
Hartmann, K.M., Cohnen Unser, I.C.: Analytical action spectroscopy with living systems: photochemical aspects and attenuance. Ber. Dtsch. Bot. Ges. *85*, 481–552 (1972)
Haupt, W.: Photodinese. In: Encyclopedia of Plant Physiology. Ruhland, W. (ed.). Berlin-Heidelberg-New York: Springer 1959, Vol. XVII, pt. 1, pp. 388–398
Haupt, W., Schönbohm, E.: Light-oriented chloroplast movements. In: Photobiology of Microorganisms. Halldal, P. (ed.). London: Wiley, 1970, pp. 283–307
Hayashi, T.: Role of the cortical gel layer in cytoplasmic streaming. In: Primitive Motile Systems in Cell Biology. Allen, R.D., Kamiya, N. (eds.). New York: Academic Press, 1964, pp. 19–29
Heilbrunn, L.V.: The viscosity of protoplasm. In: Protoplasmatologia Vol. II, C, 1. Heilbrunn, L.V., Weber, F. (eds.). Wien: Springer, 1958
Hejnowicz, Z.: Propagated disturbances of transverse potential gradient in intracellular fibrills as the source of motive forces for longitudinal transport in cells. Protoplasma *71*, 343–364 (1970)
Hepler, P.K., Palevitz, B.A.: Microtubules and microfilaments Annu. Rev. Plant Physiol. *25*, 309–362 (1974)
Hitchcock, S.E.: Regulation of motility in nonmuscle cells. J. Cell Biol. *74*, 1–15 (1977)
Hope, A.B., Walker, N.A.: The Physiology of Giant Algal Cells. Cambridge: University Press, 1975
Huxley, A.F., Simmons, R.M.: Proposed mechanism of force generation in striated muscle. Nature (London) *233*, 533–538 (1971)
Jarosch, R.: Grundlage einer Schrauben-Mechanik des Protoplasmas. Protoplasma *57*, 448–500 (1963)
Jarosch, R.: Screw-mechanical basis of protoplasmic movement. In: Primitive Motile System in Cell Biology. Allen, R.D., Kamiya, N. (eds.). New York: Academic Press, 1964, pp. 599–622
Kamitsubo, E.: A "window technique" for detailed observation of characean cytoplasmic streaming. Exp. Cell Res. *74*, 613–616 (1972a)
Kamitsubo, E.: Motile protoplasmic fibrils in cells of the Characeae. Protoplasma *74*, 53–70 (1972b)
Kamiya, N.: Protoplasmic streaming. In: Protoplasmatologia Vol. VIII, 3a. Heilbrunn, L.V., Weber, F. (eds.). Wien: Springer, 1959
Kamiya, N.: Physics and Chemistry of protplasmic streaming. Annu. Rev. Plant Physiol. *11*, 323–340 (1960)
Kamiya, N.: Protoplasmic streaming. In: Encyclopedia of Plant Physiology. Ruhland, W. (ed.). Berlin-Heidelberg-New York: Springer, 1962, Vol. XVII, pt. 2, pp. 979–1035
Kamiya, N.: Introductory remarks. In: International Cell Biology 1976–1977. Brinkley, B.R., Porter, K. (eds.). New York: Rockefeller University Press, 1977, pp. 361–366
Kamiya, N., Kuroda, K.: Rotational protoplasmic streaming in *Nitella* and some physical

properties of the endoplasm. In: Proc. 4. Int. Congr. Rheology. New York: Wiley, 1965, pp. 157–171
Kato, T., Tonomura, Y.: Identification of myosin in *Nitella* flexilis. J. Biochem. *82*, 777–782 (1977)
Kavanau, J.L.: Structure and function in biological membranes. Holden-Day, 1963
Kersey, Y.M., Hepler, P.K., Palevitz, B.A., Wessels, N.K.: Polarity of actin filaments in characean algae. Proc. Natl. Acad. Sci. USA *73*, 165–167 (1976)
Kersey, Y.M., Wessels, N.K.: Localization of actin filaments in internodal cells of characean algae. J. Cell Biol. *68*, 264–275 (1976)
Keul, M.: Das Wirkungsspektrum der Photodinese in den Wurzelhaaren der Gerste (*Hordeum vulgare* L.). Z. Pflanzenphysiol. *79*, 40–52 (1976)
Kishimoto, U., Akabori, H.: Protoplasmic streaming of an internodal cell of *Nitella flexilis* and its correlation with electric stimuli. J. Gen. Physiol. *42*, 1167–1183 (1959)
Kowallik, W.: Light effects on carbohydrate and protein metabolism in algae. In: The Photobiology of Microorganisms. Halldal, P. (ed.). London: Wiley, 1970, pp. 165–181
Läuchli, A., Lüttge, U.: Untersuchung der Kinetik der Ionenaufnahme in das Cytoplasma von *Mnium*-Blattzellen mit Hilfe der Mikroautoradiographie und der Röntgen-Mikrosonde. Planta *83*, 80–98 (1968)
Lazar-Keul, G., Keul, M., Soran, V.: Die Wirkung von Oligomycin, ATP und ADP auf die Zirkulationsströmung in den Wurzelhaaren der Tomate (*Lycopersicum esculentum*) Biochem. Physiol. Pflanz. (BPP) *166*, 255–262 (1974)
Lazar-Keul, G., Vintila, R., Keul, M.: Die Wirkung der Chelatisierung auf die ATPase-Aktivität in Beziehung zur Protoplasmaströmung in den Wurzelhaaren der Gerste (*Hordeum vulgare* L.). Rev. Roum. Biol., Bot. *17*, 113–119 (1972)
Lomagin, A.G., Antropova, T.A.: Influence of light on the resistance of cellular protoplasmic streaming in plants to 2,4 dinitrophenol. Photochem. Photobiol. *2*, 527–533 (1963)
Lüttge, U., Pitman, M.G.: Transport and energy. In: Encyclopedia of Plant Physiology N.S. 2, A. Lüttge, U., Pitman, M. (ed.). Berlin-Heidelberg-New York: Springer, 1976, pp. 251–259
Maier, J.: Lichtinduzierte Bewegungsreaktionen der Chloroplasten: Zur Beziehung zwischen der Rotationsauslösung und der Abzentrifugierbarkeit bei *Elodea canadensis*. Zulassungsarbeit, Erlangen (1975)
Marchant, H.J.: Ultrastructure, development and cytoplasmic rotation of seta-bearing cells of *Coleochaete scutata*. J. Phycol. *13*, 28–36 (1977)
Morel, J.E., Pinset-Härström, J., Gingold, M.P.: Muscular contraction and cytoplasmic streaming: A new general hypothesis. J. Theor. Biol. *62*, 17–51 (1976)
Mustacich, R.V., Ware, B.R.: Observation of protoplasmic streaming by laser-light scattering. Phys. Rev. Lett. *33*, 617–620 (1974)
Mustacich, R.V., Ware, B.R.: A study of protoplasmic streaming in *Nitella* by laser Doppler spectroscopy. Biophys. J. *16*, 373–388 (1976)
Nagai, R., Kamiya, N.: Differential treatment of *Chara* cells with cytochalasin B with special reference to its effect on cytoplasmic streaming. Exp. Cell Res. *108*, 231–237 (1977)
Nuccitelli, R., Poo, M.M., Jaffe, L.: Relations between amoeboid movement and membrane-controlled electrical currents J. Gen. Physiol. *69*, 743–763 (1977)
O'Brien, T.P., McCully, M.E.: Cytoplasmic fibres associated with streaming and saltatory-particle movement in *Heracleum mantegazzianum*. Planta *94*, 91–94 (1970)
Oda, K., Linstead, P.J.: Changes in cell length during action potentials in *Chara*. J. Exp. Bot. *26*, 228–239 (1975)
Offer, G., Elliott, A.: Can a myosin molecule bind to two actin filaments? Nature (London) *271*, 325–329 (1978)
Palevitz, B.A.: Actin cables and cytoplasmic streaming in green plants. In: Cell Motility. Goldman, R., Pollard, T., Rosenbaum, J. (eds.). Cold Spring Harbor Laboratory, 1976, pp. 601–611
Palevitz, B.A., Ash, J., Hepler, P.: Actin in the green alga *Nitella*. Proc. Natl. Acad. Sci. USA *71*, 363–366 (1974)
Palevitz, B.A., Hepler, P.K.: Identification of actin in situ at the ectoplasm-endoplasm interface in *Nitella*. J. Cell Biol. *65*, 29–38 (1975)
Pfeiffer, H.H.: Veränderungen des Strömungsdichroismus an Protoplasmasträngen pflanzlicher Dauerzellen. Ber. Dtsch. Bot. Ges. *66*, 2–7 (1953)

Pickard, W.F.: The correlation between electrical behavior and cytoplasmic streaming in *Chara braunii*. Can. J. Bot. *47*, 1233–1240 (1969)
Pickard, W.F.: Further observations on cytoplasmic streaming in *Chara braunii*. Can. J. Bot. *50*, 703–711 (1972)
Pickard, W.F.: Hydrodynamic aspects of protoplasmic streaming in *Chara braunii*. Protoplasma *82*, 321–339 (1974)
Pollard, T.D., Ito, S.: Cytoplasmic filaments of *Amoeba proteus* I. The role of filaments in consistency changes and movement. J. Cell Biol. *46*, 267–289 (1970)
Pop, E., Soran, V., Lazar, G.: The effect of ATP upon rotational streaming. Physiol. Plant *20*, 617–623 (1967)
Raven, J.A.: Cyclic and non-cyclic photophosphorylation as energy sources for active K influx in *Hydrodictyon africanum*. J. Exp. Bot. *22*, 420–433 (1971)
Rebhun, L.J.: Structural aspects of saltatory particle movement. J. Gen. Physiol. *50*, 223–239 (1967)
Sattelle, D.B., Buchan, P.B.: Cytoplasmic streaming in *Chara corallina* studied by laser light scattering. J. Cell Sci. *22*, 633–643 (1976)
Seitz, K.: Das Wirkungsspektrum der Photodinese bei *Elodea canadensis*. Protoplasma *58*, 621–640 (1964)
Seitz, K.: Wirkungsspektren für die Starklichtbewegung der Chloroplasten, die Photodinese und die lichtabhängige Viskositätsänderung bei *Vallisneria spiralis*. Z. Pflanzenphysiol. *56*, 246–261 (1967)
Seitz, K.: Die Starklichtbewegung der Chloroplasten von *Vallisneria* in Abhängigkeit von Hemmstoffen der oxydativen Phosphorylierung. Z. Pflanzenphysiol. *63*, 401–407 (1970)
Seitz, K.: Die Ursache der Phototaxis der Chloroplasten: ein ATP-Gradient? Z. Pflanzenphysiol. *64*, 241–256 (1971)
Seitz, K.: Orientation in space: plants. In: Marine Ecology Vol. II, 2. Kinne, O. (ed.). London: Wiley, 1975, pp. 451–497
Seitz, K.: Light induced changes in the centrifugability of chloroplasts: different action spectra and different influence of inhibitors in the low and high intensity range. Z. Pflanzenphysiol., in preparation (1979)
Shimmen, T., Tazawa, M.: Control of membrane potential and excitability of *Chara* cells with ATP and Mg^{2+}. J. Membr. Biol. *37*, 167–192 (1977)
Spudich, J.A., Lin, S.: Cytochalasin B, its interaction with actin and myosin from muscle. Proc. Natl. Acad. Sci. USA *69*, 442–446 (1972)
Ståfelt, M.G.: Viscosität. In: Encyclopedia of Plant Physiology II, Ruhland, W. (ed.). Berlin-Heidelberg-New York: Springer, 1956, pp. 591–606
Szent-Györgyi, A.G.: Calcium regulation of muscle contraction. Biophys. J. *15*, 707–723 (1975)
Tageeva, S.V., Kazantsev, E.N.: Regularities of cytoplasm and chloroplast motion in the cells of *Elodea canadensis* JZV. AN SSR Ser. Biol. *6*, 885–895 (1962)
Tageeva, S.V., Kazantsev, E.N., Tairbekov, M.G., Korshunova, V.S.: Elements of the mechanism of cytoplasmic structure. Mobility in plant cells. Fiz. Rast. *12*, 854–865 (1965)
Tairbekov, M.R.: Interaction between the intensity of energetic processes and the rate of cytoplasm movement in the cells of *Nitella micronata*. Biofizika (Mosk.) *15*, 629–634 (1970)
Taylor, D.L.: Dynamics of cytoplasmic structure and contractility. In: International Cell Biology 1976–1977. Brinkley, B.R., Porter, K.R. (eds.). New York: The Rockefeller University Press, 1977, pp. 367–377
Tazawa, M.: Motive force of the cytoplasmic streameng in *Nitella*. Protoplasma *65*, 207–222 (1968)
Tazawa, M.: Cytoplasmic streaming and ionic regulation in *Nitella flexilis* having artificial cell saps of low ionic strength. Bot. Mag. Tokyo *88*, 219–229 (1975)
Tazawa, M., Kishimoto, U.: Studies on *Nitella* having artificial cell sap. II. Rate of cyclosis and electric potential. Plant Cell Physiol. *5*, 45–59 (1964)
Tazawa, M., Kishimoto, U.: Cessation of cytoplasmic streaming of *Chara* internodes during action potential. Plant Cell Physiol. *9*, 361–368 (1968)
Terborgh, J.: Potentiation of photosynthetic oxygen evolution in red light by small quantities of monochromatic blue light. Plant Physiol. *41*, 1401–1410 (1966)
Thaine, R.: Surface associations between particles and the endoplasmatic reticulum in protoplasmic streaming. New Phytol. *64*, 118–130 (1965)

Tonomura, Y., Oosawa, F.: Molecular mechanism of contraction. Annu. Rev. Biophys. Bioeng. *1*, 159–190 (1972)
Vintila, R., Fabian, A., Lazar-Keul, G.: Die Kontrolle der Protoplasmaströmung in den Wurzelhaaren der Gerste (*Hordeum vulgare*) durch ATP, SH-Gruppen und Ca^{2+}-Ionen. Rev. Roum. Biol. Bot. *18*, 287–296 (1973)
Virgin, H.J.: The relation between the viscosity of the cytoplasm, the plasma flow, and the motive force. Physiol. Plant. *2*, 157–163 (1949)
Virgin, H.J.: The effect of light on the protoplasmic viscosity. Physiol. Plant. *4*, 255–357 (1951)
Virgin, H.J.: An action spectrum for the light induced changes in the viscosity of plant protoplasm. Physiol. Plant. *5*, 575–582 (1952)
Virgin, H.J.: Some effects of light on chloroplast and plant protoplasm. In: Photophysiology, Vol. I. Giese, A.C. (ed.). New York: Academic Press, 1964, pp. 273–303
Wagner, G., Klein, K.: Differential effect of Calcium on chloroplast movement in *Mougeotia*. Photochem. Photobiol. *27*, 137–140 (1978)
Weber, F., Weber, G.: Die Temperaturabhängigkeit der Plasmaviskosität. Ber. Deutsch. Bot. Ges. *34*, 836–846 (1916)
Weisenseel, M.H., Ruppert, H.K.: Phytochrome and calcium ions are involved in light-induced membrane depolarization in *Nitella*. Planta *137*, 225–229 (1977)
Williams, N.S., Terborgh, J.: Protoplasmic streaming in red light potentiated by small quantities of blue light in *Acetabularia crenulata*. Plant Physiol. *46*, (1) (1970)
Williamson, R.E.: Actin in the alga *Chara corallina*. Nature (London) *248*, 801–802 (1974)
Williamson, R.E.: Cytoplasmic streaming in *Chara*: a cell model activated by ATP and inhibited by cytochalasin B. J. Cell Sci. *17*, 655–668 (1975)
Williamson, R.E.: Actin and motility in plant cells. In: Contractile Systems in Non-muscle Tissues. Perry, S.V., Margreth, A., Adelstein, R.S. (eds.). Amsterdam: North-Holland, 1976, pp. 91–101
Wilson, J.R., Schwabe, W.W.: Effects of photoperiod on the apparent viscosity of leaf cytoplasm in *Kalanchoe blossfeldiana*. I. A method for quantitative estimation and its sources of error. Annu. Bot. *28*, 647–668 (1964)
Zurzycki, J.: Studies on the centrifugation of chloroplasts in *Lemna trisulca*. Acta Soc. Bot. Pol. *29*, 385–393 (1960)
Zurzycki, J.: The action spectrum for the light dependent movements of chloroplasts in *Lemna trisulca*. Acta Soc. Bot. Pol. *31*, 489–538 (1962a)
Zurzycki, J.: The mechanism of the movements of plastids. In: Encyclopedia of Plant Physiology Ruhland, W. (ed.). Berlin-Heidelberg-New York: Springer, 1962b, Vol. XVII, pt. 2, pp. 940–978

2.5 Chloroplast and Nuclear Migration[1]

S. J. BRITZ

1. Introduction

Organelles are capable not only of random movements within the cell, but also of directed migrations between different arrangements in response to internal and external stimuli. These directed movements are important in cell development and function (e.g., chromosome movement, vesicle translocation during endo- and exocytosis, and the dispersion and aggregation of pigment granules in animal chromatophores). A variety of mechanisms have evolved to accomplish these tasks. The present review, although restricted in scope to the migrations of chloroplasts and nuclei, adopts a comparative approach where feasible.

The intracellular distribution of chloroplasts in many cell types varies according to light intensity and direction. These light-induced movements have been studied in detail and, accordingly, they receive the greatest attention in this review. The mechanism is considered in terms of a photoreceptor-effector system assumed to comprise a means of sensing light direction (presumably through a primary physical gradient of light absorption) and intensity, an actual movement system to change chloroplast distribution, and a transducing mechanism capable of regulating the movement system. Rhythmic chloroplast movements controlled by the biological clock (see also SWEENEY, Chap. 1.4, this vol.) are discussed separately.

Nuclear migration occurs in response to wounding and to unknown internal factors. The latter may be related to developmental gradients, since nuclear migration alters the plane and position of the cell plate and is an important feature of many types of differentiation.

Light-induced chloroplast movement was reviewed twice in the last edition of the *Encyclopedia* (HAUPT, 1959b; ZURZYCKI, 1962b) and more recently elsewhere (HAUPT, 1965, 1972, 1973a; HAUPT and SCHÖNBOHM, 1970; HAUPT and WEISENSEEL, 1976; MAYER, 1971; SCHÖNBOHM, 1972b; VIRGIN, 1964, 1968). Nuclear migration was also reviewed twice in the last edition of the *Encyclopedia* (BÜNNING, 1959; GIRBARDT, 1962) as well as elsewhere (GIRBARDT, 1968; HEPLER and PALEVITZ, 1974).

[1] CIW-DPB Publication No. 633.

Abbreviations: ATP, adenosine triphosphate; CB, cytochalasin B; DCMU, dichlorophenyl dimethylurea; EGTA, ethyleneglycol-*bis*-(β-aminoethyl ether)-N,N′-tetraacetic acid; FMN, flavin mononucleotide; PCMB, p-chloromercuribenzoic acid; PCMBS, p-chloromercuribenzene sulfonic acid; Pr and Pfr, the red light and far-red light absorbing forms, respectively, of phytochrome; red light, wavelengths approximately between 600 and 700 nm; far-red light, wavelengths approximately between 700 and 800 nm; E-vector, the electric vector of linearly polarized light.

The present review does not consider nondirected organelle movements such as the spontaneous rotation of chloroplasts (GEITLER, 1961; MARCHANT, 1977) and nuclei (GIRBARDT, 1968). Neither does it consider light effects that prompt chloroplast movement irrespective of light direction (SENN, 1908) or changes in chloroplast shape, either light-induced (e.g., ZURZYCKI, 1974) or rhythmic (SWIFT and TAYLOR, 1967). The emphasis on the role of light in influencing chloroplast movement should not obscure the fact that external factors other than light (e.g., wounding in *Selaginella;* MAYER, 1964) also affect chloroplast distribution. These factors, which will not be discussed in this review, have received scant attention since the 1908 monograph of SENN.

2. Light-Induced Chloroplast Migration

2.1 Types of Response

Some typical plants and their chloroplast arrangements in response to light are shown schematically in Figure 1. In unilateral, low-intensity ("weak") light, chloroplasts generally migrate to cell walls approximately normal to the direction of illumination (epistrophe) while in high intensity ("strong") light the chloroplasts redistribute to side walls (parastrophe). Chloroplast distribution may also randomize in the dark (apostrophe). Note that both morphological terms (epistrophe, face-to-profile movement) and physiological terms (e.g., weak-light position, strong-light response) are used to describe chloroplast distribution and migration.

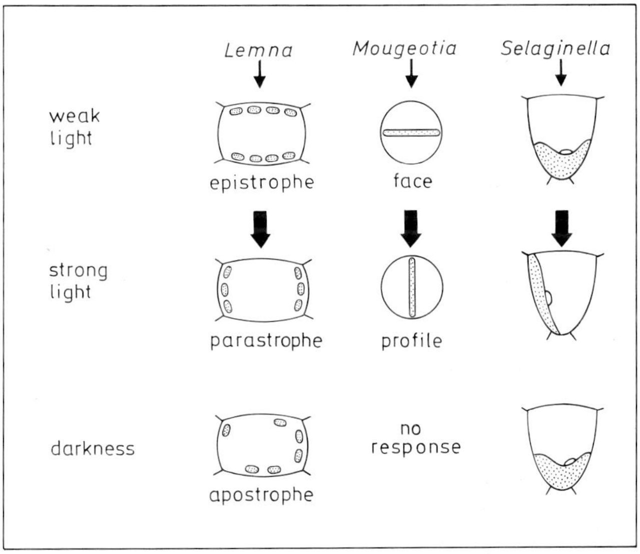

Fig. 1. Schematic representation of light-induced chloroplast arrangements in three commonly studied species. The direction of incident light is indicated by *arrows*, a *thin arrow* in the case of low-intensity light and a *thick arrow* in the case of high-intensity. *Mougeotia* chloroplasts in darkness remain in the position established by the previous light treatment. (After WEISENSEEL, 1968a)

The time required to accomplish the various changes in orientation varies from minutes to hours. The different orientations appear related to gradients of light intensity within the cells. The incident light flux is less at the sides of the cell because (1) the cross-sectional area presented is less, (2) attenuation of light by cell contents including the chloroplasts causes shading, and (3) light may be focused away from the side walls by the lens effect (see Fig. 3B). The adaptive significance of chloroplast migration is discussed below.

Light-induced chloroplast migration is widely distributed within most, if not all, plant phyla. Nonetheless, a comprehensive catalog of migration responses is still not possible, since most workers have concentrated on physiological studies in a few species of microalgae, mosses, ferns, and higher aquatic plants where direct determination of chloroplast position by light microscopy is feasible. Studies in morphologically more complex plants have been hampered by the lack of suitable investigative techniques. The recent use of indirect methods (changes in light transmission, Sect. 2.2) has identified several genera of higher plants with marked light-induced chloroplast orientation changes. Among them are *Ajuga* (ŁEŁATKO, 1970), *Begonia* (BIEBL, 1954, 1955; INOUE and SHIBATA, 1973, 1974), *Setaria* (INOUE and SHIBATA, 1973, 1974) and *Sambucus* (BIEBL, 1955; ŻURZYCKI and ŁEŁATKO, 1969). Among marine algae studied, light-induced chloroplast migration seems most widely distributed in representatives of the Phaeophyceae (NULTSCH and PFAU, 1978).

2.2 The Use of Light Transmission Changes to Study Migration

Compartmentalization of pigment in organelles causes light absorption to be less than if the same amount of pigment were dispersed homogeneously (the sieve effect; RABINOWITCH, 1951). Changes in chloroplast distribution to minimize (e.g., parastrophe) or maximize (e.g., epistrophe) the surface area presented to light will decrease or increase, respectively, the fraction of incident light absorbed by the cell. The resultant light transmission changes provide a useful indirect measure of chloroplast movement with a continuous record valuable for kinetic studies or for long-term processes such as circadian rhythms (see Fig. 7 and Sect. 3). However, if transmission is averaged over many cells, then information on the statistical distribution of response is lost.

In general, it is important to use monochromatic blue or red measuring light for studies of chloroplast migration, since the transmission changes generated by the sieve effect are largest in the blue and red spectral regions where the chloroplast pigments absorb most strongly.

Serious artifacts may result if a white measuring light is used in conjunction with a green light-sensitive photodetector. The high transmission of leaves for green light means the detector will sense mainly green light. However, the sieve effect for green light is very small, so net transmission changes may be influenced more by changes in light scattering or pigment content (BRITZ, 1975a; 1976; ZURZYCKI, 1961). This explanation may account for earlier unsatisfactory results (e.g., BIEBL, 1955). On the other hand, PFAU et al. (1974) successfully measured chloroplast migration by white light transmission changes in *Dictyota* probably because their use of a brown alga (strong light absorption in the blue-green) was complemented by a photodetector with mainly blue-green light sensitivity.

Recent studies of transmission changes have been able to minimize artifacts with measurements at wavelengths of absorption maxima and with opal glass, end-on photomultipliers and/or dual wavelength referencing techniques (BRITZ, 1976; BRITZ and BRIGGS, 1976; BRITZ et al., 1976; INOUE and SHIBATA, 1973, 1974; LEŁATKO, 1970; ZURZYCKI, 1961). In these cases the correlation between chloroplast position and light transmission has been excellent by both quantitative (BRITZ, 1976; ZURZYCKI, 1961) and qualitative (BRITZ and BRIGGS, 1976; INOUE and SHIBATA, 1973; LEŁATKO, 1970; PFAU et al., 1974) criteria. Note that unusual chloroplast orientations (ZURZYCKI, 1961) or chloroplast shape changes (BRITZ, 1976) can yield anomalous results if only transmission measurements are considered. Thus, transmission changes augment but do not replace direct visual observation.

2.3 Intensity Dependence

In *Mougeotia* (Fig. 1) the single, plate-like chloroplast turns to either face or profile position depending on intensity. In cells with many chloroplasts (e.g., *Lemna;* Fig. 1) there is a graded response. The minimum intensity change required to produce a change in chloroplast orientation is generally unknown. Measurable changes in light transmission occurred in *Dictyota* in response to increases or decreases of actinic light intensity as small as 1% to 6% (PFAU, 1977). However, it cannot be certain that chloroplast migration was actually responsible for the very small optical changes.

The intensity ranges corresponding to "weak" and "strong" light are too variable to specify. Because of the selective absorption by the photoreceptor (see Sect. 2.5.1), the effective intensities of white light may vary with the color temperature of the light source. Extrapolation of data from the recent literature to the environment is difficult because most studies have employed special techniques to sensitize plant material, as well as broad band colored or monochromatic actinic light sources.

With appropriate precautions, ZURZYCKI (1955, 1962a) found that for white light the transition between weak-light and strong-light chloroplast distributions agreed well with the approximate light saturation of photosynthesis. This correspondence suggests that under natural conditions the sensitivity of the photoreceptor for chloroplast movement can adapt according to the photosynthetic performance of the organism.

Among several terrestrial species grown outdoors, the optimal weak-light response occurs in the range of 0.08 to 1.8 W m^{-2} of blue light depending on species (INOUE and SHIBATA, 1974). The transition between the weak-light and strong-light responses occurred in the range between 1.0 and 7.3 W m^{-2}. Chloroplast migration would be expected during the normal course of the day, since the blue light component of full sunlight (150 W m^{-2}; measured through the actinic light filter) was much brighter than the intensity for the weak-to-strong light transition.

2.4 Adaptive Significance

The relation between chloroplast distribution and the sieve effect and the correlation between chloroplast position and photosynthesis (ZURZYCKI, 1955, 1962a) suggest that the displacement helps regulate light absorption in the cell. Since photosynthesis would presumably be saturated under "strong" light conditions, extra absorbed light serves no purpose and may damage the cell. A change in chloroplast arrangement which decreased absorption would serve an adaptive function. However, even in favorable cases (a one cell-layered, aquatic plant with little light scattering), the actual changes in light absorption upon chloroplast migration may be only as large as 20% at pigment absorption maxima (ZURZYCKI, 1961). In terrestrial plants with thick leaves and air spaces, light scattering will reduce or even cancel the sieve effect (ŁEŁATKO, 1970; INOUE and SHIBATA, 1974). The resultant small absorption changes are of doubtful adaptive significance. Short-term movements of algal thalli or of leaves will also obviate any value from chloroplast orientation.

Some information relating to the ecology of chloroplast migration is available. Among three intertidal species of *Fucus*, higher position in the intertidal zone (and longer exposure at low tide) correlated positively with increased sensitivity of light-induced chloroplast migration (NULTSCH and PFAU, 1978).

ZURZYCKI (1955) and LECHOWSKI (1974) compared changes in photosynthetic rates with changes in chloroplast distribution and light absorption. At intensities limiting for photosynthesis, large increases in photosynthesis were observed as chloroplasts migrated from profile or parastrophe to face or epistrophe positions. However, the increases in photosynthetic rates are much larger than can be explained by absorption increases.

In *Ajuga*, a land plant, the increased rates did not result from changes in stomatal aperture (LECHOWSKI, 1974). In *Mougeotia*, a uniseriate, filamentous alga, it was unlikely that changes in substrate or product diffusion were important (ZURZYCKI, 1955). In *Funaria* (where the dark position is quasi-parastrophe rather than random) and in *Mougeotia* (where the profile position can be stable in the dark as the face position), similar increases in photosynthesis were observed on weak-light-induced parastrophe-to-epistrophe or profile-to-face movement following either high-intensity light or darkness (ZURZYCKI, 1955). Thus, a specific effect by pre-treatment with high intensity light seems unlikely.

Conversely, plants which fail to exhibit light-induced chloroplast migration also lack long-term increases in photosynthesis upon transition from strong- to weak-light regimes.

In conclusion, it can only be said that light absorbed by profile position chloroplasts is utilized less efficiently than light absorbed by face position chloroplasts. Further insight might come from techniques which investigate the fate of absorbed light energy (e.g., in vivo fluorescence). Certainly the wide distribution of light-induced chloroplast migration requires an explanation, unless the phenomenon is a living "fossil."

2.5 Photoperception

2.5.1 Photoreceptor Pigments

Depending on species and response (weak-light or strong-light), either a blue and UV light-sensitive photoreceptor and/or the photosynthetic pigments and/or phytochrome may be involved in chloroplast migration.

2.5.1.1 Blue Light Photoreceptors

Most light-induced chloroplast migration responses are mediated by blue light. An action spectrum for the induction of the weak light response in *Funaria* (ZURZYCKI, 1967a) is shown in Figure 2a. Action maxima occur in the far UV, in the near UV from about 360 to 380 nm, and in the blue at about 454 nm. The blue band exhibits distinct fine structure including a shoulder at 480 nm. These characteristics are in best agreement with a flavoprotein as the photoreceptor (for this argument and others, see SONG and MOORE, 1974). Similar action spectra have been obtained for the blue and (where studied) UV activity in *Lemna* (ZURZYCKI, 1962a), *Vallisneria* (SEITZ, 1967a), *Vaucheria* (FISCHER-ARNOLD, 1963) and *Setaria* (INOUE and SHIBATA, 1973). Where tested, the weak- and strong-light responses had almost identical spectral dependencies. The strong-light response in *Mesotaenium* has an action maximum at 450 nm (GÄRTNER, 1970b). In *Ajuga* (LECHOWSKI, 1973) and in *Selaginella* (MAYER, 1964) the blue light action maxima are at about 470 to 480 nm. The strong-light response in *Mougeotia* is maximal at 475 nm (SCHÖNBOHM, 1963). It is not clear whether shading or the involvement of different pigment systems are responsible for the differences in the action spectra. Chloroplast movements in *Hormidium* (SCHOLZ, 1976b) and in *Avena, Begonia, Spirodela* and *Zea* (INOUE and SHIBATA, 1973) are all blue light-sensitive.

Various experiments have been attempted to test the role of flavins as blue and UV light photoreceptor pigments for chloroplast migration. For example, dark incubation of *Hormidium* decreased both FMN content and the sensitivity of the blue light response (SCHOLZ, 1976b). Added FMN partially restored light sensitivity. However, dye experiments do not establish identity between the dye and an endogenous photoreceptor (DRAWERT, 1968; ERLANGER, 1976; TAYLOR and KOSHLAND, 1975; POP et al., 1971).

KI quenches the triplet excited state of flavins (e.g., SONG and METZLER, 1967). High concentrations of iodide salts ($\geqq 0.1$ M) inhibited blue light-induced but not red light-induced chloroplast migration in *Selaginella* (MAYER, 1966) and *Vallisneria* (SEITZ, 1967b, 1970a) and, under certain conditions, also in *Hormidium* (SCHOLZ, 1976b). The authors concluded that iodide affected the blue light photoreceptor rather than the movement mechanism, and that iodide inhibition was consistent with a flavin as the photoreceptor pigment. These experiments must be interpreted cautiously, since the iodide anion is highly reactive and may have various effects. In *Mougeotia,* for example, KI inhibited the weak-light response mediated by phytochrome and this inhibition could also be achieved if KI was given after the light treatment (SCHÖNBOHM, 1967b, 1969a).

A direct effect of iodide on the movement mechanism is possible. KI depolymerizes muscle F-actin (GUBA, 1950), although concentrations higher than those effective against chloroplast migration are required. Cytoplasmic actins, which may be involved in chloroplast movement (see Sect. 2.8) may be more labile than corresponding muscle actins and perhaps more sensitive to salt depolymerization. Iodide also uncouples oxidative phosphorylation (MIDDLEBROOK and SZENT-GYORGYI, 1955). SEITZ (1970a) suggested that the lack of iodide inhibition in red light could be attributed to the occurrence of photophosphorylation. However, it is not clear why photophosphorylation by blue light would not suffice to permit chloroplast movement.

No effect on phototropic bending in *Avena* coleoptiles was found for 0.01 M KI, a concentration that inhibited photooxidation of naphthylene acetic acid (a presumed flavin-mediated triplet state reaction) in coleoptiles or coleoptile extracts (ZENK, 1967). Sensitivity to KI does not prove triplet state involvement, since iodide also quenches singlet excited states (SONG

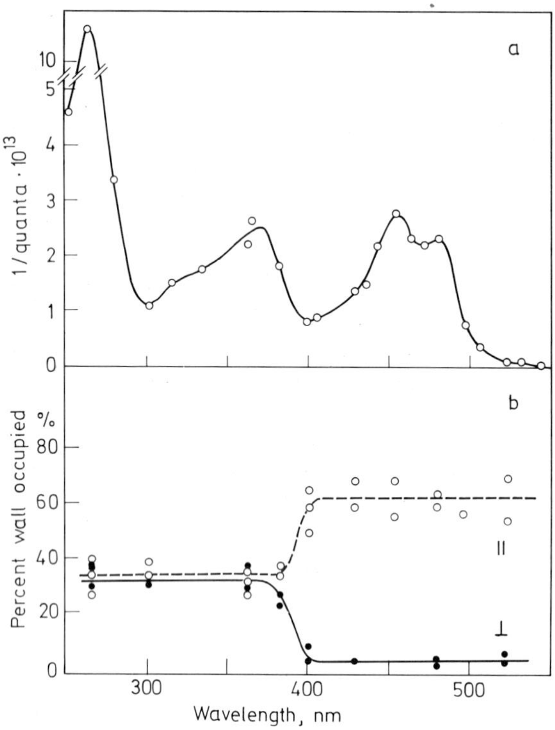

Fig. 2a and b. Spectral dependence of chloroplast migration in *Funaria*. **a** Action spectrum. The *ordinate* shows the reciprocal number of quanta required to induce 50% of the chloroplasts to assemble in epistrophe (weak-light arrangement). **b** Action dichroism. The *ordinate* shows the percentage of chloroplasts being found at the shorter anticlinal cell wall in weak-light conditions if this wall is parallel (||, ○) or perpendicular (⊥, ●) to the E-vector of low-intensity light. (After ZURZYCKI, 1967a and b)

and METZLER, 1967), and at concentrations still lower than those effective against chloroplast movement. Of course, in experiments with chloroplast migration it was not known what amount of KI penetrated the cell.

It must also be considered that more than one blue light-absorbing pigment system can mediate a single type of response in the same organism. For example, on the basis of differential KI sensitivity, FMN enhancement, and other criteria, it is likely that in blue light two separate photoreceptors are active in the weak-light response in *Hormidium* (SCHOLZ, 1976b). Moreover, the action spectra for chloroplast migration in various organisms can be segregated into two groups; in one, there is a maximum at about 450 nm and a shoulder at about 470 to 480 nm, while in the second the wavelength positions of maximum and shoulder are reversed. A detailed comparison of action spectra would be valuable but would require more resolution and uniform experimental conditions. However, it is noteworthy that the action spectrum for a light effect perhaps related to chloroplast migration (light enhancement of chloroplast centrifugation in *Elodea*, see Sect. 2.8.2) shows two separate and clearly resolved peaks, one at about 450 nm and the other at 470 to 480 nm (VIRGIN, 1952).

That one organism can possess different and independent blue light photoreceptors is indicated in *Neurospora* by the different action spectra for carotenogenesis (DEFABO et al., 1976) and suppression of circadian banding (SARGENT and BRIGGS, 1967). Separate photoreceptors for positive and negative phototaxis in *Euglena* have also been proposed (MIKOLAJCZYCK and DIEHN, 1975).

2.5.1.2 Photosynthetic Pigments

Red light absorbed by the photosynthetic pigments can cause chloroplast migration in at least three plants: the strong-light responses in *Selaginella* (MAYER, 1964) and in *Vallisneria* (SEITZ, 1967a) and, under certain conditions, the weak-light response in *Hormidium* (SCHOLZ, 1976b). The involvement of photosynthesis in the red light effect is supported by the DCMU sensitivity of the red response in *Vallisneria* (SEITZ, 1967b) and *Hormidium* (SCHOLZ, 1976b). The red and blue light effects clearly involve different photoreceptors, since blue light-induced changes in chloroplast arrangement are DCMU insensitive. In all cases, red light was much less effective than blue on a quantum basis (50-fold less in *Selaginella* and 10-fold less in *Vallisneria*). Thus, under natural light conditions chloroplast arrangement will be controlled mainly by the blue light photoreceptor.

In *Vallisneria* a small additional contribution of the photosynthetic pigments in the blue was demonstrated with enhancement maxima at about 430 and 485 nm upon simultaneous irradiation with UV light of constant wavelength and with blue light of variable wavelength (SEITZ, 1967b). The blue enhancement maxima were eliminated in the presence of DCMU.

2.5.1.3 Phytochrome

In two related green algae, *Mougeotia* (HAUPT, 1959a) and *Mesotaenium* (HAUPT and THIELE, 1961), the weak-light response is mediated by red light acting through the photochromic protein phytochrome. As is characteristic of phytochrome involvement, far-red irradiation reverses red light-induced profile-to-face movement.

The action maximum for the weak-light response in *Mougeotia* is shifted to longer wavelengths (about 680 nm) than is usual for the Pr absorption maximum (usually about 660 to 665 nm). However, *Mougeotia* phytochrome is spectrally normal, as judged by the action spectrum for another type of phytochrome response (HAUPT, 1968a). The reason for the shift lies in the orientation of the photoreceptor (see Sect. 2.5.3). *Mesotaenium,* however, may have a spectrally unusual form of phytochrome, since the Pr absorption maximum in partially purified extracts was found to be at 650 nm (TAYLOR and BONNER, 1967). An action maximum near 650 nm has been reported for a phytochrome effect in *Mesotaenium* (GÄRTNER, 1970b).

Blue light does induce a weak-light response in *Mougeotia,* but much less effectively than red light. Although Pr absorbs in the blue, HAUPT (1971), by a number of criteria, was unable to explain blue light action on the basis of phytochrome. Since a blue light photoreceptor is obviously active in the strong-light response, the possibility of a blue light photoreceptor for the weak-light response in addition to phytochrome cannot be excluded. Some type of second photoreceptor was already suggested on the basis of UV light stabilization of red light induction against far-red light reversal (HAUPT and SCHÖNFELD, 1963). Separate photoreceptors are also indicated by the different mechanisms for blue light-mediated and phytochrome-mediated phototropisms of *Mougeotia* filaments (NEUSCHELER-WIRTH, 1970). However, HARTMANN and COHNEN UNSER (1973) reexamined the data of HAUPT (1971) and concluded that they were consistent with phytochrome involvement.

While both profile and face positions in *Mougeotia* are relatively stable in the dark, profile position chloroplasts in *Mesotaenium* return partly to face position in the dark in up to 40% of the cells (DORSCHEID, 1969b). Blue or red light, acting independently of phytochrome, enhances this partial transition (DORSCHEID and WARTENBERG, 1966; DORSCHEID, 1969a; GÄRTNER, 1970a). In addition to a different spectral sensitivity, the enhanced partial reaction has a different temperature dependence from the actual phytochrome-mediated weak-light response (DORSCHEID, 1969b). GÄRTNER (1970a) suggested that endogenous physical factors may force the chloroplast from profile position in the absence of an orienting gradient established by phytochrome prior to darkening. The separate action of blue and red light would help to level such a gradient.

2.5.2 Photoreceptor Localization

The chloroplast is the obvious choice for the site of the photoreceptor in the cases where the migration response is mediated by the photosynthetic pigments. However, that is not true of the blue and UV light and phytochrome photoreceptors. Microbeam irradiation of *Mougeotia* (BOCK and HAUPT, 1961) and *Selaginella* (MAYER, 1964) cytoplasm with red and blue light, respectively, stimulates a weak-light chloroplast migration even if the chloroplast itself is not illuminated. Scattered light which might be perceived by the chloroplast is apparently not important, since in both cases neighboring cells close to the microbeams were not affected. Localization of photoreceptors in the cytoplasm was also shown in *Vaucheria* (FISCHER-ARNOLD, 1963).

In linearly polarized light, the response in almost all organisms examined depends on the angle between the electric vector of the light and some structure in the cell (e.g., the cell wall; ZURZYCKI and LEŁATKO, 1969). This action dichroism, which is detailed in Sect. 2.5.3 is best explained by a nonrandom orientation of the photoreceptor pigment transition moment. This could occur if the pigment were bound to some regular and stable cell component such as the plasma or tonoplast membranes. In contrast, an action dichroism is not observed for the red light-induced chloroplast migration in *Vallisneria* (SEITZ, 1967a) and *Selaginella* (MAYER, 1964) where the photoreceptors are most likely localized in the chloroplast.

Remarkably, the action dichroism is unaltered in plasmolysed cells of *Funaria* (ZURZYCKI, 1968) and *Lemna* (ZURZYCKI, 1969). Ultrastructural studies of *Elodea* cells plasmolysed in sucrose show the plasma membrane to be strongly wrinkled (SITTE, 1963). If these observations are not the result of artifact, and if wrinkling also occurs in plasmolysed *Lemna* and *Funaria* cells, then it seems unlikely that the photoreceptors for chloroplast migration in these plants are in the plasma membrane. Wrinkling should destroy the effects of nonrandom transition moment orientation. Experimental modification of the action dichroism is possible by other means and offers, in correlation with ultrastructural studies, a promising approach to photoreceptor localization.

2.5.3 Determination of Light Direction

Information on light direction is provided by the attenuation of light as it passes through the cell (shading mechanism, Fig. 3a) and by the focusing of light to the back of the cell (lens mechanism, Fig. 3b). Note that these mechanisms establish gradients of light intensity within the cell. In the cases where the photoreceptors are localized in the chloroplasts (MAYER, 1964; SEITZ, 1967a), migration could

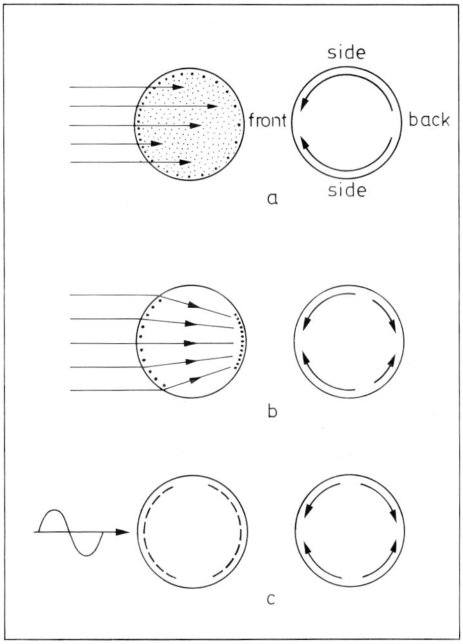

Fig. 3a–c. Mechanisms for establishing light intensity and absorption gradients in cross sections of idealized cylindrical or spherical cells. The *large dots* or *heavy dashes* around the periphery of cells on the *left* represent the distribution of "activated" photoreceptors in the cytoplasm while the *arrows* in the model cells on the *right* show the gradients established by light. The *arrows* point to regions of expected higher absorption rates while the *size of the arrows* is related to the steepness of the gradient. **a** Shading mechanism. The *stipling* represents a screening substance which attenuates the transmitted light. **b** Lens mechanism. A higher refractive index for the cell contents causes incident light to be focused to the back of the cell. **c** Pigment dichroism. Photoreceptor molecules with transition moments oriented parallel to the cell surface and in the plane of the page *(heavy dashes)* will absorb maximally at front and back if presented with linearly polarized light with its electric vector vibrating as shown in the plane of the page (sine wave). Other orientations of E-vectors and of transition moments parallel to the cell surface will not contribute to gradient formation. (After HAUPT, 1972c)

in theory proceed like a type of phototaxis with the chloroplasts sensing light intensity gradients. The blue and UV and the phytochrome photoreceptors for migration are localized in the cytoplasm apart from the chloroplasts. In these cases it is presumed that a gradient of light absorption by the cytoplasmic photoreceptors controls chloroplast orientation. The shading and lens mechanisms will still be important, since the gradient of absorption will depend on the localized light intensities.

However, the existence of an action dichroism for chloroplast migration indicates that a third mechanism, pigment dichroism (Fig. 3c) must also be considered as a means of sensing direction. If the photoreceptor pigment transition moments have a random orientation parallel to the cell surface, then more light will be absorbed at the front and back of the cell than at the sides. Note that this mechanism does not rely on a gradient of light intensity through the cell.

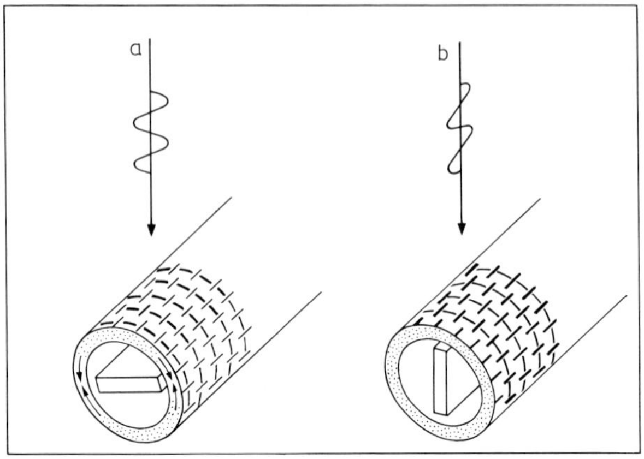

Fig. 4a and b. Postulated pattern for the absorption of linearly polarized light by the Pr form of phytochrome in *Mougeotia*. The *dashes* represent the idealized vector representations of the transition moments. The *heavier dashes* indicate orientations more likely to absorb light depending on the E-vector orientation of the incident light *(arrow with sine wave above each cell)*. The resultant absorption gradients are shown by *stipling* around the cell periphery. A *higher density of stipling* means more Pfr will be formed. **a** E-vector normal to the long axis of the cell. Light is absorbed preferentially top and bottom and results in profile-to-face chloroplast movement such that the chloroplast edges "avoid" high Pfr levels. **b** E-vector parallel to the long axis of the cell. The possibility for light absorption is more or less even around the cell, hence no gradient of Pfr forms and the chloroplast remains in its original position. (After HAUPT and SCHÖNBOHM, 1970).

In *Mougeotia,* the experimental data are consistent with the orientation of the Pr transition moment parallel to the surface of the cell (HAUPT, 1960; HAUPT and BOCK, 1962). Thus, linearly polarized red light with an E-vector normal to the long axis of the cell will be absorbed best at the front and back of the cell, the regions adjacent to the edges of the plate-like, profile position chloroplast (Fig. 4a). The higher Pfr levels established at the front and back (top and bottom in figure) somehow affect the movement mechanism and cause the chloroplast edges to move to the sides of the cell where the Pfr level is lower. On the other hand, red light linearly polarized parallel to the long axis of the cell will be absorbed more or less evenly around the cell periphery, no gradient of Pfr phototransformation will occur (Fig. 4b), and therefore no movement of the chloroplast is induced (HAUPT, 1973a). This dichroic mechanism is also operative in *Mesotaenium* (HAUPT and THIELE, 1961).

Experiments with polarized red and far-red light microbeams verified that the transition moment of Pr is parallel to the cell surface and that furthermore the transition moment for Pfr is perpendicular to the cell surface (HAUPT, 1970a; HAUPT et al., 1969). Based on the transition moment orientations and the assumption that *Mougeotia* phytochrome is spectrally "normal", HAUPT (1968a) calculated phototransformation dose response curves for the side and front of a model *Mougeotia* cell. Certain anomalies in the weak-light response could then be explained if it was further assumed that the response depended not on the total Pfr created front and back but rather on the front-to-side and back-to-side gradients in Pfr level. For example, the above-mentioned shift of the action maximum to 680 nm

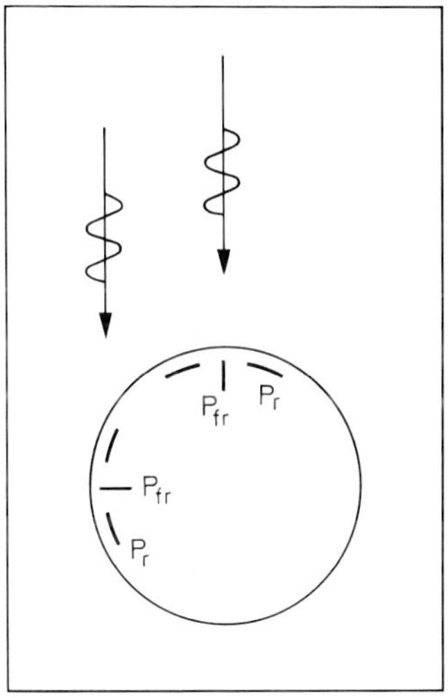

Fig. 5. Idealized Pr and Pfr transition moments at the top and sides of a *Mougeotia* cell (cross section) irradiated with unilateral, linearly polarized light with an E-vector normal to the long axis of the cell (and in the plane of the page, *single-headed arrows with sine wave*). At the top of the cell the Pr molecules are better oriented to absorb light, but at the sides Pfr is better oriented. (After HAUPT, 1972c)

(Sect. 2.5.1.3) could be predicted by the model. The empirical and theoretical results are rationalized if one considers the behavior of Pfr at the sides of the cell. In light linearly polarized perpendicular to the long axis, the orientation of Pfr transition moments, i.e., perpendicular to the cell surface, would be most favorable for light absorption at the sides (Fig. 5). The resulting back conversion to Pr, of course, is driven more effectively by 680 nm than by 660 nm light. Meanwhile at the front and back of the cell the orientation of the Pfr transition moments will be unfavorable for light absorption at both wavelengths. Hence, little back conversion will occur. Thus, 680 nm light will be more effective than 660 nm light at creating a steep gradient.

Continued investigation has suggested a number of modifications and additions to the Pfr gradient hypothesis.

1. A minimum overall level of Pfr as well as a steep gradient is important for a successful induction (HAUPT, 1968b).
2. The requirement for the steepness of the gradient may change through the course of a day (FLOHRS and HAUPT, 1971). Newly formed portions of synchronously dividing *Mougeotia* were less sensitive to red light induction (NEUSCHELER-WIRTH, 1971).
3. Pfr may transform into a less active form before its reversion back to Pr (FISCHER, 1963; HAUPT, 1966; HELLER, 1976; GRUM-HELLER, 1977).
4. Although the gradient can be established by a single pulse as short as 10^{-3} s (HAUPT, 1959a), the irradiation may have more than one effect important for the induction of chloroplast migration. Two pulses of millisecond duration but of subsaturating intensity are most effective if given with at least a 10 ms interval between them (HAUPT and BRETZ, 1976). The first pulse may be required to initiate time or intensity-dependent processes which facilitate the effect of the second pulse.
5. The physical saturation of the gradient of phytochrome phototransformation corresponds to the physiological saturation of the light-induced chloroplast migration (HAUPT, 1970b).
6. Microbeams which are much smaller than the diameter of the cell and which therefore do not transform phytochrome at the sides of the cell are able to establish an effective gradient without an enhancement by the use of polarized light (HAUPT and TRUMP, 1975).

The dichroic mechanism is also consistent with the organization of the blue light photoreceptor. In *Funaria* (ZURZYCKI, 1967b), under linearly polarized high-intensity blue light, the chloroplasts avoid all cell walls approximately parallel to the electric vector of the actinic light (not only front and back walls as is usual with unpolarized light, but also side walls). With weak light the opposite situation ensues and chloroplasts are attracted to walls parallel to the electric vector of the light, again including the side walls. This behavior is graphed in Figure 2b. Similar action dichroisms have been observed for blue light in *Selaginella* (MAYER, 1964) and in *Vallisneria* (SEITZ, 1976a).

The dichroic mechanism is not exclusive of other mechanisms for sensing light direction. GABRYS-MIZERA (1976) demonstrated with a *Funaria* model cell system that calculated photoreceptor gradients most comparable to actual orientation responses were obtained with a combination of dichroic and lens mechanisms and attenuation of light by the chloroplasts. The shading or lens mechanisms may also be used for the UV light response in *Funaria* where an action dichroism is lacking (Fig. 2b; ZURZYCKI, 1967b).

The absence of an action dichroism in the UV region (ZURZYCKI, 1967b) or the existence of different angles for the dichroism between blue and UV regions (SCHÖNBOHM, 1971b) is consistent with a calculated small angle between the blue and UV transition moments of a flavin (KURTIN and SONG, 1968).

The weak-light response of *Hormidium* combines certain aspects of dichroic, lens, and shading mechanisms (SCHOLZ, 1976a, b). The contribution of each of these mechanisms depends on light intensity, light quality, and the refractive index of the medium.

2.5.4 Light Intensity Effects

The orientation of chloroplasts in strong light is usually exactly opposite that in weak light. In the absence of more information, a simplistic interpretation is that an intracellular gradient established by unilateral weak light has reversed its polarity in strong light, or that under strong light conditions the movement mechanism responds differently to the original gradient. There are several ways in which these effects could occur (HAUPT, 1973b).

1. The photoreceptors might be inactivated above a certain intensity. A larger number of activated photoreceptors would then exist on the shaded side and the gradient would reverse in polarity.

2. The actual distribution of chloroplasts is presumed to be controlled by a secondary gradient of products from the primary physical gradient of light absorption. If, at some concentration, the effects of the secondary gradient on the movement mechanism are optimal, then at supraoptimal concentration an effective reversal of polarity would result.

3. There may be separate photoreceptors for the strong- and weak-light responses with different dose dependencies. A gradient formed by the weak-light photoreceptor would level out above saturation intensity. At this point the strong-light photoreceptor would establish a gradient with the same polarity, but with opposite effects on the movement mechanism.

4. One photoreceptor system may establish the gradient at all intensities, but above a certain intensity a tonic photoreceptor system may change the nature of the secondary products or the response of the movement mechanism to those products. Tonic photoreceptor systems separate from the direction-sensing system have been demonstrated for phototaxis in *Platymonas* (HALLDAL, 1960).

With regard to points (3) and (4) above, the two photoreceptor systems may or may not have the same spectral dependence. In the case of the fourth system, the spectral dependence of the strong light response would depend on driving both photoreceptors optimally.

Of the four mechanisms discussed above, only the tonic photoreceptor hypothesis has experimental support at present. In *Mougeotia* the strong light response utilizes phytochrome phototransformation to sense light direction, while an auxiliary blue light-sensitive tonic system senses intensity. Simultaneous irradiation with separate beams of blue and red light situated at various angles to one another showed that chloroplasts always oriented with respect to the direction of red light. The sign of response (whether face or profile) was determined by the intensity of blue light, irrespective of the angle between the blue and red light beams (SCHÖNBOHM, 1966, 1967a). Blue light alone was most effective if polarized with the E-vector at an angle of 45° to the cell axis, whereas in combination with a red light beam (to set the phytochrome gradient) blue light was most effective if polarized with an E-vector parallel to the long axis of the cell (SCHÖNBOHM, 1968). The 45° angle obviously represents a compromise between Pfr gradient formation and excitation of the tonic system.

The interaction of phytochrome and tonic receptor should also be evident in the spectral dependence of the strong-light response. Thus, the absence of the strong-light response in the red (where phytochrome absorbs well) implies the tonic photoreceptor does not absorb significantly in the red. Since phytochrome absorption is fairly featureless in the blue, it is not surprising that the action spectrum for the strong-light response represents mainly the absorption spectrum of the tonic photoreceptor system (SCHÖNBOHM, 1971b). A wavelength-dependent action dichroism for the tonic photoreceptor in the blue does indicate that several photoreceptor pigments (see Sect. 2.5.1.1) may be involved (SCHÖNBOHM, 1971b). It is unlikely that the tonic system is localized in the chloroplast (SCHÖNBOHM, 1971a).

The fortuitous involvement of clearly different photoreceptors in the direction and intensity sensing systems of *Mougeotia* (and also in *Mesotaenium*; GÄRTNER, 1970a, b) facilitated the discovery of tonic photoreceptor systems. A tonic photoreceptor might also be involved in the blue and UV light-sensitive chloroplast migration. Although in this case the action spectra for the strong- and weak-light responses are the same, that fact could be explained if the tonic system had the same absorption spectrum as the direction-sensing photoreceptor or if the tonic photoreceptor's absorption spectrum were relatively featureless in the blue. It seems doubtful that chloroplast migration action spectra have been performed with sufficient resolution to distinguish what might be subtle differences between the strong- and weak-light responses.

Enhancement experiments could be used to investigate tonic light effects in the strong-light response. In *Vaucheria*, subthreshold doses of blue light become effective if combined with red irradiations of about 100-fold greater energy (FISCHER-ARNOLD, 1963). In *Lemna* 0.1 W m^{-2} of blue light induces the weak-light response, but, if given together with 80 W m^{-2} of red light, a strong-light response is obtained (ZURZYCKI, 1964). The action spectrum for the red light enhancement in *Lemna* indicates the involvement of the photosynthetic pigments. It is unlikely that the photosynthetic pigments are involved as the normal tonic system for *Lemna* in white light, since the red light intensity required for an enhancement was more than an order of magnitude greater than the blue light intensity required to elicit a strong-light response.

2.6 Energetics

Inhibitor experiments show that metabolic energy from oxidative phosphorylation or photophosphorylation is used for light-induced chloroplast migration in *Mougeotia* (FETZER, 1963; SCHÖNBOHM, 1969a) and in *Lemna* (ZURZYCKI, 1965). Although intracellular ATP levels were not measured, inhibition of respiratory or photosynthetic gas exchange generally paralleled inhibition of movement (FETZER, 1963; ZURZYCKI, 1965). Moreover, inhibition with Atebrin, an uncoupler of photophosphorylation, was reversible with added ATP (SCHÖNBOHM, 1969a). Oxidative phosphorylation is probably sufficient for energy requirements, since DCMU inhibition of movement in *Mougeotia* can be reversed by better aeration (SCHÖNBOHM, 1969a). Glycolysis is evidently not important as an energy source for chloroplast migration (FETZER, 1963; ZURZYCKI, 1965).

Q_{10}'s for the rate of chloroplast movement during the dark, weak-light and strong-light responses in *Mougeotia, Lemna* and *Selaginella* were between 2 and 4 (WEISENSEEL, 1968b). The Q_{10} for the strong-light response in *Mesotaenium* was 2.8 (GÄRTNER, 1970a). Both authors found that a given strong-light response required higher light intensities at higher temperatures. Thus, the failure of earlier workers (cited by WEISENSEEL, 1968b) to observe a temperature dependence of movement rates may have been caused by light as the limiting factor in the migration.

Chloroplast migration is thus a temperature- and energy-dependent process. Whether the energy requirement relates to the movement mechanism, or to transduction, or both is unknown. However, the primary gradient of light absorption will presumably not be affected by the above treatments. That red light induces phytochrome phototransformation in *Mougeotia* at 4° to 5 °C (MUGELE, 1962) and under N_2 and CO gas (FETZER, 1963) is demonstrated by the onset of chloroplast migration after transfer of cells to darkness and restoration of physiological conditions.

2.7 Transduction

The primary gradient of light absorption probably influences chloroplast distribution through a secondary gradient. Candidates for the secondary gradient include ions (especially Ca), nucleotides (particularly ATP), pH, and the redox state of sulfhydryl groups. Any of these agents could influence the structure and/or function of actomyosin (see WAGNER, Chap. 2.2, this vol.) or microtubules (see FILNER and YADAV Chap. 2.1, this vol.), which may well constitute the movement-generating systems for chloroplast movement (see Sect. 2.8).

Note that actin, myosin, and tubulin are in a sense also structural proteins and that their regulation could involve not only control of their enzymatic properties, but also control of their mutual interaction, assembly or disassembly (see WAGNER, Chap. 2.2; FILNER and YADAV, Chap. 2.1, this vol.).

The secondary gradient can be established within seconds. Chloroplast accumulation is first seen 10 to 80 s after illumination in *Vaucheria* (FISCHER-ARNOLD, 1963) and possibly as soon as 15 s in *Mougeotia* (HAUPT and ÜBEL, 1975). These times are maximal for the establishment of the gradient, since they also include the activation of the movement mechanism.

The gradient can be localized in certain areas of the cell since only irradiated portions respond in *Mougeotia* (BOCK and HAUPT, 1961), *Vaucheria* (FISCHER-ARNOLD, 1963) and *Funaria* (ZURZYCKI, 1972).

Although low temperature slows the rate of chloroplast movement, eventually complete orientation responses still take place in *Mougeotia, Lemna,* and *Selaginella* at 0 ° to 5 °C (WEISENSEEL, 1968a). Thus, the gradient itself must still be established at low temperatures. As discussed above, the weak-light response in *Mougeotia* may involve different photoreceptors (HAUPT, 1971) with possibly different transduction mechanisms. In fact, the weak-light response was inhibited at 2 °C when induced by red light, but it persisted when induced by blue light (WEISENSEEL, 1968a). The possibility of various transduction mechanisms must also be considered for *Vallisneria* (SEITZ, 1967a), *Selaginella* (MAYER, 1964) and *Hormidium* (SCHOLZ, 1976b), where separate photoreceptors are thought to exist.

The secondary gradient is generally unstable. Either continuous or pulsed irradiations are required to generate a movement response in *Selaginella* (MAYER, 1964) and *Vaucheria* (FISCHER-ARNOLD, 1963). In the *Mesotaenium* weak-light response, pulses of 1 s every 5 min will suffice (HAUPT and THIELE, 1961). Although the strong-light response in *Mougeotia* requires continuous or pulsed irradiations (SCHÖNBOHM, 1963), the weak-light response in this alga can be induced with a single pulse as short as 10^{-3} s (HAUPT, 1959a). Since phytochrome once transformed to Pfr remains that way for some time, it may be able to regenerate the gradient continuously. The requirement for continuous or pulsed irradiation in the strong-light response may be a property of the tonic photoreceptor system (see Sect. 2.5.4).

Low temperatures in combination with pulsed irradiations permit longer dark intervals to be used (GÄRTNER, 1970a). Thus, temperature-dependent processes are probably involved in the decay of the gradient, an observation which may explain the ability to evoke a movement response at low temperature. Low temperature also requires less light for both the strong-light and weak-light responses, which again suggests the gradient can be maintained more easily. Weak-light response requires 0.25 W m^{-2} in *Lemna* at 20 °C but only 0.01 W m^{-2} at 5 °C (WEISENSEEL, 1968a).

2.7.1 Ions

The action dichroism for light-induced chloroplast migration indicates an ordered arrangement for the photoreceptors, possibly on a membrane. A logical suggestion is that the photoreceptors alter some membrane function such as ion flux. Calcium, known to be important in cellular motility systems, is a logical candidate. The concentration of free Ca in the cytoplasm is usually quite low, but under some conditions the ion can form temporally and spatially stable concentration gradients (ROSE and LOEWENSTEIN, 1975). Light is already known to affect Ca fluxes as, for example, in blue light-mediated phototropism in developing fucoid zygotes (ROBINSON and JAFFE, 1975).

It is generally assumed that high Ca levels are permissive for the activity of actomyosin systems (see WAGNER, Chap. 2.2, this vol.), but such need not be the case. ABRAMOWITZ et al. (1975) reported the depolymerization of an unusual form of actin by Ca. Elevated Ca levels inhibited cytoplasmic streaming in *Elodea* (FORDE and STEER, 1976) and vesicle migration along microfilament bundles in a perfused *Chara* cell system (WILLIAMSON, 1975). In the latter case, 1 mM Mg was required for movement, whereas Ca greater than 10^{-7} M was inhibitory. BARRY (1968) noted that an action potential in *Nitella* stopped streaming with Ca or Sr in the external medium but not with Mg. This observation would be consistent with the transmembrane penetration of Ca into the cell.

WAGNER and KLEIN (1978) found that prolonged culture of *Mougeotia* in Ca-free medium lowered by 50% the bound but not the soluble Ca content and inhibited chloroplast movement by 40%. While low Ca levels could affect motility in many ways (e.g., altered membrane properties), it was ascertained that ATP levels did not decrease with low Ca. It will be interesting to test whether Ca affects the rate of movement, the number of responding cells or the light requirement for a given response.

Phytochrome state did not affect fluxes into and out of *Mougeotia* of Ca (WAGNER and BELLINI, 1976) or of K and Cl (WAGNER, 1974). WEISENSEEL and SMEIBIDL (1973) found no change in membrane potential or medium conductivity upon light-induced chloroplast migration in *Mougeotia*. However, net intracellular ion levels would not change if ions cycled through an internal sequestering system.

It may be that marine algae from an ion-rich environment are more suitable for the investigation of ions (as was the case for ion involvement in *Platymonas* phototaxis; HALLDAL, 1959). Also as in *Platymonas*, it may be that changes in the ratio of several ions will be the controlling step. If a gradient of ions is postulated to control movement, then the existence of the gradient at low temperatures must be reconciled with the possibility that ion pumps will be inhibited at low temperature and that ion gradients will be leveled (VAN ROSSUM, 1970). Note, in fact, that the phytochrome-controlled weak-light response in *Mougeotia* was inhibited at low temperatures (WEISENSEEL, 1968a).

Further approaches to the role of ions would be to use the Ca-stimulated photoprotein aequorin (e.g., ETTIENNE, 1970) to record possible calcium transients upon movement or to use ionophores to disrupt ion balance. For example, a normally light-induced microfilament assembly in a marine invertebrate has been induced in darkness by the ionophore A23187 (WOOLLACOTT and PORTER, 1977) which will alter Ca, Mg and H concentrations. It may also be feasible to examine membrane preparations from plants with light-induced chloroplast migration for light effects on ion pumps or ATPases.

2.7.2 ATP

Low intensities of blue light increase 0_2 uptake (e.g., KOWALLIK, 1971) and ATP levels (KOWALLIK and SCHEIL, 1976) in nonphotosynthesizing *Chlorella*. The oxygenase activity of RUDP carboxylase is not involved. Whether blue light actually stimulates respiration and oxidative phosphorylation is not known. SEITZ (1971, 1972, 1974) speculated that blue light-stimulated differential ATP increases modulated by photosynthesis could control chloroplast arrangement. The action dichroism for light-induced chloroplast migration argues against the photoreceptor being on the mitochondria. Nonmitochondrial membrane-bound ATPases are not excluded.

If indeed a gradient of ATP functioned as a transducing signal, exogenously added ATP would be expected to cancel such an internal gradient. Accordingly, ZURZYCKI (1965) found that 5 mM ATP inhibited light-induced chloroplast migration and the maintenance of weak- and strong-light positions under constant illumination in *Lemna*. It is unlikely that ATP inhibited the movement mechanism, since, in the dark, ATP enhanced the rate of chloroplast migration. In *Vallisneria* ATP altered chloroplast distribution in darkness (SEITZ, 1971). In *Mougeotia*, however, ATP is unlikely to be involved in a gradient since exogenous ATP had no effect on light-induced chloroplast migration except to enhance the rate of movement (SCHÖNBOHM, 1969a).

2.7.3 Glycine Oxidase

Further investigations of blue light-stimulated O_2 uptake have prompted an alternative theory for transduction. Glycine oxidase catalyzes the formation of glyoxylate, H_2O_2 and NH_4^+ from glycine and O_2. This enzyme uses FMN as a coenzyme and its activity is stimulated by low intensities of blue light (SCHMID and SCHWARZE, 1969). ZURZYCKI (1972) proposed membrane-localized glycine oxidase as the photoreceptor for chloroplast migration. In fact, the chloroplasts of *Funaria* are slightly chemotactic to glyoxylate in the light or dark, whereas they are chemotactic to glycine only in the light (GODZIEMBA-CZYZ, 1973). It is not known how glyoxylate might affect a movement-generating system. NH_4^+, H_2O_2, glycolate, glyoxal, oxalate, and acetate had no effect in the light or dark.

In *Lemna* there is an excellent correlation between the dose response curves for light-enhanced O_2 uptake and light-induced chloroplast migration (ZURZYCKI, 1970). Since the gas exchange measurements consider the reaction in all parts of the cell at once, it is not surprising that an action dichroism was not observed for the rectangular *Lemna* cells. However, an action dichroism was reported for light-stimulated O_2 uptake in certain spindle-shaped cells in which the cell walls could be oriented with respect to the electric vector of light (ZURZYCKÍ, 1971). Thus, it is possible that the photoreceptors for light-stimulated O_2 uptake may be located in the cell similarly to the photoreceptors for light-induced chloroplast migration. Unfortunately, comparative dose response curves were not reported for chloroplast migration in these spindle-shaped cells. A further problem is that light stimulation of glycine oxidase stops immediately after the light is extinguished (SCHMID and SCHWARZE, 1969) while light enhanced O_2 uptake in *Chlorella* (KOWALLIK, 1971) and *Lemna* (ZURZYCKI, 1970) continues for a while in darkness.

Treatments which affect the levels of glycine and glyoxylate also affect chloroplast migration. Glyoxylate can be transaminated to glycine. Amino oxyacetate, an inhibitor of transaminases, shifts the dose requirement for chloroplast movement in *Funaria* to 100 fold higher intensities while added glycine sensitizes chloroplast migration to light (ZURZYCKI, 1972). These compounds given in the dark induce *Funaria* chloroplasts, which normally assume a quasi-parastrophe position, to randomize. Amino oxyacetate and glycine inhibit and enhance, respectively, light-stimulated O_2 uptake in *Lemna* (ZURZYCKI, 1972). It will be important to combine these studies in one organism and, also, to determine whether amino oxyacetate and glycine shift the dose response for light-stimulated O_2 uptake in the same way as they affect the chloroplast migration curves.

Although glycolate oxidase forms glyoxylate from glycolate synthesized in the chloroplasts during photosynthesis, DCMU and α-hydroxy methylsulfonate (an inhibitor of glycolate oxidase) had no effect on the rates and extent of chloroplast migration in *Funaria* at two selected light intensities (ZURZYCKI, 1972). On the other hand, in *Lemna*, added glycolate enhanced the light-stimulated O_2 uptake. Again it would be important to combine studies on the same organism. Note that blue light inhibits the activity of glycolate oxidase (SCHMID, 1969). Conditions which enhance photosynthetic glycolate formation (high O_2 and low CO_2 tension, high light intensity) should influence chloroplast distribution. Perhaps the tonic effect of high-intensity red light in *Lemna* (ZURZYCKI, 1964; see Sect. 2.5.4) was mediated through glycolate formation.

2.7.4 Photosensitization

The photoreceptor pigments could affect the movement mechanism directly through a photodynamic oxidation. Flavins are efficient photosensitizers and dye-mediated photooxidation and inhibition of myosin (EIDUS and KONDAKOVA, 1958; cited in POP et al., 1971) and actin (MARTONOSI and GOUVEA, 1961) have been reported. Both proteins are especially sensitive to oxidation. LANG-FEULNER and RAU (1975) considered the blue and UV light-mediated carotenogenesis in *Fusarium* to be consistent with an oxidative mechanism.

2.8 Mechanisms of Movement

The likely participation of actomyosin systems as the motive force for various types of intracellular movement has received widespread attention (see WAGNER, Chap. 2.2, this vol.; BRITZ, Chap. 2.3, this vol.). Actomyosin is probably also involved in at least some types of light-induced chloroplast migration. Microtubules constitute the other major class of protein involved in motility (see FILNER and YADAV, Chap. 2.1, this vol.). Although no evidence exists to link microtubules with light-induced chloroplast movement, they are probably involved in some types of rhythmic chloroplast migration (see Sect. 3.3) and therefore cannot be dismissed from consideration.

Recent work on vorticellid ciliates has uncovered a third movement system which involves the elastic, non-ATP-requiring contraction of a protein, spasmin, upon Ca-binding (AMOS et al., 1975). There is as yet no evidence to postulate the participation of a spasmin-like system in chloroplast migration.

Although actomyosin in skeletal muscle is rather highly structured, actomyosin in other cellular systems exhibits more latitude in its structural and functional organization. Evidence for possible ways in which actomyosin systems might combine to influence chloroplast distribution are discussed below. Inhibitor studies on the movement mechanism are also considered.

2.8.1 Shear Force Generation

Chloroplast movement in some plants may be related to cytoplasmic streaming (see also SEITZ, Chap. 2.4, this vol.). In *Chara*, polar microfilament bundles are bound to the chloroplasts which are anchored in the ectoplasm (KERSEY et al., 1976). Shear forces generated between the microfilaments and some form of myosin probably cause the resultant movement of the cytoplasm with respect to the chloroplasts (HUXLEY, 1973; WILLIAMSON, 1975). If the microfilaments were anchored in the ectoplasm, as is more common, then the movement of chloroplasts with respect to the ectoplasm might result. Such a mechanism may be relevant for light-induced chloroplast migration in *Vaucheria*. In this alga fibrillar bundles visible in differential interference contrast microscopy (M. Blatt, personal communication) run parallel to the channels of multistriate streaming. The chloroplasts appear to be carried passively in the streaming cytoplasm (FISCHER-ARNOLD, 1963). Under microbeam irradiation, cytoplasm and organelles accumulate in the irradiated area. Thus, the effect of light in *Vaucheria* appears to be directly on streaming, at least for the weak-light response. Between 10 ° and 20 °C the Q_{10} was equal to 2 for the speed of both the weak-light response and cytoplasmic streaming. The strong-light response is more complex, and probably involves reorientation of the channels of streaming (FISCHER-ARNOLD, 1963). Furthermore, actin-like and myosin-like proteins have been reported in chloroplast fractions isolated from spinach leaves (ONISHI, 1964).

2.8.2 Changes in Cytoplasmic Viscosity

Light may influence cytoplasmic viscosity, which in turn may influence chloroplast distribution. The centrifugal force required to displace *Elodea densa* chloroplasts, taken to be a measure of cytoplasmic viscosity, was greater in moderate light intensities (about 0.5 to 2200 lux white light) than in darkness or in bright light (VIRGIN, 1951). In *Vallisneria*, the action spectra and intensity requirements for enhanced chloroplast centrifugation, induction of cytoplasmic streaming and the strong-light chloroplast migration response were all similar (SEITZ, 1967a). A light effect on cytoplasmic viscosity remains consistent with possible actomyosin involvement because the polymerization of supramolecular structures such as actin filaments will affect cytoplasmic viscosity (e.g., POLLARD, 1976).

Since a gradient of light intensity exists around the cell periphery (Sect. 2.5.3), a gradient of viscosity might then result. Such a gradient has been shown by SEITZ (1971) in *Vallisneria*: at 0.01 W m^{-2} it was harder to centrifuge the chloroplasts from the front walls than from the side walls while at 1 W m^{-2} the situation was reversed. Chloroplasts could accumulate nonspecifically in regions of higher viscosity, analogous to a photokinetic accumulation of microorganisms in light or dark traps in which swimming rate has slowed (see HÄDER, Chap. 3.2.1, this vol.).

Studies in other organisms indicate that the light effects on chloroplast centrifugation may be more complex than just effects on cytoplasmic viscosity. In *Lemna* it was easier to centrifuge chloroplasts during migration in response to altered light conditions than it was after redistribution was completed (ZURZYCKI, 1960). The participation of a separate chloroplast fastening system was suggested. In *Elodea canadensis* both blue and red light enhanced cytoplasmic streaming rates, but the chloroplasts remained immobile in red light whereas in blue light they joined the streaming cytoplasm (SEITZ, 1964). Again, a separate blue light-sensitive fastening system is suggested.

2.8.3 Sliding Filament

Chloroplast migration also occurs in cells with irregular or no streaming, or where the chloroplasts are very large and do not move with the cytoplasm. In these cases it seems likely that a mechanism operating more specifically on the chloroplast

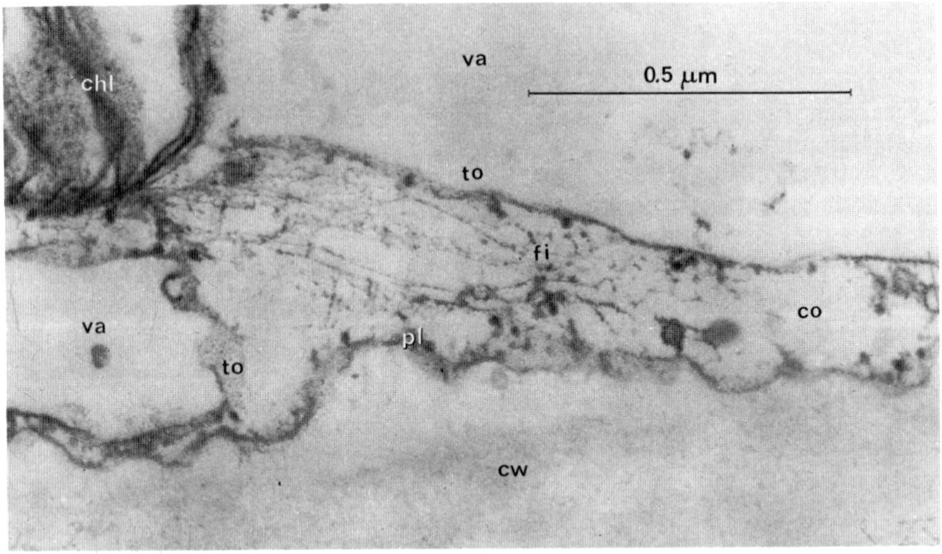

Fig. 6. Ultrastructure of a *Mougeotia* cell in cross-section including the chloroplast *(chl)*, cell wall *(cw)*, plasma membrane *(pl)*, tonoplast membrane *(to)*, and vacuole *(va)*. Cytoplasm bisecting the vacuole and connecting the chloroplast with the cortical ectoplasm *(co)* contains numerous thin filaments *(fi)* (WAGNER and KLEIN, 1978)

is involved. SCHÖNBOHM (1973a, 1975) observed with differential interference contrast microscopy the formation of fibrils between the chloroplast and ectoplasm during light-induced chloroplast migration in *Mougeotia*. An apparent active movement of particles along the fibrils was also observed. WAGNER and KLEIN (1978) reported 5 to 10 nm diameter microfilaments between the chloroplast and ectoplasm of *Mougeotia* (Fig. 6). Although the identity of these filaments as actin and their dynamic behavior during chloroplast migration must still be determined, their arrangement seems consistent with a contractile sliding filament mechanism like muscle (of course, the mutual sliding of filaments is a special case of shear force generation). The presence of actin in *Mougeotia* was demonstrated by heavy meromyosin staining of filaments isolated from ruptured cells (MARCHANT, 1976).

Light which stimulates the weak-light response in *Mougeotia* also decreases the ability of chloroplasts to be centrifuged (SCHÖNBOHM, 1972a, 1973a, b). Perhaps a light-induced filament formation serves to attach the chloroplast to the ectoplasm. A light-induced microfilament assembly has been described for a marine invertebrate (WOOLLACOTT, 1974).

Additional studies on *Mougeotia* are consistent with the above observations:

1. A chloroplast can rotate at least five times about its axis (HAUPT and HEYMANN, 1969). The limit was imposed by photoreceptor adaptation and not by some elastic retarding force. The movement mechanism must be capable of continuously forming and disaggregating.

2. The movement mechanism can span a gap between a contracted chloroplast and the ectoplasm (SCHORER-MÖRTEL, 1972). Thus, the system is highly flexible.

3. The ectoplasm does not rotate during chloroplast migration (FOOS, 1971).

2.8.4 Inhibitor Studies

Whether microtubules might be involved in chloroplast movement is unknown. However, the microtubule antagonist colchicine (WILSON and BRYAN, 1974) at 10 mM had no effect on chloroplast movement in *Mougeotia* or *Funaria* (SCHÖNBOHM, 1973a).

Cytochalasin B, on the other hand, stops various cellular movements thought to involve actin microfilaments, including cytoplasmic streaming (WESSELLS et al., 1971). At about 12 μM this drug inhibited the weak-light response of *Mougeotia* by 50% (WAGNER et al., 1972). CB did not inhibit the establishment of a gradient of phytochrome phototransformation, since the drug could be washed out in darkness after a prior red light induction and a response would still occur. CB inhibited both the strong- and weak-light response in *Mougeotia* and *Funaria* (SCHÖNBOHM, 1973a), but it did not increase chloroplast centrifugability or reverse the light-induced fibril formation (SCHÖNBOHM, 1975). However, the movement of small vesicles along these fibrils was inhibited by CB.

The mechanism of CB action is controversial, although the drug does not seem to affect microtubule-related processes (WESSELLS et al., 1971). At 50 μM, CB dramatically alters the structure of the actin polymer from animal sources and inhibits the actin-activated heavy meromyosin ATPase (SPUDICH, 1973). CB does not seem to affect plant cell actin microfilaments as strongly as animal cell microfilaments (BRADLEY, 1973; CANDE et al., 1973). However, it may affect actin-myosin interactions in plants, since, in the perfused *Chara* cell system of WILLIAMSON (1975), CB rapidly inhibited ATP-induced vesicle translocation along and release from microfilament bundles.

CB also inhibits the transport of small molecules across animal cell membranes (see COPELAND, 1974). The active concentrations of drug that inhibit transport are frequently lower than those which inhibit movement. CB does bind to membranes and, in the case of fat cells, the association constant (6×10^{-7} M) corresponds closely to the K_i for inhibition of D-glucose transport (CZECH, 1976). At very high concentrations (100 μM), CB inhibits ATP formation in isolated mitochondria (LIN et al., 1973). Although many CB-sensitive processes in plants require concentrations approaching this range, deleterious effects of CB on plant metabolism (cell wall and protein synthesis) have not been observed (BRADLEY, 1973; CANDE et al., 1973).

Sulfhydryl groups are important for chloroplast movement. Both the strong- and weak-light responses in *Mougeotia* are inhibited by p-chloromercuribenzoic acid (SCHÖNBOHM, 1969b) as well as by salyrgan and $HgCl_2$ (SCHÖNBOHM, 1972a). Consistent with an action on sulfhydryl groups, PCMB inhibition is reversed by cysteine (SCHÖNBOHM, 1969b). Unlike CB, PCMB causes the disappearance of the cytoplasmic fibrils induced by light during chloroplast migration in *Mougeotia* (SCHÖNBOHM, 1973a). Sulfhydryl reagents also increase the centrifugability of *Mougeotia* chloroplasts in the dark and inhibit the light-induced decrease in centrifugability (SCHÖNBOHM, 1972a, b).

PCMB inhibits chloroplast migration induced by either red or blue light in *Vallisneria* (SEITZ, 1970b). However, PCMB may also alter early processes following photoperception in *Vallisneria*, since a low concentration inhibited the light-enhanced chloroplast centrifugation without affecting centrifugation in the dark (SEITZ, 1974).

It has been shown for *Mougeotia* chloroplast migration that noninhibitory concentrations of PCMB become inhibitory in the presence of ADP or ATP (SCHÖNBOHM, 1969b; ATP by itself enhances chloroplast migration.) SCHÖNBOHM (1969b) cites other examples of the ATP-sulfhydryl reagent synergism, including the report by PORTZEHL (1954) that ATP enhances the Salyrgan inhibition of rabbit actomyosin thread models. These synergisms are reminiscent of the ATP- and ADP-enhanced inhibition of myosin ATPase by the sulfhydryl reagent N-ethylmaleimide (SEKINE and YAMAGUCHI, 1963). ATP-binding by myosin may cause a protein conformation change which could alter sulfhydryl sensitivity.

On the other hand, a specific effect of sulfhydryl reagents in vivo seems unlikely, especially as PCMB (GAUDEMER and LATRUFFE, 1975) and its more hydrophillic sulfonic acid derivative PCMBS (GIAQUINTA, 1976), may not penetrate membranes. The comparative effect of PCMB, PCMBS and a clearly penetrant sulfhydryl reagent such as N-ethylmaleimide on chloroplast migration and other intracellular processes (e.g., photosynthesis, respiration) should be investigated.

3. Rhythmic Chloroplast Migration

3.1 Types of Response

Internal factors regulated by the biological clock (see also SWEENEY, Chap. 1.4, this vol.) can lead to changes in chloroplast arrangement. Epidermal cell chloroplasts of *Selaginella serpens* have pronounced circadian rhythms of shape change and migration (BUSCH, 1953). Chloroplast distribution alters rhythmically in the growing tips of *Caulerpa* (DAWES and BARILOTTI, 1969). The large, parietal chloroplasts of *Ulva* migrate rhythmically (BRITZ and BRIGGS, 1976). During the day the chloroplasts of *Ulva* are arranged on the outer cell wall of the two cell-layered thallus (face position). Beginning in the latter half of the light period and continuing into the dark period, the chloroplasts move slowly to the sides of the cell (profile position, not unlike the strong-light position in *Selaginella,* Fig. 1c). Return movement begins before the start of the light period. Characteristic of circadian rhythms, the migration in *Ulva* persists for several days in constant light or darkness and constant temperature with a free-running period of about 24 h (BRITZ et al., 1976).

Since "round-the-clock" microscope studies are seldom performed, it is difficult to estimate how widespread the occurrence of rhythmic chloroplast migration is. The value of indirect photometric methods in such cases is readily apparent (see Sect. 2.2). An example of rhythmic chloroplast movement (*Ulva*) studied by a photometric method and the precision with which such measurements can be achieved is shown in Figure 7.

3.2 Adaptive Significance

Studies on rhythmic chloroplast migration fail to indicate any clear adaptive function. In *Ulva* the chloroplasts occupy face position during the middle of the day (BRITZ and BRIGGS, 1976). Thus, unlike light-induced chloroplast movement, the efficiency of light absorption is greatest at a time when natural irradiance is also greatest. Light-induced chloroplast migration in *Ulva* (strong-light response) which would modulate absorption efficiency was not observed (BRITZ and BRIGGS, 1976). Although *Ulva* also has a rhythm of photosynthetic capacity in saturating light (BRITZ and BRIGGS, 1976), there remains no evidence for a direct involvement of chloroplast movement in regulation of photosynthesis. Obviously, other adaptive

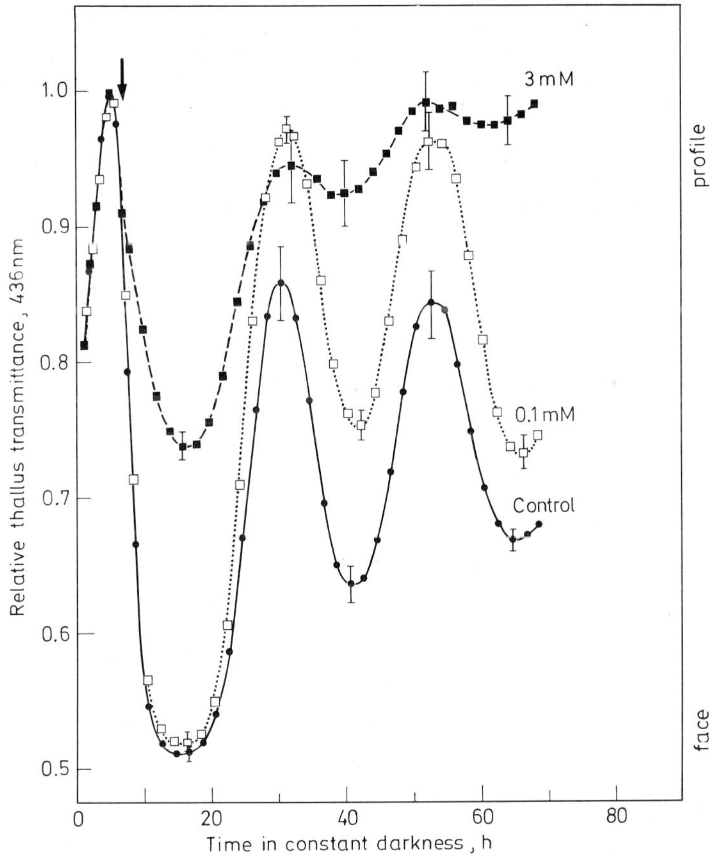

Fig. 7. The effect of colchicine on rhythmic chloroplast movement in *Ulva*. Movement was measured as change in light transmission at 436 nm during 3 days in essentially constant darkness; the dim measuring beam did not affect the rhythm. The alga had been previously cultured in a light–dark cycle; Time 0 is the end of the last light period. Each curve is the average of five samples (normalized to a transmission of 1.0 at the first maximum); the *bars* represent one standard error of the mean. *High transmission* indicates the chloroplasts are more in profile while *low transmission* indicates they are more in face position. Colchicine (3 mM, ---; or 0.1 mM, ····) or a sham treatment (——) were administered at the time indicated by the *heavy arrow*. (After BRITZ, 1976)

mechanisms must be considered. Observations of the relation between chloroplast position and cell division in the Ulvales (BRITZ and BRIGGS, 1976; SMITH, 1947; WEBER, 1959) provide circumstantial support for the suggestion (BRITZ and BRIGGS, 1976) that rhythmic chloroplast migration in *Ulva* is related to the establishment of cell polarity in the control of cell division and development.

3.3 Mechanisms

Although light-induced and rhythmic chloroplast movement share morphological similarities, it is certainly not safe to generalize about mechanism. Moreover, essentially nothing is known about the nature of the biological clock or the way in which it couples to the movement mechanism. A further difficulty inherent in rhythm work is the separation of effects on the clock from effects on the "hands" of the clock. In *Ulva,* for example, the general metabolic poison sodium azide inhibits chloroplast movement, but it also affects the free-running period and phase of the movement rhythm (BRITZ, 1976).

Certainly some types of rhythmic movement may have mechanisms similar to light-induced chloroplast migration. The rhythm of chloroplast migration in *Caulerpa,* for example, appears to result from a rhythmic change in distribution of cytoplasm with the chloroplasts carried passively in the streaming cytoplasm (DAWES and BARILOTTI, 1969). For comparison, see Sect. 2.8.1.

On the other hand, rhythmic chloroplast movement in *Ulva* may use microtubules (BRITZ, 1975b, 1976). Sub-millimolar concentrations of the microtubule antagonist colchicine shifted *Ulva* chloroplasts more into profile position without inhibiting the extent of movement, while millimolar concentrations of the drug actually stopped profile-to-face movement (Fig. 7). This inhibition could not be reversed with washing, but (consistent with an effect on microtubules) the inhibition was reversed by irradiation with UV light which forms the nontubulin binding isomer of colchicine, lumicolchicine. CB had no effect by itself on chloroplast movement in *Ulva* (BRITZ, 1975b, 1976). Thus, different mechanisms are probably involved in rhythmic chloroplast migration in *Ulva* as compared with light-induced chloroplast migration in *Mougeotia* and *Funaria* where CB but not colchicine inhibited movement (see Sect. 2.8.4).

Microtubules (see FILNER and YADAV, Chap. 2.1, this vol.) can generate movement by either polymerization-depolymerization or by an ATPase-driven sliding of tubules. Microtubules might alternatively act as guide elements, in which case their destruction would disrupt movement without actually inhibiting the movement-generating system (e.g., if an actomyosin system were involved). The particular function of mictrotubules in *Ulva* chloroplast movement is unknown, as is the mechanism for the colchicine-insensitive face-to-profile movement. Whether an actomyosin but CB-insensitive system could be involved in either direction of chloroplast movement is unknown.

Microtubules seem to be more generally involved in organelle migration in animal cell systems. Colchicine inhibits melanosome aggregation while CB inhibits both aggregation and dispersion in the fiddler crab *Uca* (LAMBERT and CROWE, 1973) and in frog skin (MALAWISTA, 1975). Ultrastructural evidence indicates both microtubules and microfilaments are involved in melanosome movement in frog skin (MCGUIRE and MOELLMANN, 1972) and in light-induced pigment epithelium migration in fish retinas (BURNSIDE, 1976). On the other hand, only the presence of microtubules has been demonstrated in chromatophores of various teleost fish (MURPHY, 1975). Interestingly, melanosome movement in the fiddler crab is also under the control of a circadian rhythm (BROWN and WEBB, 1948).

4. Nuclear Migration

4.1 Traumatotactic Nuclear Migration

Within 15 h after wounding, and preliminary to the resumption of cell division, about 85% of the nuclei in *Tradescantia* leaf cells in the first few rows bordering an incision will move to the cell wall nearest the wound. The migration is sensitive to CB and to KCN, but not to colchicine or morphactin (SCHNEPF and VON TRAITTEUR, 1973). Movement therefore involves active processes, possibly actin microfilaments. However, as reviewed by BÜNNING (1959), cytoplasmic streaming is probably not directly involved. EGTA inhibition of directed movement is partially reversible with Ca (SCHNEPF and VOLKMANN, 1974).

Inhibitor experiments indicate that transduction of the wound stimulus requires synthesis of protein and mRNA. The critical period for mRNA synthesis is between $1^1/_2$ to 2 h after wounding (SCHNEPF and VOLKMANN, 1974; SCHNEPF and KLUMP, 1975).

4.2 Premitotic Nuclear Migration

There are many examples of this type of movement. Formation of root hairs in *Hydrocharis* (CUTTER and HUNG, 1972) and guard mother cells in *Allium* (PALEVITZ and HEPLER, 1974a) are preceded by an asymmetric division which involves nuclear migration.

A relation between nuclear position and cell division plane is observed in *Ulva*. Prior to mitosis the nucleus migrates from its position under the chloroplast at the outer cell wall to a position near the inner cell wall. Division proceeds with the cleavage plane orthogonal to the thallus plane (LØVLIE and BRÅTEN, 1968). During meiosis, on the other hand, the nucleus migrates to the side wall instead, and the first cleavage plane is parallel to the plane of the thallus (BRÅTEN and NORDBY, 1973). The availability of mutants (FJELD, 1972; LØVLIE, 1968) with abnormal development and division planes may provide an opportunity to study the physiology of nuclear migration and its relation to development.

4.3 Spindle Orientation

Cell division in the guard mother cells of *Allium* proceeds with an obliquely oriented spindle and results in daughter nuclei at opposing corners of the cell (PALEVITZ and HEPLER, 1974a). The task of the developing stomatal guard cells is to change the plane of the cell plate to be parallel to the axis of the elongating cell file. During late anaphase or early telophase the spindle rotates until the proper orientation is achieved (PALEVITZ and HEPLER, 1974a, b). The cell plate ends up in the same plane as a cortical preprophase band of microtubules. The band disappears before karyokinesis, of course, but its position may represent microtubule organizing centers which later help orient the spindle. The authors point out that a very precise control mechanism seems to operate, since no movement occurs in cells where by chance the spindle forms initially in the proper orientation. Moreover, the movement is always in the correct direction without any overshoot.

Spindle migration takes about 10 to 15 min. The migration is active, since NaN_3 or dinitrophenol sufficient to inhibit streaming in neighboring cells will reversibly inhibit spindle orientation. Microtubules are probably involved, since the movement is inhibited by microtubule antagonists (colchicine, vinblastine sulfate, and isopropyl-N-phenyl carbamate) given late after karyokinesis. The dividing guard mother cells themselves have no cytoplasmic streaming and no microfilaments were observed. Cytochalasin B had variable effects on the migration.

At the start of migration the plane of the cell plate deviates from a strict 90° orientation to the spindle axis and inclines towards the plane of the preprophase band. PALEVITZ and HEPLER (1974b) suggested that some sort of force may be acting on the plate. The cell plate, however, is not necessary for orientation since it can be disrupted with caffeine and normal spindle migration will still occur.

4.4 Postmitotic Nuclear Migration

In *Acetabularia* several hundred secondary nuclei migrate to the developing cap and distribute equidistant to one another (WOODCOCK, 1971). Eventually each nucleus is apportioned into a cyst. Electron microscopy of the clear zone around each nucleus from which chloroplasts are excluded revealed a scaffolding of perinuclear microtubules extending 2 to 3 μm into the surrounding cytoplasm. Microtubules were also observed to extend radially into the cytoplasm for at least 30 μm (about one-third the distance between nuclei). The importance of the microtubular network in the spacing of nuclei was shown in experiments with colchicine and vinblastine sulfate in which disruption of microtubules corresponded to a breakdown in nuclear spacing and normal cyst development.

The elaborate bilateral symmetry of the desmids is destroyed by mitosis, but it is restored by nuclear migration after division. PICKETT-HEAPS and FOWKE (1970) describe the process for *Closterium*. In the crescent-shaped cells of this alga the nucleus is positioned centrally between two chloroplasts. After division the nucleus is located at what has become the end of the cell and it must migrate back to the middle. The nucleus becomes enclosed in a post-telophase sheath of microtubules, formed presumably from the spindle and arrayed from an organizing center in the middle of the daughter cell. The nucleus, greatly deformed, then migrates within a lobe of the chloroplast to a central cleft. After the nucleus arrives at the cleft the chloroplast divides along the cleft and the symmetry is restored. The movement has been reported to be inhibited by colchicine (unpublished data, cited in PICKETT-HEAPS, 1972).

Somewhat different examples are represented in *Micrasterias* (KIERMAYER, 1970a) and *Cosmarium* (PICKETT-HEAPS, 1972). In *Micrasterias* two half-cells are separated by a narrow isthmus in which the nucleus resides. After division at the isthmus each daughter cell forms a new half-cell. The nucleus migrates into the developing half-cell and assumes a position near the cell wall. When, after 6 h, growth is complete, the nucleus migrates back to the isthmus and the chloroplast enters the new half-cell. It is possible to inhibit the back migration of the nucleus with microtubule antagonists (colchicine, vinblastine sulfate, trifluralin and isopropyl-N-phenyl carbamate) at concentrations which do not inhibit chloroplast migration or development of the new half-cell (KIERMAYER, 1968b, 1972; KIERMAYER and HEPLER, 1970). Several different microtubule systems in *Micrasterias* have been described including a post-telophase perinuclear sheath (KIER-

MAYER, 1968a, 1970b). Nothing is known about the mechanisms of nuclear migration into the developing half-cell in the first place.

5. Conclusions

General statements about chloroplast and nuclear migrations are difficult to make. The responses exhibit many different morphological patterns. Among chloroplasts the rule is that light-induced distributions maximize the absorption efficiency in weak light and minimize it in strong light. While that relationship would seem to indicate a role for chloroplast migration in control of photosynthesis, the actual changes in percent light absorbed are usually too small to be significant. Thus, some other adaptive value must be sought. The same conclusion appears to be true for rhythmic chloroplast migrations. Nuclear migrations, on the other hand, are clearly important for normal plant development, yet much less is known about their physiology than about that of light-induced chloroplast migrations.

Light, trauma, or unknown internal factors, such as the biological clock or developmental gradients, can stimulate orientation and migration. In the case of light-induced chloroplast migration, the photosynthetic pigments, phytochrome, or a presumed flavin pigment have been identified as photoreceptors. One organism may have several photoreceptors and even a single physiological response can be mediated by more than one photoreceptor. In general, however, most chloroplast movement responses are sensitive to blue and UV light and thus seem to make use preferentially of the presumed flavin photoreceptor.

The blue–UV system (i.e., flavin) and phytochrome photoreceptors responsible for chloroplast migration are not located on the chloroplasts. Light direction is sensed mainly through the dichroic orientation of the transition moments of these pigments which results in a gradient of light absorption around the cell. This gradient is somehow able to alter chloroplast distribution. In other cases, for example when the chloroplasts are their own photoreceptors, light direction can be sensed through gradients established by other mechanisms such as shading.

Ongoing investigations are uncovering information on the mechanism for moving chloroplasts. Here many analogies with other cellular movements may exist. Actomyosin systems appear to be important in the light-induced chloroplast migration responses that have been examined. The specific mechanisms (e.g., shear force generation underlying cytoplasmic streaming, polymerization effects on viscosity, or sliding filaments) have not yet been established firmly and, in any case, vary among organisms. Microtubules are implicated in rhythmic chloroplast movement in *Ulva* and also in many types of nuclear migration. The function of the microtubules (movement-generating or guide element) likewise remains unknown.

Much less is known about transduction than about the movement mechanism, certainly in the cases of rhythmic chloroplast movement and nuclear migration. In light-induced chloroplast migrations, ions, ATP, and glyoxylate have been discussed as possible substances underlying secondary gradients which could alter chloroplast distribution.

Acknowledgments. This review was supported in part by a Fellowship to the author from the Carnegie Institution of Washington and by a National Aeronautics and Space Administration grant to Dr. A.W. Galston.

References

Abramowitz, J.W., Stracher, A., Detwiler, T.C.: A second form of actin: Platelet microfilaments depolymerized by ATP and divalent cations. Arch. Biochem. Biophys. *167*, 230–237 (1975)

Amos, W.B., Routledge, L.M., Yew, F.F.: Calcium-binding proteins in a vorticellid contractile organelle. J. Cell Sci. *19*, 203–213 (1975)

Barry, W.H.: Coupling of excitation and cessation of cyclosis in *Nitella:* Role of divalent cations. J. Cell Physiol. *72*, 153–159 (1968)

Biebl, R.: Lichttransmission und Chloroplastenbewegung. Flora *141*, 163–177 (1954)

Biebl, R.: Tagesgänge der Lichttransmission verschiedener Blätter. Flora *142*, 280–294 (1955)

Bock, G., Haupt, W.: Die Chloroplastendrehung bei *Mougeotia*. III. Die Frage der Lokalisierung des Hellrot-Dunkelrot-Pigmentsystems in der Zelle. Planta *57*, 518–530 (1961)

Bradley, M.O.: Microfilaments and cytoplasmic streaming: Inhibition of streaming with cytochalasin. J. Cell Sci. *12*, 327–343 (1973)

Bråten, T., Nordby, Ø.: Ultrastructure of meiosis and centriole behaviour in *Ulva mutabilis* Føyn. J. Cell Sci. *13*, 69–81 (1973)

Britz, S.J.: A model system to simulate chloroplast movement and accompanying transmittance changes in *Ulva*. Carnegie Inst. Washington Yearb. *74*, 794–803 (1975a)

Britz, S.J.: Inhibitor studies on the mechanism of chloroplast movement in *Ulva*. Carnegie Inst. Washington Yearb. *74*, 803–805 (1975b)

Britz, S.J.: Studies on a circadian rhythm of chloroplast movement in *Ulva*. Ph.D. Thesis, Harvard Univ., 1976

Britz, S.J., Briggs, W.R.: Circadian rhythms of chloroplast orientation and photosynthetic capacity in *Ulva*. Plant. Physiol. *58*, 22–27 (1976)

Britz, S.J., Pfau, J., Nultsch, W., Briggs, W.R.: Automatic monitoring of a circadian rhythm of change in light transmittance in *Ulva*. Plant Physiol. *58*, 17–21 (1976)

Brown, Jr., F.A., Webb, H.M.: Temperature relations of an endogenous daily rhythmicity in the fiddler crab, *Uca*. Physiol. Zool. *21*, 371–381 (1948)

Bünning, E.: Die Wirkung von Wundreizen. In: Encyclopedia of Plant Physiology. Ruhland, W. (ed.). Berlin-Göttingen-Heidelberg: Springer, 1959, Vol. XVII, pt. 1, pp. 119–134

Burnside, M.B.: Possible roles of microtubules and actin filaments in retinal pigmented epithelium. Exp. Eye Res. *23*, 257–275 (1976)

Busch, G.: Über die photoperiodische Formänderung der Chloroplasten von *Selaginella serpens*. Biol. Zentralbl. *72*, 598–629 (1953)

Cande, W.Z., Goldsmith, M.H.M., Ray, P.M.: Polar auxin transport and auxin-induced elongation in the absence of cytoplasmic streaming. Planta *111*, 279–296 (1973)

Copeland, M.: The cellular response to cytochalasin B: A critical overview. Cytologia *39*, 709–727 (1974)

Cutter, E.G., Hung, C.Y.: Symmetric and asymmetric mitosis and cytokinesis in the root tip of *Hydrocharis morsus-ranae* L. J. Cell Sci. *11*, 723–737 (1972)

Czech, M.P.: Characterization of [^{3}H]cytochalasin B binding to the fat cell plasma membrane. J. Biol. Chem. *251*, 2905–2910 (1976)

Dawes, C.J., Barilotti, D.C.: Cytoplasmic organization and rhythmic streaming in growing blades of *Caulerpa prolifera*. Am. J. Bot. *56*, 8–15 (1969)

DeFabo, E.C., Harding, R.W., Shropshire, Jr., W.: Action spectrum between 260 and 800 nanometers for the photoinduction of carotenoid biosynthesis in *Neurospora crassa*. Plant Physiol. *57*, 440–445 (1976)

Dorscheid, T.: Die Orientierung aktiver Moleküle des Chlorophyll *a* im Chloroplasten von *Mesotaenium caldariorum*. Z. Pflanzenphysiol. *61*, 46–51 (1969a)

Dorscheid, T.: Die Temperaturabhängigkeit der positiven Chloroplasten-Phototaxis bei *Mesotaenium caldariorum*. Z. Pflanzenphysiol. *61*, 52–57 (1969b)

Dorscheid, T., Wartenberg, A.: Chlorophyll als Photoreceptor bei der Schwachlichtbewegung des *Mesotaenium*-Chloroplasten. Planta *70*, 187–192 (1966)

Drawert, H.: Vitalfärbung und Vitalfluorochromierung pflanzlicher Zellen und Gewebe. Protoplasmatologia Vol. II, pt. D3, 1–749 (1968)

Erlanger, B.F.: Photoregulation of biologically active macromolecules. Annu. Rev. Biochem. *45*, 267–283 (1976)

Ettienne, E.M.: Control of contractility in *Spirostomum* by dissociated calcium ions. J. Gen. Physiol. *56*, 168–179 (1970)

Fetzer, J.: Über die Beteiligung energieliefernder Stoffwechselprozesse an den lichtinduzierten Chloroplastenbewegungen von *Mougeotia*. Z. Bot. *51*, 468–506 (1963)

Fischer, W.: Untersuchungen über die Inversion der Schwachlichtbewegung des *Mougeotia*-Chloroplasten. Z. Bot. *51*, 348–387 (1963)

Fischer-Arnold, G.: Untersuchungen über die Chloroplastenbewegung bei *Vaucheria sessilis*. Protoplasma *56*, 495–520 (1963)

Fjeld, A.: Genetic control of cellular differentiation in *Ulva mutabilis*. Gene effects in early development. Dev. Biol. *28*, 326–343 (1972)

Flohrs, H., Haupt, W.: Tagesperiodische Empfindlichkeitsschwankungen der lichtinduzierten Chloroplastenbewegung von *Mougeotia*. Z. Pflanzenphysiol. *65*, 65–69 (1971)

Foos, K.: Untersuchungen zur Feinstruktur von *Mougeotia* sp und zum Bewegungsmechanismus des Chloroplasten. Z. Pflanzenphysiol. *64*, 369–386 (1971)

Forde, J., Steer, M.W.: Cytoplasmic streaming in *Elodea*. Can. J. Bot. *54*, 2688–2694 (1976)

Gabryś-Mizera, H.: Model considerations of the light conditions in noncylindrical plant cells. Photochem. Photobiol. *24*, 453–461 (1976)

Gärtner, R.: Die Bewegung des *Mesotaenium*-Chloroplasten im Starklichtbereich. I. Zeit- und temperaturabhängige Sekundärprozesse. Z. Pflanzenphysiol. *63*, 147–161 (1970a)

Gärtner, R.: Die Bewegung des *Mesotaenium*-Chloroplasten im Starklichtbereich. II. Aktionsdichroismus und Wechselwirkungen des Photoreceptors mit Phytochrom. Z. Pflanzenphysiol. *63*, 428–443 (1970b)

Gaudemer, Y., Latruffe, N.: Evidence for penetrant and non-penetrant thiol reagents and their use in the location of rat liver mitochondrial D(-)-β-hydroxybutyrate dehydrogenase. FEBS Lett. *54*, 30–34 (1975)

Geitler, L.: Spontaneous partial rotation and oscillation of the protoplasm in *Coleochaete* and other Chlorophyceae. Am. J. Bot. *48*, 738–741 (1961)

Giaquinta, R.: Evidence for phloem loading from the apoplast. Chemical modification of membrane sulfhydryl groups. Plant Physiol. *57*, 872–875 (1976)

Girbardt, M.: Kernbewegungen. In: Encyclopedia of Plant Physiology. Ruhland, W. (ed.). Berlin-Göttingen-Heidelberg: Springer, 1962, Vol. XVII, pt. 2, pp. 920–939

Girbardt, M.: Ultrastructure and dynamics of the moving nucleus. Symp. Soc. Exp. Biol. *22*, 249–259 (1968)

Godziemba-Czyz, J.: Certain aspects of the chemotaxis reaction of chloroplasts in *Funaria hygrometrica*. Acta Soc. Bot. Pol. *42*, 453–459 (1973)

Grum-Heller, S.: Die inverse Schwachlichtbewegung des *Mougeotia*-Chloroplasten: Die Induzierbarkeit wird durch zeitabhängige Dunkelreaktionen des Phytochroms gehemmt. Z. Pflanzenphysiol. *81*, 212–225 (1977)

Guba, F.: Effect of halogen ions on F-actin. Nature (London) *165*, 439 (1950)

Halldal, P.: Factors affecting light response in phototactic algae. Physiol. Plant. *12*, 742–752 (1959)

Halldal, P.: Action spectra of induced phototactic response changes in *Platymonas*. Physiol. Plant. *13*, 726–735 (1960)

Hartmann, K.M., Cohnen Unser, I.: Carotenoids and flavins versus phytochrome as the controlling pigment for blue-UV-mediated photoresponses. Z. Pflanzenphysiol. *69*, 109–124 (1973)

Haupt, W.: Die Chloroplastendrehung bei *Mougeotia*. I. Mitteilung: Über den quantitativen und qualitativen Lichtbedarf der Schwachlichtbewegung. Planta *53*, 484–501 (1959a)

Haupt, W.: Chloroplastenbewegung. In: Encyclopedia of Plant Physiology. Ruhland, W. (ed.). Berlin-Göttingen-Heidelberg: Springer 1959b, Vol. XVII, pt. 1, pp. 278–317

Haupt, W.: Die Chloroplastendrehung bei *Mougeotia*. II. Mitteilung. Die Induktion der Schwachlichtbewegung durch linear polarisiertes Licht. Planta *55*, 465–479 (1960)

Haupt, W.: Perception of environmental stimuli orienting growth and movement in lower plants. Annu. Rev. Plant Physiol. *16*, 267–290 (1965)

Haupt, W.: Die Inversion der Schwachlichtbewegung des *Mougeotia*-Chloroplasten: Versuche zur Kinetik der Phytochrom-Umwandlung. Z. Pflanzenphysiol. *54*, 151–160 (1966)

Haupt, W.: Die Orientierung der Phytochrom-Moleküle in der *Mougeotia*zelle: Ein neues Modell zur Deutung der experimentellen Befunde. Z. Pflanzenphysiol. *58*, 331–346 (1968a)

Haupt, W.: Das Phytochromsystem der *Mougeotia*zelle: Weitere Untersuchungen über den P_{730}-Gradienten. Z. Pflanzenphysiol. *59*, 45–55 (1968b)
Haupt, W.: Über den Dichroismus von $Phytochrom_{660}$ und $Phytochrom_{730}$ bei *Mougeotia*. Z. Pflanzenphysiol. *62*, 287–298 (1970a)
Haupt, W.: Localization of phytochrome in the cell. Physiol. Veg. *8*, 551–563 (1970b)
Haupt, W.: Hellrot- und Dunkelrot-Wechselwirkungen bei der Chloroplastendrehung von *Mougeotia*. Wiss. Z. Ernst-Moritz-Arndt-Univ. Greifswald *19*, 47–54 (1970c)
Haupt, W.: Schwachlichtbewegung des *Mougeotia*-Chloroplasten im Blaulicht. Z. Pflanzenphysiol. *65*, 248–265 (1971)
Haupt, W.: Short-term phenomena controlled by phytochrome. In: Phytochrome. Mitrakos, K., Shropshire, W., Jr. (eds.). London-New York: Academic Press 1972a, pp. 349–368
Haupt, W.: Localization of phytochrome within the cell. In: Phytochrome. Mitrakos, K., Shropshire, W., Jr. (eds.). London-New York: Academic Press 1972b, pp. 553–569
Haupt, W.: Perception of light direction in oriented displacements of cell organelles. Acta Protozool. *11*, 179–188 (1972c)
Haupt, W.: Role of light in chloroplast movement. Bioscience *23*, 289–296 (1973a)
Haupt, W.: Regulation der Chloroplastenverteilung in der Zelle durch Lichtintensität und Lichtrichtung. Ber. Dtsch. Bot. Ges. *86*, 403–406 (1973b)
Haupt, W., Bock, G.: Die Chloroplastendrehung bei *Mougeotia*. IV. Die Orientierung der Phytochrom-Moleküle im Cytoplasma. Planta *59*, 38–48 (1962)
Haupt, W., Bretz, N.: Short-term reactions of phytochrome in *Mougeotia?* Planta *128*, 1–3 (1976)
Haupt, W., Heymann, N.: Versuche zur Mechanik der Chloroplastenbewegung von *Mougeotia*. Z. Pflanzenphysiol. *57*, 68–71 (1967)
Haupt, W., Mörtel, G., Winkelnkemper, I.: Demonstration of different dichroic orientation of phytochrome P_R and P_{FR}. Planta *88*, 183–186 (1969)
Haupt, W., Schönbohm, E.: Light-oriented chloroplast movements. In: Photobiology of Microorganisms. Halldal, P. (ed.). London-New York-Sydney-Toronto: Wiley-Interscience, 1970, pp. 283–307
Haupt, W., Schönfeld, I.: Die Wirkung von kurzwelliger Strahlung auf die Schwachlichtbewegung des *Mougeotia*-Chloroplasten. Z. Bot. *51*, 17–31 (1963)
Haupt, W., Thiele, R.: Chloroplastenbewegung bei *Mesotaenium*. Planta *56*, 388–401 (1961)
Haupt, W., Trump, K.: Lichtorientierte Chloroplastenbewegung bei *Mougeotia:* Die Größe des Phytochromgradienten steuert die Bewegungsgeschwindigkeit. Biochem. Physiol. Pflanz. *168*, 131–140 (1975)
Haupt, W., Übel, H.: Zum Mechanismus der Phytochromwirkung bei der Chloroplastenbewegung von *Mougeotia*. Z. Pflanzenphysiol. *75*, 165–171 (1975)
Haupt, W., Weisenseel, M.H.: Physiological evidence and some thoughts on localised responses, intracellular localisation and action of phytochrome. In: Light and Plant Development. Smith, H. (ed.). London-Boston: Butterworths, 1976, pp. 63–74
Heller, S.: Phytochromreaktionen bei *Mougeotia* mit unterschiedlicher spektraler Wirkungsverteilung. Z. Pflanzenphysiol. *79*, 8–22 (1976)
Hepler, P.K., Palevitz, B.A.: Microtubules and microfilaments. Annu. Rev. Plant Physiol. *25*, 309–362 (1974)
Huxley, H.E.: Muscular contraction and cell motility. Nature (London) *243*, 445–449 (1973)
Inoue, Y., Shibata, K.: Light-induced chloroplast rearrangements and their action spectra as measured by absorption spectrophotometry. Planta *114*, 341–358 (1973)
Inoue, Y., Shibata, K.: Comparative examination of terrestrial plant leaves in terms of light-induced absorption changes due to chloroplast rearrangement. Plant Cell Physiol. *15*, 717–721 (1974)
Kersey, Y.M., Hepler, P.K., Palevitz, B.A., Wessells, N.K.: Polarity of actin filaments in Characean algae. Proc. Natl. Acad. Sci. USA *73*, 165–167 (1976)
Kiermayer, O.: The distribution of microtubules in differentiating cells of *Micrasterias denticulata* Bréb. Planta *83*, 223–236 (1968a)
Kiermayer, O.: Hemmung der Kern- und Chloroplastenmigration von *Micrasterias* durch Colchizin. Naturwissenschaften *55*, 299–300 (1968b)
Kiermayer, O.: Causal aspects of cytomorphogenesis in *Micrasterias*. Ann. N. Y. Acad. Sci. *175*, 686–701 (1970a)

Kiermayer, O.: Elektronenmikroskopische Untersuchungen zum Problem der Cytomorphogenese von *Micrasterias denticulata* Bréb. I. Allgemeiner Überblick. Protoplasma *69*, 97–132 (1970b)
Kiermayer, O.: Beeinflussung der postmitotischen Kernmigration von *Micrasterias denticulata* Bréb. durch das Herbizid Trifluralin. Protoplasma *75*, 421–426 (1972)
Kiermayer, O., Hepler, P.K.: Hemmung der Kernmigration von Jochalgen (*Micrasterias*) durch Isopropyl-N-phenylcarbamat. Naturwissenschaften *57*, 252 (1970)
Kowallik, W.: Light stimulated respiratory gas exchange in algae and its relation to photorespiration. In: Photosynthesis and Photorespiration. Hatch, M.D., Osmond, C.B., Slatyer, R.O. (eds.). New York-London-Sydney-Toronto: Wiley-Interscience, 1971, pp. 514–522
Kowallik, W., Scheil, I.: Lichtbedingte Veränderungen des ATP-Spiegels einer chlorophyllfreien *Chlorella*-Mutante. Planta *131*, 105–108 (1976)
Kurtin, W.E., Song, P.-S.: Photochemistry of the model phototropic system involving flavins and indoles-I. Fluorescence polarization and MO calculations of the direction of the electronic transition moments in flavins. Photochem. Photobiol. *7*, 263–273 (1968)
Lambert, D.T., Crowe, J.H.: Colchicine and cytochalasin B: effects on pigment granule translocation in melanophores of *Uca pugilator* (Crustacea: Decapoda). Comp. Biochem. Physiol. *46A*, 11–16 (1973)
Lang-Feulner, J., Rau, W.: Redox dyes as artificial photoreceptors in light-dependent carotenoid synthesis. Photochem. Photobiol. *21*, 179–183 (1975)
Lechowski, Z.: The action spectrum in chloroplast translocation in multilayer leaf cells. Acta Soc. Bot. Pol. *42*, 461–472 (1973)
Lechowski, Z.: Chloroplast arrangement as a factor of photosynthesis in multilayered leaves. Acta Soc. Bot. Pol. *43*, 531–540 (1974)
Lełatko, Z.: Some aspects of chloroplast movement in leaves of terrestrial plants. Acta Soc. Bot. Pol. *39*, 453–468 (1970)
Lin, S., Lin, D.C., Spudich, J.A., Kun, E.: Inhibition of mitochondrial contraction by cytochalasin B. FEBS Lett. *37*, 241–243 (1973)
Løvlie, A.: On the use of a mutlicellular alga (*Ulva mutabilis* Føyn) in the study of general aspects of growth and differentiation. Nytt Mag. Zool. *16*, 39–49 (1968)
Løvlie, A., Bråten, T.: On the division of cytoplasm and chloroplast in the multicellular green alga *Ulva mutabilis* Føyn. Exp. Cell Res. *51*, 211–220 (1968)
Löw, I., Dancker, P.: Effect of cytochalasin B on formation and properties of muscle F-actin. Biochim. Biophys. Acta *430*, 366–374 (1976)
Malawista, S.E.: Microtubules and the movement of melanin granules in frog dermal melanocytes. Ann. N. Y. Acad. Sci. *253*, 702–710 (1975)
Marchant, H.J.: Actin in the green algae *Coleochaete* and *Mougeotia*. Planta *131*, 119–120 (1976)
Marchant, H.J.: Ultrastructure, development and cytoplasmic rotation of seta-bearing cells of *Coleochaete scutata* (Chlorophyceae). J. Phycol. *13*, 28–36 (1977)
Martonosi, A., Gouvea, M.A.: Studies on actin. V. Chemical modification of actin. J. Biol. Chem. *236*, 1338–1344 (1961)
Mayer, F.: Lichtorientierte Chloroplasten-Verlagerungen bei *Selaginella martensii*. Z. Bot. *52*, 346–381 (1964)
Mayer, F.: Lichtinduzierte Chloroplasten-Verlagerungen bei *Selaginella martensii*. Untersuchungen zur Identifizierung des Photoreceptors durch Anwendung von Quencher-Substanzen. Z. Pflanzenphysiol. *55*, 65–70 (1966)
Mayer, F.: Light-induced chloroplast contraction and movement. In: Structure and Function of Chloroplasts. Gibbs, M. (ed.). Berlin-Heidelberg-New York: Springer, 1971, pp. 35–49
McGuire, J., Moellmann, G.: Cytochalasin B: Effects on microfilaments and movement of melanin granules within melanocytes. Science *175*, 642–644 (1972)
Middlebrook, M., Szent-Györgyi, A.: The action of iodide on oxidative phosphorylation. Biochim. Biophys. Acta *18*, 407–410 (1955)
Mikolajczyck, E., Diehn, B.: The effect of potassium iodide on photophobic responses in *Euglena*: evidence for two photoreceptor pigments. Photochem. Photobiol. *22*, 269–271 (1975)
Mugele, F.: Der Einfluß der Temperatur auf die lichtinduzierte Chloroplastenbewegung. Z. Bot. *50*, 368–388 (1962)

Murphy, D.B.: The mechanism of microtubule-dependent movement of pigment granules in teleost chromatophores. Ann. N. Y. Acad. Sci. *253*, 692–701 (1975)
Neuscheler-Wirth, H.: Photomorphogenese und Phototropismus bei *Mougeotia*. Z. Pflanzenphysiol. *63*, 238–260 (1970)
Neuscheler-Wirth, H.: Die lichtorientierte Chloroplastenbewegung von *Mougeotia* in Abhängigkeit von der Vorkultur, vom Licht-Dunkel-Wechsel und vom Zellalter. Z. Pflanzenphysiol. *65*, 130–139 (1971)
Nultsch, W., Pfau, J.: Investigations on light-induced plastid movement in seaweeds. In: Proc. 8th. Int. Seaweed Symp. In Press 1978
Ohnishi, T.: Le changement de volume du chloroplaste, accompagné de photophosphorylation, et les protéines ressemblantes à l'actine et à la myosine extraites du chloroplaste. J. Biochem. *55*, 494–503 (1964)
Palevitz, B.A., Hepler, P.K.: The control of the plane of division during stomatal differentiation in *Allium*. I. Spindle Reorientation. Chromosoma *46*, 297–326 (1974a)
Palevitz, B.A., Hepler, P.K.: The control of the plane of division during stomatal differentiation in *Allium*. II. Drug Studies. Chromosoma *46*, 327–341 (1974b)
Pfau, J.: Dose response, Weber Law adherence and discrimination sensitivity of the chromatophore movement of the brown alga *Dictyota dichotoma*. Z. Pflanzenphysiol. *81*, 17–25 (1977)
Pfau, J., Throm, G., Nultsch, W.: Recording microphotometer for determination of light induced chromatophore movements in brown algae. Z. Pflanzenphysiol. *71*, 242–260 (1974)
Pickett-Heaps, J.D.: Cell division in *Cosmarium botrytis*. J. Phycol. *8*, 343–360 (1972)
Pickett-Heaps, J.D., Fowke, L.C.: Mitosis, cytokinesis, and cell elongation in the desmid, *Closterium littorale*. J. Phycol. *6*, 189–215 (1970)
Pollard, T.D.: The role of actin in the temperature-dependent gelation and contraction of extracts of *Acanthamoeba*. J. Cell Biol. *68*, 579–601 (1976)
Pop, E., Popovici, G.G., Soran, V.: The chemical and photosensitizing action of riboflavin on rotational streaming. Z. Pflanzenphysiol. *64*, 97–106 (1971)
Portzehl, H.: Gemeinsame Eigenschaften von Zell- und Muskelkontraktilität. Biochim. Biophys. Acta *14*, 195–202 (1954)
Rabinowitch, E.I.: Photosynthesis and Related Processes. New York: Interscience 1951, Vol. II., pt. 1
Robinson, K.R., Jaffe, L.F.: Polarizing fucoid eggs drive a calcium current through themselves. Science *187*, 70–72 (1975)
Rose, B., Loewenstein, W.R.: Calcium ion distribution in cytoplasm visualized by aequorin: diffusion in cytosol restricted by energized sequestering. Science *190*, 1204–1206 (1975)
Sargent, M.L., Briggs, W.R.: The effects of light on a circadian rhythm of conidiation in *Neurospora*. Plant Physiol. *42*, 1504–1510 (1967)
Schmid, G.H.: The effect of blue light on glycolate oxidase of tobacco. Hoppe-Seyler's Z. Physiol. Chem. *350*, 1035–1046 (1969)
Schmid, G.H., Schwarze, P.: Blue light enhanced respiration in a colorless *Chlorella* mutant. Hoppe-Seyler's Z. Physiol. Chem. *350*, 1513–1520 (1969)
Schnepf, E., Klump, J.: Über die traumatotaktische Bewegung der Zellkerne in *Tradescantia*-Blättern. III Wirkung von Hemmstoffen der Proteinsynthese. Biochem. Physiol. Pflanz. *168*, 123–129 (1975)
Schnepf, E., Volkmann, K.: Inhibition of traumatotactic movement of nuclei in *Tradescantia* leaf epidermis. II. Effects of heavy water, ethionine, Ca^{++}, and cyclic AMP. Protoplasma *81*, 313–321 (1974)
Schnepf, E., von Traitteur, R.: Über die traumatotaktische Bewegung der Zellkerne in *Tradescantia*-Blättern. Z. Pflanzenphysiol. *69*, 181–184 (1973)
Schönbohm, E.: Untersuchungen über die Starklichtbewegung des *Mougeotia*-Chloroplasten. Z. Bot. *51*, 233–276 (1963)
Schönbohm, E.: Der Einfluß von Rotlicht auf die negative Phototaxis des *Mougeotia*-Chloroplasten: Die Bedeutung eines Gradienten von P_{730} für die Orientierung. Z. Pflanzenphysiol. *55*, 278–286 (1966)
Schönbohm, E.: Die Bedeutung des Gradienten von $Phytochrom_{730}$ und die tonische Wirkung von Blaulicht bei der negativen Phototaxis des *Mougeotia*-Chloroplasten. Z. Pflanzenphysiol. *56*, 282–291 (1967a)

Schönbohm, E.: Die Hemmung der positiven und negativen Phototaxis des *Mougeotia*-Chloroplasten durch Jodid-Ionen. Z. Pflanzenphysiol. *56*, 366–374 (1967b)

Schönbohm, E.: Aktionsdichroismus bei der Starklichtbewegung des Chloroplasten von *Mougeotia* spec. Ber. Dtsch. Bot. Ges. *81*, 203–209 (1968)

Schönbohm, E.: Untersuchungen über den Einfluß von Photosynthesenhemmstoffen und Halogeniden auf die Starklicht- und Schwachlichtbewegung des Chloroplasten von *Mougeotia* spec. Z. Pflanzenphysiol. *60*, 255–269 (1969a)

Schönbohm, E.: Die Hemmung der lichtinduzierten Bewegung des *Mougeotia*-Chloroplasten durch p-Chlormercuribenzoat (Versuche zur Mechanik der Chloroplastenbewegung). Z. Pflanzenphysiol. *61*, 250–260 (1969b)

Schönbohm, E.: Über die Lokalisierung des Photoreceptors für den tonischen Blaulicht-Effekt bei der Verlagerung des *Mougeotia*-Chloroplasten im Starklicht. Z. Pflanzenphysiol. *65*, 453–457 (1971a)

Schönbohm, E.: Untersuchungen zum Photoreceptorproblem beim tonischen Blaulicht-Effekt der Starklichtbewegung des *Mougeotia*-Chloroplasten. Z. Pflanzenphysiol. *66*, 20–33 (1971b)

Schönbohm, E.: Die Wirkung von SH-Blockern sowie von Licht und Dunkel auf die Verankerung der *Mougeotia*-Chloroplasten im cytoplasmatischen Wandbelag. Z. Pflanzenphysiol. *66*, 113–132 (1972a)

Schönbohm, E.: Experiments on the mechanism of chloroplast movement in light oriented chloroplast arrangement. Acta Protozool. *11*, 211–223 +12 plates (1972b)

Schönbohm, E.: Kontraktile Fibrillen als aktive Elemente bei der Mechanik der Chloroplastenverlagerung. Ber. Dtsch. Bot. Ges. *86*, 407–422 (1973a)

Schönbohm, E.: Die lichtinduzierte Verankerung der Plastiden im cytoplasmatischen Wandbelag: Eine phytochromgesteuerte Kurzzeitreaktion. Ber. Dtsch. Bot. Ges. *86*, 423–430 (1973b)

Schönbohm, E.: Der Einfluß von Colchicine sowie von Cytochalasin B auf fädige Plasmastrukturen, auf die Verankerung der Chloroplasten sowie auf die orientierte Chloroplastenbewegung. Ber. Dtsch. Bot. Ges. *88*, 211–224 (1975)

Scholz, A.: Lichtorientierte Chloroplastenbewegung bei *Hormidium flaccidum*: Perception der Lichtrichtung mittels Sammellinseneffekt. Z. Pflanzenphysiol. *77*, 406–421 (1976a)

Scholz, A.: Lichtorientierte Chloroplastenbewegung bei *Hormidium flaccidum*: Verschiedene Methoden der Lichtrichtungsperception und die wirksamen Pigmente. Z. Pflanzenphysiol. *77*, 422–436 (1976b)

Schorer-Mörtel, G.: Reversible Form- und Volumenänderungen des *Mougeotia*-Chloroplasten. II. Die Kontraktion im Schwachlichtbereich und in Dunkelheit sowie Beziehungen zu den Orientierungsbewegungen. Z. Pflanzenphysiol. *68*, 193–214 (1972)

Seitz, K.: Das Wirkungsspektrum der Photodinese bei *Elodea canadensis*. Protoplasma *58*, 621–640 (1964)

Seitz, K.: Wirkungsspektren für die Starklichtbewegung der Chloroplasten, die Photodinese und die lichtabhängige Viskositätsänderung bei *Vallisneria spiralis* ssp. *torta*. Z. Pflanzenphysiol. *56*, 246–261 (1967a)

Seitz, K.: Eine Analyse der für die lichtabhängigen Bewegungen der Chloroplasten verantwortlichen Photorezeptorsysteme bei *Vallisneria spiralis* ssp. *torta*. Z. Pflanzenphysiol. *57*, 96–104 (1967b)

Seitz, K.: Zur Frage der Jodid-Wirkung auf die Starklichtbewegung der Chloroplasten von *Vallisneria spiralis* ssp. *torta*. Z. Pflanzenphysiol. *62*, 63–69 (1970a)

Seitz, K.: Die Starklichtbewegung der Chloroplasten von *Vallisneria* in Abhängigkeit von Hemmstoffen der oxydativen Phosphorylierung. Z. Pflanzenphysiol. *63*, 401–407 (1970b)

Seitz, K.: Die Ursache der Phototaxis der Chloroplasten: ein ATP-Gradient? (Versuche zum Primärprozess der Starklichtbewegung bei *Vallisneria*). Z. Pflanzenphysiol. *64*, 241–256 (1971)

Seitz, K.: Primary processes controlling the light induced movement of chloroplasts. Acta Protozool. *11*, 225–235 (1972)

Seitz, K.: Lichtabhängige Orientierungsbewegungen und ihre Regelung. Ber. Dtsch. Bot. Ges. *87*, 195–206 (1974)

Sekine, T., Yamaguchi, M.: Effect of ATP on the binding of N-ethylmaleimide to SH groups in the active site of myosin ATPase. J. Biochem. *54*, 196–198 (1963)

Senn, G.: Die Gestalts- und Lageveränderung der Pflanzen Chromatophoren. Leipzig: Wm. Engelmann, 1908

Sitte, P.: Zellfeinbau bei Plasmolyse. II. Der Feinbau der *Elodea*-Blattzellen bei Zucker- und Ionenplasmolyse. Protoplasma *57*, 304–333 (1963)
Smith, G.M.: On the reproduction of some Pacific coast species of *Ulva*. Am. J. Bot. *34*, 80–87 (1947)
Song, P.-S., Metzler, D.E.: Photochemical degradation of flavins—IV. Studies of the anaerobic photolysis of riboflavin. Photochem. Photobiol. *6*, 691–709 (1967)
Song, P.-S., Moore, T.A.: On the photoreceptor pigment for phototropism and phototaxis: is a carotenoid the most likely candidate? Photochem. Photobiol. *19*, 435–441 (1974)
Spudich, J.A.: Effects of cytochalasin B on actin filaments. Cold Spring Harbor Symp. Quant. Biol. *37*, 585–593 (1973)
Swift, E., Taylor, W.R.: Bioluminescence and chloroplast movement in the dinoflagellate *Pyrocystis lunula*. J. Phycol. *3*, 77–81 (1967)
Taylor, A.O., Bonner, B.A.: Isolation of phytochrome from the alga *Mesotaenium* and liverwort *Sphaerocarpos*. Plant Physiol. *42*, 762–766 (1967)
Taylor, B.L., Koshland, D.E. Jr.: Intrinsic and extrinsic light responses of *Salmonella typhimurium* and *Escherichia coli*. J. Bacteriol. *123*, 557–569 (1975)
van Rossum, G.D.V.: Net movements of calcium and magnesium in slices of rat liver. J. Gen. Physiol. *55*, 18–32 (1970)
Virgin, H.I.: The effect of light on the protoplasmic viscosity. Physiol. Plant. *4*, 255–357 (1951)
Virgin, H.I.: An action spectrum for the light induced changes in the viscosity of plant protoplasm. Physiol. Plant. *5*, 575–582 (1952)
Virgin, H.I.: Some effects of light on chloroplasts and plant protoplasm. In: Photophysiology. Giese, A.C. (ed.). New York-London: Academic Press, 1964, Vol. I. pp. 273–303
Virgin, H.I.: Light and chloroplast movements. Symp. Soc. Exp. Biol. *22*, 329–352 (1968)
Wagner, G.: Fluxes and compartmentation of potassium and chloride in the green alga *Mougeotia*. Planta *118*, 145–157 (1974)
Wagner, G., Bellini, E.: Light-dependent fluxes and compartmentation of calcium in the green alga *Mougeotia*. Z. Pflanzenphysiol. *79*, 283–291 (1976)
Wagner, G., Haupt, W., Laux, A.: Reversible inhibition of chloroplast movement by cytochalasin B in the green alga *Mougeotia*. Science *176*, 808–809 (1972)
Wagner, G., Klein, K.: Differential effect of calcium on chloroplast movement in *Mougeotia*. Photochem. Photobiol. *27*, 137–140 (1978)
Weber, W.: Das Polaritätsverhalten der *Enteromorpha*zelle bei Verzweigung und Schwärmerbildung. Z. Bot. *47*, 251–257 (1959)
Weisenseel, M.H.: Vergleichende Untersuchungen zum Einfluß der Temperatur auf lichtinduzierte Chloroplastenverlagerungen. I. Die Wirkung verschiedener Lichtintensitäten auf die Chloroplastenanordnung und ihre Abhängigkeit von der Temperatur. Z. Pflanzenphysiol. *59*, 56–69 (1968a)
Weisenseel, M.H.: Vergleichende Untersuchungen zum Einfluß der Temperatur auf lichtinduzierte Chloroplastenverlagerungen. II. Die statistische Bewegungsgeschwindigkeit der Chloroplasten und ihre Abhängigkeit von der Temperatur. Z. Pflanzenphysiol. *59*, 153–171 (1968b)
Weisenseel, M.H., Smeibidl, E.: Phytochrome controls the water permeability in *Mougeotia*. Z. Pflanzenphysiol. *70*, 420–431 (1973)
Wessells, N.K., Spooner, B.S., Ash, J.F., Bradley, M.O., Ludueña, M.A., Taylor, E.L., Wrenn, J.T., Yamada, K.M.: Microfilaments in cellular and developmental processes. Science *171*, 135–143 (1971)
Wieland, T.: Phallotoxins and microfilaments. In: Molecular Basis of Motility. Heilmeyer, L.M.G., Jr., Rüegg, J.C., Wieland, T. (eds.). Berlin-Heidelberg-New York: Springer 1976, pp. 203–213
Williamson, R.E.: Cytoplasmic streaming in *Chara*: a cell model activated by ATP and inhibited by cytochalasin B. J. Cell Sci. *17*, 655–668 (1975)
Wilson, L., Bryan, J.: Biochemical and pharmacological properties of microtubules. Adv. Cell Mol. Biol. *3*, 21–72 (1974)
Woodcock, C.L.F.: The anchoring of nuclei by cytoplasmic microtubules in *Acetabularia*. J. Cell Sci. *8*, 611–621 (1971)

Woollacott, R.M.: Microfilaments and the mechanism of light-triggered sperm release in ascidians. Dev. Biol. *40*, 186–195 (1974)
Woollacott, R.M., Porter, M.E.: A synchronized multicellular movement initiated by light and mediated by microfilaments. Dev. Biol. *61*, 41–57 (1977)
Zenk, M.H.: Untersuchungen zum Phototropismus der *Avena*-Koleoptile: I. Photooxydationen in vivo. Z. Pflanzenphysiol. *56*, 57–69 (1967)
Zurzycki, J.: Chloroplast arrangement as a factor in photosynthesis. Acta Soc. Bot. Pol. *24*, 27–63 (1955)
Zurzycki, J.: Studies on the centrifugation of chloroplasts in *Lemna trisulca*. Acta Soc. Bot. Pol. *29*, 385–393 (1960)
Zurzycki, J.: The influence of chloroplast displacements on the optical properties of leaves. Acta Soc. Bot. Pol. *30*, 503–527 (1961)
Zurzycki, J.: The action spectrum for the light depended movements of chloroplasts in *Lemna trisulca* L. Acta Soc. Bot. Pol. *31*, 489–538 (1962a)
Zurzycki, J.: The mechanism of the movements of plastids. In: Encyclopedia of Plant Physiology. Ruhland, W. (ed.). Berlin-Göttingen-Heidelberg: Springer, 1962b, Vol. XVII. pt. 2. pp. 940–978
Zurzycki, J.: The effects of simultaneous action of the short and long wave parts of spectrum on the movements of chloroplasts. Acta Soc. Bot. Pol. *33*, 133–139 (1964)
Zurzycki, J.: The energy of chloroplast movements in *Lemna trisulca* L. Acta Soc. Bot. Pol. *34*, 637–666 (1965)
Zurzycki, J.: Properties and localization of the photoreceptor active in displacements of chloroplasts in *Funaria hygrometrica*. I. Action Spectrum. Acta Soc. Bot. Pol. *36*, 133–142 (1967a)
Zurzycki, J.: Properties and localization of the photoreceptor active in displacements of chloroplasts in *Funaria hygrometrica*. II. Studies with polarized light. Acta Soc. Bot. Pol. *36*, 143–152 (1967b)
Zurzycki, J.: Properties and localization of the photoreceptor active in displacements of chloroplasts in *Funaria hygrometrica*. V. Studies on plasmolized cells. Acta Soc. Bot. Pol. *37*, 11–17 (1968)
Zurzycki, J.: Experimental modification of the reaction pattern of *Lemna* leaf cells to polarized light. Protoplasma *68*, 193–207 (1969)
Zurzycki, J.: Light respiration in *Lemna trisulca*. Acta Soc. Bot. Pol. *39*, 485–495 (1970)
Zurzycki, J.: Effect of linear polarized light on the O_2 uptake in leaves. Biochem. Physiol. Pflanz. *162*, 310–317 (1971)
Zurzycki, J.: Primary reactions in the chloroplast rearrangements. Acta Protozool. *11*, 189–199 (1972)
Zurzycki, J.: The effect of blue and red radiation on the volume changes of *Funaria* chloroplast *in vivo*. Photosynthetica *8*, 338–343 (1974)
Zurzycki, J., Lełatko, Z.: Action dichroism in the chloroplasts rearrangements in various plant species. Acta Soc. Bot. Pol. *38*, 493–506 (1969)

3. Locomotion in Microbial Plants

3.1 Mechanisms of Locomotion

3.1.1 Bacterial Flagella

R.M. MACNAB

For reviews of bacterial flagella and motility, see e.g., IINO (1969a), SMITH and KOFFLER (1971), DOETSCH (1971), BERG (1975a), HILMEN and SIMON (1976), IINO (1977), SILVERMAN and SIMON (1977), and earlier reviews cited therein. A less technical account is given by BERG (1975b). Reviews of bacterial chemotaxis obviously contain information on motility, and are cited in Chapter 3.2.2.

1. The Occurrence and Function of Motility in Bacteria

A very large number of bacterial species are motile. Motility of one sort or another is found among phototrophic, chemoautotrophic, and chemoheterotrophic groups; among Gram-negative and Gram-positive bacteria; among aerobes and anaerobes; among sporeformers and nonsporeformers; among filamentous as well as unicellular forms, and so on. Thus motility as a taxonomic criterion cuts across many lines, and makes any simple description of its occurrence impossible. However, a few generalizations can be made. Cocci are almost all nonmotile, whereas rods are very commonly motile. The few mycoplasma which move do so by gliding. Among the phototrophic bacteria, the cyanobacteria (formerly blue-green algae), if unicellular, are usually nonmotile, but filamentous species display gliding motility; the green bacteria are nonmotile; the purple bacteria are usually flagellated and display either helical flagellar motility or spirillar motility. Two orders of bacteria, the *Myxobacterales* and *Cytophagales*, are characterized by gliding motility. Most of the enterobacteria are peritrichously flagellated and motile. The pseudomonads are polarly flagellated and motile. Spirochetes are motile. Few coryneforms are motile. Bacteria which develop mycelia are invariably nonmotile at the mycelial stage, but may give rise to motile spores (e.g., in the genus *Actinoplanes*); this is in contrast to other spore-formers such as *Bacillus*, where the unicellular vegetative form is motile, but the spore nonmotile. *Rickettsia* are seldom motile. For information on the presence or absence of motility in a particular species, see BERGEY'S *Manual of Determinative Bacteriology* (1974).

Motility in vegetative cells presumably serves a dispersive function, reducing the likelihood of exhaustion of local food supply or excessive accumulation of excretion products, and providing access to a greater range of environments. In the case of the myxobacteria, motility permits the specific accumulation of cells to form the fruiting body. In many instances (see Chap. 3.2.2) motility is modified by environmental stimuli, allowing directed migration to a more favorable environment. However, in view of the many bacterial species which are nonmotile, without possessing any obvious compensatory features, the extent of the evolutionary advantage of motility is unclear.

2. Basic Types of Motility

All examples of motility in bacteria may be divided into two main types, those employing flagella and those employing other means. [It should be emphasized that although the term *flagellum* is used for the motor organelle of both pro- and eukaryotes, the two structures are vastly different (see the comparison on p. 244). SATIR and OJAKIAN (Chap. 3.1.2, this vol.) favor the synonymous term *cilium* for the eukaryotic organelle.]

Since the techniques for studying motility have frequently been developed in the context of chemotaxis, they are described in Chap. 3.2.2.

2.1 Flagellar Motility

Swimming and Swarming by Means of External Helical Flagella. This type of motility, particularly *swimming*, is very widespread—*Escherichia, Salmonella, Pseudomonas,* and *Bacillus* are just a few examples of common genera employing external helical flagella. The helical waveform propagates against the viscous resistance of the medium to provide the thrust for motility (the mechanism of propagation of this waveform is considered later). Whether a cell possesses one polar flagellum only (as in *Pseudomonas*) or many randomly located flagella (as in *Escherichia*), swimming in liquid media proceeds by a single coordinated helix which in multiflagellate cells is a bundled aggregate (Fig. 1a) of the individual flagella (Fig. 1b). Swimming velocities are high, typically about 20–60 $\mu m\ s^{-1}$ in water (SCHNEIDER and DOETSCH, 1974). A characteristic feature of swimming motility is its frequent interruption, by backing up in the case of monoflagellates, or bundle dispersal and tumbling (i.e., chaotic, end-over-end motion of the cell body) in multiflagellates. As will be seen later, this is the means by which tactic responses are expressed.

In contrast to swimming, which is an individual act, some flagellated genera (notably *Proteus*) move across solid surfaces in groups (rafts) of cells (Fig. 1d). This behavior is termed *swarming*. Bacteria which swarm are long (up to 80 μm, and frequently multinucleate and aseptate) and far more highly flagellated (up to 1000 flagella per cell) than swimming cells. Swarming can occur on fairly dry agar surfaces. Individual cells cannot swarm; presumably the cooperative thrust of many cells, in bullet-shaped rafts, is necessary to overcome the high viscous resistance. Until recently, it was thought that swarming behavior occurs as a chemotactic response to excretion products, but there is now good evidence against this (WILLIAMS et al., 1976). The coordination of many cells is probably a passive mechanical interaction which, because of cooperative characteristics (cf. bundle formation, below) result in substantial net thrust. Swarming velocities are much lower (ca. 1 $\mu m\ s^{-1}$) than swimming velocities of free individual cells (HENRICHSEN, 1972).

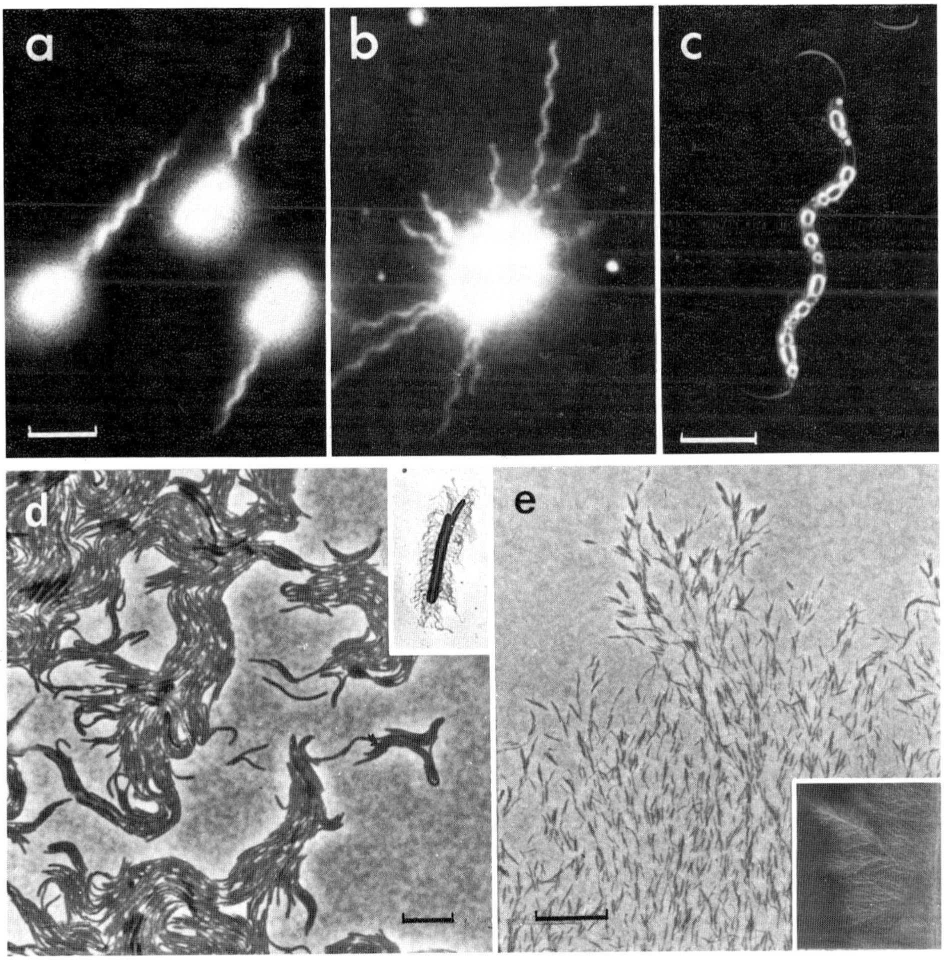

Fig. 1a and b. Swimming by means of rotating helical flagella, *Salmonella typhimurium*. **a** Swimming cell, **b** Stationary cell. The flagella originate randomly and coalesce in a bundle during swimming. In certain other species, the flagella originate polarly. Bar = 5 μm. (From MACNAB and ORNSTON, 1977). **c** Spirillar motility: *Spirillum volutans* swims by rotating polar flagellar tufts; the helical cell body, rotating in reaction, generates thrust. Bar = 10 μm. (Original); (**a**, **b** and **c**, dark-field illumination, 500 W xenon arc). **d** Swarming: *Proteus mirabilis* traveling across an agar surface in rafts of cells. The thrust is provided by prolific flagellation (stained individual cell, *inset*). Bar = 20 μm (10 μm on *inset*). **e** Gliding: Nonflagellar motility across moist surfaces, by an unknown mechanism, exemplified here by *Flexibacter columnaris*. Note delicate feathery structure of the growth pattern viewed at low magnification *(inset)*. Bar = 50 μm (13 mm on *inset*). **d** and **e** from HENRICHSEN, 1972

Swimming and swarming appear to be quite distinct behavioral manifestations of the same basic mechanism, namely propagation of a helical flagellar waveform. In fact, the same species can display either behavior depending on whether growth conditions favor a short or an extended cell shape.

Spirillar Motility. [For a review of the biology of spirilla, including their motility, see KRIEG (1976).] Found in the phototrophic genera *Rhodospirillum* and *Thiospiril-*

lum, and also in the chemoheterotrophic genus *Spirillum*, spirillar motility is characterized by opposite rotation of a *helical cell body* in reaction to rotation of flagellar tufts (typically 75 flagella per tuft) at the pole (Fig. 1c). There may be some helicity in the flagella themselves, but it is usually much less than one period in length. Judging from the nonhelical shape of the cones executed by the tufts, the function of the flagella in these organisms is to apply torque to the cell body which then, because of its helical shape, develops the thrust for motility. [WINET and KELLER (1976) consider helicity of the tufts to be significant (see Sect. 4.4)]. The swimming of spirilla is reversible and symmetric, i.e., there is no permanent head-to-tail relationship. Velocities are high, of the order of 50 μm s^{-1}. In bipolarly flagellated forms (cf. *Thiospirillum*, which is flagellated at one pole only), the two flagellar tufts are highly coordinated (head–tail) but can be made anti-coordinated (head–head or tail–tail) by certain drugs.

Spirochetal Motility. Bacteria in the order *Spirochaetales* have, like spirilla, a helical body shape. Superficially they appear nonflagellate, but in fact they possess filaments constrained within a somatic sheath. Their motility may take a variety of forms: rotation about the helical body axis (often with translation), flexing, twisting, and creeping on semi-solid surfaces. The overall helicity of the body can be substantially distorted in the process of these movements. It is believed by BERG (1976), in a development of ideas by WANG and JAHN (1972) and CHWANG et al. (1974), that motility is achieved by rotation of the helical cell within the somatic sheath, as a consequence of rotation of the contained flagella. Because the sheath conforms to the body shape, it propagates helical waves against the viscous resistance of the medium, and so thrust is developed.

2.2 Nonflagellar Motility

The categories given below follow the recommendations in the review of surface translocation by HENRICHSEN, 1972.

Gliding. Exemplified by filamentous species of cyanobacteria, the myxobacteria, and the order *Cytophagales*, gliding is a form of motility requiring a solid surface covered with an appreciable liquid film (unlike swarming, which can occur under much drier conditions). (For a detailed discussion of gliding mechanisms, see HALFEN, Chap. 3.1.3, this vol.). The cells have no external organelles and proceed axially at relatively low speeds (less than 8 μm s^{-1}, and often much less). Like swarming, gliding is sometimes a cooperative event among groups of cells (Fig. 1e). The mechanism of gliding is unknown, although PATE and ORDAL (1967) found cytoplasmic tubular rods (rhapidisomes) and peripheral fibrils in *Chondrococcus* (now *Flexibacter*) *columnaris*. SCHMITT-LORENZ and KÜHLWEIN (1968) found, in myxobacteria, cytoplasmic fibrils which they claim are the motor organelles.

In some mycoplasmas, gliding motility at speeds of ca. 0.5 μm s^{-1} is observed. Glass or a similar surface is required, and appreciable cell deformation occurs, frequently with a marked leading stalk (BREDT and RADESTOCK, 1977).

Twitching. This type of motility was not described until quite recently (LAUTROP, 1961), in an *Acinetobacter* species. It is now recognized in several Gram-negative cocci, including *Neisseria* and *Moraxella*, and also appears in a few *Pseudomonas* species. Twitching consists of intermittent jerky displacements of the entire body with quiescent periods in between. The time-averaged velocity is very low (ca. 0.03 μm s^{-1}). There is a tendency for twitching cells to displace as a group, but individual cells are also capable of this movement. Twitching seems to be confined to hydrophobic boundaries (e.g., air or oil) of aqueous media, and it is suggested that the function in twitching of the polar fimbriae (thin, straight, extracellular filaments) of these organisms (HENRICHSEN, 1975) may be to anchor the cells to the hydrophobic surface rather than to generate movement. Various mechanisms, including passive ones involving random fluctuations in parameters such as surface tension or bacterial surface charge density, have been suggested for twitching motility, but without supporting evidence. The mechanism must therefore still be regarded as unknown.

Sliding. Although not examples of motility in the usual sense, the phenomena of *sliding* and (below) *darting* do permit cell populations to sample an extended environment. Some bacteria, e.g., some *Streptococcus* and *Flavobacterium* species, when grown on solid surfaces such as agar, spread rapidly (up to 0.5 μm s^{-1}) as a monolayer as cell division proceeds, instead of spreading slowly (e.g. *E. coli*, 5 nm s^{-1}) to form compact colonies. Increasing cell number manifested in this way is termed *sliding*, and results from unusually low friction between the cell and the solid surface.

Darting. Another form of spreading resulting from growth pressure is darting, which occurs in a number of *Staphylococcus* species, where growth and cell division take place in an encapsulated cluster from which diplococci are intermittently ejected. As a means of dispersal it bears an obvious resemblance to the explosion of seed pods in higher plants.

3. Structure of Bacterial Flagella

This section and the following one on mechanism will cover only flagellar motility. Information regarding structure and mechanism of nonflagellar motility was included above in the description of basic types of motility; see also HALFEN (Chap. 3.1.3, this vol.) on gliding movement.

3.1 Flagellar Structure and Cell Morphology

[For the pattern of flagellation in any particular species, see LEIFSON'S *Atlas of Bacterial Flagellation* (1960)]. The term *flagellum* is used here to denote the entire organelle. It consists (Fig. 2) of a long slender flagellar *filament*, with a more or less pronounced helicity, a short curved connection to the cell surface (the *hook*), and the *basal body*, imbedded in the cell surface. A species may possess only one flagellum per cell, in which case it is usually located at one pole of the approximately rod-shaped cell body (polar monoflagellation, e.g., in *Pseudomonas*, Fig. 3a), although sub-polar flagellation also exists. Multiflagellate species may be polarly flagellated at one pole (as in *Chromatium okenii*) or both poles (e.g., *Spirillum* spp., Fig. 1c), or the flagella may be randomly located around

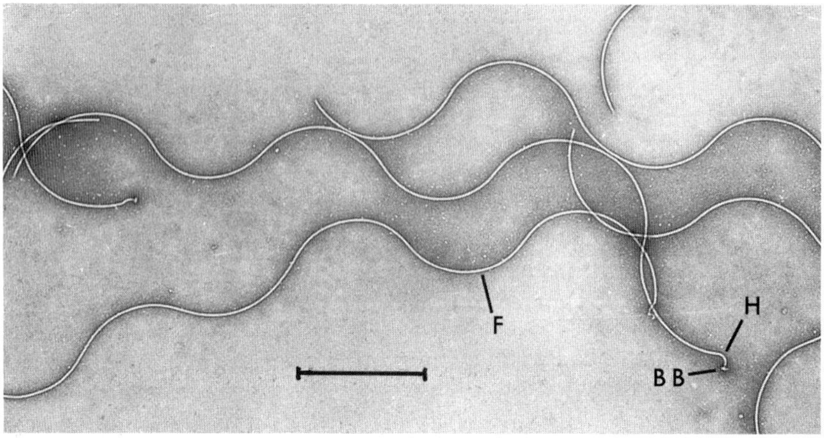

Fig. 2. Purified intact flagella from *B. subtilis*. The helical filament *F* is connected by the hook *H* to the basal body *BB*, which is anchored in the cell surface. Negatively stained with phosphotungstic acid. Bar = 1 μm. (After DIMMITT and SIMON, 1971)

the cell surface, an arrangement described as *peritrichous flagellation* (e.g., *E. coli*, Fig. 3b). Polar tufts may consist of up to 80 flagella [in *Spirillum volutans* (WILLIAMS and CHAPMAN, 1961)]; peritrichous flagellation requires a minimum of three to five flagella for an effective bundle (see below) and up to ten flagella/cell are commonly encountered in the enteric bacteria. In swarming species, the number of flagella per cell rises dramatically, partly because of an increased density per unit cell length and partly because of extra cell length.

The spirochetes represent a special case of flagellar arrangement (Fig. 3c). The flagella occur as two sub-polar fascicles or tufts (with up to 50 flagella/fascicle in some species), which are contained within an outer sheath enveloping the body. These fascicles lie back along the cell body, and overlap near the center.

3.2 The Flagellar Filament

The flagellar filament represents the great majority (about 95%) of the mass of a flagellum. Most bacteria have filaments built from a single protein, *flagellin*, [molecular weight typically 53,000 daltons (KONDOH and HOTANI, 1974; HILMEN et al., 1974)] but some species [e.g., *Rhizobium lupini* (SCHMITT et al., 1974a) and *Pseudomonas rhodos* (SCHMITT et al., 1974b)] have more complex structures. *No*

Fig. 3a–c. Morphology of flagellation: **a** Polar monotrichous flagellation. *Pseudomonas aeruginosa*, negatively stained with phosphotungstic acid. Bar = 0.5 μm. (From SUZUKI and IINO, 1977). **b** Peritrichous flagellation. *E. coli*, shadowed with platinum. Bar = 0.5 μm. (From BERG, 1975b). **c** Spirochetal flagellation. An oral spirochete, negatively stained with phosphotungstic acid. Flagella are inserted sub-polarly *(arrows)* and lie along the cell. Viewed in transverse section *(inset)* flagellar filaments *f* are seen to lie along the cell between cell membrane *cm* and somatic sheath *cw*. (From STANIER et al., 1976; after LISTGARTEN and SOCRANSKY, 1964)

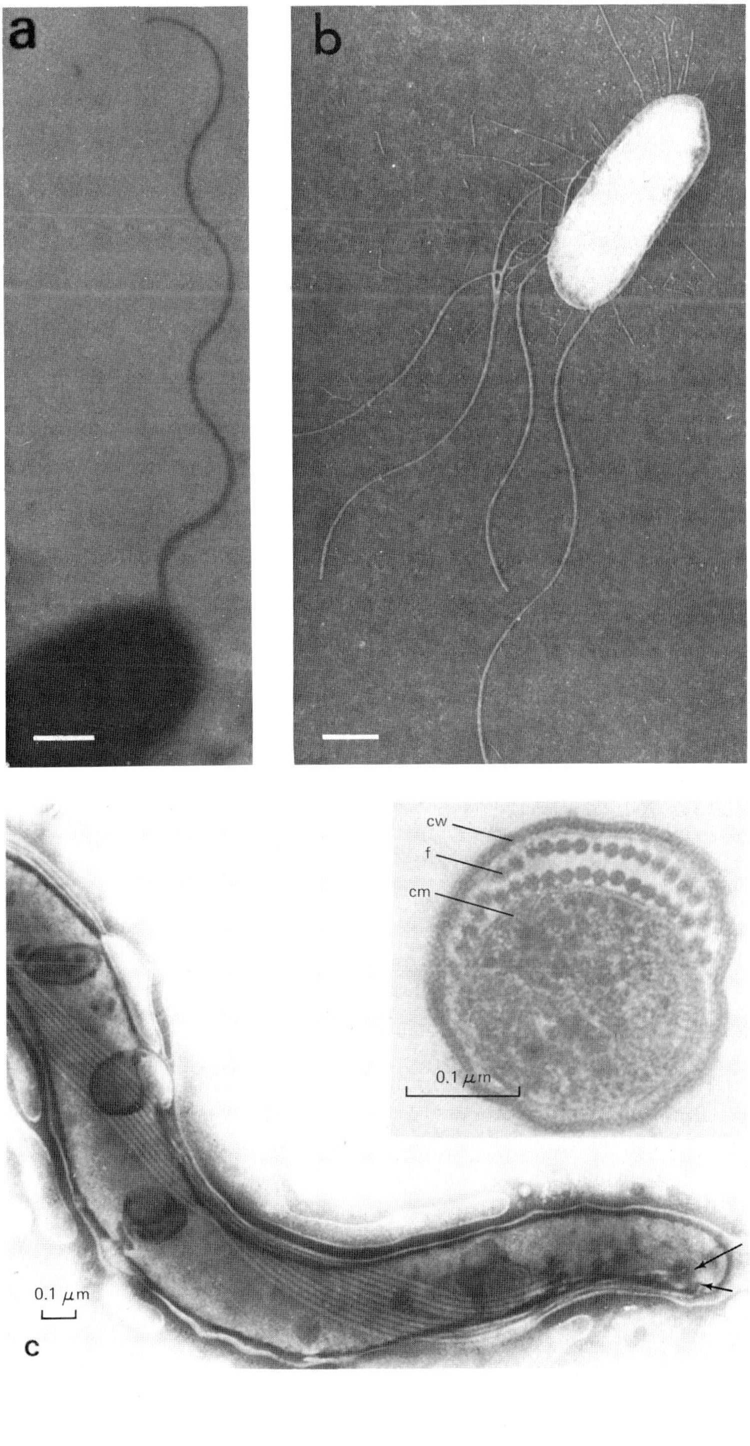
a
b
cw
f
cm
0.1 μm
0.1 μm
c

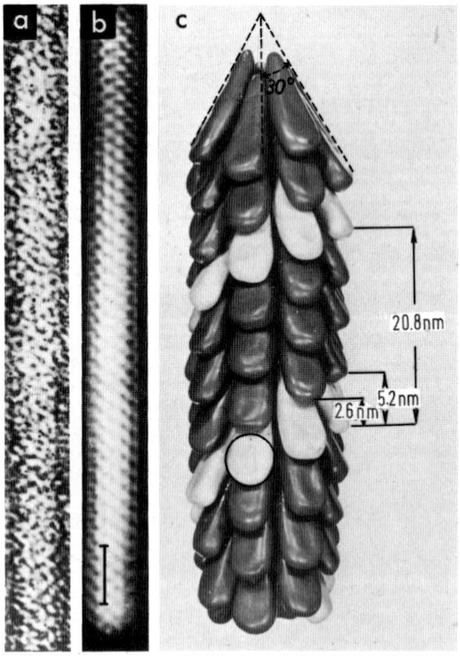

Fig. 4a–c. Structure of flagellar filaments. The filament is a hollow cylindrical structure built from subunits of the protein flagellin arranged in approximately longitudinal columns and multi-start helices. **a** and **b** are, respectively, raw and optically filtered electron micrographs of a filament from a straight flagellar mutant of *E. coli.* Bar = 20 nm. (From KONDOH and YANAGIDA, 1975). **c** Model taking into account elongated shape of subunit, and arrow-like appearance of detached filaments. (From BODE et al., 1972). [Note: The number of longitudinal columns, shown as 8 in this model, is now believed to be 11]

enzymatic activity resides in the filament (e.g., ENOMOTO, 1966). It is an extremely thin but quite rigid structure (20 nm diameter) of length 5–10 μm, which only recently has been visualized in the light microscope (MACNAB and KOSHLAND, 1974; MACNAB, 1976). The helicity, which is intrinsic, has characteristic parameters for any given species. In *Salmonella*, for example, these are: wavelength 2.3 μm, helical diameter 0.4 μm (KAMIYA and ASAKURA, 1976). The protein subunits are elongated ellipsoids (BODE et al., 1972) stacked radially in a cylindrical structure (Fig. 4) (O'BRIEN and BENNETT, 1972; KONDOH and YANAGIDA, 1975) in the case of a straight flagellar mutant. In wild-type filaments the structure possesses a slight twist and curvature, which therefore give rise to the characteristic helical shape. The helicity is left-handed in the wild-type species characterized so far (WEIBULL, 1950; MACNAB and KOSHLAND, 1974), but a variety of other stacking arrangements are possible under special circumstances, giving rise to a whole family of helical forms (Fig. 5), including ones with a right-handed helical sense (SHIMADA et al., 1975). This *polymorphism* (ASAKURA and IINO, 1972) is displayed by flagellar mutants (IINO, 1962) and, in wild-type cells, by physicochemical (e.g., pH or ionic strength) manipulation (ABRAM and KOFFLER, 1964; KAMIYA and ASAKURA, 1976) or by mechanical force (MACNAB and ORNSTON, 1977; HOTANI, unpublished observations). Mechanically induced transitions are an important feature of bacterial motility (see below).

The geometrical regularity of flagellar filaments reflects the fact that the filament is a "one-dimensional crystal," which can indeed be reconstituted in vitro from flagellin monomer; no enzymes are involved (ASAKURA, 1970; SMITH and KOFFLER, 1971; IINO, 1974). A theoretical analysis of the underlying principles of subunit

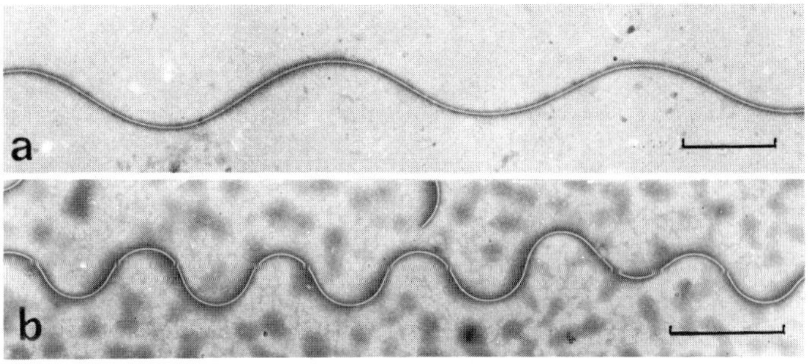

Fig. 5a and b. Polymorphism. Flagellar filaments can display a variety of discrete structures, exemplified here by **a** "normal" filament of *Salmonella* and **b** filament from "curly" mutant. Bar = 1 μm. (From ASAKURA and IINO, 1972). In wild-type cells, transitions between the two forms occur by mechanical force, and are important in expressing tactic responses

stacking has been given (CALLADINE, 1976; CALLADINE, 1978), which nicely explains the twist-curvature relationships observed within the family of polymorphic forms in *Salmonella*. Essential to the analysis is the concept of *quasi-equivalence* (KLUG, 1967), in which identical subunits occupy almost, but not quite, identical environments; with true equivalence, only straight filaments could develop.

3.3 The Hook and Basal Body

The hook is a short, approximately L-shaped structure which is slightly thicker than the filament, and is built from a single protein, molecular weight about 42,000 daltons (HILMEN et al., 1974; KAGAWA et al., 1976), which is antigenically distinct from flagellin. The function of the hook is not known, but it may be to couple the filament flexibly to the basal body.

The basal body is a complex structure, as might be expected in view of its function as the motor. Figure 6a shows hook/basal-body structures from *E. coli*, and Figure 6b is an idealized structure based on such electron micrographs (DEPAMPHILIS and ADLER, 1971). The structure comprises four rings, although the outer two (P and L) are specific to Gram-negative cells and probably serve as anchoring structures. The inner two rings (M and S), found in both Gram-positive and Gram-negative cells, are thought to be motor elements, the innermost being associated with the cell membrane. Given that the motor is a rotary device (see below), a plausible model has one ring acting as the "rotor" and the other the "stator", although conceivably the rotary action is developed between the rings and the interconnecting rod.

3.4 Flagellar Assembly

In multiflagellate species, the number of flagella per cell can be influenced by growth conditions. In *E. coli*, for example, growth above 40° C, or in the presence of metabolites such as glucose (ADLER and TEMPLETON, 1967), results in reduced

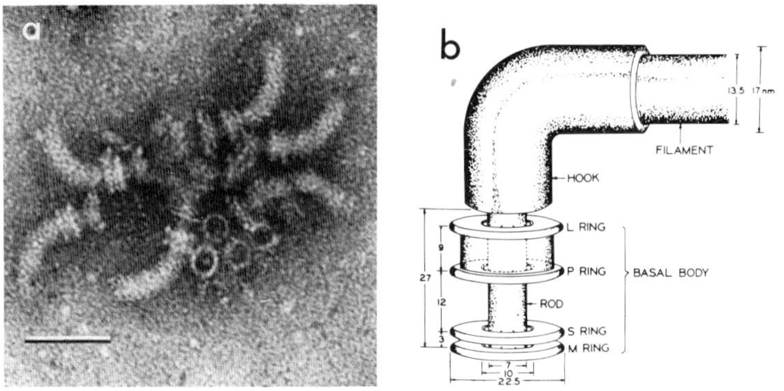

Fig. 6a and b. Structure of flagellar basal region. **a** Detached purified hook/basal-body complexes from *E. coli*, negatively stained with phosphotungstic acid. Bar = 50 nm. (From HILMEN and SIMON, 1976). **b** Model based on similar electron micrographs. The *M ring* lies in the plane of the cytoplasmic membrane, the *S ring* in the periplasmic space, the *P ring* in the peptidoglycan (cell wall) layer, and the *L ring* in the lipopolysaccharide (outer membrane) layer. (From DEPAMPHILIS and ADLER, 1971)

flagellation. The glucose effect has since been shown to be mediated by cyclic AMP (YOKOTA and GOTS, 1970). Ubiquinone deficiency, for unknown reasons, inhibits flagellar synthesis (BAR TANA et al., 1977); so also do various modifications to the outer membrane and cell wall (MCGROATY et al., 1973; AMES et al., 1974; KOMEDA et al., 1977).

Flagellar filaments are self-assembling, i.e., they require no enzymatic mediation. Growth proceeds at one end only, as was first demonstrated in vitro; in vivo, this is found to be the end *distal* to the cell (IINO, 1969b; EMERSON et al., 1970). Presumably this is achieved by extrusion of subunits through the flagellar core, but this has never been directly demonstrated. The growth rate decreases exponentially with length, indicating a decreasing rate at which monomers can be transported to the tip. Termination of filament elongation by faulty insertion of monomer is observed in vitro (HOTANI and ASAKURA, 1974), but this does not seem to be significant in vivo (IINO, 1969b).

The genetics of flagellar assembly and function (IINO, 1969a; HILMEN and SIMON, 1976) is intricate and not yet well understood. At least 20 genes are involved, excluding chemoreceptor genes. (See Sect. 7 and Fig. 5 of Chap. 3.2.2 for a summary of the genetics of motility and taxis.) The *hag* gene is a structural gene for flagellin; in *Salmonella*, two distinct flagellin genes exist, *H1* and *H2*, but only one is expressed at any given time. Oscillation between H1 and H2 occurs about 1 per 100 generations, giving rise to the phenomenon known as phase variation. A variety of mutants exists which are nonflagellate (*fla*$^-$ phenotype) in spite of normal *hag* genes; of the ten or so *fla* genes, some are structural (for components of the basal body or hook) while others have a regulatory role. A paralyzed phenotype is caused by defects in the *mot* operon. One *mot* gene product has been identified as a protein which is membrane-bound, but not part of the basal body (SILVERMAN et al., 1976). Very little is known of the assembly of the hook or basal body; one *fla* gene (*flaE*) regulates termination of hook assembly (SILVERMAN and SIMON,

1972). The structural gene for the hook protein has recently been identified as *flaK* (MATSUMARA et al., 1977).

4. Mechanism of Flagellar Motility

4.1 The Energy Source

The bacterial flagellum is driven by protons flowing across the cytoplasmic membrane (LARSEN et al., 1974a; THIPAYATHASANA and VALENTINE, 1974; MANSON et al., 1977). The protonmotive force can be obtained either directly from respiration or from ATP hydrolysis. Note that in the latter case the use of ATP to obtain a protonmotive force, which then generates mechanical force, is quite different from the direct use of ATP to generate mechanical force, in eukaryotic systems such as cilia or muscle.

WINET and KELLER (1976) estimate for *Spirillum volutans* that the energy expended in motility is small, of the order of 0.1% of the total energy available to the cell.

4.2 Evidence for a Rotary Motor

The existence of a defined helicity in the filaments, as shown by electron microscopy and dark-field light microscopy, and observations of propagation of the helical waveform by the latter technique, does not permit one to distinguish between a mechanism whereby the filament is forced to undergo true rotation at the base, and one in which it is driven through cyclic conformational changes. (The two mechanisms can be demonstrated by turning a corkscrew while holding onto a piece of elastic tubing sleeved over it. Both the corkscrew and the tubing propagate helical waves, the former by a rotational mechanism and the latter by a conformational mechanism.) A variety of evidence strongly supports the rotary mechanism for bacterial flagella (BERG and ANDERSON, 1973; SILVERMAN and SIMON, 1974; BERG, 1974; LARSEN et al., 1974b). Evidence (Fig. 7) includes the behavior of cells tethered by a flagellum or polyhook to a glass surface. Perhaps the most convincing single piece of evidence, however, is the mutual rotation of cell pairs, with flagella joined by antiflagellar antibody, of a mutant which possesses filaments which are *straight* and therefore are *incapable* of performing macroscopic conformational work. Although the conformational model has been intuitively more appealing to some workers (e.g., HARRIS, 1973), no convincing evidence has been put forward in its favor, and the rotational model is now generally accepted by most. However, a direct demonstration of rotating flagella is still needed. BERG (1974) considered the frictional and bearing properties of the cell surface, and found nothing which would rule out the idea of a rotary motor imbedded in it. ADAM (1977) has calculated that viscous streaming of the cell membrane is an energetically feasible means of applying drive to the motor.

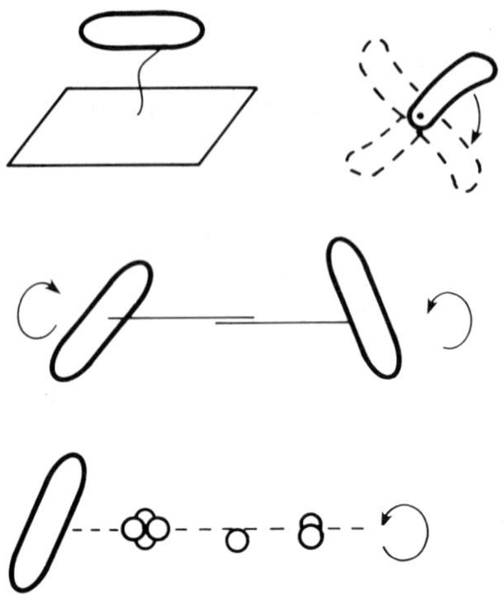

Fig. 7 a–c. Evidence for flagellar rotation. **a** Cell tethered by single filament or hook structure to glass surface. Viewed from above, cell body rotates (precessional motion is ruled out by behavior of asymmetric features). **b** Cell pairs of straight flagellar mutant, tethered to each other by antibody, rotate mutually. **c** Latex bead clusters attached by antibody to filament of straight flagellar mutant. Clusters rotate about the (inferred) axis of the filament, which could not be seen in this experiment. Drawings based on *E. coli* experiments of SILVERMAN and SIMON (1974)

4.3 Reversibility of the Motor

Another remarkable feature of the flagellar motor is its ability to rotate in either sense. In the absence of stimulation, reversals in *E. coli* and *Salmonella* occur randomly between the counterclockwise sense (viewed along the filament into the cell), which produces swimming, and the clockwise sense, which produces tumbling. The correlation between rotational sense and swimming and tumbling was made by comparing free and tethered cells of various mutants, and of cells under chemotactic stimulation (LARSEN et al., 1974b). One presumes that the protonmotive force which is the energy source for motility remains directed inward across the cytoplasmic membrane (since it is required for a variety of transport processes as well as motility), regardless of the sense of rotation of the motor. This implies the existence of a "gear" which can direct mechanical force in either of two opposing senses. The nature of this gear is obscure at the present time. One is also led to postulate a signal which instructs the gear to switch. Chemical and electrical signalling models have been proposed (see Chap. 3.2.2, Sect. 6.5).

4.4 Hydrodynamics of Propulsion

For organisms as small as bacteria, with dimensions of about 1 μm, and swimming velocities of about 10 $\mu m\ s^{-1}$, inertial forces are utterly negligible – only viscous forces need be considered (TAYLOR, 1952). Hydrodynamic analyses (HOLWILL and BURGE, 1963; CHWANG and WU, 1971; COAKLEY and HOLWILL, 1972) indicate that the body: filament diameter ratio and the wavelength: helical-amplitude ratio of bacteria are nearly optimal for efficient swimming. Experimental values of the velocities of translation, body rotation, and wave propagation are in reasonable agreement with the theoretical analysis (YOSHIDA et al., 1975; SHIMADA et al., 1976). COAKLEY and HOLWILL (1972) pointed out that hydrodynamic analysis could not distinguish between rotational and conformational models of propulsion, since they are essentially equivalent in terms of efficiency; the major work is done by lateral displacement of the filament, which applies in either case. They also noted that bundle formation results in a large increase in efficiency compared with that which would be achieved by the same number of flagella operating separately.

The hydrodynamics of spirillar motility has also been analyzed (CHWANG et al., 1972; WINET and KELLER, 1976). The latter authors conclude, partly on the basis of tuft images in cine recordings, and partly on comparison of theoretical and measured velocities, that the tufts cannot simply be a rotating cone or paraboloid, but must possess appreciable helicity. They were unable to decide whether this helicity is a result of elastic deformation, or is a structural feature of the flagella.

4.5 Bundle Formation and Function

The flagella of peritrichous bacteria are left-handed helices which therefore, when rotating counterclockwise, provide a pushing force on the cell. This results in formation of a bundle aligned more or less exactly with the long axis of the cell (Fig. 1a). Formation and synchronous function of the bundle can be explained by mechanical and hydrodynamic considerations alone (ANDERSON, 1975; MACNAB, 1977). There is no need to invoke active control by the cell, nor is there a problem of tangling among the individual rotating helices in the bundle, provided the sense of rotation is counterclockwise. Right-handed wrapping of the left-handed bundle can occur (by hydrodynamic force), to produce a very coherent bundle which can operate indefinitely. These points have been confirmed on a working model by MACNAB (1977).

4.6 Tumbling in Peritrichous Bacteria (*E. coli* and *Salmonella*)

The swimming mode of motility is, in wild-type cells, interrupted every second or two by brief chaotic motility called *tumbling* (mutants with abnormally low or high incidence of tumbling also exist). In experiments with cells tethered by a flagellum to a glass surface, LARSEN et al. (1974b) clearly showed that this correlates with clockwise rotation of the motor. However, simple clockwise rotation of a left-handed flagellar bundle should lead to jamming, not tumbling (MACNAB, 1977). Yet, without bundle dispersal, tumbling could not occur. It has now been shown that tumbling involves much more than simple reversal of rotation. The filaments, as a result of the reverse viscous torque, undergo transition from the normal (left-handed) form to an alternative right-handed crystalline form [the "curly" form observed in flagellar mutants by IINO (1962)] and, during the transition, filaments are in a heteromorphous state (Fig. 8) with a characteristic "dog-leg" between two segments of opposite helical sense (MACNAB and ORNSTON, 1977). The large resistance which such a form presents to the medium, and the presence of mixed helical sense in heteromorphous filaments, is presumed to be a major contribution to tumbling.

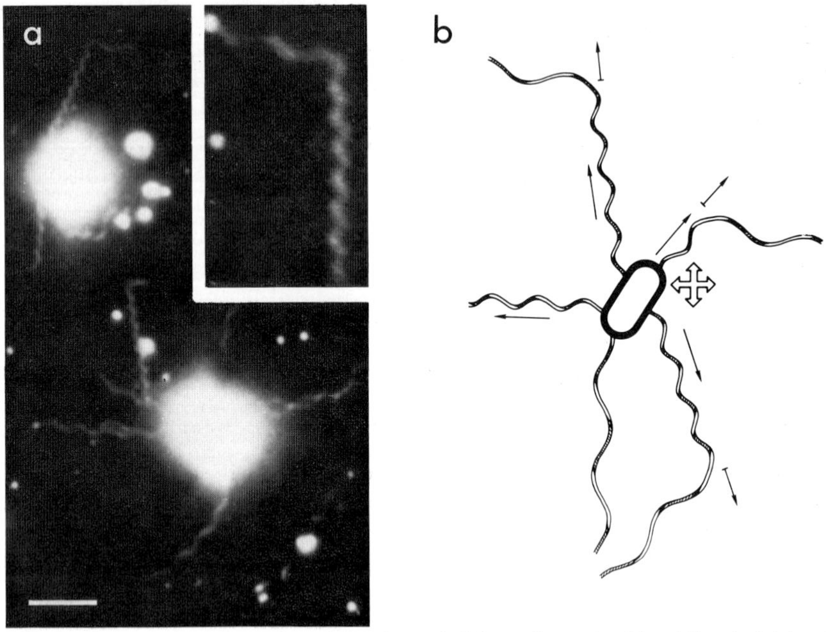

Fig. 8a and b. Mechanism of tumbling in peritrichous bacteria: Tumbling is initiated by motor reversal to clockwise rotation. The resulting right-handed torsion destabilizes the normal (left-handed) helical structure, and causes a transition to the curly (right-handed) form. The flagellar bundle (Fig. 1a) flies apart, and chaotic tumbling ensues until the motors resume counterclockwise rotation. **a** Tumbling mutant of *Salmonella*. Note presence of both waveforms, in some cases on the same filament. These heteromorphous filaments have discontinuity in their axial directions, seen more clearly in the detached filament *(inset)*, which was obtained by biochemical manipulation (*p*-fluorophenylalanine incorporation), and is stable in the heteromorphous state. Bar = 5 μm (2.5 μm on *inset*); **b** Cartoon of a tumbling cell. Direction of wave propagation is shown by *arrows; arrows with backbars* indicate propagation of the transition point, and the tumbling movement of the cell body is symbolized by *crossed open arrow*. (After MACNAB and ORNSTON, 1977)

4.7 The Basic Motility Pattern and Its Modification by Environmental Stimuli

(See Chap. 3.2.2 for a more detailed description.) The alternation between swimming and tumbling produces a trajectory of approximately straight line segments joined at random angles. Chemical gradients modify this random walk, not by introducing any specificity into the directional changes caused by tumbling, but by altering the probability that a tumble will occur. Migration therefore occurs because there is a bias in the time spent in different segments of the random walk (Fig. 3 in Chap. 3.2.2). Note that both swimming and tumbling modes of motility are essential for systematic migration up or down a chemical gradient—in fact, mutants which have either an unusually low or high incidence of tumbling are incapable of responding to environmental stimuli.

5. Concluding Remarks

The bacterial flagellum is proving to be a motor organelle with remarkable properties. It consists of a rigid, crystalline, helical filament which is driven like a "brass propeller" by a rotary motor at the base; the motor can rotate forward or in reverse; the power source is the proton potential across the membrane; other crystalline forms are induced by mechanical force and precipitate directional changes; and operation of the whole system is under the control of environmental stimuli. (See Fig. 6 in Chap. 3.2.2.) Elucidation of the molecular details of how all of this is achieved is a difficult but exciting task.

References

Abram, D., Koffler, H.: *In vitro* formation of flagella-like filaments and other structures from flagellin. J. Mol. Biol. *9*, 168–185 (1964)

Adam, G.: Rotation of bacterial flagella as driven by cytomembrane streaming. J. Theor. Biol. *65*, 713–726 (1977)

Adler, J., Templeton, B.: The effect of environmental conditions on the motility of *Escherichia coli*. J. Gen. Microbiol. *46*, 175–184 (1967)

Ames, G.F., Spudich, E.N., Nikaido, H.: Protein composition of outer membrane of *Salmonella typhimurium:* Effect of lipopolysaccharide mutations. J. Bacteriol. *117*, 406–416 (1974)

Anderson, R.A.: Formation of the bacterial flagellar bundle. In: Swimming and Flying in Nature, Vol. I. Wu, T.Y.-T., Brokaw, C.J., Brennen, C. (eds.). New York: Plenum Press, 1975, pp. 45–56

Asakura, S.: Polymerization of flagellin and polymorphism of flagella. Adv. Biophys. *1*, 99–155 (1970)

Asakura, S., Iino, T.: Polymorphism of *Salmonella* flagella as investigated by means of *in vitro* copolymerization of flagellins derived from various strains. J. Mol. Biol. *64*, 251–268 (1972)

Bar Tana, J., Howlett, B.J., Koshland, Jr., D.E.: Flagellar formation in *Escherichia coli* electron transport mutants. J. Bacteriol. *130*, 787–792 (1977)

Berg, H.C.: Dynamic properties of bacterial flagellar motors. Nature (London) *249*, 77–79 (1974)

Berg, H.C.: Bacterial behaviour. Nature (London) *254*, 389–392 (1975a)

Berg, H.C.: How bacteria swim. Sci. Am. *233*, 36–44 (1975b)

Berg, H.C.: How spirochetes may swim. J. Theor. Biol. *56*, 269–273 (1976)

Berg, H.C., Anderson, R.A.: Bacteria swim by rotating their flagellar filaments. Nature (London) *245*, 380–382 (1973)

Bergey's Manual of Determinative Bacteriology, 8th ed. Buchanan, R.E., Gibbons, N.E. (eds.). Baltimore: The Williams & Wilkins Company, 1974

Bode, W., Engel, J., Winklmair, D.: A model of bacterial flagella based on small-angle X-ray scattering and hydrodynamic data which indicate an elongated shape of the flagellin protomer. Eur. J. Biochem. *26*, 313–327 (1972)

Bredt, W., Radestock, U.: Gliding motility of *Mycoplasma pulmonis*. J. Bacteriol. *130*, 937–938 (1977)

Calladine, C.R.: Design requirements for the construction of bacterial flagella. J. Theor. Biol. *57*, 469–489 (1976)

Calladine, C.R.: Change of waveform in bacterial flagella: The role of mechanics at the molecular level. J. Mol. Biol. *118*, 457–479 (1978)

Chwang, A.T., Winet, H., Wu, T.Y.: A theoretical mechanism of spirochetal locomotion. J. Mechanochem. Cell Motil. *3*, 69–76 (1974)

Chwang, A.T., Wu, T.Y.: A note on the helical movement of micro-organisms. Proc. R. Soc. (London) Ser. B. *178*, 327–346 (1971)

Chwang, A.T., Wu, T.Y., Winet, H.: Locomotion of spirilla. Biophys. J. *12*, 1549–1561 (1972)

Coakley, C.J., Holwill, M.E.J.: Propulsion of micro-organisms by three-dimensional flagellar waves. J. Theoret. Biol. *35*, 525–542 (1972)

DePamphilis, M.L., Adler, J.: Fine structure and isolation of the hook-basal body complex of flagella from *Escherichia coli* and *Bacillus subtilis*. J. Bacteriol. *105*, 384–395 (1971)

Dimmitt, K., Simon, M.: Purification and thermal stability of intact *Bacillus subtilis* flagella. J. Bacteriol. *105*, 369–375 (1971)

Doetsch, R.N.: Functional aspects of bacterial flagellar motility. Critical Reviews in Microbiology *1*, 73–103 (1971)
Emerson, S.U., Tokuyasu, K., Simon, M.I.: Bacterial flagella: Polarity of elongation. Science *169*, 190–192 (1970)
Enomoto, M.: Genetic studies of paralyzed mutants of *Salmonella*. I. Genetic fine structure of the *mot* loci in *Salmonella typhimurium*. Genetics *54*, 715–726 (1966)
Harris, W.F.: Bacterial flagella: Do they rotate or do they propagate waves of bending? Protoplasma *77*, 477–479 (1973)
Henrichsen, J.: Bacterial surface translocation: A survey and a classification. Bacteriol. Rev. *36*, 478–503 (1972)
Henrichsen, J.: On twitching motility and its mechanism. Acta Pathol. Microbiol. Scand. Sect. B *83*, 187–190 (1975)
Hilmen, M., Silverman, M., Simon, M.: The regulation of flagellar formation and function. J. Supramol. Struct. *2*, 360–371 (1974)
Hilmen, M., Simon, M.: Motility and the structure of bacterial flagella. In: Cell Motility. Goldman, R., Pollard, T., Rosenbaum, J. (eds.). New York: Cold Spring Harbor, 1976, pp. 35–45
Holwill, M.E.J., Burge, R.E.: A hydrodynamic study of the motility of flagellated bacteria. Arch. Biochem. Biophys. *101*, 249–260 (1963)
Hotani, H., Asakura, S.: Growth-saturation *in vitro* of *Salmonella* flagella. J. Mol. Biol. *86*, 285–300 (1974)
Iino, T.: Curly flagellar mutants in *Salmonella*. J. Gen. Microbiol. *27*, 167–175 (1962)
Iino, T.: Genetics and chemistry of bacterial flagella. Bacteriol. Rev. *33*, 454–475 (1969a)
Iino, T.: Polarity of flagellar growth in *Salmonella*. J. Gen. Microbiol. *56*, 227–239 (1969b)
Iino, T.: Assembly of *Salmonella* flagellin *in vitro* and *in vivo*. J. Supramol. Struct. *2*, 372–384 (1974)
Iino, T.: Genetics of structure and function of bacterial flagella. Annu. Rev. Genet. *11*, 161–182 (1977)
Kagawa, H., Owaribe, K., Asakura, S., Takahashi, N.: Flagellar hook protein from *Salmonella* SJ25. J. Bacteriol. *125*, 68–73 (1976)
Kamiya, R., Asakura, S.: Helical transformations of *Salmonella* flagella *in vitro*. J. Mol. Biol. *106*, 167–186 (1976)
Klug, A.: The design of self-assembling systems of equal units. Symp. Int. Soc. Cell Biol. *6*, 1–18 (1967)
Komeda, Y., Icho, T., Iino, T.: Effects of *galU* mutation on flagellar formation in *Escherichia coli*. J. Bacteriol. *129*, 908–915 (1977)
Kondoh, H., Hotani, H.: Flagellin from *Escherichia coli* K12: Polymerization and molecular weight in comparison with *Salmonella* flagellins. Biochim. Biophys. Acta *336*, 117–139 (1974)
Kondoh, H., Yanagida, M.: Structure of straight filaments from a mutant of *Escherichia coli*. J. Mol. Biol. *96*, 641–652 (1975)
Krieg, N.R.: Biology of the chemoheterotrophic spirilla. Bacteriol. Rev. *40*, 55–115 (1976)
Larsen, S.H., Adler, J., Gargus, J.J., Hogg, R.W.: Chemomechanical coupling without ATP: The source of energy for motility and chemetaxis in bacteria. Proc. Natl. Acad. Sci. USA *71*, 1239–1243 (1974a)
Larsen, S.H., Reader, R.W., Kort, E.N., Tso, W.-W., Adler, J.: Change in direction of flagellar rotation is the basis of the chemotactic response in *Escherichia coli*. Nature (London) *249*, 74–77 (1974b)
Lautrop, H.: *Bacterium anitratum* transferred to the genus *Cytophaga*. Int. Bull. Bacteriol. Nomencl. *11*, 107–108 (1961)
Leifson, E.: Atlas of Bacterial Flagellation. New York: Academic Press, 1960
Listgarten, M.A., Socransky, S.S.: Electron microscopy of axial fibrils, outer envelope and cell division of certain oral spirochetes. J. Bacteriol. *88*, 1087–1103 (1964)
Macnab, R.M.: Examination of bacterial flagellation by dark-field microscopy. J. Clin. Microbiol. *4*, 258–265 (1976)
Macnab, R.M.: Bacterial flagella rotating in bundles: A study in helical geometry. Proc. Natl. Acad. Sci. USA *74*, 221–225 (1977)
Macnab, R.M., Koshland, Jr., D.E.: Bacterial motility and chemotaxis: Light-induced tumbling response and visualization of individual flagella. J. Mol. Biol. *84*, 399–406 (1974)
Macnab, R.M., Ornston, M.K.: Normal-to-curly flagellar transitions and their role in bacterial

tumbling. Stabilization of an alternative quaternary structure by mechanical force. J. Mol. Biol. *112*, 1–30 (1977)
Manson, M.D., Tedesco, P., Berg, H.C., Harold, F.M., van der Drift, C.: A protonmotive force drives bacterial flagella. Proc. Natl. Acad. Sci. USA *74*, 3060–3064 (1977)
Matsumura, P., Silverman, M., Simon, M.: Cloning and expression of the flagellar hook gene on hybrid plasmids in minicells. Nature (London) *265*, 758–760 (1977)
McGroaty, E.J., Koffler, H., Smith, R.W.: Regulation of flagellar morphogenesis by temperature: Involvement of the bacterial cell surface in the synthesis of flagellin and the flagellum. J. Bacteriol. *113*, 295–303 (1973)
O'Brien, E.J., Bennett, P.M.: Structure of straight flagella from a mutant *Salmonella*. J. Mol. Biol. *70*, 133–152 (1972)
Pate, J.L., Ordal, E.J.: The fine structure of *Chondrococcus columnaris*. III. The surface layers of *Chondrococcus columnaris*. J. Cell Biol. *35*, 37–51 (1967)
Schmidt-Lorenz, W., Kühlwein, H.: Intracellulare Bewegungsorganellen der Myxobakterien. Arch. Mikrobiol. *60*, 95–98 (1968)
Schmitt, R., Bamberger, I., Acker, G., Mayer, F.: Feinstrukturanalyse der komplexen Geißeln von *Rhizobium lupini* H 13-3. Arch. Mikrobiol. *100*, 145–162 (1974a)
Schmitt, R., Raska, I., Mayer, F.: Plain and complex flagella of *Pseudomonas rhodos*: Analysis of fine structure and composition. J. Bacteriol. *117*, 844–857 (1974b)
Schneider, W.R., Doetsch, R.N.: Velocity measurements of motile bacteria by use of a videotape recording technique. Appl. Microbiol. *27*, 283–284 (1974)
Shimada, K., Kamiya, R., Asakura, S.: Left-handed to right-handed helix conversion in *Salmonella* flagella. Nature (London) *254*, 332–334 (1975)
Shimada, K., Ikkai, T., Yoshida, T., Asakura, S.: Cinemicrographic analysis of the movement of flagellated bacteria. II. The ratio of the propulsive velocity to the frequency of the wave propagation along flagellar tail. J. Mechanochem. Cell Motil. *3*, 185–193 (1976)
Silverman, M.R., Matsumura, P., Simon, M.I.: The identification of the *mot* gene product with *Escherichia coli*-lambda hybrids. Proc. Natl. Acad. Sci. USA *73*, 3126–3130 (1976)
Silverman, M.R., Simon, M.I.: Flagellar assembly mutants in *Escherichia coli*. J. Bacteriol. *112*, 986–993 (1972)
Silverman, M.R., Simon, M.I.: Flagellar rotation and the mechanism of bacterial motility. Nature (London) *249*, 73–74 (1974)
Silverman, M.R., Simon, M.I.: Bacterial flagella. Annu. Rev. Microbiol. *31*, 397–419 (1977)
Smith, R.W., Koffler, H.: Bacterial Flagella. Adv. Microb. Physiol. *6*, 219–339 (1971)
Stanier, R.Y., Adelberg, E.A., Ingraham, J.: The Microbial World, 4th ed. Englewood Cliffs, New Jersey: Prentice-Hall, Inc., 1976, p. 135
Suzuki, T., Iino, T.: Appearance of straight flagellar filaments in the presence of *p*-fluorophenylalanine in *Pseudomonas aeruginosa*. J. Bacteriol. *129*, 527–529 (1977)
Taylor, G.: The action of waving cylindrical tails in propelling microscopic organisms. Proc. Roy. Soc. (London) Ser. A. *211*, 225–239 (1952)
Thipayathasana, P., Valentine, R.C.: The requirement for energy transducing ATPase for anaerobic motility in *Escherichia coli*. Biochim. Biophys. Acta *347*, 464–468 (1974)
Wang, C.-Y., Jahn, T.L.: A theory for the locomotion of spirochetes. J. Theor. Biol. *36*, 53–60 (1972)
Weibull, C.: Ordered aggregation of salted out and dried bacterial flagella. Arkiv Kemi *1*, 573–575 (1950)
Williams, F.D., Anderson, D.M., Hoffman, P.S., Schwarzhoff, R.H., Leonard, S.: Evidence against the involvement of chemotaxis in swarming of *Proteus mirabilis*. J. Bacteriol. *127*, 237–248 (1976)
Williams, M.A., Chapman, G.B.: Electron microscopy of flagellation in species of *Spirillum*. J. Bacteriol. *81*, 195–203 (1961)
Winet, H., Keller, S.R.: Spirillum swimming: Theory and observations of propulsion by the flagellar bundle. J. Exp. Biol. *65*, 577–602 (1976)
Yokota, T., Gots, J.S.: Requirement of adenosine 3', 5'-cyclic phosphate for flagella formation in *Escherichia coli* and *Salmonella typhimurium*. J. Bacteriol. *103*, 513–516 (1970)
Yoshida, T., Shimada, K., Asakura, S.: Cinemicrographic analysis of the movement of flagellated bacteria. I. The ratio of the propulsive velocity to the frequency of bodily rotation. J. Mechanochem. Cell Motil. *3*, 87–98 (1975)

3.1.2 Plant Cilia

P. SATIR and G.K. OJAKIAN

1. Distribution and Significance

The cilium is a widely occurring organelle of eukaryotic plant and animal cells. A cilium consists of two main interrelated and interacting parts:

1. the axoneme, an assemblage of many tens of different polypeptide chains produced under the control of chromosomal genes, and
2. the ciliary membrane.

Cilia and eukaryotic flagella are identical organelles and both are found among plants. Although both terms have long histories and are useful in specific connotations, the former has the advantage of minimizing confusion with prokaryotic cell extension, and will be used as the inclusive term here.

Since a cilium is a membrane-bounded outgrowth of a basal body found in the cytoplasm, the symmetry and construction of the axoneme are related to that primordial organelle, and the main molecular constituents of the axoneme find counterparts in other arrangements of cytoplasmic microtubules. Axonemal microtubules provide the scaffolding on which the cilium is built, as well as probably participating directly in motility itself. They are chemically similar to cytoplasmic microtubules, both being largely composed of a group of similar proteins, the tubulins.

Contemporary plant and animal cells have both made considerable use of microtubules as cytoskeletal elements in the control of cell shape, in patterning or localizing synthetic processes, such as exocytosis or cell-wall formation, and in mitosis. However, plant cells have for the most part dispensed with microtubules arranged into basal bodies unadorned by functioning cilia. In particular, in plant cells, temporal continuity in the appearance of the centriole in the cytoplasm and at the poles of the mitotic apparatus is not seen.

Despite the lack of centrioles, most plant phyla contain species whose cells are ciliated at critical points in their life cycles. To be more specific, with the notable exceptions of the red algae, certain fungi, the higher gymnosperms and all angiosperms, zoospores and male gametes in plants are generally ciliated cells, while in unicellular and colonial algae, cilia are widely found (cf. DUCKETT and RACEY, 1975).

The invention of the cilium conferred upon the newly evolving eukaryotic cell two advantages: first, motility, which is a cellular dispersion mechanism and has thermodynamic advantages as well (cf. CARLSON, 1962), and second, a potential for localized membrane differentiation, which seems a prerequisite to strain-specific membrane fusion, and therefore to gametic development. Plant cilia are utilized both in motility and in cellular, especially sexual, recognition. It is a short step to combine membrane and motile functions of cilia so that in the single-celled forms, including unicellular algae and gametes or zoospores, the organelles become the modulators of cell behavior. As such, cilia can be effectors of phototaxis and chemotaxis in plant cells possessing them. Animal cilia retain and expand this sensory aspect of ciliary function in somatic cells, but somatic cells of higher plants minimize this; generally, cilia are much more restricted on cells covered by cell walls or other permanent extracellular matrices. Nonetheless, the selective advantage of cilia to gametes and zoospores is apparently sufficient for the organelle to be retained in many higher plants, including *Gingko* and the cycads, and the organelle in these forms is not very different in detailed construction from its appearance in a unicellular green alga, such as *Chlamydomonas,* or indeed in metazoan epithelia. In what follows, detailed information, often available only for *Chlamydomonas* and a few related forms, is generally taken as representative of structure and function for all plant cilia. The reservations necessary in adopting this view appear to be, on the whole, not serious ones.

Ciliated plant cells possess one, two, or many cilia. As summarized by MANTON (1970), the cilia of the great majority of zoospores are at least paired and fall into one of two categories, isokont or heterokont. Isokont flagella, such as those of *Chlamydomonas* and many green algae, are equal in all respects and are symmetrically arranged at or near the anterior end of the cell. Heterokont flagella are different from one another in length; moreover, one flagellum is directed forward and possesses special appendages – mastigonemes (also found on *Chlamydomonas* flagella, see below) or "flimmer" – while the other lacks such appendages and is directed either backwards or sideways. In *Fucus,* for example, the flimmer flagellum is ca. 10 μm long and the trailing flagellum is about 25 μm long. Heterokont zooids occur in brown algae, yellow-green algae, some taxa of flagellates and certain water molds. The distinction seems a pragmatic one, which is useful in plant phylogeny, for which see MANTON (1970) and DUCKETT and RACEY (1975).

2. Beat Form in Plant Cilia

Although there are numerous careful recent descriptions of beat form and wave characteristics in animal cell cilia and sperm, the information on the form of motion of plant cilia is more limited. *Chlamydomonas* swims with a breast stroke, flagella first (RINGO, 1967; Fig. 1). Its two flagella beat backwards over opposite sides of the cell body in flexural motion with a classical effective stroke, and a three-dimensional recovery stroke. *Chlamydomonas* exhibits an avoidance reaction as a response to various stimuli, wherein both flagella are thrown forward, and beat with undulatory motion, generating waves that pass from base to tip. The cell body is then momentarily propelled backward, the flagella trailing.

Plant cilia, particularly in unicellular forms, seem to have the same potential for a variety of movements as animal cilia; moreover, as expected from ultrastructural considerations, beat parameters for a particular form of motion seem to be reasonably similar in plant and animal cilia. Detailed descriptions of waveform

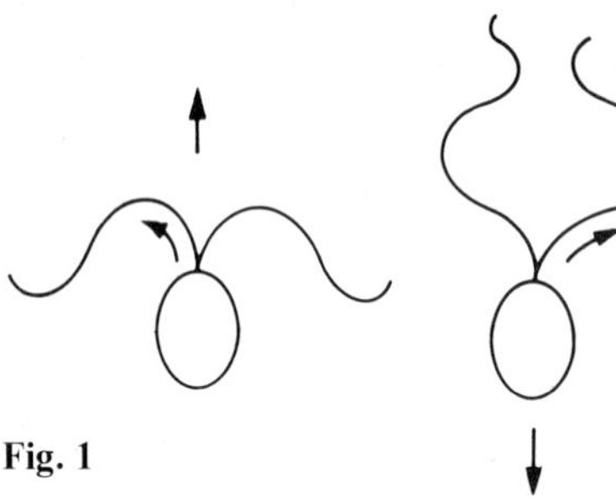

Fig. 1

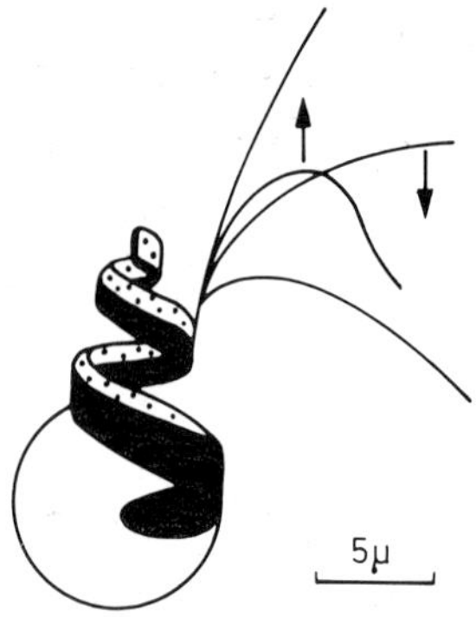

Fig. 2

Fig. 1. Diagram of beat of *Chlamydomonas* flagella. *Left:* Normal swimming. The organelles beat with a stroke characteristic of flexural motion. *Right:* Reversal. The flagella beat with undulatory motion. In each case, *arrow along flagellum* indicates observed direction of wave propagation; *arrow in medium* indicates direction of progression of the organism

Fig. 2. Diagram of beat of cilium of bracken spermatozoid. *Arrows* indicate probable effective stroke (cilium relatively straight) and recovery stroke (cilium bent) directions, respectively. (Courtesy of C.J. BROKAW, after BROKAW, 1958)

during beat are unavailable for metaphytan gametes. BROKAW (1958) has studied bracken spermatozoids, which swim at speeds of up to 300 μm/s in a helical path. The cell rolls over on its long axis three to six times a second while swimming, in the sense of a left-handed screw. Figure 2 shows the possible orientation and manner of beat of one of the cilia of this cell. A mature sperm of *Marsilea* from the studies of RICE and LAETSCH (1967) is shown in Figure 3.

Spermatozoids of *Gingko* and the cycads swim in the fluid medium of the archegonial chamber of the female gametophyte for short periods of time. Images of active spermatozoids of *Zamia* (NORSTOG, 1975) show cilia primarily with one or two bent regions on relatively long organelles (60 μm). The number of cilia on a single gamete is estimated by NORSTOG to be about 10^5; the beat of individual cilia is coordinated and is accompanied by flexing of the anterior portion of the cell. The spermatozoids swim cilia first.

Cilia are low Reynolds number systems ($<10^{-3}$) (HOLWILL, 1974; BLAKE and SLEIGH, 1974). Forces that arise from ciliary beat are primarily viscous forces, and ordinary intuitive ideas derived from inertial systems may not hold for ciliated systems, including plant cells. One consequence of these hydrodynamic considerations is that when ciliary motion ceases, the cell instantaneously stops moving forward; this is probably quite important for behavioral responses.

3. Structure and Chemistry

3.1 Axonemal Components

3.1.1 The 9+2 Pattern of Microtubules

Plant cilia were the first to be generally characterized as having a 9+2 arrangement of axonemal elements (cf. MANTON, 1952; MANTON and CLARKE, 1952; MANTON,

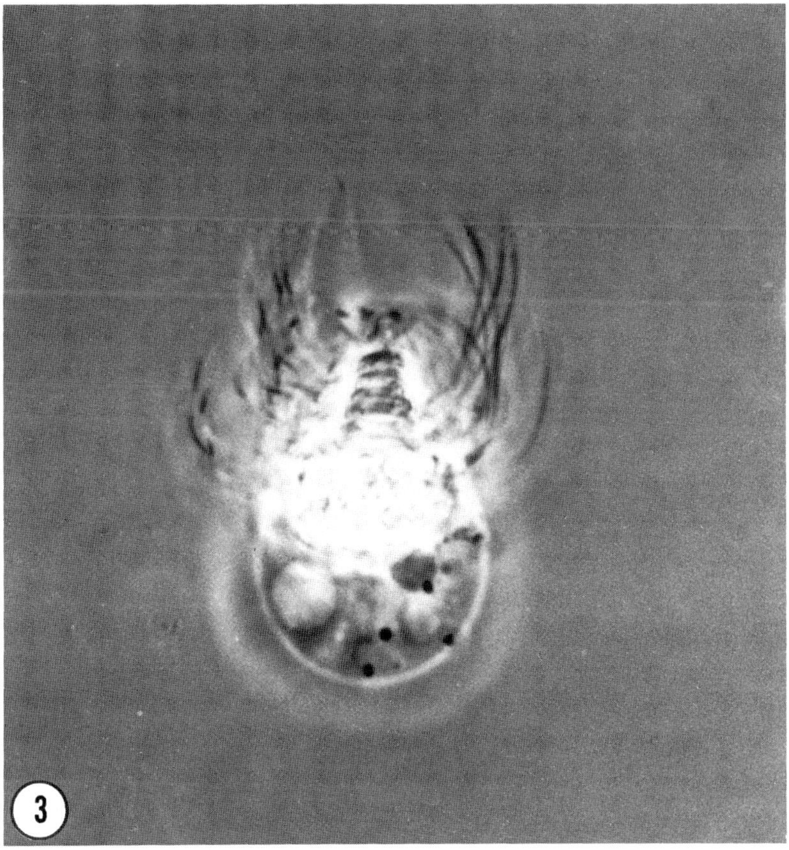

Fig. 3. A *Marsilea* sperm. In this specimen, ciliary motion has largely ceased and the arrested cilia are somewhat straighter than in actively swimming sperm. (Courtesy W.M. LAETSCH from RICE and LAETSCH, 1967)

1956). MANTON reconstructed this pattern from careful studies of whole-mounted, shadowed specimens of fern sperm, whose cilia sometimes fray apart. The morphological distinction between the central pair and the nine peripheral elements becomes apparent. Although the special geometrical packing properties of the 9+2 arrangement proposed by MANTON and CLARKE (1952) are somewhat simplified, the discernment of essential axonemal features was remarkable. The 9+2 arrangement was confirmed in animal cilia by FAWCETT and PORTER (1954) who studied cross-sections of embedded material by electron microscopy. By 1965, the 9+2 pattern, as shown in Figure 4, had been reported in zoospores and spermatozoids of many plant groups (see table in SATIR, 1965a), and is the only one known for motile ciliated plant gametes, with the exceptions of (1) certain centric diatoms (MANTON and VON STOSCH, 1966; HEATH and DARLEY, 1972) where the pattern is 9+0, the central pair being missing and of (2) the green alga *Golenkinia* which possesses a single, sinuous central microtubule (MOESTRUP, 1975). Certain nonmotile mutants of *Chlamydomonas* also have 9+0 flagella, as will be discussed further below.

Although, as mentioned above, the axonemal microtubules comprising the 9+2 pattern are similar to cytoplasmic microtubules, there are known physiological and structural differences between the two types of microtubules: (1) Axonemal microtubules are somewhat more stable than cytoplasmic microtubules, i.e., to fixatives, rising internal Ca^{2+}, extraction solutions, colchicine-induced depolymerization, etc. (2) The nine peripheral elements of the ciliary axoneme are doublet microtubules (Fig. 4b), a condition unknown in cytoplasmic microtubules. The doublet consists of subfiber A, a complete microtubule, to which is attached the incomplete, shorter element, subfiber B. The central elements of the axoneme resemble ordinary cytoplasmic microtubules to a greater degree, and are singlets. (3) There are specific structural attachments to axonemal microtubules which, if present at all on cytoplasmic microtubules, are not as regularly arranged or as prominent. One possible compositional difference between axonemal and cytoplasmic microtubules may be in the tubulins themselves, as in the multitubulin hypothesis of FULTON and SIMPSON (1976), but there may also be other unrecognized, yet important molecular differences. More complete recent reviews of microtubule structure and function, beyond the scope of this article, are provided by OLMSTED and BORISY (1973), STEPHENS and EDDS (1976) and by the recent conference series on *Cell Motility* (GOLDMAN et al., 1976; for a general discussion of microtubules in plant movements see FILNER and YADAV, Chap. 2.1, this. vol.).

The axonemal microtubules of *Chlamydomonas,* the most extensively analyzed plant microtubules, are composed of two tubulins, one of ca. 56,000 daltons, one of ca. 53,000 daltons, in approximate 1:1 ratio. On the basis of this ratio, most workers accept a heterodimer model of the tubulin subunit, the two tubulins designated α and β, respectively. This probably oversimplifies differences in primary structure between classes of tubulins, and local microheterogeneity (cf. WITMAN et al., 1972a, b). In in vitro assembly systems, *Chlamydomonas* axonemes can serve as nucleating centers for polymerization of vertebrate brain tubulin (ALLEN and BORISY, 1974a), whereby subfiber A of each doublet elongates. This indicates that despite minor differences there is substantial similarity of tubulins even across wide phylogenetic gaps.

In ciliary microtubules, the tubulin heterodimers are strung together in protofilaments. Subfiber A, as well as the central microtubules, consists of 13 protofilaments, while subfiber B contains 10 (RINGO, 1967). The protofilaments are aligned to produce a helix along subfiber A, and a somewhat different lattice along subfiber B (AMOS et al., 1976). Various intermicrotubule linkers – the arms, the spokes etc. – extend from subfiber A. More precisely, the attachment of such projections to sub-

→

Fig. 4. a–c *Chlamydomonas* flagellar fine structure. See text for detailed description. **a** Longitudinal section showing origin from basal body, ×75,000. **b** The 9+2 pattern, viewed tip to base. This is the cross-sectional appearance of the axoneme for most of its length. Note the surrounding membrane and sheath. *d*=outer dynein-1 arm attached to subfiber A of axonemal doublet, ×182,000. **c** Star pattern in transition region between basal body and cilium proper. Note goblet-like extensions from midwalls of doublets toward the membrane (*arrow*), ×168,000. **d** Schematic diagram of axonemal changes accompanying ciliary bend formation. Axonemal structure has been simplified to illustrate relative sliding of two doublet microtubules at opposite sites of the central pair as a bend is produced. Spoke groups in the developing beat region are tilted as if attached to the central sheath when sliding occurs. (See SATIR, 1968, 1976 for further details)

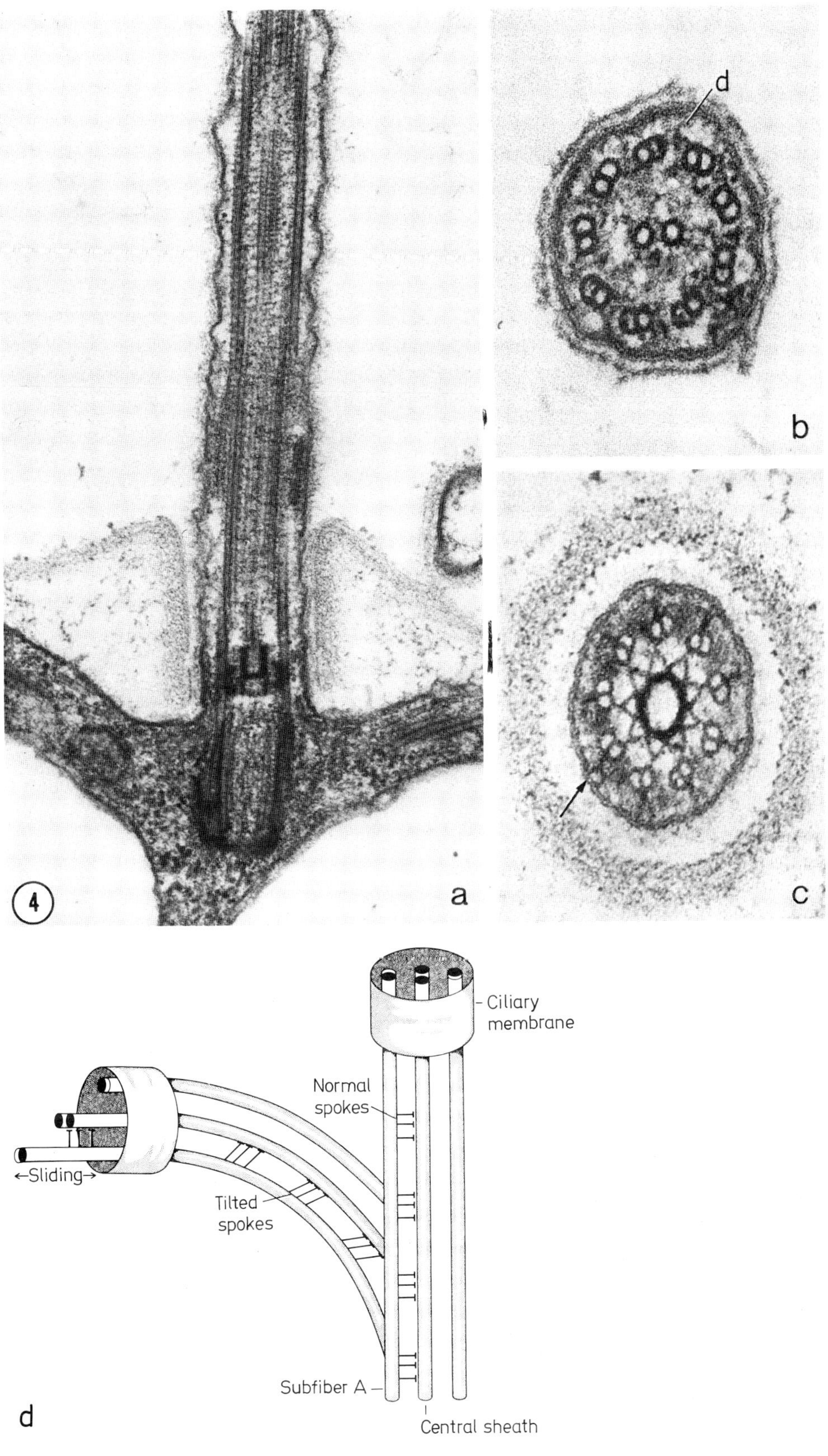
4
a
d
b
c
Ciliary
membrane
Normal
spokes
←Sliding→
Tilted
spokes
Subfiber A
Central sheath
d

fiber A generally persists upon axonemal disruption, and therefore seems more permanent than their attachments to other structures.

The ninefold symmetry of the axoneme is based on a similar symmetry of the basal body from which the axoneme extends. The central pair of microtubules confers upon the axoneme an overall bilaterality. The cilium moves perpendicular to the plane connecting the central pair (Fig. 4d), along a line often called the axis. This permits the ciliary doublets to be assigned unique numbers. By convention, the doublet transected by the axis is no. 1, and numbering increases in the direction in which the arms point (AFZELIUS, 1959; see below). All cilia thus far studied have a single enantiomorphic configuration, whereby the arms point counterclockwise when the axoneme is viewed from tip to base. The axis passes between doublets no. 5 and 6, which are connected by a special bridge in some animal cilia. The bridge is not a prominent feature of plant cilia. The main features of the 9+2 axoneme are evident in Figure 4.

The central pair is sometimes shorter than the doublets (MANTON, 1952) in a fully grown plant cilium. In *Chlamydomonas,* where the central pair is longer than the doublets, both singlets insert into a special "central pair cap" (DENTLER and ROSENBAUM, 1977). On the doublets, the B subfibers end abruptly, but each A subfiber ends in a forked fiber tail.

3.1.2 Intermicrotubule Linkage

Dynein Arms. From each subfiber A of the peripheral doublets, two rows of circumferential projections called arms (AFZELIUS, 1959) extend. Along one row, the distance between adjacent arms is about 24 nm, while the rows are probably staggered with respect to each other. The outer row of arms is slightly different in thin-section morphology and in extractability from the inner row, but both rows project toward subfiber B of the apposed doublet and are capable of attaching to sites along this doublet under rigor conditions. In 1963, using differential extraction procedures, GIBBONS showed that the arms were associated with the major ciliary ATPase which is thought to be the molecule responsible for chemical and mechanical transduction which he has since named "dynein". Recently (cf. GIBBONS et al., 1976) the term "dynein" has been extended to include all members of a class of ATPase isoenzymes that occur associated with microtubular systems, and have a very high molecular weight polypeptide subunit ($>300{,}000$ daltons in SDS-polyacrylamide gel electrophoresis). The arms apparently consist of the single isoenzyme, dynein-1. For example, extraction of half the arms of the axoneme (i.e., only the inner arms remain) removes approximately half the total dynein-1. A minor ATPase of the axoneme, dynein-2, is unaffected by this extraction.

Dynein 1 exists in two interconvertible forms. In the first, the ATPase activity is latent, while in the second it is activated (5–10 times greater activity than in the latent form). GIBBONS and GIBBONS (1976) have prepared soluble dynein-1 with latent ATPase activity and added this to demembranated sperm tails partially extracted with KCl under conditions where only about 7% of the outer arms are present on the doublets. During incubation with 1mM ATP at 23° C, the beat frequency rises from 13 Hz to about 25 Hz. Simultaneously, the percent of outer arms on the doublets rises to about 77%; that is, the soluble dynein-1 with latent ATPase activity is capable of reassembling appropriately with the arm-deficient axonemes to restore arms. Dynein-1 with activated ATPase activity is incapable of increasing the beat frequency.

Although most of these experiments have been done with sea urchin material or other animal cilia, the arms on plant cilia resemble other arms, and probably behave quite similarly. For example, Figure 5a shows the arms along *Chlamydomonas* axonemal doublets in longitudinal section. Dynein has recently been extracted from *Chlamydomonas* axonemes, and a preliminary report suggests that this dynein is also capable of recombining with the doublet microtubules to restore arms (HAIMO and ROSENBAUM, 1976).

In short, the arms are dynein-1, and the interaction of this molecule with two microtubules, the A tubule of one doublet to which it recombines and the B tubule of the adjacent doublet to which it attaches, usually transiently, is critical to powering ciliary motility. The activity of the dynein-1 arm may be thought of as analogous to the myosin cross-bridge in muscle; it is probably similarly capable of conformational change. In the absence of ATP, the arm combines with the B subfiber in a permanent rigor configuration (cf. GIBBONS and GIBBONS, 1974); in the presence of ATP, an arm is capable of performing mechanochemical work.

Radial Spokes. The second major type of projection attached to the A subfiber of each doublet is the spoke (Fig. 4d). Each spoke ends in an enlarged spoke head that abuts against and probably is capable of attaching to projections which surround the central pair of microtubules. In three dimensions, these projections form hoops, partial hoops, or perhaps a continuous helical strand, depending on the interpretation of the images seen, and are usually collectively called the *central sheath*. Spokes occur in groups of two (Fig. 5b, *Chlamydomonas*) or three (Fig. 4d). The groups appear as battlements along each doublet and were first seen and described as such by MANTON and CLARKE (1952) in *Sphagnum* cilia. Details of spoke and sheath arrangements have been studied in *Chlamydomonas*, first by HOPKINS (1970), CHASEY (1974), ALLEN and BORISY (1974b), and more recently with biochemical and functional correlations by WITMAN et al. (1976) and PIPERNO et al. (1977). The spokes within each group are conveniently labeled, beginning basally S1, S2, (S3) and the major spoke repeat (S1–S1) is about 90 nm. The normal S1–S2 distance in *Chlamydomonas* is about 29 nm. It may be that a thin fiber links the heads of S1–S2 together. Where a third spoke is present (animal cilia) the S2–S3 distance is about 21 nm. There are six sheath projections per major spoke repeat and four dynein arm repeats.

Mutants of *Chlamydomonas* flagella which systematically lack spokes (pf-14; pf = paralyzed flagellum) or where the central tubules and sheath do not assemble (pf-15A and pf-19) are paralyzed (see Sect. 4.2). Electrophoretic analysis of these and similar mutants has been undertaken using first one-dimensional and now two-dimensional gel methods. At least six polypeptides compose the spoke and six additional polypeptides compose the spoke head (PIPERNO et al., 1977). The gels as yet do not allow resolution in the molecular weight range of the dyneins.

Interdoublet Links. Near the origin of the inner row of dynein-1 arms, a further projection originates. This has a periodicity similar to that of the radial spokes, but it extends in a circumferential rather than radial direction so as to link adjacent doublets together. In some cilia, these interdoublet links have been identified with a particular gel band called "nexin" (STEPHENS, 1970). The best indication that these links are significant structures is that, after extraction of the dynein arms,

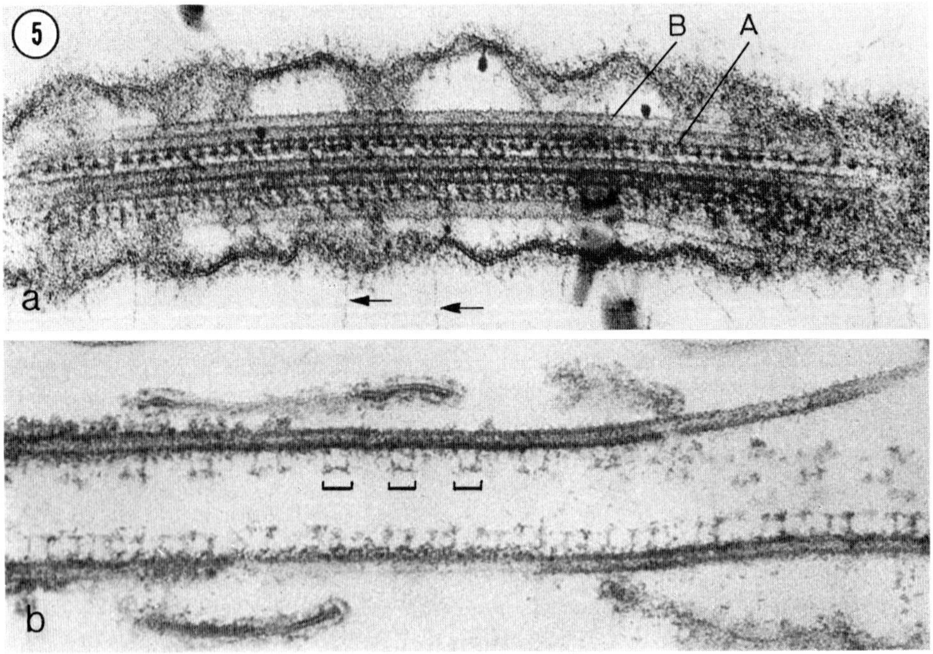

Fig. 5. a Longitudinal section of *Chlamydomonas* flagellum showing details of construction. *A* and *B* indicate subfibers *A* and *B* of one doublet respectively. Note row of dynein-1 arms along subfiber *A* of uppermost doublet. The fine hair-like extensions from the membrane are mastigonemes (*arrows*), ×136,000. **b** Paired spokes along *Chlamydomonas* axonemal subfiber As. Free spokes lie perpendicular to their doublet of origin. In the intact flagellum, the expanded head ends of the spokes abut against and may attach to the central sheath. Brackets mark spoke pairs. ×180,000

the axonemal doublets still stay together in correctly spaced closed or open chains of doublets. Nevertheless, the interdoublet links have not been characterized in great detail, and particularly are not yet well studied in plant cilia.

3.1.3 Summary

The main structures comprising the 9+2 axoneme seen with the electron microscope are now carefully described, despite minor interpretational difficulties, and are probably nearly universal in plant and animal cilia. The major protein constituents of the microtubules and arms are known: they are α and β tubulin and dynein-1 respectively. Less is known about the chemical composition of the other main structural elements: the spokes, central sheath, and interdoublet linkages. These are composed of a number of proteins which have been partially identified and characterized. There are a hundred or so more minor polypeptide constituents of the axoneme that probably play a modulating role in the function of the organelle, although they are not yet identified with or represented by specific structures.

3.1.4 Basal Body and Cytoplasmic Extensions

The basal bodies of plant cilia are similar to animal centrioles and basal bodies, as mentioned above. They have been relatively carefully examined in many plant groups because of their abrupt appearance in the cell during spermatogenesis. An early electron microscope description of the genesis of plant basal bodies was provided by MIZUKAMI and GALL (1966); more recent and more comprehensive, but essentially similar, information is available in DUCKETT and RACEY (1975) and HEPLER (1976). For example, in the pteridophyte *Equisetum* (DUCKETT, 1975), just before the last antheridial mitosis, each spermatogenous cell suddenly contains two mulberry-shaped blepharoplasts, which are multi basal-body precursors. (Depending on the organism, a single basal body or many may arise from each blepharoplast or its homolog.) Earlier precursors of the blepharoplast have been identified in *Marsilea* and followed in detail by HEPLER (1976). In the earliest stage, the organelle consists of a single, amorphous sphere in which plaques arise. It is likely that this represents a true de novo origin of this organelle. The plaques separate, producing two blepharoplasts which grow, differentiate and move from their initial position near the nuclear envelope. Individual basal bodies assemble and begin to separate from the granular matrix of the blepharoplast.

They are quickly recognizable as short cylinders ca. 0.15 μm in diameter and 0.3 μm in length composed of nine sets of triplet microtubules. The interior of the basal body contains (and perhaps assembles around) a so-called "cartwheel", from whose hub nine spokes extend, one to each triplet (cf. Fig. 3A of DUCKETT, 1975; Fig. 37 of HEPLER, 1976). On the end opposite the cartwheel in a mature basal body (which becomes the transition region between the basal body and the cilium proper), the sets of microtubules are interconnected by fibers to produce several distinctive nine-pointed stars, one of which is shown in Figure 4c. MANTON (1964) has carefully described some of these patterns in the green algae and in chrysophytes. She suggests that the star-patterned fibers may co-ordinate component doublets in the axoneme during motion, but there is as yet no evidence to support this conjecture. The basal bodies presumably migrate and attach to the membrane; the microtubule–membrane connections discussed in the next section probably also arise at this time in the transition zone, where one component of each triplet microtubule ends and the axoneme sprouts.

Fibers of various sorts and microtubule sheets often extend from or are more loosely associated with the cytoplasmic sides of basal bodies. These are structures strictly ancillary to the cilia, generally concerned with shaping the cell cortex or crown and probably stiffening and reinforcing the cytoplasm to resist the forces and distortions generated by ciliary motion and by obstacles encountered on the way to fertilization. Such extensions are reasonably conservative features in plant evolution and, like the cilia proper, seem particularly useful as indicators of plant phylogeny. PICKETT-HEAPS (1975), for example, describes two types of basal body complex: the chlamydomonad type and the bryophytan type. In *Chlamydomonas*, the two flagellar basal bodies are connected by a short striated fiber (see Fig. 7). This provides mechanical co-ordination between the flagella (HYAMS and BORISY, 1975a). This fiber seems rather like the more complex striated rootlets in the zoospore of the advanced green alga *Oedogonium* and the kinetodesmal and rootlet fibers of animal cilia. The banding of such fibers is quite striking. Limited chemical analysis of the protein composition of ciliary rootlets is available. Four sets of microtubules also originate near the flagellar bases in *Chlamydomonas* and extend around the periphery of the cell, presumably as a loose cytoskeleton.

The bryophytan type of complex seems derived from a ulotrichalean alga, similar to *Klebsormidium* for instance, and related phyletically to the main groups of green land plants. In bryophytes, two basal bodies arise after the final spermatogenous mitosis (CAROTHERS, 1975) and these are also connected by a short fibrous structure. Most importantly, simultaneous with the basal body differentiation, lying just below the basal bodies, a single flat plate or manchette of microtubules also assembles. Immediately subtending this layer of microtubules is a layer of amorphous, dark-staining material which subsequently becomes stratified. This grouping of microtubules and associated material is very characteristic and is called the multilayered structure or, since there are four distinct regions (at least in liverworts), the Vierergruppe region. A similar structure is found in spermatozoids of pteridophytes and cycads, and in zoospores of *Klebsormidium*.

3.2 The Ciliary Membrane

Examination of plant cilia by thin-section electron microscopy demonstrates very clearly that the cilium is a true cytoplasmic organelle enclosed within an extension of the plasma membrane (see Fig. 4). This ciliary membrane has the usual trilaminar appearance, approximately 8–10 nm in thickness. The detailed chemical composition of ciliary membranes of plants, with the possible exception of *Chlamydomonas*, is virtually unknown, although it is usually assumed that these membranes conform to the general pattern, with phospholipids forming a lipid bilayer in which integral proteins are embedded. There are indications, however, that in certain instances, e.g., *Ochromonas*, this picture is somewhat oversimplified. In *Chlamydomonas*, SDS polyacrylamide gel electrophoragrams show one major ciliary membrane glycoprotein with molecular weight greater than 200,000 daltons.

A considerable amount of new information on the ciliary membrane has been obtained by freeze-fracture electron microscopy. As originally proposed by BRANTON (1966), frozen membranes, when fractured, split down an internal hydrophobic plane dividing the membrane into two complementary monolayers which allow the electron microscopic examination of extensive internal membrane areas, the

Fig. 6. Freeze-fractured flagella of a *Chlamydomonas reinhardtii* gamete, indicating flagellar necklace (*n*) and bracelet (*b*). (Courtesy of R.L. WEISS and U.W. GOODENOUGH, from WEISS et al., 1977) ×67,000

Fig. 7. Thin section of *C. reinhardtii* gamete which lacks a cell wall; *tr* marks transition region between basal body and flagellum. The two flagella are connected by short striated fibers (see text). Note attachments between axoneme and membrane at the level of the necklace (*arrows*). (Courtesy of R.L. WEISS and U.W. GOODENOUGH, from WEISS et al., 1977) ×67,000

Fig. 8. Freeze-etched flagellum showing overlying cell wall (*cw*) and the relative location of the flagellar bracelet (*b*). (Courtesy of R.L. WEISS and U.W. GOODENOUGH, from WEISS et al., 1977) ×67,000

Fig. 9. Face-on view of base of cross-fractured flagellum surrounded by the flagellar bracelet. ×75,000

Fig. 10. Electron-opaque colloidal iron hydroxide stains the flagellar sheath in *C. reinhardtii*. (Axoneme unstained) ×42,000

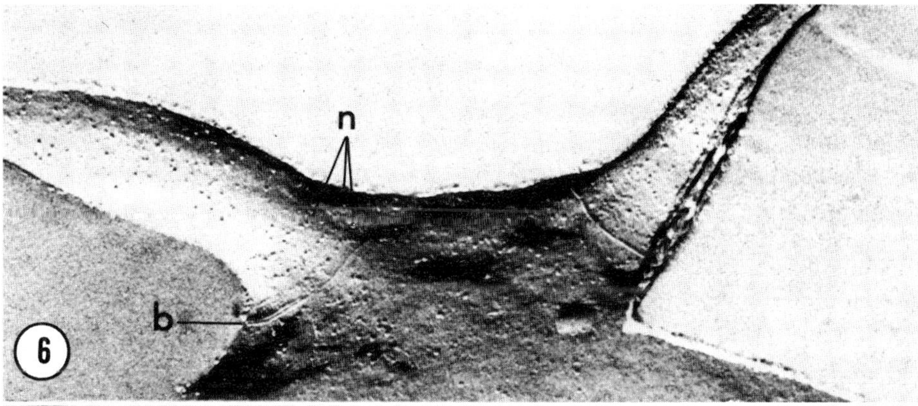
n
b
6

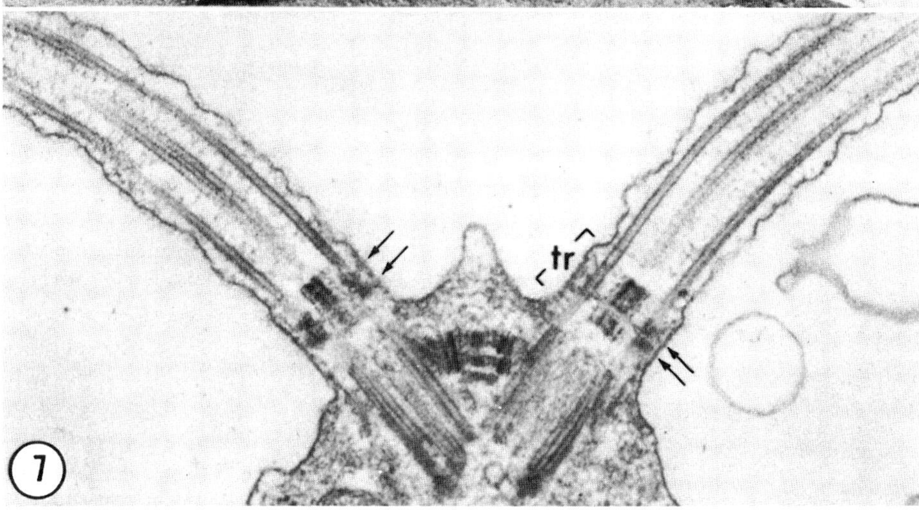
tr
7

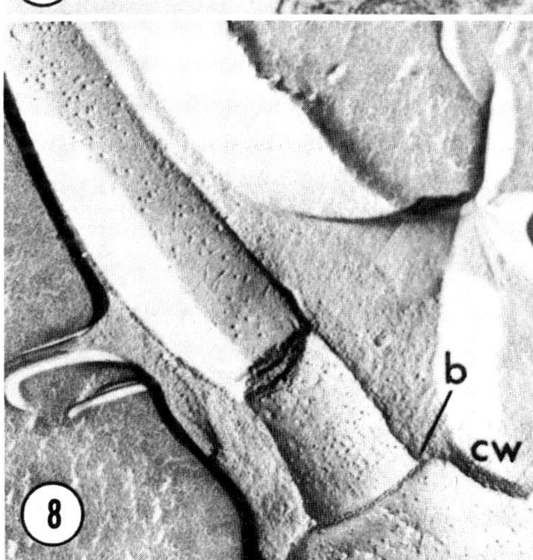
b
cw
8

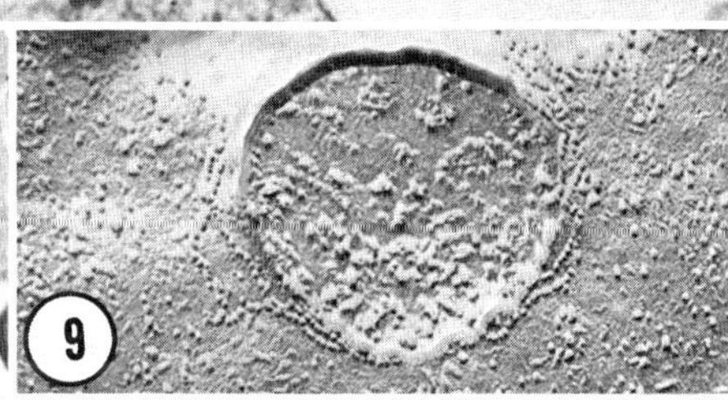
9

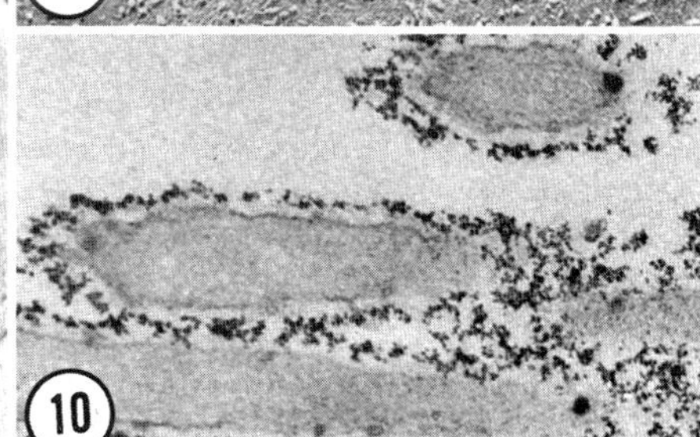
10

fracture faces. The fractured membrane has a smooth appearance, probably derived mainly from the background lamellar phase lipids, but it is interrupted by projections ca. 10 nm in diameter called intramembrane particles. There is suggestive evidence that intramembrane particles are composed of oligomeric integral membrane proteins.

In freeze-fractured flagellar membranes from vegetative cells of *Chlamydomonas reinhardtii,* as on other ciliary membranes, the particle population is sparse. The cytoplasmic, or P, fracture face contains the majority of intramembrane particles, while the extracellular, or E, fracture face has considerably fewer particles. Freeze-fracture studies on flagellar membranes of *C. eugametos* have demonstrated that, in this species, the particle density on the P fracture face is even lower than in *C. reinhardtii* (BRAY et al., 1974). The E fracture face was not shown in this study. The particles on the P fracture face are occasionally found arranged in rows running parallel to the long axis of the flagellum with complementary grooves being present on the E fracture face (BERGMAN et al., 1975; SNELL, 1976). *C. reinhardtii* gametes of both mating types have considerably more intramembrane particles on the E fracture face than do vegetative cells, although no direct correlation can yet be made between increased particle density and gametic agglutination (see below).

In many membranes, unique arrangements of intramembrane particles termed "particle arrays" have been described and linked to specific biological functions (SATIR and SATIR, 1974). A specific array called the "ciliary necklace" has been described as a common feature near the base of the ciliary membrane of various animal cells (GILULA and SATIR, 1972) (see Fig. 6 for an example of necklace arrays). The necklace represents the transmembrane portion of a structural path linking each microtubular set in the distal end of the basal body to a region of the cell surface which is specialized, at least with regard to its ability to bind substances such as cationic ferritin. Distinctive links between microtubules and membrane are visible in a characteristic cross-section of the necklace region. Recently, WEISS et al. (1977) have described a necklace which consists of three rows of particles in the flagellar membrane of *C. reinhardtii* adjacent to the transition region of the axoneme. This is shown in Figure 6. It is not known whether similar arrays are to be found in other plant cilia, but some other algae at least possess the requisite transition-region morphology. Although the function of the ciliary necklace has not yet been ascertained, the observation that there are microtubule–membrane attachments (Figs. 4c, 7) in the necklace region suggest that this array of intramembrane particles may be involved in the control of ciliary beat, possibly by regulating ion permeability. With localization techniques, Ca^{2+} precipitates have been found specifically in the necklace region, so that it is speculated further that the ion being regulated is calcium ion (see Sect. 4.3).

Using the freeze-fracture technique, WEISS et al. (1977) have recently described two new membrane modifications associated with flagella of both vegetative and gametic cells of *C. reinhardtii.* The first modification consists of one to three rings of tightly packed particles on the P fracture face located in the plasma membrane which surrounds each flagellum at its base as it emerges from the cell (Figs. 6, 8, 9). This array is more distinctive than the necklace and has been termed the "flagellar bracelet". Formation of the bracelet appears to depend upon the presence of mature basal bodies. The bracelet may function as a barrier to prevent the lateral diffusion of integral plasma membrane components into the flagellar membrane. The second modification is the existence of connections between struts, which are thicker fibrous extensions from the distal ends of basal bodies, and the plasma membrane. These arrays, termed "strut arrays", consist of small clusters of particles located in a circle on the E fracture face just outside the grooves of the flagellar necklace, 300 nm from the center of the basal body. Strut arrays are present in vegetative cells and in gametes. Probably all these arrays contribute to the anchoring of basal body to membrane which conveys mechanical stability, and perhaps is a requisite for ciliary morphogenesis.

3.2.1 Surface Features

In *Chlamydomonas* the membrane is surrounded externally by a fuzzy layer of material termed the flagellar sheath (RINGO, 1967). Interestingly, particles, such as small polystyrene microspheres, placed in the medium will attach to the flagellar sheath and are transported along the flagellum (BLOODGOOD, 1977). Since the carbohydrate stain colloidal iron hydroxide (GASIC et al., 1968) reacts strongly with the flagellar sheath (Fig. 10), it appears that this structure is composed of acid mucopolysaccharides similar to those found in the cell coat of animal cells (MARTINEZ-PALOMO, 1970).

Attached to the ciliary membranes of many plant cells are hair-like structures generally called mastigonemes (see BOUCK, 1972 for a more complete discussion on their occurrence and biogenesis). There may, in fact, be several slightly different sorts of such hairs. Figure 5a shows one example. To date, the most complete study on the structural aspects of plant mastigonemes has been done on the vegetative and gametic cells of *C. reinhardtii* using shadowing (RINGO, 1967; BERGMAN et al., 1975) and negative-staining techniques (WITMAN et al., 1972a; BERGMAN et al., 1975; SNELL, 1976). With negative staining, a mastigoneme is about 0.9 μm long and consists of a single row of subunits 16 nm wide and 20 nm in length with a center-to-center spacing of 20 nm along the long axis of the mastigoneme. Polyacrylamide gel electrophoresis indicates that *C. reinhardtii* mastigonemes are composed of a single glycoprotein of ca. 170,000 daltons (i.e., somewhat lower molecular weight than the membrane glycoprotein). There are no apparent electrophoretic differences between mastigonemes derived from vegetative cells or from gametes. Mastigonemes are first assembled from subunits within cisternae of the endoplasmic reticulum and are released to the surface of the ciliary membrane upon vesicular exocytosis. Perhaps, at the base of the cilium, the vesicle membrane itself becomes new ciliary membrane with appropriate attachment sites for mastigonemes already in place. The mastigonemes are quite regularly aligned along the membrane. Sometimes they appear associated through the membrane to the doublet microtubules (MARKEY and BOUCK, 1977). Their function is not known, although the presence of such appendages probably affects the characteristics of fluid flow around the cilium. Because the mastigonemes appear to be rigid projections, for some years it was believed that mastigonemes permit reversal of the direction of propulsive thrust of the cilium (cf. JAHN et al., 1964; HOLWILL, 1966), but this now seems rather less likely.

The surfaces of certain flagella of prasinophycean flagellates and the *Chara* group are covered with unique diamond-shaped scales (cf. MANTON et al., 1965; MANTON, 1970; MOESTRUP, 1975). These are also assembled in Golgi cisternae, sometimes together with mastigonemes, and are also deposited on the flagellum when the Golgi vesicles fuse with the plasma membrane. Again, these structures may be interesting indicators of plant phylogeny, but little is known of their function and significance.

3.2.2 Agglutination and the Mating Response

The membranes of animal cilia are often unique antigenically. Such ciliary antigens are responsible for agglutination and immobilization reactions in the ciliated protozoan *Paramecium*, for example. In plants, corresponding specificity appears in the ciliary membrane. This has been studied in several forms, including the colonial alga *Pandorina* (COLEMAN, 1963). Specificity manifests itself best during mating of ciliated gametes of opposite type in algae. For instance, when they are grown in nitrogen-free media, the vegetative cells of *Chlamydomonas* lose their cell walls and differentiate into gametes (see FRIEDMAN et al., 1968, and WIESE, 1969; for a more complete description). When the ciliated iso-gametes of opposite mating types are mixed, agglutination usually occurs through the interaction of tips of the cilia (WIESE, 1965; BERGMAN et al., 1975; GOODENOUGH and WEISS, 1975). This is a signal for the cells and is shortly followed by cell fusion via a cytoplasmic bridge to produce a zygote. Agglutination is inhibited by treatment of the gametes with trypsin, other proteases, or the plant lectin concanavalin A (WIESE and SHOEMAKER, 1970; MCLEAN and BROWN, 1974), indicating that the agglutinating cell component is probably a glycoprotein that associates with the flagellar membrane. Actually, living gametes of one mating type can be induced to agglutinate when incubated appropriately with factors derived from the opposite mating type.

The mating factor, or "gamone" (WIESE, 1965), has been isolated and characterized both structurally and biochemically by several groups. Gamone consists of membranous vesicles with attached mastigonemes and appears to be fragments of the flagellar membrane (MCLEAN and BROWN, 1974; BERGMAN et al., 1975; SNELL, 1976). When the membranes and mastigonemes are separated on either cesium chloride or sucrose density gradients, only the membrane fraction contains the agglutinating activity. Apparently, in *Chlamydomonas*, mating type recognition and agglutination resides specifically in the ciliary membrane. A protein of 70,000 daltons unique to the (+) mating type gamone of *C. reinhardtii* has been described by SNELL (1976). This protein does not however, appear to be directly involved in gamete adhesion, since it is found with the purified mastigonemes, a fraction which lacks agglutinability, as well as with the membranes.

MCLEAN and BOSMANN (1975) have proposed that the agglutination of *C. moewusii* gametes involves flagellar membrane glycosyl transferases since mixing of either flagellar membrane vesicles (gamone) or gametes of opposite mating types will cause a two- to eightfold increase in the activity of these enzymes. MCLEAN and BOSMANN suggest that their results are compatible with the general mechanism of cellular recognition involving multiglycosyltransferase systems proposed by ROSEMAN (1970) and his co-workers (ROTH et al., 1971).

4. Motile Mechanism

4.1 Sliding Microtubules in Plant and Animal Cilia

All 9+2 cilia appear to generate and control their motion by the same fundamental mechanisms. It has been known for some time that detergent-treated isolated cilia or 9+2 sperm tails will beat with apparently normal waveform and propagation specifically when ATP is added in appropriate media. Detergent removes or severely damages the ciliary membrane, leaving a membraneless axoneme. Therefore, it seems clear that the axoneme alone is responsible for generating ciliary motion, and that the power for motility comes from hydrolysis of ATP. Attention has therefore focused on the molecules that are the ciliary ATPases—dynein-1 and dynein-2—as potential mechanochemical transducer molecules of this system.

Dynein-1 is located in the inner and outer arms along the axonemal doublets, as discussed above. A major result from studies of cilia bent in various positions is that the axonemal microtubules do not change length during motion, but instead are capable of sliding with respect to one another (SATIR, 1965b, 1968) (see Fig. 4d). This sliding is produced by the dynein-1 arms of one doublet acting upon some unspecified sites—presumably of the tubulin lattice—of the adjacent doublet. As noted previously, the arms are thought to undergo a cycle of cross-bridging between doublets, somewhat similar to the cross-bridge cycle of myosin to actin.

SUMMERS and GIBBONS (1971, 1973) have demonstrated sliding of ciliary microtubules elegantly and directly. They have treated membraneless axonemes briefly with trypsin and have followed digestion as a change in turbidity. Digestion is stopped with an enzyme inhibitor and ATP is added in a reactivation medium. Such axonemes do not reactivate their typical beat; instead, they slide apart as one watches under the light microscope. Moreover, electron microscopic examination shows that, morphologically, the striking differences between untreated and trypsin-treated axonemes are the disappearance of the radial spokes and the interdoublet links as digestion proceeds. The dynein-1 arms apparently remain intact. This suggests (1) that the interaction between dynein-1 and tubulin is sufficient for generating sliding and (2) that the spokes and interdoublet links are a significant part of the mechanical or mechanochemical restraining system that limits sliding in intact axonemes and converts the sliding into local bending.

The Summers and Gibbons experiment, originally done with invertebrate sperm tails, has been repeated using mammalian sperm tails, protozoan cilia and *Chlamydomonas* flagella. Sliding in the latter case was originally described by ALLEN and BORISY (1974b) and subsequently in more extensive fashion by WITMAN et al. (1976). This limited but phylogenetically varied survey supports the general validity of the sliding microtubule mechanism of ciliary motion. Whether microtubule sliding is important in other processes central to the biology of plant and animal cells, particularly in mitosis, is still controversial, and is beyond the scope of this summary.

SALE and SATIR (1977), studying *Tetrahymena* cilia, have captured images of sliding microtubules for electron microscopic examination. They conclude that sliding always has a constant polarity, such that the dynein arms of doublet number n act effectively to push doublet # $n+1$ in a tipward direction. This immediately suggests that the doublets within the axoneme slide asynchronously or with some phase-shifting between doublets or groups of doublets; for example, doublets 1–4 may be sliding actively to produce the effective stroke of a cilium, while doublets 6–9 are inactive, and then 6–9 will be sliding actively during recovery, while 1–4 are inactive. In negative stain, dynein is a short cylinder about 9–10 nm in diameter (WARNER et al., 1977); 3–4 such cylinders comprise one arm. In *Tetrahymena*, arms are found in what appear to be two configurations, an extended configuration which is slightly tilted (40°) towards the base of the axoneme, and a more flattened configuration SALE and SATIR, 1977). Speculatively, these configurations represent positions in the dynein cross-bridge cycle.

The rate of microtubule sliding, measured by beat frequency, is directly proportional to ATP concentration and to dynein-1 concentration. Beat frequency drops to half when half the dynein-1 arms are extracted (GIBBONS and GIBBONS, 1973). Bend amplitude and general co-ordination are minimally affected.

Active microtubule sliding is usually coupled quite closely to bending in the intact cilium; the result is that the bends form circular arcs, a geometry such that the amount of sliding per spoke group is constant in a bent region. Bend angle measures active sliding of microtubules only under conditions of tight coupling. Under stringent conditions, where the radius of curvature of the bend approaches a minimum, the amount of sliding per spoke group approaches a maximum, and the spokes assume a tilted configuration (WARNER and SATIR, 1974), as seen in Figure 4d. In straight regions, above and below a bend, spokes lie normal to the axonemal microtubules. These positions are consistent with the hypothesis that bends develop and propagate in regions where spokes attach to the central sheath. WARNER and SATIR (1974) have postulated that a spoke attachment–detachment cycle may occur during a normal ciliary beat. The bend propagation mechanism is not well understood, and probably relies on mechanical and elastic features of the spokes, sheath, arms and interdoublet links. There is some suggestive evidence that the interdoublet links are somewhat flexible, perhaps acting as a sort of elastic sleeve on the bent region of the axoneme. The attachment site between spokehead and sheath is obviously a vital part of this model, and it is the potential and putative location of dynein-2.

4.2 Mutant Studies of *Chlamydomonas*

In *Chlamydomonas*, the sliding–bending mechanisms can be dissected further because of the availability of paralyzed mutants. Correlations of the specific physiological defects in such mutants with abnormal structure or proteins of the axoneme should provide an indication of the function of the abnormal or missing component. WITMAN et al. (1976) studying mutants pf-14, pf-15 A and pf-19 of *C. reinhardtii* have used this approach in a particularly illuminating way. Axonemes of wild-type cells prepared by detergent treatment can be reactivated by solutions containing 1 mM ATP. The axonemes beat rapidly with a waveform similar to in situ ciliary beat and spin around in a tight circle. Under identical conditions, the pf-14, pf-15 A and pf-19 axonemes remain paralyzed. This is the situation expected where a structural defect in the axoneme is responsible for paralysis.

As indicated in the previous section, wild-type axonemes treated with trypsin and ATP will disintegrate by microtubule sliding. Remarkably, when isolated axonemes of pf-14, pf-15 A and pf-19 are mixed appropriately with 1mM ATP and trypsin, they also disintegrate in a manner similar to wild type. The axonemal microtubules of these paralyzed mutants are in fact capable of normal sliding.

Presumably, these mutant axonemes are paralyzed because coupling between doublet sliding and the generation and propagation of bends along the axoneme is defective. Recall that structural mutant pf-14 lacks the radial spokes, while mutants pf-15 A and pf-19 lack the central pair and its sheath projections. These observations support the hypothesis that the spoke–sheath interaction is essential for the conversion of doublet sliding into axonemal bending.

The interdoublet nexin links of the paralyzed axonemes seem to be intact before trypsin treatment. Therefore, these alone would not seem to be sufficient to transduce microtubule sliding into bending. Nontrypsinized axonemes of pf-14 remain intact, neither beating nor disintegrating under reactivating conditions. The interdoublet links alone are then sufficient to maintain the integrity of the axoneme against dynein-shearing forces. Under these conditions, the nexin links must form bridges between adjacent doublets and restrain their potential displacement elastically, without inducing bend formation. The nexin links are presumably digested by trypsin, although some special peripheral links remain near the base of the axoneme in *Chlamydomonas*; these can also counterbalance doublet displacement.

4.3 Reversal and Behavioral Response of Cilia

Behavioral responses of single-celled plants and plant gametes possessing cilia normally depend upon alteration of parameters of the organellar beat, such as the amplitude and wavelength of bending, the direction in which bends propagate, the symmetry and timing of bends and so on. Among simple protists, a series of different changes may be possible at different times. The behavioral effect of such changes, regardless of the specifics, is generally an alteration in direction of swimming of the cell. The clearest responses are often called "avoiding reactions" or "ciliary reversal". Changes in beat form associated with reversal in *Chlamydomonas* have been discussed in Section 2.

The molecular mechanism of reversal has been studied in a series of cells, including *Chlamydomonas*, but is perhaps best known in *Paramecium*, which alters the direction of the effective stroke of its cilia but not their bend characteristics (cf. NAITOH and ECKERT, 1974; MACHEMER, 1974). Regardless of the way in which ciliary beat is altered in the cases so far examined, the basis of reversal seems nearly identical in terms of ionic control. In this regard, it is important to remember that membraneless ciliary models beat normally in the presence of $<10^{-7}$ M Ca^{2+} in the reactivation medium. In *Paramecium*, such models swim in a normal forward direction (NAITOH and KANEKO, 1972). In *Chlamydomonas*, connected pairs of isolated cilia can be prepared, which beat with breast strokes of the normal ciliary type but of opposite senses (i.e., left and right) with the effective stroke in each case directed posteriorly, in Ca^{2+} from 10^{-9}–10^{-6} M in buffered solutions (HYAMS and BORISY, 1975b). Above 10^{-5} M Ca^{2+}, *Chlamydomonas* flagella models alter their orientation and beat with a flagella-type motion in a reversal response, and

Paramecium models swim backwards. Since the model systems all respond to these defined changes in Ca^{2+}, it seems likely that the reversal of ciliary beat in the living cells is also controlled by the concentration of this ion in the axoneme. The suggested physiological sequence is that a sensory stimulus, e.g., mechanical deformation of the cell, leads to membrane permeability changes, possibly within the ciliary membrane itself, which open voltage-dependent Ca^{2+} channels momentarily. Ca^{2+}, which is normally in ca. millimolar concentration in the external environment, is then free to move down its concentration gradient and enter the cell (ECKERT, 1972). This causes reversal. In the absence of additional stimulation, the Ca^{2+} is then removed from the axoneme by some pumping mechanism, the location of which is still not completely defined. When Ca^{2+} drops below a critical threshold, normal beat resumes. In *Paramecium*, a series of membrane mutants exists, one of which is called pawn (after the chess piece that can move only forward). This mutant is incapable of ciliary reversal since it lacks the appropriate functional Ca^{2+} channels in the membrane (KUNG et al., 1975). The channels are largely located in the ciliary membrane itself, since amputation of the cilia nearly abolishes the Ca^{2+} conductance changes of the cell (DUNLAP and ECKERT, 1976; OGAWA and TAKAHASHI, 1976). A relationship of Ca^{2+} channels and/or pumps to intramembrane particles seen in freeze fracture, including the ciliary necklace and the patch arrays on *Paramecium*, has been suggested (cf. SATIR, 1976), partly because electron microscope methods localize Ca^{2+} deposits in these regions as mentioned above, partly because such particles are known to correspond to the proteins that are involved in calcium ion exchange across the membranes of the sarcoplasmic reticulum of vertebrate muscle. The suggestion remains speculative at this time.

The site of reversible Ca^{2+} binding to the axoneme that initiates form or directional changes in the beat is unknown. However, it is less likely to involve the basic dynein–tubulin sliding interaction than some modulator of that interaction, possibly the spokehead–central sheath attachment cycle discussed above. The reason for this is that the cilia still beat during reversal; presumably microtubule sliding is still going on. SALE and SATIR (1977) have preliminary evidence that in *Tetrahymena* axonemes digested with trypsin, addition of Ca^{2+} in millimolar concentration (presumably sufficient to produce ciliary reversal) to the ATP reactivation medium does not stop microtubule sliding nor does it alter the basic direction of dynein force generation.

4.4 Summary

Like animal cilia, plant cilia are powered by a sliding microtubule mechanism which depends on the dynein-1 arm-tubulin mechanochemical cycle, where ATP dephosphorylation accompanies sliding in a constant effective direction. In a beat cycle, microtubule sliding is asynchronous. Sliding is coupled closely to bending by mechanical and elastic feedback mechanisms of the axoneme, possibly involving a radial spoke–central sheath local attachment cycle as a prime component. Cilia beat normally with a free axonemal Ca^{2+} of $<10^{-7}$ M.

Beat parameters change to give reversal responses when Ca^{2+} momentarily rises above 10^{-6} M, which occurs physiologically as a result of membrane events following sensory stimulation in ciliated cells.

5. Morphogenesis and Growth

Many plant cilia, especially those of metaphytan gametes, grow explosively within less than one hour or so at a particular stage of the life cycle. The earlier stages of ciliogenesis in such cases involve the production of basal bodies, as described briefly in Sect. 3.14. The later stages involve elongation of the axoneme and extension of the ciliary membrane. Precursor material for the axoneme is very probably released directly from polysomes into the cytoplasmic matrix, while membrane components are assembled separately. There is a formal resemblance between ciliary morphopoiesis and other complex self-assembling systems. The sequence of ciliary assembly involves interactions of independent self-assembling structures, with controls at various levels of the overall process.

In some ciliated plant gametes, production of the cilium probably occurs in a bud at the cell surface, above a basal body that is already well differentiated and attached to the cell membrane. In other cases, such as in *Allomyces,* the basal body lies deeper in the cell and the bud protrudes into a cytoplasmic vesicle which only later fuses with the cell membrane. Ciliogenesis can sometimes be induced by simple environmental stimuli, which lead to specific gene activation. The gene products involved in triggering ciliary production are not clearly identified; in the amebo-flagellate *Naegleria,* it is suggested that a special tubulin itself is produced (FULTON and SIMPSON, 1976). Assembly is controlled locally within the elongating ciliary bud from various precursor pools; axonemes in the bud are incomplete, especially at the growing tip. Microtubule precursor material appears to be amorphous at such times. Growth of the axonemal doublet microtubules is probably unidirectional, elongation occurring at the tip. Subfiber A assembly precedes subfiber B assembly, while the central pair assembles still later and perhaps somewhat differently. The axoneme is capable of beating before it is fully grown.

A feature of assembly is the strict species-specific determination of ciliary microtubule length; how this is determined and what role the various cap and fiber structures at the microtubule ends (see Sect. 3.4) play in length determination is unclear.

Ciliary regeneration after amputation has been carefully studied. In *Chlamydomonas*, regeneration depends in part on an available pool of precursor protein and in part on new protein synthesis. Partial regeneration of cilia is found under conditions where new protein synthesis is inhibited. In single cells, when only one of the two cilia is removed, the intact organelle initially shortens, while simultaneously, even in the absence of protein synthesis, the amputated cilium elongates. When the two cilia reach identical lengths, both elongate together (cf. COYNE and ROSENBAUM, 1970). Regeneration differs from de novo morphogenesis in that the earliest processes of assembly need not occur in the former situation.

The growth of a 10 μm-long cylindrical cilium requires the production and assembly of over 6 μm^2 of ciliary membrane. The ciliary necklace is formed before the cilium proper grows. The source of ciliary membrane is not known. Where the cilium grows within a cytoplasmic vesicle, one original border of the vesicle, which is a Golgi derivative, becomes the ciliary membrane. The vesicle itself increases its circumference by further fusion; this could be a potential source for extension of the contiguous ciliary membrane.

The known genes for initiation and control of ciliary assembly are thus far all nuclear genes, although the suggestion that the basal body contains organelle-specific informational nucleic acid persists. In *Chlamydomonas* over 80 mutants with altered ciliary structure have been isolated and classified (McVITTIE, 1969), including those discussed previously, which are now the objects of intensive scrutiny.

6. Eukaryotic Cilia vs Other Motile Mechanisms

Eukaryotic plant and animal cells share a common heritage of organelles, which differ from prokaryotic structures, usually in organizational complexity or in molecular specifics. Often a particular eukaryotic cell organelle is passed down from protistan forms to higher metaphyta and metazoa without extensive modification, so that the fundamental structural plan of the organelle is preserved. Function is either preserved as well, or it is modified to serve new purposes. The cilium is an excellent example of such conservation of organellar pattern.

The essentially simple self-assembly nature of tubulin and the symmetry of basal bodies is at least reminiscent of properties of viruses, so that alternative explanations of the evolutionary origin of microtubules and cilia are (1) de novo production in primitive eukaryotes or (2) the basal body or centriole is a viral invader of the protoeukaryotic cytoplasm. The recurrent suggestion that microtubular-based organelles of ninefold symmetry are derived from prokaryotic flagella, either by direct evolution (ASTBURY, et al., 1955) or by ingestion of an endosymbiotic flagellated prokaryote (SAGAN, 1967), is much less likely. Prokaryotic flagella have neither the requisite morphology nor chemistry of ciliary microtubules (see MACNAB, Chap. 3.1.1, this vol.). Tubulin and flagellin are fundamentally different proteins. Flagellin polymerizes outside the cell proper and can be regarded as a secretory protein, while tubulin is a product of the cytoplasmic matrix. According to present concepts, secretory proteins pass through the cell membrane or its homolog at the time of formation and therefore are processed very differently from cytoplasmic proteins. In this sense, mastigonemes rather than microtubules are comparable to bacterial flagella. Moreover, bacterial flagellar motion seemingly is uncoupled from dephosphorylation of ATP. The bacterial flagellum proper, as discussed by MACNAB (Chap. 3.1.1, this vol.), is probably a rigid protein wire turned by mechanoenzymes embedded in the cell membrane—the so-called membrane motor.

On the other hand, the cryptic transmission of centriole-specific information from generation to generation in plant cells is not inconsistent with what might be expected from a well-integrated latent virus, and may be taken to support this hypothesis of basal body origin. Viral assembly may, of course, involve the production of elements of simple but striking and invariant symmetry, such as the baseplate of bacteriophage T2. SATIR and SATIR (1964) present a simple hypothesis for the origin of ninefold symmetry in assemblages of α-helical proteins. One additional advantage of the invasive hypothesis is that it highlights the intimate relationship between basal bodies and cell membranes, which has been exploited further by evolution in the cilium.

It is not clear why prokaryotic mechanisms of motility are not carried over to eukaryotes; in general, one suspects that the propulsive force generated by the bacterial flagellum is too small for the larger cell mass; however, no precise calculations seem available in this regard.

Eukaryotic plant cells also rely on actin and myosin for motility, sometimes as an alternative to ciliary motility. Actin and myosin, like tubulin and dynein, appear to be early eukaryotic features, and, as is now known, are widely distributed in both plant and animal cells (see BRITZ, Chap. 2.3; WAGNER, Chap. 2.2, in this volume, for more extensive discussion). It has sometimes been suggested on the basis of assembly characteristics, amino acid profile and so forth, that actin and tubulin are related molecules, perhaps with a common evolutionary origin; similar suggestions occur with regard to myosin vs dynein. The most careful comparisons do not support very close homologies. Regardless of the validity of such suggestions, actin and myosin are generally mobilized by the cell in different ways from tubulin and dynein. Although both sets of molecules rely on sliding interactions and ATP hydrolysis, the control mechanisms for microtubule and microfilament-based motility have obviously diverged to the point where they usually provide for antagonistic actions of the systems. In particular, actin-myosin based systems are usually turned on by local Ca^{2+} increases in the cell cytoplasm, and are inoperative at $Ca^{2+} < 10^{-6}$ M, while, as we have seen above, cilia are operative in low Ca^{2+}, and often reverse or stop in high Ca^{2+}. The eukaryotic plant cell then has a greater repertoire of molecules with which to move organelles or displace and shape itself than it would have if only one of these systems had evolved, and this may help to explain the divergence of form and subtlety of cell specialization that became possible during the evolution of such cells.

Acknowledgements. The senior author held appointment as a Miller Professor at the Univ. of California, Berkeley for much of the time during which this article was written. The work was also supported in part by grants from the USPHS. We thank Drs. Z. Cande and R.A. Bloodgood for helpful comments, P. Setzer for photographic assistance and M. Smith for secretarial assistance. Dr. Satir's present address is the Department of Anatomy at Albert Einstein College of Medicine, Bronx, New York. Dr. Ojakian's present address is SUNY Downstate Medical Center, Brooklyn, New York.

References

Afzelius, B.A. Electron microscopy of the sperm tail. Results obtained with a new fixative. J. Biophys. Biochem. Cytol. *5*, 269–278 (1959)

Allen, C., Borisy, G.G.: Structural polarity and directional growth of microtubules of *Chlamydomonas* flagella. J. Mol. Biol. *90*, 381–402 (1974a)

Allen, C., Borisy, G.G.: Flagellar motility in *Chlamydomonas*: reactivation and sliding in vitro. J. Cell. Biol. *63*, 5a (1974b)

Amos, L.A., Linck, R.W., Klug, A.: Molecular structure of flagellar microtubules. In: Cell Motility. Goldman, R., Pollard, T., Rosenbaum, J.L. (eds.). New York: Cold Spring Harbor Lab, 1976, pp. 847–867

Astbury, W., Beighton, E., Werbull, C.: The structure of bacterial flagella. Symp. Soc. Exp. Biol. *9*, 282–305 (1955)

Bergman, K., Goodenough, U.W., Goodenough, D.A., Jawitz, J., Martin, H.: Gametic differentiation in *Chlamydomonas reinhardtii* II. Flagellar membranes and the agglutination reaction. J. Cell Biol. *67*, 606–622 (1975)

Blake, J.R., Sleigh, M.A.: Hydromechanical aspects of ciliary propulsion. In: Swimming and Flying in Nature. Wu, T. Y-T., Brokaw, C.J., Brennen, C. (eds.). New York: Plenum Press, 1974, Vol. I pp. 133–141

Bloodgood, R.A.: Motility occurring in association with the surface of the *Chlamydomonas* flagellum. J. Cell Biol. *75*, 983–989 (1977)

Bouck, G.B.: Architecture and assembly of mastigonemes. In: Advances in Cell and Molecular Biology. DuPraw, E.J. (ed.). New York: Academic Press, 1972, Vol II, pp. 237–271

Branton, D.: Fracture faces of frozen membranes. Proc. Natl. Acad. Sci. USA *55*, 1048–1056 (1966)

Bray, D.F., Nakamura, K., Costerton, J.W., Wagenaar, E.B.: Ultrastructure of *Chlamydomonas eugametos* as revealed by freeze-etching: cell wall, plasmalemma and chloroplast membrane. J. Ultrastruct. Res. *47*, 125–141 (1974)

Brokaw, C.J.: Chemotaxis of Braken Sperm. Ph. D. thesis, Cambridge, 1958, 133 pp.

Carlson, F.D.: A theory of the survival value of motility. In: Spermatozoon Motility. Bishop, D.W. (ed.). Washington, D.C.: Am. Assoc. Adv. Sci., 1962, pp. 137–146

Carothers, Z.B.: Comparative studies on spermatogenesis in bryophytes. In: The Biology of the Male Gamete. Duckett, J.G., Racey, P.A. (eds.). Biol. J. Linnean Soc. *7*, suppl. no 1. London: Academic Press, 1975, pp. 71–84

Chasey, D.: The three-dimensional arrangement of radial spokes in the flagella of *Chlamydomonas reinhardtii*. Exp. Cell Res. *84*, 374–380 (1974)

Coleman, A.: Immobilization, agglutination and agar precipitin effects of antibodies to flagella of *Pandorina* mating types. J. Protozool. *10*, 141–148 (1963)

Coyne, B., Rosenbaum, J.L.: Flagellar elongation and shortening in *Chlamydomonas*. J. Cell Biol. *47*, 777–781 (1970)

Dentler, W.L., Rosenbaum, J.L.: Flagellar elongation and shortening in *Chlamydomonas* III. Structures attached to the tips of flagellar microtubules and their relationship to the directionality of flagellar microtubule assembly. J. Cell. Biol. *74*, 747–759 (1977)

Duckett, J.G.: Spermatogenesis in pteridophytes. In: The Biology of the Male Gamete. Duckett, J.G., Racey, P.A. (eds.). Biol. J. Linnean Soc. *7* suppl. no. 1 London: Academic Press, 1975, pp. 97–127

Duckett, J.G., Racey, P.A. (eds.): The Biology of the Male Gamete. Biol. J. Linnean Soc. *7* suppl. No. 1. London: Academic Press, 1975

Dunlap, K., Eckert, R.: Calcium channels in the ciliary membrane of *Paramecium*. J. Cell Biol. *70*, 245a (1976)

Eckert, R.: Bioelectric control of ciliary activity. Science *176*, 473–481 (1972)

Fawcett, D.W., Porter, K.R.: A study of the fine structure of ciliated epithelia. J. Morphol. *94*, 221–281 (1954)

Friedman, I., Colwin, A.L., Colwin, L.H.: Fine structural aspects of fertilization in *Chlamydomonas reinhardtii*. J. Cell. Sci. *3*, 115–128 (1968)

Fulton, C., Simpson, P.A.: Selective synthesis and utilization of flagellar tubulin. The multitubulin hypothesis. In: Cell Motility. Goldman, R., Pollard, T., Rosenbaum, J.L. (eds.). New York: Cold Spring Harbor Lab., 1976, pp. 987–1005

Gasic, G.J., Berwick, L., Sorrentino, M.: Positive and negative colloidal iron as cell surface electron stains. Lab. Invest. *18*, 63–71 (1968)

Gibbons, B.H., Gibbons, I.R.: The effect of partial extraction of dynein arms on the movement of re-activated sea urchin sperm. J. Cell. Sci. *13*, 337–357 (1973)

Gibbons, B.H., Gibbons, I.R.: Properties of flagellar "rigor waves" produced by abrupt removal of adenosine triphosphate from actively swimming sea urchin sperm. J. Cell. Biol. *63*, 970–985 (1974)

Gibbons, B.H., Gibbons, I.R.: Functional recombination of dynein-1 with demembranated sea urchin sperm partially extracted with KCl. Biochem. Biophys. Res. Comm. *73*, 1–6 (1976)

Gibbons, I.R., Fronk, E., Gibbons, B.H., Ogawa, K.: Multiple forms of dynein in sea urchin sperm flagella. In: Cell Motility. Goldman, R., Pollard, T., Rosenbaum, J.L. (eds.). New York: Cold Spring Harbor Lab., 1976, pp. 915–932

Gilula, N.B., Satir, P.: The ciliary necklace. A ciliary membrane specialization. J. Cell Biol. *53*, 494–509 (1972)

Goldman, R., Pollard, R., Rosenbaum, J.R. (eds.): Cell Motility (esp. Book C) New York: Cold Spring Harbor Lab, 1976

Goodenough, U., Weiss, R.L.: Gametic differentiation in *Chlamydomonas reinhardtii* III. Cell wall lysis and microfilament associated mating structure activation in wild type and mutant strains. J. Cell Biol. *67*, 623–637 (1975)
Haimo, L., Rosenbaum, J.: Recombination of dynein with outer fiber of *Chlamydomonas reinhardtii*. J. Cell Biol. *70*, 216a (1976)
Heath, I.B., Darley, W.M.: Observations on the ultrastructure of the male gametes of *Biddulphia levis*. Eur. J. Phycol. *8*, 51–59 (1972)
Hepler, P.K.: The blepharoplast of *Marsilea*; its de novo formation and spindle association. J. Cell Sci. *21*, 361–390
Holwill, M.E.J.: The role of body oscillation in the propulsion of microorganisms. In: Swimming and Flying in Nature. Wu, T. Y-T., Brokaw, C.J., Brennen, C. (eds.). New York: Plenum Press, 1974, Vol. I pp. 133–141
Holwill, M.E.J.: Physical aspects of flagellar movement. Physiol. Rev. *46*, 696–785 (1966)
Hopkins, J.M.: Subsidiary components of the flagella of *Chlamydomonas reinhardtii*. J. Cell Sci. *7*, 823–839 (1970)
Hyams, J.S., Borisy, G.G.: Flagellar co-ordination in *Chlamydomonas reinhardtii*: isolation and reactivation of the flagellar apparatus. Science *189*, 891–893 (1975a)
Hyams, J.S., Borisy, G.G.: The dependence of the waveform and direction of beat of *Chlamydomonas* flagella on calcium ions. J. Cell Biol. *67*, 186a (1975b)
Jahn, T.L., Landman, M., Fonseca, J.R.: The mechanism of locomotion in flagellates. II. Function of the mastigonemes of *Ochromonas*. J. Protozool. *11*, 291–296 (1964)
Kung, C., Chang, S.Y., Satow, Y., Van Houten, J., Hansma, H.: Genetic dissection of behavior in *Paramecium*. Science *188*, 898–904 (1974)
Machemer, H.: Ciliary activity and metachronism in protozoa. In: Cilia and Flagella. Sleigh, M.A. (ed.). New York: Academic Press, 1974, pp. 199–286
Manton, I.: The fine structure of plant cilia. Symp. Soc. Exp. Biol. *6*, 306–319 (1952)
Manton, I.: Plant cilia and associated organelles. In: Cellular Mechanisms in Differentiation and Growth. Rudnick, D. (ed.). Princeton, N.J.: Princeton Univ. Press, 1956, pp. 61–71
Manton, I.: The possible significance of some details of flagellar bases in plants. J. Roy. Micro. Soc. *82*, 279–285 (1964)
Manton, I.: Plant spermatozoids. In: Comparative Spermatology. Baccetti, B. (ed.). New York: Academic Press, 1970, pp. 143–158
Manton, I., Clarke, B.: An electron microscope study of the spermatozoid of *Sphagnum*. J. Exp. Bot. *3*, 265–275 (1952)
Manton, I., Rayns, D.G., Ettl, H., Parke, M.: Further observations on green flagellates with scaly flagella: the genus *Heteromastix korshikov* J. Mar. Biol. Assoc. UK *45*, 241–255 (1965)
Manton, I., von Stosch, H.A.: Observations on the fine structure of the male gamete of the marine centric diatom *Lithodesmium undulatum* J. Roy. Microsc. Soc. *85*, 119–134 (1966)
Markey, D.R., Bouck, B.: Mastigoneme attachment in *Ochromonas*. J. Ultrastruct. Res. *59*, 173–177 (1977)
Martinez-Palomo, A.: The surface coat of animal cells. Int. Rev. Cytol. *29*, 29–75 (1970)
McLean, R.J., Bosmann, H.B.: Cell-cell interactions: enhancement of glycosyl transferase ectoenzyme systems during *Chlamydomonas* gametic contact. Proc. Natl. Acad. Sci. USA *72*, 310–313 (1975)
McLean, R.J., Brown, R.M.: Cell surface differentiation of *Chlamydomonas* during gamete genesis. Dev. Biol. *36*, 279–285 (1974)
McVittie, A.: Studies on flagella-less, stumpy and short flagellum mutants of *Chlamydomonas reinhardi*. Proc.Roy. Soc. London ser. B *173*, 59–60 (1969)
Mizukami, I., Gall, J.: (1966) Centriole replication II. Sperm formation in the fern *Marsilea* and the cycad *Zamia* J. Cell Biol. *29*, 97–111 (1966)
Moestrup, Ø.: Some aspects of sexual reproduction in eucaryotic algae. In: The Biology of the Male Gamete. Duckett, J.G., Racey, P.A. (eds.). Biol. J. Linnean Soc. 7 suppl. no. 1. London: Academic Press, 1975, pp. 23–35
Naitoh, Y., Eckert, R.: The Control of Ciliary Activity in Protozoa. In: Cilia and Flagella. Sleigh, M.A. (ed.). New York: Academic Press, 1974, pp. 305–352

Naitoh, Y., Kaneko, H.: Reactivated triton-extracted models of *Paramecium*: Modification of ciliary movement by calcium ions. Science *176*, 523–524 (1972)
Norstog, K.: The motility of cycad spermatozoids in relation to structure and function. In: The Biology of the Male Gamete. Duckett, J.G., Racey, P.A. (eds.). Biol. J. Linnean Soc. *7* suppl. no. 1 London: Academic Press, 1975, pp. 135–142
Ogawa, A., Takahashi, K.: Artificial deciliation causes loss of calcium-dependent responses in *Paramecium*. Nature (London) *264*, 170–172 (1976)
Olmsted, J.B., Borisy, G.G.: Microtubules. Ann. Rev. Biochem. *42*, 507–540 (1973)
Pickett-Heaps, J.: Structural and phylogenetic aspects of microtubular systems in gametes and zoospores of certain green algae. In: The Biology of the Male Gamete. Duckett, J.G., Racey, P.A. (eds.). Biol. J. Linnean Soc. *7* suppl. no. 1 London: Academic Press, 1975, pp. 37–44
Piperno, G., Huang, B., Luck, D.L.: Two-dimensional analysis of flagellar proteins from wild-type and paralyzed mutants of *Chlamydomonas reinhardtii*. Proc. Natl. Acad. Sci. USA *74*, 1600–1604 (1977)
Rice, H.V., Laetsch, W.M.: Observations on the morphology and physiology of *Marsilea* sperm. Am. J. Bot. *54*, 856–866 (1967)
Ringo, D.L.: Flagellar motion and fine structure of the flagellar apparatus in *Chlamydomonas*. J. Cell. Biol. *33*, 543–571 (1967)
Roseman, S.: The synthesis of complex carbohydrates by multiglycosyl transferase systems and their potential function in intercellular adhesion. Chem. Phys. Lipids *5*, 270–297 (1970)
Roth, S., McGuire, E.J., Roseman, S.: Evidence for cell-surface glycosyl transferases: their potential role in cell recognition. J. Cell. Biol. *51*, 536–547 (1971)
Sagan, L.: On the origin of mitosing cells. J. Theoret. Biol. *14*, 225–274 (1967)
Sale, W.S., Satir, P.: The direction of active sliding of microtubules in *Tetrahymena*. Proc. Natl. Acad. Sci. USA *74*, 2045–2049 (1977)
Satir, P.: Structure and function in cilia and flagella. Facts and Problems. Protoplasmatologia III E 52 pp (1965a)
Satir, P.: Studies on cilia. II. Examination of the distal region of the ciliary shaft and the role of the filaments in motility. J. Cell. Biol. *26*, 805–834 (1965b)
Satir, P.: Studies on cilia III. Further studies on the cilium tip and a "sliding filament" model of ciliary motility. J. Cell Biol. *39*, 77–94 (1968)
Satir, P.: Production and arrest of ciliary motility. In: Contractile Systems in Non-Muscle Tissues. Perry, S.V., Margreth, A., Adelstein, R.S. (eds.). Amsterdam: Elsevier North Holland, 1976, pp. 263–273
Satir, P., Satir, B.: A model for ninefold symmetry in α-keratin and cilia. J. Theoret. Biol. *7*, 123–128 (1964)
Satir, P., Satir, B.: The design and function of site-specific particle arrays in the cell membrane. In: Control of Cell Proliferation. Clarkson, B., Baserga, R. (eds.). New York: Cold Spring Harbor Lab, 1974, pp. 233–249
Snell, W.J.: Mating in *Chlamydomonas*: a system for the study of specific cell adhesion I. Ultrastructural and electrophoretic analysis of flagellar surface components involved in adhesion. J. Cell Biol. *68*, 48–69 (1976)
Stephens, R.E.: Isolation of nexin—the linkage protein responsible for the maintenance of the nine-fold configuration of flagellar axonemes. Biol. Bull. *139*, 438 (1970)
Stephens, R.E., Edds, K.T.: Microtubules: Structure, Chemistry and Function. Biophys. Rev. *56*, 709–778 (1976)
Summers, K.E., Gibbons, I.R.: ATP-induced sliding of tubules in trypsin-treated flagella of sea urchin sperm. Proc. Natl. Acad. Sci. USA *68*, 3092–3096 (1971)
Summers, K.E., Gibbons, I.R.: Effects of trypsin digestion on flagellar structures and their relationship to motility. J. Cell Biol. *58*, 618–629 (1973)
Warner, F.D., Mitchell, D.R., Perkins, C.R.: Structural conformation of the ciliary ATPase dynein. J. Mol. Biol. *114*, 367–384 (1977)
Warner, F.D., Satir, P.: The structural basis of ciliary bend formation. J. Cell Biol. *63*, 35–63 (1974)
Weiss, R.L., Goodenough, D.A., Goodenough, U.W.: Membrane particles associated with the basal body and contractile vacuole secretion in *Chlamydomonas*. J. Cell Biol. *72*, 133–142 (1977)

Wiese, L.: On sexual agglutination and mating-type substances (gamones) in isogamous heterothallic Chlamydomonads I. Evidence of the identity of the gamones with the surface components responsible for sexual flagellar contact. J. Phycol. *1*, 46–54 (1965)
Wiese, L.: Algae. In: Fertilization. Metz, C.B., Monroy, A. (eds.). New York: Academic Press, 1969, Vol. II, pp. 135–188
Wiese, L., Shoemaker, D.W.: On sexual agglutination and mating-type substances (gamones) in isogamous heterothallic Chlamydomonads. II. The effect of concanavalin A upon the mating type reaction. Biol. Bull. *138*, 88–95 (1970)
Witman, G.B., Carlson, K., Berliner, J., Rosenbaum, J.L.: *Chlamydomonas* flagella. I. Isolation of electrophoretic analysis of microtubules, matrix, membranes and mastigonemes. J. Cell Biol. *54*, 507–539 (1972a)
Witman, G.B., Carlson, K., Rosenbaum, J.L.: *Chlamydomonas* flagella. II. The distribution of tubulins 1 and 2 in the outer doublet microtubules. J. Cell Biol. *54*, 540–555 (1972b)
Witman, G.B., Fay, R., Plummer, J.: *Chlamydomonas* mutants: evidence for the roles of specific axonemal components in flagellar movement: In: Cell Motility. Goldman, R., Pollard, T., Rosenbaum, J.L. (eds.). New York: Cold Spring Harbor Lab, 1976, pp. 969–986

3.1.3 Gliding Movements

L.N. HALFEN

1. Introduction

1.1 Terminology and General Aspects

The term "gliding movements" is used to describe active displacement of an organism in contact with a solid substrate, where there is no visible structure responsible for the movement, and no change in the shape of the organism (JAROSCH, 1962). Light microscopy has generally not been able to reveal any morphological aspect of gliding organisms which is capable of producing movement. Gliding motility requires contact with a reasonably firm surface, and should not be confused with the ability to swim through a liquid. Gliding rates are also generally slower than the rates of locomotion displayed by swimming forms. Reported rates of gliding range from 2 μm min^{-1} up to 11.1 μm s^{-1} (GEITLER, 1936; DREWS and NULTSCH, 1962; HALFEN and CASTENHOLZ, 1971a). In the genus *Oscillatoria*, motility is often a strikingly uniform process, while unicellular gliding organisms tend to be much more irregular in their pattern and rate of movement. No one in this century has seriously suggested that flagella or cilia are implicated in gliding with the few exceptions noted by JAROSCH (1962). From the outset, motility hypotheses which require the rapid discharge of cell contents to induce motility are dismissed. While gliding *Oscillatoria* trichomes may display rates of several μm s^{-1}, India ink particles near the surface of these trichomes are not disturbed in a fashion which suggests a discharge process (HALFEN, 1970).

Motile cyanophytes may exhibit a wide range of trichome conformations, including the straight hormogonia of *Symploca*, the well-developed curved and attenuated tip regions of *Oscillatoria* and the extreme helical morphology of *Spirulina*. In all these cases, the trichome conformation remains essentially unaltered during the movement process, and certainly is not responsible for the observed rates of movement. Trichome distortion may result from impeded movement activity, particularly in flexible smaller organisms (CASTENHOLZ, 1973).

Gliding eukaryotic organisms display movement which is invariably associated with the production of a substantial slime trail (NULTSCH, 1974). Considering desmids and diatoms, the directed secretion of this trail is generally accepted as the mechanism responsible for movement in these forms, and this view is consistent with their slow rates of movement (JAROSCH, 1962; DREWS and NULTSCH, 1962; NULTSCH, 1974). No conformational change or motility organelle is visible in these organisms using light microscopy.

Central to the production of gliding movements is the establishment of a shear zone between the organism and the underlying substrate. In evaluating how an organism is able to move, it is necessary to postulate that the organism is responsible for exerting a force against the substrate sufficient to effect the displacement of the organism. Thus, a velocity gradient is created between the organism and the substrate. Modern attempts to illucidate gliding have focused upon this shear zone and its components (e.g., SCHULZ, 1955; WALSBY, 1968; HALFEN and CASTENHOLZ, 1971a; RIDGEWAY et al., 1975).

1.2 Occurrence

Gliding movements have been noted in several photo-autotrophic blue-green algae including members of the Oscillatoriaceae, Nostocaceae and Chroococcales. Gliding motility is also found in the thiotrophic analogs of cyanophytes: the Beggiatoales, and in heterotrophic forms: the Vitreoscillaceae. There are gliding myxobacteria, spirochaetes and carotenoid-containing and photosynthetic flexibacteria. (For reviews see WEIBULL, 1960; DOETSCH and HAGEAGE, 1968; CASTENHOLZ, 1973; NULTSCH, 1974.)

Gliding motility is readily exemplified in the genus *Oscillatoria*, where most healthy trichomes display remarkably smooth and sustained rates of locomotion (LAMONT, 1969; HALFEN and CASTENHOLZ, 1971a). In contrast, the gliding movements of the Nostocaceae are generally slower and often restricted to hormogonial stages (LAZAROFF and VISHNIAC, 1961; DREWS and NULTSCH, 1962; CASTENHOLZ, 1973). Branched or heavily sheathed cyanophytes are generally not motile except when hormogonia are liberated (CASTENHOLZ, 1973; FOGG et al., 1973). Unicellular blue-green algae are reported to display movements which are jerky and quite slow (GEITLER, 1936; STANIER et al., 1971). Motile bacteria that are morphological analogs to blue-green algae move at rates comparable to their cyanophyte equivalent (DOETSCH and HAGEAGE, 1968).

Eukaryotic organisms capable of gliding include several genera of pennate diatoms that possess a true raphe, which is apparently essential for motility (JAROSCH, 1962; GORDON and DRUM, 1970; NULTSCH, 1974). Some green algae of the order Conjugales also glide, especially desmids of the genera *Closterium* and *Micrasterias* (JAROSCH, 1962; NEUSCHELER, 1967; PAO-ZUN YEH and GIBOR, 1970; NULTSCH, 1974). Gliding motility may also be observed occasionally in certain Euglenoids, protozoans, flagellated cells in contact with solid substrates, and specialized reproductive cells of red algae and fungi (JAROSCH, 1962; NULTSCH, 1974).

1.3 Historical Perspectives

The literature dealing with gliding movements spans more than two centuries and has been regularly and recently reviewed (DOETSCH and HAGEAGE, 1968; CASTENHOLZ, 1973; FOGG et al., 1973; NULTSCH, 1974). Beginning with ADANSON (1767), dozens of investigators have described gliding movements, have discussed their occurrence in nature and have offered explanations of the mechanism responsible for motility. The clear distinctions among these explanations allowed BURKHOLDER (1934) to divide them into four categories listed on the following page:

1. Osmotic forces.
2. Surface tension activities.
3. Slime secretion.
4. Contractile waves.

These hypotheses were based upon direct observations of gliding organisms without the aid of ultrastructural techniques. The first two groups of hypotheses were poorly supported by experimental evidence and have not been seriously considered in the literature in the last four decades (BURKHOLDER, 1934; DOETSCH and HAGEAGE, 1968; HALFEN and CASTENHOLZ, 1970; FOGG et al., 1973). In contrast, the remaining two groups of motility explanations have been repeatedly employed, particularly after shear-zone evaluation by electron microscopy (e.g., SCHULZ, 1955; PATE and ORDAL, 1967; HALFEN, 1973; RIDGEWAY et al., 1975).

Sheath extrusion (slime secretion) mechanisms for inducing gliding movements are based upon observations that gliding organisms produce a slime-like sheath (or trail) as they move (NIKLITSCHEK, 1934; FRITSCH, 1945). This sheath is displaced over cyanophyte trichomes as they move, and will accumulate with time on immobilized trichomes (HOSOI, 1951; SCHULZ, 1955; DODD, 1960). SIEBOLD (1849) and FECHNER (1915) suggested that movement was caused by streaming of the mucilage composing the sheath layer. Later, HARDER (1918) modified this proposal, indicating that motility may result from swelling of the mucilage once it has left the cell. PRELL (1921) developed an elaborate and elegant explanation of *Oscillatoria* motility based upon the directed extrusion of sheath strands through helically arranged pores in the cell wall. PRELL's work was noteworthy because his model was one of the first to incorporate other cyanophyte motility characteristics in addition to the obvious displacement parallel to the long axis of the trichome. Mucilage extrusion proposals attempting to explain gliding in filamentous prokaryotic organisms have surfaced repeatedly in the literature in recent times (HOSOI, 1951; SCHULZ, 1955; WALSBY, 1968; RIDGEWAY et al., 1975). However, ultrastructural studies failed to reveal cellular structures in *Oscillatoria* which could function in extruding mucilage to induce movement (RIS and SINGH, 1961; HALFEN and CASTENHOLZ, 1971a).

Sheath extrusion as a basis for motility was countered by the observation that this mechanism was unable to explain all the behavior associated with cyanophyte locomotion, and could not account for the rates of linear translation recorded (DOETSCH and HAGEAGE, 1968; HALFEN, 1970). Neither of these criticisms apply, however, to the gliding movements of pennate diatoms and certain green algae. Here, the best proposals to date involve slime extrusion and possible associated swelling to displace these cells in their slow and frequently erratic paths (GORDON and DRUM, 1970; PAO-ZUN YEH and GIBOR, 1970; NULTSCH, 1974; PICKET-HEAPS, 1975).

Motility hypotheses based upon contractile-wave mechanisms were introduced by PHILLIPS (1904), SCHMID (1918, 1921) and ULLRICH (1926). ULLRICH produced cinematographic records of longitudinal waves in *Oscillatoria jenensis* which were accepted for three decades until SCHULZ (1955) challenged them as optical artifacts. JAROSCH (1962, 1964) initially postulated the presence of fibrillar elements in motility systems of gliding organisms at the molecular level. He believed that a type of protein fibril was helically wrapped around the cell wall of *Oscillatoria* which, when twisted or rotated, would create higher-order helices that moved the length of the fibril, creating submicroscopic waves at the trichome surface (JAROSCH, 1964). If the direction of propagation of these torsion-induced waves was the same in a majority of the fibrils, the cells were supposedly displaced in the opposite direction over a firm substrate. Aspects of this proposal were supported by electron microscopy of three *Oscillatoria* species which displayed a fibrillar array as a component of the cell wall (HALFEN and CASTENHOLZ, 1971a; HALFEN, 1973). However, no evidence was forthcoming which indicated that these fibrils were twisted or rotated to produce waves on the cell surface. These fibrils were found in the shear zone lying on the outside of the murein (peptide-glycan) layer of the cell wall.

The waves-of-contraction concept found less support in diatoms, where recent hypotheses center about the secretion of fibrous slime. This material is apparently directed out of the raphe of the cell. Subsequent hydration of the slime produces sufficient swelling to move the organism (DRUM and HOPKINS, 1966; HOPKINS and DRUM, 1966).

2. Behavior and Functional Morphology

2.1 Individual Organisms

Linear translation is the component common to all gliding systems which involve the linear movement of the organism from one location to another over (or, in some cases, through) a solid substrate. With filamentous forms, linear translation is a coordinated trichome displacement in a direction parallel to the long axis of that trichome. This displacement establishes anterior and posterior ends in what may be an otherwise undifferentiated linear array of cells. Table 1 presents a listing of published rates of linear translation, revealing the range of rates which exists for different species.

Most gliding organisms have the capacity for rapid cessation of linear translation and for initiation of movement in the opposite direction. This change in polarity apparently does not impair the rate of linear translation in the new direction (HALFEN and CASTENHOLZ, 1971a). The frequency of reversals of this type varies with conditions and also with the size of the gliding entity; unicellular forms display this behavior more frequently than organisms existing as trichomes (CASTENHOLZ, 1973; FOGG et al., 1973). Reversals have been seen to result from environmental changes (e.g., light, temperature) or simply because of physical obstructions in the original line of advance (BURKHOLDER, 1934; JAROSCH, 1962; HALFEN, 1970; NULTSCH, 1974).

Using the capacity for reversals, a trichome moving through an environmental gradient may ultimately be successful in positioning itself in a favorable location in that gradient. This eliminates (in organisms with this type of behavior) the necessity of postulating the existence of a "steering" mechanism to accomplish this end (BURKHOLDER, 1934; NULTSCH, 1962a, 1962c; CASTENHOLZ, 1973; see HÄDER Chap. 3.2.1, this vol.). A functional implication of reversal behavior is the demonstrated ability of the mechanism responsible for linear translation to operate with comparable effectiveness in either direction, following rapid and possibly repetitive

Table 1. Motility rates for gliding organisms

Organism	Rate μm s⁻¹	Reference
Cyanophyta		
Anabaena cylindrica	1	Walsby (1968)
A. variabilis	0.33	Drews (1959)
Anabaena sp.	2.5	Castle (1926)
Cylindrospermum licheniformis	0.25	Drews (1959)
Lyngbya sp.	5	Lamont (1969)
Oscillatoria amoena	1.8	Halfen and Castenholz (1971 a)
O. animalis	1.8	Halfen and Castenholz (1971 a)
O. chalybea	5	Lamont (1969)
O. curviceps	1.1–7.5	Schmid (1918)
O. formosa	4.2	Castle (1926)
O. limosa	2–3	Jarosch (1958)
O. mougeotii	3.3	Drews and Nultsch (1962)
O. princeps	5	Castle (1926)
O. princeps	11.1	Halfen (1970)
O. sancta	0.5–2	Schulz (1955)
O. terebriformis	8.3	Castenholz (1968)
Oscillatoria sp.	4.1	Crozier and Federighi (1924)
Oscillatoria sp.	2.5–3	Abe (1963)
Oscillatoria sp.	4.3	Halfen and Castenholz (1971 a)
Phormidium autumnale	2.2–2.8	Nultsch (1961)
P. uncinatum	2.5	Drews and Nultsch (1962)
Scytonema sp.	0.004–0.058	Castle (1926)
Spirulina sp.	1–1.8	Halfen (1970)
Symploca muscorum	0.006–0.03	Halfen and Castenholz (1971 a)
Diatoms		
Amphora montana	2–5	Nultsch (1956)
Bacillaria paradoxa	8	Jarosch (1958)
Cymatopleura solea	4–6	Jarosch (1958)
Donkinia carinata	22.5	Nultsch (1962 b)
Navicula Buderi	4–6	Nultsch (1956)
N. radiosa	14–20	Jarosch (1962)
Navicula sp.	4–6	Jarosch (1958)
Nitzschia closterium	5–9	Jarosch (1962)
N. communis	5–8.5	Nultsch (1956)
N. longissima	25	Abe (1963)
N. palea	8–14	Nultsch (1956)
N. putrida	2.7	Jarosch (1962)
N. stagnorum	5–9	Nultsch (1956)
Pleurosigma angulatum	17.5	Nultsch (1962 b)
Tropidoneis gibberula	25	Nultsch (1962 b)
Desmids		
Closterium acerosum	3.7	Klebs (1885)
Micrasterias denticulata	0.5–1	Neuscheler (1967)
Other Organisms		
Bacterium anitratum	0.032	Hageage (1963)
Beggiatoa sp.	6–8	Jarosch (1958)
Cytophaga sp.	2.5	Stanier (1942)
Euglena deses	2–3	Jarosch (1958)
Saprospira sp.	2–3	Lewin (1965)
Spirochaeta plicatilis	1–5	Jarosch (1958)
Vitreoscilla sp.	0.4	Costerton et al. (1961)
Vitreoscilla sp.	0.16–0.32	Doetsch and Hageage (1968)

changeovers. This capacity is a fundamental consideration for any explanation of the basis for gliding movement. Linear translation is usually associated with secretion of a sticky mucilage sheath layer or trail (JAROSCH, 1962). DODD (1960) suggested that the role of this mucilage was to facilitate contact between the gliding organism and the substrate. Thus, the apparatus producing motility would be assured of a closely associated physical structure against which to act. Diatom movement was observed to vary markedly depending upon the nature of the substrate and the degree of contact with it (JAROSCH, 1962; HOPKINS and DRUM, 1966). This relationship is fundamental to understanding the probable mechanisms responsible for movement (see Sect. 3).

Mucilage sheath is produced on the entire length of trichomes of *Oscillatoria princeps* (Fig. 1) and is shed as a collapsed twisted strand behind the advancing trichome (HALFEN and CASTENHOLZ, 1971a). Ultrastructural examination of this organism revealed a system of small junctional pores in the cell wall at the site of the crosswall intersection (HALFEN, 1970; HALFEN and CASTENHOLZ, 1971a). Large membrane-lined blind-ended pores (Fig. 2) were observed penetrating the murein (peptido-glycan, L-2) layer of the longitudinal cell wall (RIS and SINGH, 1961; HALFEN and CASTENHOLZ, 1971a). Neither of these pore systems provides ultrastructural detail consistent with direct translocation of mucilage or its precursor to the exterior of the cells. LAMONT (1969) offered an alternative hypothesis that mucilage was shear-polymerized on the exterior of Oscillatoriacean trichomes, following the diffusion of components through the layers of the cell wall.

In the case of *O. princeps*, the existence of an array of large cytoplasm-filled pores in the exceedingly thick L-2 cell-wall layer may readily facilitate the process of diffusion of sheath precursor (HALFEN and CASTENHOLZ, 1971a). Smaller trichomes with much thinner cell walls may be able to export sheath precursor more readily, reducing the need for this type of pore system.

The mucilage secreted by gliding organisms is generally thought to be composed of polysaccharides. (See NULTSCH, 1974.) This mucilage material stains with ruthenium red (LEAK, 1968; GORDON and DRUM, 1970) and will trap suspended carmine or India ink particles (SCHULZ, 1955; WALSBY, 1968; HALFEN, 1970). The trail of mucilage deposited by motile desmids is degraded by lysozyme (PAO-ZUN YEH and GIBOR, 1970). Natural populations of gliding organisms which produce expendable mucilage probably support communities of heterotrophic microorganisms that exploit this material (CASTENHOLZ, 1967). Axenic aerated cultures of some cyanophytes such as *Oscillatoria terebriformis* develop accumulations of expended sheath dispersed in the culture which significantly increase the viscosity of the medium. *O. princeps* collects expended sheath in the center of an aggregation of motile trichomes, eventually forming a gel-like ball of dead trichomes and mucilage (HALFEN, 1970). NIKLITSCHEK (1934) contended that no gliding motility was possible in cyanophytes without sheath production. Certainly, motility in diatoms (DRUM and HOPKINS, 1966) and desmids (PAO-ZUN YEH and GIBOR, 1970) is questionable without production of a slime trail, which these authors have proposed as a basis for the observed movement.

The motility of many Oscillatoriacean cyanophytes is characterized by obligate rotation associated with the advance of the trichome, causing a surface point on a cell to trace a regular helix (FRITSCH, 1945; LAMONT, 1969; HALFEN and CASTENHOLZ, 1970, 1971a). Immobilized trichomes displace sheath material in

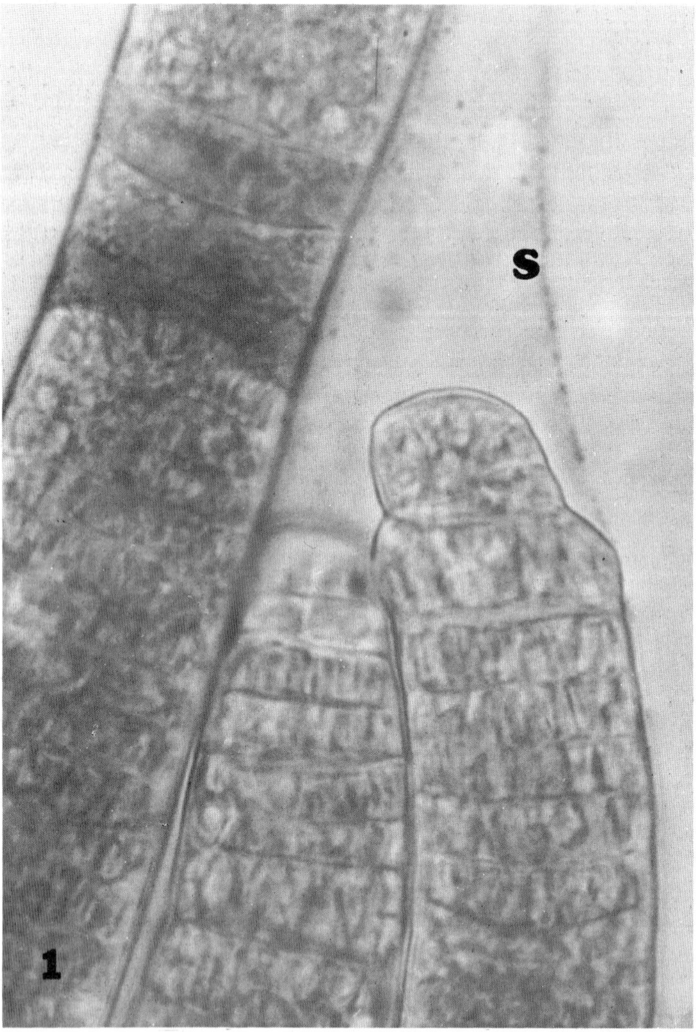

Fig. 1. A light micrograph of gliding trichomes of *O. princeps* which have been exposed to India ink. The sheath *(S)* is being shed as a sticky strand behind the trichomes. The trichomes are 35 μm in diameter at the bottom of this view

a helical fashion over the trichome surface (HOSOI, 1951). FRITSCH (1945) has suggested that the direction of rotation is a species-specific characteristic.

The obligate nature of trichome rotation during linear translation implies that the motility system may be itself helically distributed or organized over the cell surfaces to produce the helical advance of the organism (PRELL, 1921; HALFEN and CASTENHOLZ, 1971a). Close observation of impeded *O. princeps* trichomes revealed little or no slippage of the helix when rotation was prevented without hindering linear translation. In all cases, without free rotation, linear translation ceased until torsions were sufficient to snap the trichome free (HALFEN, 1970). The

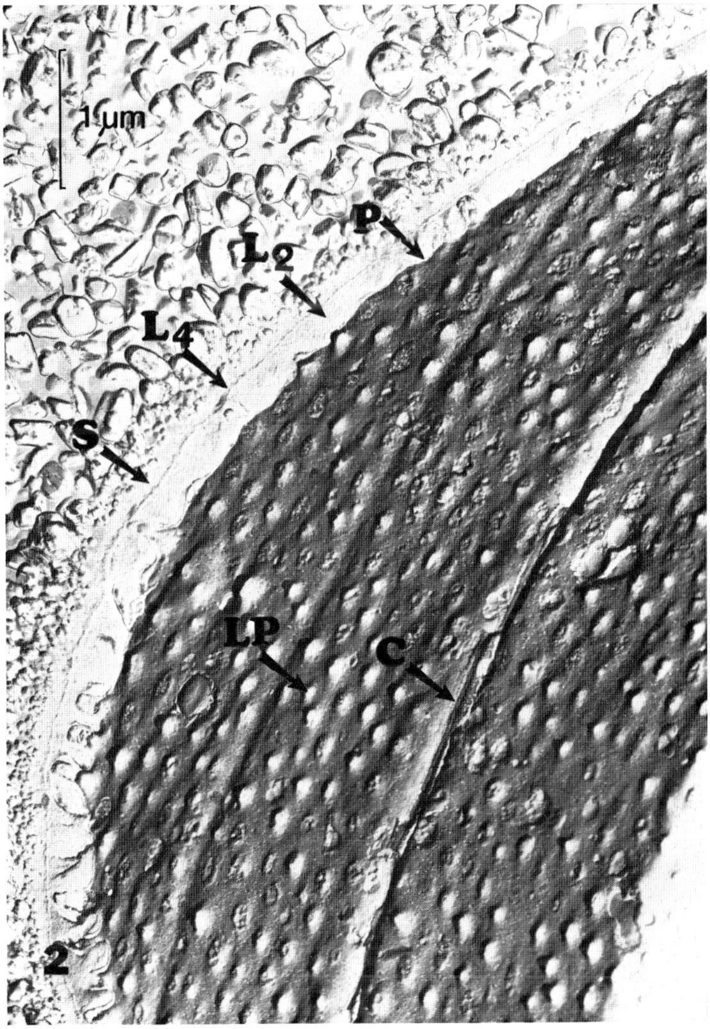

Fig. 2. A transmission electron micrograph of a freeze-etched cell wall of *O. princeps* revealing a sheath region *(S)*, the *L-4* and *L-2* layers of the cell wall with an underlying plasma membrane *(P)*. This view displays the internal aspect of the longitudinal cell wall with a fracture face revealing cross sectional detail of the cell wall. The large pores *(LP)* in the cell wall and a cross wall *(C)* are indicated. The marker is 1 µm

right- or left-handedness of the helix is preserved during trichome reversal and is merely retraced in the opposite sense (HALFEN and CASTENHOLZ, 1970, 1971a).

Those organisms which rotate as they glide frequently form circles of trichomes when allowed to glide on a smooth surface. The entry into these circles is clockwise for an organism that rotates to the right while it is counterclockwise for those that show a left-handed rotation (HALFEN, 1970; THOMAS, 1970). This behavior is the result of a lateral force created by the rotating trichome by virtue of a single line of contact with the underlying substrate. Alternatively, *O. princeps* may also glide through solidified agar, in which case a straight course is usually followed. In these situations, the motility apparatus is in contact with the substrate around the entire cirumference of the trichome, resulting in no uncompensated lateral forces to deflect the trichome into a curved path (HALFEN, 1970).

The degree of path curvature for organisms gliding on the surface of a substrate may be decidedly increased by the presence of a curved tip region such as described for many *Oscillatoria* (see DESIKACHARY, 1959; SHUKOVSKY and HALFEN, 1976). This tip region describes a cone in space as it rotates during trichome linear translation. Trichomes will circle more quickly when the curved tip regions strike the substrate, thereby displacing the trichomes laterally (HALFEN, 1970). The rotation of curved tip regions also facilitates snagging other trichomes, thus promoting aggregations even in rigid forms, by effectively increasing the size of the sweep through space made by the trichome as it rotates.

Some filamentous cyanophytes, including members of the Nostocaceae, are able to glide without rotation (BURKHOLDER, 1934; DREWS and NULTSCH, 1962; WEIBULL, 1960; FOGG et al., 1973). Interestingly, some of these organisms are still able to form circular accumulations of trichomes on an agar or glass surface (LAZAROFF and VISHNIAC, 1964). These nonrotating trichomes will displace mucilage in a nonhelical fashion over the surface of the trichome. The mucilage accumulates in the depressions existing between individual cells of the trichome (WALSBY, 1968). Additional investigation is needed to define the differences in motility mechanisms existing between these nonrotating organisms and the rotating forms discussed earlier.

Flexional activity is also a component of gliding behavior in many filamentous species (BURKHOLDER, 1934; CASTENHOLZ, 1973; NULTSCH, 1974). These movements often result from rotation of a curved region of a trichome as is the usual case with flexible *Oscillatoria* and *Spirulina* (HALFEN and CASTENHOLZ, 1971a; CASTENHOLZ, 1973). Flexions have been reported in nonrotating cyanophytes (DREWS and NULTSCH, 1962), but have not received much corroboration (CASTENHOLZ, 1973). Flexibacteria, on the other hand, do exhibit active flexional activity (DOETSCH and HAGEAGE, 1968; LEWIN and LOUNSBERY, 1969). It is assumed that flexional activity is associated with and derived from the function of the gliding mechanism (HALFEN and CASTENHOLZ, 1971a; FOGG et al., 1973).

NULTSCH (1956) has classified diatom movements into three groupings based upon raphe shapes. These are *Navicula*-type (straight path), *Amphora*-type (usually curved) and *Nitzschia*-type (always curved, with variable radii). NULTSCH explained the "swaying" movements of some diatoms as pivoting about the point of mucilage attachment at the end of the cell. In contrast, "rotating" movements resulted when the point of substrate attachment was at the middle of the cell. These types of diatom movement probably are the result of inadequate substrate contact for effective gliding motility (HOPKINS and DRUM, 1966).

Movements of the motile desmid *Closterium* observed by PAO-ZUN YEH and GIBOR (1970) were jerky with abrupt stops and direction changes. They note that desmid movement is not accomplished in a straight line, but rather is composed of wide undulations while leaving a clear trail of mucilage behind the advancing cell.

In *Spirogyra*, these authors observed slow filament rotation and linear translation in some filaments. *Spirogyra* cultures produced mucilage, with growth characterized by ring formation, which was attributed to filament rotation (PAO-ZUN YEH and GIBOR, 1970).

2.2 Associations

Both WALSBY (1968) and CASTENHOLZ (1967) have described behavior mechanisms whereby masses of motile filamentous cyanophytes are able to form dynamic aggregations or clumps. CASTENHOLZ (1967, 1969) has reviewed the ecological aspects of this behavior for survival of hot-spring organisms. HALFEN and MCCANN (1975) discuss the dynamics of clump behavior of benthic *Oscillatoria* as a dispersal strategy. Rafts of benthic *Oscillatoria* trichomes may become buoyant because of photosynthetic gas accumulations. These clumps may drift about in lentic habitats, eventually colonizing new benthic substrates by extending ropes of motile trichomes down from the floating mass.

The clumping process is thought to be the result of flexional movements of individual trichomes which allows contact between trichomes which eventually glide over each other. Masses of intertwined trichomes continue flexions while forming a "rope" or "clump" as they glide over each other, thus producing an increasingly compact mass of actively moving trichomes. Trichomes with reduced flexional activity (e.g., *O. princeps*) are still able to aggregate but here, reduced flexional activity is replaced to some degree by swings of the curved tip regions induced by trichome rotation (HALFEN, 1970; SHUKOVSKY and HALFEN, 1976). Individual trichomes in these clumps are motile, and shift position in the mass producing several strands composed of many trichomes which radiate from the original clump (CASTENHOLZ, 1967).

LAZAROFF and VISHNIAC (1961, 1964) have described developmental stages in *Nostoc* trichomes where associations of motile hormogonia are involved. LAZAROFF (1973) has discussed the behavior of masses of these hormogonia gliding on flat surfaces while producing circular swarms of many trichomes. While trichome rotation is apparently not involved, the formation of these circular swarms has also been noted in *Anabaena* (HALFEN, unpublished) and occasionally in *Mastigocladus* (CASTENHOLZ, 1973). These motile hormogonia are undoubtedly a dissemination mechanism for the organism (LAZAROFF, 1973; CASTENHOLZ, 1973).

Closterium cultures allowed to age produced mucilage-embedded aggregations of cells which were regularly spaced in the cluster (PAO-ZUN YEH and GIBOR, 1970). There was a network of trails in the mass through which the cells were able to glide. The mucilage was thought to be a means of trapping soluble nutrients for the colony.

3. Mechanisms

3.1 Shear Zone Evaluation

This section will emphasize the motility mechanism of oscillatoriacean trichomes. This decision is based upon the author's experience with these organisms, and also upon the paucity of experimentally substantiated hypotheses attempting to explain the mechanism of motility in other groups.

Any mechanism capable of inducing motility must establish a shear zone or velocity gradient between the organism and the substrate. Gliding requires the shear zone to be very small to allow the sub-microscopic motility apparatus to displace the cells by acting against the substrate. No motility organelle or shape changes by the cell wall in this shear zone are visible in the light microscope (JAROSCH, 1962; HALFEN, 1970; CASTENHOLZ, 1973; FOGG et al., 1973).

3.1.1 Blue-Green Algae

Ultrastructural examination of gliding cyanophyte trichomes has confirmed the existence of a fibrous mucilage sheath which lies to the exterior of the cell wall layers (LEAK, 1968; LAMONT, 1969; HALFEN and CASTENHOLZ, 1971a). This layer is, in effect, the shear zone (NIKLITSCHEK, 1934; HALFEN and CASTENHOLZ, 1970, 1971a) and the direction of movement may cause individual sheath fibrils to become oriented in the same direction (LAMONT, 1969). The cyanophyte motility apparatus has been postulated to lie beneath this expendable sheath layer and to operate on the mucilage to induce movement (HALFEN, 1970; HALFEN and CASTENHOLZ, 1971a).

The use of electron microscopy for evaluation of motility provides high resolution, but is problematic because of the inability to observe directly the function of living systems. This forces the investigator to deal in functional inferences derived from observations of fine structure which may bear a relationship to behavioral characteristics in living organisms. Considering *Oscillatoria*, ultrastructural detail has not provided significant support for mucilage extrusion hypotheses of gliding motility (LAMONT, 1969; HALFEN and CASTENHOLZ, 1971a; FOGG et al., 1973; NULTSCH, 1974). There is an absence of cellular components which could effectively induce the extrusion or export of material to produce observed rates of linear translation (HOLTON and FREEMAN, 1965; JOST, 1965; HALFEN and CASTENHOLZ, 1971a). Although junctional pores exist at crosswall intersections with the cell wall, they do not appear to penetrate every layer of the cell wall (PANKRATZ and BOWEN, 1963; JOST, 1965; LAMONT, 1969; HALFEN and CASTENHOLZ, 1971a). Indeed, no unique accumulations or attachments of mucilage were seen to be associated with junctional pores in Oscillatoriacean trichomes.

The cell wall of many cyanophytes has a structural component composed of lysozyme-sensitive murein (peptidoglycan) which is termed the L-2 layer (Fig. 2) (JOST, 1965; LANG, 1968; WEISE et al., 1970). The motility system must lie to the exterior of the L-2 layer or operate through it in a way which does not alter its conformation (JAROSCH, 1962; HALFEN and CASTENHOLZ, 1971a). The outer layer of the *Oscillatoria* cell wall is the so-called L-4 membrane which is continuous over the surface of the trichome (JOST, 1965; HALFEN and CASTENHOLZ, 1971a). In chemically fixed preparations, this outer membrane frequently displays wave-like undulations which led COSTERTON et al. (1961) to suggest that the L-4 layer was able to create waves of contraction. Subsequently, JOST (1965) published freeze-etched views of the L-4 layer lacking undulations, which support the contention that the undulations in chemically fixed material were preparation artifacts. HALFEN and CASTENHOLZ (1970, 1971a) identified a parallel array of fibrillar ele-

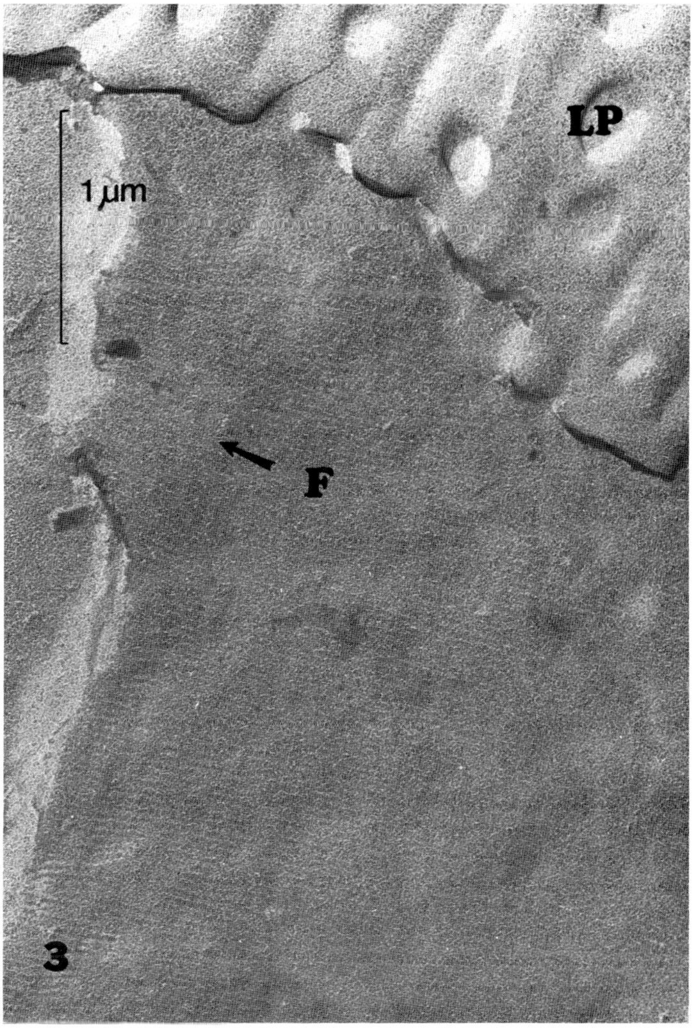

Fig. 3. A transmission electron microscope view of the surface of a platinum-carbon replica of a cell wall fragment of *O. princeps* prepared by ultrasonic disruption. The large pores *(LP)* are visible on the inner surface of the wall while the fibril layer *(F)* is labeled on the outer surface. The wall fragment is folded in this replica to reveal both the inner and outer surfaces. *Arrow* indicates the orientation of the fibrillar array. The marker is 1 µm

ments composing the L-3 layer (on the surface of the L-2 layer beneath the L-4 membrane) which were 5 to 8 nm in diameter. Ultrasonic disruption of *O. princeps* trichomes has been used to isolate cell wall fragments. This technique usually results in the loss of the L-4 layer from the cell wall, but the fibrillar array remains essentially intact (Fig. 3; HALFEN, 1973). This array of fibrils is continuous, and is immediately approached, but not interrupted, by the large pores in the L-2 layer.

HALFEN and CASTENHOLZ (1971a) have postulated that the fibrillar array on the surface of the L-2 layer is responsible for gliding motility in *Oscillatoria* (Fig. 4).

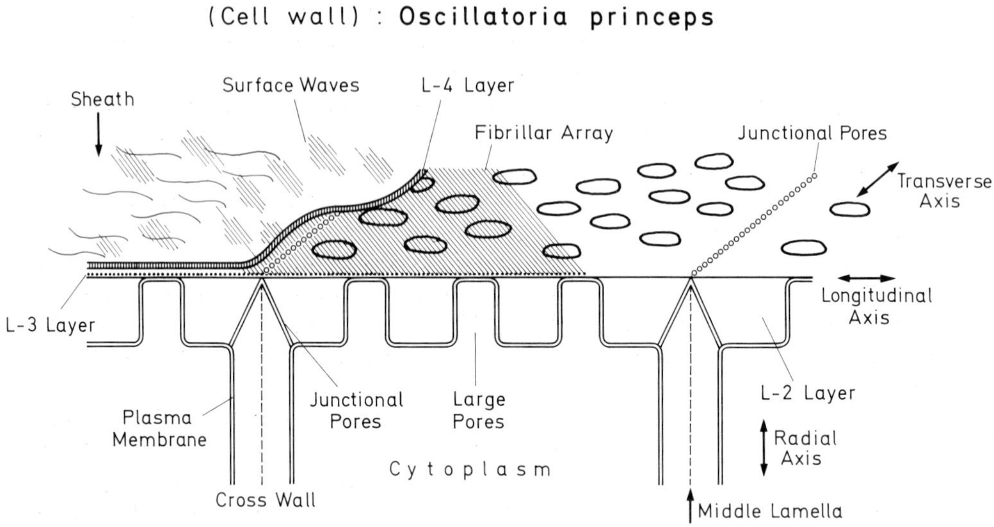

Fig. 4. A diagramatic representation of the organization of the motility apparatus of *O. princeps*. See the text for details

They contend that these protease-sensitive fibrils are capable of producing coordinated sub-microscopic waves of bending which travel the length of this continuous fibrillar array on the cell surface (see also, HALFEN, 1973). The amplitude of these waves is thought to be sufficient to act through the L-4 layer to gain a purchase against the close-fitting elastic mucilage of the sheeath, thereby inducing the uniform gliding movements seen in these trichomes. The role of the L-4 layer overlying the motility apparatus may be to protect the system from extra-cellular chemical variations while providing the proper physiological environment for fibrillar function (HALFEN, 1970; HALFEN and CASTENHOLZ, 1971a).

This explanation of the basis for motility in *Oscillatoria* is the only proposal which incorporates all the behavioral observations discussed earlier. It is assumed that the fibrillar array would have to produce rapid linear translation. Furthermore, ultrastructural evaluation of the fibrils has revealed no polarity which would prevent the propagation of waves in either direction along the fibrillar array. To accomplish a reversal, it is necessary to postulate a change in the direction of wave propagation in the fibrillar layer which rapidly becomes effective throughout the layer. Trichome rotation associated with linear translation in *O. princeps* necessitates the orientation of the fibrillar array in a right-handed helix with a pitch of 60°, a feature which has been confirmed ultrastructurally (HALFEN, 1970; HALFEN and CASTENHOLZ, 1971a; HALFEN, 1973). The orientation of the fibrillar array in *O. princeps* corresponds exactly to the rotation characteristics of that species; this correlation was also found in two other *Oscillatoria* species (HALFEN, 1970, 1973). Little flexional activity is noted in *O. princeps* because of its thick cell wall.

A fibrillar layer was also found in *O. animalis* and in *O. terebriformis* (HALFEN und CASTENHOLZ, 1971a; HALFEN, 1973). These thinner trichomes are more flexible than the thick-walled *O. princeps*, and, unlike that species, display a considerable amount of flexional activity. Such flexional activity is probably the result of rotation-induced bending of these flexible trichomes. In very small forms, independent flexional activity may result from a loss of coordination of wave propagation in the fibrillar array, thus bending the trichome.

Sheath production is essential to this motility model because it would be impossible for sub-microscopic waves at the trichome surface to induce movement without the facilitated substrate contact provided by a close-fitting mucilage sheath. Since the sheath completely surrounds the trichome, it is possible to bring the entire fibrillar array into action against the sheath–substrate complex, even though the trichome may only be resting on the surface of a substrate rather than moving through it (HALFEN, 1970). The role of mucilage in enhancing substrate contact is not unique to this model (JAROSCH, 1962; FOGG et al., 1973), but the sheath has been assigned an essential role distinctly different from inducing movement itself.

The motility model presented by HALFEN and CASTENHOLZ (1971a) applies to *Oscillatoria* and probably also to some of the morphological analogs of the Oscillatoriaceae among the bacteria. However, there remain a number of gliding organisms which display movements which are not entirely the same as in *Oscillatoria*. Unfortunately, the motile Nostocacean trichomes and the gliding unicellular forms found in the Chroococcales have not been closely examined with respect to the location, organization and function of their motility system (WALSBY, 1968; STANIER et al., 1971; LAZAROFF, 1973). If these organisms utilize an *Oscillatoria*-like gliding mechanism, the system is undoubtedly modified to some degree, forming the basis for a potentially interesting comparative investigation (see FOGG et al., 1973).

3.1.2 Other Groups of Gliding Organisms

Recently RIDGEWAY et al. (1975) have implicated adhesive fiber extrusion to the movement of flexibacteria. Goblet-like structures, seen on the cell wall, had fibers of what was interpreted to be adhesive material extending from them. These authors suggest that these goblets may be involved in the directed export of polymers from the cell in connection with motility. However, they do not use this mechanism to address the entire motility–behavior complex which provides this group of organisms with their name.

The motility of desmids is attributed to the extrusion of a mucilage trail by the cells, using numerous pores in the cell wall (JAROSCH, 1962; MIX, 1966, 1969; NULTSCH, 1974). The presence of the slime trail which is required for motility, and the irregular slow movements of these organisms are consistent with this view. Subsequent swelling of desmid mucilage during hydration may be responsible for cell displacement (NULTSCH, 1974).

Diatom motility was first thought to be the result of raphe contact with the gliding substrate, allowing motile forces from cytoplasmic streaming to induce movement (see NULTSCH, 1962b, 1974). JAROSCH (1962) maintained that diatom movements were the result of extramembranous fibrils, which propelled either mucilage over the cell or the cell itself over a substrate. More recent proposals center about deposition of slime precursors in the region of the terminal pore at the anterior end of the raphe, with subsequent propulsion down the raphe by a cytoplasmic fibrillar bundle overlying the raphe (HOPKINS and DRUM, 1966;

DRUM and HOPKINS, 1966). Swelling by hydration has also been implicated in cell displacement, when the mucilage is ejected from the raphe at the posterior end of the cell (GORDON and DRUM, 1970).

This explanation of diatom movement meets behavioral criteria, and is further supported by the presence of crystalloid bodies and cytoplasmic fibrils in the raphe region of all motile diatoms examined so far (NULTSCH, 1974).

3.2 Energy Supply

The supply of ATP for motility may originate either from photo- or from oxidative phosphorylation mechanisms which in the cyanophyta may come either from photosynthesis or from respiration. CASTENHOLZ (1973) has reported the loss of motility in *O. terebriformis* following 24 h of darkness, or within minutes when oxidative phosphorylation was inhibited in the dark. However, *O. terebriformis* had its movement suppressed by 2.10^{-4} M DNP to only 75% of the level in the light, suggesting the involvement of photophosphorylation as well. NULTSCH (1962a, 1968) has evidence that ATP supply is almost completely derived from photophosphorylation in *Phormidium*. Administration of DCMU to *P. autumnale* reduced motility to the very slow rates seen in darkness, where oxidative phosphorylation presumably empowered the motility system (NULTSCH, 1965). *O. princeps*, on the other hand, was able to remain motile after several days of darkness and was not slowed by DCMU (HALFEN, 1970; HALFEN and CASTENHOLZ, 1971a). These trichomes were halted by treatment with DNP or NaCN, indicating that energy for movement in this organism is derived almost completely from respiration (HALFEN, 1970). These findings are similar to results obtained from studies of photokinesis (see HÄDER, Chap. 3.2.1, this vol.).

Smaller organisms, including *O. terebriformis* and species of *Phormidium*, are certainly not able to store large amounts of reserve carbohydrates (see GIESY, 1964) in their cells, which forces their heavy dependence on photophosphorylation as an ATP source (CASTENHOLZ, 1973). The large trichomes of *O. princeps* provide space for considerable deposition of polyglucosidic bodies. These probably serve as a substrate for respiratory activity during dark periods, thus permitting prolonged motility in the dark.

HALFEN and CASTENHOLZ (1971b) evaluated the rate of gliding movement of *O. princeps* trichomes in a medium of known viscosity and calculated the amount of ATP necessary to produce the observed movement. The rate of ATP utilization was compared with the assumed rate of ATP production as a result of cellular respiration. Their calculations showed that a maximum of 5% of the available respiratory energy would be required to produce the observed rates of gliding.

4. Ecological Aspects

The ability to glide, allowing an organism to change location in a habitat, is of distinct survival value. The movement may be unrelated to any specific variable but the organism may still derive a benefit, for instance from exposure to new supplies of nutrient. Another example involves gliding by desmids and certain diatoms to remain on the surface of benthic sediments.

Some individual organisms exhibit motility reactions to environmental factors such as temperature, light intensity, and/or chemical changes (see CASTENHOLZ, 1968, 1969, 1973). Frequently, the direction of gliding is related to the direction of the applied variable qualifying the motile reaction to be treated as a tactic response (see HÄDER, Chap. 3.2.1, this vol.).

The value of gliding is not limited to an environmental compensation or reaction. Heterocyst-forming cyanophytes produce motile hormogonia during a single stage in their life cycle, and these trichomes serve to spread and increase the population of that organism in that environment (see LAZAROFF, 1973). HALFEN and MCCANN (1975) have described a mechanism whereby benthic communities of motile *Oscillatoria* are able to produce floating rafts of trichomes, which move by wind or current to new areas, where strings of trichomes glide down to the substrate and colonize the area. The incorporation of gliding capability into this mechanism ensures the success of this method of dissemination and helps to explain the rapid spread of benthic communities once they become established in a body of water.

References

Abe, S.: The effect of p-chloromercuribenzoate on amoeboid movement, flagellar movement and gliding movement. Biol. Bull. *124*, 107–114 (1963)

Adanson, M.: Un mouvement particulier découvert dans une plante appellée tremella. Mem. Acad. Sci. Paris 415–431 (1967)

Burkholder, P.R.: Movement in the Cyanophyceae. Quart. Rev. Biol. *9*, 438–459 (1934)

Castenholz, R.W.: Aggregation in a thermophilic *Oscillatoria*. Nature (London) *215*, 1285–1286 (1967)

Castenholz, R.W.: The behavior of *Oscillatoria terebriformis* in hot springs. J. Phycol. *5*, 123–139 (1968)

Castenholz, R.W.: Thermophilic blue-green algae and the thermal environment. Bact. Rev. *33*, 476–504 (1969)

Castenholz, R.W.: Movements. In: The Biology of the Blue-Green Algae. Carr, N.G., Whitton, B.A. (eds.). Berkeley: Univ. California Press, 1973 pp. 320–339

Castle, E.S.: Observations on the motility in certain Cyanophyceae. Biol. Bull. *51*, 69–72 (1926)

Costerton, J.W.F., Murray, G.E., Robinow, C.F.: Obervations on the motility and structure of *Vitreoscilla*. Can. J. Microbiol. *7*, 329–339 (1961)

Crozier, W.N., Federighi, H.: Critical thermal increment for the movement of *Oscillatoria*. J. Gen. Physiol. *7*, 137–150 (1924)

Desikachary, T.V.: Cyanophyta. New Dehli: Indian Council Agri. Res., 1959, pp. 686

Dodd, J.D.: Filament movement in *Oscillatoria sancta* (Kuetz.) Gomont. Trans. Am. Microsc. Soc. *79*, 480–485 (1960)

Doetsch, R.N., Hageage, G.J.: Motility in prokaryotic organisms: problems, points of view, and perspectives. Biol. Rev. 317–362 (1968)

Drews, G.: Beiträge zur Kenntnis der phototaktischen Reaktionen der Cyanophyceen. Arch. Protistenk. *104*, 389–430 (1959)

Drews, G., Nultsch, W.: Spezielle Bewegungsmechanismen von Einzellern. In: Handbuch der Pflanzenphysiologie, B. XVII, Teil 2. Bünning, E. (ed.). Berlin-Heidelberg-New York: Springer 1962, pp. 876–919

Drum, R.W., Hopkins, J.T.: Diatom motility: an explanation. Protoplasma *62*, 1–33 (1966)

Fechner, R.: Die Chemotaxis der Oscillatorien und ihre Bewegungserscheinungen überhaupt. Z. Bot. *7*, 289–362 (1915)

Fogg, G.E., Stewart, W.D.P., Fay, P., Walsby, A.E.: The Blue-Green Algae. London: Academic Press, 1973, pp. 459
Fritsch, F.E.: The Structure and Reproduction of the Algae. Vol. II. Cambridge: Cambridge Univ. Press, 1945, pp. 939
Geitler, L.: Schizophyceen. In: Handbuch der Pflanzenanatomie, Vol. VI, Part 1. Zimmerman, W., Ozenda, O. (eds.). Berlin: Bornträger, 1936
Giesy, R.M.: A light and electron microscopic study of interlamellar polyglucosidic bodies in *Oscillatoria chalybia*. Am. J. Bot. *51*, 388–396 (1964)
Gordon, R., Drum, R.W.: A capillarity mechanism for diatom gliding locomotion. Proc. Natl. Acad. Sci. USA *67*, 338–344 (1970)
Hageage, G.J.: A Cytological Study of Gliding Microorganisms. College Park: Ph. D. thesis, Univ. Maryland, 1963
Halfen, L.N.: A Study of Gliding Motility in the Blue-Green Alga *Oscillatoria princeps*. Eugene: Ph. D. thesis, Univ. Oregon, 1970, pp. 167
Halfen, L.N.: Gliding motility of *Oscillatoria*: ultrastructural and chemical characterization of the fibrillar layer. J. Phycol. *9*, 248–253 (1973)
Halfen, L.N., Castenholz, R.W.: Gliding in a blue-green alga: a possible mechanism. Nature (London) *225*, 1163–1165 (1970)
Halfen, L.N., Castenholz, R.W.: Gliding in the blue-green alga, *Oscillatoria princeps*. J. Phycol. *7*, 133–145 (1971a)
Halfen, L.N., Castenholz, R.W.: Energy expenditure for gliding motility in a blue-green alga. J. Phycol. *7*, 258–260 (1971b)
Halfen, L.N., McCann, M.T.: Behavioral aspects of benthic communities of filamentous blue-green algae in lentic habitats. Mich. Bot. *14*, 49–56 (1975)
Harder, R.: Über die Bewegung der Nostocaceen. Z. Bot. *10*, 177–244 (1918)
Holton, R.W., Freeman, A.W.: Some theoretical and experimental considerations of the gliding movement of blue-green algae. Am. J. Bot. *52*, 640 (1965)
Hopkins, J.T., Drum, R.W.: Diatom motility: an explanation and a problem. Brit. Phycol. Bull. *3*, 63–67 (1966)
Hosoi, A.: Secretion of a slime substance in *Oscillatoria* in relation to its movement. Bot. Mag. *64*, 14–17 (1951)
Jarosch, R.: Zur Gleitbewegung der niederen Organismen. Protoplasma *50*, 177–189 (1958)
Jarosch, R.: Gliding. In: Physiology and Biochemistry of Algae. Lewin, R.A. (ed.) New York: Academic Press, 1962, pp. 573–581
Jarosch, R.: Gleitbewegung und Torsion von Oscillatorien. Öst. Bot. Z. *111*, 143–148 (1964)
Jost, M.: Die Ultrastruktur von *Oscillatoria rubescens* D.C. Arch. Mikrobiol. *50*, 211–245 (1965)
Klebs, G.: Über Bewegung und Schleimbildung der Desmidiaceen. Biol. Zbl. *5*, 353–367 (1885)
Lamont, H.C.: Shear-oriented microfibrils in the mucilage investments of two motile Oscillatoriacean blue-green algae. J. Bact. *97*, 350–361 (1969)
Lang, N.J.: The fine structure of blue-green algae. Annu. Rev. Microbiol. *22*, 15–46 (1968)
Lazaroff, N.: Photomorphogenesis and Nostocacean development. In: The Biology of the Blue-Green Algae. Carr, N.G., Whitton, B.A. (eds.). Berkeley: Univ. California Press, 1973, pp. 279–319
Lazaroff, N., Vishniac, W.: The effect of light on the developmental cycle of *Nostoc muscorum* a filamentous blue-green alga. J. Gen. Microbiol. *25*, 365–374 (1961)
Lazaroff, N., Vishniac, W.: The relationship of cellular differentiation to colonial morphogenesis of the blue-green alga *Nostoc muscorum*. A. J. Gen. Microbiol. *35*, 447–457 (1964)
Leak, L.V.: Fine structure of the mucilage sheath of *Anabaena*. J. Ultrastruct. Res. *21*, 61–74 (1968)
Lewin, R.A.: *Saprospira grandis* Gross, and suggestions for reclassifying helical apochlorotic gliding organisms. Can. J. Microbiol. *8*, 555–563 (1961)
Lewin, R.A.: Freshwater species of *Saprospira*. Can. J. Microbiol. *11*, 135–139 (1965)
Lewin, R.A., Lounsbery, D.M.: Isolation, cultivation and characterization of Flexibacteria. J. Gen. Microbiol. *58*, 145–170 (1969)
Mix, M.: Licht- und elektronen-mikroskopische Untersuchungen an Desmidiaceen XII. Zur Feinstruktur der Zellwände und Mikrofibrillen einiger Desmidiaceen von Cosmarium-Typ. Arch. Mikrobiol. *55*, 116–133 (1966)

Mix, M.: Zur Feinstruktur der Zellwände in der Gattung *Closterium* (Desmidiaceae) unter besonderer Berücksichtigung des Porensystem. Arch. Mikrobiol. *68*, 306–325 (1969)

Neuscheler, W.: Bewegung und Orientierung bei *Micrasterias denticulata* Bréb. im Licht. I. Zur Bewegungs- und Orientierungsweise. Z. Pflanzenphysiol. *57*, 46–59 (1967)

Niklitschek, A.: Das Problem der Oscillatorien-Bewegung. Beih. Bot. Zbl. Abt. A. *52*, 205–254 (1934)

Nultsch, W.: Studien über die Phototaxis der Diatomeen. Arch. Protistenk. *101*, 1–68 (1956)

Nultsch, W.: Der Einfluß des Lichts auf die Bewegung der Cyanophyceen. I. Mitteilung: Photopotaxis von *Phormidium autumnale*. Planta *56*, 632–647 (1961)

Nultsch, W.: Der Einfluß des Lichtes auf die Bewegung der Cyanophyceen. II. Photokinesis bei *Phormidium autumnale*. Planta *57*, 613–623 (1962a)

Nultsch, W.: Über das Bewegungsverhalten der Diatomeen. Planta *58*, 22–33 (1962b)

Nultsch, W.: Phototaktische Aktionsspektren von Cyanophyceen. Ber. Dtsch. Bot. Ges. *75*, 443–453 (1962c)

Nultsch, W.: Light reaction systems in Cyanophyceae. Photochem. Photobiol. *4*, 613–619 (1965)

Nultsch, W.: Einfluß von Redoxsystemen auf die Bewegungsaktivität und das phototaktische Reaktionsverhalten von *Phormidium uncinatum*. Arch. Mikrobiol. *63*, 295–320 (1968)

Nultsch, W.: Movements. In: Algal Physiology and Biochemistry. Stewart, W.D.P. (ed.) Berkeley: Univ. California Press, 1974, pp. 864–893

Pankratz, H.S., Bowen, C.C.: Cytology of blue-green algae. I. The cells of *Symploca muscorum*. Am. J. Bot. *50*, 387–399 (1963)

Pao-Zun Yeh, Gibor, A.: Growth pattern and motility of *Spirogyra sp.* and *Closterium acerosum*. J. Phycol. *6*, 44–48 (1970)

Pate, J.L., Ordal, E.J.: The fine structure of *Chondrococcus columnaris*. III. The surface layers of *Chondrococcus columnaris*. J. Cell. Biol. *35*, 37–51 (1967)

Phillips, O.P.: A comparative study of the cytology and movements of the cyanophyceae. Cont. Bot. Lab. Univ. Penn. *2*, 237–335 (1904)

Pickett-Heaps, J.D.: Green Algae. Sunderland: Sinaur Associates, 1975, pp. 606.

Prell, H.: Zur Theorie der sekretorischen Ortsbewegung I. Die Bewegung der Cyanophyceen. Arch. Protistenk. *42*, 99–156 (1921)

Ridgeway, H.F., Wagner, R.M., Dawsey, W.T., Lewin, R.A.: Fine structure of the cell envelope layers of *Flexibacter polymorphus*. Can. J. Microbiol. *21*, 1733–1750 (1975)

Ris, H., Singh, R.N.: Electron microscope studies on blue-green algae. J. Biophys. Biochem. Cytol. *9*, 63–80 (1961)

Schmid, G.: Zur Kenntnis der Oscillarienbewegung. Flora *111*, 327–379 (1918)

Schmid, G.: Über Organisation und Schleimbildung bei *Oscillatoria jenensis*. Jb. Wiss. Bot. *60*, 572–627 (1921)

Schulz, G.: Bewegungsstudien an Cyanophyceen. Arch. Mikrobiol. *21*, 335–370 (1955)

Shukovsky, E.S., Halfen, L.N.: Cellular differentiation of terminal regions of trichomes of *Oscillatoria princeps* (Cyanophyceae) J. Phycol. *12*, 336–342 (1976)

Siebold, C.T., von: Über einzellige Pflanzen und Tiere. Z. Wiss. Zool. Abt. A. *1*, 270–294 (1849)

Stanier, R.Y.: The cytophaga group: A contribution to the biology of Myxobacteria. Bact. Rev. *6*, 143–196 (1942)

Stanier, R.Y., Kanisawa, R., Mandel, M., Cohen-Bazire, G.: Purification and properties of unicellular blue-green algae (Order chroococcales). Bact. Rev. *35*, 171–205 (1971)

Thomas, E.A.: Beobachtungen über das Wandern von *Phormidium autumnale* und *Oscillatoria princeps* auf Agar. Schweiz. Z. Hydrol. *32*, 523–531 (1970)

Ullrich, H.: Über die Bewegung von *Beggiatoa mirabilis* und *Oscillatoria jenensis*. Planta *2*, 295–324 (1926)

Walsby, A.E.: Mucilage secretion and the movements of blue-green algae. Protoplasma *65*, 223–238 (1968)

Weibull, C.: Movement. In: The Bacteria, Vol. I. Gunsalivs, I.C., Stanier, R.Y. (eds.). New York: Academic Press, 1960, pp. 153–198

Weise, G., Drews, G., Jann, B., Jann, K.: Identification and analysis of a lipopolysaccharide in the cell walls of the blue-green alga *Anacystis nidulans*. Arch. Mikrobiol. *71*, 89–98 (1970)

3.2 Control of Locomotion

3.2.1 Photomovement

D.-P. Häder

1. Introduction

The term photomovement is used in this chapter to describe all light-influenced motor responses of free-moving microorganisms, unicellular or multicellular. (Light here includes near UV and IR). The organisms dealt with range from bacteria to all groups of algae which are motile, or in which motile stages occur.

1.1 General Characteristics of the Responses and Terminology

Although most of the basic phenomena described in this review have been studied since the nineteenth century, no general agreement on terminology has been reached. Different schools have used divergent terms and definitions, which has led to mutual misunderstanding and caused a considerable amount of confusion. In this article we follow the terminology and definitions of Diehn et al. (1977), which have been widely accepted. There are three main motor reactions in microorganisms elicited by light.

1.1.1 Photokinesis

Photokinesis is a response in which the steady-state speed of an organism is governed by light intensity (I), independent of light direction. The response is called positive when the speed is higher in the presence of a stimulus (light of a given intensity) than in its absence. In the dark control, the speed may even be zero (Dunkelstarre) (Fig. 1). A negative photokinesis means a decrease of the steady-state speed compared with the dark control—in some cases to a complete stop (Lichtstarre).

Fraenkel and Gunn (1961) made a distinction between orthokinesis, which they defined as a change in linear velocity, and klinokinesis, which means an alteration in the frequency of directional change, or angular velocity. In this review, however, the term photokinesis is used always in the sense of orthophotokinesis.

1.1.2 Photophobic Reactions

Photophobic reactions are mediated by sudden changes in light intensity (dI/dt), independent of light direction, as in the case of photokinesis. However, unlike photokinesis, photophobic response is a transient phenomenon, after which the

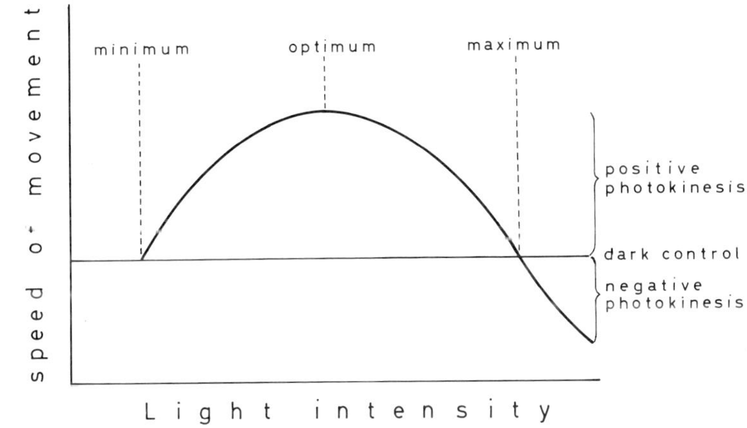

Fig. 1. Schematic diagram of the relation between speed of movement and light intensity

original steady-state speed is resumed. The mode of reaction is a characteristic of the species and shows wide variation including a sudden or delayed stop of motion, backward movement along the original, or a different path, or a rotation of the cell. Other terms used for this phenomenon are shock response, stop response, motor response or skotophobic response (NULTSCH, 1975). In this chapter, distinction will be made between step-down and step-up photophobic response as proposed by HAUPT (1974b). These attributes indicate that the response is triggered either by a sudden decrease or by an increase in light intensity. Thus step-down and step-up refer to the stimulus rather than to the reaction, which, for a given organism, is always the same.

This terminology is much more logical than the adjectives positive and negative, which were used earlier, and which derived from the particular set-up of the light trap (see Sect. 3.2.2). Accumulation in a light trap was considered, at that early time, as a phenomenon of phototaxis and named, therefore, photophobotaxis, to distinguish it from phototopotaxis (PFEFFER, 1904). Because only the latter fits the definition of an orientation response, the terms phobotaxis and topotaxis should no longer be used.

1.1.3 Phototaxis

Phototaxis is a movement oriented in relation to the light direction. Oriented movement toward the light source is called positive, and movement away from the stimulus source, parallel to the actinic light rays, negative. Perpendicular orientation is expressed by the terms "transverse" or "dia-phototaxis" (PFEFFER, 1904; BURKHOLDER, 1934). Phototactic orientation can be brought about by active steering or by trial and error.

1.2 Systematic Groups in Which Photomotion Occurs

Bacteria show photokinesis, phobic reactions, and weak phototactic orientation. Photomovement has been studied mostly in photosynthetic purple bacteria, but

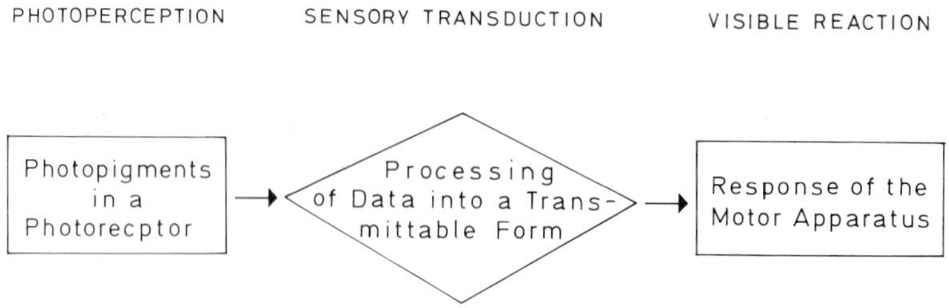

Fig. 2. Reaction chain of light-induced motor responses performed by motile microorganisms

has recently been studied in nonphotosynthetic bacteria as well. In blue-green algae, photomovement is exhibited primarily by the filamentous Hormogonales; all three reaction types are clearly discernible. Phototaxis has been the main response studied in flagellates, including flagellated stages of algae and fungi, although photokinesis and phobic reactions are known in these groups as well. Other motile species which show photomotion have been found in desmids, diatoms and red algae. Phototactic orientation in slime molds will be completely omitted from this article, since this topic is covered by POFF and WHITAKER, Chap. 3.2.4, this volume.

1.3 Reaction Chain

A photoresponse in an organism can be divided into three successive steps which may be arranged in a reaction chain (Fig. 2). The primary photoperception, the absorption of quanta, triggers the response of the motor apparatus. The "telephone line" which links the two processes has been called sensory transduction.

It is a truism that radiation must be absorbed by a photopigment to become effective. Much work has been invested in the study of light perception, starting with ENGELMANN (1883), who measured crude action spectra of photokinesis. Comparison of action spectra with in vivo absorption spectra and absorption spectra of isolated pigments in vitro, corrected for wavelength shift, can yield information on the photopigments involved in the primary process of photoperception. However, only in a few cases has identification been successful, and hardly anything is known about the photoevents on a molecular basis.

Sensory transduction within the reaction chain has often been called the "black box" or "internal mechanism" (HAND and DAVENPORT, 1970) and indeed very little material has been published yet. There are only a few recent investigations in blue-green algae and flagellates which attack this problem (cf. Sects. 3.3.3 and 4.4.1).

Since the final reaction of the motor apparatus is accessible by microscopic observation, abundant descriptions of photoreactions in all sorts of organisms are available. However, especially in rapidly moving organisms further investigations using new techniques like high-speed microcinematography might be promising.

2. Photokinesis

Photokinetic behavior was first observed a century ago (STRASBURGER, 1878). The numerous older reports are summarized, for example, in the recent review articles by HAUPT (1959), HAND and DAVENPORT (1970) and NULTSCH (1970, 1974b).

2.1 Methodology

According to the definition of photokinesis, all quantitative assessment of kinetic reactions must necessarily measure the speed of movement. This can be determined either by population or by individual cell methods (DAVENPORT, 1973; NULTSCH, 1975; SCHUCHART, 1975). While single cell observations are rather time-consuming and demand a high degree of repetition to ensure statistical significance, population experiments provide an averaged result of many reacting organisms (cf. MACNAB, Chap. 3.2.2, this vol). On the other hand, controls are indispensable to make sure that only photokinetic behavior is being measured.

2.1.1 Individual Cell Methods

Most earlier investigators assessed speed of movement under a microscope, by measuring the time needed for an organism to traverse a given distance. This method was improved by FEINLEIB and CURRY (1967), who measured the lengths of tracks on photomicrographs taken from moving *Chlamydomonas* cells with a constant exposure time. Similar techniques have been used by KUNG (1971) and GIBBONS and GIBBONS (1972). Video recording and microcinematography of swimming cells have been successfully performed by PHILLIPS (1972), IWASA and SHIMIZU (1972), OJAKIAN and KATZ (1973) and SCHNEIDER and DOETSCH (1974). NEUSCHELER (1967a, b) stained and measured the slime tracks left by creeping desmids.

Recently the speed of single *Euglena* cells was measured by means of the Doppler effect based on laser light-scattering (ASCOLI, 1975; ASCOLI et al., 1971; CHECCUCCI, 1973).

2.1.2 Population Methods

NULTSCH (1962c) estimated photokinetic reaction values by measuring the diameter of a circular area that had spread out from a central inoculation of blue-green algae on a semi-solid agar layer, after a given time of illumination. The relation between this diameter (minus that of the initial inoculation area) and that of the dark control was used as photokinetic reaction value. A technique introduced by ADLER and DAHL (1967) and further used by STAVIS and HIRSCHBERG (1973) to measure random motility in bacteria and green flagellates could also be used–but has not yet been–for kinetic experiments: after application from one side, they allowed organisms to distribute themselves in a capillary for several hours, then broke the capillary into equal pieces and estimated the number of organisms in each compartment. In each such experiment it has to be taken into account that the actual path length traversed by the individual cell is much greater than the distance which the whole population migrates, since the track of an individual is far from being straight. Thus a population of *Phormidium autumnale* spreads with a speed of about 1.67 mm/h under 1000 lx white light (NULTSCH, 1962c), while the average speed of an individual filament is 7.8 mm/h (NULTSCH, 1961). Indirect estimation of motility and photokinetic effects has been carried out by THROM (1968) in *Rhodospirillum rubrum* and by NULTSCH et al. (1971) in *Chlamydomonas*, by measuring the kinetics of phobic and phototactic reactions respectively, since the time needed for performing these reactions depends on the speed of movement.

2.2 Prokaryotic Organisms

2.2.1 White Light Behavior

In all prokaryotes investigated the photokinetic response follows an optimum curve. The minimal light intensity producing a photokinetic effect is 0.02 lx in *Phormidium autumnale* and *Ph. ambiguum* (NULTSCH, 1962c; NULTSCH and HELLMANN, 1972), while in *Rhodospirillum* it is >100 lx (THROM, 1968). The optima found in the two *Phormidium* species are 2000 and 1000 lx, respectively, while a maximum is reached at about 90,000 lx (NULTSCH, 1974a). Higher light intensities may produce negative photokinetic reactions (NULTSCH, 1962c). After decrease or increase of light intensity, the new speed is reached within a few minutes.

2.2.2 Correlations Between Photokinesis and Photosynthesis

Action Spectra. In the photosynthetic purple bacterium *Rhodospirillum rubrum*, the action spectrum for photokinesis (THROM, 1968) shows peaks at 510 and 886 nm, indicating the involvement of spirilloxanthin and bacteriochlorophyll (Fig. 3). Thus it resembles the absorption spectrum in vivo and the action spectrum of photosynthesis (THOMAS, 1950; DUYSENS, 1952); but it is interesting to notice that, in contrast to photosynthesis, in photokinesis the quantum yield of the carotenoids is more effective than that of bacteriochlorophyll.

Action spectra of *Phormidium uncinatum* (NULTSCH, 1962b) and *Ph. ambiguum* (NULTSCH and HELLMANN, 1972) indicated the participation of chlorophyll a, while the accessory pigments of photosystem II were not photokinetically active (Fig. 4). However, in *Anabaena variabilis*, the maximum in the photokinetic action spectrum coincided with the absorption maximum of C-phycocyanin (Fig. 4). Thus, two types of photokinetic mechanism have been found in blue-green algae (1) the *Phormidium* type, which seems to be mediated by cyclic photophosphorylation and (2) the *Anabaena* type, driven by noncyclic photophosphorylation. Further investigations on other organisms showed that a very strict categorization cannot be made, since there are also intermediate types (NULTSCH, 1973, 1974a).

Inhibitor Studies. Uncouplers of photophosphorylation block ATP production, but do not inhibit (and sometimes even stimulate) photosynthetic electron transport (HORIO et al., 1965). Application of desaspidin, which, according to TSUJIMOTO et al. (1966), uncouples photophosphorylation, specifically inhibited photokinesis in

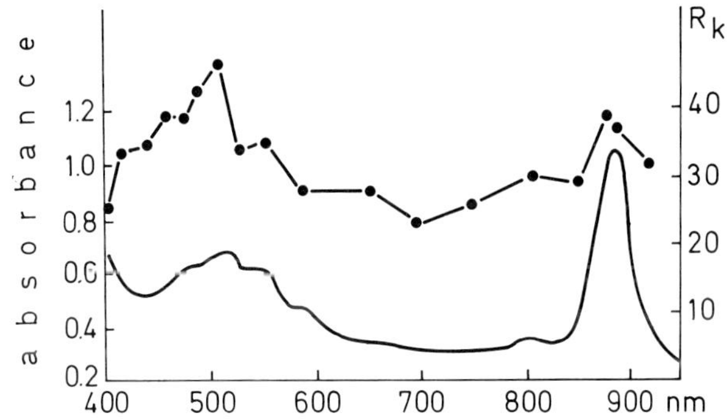

Fig. 3. Action spectrum of photokinetic reaction in *Rhodospirillum rubrum* (*dots, right hand ordinate* with a photokinetic reaction value R_k) compared with the absorption spectrum (*solid curve, left hand ordinate*). (After THROM, 1968)

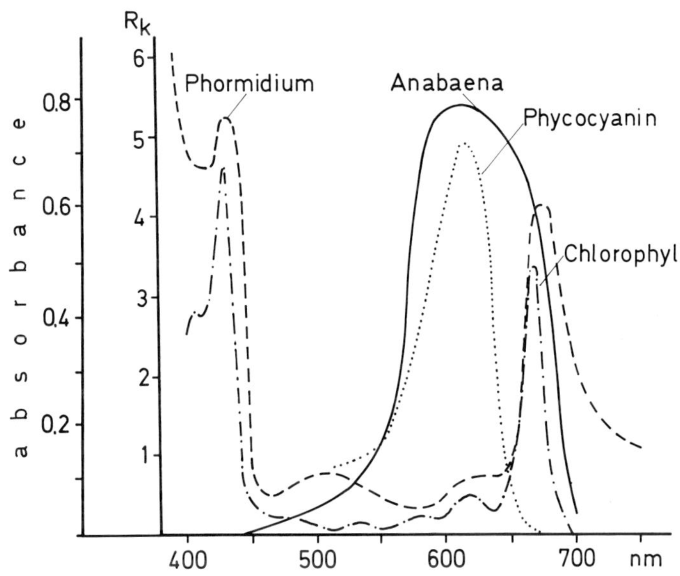

Fig. 4. Photokinetic action spectra of *Phormidium ambiguum* (*broken curve*) and *Anabaena variabilis* (*solid curve*) in comparison with the absorption spectra of C-phycocyanin (*dotted curve*) and chlorophyll a (*dot-dash curve*). (After NULTSCH and HELLMANN, 1972)

Phormidium uncinatum (NULTSCH, 1974a) and *Rhodospirillum rubrum* (THROM, 1968). The same was found for DNP (dinitrophenol), CCCP (carbonylcyanide-m-chlorophenyl-hydrazone), PMS (N-methylphenazonium-methosulfate) and other uncouplers (NULTSCH, 1966, 1967; THROM, 1968), suggesting that photokinesis in bacteria and blue-green algae is evoked by the additional ATP supply from photophosphorylation. This was supported further by the fact that, in *Rhodospirillum rubrum,* externally applied ATP increased the speed of movement by 500% compared with the dark control. Unfortunately almost all uncouplers are quite unspecific in that they uncouple all types of photophosphorylation and often oxidative phosphorylation as well (BALTSCHEFFSKY and DE KIEWIET, 1964; RENGER, 1969).

Better results were obtained with inhibitors of the photosynthetic electron transport chain such as SAO (salicylaldoxime), DCMU [3-(3,4-dichlorophenyl)-1,1-di-

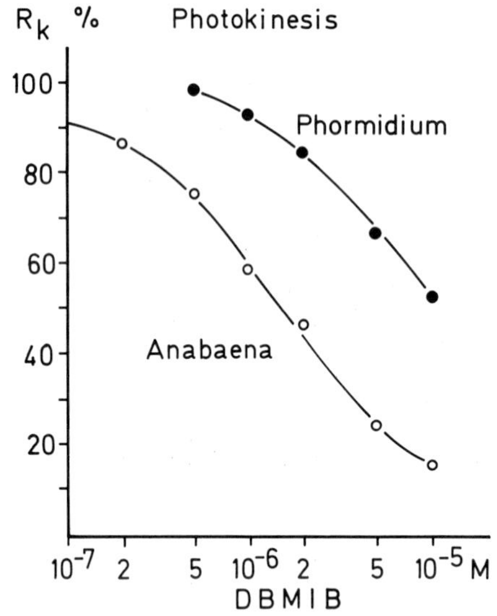

Fig. 5. Effect of DBMIB (molarity in logarithmic units on abscissa) on photokinetic reactions in *Phormidium ambiguum* (*dots*) and *Anabaena variabilis* (*circles*). *Ordinate,* reaction value in percentage of the uninhibited control. (After NULTSCH and HELLMANN, 1972)

methylurea] and DBMIB (dibromomethyl isopropyl benzoquinone; NULTSCH and JEEJI-BAI, 1966; TREBST et al., 1970). The effects of DBMIB on photokinesis were quite different in *Anabaena* and in *Phormidium*: DBMIB impairs photokinesis in *Anabaena* at considerably lower concentrations than in *Phormidium* (Fig. 5). This confirmed the conclusions drawn from the action spectra (see above; NULTSCH and JEEJI-BAI, 1966; NULTSCH and HELLMANN, 1972).

Application of artificial redox systems blocked photokinesis in *Phormidium uncinatum* maximally if the substances had midpoint potentials, enabling them to remove electrons from the primary acceptor of photosystem I, thus interfering with cyclic electron transport in vivo (NULTSCH, 1968).

2.3 Gliding Eukaryotic Organisms

2.3.1 Diatoms

NULTSCH (1957, 1971) studied the wavelength dependence of photokinetic behavior in various species of diatoms. The action spectrum of *Nitzschia communis* has a prominent peak in the red absorption band of chlorophyll a (Fig. 6). Although, on the other hand, blue light absorbed by chlorophyll a is not very effective, at least in some diatoms, photokinesis is energized by photophosphorylation, as it is in prokaryotes. Mainly photosystem I seems to be involved, since wavelengths between 500 and 550 nm, which are absorbed by the main auxiliary pigment, fucoxanthin, are of minor effectiveness.

2.3.2 Red Algae

In the red alga *Porphyridium cruentum,* wavelengths absorbed by the biliprotein R-phycoerythrin are even more effective in evoking photokinesis than those

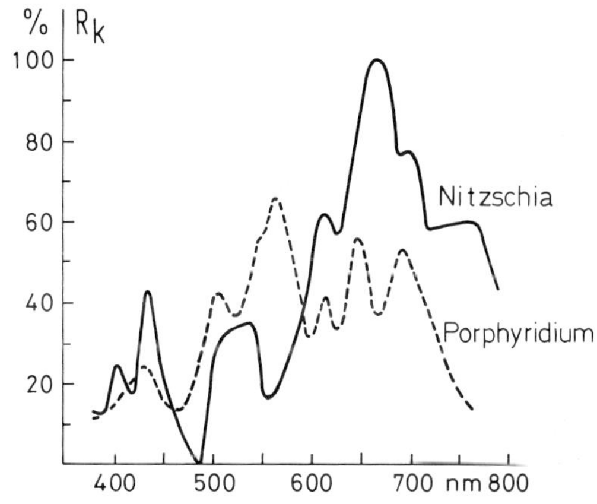

Fig. 6. Photokinetic action spectra of *Nitzschia communis* (*solid curve;* after NULTSCH, 1971) and *Porphyridium cruentum* (*broken curve;* after NULTSCH, 1974a)

absorbed by chlorophyll a (NULTSCH, 1974a, 1975). This indicates the participation of noncyclic photophosphorylation involving both photosystem II and I (Fig. 6). Similar action spectra have been measured for the photosynthetic oxygen production by BLINKS (1960) and by HAXO (1960).

2.3.3 Desmids

NEUSCHELER (1967a) found that in *Micrasterias* the position of cells with respect to the substrate depends on light intensity and is also influenced by a circadian rhythm: at midnight, vertical irradiation of 700 lx white light induced most of the cells to lie almost motionless, flat on the substrate, whereas, under the same irradiation at noon, many stood on their edges, creeping at a comparatively high speed (2–3 mm/h). During the night, the zero threshold was 100 times higher than in the daytime. As in blue-green algae, the dose-response curves in white light and in monochromatic light are optimum curves. In monochromatic light (434 and 666 nm) the minimum was found between 10^{-5} and 10^{-4} Wm^{-2}, and the optimum between 10^{-2} and 10^{-1} Wm^{-2}. Beyond the optimum there is a decrease and an additional second increase. Although no action spectrum has been determined, photokinesis in desmids seems to be mediated by near UV, blue, and red light, while green light has only a weak effect.

2.4 Flagellates

2.4.1 Positive and Negative Photokinesis in White Light

Photokinesis in motile cells of the green algae *Ulothrix, Botrydium* and *Haematococcus* was first described by STRASBURGER (1878). Most organisms show a positive photokinetic behavior, while others were reported to be negatively photokinetic. A third group was found to be photokinetically indifferent (BOLTE, 1920). FEINLEIB and CURRY (1967) failed to demonstrate photokinesis in *Chlamydomonas* and *Ulva* gametes, while WOLKEN and SHIN (1958) found a saturation level of photokinesis in *Euglena gracilis* at about 300 lx (cf. Fig. 7). A lag period of 10–15 min is

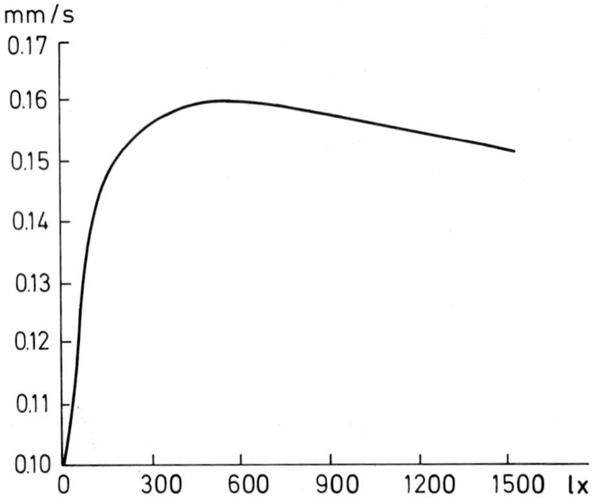

Fig. 7. Effect of white light intensity (in lx) on the speed of movement in *Euglena gracilis*. (After WOLKEN and SHIN, 1958)

necessary for the organisms to reach a new steady-state speed at the new light intensity (MAST, 1911). According to ASCOLI (1975), the acceleration in *Euglena* is due to an increase of the flagellar beat frequency. In *Chlamydomonas* NULTSCH and THROM (1975) could demonstrate a positive photokinesis only after 72 h of darkening.

2.4.2 Wavelength Dependence

Rough action spectra of photokinesis were measured in *Eudorina*, *Volvox* and *Chlamydomonas* (cf. HAUPT, 1959). Later an action spectrum was measured by WOLKEN and SHIN (1958) in *Euglena;* it showed a prominent peak at 465 nm and a broader, but lower maximum at 630 nm; the authors ascribed the 465 nm peak to β-carotene absorption, but a more fit candidate for the photoreceptor might be chlorophyll b (NULTSCH, 1975). ASCOLI (1975) reported an increase of motility in *Euglena* in blue (431 nm) and red (676 nm) light, while green (521 nm) had no effect. However, at high light intensities (30 Wm^{-2}) DIEHN (1973b) observed a complete immobilization in the blue region. Thus, it is still an open question whether photokinesis in flagellates is driven by photosynthesis, as in prokaryotic organisms (cf. Sect. 2.2.2), or whether a specialized blue-light receptor is involved.

3. Photophobic Responses

3.1 Characteristics of Phobic Reactions

Photophobic behavior was described by ENGELMANN (1883), who called it shock movement (Schreckbewegung). The investigations of the numerous earlier authors have been summarized by HAUPT (1959) and NULTSCH (1975).

To produce a phobic reaction, the intensity change dI must take place in a rather short time (dt); furthermore it must exceed a discrimination threshold, which denotes the minimal difference between the light intensity before and after induction of a phobic reaction (Clayton, 1959). This threshold value either follows the Weber law, as in the case of purple bacteria (Clayton, 1953), at least over a certain range of intensities, or it exhibits a dependency according to an optimum curve (Nultsch and Häder, 1970). Furthermore, as in the case of photokinesis and phototaxis, a zero threshold, i.e., a minimal light intensity, must be exceeded (Nultsch, 1962c).

3.2 Methodology

3.2.1 Microscopic Observation

Phobic reactions can easily be evoked in many organisms under a microscope by interrupting the light path between light source and microsope stage for a short while. Microscopic methods have been used, for example by Hildebrand and Dencher (1974, 1975) and Häder (1977). Improvements have been obtained by analysis of microcinematographic (Schmidt et al., 1974) or video recording (Feinleib and Curry, 1971a) or with the flying spot scanning apparatus (Davenport et al., 1962; Hand et al., 1967). Partial shading of single bacteria, diatoms and trichoms of blue-green algae were already used by Buder (1915) and recently by Nultsch and Wenderoth (1973) and Wenderoth (1975) to determine the amount and location of area which must be shaded to induce a phobic reaction. Wenderoth (1975) described an automatic devicc in which transmission changes, produced by a diatom moving in a capillary, are automatically registered by a photomultiplier mounted on top of a microscope. Schletz (1976) observed phobic responses induced by changes in the wavelength of the actinic light.

3.2.2 Light-Trap Methods

If an organism is allowed to move at random in a liquid or a semi-solid agar, it will undergo a phobic reaction when it hits a light–dark border, because by traveling along the spatial light gradient it experiences a change in light intensity. Homogeneously distributed microorganisms which react upon a step-down signal will tend to accumulate in a light field. On this basis Engelmann (1883) devised a light-trap method which, in many variations, is still in use now. In slowly moving organisms, such as diatoms, desmids and blue-green algae, accumulations in the light trap, resulting from repetitive phobic responses, remain stable for some minutes (Fig. 8). This allows counting or photographing the organisms, or estimating their number densitometrically (Nultsch, 1962d; Häder and Nultsch, 1971). In bacteria and flagellates, the accumulation rate has to be measured continuously. This may be done in a single-beam apparatus (Bruce and Pittendrigh, 1960; Brinkmann, 1966; Bruce, 1970). On the other hand, actinic beam and measuring beam may be separate, as in the phototaxigraph devised by Lindes et al. (1965) and improved by Diehn (1969c), originally built to measure phototactic reactions (cf. Sect. 4.1.2).

A drawback of light-trap methods is that organisms which have already entered the light trap may serve as a secondary light source according to the Tyndall

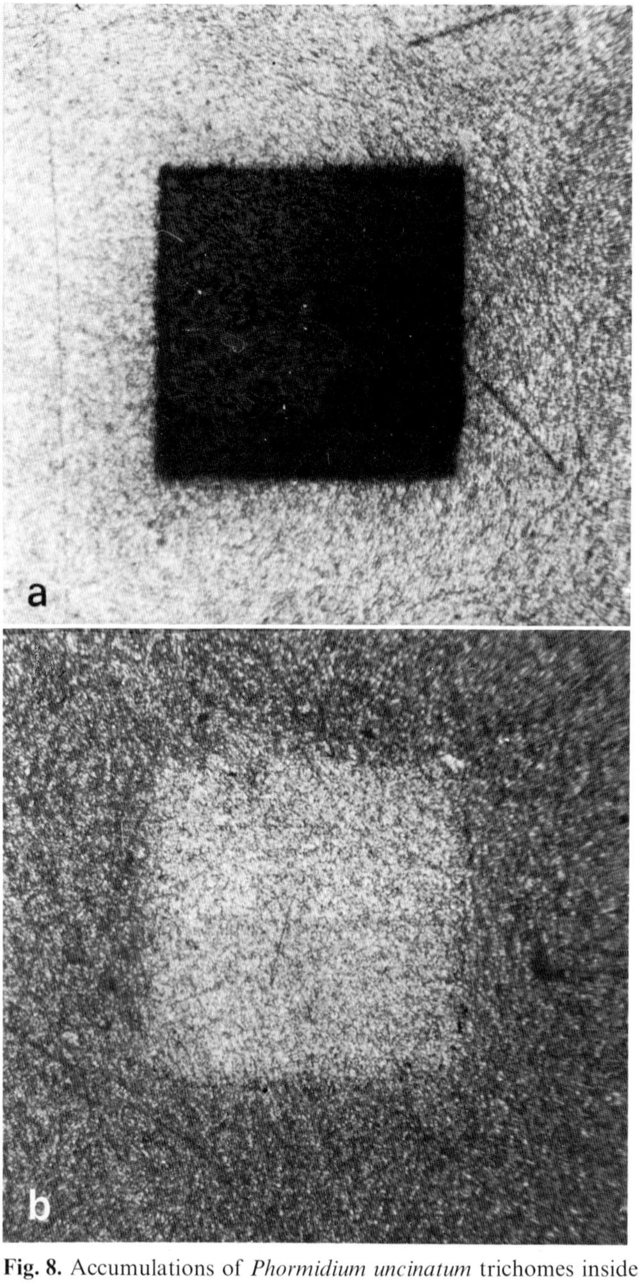

Fig. 8. Accumulations of *Phormidium uncinatum* trichomes inside **a** and outside **b** a light trap. (After HÄDER, 1975)

effect, and in this way phototactically attract organisms from outside the light trap. To minimize this disadvantage, HÄDER and NULTSCH (1973) and HÄDER (1974) devised a double irradiation technique which allows the projection of a light trap of a certain wavelength onto a homogeneous background of the same or a different wavelength.

HÄDER (1976a) estimated the effect of the two photosynthetic photosystems in the phobic response of blue-green algae by establishing a boundary between two adjacent monochromatic light fields of different wavelengths. This method was modified from a technique originally developed by MANTEN (1948).

Although all light-trap experiments incorporate the advantages of population methods, they may be subject to serious misinterpretation, as CLAYTON (1964) pointed out. Accumulations can be brought about by negative photokinesis, when organisms, which are motile outside, slow down or become motionless in the light trap; an emptying of the light trap can be brought about by the contrary effect. Therefore the analysis of phobic accumulations in a light trap should never be limited to macroscopic determination, but should include microscopic observation of single organism behavior (cf. Sect. 2.1).

3.3 Prokaryotic Organisms

3.3.1 Purple Bacteria

Earlier light-trap experiments on *Rhodospirillum rubrum* (ENGELMANN, 1883; MANTEN, 1948; MILATZ and MANTEN, 1953) revealed a close resemblance between the action spectrum for step-down photophobic response and the absorption spectrum of bacteriochlorophyll and rhodopin, suggesting that phobic responses are triggered by some photosynthetic product. THROM (1968) studied the influence of uncouplers and inhibitors on photophobic responses. The uncouplers dinitrophenol (DNP), desaspidin and imidazol had no effect on the phobic response, although they impaired photokinesis quite drastically (Sect. 2.2.2). On the other hand, inhibitors of photosynthetic electron transport (antimycin A, o-phenanthroline and atebrine) specifically blocked the phobic response. It was concluded that phobic reactions in purple bacteria are triggered by changes in the photosynthetic electron transport chain, rather than by differences in the ATP pool, as had earlier been suggested by THOMAS (1950) and LINKS (1955). The magnitude of photophobic accumulation also depends on pH and O_2 concentration.

In *Rhodospirillum rubrum* phobic responses are usually found only in anaerobically cultivated organisms. However, HARAYAMA and IINO (1976) demonstrated the existence of step-down reactions in aerobic cultures at very high light intensity ($3 \cdot 10^4$ Wm^{-2}).

3.3.2 Nonphotosynthetic Bacteria

Recently photophobic reactions have been demonstrated in the nonphotosynthetic *Halobacterium halobium* by HILDEBRAND and DENCHER (1974). Step-up and step-down reactions occur after a lag phase of about a second; the organism stops for another second and then reverses direction along its old path. The action spectrum of the step-down response shows a maximum at about 565 nm, which coincides with the in vitro absorption maximum of bacteriorhodopsin (Fig. 9). This pigment is located in distinct patches in the cell membrane BLAUROCK and STOECKENIUS, 1971; OESTERHELT and STOECKENIUS, 1971, 1973; DRACHEV et al., 1974). The chromophore seems to be reversibly dissociable from the protein (OES-

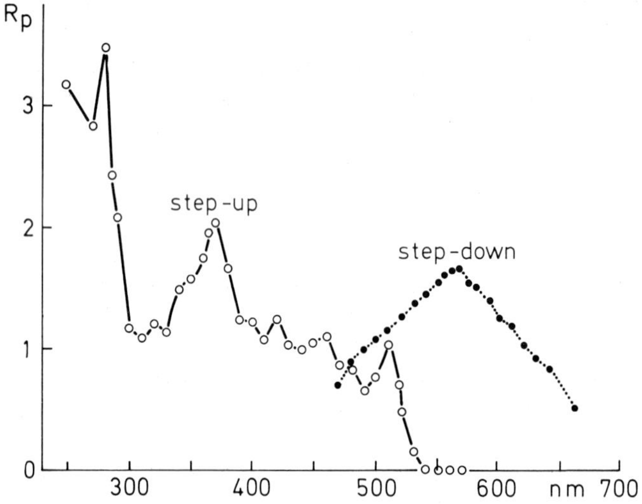

Fig. 9. Action spectra of step-down (*dots*) and step-up (*circles*) photophobic reactions in *Halobacterium halobium*. The ordinate (R_p) gives the photophobic reaction value. (After HILDEBRAND and DENCHER, 1975)

TERHELT et al., 1973), and to act as a light-driven proton pump according to the chemiosmotic hypothesis. The action spectrum of the step-up response has two distinct peaks, at 280 and 370 nm, and several smaller maxima and shoulders up to 530 nm (Fig. 9). This spectrum suggested the involvement of either a retinyliden protein or a flavoprotein (HILDEBRAND and DENCHER, 1975). Thus, step-up and step-down photophobic reactions are mediated by different photoreceptor pigments in the same organism, a rather exceptional behavior (HILDEBRAND, 1977).

MACNAB and KOSHLAND (1974) found a photophobic step-up response in *Salmonella typhimurium*: high light-intensity pulses in the presence of high oxygen concentrations induced tumbling of the bacteria. Since only wavelengths lower than 510 nm were effective, the authors proposed flavins as photoreceptors for this response (cf. MACNAB, Chap. 3.2.2 this vol.).

3.3.3 Blue-Green Algae

White Light Behavior. Characterization of stimulus and description of phobic responses as studied by earlier authors have been extensively reviewed by HAUPT (1974), NULTSCH (1973, 1975) and WOLK (1973). Almost all organisms show a step-down reaction, even at very high light intensities–at least under white light conditions; only *Oscillatoria jenensis* has been reported to display step-up responses (SCHMID, 1923). The photoreceptive zone of the organisms is evidently found in the front, since photophobic reactions can be induced by darkening the front 5%–10% of the trichom length, while darkening the rear or middle portion of the filament up to 50% of its length has no effect at all (SCHMID, 1923; HÄDER, 1977).

In *Phormidium uncinatum*, weak phobic reactions can be induced in a light trap of less than 0.1 lx white light (DREWS, 1959; NULTSCH, 1962d). The optimum was found at 5000 lx, and at 50,000 lx accumulations in the light trap still occur; higher light intensities have not been tested. Related species with higher zero thresholds (1 or 20 lx, respectively) nevertheless show nearly the same optimum intensity (NULTSCH, 1962b). The accumulation density is linearly related to exposure time up to 4–5 h in light of 1000 lx, after which a saturation level is approached.

Action Spectra. Crude action spectra have been measured by DANGEARD (1928) and DREWS (1957). NULTSCH (1962d) measured an action spectrum of the phobic response in *Phormidium uncinatum* on an equal energy basis (5 Wm^{-2}). Maximal activity was found in the range between 550 and 560 nm, coinciding with the absorption maximum of C-phycoerythrin. C-phycocyanin contributes a minor peak near 620 nm. The maximum at 665 nm can be correlated with the absorption peak of chlorophyll a.

It is interesting to note that the major maximum in the photophobic action spectrum shows a double peak which has also been demonstrated in the absorption spectrum of phycoerythrin in vitro (PECCI and FUJIMORI, 1968; GRAY and GANTT, 1975) and in the in situ low temperature spectrum (–197° C; HÄDER and NULTSCH, 1971).

As in photosynthetic action spectra of blue-green and red algae (HAXO, 1960; NULTSCH and RICHTER, 1963) the quantum efficiency of the main accessory pigment, phycoerythrin, is markedly higher than that of chlorophyll a (both in the red and blue absorption bands). A similar action spectrum has been obtained for phobic reactions in *Phormidium ambiguum* (NULTSCH, 1962d).

Inhibitor Studies. The close resemblance between photosynthetic and photophobic action spectra could be interpreted at first sight as confirming the hypothesis of LINKS (1955), according to which phobic reactions are mediated by sudden changes in the ATP supply (see Sect. 3.3.1). Since externally applied ATP or glucose had no effect at all (NULTSCH, 1962d), inhibitor and uncoupler studies became necessary. If the hypothesis of LINKS (1955) is correct, uncouplers of photophosphorylation would be expected to block phobic reactions. NULTSCH (1966, 1967) studied the effect of the uncouplers DNP (dinitrophenol), NaN_3, imidazole, atebrin and others. Compared to the influence on photokinesis the effect on phobic reactions was small (Table 1a). Since photophobic reactions are not specifically blocked by uncouplers, they obviously cannot be triggered by changes in the ATP production. Therefore NULTSCH (1966) and NULTSCH and JEEJI-BAI (1966) studied the effect of inhibitors of photosynthetic electron transport (Table 1b). Most of the tested inhibitors (e.g., antimycin A, DCMU, SAO, HQNO, o-phenanthroline) impaired the photophobic reaction in significantly lower concentrations than they impaired photokinesis. These results indicated that photophobic reactions are coupled to photosynthetic electron transport as in photosynthetic bacteria (Sect. 3.3.1).

The application of artificial redox systems (NULTSCH, 1968) with redox potentials E'_0 between -0.44 V and $+0.356$ V yielded a pronounced inhibition between 0.0 V and +0.1 V (Fig. 10). Redox systems with an E'_0 value in this region (methylene blue, toluidine blue and especially thionine) are able to draw electrons from the natural electron acceptor of photosystem II, thus inhibiting noncyclic electron transport.

Table 1. Effect of uncouplers and inhibitors of electron transport on photophobic reactions and photokinesis in *Phormidium uncinatum*

Substance	Concentration to produce 50% inhibition (in M)			
	Phobic reaction	Photo-kinesis	Q	Reference
a. Uncouplers				
NaN_3	$3\cdot10^{-3}$	$9\cdot10^{-4}$	3.3	NULTSCH (1967)
Imidazole	$3.8\cdot10^{-2}$	$4.6\cdot10^{-3}$	8.3	NULTSCH (1967)
Ammonium ions	$>10^{-1}$	$>2\cdot10^{-2}$	5.0	NULTSCH (1967)
b. Inhibitors				
Hydroxylamin	$1.3\cdot10^{-4}$	$2.6\cdot10^{-4}$	0.5	NULTSCH and JEEJI-BAI (1966)
o-Phenanthroline	$1.2\cdot10^{-5}$	$4\cdot10^{-5}$	0.3	NULTSCH and JEEJI-BAI (1966)
Salicyladoxim	$5\cdot10^{-5}$	10^{-3}	0.05	NULTSCH and JEEJI-BAI (1966)

Q is a quotient which is obtained by dividing the molarity needed to inhibit the phobic response to 50% by that which is necessary to produce a 50% inhibition of photokinesis

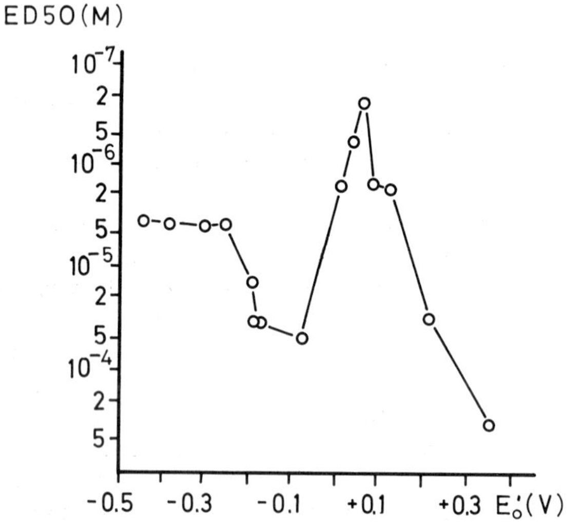

Fig. 10. Effect of externally applied redox systems on photophobic reactions in *Phormidium uncinatum*. *Abscissa*, redox potential, *ordinate*, molar concentration of the tested substance to produce 50% inhibition. (After NULTSCH, 1968)

Electron Pool Hypothesis. HÄDER (1973, 1974) used a double-irradiation technique (Sect. 3.2.2) to elucidate the roles of photosystems II and I. Light traps of a certain wavelength were projected into a homogeneous background light. Thus organisms were illuminated by the background light when outside the light trap, and by both background and trap light within. Photosystem II wavelengths in the trap (e.g., 563 nm) induced the organisms to react upon leaving the light trap, resulting in accumulations. With photosystem I light in the trap (≧700 nm) the organisms displayed phobic reactions, either upon entering the light trap (yielding empty traps, cf. Fig. 8), or upon leaving the light trap (yielding accumulations), depending on background wavelength and on the relative energy of the two light sources (HÄDER and NULTSCH, 1973; NULTSCH and HÄDER, 1974). Constant irradiation of the background with photosystem I light and variation of the trap wavelength produced an action spectrum of photosystem II and vice versa (Fig. 11; NULTSCH and HÄDER, 1974). These results gave rise to the electron pool hypothesis: In the noncyclic electron transport chain there is an electron pool, localized between photosystems II and I, which measures the electron flow rate (Fig. 12). This flow rate can be unbalanced when the number of electrons entering the pool decreases, or when the pool is drained via photosystem I. Both conditions result in photophobic reactions, although causing different patterns of accumulation in light traps.

The electron pool must be localized after the site of DCMU blocking, which has been reported to be near the primary acceptor of photosystem II (AMESZ, 1975): DCMU inhibits the photophobic reaction dependent on photosystem II much more effectively than the photosystem I-dependent reaction. On the other hand DBMIB, which prevents the reoxidation of plastoquinone (TREBST and BURBA, 1967; WARDEN, 1976) inhibits the photosystem I-dependent reaction more strongly than the photosystem II-dependent reaction (HÄDER, 1975).

KCN and DSPD (disalicylidenpropandiamin) which inhibit redox systems further down the electron transport chain produced similar results as DBMIB (HÄDER, 1976b).

These results, and the findings that plastoquinone is involved in cyclic as well as in noncyclic electron transport (SUNDBERG et al., 1976; SIGGEL, 1976) favor plastoquinone as the most promising candidate for the site of the electron pool. Furthermore, plastoquinone has been demonstrated to be the only greater pool of redox systems (about 10 times the amount of P 700) within the photosynthetic electron transport chain (AMESZ, 1975).

Sensory Transduction. In principle, sensory transduction could be brought about by various devices, e.g., piezoelectric effects (ZIMMERMAN, 1976), chemical transmitters, mechanical transmission or electrical effects. The involvement of chemical transmitters can be ruled out for photophobic reactions in blue-green algae, since diffusion of even small molecules is too slow to account for the reaction time (LEE and FITZSIMONS, 1976). It is more likely that sensory transduction is brought about by changes in the membrane potential. In photosynthetic organisms, the membrane potential may be subject to changes due to alterations in light intensity (THROM, 1970; BENTRUP, 1974; FELLE and BENTRUP, 1974). According to the Mitchell chemiosmotic hypothesis, the electron transport across the thylakoid membrane produces a proton gradient and an electrical potential (MITCHELL, 1975). In blue-green algae this photopotential must alter the plasma membrane potential, since the thylakoids are not enclosed in a chloroplast membrane, but are connected to the plasmalemma (LANG, 1968).

Light-induced changes in the membrane potential can be measured in *Phormidium* with the aid of externally applied platinum electrodes (HÄDER, 1976c); the amplitude of the potential change depends on the light intensity. The action spectrum of the phenomenon shows peaks which correlate with the absorption maxima of photosystem II and I pigments, biliproteins and chlorophyll a; thus, it resembles the action spectrum of the photophobic reaction. By penetration of the cell with microelectrodes (0.1 μm tip diameter), a membrane potential of about 30–40 mV can be measured (HÄDER, 1976c); light-induced potential changes have been obtained in this system as well.

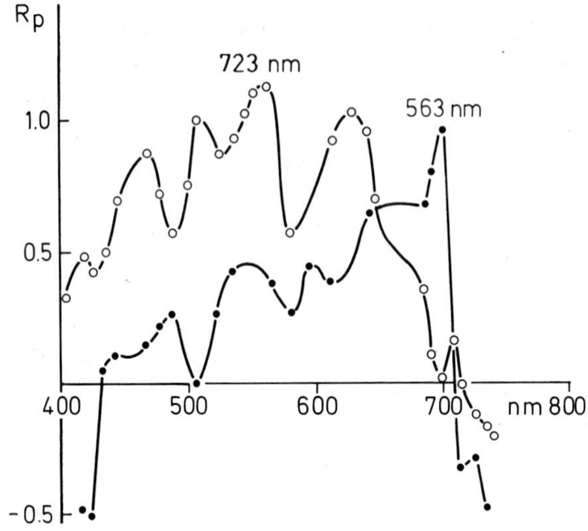

Fig. 11. Photophobic action spectra of *Phormidium uncinatum* under a background illumination of 563 nm (*dots*) and 723 nm (*circles*), respectively. (After NULTSCH and HÄDER, 1974). *Abscissa:* trap wavelength

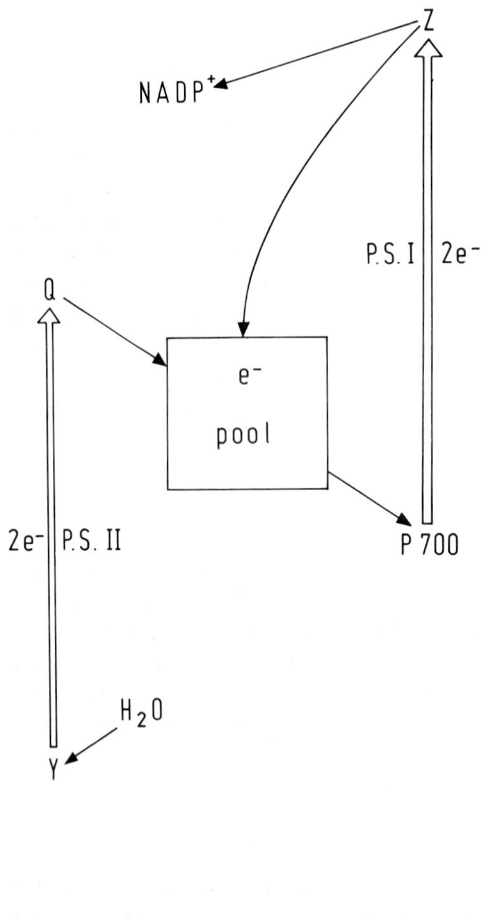

Fig. 12. Schematic diagram explaining the electron pool hypothesis. (After HÄDER, 1973)

External direct current fields, applied to suspensions of *Phormidium* trichomes, inhibited accumulations in a light trap when the voltages induced along the length of an organism exceeded 10 mV (HÄDER, 1977). The external potentials apparently disrupted the internal photopotential and thus inhibited photophobic reactions. However, phobic reactions were suppressed only in those trichomes which were oriented more or less parallel to the electric field lines. In *Chlamydomonas,* a similar experiment was carried out by MARBACH and MAYER (1971); induced external potentials of about the same magnitude as in *Phormidium* reversibly suppressed phototactic orientation.

These results have been taken as hints that sensory transduction in photophobic reactions of blue-green algae is mediated by light-induced membrane potential changes. Changes in membrane potentials have been proposed as triggering phobic reactions in bacteria as well (DOETSCH, 1972).

3.4 Gliding Eukaryotic Organisms

3.4.1 Diatoms

Earlier reports on photophobic reactions in diatoms have been reviewed by e.g., HAUPT (1959, 1974) and NULTSCH (1970, 1974b, 1975). HEIDINGSFELD (1940) was the first to conduct a detailed study of photophobic reactions in the diatom *Navicula radiosa* by means of the light-trap method.

Initially there was some doubt about the true nature of red light accumulations, since they lack the sharp edges found in blue light and generally elsewhere in light-trap experiments. This could mean in fact a chemotactic orientation in response to oxygen produced photosynthetically by organisms in the red light field (NULTSCH, 1962a, 1971; NULTSCH and WENDEROTH, 1973). However, WENDEROTH (1975), who measured the action spectrum of *Navicula peregrina* (Fig. 13), was able to induce phobic reactions by partial illumination with red light fields. These experiments further revealed that the photoreceptor pigments are localized in the phaeoplast. From the similarity between action spectra and absorption spectra, it can be concluded that phobic reactions in diatoms are mediated by the photosynthetic apparatus, as in prokaryotic microorganisms.

3.4.2 Desmids

NEUSCHELER (1967a, b) failed to demonstrate phobic reactions in desmids. Recently HÄDER and WENDEROTH observed accumulations with the light-trap method in *Cosmarium botrytis, C. cucumis* and *C. margaritiferum.* The zero threshold in white light was 5 lx, 0.25 lx and 0.5 lx respectively. There are two types of phobic responses in *Cosmarium* and other desmids: either the cells swing sideways into the new direction, or they reverse their movement by expulsion of a new slime stream from the former front end.

3.4.3 Red Algae

The unicellular red alga *Porphyridium cruentum* moves by means of a gliding mechanism involving extrusion of slime. Comparison between photophobic action

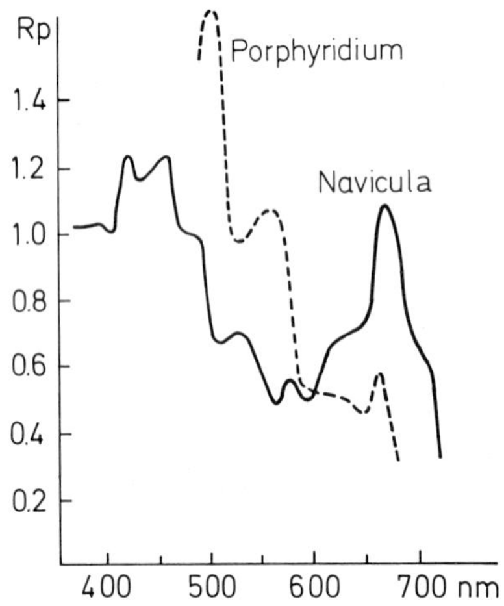

Fig. 13. Photophobic action spectra of *Navicula peregrina* (*solid curve;* after WENDEROTH, 1975) and *Porphyridium cruentum* (*broken curve;* after NULTSCH, 1975)

spectrum and in vivo absorption spectrum reveals the participation of B-phycoerythrin (560 nm) and chlorophyll a (665 nm) (NULTSCH, 1975; Fig. 13). The peak near 500 nm seems to be an artifact due to phototactic orientation, and coincides with the maximum in the action spectrum of phototaxis.

3.5 Flagellates

3.5.1 Behavioral Aspects

Among photoresponsive microorganisms, flagellates seem to be the exception to the rule. While in all other microorganisms the photophobic reaction is linked to photosynthesis, in most flagellates a different mechanism, involving a blue light-absorbing photoreceptor, seems to exist.

The numerous reports on photophobic reactions in flagellates by earlier authors have been reviewed by FEINLEIB and CURRY (1971b), DIEHN (1973a) and NULTSCH (1975). The reaction in *Euglena, Phacus, Ceratium,* and others consists in a sudden ventral beat which forces the organism into a new direction. Commonly, a given organism shows step-down responses at low light intensities and step-up responses at high intensities. Both these intensity ranges are separated by an "indifference zone" in which no response is evident.

3.5.2 *Euglena*

Many investigations in *Euglena* have been carried out using the "phototaxigraph" (cf. Sect. 4.1.2; BOUND and TOLLIN, 1967; DIEHN, 1969a; CHECCUCCI et al., 1974, 1975, 1976). The action spectrum of step-up phobic reactions in *Euglena* shows peaks at 365, 412, 450 and 480 nm (DIEHN, 1969a), and is interpreted to represent a flavin pigment. The action spectrum of the step-down reaction peaks at 375 and 480 nm and resembles that of positive phototaxis (Fig. 14). This spectrum

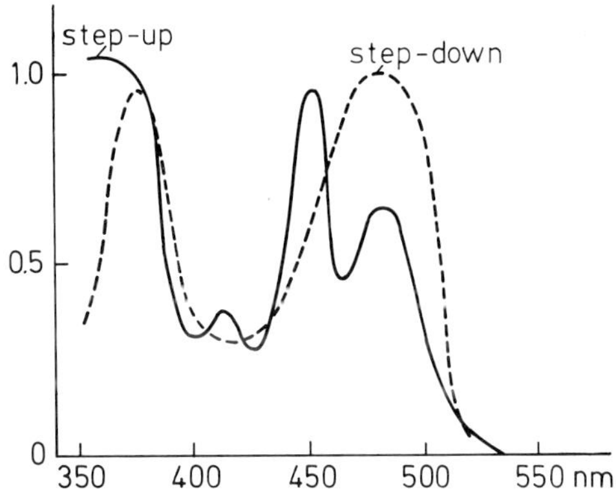

Fig. 14. Action spectra of step-down (*broken curve*) and step-up (*solid curve*) photophobic reactions in *Euglena gracilis*. The latter is composed of the action spectra obtained by polarized light vibrating parallel and perpendicular to the long axis of the cell. (After DIEHN, 1969a)

supposedly represents the modified action of photoreceptor and screening pigments. However, potassium iodide, a quencher of flavin fluorescence, fails to inhibit step-up responses and negative phototaxis in *Euglena*, whereas it does inhibit the step-down response, although at rather high concentrations (150 mM; DIEHN and KINT, 1970; MIKOLAJCZYK and DIEHN, 1975).

Another argument for the flavin nature of the photoreceptor has been found by DIEHN (1973a): in a reducing environment, phobic reactions can be induced by red light. Consistent with this observation, some flavins absorb in this region when they are in a more reduced state than usual.

It is now widely accepted that the stigma serves as a shading device for the photoreceptor in the phototactic orientation mechanism (cf. Sect. 4.4.1). The photoreceptor for photophobic response and for phototaxis is thought to be localized in a swelling of the emergent flagellum called the paraflagellar body (DIEHN, 1969b; CHECCUCCI et al., 1975; CHECCUCCI, 1976). GÖSSEL (1957) demonstrated that a stigmaless *Euglena* which possessed the paraflagellar body performed phobic reactions, while *Astasia*, lacking this organelle, did not. *Euglena* mutants lacking the paraflagellar body also lack the phobic response (DIEHN, 1973a). Recent evidence indicates that the paraflagellar body consists of flavoproteins (FROEHLICH and DIEHN, 1974; COLOMBETTI et al., 1975; BENEDETTI and CHECCUCCI, 1975; MCKELLAR et al., 1975). Fluorescence studies have indicated that the flavin molecules are highly ordered within the paraflagellar body (BENEDETTI and CHECCUCCI, 1975). However, fluorometric evidence has recently been found that flavins are also present in the stigma (SPERLING et al., 1976).

The stigma is composed of a layer of thirty or more globules (LEEDALE, 1967; DODGE, 1973; KIVIC and VESK, 1974), filled with unoriented carotenoids (WOLKEN, 1960, 1961; BUETOW, 1968; BENEDETTI et al., 1976). Recently, absorption spectra of the stigma in green and dark-grown bleached *Euglena* and of isolated stigma globules have been measured microspectrographically by BENEDETTI et al. (1976; Fig. 15). Since the absorption spectrum of the stigma coincides with that of the photoreceptor (as judged from phototactic action spectra), the stigma acts as a neutral attenuator. It absorbs about 50% of the incident light in green cells and about 30% in the dark-grown cells. Absorbance of isolated globules is even higher, because of swelling of globules outside the cell.

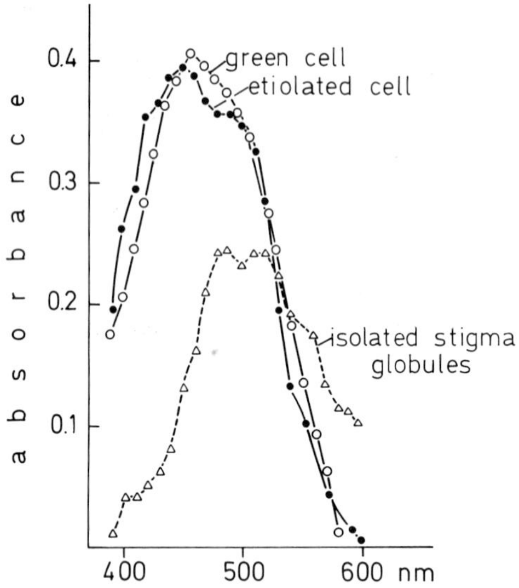

Fig. 15. Absorption spectra of the stigma in *Euglena gracilis*. *Circles*, stigma inside a green cell; *dots*, stigma inside an etiolated cell; *triangles*, isolated stigma globules. (After BENEDETTI et al., 1976)

Recently MIKOLAJCZYK and DIEHN (1976) studied the effect of viscosity on photophobic reactions. In media of high viscosity, organisms stop their flagellar beat and respond to changes in light intensity by body contractions.

3.5.3 Other Flagellates

In *Gyrodinium dorsum* the step-up photophobic response consists of a complete stop for 1 to 5 s followed by a rapid rotation of the cell, seldom exceeding 0.5 s in duration (HAND, 1970). Again the photoreceptor is maximally stimulated by blue light, which is attributed by the author to the action of a carotene-type pigment. The action spectrum for phobic responses shows maxima at 280 and 470 nm (FORWARD, 1974). It is, however, strange to notice that preillumination with 620 nm increases the sensitivity towards blue light of 470 nm, while, after preillimination with 700 nm, the sensitivity at 490 nm is increased (FORWARD and DAVENPORT, 1968, 1970; FORWARD, 1970). This behavior has been attributed to the action of a phytochrome system (HAND et al., 1967; FORWARD and DAVENPORT, 1968).

Volvox and *Pleodorina* exhibit a different type of photophobic response. Upon a change in light intensity the frequency of the flagellar beat decreases. If this change in intensity is strong enough, the flagella are arrested in a characteristic position. Depending on the overall intensity, this phobic response is brought about by either a sudden decrease or increase, and normal movement can be restored immediately by increasing or decreasing the light intensity to the former value (GERISCH, 1959). The stigma has been postulated by the author to be the photoreceptor organelle for the photophobic response in these organisms. Thus, there is no general agreement on a single photoreceptor pigment in the photophobic responses of flagellates: while the photoreceptor pigment is supposed to be a flavin in *Euglena*, both carotenoids and flavins are being discussed for other flagellates.

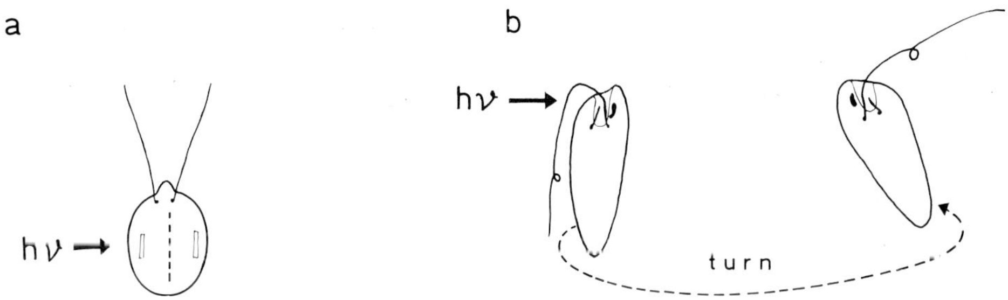

Fig. 16. Schematic drawings explaining one-instant **a** and two-instant **b** mechanisms. (After BOSCOV, 1974, and FEINLEIB, 1975)

4. Phototaxis

Phototaxis describes the phenomenon of oriented movement in relation to the direction of the impinging light rays. The question of whether phototaxis is brought about by a directional movement towards the light source, or an orientation in a spatial gradient of light intensity, was answered by an ingenious experiment by BUDER (1919): in a convergent light beam, organisms still move toward the light source, although light intensity decreases with increasing distance from the focal point. Another problem is the determination of light direction by the organism, which seems to be solved by one of two basic types of mechanism (HAUPT, 1966; FEINLEIB and CURRY, 1971a; BOSCOV, 1974). In the "one-instant" mechanism (FEINLEIB, 1975b), at least two photoreceptors detect an internal light-intensity gradient by comparing light absorption in two regions of the cell at the same time. The alternative to this is the "two-instant" mechanism: successive measurement of light intensity is performed by one photoreceptor as the cell changes its position in relation to the light source (Fig. 16). This latter device seems to be verified in the flagellates *Euglena* (MAST, 1911; DIEHN, 1973a), *Gyrodinium* (HAND and SCHMIDT, 1975), and *Volvox* (SCHLETZ, 1976), while gliding organisms seem to use the one-instant mechanism.

For evaluation of pre-sixties authors see reviews by BENDIX (1960a), HAUPT (1965, 1972), FEINLEIB and CURRY (1971a), DIEHN (1973a) and NULTSCH (1975). Since phototactic behavior is far from being uniform in the various systematic groups, no attempts will be made to encompass it in a comparative description. Only a few comments will be made on the light-intensity dependence of various organisms (Table 2). While the desmid *Micrasterias* is extremely sensitive to phototactic stimulation, the zero thresholds of other organisms are in the range of several lx. Comparing the optima, the flagellates are again the exception. Supraoptimal intensities decrease the sensitivity, until in most cases the response becomes negative at high light intensities; only diatoms fail to perform negative phototactic reactions (NULTSCH, 1956, 1971). At the inversion intensity between positive and negative phototaxis, unoriented movement has been observed in *Volvox* (cf. HAUPT, 1959), while in *Chlamydomonas* the path of any given cell is always directed either toward or away from the light source (FEINLEIB and CURRY, 1971b).

Table 2. Light intensity dependence of phototaxis in various microorganisms (after NULTSCH, 1975)

Systematic group	Zero threshold[a]	Optimum	Inversion intensity[b]	Reference
Photosynthetic bacteria (Rhodospirillum)	< 30 lx	60 lx	18,000 lx	THROM (1968)
Blue-green algae (Phormidium)	1–30 lx	50–200 lx	1,000–10,000 lx	DREWS (1959), NULTSCH (1962b)
Diatoms (Nitzschia)	1–30 lx	200 lx	—	HEIDINGSFELD (1940), NULTSCH (1956, 1971)
Desmids (Micrasterias)	10^{-4}–10^{-5} lx 10^{-6}–10^{-7} W m^{-2} (666 nm)	— 10^{-2}–10^{-3} W m^{-2} (666 nm)	4,000 lx –	NEUSCHELER (1967a, b)
Flagellates (Chlamydomonas)	< 1 mW m^{-2} (blue-green light)	1,000 lx	20 W m^{-2} (blue-green light)	FEINLEIB and CURRY (1971b) NULTSCH et al. (1971a)

[a] The lowest intensity at which positive phototaxis is detected.
[b] The intensity at which positive phototaxis changes to negative.

4.1 Methodology

As in photokinesis and photophobic reactions, the methods used to study phototactic behavior can be grouped into two categories: (1) individual cell methods and (2) mass movement or population methods.

4.1.1 Individual Cell Methods

The accumulation rate of phototactically reacting organisms within a given time depends not only on their speed, but also on the straightness of their path. The details of movements by which the cells orient themselves toward the source of the stimulus can be resolved only by individual-cell methods.

Some of the techniques used in phototaxis research have also been applied to photokinesis experiments (cf. Sect. 2.1.1). NULTSCH (1957) traced the tracks of diatoms after microscopic observation. All microscopic techniques demand high light intensities, which may have a direct or indirect impact on the photobehavior (HAND and DAVENPORT, 1970). Photomicrographic techniques have been applied for example by FEINLEIB and CURRY (1967, 1971b) and FEINLEIB (1975a). The path of organisms can be followed by the flying spot scanning device of DAVENPORT et al. (1962) and HAND et al. (1967). This device allows for continuous registration, and yields images of the organisms several times per second. A further improvement was the introduction of a video camera mounted over a microscope connected to a video recorder. All these techniques have two drawbacks: one is the limitation to two dimensions, which demands a restriction of space available for the moving organisms, otherwise the organisms swim out of the focal plane. Furthermore experiments are rather tedious, since reactions of one or only a few organisms can be followed at a time.

The method of computerized television combines a video camera mounted on a microscope, with a "bug-watcher" (a unit which processes the video signal for quantification) and a digital computer (DAVENPORT et al., 1970; DAVENPORT, 1973). This apparatus overcomes one of the drawbacks of other individual cell methods, since traces of a number of organisms can be followed at the same time.

4.1.2 Population Methods

All automatic devices used to estimate phototactic reactions utilize absorbance changes caused by the accumulating organisms. BRUCE and PITTENDRIGH (1956) designed an instrument in which a silicon solar cell measures, at 2 h intervals, the density of organisms heading toward the actinic white light.

LINDES et al. (1965) designed the so-called phototaxigraph, which was used and improved by DIEHN (1970), TOLLIN and ROBINSON (1969) and STAVIS (1974). This apparatus consists of a rotating cylindrical cuvette, with the actinic beams entering perpendicular to the long axis from above and below. Phototactic response is elicited by light scattered by organisms already inside the actinic zone. The degree of phototactic accumulation is estimated by continuously comparing the absorbance of the nonactinic measuring beam (800 nm) with one or two reference beams, which pass through the suspension outside the actinic zone. This method has several complications, which may lead to serious misinterpretations (NULTSCH, 1975; SCHUCHART, 1975): the phototactic stimulus cannot be quantified and it changes during the experiment, phobic reactions occur along the border of the actinic zone, and photokinetic effects cannot be excluded. Thus, this instrument has only limited use in phototaxis research.

In the apparatus designed by FEINLEIB and CURRY (1967), the actinic beam enters the cylindrical rotating (60 rpm) cuvette from one end, parallel to its long axis. Two measuring beams (>620 nm) pass through the cuvette, perpendicular to the long axis, and strike two photocells. A movement of organisms toward the light source causes an asymmetrical absorption, and thus a difference in the photocurrents.

In the instrument devised by THROM (1968) and improved by NULTSCH and THROM (1975), the light beam enters the cuvette at 45°; this reduces the light gradient resulting from absorption by the organisms. Since the phototactic response is governed by a circadian rhythm in several organisms (FELDMAN, 1967; BRINKMANN, 1976), this instrument has been linked to a homocontinuous culture (NULTSCH and THROM, 1975; NULTSCH, 1977). This randomizes the individual phases of cell cycles.

Assessment of phototaxis in diatoms and blue-green algae can be quantified by a method used by DREWS (1959) and NULTSCH (1961). Organisms inoculated in the center of an agar surface form a spreading area directed toward the unilateral light source.

4.2 Prokaryotic Organisms

4.2.1 Bacteria

Although phototactic reactions were discovered quite early in bacteria, phototaxis has not yet been investigated extensively, while chemotaxis has been thoroughly studied (e.g., SEYMOUR and DOETSCH, 1973; ADLER, 1975, 1976; cf. MACNAB, Chap. 3.2.2, this vol.). ENGELMANN (1883) observed a weak phototactic response in the purple *Bacterium photometricum* (=*Chromatium?* CLAYTON, 1959) in red light. CLAYTON (1953, 1958) presumed that all phototactic reactions reported in *Rhodospirillum rubrum* are based on artifacts due to photophobic reactions. However, THROM (1968) could demonstrate true, although weak phototactic orientation in this bacterium in white light; the zero threshold was 30 times lower than in the phobic response.

4.2.2 Blue-Green Algae

Phototactic Orientation. Most blue-green algae show both positive and negative phototaxis (FAMINTZIN, 1867), while *Beggiatoa* and *Oscillatoria jenensis* display only a negative response. Most of the phototactic genera are filamentous, but unicellular Chroococcales have also been reported as showing phototaxis (STANIER et al., 1971).

In blue-green algae two different types of phototactic orientation are known: (1) In *Anabaena* there is an active steering mechanism; trichomes, which are either straight or U-shaped, turn actively into the new light direction (DREWS, 1959); (2) A presumably more primitive mechanism is found in *Oscillatoria* and *Phormidium*: Unstimulated trichomes move back and forth according to an autonomous rhythm. Upon unilateral illumination, the movement towards the light source is extended in those organisms which glide more or less parallel to the incident light rays, while the backward movement is shortened (NULTSCH, 1975). This results in a net movement toward the light direction. Whether this difference in mode of response is related to the fact that *Oscillatoria*-type blue-green algae rotate during their locomotion, while *Anabaena* does not, has not yet been revealed (see HALFEN, Chap. 3.1.3, this vol.).

Action Spectra. There are quite early observations about wavelength dependency. DANGEARD (1928) found main activity in the red and infrared. DREWS (1959) discovered considerable differences among several species, but never observed phototaxis in the IR >720 nm. In *Phormidium autumnale* the action spectrum for phototaxis coincides with the absorption maxima of the photosynthetic accessory pigments C-phycoerythrin and C-phycocyanin; carotenoids might be active as well (NULTSCH, 1961). Chlorophyll a does not seem to participate in the reaction. In *Phormidium uncinatum* the spectrum resembles that of *Ph. autumnale* (NULTSCH, 1962b). These results seem to exclude any relation between photosynthesis and phototaxis, but since no uncoupler or inhibitor studies have been reported, this possibility cannot yet be ruled out definitely.

4.3 Gliding Eukaryotic Organisms

4.3.1 Desmids

Phototaxis in desmids was described as early as 1861 (PRITCHARD). Both positive (low light intensities) and negative (high light intensities) reactions have been observed. NEUSCHELER (1967a, b) found that in *Micrasterias denticulata* phototaxis is governed by a circadian rhythm: the thresholds both for phototaxis and for photokinesis were 100 times higher by night. During the day, the cell tends to align its longitudinal axis parallel to the incident light rays, i.e., towards the light at low intensities and away from the light at high intensities, and the meandering path becomes straight. In the indifference range, the cell turns into a position perpendicular to the light beam without moving forward. Obviously desmids are able to perceive light direction by means of a one-instant mechanism (cf. p. 289).

The action spectrum shows a broad maximum between 370 and 470 nm and a peak at 660 nm, which could represent chlorophyll a absorption (NEUSCHELER, 1967b). In her action spectrum BENDIX (1957, 1960a, b) found two sharp peaks at 650 and 690 nm in *M. rotata*, but low activity in the blue. These differences between the action spectra in various species of *Micrasterias* may be explained by the investigations by NEUSCHELER (1967b). The relative efficiency of blue and red light depended both on light intensity and on phase of diurnal rhythm. Chlorophyll a has been suggested to be the photoreceptor pigment for red light-induced phototactic orientation, while flavins, carotenoids or chlorophyll may be involved in blue light responses. Blue light and red light systems may operate separately from one another.

4.3.2 Diatoms and Red Algae

Phototaxis in diatoms is brought about by the same mechanism as in the blue-green algae of the *Phormidium*-type (p. 292) (HEIDINGSFELD, 1940; NULTSCH, 1956, 1957). Blue light (400–550 nm) has been found to be most effective in *Navicula* and other diatoms (HEIDINGSFELD, 1940; NULTSCH, 1959).

In red algae, on the other hand, a true steering mechanism seems to be involved in phototactic orientation. The primitive unicellular *Porphyridium cruentum* is one of only a few motile organisms in this group. The cells move toward the light source, leaving a track of mucilage, which may be forked in places where cells had divided (Fig. 17; PRINGSHEIM, 1968). Neither action spectra nor inhibitor studies have yet been published.

4.4 Flagellates

4.4.1 Phototaxis in *Euglena*

Mechanisms of Phototactic Orientation. Most of the investigations in phototaxis have been done on flagellates and especially on *Euglena*. Today a widely accepted theory of phototactic orientation is the periodic shading mechanism (BUDER, 1919; LEEDALE, 1967; DIEHN, 1973a). This mechanism incorporates the photoreceptor (a swelling in the longer flagellum inside the reservoir, called paraflagellar body;

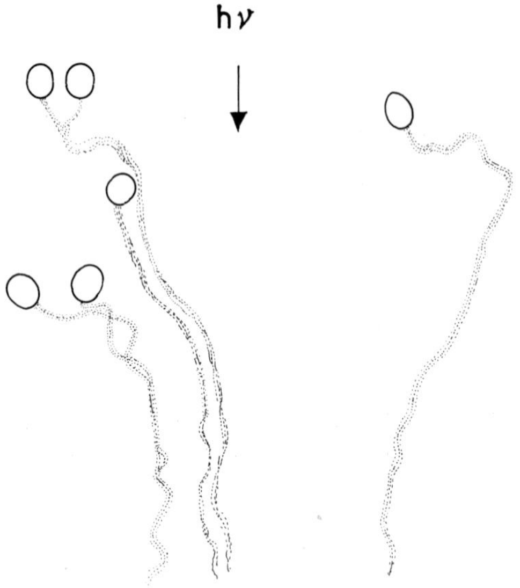

Fig. 17. Slime traces left by phototactically oriented moving *Porphyridium cruentum* cells. Traces are forked where cells had divided. (Drawn after a microphotograph by PRINGSHEIM, 1968)

KIVIC and VESK, 1972a, b; KRONESTEDT and WALLES, 1975; cf. also Sect. 3.5.2) and the stigma. In lateral illumination the stigma periodically shades the photoreceptor due to rotation of the organism around its long axis (WOLKEN, 1960; DIEHN, 1973a). This sudden shading mediates a comparatively small motor response, consisting of a ventral beat of the flagellum, which slightly corrects the course of the organism toward the light source (Fig. 18). CHECCUCCI (1976) called this an elementary trajectory correction. This procedure is repeated until the front end faces the light source. In that position, further shading is morphologically prevented. An analogous mechanism has been suggested for negative phototaxis, with the modification that in this case the whole rear end, including chloroplasts, acts as a shading device (DIEHN, 1973a; DIEHN et al., 1975).

There are several approaches to confirm this hypothesis. DIEHN (1969c) simulated periodic shading by using repetitive flashes of actinic light. Maximal effect on phototaxis was obtained at a pulse rate of 1.05 s, which indeed is the time *Euglena* takes for one rotation. *Euglena* cells lacking both stigma and paraflagellar body do not orient themselves phototactically (PRINGSHEIM, 1948; VAVRA, 1962). Dark-bleached cells containing stigma and paraflagellar body perform positive phototaxis, while organisms with paraflagellar body, but without an absorbing stigma, show only negative phototaxis (VAVRA, 1962; CHECCUCCI et al., 1976; FERRARA, 1975).

The observations described above may also be explained by a more recent model, which has been derived from experiments with polarized light (BOUND and TOLLIN, 1967; DIEHN, 1969b; CREUTZ and DIEHN, 1976). Single cell observation has revealed that at intensities ranging from 3 mW m^{-2} to 3 W m^{-2} *Euglena* prefers to swim perpendicular to the plane of vibration. Since no evidence of dichroism was found in the stigma, but a remarkable birefringence appeared in the paraflagellar body, the authors concluded that the photoreceptor molecules in the paraflagellar body are oriented perpendicular to the long axis of the organism.

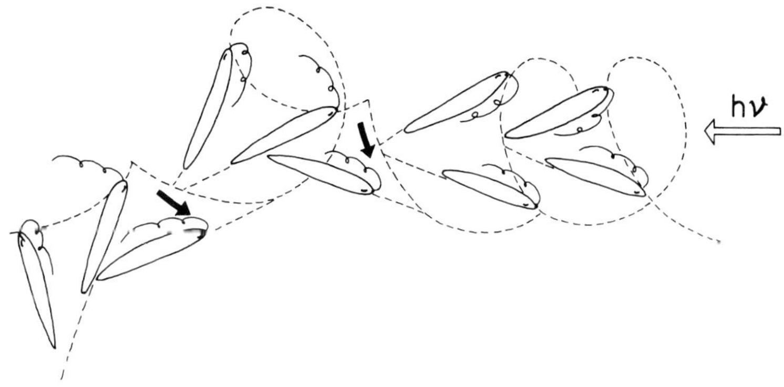

Fig. 18. Schematic diagram of phototactic orientation by means of photophobic course corrections (*solid arrows*) during rotation of *Euglena gracilis*. (After CHECCUCCI, 1976)

Shading by the rear end of the cell prevents excitation of photoreceptor pigments from a posterior direction. If a cell moves transversely to the direction of an unpolarized actinic light beam, only part of the photoreceptor molecules absorb light, while, when the cell moves toward the light source, all molecules are excited. Course corrections are achieved in this model by step-down photophobic responses as in the periodic shading mechanism.

Action Spectra of Photoresponses in Euglena. In 1917, OLTMANNS measured a crude action spectrum in *Euglena*, with a broad maximum between 450 and 500 nm. Since then, numerous action spectra were measured. A number of factors have been found to influence phototaxis in *Euglena*, including circadian rhythm (BÜNNING and SCHNEIDERHÖHN, 1956; BRUCE and PITTENDRIGH, 1960; BRUCE, 1973), dark or light adaptation (BÜNNING and GÖSSEL, 1959; WOLKEN and SHIN, 1958), preillumination with bright red light (TOLLIN and ROBINSON, 1969) and presence or absence of chloroplasts and stigma (GÖSSEL, 1957; VAVRA, 1962; COOK et al., 1974; CHECCUCCI et al., 1975).

All authors agree on the fact that true light-mediated orientation is limited to short wavelengths up to about 530 nm. Red light-induced responses are due to aerotactic accumulation (CHECCUCCI et al., 1974). Furthermore phototactic experiments with population methods may be disturbed by pattern formation, which occurs independent of any photoeffect (LEVANDOWSKY et al., 1975).

From the similarity of phototactic action spectra in organisms with and without stigma, CHECCUCCI et al. (1976) concluded that the action spectra are identical with the absorption spectrum of the photoreceptor and are not distorted by the screening pigments. The photoreceptor seems to consist of flavoproteins (cf. Sect. 3.5.2). Potassium iodide, which inhibits positive phototaxis and step-down photophobic reactions (DIEHN and KINT, 1970; MIKOLAJCZYK and DIEHN, 1975), does not impair step-up photophobic reactions and negative phototaxis (MIKOLAJCZYK and DIEHN, 1975). Therefore the authors concluded that these responses are mediated by a different, nonflavin pigment system.

Sensory Transduction in Euglena. The transmission of signals from the photoreceptor to the motor of the cell is still a "black box". DIEHN (1973b) offered a model incorporating a receptor (paraflagellar body), a processor (morphological correlate unknown) and an effector, which controls the orientation of the motor (flagellum). By means of a computer simulation, the reactions of the cell under different light programs could be predicted. The coupling between the energizers of the organism (mitochondria and chloroplasts) and the phototactic machinery cannot be simply by the ATP flux, since inhibitors and uncouplers had only minor effects (DIEHN and TOLLIN, 1967).

It has been proposed that differences in the membrane potential might be used for sensory transduction (FROEHLICH and DIEHN, 1974; HILDEBRAND, 1974) as in *Paramaecium* (ECKERT, 1972). Indeed, flagellar activity can be controlled by injecting negative current pulses with the aid of inserted microelectrodes (NICHOLS and RIKMENSPOEL, 1977).

BOVEE and JAHN (1972) proposed a theory involving a structure within the flagellum that had piezoelectric properties. Whether the paraflagellar rod, a structure located between the paraflagellar body and the flagellar microtubules in a number of flagellates, plays a part in this process is also an open question (WALNE and ARNOTT, 1967; PICCINNI et al., 1975; PICCINNI and OMODEO, 1975).

KRISTIANSEN and WALNE (1976) found in *Dinobryon, Chromulina* and other organisms a structural connection between flagellar base and stigma to which they ascribed the function of sensory transduction in phototaxis, since the stigma has been proposed to be the photoreceptor in these organisms. Similar structures have not been found to date in *Euglena*.

4.4.2 Other Algal Flagellates

Action Spectra and Inhibitor Studies. Action spectra similar to those in *Euglena,* with main sensitivity in blue light, have been found in *Chlamydomonas* (NULTSCH et al., 1971), *Gyrodinium* (HAND et al., 1967; FORWARD, 1973, 1974), *Peridinium, Dunaliella* and *Platymonas* (HALLDAL, 1960, 1963). The main candidates for photoreceptor pigments have been carotenoids (FERRARA, 1976).

There are, however some interesting exceptions to the rule: WATANABE and FURUYA (1974) published an action spectrum for the cryptoflagellate *Cryptomonas* which shows that light between 400 and 640 nm is phototactically effective, with a distinct maximum at 560 nm. This maximum corresponds to the absorption peak of phycoerythrin, which is the photoreceptor pigment for phototaxis in blue-green algae as well (cf. Sect. 4.2.2). The other exception is found in the dinoflagellate *Prorocentrum,* with a maximum at 570 nm due to absorption by an unknown pigment (HALLDAL, 1964). Figure 19 summarizes action spectra of several species.

The effects of uncouplers and inhibitors of the photosynthetic electron transport chain on phototaxis in *Chlamydomonas* clearly revealed that there is no direct connection between phototaxis and photosynthesis in that organism (SACHS and MAYER, 1961; MARBACH and MAYER, 1970; STAVIS and HIRSCHBERG, 1973; STAVIS, 1974). The same is true for *Cryptomonas* (WATANABE et al., 1976).

Photoreceptor Organelles and Mechanisms Involved in Phototactic Orientation. For some unicellular flagellates, e.g., *Tetracystis* (ARNOTT and BROWN, 1967) and *Gyrodinium* (HAND, 1970), as well as for colonial forms, e.g., *Volvox* (SCHLETZ, 1976) and other Volvocales (HALLDAL, 1964), a periodic shading mechanism has been proposed, as for *Euglena*.

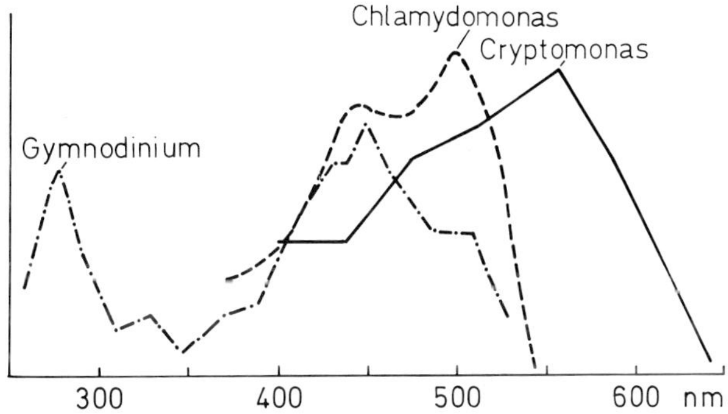

Fig. 19. Phototactic action spectra of *Gymnodinium splendens* (*dot-dash curve;* after FORWARD, 1974), *Chlamydomonas reinhardii* (*broken curve;* after NULTSCH et al., 1971) and *Cryptomonas sp.* (*solid curve;* after WATANABE and FURUYA, 1974)

In the marine dinoflagellate *Gyrodinium*, the receptor is presumably located at the sulcus–girdle junction. Contrary to the situation in *Euglena*, the receptor is continuously shaded by the front half of the cell, when the organism has become oriented toward the light (HAND and SCHMIDT, 1975).

In a number of flagellates the stigma has been proposed to be involved in the light-perception process. Electron microscopy of algal stigmata and presumed photoreceptors revealed a wide variety of ultrastructure (DODGE, 1969; DODGE and CRAWFORD, 1969). The stigma may be part of the plastid or independent of it. It may be located in the vicinity of the paraflagellar body or at a different place. Some dinoflagellates and peridins have been found to possess a rather complicated photoreceptor apparatus including a lens-like structure and a cup-shaped pigmented zone which contains a retinoid (FRANCIS, 1967; GREUET, 1968; MESSER and BEN-SHAUL, 1969).

In *Chlamydomonas*, the location of the photoreceptor is unknown, but the stigma does not seem to be a necessary requisite for phototaxis, since stigmaless mutants are able to perform photoorientation (HARTSHORNE, 1953). Phototactic reactions are sometimes preceded by a phobic response, which is followed by positive or negative reorientation, depending on light intensity (FEINLEIB and CURRY, 1971b; FEINLEIB, 1975a). The same is true for other flagellates (HAND, 1970; FORWARD, 1973, 1974).

Responses of *Chlamydomonas* to single and repetitive flashes of light (6 μs) have been studied by videomicroscopic recording (BOSCOV, 1974; FEINLEIB, 1975a, b). About 50%–60% of the cells respond to a single flash of high intensity: 70%–90% of these turn away from the flash source. The authors suggest that the only cells to respond may be those in which the photoreceptor is directed toward the light source at the time of the flash. If so, the cells may detect light direction by a two-instant mechanism: measurements would be taken before and during the flash.

In *Volvox* only the front end is sensitive to light, and a periodic shading mechanism has been proposed (HUTH, 1970a, b; SCHLETZ, 1976). HAND and HAUPT (1971) studied the flagellar activity by adding latex particles (1 μm ∅) to the swimming medium. Photomicrographic analysis revealed that, upon lateral illumination, the flagella, in a step-up photophobic response, stop beating on the illuminated side, which causes a turning into the new direction. This "rowboat mechanism" has been confirmed by SCHLETZ (1976).

During rotation *Volvox* deviates from the light direction by a certain angle, which is a function of the efficiency of the shading mechanism, because it is the resultant of a physical tendency to swim in a left-handed curve and the phototactic orientation. Therefore, the dependence of the deviation angle on the wavelength of the incident light has been used to obtain an action spectrum of phototaxis in *Volvox* (SCHLETZ, 1976).

There are protoplasmic connections between the single cells in *Volvox*, which have been proposed to be involved in sensory transduction in phototaxis (DOLZMANN and DOLZMANN, 1964; IKUSHIMA and MARUYAMA, 1968). This hypothesis has been questioned by HIATT and HAND (1972), since the protoplasmic connections are not found in all *Volvox* species; moreover, differential motor responses of the different cells, which are necessary for phototactic orientation, would be impossible if signals were transmitted from cell to cell all over the colony.

4.4.3 Zoospores of Lower Fungi

ROBERTSON (1972) reported that the action spectrum of phototaxis in zoospores of *Allomyces* shows a broad maximum between 470 and 525 nm. A possible photoreceptor for this response has not been suggested. Some species of Phycomycetes contain an organelle, called rumposoma, consisting of radially arranged tubuli, which are thought to guide the light to the photopigment: the photoreceptor is thought to lie in an electron-dense mass beneath this structure (KAZAMA, 1972). In all Chytridiomycetes the rumposoma is connected with a lipid drop, which may act as a lens (OMODEO, 1975).

4.4.4 Influence of External Factors

HALLDAL (1959, 1963) found evidence that the sense of the phototactic reaction – positive or negative – may be influenced by the concentration and ratio of ions in the medium. NULTSCH (1977) demonstrated that the CO_2 concentration has an influence on the inversion intensity. This is only partly due to changes in the pH. Oxygen, on the other hand, has only a slight effect.

Apart from eliciting phototactic orientation, light has indirect effects on the light-induced reaction (NULTSCH and THROM, 1975): predarkening of the organisms raises the phototactic sensitivity, while preillumination has no significant effect.

In a number of flagellates, phototactic responses are governed by an endogenous rhythm. In marine dinoflagellates this causes daily movements (HAND et al., 1967; FORWARD and DAVENPORT, 1968). BRINKMANN (1966, 1976) studied the circadian rhythm in *Euglena*. Experiments with alcohols indicated that the increase in period length upon application of ethanol or butanol might be due to the activation of mitochondrial enzymes. The circadian rhythm in phototactic orientation of *Chlamydomonas* can be induced by a single period of darkness or low light intensity, but not by high light intensity (NULTSCH and THROM, 1975).

5. Conclusions

5.1 Phobic Reactions: Prokaryotes versus Eukaryotes?

Although all prokaryotes investigated so far make use of the photosynthetic apparatus for their photophobic responses, the flagellates, which are sensitive only to wavelengths shorter than 530 nm, must have developed a completely different mechanism. Contrary to the earlier suggestion by NULTSCH (1974), this is not a characteristic of prokaryotes versus eukaryotes, because all gliding eukaryotes seem to behave, in this respect, like the prokaryotes. Thus, the flagellates seem to be the exception to the rule, possibly due to their highly specialized mode of propulsion.

5.2 Ecological Aspects

It could be speculated that photophobic reactions were "invented" to prevent an organism from entering unfavorable light conditions; but this seems to be an inefficient mechanism for those organisms that tumble or rotate upon phobic stimulation, and then continue along their way in a randomly determined direction. This kind of mechanism might be more efficient in organisms which usually return along their own track (e.g., blue-green algae).

Phototactic orientation seems to be even more effective for light-dependent plants, since it guides the individual or the population toward a favorable light intensity or away from a harmful radiation. In some marine dinoflagellates a circadian vertical movement of about 10 m has been found to result from phototactic orientation (HASLE, 1950).

The mechanism in diatoms, some blue-green algae, and possibly bacteria has to be regarded as primitive, since the organism – provided its course is more or less parallel to the impinging light rays – merely prolongs those movements that happen to be directed toward the light source. Other blue-green algae (Nostocaceae) have developed a true steering mechanism. The periodic shading mechanism, in which phototactic orientations is achieved by means of small corrections, has already been discussed (Sect. 4.4.1). Of course, "searching" for energetically useful light conditions cannot be claimed as a sensible aim for motile fungi.

Positive photokinesis does not seem to fit into this concept, since increase in speed in fact shortens the time in high light intensities, and often results in an accumulation outside the light field (HALLDAL, 1964). Theoretically, negative photokinesis could bring about accumulations in a light field; but in fact no organism is known to use photokinesis as its major photomovement strategy. In most cases this is outweighed by phototaxis or photophobic responses, by which almost all organisms capable of light-induced reactions accumulate in zones of favorable light conditions.

5.3. Photoreceptor Pigments and Structures

The pigments involved in primary photoevents in photomovement display considerable diversity; thus a general statement cannot be formulated. Even within one type of photomovent (e.g. photokinesis) there are two classes of photoreceptor pigments. On the one hand there are the photosynthetic pigments: bacteriochlorophyll, chlorophyll a, and different accessory pigments in various systematic groups, and, on the other hand, the blue light-absorbing pigments. In phototaxis, the concept of flavin absorption seems to be well established for *Euglena,* while carotenoids are being discussed as possible receptor pigments for other flagellates *(Chlamydomonas, Volvox, Gymnodinium).* No mechanism has been proposed for responses in which accessory photosynthetic pigments are involved but chlorophyll a is not, as in the phototaxis of *Cryptomonas* or in the photophobic reaction of *Nitzschia.* The opposite problem arises in *Micrasterias,* in which chlorophyll a seems to be active, while the accessory pigments, chlorophyll b and carotenoids, are not. No hypothesis has as yet been offered to explain these mechanisms.

Several possible photochemical processes have been proposed for carotenoid and flavin pigments, but none has yet been shown to take place in the living cell. For carotenoids cis–trans isomerization, energy transfer and polarization have been discussed, any of which might enable the molecule to function as photoreceptor pigment (Bensasson, 1975; Bonačić-Koutecký et al., 1975). On the other hand, theoretical considerations by Song and Moore (1974) make carotenoid involvement highly improbable for photoperception.

For flavins, on the other hand, photoreduction, photobleaching and dark-reversible light-induced fluorescence have been discussed (Lenci, 1975).

Since in bacteria and blue-green algae the photosynthetic pigments, localized in the thylakoids, are involved in photokinesis and photophobic responses, no separate photoreceptors are to be expected. In diatoms, desmids, and red algae the plastids are responsible for photoperception. The situation is different in flagellates. In *Euglena* the photoreceptor seems to be located in the paraflagellar body, while the stigma modulates the lateral actinic light beam. In many flagellates no photoreceptor has yet been identified, although some electron microscope investigations revealed structures which might be responsible for light perception.

In *Chlamydomonas* several phototaxis mutants have been isolated (Smyth and Ebersold, 1970; Hudock and Hudock, 1973; Hirschberg and Stavis, 1977); these might prove useful tools in analyzing the reaction chain step by step.

Last but not least, theoretical considerations and model building may well prove useful, but will of course have to be checked against the accessible experimental data.

References

Adler, J.: Chemotaxis in bacteria. In: Primitive Sensory and Communication Systems. Carlile, M.J. (ed.). London-New York-San Francisco: Academic Press, 1975, pp. 91–100

Adler, J.: The sensing of chemicals by bacteria. Sci. Am. *234*, 40–47 (1976)

Adler, J., Dahl, M.M.: A method for measuring the motility of bacteria and for comparing random and non-random motility. J. Gen. Microbiol. *46*, 161–173 (1967)

Allen, M.B., Bendix, S.A., Murchio, J.C.: Concerning "long wave length" pigments in algae. Arch. Mikrobiol. *42*, 36–39 (1962)

Amesz, J.: Photosynthesis. Biophysical aspects. Fortschr. Bot. *37*, 107–120 (1975)

Arnott, H.J., Brown, R.M. jr.: Ultrastructure of the eyespot and its possible significance in phototaxis of *Tetracystis excentrica*. J. Protozool. *14*, 529–539 (1967)

Ascoli, C.: New techniques in photomotion methodology. In: Biophysics of Photoreceptors and Photobehavior of Microorganisms, Colombetti, G. (ed.). Pisa: LitoFelici, 1975, pp. 109–120

Ascoli, C., Barbi, M., Frediani, C., Petracchi, D., Trimarco, C.: Osservazioni sul moto di *Euglena* gracilis mediante diffusione di luce laser. Atti del I. Congresso di Cibernetica. Vol. I. Pisa: LitoFelici, 1971, pp. 296–300

Baltscheffsky, H., de Kiewiet, D.Y.: Existence and localization of two phosphorylation sites in photophosphorylation of isolated spinach chloroplasts. Acta Chem. Scand. *18*, 2406–2408 (1964)

Batra, P.P., Tollin, G.: Phototaxis in *Euglena*. I. Isolation of the eye-spot granules and identification of the eye-spot pigments. Biochim. Biophys. Acta *79*, 371–378 (1964)

Bendix, S.: Phototaxis in the desmid *Micrasterias rotata*. Thesis, Univ. Calif., 1957

Bendix, S.: Pigments in phototaxis. In: Comparative Biochemistry of Photoreactive Systems. Allen, M.B. (ed.). London-New York: Academic Press, 1960a, pp. 107–127

Bendix, S.: Phototaxis. Bot. Rev. *26*, 145–208 (1960b)

Benedetti, P.A., Bianchini, G., Checcucci, A., Ferrara, R., Grassi, S., Percival, D.: Spectroscopic properties and related functions of the stigma measured in living cells of *Euglena gracilis*. Arch. Microbiol. *111*, 73–76 (1976)

Benedetti, P.A., Checcucci, A.: Paraflagellar body (PFB) pigments studied by fluorescence microscopy in *Euglena gracilis*. Plant Sci. Lett. *4*, 47–51 (1975)

Bensasson, R.V.: Spectroscopic and biological properties of carotenoids. In: Biophysics of Photoreceptors and Photobehaviour of Microorganisms. Colombetti, G. (ed.). Pisa: LitoFelici, 1975, pp. 146–163

Bentrup, F.-W.: Lichtabhängige Membranpotentiale bei Pflanzen. Ber. Dtsch. Bot. Ges. *87*, 515–528 (1974)

Blaurock, A.E., Stoeckenius, W.: Structure of the purple membrane. Nature (New Biol.) *233*, 152–155 (1971)

Blinks, L.R.: Chromatic transients in the photosynthesis of green, brown and red algae. In: Comparative Biochemistry of Photoreactive Systems. Allen, M.B. (ed.). New York-London: Academic Press, 1960, pp. 367–376

Bolte, E.: Über die Wirkung von Licht und Kohlensäure auf die Beweglichkeit grüner und farbloser Schwärmzellen. Jb. Wiss. Bot. *59*, 287–324 (1920)

Bonačić-Koutecký, V., Bruckman, P., Hiberty, P., Koutecký, J., Leforestier, C., Salem, L.: Plötzliche Polarisation im zwitterionischen angeregten Zustand Z_1 organischer Zwischenstufen — photochemische Auswirkungen. Angew. Chem. *87*, 599–601 (1975)

Boscov, J.S.: Responses of *Chlamydomonas* to single flashes of light. Thesis, Tufts Univ., 1974

Bound, K.E., Tollin, G.: Phototactic response of *Euglena gracilis* to polarized light. Nature (London) *216*, 1042–1044 (1967)

Bovee, E.C., Jahn, Th.L.: A theory of piezoelectric activity and ion-movements in the relation of flagellar structures and their movements to the phototaxis of *Euglena*. J. Theor. Biol. *35*, 259–276 (1972)

Brinkmann, K.: Temperatureinflüsse auf die circadiane Rhythmik von *Euglena gracilis* bei Mixotrophie und Autotrophie. Planta *70*, 344–389 (1966)

Brinkmann, K.: The influence of alcohols on the circadian rhythm and metabolism of *Euglena gracilis*. J. Interdiscipl. Cycle Res. *7*, 149–170 (1976)

Bruce, V.G.: The biological clock in *Chlamydomonas reinhardi*. J. Protozool. *17*, 328–334 (1970)

Bruce, V.G.: The role of the clock in controlling phototactic rhythms. In: Behaviour of Micro-organisms. Pérez-Miravete, A. (ed.). London-New York: Plenum Press, 1973, pp. 257–266

Bruce, V.G., Pittendrigh, C.S.: Temperature independence in a unicellular "clock". Proc. Natl. Acad. Sci. USA *42*, 676–682 (1956)

Bruce, V.G., Pittendrigh, C.S.: An effect of heavy water on the phase and period of the circadian rhythm in *Euglena*. J. Cell. Comp. Physiol. *56*, 25–31 (1960)

Brucker, W.: Beiträge zur Kenntnis der Phototaxis grüner Schwärmzellen. 1. Die Lichtempfindlichkeit von *Lepocinclis texta* und ihre Abhängigkeit von der Vorbelichtung und vom Kohlensäuregehalt des Mediums. Arch. Protistenk. *99*, 294–327 (1954)

Buder, J.: Zur Kenntnis des *Thiospirillum jenense* und seiner Reaktionen auf Lichtreize. Jb. Wiss. Bot. *56*, 529–584 (1915)
Buder, J.: Zur Kenntnis der phototaktischen Richtungsbewegungen. Jb. Wiss. Bot. *58*, 105–220 (1919)
Buetow, D.E.: The Biology of *Euglena*, Vol. I. General Biology and Ultrastructure. New York-London: Academic Press, 1968
Bünning, E., Gössel, I.: Ergänzende Versuche über die phototaktischen Aktionsspektren von *Euglena*. Arch. Mikrobiol. *32*, 319–321 (1959)
Bünning, E., Schneiderhöhn, G.: Über das Aktionsspektrum der phototaktischen Reaktionen von *Euglena*. Arch. Mikrobiol. *24*, 80–90 (1956)
Burkholder, P.R.: Movement in the Cyanophyceae. Quart. Rev. Biol. *9*, 438–459 (1934)
Callame, B., Debyser, J.: Observations sur les mouvements des diatomées à la surface des sédiments marins de la zone intercotidale. Vie et Milieu *5*, 242–249 (1954)
Checcucci, A.: Photomotion methodology in flagellates. In: Primary Molecular Events in Photobiology. Checcucci, A., Weale, R.A. (eds.). Amsterdam-London-New York: Elsevier, 1973, pp. 217–244
Checcucci, A.: Molecular sensory physiology of *Euglena*. Naturwissenschaften *63*, 412–417 (1976)
Checcucci, A., Colombetti, G., del Carratore, G., Ferrara, R., Lenci, F.: Red light-induced accumulation of *Euglena gracilis*. Photochem. Photobiol. *19*, 223–226 (1974)
Checcucci, A., Colombetti, G., Ferrara, R., Lenci, F.: Action spectra for photoaccumulation of green and colorless *Euglena*: Evidence for identification of receptor pigments. Photochem. Photobiol. *23*, 51–54 (1976)
Checcucci, A., Favati, L., Grassi, S., Piaggesi, T.: The measurement of photoactivity in *Euglena gracilis* Klebs. Monitore Zool. Ital. N. S. *9*, 83–98 (1975)
Clayton, R.K.: Studies in the phototaxis of *Rhodospirillum rubrum*. Arch. Mikrobiol. *19*, 107–165 (1953)
Clayton, R.K.: On the interplay of environmental factors affecting taxis and motility in *Rhodospirillum rubrum*. Arch. Mikrobiol. *29*, 189–212 (1958)
Clayton, R.K.: Phototaxis in purple bacteria. In: Encyclopedia of Plant Physiology. Vol. XVII/1. Ruhland, W. (ed.). Berlin-Heidelberg-New York: Springer 1959, pp. 371–387
Clayton, R.K.: Phototaxis in microorganisms. In: Photophysiology. Vol. II. Giese, A.C. (ed.). New York-London: Academic Press, 1964, pp. 51–77
Colombetti, G., Lenci, F., McKellar, J.F., Phillips, G.O.: Light-induced effects on a flavoprotein, d-amino-acid oxidase. Photochem. Photobiol. *21*, 303–306 (1975)
Cook, J.R., Harris, P., Nachtwey, D.S.: Irreversible plastid loss in *Euglena gracilis* under physiological conditions. Plant. Physiol. *53*, 284–290 (1974)
Creutz, C., Diehn, B.: Motor responses to polarized light and gravity sensing in *Euglena gracilis*. J. Protozool. *23*, 552–556 (1976)
Dangeard, M.P.-A.: Le déterminisme des mouvements chez les organismes inférieurs. Ann. Protistol. (Paris) *1*, 3–10 (1928)
Dangeard, M.P.-A.: Sur la mobilité des certaines cellules du *Porphyridium cruentum* Naegeli. C. R. Ser. D *190*, 819–821 (1930)
Davenport, D.: Studies in microorganismal behaviour by computerized television. In: Behaviour of Micro-organisms. Pérez-Miravete, A. (ed.). London-New York: Plenum Press, 1973, pp. 106–116
Davenport, D., Culler, G.J., Greaves, J.O.B., Forward, R.B., Hand, W.G.: The investigation of the behavior of microorganisms by computerized television. IEEE Transactions Biomedical Engeneering *17*, 230–237 (1970)
Davenport, D., Wright, C.A., Causley, D.: Technique for the study of the behavior of motile microorganisms. Science *135*, 1059–1060 (1962)
Diehn, B.: Action spectra of the phototactic responses in *Euglena*. Biochim. Biophys. Acta *177*, 136–143 (1969a)
Diehn, B.: Two perpendicularly oriented pigment systems involved in phototaxis of *Euglena*. Nature (London) *221*, 366–367 (1969b)
Diehn, B.: Phototactic response of *Euglena* to single and repetitive pulses of actinic light: orientation time and mechanism. Exp. Cell Res. *56*, 375–381 (1969c)
Diehn, B.: Phototaxis in *Euglena*. 1.: Physiological basis of photoreception and tactic orienta-

tion. In: Behaviour of Micro-organisms. Pérez-Miravete, A. (ed.). London-New York: Plenum Press, 1973a, pp. 83–90

Diehn, B.: Phototaxis and sensory transduction in *Euglena*. Science *181*, 1009–1015 (1973b)

Diehn, B., Feinleib, M.E., Haupt, W., Hildebrand, E., Lenci, F., Nultsch, W.: Terminology of behavioral responses of motile microorganisms. Photochem. Photobiol. *26*, 559–560 (1977)

Diehn, B., Fonseca, J.R., Jahn, Th.L.: High speed cinemicrography of the direct photophobic response of *Euglena* and the mechanism of negative phototaxis. J. Protozool. *22*, 492–494 (1975)

Diehn, B., Kint, B.: The flavin nature of the photoreceptor molecule for phototaxis in *Euglena*. Physiol. Chem. and Physics. *2*, 483–488 (1970)

Diehn, B., Tollin, G.: Phototaxis in *Euglena*. IV. Effect of inhibitors of oxidative and photophosphorylation on the rate of phototaxis. Arch. Biochem. Biophys. *121*, 169–177 (1967)

Dodge, J.D.: A review of the fine structure of algal eyespots. Brit. Phycol. J. *4*, 199–210 (1969)

Dodge, J.D.: The fine structure of algal cells. London-New York: Academic Press 1973

Dodge, J.D., Crawford, R.M.: Observations on the fine structure of the eyespot and associated organelles in the dinoflagellate *Glenodinium foliaceum*. J. Cell Sci. *5*, 479–493 (1969)

Doetsch, R.N.: A unified theory of bacterial motile behavior. J. Theor. Biol. *35*, 55–66 (1972)

Dolzmann, R., Dolzmann, P.: Untersuchungen über die Feinstruktur und die Funktion der Plasmodesmen von *Volvox aureus*. Planta *61*, 332–345 (1964)

Drachev, L.A., Kaulen, A.D., Ostroumov, S.A., Skulachev, V.P.: Electrogenesis by bacteriorhodopsin incorporated in a planar phospholipid membrane. FEBS Lett. *39*, 43–45 (1974)

Drews, G.: Die phototaktischen Reaktionen einiger Cyanophyceen. Ber. Dtsch. Bot. Ges. *70*, 259–262 (1957)

Drews, G.: Beiträge zur Kenntnis der phototaktischen Reaktionen der Cyanophyceen. Arch. Protistenk. *104*, 389–430 (1959)

Duysens, L.N.M.: Transfer of excitation energy in photosynthesis. Diss. Univ. Utrecht 1952

Eckert, R.: Bioelectric control of ciliary activity. Science *176*, 473–481 (1972)

Engelmann, Th.W.: *Bacterium photometricum*. Ein Beitrag zur vergleichenden Physiologie des Licht- und Farbensinnes. Pflügers Arch. Ges. Physiol. *30*, 95–124 (1883)

Famintzin, A.: Die Wirkung des Lichtes auf Algen und einige andere ihnen nahe verwandte Organismen. Jb. Wiss. Bot. *6*, 1–48 (1867)

Feinleib, M.E.: Phototactic response of Chlamydomonas to flashes of light. I. Response of cell populations. Photochem. Photobiol. *21*, 351–354 (1975a)

Feinleib, M.E.: Behaviour models in flagellates. In: Biophysics of Photoreceptors and Photobehaviour of Microorganisms. Colombetti, G. (ed.). Pisa: LitoFelici, 1975b, pp. 68–84

Feinleib, M.E., Curry, G.M.: Methods for measuring phototaxis of cell populations and individual cells. Physiol. Plant. *20*, 1083–1095 (1967)

Feinleib, M.E., Curry, G.M.: The nature of the photoreceptor in phototaxis. In: Handbook of Sensory Physiology. Vol. I. Receptor Mechanisms. Berlin-Heidelberg-New York: Springer, 1971a, pp. 366–395

Feinleib, M.E., Curry, G.M.: The relationship between stimulus intensity and oriented phototaxis response (topotaxis) in *Chlamydomonas*. Physiol. Plant. *25*, 346–352 (1971b)

Feldman, J.F.: Lengthening the period of a biological clock in *Euglena* by cycloheximide, an inhibitor of protein synthesis. Proc. Natl. Acad. Sci. USA *57*, 1080–1087 (1967)

Felle, H., Bentrup, F.-W.: Lichtabhängigkeit des Membranpotentials der Rhizoidzelle von *Riccia fluitans*: Versuche zur Signalübertragung. Ber. Dtsch. Bot. Ges. *87*, 223–228 (1974)

Ferrara, R.: General review on phototactic action spectra. In: Biophysics of Photoreceptors and Photobehaviour of Microorganisms. Colombetti, G. (ed.). Pisa: LitoFelici, 1975, pp. 121–145

Forward, R.B. jr.: Change in the photoresponse action spectrum of the dinoflagellate *Gyrodinium dorsum* Kofoid by red and far-red light. Planta *92*, 248–258 (1970)

Forward, R.B. jr.: Phototaxis in a dinoflagellate: action spectra as evidence for a two-pigment system. Planta *111*, 167–178 (1973)

Forward, R.B. jr.: Phototaxis by the dinoflagellate *Gymnodinium splendens* Lebour. J. Protozool. *21*, 312–315 (1974)

Forward, R.B. jr., Davenport, D.: Red and far-red light effects on a short term behavioral response in a dinoflagellate. Science *161*, 1028–1029 (1968)
Forward, R.B. jr., Davenport, D.: The circadian rhythm of a behavioral photoresponse in the dinoflagellate *Gyrodinium dorsum*. Planta *92*, 259–266 (1970)
Francis, D.: On the eyespot of the dinoflagellate, *Nematodinium*. J. Exp. Biol. *47*, 495–501 (1967)
Fraenkel, G.S., Gunn, D.L.: The Orientation of Animals. New York: Dover, 1961
Froehlich, O., Diehn, B.: Photoeffects in a flavin-containing lipid bilayer membrane and implications for algal phototaxis. Nature (London) *248*, 802–804 (1974)
Geitler, L.: Das Bewegungsverhalten von *Porphyridium cruentum* am natürlichen Standort. Öster. Bot. Z. *100*, 672–676 (1953)
Gerisch, G.: Die Zelldifferenzierung bei *Pleodorina californica* Shaw und die Organisation der Phytomonadinenkolonien. Arch. Protistenk. *104*, 292–358 (1959)
Gibbons, B.H., Gibbons, I.R.: Flagellar movement and adenosine triphosphatase activity in sea urchin sperm extracted with triton X-100. J. Cell Biol. *54*, 75–97 (1972)
Gössel, I.: Über das Aktionsspektrum der Phototaxis chlorophyllfreier Euglenen und über die Absorption des Augenflecks. Arch. Mikrobiol. *27*, 288–305 (1957)
Gray, B.H., Gantt, E.: Spectral properties of phycobilisomes and phycobiliproteins from the blue-green alga-*Nostoc* sp.. Photochem. Photobiol. *21*, 121–128 (1975)
Greuet, C.: Organisation ultrastucturale de l'ocelle de deux péridiniens Warnowiidae, *Erythropsis pavillardi* Kofoid et Swezy et *Warnowia pulchra* Schiller. Protistologica *4*, 209–230 (1968)
Häder, D.-P.: Untersuchungen zur Photo-phobotaxis bei *Phormidium uncinatum*. Diss. Univ. Marburg 1973
Häder, D.-P.: Participation of two photosystems in the photo-phobotaxis of *Phormidium uncinatum*. Arch. Microbiol. *96*, 255–266 (1974)
Häder, D.-P.: The effect of inhibitors on the electron flow triggering photo-phobic reactions in Cyanophyceae. Arch. Microbiol. *103*, 169–174 (1975)
Häder, D.-P.: Phobic reactions between two adjacent monochromatic light fields. Z. Pflanzenphysiol. *78*, 173–176 (1976a)
Häder, D.-P.: Further evidence for the electron pool hypothesis. The effect of KCN and DSPD on the photophobic reaction in the filamentous blue-green alga *Phormidium uncinatum*. Arch. Microbiol. *110*, 301–303 (1976b)
Häder, D.-P.: Speculations about sensory transduction. Proc. 7 th Int. Congr. Photobiol. Rome, Aug. 29th–Sept. 3rd (1976c)
Häder, D.-P.: Influence of electric fields on photophobic reactions in blue-green algae. Arch. Microbiol. *114*, 83–86 (1977)
Häder, D.-P., Nultsch, W.: Untersuchungen zur Abgrenzung der Arten *Phormidium uncinatum* und *Phormidium autumnale*. Schweiz. Z. Hydrol. *33*, 566–577 (1971)
Häder, D.-P., Nultsch, W.: Negative photo-phobotactic reactions in *Phormidium uncinatum*. Photochem. Photobiol. *18*, 311–317 (1973)
Häder, D.-P., Wenderoth, K.: Role of three basic light reactions in photomovement of desmids. Planta, *137*, 207–214 (1977)
Halldal, P.: Factors affecting light response in phototactic algae. Physiol. Plant. *12*, 742–752 (1959)
Halldal, P.: Action spectra of induced phototactic response changes in *Platymonas*. Physiol. Plant. *13*, 726–735 (1960)
Halldal, P.: Zur Frage des Photoreceptors bei der Topophototaxis der Flagellaten. Ber. Dtsch. Bot. Ges. *76*, 323–327 (1963)
Halldal, P.: Phototaxis in protozoa. In: Biochemistry and Physiology of Protozoa. Vol. III. Hutner, S.H. (ed.). New York-London: Academic Press 1964, pp. 277–296
Hand, W.G.: Phototactic orientation by the marine dinoflagellate *Gyrodinium dorsum* Kofoid. I. A mechanism model. J. Exp. Zool. *174*, 33–38 (1970)
Hand, W.G., Davenport, D.: The experimental analysis of phototaxis and photokinesis in flagellates. In: Photobiology of Microorganisms. Halldal, P. (ed.). London-New York-Sidney-Toronto: Wiley, 1970, pp. 253–282
Hand, W.G., Forward, R., Davenport, D.: Short-term photic regulation of a receptor mechanism in a dinoflagellate. Biol. Bull. *133*, 150–165 (1967)

Hand, W.G., Haupt, W.: Flagellar activity of the colony members of *Volvox aureus* Ehrbg. during light stimulation. J. Protozool. *18*, 361–364 (1971)
Hand, W.G., Schmidt, J.A.: Phototactic orientation by the marine dinoflagellate *Gyrodinium dorsum* Kofoid. II. Flagellar activity and overall response mechanism. J. Protozool. *22*, 494–498 (1975)
Harayama, S., Iino, T.: Phototactic response of aerobically cultivated *Rhodospirillum rubrum*. J. Gen. Microbiol. *94*, 173–179 (1976)
Hartshorne, J.N.: The function of the eyespot in Chlamydomonas. New Phytol. *52*, 292–297 (1953)
Hasle, R.G.: Phototactic vertical migration in marine dinoflagellates. Oikos *2*, 162–175 (1950)
Haupt, W.: Die Phototaxis der Algen. In: Encyclopedia of Plant Physiology. Vol. XVII/1. Ruhland, W. (ed.). Berlin-Heidelberg-New York: Springer, 1959, pp. 318–370
Haupt, W.: Perception of environmental stimuli orienting growth and movement in lower plants. Annu. Rev. Plant Physiol. *16*, 267–290 (1965)
Haupt, W.: Phototaxis in plants. Int. Rev. Cytol. *19*, 267–299 (1966)
Haupt, W.: Bewegungen. Fortschr. Bot. *34*, 227–235 (1972)
Haupt, W.: Locomotion. Fortschr. Bot. *37*, 177–185 (1974)
Haxo, F.T.: Photosynthesis in algae containing special pigments. In: Encyclopedia of Plant Physiology. Vol. V/2. Ruhland, W. (ed.). Berlin-Heidelberg-New York: Springer, 1960, pp. 349–363
Heidingsfeld, I.: Phototaktische Untersuchungen bei Diatomeen. Diss. Univ. Breslau 1940
Hiatt, J.D.F., Hand, W.G.: Do protoplasmic connections function in the phototactic coordination of the *Volvox* colony during light stimulation? J. Protozool. *19*, 488–489 (1972)
Hildebrand, E.: Effects of some inhibitors in *Euglena* photomotion. In: Progress in Photobiology, Proc. 6th. Int. Congr. Photobiol. Schenk, G. (ed.). Frankfurt: Dtsch. Ges. Lichtforsch. e.V. 1974, p. 358
Hildebrand, E.: What does *Halobacterium* tell us about photoreception? Biophys. Struct. Mechanism. *3*, 69–77 (1977)
Hildebrand, E., Dencher, N.: Photophobische Reaktionen von *Halobacterium halobium* und ihre Beziehung zum Retinal-Protein-Komplex der Zellmembran. Ber. Dtsch. Bot. Ges. *87*, 93–99 (1974)
Hildebrand, E., Dencher, N.: Two photosystems controlling behavioural responses of *Halobacterium halobium*. Nature (London) *257*, 46–48 (1975)
Hirschberg, R., Stavis, R.: Phototaxis mutants of *Chlamydomonas reinhardtii*. J. Bacteriol. *129*, 803–808 (1977)
Horio, T., Nishikawa, K., Katsumata, M., Yamashita, I.: Possible partial reactions of the phosphorylation process in chromatophores from *Rhodospirillum rubrum*. Biochim. Biophys. Acta *94*, 371–382 (1965)
Hudock, G.A., Hudock, M.O.: Phototaxis: Isolation of mutant strains of *Chlamydomonas reinhardi* with reversed sign of response. J. Protozool. *20*, 139–140 (1973)
Huth, K.: Bewegung und Orientierung bei *Volvox aureus* Ehrb.. I. Mechanismus der phototaktischen Reaktion. Z. Pflanzenphysiol. *62*, 436–450 (1970a)
Huth, K.: Bewegung und Orientierung bei *Volvox aureus* Ehrb.. II. Richtungsabweichung bei taktischen Reaktionen. Z. Pflanzenphysiol. *63*, 344–351 (1970b)
Ikushima, N., Maruyama, S.: The protoplasmic connections in *Volvox*. J. Protozool. *15*, 136–140 (1968)
Iwasa, K., Shimizu, A.: Motility of the diatom *Phaeodactylum tricornutum*. Exp. Cell Res. *74*, 552–558 (1972)
Kazama, F.Y.: Ultrastructure and phototaxis of the zoospores of *Phlyctochytrium* sp., an Estuarine Chytrid. J. Gen. Microbiol. *71*, 555–566 (1972)
Kennedy, J.S.: Classification and nomenclature of animal behaviour. Nature (London) *156*, 754 (1945)
Kivic, P.A., Vesk, M.: Structure and function in the euglenoid eyespot apparatus: the fine structure, and response to environmental changes. Planta *105*, 1–14 (1972a)
Kivic, P.A., Vesk, M.: Structure and function of the euglenoid eyespot. The probable location of the phototaxis photoreceptor. J. Exp. Bot. *23*, 1070–1075 (1972b)
Kivic, P.A., Vesk, M.: The structure of the eyespot apparatus in bleached strains of *Euglena gracilis*. Cytobiologie *10*, 88–101 (1974)

Kristiansen, J., Walne, P.L.: Structural connections between flagellar base and stigma in *Dinobryon*. Protoplasma *89*, 371–374 (1976)
Kronestedt, E., Walles, B.: On the presence of plastids and the eyespot apparatus in a porfiromycin-bleached strain of *Euglena gracilis*. Protoplasma *84*, 75–82 (1975)
Kung, Ch.: Genic mutants with altered system of excitation in *Paramecium aurelia*. I. Phenotypes of the behavioral mutants. Z. Vergl. Physiol. *71*, 142–164 (1971)
Lang, N.J.: The fine structure of blue-green algae. Annu. Rev. Microbiol. *22*, 15–46 (1968)
Lee, A.G., Fitzsimons, J.T.R.: Motility in normal and filamentous forms of *Rhodospirillum rubrum*. J. Gen. Microbiol. *93*, 346–354 (1976)
Leedale, G.F.: Euglenida/Euglenophyta. Annu. Rev. Microbiol. *21*, 31–48 (1967)
Lenci, F.: Photochemistry of flavins. In: Biophysics of Photoreceptors and Photobehaviour of Microorganisms. Colombetti, G. (ed.) Pisa: LitoFelici, 1975, pp. 164–183
Levandowsky, M., Childress, W.S., Spiegel, E.A., Hutner, S.H.: A mathematical model of pattern formation by swimming microorganisms. J. Protozool. *22*, 296–306 (1975)
Lindes, D.A., Diehn, B., Tollin, G.: Phototaxigraph: recording instrument for determination of rate of response of phototactic microorganisms to light of controlled intensity and wavelength. Rev. Sci. Instr. *36*, 1721–1725 (1965)
Links, J.: I. Een hypothese over het mechanisme van de (phobo-) chemotaxis. II. De carotenoiden, steroiden en vetzuren van *Polytoma uvella*. Diss. Univ. Leiden 1955
Macnab, R., Koshland, D.E. jr.: Bacterial motility and chemotaxis: Light-induced tumbling response and visualization of individual flagella. J. Mol. Biol. *84*, 399–406 (1974)
Manten, A.: Phototaxis, phototropism and photosynthesis in purple bacteria and blue-green algae. Diss. Univ. Utrecht 1948
Marbach, I., Mayer, A.M.: Direction of phototaxis in *Chlamydomonas reinhardii* and its relation to cell metabolism. Phycologia *9*, 255–260 (1970)
Marbach, I., Mayer, A.M.: Effect of electric field on the phototactic response of *Chlamydomonas reinhardii*. Israel J. Bot. *20*, 96–100 (1971)
Mast, S.O.: Light and Behavior of Organisms. New York: Wiley, 1911
McKellar, J.F., Phillips, G.O., Checcucci, A., Lenci, F.: Excited states of flavins in photoreception processes. In: Excited States of Biological Molecules. Birks, J.B. (ed.). New York: Wiley, 1975
Messer, G., Ben-Shaul, Y.: Fine structure of *Peridinium westii* Lemm., a freshwater dinoflagellate. J. Protozool. *16*, 272–280 (1969)
Mikolajczyk, E., Diehn, G.: The effect of potassium iodide on photophobic responses in *Euglena:* Evidence for two photoreceptor pigments. Photochem. Photobiol. *22*, 269–271 (1975)
Mikolajczyk, E., Diehn, G.: Light-induced body movement of *Euglena gracilis* coupled to flagellar photophobic responses by mechanical stimulation. J. Protozool. *23*, 144–147 (1976)
Milatz, J.M.W., Manten, A.: The quantitative determination of the spectral distribution of phototactic sensitivity in the purple bacterium *Rhodospirillum rubrum*. Biochim. Biophys. Acta *11*, 17–27 (1953)
Mitchell, P.: Protonmotive redox mechanism of the cytochrome b-c_1 complex in the respiratory chain: protonmotive ubiquinone cycle. FEBS Lett. *56*, 1–6 (1975)
Neuscheler, W.: Bewegung und Orientierung bei *Micrasterias denticulata* Bréb. im Licht. I. Zur Bewegungs- und Orientierungsweise. Z. Pflanzenphysiol. *57*, 46–59 (1967a)
Neuscheler, W.: Bewegung und Orientierung bei *Micrasterias denticulata* Brèb. im Licht. II. Photokinesis und Phototaxis. Z. Pflanzenphysiol. *57*, 151–172 (1967b)
Nichols, K.M., Rikmenspoel, R.: Mg^{2+}-dependent electrical control of flagellar activity in *Euglena*. J. Cell. Sci. *23*, 211–225 (1977)
Nultsch, W.: Studien über die Phototaxis der Diatomeen. Arch. Protistenk. *101*, 1–68 (1956)
Nultsch, W.: Die Bewegung der Diatomeen. Mikrokosmos *46*, 220–227 (1957)
Nultsch, W.: Der Einfluß des Lichtes auf die Bewegung der Cyanophyceen. I. Phototopotaxis von *Phormidium autumnale*. Planta *56*, 632–647 (1961)
Nultsch, W.: Über das Bewegungsverhalten der Diatomeen. Planta *58*, 22–33 (1962a)
Nultsch, W.: Phototaktische Aktionsspektren von Cyanophyceen. Ber. Dtsch. Bot. Ges. *75*, 443–453 (1962b)
Nultsch, W.: Der Einfluß des Lichtes auf die Bewegung der Cyanophyceen. II. Photokinesis bei *Phormidium autumnale*. Planta *57*, 613–623 (1962c)

Nultsch, W.: Der Einfluß des Lichtes auf die Bewegung der Cyanophyceen. III. Photophobotaxis von *Phormidium uncinatum*. Planta *58*, 647–663 (1962d)
Nultsch, W.: Über den Antagonismus von Atebrin und Flavinnucleotiden im Bewegungs- und Lichtreaktionsverhalten von *Phormidium uncinatum*. Arch. Mikrobiol. *55*, 187–199 (1966)
Nultsch, W.: Untersuchungen über den Einfluß von Entkopplern auf die Bewegungsaktivität und das phototaktische Reaktionsverhalten blaugrüner Algen. Z. Pflanzenphysiol. *56*, 1–11 (1967)
Nultsch, W.: Einfluß von Redox-Systemen auf die Bewegungsaktivität und das phototaktische Reaktionsverhalten von *Phormidium uncinatum*. Arch. Mikrobiol. *63*, 295–320 (1968)
Nultsch, W.: Photomotion of microorganisms and its interaction with photosynthesis. In: Photobiology of Microorganisms. Halldal, P. (ed.). London-New York-Sidney-Toronto: Wiley, 1970, pp. 213–251
Nultsch, W.: Phototactic and photokinetic action spectra of the diatom *Nitzschia communis*. Photochem. Photobiol. *14*, 705–712 (1971)
Nultsch, W.: Relation between photomotion and photosynthesis. In: Primary Molecular Events in Photobiology. Checcucci, A., Weale, R.M. (eds.). Amsterdam-London-New York: Elsevier 1973, pp. 245–273
Nultsch, W.: Der Einfluß des Lichtes auf die Bewegung phototropher Microorganismen. I. Photokinesis. Abhandlungen der Marburger Gelehrten Gesellschaft. Jg. 1972 Nr. 2. München: Fink, 1974a, pp. 143–213
Nultsch, W.: Movements. In: Algal Physiology and Biochemistry. Stewart, W.D.P. (ed.). Oxford-London-Edinburgh-Melbourne: Blackwell Scientific Publications, 1974b, pp. 864–893
Nultsch, W.: Phototaxis and photokinesis. In: Primitive Sensory and Communication Systems. Carlile, M.J. (ed.). London-New York-San Francisco: Academic Press 1975, pp. 29–90
Nultsch, W.: Effect of external factors on phototaxis of *Chlamydomonas reinhardtii*. II. Carbon dioxide, oxygen and pH. Arch. Microbiol. *112*, 179–185 (1977)
Nultsch, W., Häder, D.-P.: Bestimmung der photo-phobotaktischen Unterschiedsschwelle bei *Phormidium uncinatum*. Ber. Dtsch. Bot. Ges. *83*, 185–192 (1970)
Nultsch, W., Häder, D.-P.: Über die Rolle der beiden Photosysteme in der Photo-phobotaxis von *Phormidium uncinatum*. Ber. Dtsch. Bot. Ges. *87*, 83–92 (1974)
Nultsch, W., Hellmann, W.: Untersuchungen zur Photokinesis von *Anabaena variabilis* Kützing. Arch. Mikrobiol. *82*, 76–90 (1972)
Nultsch, W., Jeeji-Bai: Untersuchungen über den Einfluß von Photosynthese-Hemmstoffen auf das phototaktische und photokinetische Reaktionsverhalten blaugrüner Algen. Z. Pflanzenphysiol. *54*, 84–98 (1966)
Nultsch, W., Richter, G.: Aktionsspektrum des photosynthetischen $^{14}CO_2$-Einbaus von *Phormidium uncinatum*. Arch. Mikrobiol. *47*, 207–213 (1963)
Nultsch, W., Throm, G.: Effect of external factors on phototaxis of *Chlamydomonas reinhardtii*. Arch. Microbiol. *103*, 175–179 (1975)
Nultsch, W., Throm, G., v. Rimscha, I.: Phototaktische Untersuchungen an *Chlamydomonas reinhardii* Dangeard in homokontinuierlicher Kultur. Arch. Mikrobiol. *80*, 351–369 (1971)
Nultsch, W., Wenderoth, K.: Phototaktische Untersuchungen an einzelnen Zellen von *Navicula peregrina* (Ehrenberg) Kützing. Arch. Mikrobiol. *90*, 47–58 (1973)
Oesterhelt, D., Meentzen, M., Schuhmann, L.: Reversible dissociation of the purple complex in bacteriorhodopsin and identification of 13-cis and all-trans-retinal as its chromophores. Eur. J. Biochem. *40*, 453–463 (1973)
Oesterhelt, D., Stoeckenius, W.: Rhodopsin-like protein from the purple membrane of *Halobacterium halobium*. Nature (New Biol.) *233*, 149–152 (1971)
Oesterhelt, D., Stoeckenius, W.: Functions of a new photoreceptor membrane. Proc. Natl. Acad. Sci. USA. *70*, 2853–2857 (1973)
Ojakian, G.K., Katz, D.F.: A simple technique for the measurement of swimming speed of *Chlamydomonas*. Exp. Cell Res. *81*, 487–491 (1973)
Oltmanns, F.: Über Phototaxis. Z. Bot. *9*, 257–338 (1917)
Omodeo, P.: Morphology of the phototactic apparatus in eucariotic flagellated cells. In: Biophysics of Photoreceptors and Photobehaviour of Microorganisms. Colombetti, G. (ed.) Pisa: LitoFelici, 1975, pp. 24–48
Pecci, J., Fujimori, E.: Spectral change of phycoerythrin from *Hydrocoleum* species and

its relationship to protein dissociation. Effect of mercurials on single- and double-peaked forms. Biochim. Biophys. Acta *154*, 332–341 (1968)
Pfeffer, W.: Pflanzenphysiologie. Leipzig 1904
Phillips, D.M.: Comparative analysis of mammalian sperm motility. J. Cell Biol. *53*, 561–573 (1972)
Piccinni, E., Albergoni, V., Coppellotti, O.: ATPase activity in flagella from *Euglena gracilis*. Localization of the enzyme and effects of detergents. J. Protozool. *22*, 331–335 (1975)
Piccinni, E., Omodeo, P.: Photoreceptors and phototactic programs in protista. Boll. Zool. *42*, 57–79 (1975)
Pringsheim, E.G.: The loss of chromatophores in *Euglena gracilis*. New Phytol. *47*, 52–87 (1948)
Pringsheim, E.G.: Kleine Mitteilungen über Flagellaten und Algen. XV. Zur Kenntnis der Gattung *Porphyridium* (Rhodophyceae). Arch. Mikrobiol. *61*, 169–180 (1968)
Pringsheim, E.G., Pringsheim, O.: The growth requirements of *Porphyridium cruentum*: with remarks on the biology of brackish water algae. J. Ecol. *37*, 57–64 (1949)
Pritchard, A.: A History of Infusoria. London: Whittaker, 1861
Renger, G.: Reaction of CCCP in photosynthesis on an intermediate between chlorophyll-a_{II} and water. Naturwissenschaften *56*, 370 (1969)
Robertson, J.A.: Phototaxis in a new allomyces. Arch. Mikrobiol. *85*, 259–266 (1972)
Sachs, T., Mayer, A.M.: Studies on the relation between metabolism and phototaxis of *Chlamydomonas snowiae*. Phycologia *1*, 149–159 (1961)
Schletz, K.: Phototaxis bei *Volvox* – Pigmentsysteme der Lichtrichtungsperzeption. Z. Pflanzenphysiol. *77*, 189–211 (1976)
Schmid, G.: Das Reizverhalten künstlicher Teilstücke, die Kontraktilität und das osmotische Verhalten der *Oscillatoria jenensis*. Jb. Wiss. Bot. *62*, 328–419 (1923)
Schmidt, J.A., Creutz, C., Diehn, B.: Factors controlling the expression of phototaxis in *Euglena*. In: Progress in Photobiology, Proceedings of the VI. Int. Congr. Photobiol. Schenck, G. (ed.). Frankfurt: Dtsch. Ges. Lichtforsch. e.V. 1974, p. 359
Schneider, W.R., Doetsch, R.N.: Velocity measurements of motile bacteria by use of a videotape recording technique. Appl. Microbiol. *27*, 283–284 (1974)
Schuchart, H.: Phototaxis methodology. In: Biophysics of Photoreceptors and Photobehaviour of Microorganisms. Colombetti, G. (ed.). Pisa: LitoFelici, 1975, pp. 85–108
Seymour, F.W.K., Doetsch, R.N.: Chemotactic responses by motile bacteria. J. Gen. Microbiol. *78*, 287–296 (1973)
Siggel, V.: The function of plastoquinone as electron and proton carrier in photosynthesis. Bioelectrochem. Bioenergetics *3*, 302–318 (1976)
Smyth, R.D., Ebersold, W.T.: A *Chlamydomonas* mutant with altered phototactic response. Genetics *64*, 62 (1970)
Song, P.-S., Moore, T.A.: On the photoreceptor pigment for phototropism and phototaxis: is a carotenoid the most likely candidate? Photochem. Photobiol. *19*, 435–441 (1974)
Sperling Pagni, P.G., Walne, P.L., Wehry, E.L.: Fluorometric evidence for flavins in isolated eyespots of *Euglena gracilis* var. *bacillaris*. Photochem. Photobiol. *24*, 373–375 (1976)
Stanier, R.Y., Kunisawa, R., Mandel, M., Cohen-Bazire, G.: Purification and properties of unicellular blue-green algae (order Chroococcales). Bacteriol. Rev. *35*, 171–205 (1971)
Stavis, R.L.: The effect of azide on phototaxis in *Chlamydomonas reinhardi*. Proc. Natl. Acad. Sci. USA *71*, 1824–1827 (1974)
Stavis, R.L., Hirschberg, R.: Phototaxis in *Chlamydomonas reinhardtii*. J. Cell Biol. *59*, 367–377 (1973)
Strasburger, E.: Wirkung des Lichtes und der Wärme auf Schwärmsporen. Jena 1878
Strother, G.K., Wolken, J.J.: Microspectrophotometry of *Euglena*. Chloroplast and eyespot. Nature (London) *188*, 601–602 (1960)
Strother, G.K., Wolken, J.J.: In vivo absorption spectra of *Euglena*: chloroplast and eyespot. J. Protozool. *8*, 261–265 (1961)
Sundberg, I., Tillberg, J.-E., Björklund, G.: Effects of 2,5-dibromo-3-methyl-6-isopropyl-p-benzoquinone (DBMIB) on light-induced and glycolytic phosphate uptake in *Scenedesmus*. Physiol. Plant. *38*, 141–146 (1976)
Thomas, J.B.: On the rôle of the carotenoids in photosynthesis in *Rhodospirillum rubrum*. Biochim. Biophys. Acta *5*, 186–196 (1950)

Throm, G.: Untersuchungen zum Reaktionsmechanismus von Phobotaxis und Kinesis an *Rhodospirillum rubrum*. Arch. Protistenk. *110*, 313–371 (1968)
Throm, G.: Die lichtabhängige Änderung des Membranpotentials bei *Griffithsia setacea*. Z. Pflanzenphysiol. *63*, 162–180 (1970)
Tollin, G., Robinson, M.J.: Phototaxis in *Euglena*. V. Photosupression of phototactic activity by blue light. Photochem. Photobiol. *9*, 411–416 (1969)
Trebst, A., Burba, M.: Über die Hemmung photosynthetischer Reaktionen in isolierten Chloroplasten und in *Chlorella* durch Disalicylidenpropandiamin. Z. Pflanzenphysiol. *57*, 419–433 (1967)
Trebst, A., Harth, E., Draber, W.: On a new inhibitor of photosynthetic electron-transport in isolated chloroplasts. Z. Naturforsch. *25b*, 1157–1159 (1970)
Tsujimoto, H.Y., McSwain, B.D., Arnon, D.I.: Differential effects of desaspidin on photosynthetic phosphorylation. Plant Physiol. *41*, 1376–1380 (1966)
Vavra, J.: Instability of the stigma in apochlorotic *Euglena gracilis* var. *bacillaris*. J. Protozool. *9*, Suppl., 28–29 (1962)
Walne, P.L., Arnott, H.J.: The comparative ultrastructure and possible function of eyespots: *Euglena granulata* and *Chlamydomonas eugametos*. Planta *77*, 325–353 (1967)
Warden, J.T.: Flash photolysis-electron spin resonance studies of photosystem I. A fast reduction component of $P–700^+$. Biochim. Biophys. Acta *440*, 89–97 (1976)
Watanabe, M., Furuya, M.: Action spectrum of phototaxis in a cryptomonad alga, *Cryptomonas sp*. Plant Cell Physiol. *15*, 413–420 (1974)
Watanabe, M., Miyoshi, Y., Furuya, M.: Phototaxis in *Cryptomonas* sp. under conditions suppressing photosynthesis. Plant Cell Physiol. *17*, 683–690 (1976)
Wenderoth, K.: Untersuchungen der photo-phobotaktischen Reaktionen einzelner Diatomeenzellen. Diss. Univ. Marburg 1975
Wolk, C.P.: Physiology and cytological chemistry of blue-green algae. Bacteriol. Rev. *37*, 32–101 (1973)
Wolken, J.J.: Photoreceptors: Comparative Studies. In: Comparative Biochemistry of Photoreactive Systems. Allen, M.B. (ed.). New York-London: Academic Press 1960, pp. 145–167
Wolken, J.J.: *Euglena* – An Experimental Organism for Biochemical and Biophysical Studies. Pittsburgh: Rutgers, The State University, 1961
Wolken, J.J., Shin, E.: Photomotion in *Euglena gracilis*. I. Photokinesis, II. Phototaxis. J. Protozool. *5*, 39–46 (1958)
Zimmerman, R.L.: Induced piezoelectricity in isotropic biomaterial. Biophys. J. *16*, 1341–1348 (1976)

3.2.2 Chemotaxis in Bacteria

R.M. MACNAB

For reviews of bacterial chemotaxis, see e.g., WEIBULL (1960), ZIEGLER (1962), ADLER (1975, 1976), BERG (1975), PARKINSON (1975), KOSHLAND (1977a, b), HAZELBAUER and PARKINSON (1977), PARKINSON (1977), MACNAB (1978). For reviews of phototaxis, see CLAYTON (1964), NULTSCH (1970), and HÄDER (Chap. 3.2.1, this vol.).

1. Introduction[1]

Motility in bacteria is very common and it is probably safe to say that, wherever motility exists, it is capable of modification by external stimuli to give net migration (various illustrations are given in Fig. 1) to environments which are more beneficial (elevated nutrient concentrations) or less harmful (reduced concentration of toxic materials). Chemotactic responses have been demonstrated in a wide range of genera (WEIBULL, 1960).

Responses to oxygen and food particles were noted from the beginnings of microscopy and, along with photoresponses, were studied in some detail in the late nineteenth and early twentieth centuries by ENGELMANN, PFEFFER, ROTHERT, and others. PFEFFER in 1884 developed a rough assay where a capillary tube was filled with an attractant such as meat extract, and the tube was then immersed in a bacterial suspension. Visual estimates of the accumulation of cells at the capillary mouth (Fig. 1a) provided a measure of the potency of the stimulus. He was able to show (PFEFFER, 1888) that the cells utilized a concentration-ratio sensing mechanism (see Sect. 5.3) and that there was considerable specificity in the attractant/cell interaction. After a long hiatus, a great resurgence of interest in bacterial sensing of chemicals was initiated by ADLER (1969) and his co-workers, when they demonstrated in a detailed way that there were specific chemoreceptors for a variety of chemicals, and that the response is *not* dependent on transport or metabolism of the chemicals. In the years since, ADLER's efforts have been joined by those of a number of other laboratories, and much new fundamental knowledge has been obtained.

Following the example of PFEFFER and others, and in keeping with current practice, we use here the term *chemotaxis* to refer to migration in chemical gradients, without connoting a particular mechanism. For those who are accustomed to a more rigid terminology, the bacterial response of *Escherichia coli* and *Salmonella typhimurium* can be denoted as *chemoklinokinesis with adaptation* (see below).

---

[1] The reader may find it helpful to refer from time to time to the schematic diagram of bacterial chemotaxis in Figure 6.

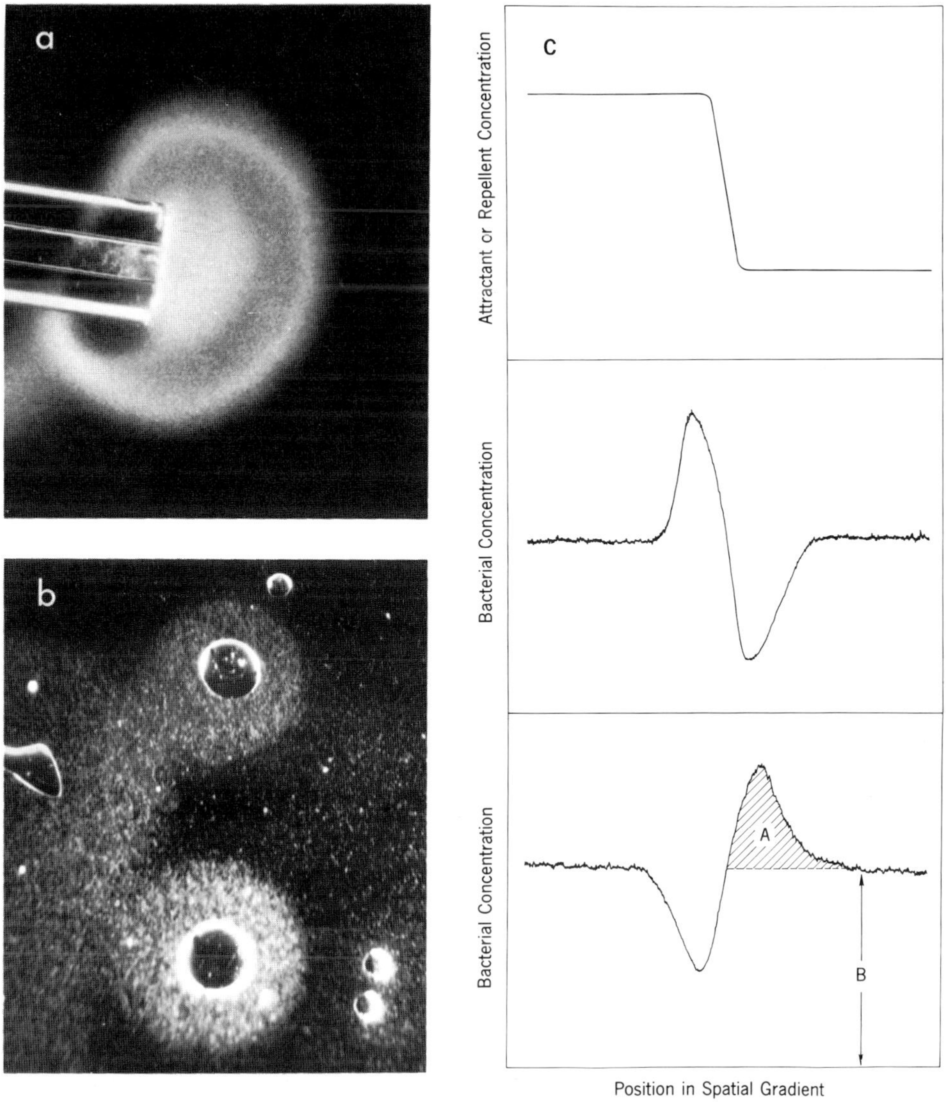

Fig. 1. Tactic responses in a peritrichous bacterium, *Salmonella typhimurium*. **a** accumulation of cells around the mouth of a capillary containing an attractant, L-serine, at 10^{-2} M; **b** aerotactic bands of cells around air bubbles trapped under a microscope coverslide; **c** transfer of cells across an essentially discontinuous change in concentration of a chemotactic agent (*above*). If the agent is an attractant, such as serine (10^{-3} M/10^{-4} M) the transfer is to the zone of higher concentration (*center*); if a repellent, such as phenol (10^{-4} M/10^{-5} M), the transfer is to the lower concentration (*below*). Cell concentrations were measured by light scattering; the response can be quantified by normalizing the transfer area A to the mean signal height B. (**c** from TSANG et al., 1973)

Responses to oxygen, magnesium ion, and pH may be mechanistically distinct from responses to larger molecules and therefore are considered separately. Before considering chemoresponses in detail, a brief survey of responses to a variety of physical stimuli will also be given.

2. Survey of Bacterial Motility Responses to Environmental Stimuli

2.1 Responses to Chemical Stimuli

Chemotactic responses have been demonstrated in a wide variety of genera, such as *Bacillus* (ORDAL and GOLDMAN, 1976; ORDAL and GIBSON, 1977), *Bdellovibrio* (STRALEY and CONTI, 1974), *Clostridium* (ROTHERT, 1904), *Escherichia* (see below), *Proteus, Pseudomonas,* and *Spirillum* (SEYMOUR and DOETSCH, 1973), *Salmonella* (see below), *Spirochaeta* (GREENBERG and CANALE-PAROLA, 1977), and *Streptococcus* (VAN DER DRIFT et al., 1975). For any given species, the responses are specific to certain chemicals, but there is considerable variation between bacteria in different genera (e.g., phenol is a repellent for *Salmonella*, but not *Escherichia*, and glycerol has no effect on either of these genera, but attracts *Proteus*). Since almost all of the fundamental investigation of the nature of bacterial chemotaxis in recent years has been on *Escherichia*, *Salmonella*, or *Bacillus*, Sections 4–7 will be confined to these genera.

2.2 Responses to Other Stimuli

Oxygen. Tactic responses occur in many aerobic species (Fig. 1b). In 1881 ENGELMANN demonstrated the accumulation of *Bacterium termo* in regions of high oxygen concentration. The studies were extended to other species by BEIJERINCK in 1893, and more recently by BARACCHINI and SHERRIS (1959) and ADLER (1966a). These studies were phenomenological in character, but recently TAYLOR et al. (1978) have provided some insight into the mechanism by the demonstration that, under anaerobic conditions, taxis occurs toward alternative electron acceptors such as nitrate or fumarate, but only if the corresponding reductase is functional. They suggest, but have so far been unable to demonstrate directly, that the "receptor" for oxygen may be cytochrome oxidase, a suggestion which is in agreement with the LINKS-CLAYTON hypothesis referred to in Section 6.5.

pH. Negative tactic responses to both high and low pH occur in many species, with obvious selective advantages. Since the proton potential across the membrane is the energy source for bacterial motility (see below), and since perturbation of this potential elicits tumbling or reversal responses, there is probably not a "proton receptor" in the same sense as there are receptors for other chemicals.

Divalent Cations. Magnesium, calcium, manganese, and zinc ions all act as attractants for *Salmonella* and *E. coli* (ZUKIN and KOSHLAND, 1976). The effect requires a functional membrane-bound magnesium/calcium-dependent ATPase. ATP levels and, indirectly, membrane potential can therefore presumably be perturbed by changes in divalent cation levels. With perturbation of the energy source for motility (protonmotive force across the membrane) and a substrate (ATP) for the chemotactic methylation reaction (see below), it is perhaps not surprising that tactic responses occur. ORDAL (1976a) found permanent effects on tumbling frequency in *B. subtilis* upon manipulation of magnesium concentration in the presence of an ionophore specific to it.

Potassium was also found to attract *Salmonella* and *E. coli*, but only at very elevated levels (ZUKIN and KOSHLAND, 1976).

Light. Photomovement has been demonstrated in the swimming phototrophic bacteria *Chromatium, Thiospirillum,* and *Rhodospirillum*, and in the gliding cyanobacteria (CLAYTON, 1964; NULTSCH, 1970). Since light is an energy source for motility in these species, population analysis of these movements is complicated by effects on translational velocity. However, by analyzing behavior of single cells, it has been established that in the swimming bacteria, the characteristic response is an increased probability of reversal on reduction of spatially uniform light intensity, with subsequent adaptation: a so-called "photophobic" response (BUDER, 1915; METZNER, 1920). Thus the photo- and chemotactic responses (see below) are similar in that they involve temporal comparison. Note however that the photophobic response in bacteria usually is an *avoidance* response to *decreases* in stimulus intensity, whereas chemotaxis to attractants results from a *suppression* of spontaneous "avoidances" (tumbles) in response to *increases* in stimulus intensity, as will be discussed in detail later. The photoreceptors for the phobic response are, as judged by the action spectrum, identical to those of the photosynthetic apparatus, namely bacteriochlorophyll and the carotenoids (MANTEN, 1948; CLAYTON, 1953; VAN NIEL et al., 1956).

In the gliding filamentous cyanobacteria the response is a true phototaxis, i.e., it is a response to the *direction* of the light stimulus (NULTSCH, 1970). The mechanism for this response is distinct from that of phobic responses in the phototrophic bacteria, and chemotaxis in bacteria generally, because it involves a detection of stimulus vectoriality by the sensory apparatus.

A response to high-intensity light occurs in nonphotosynthetic bacteria, and this will be discussed in Section 6.5.

Temperature. Accumulation around a heated wire was observed in various bacilli and cocci by SCHENK in 1893 and studied in more detail in *Spirillum volutans* by METZNER (1920), who noted that high probability of reversal correlated with movement down a thermal gradient. Similarly, the negative temporal temperature gradient created when a heated suspension of cells was allowed to cool resulted in a high incidence of reversals.

MAEDA et al. (1976) showed that *E. coli* grown at 34 °C accumulate in a band at 30 °C when placed in a thermal gradient, but are careful to point out the possible participation of oxygen artifacts. It is hard to see how this objection would apply to their demonstration that a maximum in tumbling frequency occurs at the growth temperature of 34 °C, and that a sudden increase in temperature elicits transient smooth swimming. MILLER and KOSHLAND (1977a) confirmed over a considerable temperature range that tumble frequency is maximized at the growth temperature. [They also showed that tactic responses could occur below the membrane phase transition temperature, in contrast to a previous report (LOFGREN and FOX, 1974).]

The effects of temperature on motility are thus seen to be complex, but in some way they combine to give migration in thermal gradients, a response which may well have evolved to ensure migration to a favorable temperature for growth.

Gravity. One early report exists of geotaxis in *Spirillum*, by MASSART in 1891. The possibility of geotaxis as a hydrodynamic phenomenon was pointed out by ROBERTS (1970) and has now been observed in smooth-swimming mutants of *Salmonella* (ASWAD and KOSHLAND, 1975a). This is probably a "parachute" effect—a cell initially swimming horizontally will tend to curve down as the body sediments. The phenomenon appears to be insignificant in wild-type cells, where active directional changes by tumbling occur continually.

Electrical and Magnetic Fields. Electrophoretic effects can occur because of net charge on the cell, but this is unlikely to be of biological significance. *S. volutans* exhibits flagellar reversal induced by electrical field transients (CARAWAY and KRIEG, 1972). This is probably a consequence of membrane potential perturbation, which is discussed further in Section 6.5.

Bacterial *magnetotaxis* has been demonstrated in a species possessing iron-rich vesicles (BLAKEMORE, 1975). The author suggests a possible selective advantage for this marsh-dwelling anaerobe, in that the earth's magnetic field at any appreciable distance from the poles has a vertical component, and so cells would be driven away from the oxygen-rich water surface, but this suggestion must be regarded as speculative at the present time.

Sex. BEZDEK and SOSKA (1972) demonstrated that Hfr strains of *Salmonella* preferentially invade colonies of F^- strains. Nothing is known regarding mechanism.

3. Experimental Techniques

A variety of assays depend on migration or accumulation of populations of cells. These include the capillary assay of PFEFFER, as modified by ADLER (1973) to include plating and colony counting of the capillary contents, visual assessment of chemotactic bands on semi-solid agar plates (ADLER, 1966b), and a quantitative assay of response in calibrated unidimensional gradients, as described by DAHLQUIST et al. (1972). Since no microscopic observation of individual cells takes place in these assays, inference as to mechanism can be hazardous; for example, motility may be inhibited by a test substance or, in crowded regions, tactic responses to oxygen gradients may interfere. In spite of these *caveats*, assays of this type, especially the capillary assay of ADLER, have proved invaluable.

Observation of individual cells, though time-consuming, yields most information. This was done subjectively in the rapid-mixing technique of MACNAB and KOSHLAND (1972); cells were exposed to abrupt changes in the concentration of an attractant or repellent and the response of the cells in terms of enhancement or suppression of tumbling was assessed visually. This simple approach has proved very useful for qualitative and semiquantitative determinations. A refined photographic version was introduced by SPUDICH and KOSHLAND (1975). Detailed studies of the motion of swimming cells were made possible by the development of an automated tracking microscope (BERG, 1971; BERG and BROWN, 1972) which accumulated data on the three-dimensional trajectory of a cell, as a function of time, for subsequent computer analysis. Tracking techniques have also been employed by LOVELY et al. (1974). A technique was devised by SILVERMAN and SIMON (1974) for tethering cells by a single flagellum or by the hook structure to a microscope slide by means of antibody (Fig. 7a of Chap. 3.1.1); this technique permits indefinite viewing of an individual cell's behavior, under a variety of stimulus conditions. A suitably designed high-intensity dark-field microscopic arrangement allows individual functioning flagella on a cell to be studied (MACNAB and KOSHLAND, 1974), and can be combined with tethering of the body to the microscope slide to eliminate cell motion (MACNAB, unpublished). Video recording of individual flagella is now feasible, with an ultra-sensitive silicon intensifying television camera (HOTANI, unpublished).

4. The Nature of the Stimulus in Chemotaxis

(This section covers chemical agents which initiate a response as a result of recognition by specific receptors. Oxygen, pH, and divalent cations, which were discussed in Section 2, probably do not utilize the specific chemotactic machinery, but instead directly affect the metabolic state of the cell.)

4.1 Chemical Nature and Specificity of Chemotactic Agents

The most detailed surveys of chemicals for their tactic effectiveness have been those of ADLER and co-workers in *E. coli* (Table 1). The principal classes of attractants are sugars (ADLER et al., 1973) and amino acids (MESIBOV and ADLER, 1972). All strong attractant sugars are either D-pentoses or D-hexoses or derivatives thereof such as glycosamines, sugar alcohols, sugar phosphates, alkyl glycosides, and disaccharides.

Eight of the common protein amino acids are attractants: alanine, asparagine, aspartate, cysteine, glutamate, glycine, serine, and (less potent) methionine. The corresponding D-amino acids are also fairly strong attractants, as are some α- and β-methyl derivatives. Removal or blocking of the amino group greatly reduces potency, and in fact ammonium ion itself is an attractant, utilizing the serine receptor.

In contrast to chemical signaling in higher systems, adenosine-3′,5′-cyclic monophosphate (cyclic AMP) and other nucleotides are ineffective as chemotactic agents for *E. coli*.

In *Salmonella*, an exhaustive survey of attractants has not been undertaken, but they include a number of the demonstrated attractants for *E. coli*, such as ribose (AKSAMIT and KOSHLAND, 1972), galactose and glucose (STRANGE and KOSHLAND, 1976), aspartate (MACNAB and KOSHLAND, 1972), and serine (DAHLQUIST et al., 1972). Citrate is an attractant for *Salmonella* also (MACNAB and KOSHLAND, unpublished observations). *B. subtilis* shows taxis to all 20 common amino acids (ORDAL and GIBSON, 1977).

Detailed studies of negative chemotaxis in *E. coli* by TSO and ADLER (1974) show that the following are repellents: short-chain saturated and unsaturated fatty

Table 1. Principal attractants, repellents, and receptor classes[a] in *E. coli* (excluding O_2, pH, and inorganic ions)

Overall class	Receptor name	Receptor location or identity	Compounds detected
Attractant amino acids (MESIBOV and ADLER, 1972)	Aspartate	?	Aspartate Glutamate Asparagine Cysteine Methionine
	Serine	?	Serine Glycine Alanine Asparagine Cysteine
Attractant sugars (ADLER et al., 1973 and refs. in Sec. 6.1)	Galactose	Periplasmic	Galactose Glucose Glycerol-β-galactoside Methyl-β-galactoside Methyl-β-glucoside
	Maltose	Periplasmic	Maltose
	Ribose	Periplasmic	Ribose
	Trehalose	?	Trehalose
	Glucose	Corresponding enzyme II of phosphotransferase system	Glucose Glucose-1-P (possibly hydrolyzed to glucose first)
	Mannose	Corresponding enzyme II of phosphotransferase system	Mannose Glucose
	Fructose	Corresponding enzyme II of phosphotransferase system	Fructose
	Sorbitol	Corresponding enzyme II of phosphotransferase system	Sorbitol
	Galactitol	Corresponding enzyme II of phosphotransferase system	Galactitol
	Mannitol	Corresponding enzyme II of phosphotransferase system	Mannitol
	N-acetyl-glucosamine	Corresponding enzyme II of phosphotransferase system	N-acetylglucosamine
Repellents (TSO and ADLER, 1974)	Fatty acid	?	C_2-C_6 (saturated and unsaturated)
	Aliphatic alcohol	?	C_2-C_4
	Hydrophobic amino acid	?	Leucine, Isoleucine Valine, Tryptophan
	Indole	?	Indole, Skatole
	Aromatic acid	?	Benzoate Salicylate

[a] Where the receptor has not been identified, receptor classes are based on competition studies of chemotaxis.

acids; aliphatic alcohols; hydrophobic amino acids; indole (effective at 10^{-6} M, a much lower threshold than for most repellents) and analogs; aromatic acids. A number of other compounds also repel, but not as strongly as those listed above.

Salmonella shows a marked negative response to phenol [unlike *E. coli* (LEDERBERG, 1956)], as well as to the fatty acids, the aliphatic amino acids, and indole (TSANG et al., 1973).

In these surveys of attractants and repellents, competition experiments have revealed classes of chemicals; within a class, a high uniform concentration of one compound (e.g., galactose at 10^{-2} M) inhibits or abolishes the response to a gradient of a second analogous compound (e.g., fucose at any concentration). From this, the concept of receptors has emerged, where the number of types of receptor is considerably less than the number of chemotactic agents. The subject of receptors will be taken up in more detail in Section 6.1.

Specificity of chemotactic agents for the receptors is comparable to substrate/enzyme specificities, and seems to be less related to chemical reactivity than to steric factors. As examples of specificity, the taxis threshold of galactose (the primary agent) is 10^{-6} M, 1000-fold lower than that of an analog, 2-deoxy-D-galactose (ADLER et al., 1973); for aspartate and β-methyl aspartate, the thresholds are 6×10^{-8} and 5×10^{-6} M respectively (MESIBOV and ADLER, 1972).

The existence in *E. coli* or *Salmonella* of certain compounds as attractants and others as repellents can be roughly rationalized. The advantage to a cell of migration to a sugar like glucose or an amino acid like aspartate is obvious. Yet many excellent carbon sources such as glycerol are not attractants, nor are many of the protein amino acids. Perhaps a representative selection of agents can normally act as an adequate signal that a biological source of nutrition is nearby—aspartate probably indicates hydrolyzing protein and hence the availability of the other amino acids, and so on.

Repellents are somewhat harder to explain. With the exception of indole, they only operate at considerably higher concentrations than attractants, although still well below toxic levels. Some (e.g., acetate) can be excretion products, indicative of overcrowding; the aliphatic amino acids are subject to cross-inhibition which can prevent growth; phenol is toxic. Such rationales are not wholly satisfactory: for example, phenol is equally toxic to *E. coli* and *Salmonella*, yet repels only the latter, and many other toxic materials, although they may inhibit motility, do not cause tactic responses.

4.2 The Fate of Chemotactic Agents

Bearing in mind that photomovement and aerotaxis are controlled *via* metabolism, it is natural to suppose that chemotaxis is a consequence of some aspect of the *metabolism* of the relevant agent. This was clearly shown by ADLER and co-workers to be incorrect (ADLER, 1969), since (a) nonmetabolizable analogs (e.g., 6-deoxy-D-galactose) of the primary agent can be attractants (b) mutants unable to metabolize, or in some cases even transport the agent are still chemotactically competent (c) mutants which are chemotactically defective to a chemical can be capable of metabolizing it, and (d) many metabolizable compounds, including metabolic products of attractants (e.g., glucose-6-phosphate) are not attractants.

Some connection between transport and chemotaxis exists, in that the chemoreceptors which have been identified so far are also elements of the corresponding transport systems. However, as has been demonstrated in the case of galactose-binding protein (*mglB* gene product), loss of function as a galactose chemoreceptor does not necessarily imply loss of galactose transport function or vice versa (ORDAL and ADLER, 1974a, b).

The role of the chemotactic agent is therefore confined to the act of binding to a surface receptor. It does not have to be either transported or metabolized to provide the stimulus, although of course both of these processes do frequently accompany recognition of a nutrient.

4.3 Gradient Nature of the Stimulus

The possibility that chemotactic agents act by affecting motility differently at different concentrations rather than by a true gradient-sensing mechanism has been ruled out [MACNAB and KOSHLAND, 1972; BERG and BROWN, 1972 (aspartate results)]. Provided an adequate energy supply is available and provided the agent in question is not present at toxic levels, neither velocity nor tumbling frequency is affected by its presence. Only chemical *gradients* are effective in eliciting any sort of motility response. Gradient measurement implies comparison, but the small size of a bacterium (1 or 2 μm) makes sensing of quite shallow chemical gradients by a spatial (e.g., head–tail) comparison unlikely, yet *Salmonella*, for example, shows distinct migration in gradients where the mean concentration difference across the body is only one part in 10^4 (DAHLQUIST et al., 1972). A cell swimming in a spatial gradient is generating (in its own frame of reference) a temporal gradient, which could be the basis for an extended comparison of concentration, provided the comparison occurs on a time scale which permits the cell to translate through many body lengths. That bacteria do indeed use a temporal gradient-sensing mechanism in chemotaxis was demonstrated (Fig. 2) by MACNAB and KOSHLAND (1972) who subjected bacteria to concentration jumps in a rapid-mixing apparatus. These studies were later extended to include repellents (TSANG et al., 1973). "Favorable" stimuli (increase of attractant or decrease of repellent) suppress tumbling, whereas "unfavorable" stimuli (decrease of attractant or increase of repellent) enhance tumbling but, in either case, the response is temporary as would be expected from a gradient-sensing mechanism subjected to a "one-shot" gradient stimulus (an indefinite response, i.e., failure to adapt to a new concentration, would be counter-productive). BROWN and BERG (1974) later confirmed temporal-gradient sensing by enzymatically creating or destroying chemotactic agents; while this was happening, the cells responded. Combined with the concept of receptors, chemotactic stimuli can be seen to operate *via* a time-dependent change in the state of occupancy of these receptors by the chemotactic agent.

5. The Nature of the Response

5.1 Motility–The Random Walk

Peritrichous bacteria alternately swim (by counterclockwise rotation of the flagellar bundle) and tumble (by clockwise flagellar rotation which, by a polymorphic transi-

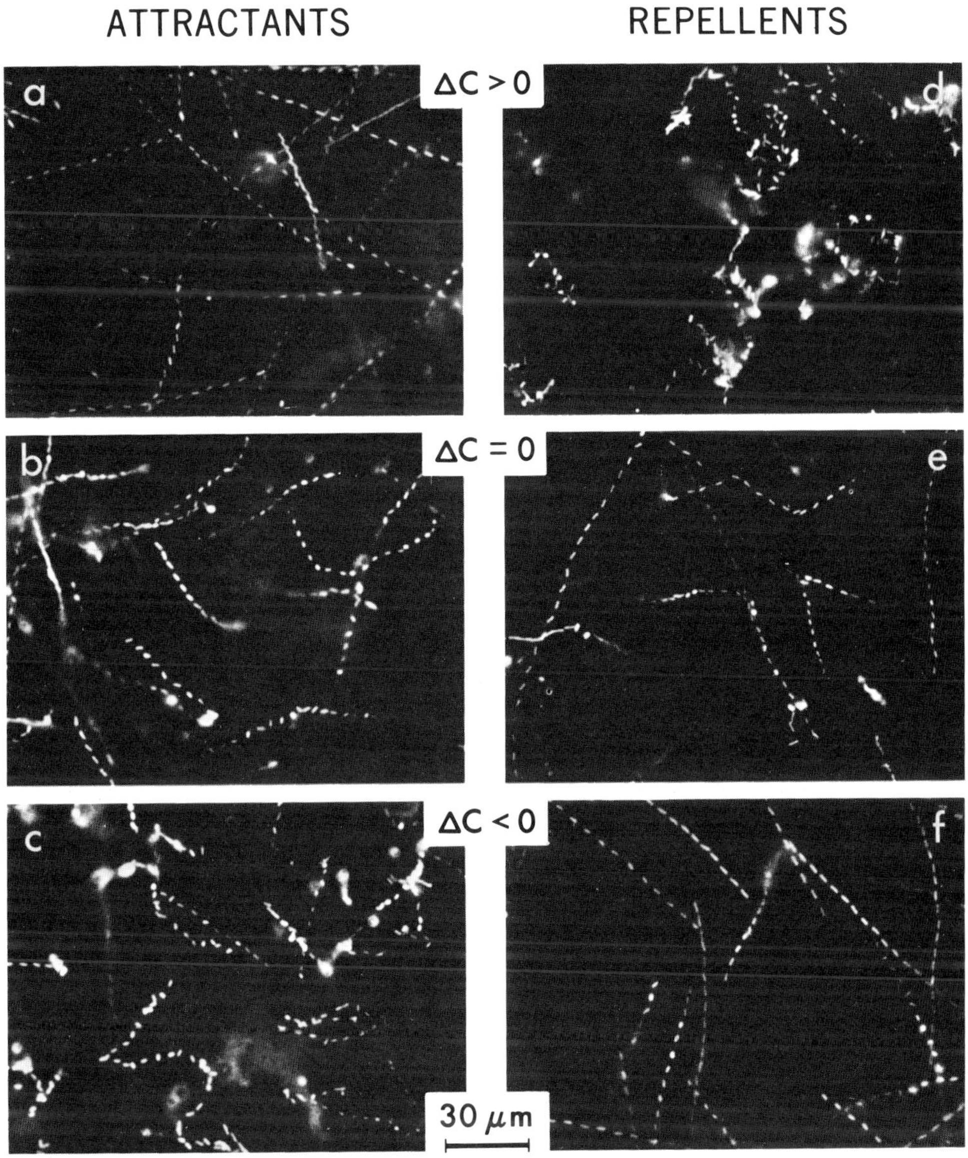

Fig. 2 a–f. Motility tracks of *Salmonella* in the interval 2 to 7 s after subjection to temporal gradients of attractants and repellents. Photographs taken in dark field with stroboscopic illumination operating at 5 pulses per s. **a**, **b**, **c** attractants; **a** serine increase from zero to 7.5×10^{-4} M; **b** no concentration change (control); **c** serine decrease from 10^{-3} to 2.4×10^{-4} M; **d**, **e**, **f** repellents; **d** phenol increase from zero to 7.5×10^{-4} M; **e** no concentration change (control); **f** phenol decrease from 3×10^{-4} to 7.5×10^{-5} M. The smooth motility response to "favorable" gradients (**a** and **f**) and the tumbling response to "unfavorable" gradients (**c** and **d**) eventually give way to the normal motility pattern (**b** and **e**). (From TSANG et al., 1973)

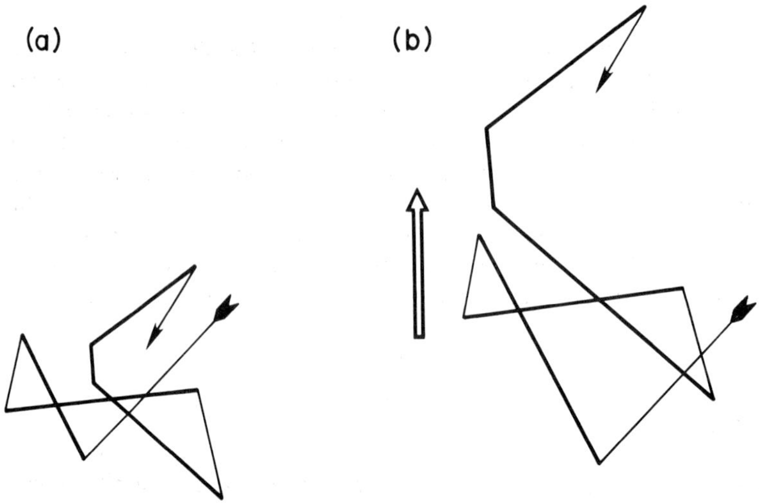

Fig. 3a and b. Idealized trajectories of a swimming bacterium such as *E. coli* or *Salmonella*: **a** in an isotropic medium; **b** in a unidirectional attractant gradient (*open arrow*). In either case, the cell swims in a straight line, randomizes its direction by tumbling, swims again, tumbles, etc., yielding a three-dimensional random walk. The effect of an attractant gradient is to extend (by tumble suppression) the mean duration of swimming segments in an up-gradient direction (*heavy lines*); (segments in the down-gradient direction (*fine lines*) are not appreciably shortened). To illustrate this, the cells **a** and **b** start at the same point (*arrow tail*), and execute the same number of tumbles with the same resulting directional changes, but the cell **b** in the gradient has up-gradient segments extended by a factor cosθ where θ is the angle with respect to the gradient direction. It therefore displays migration in the gradient direction by executing a time-biased random walk

tion, causes bundle dispersal) (see Chap. 3.1.1). The swimming intervals generate approximately linear displacements of the cell, whereas the tumbles effectively ensure that successive swimming displacements ("runs") are directionally uncorrelated (there is actually a slight forward bias but this is not important to the argument). The mean duration of tumbles (ca. 10^{-1} s) is much less than that of runs (1 or 2 s). In the absence of stimulation, termination of both runs and tumbles in *E. coli* occurs with constant probability (cf. disintegration of a radioactive nucleus), and so the lengths of runs, as well as their directions, are randomly distributed (BERG and BROWN, 1972). [This description has been confirmed in *Salmonella* (MACNAB and KOSHLAND, 1973), and unpublished results; it is probably valid for all peritrichs.] The cell trajectory is therefore a random zig-zag in three-dimensional space (Fig. 3a).

5.2 Gradient-induced Bias

As was described in Section 4.3, the effect of a positive temporal gradient (concentration increase) of attractant or a negative temporal gradient (concentration decrease) of repellent is to suppress tumbles. This applies whether the temporal gradient in the cell's frame of reference results from a temporal gradient per se, or from swimming through a spatial gradient. In the latter case, however, the magnitude and sign of the gradient sensed depends on the direction in which the cell is swimming: a cell swimming directly up-gradient will experience a positive temporal gradient; a cell swimming at an angle up-gradient will experience a smaller positive temporal gradient, and so on. The run length therefore no longer has a constant probability of being terminated by a tumble, but instead has a probability which correlates with the current direction of travel. A bias in the random walk has now been introduced by (on the average) extending runs in which the cell is sensing a "favorable" gradient (Fig. 3b; BERG and BROWN, 1972). Note, however, that the cell does not "turn in the right direction", as do many higher organisms.

5.3 Quantitative Aspects

To give some idea of the magnitude of chemotactic responses, *Salmonella* placed in an exponential gradient of serine, with concentration increasing by a factor of e in 6.4 mm, migrate up-gradient at 2.8 $\mu m\, s^{-1}$, i.e., at about 10% of the instantaneous swimming speed (DAHLQUIST et al., 1972). [Note how shallow this gradient is on the scale of a bacterial cell (2 µm); the concentration difference across the cell would be only 3 parts in 10^4.] As an indication of how successfully a responding population can invade a favorable region, the entry over a 1-h period of bacteria into a capillary containing 10^{-3} M aspartate (4×10^5 cells), is 40-fold higher than the control value (10^4 cells) (ADLER, 1973).

In early studies, PFEFFER (1888) noted that accumulation of bacteria in a capillary occurred when the capillary attractant concentration equalled or exceeded a constant multiple of the external concentration. This adherence to the WEBER law of sensory perception and the related law of FECHNER (which states that magnitude of response is proportional to relative increase in stimulus intensity) has been further explored in recent years. DAHLQUIST et al. (1972) found, in agreement with the above, that over several orders of magnitude of serine concentration, *Salmonella* displayed fairly constant migrational velocity in an exponential serine gradient. Serine does not seem to be typical in this regard however (see below); MESIBOV et al. (1973) showed that *E. coli* responses to α-methylaspartate and galactose adhere only approximately to Weber-Fechner behavior, and only over a rather narrow concentration range. They postulated that size of chemotactic response (as measured by accumulation in a capillary) is proportional to the change in extent of receptor bound to attractant. For simple noncooperative binding, this postulate leads to approximate Weber-Fechner behavior in the immediate concentration range of the receptor dissociation constant, and dissociation constants measured from chemotaxis assays in this way are in good agreement with the values measured directly on purified receptors (MESIBOV et al., 1973; SPUDICH and KOSHLAND, 1975; BERG and TEDESCO, 1975). The extended range of serine concentrations to which bacteria respond indicates either a receptor system with negative cooperativity, more than one receptor system, or a nonlinear relationship between change of receptor occupancy and response (DAHLQUIST et al., 1976).

In temporal stimulus studies, a sufficiently large attractant increase (or repellent decrease) can suppress tumbling for periods of the order of minutes, where a corresponding attractant decrease (repellent increase) causes continuous tumbling for much shorter times, of the order of 10 s (MACNAB and KOSHLAND, 1972; TSANG et al., 1973). This asymmetry is sufficiently great that, in the shallow temporal gradients that a cell generates by its own swimming, tumble *suppression* by "favorable" stimuli (attractant increases or repellent decreases) is the major element of the response (BERG and BROWN, 1972). This is in contrast to the phobic response of photoresponsive bacteria, and protozoans.

The time scale of the temporal comparison is still unclear, at least in part because it is not obvious how to define such a scale. *Salmonella* persist (i.e., have a positive velocity component) up a unidirectional spatial gradient of serine for approximately 20 s before turning into the down-gradient direction (MACNAB and KOSHLAND, unpublished) compared with a persistence time[2] of 12 s in any given direction, in the absence of a gradient (MACNAB and KOSHLAND, 1973). BERG and TEDESCO (1975), studying the response of tethered cells to concentration jumps, found that adaptation to attractant increases (repellent decreases) is slow, while adaptation to attractant decreases (repellent increases) is fast. A mechanism can therefore be invoked which has a relatively long time constant, without this resulting in carry-over of "false" information when the swimming cell changes to the down-gradient direction. Current estimates of the time scale range from less than one second to several seconds.

TSANG et al. (1973) and ADLER and TSO (1974) found that if more than one kind of chemical stimulus was given to a bacterial population simultaneously, the response was approximately the algebraic sum of the responses to the stimuli given individually; this was later demonstrated quantitatively by SPUDICH and KOSHLAND (1975) and BERG and TEDESCO (1975), and leads to the conclusion that simultaneous access to a common signalling system is possible.

6. The Transduction Process Between Stimulus and Response

As was indicated above, the initial stimulus is a change in attractant (or repellent) concentration leading to change in the state of occupancy of a set of chemoreceptors at the cell surface. The ultimate response is a transient modification of the pattern of spontaneous reversal of the flagella. This section will describe what is currently known about the events linking the stimulus to the response.

6.1 The Chemoreceptors (see Table 1)

Although a considerable number of classes of attractants and repellents have been described on the basis of competition studies, and there is a prevailing assumption that to each class there is a chemoreceptor, isolation or other characterization of the chemoreceptors is still confined to a few cases (receptor mutants have

[2] Run time and persistence time are distinct quantities. *Run time* is the time interval between two successive tumbles, whereas *persistence time* in any defined direction is the time interval during which the *sign* of the velocity component with respect to that direction remains unchanged, even though tumbles may have occurred (e.g., the duration of the penultimate three runs in Fig. 3b would constitute a persistence time up the gradient). Persistence time therefore is generally longer than run time.

been found in some cases, without identification of the corresponding gene product). In fact, all chemotactic receptors isolated so far are for sugars; it remains to be seen whether the receptors for other types of agents (attractant amino acids, and repellents) have similar properties. (Since amino acid attractants elicit much stronger responses than do sugars, there may be substantial differences in the receptors.)

Sugar receptors for galactose and ribose (*E. coli* and *Salmonella*) and maltose (*E. coli*) are periplasmic binding proteins, loosely associated with the outer surface of the cytoplasmic membrane and contained by the cell wall (ANRAKU, 1968; WILLIS and FURLONG, 1974; STRANGE and KOSHLAND, 1976; AKSAMIT and KOSHLAND, 1972; KELLERMAN and SZMELCMAN, 1974). They can be released by osmotic shock, with the number of copies released per cell ranging from 10^3 to 6×10^4. They have molecular weights in the range 30–40,000 daltons, are soluble as monomers in aqueous buffers, are heat-stable, and lacking cysteine; are insensitive to sulfhydryl reagents. The percentage of nonpolar amino acids (approximately 50%) is not unusually high. Initial reports (BOOS et al., 1972) that the galactose-binding protein of *E. coli* has two binding sites for galactose (a feature which could have mechanistic implications) now appear to be incorrect (ZUKIN et al., 1977a). Each galactose- and ribose-binding protein of *Salmonella* has only a single site (AKSAMIT and KOSHLAND, 1972; STRANGE and KOSHLAND, 1976). Dissociation constants are about 10^{-7} to 10^{-6} M. The binding constants for the primary agent and various analogs are in agreement with their relative effectiveness as chemotactic agents. Mutants lacking a given binding protein lack the corresponding tactic response; this, together with the binding property correlations just described, is the basis of their identification as receptors. The receptor proteins are also components of transport systems for the corresponding substrate.

The other class of sugar chemoreceptor which has been found to mediate chemotaxis is the sugar-specific enzyme IIA moiety of the phosphotransferase system (ADLER and EPSTEIN, 1974). Each enzyme IIA protein is a hydrophilic, but membrane-bound, component of the corresponding transport system, in which its function appears to be the recognition of the chemical prior to its phosphorylation and transport. Specific enzyme IIA proteins for glucose, fructose, mannose, N-acetylglucosamine, and three hexitols have been described and correlated with tactic responses (ADLER and EPSTEIN, 1974; LENGELER, 1975). The separability of function in taxis and transport is less clear-cut in the phosphotransferase system than in the case of the periplasmic proteins described above—components of the phosphotransferase systems which are not sugar-specific *are* nevertheless required for specific sugar taxes. A probable explanation is that tactic function requires the enzyme IIA to be integrated into the complete transport complex—there is no evidence that phosphorylation or transport is required for chemotaxis and indeed the corresponding phosphorylated sugars (e.g., glucose-6-phosphate) are poor attractants.

A number of attempts have been made to determine what changes, if any, take place in isolated receptor preparations upon substrate binding. Boos and co-workers (see BOOS, 1974 for a discussion) demonstrated absorption and fluorescence changes on binding of galactose to the corresponding periplasmic binding protein, a result which they interpreted in terms of an active site tryptophan rather than an allosteric effect. They also observed electrophoretic changes upon agent binding which they attribute to surface charge changes. ZUKIN et al. (1977b), taking advantage of the fact that the galactose-binding protein of *Salmonella* has only one tryptophan residue, covalently bound a second fluorophor (5-iodoacetamidofluorescein). Fluorescence energy transfer measurements indicated that the two fluorophors were at least 3 nm apart, yet they were both affected by agent binding, providing strong evidence for a long-range conformational change. No corresponding in vivo experiments have been described.

6.2 Transmission of Information from the Receptor

Evidence has been presented above for receptor changes on binding. No direct evidence has been obtained yet for other cellular components in physical association with the chemoreceptor, but certain examples of competition and of genetic lesions in taxis cannot be explained at the level of the primary receptor alone. Both *E. coli* and *Salmonella* have different receptors for ribose and galactose, yet high uniform concentration of one substrate reduces or eliminates tactic responses to the other (ORDAL and ADLER, 1974a, b; STRANGE and KOSHLAND, 1976). Similar effects have been observed among the agents of the sugar-specific enzyme IIA receptors of the phosphotransferase system (ADLER et al., 1973; LENGELER, 1975) and among many amino acids in *B. subtilis* (although the receptors have not been isolated in this case; ORDAL et al., 1977). Inhibition operates via the specific receptor for the inhibiting species – in a ribose-binding protein mutant, ribose has no effect on the galactose response. The results just described could be explained by an interaction between the receptors, but the existence of a genetic locus, *trg*, which affects ribose and galactose taxis simultaneously, suggests that the corresponding gene product is part of a second stage of the information transmission process, shared by the ribose and galactose receptors.

Another example of a shared component is the *tsr* gene product (HAZELBAUER et al., 1969; TSO and ADLER, 1974). Lesions in this gene abolish responses to a variety of repellents and to the attractant serine. [Serine is an unusual attractant in other regards – it is effective over an extremely wide concentration range (see Sect. 5.3), and can affect swimming behavior under nongradient conditions (BERG and BROWN, 1972.] The *tsr* product affects central aspects of the chemotactic apparatus. One class of general chemotactically defective mutants (*cheD*) is now believed to be in the *tsr* gene (PARKINSON, 1974, and unpublished). In this class, not only has response to serine and repellents been abolished, but the phenotype is one of smooth swimming, which prevents chemotactic responses of any sort.

6.3 The Methyl-Accepting Chemotaxis Protein (MCP)

For many years it has been known that methionine is necessary for tactic responses to other chemicals (ADLER and DAHL, 1967; ASWAD and KOSHLAND, 1974; ORDAL, 1976b), as well as being a weak attractant itself. Cells noninduced for, or incapable of, methionine synthesis swim smoothly unless methionine is supplied. The methionine requirement is for the synthesis of S-adenosylmethionine, with ATP as the adenosyl donor (ARMSTRONG, 1972; ASWAD and KOSHLAND, 1975b). KORT et al.

(1975) fractionated *E. coli* cells which had been pulse-labeled with [methyl-^{3}H]-methionine, and found, by gel electrophoresis, a labeled component of the cytoplasmic membrane fraction. Because of its correlation with chemotaxis (see below), and because only the methyl group of methionine was incorporated, they named the molecule MCP (for *methyl-accepting chemotaxis protein*). It has a molecular weight of 62,000 daltons; there are at least 700 copies/cell. Its implication in chemotaxis was based on several lines of evidence. (1) General chemotaxis mutants (*che*$^-$) had abnormal methylation levels, and reversion to wild-type tactic behavior was accompanied by reversion to normal methylation levels. (However, the abnormal levels of the mutants did not fit any obvious pattern—examples of below-normal and above-normal levels were found in various *che* classes, and among smooth and tumbling phenotypes.) (2) Chemotactic stimulation also stimulates methylation. (3) Methylation levels in various *fla* mutants (nonflagellate phenotype) were extremely low. (4) The time scales of methylation and demethylation are fairly short (ca. 10 min), in the same general range as the time scales of gain or loss of chemotactic competence by manipulation of methionine concentration.

ASWAD and KOSHLAND (1974) showed that a tumbling, methionine-auxotrophic mutant of *Salmonella* continued to tumble in the absence of methionine, but its smooth swimming response to large increases in concentration of attractant was greatly extended. From this they concluded that while methionine participates in the regulation of tumbling, it is *not* a prerequisite for tumbling. M. SPRINGER et al. (1975), in analogous studies with wild-type and four tumbling mutants of *E. coli*, found that methionine starvation eliminates tumbling in the wild-type and in two of the mutants, and that the presence of *constant* levels of attractants accelerates this process and enables its occurrence in a third mutant. Methionine starvation extended smooth swimming responses to chemotactic stimuli given prior to the final loss of the ability to tumble. Arsenate (which depletes ATP levels) eliminated tumbling in the fourth mutant which had been unaffected by methionine starvation. The authors use the latter result to suggest (contrary to ASWAD and KOSHLAND) that S-adenosylmethionine (and therefore methionine) *is* necessary for tumbling—but at low levels, so that methionine derived from protein can suffice in many strains.

In a subsequent paper (M. SPRINGER et al., 1977), they found that only the process of adaptation to a new (higher) attractant concentration requires methionine; maintenance of the adapted state, and de-adaptation to the original state does not. As they stand, these conclusions are not compatible with the observation that constant attractant levels cause rapid methyl turnover; methionine starvation should then cause net demethylation and prevent maintenance of the adapted state, which it does not. M. SPRINGER et al. therefore make an additional assumption, namely that not only the methylation, but also the *de-methylation* aspects of methyl turnover requires the presence of methionine. The enzyme responsible for methylation of MCP (*chemotaxis methyl transferase*, CMT) has recently been identified in *Salmonella* by W. SPRINGER and KOSHLAND (1977) as the *cheR* gene product. It is a cytoplasmic enzyme, molecular weight 38,000 daltons, and is substrate-specific; it did not methylate a variety of proteins or membranes which were tested (but, interestingly, did methylate the *E. coli* MCP). The methylated moiety of MCP is a γ-glutamyl methylester (VAN DER WERF and KOSHLAND, 1977, KLEENE et al., 1977). *CheR*$^-$ mutants tumble upon negative chemotactic stimulation with phenol, lending support to the idea that methylation is not an absolute requirement for tumbling.

To summarize the present state of knowledge on this topic, a membrane-bound protein undergoes reversible methylation mediated by a soluble methyltransferase and an unidentified demethylase. Depletion of the methylated form of the protein, whether because of methionine deprivation or other reasons, suppresses tumbling. Chemotactic stimulation is one of the regulatory inputs to the methylation system, and hence to the control of tumbling.

6.4 Gradient Inference

As was described earlier, chemotactic responses are proportional to the change in occupancy of the corresponding receptors. How does the cell measure this change in occupancy? Some type of memory device is called for which can compare past with present. Essential to such a device is a fast process to measure the present, a slow process to keep a record of the past, and a comparator. MACNAB and KOSHLAND (1972) used as an illustrative example the enzymatic regulation model shown in Figure 4, which has since been elaborated to include the methylation aspects of regulation. BERG and TEDESCO (1975) present an abstract model involving a (fast) signal reservoir, a (slow) adaptation reservoir, and a comparator to instruct the motor and control the adaptation process. An important feature of their model is that the adaptation reservoir has slow "filling" but rapid "draining" kinetics, in order to account for the observed asymmetry between responses to positive and negative gradient stimuli.

6.5 The Signal to the Motor

The motor is energized by a protonmotive force across the membrane, and is a spontaneously reversible device (Chap. 3.1.1). The effect of "favorable" (i.e., positive attractant or negative repellent) gradient stimuli is to temporarily suppress spontaneous tumbling. LARSEN et al. (1974) in tethered cell experiments with smooth and tumbling mutants, and with wild-type cells under chemotactic stimulation, concluded that swimming in free cells occurs during counterclockwise flagellar rotation (viewed along the filament into the cell) and tumbling in free cells occurs during clockwise rotation (note that tumbling also involves structural transitions in the filaments; see Chap. 3.1.1). TAYLOR and KOSHLAND (1974) confirmed this in *Pseudomonas*. Therefore chemotactic responses are achieved by modulating motor reversibility. A variety of other stimuli affect the reversibility of the motor and, since many of these [such as pulses of the proton ionophore FCCP (ORDAL and GOLDMAN, 1975); high-intensity blue light with an action spectrum suggesting flavin excitation (MACNAB and KOSHLAND, 1974; TAYLOR and KOSHLAND, 1975); and oxygen gradients (Sect. 2.2)] can be related to transient changes in membrane potential, it is natural to consider membrane potential as a candidate for the chemotactic signal to the motor. The well-substantiated LINKS-CLAYTON hypothesis for phototaxis in fact has energy perturbation as its basis (CLAYTON, 1958). Does chemotaxis operate by a membrane potential signal or by another means, such

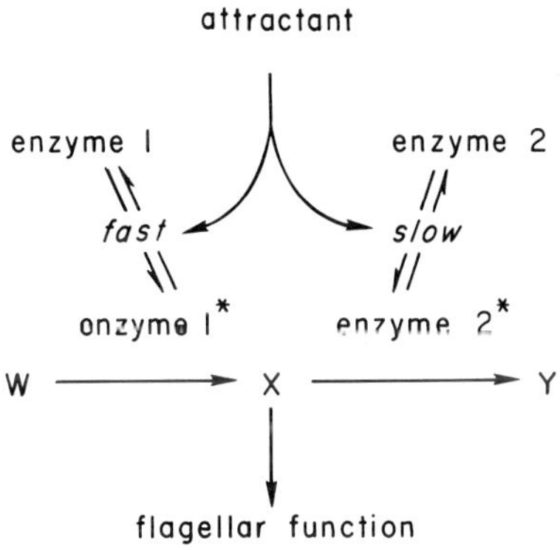

Fig. 4. Temporal gradient-sensing requires a fast process, a slow process, and a comparator. In this illustrative example (from MACNAB and KOSHLAND, 1972) of a possible enzymatic mechanism, attractant activates (fast process) enzyme 1 which mediates the production of X, and also activates (slow process) enzyme 2 which mediates the destruction of X. The pool size of X (comparator), which modulates flagellar function, increases during positive temporal gradient stimulation, decreases during negative temporal gradient stimulation, but remains unchanged during nongradient conditions

as a chemical transmitter? Conflicting results have been obtained. SZMELCMAN and ADLER (1976) using a radioactively labeled permeant cation to measure membrane potential, found transient hyperpolarization specifically accompanying chemotactic stimulation. MANSON et al. (1977) on the other hand, found that a motile strain of *Streptococcus faecalis* was chemotactically competent even when the protonmotive force was clamped by valinomycin or gramicidin C, and MILLER and KOSHLAND (1977b) found no changes in the voltage-sensitive fluorescence of a cyanine dye probe bound to cells which were under chemotactic stimulation. Both results would seem to indicate that neither overall membrane potential nor protonmotive force is the signal for tumble regulation. It should, however, be pointed out that, in at least one bacterial species (*S. volutans*), synchrony of reversal of the two polar tufts is too precise to be mediated by diffusion-limited chemical signaling along the 40 μm body length.

7. Genetics of Taxis and Motility

(See HILMEN and SIMON, 1976; TSUI COLLINS and STOCKER, 1976; WARRICK et al., 1977; SILVERMAN and SIMON, 1977; HAZELBAUER and PARKINSON, 1977, for recent summaries.)

Reference has been made to various genetic aspects throughout the text; for convenience, a summary is given here and in Figure 5 of the relevant genes and their functions, where known. Unless otherwise indicated, genes refer to *E. coli*. In both *E. coli* and *Salmonella typhimurium*, a massive cluster of around 20 genes (*hag, mot, che,* and *fla*) affecting motility and/or taxis exist near the *his* region of the map; in *E. coli*, the cluster is split into two regions (regions II and III)

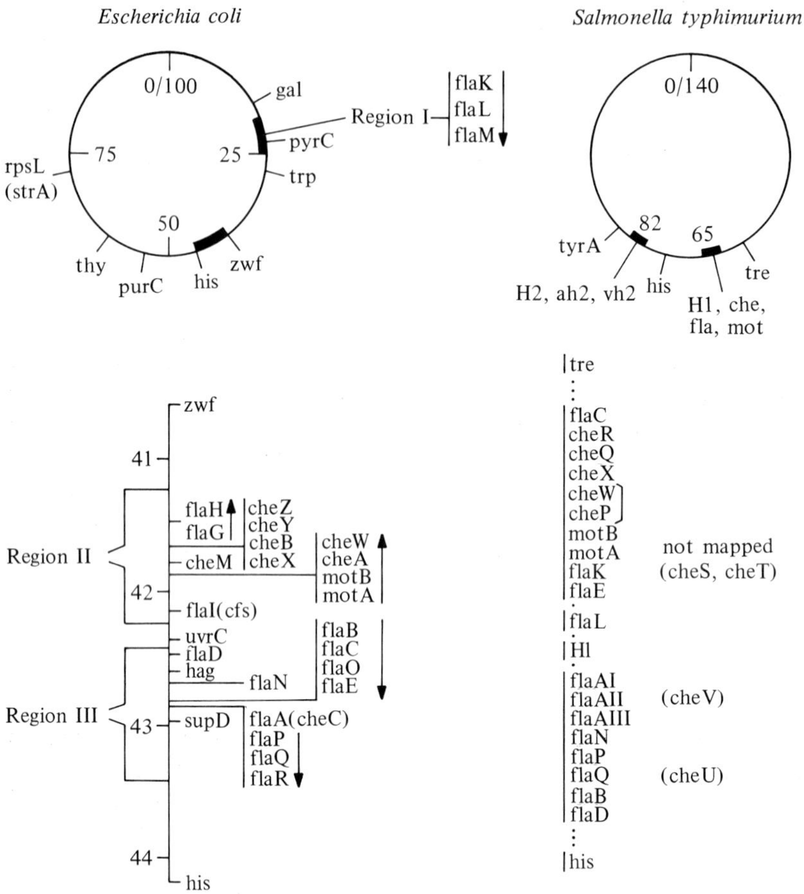

Fig. 5. Genes affecting motility and taxis in *E. coli* and *Salmonella typhimurium*. Both species have a large cluster of these genes on the early side of *his*, as detailed in the lower figures. In addition, *E. coli* has a small group of *fla* genes in region I, and *Salmonella* has a separate group of genes regulating phase-2 flagellin production. [*E. coli*, from SILVERMAN and SIMON (1977); *Salmonella* (*upper*), after SANDERSON (1972); (*lower*), after WARRICK et al. (1977).] Chemoreceptor genes, and the *trg* and *tsr* genes are dispersed throughout the genetic map

by the *uvrC* gene. Another *fla* region (region I) with at least four genes exists near *pyrC*. Receptor genes, and secondary signaling components such as *trg*, are located elsewhere on the map, as is the second flagellin production system of *Salmonella*.

7.1 Chemoreceptor Genes

Fructose (*ptsF*), glucose (*ptsG*), mannose (*ptsM*) galactitol (*gatA*), glucitol (*gutA*), mannitol (*mtlA*), galactose (*mglB*), ribose (*rbsP*), and maltose (*malE*). *LamB*, the phage-lambda receptor in the outer membrane of *E. coli*, participates in periplasmic accumulation of maltose, and therefore also affects maltose taxis.

7.2 Secondary Signaling Components

Genes coding for a shared component of ribose and galactose taxis (*trg*) and of serine and repellents (*tsr*) have been identified. Certain mutations in the latter gene can also affect chemotaxis generally and in this context it has been named *cheD*.

7.3 The Motor Organelle

At least 20 genes participate, and at least 11 different polypeptides are present, in the flagellum (comprising filament, hook, and basal body—see Chap. 3.1.1). Genes for flagellin (in *E. coli*, *hag*; in *Salmonella*, *H1* and *H2* and, for regulation of phase, *ah1*, *ah2*, and *vh2*). Hook protein (one of *flaK*, *L*, or *M*, in region I); regulation of hook length (*flaE*). Genes resulting in total absence of the organelle (*fla* genes); the exact function of most of these genes is not known; some (*flaI*, *flaH*) are known to be regulatory. One (*flaA*) is involved in regulation of motor reversibility, since some missense mutants display *che*$^-$ phenotype (*cheC*) instead of absence of flagella. The *mot* genes are essential for flagellar rotation.

7.4 Chemotactic Competence

In *E. coli*, at least nine genes affect chemotactic competence by affecting reversibility of the motor. The classical *cheA* has recently been found to consist of two genes (*cheA* and *cheW*) co-transcribed with *mot* (*mocha* operon). The classical *cheB* is now replaced by four genes (*cheX*, *cheB*, *cheY*, and *cheZ*). A seventh gene (*cheM*) has not been mapped precisely, but is believed to code for the membrane-bound methyl-accepting chemotaxis protein (MCP). The other *che* genes yield cytoplasmic products (SILVERMAN and SIMON, 1977). *CheC* is the same gene as *flaA*, and *cheD* as *tsr*.

In *Salmonella*, there are also at least nine such genes (*cheP*–*cheX*) (WARRICK et al., 1977). *CheR* has been identified as the gene for the chemotactic methyltransferase (W. SPRINGER and KOSHLAND, 1977). As with *E. coli*, some *che* genes affect *fla* phenotype.

Two *che*$^-$ phenotypes are encountered; mutants either never tumble (smooth swimming phenotype) or tumble constantly (tumbling phenotype).

8. Concluding Remarks

The last decade or so has seen major advances in our understanding of the phenomenon of bacterial motility and chemotaxis. These are summarized schematically in Figure 6. Genetic, biochemical, and biophysical approaches have all been used to advantage. The mechanism and energy source for motility are now known, as is the motor response to tactic stimuli; some chemoreceptors have been isolated,

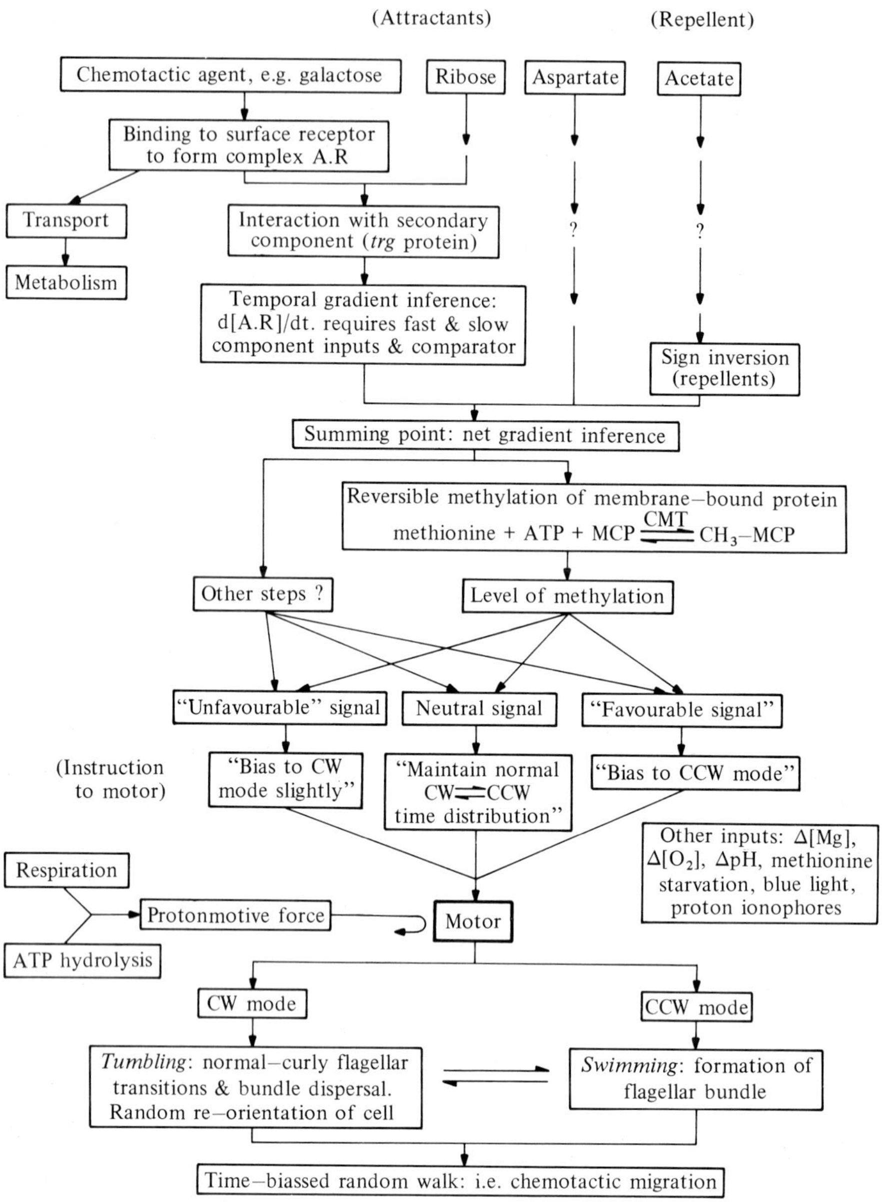

Fig. 6. Chemotaxis in *E. coli* and *Salmonella*. A schematic representation of present knowledge. The exact point at which gradient inference and summing occurs is not clear, nor is the interplay between methylation and "other steps". A protonmotive force drives the motors, in either counterclockwise (*ccw*) or clockwise (*cw*) rotational modes; the "gear" receives gradient signals from the chemoreceptor system, and a time-bias arises in the cell's motion, producing migration

purified, and characterized, and certain aspects of the cellular biochemistry related to chemotaxis are beginning to emerge; the intricate genetic basis for the phenomenon is progressively yielding to molecular cloning and other techniques. In spite of this, much is still unknown, including such fundamental questions as the molecular basis of gradient inference, of conversion of protonmotive force into the mechanical force of the rotary motor, of the reversibility of the device, and of the switch. The whole system is proving to be remarkably complex and interesting, and if, as seems quite possible, it can eventually be understood in detail, it should provide some valuable organizational and mechanistic principles which may have a much wider validity.

References

Adler, J.: Effect of amino acids and oxygen on chemotaxis in *Escherichia coli*. J. Bacteriol. *92*, 121–129 (1966a)

Adler, J.: Chemotaxis in bacteria. Science *153*, 708–716 (1966b)

Adler, J.: Chemoreceptors in bacteria. Science *166*, 1588–1597 (1969)

Adler, J.: A method for measuring chemotaxis and use of the method to determine optimum conditions for chemotaxis by *Escherichia coli*. J. Gen. Microbiol. *74*, 77–91 (1973)

Adler, J.: Chemotaxis in bacteria. Annu. Rev. Biochem. *44*, 341–356 (1975)

Adler, J.: The sensing of chemicals by bacteria. Sci. Am. *234*, 40–47 (1976)

Adler, J., Dahl, M.M.: A method for measuring the motility of bacteria and for comparing random and non-random motility. J. Gen. Microbiol. *46*, 161–173 (1967)

Adler, J., Epstein, W.: Phosphotransferase-system enzymes as chemoreceptors for certain sugars in *Escherichia coli* chemotaxis. Proc. Natl. Acad. Sci. USA *71*, 2895–2899 (1974)

Adler, J., Hazelbauer, G.L., Dahl, M.M.: Chemotaxis toward sugars in *Escherichia coli*. J. Bacteriol. *115*, 824–847 (1973)

Adler, J., Tso, W.-W.: "Decision"-making in bacteria: Chemotactic response of *Escherichia coli* to conflicting stimuli. Science *184*, 1292–1294 (1974)

Aksamit, R., Koshland, Jr., D.E.: A ribose binding protein of *Salmonella typhimurium*. Biochem. Biophys. Res. Commun. *48*, 1348–1353 (1972)

Anraku, Y.: Transport of sugars and amino acids in bacteria. I. Purification and specificity of the galactose- and leucine-binding proteins. J. Biol. Chem. *243*, 3116–3122 (1968)

Armstrong, J.B.: An S-adenosylmethionine requirement for chemotaxis in *Escherichia coli*. Can. J. Microbiol. *18*, 1695–1701 (1972)

Aswad, D., Koshland, Jr., D.E.: Role of methionine in bacterial chemotaxis. J. Bacteriol. *118*, 640–645 (1974)

Aswad, D., Koshland, Jr., D.E.: Isolation, characterization and complementation of *Salmonella typhimurium* chemotaxis mutants. J. Mol. Biol. *97*, 225–235 (1975a)

Aswad, D., Koshland, Jr., D.E.: Evidence for an S-adenosylmethionine requirement in the chemotactic behavior of *Salmonella typhimurium*. J. Mol. Biol. *97*, 207–223 (1975b)

Baracchini, O., Sherris, J.C.: The chemotactic effect of oxygen on bacteria. J. Pathol. Bacteriol. *77*, 565–574 (1959)

Berg, H.C.: How to track bacteria. Rev. Sci. Instrum. *42*, 868–871 (1971)

Berg, H.C.: Chemotaxis in bacteria. Annu. Rev. Biophys. Bioeng. *4*, 119–136 (1975)

Berg, H.C., Brown, D.A.: Chemotaxis in *Escherichia coli* analysed by three-dimensional tracking. Nature (London) *239*, 500–504 (1972)

Berg, H.C., Tedesco, P.M.: Transient response to chemotactic stimuli in *Escherichia coli*. Proc. Natl. Acad. Sci. USA *72*, 3235–3239 (1975)

Bezdek, M., Soska, J.: Sex-determined chemotaxis in *Salmonella typhimurium* LT2. Folia Microbiol. (Prague) *17*, 366–369 (1972)

Blakemore, R.: Magnetotactic bacteria. Science *190*, 377–379 (1975)

Boos, W.: The properties of the galactose-binding protein, the possible chemoreceptor for galactose chemotaxis in *Escherichia coli*. Antibiot. Chemother. *19*, 21–54 (1974)
Boos, W., Gordon, A.S., Hall, R.E., Price, H.D.: Transport properties of the galactose-binding protein of *Escherichia coli*. Substrate-induced conformational change. J. Biol. Chem. *247*, 917–924 (1972)
Brown, D.A., Berg, H.C.: Temporal stimulation of chemotaxis in *Escherichia coli*. Proc. Natl. Acad. Sci. USA *71*, 1388–1392 (1974)
Buder, J.: Zur Kenntnis des *Thiospirillum jenense* und seiner Reaktion auf Lichtreize. Jahrb. Wiss. Bot. *56*, 529–584 (1915)
Caraway, B.H., Krieg, N.R.: Uncoordination and recoordination in *Sprillum volutans*. Can. J. Microbiol. *18*, 1749–1759 (1972)
Clayton, R.K.: Studies in the phototaxis of *Rhodospirillum rubrum*. I. Action spectrum, growth in green light, and Weber Law adherence. Arch. Mikrobiol. *19*, 107–124 (1953)
Clayton, R.K.: On the interplay of environmental factors affecting taxis and motility in *Rhodospirillum rubrum*. Arch. Mikrobiol. *29*, 189–212 (1958)
Clayton, R.K.: Phototaxis in microorganisms. Photophysiology *2*, 51–77 (1964)
Dahlquist, F.W., Elwell, R.A., Lovely, P.S.: Studies of bacterial chemotaxis in defined concentration gradients. A model for chemotaxis toward L-serine. J. Supramol. Struct. *4*, 329–342 (1976)
Dahlquist, F.W., Lovely, P., Koshland, Jr., D.E.: Quantitative analysis of bacterial migration in chemotaxis. Nature (London) *236*, 120–123 (1972)
Drews, G.: Beiträge zur Kenntnis der phototaktischen Reaktionen der Cyanophyceen. Arch. Protistenk. *104*, 389–430 (1959)
Drift, C. van der, Duiverman, J., Bexkens, H., Krijnen, A.: Chemotaxis of a motile streptococcus towards sugars and amino acids. J. Bacteriol. *124*, 1142–1147 (1975)
Greenberg, E.P., Canale-Parola, E.: Chemotaxis in *Spirochaeta aurantia*. J. Bacteriol. *130*, 485–494 (1977)
Hazelbauer, G.L., Mesibov, R.E., Adler, J.: *Escherichia coli* mutants defective in chemotaxis toward specific chemicals. Proc. Natl. Acad. Sci. USA *64*, 1300–1307 (1969)
Hazelbauer, G.L., Parkinson, J.S.: Bacterial chemotaxis. In: Receptors and Recognition: Microbial Interactions. Reissig, J. (ed.). Chapman and Wall, London, 1977, pp. 60–98
Hilmen, M., Simon, M.: Motility and the structure of bacterial flagella. In: Cell Motility. Goldman, R., Pollard, T., Rosenbaum, J. (eds.). Cold Spring Harbor, New York, 1976, pp. 35–45
Kellermann, O., Szmelcman, S.: Active transport of maltose in *Escherichia coli* K12. Involvement of a "periplasmic" maltose binding protein. Eur. J. Biochem. *47*, 139–149 (1974)
Kleene, S.J., Toews, M.L., Adler, J.: Isolation of glutamic acid methyl ester from an *Escherichia coli* membrane protein involved in chemotaxis. J. Biol. Chem. *252*, 3214–3218 (1977)
Kort, E.N., Goy, M.F., Larsen, S.H., Adler, J.: Methylation of a membrane protein involved in bacterial chemotaxis. Proc. Natl. Acad. Sci. USA *72*, 3939–3942 (1975)
Koshland, Jr., D.E.: A response regulator model in a simple sensory system. Science *196*, 1055–1063 (1977a)
Koshland, Jr., D.E.: Sensory response in bacteria. In: Advances in Neurochemistry 2. Agranoff, B.W., Aprison, M.H. (eds.). New York: Plenum Press 1977b, pp. 277–341
Larsen, S.H., Reader, R.W., Kort, E.N., Tso, W.-W., Adler, J.: Change in direction of flagellar rotation is the basis of the chemotactic response in *Escherichia coli*. Nature (London) *249*, 74–77 (1974)
Lederberg, J.: Linear inheritance in transductional clones. Genetics *41*, 845–871 (1956)
Lengeler, J.: Mutations affecting transport of the hexitols D-mannitol, D-glucitol, and galactitol in *Escherichia coli* K-12: Isolation and mapping. J. Bacteriol. *124*, 26–38 (1975)
Lofgren, K.W., Fox, C.F.: Attractant-directed motility in *Escherichia coli*: Requirement for a fluid lipid phase. J. Bacteriol. *118*, 1181–1182 (1974)
Lovely, P., Dahlquist, F.W., Macnab, R., Koshland, Jr., D.E.: An instrument for recording the motions of microorganisms in chemical gradients. Rev. Sci. Instrum. *45*, 683–686 (1974)
Macnab, R.M.: Bacterial Motility and Chemotaxis. Crit. Rev. Biochem.*5*, 291–341 (1978)
Macnab, R.M., Koshland, Jr., D.E.: The gradient-sensing mechanism in bacterial chemotaxis. Proc. Natl. Acad. Sci. USA *69*, 2509–2512 (1972)

Macnab, R.M., Koshland, Jr., D.E.: Persistence as a concept in the motility of chemotactic bacteria. J. Mechanochem. Cell Motil. *2*, 141–148 (1973)
Macnab, R.M., Koshland, Jr., D.E.: Bacterial motility and chemotaxis: Light-induced tumbling response and visualization of individual flagella. J. Mol. Biol. *84*, 399–406 (1974)
Maeda, K., Imae, Y., Shioi, J.-I., Oosawa, F.: Effect of temperature on motility and chemotaxis of *Escherichia coli*. J. Bacteriol. *127*, 1039–1046 (1976)
Manson, M.D., Tedesco, P., Berg, H.C., Harold, F.M., Drift, C van der.: A protonmotive force drives bacterial flagella. Proc. Natl. Acad. Sci. USA *74*, 3060–3064 (1977)
Manten, A.: Phototaxis in the purple bacterium *Rhodospirillum rubrum*, and the relation between phototaxis and photosynthesis. Antonie van Leeuwenhoek *14*, 65–86 (1948)
Mesibov, R., Adler, J.: Chemotaxis toward amino acids in *Escherichia coli*. J. Bacteriol. *112*, 315–326 (1972)
Mesibov, R., Ordal, G.W., Adler, J.: The range of attractant concentrations for bacterial chemotaxis and the threshold and size of response over this range. Weber law and related phenomena. J. Gen. Physiol. *62*, 203–223 (1973)
Metzner, P.: Die Bewegung und Reizbeantwortung der bipolar gegeisselten *Spirillen*. Jahrb. Wiss. Bot. *59*, 325–412 (1920)
Miller, J.B., Koshland, Jr., D.E.: Membrane fluidity and chemotaxis: Effects of temperature and membrane lipid composition on the swimming behavior of *Salmonella typhimurium* and *Escherichia coli*. J. Mol. Biol. *111*, 183–201 (1977a)
Miller, J.B., Koshland, Jr., D.E.: Bacterial electrophysiology: Relationship of the membrane potential to motility and chemotaxis in *Bacillus subtilis*. Proc. Natl. Acad. Sci. USA *74*, 4752–4756 (1977b)
Niel, C.B. van, Goodwin, T.W., Sissins, M.E.: The nature of the changes in carotenoid synthesis in *Rhodospirillum rubrum* during growth. Biochem. J. *63*, 408–412 (1956)
Nultsch, W.: Photomotion of microorganisms and its interactions with photosynthesis. In: Photobiology of Microorganisms. Halldal, P. (ed.). New York: Wiley, 1970, pp. 213–251
Ordal, G.W.: Control of tumbling in bacterial chemotaxis by divalent cation. J. Bacteriol. *126*, 706–711 (1976a)
Ordal, G.W.: Effect of methionine on chemotaxis by *Bacillus subtilis*. J. Bacteriol. *125*, 1005–1012 (1976b)
Ordal, G.W., Adler, J.: Isolation and complementation of mutants in galactose taxis and transport. J. Bacteriol. *117*, 509–516 (1974a)
Ordal, G.W., Adler, J.: Properties of mutants in galactose taxis and transport. J. Bacteriol. *117*, 517–526 (1974b)
Ordal, G.W., Gibson, K.J.: Chemotaxis toward amino acids by *Bacillus subtilis*. J. Bacteriol. *129*, 151–155 (1977)
Ordal, G.W., Goldman, D.J.: Chemotaxis away from uncouplers of oxidative phosphorylation in *Bacillus subtilis*. Science *189*, 802–805 (1975)
Ordal, G.W., Goldman, D.J.: Chemotactic repellents of *Bacillus subtilis*. J. Mol. Biol. *100*, 103–108 (1976)
Ordal, G.W., Villani, D.P., Gibson, K.J.: Amino acid chemoreceptors of *Bacillus subtilis*. J. Bacteriol. *129*, 156–165 (1977)
Parkinson, J.S.: Data processing by the chemotaxis machinery of *Escherichia coli*. Nature (London) *252*, 317–319 (1974)
Parkinson, J.S.: Genetics of chemotactic behavior in bacteria. Cell *4*, 183–188 (1975)
Parkinson, J.S.: Behavioral genetics in bacteria. Annu. Rev. Genet. *11*, 397–414 (1977)
Pfeffer, W.: Über chemotaktische Bewegungen von Bakterien, Flagellaten und Volvocineen. Unters. Botan. Inst. Tübingen *2*, 582–663 (1888)
Roberts, A.M.: Geotaxis in motile micro-organisms. J. Exp. Biol. *53*, 687–699 (1970)
Rothert, W.: Über die Wirkung des Aethers und Chloroforms auf die Reizbewegungen der Mikroorganismen. Jahrb. Wiss. Bot. *39*, 1–70 (1904)
Sanderson, K.E.: Linkage map of *Salmonella typhimurium*, Edition IV. Bacteriol. Rev. *36*, 558–586 (1972)
Schenk, S.L.: Zentralbl. Bakteriol. Parasitenk. *14*, 33 (1893)
Seymour, F.W.K., Doetsch, R.N.: Chemotactic responses by motile bacteria. J. Gen. Microbiol. *78*, 287–296 (1973)

Silverman, M., Simon, M.: Flagellar rotation and the mechanism of bacterial motility. Nature (London) *249*, 73–74 (1974)

Silverman, M., Simon, M.: Identification of polypeptides necessary for chemotaxis in *Escherichia coli*. J. Bacteriol. *130*, 1317–1325 (1977)

Springer, M.S., Goy, M.F., Adler, J.: Sensory transduction in *Escherichia coli*: A requirement for methionine in sensory adaptation. Proc. Natl. Acad. Sci. USA *74*, 183–187 (1977)

Springer, M.S., Kort, E.N., Larsen, S.H., Ordal, G.W., Reader, R.W., Adler, J.: Role of methionine in bacterial chemotaxis: Requirement for tumbling and involvement in information processing. Proc. Natl. Acad. Sci. USA *72*, 4640–4644 (1975)

Springer, W.R., Koshland, Jr., D.E.: Identification of a protein methyltransferase as the *cheR* gene product in the bacterial sensing system. Proc. Natl. Acad. Sci. USA *74*, 533–537 (1977)

Spudich, J.L., Koshland, Jr., D.E.: Quantitation of the sensory response in bacterial chemotaxis. Proc. Natl. Acad. Sci. USA *72*, 710–713 (1975)

Straley, S.C., Conti, S.F.: Chemotaxis in *Bdellovibrio bacteriovorus*. J. Bacteriol. *120*, 549–551 (1974)

Strange, P.G., Koshland, Jr., D.E.: Receptor interactions in a signalling system: Competition between ribose receptor and galactose receptor in the chemotaxis response. Proc. Natl. Acad. Sci. USA *73*, 762–766 (1976)

Szmelcman, S., Adler, J.: Change in membrane potential during bacterial chemotaxis. Proc. Natl. Acad. Sci. USA *73*, 4387–4391 (1976)

Taylor, B.L., Koshland, Jr., D.E.: Reversal of flagellar rotation in monotrichous and peritrichous bacteria: Generation of changes in direction. J. Bacteriol. *119*, 640–642 (1974)

Taylor, B.L., Koshland, Jr., D.E.: Intrinsic and extrinsic light responses of *Salmonella typhimurium* and *Escherichia coli*. J. Bacteriol. *123*, 557–569 (1975)

Taylor, B.L., Miller, J.B., Warrick, H.M., Koshland, Jr., D.E.: Electron transfer pathways are essential for tactic responses to light and electron acceptors by *Salmonella typhimurium* and *Escherichia coli*. (ms. in preparation, 1978)

Tsang, N., Macnab, R., Koshland, Jr., D.E.: Common mechanism for repellents and attractants in bacterial chemotaxis. Science *181*, 60–63 (1973)

Tso, W.-W., Adler, J.: Negative chemotaxis in *Escherichia coli*. J. Bacteriol. *118*, 560–576 (1974)

Tsui Collins, A.L., Stocker, B.A.D.: *Salmonella typhimurium* mutants generally defective in chemotaxis. J. Bacteriol. *128*, 754–765 (1976)

Warrick, H.M., Taylor, B.L., Koshland, Jr., D.E.: The chemotactic mechanism of *Salmonella typhimurium*: Mapping and characterization of mutants. J. Bacteriol. *130*, 223–231 (1977)

Weibull, C.: Movement. In: The Bacteria, Vol. 1. Gunsalus, I.C., Stanier, R.Y. (eds.). New York: Academic Press, 1960, pp. 153–205

Werf, P. van der, Koshland, Jr., D.E.: Identification of a γ-glutamyl methyl ester in bacterial membrane protein involved in chemotaxis. J. Biol. Chem. *252*, 2793–2795 (1977)

Willis, R.C., Furlong, C.E.: Purification and properties of a ribose-binding protein from *Escherichia coli*. J. Biol. Chem. *249*, 6926–6929 (1974)

Ziegler, H.: Chemotaxis. In: Handbuch der Pflanzenphysiologie, Vol. 17, Ruhland, W. (ed.). Berlin, Göttingen, Heidelberg: Springer, 1962, pp. 484–532

Zukin, R.S., Hartig, P.R., Koshland, Jr., D.E.: Use of a distant reporter group as evidence for a conformational change in a sensory receptor. Proc. Natl. Acad. Sci. USA *74*, 1932–1936 (1977b)

Zukin, R.S., Koshland, Jr., D.E.: Mg^{2+}, Ca^{2+}-dependent adenosine triphosphatase as receptor for divalent cations in bacterial sensing. Science *193*, 405–408 (1976)

Zukin, R.S., Strange, P.G., Heavey, L.R., Koshland, Jr., D.E.: Properties of the galactose binding protein of *Salmonella typhimurium* and *Escherichia coli*. Biochemistry *16*, 381–386 (1977a)

3.2.3 Chemotaxis in Unicellular Eukaryotes

B. BEAN

1. The Nature and Diversity of Eukaryotic Chemoresponsiveness

Chemosensitive behavior is widespread among both multicellular and unicellular eukaryotes. In contrast to recent progress in understanding of bacterial chemotaxis, few eukaryote chemoresponses have been analyzed or even described in detail. The analysis of eukaryotic chemoresponses is complicated conceptually and technically by the diversity and complexity of the systems involved; but the very features that make these systems seem bewildering also make them interesting.

Chemotaxis has both immediate and long-term adaptive consequences. Eukaryotic chemoresponses reflect evolutionary strategies that vary with the life style of the species. The adaptive advantage of chemotaxis is usually revealed in migration: (1) toward increased nutrient supplies, (2) away from chemically harsh regions, or (3) toward reproductive targets. Few biological phenomena offer such simple and compact models of "goal-oriented" processes. Such systems constitute interesting models in biological cybernetics, and merit analysis at both the molecular and evolutionary levels.

In cases where chemoresponses directly influence reproduction, evolutionary changes in these systems may have played a key role in the reproductive isolation of incipient new species. This view is supported by the existence of groups of organisms in which closely related species are distinguished by generally similar chemoresponse systems that show somewhat different chemical sensitivities (e.g., TSUBO, 1961; MÜLLER, 1976).

Several notable reviews have appeared on chemotaxis and chemotropism of several groups of organisms that are favorites for laboratory study. ZIEGLER'S (1962) general review provides a substantial coverage of the classical observations on chemotaxis in plants and is recommended as a guide to the earlier papers in this field. A recent monograph by VAN DEN ENDE (1976) reviews the chemical coordination of sexual reproduction in genera *Allomyces*, *Achlya*, *Mucor*, *Chlamydomonas*, *Volvox*, and *Oedogonium*, and in the yeasts, brown algae, ferns, and flowering plants. A valvable recent collection of reviews edited by CARLILE (1975a) includes four on chemotaxis: in bacteria (ADLER, 1975), cellular slime molds (KONIJN, 1975), fungi and algae (GOODAY, 1975), and leukocytes (WILKINSON, 1975). MACHLIS and RAWITSCHER-KUNKEL (1963) have reviewed the chemical orientation systems of some plant gametes. Two reviews by WIESE (1969; 1976) cover the events of sexual reproduction, including chemotaxis, in algae. Useful related material appears in reviews by MCCUTCHEON (1946), HALLDALL (1962), HICKMAN and HO (1966), JAHN and BOVEE (1967), JAHN and VOTTA (1972), GOODAY (1974), NULTSCH (1974) and in recent collections of research reports edited by PÉREZ-MIRAVETE (1973), SLEIGH (1973), SORKIN (1974), DUCKET and RACEY (1975), WU et al. (1975), INOUE and STEPHENS (1975), and GOLDMAN et al. (1976). Chemotaxis in slime molds is covered in this volume (POFF and WHITAKER, Chap. 3.2.4).

It is difficult to determine what importance chemoresponse systems have in nature. In the natural environment, individual organisms encounter, and may respond to, a complex of stimuli, and hence it is frequently difficult to identify the actual stimulus.

This treatment will attempt to focus on common features and fundamental processes that operate in eukaryote chemotaxis systems. In view of the previously cited recent reviews on various facets of this field, it seems inappropriate to attempt here a comprehensive cataloging of phenomena.

2. Classification and Mechanics of Chemoresponsive Behavior

Chemoresponsive behavior includes any change in movement (response) of an organism that is mediated with some specificity by a particular external chemical compound (stimulus). This general formulation postulates an interplay of a chemical receptor with an effector; the linkage between these two key elements is an interesting but poorly understood area.

FRAENKEL and GUNN (1961) and others (CARLILE, 1975b; DIEHN et al., 1977) have classified and discussed orientation behaviors in depth. The terms are based mainly on the mechanics of the response. The terms kinesis or orthokinesis denote reactions that involve changes in rate of movement. Klinokinesis defines that general set of responses that result from changes in the frequency of turning (reorientation maneuvering). A phobic response is a sudden change in movement elicited by a change in stimulus intensity. The term taxis (from the Greek ταξις; position, order, arrangement) denotes motion that is oriented relative to a stimulus, where the achievement (and perhaps maintenance) of orientation requires perception of the direction of the stimulus. Bacterial chemotaxis is a klinokinesis; however, the more classic term chemotaxis has the sanction of common usage. The general aspects of motile behavior are further discussed in the introduction to this volume (HAUPT and FEINLEIB, Chap. 1.1).

Unfortunately, investigators often do not distinguish the orientation behavior categories, or they adopt assays that obscure the mode of response, and characterization of cell movement is lacking. In general the term chemotaxis in common use can only be taken as an indication that redistribution of population takes place.

In many cases, responses have proven to be quite complex. The mechanism for maintaining an asymmetric distribution may be quite different from the mechanism employed in the original orientation event. Examples of sensory adaptation, alteration of swimming style, and combination responses (e.g., taxis plus kinesis) are known. Investigators should be careful and complete in describing patterns of movement.

3. Assessment of Chemotaxis

3.1 Assays of Free-Swimming Cells

A number of methods have been used for the evaluation of chemotactic behavior of microorganisms and other motile cells. PFEFFER (1884) and other pioneers in the study of chemotaxis inserted capillary tubes containing the test solution into a suspension of cells, and directly observed migration of cells away from, toward, or into the capillary tube. Such simple assays have been used widely, but they present some difficulties. Detection of chemotaxis often depends upon the persistence of a chemical concentration gradient, which is difficult to maintain if an open capillary tube is immersed amid a swarm of highly motile cells. Capillary tubes can function as light pipes, and thereby promote phototactic redistribution of photosensitive cells. Tubes might also act as traps for cells that cannot turn effectively within them, or for negatively geotactic cells that originally entered by purely random swimming, but cannot escape randomly. Direct observation of chemotaxis by such methods must be performed with appropriate controls and interpreted carefully.

In spite of these problems, capillary tube assays can be very sensitive and qualitatively accurate. The assay starts with a sizeable population of randomly distributed motile cells; the formation of a nonrandom array is readily apparent, and is usually taken as a criterion of response.

In all capillary assay systems the time course of the response is important. Either physiological or physical factors may limit detection of an orientation response. Chemosensory adaptation is known in some microbial systems (MACHLIS, 1973b; BROKAW 1974; SPRINGER et al., 1977), suggesting that possible *initial* responses may be the most important in some systems, and may be observable only briefly. In some cases, responsive cells are able to take up and metabolize or otherwise inactivate chemoactive substances (MACHLIS, 1973a; MUSGRAVE and NIEUWENHUIS, 1975), or to inhibit further production of attractants by the target sexual cells.

Many interesting modifications of the capillary tube method have been introduced. ROTHSCHILD (1952, 1956) added 1% agar to his test solution, thus limiting diffusion into and out of the capillary tube, and blocking entry into the capillary. Cinematography was used to record cell movements near the mouth of the capillary, for subsequent analysis. The labor of frame-by-frame analysis can be avoided in systems where dark-field time-exposure or stroboscopic photography can be used to record cell movements. Such systems have been used, e.g., by BROKAW (1958b, 1959, 1974), KATSURA and MIYATA (1971), and by ALLEN and NEWHOOK (1973).

The availability of records of the details of the paths of cell movements makes it possible to study the modes of behavior in responsive and nonresponsive cells. Depending on the methods employed, cell tracks contain information on: (1) rates of movement of cells, (2) frequencies of turning, and (3) the rate and character of the turning maneuvers. MCCUTCHEON (1923, 1946) described the orientation maneuvers of individual cells in terms of a chemotaxis ratio, where the numerator is the net progress toward the target and the denominator is the actual path length through which the cell traveled. Ratios close to +1.0 indicate paths straight toward the target, while lower ratios reflect circuitous routes. Such values can be determined for different subpopulations, or for cells influenced by different experimental variables. NOSSAL and ZIGMOND (1976) recently analyzed the turn-angle probabilities for chemotactic polymorphonuclear leukocytes, and developed a mathematical model (NOSSAL, 1976) to represent their behavior.

Many other chemotaxis assays have been used. One method involves spotting or streaking of a solution to be tested on a microscope slide, and allowing the material to evaporate to dryness. A sample of uniformly suspended cells, spotted on the slide to cover part or all of the streak of dried material is monitored microscopically for formation of an asymmetric distribution (McCUTCHEON, 1946). In other assay systems, glass beads or other objects are coated with the material to be tested. When appropriately treated glass microbeads are added to a suspension of chemotactic male gametes of the fungus *Allomyces*, clouds of cells gather about the beads within a few minutes (CARLILE and MACHLIS, 1965a). Similar observations have been reported for an assay system in which cells accumulate in the vicinity of fragments of agar, or oil droplets, (or other material) into which the chemotactic agent has been incorporated (MÜLLER, 1972; JAENICKE et al., 1974). In other cases, natural structures to which motile cells are responsive are used directly as particulate chemotaxis targets. For example, zoospores of the fungus *Rozella* accumulate about hyphae of its host *Allomyces* (HELD, 1973), root fragments attract zoospores of the parasitic fungus *Phytophthora* (DUKES and APPLE, 1961), and oogonial mother cells and oogonia of the green alga *Oedogonium* attract motile male cells (HOFFMAN, 1960; RAWITSCHER-KUNKEL and MACHLIS, 1962).

A very sensitive qualitative assay for chemotaxis of free-swimming cells has been used by POMMERVILLE (1977) in studies on *Allomyces*. The assay situation resembles that used in immunodiffusion studies—small wells (3 mm diameter) are cut in a thin layer of agar to give a symmetric array of four peripheral wells surrounding a single central well. Motile cells are placed in the peripheral wells, compounds or structures to be tested (or appropriate controls) are placed in the central well. Redistribution of cells within the peripheral wells is monitored microscopically for 15 min, and cell distributions are photographed. In cases of attraction, male gametes of *Allomyces* accumulate at or near the side of their well that is closest to the source of the diffusing attractant.

3.2 Assays Involving Immobilization

There are several assays that exploit the property of some species to attach physically to certain types of substrates. Machlis and colleagues developed a chemotaxis assay for *Allomyces* that used special chambers with bottoms of cellophane dialysis membrane. These wells are suspended in a reservoir containing a population of motile *Allomyces*, and the wells are loaded with solutions to be tested. The chambers are incubated for a measured period, and the number of cells that become attached per unit area of membrane is determined (MACHLIS, 1969a, 1973a).

KHEW and ZENTMYER (1973) recently described a quantitative assay for chemotaxis of *Phytophthora* zoospores (see Fig. 1). The zoospore suspension is placed in a reservoir on a microscope slide, and from either end, glass capillary micropipettes (pre-filled with agar containing the experimental and control substances) are introduced. The distribution of fungal cells is monitored at intervals during the subsequent 3–5 h, by which time encystment is complete. Counts are then made of the numbers of encysted zoospores at the mouth of the capillary containing the attractant and at a distance 6–8 mm away, where "randomly encysted" zoospores are located. The ratio of the number near the capillary relative to the number randomly encysted is taken as the chemotactic index, i.e., values significantly different from 1.0 reflect chemotactically asymmetric distributions. HALSALL (1976) used a similar assay system with some refinements including photography

Fig. 1. Microscope slide chemotaxis assay chamber. (After KHEW and ZENTMYER, 1973)

of the area at the mouth of the capillary, and MÜLLER (1976) used a similar ratio to characterize chemotaxis in brown algae.

Levandowsky and colleagues also have used attachment at an interface in a chemoaccumulation assay. Two opposite quadrants of a divided petri dish are prefilled with agar containing the chemical to be tested, while the remaining quadrants are filled with unsupplemented agar as a control. A suspension of the motile dinoflagellate *Crypthecodinium* is added above the level of the agar substrate. During a 3-h incubation period, the organisms become attached to the agar substrate in an asymmetric array (HAUSER et al., 1975a, 1975b; LEVANDOWSKY et al., 1975b). It is not clear what the behavioral basis of this asymmetric attachment phenomenon is (e.g., chemotaxis, chemophobic response, chemokinesis), and the authors correctly use a general term "chemosensory responses" to characterize the phenomenon. This assay again permits quantification: the authors test the statistical significance of asymmetry in the distribution of cells, and report specific concentration ranges over which significant positive or negative responses are observed. This method provides a very sensitive measure of the differential distribution of individuals, but quantifies a complex response without distinguishing component processes. This particular case is further complicated by the fact that reproducible results could not be obtained for chemotaxis to capillary sources (HAUSER et al., 1975b). Furthermore, the species was also reported to be phototactic, yet the petri dish chemoresponse assay was carried out under illumination for long periods (HAUSER et al., 1975b). While the results reported for this system are indeed interesting, it is important in such cases to develop techniques that provide better resolution of responses.

In cases that entail attachment or immobilization, it is not always apparent whether accumulation is the result of chemotaxis, chemokinesis, or simply of selective removal of individuals from the population by immobilization at that particular location (see discussions by MACHLIS and RAWITSCHER-KUNKEL, 1963; WIESE, 1969; CHANG-HO and HICKMAN, 1970; HICKMAN and HO, 1977). In *Allomyces,* cells orient rapidly and move directly toward the membrane of the test chamber prior to attachment (CARLILE and MACHLIS, 1965a, 1965b): a clear case of chemotaxis. Related species may show different characteristics or combinations of chemoresponses, e.g., zoospores of different phycomycetes possess different capabilities for chemotaxis and chemokinesis (ALLEN and HARVEY, 1974; GOODE, 1956; KHEW and ZENTMYER, 1973; ROYLE and HICKMAN, 1964a, 1964b; ZENTMYER, 1961).

4. Mechanisms of Eukaryotic Chemotaxis

Analysis of eukaryotic chemotaxis at the subcellular and molecular levels is now a feasible objective. What are the biological structures, processes, and regulatory phenomena that make taxis possible? While the goal of comprehensive understanding remains elusive, some important ideas have recently emerged.

Chemoresponsive behavior may be viewed as including three operational levels: (1) chemosensory or receptor processes (2) transduction events, and (3) effector action, usually a motor response. (The term transduction is intended to include transmission of signals as well as processing or modification or interpretation of those signals relative to the current or past physiological condition or location of the organism.) Biochemical structures, whether in soluble form or in macromolecular assemblies, are ultimately responsible for the encoding of information.

4.1 The Receptor Level

Chemotaxis systems show remarkable diversity in both the qualitative and quantitative aspects of their sensitivity. Some systems are exacting in their specificity. Known agents include both high and low molecular weight structures; cationic, anionic and nonionic compounds; acids, bases, and salts; lipid-soluble, and water-soluble materials; volatile and nonvolatile forms; corrosive and noncorrosive substances; synthetic and natural products. The ranges of sensitivity shown are wide in some cases, and narrow in others. The threshold concentration (the minimum to which a response is detected) is low in some instances, and relatively high in others. Responses may be conditioned by, or influenced by other chemical or physical features of the organisms' environment.

In several experimental systems it has been possible to challenge an organism with a family of chemically similar substances to delineate stimulus-receptor specificities (cf. ZIEGLER, 1962; HICKMAN and HO, 1966). The sensitivities and specificities of reception, at least in several cases (see 4.1.2 below), suggest that highly specific chemical recognition interactions take place at or near the cell surface—probably involving specific receptor proteins such as those discussed for bacterial chemotaxis (cf. MACNAB, Chap. 3.2.2; VAN DEN ENDE, 1976). Systems that involve specific interactions between the chemical stimulus and a receptor protein could at once reconcile chemotaxis concentration sensitivity ranges, chemical specificities, and evolutionary diversity.

4.1.1 Nonspecific Responses

On the other hand, systems are known in which responses are relatively generalized or nonspecific. Some organisms show apparent generalized response to pH, ionic strength, harsh chemicals, inorganic ion concentrations, etc. Broad spectra or generalized responses of this type suggest the existence of relatively nonspecific sensors of the chemical environment. Some of these systems may rely on general structural or functional changes in the plasma membrane for detection of chemical conditions. In *Chlamydomonas reinhardtii,* for instance, a general avoidance of conditions

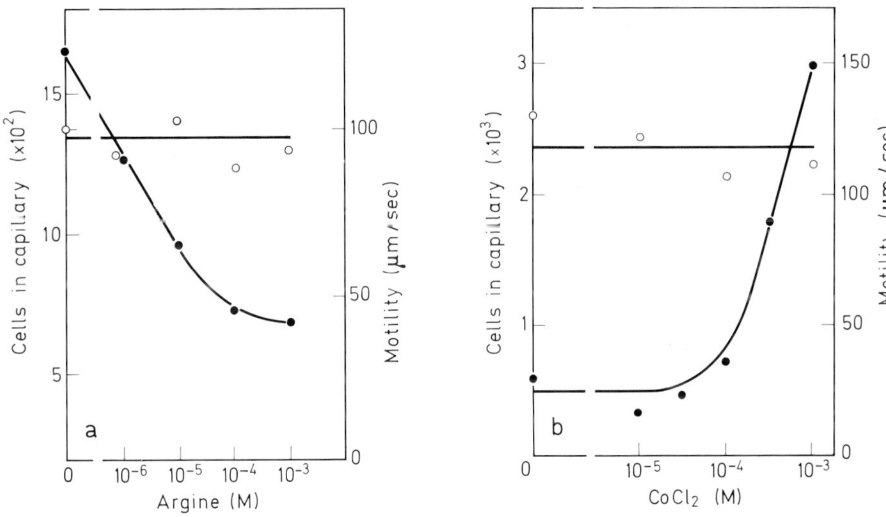

Fig. 2. Negative and positive responses of *Chlamydomonas reinhardtii* (●——●) to **a** arginine and **b** $CoCl_2$, respectively, assayed by Adler's capillary method. Motility (○——○) given for comparison. (From HIRSCHBERG and RODGERS, 1978)

of high ionic strength is observed (JEFFERS and BEAN, unpublished observations). With many compounds, this alga successfully avoids damaging conditions, while with other compounds cells become immobilized (see Figs. 2 and 3). Considering the diverse ionic compounds to which such response is shown, and the high concentration necessary to evoke the response, it seems safe to conclude that no specific receptors are involved. However, genuine avoidance behavior is involved, since the cells effectively avoid entering the region where adverse conditions would result in immobilization, accumulation, and death of the cells. The sensing mechanism probably entails local damage or depolarization in the plasma membrane or in the flagellar membrane, coupled with a consequent reorientation of the cell. Nonspecific responses of this type have been identified in several groups among the lower plants (cf. ZIEGLER, 1962; GOODAY, 1975).

4.1.2 Specific Responses

MACHLIS and CARLILE observed specific chemoresponses in the water-mold *Allomyces*. There are five distinct flagellated unicellular stages in the life cycle of *Allomyces*, and cells at different stages show different chemoresponse sensitivities (CARLILE and MACHLIS, 1965b; MACHLIS, 1969a). "Chemotaxis" is evaluated after attachment of cells to a membrane, as previously discussed. Zygotes, mitospores, and meiospores, but not male or female gametes, exhibit positive chemotaxis toward certain amino acids at concentrations that are thought to be of nutritional significance. These cells respond to 0.4% casein hydrolysate, or to specific combinations of certain amino acids at 2×10^{-4} M each (CARLILE and MACHLIS, 1965b; MACHLIS, 1969a, 1969b). There are differences in responsiveness between closely related species. The mitospores and meiospores of *A. macrogynus* respond to a mixture

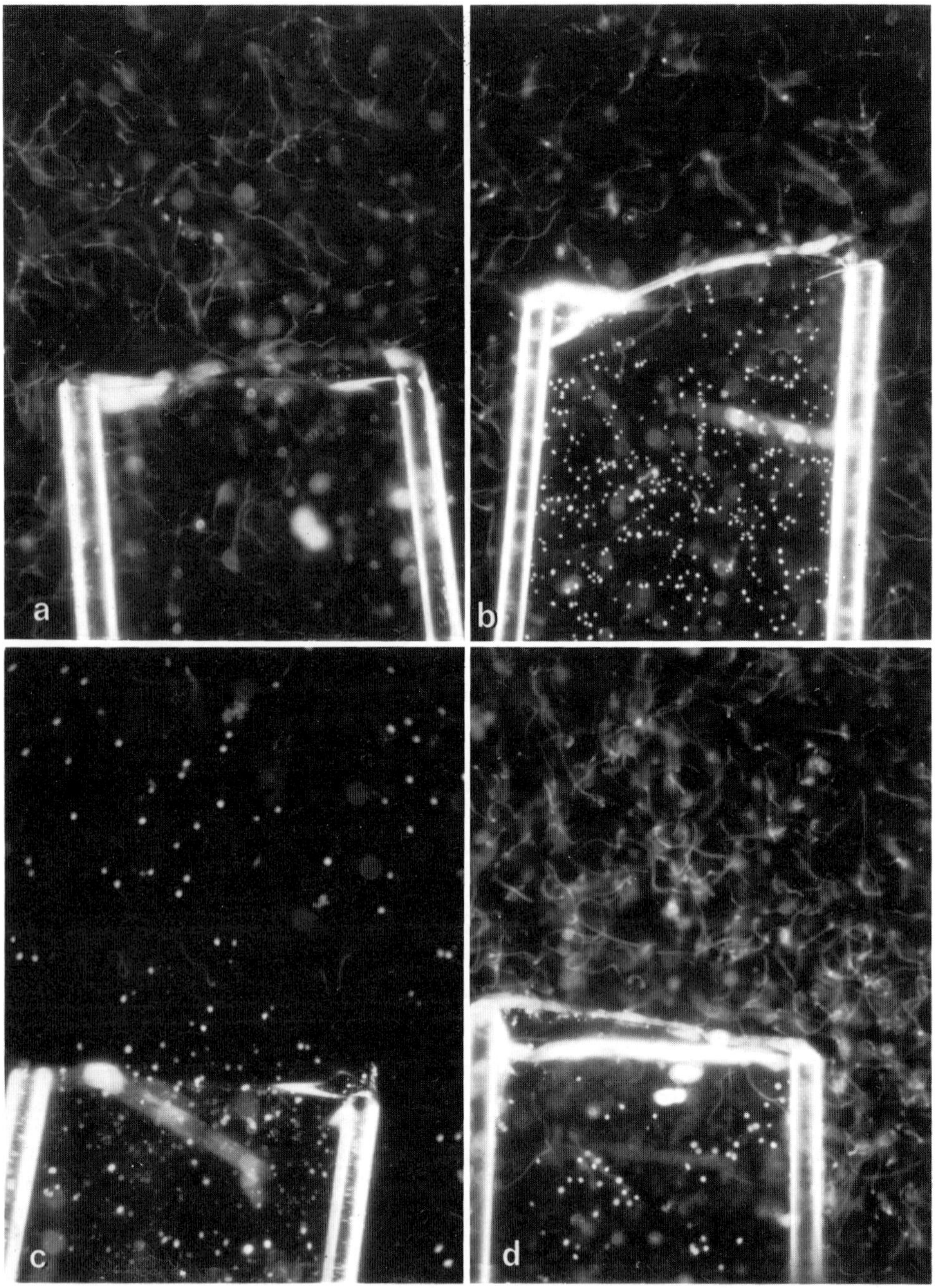

Fig. 3a–d. Complexity of responses: nonphototactic mutant of *Chlamydomonas* examined in the assay of KHEW and ZENTMYER by dark field microscopy (JEFFERS and BEAN, unpublished observations). **a** Control (1.7×10^6 cells/ml, $1^1/_2$ s. exposure); **b** immobilization and adhesion of cells to capillary (containing 0.5 g/ml sucrose); **c** avoidance of arginine (0.5 g/ml in capillary), **d** attraction to ammonium tartrate (0.38 g/ml in capillary)

of L-leucine and L-lysine, but do not respond to either alone, or to most other amino acids or combinations thereof (MACHLIS, 1969a). The response is enhanced by proline, although this species will not respond to proline alone. *A. arbusculus* shows a similar response, but is also responsive to proline alone (MACHLIS, 1969a). Neither male nor female gametes of either species show sensitivity to amino acids, but zygotes, within 5 min after fusion of the gametes, develop the ability for the amino acid response (MACHLIS, 1969b).

The specificities of chemotaxis in *Allomyces* suggest that responsiveness is dependent upon the state of biological differentiation in the cell. Since the gametes are motile and fully capable of chemotaxis to another attractant (viz., sirenin, see the paragraph after next), it seems likely that the difference in amino acid sensitivity is due to differences at the receptor level. The data strongly suggest that the primary sensory event involves the recognition of particular amino acids by specific receptor molecules. Chemotaxis toward amino acids in *Allomyces* would make an excellent system for further experimental analysis. The binding and incorporation of specific radiolabeled amino acids should be studied in relation to chemical conditions that elicit different behaviors (i.e., with and without the appropriate combination of amino acids for attraction; in the presence or absence of an attractant gradient). The fate of labeled molecules during the developmentally distinct stages should be compared. The effects and fate of nonmetabolizable analogs should be contrasted with those for the natural attractants.

DILL and FULLER (1971) have used treatment with amino acids (under conditions similar to those used by MACHLIS and colleagues) to immobilize cells and to synchronize germination in a third species, *Allomyces neomoniliformis*. They observed that the combination of leucine and lysine arrested flagellar action in zoospores. In further work on these systems, it would be desirable to distinguish effects on motility, orientation, and attachment, and to investigate the sequence and regulation of events during the response.

MACHLIS also led investigations into the chemotactic response of male gametes of *Allomyces* to the sex-attractant sirenin. Sirenin is a sex hormone or erotactin that is specifically produced by female gametes; it elicits a specific positive chemotactic response of male gametes (CARLILE and MACHLIS 1965a; MACHLIS 1958a, 1972; MACHLIS and RAWITSCHER-KUNKEL, 1963). The effective concentrations are much lower than those of the amino acid attractants, and span a broad concentration range. A significant attraction toward sirenin is evident at concentrations as low as 10^{-10} M, and as high as 10^{-4} M, with an optimum around 10^{-6} M (CARLILE and MACHLIS 1965a). Higher concentrations of sirenin cause loss of orientation, suggesting that these higher concentrations result in saturation of receptor sites (CARLILE and MACHLIS, 1965a).

Careful studies were undertaken to purify, characterize, and synthesize sirenin (MACHLIS 1958b; MACHLIS et al., 1966; MACHLIS et al., 1968; NUTTING et al., 1968; PLATTNER et al., 1969; COREY et al., 1969; GRIECO, 1969; BHALERAO et al., 1970). These efforts established the identity of sirenin as 9-[(E)-4-hydroxy-3-methyl-2-butenyl]-10-hydroxy-2-carene, a sesquiterpene with a molecular formula $C_{15}H_{24}O_2$ (see Fig. 4). The response of male gametes is highly selective for 1-sirenin; closely related compounds, including stereoisomers, positional isomers, and the corresponding deoxy compounds, do not stimulate chemotaxis as effectively (BHALERAO et al., 1970; MACHLIS, 1973b). Different species and interspecies hybrids

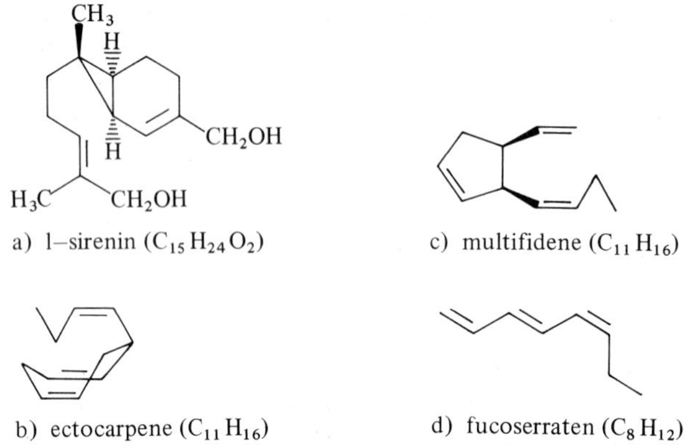

Fig. 4a–d. Structures of some known sexual attractants

of *Allomyces* respond differently to sirenin (MACHLIS, 1968) and to some other compounds tested (MACHLIS, 1973b).

CARLILE and MACHLIS (1965a; MACHLIS, 1973b) showed that sirenin is inactivated by male gametes of *Allomyces*. Sirenin is also subject to some inactivation by nonresponsive or heterologous cells such as mitospores of *Allomyces* or yeast cells. After exposure to moderate amounts of sirenin, furthermore, male gametes are temporarily unable to orient chemotactically to a gradient of sirenin, although they can still inactivate the compound. Studies on the rate of disappearence of sirenin reveal that the attractant is taken up by male gametes, with first order kinetics. Once taken up, sirenin cannot be recovered from the cells, suggesting that it must be rapidly converted to a chemotactically inactive form.

Both the inactivation of sirenin and the post-exposure period of insensitivity to it may play significant roles in the mechanism of orientation in a sirenin gradient. The rapid inactivation may account for the very broad concentration range over which these cells respond. Inactivation could serve to maintain and accentuate a gradient of sirenin concentration, and perhaps act to "amplify" differences in the region surrounding responsive cells. Considering the possible competitiveness among male gametes and among female gametes for participation in fertilization, biochemical inactivation of the stimulus would appear to have adaptive value. The first cells to travel a gradient would increase their own prospects for sexual success, and simultaneously redirect later arrivals toward more fertile adventures.

The continued ability of male gametes of *Allomyces* to inactivate sirenin while they are unable to orient (MACHLIS, 1973b) indicates that the refractory period is not due to a simple substrate overload in the enzymatic system that inactivates sirenin, but rather to saturation of a specific recognition or receptor protein in the male cell.

A refractory period such as that reported by MACHLIS (1973b) could contribute importantly to the orientation mechanism through its influence on the frequency of reorientation maneuvers. Recently, POMMERVILLE (1978) reported that randomly swimming male gametes of *Allomyces* move in short, smooth arcs, and that oriented swimming is characterized by a more steady spiral swimming toward the female gametes, with a suppression of randomizing reorientation maneuvers. Thus, a cell swimming away from a sirenin source is likely to change its orientation; whereas a cell swimming toward a sirenin source experiences a refractory period during which reorientation does not occur, and continues to move toward the source.

Transient changes in behavior of *Allomyces* male gametes following the rapid introduction of sirenin are consistent with those seen in the vicinity of female cells, and provide evidence for a temporal sensing mechanism (POMMERVILLE, 1978, cf. also MACNAB, Chap. 3.2.2, Sect. 4.3, this vol.). The state of binding sites, turnover rates of the receptor molecules, and activities of enzymes for sirenin transport and metabolism might interact in determining the fundamental timing of comparator events.

The specificities of sirenin chemotaxis may have evolved through a series of simple changes in the regulation of gene expression or modifications of enzyme characteristics in a nonessential pathway for the production, transport, and/or metabolism of sirenin and closely related compounds.

POMMERVILLE (1977) recently reported reciprocal sexual orientation in *Allomyces macrogynus:* male gametes produce a substance that attracts the motile female gametes. This attraction is weaker than that of male gametes toward sirenin in the same species.

Sex-taxis of the type observed for *Allomyces* is reasonably widespread among the lower plants, but only a few of these systems have been studied in detail. MÜLLER et al., (1971) studied the compound ectocarpene that is produced by female gametes of the brown alga *Ectocarpus siliculosus* and attracts male gametes. The compound has the formula $C_{11}H_{16}$, allo-cis-1-(cycloheptadien-2′, 5′-yl)-butene-1 (Fig. 4b; MÜLLER et al., 1971). JAENICKE et al. (1974) found that the attractant produced by the gynogametes of *Cutleria multifida* is a different $C_{11}H_{16}$ hydrocarbon (multifidene Fig. 4c). Recent work has also led to the identification of the chemoattractant produced by eggs of the alga *Fucus serratus* as 1, 3-trans-5-cis-octatriene (fucoserraten, Fig. 4d; MÜLLER, 1972; JAENICKE and SEFERIADIS, 1975).

These compounds act with considerable chemical selectivity in the attraction of male gametes toward conspecific female gametes. Related compounds are not as effective, although there does seem to be some general capacity for chemotaxis to hydrocarbons in many algae (MÜLLER, 1976). Unfortunately there has been little work yet on the details of the behavioral nature of the response.

In members of the green algal genus *Chlamydomonas*, events of mating involve specificities that are behavioral as well as structural. These interesting interactions include stages of general clumping and specific pairwise agglutination and signalling (cf. VAN DEN ENDE, 1976; SOLTER and GIBOR, 1977; WIESE, 1969). There is at least one case (TSUBO, 1957, 1961) in which cells of opposite mating type locate each other by means of chemotaxis, and it seems likely that some other volvocales also utilize chemotaxis. In the strain *Chlamydomonas moewusii* var. *rotunda*, studied by Tsubo, there is unidirectional chemotaxis by gametes of the plus mating type toward products secreted by gametes of the minus mating type. This response does not occur in vegetative cells. Tsubo showed that the attractant was volatile, and that the response could be simulated by low molecular weight hydrocarbons, such as ethylene or acetylene. In tests of the species specificity of response, TSUBO (1961) found that gametes of two other varieties of *C. moewusii* and *C. eugametos* were attracted to supernatant from minus mating type *C. moewusii* var. *rotunda*, but those of *C. reinhardtii* were not. Of the species studied, Tsubo found intraspecific sexual chemotaxis only in *C. moewusii* var. *rotunda* and in *Chlamydomonas* sp. 24 (TSUBO, 1957, 1961).

In addition to specific chemotaxis systems, many species of lower plants exhibit chemotropisms. While the effector aspects are quite different for chemotropism than they are for chemotaxis, the chemoreception mechanisms may be very similar. Tropisms and taxes clearly operate within different time frames, but the difference may not be at the receptor level. Further discussions of tropisms appear in CARLILE (1975), and VAN DEN ENDE (1976), and elsewhere in this volume (Part 5, Growth Movements).

4.1.3 The Nature of the Receptor

As discussed above, the biological specificity usually associated with eukaryotic chemotaxis depends on specific receptors that recognize and bind particular chemicals. While there is no direct evidence, it seems likely that the receptors are protein molecules of the cell surface. The effective concentration range for a chemotactic agent would reflect characteristics of interaction between the receptor protein and its "substrate"—including the dissociation constant, the number, location, and rate of turnover of active receptor molecules, as well as competition with other proteins for the substrate (before or after binding).

The receptor proteins are probably rather sophisticated molecules. In addition to their primary binding site for the chemotactic agent, they may also interact specifically with other proteins involved in the metabolism or transport of the chemotactic agent. They must also have specificity for interaction with other cell structures such that information flow can take place—some biochemical process must be modulated by the presence or absence of a bound ligand at the primary receptor site of the receptor protein; in addition, they may have recognition sites for regulatory agents.

It may be possible to study the subcellular distribution and physiology of receptors in at least some cases. Experimental approaches might include: (1) in situ autoradiography of radioactive attractant molecules (2) substitution of analogs of natural attractants (3) blocking of transport and/or metabolism of attractants with antimetabolites or by using the appropriate mutants (4) localization studies using fluorescent or electron-opaque markers. Studies on mutant strains that are unable to sense the chemotactic agent could be of particular analytic importance.

BROKAW (1974) has outlined a possible mechanism for chemoreception that reconciles (and reinterprets) many observations (BROKAW, 1957, 1958a, 1958b, 1959), on the orientation behavior of spermatozoids of the bracken fern *Pteridium aquilinum*. Individuals are attracted toward acid, calcium ion, and malate, and also exhibit anodic galvanotaxis. Importantly, these orientation responses require low concentrations of both bimalate ion and Ca^2. Brokaw envisions a transport-limited receptor membrane in which the passage of both malate and Ca^{2+} is restricted. If either species is absent, the other cannot pass the membrane: orientation behavior does not occur, and random swimming is observed. When both ions are present, however, coupled transport can occur via carrier molecules in the receptor membrane. When fluxes of both ions occur, modulation of flagellar activity takes place, which changes the orientation of the spermatozoid.

BROKAW'S (1974) mechanism involves temporal discrimination of environmental conditions: increase in concentration of one ion species in the medium generates a period of increased influx for both, and thus a transiently higher concentration of ions in the intraflagellar space. After a lag period, the ion balance across the membrane is restored. The length of this period is determined by the physical character of the gradient and by the path of the cell in the gradient. The period required for ion flux and restoration of a balanced state constitutes a time frame for the response. Thus, by virtue of its movement, the cell transforms a spatial gradient into a period of changing intraflagellar ion concentrations. Such concentrations are responsible for the character of flagellar activity and the nature of the maneuver executed. It appears that this single mechanism accounts for both the achievement and maintainance of correct orientation relative to the stimulus.

BROKAW's ideas are attractive in many ways. The coupled diffusion model of chemoreception could operate in other chemotaxis systems as well. Different protein carrier molecules could account for the different responses. At present, Brokaw's hypothesis represents a stimulating and relatively simple model for chemoreceptor function. Its correctness remains to be determined, but its heuristic value is clear.

4.2 Signal Transduction and Effector Action

In recent years remarkable progress has been made in understanding the mechanisms of eukaryotic motility. In these effector systems specific protein molecules catalyze the hydrolysis of ATP, and trap the energy in the form of specific interactions between protein molecules; the structure, assembly, and biochemical activity of these proteins is influenced by nucleotides, ions, and other specific proteins. Some recent perspectives on basic effector function are presented in this volume (SATIR and OJAKIAN, Chap. 3.1.2) and elsewhere (SLEIGH, 1973; INOUE and STEPHENS, 1975; GOLDMAN et al., 1976).

Most plant chemotaxis systems employ cilia or flagella as effectors. Recent studies in several organisms have implicated Ca^{2+} as one important modulator of axonemal action (BROKAW, 1974; BROKAW et al., 1974; HYAMS and BORISY, 1975; KUNG et al., 1975; BROWNING et al., 1976; ORDALL, 1977; BEAN and HARRIS, 1979). There are several interesting ideas about the mechanisms by which Ca^{2+} influences axonemal activity e.g., through direct or indirect modulation of an ATPase activity of the flagellum (cf. SATIR and SALE, 1977). In *Chlamydomonas* both Ca^{2+}-activated and Mg^{2+}-activated flagellar ATPases have been identified (WATANABE and FLAVIN, 1973; WITMAN et al., 1976), and probably exist in other organisms as well (GIBBONS, 1966; KUNG and NAITOH, 1973; BROKAW et al., 1974; HAYASHI, 1974; ECKERT and MACHEMER, 1975).

BROKAW (1974) observed differences in the swimming style of multiflagellate bracken fern spermatozoids that depended upon their orientation, and he related the differences to the intracellular calcium ion concentration (discussed in Section 4.1.3). Individuals moving up an attractant gradient swim in tight helices relative to randomly swimming spermatozoids, while those heading down-gradient swim with a helical path that has a large radius. Individuals heading away from an attractant therefore (1) make slow net progress away from the chemical, and (2) exhibit a path that is more irregular and subject to directional changes. BROKAW suggests that these movements result from greater asymmetry in the activity of the several dozen flagella of cells heading away from an attractant, compared with those heading toward the attractant (BROKAW, 1957, 1958a, 1958b, 1959, 1974). BROKAW (1974) rejects his own earlier (1958b) suggestion that the mechanism of orientation in galvanotaxis involves binding of malate to the sperm in an asymmetric array that makes them turn passively in a voltage gradient. Observations on the swimming speed and turning rates of orienting individuals suggest instead a mechanism that entails active flagellar response.

Paramecium has been an excellent subject for analysis since it is amenable to electrophysiological and genetic study. In this and other organisms, influx of calcium into the axoneme is controlled by biochemical and biophysical properties of the cortical membrane. The characteristics of ion flow correlate neatly with the arrival of appropriate sensory stimuli, and with the triggering of reorientation maneuvers by the responding cell (cf. ECKERT and MACHEMER, 1975; KUNG et al., 1975). Recent studies using electron microscopy and electron microprobe analysis have revealed the distribution of local concentrations of ions in the cell after fixation. In *Paramecium* and other species, deposits of Ca^{2+} (or other divalent cations) are most frequently localized near the cell surface in association with the ciliary membranes, basal bodies, and axonemes (FISHER et al., 1976; PLATTNER, 1975; PRZELECKA et al., 1977; TSUCHIYA and TAKAHASHI, 1976). ECKERT and MACHEMER (1975) suggest a generalization that, in protozoa, Ca^{2+} concentrations in the axoneme directly govern the direction and frequency of beat in cilia.

While there has been significant progress in achieving an understanding of the basic mechanisms of axonemal activity, attention to other levels of effector organization has been irregular. There are many different patterns of physical organization and modes of stroking of flagella (cf. JAHN and BOVEE, 1967). Motile behavior is complex, particularly during orientation, involving coordinated movements of two or more axonemes, or employing several possible modes of axoneme movement or turning maneuvers (BEAN, 1977; BROKAW, 1963, 1974; GITTLESON et al., 1974).

Another complexity that arises at the effector level is the possibility for intercellular interaction. In cultures of many species, for instance, cells have the ability to form and maintain characteristic patterns of aggregation, a phenomenon often called pattern swimming. Pattern swimming is thought to involve geotaxis and/or chemotaxis, coupled with the behavioral effects of intercellular interactions (GITTLESON and JAHN, 1968; LEVANDOWSKY et al., 1975a; PLATT, 1961; PLESSET et al., 1975; WAGER, 1910).

From the viewpoint of biological regulation, stimulus transduction is probably the most interesting realm of behavioral organization, yet the least approachable and certainly the least well understood. It is now clear that several genes specify transduction in bacterial chemotaxis (MACNAB, Chap. 3.2.2, this vol.).

One type of signal transduction system, the membrane-controlled ion-gating system of the ciliates, is perhaps the best understood (cf. KUNG et al., 1975). In the thigmotactic and galvanotactic responses of *Paramecium* it may be that the polarized membrane is itself the receptor—and the same may be true of nonspecific chemotactic responses. However, it is also easy to envision response systems that could exploit ion-gating as a transduction level process. In such a system, specific receptor proteins at the cell surface would be coupled with an ion pump or with some other function of the plasma membrane. This would introduce the element of specificity that is well known in chemotaxis systems, and permit triggering of membrane depolarization under conditions when depolarization would not otherwise occur.

Further analysis of behavioral mutants of *Paramecium* should make it possible to detect response systems in which specific receptors operate, and the membrane functions in a transduction capacity. Indeed, VAN HOUTEN (1977) has recently characterized a chemotaxis mutant of *Paramecium* that appears to carry a defect in the degree of hyperpolarization in the ciliary membrane under certain chemical conditions.

There have been several experimental approaches to signal transduction in the case of phototaxis in unicellular eukaryotes (cf. STAVIS, 1974; NULTSCH, 1975; HIRSCHBERG and STAVIS, 1977), although there is no consensus on the biochemical nature of these processes. The photoresponses of *Chlamydomonas* have been well studied (FEINLEIB and CURRY, 1971a, b; BRUCE, 1972; HUDOCK and HUDOCK, 1973; STAVIS and HIRSCHBERG, 1973; STAVIS, 1974; FEINLEIB, 1975; HIRSCHBERG and STAVIS, 1977), and chemotaxis has also been found (A. GIBOR, personal communication; TSUBO, 1957, 1961; HIRSCHBERG and RODGERS, 1977, 1978; JEFFERS and BEAN, unpublished observations). Experimentally versatile genetic systems are known for this genus. Given these advantages, it should be possible to detect the possible sharing of transduction level steps in the pathways for processing of information for phototaxis and chemotaxis in *Chlamydomonas*. It would seem likely from evolutionary considerations that such steps should be shared—this could result in more efficient use of the relevant biochemical machinery, and could provide for behavioral coordination between phototactic and chemotactic responses.

Selective inhibitors should prove important in the analysis of regulation at the level of signal transmission. It is difficult to pinpoint the specific actions of new inhibitors, or to find inhibitors that have a selectivity coincident with the wishes of the investigator. Nevertheless, some promising selective effects are known. In cases where a given metal ion is important in the regulation of behavioral activity, the effects of other ions are often inhibitory (KUZNICKI, 1963; SCHEIN, 1976; BEAN and HARRIS, 1978). Ionophores and transport blocking agents may generate abnormal behavior. In *Chlamydomonas* the inhibitors antimycin A and sodium azide may be selective toward processes involved in taxis behavior (STAVIS and HIRSCHBERG, 1973; STAVIS, 1974; BEAN, 1977).

It is probable that current speculations and model systems for eukaryotic transduction processes are simplistic. The most promising experimental avenue for elucidating these processes appears to entail comparative analysis of behavioral mutants as has been done for bacterial chemotaxis. The eukaryotic mechanisms may reveal more complexity, diversity, and beauty.

Acknowledgments. I am grateful to Drs. R.M. MACNAB and M.E.H. FEINLEIB for careful criticism, and to A.M. STERN for expert secretarial assistance.

References

Adler, J.: Chemotaxis in bacteria. In: Primitive Sensory and Communications Systems: The Taxes and Tropisms of Microorganisms and Cells. Carlile, M.J. (ed.). London-New York-San Francisco: Academic Press, 1975, pp. 91–100

Allen, R.N., Harvey, J.D.: Negative chemotaxis of zoospores of *Phytophthora cinnamomi*. J. Gen. Microbiol. *84*, 28–38 (1974)
Allen, R.N., Newhook, F.J.: Chemotaxis of zoospores of *Phytophthora cinnamomi* to ethanol in capillaries of soil pore dimensions. Trans. Brit. Mycol. Soc. *61*, 287–302 (1973)
Bean, B.: Geotactic Behavior of *Chlamydomonas*. J. Protozool. *24*, 394–401 (1977)
Bean, B., Harris, A.: Selective inhibition of flagellar activity in *Chlamydomonas* by Ni^{++}. J. Protozool. in press (1979)
Bhalerao, U.T., Plattner, J.J., Rapoport, H.: Synthesis of dl-sirenin and dl-isosirenin. J. Am. Chem. Soc. *92*, 3429–3433 (1970)
Brokaw, C.J.: "Electro-chemical" orientation of bracken spermatozoids. Nature (London) *179*, 525 (1957)
Brokaw, C.J.: Chemotaxis of bracken spermatozoids: The role of bimalate ions. J. Exp. Biol. *35*, 192–196 (1958a)
Brokaw, C.J.: Chemotaxis of bracken spermatozoids: Implications of electrochemical orientation. J. Exp. Biol. *35*, 197–212 (1958b)
Brokaw, C.J.: Random and oriented movements of bracken spermatozoids. J. Cellular Comp. Physiol. *54*, 95–101 (1959)
Brokaw, C.J.: Movement of the flagella of *Polytoma uvella*. J. Exp. Biol. *40*, 149–156 (1963)
Brokaw, C.J.: Calcium and flagellar response during the chemotaxis of bracken spermatozoids. J. Cell. Physiol. *83*, 151–158 (1974)
Brokaw, C.J., Josslin, R., Bobrow, L.: Calcium ion regulation of flagellar beat symmetry in reactivated sea urchin spermatozoa. Biochem. Biophys. Res. Comm. *58*, 795–800 (1974)
Browning, J.L., Nelson, D.L., Hansma, H.G.: Ca^{2+} influx across the excitable membrane of behavioural mutants of *Paramecium*. Nature *259*, 491–494 (1976)
Bruce, V.: Mutants of the biological clock in *Chlamydomonas reinhardi*. Genetics *70*, 537–548 (1972)
Carlile, M.J., (ed.).: Primitive Sensory and Communication Systems: The Taxes and Tropisms of Microorganisms and Cells. London-New York-San Francisco: Academic Press, 1975a, 258 pp.
Carlile, M.J.: Taxes and tropisms: Diversity, biological significance and evolution. In: Primitive Sensory and Communication Systems. Carlile, M.J. (ed.). London-New York-San Francisco: Academic Press, 1975b, pp. 1–28
Carlile, M.J., Machlis, L.: The response of male gametes of *Allomyces* to the sexual hormone sirenin. Am. J. Bot. *52*, 478–483 (1965a)
Carlile, M.J., Machlis, L.: A comparative study of the chemotaxis of the motile phases of *Allomyces*. Am. J. Bot. *52*, 484–486 (1965b)
Chang-Ho, Y., Hickman, C.J.: Some factors involved in the accumulation of phycomycete zoospores on plant roots. In: Root Diseases and Soil-born Pathogens. Tousson, T.A., Bega, R.V., Nelson, P.E. (eds.). Berkeley-Los Angeles-London: Univ. Calif. 1970, pp. 103–108
Corey, E.J., Achiwa, K., Katzenellenbogen, J.A.: Total synthesis of dl-sirenin. J. Am. Chem. Soc. *91*, 4318–4320 (1969)
Diehn, B., Feinleib, M., Haupt, W., Hildebrand, E., Lenci, F., Nultsch, W.: Terminology of behavioral responses of motile microorganisms. Photochem. Photobiol. *26*, 559–560 (1977)
Dill, B.C., Fuller, M.S.: Amino acid immobilization of fungal motile cells. Arch. Microbiol. *78*, 92–98 (1971)
Duckett, J.G., Racey, P.A. (eds.): The Biology of the Male Gamete. London: Academic Press, 1975, 460 pp.
Dukes, P.D., Apple, J.L.: Chemotaxis of zoospores of *Phytophthora parasitica* var. nicotinae by plant roots and certain chemical solutions. Phytopathology *51*, 195–197 (1961)
Eckert, R., Machemer, H.: Regulation of ciliary beating frequency by the surface membrane. In: Molecules and Cell Movement. Inoue, S., Stephens, R.E. (eds.). New York: Raven Press, 1975, pp. 151–164
Ende van den, H.: Sexual Interactions in Plants: The Role of Specific Substances in Sexual Reproduction. London-New York-San Francisco: Academic Press, 1976, 186 pp.
Feinleib, M.E.: Phototactic responses of *Chlamydomonas* to flashes of light. I. Responses of cell populations. Photochem. Photobiol. *21*, 351–354 (1975)

Feinleib, M.E., Curry, G.M.: The nature of the photoreceptor in phototaxis. In: Handbook of Receptor Physiology I. Principles of Receptor Physiology. Lowenstein, W.R. (ed.). Berlin: Springer, 1971a, pp. 365–395
Feinleib, M.E., Curry, G.M.: The relationship between stimulus intensity and oriented phototactic response (topotaxis) in *Chlamydomonas*. Physiol. Plant. *25*, 346–352 (1971b)
Fisher, G., Kaneshiro, E.S., Peters, P.D.: Divalent cation affinity sites in *Paramecium aurelia*. J. Cell Biol. *69*, 429–442 (1976)
Fraenkel, G.S., Gunn, D.L.: The Orientation of Animals, 2nd ed. New York: Dover Press, 1961, 376 pp.
Gibbons, I.R.: Studies on the adenosine triphosphatase activity of 14S and 30S dyncin from cilia of *Tetrahymena*. J. Biol. Chem. *241*, 5590–5596 (1966)
Gittleson, S.M., Hotchkiss, S.K., Valencia, F.G.: Locomotion in the marine dinoflagellate *Amphidinium carterae* (Hulburt). Trans. Am. Micros. Soc. *93*, 101–105 (1974)
Gittleson, S.M., Jahn, T.L.: Pattern swimming by *Polytomella agilis*. Am. Nat. *103*, 413–25 (1968)
Goldman, R., Pollard, T., Rosenbaum, J. (eds.): Cell Motility. New York: Cold Spring Harbor Laboratory, 1976. 3 vols.
Gooday, G.W.: Fungal sex hormones. Annu. Rev. Biochem. *43*, 35–49 (1974)
Gooday, G.W.: Chemotaxis and chemotropism in fungi and algae. In: Primitive Motile and Communication Systems: The Taxes and Tropisms of Microorganisms and Cells. Carlile, M.J. (ed.). London-New York-San Francisco: Academic Press, 1975, pp. 155–204
Goode, P.M.: Infection of strawberry roots by zoospores of *Phytophthora fragariae*. Trans. Brit. Mycol. Soc. *39*, 367–377 (1956)
Grieco, P.A.: The total synthesis of sirenin. J. Am. Chem. Soc. 91, 5660–5661 (1969)
Halldal, P.: Taxes. In: Physiology and Biochemistry of Algae. Lewin, R.A. (ed.). New York: Academic Press, 1962, pp. 583–592
Halsall, D.M.: Zoospore chemotaxis in Australian isolates of *Phytophthora* species. Can. J. Microbiol. *22*, 409–422 (1976)
Hauser, D.C.R., Levandowsky, M., Glassgold, J.M.: Ultrasensitive chemosensory responses by a protozoan to epinephrine and other neurochemicals. Science *190*, 285–286 (1975a)
Hauser, D.C.R., Levandowsky, M., Hutner, S.H., Chunosoff, L., Hollwitz, J.S.: Chemosensory responses by the heterotrophic marine dinoflagellate *Crypthecodinium cohnii*. Microbial Ecology *1*, 246–254 (1975b)
Hayashi, M.: Kinetic analysis of axoneme dynein ATPase from sea urchin sperm. Arch. Biochem. Biophys. *165*, 288–296 (1974)
Held, A.A.: Encystment and germination of the parasitic chytrid *Rozella allomycis* on host hyphae. Can. J. Bot. *51*, 1825–1836 (1973)
Hickman, C.J., Ho, H.H.: Behaviour of zoospores in plant-pathogenic Phycomycetes. Ann. Rev. Phytopathol. *4*, 195–220 (1966)
Hirschberg, R., Rodgers, S.: Chemotaxis in *Chlamydomonas reinhardtii*. Abst. Ann. Meeting Am. Soc. Microbiol. 1977, p. 162 (1977)
Hirschberg, R., Rodgers, S.: Chemoresponses of *Chlamydomonas reinhardtii*. J. Bacteriol. *130*, 671–673 (1978)
Hirschberg, R., Stavis, R.: Phototaxis mutants of *Chlamydomonas reinhardtii*. J. Bacteriol. *129*, 803–808 (1977)
Hoffman, L.R.: Chemotaxis of *Oedogonium* sperms. Southwestern Naturalist *5*, 111–116 (1960)
Houten van, J.: A mutant of *Paramecium* defective in chemotaxis. Science *198*, 746–748 (1977)
Huang, B., Rifkin, M.R., Luck, D.J.L.: Temperature-sensitive mutations affecting flagellar assembly and function in *Chlamydomonas reinhardtii*. J. Cell Biol. *72*, 67–85 (1977)
Hudock, G.A., Hudock, M.O.: Phototaxis: isolation of mutant strains of *Chlamydomonas reinhardi* with reversed sign of response. J. Protozool. *20*, 139–140 (1973)
Hyams, J.S., Borisy, G.G.: The dependence of the waveform and direction of beat of *Chlamydomonas* flagella on calcium ions. J. Cell Biol. *67*, 186a (1975)
Inoue, S., Stephens, R.E. (eds.): Molecules and Cell Movement. New York. Raven, 1975, 450 pp.
Jaenicke, L., Müller, D.G., Moore, R.E.: Multifidene and aucantene, C_{11} hydrocarbons in the male-attracting essential oil from the gynogametes of *Cutleria multifida* (Smith) Gren. (Phaeophyta). J. Am. Chem. Soc. *96*, 3324–3325 (1974)

Jaenicke, L., Seferiadis, K.: Die Stereochemie von Fucoserraten, dem Gametenlockstoff der Braunalge *Fucus serratus* L. Chem. Ber. *108*, 225–232 (1975)
Jahn, T.L., Bovee, E.C.: Motile behavior of Protozoa. In: Research in Protozoology, Vol. 1. Chen, T-T. (ed.). Oxford: Pergamon Press, 1967, pp. 41–200
Jahn, T.L., Votta, J.J.: Locomotion of protozoa. Ann. Rev. Fluid Mech. *4*, 93–116 (1972)
Jennings, H.S.: Behaviour of the Lower Organisms. New York: Columbia University Press, 1906. Republication: Bloomington: Indiana Univ. Press, 1962. 366 pp.
Katsura, K., Miyata, Y.: Swimming behavior of *Phytophthora capsici* zoospores. In: Morphological and Biochemical Events in Plant-Parasite Interaction. Akai, S., Ouchi, S. (eds.). Tokyo, Phytopathol. Soc. Japan, 1971
Khew, K.L., Zentmyer, G.A.: Chemotactic response of zoospores of seven species of *Phytophthora*. Phytopathology 63, 1511–7 (1973)
Konijn, T.M.: Chemotaxis in the cellular slime moulds. In: Primitive Sensory and Communication Systems: The Taxes and Tropisms of Micro-organisms and Cells. Carlile, M.J. (ed.). London-New York-San Francisco: Academic Press, 1975, pp. 101–153
Kung, C., Chang, S.-Y., Satow, Y., van Houten, J., Hansma, H.: Genetic dissection of behavior in *Paramecium*. Science *188*, 898–904 (1975)
Kung, C., Naitoh, Y.: Calcium-induced ciliary reversal in the extracted models of "Pawn", a behavioral mutant of *Paramecium*. Science *179*, 195–196 (1973)
Kuznicki, L.: Reversible immobilization of *Paramecium caudatum* evoked by nickel ions. Acta Protozool. *1*, 301–312 (1963)
Levandowsky, M., Childress, W.S., Spiegel, E.A., Hutner, S.H.: A mathematical model of pattern formation by swimming microorganisms. J. Protozool. *22*, 296–306 (1975a)
Levandowsky, M., Hauser, D.C.R., Glassgold, J.M.: Chemosensory responses of a protozoan are modified by antitubulins. J. Bacteriol. *124*, 1037–1038 (1975b)
Machlis, L.: Evidence for a sexual hormone in *Allomyces*. Physiol. Plant. *11*, 181–192 (1958a)
Machlis, L.: A procedure for the purification of sirenin. Nature (London) *181*, 1790–1791 (1958b)
Machlis, L.: The response of wild type male gametes of *Allomyces* to sirenin. Plant Physiol. *43*, 1319–1320 (1968)
Machlis, L.: Zoospore chemotaxis, in the watermold *Allomyces*. Physiol. Plant. *22*, 126–139 (1969a)
Machlis, L.: Fertilization-induced chemotaxis in the zygotes of the watermold *Allomyces*. Physiol. Plant. *22*, 392–400 (1969b)
Machlis, L.: The coming of age of sex hormones in plants. Mycologia *64*, 235–247 (1972)
Machlis, L.: Factors affecting the stability and accuracy of the bioassay for the sperm attractant sirenin. Plant Physiol. *52*, 524–526 (1973a)
Machlis, L.: The chemotactic activity of various sirenins and analogues and the uptake of sirenin by the sperm of *Allomyces*. Plant Physiol. *52*, 527–530 (1973b)
Machlis, L., Nutting, W.H., Rapoport, H.: The structure of sirenin. J. Am. Chem. Soc. *90*, 1674–1676 (1968)
Machlis, L., Nutting, W.H., Williams, M.W., Rapoport, H.: Production, isolation and characterization of sirenin. Biochemistry *5*, 2147–2152 (1966)
Machlis, L., Rawitscher-Kunkel, E.: Mechanisms of gametic approach in plants. Int. Rev. Cytol. *15*, 97–138 (1963)
McCutcheon, M.: Studies on the locomotion of leukocytes. Am. J. Physiol. *66*, 180–184 (1923)
McCutcheon, M.: Chemotaxis in leucocytes. Physiol. Rev. *26*, 319–336 (1946)
Müller, D.G.: Detection and isolation of the attractant released by eggs of *Fucus serratus* L. Naturwissenschaften *59*, 166 (1972)
Müller, D.G.: Quantitative evaluation of sexual chemotaxis in two marine brown algae. Z. Pflanzenphysiol. *80*, 120–130 (1976)
Müller, D.G., Jaenicke, L., Donike, M., Akintobi, T.: Sex attractant in a brown alga: Chemical structure. Science *191*, 815–817 (1971)
Musgrave, A., Nieuwenhuis, D.: Metabolism of radioactive antheridiol by *Achyla* species. Arch. Microbiol. *105*, 313–317 (1975)
Nossal, R.: Directed cell locomotion arising from strongly biased turn angles. Math. Biosci. *31*, 121–129 (1976)

Nossal, R., Zigmond, S.H.: Chemotropism indices for polymorphonuclear leukocytes. Biophys. J. *16*, 1171–1182 (1976)
Nultsch, W.: Movements. In: Algal Physiology and Biochemistry. Stewart, W.D.P. (ed.). Berkeley Los Angeles: Univ. Calif. 1974, pp. 864–893
Nultsch, W.: Phototaxis and photokinesis. In: Primitive Sensory and Communication Systems: The Taxes and Tropisms of Micro-organisms and Cells. Carlile, M.J. (ed.). London-New York-San Francisco: Academic Press, 1975, pp. 29–90
Nutting, W.H., Rapoport, H., Machlis, L.: The structure of sirenin. J. Am. Chem. Soc. *90*, 6434–6438 (1968)
Ordall, G.W.: Calcium ion regulates chemotactic behaviour in bacteria. Nature (London) *270*, 66–67 (1977)
Pérez-Miravete, A., (coordinator): Behaviour of Micro-organisms. London-New York: Plenum Press, 1973, 301 pp.
Pfeffer, W.: Locomotorische Richtungsbewegungen durch chemischen Reize. Untersuchungen Bot. Inst., Tübingen *1*, 364–482 (1884)
Platt, J.R.: "Bioconvection patterns" in cultures of free-swimming organisms. Science *133*, 1766–1767 (1961)
Plattner, H.: Ciliary granule plaques: membrane-intercalated particle aggregates associated with Ca^{2+}-binding sites in *Paramecium*. J. Cell Sci. *18*, 257–269 (1975)
Plattner, J.J., Bhalerao, U.T., Rapoport, H.: Synthesis of dl-sirenin. J. Am. Chem. Soc. *91*, 4933–4950 (1969)
Plesset, M.S., Whipple, C.G., Winet, H.: Analysis of the steady state of the bioconvection in swarms of swimming microorganisms. In: Swimming and Flying in Nature, Vol. I. Wu. T.Y., Brokaw, C.J., Brennan, C. (eds.). New York-London: Plenum Press, 1974, pp. 339–360
Pommerville, J.: Chemotaxis of *Allomyces* gametes. Exp. Cell Res. *109*, 43–51 (1977)
Pommerville, J.: Analysis of gamete and zygote motility in *Allomyces*. Exp. Cell Res. 161–172 (1978)
Przelecka, A., Sobota, A., Brutkowska, M.: Ca-dependent deposits in cilia of *Tetrahymena pyriformis* GL. J. Protozool. *24*, 24a–25a (1977)
Randall, J., Starling, D.: Genetic determinants of flagellum phenotype in *Chlamydomonas reinhardtii*. In: Proc. Int. Symp. The Genetics of the Spermatozoon. Beatty, R.A., Glueck-sohn-Waelsch, S. (eds.). Edinburgh-New York: Univ. Edinburgh, 1972, pp. 13–36
Rawitscher-Kunkel, E., Machlis, L.: The hormonal integration of sexual reproduction in *Oedogonium*. Am. J. Bot. *49*, 177–183 (1962)
Rothschild, L.: The behavior of spermatozoa in the neighborhood of eggs. Int. Rev. Cytol. *1*, 257–263 (1952)
Rothschild, L.: Fertilization. London: Methuen, and New York: Wiley, 1956, 170 pp.
Royle, D.J., Hickman, C.J.: Analysis of factors governing *in vitro* accumulation of zoospores of *Phythium aphanidermatum* on roots. I. Behaviour of zoospores. Can. J. Microbiol. *10*, 151–162 (1964a)
Royle, D.J., Hickman, C.J.: Analysis of factors governing *in vitro* accumulation of zoospores of *Phythium aphanidermatum* on roots. II. Substances causing the response. Can. J. Microbiol. *10*, 201–219, (1964b)
Satir, P., Sale, W.: Tails of *Tetrahymena*. J. Protozool. *24*, 498–501 (1977)
Schein, S.J.: Nonbehavioral selection for pawns, mutants of *Paramecium aurelia* with decreased excitability. Genetics *84*, 453–468 (1976)
Sleigh, M.A. (ed.): Cilia and Flagella. London: Academic Press, 1973, 500 pp.
Solter, K., Gibor, A.: Evidence for role of flagella as sensory transducers in mating of *Chlamydomonas reinhardi*. Nature (London) *265*, 444–445 (1977)
Sorkin, E. (ed.): Chemotaxis: Its Biology and Biochemistry. Basel: Karger, 1974, 466 pp.
Springer, M.S., Goy, M.F., Adler, J.: Sensory transduction in *Escherichia coli:* A requirement for methionine in sensory adaptation. Proc. Natl. Acad. Sci. U.S. *74*, 183–187 (1977)
Stavis, R.L.: The effect of azide on phototaxis in *Chlamydomonas reinhardi*. Proc. Natl. Acad. Sci. U.S. *71*, 1824–1827 (1974)
Stavis, R.L., Hirschberg, R.: Phototaxis in *Chlamydomonas reinhardtii*. J. Cell Biol. *59*, 367–377 (1973)

Tsubo, Y.: On the mating reaction of a *Chlamydomonas,* with special references to clumping and chemotaxis. Botanical Magazine (Tokyo) *70*, 327–334 (1957)
Tsubo, Y.: Chemotaxis and sexual behavior in *Chlamydomonas.* J. Protozool. *8*, 114–121 (1961)
Tsuchiya, T., Takahashi, K.: Localization of possible calcium-binding sites in the cilia of *Paramecium caudatum.* J. Protozool. *23*, 523–526 (1976)
Wager, H.: On the effects of gravity upon the movements and aggregation of *Euglena viridis,* Ehrb., and other microorganisms. Phil. Trans. Roy. Soc. London *B201*, 333–390 (1910)
Watanabe, T., Flavin, M.: Two types of adenosine triphosphatase from flagella of *Chlamydomonas reinhardi.* Biochem. Biophys. Res. Comm. *52*, 195–201 (1973)
Wiese, L.: Algae. In: Fertilization, Vol. II. Metz, C.B., Monroy, A. (eds.). New York: Academic Press, 1969, pp. 135–188
Wiese, L.: Genetic aspects of sexuality in Volvocales. In: The Genetics of Algae. Lewin, R.A. (ed.). Oxford: Blackwell, 1976, pp. 174–197
Wilkinson, P.C.: Chemotaxis of leucocytes. In: Primitive Motile and Communication Systems: The Taxes and Tropisms of Micro-organisms and Cells. Carlile, M.J. (ed.). London-New York-San Francisco: Academic Press, 1975, pp. 205–243
Witman, G.B., Fay, R., Plummer, J.: *Chlamydomonas* mutants: Evidence for the roles of specific axonemal components in flagellar movement. In: Cell Motility (Book C). Goldman, R., Pollard, T., Rosenbaum, J. (eds.). Cold Spring Harbor, Cold Spring Harbor Laboratory, 1976, pp. 969–986
Wu, T.Y., Brokaw, C.J., Brennen, C. (eds.): Swimming and Flying in Nature, Vol. I. New York-London: Plenum Press, 1974, 421 pp.
Zentmyer, G.A.: Chemotaxis of zoospores for root exudates. Science *133*, 1595–1596 (1961)
Ziegler, H.: Chemotaxis. In: Handbuch der Pflanzenphysiologie, Vol. XVII, pt. 2. Ruhland, W. (ed.). Berlin-Heidelberg-New York: Springer, 1962, pp. 484–532

3.2.4 Movement of Slime Molds

K.L. Poff and B.D. Whitaker

1. Introduction

Much of the work which has been reported in the past fifteen years on the mycetozoans is clearly related to the trend in biology for increased research on fewer "model" systems. Consistent with this trend, the majority of the work on the mycetozoans has been with two species, *Dictyostelium discoideum* and *Physarum polycephalum*. These two species are representatives of two distinctly different groups of mycetozoans, the cellular and the acellular slime molds. The reader is referred to a monograph by Olive (1975) for a thorough discussion of the heterogeneous mycetozoans.

The mycetozoans are perhaps the most primitive eukaryotic organisms, and this accounts at least in part for their use as model systems with many results which may be extrapolated both to higher plants and to animals. Ease of culture in the laboratory and a well-defined morphogenesis also contribute to the popularity of *D. discoideum* and *P. polycephalum*, particularly in studies on growth and development. The different types of movement by the mycetozoans and the rich variety of environmental stimuli to which these organisms will respond strongly encourage their use in studies on movement and sensory transduction[1]. Because of the use of *D. discoideum* and *P. polycephalum* as model systems, this review will discuss movement mainly in these two organisms. Information concerning other organisms will be included when available.

2. Cellular Slime Molds

The vegetative stage of *D. discoideum* is cellular, consisting of myxamoebae (Fig. 1a, b) which feed on bacteria in the mulch on the forest floor (several axenic mutants are available). The amoebae are motile by means of pseudopodia and will move toward food bacteria. Upon starvation, the amoebae enter a quiescent phase (termed interphase) for 6–8 h, and then aggregate to form a multicellular mass of 10^3 to 10^5 cells surrounded by a noncellular slime sheath. The mechanism for aggregation is chemotaxis by the amoebae toward pulses of cyclic 3′,5′ adenosine monophosphate (cAMP) released by other *D. discoideum* amoebae (Fig. 1c).

---

[1] Sensory transduction is used in this chapter in its broadest sense including the complete sequence of biophysical and biochemical events from reception of the stimulus through the processing of information to the behavioral response.

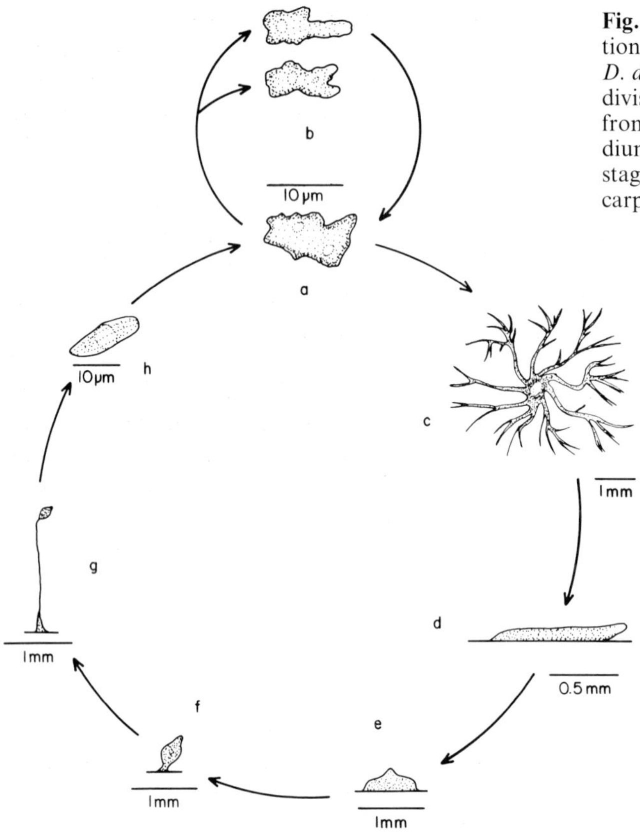

Fig. 1. Diagrammatic representation of the developmental cycle of *D. discoideum*. *a*, amoebae; *b*, cell division; *c*, aggregation viewed from above; *d*, pseudoplasmodium viewed from the side; *e*+*f*, stages of culmination; *g*, sorocarp; *h*, spore

The first multicellular stage, called a pseudoplasmodium, slug, or grex (Fig. 1d), may migrate for some period of time (toward the surface of the forest mulch in nature) and then culminates in the formation of a sorocarp (Fig. 1g), composed of spores supported by a stalk of vacuolated cells with cellulose walls. The spores (Fig. 1h) may be disseminated by wind, rain, or insects, and germinate under suitable conditions to liberate amoebae, thus reinitiating the developmental cycle.

Several reviews concerned with *Dictyostelium* are available (ASHWORTH, 1971; BONNER, 1971; GERISCH, 1968; GERISCH et al., 1974; GREGG, 1966; SUSSMAN and BRACKENBURG, 1976), as well as a recent monograph by LOOMIS (1975a), and the reader is referred to these for further general information. Movement of the amoebae and of the pseudoplasmodia probably are based on different mechanisms and controlled in completely different ways, and therefore will be discussed in separate sections.

2.1 Movement of the Amoeba

2.1.1 Mechanism of Movement

2.1.1.1 The Mechanism for Pseudopodium Formation

Despite the inherent genetic and dimensional differences between giant protozoan amoebae (~100 µm across) and myxamoebae of the acrasiales (~10 µm across), it is quite probable that the underlying molecular and ultrastructural basis for the generation of the driving force, if not the gross mechanism of movement itself, is essentially the same for the two groups. However, no single model completely consistent with all the features of movement for any one amoeboid organism has yet been agreed upon (ALLEN, 1972; BOVEE and JAHN, 1972; KOMNICK et al., 1973). Furthermore, it is frequently observed that a single species of amoeba will exhibit several distinctly different kinds of movement depending on its environment or stage of development (BOVEE, 1964; BOVEE and JAHN, 1972; KOMNICK et al., 1973). The one feature of amoeboid movement which has received general acceptance is the contractile basis for the generation of a driving force. It is reasonably certain that the driving force for amoeboid locomotion is produced by local contractions involving supramolecular filaments of the proteins actin and myosin within the cell (POLLARD, 1972; POLLARD and WEIHING, 1974; TAYLOR, 1976). Species of these contractile proteins have been isolated from or demonstrated within a wide variety of eukaryotic cells, and are believed to participate in many different types of cell movements (DURHAM, 1974; POLLARD, 1972; POLLARD and WEIHING, 1974; see also WAGNER, Chap. 2.2, this vol.).

The majority of the pioneering work on amoeboid movement has focused on the large protozoan amoebae, *Amoeba proteus* and *Chaos chaos*. The major aspects of this work with large protozoan amoebae will be outlined where pertinent to provide a framework for discussion of amoeboid movement in the cellular slime molds.

Microtubules[2]. There is little evidence for a role of microtubules in the locomotion of slime mold amoebae. The number of microtubules is low in the highly motile vegetative cells and also in the interphase cells, but increases following interphase. The bundles of microtubules which appear in the cytoplasm following interphase persist after aggregation is complete and most probably are important in post-aggregation development (ECKERT et al., 1977). Amoebae of *D. discoideum* treated with high concentrations of the inhibitors of tubulin synthesis and assembly, griseofulvin and isopropyl N-phenyl carbamate, or the inhibitors of microtubule function, colchicine and vinblastine, continue to aggregate normally although sometimes after a delay (CAPPUCCINELLI and ASHWORTH, 1976). Cytoplasmic particles are normally observed to move about the cytoplasm of an amoeba independently of the direction of cell migration. In the presence of any of the "anti-tubulins", these saltatory movements of cytoplasmic particles cease, with the subsequent accumulation of these particles in the posterior of migrating amoebae (ECKERT et al., 1977). Thus, microtubules are involved in some intracellular movements but are probably not the basis of amoeboid movement itself in *D. discoideum*.

Actomyosin. It is now widely accepted that the driving force for amoeboid movement is provided by the cytoplasmic contractions produced by the interactions of actin and myosin. Bulk contractility and streaming in extracts of cytoplasm prepared from *Amoeba proteus* were first demonstrated by THOMPSON and WOLPERT (1963) and later by POLLARD and ITO (1970). These cytoplasmic extracts were shown to contain actin filaments and thick nonactin filaments that interacted in the presence of ATP at room temperature (POLLARD and KORN, 1971). More

[2] For a general discussion of microtubules in plant movements, see FILNER and YADAV, Chap. 2.1, this vol.

recently, species of actin and myosin have been isolated from amoebae of *D. discoideum* and purified to electrophoretic homogeneity (CLARKE and SPUDICH, 1974; SPUDICH, 1974; SPUDICH and COOKE, 1975). The concentration of both actin and myosin remains high during the vegetative and aggregation phases of the life cycle of *D. discoideum*, corresponding well with the high motility during these periods. Actin is in fact the most abundant protein present at the onset of aggregation. Finally, during the interphase of development, actomyosin content and motility both decrease (ECKERT et al., 1977).

Intracellular Localization of Cytoplasmic Contractions. The principal intracellular location of the cytoplasmic contractions produced by the interactions of actin and myosin is a current subject of controversy (ALLEN, 1972; ALLEN and TAYLOR, 1975; BOVEE and JAHN, 1972; KOMNICK et al., 1973). Two main opposing theories of the gross mechanism of "simple" amoeboid movement have evolved from the work on protozoan amoebae: (1) the hydraulic pressure flow (ectoplasmic tube contraction) theory, and (2) the front zone contraction (fountain streaming) theory (ALLEN, 1972; BOVEE and JAHN, 1972; KOMNICK et al., 1973). The first theory states that local contractions of the ectoplasmic cortex, predominantly in the uroid (tail) region, cause an increased pressure on the central liquid endoplasm which is pushed into the front end of an extending pseudopod where pressure is relatively low. Thus, a gradient of decreasing pressure exists from the rear to the front of a moving amoeba, the endoplasmic core playing a passive role in the process (JAHN and BOVEE, 1969; MAST, 1926; PANTIN, 1923). The second theory states that contractions occurring during the transformation of frontal endoplasm to ectoplasm (sol to gel state) serve to pull the liquid endoplasmic core forward. Thus the contracting endoplasm within a pseudopodium plays an active role, generating a reduced pressure or tension on the distal uncontracted endoplasm to draw it forward into the "fountain zone" (ALLEN, 1961, 1962). The first theorized mechanism is similar to the extrusion of tooth paste from a tube through pressure applied to the tube; the latter mechanism is similar to drawing toothpaste from the tube by means of a vacuum pump. These two proposed mechanisms are not necessarily mutually exclusive; a combination or alternation of the two could account for the observed features of movement by an amoeba of the proteus type (WOLPERT and GINGELL, 1968).

SHAFFER'S (1964) description of the process of movement by individual *D. discoideum* amoebae most closely approximates the hydraulic pressure flow mechanism, with a few modifications. SHAFFER (1964) cites contractions of a relatively rigid posterior cell surface region as the source of the hydraulic pressure gradient. The lateral surfaces of the cell are reportedly rigid and stationary, with recycling of cell surface material occurring by removal at the rear and redeposition at the front of the amoeba. In this respect, slime mold amoebae differ from the larger protozoan amoebae. Presumably, this greater lateral rigidity is important during the multicellular stages of development when the myxamoebae must maintain traction for movement while stacked many cells high as in a migrating pseudoplasmodium (see Sect. 2.2).

Slime mold amoebae also differ from giant protozoan amoebae in their ability to form "filopodia" (thin cytoplasmic extensions which may precede pseudopodium formation) in addition to the more prominent pseudopodia (Fig. 2; SHAFFER, 1964). These structures appear to play an integral part in the rapid responses of slime

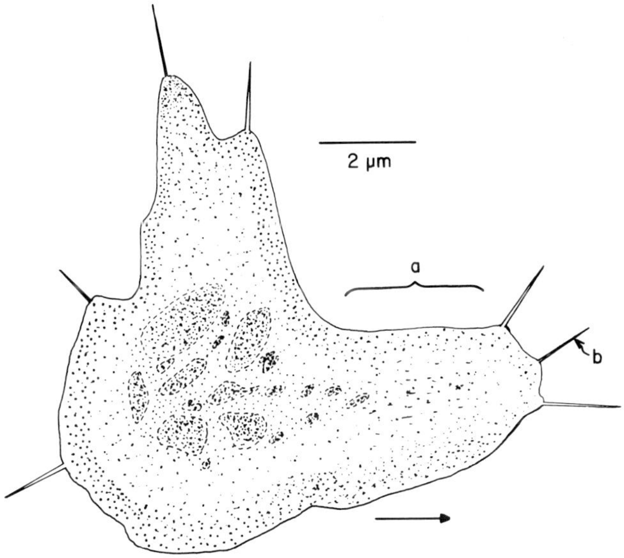

Fig. 2. Drawing of an amoeba of *D. discoideum* as seen using phase contrast microscopy. *a*, pseudopodium; *b*, filopodium

mold amoebae to mechanical or chemical stimuli (GERISCH et al., 1975; SHAFFER, 1964). A filopodium receiving a positive stimulus is "rewarded", enlarging to form a new pseudopodium, and thereby altering the direction of movement. This phenomenon may involve a weakening of the filopodium surface structure, permitting pressure-driven enlargement (BOVEE and JAHN, 1972; DURHAM, 1974).

2.1.1.2 Control of Amoeboid Movement

Calcium. The parallels between the molecular mechanism of amoeboid locomotion and that of contraction and relaxation of striated muscle have become more and more striking with the recent advances of research on nonmuscle motile systems (DURHAM, 1974; KOMNICK et al., 1973; POLLARD, 1972; POLLARD and WEIHING, 1974). It is no surprise, then, that a large body of evidence now points to Ca^{2+} as the mediator of the actomyosin contractions that provide the motive force for amoeboid movement (DURHAM, 1974; TAYLOR et al., 1973; 1976). TAYLOR's group (ALLEN and TAYLOR, 1975; TAYLOR, 1976; TAYLOR et al., 1973; 1976) has reported a series of interrelated experiments on calcium-induced contraction with motile extracts and demembranated, intact cell models of *Chaos carolinensis* and *A. proteus*. They have demonstrated that a threshold calcium ion concentration ($\simeq 7.0 \times 10^{-7}$ M) in the presence of a Mg-ATP energy source can elicit the extension of membraneless pseudopodia from a single-cell cytoplasmic model. Microinjection of the same threshold Ca^{2+} concentration into intact amoebae produces local contractions of the endoplasm, both in the uroid region and in the pseudopodia. Injection of suprathreshold Ca^{2+} ($\simeq 10^{-6}$ M) produces extensive contractions of the ectoplasmic cortex as well as the local endoplasm. The ability of intact cells to recover from such calcium-induced contractions, and the inhibition of "flare streaming" in cytoplasmic models by Ca^{2+} below 10^{-7} M or in excess of 10^{-6} M, indicate that pseudopodium formation is a highly controlled process requiring an efficient calcium-sequestering mechanism. REINOLD and STOCKEM (1972) have reported a salyrgan-sensitive calcium pump (ATP-ase) in cytoplasmic membranes of *A. proteus* and *C. chaos* capable of maintaining the concentration of free cytoplasmic Ca^{2+} below 10^{-6} M. Based on Ca^{2+} localization in histochemical preparations, they conclude that the endoplasmic reticulum in the uroid of proteus amoebae serves a function analogous to that of the sarcoplasmic reticulum in striated muscle—sequestering Ca^{2+} such that the concentration of free Ca^{2+} in the cytoplasm is maintained below the contraction threshold. Reversible transformations of actin and myosin from less structured states to filamentous states, under the control of Ca^{2+} and several accessory proteins, could also play a role in normal cell movements (see also BRITZ, Chap. 2.3 and 2.5, this vol.; WAGNER, Chap. 2.2, this vol.). This is supported by the observation that Ca^{2+} elicits the formation of many separate F-actin filaments from "amorphous aggregates", in addition to triggering Mg-ATP mediated contraction of actomyosin (TAYLOR, 1976; TAYLOR et al., 1975).

Calcium appears to be equally important for the control of movement by slime mold amoebae. An extracellular concentration $\geqq 10^{-6}$ M is essential for aggregation of *D. discoideum* (MASON et al., 1971). Primarily because of the size limitation, microinjection experiments have not been performed with myxamoebae. However, the relative ease of culturing cellular slime molds facilitates the collection of large volumes of amoebae, and thus the characterization of the molecular components of motility. TAYLOR et al. (1975) have prepared motile extracts from amoebae of *D. discoideum* by the same procedures employed in their isolation of contractile proteins from *A. proteus*. The extracts from *D. discoideum* and those of *A. proteus* behave in an essentially identical manner upon the addition of micromolar quantities of calcium in the presence of Mg-ATP: i.e., the induction of actomyosin contraction is preceded by the formation of numerous F-actin filaments. SPUDICH and COOKE (1975) have prepared highly purified actin from amoebae of *D. discoideum*, and have observed extensive polymerization into filaments at an optimal KCl concentration of 0.1 M. This KCl-dependent polymerization of actin from *D. discoideum*, and the apparent equilibrium constant of 1.3 μM for the transition from monomer to polymer, is shared by purified actin from striated muscle. Actin from *D. discoideum* differs from muscle actin in its ability to form two-dimensional nets in 10 mM $CaCl_2$, and filament bundles 0.1 μM in diameter in the presence of 5 mM $MgCl_2$. These bundles assemble only in partially purified preparations of actin from *D. discoideum*, indicating that the assembly may be dependent on the participation of an additional component accessory to actin. The actin-filament bundles reconstituted in vitro reportedly resemble the actin-containing bundles found in situ in the cortical region of *D. discoideum* amoebae by electron microscopy (CLARKE et al., 1975).

Evidence for one further major homology between Ca^{2+} regulation of contractions in striated muscle and in amoebae of *D. discoideum* has recently been presented. In striated muscle an accessory protein, troponin, is known to confer Ca^{2+} sensitivity upon the ATPase of myosin in the actomyosin complex, thus allowing for the control of muscular contractions through the regulation of the free Ca^{2+} concentration in the sarcoplasm. MOCKRIN and SPUDICH (1976) have shown that a troponin-like protein fraction from *D. discoideum* confers Ca^{2+} sensitivity on the activation of the ATPase of purified myosin from *D. discoideum* by actin purified from either *D. discoideum* or striated muscle. These findings are clearly consistent with the view that slime mold amoebae control their movements through local regulation of the concentration of free calcium ions in the cytoplasm.

Physical Properties of the Plasma Membrane. The deformability of the plasma membrane and its adhesion to the substratum at the lower surface of an extending pseudopodium may also control amoeboid movement (KOMNICK et al., 1973; WOLPERT and GINGELL, 1968). The rate of turnover of the cell membrane is not sufficient to allow for continuous deposition at the front and resorption at the rear of a migrating amoeba (CZARSKA and GREBECKI, 1966; GARROD and WOLPERT, 1968; WOLPERT and GINGELL, 1968). It has been suggested that the membrane moves forward by a sliding mechanism involving the unfolding of pleats, predominantly in the uroid, to provide slack for the advancing pseudopod (CZARSKA and GREBECKI, 1966; DURHAM, 1974; KOMNICK et al., 1973). Thus, the total area of the plasma membrane remains constant, but the convolution of the membrane surface varies greatly (Fig. 3). It is noteworthy that a complex network of actin filament bundles appears to be directly associated with the cytoplasmic face of *D. discoideum* cell membranes (CLARKE et al., 1975). Local breakdown of this network could result in the increased deformability of the plasma membrane required for the extension of a pseudopod by either of the mechanisms proposed above. This interaction of the plasma membrane and the ectoplasm could produce the rigidity of the lateral cell surfaces required for the synchronous movement of multicellular masses in aggregation streams or pseudoplasmodia (SHAFFER, 1964; see Sect. 2.1.1.1). Fine control of contractility and deformability at the cytoplasmic face of the cell membrane could also account for the formation of filopodia by slime mold cells.

2.1.2 Chemotaxis

The responses of slime mold amoebae to mechanical or chemical stimuli appear to entail the perception of the stimulus at the plasma membrane surface, followed by the activation of the contractile machinery within the cell. Exploratory filopodia are "rewarded" through this process and consequently perpetuate or change the direction of cell movement. Myxamoebae of the cellular slime molds are capable of directed movement in response to two separate classes of chemoattractants during different phases of the life cycle. Vegetative amoebae are chemotactically sensitive to folic acid and several other pterin derivatives (PAN et al., 1972; WURSTER and SCHUBIGER, 1977). This sensitivity apparently serves as a food-seeking mechanism, enabling amoebae to locate bacteria that release folate (PAN et al., 1975). The second class of chemoattractants is called "acrasins", a generic term coined by BONNER (1947) for the group of low molecular weight compounds which serve to entrain aggregation fields during development. The acrasin of *D. discoideum* and three closely related species (*D. mucoroides, D. rosarium* and *D. purpureum*) is almost certainly cAMP. The acrasins of other slime mold species, e.g., *D. minutum, Polysphondylium pallidum* and *P. violaceum*, are as yet unidentified (KONIJN, 1973; PAN et al., 1975). The possibility that folic acid doubles as an acrasin for any of these species has been ruled out by PAN et al. (1975).

During the interphase period of starvation preceding aggregation, amoebae of *D. discoideum* acquire the competence for chemotactic migration toward cAMP

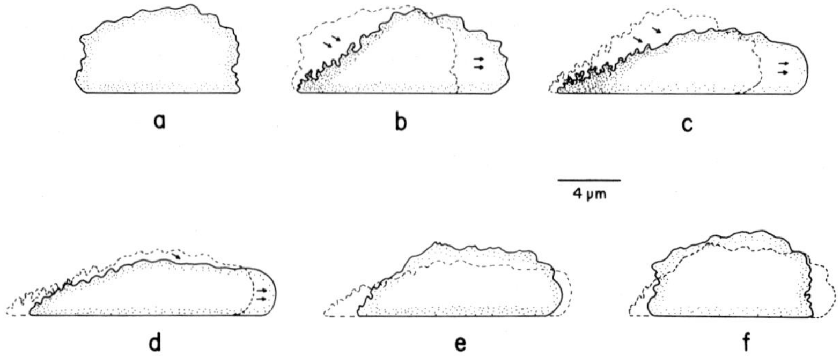

Fig. 3 a–f. Diagrammatic representation of chemotactic amoeboid movement in *D. discoideum*. **a** Outline of a resting amoeba. *Stippling* represents the extent of contraction in the ectoplasmic cortex adjoining the plasma membrane (e.g., greater stippling means greater contraction). **b** A wave of cAMP diffusing from right to left initiates a step of chemotactic movement by the amoeba. The *dashed outline* represents the position of the cell in the preceding figure. Contraction at the rear of the cell drives the ectoplasmic core forward to form an extending pseudopodium. *Arrows* indicate the direction of bulk cytoplasmic flow. The increased convolution of the plasma membrane at the rear of the cell results from contraction of the adjoining endoplasmic cortex. **c** Continued contraction at the rear of the amoeba maintains the forward pressure-driven flow of the endoplasmic core. Local relaxation of the anterior ectoplasm facilitates the flow of endoplasm at the front of the pseudopodium and enables the anterior plasma membrane pleats to unfold. **d** The highly contracted posterior ectoplasm begins to relax. The subsiding flow of endoplasm into the pseudopodium continues to stretch the membrane forward, unfolding pleats at the rear of the amoeba. **e** The migrating amoeba comes to rest and begins to round up as adhesion to the substratum diminishes. **f** The amoeba is ready to begin another "step" of movement

(ROBERTSON and DRAGE, 1975) and become less responsive to folic acid (WURSTER and SCHUBIGER, 1977). Several hours later the amoebae develop the competence for signal relaying—the ability to synthesize and secrete an amplified pulse of cAMP in response to a transient exogenous cAMP stimulus (GINGLE and ROBERTSON, 1976; ROBERTSON and DRAGE, 1975). Aggregation begins when a small percentage of the total cell population begins to release pulses of cAMP autonomously (COHEN and ROBERTSON, 1974; ROBERTSON and DRAGE, 1975; ROBERTSON and GRUTSCH, 1974). The signal relay mechanism serves to amplify and relay the autonomous pulses to successive bands of neighboring amoebae. The lifetime of a wave of cAMP, and therefore the spatial limits for the centrifugal diffusion of that wave, are set by two phosphodiesterases, one extracellular and one plasma membrane-bound, which cleave 3′,5′-cAMP to yield 5′-AMP (GERISCH et al., 1974; MALCHOW and GERISCH, 1974). Once stimulated to signal relay, an amoeba becomes refractory to further stimulation for a period of 2 to 7 min (COHEN and ROBERTSON, 1974; ROBERTSON and DRAGE, 1975; ROBERTSON and GRUTSCH, 1974). This refractory phase prevents the perpetual reflection of the cAMP signal back and forth between adjacent amoebae. There is no refractory period per se for the chemotactic response to cAMP, but the movement step of a band of amoebae elicited by an outwardly diffusing wave of cAMP does represent a relatively long-term (ca. 100 s) "all or nothing" response produced by a very brief (ca. 0.2 s) stimulus (ALCANTARA and MONK, 1974; COHEN and ROBERTSON, 1971).

The central question concerning chemotaxis of slime mold amoebae in response to folic acid or cAMP is how the basic phenomenon of amoeboid movement is translated into directed movement. Experiments with tritiated cAMP indicate that the chemoattractant does not enter the cell, but, rather, is briefly bound to receptors on the plasma membrane surface before inactivation by membrane-bound phosphodiesterase (GERISCH et al., 1974). Two species of cAMP receptors have been reported. Their relative numbers and developmental times of appearance correlate well with the thresholds and times of acquisition of the competences for chemotaxis and signal relaying (GREEN and NEWELL, 1975; HENDERSON, 1975; ROBERTSON and GRUTSCH, 1974). COHEN and ROBERTSON (1971) have reported that, during the interphase preceding aggregation, amoebae of *D. discoideum* establish a definite polarity not evident in vegetative amoebae, and subsequently extend pseudopodia from only the anterior $^1/_3$ of the cell. Such cell polarity could account for the almost linear migration of individual amoebae once oriented in the direction of an increasing cAMP gradient.

The next question is: What is the link joining the perception of the cAMP stimulus at the surface of the cell membrane with the activation of the actomyosin network in the cortical region of the surface cytoplasm? Actomyosin interactions may be governed by cAMP through controlled movement of free calcium ions. The binding of cAMP to the cell membrane surface may regulate local levels of free Ca^{2+} at the membrane-cytoplasm interface (DURHAM, 1974). Such a regulation could be mediated either by an alteration of membrane permeability to Ca^{2+} or by the activation of a membrane-associated calcium ATPase. Given a local efflux of Ca^{2+} across the cell membrane, reducing the level of free Ca^{2+} in a specific region of the cytoplasmic cortex below the contraction threshold, one would anticipate a relaxation of the actomyosin network in that area (MOCKRIN and SPUDICH, 1976; TAYLOR, 1976). This would permit a pressure-driven flow of cytoplasm into the relaxed region to form an extending pseudopodium (BOVEE and JAHN, 1972; DURHAM, 1974; SHAFFER, 1964). Concomitantly, binding of effluxed Ca^{2+} to the outer membrane surface of the pseudopodium would result in the neutralization of negative charges on this outer membrane surface and could thereby account for increased adhesion to a negatively charged substrate (DURHAM, 1974; KOMNICK et al., 1973; MANERY, 1966).

The effects of the calcium ionophore, A 23187, on the rhythm of light scattering in dense suspensions of aggregation-competent *D. discoideum* amoebae have recently been studied. Such a cell suspension will exhibit an autonomous rhythm in light scattering with oscillations at 5–12 min intervals. These changes in light scattering appear to be a manifestation of the chemotactic responses to cAMP released in pulses by the amoebae (GERISCH and HESS, 1974). Transient changes in light scattering also may be induced by exogenous cAMP. WHITAKER and POFF (unpublished data) have observed that A 23187 at $2–6 \times 10^{-5}$ M abolishes both the autonomous rhythm and the transient light scatter changes induced by exogenous cAMP.

These data clearly support the hypothesis that Ca^{2+} exchange is intimately involved in chemotactic movement. However cAMP itself may not directly trigger chemotactic movement but may act through cGMP (3′-5′-guanosine monophosphate). MATO et al. (1977a, b, c) MATO and KONIJN, 1977 have recently reported data indicating that intracellular cGMP levels increase rapidly to a maximum within 10 s following stimulation of cells with cAMP. This increase in internal cGMP concentration is the most rapid event to have been measured following cAMP reception, and is consistent with the hypothesis that cGMP is an intracellular mediator of the response by the myxamoebae to cAMP [3].

2.2 Movement of the Pseudoplasmodium

2.2.1 Mechanism of Movement

The multicellular pseudoplasmodium exhibits a gliding motion, and is positively phototactic and positively thermotactic (BONNER et al., 1950). Three partially separable aspects of migration by pseudoplasmodia will be discussed: the control of polarity, the control of turning, and the driving mechanism for migration. At present, much more is known about the control of polarity and turning than about the driving mechanism itself.

2.2.1.1 Control of Polarity

Under natural circumstances, differentiation into spore cells and stalk cells begins in the early pseudoplasmodium and is not subsequently reversed (RAPER, 1940), with the cells in the anterior ~20% portion of the pseudoplasmodium developing into stalk cells, and the cells in the posterior ~80% of the pseudoplasmodium developing into spore cells (LOOMIS, 1975b). Thus, following aggregation, a polarity is established which controls the direction of pseudoplasmodium migration and is related to subsequent cell differentiation.

One well-defined theory proposed by LOOMIS (1972, 1975b) attributes the control of polarity to a gradient in the thickness of the slime sheath covering the pseudoplasmodium. The slime sheath is first formed following aggregation, as a small mound of cells (termed a conus) is raised from the substratum. As the conus rises, slime sheath deposition proceeds until the sheath surrounds the mass of cells and adheres to the substrate. The conus continues to rise, forming a worm-like projection which eventually bends over, contacts the substrate and develops into the pseudoplasmodium. LOOMIS (1972, 1975b) suggests that the physical properties of the slime sheath limit the size of the pseudoplasmodium and also set the polarity—the anterior end is derived from the tip of the conus, which is the area of initial slime sheath deposition, and the posterior end is defined by the point where the slime sheath adheres to the substrate.

The amoebae composing the pseudoplasmodium migrate through the tube formed by the slime sheath, which is produced throughout the length of the pseudoplasmodium and is left behind as a collapsed tube. Thus, the thickness of the slime sheath increases linearly along the pseudoplasmodium from anterior to poste-

---

[3] More recent data from MATO et al. [MATO, J.M., VAN HAASTERT, P.J.M., KRENS, F.A., KONIJN, T.M.: Chemotaxis in *Dictyostelium discoideum*: effect of concanavalin A on chemoattractant mediated cyclic GMP accumulation and light scattering decrease. Cell Biol. Int. Repts. 2, 163–170 (1978)] have shown that light scatter changes and rising intracellular cGMP elicited by a pulse of cAMP in a dense cell suspension are independent events. Their data are consistent with the hypothesis that regulation of the intracellular cGMP concentration is related to the signal relaying mechanism rather than to the chemotactic response.

rior (FARNSWORTH and LOOMIS, 1975a). The slime sheath appears to polymerize with time (FREEZE and LOOMIS, 1977a, b; SHAFFER, 1965) thereby inhibiting lateral movements by the cells. Furthermore, the tendency for the slime sheath tube to collapse behind the pseudoplasmodium (perhaps as a result of elasticity) undoubtedly impedes any posterior movement by the cells (FREEZE and LOOMIS, 1977b). The only direction left for movement by the cells is through the freshly deposited thin layer of slime sheath at the anterior of the pseudoplasmodium.

LOOMIS (1972) also suggests that differentiation of the anterior and posterior cells into stalk cells and spore cells, respectively, is controlled by an effector, the diffusion of which is limited by the sheath. Thus, positional information would be obtained from a gradient in effector concentration. This gradient would result in turn from slow diffusion of the effector through the thickened sheath at the posterior end of the pseudoplasmodium, and a rapid diffusion through the thin sheath at the anterior end. The gradient of effector concentration then could control spore differentiation[4]. This sheath-effector hypothesis implicates the topography of sheath deposition in the control of polarity and cell differentiation. The major alternative hypothesis is that cell sorting, based on some subtle cell characteristic such as size or ploidy, occurs prior to pseudoplasmodium formation (BONNER, 1971). Polarity would then be a manifestation of the established cell types and the sheath merely a convenient packaging. Although it is difficult to refute such a model, one can conclude that any cell apportioning in the pseudoplasmodium is incomplete, and that each cell type can rapidly de-differentiate and re-differentiate into the opposite cell type (BONNER, 1971).

2.2.1.2 Control of Turning

Turning is discussed separately from the driving mechanism in migration to emphasize that these two aspects of migration may be separate phenomena.

A turn of a pseudoplasmodium could result from directed movement of the anterior cells of the pseudoplasmodium or from a differential rate of movement (differential cell movement or differential deposition of slime sheath) on the two sides of the pseudoplasmodium. An analogy for the first mechanism is the turning of an automobile by means of a change of direction of the front wheels whereas the second mechanism is similar to the turning of a caterpillar tractor through a change in the relative speed of the two tracks. A distinction between the two modes of turning ultimately depends on a demonstration of directed movement by individual cells taken from the tip of the pseudoplasmodium. Phototactic turns by a pseudoplasmodium can be more easily explained on the basis of the second mechanism since directed movement by individual cells has not been observed (see Sect. 2.2.2.1). Insufficient data are currently available to attribute thermotactic turns to either of these two mechanisms (see Sect. 2.2.2.2).

2.2.1.3 The Driving Mechanism for Migration of Pseudoplasmodia

Migration by pseudoplasmodia is a relatively smooth phenomenon (however see Sect. 2.2.1.1), and thus requires the continuous application of a driving force.

---

[4] SUSSMAN and SCHINDLER (SUSSMAN, M., SCHINDLER, J.: A possible mechanism of morphogenetic regulation in *Dictyostelium discoideum*. Differentiation *10*, 1–5 (1978)) have recently presented data to support a model of differentiation in *D. discoideum* based on NH_3 as a diffusible effector.

This driving force appears not to result from contraction forces in the pseudoplasmodium itself. (The slime sheath, given the property of contraction increasing with age, could constrict the posterior section of the pseudoplasmodium and thus propel the organism anteriorly. Such a contraction should however result in a decreased width of the posterior pseudoplasmodium and this is not observed.) Alternatively, one must look to some property of the individual cells to explain the driving mechanism within the pseudoplasmodium. Two reasonable hypotheses have been suggested. These are based on two different methods for directing amoeboid movement within the pseudoplasmodium.

Amoeboid Movement Directed by Chemotaxis. A gradient of cAMP has been described in the pseudoplasmodium of *D. discoideum*, with the highest concentration in the anterior tip of the organism. One potential mechanism for migration of the pseudoplasmodium would be chemotaxis by the individual cells toward some substance (such as cAMP) diffusing from the tip (Cohen and Robertson, 1972). The major support for this mechanism is the observation by Cohen and Robertson (1972) of pulsatile movement in pseudoplasmodia. This observation was in substantial agreement with a model for migration of pseudoplasmodia based on chemotactic migration by amoebae to pulses of cAMP released in the anterior of the pseudoplasmodium. There are, however, several difficulties with this model. There is disagreement concerning the existence of the gradient of cAMP along the length of the pseudoplasmodium. Garrod and Malkinson (1973) and, more recently, Brenner (1977) report that the concentration of cAMP in the pseudoplasmodium tip is not more than twice the concentration in the rear. Moreover, the cAMP binding sites on the amoebae are developmentally controlled and are at a minimum in the pseudoplasmodium (Gerisch et al., 1974; Green and Newell, 1975). Thus, even if there is a gradient of cAMP in the pseudoplasmodium, the amoebae might not be capable detecting this gradient and responding chemotactically. Indeed, cells taken from pseudoplasmodia will not migrate chemotactically toward an undisturbed pseudoplasmodium, whereas cells taken from aggregation streams (before pseudoplasmodium formation) will exhibit normal chemotactic migration toward such a pseudoplasmodium (Garrod and Malkinson, 1973). Cohen and Robertson (1972) have suggested, however, that the chemotactic stimulus may be transmitted from cell to cell via cell contact sites. There is also of course no requirement that the chemotactic stimulus proposed for cells in pseudoplasmodia be cAMP.

Amoeboid Movement Directed by Cell Polarity. Shaffer (1964) and Garrod (1969) have reported that pseudopodium formation by one amoeba will induce movement in the same direction by any second amoeba in contact with the first. Coined "contact following" by Shaffer (1964), this intercellular transfer-of-movement stimulus could account for a smooth migration by an entire population of amoebae in a single direction. (Could contact following be considered a special case of chemotaxis where the stimulus is transferred only upon contact?).

Garrod (1969) has described an alternative theory for pseudoplasmodium migration based on contact following by the amoebae and on the mechanical properties of the surface sheath. He suggests that pseudopodium extension is inhibited more by the surface sheath at the sides and rear of the pseudoplasmodium than at the anterior tip, where the surface sheath is thin, and therefore, pseudopodia are extended only in a forward direction. He further suggests an intercellular transfer of the movement stimulus resulting in forward migration by the entire mass of amoebae. Since new surface sheath is deposited at the anterior tip of the pseudoplasmodium, and since the tube of surface sheath collapses behind the multicellular mass, the postulated conditions for migration are continually regenerated.

We suggest an additional hypothesis for the driving mechanism in pseudoplasmodia based on amoeboid movement, which may be random with respect to time, but is spatially polarized. During the interphase period following starvation and preceding aggregation, the individual amoebae are reported to acquire a polarity such that there is an evident front and rear end of each amoeba (COHEN and ROBERTSON, 1971; ROBERTSON and GRUTSCH, 1974). During aggregation, front-to-rear cell contacts are established (SHAFFER, 1964) such that the polarity of the entire population of cells is similar. This cellular polarity, if maintained in the pseudoplasmodium, would result in the pseudopodium-forming end of each amoeba being directed toward the anterior tip of the pseudoplasmodium. Given the polarity of the pseudoplasmodium as a result of the constraints imposed by the slime sheath (see Sect. 2.2.1.1), and a similar polarity of the individual amoebae, movement by the amoebae would directly cause a forward migration by the pseudoplasmodium. The advantage of a hypothesis of movement by pseudoplasmodia based on the movement of polarized cells is that chemotactic stimuli and synchronous amoeboid motion become unnecessary to explain migration by pseudoplasmodia. However, the major basis for this hypothesis of movement by pseudoplasmodia is the existence of a cell polarity following interphase. The existence of such a polarity has been questioned by GERISCH et al. (1975).

An observation which may be inconsistent with this last hypothesis is derived from the microsurgical experiments reported by LOOMIS (1975a). The central section of nineteen separate pseudoplasmodia was freed from the anterior and posterior cells and from any slime sheath on the cut surface. Seventeen of these pseudoplasmodia resumed migration after some delay period. Ten of these continued migrating in the same direction as that before surgery. The remaining seven migrated in a direction opposite to that before surgery. This observation is difficult to reconcile with an established polarity of cell movement. Perhaps, during the quiescent period following surgery, chemotactic competence is regained. The amoebae could then reorient to the newly developing pseudoplasmodium tip. This possibility is supported by GARROD's (1974) report that amoebae re-acquire chemotactic competence approximately one hour after dispersal from a pseudoplasmodium.

Amoebae in newly formed pseudoplasmodia may still be capable of cAMP-directed chemotaxis, and this might account for the observation of pulsatile movement in pseudoplasmodia. Upon the gradual loss of chemotactic competence by the amoebae in the pseudoplasmodium, chemotaxis would have a decreased importance in migration, and would be replaced by a different mechanism. An assessment of the role of chemotaxis in migration should be preceded by measurements of chemotactic competence by tip and rear amoebae at various times following pseudoplasmodium formation. Similarly, amoebae at various times in the developmental sequence should be examined for cell polarity. It is difficult at present to assess the relative roles of cell polarity and contact following in the mechanism for migration of pseudoplasmodia.

2.2.2 Control of Movement by Light and Heat

Thermotaxis was orginally proposed by BONNER et al. (1950) as a general sensory response by pseudoplasmodia of *D. discoideum* with phototaxis considered a specialized aspect of thermotaxis. It was visualized that unilateral light was absorbed on the proximal side of the pseudoplasmodium and dissipated as heat. The pseudoplasmodium would then turn in response to the resultant heat gradient. This has been shown not to be the case by the following lines of evidence. (1) FRANCIS (1964) has shown that pseudoplasmodia will exhibit phototaxis toward unilateral light which, even if it were totally absorbed on the proximal side [which clearly has been shown with microbeams not to be the case (FRANCIS, 1964; POFF and LOOMIS, 1973)] and dissipated as heat, would produce a temperature increase too small to elicit a thermotactic response. (2) Vertical microbeams have been used to show that unilateral visible light is absorbed primarily on the distal side of the pseudoplasmodium (FRANCIS, 1964; POFF and LOOMIS, 1973). In thermotaxis, by contrast, the pseudoplasmodium turns toward the warmer side of the temperature gradient (POFF and SKOKUT, 1977). (3) LOOMIS (1970) has described a mutant which lacks phototaxis but shows unimpaired thermotaxis (POFF and SKOKUT, 1977). Ideally, one would also like a mutant lacking thermotaxis but with normal phototaxis, but such a mutant has not yet been reported.

2.2.2.1 Phototaxis

Much more is known at present about phototaxis than about thermotaxis in *D. discoideum*, although they were described at the same time (BONNER et al., 1950). Action spectra for phototaxis show two maxima, one in the blue and one in the green region of the spectrum (FRANCIS, 1964; POFF et al., 1973). Of the two action spectra available, that with the higher resolution shows the blue maximum at about 430 nm and the green maximum as a broad peak from about 540 nm to about 620 nm (POFF et al., 1973).

A spectrophotometrically measurable light-induced absorbance change has been described with an action spectrum similar to the action spectrum for phototaxis (POFF et al., 1973). Using this light-induced absorbance change as an assay, a pigment has been isolated and characterized. The pigment, called phototaxin, is a heme protein of molecular weight 240,000, and is present in the reduced form in vivo in darkness. The pigment is oxidized in light, and re-reduced in darkness by an unknown reductant in vivo, and by a low potential reductant such as sodium dithionite in vitro (POFF et al., 1974). This pigment has been considered the photoreceptor pigment for phototaxis on the basis of the similarity between the absorption spectrum of the purified pigment and the action spectrum for phototaxis (POFF and BUTLER, 1974). Although phototaxin was partially purified in its isolation by collecting a mitochondrial fraction (POFF et al., 1974), this fraction was undoubtedly contaminated with other subcellular components. Therefore, until a rigorous subcellular isolation of the pigment has been conducted, the localization of the pigment in the mitochondrion must be open to question. Other subcellular components which would have sedimented along with the mitochondria and which should be viewed as possible sites of phototaxin attachment are the plasma membrane, and the endoplasmic reticulum.

Table 1. Distinguishing characteristics of the two photoreceptor pigment systems described in *Dictyostelium discoideum*

Characteristic	Pigment system	
	Phototaxin	Blue light system
Physiological response	Phototaxis	No known response in *D. discoideum*
Action spectrum		
Blue region (400–500 nm)	Single maximum at about 430 nm	Broad maximum centered about 465 split in some systems into 3 peaks at 450, 465 nm and 485 nm
Green–red region (500–700)	Broad action maximum from ~540 to ~610 nm	No known physiological action
Phototeceptor pigment	240,000 daltons heme protein	Flavin
Action of light	Photooxidation of photoreceptor pigment	Photoreduction of flavin which in turn reduces cyt b-559

A second light-induced absorbance change has been described in *D. discoideum*. This absorbance change results from the flavin-mediated reduction of a b-type cytochrome. The action spectrum for this light-induced absorbance change and the absorption spectrum for the flavin photoreceptor pigment is similar to the action spectra for the so-called blue light responses in plants and animals (e.g., phototropism in corn and *Phycomyces*, induction of carotene formation, phase shifts in circadian rhythms; POFF and BUTLER, 1974, 1975), and *D. discoideum* has been employed as a convenient source from which the components of this pigment system may be purified and characterized (POFF and BUTLER, 1975). Unfortunately, this flavin-cytochrome-b system has been confused with the heme protein, phototaxin, related to phototaxis (SUSSMAN and BRACKENBURG, 1976) with which it has no known connection.

The two pigment systems are contrasted in Table 1.

The mechanism whereby phototaxin affects cellular metabolism is not known, but the ultimate effect of light is to increase the rate of migration by the pseudoplasmodia (POFF and LOOMIS, 1973). Unilateral light is focused by refraction at the proximal pseudoplasmodium-air interface such that there is a greater absorption of light in the distal half than in the proximal half of the pseudoplasmodium. The light energy absorbed by phototaxin causes the photooxidation of the pigment, and by some unknown mechanism, the oxidized phototaxin is presumed to cause an increase in migration rate on that side of the pseudoplasmodium absorbing the greatest amount of light. The pseudoplasmodium must then inexorably turn. Since the increase in migration rate is on the side away from the light, the turn is toward the light source. The two sides of a pseudoplasmodium migrating either directly toward or directly away from the light move at equal rates and no phototactic turns result until the pseudoplasmodium by chance deviates slightly from its path giving a focusing advantage to the side away from the light. Thus phototaxis may be conceived as a null-seeking system, with light as an input and control of migration on the two sides of the pseudoplasmodium as an output.

Phototaxis by individual amoebae has never been observed. It has therefore been reasoned that phototaxis results from the stimulation of some rate-limiting process in migration (POFF and LOOMIS, 1973). The formation of surface sheath may be such a process. If the rate of migration of pseudoplasmodia is limited by slime sheath production, light could control migration rate through a stimulation of slime sheath production. A turn could also result from a stimulation of the migration rate of individual amoebae composing the pseudoplasmodium.

2.2.2.2 Thermotaxis

The three major characteristics of thermotaxis by *D. discoideum* are high sensitivity, narrow response range and adaptability (POFF and SKOKUT, 1977). Of these, the single most startling characteristic is the extreme sensitivity to temperature. BONNER et al. (1950) originally described thermotactic orientation to gradients of 0.05° C/cm or 0.0005° C across a large (1 mm × 0.1 mm) pseudoplasmodium. This measurement of sensitivity has been repeated and substantiated recently (POFF and SKOKUT, 1977). Theoretically, such a high sensitivity could be accounted for by the organism measuring the environmental parameter at different times during movement (temporal measurement), rather than by simultaneous measurements at different locations (spatial measurement). The temporal method of measurement, employed by many motile organisms such as bacteria, seems not to be used by *D. discoideum* in thermotaxis. In a temporal measurement, one would expect an initial random directional response by the organism and this is not observed in *D. discoideum* where ~90% of the pseudoplasmodia give a response of positive thermotaxis as the first directional movement following the imposition of a thermal gradient (POFF and SKOKUT, 1977).

Pseudoplasmodia formed from cells grown at 23.5° C exhibit thermotaxis only over the temperature range from 21.5° to 28.5° C, while migration is possible over a much wider range (Fig. 4). For pseudoplasmodia formed from cells grown at 20° C, this temperature range of thermotaxis is shifted toward lower temperatures (POFF and SKOKUT, 1977). On the basis of these observations, POFF and SKOKUT (1977) have suggested that a specific "bio-thermometer" is used for measuring temperature (i.e., that the measurement of temperature is not simply via the effects of temperature on general metabolism) and that the narrow range of thermotaxis results from the narrow temperature range over which the "bio-thermometer" is active. Adaptation is then the ability of the organism to modify its "bio-thermometer" dependent upon the temperature of growth. Furthermore, it has been suggested that the thermo transducer mechanism is a specific substance undergoing a temperature-dependent phase transition. This specific substance could be a lipid in a membrane matrix (POFF and SKOKUT, 1977). This hypothesis has the advantage of combining a relatively narrow phase transition, such as those exhibited by lipids, with an obvious mechanism for adaptation (many cases have been described in which the membrane lipid composition of an organism is dependent upon growth temperature [CHAPMAN, 1975; HUANG et al., 1974; SINENSKY, 1974]). It is difficult at present to explain the high sensitivity, although it is clear that one

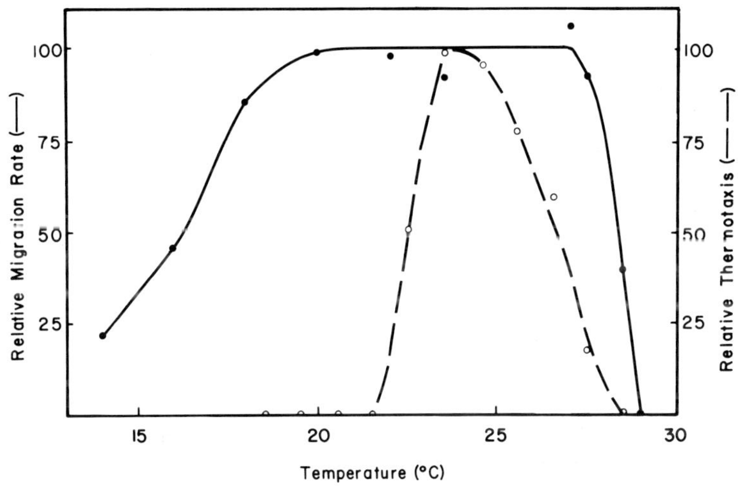

Fig. 4. A comparison of the temperature dependent capacities of pseudoplasmodia of *D. discoideum* for migration (*solid line*) and for thermotaxis to a 0.11°/cm temperature gradient (*broken line*) (data from POFF and SKOKUT, 1977)

or more amplification steps must be present. The hypothesis that a thermal gradient might control the direction of migration through a temperature effect on the local relative humidity (LOOMIS, 1975a) is unlikely. Such an effect on relative humidity should be temperature-dependent over a wide range of temperatures, while thermotaxis is exhibited only over a narrow range of temperatures.

Answers to the following questions would assist in an understanding of the mechanism of thermotaxis. (1) Is the "bio-thermometer" a portion of a cellular membrane, and if so, what portion of which membrane? (2) What causes adaptation to occur at a rate much slower than thermotaxis itself? (3) If the "bio-thermometer" is a substance exhibiting a temperature-dependent phase transition, what mechanism is employed by the cell to detect that transition? (4) By what mechanism is the direction of the pseudoplasmodium controlled following the temperature measurement?

3. Acellular Slime Molds

The true plasmodium—the distinguishing feature of the acellular slime mold—is a motile, multinucleate, acellular mass of protoplasm laced with veinlike strands up to 1 mm in diameter. These strands consist of an ectoplasmic gel sheath surrounding a liquid endoplasmic core, which flows vigorously (at speeds up to 1.3 mm/s) back and forth through the plasmodium. This cytoplasmic movement, called "shuttle streaming" (see BRITZ, Chap. 2.3, this vol.), will move the entire mass of protoplasm when the endoplasmic flow in one direction exceeds the flow in the reverse direction over an extended period of time (KOMNICK et al., 1973). In a migrating plasmodium, the "veins" are unevenly distributed, with a few

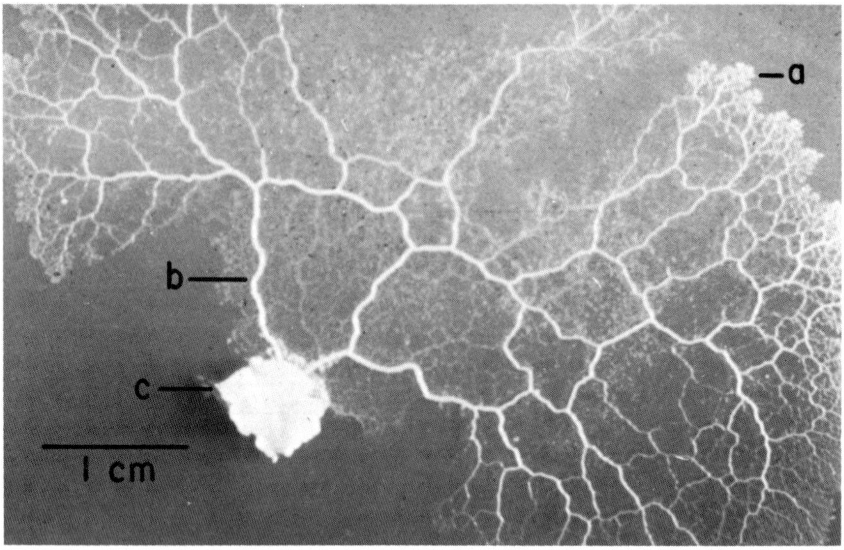

Fig. 5. Migrating plasmodium of *Physarum polycephalum* radiating on agar from the inoculum block. *a* Margin of advancing plasmodium; *b* "vein"; *c* inoculum block

strands of the greatest diameter located in the rear, extending forward into a highly branched network of very fine strands in the front (Fig. 5). The anterior veins originate from hyaloplasmic lobes, often with filose pseudopodia, that develop along the advancing margin. The initially hyaline lobes slowly become granular and eventually form reticulate strands which anastomose with the veinal network. The plasma membrane surrounding a plasmodium is covered by a thin slime sheath composed predominantly of a sulfated galactose polymer. Like the slime sheath which envelops pseudoplasmodia of the cellular slime molds, the plasmodial sheath is thinnest at the anterior end and collapses behind the migrating plasmodium, leaving a slime trail (Olive, 1975).

The natural habitat for the plasmodium of the acellular slime mold is typically on or beneath moist, shady, decaying vegetation where a plentiful food supply (bacteria and simple organic nutrients) might be expected (Olive, 1975). Plasmodia may be found ranging from microscopic up to 0.2 m^2 in surface area. Under the proper conditions, the acellular slime molds form fruiting bodies which typically are more exposed than are the plasmodia (e.g., on the bark surface rather than beneath the bark). One may presume that this exposed habit of the fruiting body increases the probability of spore dispersal by wind, water, or insects.

3.1 Movement of the Plasmodium

The pattern of protoplasmic streaming exhibited by slime mold plasmodia is too complex to permit analysis of movement of the entire plasmodium. Thus, studies on streaming and migration have frequently used models such as isolated strand segments, protoplasmic drops from punctured veins, or "microplasmodia". The latter can be generated by separation of a protoplasmic drop from the vein (WOHLFARTH-BOTTERMANN, 1962), by fragmentation of a plasmodium by brief treatment with 10 mM caffeine (HATANO and OOSAWA, 1971) or by growth of the organism in liquid shake cultures (DANIEL and BALDWIN, 1964).

KAMIYA (1959) used isolated strand segment models to study the basis of motive force generation in myxomycete plasmodia. He found that the liquid endoplasm streams passively, forced along by hydraulic pressure produced by periodic contractions of the ectoplasm surrounding the stream. LOEWY (1952) suggested that ectoplasmic contractions resulting in shuttle streaming are generated by contractile proteins analogous to actin and myosin in muscle. This hypothesis was supported by the observation that glycerinated whole plasmodia (KAMIYA and KURODA, 1965) or protoplasmic drops (ACHTERRATH, 1969) contract upon the addition of ATP. Further substantiation came with the isolation of *Physarum* actin by specific binding to muscle myosin (HATANO and OOSAWA, 1966).

3.1.1 The Structural Basis of Movement

Microscopically observable plasma fibrils appear for a brief time in contracting ectoplasm. In isolated strand segments, an ectoplasmic wall composed of a random three-dimensional fibrillar network forms around the endoplasmic core in the region of greatest contraction. Peripheral fibrils also attach at invaginations of the plasma membrane of intact plasmodia (WOHLFARTH-BOTTERMANN, 1965). These connections between the fibrillar network and the plasma membrane permit contraction of the ectoplasm-plasma membrane complex at the perimeter of the plasmodium (KOMNICK et al., 1973). These plasma fibrils have been identified as bundles of 7 nm F-actin filaments (ALLÉRA et al., 1971). Both actin and a species of myosin have been isolated from *Physarum* (ADELMAN and TAYLOR, 1969a, b; HATANO and OHNUMA, 1970; HATANO and TAZAWA, 1968), and both contractile proteins have been purified and extensively characterized (HATANO and OWARIBE, 1976; JACOBSON et al., 1976; KOMNICK et al., 1973; POLLARD and WEIHING, 1974). The physical and chemical properties of highly purified isolates of these proteins are similar to those of their counterparts from striated muscle. Either *Physarum* actin and muscle myosin or muscle actin and *Physarum* myosin can be cross "hybridized" to yield a functional actomyosin complex (HATANO and OHNUMA, 1970; HATANO and OOSAWA, 1966; KATO and TONOMURA, 1975a; cf. also BRITZ, Chap. 2.3, this vol.)

The possibility that microtubules take part in the shuttle-streaming process can probably be ruled out because of their very sparse distribution in the plasmodium. Extracts of *Physarum* show no appreciable colchicine or colcemid binding. Extracts from plasmodia contain two polypeptides which migrate similarly to authentic tubulin on SDS-gels, but these polypeptides are not precipitated by vinblastine, do not polymerize under conditions conducive to in vitro microtubule assembly, and do not specifically bind to DEAE-cellulose. These data indicate that the levels of tubulin are either very low, or that slime mold tubulin is quite different from mammalian tubulin (JACOBSON et al., 1976).

3.1.2 The Molecular Basis of Movement

Although actomyosin from *Physarum* and from striated muscle are similar in vitro, the two contractile systems are quite different in vivo. Protoplasmic streaming in plasmodia of *Physarum* requires a highly dynamic contractile apparatus, analogous to the cytoplasm of amoebae which, unlike muscle, is highly variable in structure and location (KOMNICK et al., 1973). The assembly and disassembly of plasma fibrils in a migrating plasmodium probably involves a number of accessory proteins, associated with actin and myosin, which confer lability or stability to the complex, depending upon local concentrations of ATP and divalent cations (JACOBSON et al., 1976; KOMNICK et al., 1973; POLLARD and WEIHING, 1974). HATANO and OWARIBE (1976) report one such accessory protein, dubbed "plasmodium actinin", which can be removed from *Physarum* actomyosin by column chromatography (ADELMAN and TAYLOR, 1969a, b). Purified plasmodium G-actin polymerizes to F-actin with physical properties identical to those of purified muscle F-actin. However, when combined with 10%–15% plasmodium actinin, plasmodium G-actin forms a nearly globular "Mg-polymer" quite different from the normal filamentous form of F-actin (HATANO and OWARIBE, 1976). Local levels of free Ca^{2+} and Mg^{2+} appear to regulate the extent of fibrillar differentiation of the plasmodial ground plasm, but the basis for this regulation is as yet not understood (JACOBSON et al., 1976; KOMNICK et al., 1973). Calcium could conceivably initiate fibril assembly either by direct interaction with *Physarum* actomyosin or in combination with one or more accessory proteins. An actin-linked calcium control system in *Physarum*, remarkably homologous to the troponin-tropomyosin complex of striated muscle, has recently been described by KATO and TONOMURA (1975a, b). Their most convincing evidence is the demonstration that the troponin-tropomyosin complex from *Physarum* confers calcium sensitivity on a purified actomyosin complex from rabbit skeletal muscle (KATO and TONOMURA, 1975b).

The protoplasmic contractions providing the motive force for shuttle streaming and migration of plasmodia are probably induced by a transient, localized increase in the level of free Ca^{2+} ions. Thus, one would anticipate the presence of a Ca^{2+} sequestering system in *Physarum*. BRAATZ and KOMNICK (1970) have reported a Ca^{2+} pump in vacuoles of slime mold plasmodia. Like the pump associated with cytoplasmic membranes of giant amoebae (see Sect. 2.1.1.2), the *Physarum* vacuolar enzyme requires ATP and can accumulate Ca^{2+} from an external concentration of 10^{-6} M. MATTHEWS (1977) has also reported a Ca^{2+} sequestering system in cytoplasmic membranes of *Physarum* microplasmodia which, like the sarcoplasmic reticulum of striated muscle, release Ca^{2+} upon stimulation with caffeine.

Regulation of the periodic reversal of protoplasmic streaming may be an electrophysiological phenomenon (KOMNICK et al., 1973). Over 25 years ago KAMIYA and ABE (1950) demonstrated a correlation between protoplasmic streaming and the voltage potential along a contractile strand segment. RIDGWAY and DURHAM (1976) have recently shown a similar correlation between the voltage potential and an oscillation in the concentration of free Ca^{2+} within the strand of plasmodium—Ca^{2+} was measured by the luminescence of the photoprotein aequorin, microinjected into isolated strand segments.

3.1.3 Synchronization of Movement

A type of neuromotor control of shuttle streaming was suggested by NAKAJIMA and HATANO (1962) with their discovery of acetyl choline sensitivity and acetyl cholinesterase activity in *Physarum*. A subsequent histochemical study demonstrated esterase activity associated with invaginations of the plasma membrane in *P. confertum* (BAUER, 1967). These invaginations are arranged in rows, and appear to be linked to plasma fibrils, forming a structure reminiscent of the T system of cross-striated muscle. How such a system could accomplish the rhythmic control of bioelectrical membrane phenomena ultimately resulting in shuttle streaming remains a mystery in view of the absence of a nervous control center in *Physarum* (KOMNICK et al., 1973).

3.2 Control of Movement by External Stimuli

Although sensory transduction by the true slime molds has been studied much less thoroughly than has sensory transduction by the cellular slime molds, the available reports concerning taxes in the Myxomycetes are more than sufficient to stimulate the interest of the student of perception.

3.2.1 Chemosensory

Positive and negative chemotaxis in Myxomycetes occurs in response to various substances including nutrients, inorganic salts and diffusates from several Basidiomycetes. COMAN (1940) demonstrated positive chemotaxis to glucose by plasmodia of *Physarum polycephalum*. More recently, using axenic cultures, CARLILE (1970) has shown that *P. polycephalum* will also respond to maltose, mannose, galactose, and peptone, but will not respond to sucrose, fructose, or ribose. CARLILE (1970) has pointed out that "... chemotactic effectiveness of sugars paralleled their ability to support growth". Chemotactic stimuli are not limited to potential nutrients although the chemotactic response to a given chemical may depend upon the nutritional status of the plasmodium (KINCAID, 1977). *P. polycephalum* responds with a positive chemotaxis to the phosphate compounds cyclic AMP, sodium-ATP, sodium pyrophosphate, and potassium phosphate (KH_2PO_4) in addition to thorium nitrate [$Th(NO_3)_4$], and negatively to several chloride salts (KCl, $CaCl_2$), and ($CaCl_3$) (UEDA et al., 1975). Moreover, the Myxomycete *Badhamia utricularis* will move toward extracts of the Basidiomycetes *Stereum hirsutum* and *Metarrhizum anisopline* (MADELIN et al., 1975). The active principle from *S. hirsutum* has been partially purified and characterized as a relatively low molecular weight, thermostable molecule which diffuses either through the substrate or through the atmosphere (KNOWLES and MADELIN, 1975).

UEDA et al. (1965) have measured the motive force of movement by *P. polycephalum* and the differences in membrane potential in response to different concentrations of a variety of nutrients and inorganic salts. For each substance, a chemotactic response occurs only when the compound is present above some concentration threshold specific for that compound at a particular temperature. Exposure of one end of a plasmodium to a suprathreshold concentration of such a compound causes a change in motive force (measured as the force necessary to stop net migration) of about 10 cm H_2O, and a change in potential along the length of the plasmodium of up to 40 mV. UEDA et al. (1975) suggest that this change in potential is consistent with a depolarization of the membrane and may be used to measure the difference in concentration of the chemoattractant along the plasmodium. The data clearly support a central role for the plasma membrane both in chemoreception and in signal transmission throughout the plasmodium.

3.2.2 Galvanosensory

Physarum polycephalum also exhibits directed movement in an electric field. ANDERSON (1951) observed that migration is inhibited toward the positive side of an electric field. This galvanotaxis might result from an interaction between the electric field and the oscillations in electrical potential along individual strands of protoplasm (see Sect. 3.1.2), or from electrophoresis of inorganic ions affecting movement of the organism.

3.2.3 Thermosensory

Thermotaxis in *P. polycephalum*, described by TSO and MANSOUR (1975), is a directed movement in a temperature gradient toward an optimum temperature of 29° ± 1° C, independent of the growth temperature of the organism. The maximum sensitivity by the plasmodium is to a temperature difference of 3° C (i.e., directed migration will not occur between the two halves of a Petri dish if they differ by less than 3° C.) This thermotactic behavior toward an optimum temperature is suggested by TSO and MANSOUR (1975) to result from the combined effects of temperature on various physiological processes such as migration rate. Thus, thermotaxis in *P. polycephalum* and *D. discoideum* appear to involve radically different mechanisms.

3.2.4 Photosensory

BIALCZYK and RAKOCZY (1974, 1975) have studied the reactions by plasmodia of *P. polycephalum* and *P. nudum* when half of the plasmodium is illuminated and the other half in darkness. Young (4-day) plasmodia of *P. polycephalum* and *P. nudum* consistently migrate away from the illuminated area into darkness (BIALCZYK and RAKOCZY, 1974). The reaction of old (≧8-day) plasmodia is less consistent, with some plasmodia migrating into the illuminated area. The proportion of old plasmodia reacting positively to light increases with light intensity to ca. 250 μW cm^{-2}, and decreases again at higher light intensities (BIALCZYK and RAKOCZY, 1975). Thus, the young plasmodium migrates away from light which inhibits growth, while the older plasmodium migrates toward light which is required for sporulation. RAKOCZY and BIALCZYK (1976) have also reported that the action spectrum for the negative photoresponse by young plasmodia shows two action maxima, one at 375 nm and one at 452 nm. Such an action spectrum suggests that this is one further process regulated by the virtually ubiquitous blue light photoreceptor pigment(s) (See Sect. 2.2.2.1).

4. Summary

4.1 Movement

Present lines of evidence indicate that movement in both the cellular and acellular slime molds is ultimately dependent upon contractions of actomyosin which are probably controlled by the cell (organism) through the regulation of localized intracellular concentrations of Ca^{2+}. Cellular slime mold myxamoebae may use the contractions of actomyosin to fold the plasma membrane, developing the pressure necessary for the extension of filopodia or pseudopodia. Similarly, actin fibrils attached to the plasma membrane of the acellular slime mold permit contractions of the ectoplasm of the plasmodium, developing the pressure necessary for shuttle streaming and thereby gross movement of the organism. The mechanism for migration of the cellular slime mold pseudoplasmodium probably involves amoeboid movement directed by some presently unknown method toward the anterior of the pseudoplasmodium.

Important features of slime mold movements which need to be clarified include: the mechanism whereby local intracellular calcium concentrations are controlled in cellular and acellular slime molds; the method of synchronization of contractions through the acellular slime mold plasmodium; and the basis for the polarity of cell movement in the cellular slime mold pseudoplasmodium.

4.2 Sensory Transduction

Cellular slime mold myxamoebae respond chemotactically to folic acid and to cAMP, extending pseudopodia toward the source of the stimulus. Major unknowns for both chemical attractants are the chemoreceptor molecules and the mechanism that translates chemoreception into directed movement. This mechanism appears to involve the transport of Ca^{2+}.

Cellular slime mold plasmodia migrate toward light and toward heat. Reception of light by the photoreceptor pigment leads to a stimulation of migration rate through an unknown mechanism. This increased migration rate localized on the distal side of the pseudoplasmodium (with respect to the light), results in a turn toward the light source. The measurement of temperature by the pseudoplasmodium is extremely sensitive and operates over a narrow temperature range, which depends on the growth temperature of the organism. The bio-thermometer itself is unknown, as is the mechanism by means of which the measurement of a temperature differential affects the direction of migration of the pseudoplasmodium.

Acellular slime mold plasmodia exhibit responses of directed movement to a rich variety of stimuli including chemicals, light, heat, and electrical potential. In no case is the receptor known, but many of the data are consistent with the receptors being localized at or near the plasma membrane and affecting shuttle streaming.

We suggest that the control of Ca^{2+} transport and the interaction of this control with membrane-bound receptors of sensory stimuli is the arena for exciting advances both in movement and in sensory transduction in slime molds.

Acknowledgements. The preparation of this review was supported in part by the U.S. Energy Research and Development Administration (now Department of Energy) under Contract EY-76-C-02-1338.

References

Achterrath, M.: Reaction of glycerinated protoplasma drops of *Physarum polycephalum* on addition of ATP. Cytobiology *1*, 169–183 (1969)

Adelman, M.R., Taylor, E.W.: Isolation of an actomyosin-like protein complex from slime mold plasmodium and separation of the complex into actin- and myosin-like fractions. Biochemistry *8*, 4964–4975 (1969a)

Adelman, M.R., Taylor, E.W.: Further characterization of slime mold myosin and slime mold actin. Biochemistry. *8*, 4976–4988 (1969b)

Alcantara, F., Monk, M.: Signal propagation during aggregation in the slime mold *Dictyostelium discoideum*. J. Gen. Microbiol. *85*, 321–334 (1974)

Allen, R.D.: A new theory of amoeboid movement and protoplasmic streaming. Exp. Cell Res. Suppl. *8*, 17–31 (1961)

Allen, R.D.: Amoeboid movement. Sci. Am. *206*, 112–122 (1962)

Allen, R.D.: Biophysical aspects of pseudopodium formation and retraction. In: The Biology of Amoeba. Jeon, K. (ed.). New York: Academic Press, 1972, pp. 201–247

Allen, R.D., Taylor, D.L.: The molecular basis of amoeboid movement. In: Molecules and Cell Movement. Inoué, S., Stephens, R.E. (eds.). New York: Raven Press, 1975, pp. 239–258

Alléra, A., Beck, R., Wohlfarth-Bottermann, K.E.: Extensive fibrillar protoplasmic differentiations and their significance for protoplasmic streaming. VIII. Identification of the plasmafilaments in *Physarum polycephalum* as F-actin by *in situ* binding of heavy meromyosin. Cytobiology *4*, 437–449 (1971)

Anderson, J.D.: Galvanotaxis of slime mold. J. Gen. Physiol. *35*, 1–16 (1951)

Ashworth, J.M.: Cell development in the cellular slime mould *Dictyostelium discoideum*. Symp. Soc. Exp. Biol. *25*, 27–49 (1971)

Bauer, L.G.: On the similar orientation of fibrillar structures and rows of esterolytic invaginations in plasmodia of the slime mold, *Physarum confertum* Macbr. J. Exp. Zool. *164*, 69–80 (1967)

Bialczyk, J., Rakoczy, L.: Phototaxis in the *Myxomycetes*. Migration of the young plasmodia in white light. Bull. Acad. Pol. Sci. Sér. Sci. Biol. *22*, 871–874 (1974)

Bialczyk, J., Rakoczy, L.: Phototaxis of the *Myxomycetes*. Behavior of old plasmodia in white light. Bull. Acad. Pol. Sci. Sér. Sci. Biol. *23*, 571–575 (1975)

Bonner, J.T.: Evidence for the formation of cell aggregates by chemotaxis in the development of the slime mold *Dictyostelium discoideum*. J. Exp. Zool. *106*, 1–26 (1947)

Bonner, J.T.: Aggregation and differentiation in the cellular slime molds. Annu. Rev. Microbiol. *25*, 75–92 (1971)

Bonner, J.T., Clarke, W.W. Jr., Neeley, C.L. Jr., Slifkin, M.K.: The orientation to light and the extremely sensitive orientation to temperature gradients in the slime mold *Dictyostelium discoideum*. J. Cell. Comp. Physiol. *36*, 149–158 (1950)

Bovee, E.C.: Morphological differences among pseudopodia of various small amoebae and their functional significance. In: Primitive Motile Systems in Cell Biology. Allen, R.D., Kamiya, N. (eds.). New York: Academic Press, 1964, pp. 189–219

Bovee, E.C., Jahn, T.L.: Locomotion and behavior. In: The Biology of Amoeba. Jeon, K., (ed.). New York: Academic Press, 1972, pp. 249–290

Braatz, R., Komnick, H.: Histochemischer Nachweis eines calciumpumpenden Systems in Plasmodien von Schleimpilzen. Cytobiol. *2*, 457–463 (1970)

Brenner, M.: Cyclic AMP gradient in migrating pseudoplasmodia of the cellular slime mold *Dictyostelium discoideum*. J. Biol. Chem. *252*, 4073–4077 (1977)

Cappuccinelli, P., Ashworth, J.M.: The effect of inhibitors of microtubule and microfilament function on the cellular slime mold *Dictyostelium discoideum*. Exp. Cell. Res. *103*, 387–393 (1976)

Carlile, M.J.: Nutrition and chemotaxis in the myxomycete *Physarum polycephalum*: the effect of carbohydrates on the plasmodium. J. Gen. Microbiol. *63*, 221–226 (1970)

Chapman, D.: Phase transitions and fluidity characteristics of lipids and cell membranes. Q. Rev. Biophys. *8*, 185–235 (1975)

Clarke, M., Schatten, G., Mazia, G., Spudich, J.A.: Visualization of actin fibers associated with the cell membrane in amoebae of *Dictyostelium discoideum*. Proc. Natl. Acad. Sci. U.S.A. *72*, 1758–1762 (1975)

Clarke, M., Spudich, J.A.: Biochemical and structural studies of actomyosin-like proteins from non-muscle cells: Isolation and characterization of myosin from amoebae of *Dictyostelium discoideum*. J. Mol. Biol. *86*, 209–222 (1974)
Cohen, M.H., Robertson, A.: Wave propagation in the early stages of aggregation of cellular slime molds. J. Theor. Biol. *31*, 101–118 (1971)
Cohen, M.H., Robertson, A.: Chemotaxis and the early stages of aggregation in cellular slime molds. J. Theor. Biol. *31*, 119–130 (1971)
Cohen, M.H., Robertson, A.: Cell migration and the control of development. In: Statistical Mechanics. Rice, S.A., Freed, K.F., Light, J.C. (eds.). Chicago: Univ. of Chicago Press 1972, pp. 131–157
Cohen, M.H., Robertson, A.: Cellular organization and communication in *Dictyostelium discoideum* and other cellular slime molds. In: Proc. 4th Int. Symp. Theor. Phys. Biol., Institut de la Vie, Paris, France. 1974
Coman, D.R.: Additional observations on positive and negative chemotaxis. Experiments with a myxomycete. Arch. Pathol. *29*, 220–228 (1940)
Czarska, L., Grebecki, A.: Membrane folding and the plasma-membrane ratio in the movement and shape transformation in *Amoeba proteus*. Acta Protozool. *4*, 201–240 (1966)
Daniel, J.W., Baldwin, H.: Methods of culture for plasmodial myxomycetes. In: Methods in Cell Physiology. Prescott, D.M. (ed.). New York: Academic Press, 1964, Vol. I, pp. 9–41
Durham, A.C.H.: A unified theory of the control of actin and myosin in nonmuscle movements. Cell *2*, 123–136 (1974)
Eckert, B.S., Warren, R.H., Rubin, R.W.: Structural and biochemical aspects of cell motility in amoebas of *Dictyostelium discoideum*. J. Cell Biol. *72*, 339–350 (1977)
Farnsworth, P.A., Loomis, W.F., Jr.: A gradient in the thickness of the surface sheath in pseudoplasmodia of *Dictyostelium discoideum*. Dev. Biol. *46*, 349–357 (1975)
Francis, D.W.: Some studies on phototaxis of *Dictyostelium*. J. Cell. Comp. Physiol. *64*, 131–138 (1964)
Freeze, H., Loomis, W.F., Jr.: The role of the fibrillar component of the surface sheath in the morphogenesis of *Dictyostelium discoideum*. Dev. Biol. *56*, 184–194 (1977a)
Freeze, H., Loomis, W.F., Jr.: Isolation and characterization of a component of the surface sheath of *Dictyostelium discoideum*. J. Biol. Chem. *252*, 820–824 (1977b)
Garrod, D.R.: The cellular basis of movement of the migrating grex of the slime mold *Dictyostelium discoideum*. J. Cell Sci. *4*, 781–798 (1969)
Garrod, D.R.: The cellular basis of movement of the migrating grex of the slime mould *Dictyostelium discoideum*: chemotactic and reaggregation behaviour of grex cells. J. Embryol. Exp. Morphol. *32*, 57–68 (1974)
Garrod, D.R., Malkinson, A.M.: Cyclic AMP, pattern formation and movement in the slime mould, *Dictyostelium discoideum*. Exp. Cell Res. *81*, 492–495 (1973)
Garrod, D.R., Wolpert, L.: Behavior of the cell surface during movement of the slime mold *Dictyostelium discoideum* studied with fluorescent antibody. J. Cell Sci. *3*, 365–372 (1968)
Gerisch, G.: Cell aggregation and differentiation in *Dictyostelium*. Curr. Top. Dev. Biol. *3*, 157–197 (1968)
Gerisch, G., Hess, B.: Cyclic-AMP-controlled oscillations in suspended *Dictyostelium* cells: their relation to morphogenetic cell interactions. Proc. Natl. Acad. Sci. U.S.A. *71*, 2118–2122 (1974)
Gerisch, G., Hülser, D., Malchow, D., Wick, U.: Cell communication by periodic cyclic-AMP pulses. Phil. Trans. R. Soc. (London) B. *272*, 181–192 (1975)
Gerisch, G., Malchow, D., Hess, B.: Cell communication and cyclic-AMP regulation during aggregation of the slime mold *Dictyostelium discoideum*. In: Biochemistry of Sensory Functions. Jaenicke, A. (ed.). Berlin-Heidelberg-New York: Springer, 1974, pp. 279–298
Gingle, A.R., Robertson, A.: The development of the relaying competence in *Dictyostelium discoideum*. J. Cell Sci. *20*, 21–27 (1976)
Green, A.A., Newell, P.C.: Evidence for the existence of two types of cAMP binding sites in aggregating cells of *Dictyostelium discoideum*. Cell *6*, 129–136 (1975)
Gregg, J.H.: Organization and synthesis in the cellular slime molds. In: The Fungi, an Advanced Treatise Ainsworth, G.C., Sussman, A.S. (eds.). New York: Academic Press, 1966 Vol. II, pp. 235–281

Hatano, S., Ohnuma, J.: Purification and characterization of myosin A from the Myxomycete plasmodium. Biochim. Biophys. Acta *205*, 110–120 (1970)
Hatano, S., Oosawa, F.: Extraction of an actin-like protein from the plasmodium of a myxomycete and its interaction with myosin A from rabbit striated muscle. J. Cell Physiol. *68*, 197–202 (1966)
Hatano, S., Oosawa, F.: Movement of cytoplasm in plasmodial fragments obtained by caffeine treatment. I. Its Ca^{++} sensitivity. J. Physiol. Soc. Jap. *33*, 589–590 (1971)
Hatano, S., Owaribe, K.: Actin and actinin from myxomycete plasmodia. In: Cell Motility. Book B. Actin, Myosin, and Associated Proteins. Goldman, R., Pollard, T., Rosenbaum, J. (eds.). Cold Spring Harbor, New York: Cold Spring Harbor, 1976, pp. 499–511
Hatano, S., Tazawa, M.: Isolation, purification and characterization of myosin B from Myxomycete plasmodium. Biochim. Biophys. Acta *154*, 507–519 (1968)
Henderson, E.J.: The cyclic adenosine 3′, 5′-monophosphate receptor of *Dictyostelium discoideum*. J. Biol. Chem. *250*, 4730–4736 (1975)
Huang, L., Lorch, S.K., Smith, G.G., Haug, A.: Control of membrane lipid fluidity in *Acholeplasma laidlawii*. FEBS Lett. *43*, 1–5 (1974)
Jacobson, D.N., Johnke, R.N., Adelman, M.R.: Studies on motility in *Physarum polycephalum*. In: Cell Motility. Book B. Actin, Myosin and Associated Proteins. Goldman, R., Pollard, T., Rosenbaum, J. (eds.). Cold Spring Harbor, New York: Cold Spring Harbor, 1976, pp. 749–770
Jahn, T.L., Bovee, E.C.: Protoplasmic movements within cells. Physiol. Rev. *49*, 793–862 (1969)
Kamiya, N.: Protoplasmic Streaming. In: Protoplasmatologia. Wien: Springer, 1959, Vol. VIII, pt. 3a, pp. 1–199
Kamiya, N., Abe, S.: Bioelectric phenomena in the myxomycete plasmodium and their relation to protoplasmic flow. J. Colloid Sci. *5*, 149–163 (1950)
Kamiya, N., Kuroda, K.: Movement of the myxomycete plasmodium. I. A study of glycerinated models. Proc. Jpn. Acad. *41*, 837–841 (1965)
Kato, T., Tonomura, Y.: Ca^{2+}-sensitivity of actinomyosin ATPase purified from *Physarum polycephalum*. J. Biochem. *77*, 1127–1134 (1975a)
Kato, T., Tonomura, Y.: *Physarum* tropomyosin-troponin complex isolation and properties. J. Biochem. *78*, 583–588 (1975b)
Kincaid, R.: Chemotaxis in the acellular slime mold, *Physarum polycephalum*. Ph.D. thesis, Stanford Univ. 184 p. 1977
Knowles, D., Madelin, M.F.: Chemotactic and other responses of plasmodia of *Badhamia utricularis* to an extract of *Stereum hirsutum* and to certain other substances. J. Gen. Microbiol. *89*, 235–244 (1975)
Komnick, H., Stockem, W., Wohlfarth-Bottermann, K.E.: Cell motility: mechanisms in protoplasmic streaming and amoeboid movement. Int. Rev. Cytol. *34*, 169–249 (1973)
Konijn, T.M.: Chemotaxis and aggregation in the slime molds. In: Proc. 10th Int. Cong. Microbiol. Mexico City, Mexico. Pérez-Mirauete, A. (ed.). New York: Plenum 1973, pp. 48–61
Loewy, A.G.: An actomyosin-like substance from the plasmodium of a myxomycete. J. Cell. Comp. Physiol. *40*, 127–156 (1952)
Loomis, W.F.,Jr.: Mutants in phototaxis of *Dictyostelium discoideum*. Nature (London) *227*, 745–746 (1970)
Loomis, W.F.,Jr.: Role of the surface sheath in the control of morphogenesis in *Dictyostelium discoideum*. Nature (New Biol.) *240*, 6–9 (1972)
Loomis, W.F., Jr.: *Dictyostelium discoideum*, a developmental system. New York: Academic Press, 1975a, p. 214
Loomis, W.F., Jr.: Polarity and pattern in *Dictyostelium*. Mol. Cell Biol. *2*, 109–128 (1975b)
Madelin, M.F., Audus, F., Knowles, D.: Attraction of plasmodia of the Myxomycete, *Badhamia utricularis*, by extracts of the Basidiomycete, *Stereum hirsutum*. J. Gen. Microbiol. *89*, 229–234 (1975)
Malchow, D., Gerisch, G.: Short term binding and hydrolysis of cyclic 3′:5′-adenosine monophosphate by aggregating *Dictyostelium* cells. Proc. Natl. Acad. Sci. U.S.A. *71*, 2423–2427 (1974)
Manery, J.F.: Effects of Ca ions on membranes. Fed. Proc. *25*, 1804–1810 (1966)

Mason, J.W., Rasmussen, H., DiBella, F.: 3′:5′ cAMP and Ca^{2+} in slime mold aggregation. Exp. Cell Res. *67*, 156–160 (1971)
Mast, S.O.: Structure, movement, locomotion and stimulation in *Amoeba*. J. Morphol. *41*, 347–425 (1926)
Matthews, L.M., Jr.: Ca^{++} regulation in caffeine-derived microplasmodia of *Physarum polycephalum*. J. Cell Biol. *72*, 502–505 (1977)
Mato, J.M., van Haastert, P.J.M., Krens, F.A., Rhijnsburger, E.H., Dobbe, F.C.P.M., Konijn, T.M.: Cyclic AMP and folic acid mediated cyclic GMP accumulation in *Dictyostelium discoideum*. FEBS Lett. *79*, 331–336 (1977a)
Mato, J.M., Konijn, T.M.: Chemotactic signal and cyclic GMP accumulation in *Dictyostelium*. In: Developments and Differentiation in the Cellular Slime Moulds. Cappuccinelli, P., Ashworth, J.M. (eds.). Elsevier/North Holland Biomedical Press 1977, pp. 93–103
Mato, J.M., Krens, F.A., van Haastert, P.J.M., Konijn, T.M.: 3′:5′-Cyclic AMP-dependent 3′:5′-cyclic GMP accumulation in *Dictyostelium discoideum*. Proc. Natl. Acad. Sci. U.S.A. *74*, 2348–2351 (1977b)
Mato, J.M., Krens, F.A., van Haastert, P.J.M., Konijn, T.M.: Unified control of chemotaxis and cAMP mediated cGMP accumulation by cAMP in *Dictyostelium discoideum*. Biochim. Biophys. Res. Commun. *77*, 399–402 (1977c)
Mockrin, S.C., Spudich, J.A.: Calcium control of actin-activated myosine adenosine triphosphatase from *Dictyostelium discoideum*. Proc. Natl. Acad. Sci. U.S. *73*, 2321–2325 (1976)
Nakajima, H., Hatano, S.: Acetylcholinesterase in the plasmodium of the Myxomycete, *Physarum polycephalum*. J. Cell. Comp. Physiol. *59*, 259–263 (1962)
Olive, L.S.: The Mycetozoans. New York: Academic Press 1975, p. 293
Pan, P., Hall, E.M., Bonner, J.T.: Folic acid as the second chemotactic substance in the cellular slime molds. Nature (New Biol.) *237*, 181–182 (1972)
Pan, P., Hall, E.M., Bonner, J.T.: Determination of the active portion of the folic acid molecule in cellular slime mold chemotaxis. J. Bacteriol. *122*, 185–191 (1975)
Pantin, C.F.A.: On the physiology of amoeboid movement. J. Mar. Biol. Assoc. (U.K.) *13*, 24–69 (1923)
Poff, K.L., Butler, W.L.: Absorbance changes induced by blue light in *Phycomyces blakesleeanus* and *Dictyostelium discoideum*. Nature (London) *248*, 799–801 (1974a)
Poff, K.L., Butler, W.L.: Spectral characteristics of the photoreceptor pigment of phototaxis in *Dictyostelium discoideum*. Photochem. Photobiol. *20*, 241–244 (1974b)
Poff, K.L., Butler, W.L.: Spectral characterization of the photoreducible *b*-type cytochrome of *Dictyostelium discoideum*. Plant Physiol. *55*, 427–429 (1975)
Poff, K.L., Butler, W.L., Loomis, W.F., Jr.: Light-induced absorbance changes associated with phototaxis in *Dictyostelium*. Proc. Natl. Acad. Sci. U.S.A. *70*, 813–816 (1973)
Poff, K.L., Loomis, W.F., Jr.: Control of phototactic migration in *Dictyostelium discoideum*. Exp. Cell. Res. *82*, 236–240 (1973)
Poff, K.L., Loomis, W.F., Jr., Butler, W.L.: Isolation and purification of the photoreceptor pigment associated with phototaxis in *Dictyostelium discoideum*. J. Biol. Chem. *249*, 2164–2167 (1974)
Poff, K.L., Skokut, M.: Thermotaxis by pseudoplasmodia of *Dictyostelium discoideum*. Proc. Natl. Acad. Sci. U.S.A. *74*, 2007–2010 (1977)
Pollard, T.D.: Progress in understanding amoeboid movement at the molecular level. In: The Biology of Amoeba. Jeon, K. (ed.). New York: Academic Press, 1972, pp. 291–317
Pollard, T.D., Ito, S.: Cytoplasmic filaments of *Amoeba proteus*. I. The role of cytoplasmic filaments in consistency changes and movement. J. Cell Biol. *46*, 267–289 (1970)
Pollard, T.D., Korn, E.D.: Filaments of *Amoeba proteus*. II. Binding of heavy meromyosin by thin filaments in motile cytoplasmic extracts. J. Cell. Biol. *48*, 216–219 (1971)
Pollard, T.D., Weihing, R.R.: Actin and myosin and cell movement. CRC Crit. Rev. Biochem. *2*, 1–65 (1974)
Rakoczy, L., Bialczyk, J.: Phototaxis of the Myxomycete *Physarum nudum* plasmodia. In: Proc. 7th Int. Cong. Photobiol. 1976, p. 108
Raper, K.B.: Plasmodium formation and organization in *Dictyostelium discoideum*. J. E. Mitchell Sci. Soc. *56*, 241–282 (1940)
Reinold, M., Stockem, W.: Darstellung eines ATP-sensitiven Membransystems mit Ca^{2+}-transportierender Funktion bei Amöben. Cytobiol. *6*, 182–194 (1972)

Ridgway, E.B., Durham, A.C.H.: Oscillations of calcium ion concentrations in *Physarum polycephalum*. J. Cell Biol. *69*, 223–226 (1976)
Robertson, A., Drage, D.J.: Stimulation of late interphase *Dictyostelium discoideum* amoebae with an external cyclic AMP signal. Biophys. J. *15*, 765–775 (1975)
Robertson, A., Grutsch, J.: The role of cyclic AMP in slime mold development. Life Sci. *15*, 1031–1043 (1974)
Shaffer, B.M.: Intracellular movement and locomotion of cellular slime-mold amebae. In: Primitive Motile Systems in Cell Biology. Allen, R.D., Kamiya, N. (eds.). New York, London: Academic Press, 1964, pp. 387–405
Shaffer, B.M.: Cell movement within aggregates of the slime mold *Dictyostelium discoideum* revealed by surface markers. J. Embryol. Exp. Morphol. *13*, 97–117 (1965)
Sinensky, M.: Homeoviscous adaptation—a homeostatic process that regulates the viscosity of membrane lipids in *Escherichia coli*. Proc. Natl. Acad. Sci. U.S.A. *71*, 522–525 (1974)
Spudich, J.A.: Biochemical and structural studies of actomyosin-like proteins from non-muscle cells. II. Purification, properties and membrane association of actin from amoebae of *Dictyostelium discoideum*. J. Biol. Chem. *249*, 6013–6022 (1974)
Spudich, J.A., Cooke, R.: Supramolecular forms of actin from amoebae of *Dictyostelium discoideum*. J. Biol. Chem. *250*, 7485–7491 (1975)
Sussman, M., Brackenbury, R.: Biochemical and molecular-genetic aspects of cellular slime mold development. Annu. Rev. Plant. Physiol. *27*, 229–265 (1976)
Taylor, D.L.: Motile model systems of amoeboid movement. In: Cell Motility. Book B. Actin, Myosin and Associated Proteins. Goldman, R., Pollard, T., Rosenbaum, J. (eds.). Cold Spring Harbor, New York: Cold Spring Harbor, 1976, pp. 797–821
Taylor, D.L., Condeelis, J.S., Moore, P.L., Allen, R.D.: The contractile basis of amoeboid movement. I. The chemical control of motility in isolated cytoplasm. J. Cell. Biol. *59*, 378–394 (1973)
Taylor, D.L., Rhodes, J.A., Hammond, S.A.: Calcium sensitive motile extracts. J. Cell Biol. *67*, 427a (1975)
Taylor, D.L., Rhodes, J.A., Hammond, S.A.: The contractile basis of amoeboid movement. II. Structure and contractility of motile extracts and plasmalemma-ectoplasm ghosts. J. Cell Biol. *70*, 123–143 (1976)
Thompson, C.M., Wolpert, L.: The isolation of motile cytoplasm from *Amoeba proteus*. Exp. Cell Res. *32*, 156–160 (1963)
Tso, W., Mansour, T.E.: Thermotaxis in a slime mold, *Physarum polycephalum*. J. Behav. Biol. *14*, 499–504 (1975)
Ueda, T., Terayama, K., Kurihara, K., Kobatake, Y.: Threshold phenomena in chemoreception and taxis in slime mold *Physarum polycephalum*. J. Gen. Physiol. *65*, 223–234 (1975)
Wohlfarth-Bottermann, K.E.: Weitreichende, fibrilläre Protoplasmadifferenzierungen und ihre Bedeutung für die Protoplasmaströmung. Protoplasma *54*, 514–539 (1962)
Wohlfarth-Bottermann, K.E.: Weitreichende, fibrilläre Protoplasmadifferenzierungen und ihre Bedeutung für die Protoplasmaströmung. III. Entstehung und experimentell induzierbare Musterbildungen. Wilhelm Roux' Arch. Entwicklungsmech. Organism. *156*, 371–403 (1965)
Wolpert, L., Gingell, D.: The cell surface membrane and amoeboid movement. In: Aspects of Cell Motility. Miller, P.L. (ed.). Cambridge: University Press, 1968, pp. 169–198
Wurster, B., Schubiger, K.: Oscillations and cell development in *Dictyostelium discoideum* stimulated by folic acid pulses. J. Cell Sci. *27*, 105–114 (1977)

4. Movements Using Turgor Mechanisms

4.1 Movements of Stomata

K. RASCHKE

1. Function

1.1 Bias in Research on Stomatal Movement

Stomata move in response to changes within the plant and in the plant's environment. Stomatal movement affects rates of uptake of CO_2 by the plant and of loss of water and heat from the plant. We would like to see a purpose in stomatal movement and reason that it must be the relationship between water loss and CO_2 uptake which is optimized by the stomata. Stomata may play an additional role in the prevention of overheating of leaves. This teleological attitude has stimulated research and has helped to bring order into the description of what has been observed on stomata during the past 200 years. I will not make attempts to avoid teleology in this chapter; on the other hand I feel a responsibility to point out bias. Bias was present right from the early studies of stomatal movement, although more for reasons of convenience and excitement than for teleology. Knowingly or unknowingly, species were chosen for observation and experimentation whose stomata responded rapidly and with wide amplitude, or whose epidermis could be peeled easily. Obviously, caution is indicated if one wishes to extrapolate from results obtained with selected material to stomatal behavior in the rest of the plant world.

The prepossession that stomata have an optimizing function does have an advantage: if stomata indeed participate in resolving conflicts of interest we are prepared to accept contradiction in stomatal behavior. We expect to see stomata respond to similar perturbations at different times in different ways, depending on the state in which the plant happens to be when the perturbation occurs and on the control law that evolution may have selected and deposited in the genetic memory of the plant for that state.

The purpose of this chapter is to describe the mechanisms involved in stomatal regulation of gas exchange. But since these mechanisms represent the "hardware" serving a complicated control system, it is necessary to describe briefly what we think is the aim of the control system and which control methods we might find applied.

1.2 Water Loss During Carbon Dioxide Uptake

The evolution of land plants began several hundred million years ago but to this date no plant substance (nor a man-made one) has emerged that has a higher permeability for CO_2 than for water vapor. Plants evolved instead an epidermis covered with a cuticle of low permeability for both, CO_2 and water vapor, and evolved also stomata, turgor-operated valves in the epidermis which allow gas exchange between the atmosphere and the intercellular spaces to be established or interrupted, depending on the needs of the plant.

Since the surfaces within the leaf are virtually completely wet (FARQUHAR and RASCHKE, 1978) and the water vapor gradient between leaf and air is under most circumstances considerably greater than the CO_2 gradient, frequently several hundred molecules of H_2O are lost from the plant for each CO_2 molecule taken up. A molar evaporation ratio (transpiration ratio) can be estimated: $\frac{E}{A}=\frac{(w_i-w_a)}{(c_i-c_a)}\cdot\frac{g^w}{g^c}$, where E ("evaporation") and A ("assimilation") are the molar fluxes of water vapor and CO_2, respectively, (w_i-w_a) and (c_i-c_a) are the differences in the molar fractions of water vapor and CO_2 between intercellular spaces (subscript i) and atmosphere (subscript a), and g^w and g^c are the conductances of the diffusion paths for water vapor and CO_2 through the boundary layer of the air surrounding the plant, through the stomata and through the intercellular spaces. (Gas exchange through the cuticle is ignored in this estimation; it could be easily added). The ratio g^w/g^c is equal to the ratio of the diffusion constants of water vapor and CO_2 in air ($D^w/D^c=1.56$) for diffusion through still air (as for instance through the stomatal pore); for diffusion through a laminar boundary layer $g^w/g^c=(D^w/D^c)^{2/3}=1.35$; with turbulence occurring, g^w/g^c may approach the value of 1.

A change in stomatal aperture and hence a change in g^c and g^w affects transpiration (=evaporation through a multiperforate septum) and CO_2 assimilation in different ways. Although E exerts a negative feedback effect on itself through transpirational cooling (which lowers the water vapor pressure in the leaf) the relationship between transpiration and conductance usually does not deviate very much from linearity (Fig. 1, left diagram[1]). In contrast, the rate of assimilation of CO_2 exerts strong negative feedback on itself because it lowers the intercellular CO_2 concentration which in turn reduces the rate of assimilation. (This effect has been marked "a" in the flow chart on p. 388). The decrease in CO_2 concentration across the epidermis can be described in analogy to Ohm's Law: $c_a-c_i=A/g^c$. Rearrangement results in a linear "supply function" $A=(c_a-c_i)g^c$ in which the conductance g^c is the slope. In the steady state supply must balance demand. The latter is represented by the photosynthetic saturation curve $A'(c_i, I, T_L)$, where I stands for quantum flux and T_L for the leaf temperature. Figure 2 shows assimilation curves at constant T_L and supply functions for three C-3 and two C-4 species. The intersections of the saturation curves with the linear supply functions indicate rates of CO_2 assimilation and intercellular CO_2 concentrations that will occur if a certain conductance g^c is established. One can see that conditions exist in C-4 plants, under which a reduction in conductance will not reduce assimilation. This lack of dependence of assimilation on conductance will occur when photosyn-

[1] Following COWAN (1977) and FARQUHAR et al. (1978) I use molar fractions, molar fluxes and molar conductances for the sake of simplicity in the demonstration of relationships and in order to obtain a measure of conductance of low temperature dependence and independent of atmospheric pressure. The new conductance unit has the dimension of a permeability (flux per unit potential difference). At a temperature of 25 °C and atmospheric pressure, a conductance of 0.4 mol m^{-2} s^{-1} corresponds to 1 cm s^{-1}.

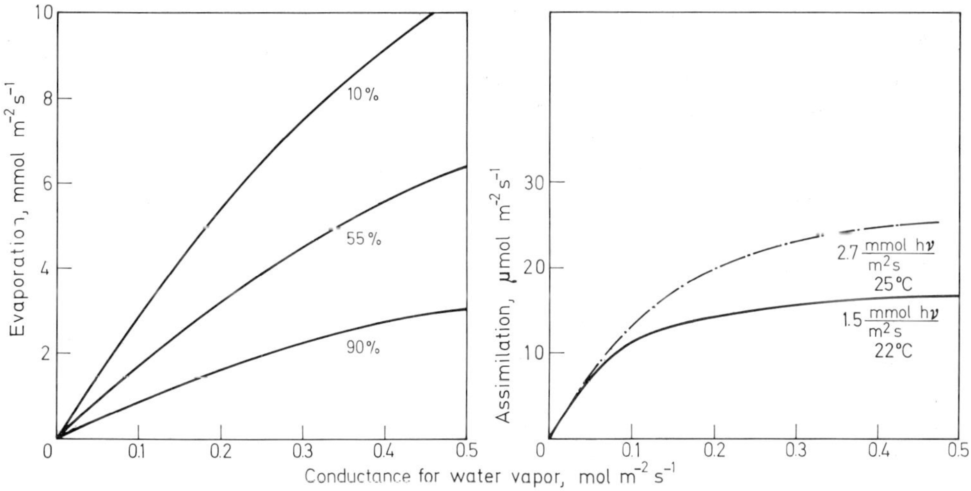

Fig. 1. The dependences of evaporation and assimilation of CO_2 on the conductance for gases diffusing through the stomata and the boundary layer of a leaf of *Xanthium strumarium*. Leaf width of 5 cm and wind speed of 2 m s^{-1} result in a boundary layer resistance of 0.25 s cm^{-1} (corresponding to a conductance of 1.6 mol m^{-2} s^{-1}) on each leaf side. Transpiration (evaporation) was computed for a net radiation of 350 W m^{-2}, air temperature of 25° C and relative humidities of 10%, 55%, and 90%. Assimilation was measured at two quantum fluxes and air temperatures. Dependence of assimilation on conductance was determined graphically as shown in Figure 2. Note the 200-fold difference in scales between evaporation and assimilation. Conductances for CO_2 are 0.64 times those for water vapor. The conductance of the boundary layer for water vapor may vary between 0.4 and more than 10 mol m^{-2} s^{-1} for leaves between 1 and 20 cm wide and exposed to wind speeds between 0.1 and 4 m s^{-1}

thesis is saturated with respect to CO_2, or using an expression of COWAN (1977), above the "breakpoint" of the relationship between A and c_i. In C-3 plants saturation of assimilation with respect to CO_2 will never occur under natural conditions. (Nevertheless, it can be seen that also in C-3 species a reduction of conductance will affect assimilation more strongly when stomata are narrow i.e., when conductance is low, than when they are wide.) We conclude that under natural conditions, stomata of C-3 species should be at their maximal aperture if maximal photosynthesis is the goal. In C-4 species, this goal is reached once the breakpoint is exceeded.

Since a reduction in conductance affects transpiration and assimilation to different degrees, the ratio between the two fluxes ("transpiration ratio") usually declines when stomata close. Using both diagrams of Figure 1 we can determine that at a quantum flux of 1.5 mmol $m^{-2}s^{-1}$ and a relative humidity of 55% leaves of *Xanthium strumarium* lose 336 molecules of water for each molecule of CO_2 assimilated when g^w is 0.4 mol m^{-2} s^{-1}. If g^w is reduced to 0.1 mol m^{-2} s^{-1} the transpiration ratio declines to 156 H_2O/CO_2.

1.3 Feedback and Feedforward

Negative Feedback. If stomata help to solve a conflict between provision of CO_2 and prevention of water loss they must be able to observe both processes simultaneously. We believe that this is accomplished by the presence of two major feedback loops (see diagram p. 388). One of them was recognized by SCARTH (1932), who interpreted stomatal opening in response to low CO_2 concentrations as a regulation against a CO_2 deficiency caused by photosynthesis (loop b) (more in Sect. 6.1.2). A few years earlier, STÅLFELT (1929) had already described how a decline in the water content of leaves triggered a then unidentified metabolic event that led to the loss of turgor in guard cells, stomatal closure and reduction of water loss (loop c). As a result, leaf turgor recovered and after a while stomata opened again in illuminated leaves. STÅLFELT also observed a tendency of the stomata to overshoot and recognized that this behavior may result in an oscillatory equilibration of opposing forces acting in the directions of "photoactive" opening and "hydroactive" closure (STÅLFELT's terms). Forty years later, WRIGHT and HIRON (1969) suggested on the basis of their experimental evidence that (+)-abscisic acid which accumulates in water-stressed leaves and was known to close stomata might close a feedback loop regulating the water relations of leaves (Sect. 6.2.1).

As a plant proceeds through its development, environmental conditions vary over a wide range. The optimal relationship between CO_2 uptake and water loss will therefore vary. The degree of feedback in the two opposing major loops can therefore be expected also to vary in order to provide flexibility. Such relative changes in gain have indeed been observed and will be described when interactions between stomatal responses to CO_2 and ABA are discussed in Sections 5.2 and 6.1.2.

Feedforward and Positive Feedback. Negative feedback alone is insufficient optimize gas exchange. We remember that for instance in *Zea mays* CO_2 assimilation does not increase if the intercellular CO_2 concentration is increased beyond the breakpoint (Fig. 2). If stomata optimize gas exchange they should keep the intercellular CO_2 concentration constant (near the breakpoint) as long as light intensity is constant. But if light intensity increases, they should allow an increase in the intercellular CO_2 concentration, corresponding to the rise in the breakpoint. Feedback of CO_2 (loop b) is inadequate for making this adjustment because it would tend to stabilize the intercellular CO_2 concentration when light intensity increases, not raise it. (Actually, an increase in CO_2 assimilation would even *lower* the intercellular CO_2 concentration because the feedback gain is relatively low, 6.1.2). An additional signal is required in the direction of stomatal opening: either by positive feedback of assimilation on stomata (d) or by a direct feedforward effect of light on guard cells (e) superimposed on the response to CO_2 (loop b).

Plants have also evolved mechanisms aiding and overriding negative feedback in the regulation of their water content. Since the bulk of the plant consists of water it takes transpiration much longer to reduce this amount by a certain fraction than it takes assimilation to remove an equal fraction of the CO_2 contained in the tissue. In a leaf the ratio of time constants for these two processes may be 10^3 to 10^4. This difference in time constants could theoretically be compensated for by a high sensitivity of the water-feedback loop to changes in water content. But a high gain in the water loop would counteract corrective actions of the CO_2 loop; flexibility in the simultaneous action of the two major feedback loops would be lost. The problem of reconciling widely differing time constants seems to have been solved by the plant in several ways. Firstly, by tolerating reductions in water content over a wide range and responding to them only after a threshold

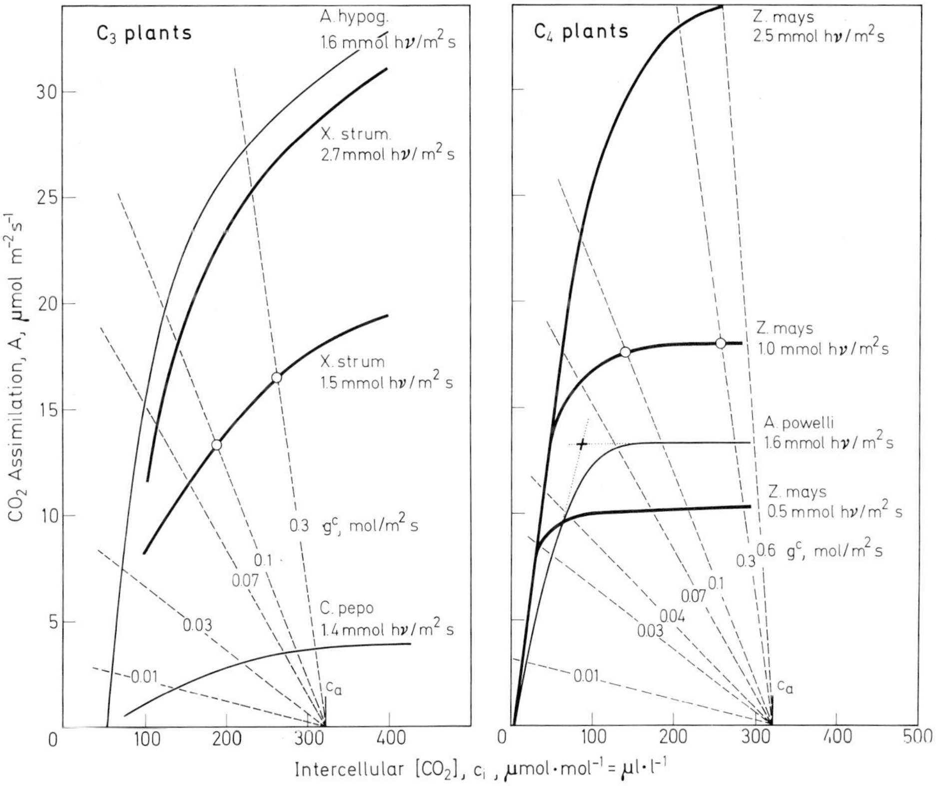

Fig. 2. Saturation curves of assimilation with respect to intercellular $[CO_2]$ ("demand functions", $A=A'(c_i, I)$), *solid curves*, and lines representing rates of diffusion of CO_2 into leaves ("supply functions", $A=(c_a-c_i)g^c$), *broken curves*, at various conductances for CO_2 (=slopes of lines). Supply functions are drawn for an atmospheric $[CO_2]$ of 320 μl l^{-1} (c_a). Intersections between demand and supply functions indicate intercellular $[CO_2]$s (*abscissa*) and rates of CO_2 assimilation (*ordinate*) which will occur when a conductance g^c is set by the stomata (and the boundary layer) of a particular leaf in a particular species. Example (*circled intersections*): lowering g^c from 0.3 to 0.1 mol m^{-2} s^{-1} will reduce c_i in a leaf of *Zea mays* (exposed to 1 mmol quanta m^{-2} s^{-1}) from 260 to 140 μl l^{-1} and in a leaf of *Xanthium strumarium* (at 1.5 mmol quanta m^{-2} s^{-1}) from 265 to 190 μl l^{-1}. In *Z. mays*, assimilation will decline from 18.0 to 17.6 μmol m^{-2} s^{-1}, in *X. strumarium* from 16.5 to 13.3 μmol m^{-2} s^{-1}. The + indicates the position of the "breakpoint" for *Amaranthus powelli*. Leaf temperatures: *Amaranthus powelli, Arachis hypogaea, Xanthium strumarium*, and *Cucurbita pepo*: 22°–25° C; *Zea mays*: 28°–32° C

value has been passed. At the threshold some or most of the leaf cells lose turgor and accumulation of the presumed feedback messenger ABA sets in (in loop c). ABA causes an effective counteraction (closure) by the stomata (Sect. 6.2.1). Secondly, by circumventing the slow negative feedback path involving the whole water content of the tissue. This is accomplished by a more or less effective separation of the water relations of the epidermis from those of the rest of the leaf. This enables stomata to respond to atmospheric dryness directly (open loop f), in what could be looked upon as an anticipating reaction (cf. Sect. 6.2.2). Since in this case the homeostatic success of a stomatal closure is not reported back to the stomata, overreactions may occur and leaves in dry air may contain more water than those exposed to moist air (SCHULZE et al., 1972). Thirdly, by using the mechanical antagonism between guard cells and their neighboring cells. When water supply declines or evaporation increases, pressure exerted by the ordinary epidermal cells on the stomata decreases; stomata may respond by further opening, temporarily or even for longer duration. COWAN (1972) showed that this response in the "wrong" direction (positive feedback, g) can accelerate stomatal opening or closing.

Interaction of the Feedback and Feedforward Paths. In summary, stomatal movements can be considered manifestations of a control system optimizing gas exchange and employing feedback (closed loop) and feedforward (open loop) methods of control:

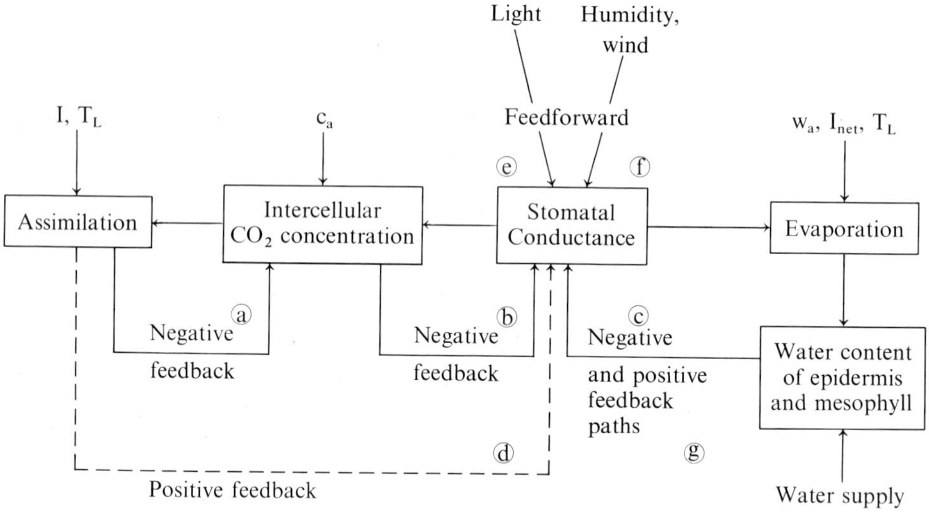

(Fig. 15 will present a further development of this network).

This introduction was intended as a reminder that the experimental isolation of a partial stomatal mechanism may take this mechanism out of its context in the complete feedback system. The observed responses may only remotely resemble those occurring in the complete system.

The possible role of stomatal movement in the dynamic optimization of gas exchange has been discussed by FARQUHAR (1973) and COWAN (1977). The article by COWAN is far more than a summary of our knowledge. It is an admirable exploration of the possible functioning of stomata in relation to their presumed control task. In another article, COWAN and FARQUHAR (1977) develop the idea that stomata may optimize gas exchange by keeping the ratio between cost (of water) and benefit (of carbon) constant: $(\partial E/\partial g)/(\partial A/\partial g)=\text{const.}$

1.4 Transpirational Cooling

Stomata may also be involved in temperature regulation. There seems to be a general tendency of stomata to open more fully with increasing temperature as long as water supply keeps pace with water loss. This has been demonstrated, for example, in *Xanthium strumarium* above 35 °C (DRAKE et al., 1970; RASCHKE, 1975a). Such a response can be of great benefit to the plant, as it not only reduces the danger of overheating if sufficient water is available, but also because it may keep the leaf near optimum temperature for photosynthesis.

We do not know yet, however, whether this tendency is the result of the general dependence on temperature of the chemical reactions proceeding in guard cells, or whether the positive temperature dependence can be ascribed to one dominating process. Great flexibility in stomatal behavior could be expected if *two* processes opposing each other determined stomatal behavior. In the absence of definitive evidence for a contribution of stomatal movement to temperature regulation of leaves, temperature regulation will not be discussed further in this chapter.

2. Mechanics of Stomatal Movement

2.1 The Basic Design of the Stomatal Valve

2.1.1 Anatomy

In its simplest form, the stoma consists of two kidney-shaped guard cells surrounding a pore (Fig. 3, *Vicia faba*). When the guard cells inflate they push each other apart and the pore opens. In grasses and sedges, guard cells have been modified into dumbbell shape. The inflatable ends of these cells are separated by a narrow, stiff middle portion (Fig. 3, *Zea mays*). Again, inflation of the guard cells produces a separation of their middle portions from each other. The slit formed between a pair of dumbbell-shaped guard cells is generally hexagonal, that between a pair of kidney-shaped guard cells elliptical. The term stoma (=mouth) usually includes the guard cells, much as lips are considered parts of the mouth.

One or several epidermal cells adjoining guard cells may be morphologically distinct from the ordinary epidermal cells (Fig. 3). These accessory or *subsidiary cells*, together with the guard cells, are considered parts of the *stomatal complex*. Subsidiary cells participate in stomatal movement mechanically (Sect. 2.3) and by serving as ion reservoirs (Sect. 3.2.2).

Much variation exists in morphology and size of the stomatal cells and how they are embedded in the tissue (VON GUTTENBERG, 1959, 1971; NAPP-ZINN, 1966, 1973; OGURA, 1972). Sizes and frequencies of stomata depend very much on the conditions which prevailed during the ontogeny of a plant. We do not know the functional significance of this great variability.

2.1.2 Observations on Guard-Cell Walls

When epidermis is viewed under a microscope with crossed polarizers, guard cells stand out. If they are kidney-shaped a diagonal pattern of reinforcement and extinction appears and remains virtually static while the stoma is being rotated, indicating a radial orientation of cellulose microfibrils in the cell walls (Fig. 4a; ZIEGENSPECK, 1938). In dumbbell-shaped guard cells the pattern is modified with apparently no radial micellation in the middle portions of the cells (Fig. 4b). Electron microscopy supports the notion of hoop-like orientation of microfibrils in the walls of kidney-shaped guard cells (PALEVITZ and HEPLER, 1976; SINGH and SRIVASTAVA, 1973). Guard cells appear to be constructed like radial tires: the incorporation of inelastic reinforcements into the wall allows a change in shape of

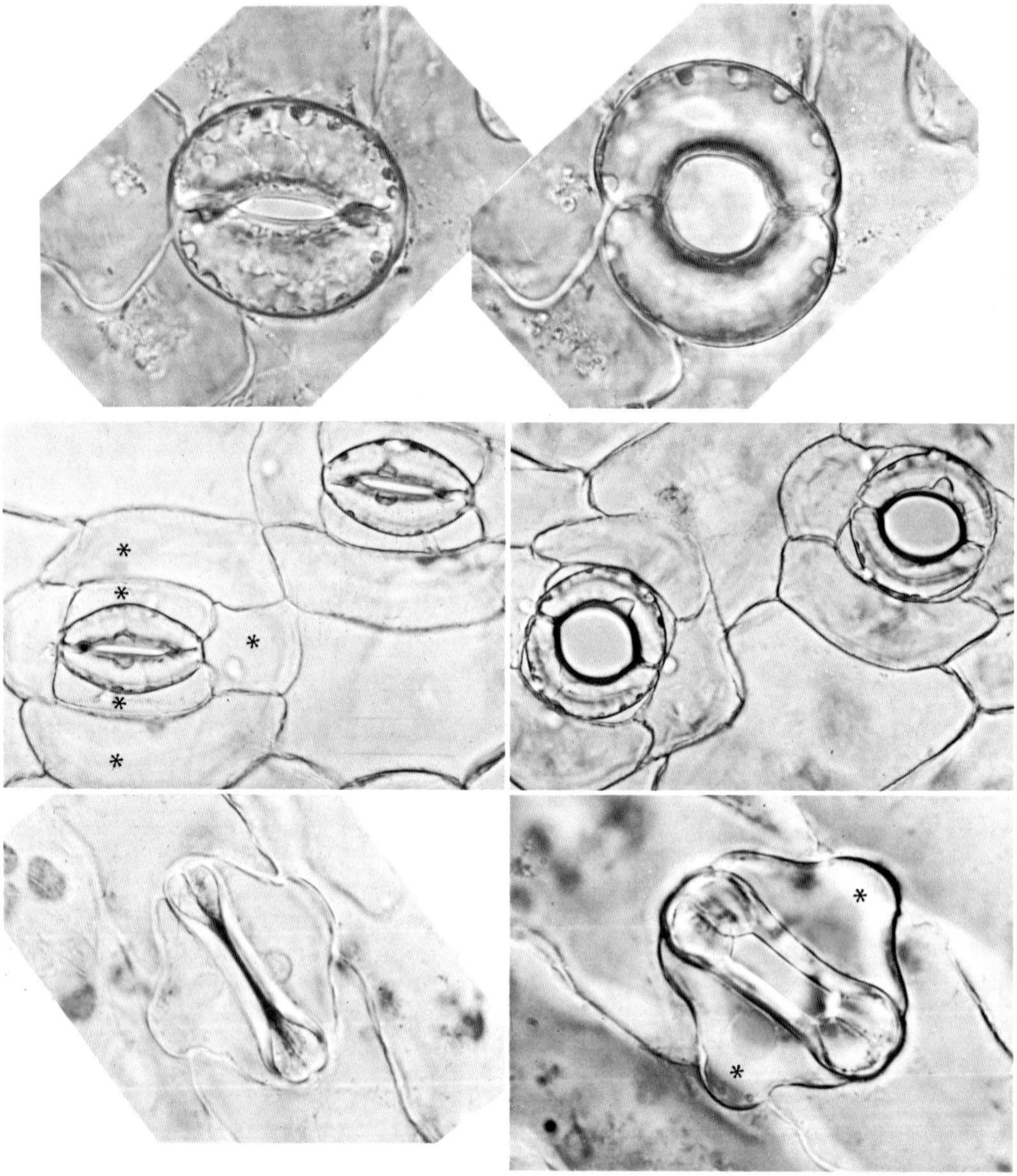

Fig. 3. Stomata of *Vicia faba* (*top*), *Commelina communis* (*middle*), and *Zea mays* (*bottom*) in two stages of opening. Stomatal complexes of *C. communis* possess six subsidiary cells* (the inner lateal subsidiary cells are well visible only when stomatal apertures are narrow, *left*); stomata of *Z. mays* possess two subsidiary cells*. The example from *V. faba* is the same stoma shown in the inflated state in water (*right*) and in a deflated state in 0.2 M sucrose (*left*)

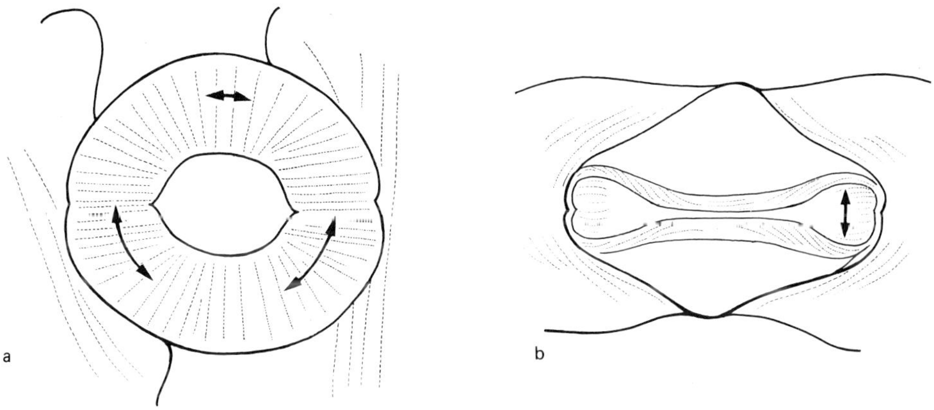

Fig. 4. Micellation in the cell walls of stomata of *Vicia faba* (**a**) and *Zea mays* (**b**). Double arrows indicate the direction of elastic stretching (in the case of *Z. mays* only presumed)

the cross section of the tire or cell but not an expansion in the circumference of that cross section.

The walls of guard cells are peculiar in another way. They stain deeply with ruthenium red (ARENS, 1968; SINGH and SRIVASTAVA, 1973; SRIVASTAVA and SINGH, 1972), are digested by snail pectinase (KELLE, 1934) or pectinases of fungal origin, and adsorb heavy metal ions (unpublished observations). These observations indicate that guard cell walls are filled with pectins. In consequence the walls must be hydrophilic, probably elastic in directions normal to the direction of the cellulose microfibrils, and anionic (their capacity to exchange cations will be discussed in Sect. 3.3.2). A considerable osmotic pressure must exist in the cell walls; it should decline with decreasing pH and the number of bound cations.

The hydrophilic character of the guard-cell walls leads to serious artefacts during preparation of epidermal samples for conventional electron microscopy. We found that guard-cell walls of *Vicia faba* shrank to one-half of their original thickness (as determined by light microscopy) during fixation with glutaraldehyde, desiccation, and embedding. Changes occurred also in cell shape, particularly in the shape of the cross sections.

2.1.3 Mechanical Work in Opening

Two kinds of work have to be done to open a stomatal pore. Epidermal tissue has to be displaced by a volume at least equal to that of the pore (COWAN, 1977), and the walls of the guard cells have to be flexed and stretched in order to produce inflation. Work done against epidermal pressure again has two components: one is the turgor in the epidermal cells, the other the rigidity of their walls. If the epidermis is closely attached to the underlying mesophyll, then the latter may also limit stomatal movement as STÅLFELT (1929) already recognized. The effects of epidermal pressure and mechanical constraints set by connections between epidermis and mesophyll on stomatal aperture become very obvious when stomatal apertures are measured before and after peeling epidermal samples. Depending on species and growing conditions, many or most of the ordinary epidermal cells rupture during stripping. In virtually all cases apertures are larger after peeling than before.

Guard cells may undergo changes in volume and shape while the stomatal pore is closed. Particularly in plants under stress and in the upper epidermis guard cells can lose turgor to a degree that their projection is no longer kidney-shaped and that their outer cuticular ledges are pressed against each other. In some species this contact of the ledges always occurs when stomata close.

When stomata are closed, forces acting in the direction of closure often exceed those acting in the direction of opening rather than being in equilibrium with each other. Pressure and volume of guard cells have to increase before opening can occur: the "motorische Phase" of stomatal opening is frequently preceded by a "Spannungsphase" (tension phase; STÅLFELT, 1927, quoted 1956).

For the sake of simplicity, all mechanical effects of epidermis and mesophyll on stomatal opening will be ignored in the following paragraphs. They will be reintroduced later in Section 2.3.

2.2 Changes in Size, Shape and Volume of Guard Cells

2.2.1 Changes in the Plane of the Epidermis

The kidney-shaped stomata of *Vicia faba* open because the ends of the guard cells bulge while the length of the stomatal apparatus remains unchanged and the width of the middle portion of each guard cell decreases even slightly (Fig. 5a). Thus expansion must occur outward, into the surrounding epidermal cells. The projected perimeter of the open pore is larger than that of the closed one and the perimeter of the whole stomatal apparatus is longer when the stoma is open than when it is closed. Similar observations were made on stomata of *Commelina communis, Tulipa gesneriana* and other species, irrespective of whether movement was brought about by physiological mechanisms or by exposure of open stomata to nonionic plasmolytica of various osmotic pressures.

Since the length of the stomatal pore of *V. faba* does not change during opening the area of the elliptical pore is linearly related to its aperture. Upon evaluating photomicrographs we found: Area (in μm^2) $= 14.3 \cdot$ aperture (in μm) with a correlation coefficient $r = 0.978$ and $n = 346$. Moreover, although the shape of the stomatal throat changes during stomatal opening, one expects a fairly linear relationship between stomatal aperture and the conductance for the diffusion of gases through stomata of *V. faba* and other species with elliptical pores.

In grasses, the length of stomata slightly decreases when the bulbous ends of their guard cells inflate (Fig. 5b). The increase in width of the stomatal slit is usually equal to the increase in width of the bulbous portions. These observations indicate that the latter act as bellows, moving apart the rigid middle portions of the guard cells, in parallel to each other or, as SCHWENDENER (1889) described it, "like the knife edges of a spectral apparatus". The length of the pore, as well as the length of the cuticular ledges, seems to increase in grasses during stomatal opening. These changes in length are probably only changes in projection because scanning electron micrographs indicate that the walls of the guard-cells

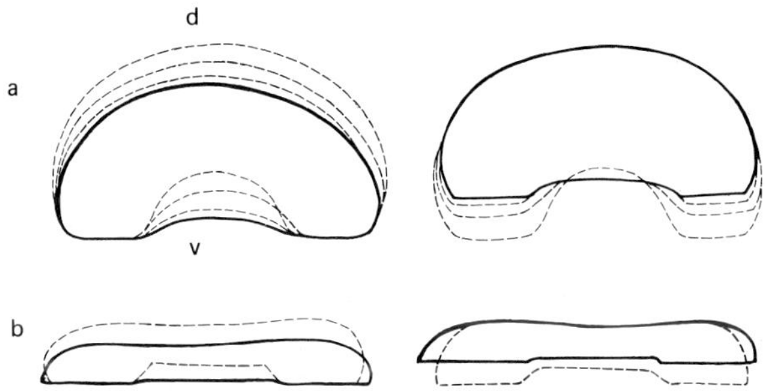

Fig. 5. Tracings of photographs of a guard cell of *Vicia faba* (**a**) and of *Zea mays* (**b**) taken during stomatal movement. *Solid line*, closed position. Reference line was either the axis of symmetry of the stomatal apparatus (*left*) or the dorsal wall of the guard cells (*right*). The left part demonstrates the displacement of epidermis when a stoma opens, the right drawings show that opening occurs because the ends of the guard cells bulge. No change in the curvature of the dorsal wall occurs in *V. faba* while the middle portion of the cells contracts slightly. In *Z. mays*, the cell length decreases during stomatal opening and the middle portion widens slightly. *d* and *v* denote dorsal and ventral wall, respectively

unfold in the corners at either end of the pore and that the cuticular ledges form arcs which swing open into the plane of the epidermis when the pore opens.

2.2.2 Changes Normal to the Plane of the Epidermis

Observations of changes in stomatal dimensions perpendicular to the epidermal plane are difficult to make: sectioning of fresh material collapses cells; preparation for conventional electron microscopy causes serious distortions of guard-cell walls (see Sect. 2.1.2). But with some skill and particularly with great patience it is possible to view turgid cells in cross section in epidermal samples put upright under the light microscope. By comparison of photographs of optical sections obtained this way with transmission and scanning electron micrographs I arrive at the following view (unpublished): The familiar triangular cross section of the lumen of the middle portion of a kidney-shaped guard cell is typical for low or no turgor (Fig. 6c′, shaded area). When turgor increases, the middle portion rounds off and becomes almost circular in cross section (Fig. 6c).

If the cellulose microfibrils form hoops in the cell walls and do not allow an expansion of the circumference of the cross section, a transition from the triangular to the circular cross section may lead to a decrease, an increase, or no change at all in the width of a guard cell as viewed in the plane of the epidermis. If the cross section approaches an equilateral triangle in the deflated state with the dorsal wall normal to the surface of the leaf, the projection of the guard cell in the epidermal plane will increase in width during inflation by 10%. However, if the cross section resembles an isosceles triangle no change will occur if the dorsal wall is the base of the triangle and measures 0.85 of the top or bottom wall. Contraction, as observed in *V. faba*, will occur if the ratio is <0.85. VON MOHL discovered as early as in 1856 that the width of guard cells either remained constant during opening or declined. Theoretically, therefore, stomata could open merely by a transition from a triangular through an elliptical to a circular cross section of the guard cells. KRAUS suggested that such a mechanism operated in mosses (VON GUTTENBERG, 1959).

More important, probably, are changes in shape and cross-sectional area in the polar regions of the guard cells during stomatal opening. Folds at the inner

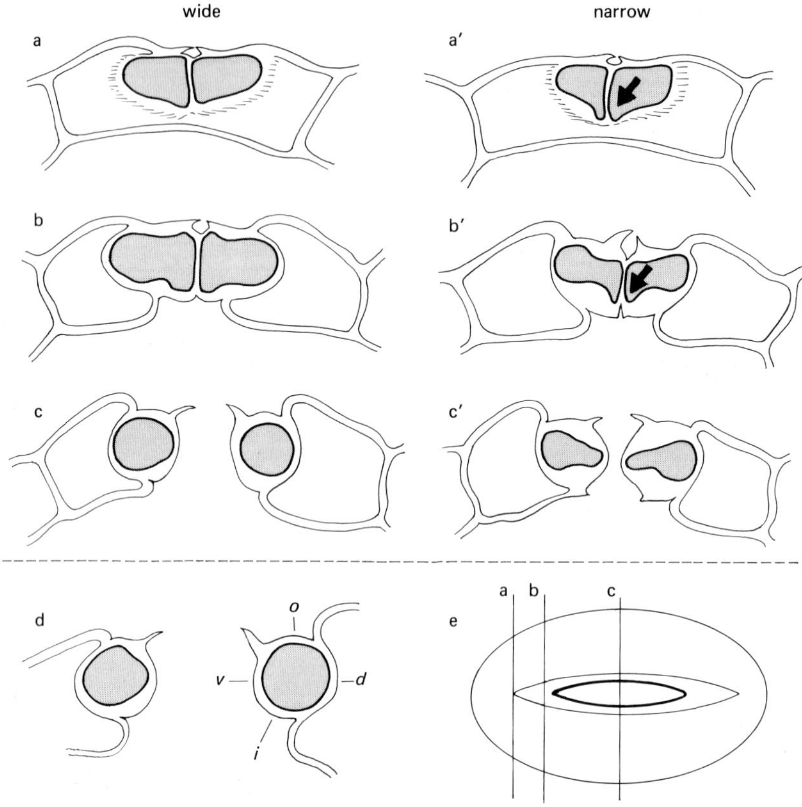

Fig. 6 a–c Tracings of optical cross sections through an intact stoma of *Vicia faba* in a wide and a narrow state of aperture. The stoma was first photographed in the open state and then plasmolyzed with a sucrose solution. Approximate position of the focal planes is shown in **e**. Stomatal apertures: open: 10.5 μm, closed: 1.5 μm. **d** is a median section through another stoma which had opened to an aperture as wide as 21.2 μm. *o, i, d, v* denote outer, inner, dorsal, and ventral wall of guard cell. The lumen of the guard cell has been *shaded*. *Arrows* point to the fold that forms in the polar parts of the guard cell when the stoma closes

wall of each guard cell characterize the closed stoma (Fig. 6a′, b′). When the guard cell expands the folds are stretched out and the wall gets thinner (Fig. 6a, b). This occurs because the cell bulges at its ends and increases in length along its long curved axis (Fig. 5a).

From the observations reported so far it can be expected that, during stomatal opening, the depth of kidney-shaped guard cells increases slightly in the middle portions of the cells and hardly at all in their polar regions. In grasses, the difference in depth between middle portions and bulbous ends is very striking (Fig. 7). In *Zea mays*, the middle portion is only about 3 to 4 μm deep. Its inner and outer walls are thick; the wall facing the subsidiary cell is thin, as if for easy transfer of solutes and water. During inflation, the diameter of the middle portion increases slightly. Most of the volume change occurs in the ends of the guard cells, which are about 12 μm deep; when inflated, they become almost perfectly spherical.

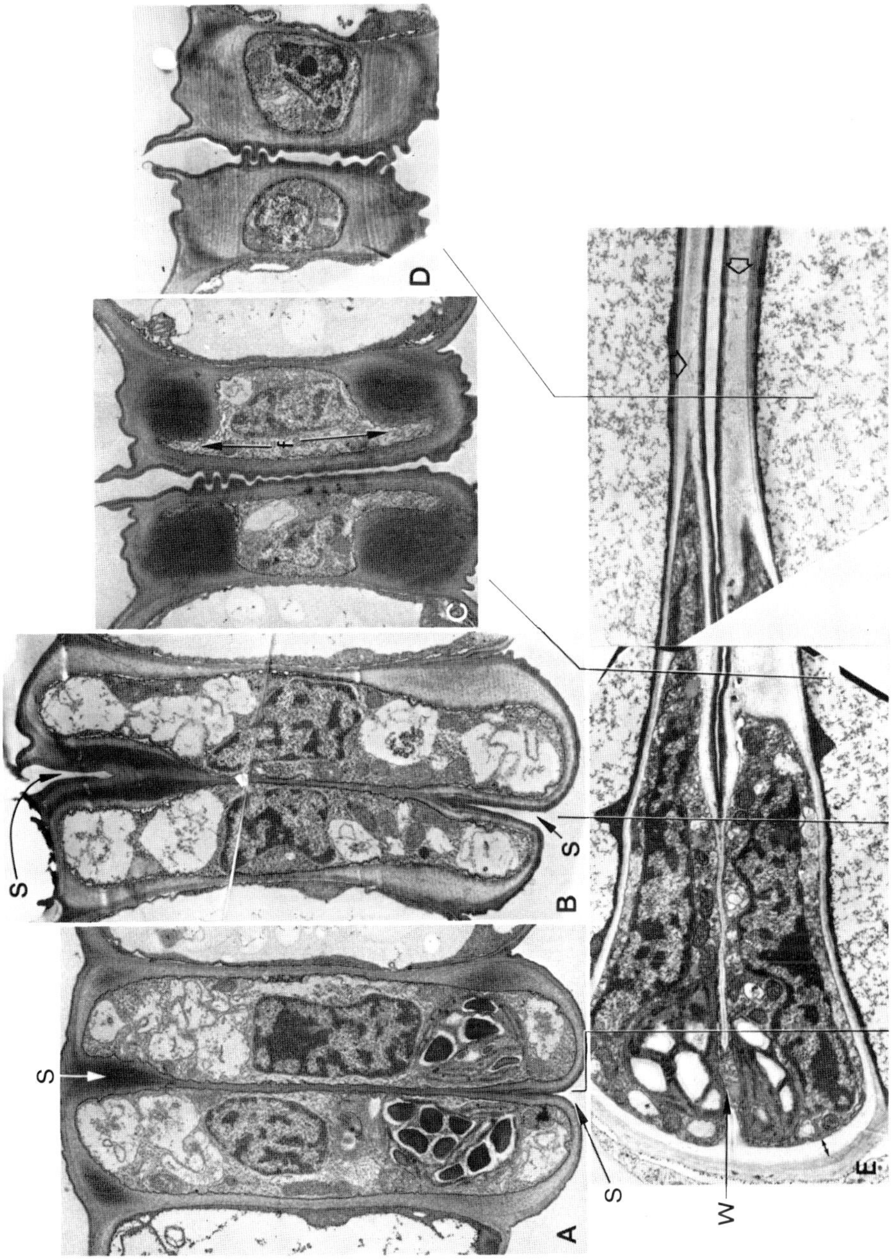

Fig. 7 A–E. Electron micrographs of sections of guard cells of a closed stoma of *Zea mays* (all ×9000). Walls may have shrunk and shapes may have changed during preparation of the samples for electron microscopy. **E** Section in the plane of the epidermis. Note the incomplete wall between the bulbous ends of the guard cells (*arrow w*). **A**, **B**, **C**, **D** Sections normal to the epidermal plane across a stoma at the locations indicated in **E**. **A**, **B** Section through the bulbous ends show that the two guard cells are not joined over all their depth but allow a partial separation when they inflate (*arrows s*) **C**, **D** Sections through the middle portions show the wall thickening making it rigid. Note the folds in **C**, reminiscent of a similar fold in kidney-shaped guard-cells (Fig. 6) which permits an expansion in the corners of the stoma when the pore opens (*arrows f*) (rearranged from SRIVASTAVA and SINGH, 1972)

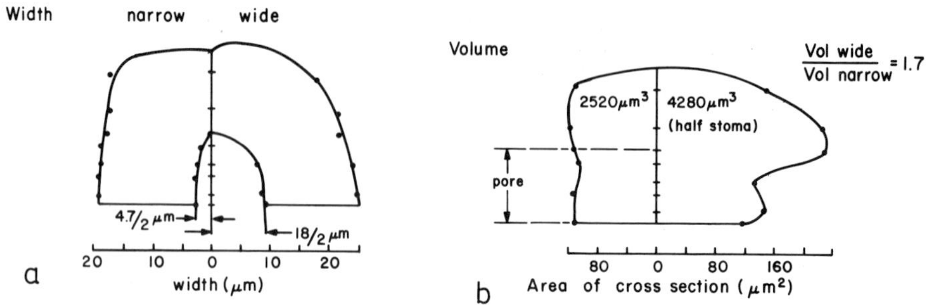

Fig. 8a and b. Projection and volume of half guard cells of *Vicia faba*. **a** Drawing of the projection in the epidermal plane with a narrow (2.3 μm) and wide (9 μm) aperture, respectively. From the long axis in the middle of the figure, the width is plotted to the left and to the right for the different positions along the axis. **b** Cross sections of the living guard cell were photographed. The areas obtained by planimetry were plotted, for the different positions along the axis, to the left and to the right. Integration by planimetry of the distribution of the cross-sectional areas over the length of the axis results in the volume of the half guard cell in the closed (*left*) and open (*right*) position, respectively. The extent of the pore is indicated on the ordinate. Note that the volume change occurs mainly in the polar portion of the half cell

Volume change appears to be brought about mainly through shape change. Deep folds inside and outside the guard cells appear to facilitate the expansion (Fig. 7).

2.2.3 Volume Changes

Optical sectioning of guard cells of *V. faba* allowed an estimation of cell volumes by integration (Fig. 8 gives an example). Determinations of volumes of several stomata of the same species led to the result that the relationship between stomatal aperture (A, μm) and guard-cell volume (V, per pair of guard cells, pl) is linear: $V=6.3+0.43$ A. Thus, the volume change ($\Delta V/\Delta A$) occurring in guard cells during stomatal opening is 0.43 $pl \cdot \mu m^{-1}$, and the variation of this value among individual cells (from 0.42 to 0.44 $pl \cdot \mu m^{-1}$) is very small, in contrast to the marked variation in V for closed stomata (from 5.0 to 7.3 pl). The volume V does *not* include that of the cell walls. The latter was quite variable and difficult to determine. It was of the same magnitude as that of the lumen. Conversion of data quoted by ALLAWAY and MILTHORPE (1976) yields $V=6.4+0.8$ A for pairs of guard cells of *Commelina cyanea* and $V=3.0+0.26$ A for *Vicia faba*. Mesophyll cells were estimated to contain 40 to 60 pl. A pair of guard cells therefore has a volume of between 1/20 and 1/4 of that of a mesophyll cell, depending on species and state of stomatal opening.

The volume change for a unit change in stomatal opening ($\Delta V/\Delta A$) required by the guard cells of grasses (Table 1) is much smaller than in kidney-shaped guard cells: these changes are only about one-third of those determined for guard cells of *Vicia faba*. It is interesting to see the volume of the subsidiary cells decline during stomatal opening, indicating a movement of water from the subsidiary to the guard cells. That the observed volume changes in guard and subsidiary cells do not compensate each other is to be ascribed to the inaccuracy of the volume determination of the subsidiary cells.

Table 1. Volumes of the components of a stomatal complex of *Zea mays* at two stages of stomatal opening (RASCHKE and DICKERSON, unpublished)

	Aperture µm	Volume of guard cells pl			Volume of subsidiary cells pl
		bulbs	cylinders	total	
	5.2	2.33	0.30	2.63	7.10
	1.7	1.92	0.25	2.17	7.80
Difference:	$\Delta A = 3.5$	0.41	0.05	$\Delta V = 0.46$	−0.70

$\Delta V/\Delta A = 0.13$ pl μm^{-1}

2.2.4 Notions on the Mechanics of Stomatal Movement. Summary and Remarks on Models

Stomata with kidney-shaped guard cells appear to operate on the same principle as those with dumbbell-shaped cells: Inflation of the ends of the cells pushes the guard cells apart and the pore opens. A difference may exist in the way the cell walls behave. In a kidney-shaped guard cell, walls appear to expand elastically along the curved axis of the cell, an anisotropy of the walls leads to curling, as observations on extremely swollen guard cells indicate (Fig. 9). Individual guard cells separated by pectinases also curl until all wall material has been completely digested. Naked guard cell protoplasts are spherical (ZEIGER and HEPLER, 1977).

Dumbbell-shaped guard cells could function solely by shape change without wall expansion (as was suggested in the previous section) although the orientation of the microfibrils in the bulbous ends indicates the possibility of an additional elastic expansion (Fig. 4b). As shown in the preceding section dumbbell-shaped guard cells require a much smaller volume change for a unit change in stomatal opening than do kidney-shaped cells.

No asymmetric wall thickening adjacent to the stomatal pore is necessary to bring about stomatal opening when guard cells inflate (SCHWENDENER, 1882, and many later texts) although asymmetry may assist stomatal opening, nor do guard cells buckle because their longitudinal expansion is constrained. Direct observation of stomatal movement through the microscope shows that guard cells do not behave like bending beams, as was suggested by DEMICHELE and SHARPE (1973) and SHOEMAKER and SRIVASTAVA (1973) but rather like sausage-shaped balloons with an anisotropic membrane, as AYLOR et al. (1973) demonstrated with a rubber model whose walls had been reinforced radially with adhesive tape. The model of AYLOR et al. does however require a longitudinal constraint of movement which may not be necessary in the real world (cf. Fig. 9).

The most sophisticated attempt so far to simulate the behavior of kidney-shaped guard cells is the finite element shell analysis performed by COOKE et al. (1976). By treating the guard cell as a sector of an elliptical torus and dividing up its walls into doubly-curved elastic triangles of variable thickness, these authors were able to simulate guard cell deformations in response to pressure changes in guard cells and epidermis. COOKE et al. concluded that stomatal opening was the result mainly of a contraction of the width of the guard cells and deformation out of the plane, in agreement with the mechanism that has been proposed to apply to moss stomata (Sect. 2.2.2), but in disagreement with the observed impor-

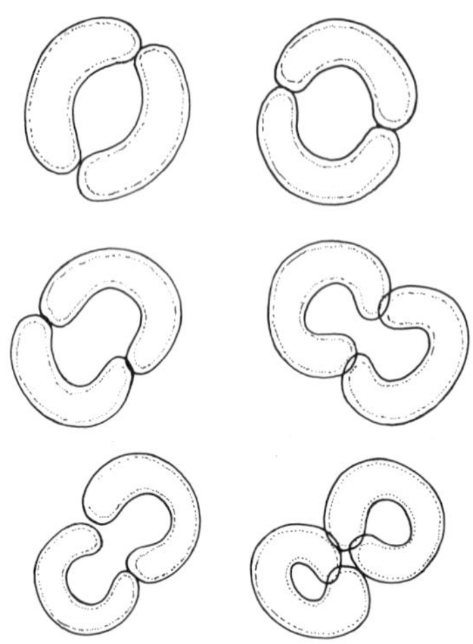

Fig. 9. Extremely wide stomatal opening and curling of the guard cells of *Rumex patientia* after rehydration of wilted leaf sections (ILJIN, 1932)

tance of the expansion of the polar portions of the guard cells. However the disagreement may only be apparent; with modification of dimensions and coefficients representing the mechanical properties of guard cell walls, an improved agreement between simulation and observation may be achievable. Finite element shell analysis appears to be appropriate for simulation and understanding of stomatal mechanics. The almost total ignorance of the mechanical properties of guard cell walls has, however, greatly hampered the application of methods of analysis available to the engineer. (For an estimate of the volume modulus of elasticity of guard cells see the following section.)

2.3 Pressure in the Stomatal Complex

2.3.1 Estimates of Pressures in Elastic Guard Cells

If guard cells behave like ideally elastic bodies (which, within limits, kidney-shaped guard cells appear to do) then pressure, volume, and modulus of elasticity are related to each other according to

$$dp = \varepsilon \, dV/V \tag{2.1}$$

or, in integrated form, $p = \varepsilon \ln (V/V_o)$, where V_o is the volume at $p = 0$. The pressure change required to produce a change in volume declines with the volume of the guard cell: $dp/dV = \varepsilon/V$ which explains the observed "giving" of the guard cell walls during stomatal opening [RASCHKE, 1976; the empirically derived relationship given there approximates the logarithmic relationship of Eq. (2.1)]. For large stomatal openings and guard cell volumes, guard cell walls must be expected to stiffen and ε should rise. From the linear relationships between stomatal aperture and guard cell volume in *Vicia faba* (Sect. 2.2.3) and between stomatal aperture and solute content in the same species (Sect. 3.1.2), I estimate ε to be between 56 and 80 atm, with 70 atm as the most likely value. No indication of wall stiffening occurred at apertures between 0 and 16 μm.

Since stomatal aperture is approximately linearly related to guard cell volume we can also expect a direct proportionality with pressure for small volume changes, ΔV, because then $\ln(V/V_o) \approx (\Delta V)/V_o$.

Data of BEARCE and KOHL (1970) indicate that guard cells of *Chrysanthemum* and *Pelargonium* require 4.2 atm per μm increase in aperture. On the other hand, EDWARDS and MEIDNER (1977) found values between 0.34 atm μm^{-1} for *Tradescantia virginica* and 1.7 atm μm^{-1} for *Triticum vulgare*.

The former values were obtained by cryoscopic determination of osmotic pressure, π, and calculation of turgor, p, according to the equation for water potential, $\psi = p - \pi$. The latter values, in contrast, were measured directly by impaling subsidiary cells with capillaries connected to pressure sensors; see Section 2.3.2, below.

2.3.2 Antagonism Between Guard Cells and Their Neighbors

Antagonism Ratio. We return to the pair of guard cells embedded in an epidermis; we consider again the additional work required to displace epidermal tissue. As early as 1856, VON MOHL stated that stomatal movement is the result of an antagonism between guard cells and epidermal cells. HEATH (1938), MONZI (1939), IMAMURA (1943), STÅLFELT (1956), and many others provided further evidence for the opposing action of guard cell and epidermal pressures. However, it would be wrong to conclude that stomatal opening is a unique function of the pressure difference between guard cells and their neighbors. Epidermal cells are usually deeper than guard cells; the pressure, p_e, in epidermal cells surrounding guard cells acts on a larger area of attack, F_e, than the area, F_g, acted upon by the pressure in the guard cells, p_g. In equilibrium $p_e F_e = p_g F_g$; i.e. epidermal cells have a "mechanical advantage", $m = F_e/F_g$, over the guard cells: $p_g = m p_e$. VON MOHL (1856) recognized this relationship and DEMICHELE and SHARPE (1973) pointed it out again. COWAN (1977) drew our attention to another aspect of m. The work done during pore opening is $p_g \Delta V_g - p_e(\Delta V_g + \Delta V_p)$, with V_g being the internal volume of the guard cell and V_p the volume of the pore. It follows that $m = 1 + dV_p/dV_g$; m obviously will depend greatly on the "volume efficiency" dV_p/dV_g of the particular stomatal design.

COOKE et al. (1976) recognized that the relationship between stomatal aperture, A, and the pressures in the guard cells and the cells adjoining them is determined by the ratio between the change in A caused by a change in p_e at constant p_g, and the change in A caused by a change in p_g at constant p_e:

$$A = A_0 + \left(\frac{\partial A}{\partial p_g}\right)_{p_e} p_g + \left(\frac{\partial A}{\partial p_e}\right)_{p_g} p_e,$$

or simplified $A = A_0 + b_g p_g + b_e p_e = A_0 + b_g(p_g - a p_e)$, (2.2)

where b_g and b_e are the partial differentials and where a is the "antagonism ratio" $= -\frac{b_e}{b_g}$ (so named by COOKE et al. to honor VON MOHL). Normally b_g is positive and b_e negative.

The presence of A_o allows for the possibilities that the pore may be slightly open ($A_o > 0$) or tightly closed ($A_o < 0$) when $p_g = p_e = 0$. Equation (2.2) includes the following statements: (1) Stomatal aperture is not a function of the pressure difference between guard cells and epidermal cells (or subsidiary cells), but depends on the difference between two functions of the pressures in these cells. One and the same aperture can be produced by various combinations of p_g and p_e, and maximal aperture does not necessarily occur at maximal p_g. (2) Since $(\partial A/\partial p_e)_{p_g} = b_e$ and $(\partial A/\partial p_g)_{p_e} = b_g$ need not be constant, b_e, b_g, and their quotient, a, cannot be expected to be constant when stomata move. The little experimental evidence we have so far however indicates that the antagonism ratio may be surprisingly constant (COOKE et al., 1976). (3) Depending on the magnitude of a and sign and magnitude of A_o, the pressure in the guard cells has to build up above that in the epidermal (or subsidiary) cells until stomatal opening can occur: $p_g > A_o/b_g + ap_e$. The phase of pressure increase which precedes stomatal opening represents STÅLFELT's "Spannungsphase".

The relationship between stomatal aperture, and volume and pressure of guard cells is complex. It is difficult to predict whether a perturbation of the water potential in a leaf will lead to stomatal opening or closure, unless one has all the necessary information on the particular stomatal mechanism under consideration. The complexity of the mechanism also indicates a possibility that certain patterns of regulatory behavior may be built into the anatomy of the epidermis or the whole leaf. See COWAN (1977) for more detail on the mechanical interaction between guard cells and epidermis and DELWICHE and COOKE (1977) for an incorporation of this interaction into a model describing the dynamics of stomatal behavior.

Determination of Antagonism Ratios and Pressures in Subsidiary Cells that Close Open Stomata. EDWARDS et al. (1976) were able to insert capillaries into guard cells and subsidiary cells of *Tradescantia virginiana* and to measure turgor simultaneously in the two cell types. They were also able to close stomata by forcing liquid through the capillaries and thereby increase the pressure in the subsidiary cells, demonstrating very convincingly the presence of an antagonism between the two cell types. COOKE et al. (1976) computed an antagonism ratio of about 1.6 from the data of EDWARDS et al. Meanwhile results of new measurements became available (EDWARDS and MEIDNER, 1977), demonstrating that pressure-aperture relationships similar to those in *T. virginiana* exist also in *Commelina communis, Triticum vulgare,* and *Vicia faba*. The reduction in stomatal aperture caused by an increase in the pressure of a subsidiary cell or an epidermal cell adjoining a guard cell was fairly linear. The slope $\Delta p_e/\Delta A$ was steepest in *Triticum vulgare* (with ~ 1.1 atm μm^{-1}) and shallowest in *Tradescantia virginiana* (with ~ 0.23 atm μm^{-1}). Unfortunately it was not possible to measure the turgor of the guard cells in these experiments, so that antagonism ratios cannot be given for these species.

2.3.3 Speculation on Pressures in Inelastic But Flexible Guard Cells

It is conceivable that grass stomata open by the bellows action of inelastic walls (Sect. 2.2, Figs. 3, 5, 7), in the way that a car is lifted by its tires (Fig. 10). The subsidiary cells exert a load 2L on the guard cells; L is balanced by ap_g, and the length (a) varies with the load.

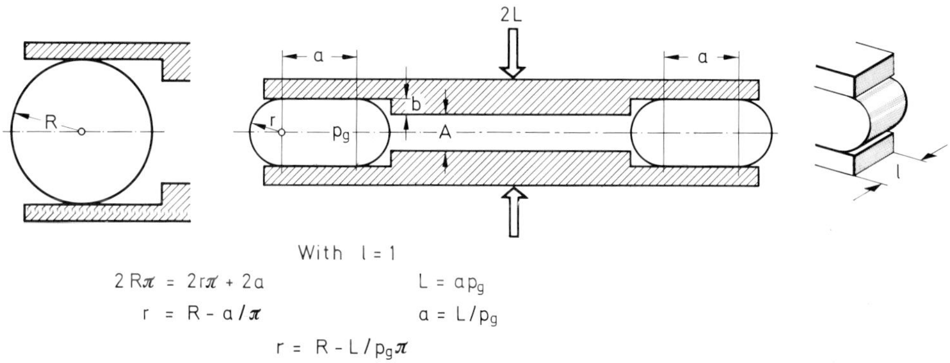

Fig. 10. Model of a grass stoma with inelastic but flexible guard-cell walls. Subsidiary cells exert a load 2L on the guard cells which is balanced by the product of guard cell pressure p_g and the area of contact $a \cdot l$, assuming cell to extend normally to the plane of the diagram by the length *l*. *R* is the radius of a guard cell head fully inflated to a cylinder (i.e., circular in cross section) whereas *r* is the radius of the curved part of the partially inflated head

Since aperture $A = 2(r - b)$ and $r = R - L/\pi p_g$, the width of the stomatal slit is in this model determined by the *ratio* between the load of the subsidiary cells on the guard cells and the pressure in the guard cells. In this arrangement, stomatal aperture is highly sensitive to changes in the hydrostatic pressure in the subsidiary cells. No information is available allowing us to determine whether dumbbell-shaped guard cells behave like elastic bodies, or inelastic but flexible ones, or a combination of both.

3. Osmotics

3.1 Solute Requirement for Stomatal Opening

3.1.1 Theory

The solute requirement for a volume increase of an ideally elastic guard cell can be estimated. Equation (2.1) defined

$$\frac{dV}{dp} = \frac{V}{\varepsilon}. \tag{3.1}$$

Turgor, p, is produced in guard cells, if the osmotic pressure, π, exceeds the negative value of the water potential, ψ, in the tissue:

$$p = \pi - (-\psi) = \frac{s\,RT}{V} + \psi, \tag{3.2}$$

where s is the solute content of the guard cell expressed in osmoles. Differentiation of Eq. (3.2) results in

$$\frac{dp}{ds} = \frac{RT}{V}, \tag{3.3}$$

i.e., the pressure change caused by a change in solute content is inversely proportional to the volume of the cell and only indirectly related to the water potential of the tissue (through V). Multiplication of Eqs. (3.1) and (3.3) yields

$$\frac{dV}{ds} = \frac{RT}{\varepsilon}. \tag{3.4}$$

Thus, the solute requirement for an increment in guard-cell volume depends solely on the modulus of elasticity of the cell and is independent of the volume of the cell and the water potential. Since aperture was found to be linearly related to guard cell volume (in *Vicia faba*) we can expect linear relationships between stomatal aperture, guard cell volume and the osmolarity of the solute content of the cell. This is true as long as the cell behaves like a perfectly elastic body. Stiffening of the cell walls at large apertures could lead to an increase of ε and a decrease of the volume increment produced by an increment in solute content.

Since nearly nothing is known of the relationship among aperture, volume and hydrostatic pressure of grass stomata no theory of the solute requirement of dumbbell-shaped guard cells will be given here.

3.1.2 Measurements

Approximate linearity between solute content, or osmotic pressure, and stomatal aperture has indeed been determined by different methods. Applying the technique of incipient plasmolysis to guard cells of *Vicia faba*, FISCHER (1972) found that the relationship between osmotic pressure and aperture had a slope of 2 atm μm^{-1}. I found over an aperture range from 0 to 16 µm in the same species a slope of 4.8 atm μm^{-1}. Since salts of K, including K malate, serve as osmotica in guard cells (Sect. 3.2), determinations of K and malic acid contents can yield information on solute levels in guard cells. Data obtained by analysis of epidermal samples are useful only if all ordinary epidermal cells are ruptured. Linear dependence of stomatal aperture on relative K contents of guard cells emerged from the results of microprobe analyses performed by SAWHNEY and ZELITCH (1969) on *Nicotiana tabacum* and TURNER (1973) on *Phaseolus vulgaris*. When malate functioned as a major anion in guard cells, malate content of epidermis whose ordinary cells had ruptured increased in direct proportion with stomatal aperture, as TRAVIS and MANSFIELD (1977) demonstrated with *Commelina communis*, and VAN KIRK and RASCHKE (1978a) with *Vicia faba*. The malate data obtained with *Vicia faba* lead to an estimated slope of 2.4 atm μm^{-1} which is one-half that obtained by us earlier by incipient plasmolysis of stomata of the same species but about equal to that determined by FISCHER. This variation in slope indicates a variation in the elastic modulus of the guard cells which probably is influenced by growing conditions and age of the plant.

Osmotic pressure of guard cells has most commonly been determined by the simple method of incipient plasmolysis, but the results are uncertain because the cells change in volume and lose solutes by leakage during exposure to the plasmolyticum. The half-time for equilibration between the water potentials of epidermal

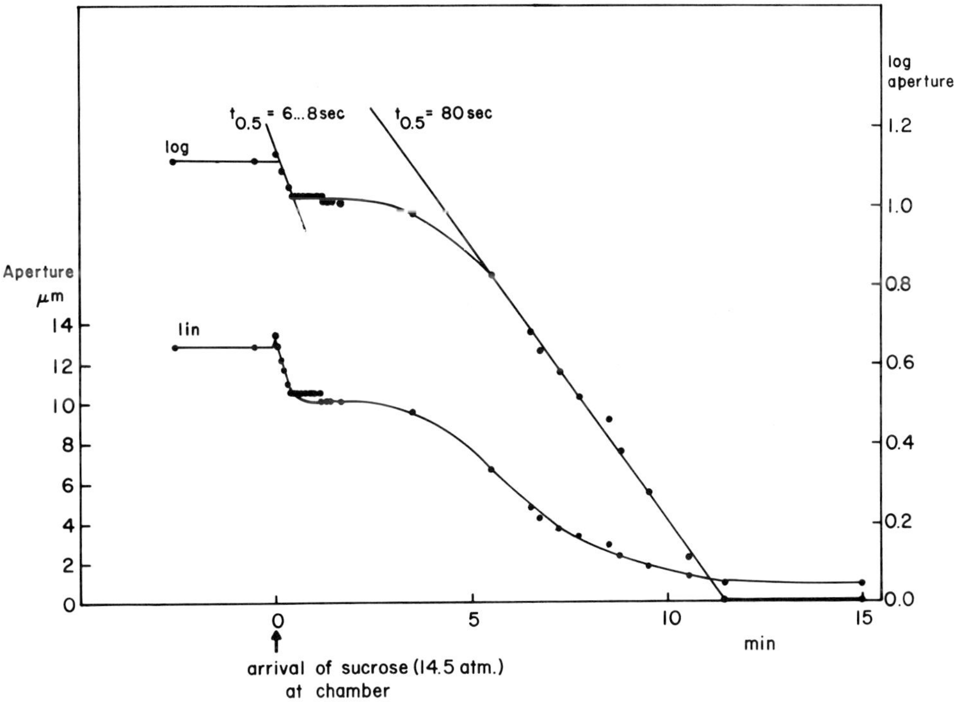

Fig. 11. Linear and semilogarithmic plots of change in stomatal aperture with time in the presence of sucrose as a plasmolyzing agent. *Arrow* indicates the time when the water potential of the liquid passing by an epidermal strip with open stomata of *Vicia faba* was changed from 0 to −14.5 atm. Two phases of equilibration with their half times $t_{0.5}$ indicated

tissue and the plasmolyzing solution is of the order of 10 s (FISCHER, 1973). After water potentials have equilibrated, a second phase of closing is often observed (Fig. 11); the rapidity of its onset depends on the strength of the plasmolyzing solution. This second phase of closure is caused by solute loss. Loss can be retarded by the addition of K to the solution (FISCHER, 1973; see also IMAMURA, 1943). Guard cells of *Vicia faba* may lose one-half of their solutes within 10 min of exposure to low water potential; within the same time span, solute loss in *Zebrina pendula* is even greater (IMAMURA, 1943).

The problem of solute leakage can be minimized if the time of exposure of the tissue to the plasmolyzing solution is kept short; i.e., 0.5 to 1.5 min. Short-term determinations on *Zebrina pendula* (IMAMURA, 1943) and on *Vicia faba* (RASCHKE, unpublished) yielded mean osmotic pressures of 70 to 90 atm in guard cells of open stomata. However, volume changes have to be considered as well. The measured values represent osmotic pressures at incipient plasmolysis, i.e., at V_0 of the guard cells. Since the volume of fully inflated guard cells is about twice that of the resting-stage cells (see Sect. 2.2.3, Fig. 8), the *actual* osmotic pressures in the inflated cells should be about half of the measured values; i.e., ca. 40 atm.

These corrected values are in good agreement with those determined by the more difficult but more accurate method of microcryoscopy. Using this technique, osmotic pressures as high as ca. 50 atm were measured for guard cells in *Chrysan-*

themum morifolium and *Pelargonium hortorum* (when stomatal apertures were 6 μm wide; BEARCE and KOHL, 1970) as well as in *Hordeum vulgare* (AYRES, 1972). Clearly, solute content and osmotic pressure can be quite high in guard cells of open stomata.

Osmotic pressures in the surrounding epidermal cells are much lower than in the guard cells: in *Vicia faba* they are about 6 atm (RASCHKE, unpublished), in *Chrysanthemum* and *Pelargonium* between 10 and 20 atm (BEARCE and KOHL, 1970) and in *Hordeum vulgare* between 22 and 30 atm (AYRES, 1972). Unlike guard cells, epidermal cells do not undergo significant changes in osmotic pressure during stomatal movement. That means that considerable differences in hydrostatic pressure build up between the two cell types during stomatal opening. Plasmodesmata can probably not stand these pressures and are therefore not (or only rarely) found to connect mature and functional guard cells with the surrounding tissue. If plasmodesmata are present they are convoluted (review by ALLAWAY and MILTHORPE 1976). CHABOT and CHABOT (1977), SANCHEZ (1977), and VASSILYEV and VASSILYEVA (1976) confirm absence of plasmodesmata from guard cells.

3.2 Osmotica in Guard Cells

HSIAO (1976) summarized the work which ultimately led to the conclusion that stomatal opening was the osmotic effect of an accumulation of salts of K (and perhaps Na) in the guard cells. Here it may suffice to state that K has been discovered in guard cells of open stomata of more then 50 species, including ferns, ginkgo, conifers, and CAM plants whose stomata open in the night (e.g. DAYANANDAN and KAUFMAN, 1975; WILLMER and PALLAS, 1973)[2]. Sodium has been found in addition to K^+ in two species grown under saline conditions (ESHEL et al., 1974; TAL et al., 1976).

3.2.1 Quantity of K Salts in Guard Cells Sampled from Leaves

Simultaneous determinations of osmotic pressures, volumes, and K contents of guard cells of *Vicia faba* led to the conclusion that accumulation and discharge of salts of K can fully account for the turgor changes in guard cells that give rise to stomatal movement. The data in Table 2 show that an increase in stomatal aperture from 2 to 12 μm requires an increase in the solute content of a pair of guard cells by 5.4 posmol. The measured increase in K^+ content was 4 pEq, which, if K^+ was associated with a monovalent anion, would give rise to an increase of the solute content of about 7 posmol. No large change in the Cl content was measured, and none in the contents of S and P. The majority of the counter ions to K^+ must therefore have been organic and most probably divalent. Indeed ALLAWAY (1973) found that the malate content of epidermis[3]

[2] NELSON and MAYO (1977) did not see K^+ accumulated in guard cells of *Paphiopedilum leeanum* sampled from leaves exposed to light and CO_2-free air. Since stomatal opening was not confirmed by direct observation the absence of K^+ from guard cells does not prove that stomata of *P. leeanum* open without an involvement of K^+ salts as osmotica.

[3] "Epidermis", "epidermal strip" and "epidermal sample" as used here and in the remainder of Section 3 should be understood to consist of guard cells plus surrounding epidermal cells.

Table 2. K and Cl content, volume, osmotic pressure and solute content of pairs of guard cells from open (after 3h light) and closed stomata of *Vicia faba*. (Data from HUMBLE and RASCHKE, 1971)

Stomata	Aperture μm	K content[a] pEq $stoma^{-1}$	Cl content[a] pEq $stoma^{-1}$	Volume[b] pl $stoma^{-1}$	π_0^c atm	π^d atm	Osmolarity osmol l^{-1}	Solute content posmol $stoma^{-1}$
Open	12	4.24	0.22	4.8	80	37	1.5	7.2
Closed	2	0.20	0.00	2.6	20	17	0.7	1.8
Difference between open and closed	10	4.04	0.22	2.2	60	20	0.8	5.4

[a] From electron probe microanalysis.
[b] Volume of a pair of guard cells at incipient plasmolysis: 2.2 pl.
[c] Osmotic pressure at incipient plasmolysis, corrected for solute loss during plasmolysis.
[d] Osmotic pressure at volume of the nonplasmolyzed cells.

of *V. faba* increased with stomatal opening by an amount that would have balanced about one-half of the K^+ in the guard cells. (ALLAWAY rolled the epidermis before analysis, with the result that all epidermal cells except the guard cells were ruptured. Thus, his "epidermis" in fact consisted of guard cells.). Recent determinations of malate contents of epidermis from leaves of *Commelina communis* (TRAVIS and MANSFIELD, 1977) and *V. faba* (VAN KIRK and RASCHKE, 1978a) resulted in values that could be sufficient to balance all the K^+ in guard cells. At an average stomatal aperture of 14 μm, epidermis of *C. communis* contained about 180 pmol mm^{-2} malate and that of *V. faba* 390 pmol mm^{-2}. Since the stomatal density was 60 mm^{-2} in *V. faba*, a pair of guard cells of this species contained 6.5 pmol malate, corresponding to about 15.6 posmol potassium plus malate ions and an osmotic pressure of 30 atm. These values agree well with what would have been expected from the relationships between aperture, volume and pressure of pairs of guard cells estimated earlier (Sect. 3.1.2). OUTLAW and LOWRY (1977) succeeded in isolating pairs of guard cells from freeze-dried tissue of *V. faba* and determining their contents of K and organic acids. They found that, during stomatal opening, each pair accumulated on the average 3.2 pEq of K^+, and 1 pEq each, of malic and citric acids, leaving 1.2 pEq of anion unaccounted for. Although the stomatal apertures attained are not known, the order of magnitude of these findings agrees with the results of the earlier determinations by electron microprobe analysis and malate analysis of epidermal strips.

The relationship between stomatal aperture and K salt content has been determined also for *Allium cepa* (SCHNABL and RASCHKE, in prep.). Stomata opened on illuminated leaves by 5.1 μm within 1 h. During this time guard cells accumulated on the average 0.30 pEq K^+ per pair and 0.27 pEq Cl^-: K^+ was virtually completely balanced by Cl^-. There was no evidence for a participation of malate (Sect. 3.3.4). The absolute salt contents were smaller than in guard cells of *V. faba* and *C. communis* because onion guard cells are smaller than those of the other two species.

We conclude that either K malate (and salts of other acids of the tricarboxylic acid cycle) or KCl appear as dominating osmotica in guard cells of open stomata;

which salt dominates depends on the species. However, analysis of guard cell contents of *Zea mays* and *Commelina communis* indicates that under natural conditions (stomata opened on illuminated leaves) Cl^- and organic anions may appear simultaneously in *Z. mays* in varying ratios (and probably also in *C. communis*). Participation of Cl^- in stomatal movement has been demonstrated in several species (DAYANANDAN and KAUFMAN, 1975; WILLMER and PALLAS, 1973). Pairs of guard cells of *Z. mays* contained on the average 0.3 pEq K^+ and 0.1 pEq Cl^- when stomata were closed, and 0.7 pEq K^+ and 0.3 pEq Cl^- when they were open (RASCHKE and FELLOWS, 1971). In guard cells of open stomata the average Cl/K ratio was 0.4, but the ratio varied widely between stomata. In individual pairs of guard cells complete balance between K^+ and Cl^- did occur. The estimated concentrations in guard cells of open stomata were 0.4 Eq K^+ l^{-1} and about 0.15 Eq Cl^- l^{-1}. PENNY and BOWLING (1974) and PENNY et al. (1976) measured (in separate investigations) the concentrations of K^+ and Cl^- in stomatal complexes of *Commelina communis* with ion-selective electrodes and found that guard cells of closed stomata were 0.095 M with respect to K^+ and 0.033 M with respect to Cl^-; guard cells of open stomata (aperture not stated) were 0.448 M with respect to K^+ and 0.121 M with respect to Cl^- (very similar to the concentrations determined for *Z. mays*). The Cl/K ratio was about 0.3. Additional data on ion contents of guard cells can be found in HSIAO (1976).

In guard cells of closed stomata the residual osmotic pressure is produced by solutes *other* than salts of K (at least in *V. faba*). Therefore, the possibility exists that under unknown circumstances stomatal opening could be caused by osmotica normally providing only the resting pressure. If evolution has led to two kinds of osmotica in guard cells one can speculate that only the accumulation and dispersal of salts of K is regulated with the purpose of affecting stomatal movement.

The concentration of K^+ in guard cells can reach 0.9 Eq l^{-1}. We do not know whether the K^+ is distributed over cytoplasm, vacuole and other organelles, or whether it is all concentrated in the vacuole. If the latter is the case, we do not know whether special compensatory and enzyme-compatible osmotica are formed in the cytoplasm and organelles to prevent compression during stomatal opening.

3.2.2 Availability of K and Cl to Guard Cells

Since plasmodesmata are virtually absent from functional guard cells, guard cells must obtain K^+ and Cl^- through the apoplast; malate can be produced within the guard cells (Sect. 3.3.4 and 4.1). It is not known whether the ion exchange capacity of the apoplast suffices to supply the guard cells with K^+; Cl^- cannot be stored in the apoplast in large amounts because epidermal cell walls are negatively charged. Species with rapidly responding stomata, such as *Zea mays*, possess subsidiary cells that evolved as effective ion reservoirs in close proximity to the guard cells (PALLAGHY, 1971; RASCHKE and FELLOWS, 1971). During stomatal movement, there is no change in the total K^+ and Cl^- content of a stomatal complex, but only in the distribution of the two ions between guard and subsidiary cells (Fig. 12). Since the volume of the subsidiary cells of *Z. mays* is 3 to 5 times larger than that of the guard cells, pressure changes resulting from the ion shuttle occur mainly in the guard cells.

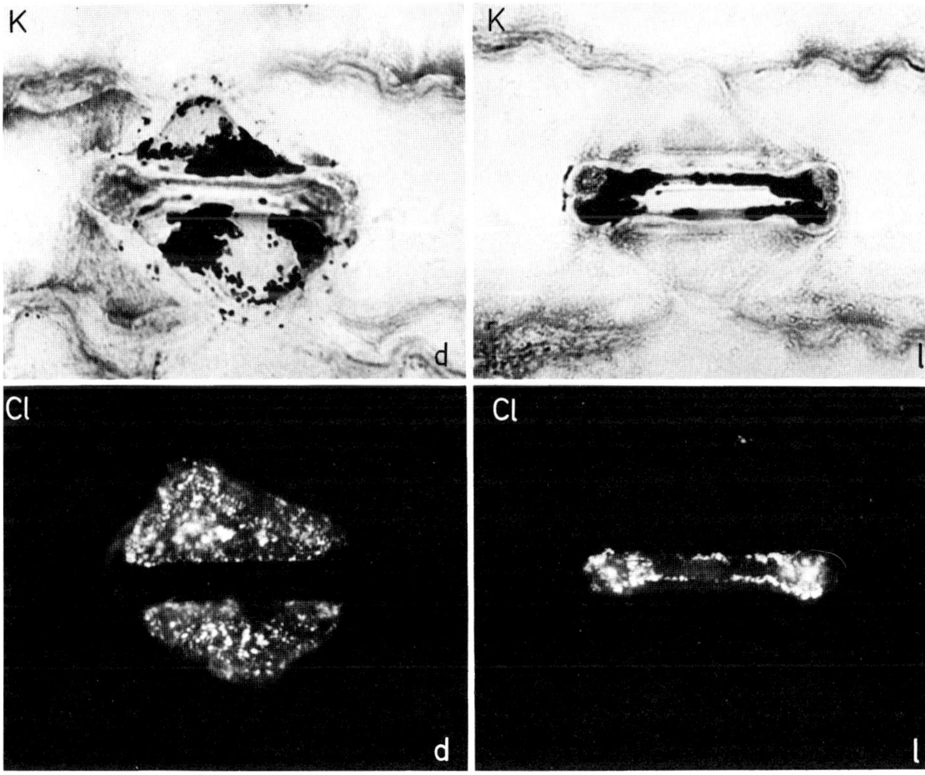

Fig. 12. Shuttle of K (*top*) and Cl (*bottom*) between guard cells and subsidiary cells of *Zea mays*. In the dark (*left*), K and Cl reside in the subsidiary cells; in the light (*right*), both elements migrate into the guard cells. (Histochemical reactions with cobaltinitrite or Ag^+; RASCHKE and FELLOWS, 1971)

The ability of subsidiary cells to accumulate K^+ and Cl^- appears to be universal. The horse-shoe shaped subsidiary cells of ferns, for instance, show the histochemical reaction for K^+ (WILLMER and PALLAS, 1973). If stomatal complexes include several subsidiary cells, as in *Commelina communis*, gradients in the concentrations of K^+ and Cl^- appear between these cells (PENNY and BOWLING, 1974; PENNY et al., 1976). Even in species without morphologically differentiated subsidiary cells, one or several epidermal cells in contact with a pair of guard cells will often contain more K^+ or Cl^- when the stoma is closed than when it is open (DAYANANDAN and KAUFMAN, 1975; WILLMER and PALLAS, 1973). In some species there are individual epidermal cells that contain large amounts of K^+ and Cl^- but are not in direct contact with a stomatal complex (RASCHKE and FELLOWS, 1971; WILLMER and PALLAS, 1973). We do not know whether the ions in such "storage cells" are available to guard cells. The presence of many mitochondria in subsidiary cells of *Zea mays* (SRIVASTAVA and SINGH, 1972; communication by H. ZIEGLER, and own observations) indicates an active participation of these cells in the ion shuttle. However we do not know how uptake of ions by guard cells is synchronized with release of ions from subsidiary cells. The observed variation in the Cl/K ratio has to be taken into consideration. In *Polypodium vulgare* and *Tradescantia pallidus* special structures of the guard cell walls may function in ion storage (STEVENS and MARTIN, 1977a, b).

3.3 Ion Transfer and Acid Production in Isolated Epidermis

The presumed function of the stomata is related to the metabolic activity of the mesophyll: its assimilation of CO_2 and its water relations. Accordingly, the possible optimization of the gas exchange of the plant by stomatal activity may depend on signals from the mesophyll to the stomata concerning not only the physiological state of the tissue but also the state of the ontogeny of the organism. An analysis of the operation of the stomatal mechanism itself therefore requires separation of the epidermis from the mesophyll. Still better would be the investigation of isolated functioning pairs of guard cells. But since MOURAVIEFF's (1956) experiments on the response of excised stomata to CO_2 nobody has had the patience to isolate living pairs of guard cells. However ZEIGER and HEPLER's (1977) recent success in preparing living guard cell protoplasts by enzyme action may open a new avenue for studying transport and metabolic capabilities of guard cells without interference by other tissues (or by guard-cell walls for that matter).

3.3.1 The Peculiar Situation of Guard Cells in an Isolated Epidermis

In an isolated epidermis floating on a solution, guard cells are no longer mechanically and metabolically supported or constrained by the presence of the mesophyll. Moreover, the water potential of the solution is usually higher than that prevailing in the leaf—in fact, often close to that of free water. Turgor in the cells increases as a consequence. Epidermal cells that survive peeling, rupture sooner or later on the solution. In the whole leaf, guard cells are in contact with the rest of the leaf through the relatively narrow connections of their cell walls and possibly through selective mediation by cells adjoining them. Guard cells in epidermal strips, on the other hand, appear to have a much larger surface available for gain and loss of solutes. Consequently, guard cells in isolated epidermal strips floating on water are exposed to a condition of depletion of virtually everything soluble, even more so if the medium is stirred. Only substances added to the medium in a concentration sufficient to prevent net loss and substances effectively sequestered within the guard cells will be retained. As will be indicated in the next paragraph, the electrical potential distributions in epidermal strips may also differ from those occurring in leaves.

3.3.2 Guard Cell Wall as Cation Exchanger

Microelectrodes inserted into walls of guard cells indicate electrical potentials that respond to a 10 fold change in the external alkali ion solution with a change by 59 mV. The wall is negative with respect to the outside solution. For guard cells of *Commelina communis*, the potential difference changes linearly with the logarithm of the alkali activity from 1 to 300 mEq K^+. The potential difference is not affected by the presence of metabolic inhibitors and is measurable in walls of cells which were treated with detergents or were frozen. The walls of guard cells behave like Donnan spaces with fixed negative charges at a density between 0.3 and 0.5 M (SAFTNER and RASCHKE, unpublished). The charge density declines with pH; half-maximal density occurs at pH 4 in the medium, corresponding to an estimated pH of 3.3 in the wall. The pattern of the dependence on pH is almost identical with the titration curve of polygalacturonic acid, consistent with the view that guard cell walls are pectinic (Sect. 2.1.2).

Some peculiar observations could be related to Donnan exclusion of anions in cell walls (and membranes?). For instance, stomata of *Zea mays* did not open on solutions containing salts of K^+, but did open on distilled water (see also PALLAGHY, 1971). Open stomata of epidermal strips of *Allium cepa* close when in contact with solutions containing K^+ in combination with iminodiacetate or with Cl^-, if the latter is present at <0.1 Eq l^{-1} (SCHNABL and RASCHKE, unpublished). Possibly cations supplied from outside neutralize fixed negative charges in walls and membranes and guard cell salts begin to leak because anions begin to leak.

3.3.3 Exchange of Protons for Alkali Ions

Guard cells can take up K^+ without accompanying anion; electroneutrality is maintained by excretion of H^+ (RASCHKE and HUMBLE, 1973). The amount of H^+ released by epidermal strips of *V. faba* during stomatal opening was titrated and found approximately equal to the estimated amount of K^+ required to produce the observed stomatal opening.

Guard cell vacuoles possess a negative electrical potential, which normally is more negative than that in epidermal cells (Table 3); it appears to be a diffusion potential. If the concentration of alkali salts (except Li^+) in the bathing medium of epidermal strips is changed, the internal potential responds in agreement with the Nernst equation. This was found by PALLAGHY (1968) with *Nicotiana tabacum* and K^+ (between 0.1 and 10 mEq l^{-1}), and by SAFTNER and RASCHKE (unpublished) with *Commelina communis* exposed to the chlorides of Na^+, K^+, Rb^+, Cs^+, $NH_4{}^+$, and choline (between 3 and 30 mEq l^{-1}); Li^+ caused hyperpolarization. PENNY and BOWLING (1974) did not discover a potential difference between guard cells and epidermal cells. But they reported that guard cells collapsed during impalement. R. SAFTNER verified with blunt electrodes that the potential difference disappears when cell contents leak out and guard cells deflate.

Fusicoccin, the fungal toxin that causes accumulation of K^+ in guard cells and stomatal opening (TURNER, 1973), is known to hyperpolarize the electrical potential and to enhance H^+ excretion in many types of plant cell (MARRÈ, 1977). Fusicoccin also enhances release of H^+ from epidermal strips of *Tulipa gesneriana*, *V. faba*, and *C. communis* if K^+ is present (RASCHKE, 1977). The enhancement of stomatal opening by fusicoccin can be overcome by abscisic acid (SQUIRE and MANSFIELD, 1972a), which appears to inhibit the expulsion of H^+ or to activate a mechanism that stimulates H^+ influx into guard cells (Sect. 5.2).

A mechanism of H^+-alkali ion exchange[4] appears to act in which proton transfer sets up an electrical gradient superimposed on existing gradients of ion

[4] The term "exchange" refers to the net effect of ion transfer and implies nothing about the mechanisms involved.

Table 3. Electrical potentials in epidermal cells (presumably vacuolar potentials), mean in mV ± s.e. (n)

Species	Tissue	Bathing solution	Potential in: Ordinary epidermal cell	Potential in: Subsidiary cell if present	Potential in: Guard cell	Reference
Nicotiana tabacum	epidermal strip	10 mEq l^{-1} KCl; 1 mEq l^{-1} NaCl	–		−76 ±1.5(20)	PALLAGHY (1968)
Tradescantia albiflora	leaf	Knop's solution	−133 ±4 [a]	−135±5	−166 ±4 [a]	GUNAR et al. (1975)
Allium cepa	paradermal section	1 mM KCl; 1 mM $CaCl_2$; 0.23 M mannitol	–		−72 ±29(45)	MOODY and ZEIGER (1978)
Allium cepa	epidermal strip	30 mM KCl; pH 7.0	−26.5 ±1.9(8)		−39.0 ±2.5(10)	SAFTNER and RASCHKE, unpubl.
Commelina communis	epidermal strip	30 mM KCl; pH 7.0	−20.1 ±2.7(10)	−24.8 ±2.1(10)	−42.1 ±2.6(10)	SAFTNER and RASCHKE, unpubl.
Nicotiana glauca	epidermal strip	30 mM KCl; pH 7.0	−23.4 ±5.5(11)		−50.3 ±2.3(11)	SAFTNER and RASCHKE, unpubl.
Vicia faba	epidermal strip	30 mM KCl; pH 7.0	−26.9 ±3.1(10)		−42.1 ±2.4(10)	SAFTNER and RASCHKE, unpubl.
Zea mays	epidermal strip	30 mM KCl; pH 7.0	−26.5 ±1.9(8)	−28.9 ±1.5(10)	−50.5 ±1.9(6)	SAFTNER and RASCHKE, unpubl.

[a] n = 7 or 8.

activity. Redistribution of ions follows between guard cells and their neighbors and between compartments within these cells. Ion permeability between cells appears to be high. The electrical resistance of guard cells is about one-third of that of other plant cells (< 1.0 versus 3 kΩ cm^2; R. SAFTNER, unpublished). The contents of guard cells easily leak into stirred water.

The specificity of guard cells for particular alkali ions appears to be low. For instance, ARENDS observed as early as in 1926 that stomata of *Allium* species opened in paradermal leaf sections almost as wide as under optimal conditions on the plant if the sections were supplied with 0.5 M NaCl. WILLMER and MANSFIELD (1969) showed that guard cells in epidermal strips of *Commelina communis* can function at least as well with Na^+ as they do with K^+. Tl^+ can also substitute for K^+ (PALLAGHY, 1972).

HUMBLE and HSIAO (1969) provided evidence for an additional uptake mechanism specifically transporting K^+ and Rb^+ into guard cells of epidermal strips of *V. faba*. While in darkness all alkali ions stimulated stomatal opening in proportion to the concentration at which they were offered (between 0.1 and 100 mEq l^{-1}), in the light K^+ and Rb^+ produced almost full opening when present at 1 mEq l^{-1}, while the approximately linear relationship between the concentrations of Li^+, Na^+, and Cs^+ and stomatal opening was maintained in the range from 0.1 to 100 mEq l^{-1}.

3.3.4 Transfer and Production of Anions

When epidermal strips of *Vicia faba* are incubated in the light on solutions free of Cl^-, Cl^- contributes only about 6% of the total anion equivalents in the guard cells of open stomata (RASCHKE and SCHNABL, 1978). About the same fraction was found in guard cells of stomata that opened on leaf samples from plants which had been fertilized with Hoagland solution (HUMBLE and RASCHKE, 1971); Hoagland solution contains Cl^- as a trace element only. When Cl^- is included in the incubation medium, guard cells of *V. faba* may satisfy 33% (PALLAGHY and FISCHER, 1974) to 50% (RASCHKE and SCHNABL, 1978) of their anion requirement by an uptake of Cl^-. When NaCl was offered to guard cells of *V. faba* in the absence of Ca^{2+} and CO_2, the Cl^- fraction was even as high as 73% (RASCHKE and PIERCE, unpublished). Apparently, availability determines the use of Cl^-. Analysis of epidermal samples for their malate contents confirms this view: in epidermal samples of *V. faba* exposed to light on solutions containing K iminodiacetate,[5] a linear relationship between malate content and stomatal aperture was found (VAN KIRK and RASCHKE, 1978a) The positive correlation between malate content and aperture disappeared when Cl^- was added to the solutions (to give $>10^{-5}$ Eq l^{-1}). Microprobe analysis combined with malate analysis showed that malate and Cl^- complemented each other in balancing K^+ in guard cells of isolated epidermal strips of *V. faba* (Table 4). The malate content of epidermis with open stomata declined very roughly with the logarithm of the concentration of Cl^- in the medium (between 10^{-5} and 10^{-1} Eq l^{-1}; RASCHKE and SCHNABL, 1978).

WILLMER et al. showed in 1973 that PEP carboxylase activity of an extract of epidermis of *Commelina communis* increased with pH up to an optimum between pH 8 and 9. When this became known, regulation of pH by modulation of PEP carboxylase activity could be invoked as a mechanism by which guard cells tune malate production to Cl^- import. The swelling of guard cells possibly begins with an expulsion of H^+, followed by an uptake of alkali ions; OH^- remaining in the guard cell can then be exchanged for Cl^-. If no Cl^- is available, excess OH^- will make the cytoplasm alkaline. The increase in pH in turn will activate PEP carboxylase. Malic and other acids will be formed, providing H^+ for pH adjustment and ion exchange, and anions needed as counterions for K^+ in the vacuole. This proposed sequence of events is analogous to that proposed to operate in roots (HIATT and HENDRICKS, 1967); it receives support from the observation that the initial rates of H^+ excretion by epidermal strips of *V. faba* were smaller in the presence of KCl than when K iminodiacetate was offered (RASCHKE and HUMBLE, 1973). With Cl^- present, an antiport of Cl^- and OH^- would occur (or a symport of H^+; MITCHELL, 1966), reducing the net efflux of H^+. But a net efflux of H^+ would ultimately raise the cytoplasmic pH and, as a consequence, malate production. (More of the possible involvement of pH regulation in the mechanism of stomatal movement in Sects. 5.1 and 5.2).

Guard cells of *Allium cepa* have an absolute requirement of Cl^- for stomatal opening (Sect. 4.1.2); Br^- can substitute for Cl^-. The epidermal malate content

[5] Iminodiacetate was used in place of Cl^- because cell membranes are presumed to be impermeable to this zwitterion.

Table 4. Changes with stomatal opening of contents of K and Cl of pairs of guard cells and malate contents of isolated epidermis of *Vicia faba* and *Allium cepa*. Guard cells of *V. faba* contain starch, those of *A. cepa* do not. Epidermal strips were floated for 4 h on 100 mM solutions of K iminodiacetate = KIDA (no Cl) or KCl. K and Cl were determined by electron probe microanalysis, malate was measured enzymically. Data from RASCHKE and SCHNABL (1978) and SCHNABL and RASCHKE (unpublished data on *A. cepa*)

Species	Salt	Initial aperture μm	Final aperture μm	ΔK pEq $stoma^{-1}$	ΔCl pEq $stoma^{-1}$	ΔMalate pEq $stoma^{-1}$
Vicia faba	KIDA	6.4	11.7	2.30	0.24	2.10
	KCl	6.4	13.9	1.41	0.64	0.70
Allium cepa	KIDA	4.7	2.1	−0.25	0.04	−0.99
	KCl	4.7	8.7	0.44	0.42	−0.74

declines while stomata open (Table 4). Therefore, the participation of malate as stomatal osmoticum appears not to be likely in guard cells of onion.

The wall and the membranes of guard cells appear to be considerably less permeable to anions than to alkali ions. This difference may not be due solely to Donnan exclusion in the cell walls. Regardless of whether the external solution contained KCl or K iminodiacetate, the electrical potential in the vacuoles of guard cells of *Commelina communis* changed in virtually the same way in response to concentration changes of the salt. This was true, even when the concentration was high enough to neutralize the fixed charges in the wall (R. SAFTNER, unpublished). Chloride thus had the same effect as iminodiacetate which presumably does not penetrate membranes. Only at an external concentration of 3 mM was the vacuolar potential about 10 mV more negative with KCl than with K iminodiacetate. This effect could be ascribed to an active uptake of Cl^- (into the vacuole?). Strong evidence for an active transport of Cl^- across the cells of the stomatal complex of *C. communis* comes from the measurements of Cl^- concentrations made by PENNY et al. (1976), particularly if one recognizes that the gradient in electrical potential is not zero, as PENNY et al. believe, but opposes the activity gradient of Cl^- when stomata are open (Tables 3, 5).

Interpreting the evidence listed in this section: Transfer of inorganic anions (chiefly Cl^-) and production of organic anions (chiefly $malate^{2-}$) are consequences of an expulsion of H^+ at the plasmalemma of guard cells. Turgor is probably produced by transport of anions into the vacuole; cations follow the electrical gradient into the vacuole.

It is tempting to explain ion transfers across plasmalemma and tonoplast of guard cells as chemiosmotic effects of proton transfer across these membranes. However the ease with which all ion movements into and out of guard cells can be explained by applying MITCHELL's (1966) theory to guard cells raises my suspicion. It is by postulate and not by knowledge that we see the requirements for an operation of chemiosmotic mechanisms in plasmalemma and tonoplast of guard cells fulfilled. Nevertheless, basing the design of new experiments on MITCHELL's theory should be very stimulating.

3.3.5 Effects of pH and Ca

If proton extrusion is essential for stomatal opening, high external pH should enhance the process. This was not borne out in all experiments. FUJINO (1959)

Table 5. Gradients in ion concentrations and electrical potential across stomatal complexes of *Commelina communis*. Data from PENNY and BOWLING (1974, 1975); PENNY et al. (1976); R. SAFTNER (unpublished electrical potential measurements)

Cell type	Ion concentrations ($mEq\ l^{-1}$)					
	K^+		Cl^-		K^+	Cl^-
	Closed	open	closed	open	open minus closed	
Guard cell	95	448	33	121	353	88
Inner lateral subsid. cell	156	293	36	62	137	26
Outer lateral subsid. cell	199	98	55	47	−101	− 8
Epidermal cell	448	73	117	86	−375	−31
	Vacuolar pH		Potential (mV)[a]			
	closed	open	closed	open		
Guard cell	5.19	5.60	−30	−45		
Inner lateral subsid. cell	5.60	–	−30	−30		
Outer lateral subsid. cell	5.78	5.56	−32	−31		
Epidermal cell	5.74	5.11	−34	−31		

[a] Epidermal samples with open or closed stomata were taken from illuminated or darkened leaves, respectively, and bathed in 30 $mEq\ l^{-1}$ K^+, 29.8 $mEq\ l^{-1}$ Cl, 1.0 $mEq\ l^{-1}$ MOPS, pH 7.

observed maximal opening in epidermis of *Commelina communis* between pH 8 and 9 (after 15 h of exposure). IMAMURA (1943) made similar observations on the same species and on the related *Zebrina pendula*, but he discovered that stomata opened also at pH < 5.5. That stomata of *C. communis* opened at low pH was confirmed by SQUIRE and MANSFIELD (1972a). They could however not see any opening at pH > 5. Wide opening at a pH < 5 may simply be ascribed to the fact that surrounding epidermal cells are much more sensitive to acidity than guard cells, and are irreversibly damaged at low pH. As a matter of fact, SQUIRE and MANSFIELD proposed to expose epidermis to solutions of pH 4.5 for 1.5 h in order to "isolate" functional guard cells. However, it is possible that a yet unexplained response of guard cells to high acidity contributes to the widening of stomatal aperture that is sometimes observed at low pH. Future investigations will also have to consider the high ion exchange capacity of the guard cell walls and the identity of the cations acting as counterions to the fixed charges in the walls.

In general, Ca^{2+} is known to reduce or inhibit stomatal opening: $CaCl_2$, as well as $MgCl_2$, when given at 1 mM, suppressed stomatal opening in strips of *Commelina communis* and *C. sikkimensis* and, to a lesser extent, in strips of *V. faba* (WILLMER and MANSFIELD, 1969). Even a $CaCl_2$ concentration as low as 0.1 mM reduced apertures in *V. faba* (FISCHER, 1972) and *C. communis* (FUJINO, 1967). Conversely, the chelator EDTA enhanced stomatal opening (FUJINO, 1967). In addition one wonders whether the observed stimulation of stomatal opening by ATP, ADP (RAGHAVENDRA et al., 1976) or citric acid is caused by sequestering of Ca^{2+}. Apparently, Ca^{2+} can control ion transfer between guard cells and their environment, and indirectly, the production of organic acids in guard cells. For a possible interpretation, it is interesting to note that placement of epidermis on solutions of salts of Ca^{2+} acidified the medium: this is presumed to result from displacement of H^+ in the walls by Ca^{2+} (RASCHKE and HUMBLE, 1973). It is, however, unlikely that Ca^{2+} inhibition of stomatal opening is mediated by the liberation of H^+; the main effect of Ca^{2+} seems to be direct. A new effort should be made to examine the involvement of Ca^{2+} in stomatal movement.

3.4 Osmotic Efficiency of Permeant Ions

The evidence presented in Section 3.3. indicates that alkali ions permeate walls and membranes of guard cells with relative ease. Plant cells in general have a high permeability for K^+, and this fact has led GRADMANN (1977) to question the contribution of K ions to the production of turgor pressure in guard cells and other motor tissues altogether. But he seems not to be aware of the fact that the reflection coefficient of an ionic solute is determined by the least permeant ion, a result of tight electrical coupling between anions and cations. If the permeability of guard cell membranes is low for Cl^- or malate, then the reflection coefficient for KCl and K malate is high and K^+ will contribute to the osmolality of the cells as much as Cl^- or malate does (HASTINGS and GUTKNECHT, 1978). To paraphrase the theme: the vacuole of a guard cell can be considered a Donnan space. The osmotic pressure or swelling pressure of a Donnan space is determined by the activity of all charges present in the space, the fixed ones as well as the freely diffusible ones. If the reflection coefficient for K salts in guard cells is determined primarily by anion permeability, an increase in the latter is sufficient to initiate stomatal closure.

4. Metabolism and Energetics

4.1 Origin and Fate of Malate

4.1.1 Fixation of CO_2

The discussion on transfer and production of anions (Sect. 3.3.4) already noted that malate is produced in guard cells by carboxylation of PEP. Using microautoradiography, SHAW and MACLACHLAN (1954) proved that isolated epidermis of *Allium, Hordeum, Nicotiana, Sedum, Tradescantia, Tulipa* and *Vicia* is able to fix $^{14}CO_2$ in the light and epidermis of *Allium, Sedum, Tulipa* and *Vicia* also in darkness and that this ability resides in the guard cells. In epidermal extracts of *Commelina communis* and *Tulipa gesneriana* WILLMER et al. (1973) found the activities of PEP carboxylase, malic enzyme, and the malate dehydrogenases. PEP carboxylase activity was proportional to stomatal frequency. OUTLAW and KENNEDY (1978) succeeded in determining that the PEP carboxylase activity in single pairs of guard cells of *Vicia faba* was sufficient to produce oxaloacetate at a rate of 650 mmol per kg dry weight per hour. Since a pair of guard cells of

V. faba has a dry weight of 6 ng (OUTLAW and LOWRY, 1977), this rate would correspond to nearly 4 pmol malate stoma^{-1}h^{-1} or about 9.4 posmol stoma^{-1}h^{-1}. This capacity exceeds considerably the rate of 5.4 posmol per *three* hours required to open stomata of *V. faba* by 10 μm (Table 2).

However, enhancement of stomatal opening by decreasing CO_2 concentration conflicts with the requirement of CO_2 for malate production, and doubts arise as to whether CO_2 supply to guard cells suffices when stomata open in response to low CO_2 tensions. Yet we assume that sufficient malate is produced, since it is only in exceptional cases that chloride satisfies the anion requirement of guard cells. Perhaps endogenous CO_2 evolution suffices to meet the CO_2 requirement for malate synthesis. If one ATP is needed to exchange one H^+ for K^+, and if 36 ATP molecules are produced during oxidation of one glucose molecule, then the latter would provide enough ATP for uptake of 36 K^+. This in turn would require formation of 18 malate molecules, assuming complete dissociation of malate in guard cells. The 6 CO_2 resulting from oxidation of one glucose would meet one third of that requirement. One could readily suggest possible metabolic sources, like an intensive maintenance metabolism of the guard cells and loading of ions into the vacuole, for the remainder of the CO_2 requirement. Nevertheless, CO_2 supply may limit stomatal opening at low external CO_2 concentrations; in many species maximal stomatal opening has not been observed in CO_2-free air but at about 100 μl CO_2 l^{-1}.

The main products of the assimilation of $H^{14}CO_3^-$ and $^{14}CO_2$ into isolated epidermis are malic and aspartic acids, other acids of the tricarboxylic acid cycle and their transamination products also appear (*C. communis, C. diffusa, T. gesneriana*: WILLMER and DITTRICH, 1974; *C. communis* and *T. gesneriana*: RASCHKE and DITTRICH, 1977; *Allium cepa*: SCHNABL, 1977; *Vicia faba*: H. SCHNABL, unpublished). If the epidermal samples are not contaminated by mesophyll cells and chloroplasts, then the labeling pattern obtained in the light is similar to that in darkness. No phosphoglyceric acid or sugar phosphates appear as early fixation products: guard cell chloroplasts are not capable of photosynthetic reduction of CO_2, at least not in the four species tested so far. Epidermis of *C. communis* does possess some RuDP carboxylase activity but cannot catalyze the conversion of exogenous R-5-P to RuDP (RASCHKE and DITTRICH, 1977). CO_2 assimilation into organic acids was not enhanced by light in *A. cepa* (SCHNABL, 1977), was enhanced between 1.2- and 1.4-fold in *Commelina diffusa* (WILLMER and DITTRICH, 1974) and *V. faba* (SCHNABL, unpublished) and 2.6-fold in *T. gesneriana* (SHAW and MACLACHLAN, 1954; RASCHKE and DITTRICH, 1977). The rate of CO_2 fixation did not depend on the state of stomatal opening (RASCHKE and DITTRICH, 1977). Guard cells of lady slipper orchids (*Paphiopedilum* spp.) do not contain chlorophyll, but do open and close (NELSON and MAYO, 1975), demonstrating that ability to photosynthesize is not an essential component of the stomatal mechanism.

4.1.2 Carbon Skeletons From Starch

With the exception of a few species (of the genus *Allium* for instance) all plants possess starch in their guard cells. In general, starch grains decrease in size when stomata open and increase when stomata close. LLOYD (1908) therefore concluded that in guard cells "in the morning starch quickly produced powerful osmotic substances" among which he explicitly included organic acids. If one feeds ^{14}C-labeled glucose-l-phosphate or sugars to epidermal strips one sees radioactivity appear in malate as well as in water-soluble carbohydrates and in starch (DITTRICH and RASCHKE, 1977b). Guard cells therefore possess the enzymes required to derive PEP from sugars and starch. The two species investigated (*C. communis* and *T. gesneriana*) differed greatly, however, in the relative amounts of radioactive carbon channeled into malate and starch.

That starch is required as source of carbon atoms 1 through 3 of malate is strikingly demonstrated by the fact that *Allium* epidermis has an absolute requirement for Cl^- as the counterion for K^+ in stomatal opening and thus presumably contains little malate in the guard cells. Species of the genus *Allium*, like other species of the Liliaceae, Iridaceae, and Amarylladeae families, possess only poorly developed chloroplasts in the guard cells; starch is often completely absent (STEINBERGER, 1922; HEATH, 1947; ALLAWAY and SETTERFIELD, 1972; SCHNABL and ZIEGLER, 1977). Epidermis of *A. cepa* does not possess phosphorylase activity (ONO, 1953; SCHNABL and ZIEGLER, 1977), in contrast to epidermis of *V. faba*, which has starch grains in the guard cells (SCHNABL, 1976). Phosphorylase activity and stomatal opening can be inhibited by Al^{3+} in *V. faba* but not in *A. cepa*.

4.1.3 Transfer of Carbohydrates to Guard Cells

If the reductive pentose phosphate pathway does not function in guard cells they have to rely on import for a supply of carbohydrates. In some plants (ferns, for instance) chloroplasts occur in the ordinary epidermal cells, but we do not know whether they supply guard cells with assimilates. In most other plants, ordinary epidermal cells are chloroplast-free. In these instances, we must assume that guard cells receive assimilates from the underlying mesophyll. Epidermis of *C. communis* contained 16 times more radioactivity if it was still attached to the leaf during a 10-min exposure to $^{14}CO_2$ than if it was isolated and floated on water. In the first case, sugars dominated the labeling pattern, which resembled that of the mesophyll; in the second case, only organic acids appeared (mostly malate and aspartate) and no sugars. If isolated epidermis was placed on prelabeled mesophyll (obtained by exposure of peeled leaves to $^{14}CO_2$ for 10 min) radioactive sugars appeared in the epidermis; they must have been received from the mesophyll (DITTRICH and RASCHKE, 1977b). In further experiments, isolated epidermis of *C. communis* and *T. gesneriana* took up ^{14}C-labeled glucose-1-phosphate (without dephosphorylation), glucose, sucrose, and maltose, and transformed these substances into other sugars and, simultaneously, into malic acid. Radioactivity appeared also in starch in the guard cells (made visible by microautoradiography) and in compounds of the tricarboxylic acid cycle, including their transamination products. The results described in this section support LLOYD's (1908) early view "that guard cells derive their starch chiefly from the underlying tissues rather than through the photosynthetic activity of their own plastids".

Disappearance and reappearance of starch in guard cells is not always well synchronized with transfer of K^+, change in osmotic pressure and stomatal movement (IMAMURA, 1943; YAMASHITA, 1952). In view of the possibility of simultaneous consumption and replenishment of carbohydrates in guard cells, strict synchrony would be unexpected.

4.1.4 The Fate of Malate

When stomata close, a large portion of the malate in the guard cells is released into the apoplast. When epidermal strips are floated on water, the malate appears in the liquid in amounts complementing the residual amount in the epidermis and correlating negatively with the extent of stomatal aperture (VAN KIRK and RASCHKE, 1978b). It is not known whether in the intact leaf, the released malate is reused by other cells (The absolute amounts of malate involved are relatively small; guard cells occupy only a few percent of the total volume of the epidermis.) In addition, some of the malate in guard cells is metabolized in the tricarboxylic acid cycle, and some enters gluconeogenesis. Radioactivity from uniformly labeled exogenous malate appears in guard cell starch, as was demonstrated by microautoradiography (DITTRICH and RASCHKE, 1977a). Malate labeled in 4-C position only does not give rise to radioactive carbohydrates, confirming decarboxylation during gluconeogenesis.

Malic enzyme and malate dehydrogenase activities of epidermal extracts possess pH optima lower than the optimum for PEP carboxylase (WILLMER et al., 1973) and therefore, in conjunction with PEP carboxylase, form a push–pull system stabilizing intracellular pH according to a general mechanism described in more detail by DAVIES (1973). This system may enable guard cells to sense [CO_2] (Sect. 5.1).

4.2 Requirement for and Provision of Energy

4.2.1 Energy Requirement for Stomatal Opening

The work, W, done during inflation of a pair of guard cells can be estimated: $W = V\Delta p + p_o \Delta V$, where ΔV is the increase in volume, V the final volume, p_o the initial pressure and Δp the increase in pressure. Using data supplied in Sections 2.2.3 and 2.3.1 for *Vicia faba*, one computes a requirement of 51 nJ or 12 ncal for a stoma to open from 0 to 18 μm. This work corresponds to 50 fmol or 10 pg of glucose, if used at 30% efficiency. Since a pair of guard cells of *V. faba* has a dry weight of about 6 ng (OUTLAW and LOWRY, 1977), less than $2 \cdot 10^{-3}$ of the dry matter of a pair of guard cells needs to be oxidized to provide the energy for opening a stoma wide. This estimate may however be too favorable with respect to efficiency. It may be more reasonable to base the estimation on the number of ATP needed for stomatal opening. Starting from a hypothetical requirement of 1 ATP for each K^+ exchanged and a requirement of 23 posmol to achieve an aperture of 18 μm one obtains an ATP requirement of 19 pmol under the assumption that K malate is the osmoticum[6]. This amount of ATP would correspond to 126 ncal or 0.5 pmol glucose, which is less than 2% of the dry matter of a pair of guard cells. The energy requirement for stomatal opening is small and probably easily met unless it has to be met quickly.

[6] With a solute requirement of 4.8 atm μm^{-1}, as determined by incipient plasmolysis (high value from Sect. 3.1.2), and a guard cell volume of 6.3 pl at incipient plasmolysis (Sect. 2.2.3) one obtains a solute requirement of 23 posmol at 20 °C. The activity coefficient of K_2 malate with respect to osmolarity was estimated to be 0.8.

4.2.2 Energy Supply and Effects of Metabolic Inhibition

Stomatal opening and closure are both energy-requiring processes. Several lines of indirect evidence suggest that power for these movements is provided chiefly by oxidative phosphorylation. Firstly, the ratio of mitochondria to chloroplasts is, on the average, four times as high in guard cells as it is in mesophyll cells (ALLAWAY and SETTERFIELD, 1972; ALLAWAY and MILTHORPE, 1976). Secondly, activity of dehydrogenases of the tricarboxylic acid cycle has been demonstrated in guard cells (FUJINO, 1960). Finally, conditions under which respiration is inhibited also tend to inhibit stomatal movement.

Inhibitors of respiration like azide, cyanide, and DNP prevent stomatal opening in epidermal strips (PALLAGHY and FISCHER, 1974; TURNER, 1973), but sometimes high concentrations have to be applied if the inhibitors are acids, probably in order to overcome Donnan exclusion in the cell walls. Most effective are nonionic uncouplers, like FCCP and CCCP (PALLAGHY and FISCHER, 1974). Uncouplers, however, may act not only on mitochondria but also directly on plasmalemma and tonoplast, if hydrogen ion gradients across plasmalemma and tonoplast drive transport of ions into guard cells and out of them.

Stomatal closing was arrested in whole leaves by anaerobic conditions (if $[O_2] < 50\ \mu l\ l^{-1}$; AKITA and MOSS, 1973; LOUGUET, 1972) and in epidermal strips by NaF as well as NaN_3 (FUJINO, 1967; WALKER and ZELITCH, 1963).

Inhibitors of protein synthesis and ion transport prevent stomatal opening; they are oligomycin, chloramphenicol, and cycloheximide (PALLAGHY and FISCHER, 1974; RAGHAVENDRA et al., 1976). Sulfhydryl blockers, like phenylmercuric acetate (ZELITCH and WAGGONER, 1962) and detergents, like Triton X-100 may also close stomata effectively.

4.2.3 Light as Energy Source?

Whether or not chloroplasts in guard cells can contribute to the energy supply for stomatal movement, cannot be stated yet in general.

Plastids in guard cells may be few or numerous, green or entirely free of chlorophyll (as in *Paphiopedilum* species; NELSON and MAYO, 1975). Their role as photoreceptors essential for stomatal functioning appears to vary between species. Stomata of CAM plants open in darkness in response to a reduction of the intercellular CO_2 level, yet in other plants light does cause an increase in stomatal aperture at constant intercellular CO_2 (HEATH and RUSSEL, 1954; Sect. 6.1.3). Light was necessary to maintain stomatal opening in epidermal strips of *Senecio odoris*. The action spectrum of this effect resembled that of photosynthesis with a blue light enhancement superimposed (KUIPER, 1964). Action spectra of stomatal opening and uptake of Rb^+ in epidermal strips of *Vicia faba* proved the involvement of a blue-absorbing photoreceptor but not of photosynthesis, except at high light intensities where a response to red light appeared which was attributed to chlorophyll (HSIAO et al., 1973). The inhibitor of photosynthetic electron flow, DCMU, had only a slight inhibitory effect on stomata opening in epidermal strips in one case (HUMBLE and HSIAO, 1970) and caused open stomata to close in another (KUIPER, 1964). The participation of the guard cells' photosynthetic apparatus in stomatal movement appears to be facultative. The variation in the importance of photosynthesis could be explained if guard cell chloroplasts functioned as CO_2 scrubbers (RASCHKE, 1975a). But after the inability of guard cell chloroplasts to reduce CO_2 had been demonstrated in five species (Sect. 4.1.1) new explanations for the variability in the involvement of photosynthesis in the stomatal mechanism have to be sought. PALLAS and DILLEY (1972) considered the possibility of photophosphorylation supplying ATP for the transport of K^+ into guard cells. They concluded that guard-cell chloroplasts of *V. faba* had sufficient ATP-generating capacity (2.25 pmol ATP $stoma^{-1}h^{-1}$) to support the observed fluxes of K^+. (If according to Sect. 4.2.1 stomatal opening by 1 µm requires 1 pmol ATP, the estimated capacity would not suffice to meet the total energy requirement but would substantially contribute to it.) JACOBY (1975) has evidence indicating that the uptake of $^{86}Rb^+$ and $^{42}K^+$ from 0.1 mM solutions into leaf slices of *Phaseolus vulgaris* is caused by the activity of the guard cells. This ion absorption is inhibited by DCMU. If the site of action of DCMU is indeed within the guard cells, the alkali-importing mechanism operating at low external concentrations (Sect. 3.3.3) may be supported by photosynthetic electron transport.

Another possible function of guard-cell chloroplasts would be similar to that of chloroplasts in the mesophyll cells of C-4 plants. In the latter, the reduction of oxaloacetic acid to malic acid is coupled to the presumed photosynthetic production of NADPH (SALIN et al., 1973). In many cases, malate production by guard cells was higher in the light than in darkness (Sect. 4.1.1) but it remains to be shown that the enhancement was the effect of an increased availability of reducing equivalents.

ZEIGER and HEPLER (1977) propose that the stimulation of stomatal opening by blue light and in particular the swelling of guard cell protoplasts occurring in blue light are results of a *direct* establishment of a proton-motive force by the photoreceptor. This contention needs more experimental evidence than has been presented so far. Blue light as a controlling factor will be discussed in the following section.

If stomata optimize gas exchange, they may possess an ability to weigh signals given by light (Sect. 1.3, 6.1.3). Not knowing the game stomata are playing at each particular instance makes an experimental analysis of stomatal responses to light difficult.

5. Control by CO_2, ABA, Light

5.1 Carbon Dioxide

In general, absence of CO_2 enhances stomatal opening; increases in the partial pressure of CO_2 cause closure (see the summaries by MEIDNER and MANSFIELD, 1968; RASCHKE, 1975a). The magnitude of the response to CO_2 varies greatly among species and with the degree of stress the individual plant has suffered or is suffering (Sect. 6.1.2). Stomata in epidermal strips floating on solutions do respond to CO_2 (e.g., PALLAGHY, 1971) but the response is often smaller than on the intact leaf and can be further reduced by increasing the concentration of K^+ in the solution (communication by C.A. VAN KIRK). Acceleration of the uptake of K by guard cells through the action of fusicoccin (which enhances H^+ expulsion; MARRÈ, 1977) completely abolishes the effect of CO_2 (SQUIRE and MANSFIELD, 1974).

The site of sensing $[CO_2]$ is located within the guard cell, as MOURAVIEFF (1956) elegantly demonstrated with isolated pairs of guard cells. The sensitivity is specific for CO_2 (or HCO_3^-); application of the CO_2 analogs ethylene or allene (C_3H_4) (PALLAGHY and RASCHKE, 1972) or SO_2 at low concentrations (MAJERNIK and MANSFIELD, 1972) does not abolish the effect of CO_2. However BONTE et al. (1977) did record temporary and KONDO and SUGAHARA (1978) lasting reductions in stomatal aperture in response to 2 μl SO_2 l^{-1}.

The velocity of stomatal closure follows saturation kinetics with respect to CO_2 ind *Zea mays* in light and in darkness; half saturation occurs near 200 μl l^{-1} (RASCHKE, 1972). Nocturnal stomatal opening of the CAM plant *Agave americana* can be reduced by increases in $[CO_2]$, and this effect saturates (as it does in maize) between 1000 and 2000 μl l^{-1} (NEALES, 1970).

Strictly anaerobic conditions prevent stomatal closure in response to CO_2 (AKITA and MOSS, 1973; LOUGET, 1972). This may be related to the observation that dark fixation of CO_2 and malate formation by CAM plants also requires oxygen (RANSON and THOMAS, 1960).

All these observations taken together led to a hypothesis according to which malate formation is not only necessary to meet an anion requirement for stomatal opening but is also responsible for stomatal closure (RASCHKE, 1975a, b). I propose that the malate level in the cytoplasm of guard cells reflects the level of CO_2 in the environment of these cells and that a high level of malate in the cytoplasm causes ion leakage from the vacuoles by reduction of the semipermeability of tonoplast and plasmalemma. The malate level in the cytoplasm reflects the balance between malate formation, malate removal into the vacuole, and deacidification (Sect. 4.1.4). The rates of malate formation and decarboxylation are determined by the pH dependence of the activities of PEP carboxylase and malic enzyme. The activity of PEP carboxylase is in addition strongly affected by the presence of malate. High malate and low pH inhibit PEP carboxylase while the activity of malic enzyme increases as the cytoplasmic pH falls. The enzyme pair PEP carboxylase-malic enzyme could thus cause cytoplasmic malate and hydrogen ion levels to fluctuate in response to changes in the intercellular CO_2 concentration. Similar processes have been described to occur in roots (HIATT, 1967; HIATT and HENDRICKS, 1967). Aliphatic and aromatic acids of relatively short chain length, including malic acid, cause leakage of ions from roots (JACKSON and TAYLOR, 1970; LEE, 1977). The effect depends on the concentration of the undissociated form of these acids and on the degree of their lipophilic character. These properties are probably important for passage of the plasmalemma only because for action an ionizable carboxyl group has to be present (LEE, 1977).

The possibility that high levels of organic acids in the cytoplasm of guard cells cause closure was tested on epidermal strips with open stomata taken from plants of *Commelina communis* and *Vicia faba*. These strips were exposed to solutions of organic acids at a low pH (3.4 in the case of malic acid): Stomata closed. They did not close, when at pH 3.4 the solution contained a strong acid, like HCl, or a nonabsorbable one, like 2-(N-morpholino) ethanesulfonic acid, instead of a short-chain organic acid, like malic acid. This indicates that an exogenous acid has to be in its undissociated form in order to be able to enter guard cells and cause closure (RASCHKE, unpublished).

Guard cells continuously incorporate $^{14}CO_2$ into malate, independent of the state of stomatal opening and in the light as well as in darkness (Sect. 4.1.1).

Continuous acid metabolism in guard cells may therefore serve as a sensing system for CO_2. This notion may even apply to stomata of *Allium cepa* whose guard cells cannot produce malate in quantities sufficient for osmotic purposes but may be able to make malate for the purpose of control. SCHNABL (1977) demonstrated that epidermal samples of onion do incorporate exogenous $^{14}CO_2$ into malate and we know from HEATH and MEIDNER's (1957) work that onion stomata do respond to CO_2.

5.2 Abscisic Acid

In order to understand the significance of the role of ABA in stomatal regulation of gas exchange, we first have to consider how strongly a change in water potential of a leaf would affect stomatal aperture. Differentiation of Eq. (3.2) with respect to V allows to estimate the magnitude of the volume change that would occur in a guard cell if its solute content remained constant. With $\frac{dp}{dV}=\frac{\varepsilon}{V}$ and $\frac{d\pi}{dV}=-\frac{sRT}{V\cdot V}=-\frac{\pi}{V}$, the differentiation yields

$$\frac{d\psi}{dV}=\frac{dp}{dV}-\frac{d\pi}{dV}=\frac{\varepsilon+\pi}{V};\quad \text{hence}\quad \frac{dV}{V}=\frac{d\psi}{\varepsilon+\pi}. \tag{5.1}$$

Since aperture was found to be proportional to guard cell volume, Eq. (5.1) also describes sensitivity of stomatal aperture to changes in leaf water potential. The modulus of elasticity (ε) of guard cells of *Vicia faba* was estimated to be of the order of 70 atm (Sect. 2.3.1), the osmotic pressure (π) in guard cells of open stomata about 30 atm. Substituting these values into Eq. (5.1) results in a denominator of 100 and tells us that a reduction of the water potential by 10 atm will produce a reduction of stomatal aperture by not more than 10% of the initial value. The magnitude of this response appears to be insufficient for the purpose of regulation of water loss but has been experimentally confirmed by us. Obviously, guard cells may have to release solutes in order to close the stomatal pore. When epidermis of *V. faba* with open stomata was exposed to solutions of decreasing water potentials, solute loss set in when the water potential was under −7 atm, indicating the possible operation of a sensing mechanism that is located in the guard cells themselves. We are however not sure whether this mechanism operates in whole leaves. However there is good evidence that solute loss from guard cells begins when ABA produced in the mesophyll arrives at the stomata and reports that mesophyll cells have sensed stress (see the reviews by HSIAO, 1973, 1976; and RASCHKE, 1975a).

ABA prevents stomatal opening and causes closure. The response can occur within a few minutes of application (MITTELHEUSER and VAN STEVENINCK, 1969; CUMMINS et al., 1971; KRIEDEMANN et al., 1972) and is reversible within a short time span (CUMMINS et al., 1971). ABA may be sequestered in a compartment of the leaf which does not communicate with the stomata (CUMMINS, 1973). A unique correlation between ABA content of a leaf and the degree of stomatal opening should therefore not be expected (Sect. 6.2.1).

Stomatal responses are specific for (+)-ABA (CUMMINS and SONDHEIMER, 1973; KRIEDEMANN et al., 1972). Only a few analogs are active; they have in common an oxygen atom on C-1 (ORTON and MANSFIELD, 1974). The complete 2,4-pentadiene side chain appears not to be essential for activity but the ring double bond and perhaps the ring carbonyl and hydroxyl groups are necessary; (+)-vomifoliol turned out to be as active as ABA on stomata in isolated epidermis of *Eichhornia crassipes* (STUART and COKE, 1975; COKE et al., 1975). Xanthoxin does not affect guard cells if applied to epidermal strips, but closes stomata within minutes if fed into the transpiration stream (RASCHKE et al., 1975). Probably in the leaf xanthoxin is rapidly converted into ABA.

ABA is possibly not the only substance formed in plants under stress that can serve as messenger of stress to stomata. WELLBURN et al. (1974) found all-trans-farnesol to increase with water stress in plants of *Sorghum sudanense*. This sesquiterpenoid closed open stomata in epidermal strips of *C. communis* and acted as an antitranspirant on leaves of *Sorghum bicolor* that had been treated by temporary immersion into emulsions of the compound (FENTON et al., 1977).

LANCASTER et al. (1977) reported that stomata of *Lupinus luteus* were insensitive to ABA but closed readily in wilting leaves. This observation raises the question whether stomata of lupin can be in a state of insensitivity to ABA, or whether a messenger other than ABA appears in wilting leaves of lupin.

We do not know how ABA acts on guard cells and why its action is easily reversible. Reversibility would be facilitated by binding with low affinity at the surface only of guard cells and by rapid removal to sinks in which ABA is either sequestered or metabolized to phaseic acid. Speculation on the mechanism of action is based on two findings, (a) on the ability of fusicoccin to overcome the effects of ABA and CO_2 (SQUIRE and MANSFIELD, 1972a, 1974) and (b) on the requirement for the simultaneous presence of ABA and CO_2 to cause stomatal closure in unstressed plants of *Xanthium strumarium* (and other species; Sect. 6.1.2). Abscisic acid sensitizes stomata to CO_2 and, vice versa, CO_2 enhances stomatal responses to ABA (Fig. 13). This enhancement follows saturation kinetics with respect to CO_2; half saturation occurs at about 200 μl CO_2 l^{-1} (RASCHKE, 1975b). The obvious linkage between effects of ABA and CO_2 on stomata can be explained by an extension of the hypothesis that an excess of malic acid in the cytoplasm causes stomatal closure (Sect. 5.1). ABA appears to facilitate acidification either by blocking the expulsion of H^+ or by activating an H^+ pump directed into the cytoplasm. The observations on simultaneous effects of CO_2 and ABA on guard cells are most easily explained by the first possibility because it requires the production of an acid, for instance malic from CO_2. Nevertheless, no direct evidence exists for an acidification of the guard cells' cytoplasm after an application of CO_2 or ABA, or both simultaneously. Alternative explanations for the effects of CO_2 and ABA should therefore be sought and tested.

5.3 Light

The possibility of guard cells using light directly as an energy source was discussed in Section 4.2.3. Light may stimulate stomatal opening at very low irradiances, too low to consider them as flows of supplementary energy. In stomata of *Helianthus annuus*, wild type and xantha mutant, HABERMANN (1973) measured responses to fluxes of blue, green, red and far-red light in a range between 2 and 8 μmol quanta m^{-2} s^{-1} (full sunlight supplies about 2000 μmol photosynthetically active quanta m^{-2} s^{-1}). Her conclusion that phytochrome might be involved could not be supported by results EVANS and ALLAWAY (1972) obtained with *Vicia faba*.

Clear evidence exists for the presence in the guard cells of a receptor absorbing blue light (MEIDNER and MANSFIELD, 1968; RASCHKE, 1975a). Blue light stimulates stomatal opening and uptake of ^{86}Rb as tracer for K (HSIAO et al., 1973); blue light also induces swelling of guard cell protoplasts of onion in the presence of K (ZEIGER and HEPLER, 1977). HSIAO et al. determined a threshold quantum flux of 7 μmol quanta m^{-2} s^{-1} above which stimulation occurred and prepared action spectra of stomatal opening and epidermal Rb uptake for quantum fluxes of 12 and 63 μmol quanta m^{-2} s^{-1}. At both light levels, the action spectra of stomatal opening and Rb uptake were very well correlated. They differed, however, between the two levels: at the low quantum flux, action was confined to wavelengths <500 nm, with a broad maximum around 460 nm, whereas at the higher quantum flux, a red light effect appeared in addition, indicating a possible contribution from chlorophyll (Sect. 4.2.3). In epidermal strips, the effect of blue light saturated at about 50 μmol quanta m^{-2} s^{-1}, in leaf sections at about

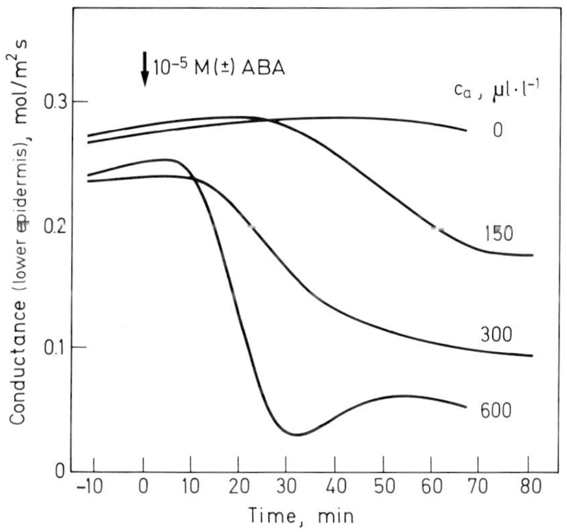

Fig. 13. Enhancement by CO_2 of stomatal responses to ABA in *Xanthium strumarium*. ABA was added to the irrigation water of detached leaves that were exposed to the indicated external CO_2 concentrations, c_a. Change in conductance is the index of stomatal response (RASCHKE, 1975b)

100 μmol m^{-2} s^{-1} which would correspond to about 1/40 and 1/20, respectively, of the full photosynthetically active radiation from the sun.

The tonoplast of isolated guard cell protoplasts from onion cotyledons fluoresces green (ZEIGER and HEPLER, 1977). ZEIGER and HEPLER think that fluorescence and swelling of protoplasts in blue light indicate the presence of a flavin or flavoprotein in the tonoplast, and perhaps also plasmalemma, that mediates responses to blue light through separation of H^+ and OH^- across the membranes. Fluorescence from guard cell vacuoles was so far seen only in onion.

Blue light causes rapid transient opening responses in grasses and sedges (BROGÅRDH, 1975; JOHNSSON et al., 1976). Stomata of species with dumbbell-shaped guard cells respond to increases in light intensity in a biphasic manner: a rapid opening movement is followed by a less rapid, often oscillatory adjustment to an equilibrium aperture which in magnitude will depend on irradiance and which may even be considerably lower than the maximum aperture reached during the transient (Fig. 14a); the equilibrium, furthermore, depends on the CO_2 content of the air (BROGÅRDH, 1975). The first, rapid phase does not occur in plants with kidney-shaped guard cells, irrespective of light quality, and does not occur in plants with grass-type stomata if red light is used (Fig. 14c, b). An action spectrum of the rapid response has meanwhile been determined (communication by A. JOHNSSON). It resembles the action spectra of Rb uptake and enhancement of stomatal opening in epidermis of *Vicia faba* as determined by HSIAO et al. (1973) for the lower light level. Action spectra of the light response serving the presumed optimization of the relationship between light intensity, intercellular CO_2 content, and assimilation, particularly in C-4 plants, have not yet been measured. Discussion of stomatal responses to light in the whole leaf will be continued in Section 6.1.3.

Rapid opening responses like the ones initiated by blue light occur also at constant light levels if the CO_2 concentration around leaves of *Zea mays* is suddenly lowered. This raises the question whether blue light and removal of CO_2 produce a sudden deacidification of the cytoplasm in the guard cells.

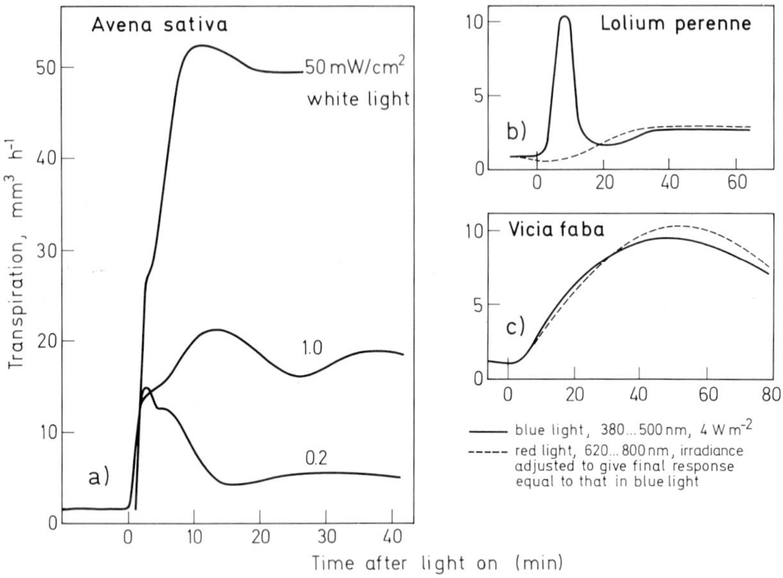

Fig. 14. Stomatal opening, as indicated by transpiration rate, in response to illumination. **a** and **b** plants with dumbbell-shaped guard cells. **c** plant with kidney-shaped guard cells. Dumbbell-shaped guard cells respond in a biphasic manner. In *Avena*, the fast response is particularly prominent at the lowest irradiance, when the slow response is of too small a magnitude to sustain a high rate of transpiration. Species with kidney-shaped guard cells (*V. faba*) do not show the fast initial opening in blue light. (Redrawn from BROGÅRDH and JOHNSSON, 1975 and JOHNSSON et al., 1976)

Responses of guard cells to CO_2, ABA, and light play a key role in the regulation of the gas exchange of leaves; our ignorance of the mechanisms involved is therefore most disturbing.

6. Integration: Stomatal Behavior in Response to Signals from the Plant and the Environment

We return to the function of stomata as moderators, regulators, optimizers of gas exchange; Figure 15 depicts the signal flow producing feedback and feedforward. Numbers following subheadings of the following section refer to diagrammatic representations of the corresponding relationships in Figure 15. The discussion can be brief because the role of stomata in the optimization of water loss and carbon assimilation will be treated in more detail in part B on *Physiological Plant Ecology* of this encyclopedia.

6.1 Movements Related to the Assimilation of CO_2 (1, 2, 4, 5, 8, 9, 12)

We can think of at least three mechanisms by which a plant can synchronize the requirement of the assimilatory apparatus for CO_2 with the supply of CO_2

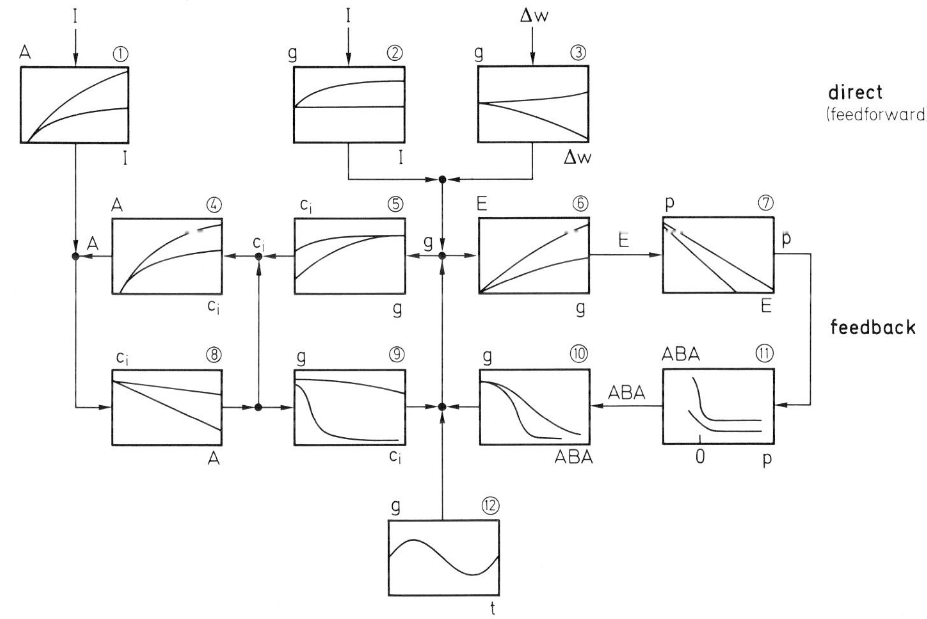

I Light intensity, modified by quality
A Rate of CO_2 assimilation
c_i Intercellular $[CO_2]$
g Stomatal conductance
t Time of day
Δw Difference in molar fraction of water vapor
E Rate of evaporation
ABA Amount or concentration of abscisic acid
p Turgor

Fig. 15. Interrelationships between variables participating in the adjustment of stomatal conductance to the requirements of CO_2 supply and water economy (to be used in conjunction with the flow sheet on p. 388). Two curves in each panel. They indicate the range of variation of the depicted relationship; in most cases the traces entered are quite arbitrary.
① Light dependence of CO_2 assimilation. ② Direct effect of light on guard cells as measured by stomatal conductance; may be unimportant or even lacking (Sects. 5.3, 6.1.3). ③ Direct effect of water vapor pressure difference between leaf and air on stomatal conductance. Conductance may increase slightly with Δw (resulting from release of epidermal pressure; Sect. 2.3.2) or may decrease (Sect. 6.2.2). ④ Saturation curve for CO_2 assimilation with respect to $[CO_2]$ (Sect. 1.2; Figs. 1, 2). ⑤ Drop of $[CO_2]$ across the epidermis (along the stomatal pores): $c_i = c_a - A/g$ (Sects. 1.2, 6.1.2). ⑥ Evaporation from leaf as function of conductance (Sect. 1.2; Fig. 1). ⑦ Decline of leaf turgor with evaporation rate. Turgor is also a function of water supply, modulus of elasticity and solute content of the tissue. ⑧ Depletion of intercellular spaces of CO_2 by assimilation: $c_i = c_a - A/g$ (Sects. 1.2, 6.1.2). ⑨ Stomatal conductance as a function of $[CO_2]$ (5.1, 6.1.2, 6.1.3); stomata may be sensitive or insensitive to CO_2 (Sects. 5.1, 6.1.2, Fig. 17). ⑩ Stomatal conductance as a function of ABA; stomata may be sensitive or insensitive to ABA (Sects. 5.2, 6.2.1). ⑪ ABA accumulation in the leaf as a function of turgor (ABA accumulates if turgor approaches zero. Water potential at zero turgor depends on osmotic pressure. $\psi = -\pi$ for $p = 0$; Sect. 6.2.1). ⑫ Circadian rhythm of conductance (Sect. 6.1.2). **Not entered are:** (a) Direct effects of turgor of leaf tissue and epidermis on conductance (Sects. 2.1.3, 2.3.2). (b) Interrelationship between effects of CO_2 and ABA on guard cells (Sects. 5.1, 5.2, 6.1.2, Fig. 13). (c) Possible direct signal from assimilating tissue to guard cells (Fig. 18). (d) Signals related to the plant's ontogeny. (e) Involvement of heat transfer and temperature (Sect. 1.4)

through the stomata. The first would couple stomatal activity to a circadian clock. The second would employ signals emitted from the assimilatory apparatus when the mesophyll consumes CO_2 and produces assimilates, and the third mechanism would involve a direct effect of light on guard cells. All three mechanisms appear to exist, although their relative importance varies.

6.1.1 Endogenous Rhythm (12)

Stomatal activity is linked to a circadian clock in many species, but control by endogenous rhythmicity does not go beyond a shortening or elimination of the Spannungsphase (Sects. 2.1.3, 2.3.2). Often turgor in the guard cells of closed stomata begins to increase after midnight and reaches its maximum at a time when the sun would normally rise. Stomata can then be slightly open even if leaves are kept in darkness. At this stage stomata respond strongly to reductions in intercellular CO_2 (MANSFIELD, 1976a). Consequently, circadian rhythmicity occurs in the rapidity to respond to light steps (BROGÅRDH and JOHNSSON, 1975) and in the stomatal tendency to oscillate (HOPMANS, 1971). Previous light treatments will shift the phase of the cycle (MANSFIELD and HEATH, 1963; MARTIN and MEIDNER, 1972; BROGÅRDH and JOHNSSON, 1975). The clock is within the tissue of the individual leaf (MANSFIELD and HEATH, 1963; KANA and MILLER, 1977). In *Xanthium strumarium*, stomatal ability to open increased also with photoperiodic induction (KRIZEK and MILTHORPE, 1973).

6.1.2 Feedback Involving CO_2 (5→9→5 and 4→8→4)

Stomatal sensitivity to changes in intercellular CO_2 concentration varies greatly between species and within a species between plants that are well supplied with water, and those which have been stressed by drought or low temperature. The magnitude of stomatal responses to CO_2 and the effects of these responses on conductance and the finally established intercellular CO_2 concentration, c_i, can be estimated, if the gain of the feedback loop is known. The gain of a loop is a measure of the strength of feedback; it consists of a physical and a physiological component (Fig. 16). The physical component of the loop involving CO_2 and stomata describes how much c_i changes when a stoma moves (5); it is derived from the equation $c_i = c_a - A/g$, with $g = g^c$ (1.2): $(\partial c_i/\partial g)_A = A/g^2$. The physiological component represents the stomatal response to a change in c_i; values of the differential dg/dc_i can be estimated from determinations of the relationship between g and c_i; the latter relationship is conveniently obtained from experiments in which the CO_2 concentration in the atmosphere was varied (Fig. 17; FARQUHAR et al., 1978). The product $(\partial c_i/\partial g)_A$ (dg/dc_i) is the gain G_g of the open loop involving stomata and c_i. Since g decreases when c_i increases, dg/dc_i and G_g are negative. The closed loop gain $1/(1-G_g)$ is the factor by which a perturbation is reduced (when negative feedback occurs) or increased (in the case of positive feedback) through the action of feedback (Fig. 16). The same factor applies if a perturbation occurs in c_i instead of g: $\Delta c_i = \overrightarrow{\Delta c_i}/(1-G_g)$, as can be derived easily by using the approach outlined in Figure 16.

So far we have considered only the loop involving CO_2 and conductance (9→5→9 or 5→9→5). Changes in c_i will however affect also the rate of CO_2

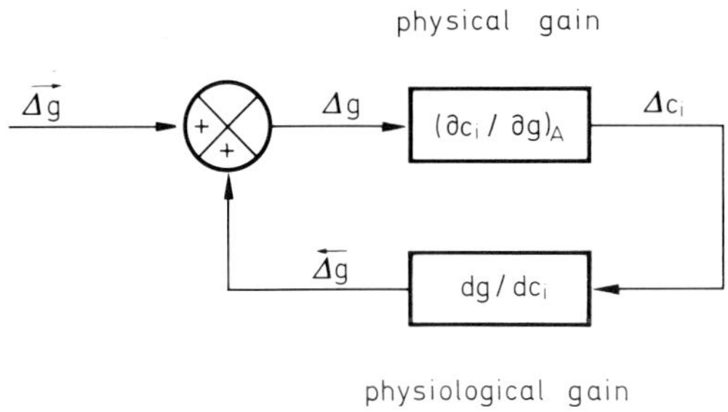

$$\Delta c_i = \Delta g \, (\partial c_i / \partial g)_A \qquad (1)$$

$$\overleftarrow{\Delta g} = \Delta c_i \, (dg / dc_i) \qquad (2)$$

$$\Delta g = \overrightarrow{\Delta g} + \overleftarrow{\Delta g} \qquad (3)$$

Substitution of (1) into (2) and definition of an open-loop gain

$$G_g = (\partial c_i / \partial g)_A \, (dg / dc_i) \qquad (4)$$

results in

$$\overleftarrow{\Delta g} = \Delta g \, G_g \qquad (5)$$

Substitution of (5) into (3) and rearrangement gives the closed-loop gain

$$\frac{\Delta g}{\overrightarrow{\Delta g}} = \frac{1}{1 - G_g} \qquad (6)$$

Fig. 16. Diagrammatic representation of the feedback loop involving CO_2 and stomatal conductance. The conductance change, $\overrightarrow{\Delta g}$, is a direct effect on stomata imposed from outside the loop, a perturbation; $\overleftarrow{\Delta g}$ is the response within the loop. Since in the steady state dg/dc_i has a negative sign, $\overleftarrow{\Delta g}$ opposes $\overrightarrow{\Delta g}$. The closed loop gain $1/(1-G_g)$ is the ratio, $\Delta g/\overrightarrow{\Delta g}$, of the response occurring in the presence of feedback to the response that would occur if feedback did not exist

assimilation, A, (4) and A will affect c_i (8). CO_2 is involved in a second feedback loop whose gain again consists of a physical component $(\partial c_i/\partial A)_g$ and a physiological one (dA/dc_i). The physical component describes the reduction in c_i caused by the assimilatory activity; again based on the equation $c_i = c_a - A/g$, $(\partial c_i/\partial A)_g = -1/g$. The physiological component dA/dc_i is the slope of the saturation curve of A with respect to c_i. Again an open loop gain $G_A = (\partial c_i/\partial A)_g \, (dA/dc_i)$ and a closed loop gain $1/(1-G_A)$ can be defined. The simultaneous equations applying to both, the conductance loop as well as the assimilation loop, can be solved together; the solution results in a combined closed loop gain of $1/(1-G_g-G_A)$ (see FARQUHAR et al., 1978). This closed loop gain describes to what extent perturbations in g or c_i will become effective in changing g and c_i. If the open loop gains G_g and G_A are large, the closed loop gain will be small and so will be the effect of a perturbation on g and c_i. The closed loop gain, of course, also determines the magnitude of a stomatal response to a change in the assimilation rate caused, for instance, by a change in light intensity, as will be shown in the following Section [Eq. (6.1)].

Figure 17 gives examples for variations in the dependence of stomatal conductance on c_i. Stomata of *Xanthium strumarium* can be insensitive to CO_2 but are sensitized to CO_2 by exogenous ABA or stress. Wilting and chilling are known to induce accumulation of ABA. In *Zea mays* differences occur in stomatal sensitivity to CO_2 between plants grown in the field and in growth chambers; the cause of the difference is not known. In field-grown plants of *Z. mays*, stomatal conductance of the upper epidermis may be equal to that of the lower or even exceed it; in growth-chamber grown plants, the lower epidermis has virtually always the higher conductance.

From curves like those shown in Figure 17 and corresponding ones describing the relationship between A and c_i (Fig. 2), gains of the feedback loops involving

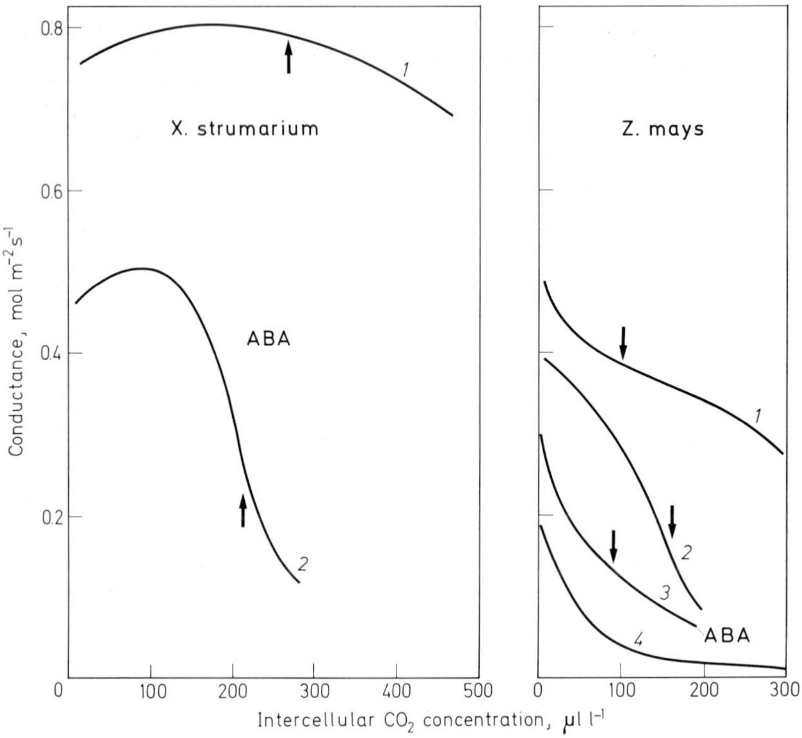

Fig. 17. Stomatal conductances for water vapor in relation to intercellular CO_2 concentration in leaves of *Xanthium strumarium* and *Zea mays*. *Arrows* indicate operating points in normal air at a quantum flux of about 1.5 mmol quanta m^{-2} s^{-1}. Stomata of greenhouse grown *X. strumarium* are virtually insensitive to CO_2 (*curve 1*) but become sensitive if the transpiration stream is 10^{-5} M (±)-ABA (*curve 2*). *Curve 1* for *Z. mays* is from a field grown plant, the others are from plants in growth chambers. *Curves 2* and *3* have been obtained with comparable leaves, leaf 3 received 10^{-5} M (±)-ABA. *Curve 4* is a composite of measurements on many leaves of a different batch of growth chamber-grown plants characterized by low stomatal sensitivity to light. Data on *X. strumarium* and *Z. mays 2, 3* from DUBBE et al. (1978), *Z. mays 4* from RASCHKE et al. (1978), *Z. mays 1*: original

CO_2, conductance, and assimilation can be calculated (FARQUHAR et al., 1978). A few results have been compiled in Table 6. If for instance a decrease in assimilation caused an increase in c_i by, say, 20 µl l^{-1} in a leaf with locked stomata. release of stomatal movement and feedback of CO_2 would cause a corrective action of a magnitude indicated by the closed loop gain. In unstressed plants of *X. strumarium* (Table 6), stomata would respond hardly at all, the final increase in c_i would be $0.93 \cdot 20 = 19$ µl l^{-1}; in "stressed" *X. strumarium* (i.e., with ABA) stomatal narrowing would reduce the increase to $0.40 \cdot 20 = 8$ µl l^{-1} and in growth-chamber grown maize stomata would close so that c_i increases just by $0.21 \cdot 20 = 4$ µl l^{-1}. Obviously, stomatal closure associated with increased sensitivity to CO_2 leads to savings in water which can be estimated for particular cases if the meteorological conditions are known.

The variation in gains gives the plant an opportunity to weigh the importance of signals in the attempt to reconcile the opposing priorities of CO_2 supply and

Table 6. Open and closed loop gains of the feedback loops involving CO_2, stomatal conductance, and assimilation of CO_2 for leaves in a normal atmosphere ($c_a = 300$ to $320\,\mu l\ l^{-1}$)

Species	Origin	Treatment	Open loop gains		Closed loop gain
			G_g	G_A	$(1/1 - G_g - G_A)$
Zea mays[a]	Field	–	−1.16	−0.24	0.42
	Growth chamber	–	−3.59	−0.21	0.21
Xanthium[b] *strumarium*	Greenhouse	−ABA	0.00	−0.08	0.93
		+ABA[c]	−1.14	−0.33	0.40
Amaranthus[b] *powelli*	Growth chamber	−ABA	−0.46	0.00	0.68
		+ABA[c]	−2.00	−0.19	0.31

[a] FARQUHAR et al., 1978.
[b] DUBBE et al., 1978.
[c] 10^{-5} M(±)-ABA in the transpiration stream of detached leaves.

water husbandry. For the same reasons the gains are not large, because precise homeostasis of c_i would make the control of water loss imprecise and, as was already indicated earlier, would make it impossible for the plant to optimize c_i (Sect. 1.3).

6.1.3 Direct Stomatal Responses to Light (2)

In Section 5.3 evidence was presented for the ability of guard cells to respond directly to light. On the other hand data exist that show that stomata can be in a state in which they do not respond to light. In the study of RASCHKE et al. (1978), stomatal conductance in leaves of *Zea mays* was clearly related to the intercellular CO_2 concentration and not to the light intensity impinging on the guard cells. In other experiments, again conducted with leaves of *Zea mays*, stomata responded strongly to changes in light intensity (Fig. 18). This response cannot have been mediated by c_i because c_i was kept constant at $100\,\mu l\ l^{-1}$ by manipulation of the external CO_2 concentration. The increase also cannot have been caused by an increase in leaf temperature because leaf temperature was kept constant. Clearly, stomatal sensitivity to light can be strong or virtually absent in the same species. We do not know whether this variation is used by the plant for optimization of gas exchange. We also do not know which conditions favor and which suppress direct responses to light. At present we can only state that stomata of *Z. mays* that did not respond to light were narrow and extremely sensitive to CO_2 at low CO_2 concentrations (Fig. 17, curve 4 is from RASCHKE et al., 1978), and those which responded to variations in strong light were wide (had relatively high conductances, Fig. 18).

We think that direct effects of light on stomata are mediated by the blue light receptor in guard cells. Enhancement of CO_2 uptake of leaves by blue light was correlated with increased transpiration (VOSKRESENSKAYA and POLYAKOV, 1975). But the fact that blue light effects saturated in leaf sections of *Vicia faba* at lower quantum fluxes (0.1 mmol quanta $m^{-2}\ s^{-1}$; Sect. 5.3) than those which saturated light enhanced opening in *Zea mays* (about 1.4 mmol $m^{-2}\,s^{-1}$; Fig. 18) calls for the determination of saturation curves and action spectra on leaves of

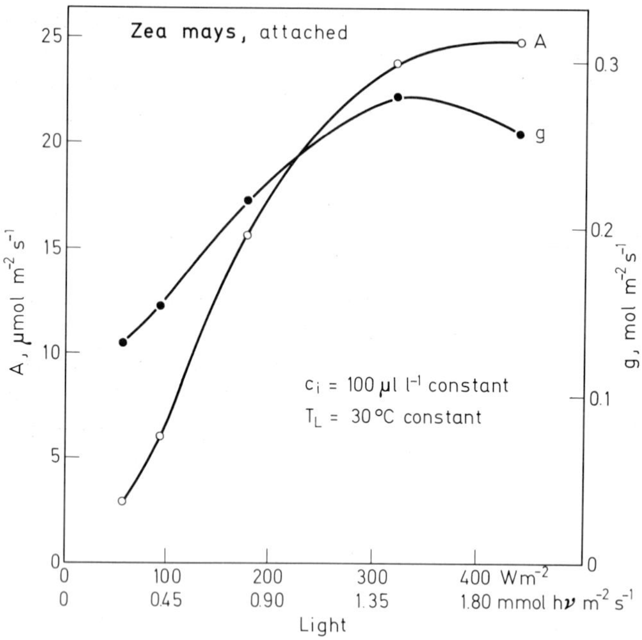

Fig. 18. CO_2 assimilation (*A*) and conductance for CO_2 (*g*) of an attached leaf of *Zea mays* exposed to increasing quantum fluxes. Intercellular CO_2 concentration and leaf temperature were kept constant at 100 µl l^{-1} and 30 °C, respectively. White light from a xenon-arc lamp

several species. Answering the question why stomata in the upper epidermis need higher light intensities than stomata in the lower epidermis to produce the same relative degree of opening (TURNER, 1970) should be made part of the investigation.

There remains the possibility that stimulation by light is actually a response to an assimilate. This could be tested, using the mesophyll as shade (TURNER, 1970; RASCHKE et al., 1978) in combination with inhibitors of photosynthesis. So far, no assimilate has been reported to stimulate stomatal opening.

Analysis of feedback in the leaf shows that stomatal responses to light will be modified considerably by the gains of the feedback loops. The approach outlined in Section 6.1.2 and Figure 16 can be used to assess the magnitude of the effects direct and indirect stomatal responses to light have on stomatal conductance (and c_i). FARQUHAR et al. (1978) derived the following equation in which $(\partial g/\partial I)_{c_i}$ represents the direct response of guard cells to light[7] (Since c_i was kept constant in the experiment of Figure 18, tangents applied to its conductance curve are $(\partial g/\partial I)_{c_i}$).

$$\frac{dg}{dI}=\frac{1}{1-\hat{G}_g-\hat{G}_A}\left\{\left(\frac{\partial g}{\partial I}\right)_{c_i}\left(1-\left(\frac{\partial A}{\partial c_i}\right)_I\left(\frac{\partial c_i}{\partial A}\right)_g\right)+\left(\frac{\partial g}{\partial c_i}\right)_I\left(\frac{\partial c_i}{\partial A}\right)_g\left(\frac{\partial A}{\partial I}\right)_{c_i}\right\} \quad (6.1)$$

closed loop gain — direct stomatal response to light — indirect stomatal response to light

[7] The definition of the gains differs from that given in Section 6.1.2. because light has been introduced as additional variable: $\hat{G}_g=(\partial c_i/\partial g)_A(\partial g/\partial c_i)_I$ and $\hat{G}_A=(\partial c_i/\partial A)_g(\partial A/\partial c_i)_I$; see FARQUHAR et al. (1978).

We recognize that the closed loop gain determines the magnitude of direct as well as indirect stomatal responses to light. High stomatal sensitivity to CO_2 is expressed in a large value of $(\partial g/\partial c_i)_I$, a large open-loop gain, $\hat{G}_g$, and a low closed-loop gain, $1/(1-\hat{G}_g-\hat{G}_A)$; (Sect. 6.1.2). Direct effects of light on stomatal conductance will therefore be reduced by high stomatal sensitivity to CO_2. Indirect responses of stomata to light are mediated through responses to changes in c_i brought about by the light-dependence of the rate of CO_2 assimilation. The magnitude of these indirect responses should therefore increase with stomatal sensitivity to CO_2. But since the latter determines also the closed loop gain, predictions of the stomatal responses to light can only be made if the relative magnitudes of all the components of Eq. (6.1) are known. The discussion could be carried further. It may have sufficed to indicate the richness plants possess in means to modify responses to light for the presumed purpose of optimization. At the same time it may also have become clear that approaches exist to an experimental analysis of the complicated relationship between stomatal conductance and light.

6.2 Movements Related to the Loss of Water (3, 6, 7, 10, 11)

6.2.1 Feedback Mediated Through Accumulation of ABA (6→7→11→10→6)

Sudden changes in water potential of a leaf or in atmospheric humidity may cause transient stomatal responses; but in general, steady-state apertures are amazingly insensitive to changes in the water contents of the plant or the atmosphere until a threshold in water content of the plant is reached beyond which stomata close rather abruptly (Sect. 1.3, reviews by HSIAO, 1973; RASCHKE, 1975a; COWAN, 1977; and the paper by BEARDSELL and COHEN, 1975). Insensitivity may to some extent be caused by compensatory reactions in guard cells and the rest of the epidermis (Sect. 2.3); insensitivity results also from high solute content and modulus of elasticity of guard cells (Sect. 5.2). The latter properties of guard cells make a solute loss necessary in order to achieve effective closure (Sect. 5.2). (Admittedly, not all species may have this requirement. Stomata of *Polypodium vulgare* appear to open and close reversibly without obvious solute loss; LÖSCH, 1977). The discovery that a reduction of the water potential of a plant below a threshold value causes the ABA level in the plant to rise (reviews by HSIAO, 1973; MILBORROW, 1974; papers by ZABADAL, 1974; BEARDSELL and COHEN, 1975; WALTON et al., 1977) in combination with the ability of guard cells to close rapidly in response to ABA (Sect. 5.2) led to the contention that ABA serves as messenger in a feedback loop controlling water loss. HIRON and WRIGHT's (1973) simultaneous measurements of stomatal resistances and ABA levels in wilting and recovering bean plants support the contention convincingly. However, other experiences appear to be in conflict with this view. BEARDSELL and COHEN (1975) noticed that stomata of *Zea mays* and *Sorghum bicolor* closed before ABA level increased measurably, and after rewatering stressed plants, normal stomatal functioning returned 2 days *after* the ABA level had fallen to control values. WALTON et al. (1977) also noticed stomatal closure before increases in ABA levels were detected. WALTON et al. suggest that it is the rate of ABA synthesis and not the ABA content that determines the status of stomatal aperture. Obviously, only ABA in the vicinity of the guard cells will act on them. It should therefore not come as a surprise that an increase of the ABA content of a leaf by the addition of ABA through the transpiration stream may cause stomatal closing if the addition is two orders of magnitude smaller than the initial ABA content of that leaf (RASCHKE, 1975b). The transpiration stream probably terminates near the guard cells (MEIDNER, 1975; BYOTT and SHERIFF, 1976; COWAN, 1977).

The threshold water potential below which ABA accumulation begins and stomata close varies between species and between conditions under which plants were grown. It was suspected that turgor rather than water potential determines when stomata close (TURNER, 1974; COWAN, 1977) and ABA rises (BEARDSELL and COHEN, 1975). Indeed, accumulation of ABA begins in leaf tissue when turgor, p, of the leaf approaches or reaches zero (PIERCE and RASCHKE, in preparation). Since $\psi = -\pi$ for $p = 0$, the point of zero turgor will depend on the solute content of the tissue. ABA accumulation will therefore set in at lower water potentials in tissue that had increased its solute content in response to drought than in tissue that had not adapted osmotically. We assume that the site of ABA formation or liberation in stressed leaves is the mesophyll. LOVEYS (1977) reported that ABA appeared only in epidermis of *Vicia faba* that had been in contact with stressed mesophyll; the ABA content did not increase in epidermis stressed separately. Steady-state gains of the feedback loop involving evaporation, turgor, ABA, and conductance have not been determined yet. For the time being, an on–off response near the point of zero turgor may suffice to describe the relationship between water content of a leaf and stomatal aperture. The sharpness will be blurred by the variation in osmotic pressures among the leaf cells that release ABA and by the variation in stomatal sensitivity to ABA. In general, but not always, guard cells of the upper epidermis respond more strongly to closing stimuli than those of the lower epidermis.

6.2.2 Direct Stomatal Responses to Humidity (3)

Optimization of the relationship between water loss and CO_2 uptake may require that the plant recognizes dry air and high wind speed as causes of high evaporative demand before large water loss occurs (Sect. 1.3). Direct water loss from the epidermis was seen to be the reason why at the same water content stomatal conductance was much lower in leaves of *Zea mays* exposed to dry air than in leaves that were in moist air (RASCHKE, 1970b). LANGE et al. (1971) observed directly through a microscope reversible stomatal closure by alternating local applications of dry or moist air to isolated epidermis of *Polypodium vulgare*. SCHULZE et al. (1972) demonstrated strikingly that stomata can close in response to increases in the evaporative demand of the air, even if transpiration decreases and leaf water content increases (HALL et al., 1976, summarized recent relevant observations). Whether direct, feedforward responses to evaporative demand occur will depend on the relative contributions of transpiration through the stomata and through the cuticle to the water relations of the epidermis. Direct stomatal responses to atmospheric water content therefore need not always occur (RAWSON et al., 1977) and can be partially overridden by responses to a depletion of CO_2 (WEST and GAFF, 1976). Reference to Figure 15 will indicate that a stomatal response will depend on the relative gains in the various feedback paths of the network involved.

6.3 Nonstationary Behavior

So far, the network depicted in Figure 15 was considered to be in the steady state; steady-state gains were applied to describe the relationships between variables. But movement of stomata involves transfer of mass; movement of ions, water; and involves biochemical reactions. Resistance occurs to transfer, and each compartment (in the broadest sense) of the stomatal mechanism has a capacitance, an ability to store matter in relation to existing potential differences. Stomatal movements therefore do not occur instantaneously but within time and are characterized by the time constants of the tissues, cells, or organelles, and by the time constants of the biochemical processes involved (time constant τ = resistance · capacitance). With respect to water relations, time constants of cells have a magnitude of the order of ten seconds, those of tissues of a few minutes, those of herbaceous plants of an hour of less.

The assumption of time-independent inputs, gains, and outputs in the previous sections was a simplification. In reality input, gain, and output are functions of frequency and time. The frequency dependence of each gain can in the simplest case be expressed by the angular velocity ω, which is the reciprocal of the time constant, $\omega = 1/\tau$. A signal travelling in a feedback loop experiences a delay. The response to a step change in the input becomes blurred, and responses to periodic changes in the input experience various phase shifts. If the phase shift is 180°, the feedback signal reinforces the originating perturbation. The sign of the gain becomes positive; oscillations may be the result. Oscillating stomata do not necessarily require an oscillating input. Noise in the system contains many frequencies and may be sufficient to trigger cycling. Cycling will continue if, for instance, the gain of the physical component in a feedback loop is large. FARQUHAR and COWAN (1974) started oscillations in the feedback loop involving stomata and evaporation from a cotton plant by lowering humidity. This manipulation increased the physical component of the loop gain which led to sustained oscillations with a period of about 40 min.

The antagonism between guard and subsidiary cells (Sect. 2.3.2) probably participates in producing an 180° phase shift (COWAN, 1972; DELWICHE and COOKE, 1977). Oscillatory transients occurring in response to step changes in light intensity (APEL, 1967; BROGÅRDH and JOHNSSON, 1975; Fig. 14), CO_2 concentration (APEL, 1967; GULJAEV, 1977), ABA concentration (RASCHKE, 1975b), water supply (RASCHKE, 1970b; GULJAEV, 1977), humidity (APEL, 1967) and temperature (RASCHKE, 1970a) indicate that the behavior of the stomatal control system can be described by a second or higher order equation. (For information on oscillations in biological control systems in general see for instance JONES, 1973, and on oscillations in the stomatal regulation of gas exchange see COWAN, 1977). Several efforts have been made to simulate the oscillatory behavior of stomata. But since very similar simulations can be achieved by different models, models need to be tested to prove their credibility.

COWAN (1972) has reflected on possible advantages of oscillatory behavior of stomata and concluded that oscillations may affect favorably the ratio between CO_2 assimilation and transpiration and may help to optimize the relationship between assimilation and growth. In particular, stomatal oscillations may be looked upon as exploratory excursions in an approach to an optimal state.

Acknowledgments. My research and the preparation of this contribution were supported by the U. S. Department of Energy under Contract EY-76-C-02-1338. I thank Margaret L. Pierce for criticism and suggestion of improvements.

References

Akita, S., Moss, D.N.: The effect of an oxygen-free atmosphere on net photosynthesis and transpiration of barley (*Hordeum vulgare* L.) and wheat (*Triticum aestivum* L.) leaves. Plant Physiol. *52*, 601–603 (1973)

Allaway, W.G.: Accumulation of malate in guard cells of *Vicia faba* during stomatal opening. Planta *110*, 63–70 (1973)

Allaway, W.G., Milthorpe, F.L.: Structure and functioning of stomata. In: Water Deficits and Plant Growth. Kozlowski, T.T. (ed.). New York: Academic Press, 1976, Vol. IV, pp. 57–102

Allaway, W.G., Setterfield, G.: Ultrastructural observations on guard cells of *Vicia faba* and *Allium porrum*. Can. J. Bot. *50*, 1405– 1413 (1972)

Apel, P.: Über rhythmisch verlaufende Änderungen in der CO_2-Aufnahme von Blättern. Ber. Dtsch. Bot. Ges. *80*, 3–9 (1967)

Arends, J.: Über den Einfluß chemischer Agenzien auf Stärkegehalt und osmotischen Wert der Spaltöffnungsschließzellen. Planta *1*, 84–115 (1926)

Arens, T.: Radialstrukturen in den Stomata von *Ouratea spectablis* (mart.) Engl. Protoplasma *66*, 403–411 (1968)

Aylor, D.E., Parlange, J.-Y., Krikorian, A.D.: Stomatal mechanics. Am. J. Bot. *60*, 163–171 (1973)

Ayres, P.G.: Abnormal behaviour of stomata in barley leaves infected with *Rynchosporium secalis* (Oudem.) J.J. Davis. J. Exp. Bot. *23*, 683–691 (1972)

Bearce, B.C., Kohl, H.C., Jr.: Measuring osmotic pressure of sap within live cells by means of a visual melting point apparatus. Plant Physiol. *46*, 515–519 (1970)

Beardsell, M.F., Cohen, D.: Relationships between leaf water status, abscisic acid levels, and stomatal resistance in maize and sorghum. Plant Physiol. *56*, 207–212 (1975)

Bonte, J., de Cormis, L., Louget, P.: Inhibition, en anaerobiose, de la reaction de fermeture des stomates du *Pelargonium* en presence de dioxyde de soufre. Environ. Pollut. *12*, 125–133 (1977)

Brogårdh, T.: Regulation of transpiration in *Avena*. Responses to red and blue light steps. Physiol. Plant. *35*, 303–309 (1975)

Brogårdh, T., Johnsson, A.: Regulation of transpiration in *Avena*. Responses to white light steps. Physiol. Plant. *35*, 115–125 (1975)

Byott, G.S., Sheriff, D.W.: Water movement into and through *Tradescantia virginiana* (L.) leaves: II: Liquid flow pathways and evaporative sites. J. Exp. Bot. *27*, 634–639 (1976)

Chabot, J.F., Chabot, B.F.: Ultrastructure of the epidermis and stomatal complex of balsam fir (*Abies balsamea*). Can. J. Bot. *55*, 1064–1075 (1977)

Coke, L.B., Stuart, K.L., Whittle, Y.G.: Further effects of vomifoliol on stomatal aperture and on the germination of lettuce and the growth of cucumber seedlings. Planta *127*, 21–25 (1975)

Cooke, J.R., de Baerdemaeker, J.-G., Rand, R.H., Mang, H.A.: A finite element shell analysis of guard cell deformations. Trans. ASAE *19*, 1107–1121 (1976)

Cowan, I.R.: Oscillations in stomatal conductance and plant functioning associated with stomatal conductance. Observations and a model. Planta *106*, 185–219 (1972)

Cowan, I.R.: Stomatal behaviour and environment. Adv. Bot. Res. *4*, 117–228 (1977)

Cowan, I.R., Farquhar, G.D.: Stomatal function in relation to leaf metabolism and environment. Symp. Soc. Exp. Biol. *31*, 471–505 (1977)

Cummins, W.R.: The metabolism of abscisic acid in relation to its reversible action on stomata in leaves of *Hordeum vulgare* L. Planta *114*, 159–167 (1973)

Cummins, W.R., Kende, H., Raschke, K.: Specificity and reversibility of the rapid stomatal response to abscisic acid. Planta *99*, 347–351 (1971)

Cummins, W.R., Sondheimer, E.: Activity of the asymmetric isomers of abscisic acid in a rapid bioassay. Planta *111*, 365–369 (1973)

Davies, D.D.: Control of and by ph. Symp. Soc. Exp. Biol. *27*, 513–529 (1973)

Dayanandan, P., Kaufman, P.B.: Stomatal movements associated with potassium fluxes. Am. J. Bot. *62*, 221–231 (1975)

Delwiche, M.J., Cooke, J.R.: An analytical model of the hydraulic aspects of stomatal dynamics. J. Theor. Biol. *69*, 113–141 (1977)

DeMichele, D.W., Sharpe, P.J.H.: An analysis of the mechanics of guard cell motion. J. Theor. Biol. *41*, 77–96 (1973)

Dittrich, P., Raschke, K.: Malate metabolism in isolated epidermis of *Commelina communis* L. in relation to stomatal functioning. Planta *134*, 77–81 (1977a)

Dittrich, P., Raschke, K.: Uptake and metabolism of carbohydrates by epidermal tissue. Planta *134*, 83–90 (1977b)

Drake, B.G., Raschke, K., Salisbury, F.B.: Temperatures and transpiration resistances of *Xanthium* leaves as affected by air temperature, humidity, and wind speed. Plant Physiol. *46*, 324–330 (1970)

Dubbe, D.R., Farquhar, G.D., Raschke, K.: The effect of abscisic acid on the gain of the feedback loop involving CO_2 and stomata. Plant Physiol. *62* (1978), in press

Edwards, M., Meidner, H.: Direct measurements of turgor pressure potentials III (A) injections of solutions into stomatal cells, and (B) comparisons of pressure potentials on stomatal cells of different species. J. Exp. Bot. *28*, 669–677 (1977)

Edwards, M., Meidner, H., Sheriff, D.W.: Direct measurements of turgor pressure potentials of guard cells II. The mechanical advantage of subsidiary cells, the Spannungsphase, and the optium leaf water deficit. J. Exp. Bot. *27*, 163–171 (1976)

Eshel, A., Waisel, Y., Ramati, A.: The role of sodium in stomatal movements of a halophyte: a study by x-ray microanalysis. In: 7th Intl. Colloquium on Plant Analysis and Fertilizer Problems. Wehrmann, J. (ed.). Hannover, Germany: German Soc. Plant Nutrition, 1974

Evans, L.T., Allaway, W.G.: Action spectrum for the opening of *Albizzia julibrissin* pinnules, and the role of phytochrome in the closing movements of pinnules and of stomata of *Vicia faba*. Aust. J. Biol. Sci. *25*, 885–893 (1972)

Farquhar, G.D.: A study of the responses of stomata to perturbations of environment. PhD Thesis, Australian National University, 1973

Farquhar, G.D., Cowan, I.R.: Oscillations in stomatal conductance. Plant Physiol. *54*, 769–772 (1974)

Farquhar, G.D., Raschke, K.: On the resistance to transpiration of the sites of evaporation within the leaf. Plant Physiol. *61*, 1000–1005 (1978)

Farquhar, G.D., Dubbe, D.R., Raschke, K.: The gain of the feedback loop involving CO_2 and stomata: theory and measurement. Plant Physiol. *62* (1978), in press

Fenton, R., Davies, W.J., Mansfield, T.A.: The role of farnesol as a regulator of stomatal opening in *Sorghum*. J. Exp. Bot. *28*, 1043–1053 (1977)

Fischer, R.A.: Aspects of potassium accumulation by stomata of *Vicia faba*. Aust. J. Biol. Sci. *25*, 1107–1123 (1972)

Fischer, R.A.: The relationship of stomatal aperture and guard-cell turgor pressure in *Vicia faba*. J. Exp. Bot. *24*, 387–399 (1973)

Fujino, M.: Correlation among the stomatal movement, pH, salt content, starch content and osmotic value in *Commelina*. Kagaku *29*, 147–148 (1959) (In Japanese)

Fujino, M.: Stomatal movement and enzyme activity in guard cells. Kagaku *30*, 89–90 (1960) (In Japanese)

Fujino, M.: Role of adenosinetriphosphate and adenosinetriphosphatase in stomatal movement. Science Bull. Fac. Educ., Nagasaki Univ. *18*, 1–47 (1967)

Gradmann, D.: Potassium and turgor pressure in plants. J. Theor. Biol. *65*, 597–599 (1977)

Guljaev, B.I.: Self-excited relaxation oscillation of gas exchange in leaves. Fiziol. Biochim. Kult. Rast. *9*, 520–526 (1977) (In Russian)

Gunar, I.I., Zlotnikova, I., Panichkin, L.A.: Electrophysiological investigation of cells of the stomate complex in spiderwort. Fiziol. Rast. *22*, 810–813 (1975) (In Russian)

Guttenberg, H., von: Die physiologische Anatomie der Spaltöffnungen. In: Handbuch der Pflanzenphysiologie. Bünning, E. (ed.). Berlin, Göttingen, Heidelberg: Springer Verlag, 1959, Vol. XVII/1, pp. 399–314

Guttenberg, H., von: Bewegungsgewebe und Perzeptionsorgane. In: Handbuch der Pflanzenanatomie. Carlquist, S., Zimmermann, W., Ozenda, P., Wulff, H.D. (eds.) Berlin, Stuttgart: Gebrüder Borntraeger, 1971, Vol. V/5, pp. 203–219

Habermann, H.M.: Evidence for two photoreactions and possible involvement of phytochrome in light-dependent stomatal opening. Plant Physiol. *51*, 543–548 (1973)

Hall, A.E., Schulze, E.D., Lange, O.L.: Current perspectives of steady-state stomatal responses to environment. In: Ecological Studies. Analysis and Synthesis. Water and Plant Life. Lange, O.L., Kappen, L., Schulze, E.D. (eds.). Berlin-Heidelberg-New York: Springer, 1976, Vol. 19, pp. 169–188

Hastings, D.F., Gutknecht, J.: Potassium and turgor pressure in plants. J. Theor. Biol. *72*, in press (1978)

Heath, O.V.S.: An experimental investigation of the mechanism of stomatal movement, with some preliminary observations upon the response of the guard cells to "shock". New Phytol. *37*, 385–395 (1938)

Heath, O.V.S.: Role of starch in light-induced stomatal movement, and a new reagent for staining stomatal starch. Nature (London) *159*, 647 (1947)

Heath, O.V.S., Meidner, H.: Effects of carbon dioxide and temperature on stomata of *Allium cepa* L. Nature (London) *180*, 180–182 (1957)

Heath, O.V.S., Russell, J.: Studies in stomatal behaviour. VI. An investigation of the light responses of wheat stomata with the attempted elimination of control by the mesophyll. Part 1. Effects of light independent of carbon dioxide and their transmission from one part of the leaf to another. J. Exp. Bot. *5*, 1–15 (1954)

Hiatt, A.J.: Reaction in vitro of enzymes involved in CO_2 fixation accompanying salt uptake by barley roots. Z. Pflanzenphysiol. *56*, 233–245 (1967)

Hiatt, A.J., Hendricks, S.B.: The role of CO_2 fixation in accumulation of ions by barley roots. Z. Pflanzenphysiol. *56*, 220–232 (1967)

Hiron, R.W.P., Wright, S.T.C.: The role of endogenous abscisic acid in the response of plants to stress. J. Exp. Bot. *24*, 769–781 (1973)

Hopmans, P.A.M.: Rhythms in stomatal opening of bean leaves. Meded. Landbouwhogesch. Wageningen *71*-(3), 1–86 (1971)

Hsiao, T.C.: Plant responses to water stress. Annu. Rev. Plant Physiol. *24*, 519–570 (1973)

Hsiao, T.C.: Stomatal ion transport. In: Ency. Plant Physiol. Part B. Lüttge, U., Pitman, M.G. (eds.). Berlin-Heidelberg-New York: Springer, 1976, Vol. II, pp. 195–221

Hsiao, T.C., Allaway, W.G., Evans, L.T.: Action spectra for guard cell Rb^+ uptake and stomatal opening in *Vicia faba*. Plant Physiol. *51*, 82–88 (1973)

Humble, G.D., Hsiao, T.C.: Specific requirement of potassium for light-activated opening of stomata in epidermal strips. Plant Physiol. *44*, 230–234 (1969)

Humble, G.D., Hsiao, T.C.: Light-dependent influx and efflux of potassium of guard cells during stomatal opening and closing. Plant Physiol. *46*, 483–487 (1970)

Humble, G.D., Raschke, K.: Stomatal opening quantitatively related to potassium transport. Evidence from electron probe analysis. Plant Physiol. *48*, 447–453 (1971)

Iljin, W.S.: Über Öffnen der Stomata bei starkem Welken der Pflanzen. Jahrb. Wiss. Bot. *77*, 220–251 (1932)

Imamura, S.: Untersuchungen über den Mechanismus der Turgorschwankung der Spaltöffnungsschliesszellen. Jpn. J. Bot. *12*, 251–346 (1943)

Jackson, P.C., Taylor, J.M.: Effects of organic acids on ion uptake and retention in barley roots. Plant Physiol. *46*, 538–542 (1970)

Jacoby, B.: The effect of dichlorophenyldimethylurea on light-stimulated $^{22}Na^+$, $^{42}K^+$ and $^{86}Rb^+$ absorption in different tissues of *Phaseolus vulgaris* leaves. Physiol. Plant. *35*, 1–4 (1975)

Jones, R.W.: Principles of Biological Regulation. New York-London: Academic Press, 1973

Johnsson, M., Issaias, S., Brogårdh, T., Johnsson, A.: Rapid, blue-light-induced transpiration response restricted to plants with grass-like stomata. Physiol. Plant. *36*, 229–232 (1976)

Kana, T.M., Miller, J.H.: Effect of photoperiod on stomatal opening in *Vicia faba*. Plant Physiol. *60*, 803–804 (1977)

Kelle, A.: Zur Physiologie der Nebenzellen des Spaltöffnungsapparates. Doktor-Dissertation, Universität Münster 1934

Kondo, N., Sugahara, K.: Changes in transpiration rate of SO_2-resistant and -sensitive plants with SO_2 fumigation and the participation of abscisic acid. Plant Cell Physiol. *19*, 365–373 (1978)
Kriedemann, P.E., Loveys, B.R., Fuller, G.L., Leopold, A.C.: Abscisic acid and stomatal regulation. Plant Physiol. *49*, 842–847 (1972)
Krizek, D.T., Milthorpe, F.L.: Effect of photoperiodic induction on the transpiration rate and stomatal behaviour of debudded *Xanthium* plants. J. Exp. Bot. *24*, 76–86 (1973)
Kuiper, P.J.C.: Action spectrum of stomatal movement in epidermis tissue of *Senecio odoris* Defl. Plant Physiol. *39*, 952–955 (1964)
Lancaster, J.E., Mann, J.D., Porter, N.G.: Ineffectiveness of abscisic acid in stomatal closure of Yellow Lupin, *Lupinus luteus* var. Weiko III. J. Exp. Bot. *28*, 184–191 (1977)
Lange, O.L.: Untersuchungen über Wärmehaushalt und Hitzeresistenz mauretanischer Wüsten- und Savannenpflanzen. Flora *147*, 595–651 (1959)
Lange, O.L., Lösch, R., Schulze, E.-D., Kappen, L.: Responses of stomata to changes in humidity. Planta *100*, 76–86 (1971)
Lee, R.B.: Effects of organic acids on the loss of ions from barley roots. J. Exp. Bot. *28*, 578–587 (1977)
Linacre, E.T.: A note on a feature of leaf and air temperatures. Agric. Meteorol. *1*, 66–72 (1964)
Lloyd, F.E.: The physiology of stomata. Carnegie Inst. Washington, Publ. No. *82* (1908)
Louguet, P.: Influence de la pression partielle d'oxygène sur la vitesse d'ouverture et de fermeture des stomates de *Pelargonium* × *hortorum* à l'obscurité. Physiol. Veg. *10*, 515–528 (1972)
Lösch, R.: Responses of stomata to environmental factors – experiments with isolated epidermal strips of *Polypodium vulgare*. I. Temperature and humidity. Oecologia *29*, 85–97 (1977)
Loveys, B.R.: The intracellular location of abscisic acid in stressed and non-stressed leaf tissue. Physiol. Plant. *40*, 6–10 (1977)
Majernik, O., Mansfield, T.A.: Stomatal responses to raised atmospheric CO_2 concentrations during exposure of plants to SO_2 pollution. Environ. Pollut. *3*, 1–7 (1972)
Mansfield, T.A.: Mechanisms involved in turgor changes of guard cells. In: Perspectives in Experimental Biology. Sunderland, N. (ed.). Oxford, New York: Botany, Pergamon Press, 1976, Vol. 2, pp. 453–462
Mansfield, T.A., Heath, O.V.S.: Studies in stomatal behaviour. IX. Photoperiodic effects on rhythmic phenomena in *Xanthium pennsylvanicum*. J. Exp. Bot. *14*, 334–352 (1963)
Marrè, E.: Effects of fusicoccin and hormones on plant cell membrane activities: observations and hypotheses. In: Regulation of Cell Membrane Activities in Plants. Marrè, E., Ciferri, O. (eds.). Amsterdam, Oxford, New York: North-Holland Publ. Comp., 1977, pp. 185–202
Martin, E.S., Meidner, H.: The phase-response of the dark stomatal rhythm in *Tradescantia virginiana* to light and dark treatments. New Phytol. *71*, 1045–1054 (1972)
Meidner, H.: Water supply, evaporation, and vapour diffusion in leaves. J. Exp. Bot. *26*, 666–673 (1975)
Meidner, H., Mansfield, T.A.: Physiology of Stomata. New York: McGraw Hill, 1968
Milborrow, B.V.: The chemistry and physiology of abscisic acid. Annu. Rev. Plant Physiol. *25*, 259–307 (1974)
Mitchell, P.: Chemiosmotic coupling in oxidative and photosynthetic phosphorylation. Biol. Rev. *41*, 445–502 (1966)
Mittelheuser, C.J., van Steveninck, R.F.M.: Stomatal closure and inhibition of transpiration induced by (RS)-abscisic acid. Nature (London) *221*, 281–282 (1969)
Mohl, H., von: Welche Ursachen bewirken Erweiterung und Verengung der Spaltöffnungen? Bot. Z., *14*, 697–704, 713–721 (1856)
Moody, W., Zeiger, E.: Electrophysiological properties of onion guard cells. Planta *139*, 159–165 (1978)
Mouravieff, P.J.: Action du CO_2 et de la lumière sur l'appareil stomatique séparé du mésophylle. Botaniste *40*, 195–212 (1956)
Monzi, M.: Die Mitwirkung der Stomata-Nebenzellen auf die Spaltöffnungsbewegung. Japan. J. Bot. *9*, 373–394 (1939)
Napp-Zinn, K.: Anatomie des Blattes. I. Blattanatomie der Gymnospermen. In: Handbuch

der Pflanzenanatomie. Part 1. Zimmermann, W., Ozenda, P., Wulff, H.D. (eds.) Berlin-Nikolassee: Gebrüder Borntraeger, 1966, Vol. VIII, pp. 48–87

Napp-Zinn, K.: Anatomie des Blattes. II. Blattanatomie der Angiospermen. A. Entwicklungsgeschichtliche und topographische Anatomie des Angiospermenblattes. In: Handbuch der Pflanzenanatomie. Part 2A, Carlquist, S., Ozenda, P., Wulff, H.D. (eds.). Berlin, Stuttgart: Gebrüder Borntraeger, 1973, Vol. VIII, pp. 221–319

Neales, T.F.: Effect of ambient carbon dioxide concentration on the rate of transpiration of *Agave americana* in the dark. Nature (London) *228*, 880–882 (1970)

Nelson, S.D., Mayo, J.M.: The occurrence of functional non-chlorophyllous guard cells in *Paphiopedilum* spp. Can. J. Bot. *53*, 1–7 (1975)

Nelson, S.D., Mayo, J.M.: Low K^+ in *Paphiopedilum leeanum* leaf epidermis: implications for stomatal functioning. Can. J. Bot. *55*, 489–495 (1977)

Ogura, Y.: Comparative anatomy of vegetative organs of the pteridophytes. In: Handbuch der Pflanzenanatomie. Part 3. Carlquist, S., Ozenda, P., Wulff, H.D. (eds.) Berlin, Stuttgart: Gebrüder Borntraeger, 1972, Vol. VII

Ono, H.: On the mechanism of the formation of starch in plants. 1. Especially in guard cells. Bot. Mag. *66*, 103–111 (1953)

Orton, P.J., Mansfield, T.A.: The activity of abscisic acid analogues as inhibitors of stomatal opening. Planta *121*, 263–272 (1974)

Outlaw, W.H., Jr., Kennedy, J.: Enzymic and substrate basis for the anapleurotic step in guard cells. Plant Physiol. *62*, in press (1978)

Outlaw, W.H., Jr., Lowry, O.H.: Organic acid and potassium accumulation in guard cells during stomatal opening. Proc. Natl. Acad. Sci. U.S.A. *74*, 4434–4438 (1977)

Palevitz, B.A., Hepler, P.K.: Cellulose microfibril orientation and cell shaping in developing guard cells of *Allium*: The role of microtubules and ion accumulation. Planta *132*, 71–93 (1976)

Pallaghy, C.K.: Electrophysiological studies in guard cells of tobacco. Planta *80*, 147–153 (1968)

Pallaghy, C.K.: Stomatal movement and potassium transport in epidermal strips of *Zea mays*: The effect of CO_2. Planta *101*, 287–295 (1971)

Pallaghy, C.K.: Localization of thallium in stomata is independent of transpiration. Aust. J. Biol. Sci. *25*, 415–417 (1972)

Pallaghy, C.K., Fischer, R.A.: Metabolic aspects of stomatal opening and ion accummulation by guard cells in *Vicia faba*. Z. Pflanzenphysiol. *71*, 332–344 (1974)

Pallaghy, C.K., Raschke, K.: No stomatal response to ethylene. Plant Physiol. *49*, 275–276 (1972)

Pallas, J.E., Jr., Dilley, R.A.: Photophosphorylation can provide sufficient adenosine 5′-triphosphate to drive K^+ movements during stomatal opening. Plant Physiol. *49*, 649–650 (1972)

Penny, M.G., Bowling, D.J.F.: A study of potassium gradients in the epidermis of intact leaves of *Commelina communis* L. in relation to stomatal opening. Planta *119*, 17–25 (1974)

Penny, M.G., Bowling, D.J.F.: Direct determination of pH in the stomatal complex of *Commelina*. Planta *122*, 209–212 (1975)

Penny, M.G., Kelday, L.S., Bowling, D.J.F.: Active chloride transport in the leaf epidermis of *Commelina communis* in relation to stomatal activity. Planta *130*, 291–294 (1976)

Raghavendra, A.S., Rao, I.M., Das, V.S.R.: Adenosin triphosphatase in epidermal tissue of *Commelina benghalensis*; possible involvement of isozymes in stomatal movement. Plant. Sci. Lett. *7*, 391–396 (1976)

Ranson, S.L., Thomas, M.: Crassulacean acid metabolism. Annu. Rev. Plant Physiol. *11*, 81–110 (1960)

Raschke, K.: Temperature dependence of CO_2 assimilation and stomatal aperture in leaf sections of *Zea mays*. Planta *91*, 336–363 (1970a)

Raschke, K.: Stomatal responses to pressure changes and interruptions in the water supply of detached leaves of *Zea mays* L. Plant Physiol. *45*, 415–423 (1970b)

Raschke, K.: Saturation kinetics of the velocity of stomatal closing in response to CO_2. Plant Physiol. *49*, 229–234 (1972)

Raschke, K.: Stomatal action. Annu. Rev. Plant Physiol. *26*, 309–340 (1975a)

Raschke, K.: Simultaneous requirement of carbon dioxide and abscisic acid for stomatal closing in *Xanthium strumarium* L. Planta *125*, 243–259 (1975b)

Raschke, K.: Transfer of ions and products of photosynthesis to guard cells. In: Transport and Transfer Processes in Plants. Wardlaw, I.F., Passioura, J.B. (eds.). New York: Academic Press 1976, pp. 203–215

Raschke, K.: The stomatal turgor mechanism and its responses to CO_2 and abscisic acid: observations and a hypothesis. In: Regulation of Cell Membrane Activities in Plants. Marrè, E., Ciferri, O. (eds.). Amsterdam: North Holland Publ. Comp. 1977, pp. 173–183

Raschke, K., Dittrich, P.: [^{14}C] carbon-dioxide fixation by isolated epidermis with stomata closed or open. Planta *134*, 69–75 (1977)

Raschke, K., Fellows, M.P.: Stomatal movement in *Zea mays*. Shuttle of potassium and chloride between guard cells and subsidiary cells. Planta *101*, 296–316 (1971)

Raschke, K., Firn, R.D., Pierce, M.: Stomatal closure in response to xanthoxin and abscisic acid. Planta *125*, 149–160 (1975)

Raschke, K., Hanebuth, G.D., Farquhar, G.D.: Relationship between stomatal conductance and light intensity in leaves of *Zea mays* L., derived from experiments using the mesophyll as shade. Planta *139*, 73–77 (1978)

Raschke, K., Humble, G.D.: No uptake of anions required by opening stomata of *Vicia faba*: guard cells release hydrogen ions. Planta *115*, 47–57 (1973)

Raschke, K., Pierce, M., Popiela, C.C.: Abscisic acid content and stomatal sensitivity to CO_2 in leaves of *Xanthium strumarium* L. after pretreatments in warm and cold growth chambers. Plant Physiol. *57*, 115–121 (1976)

Raschke, K., Schnabl, H.: Availability of chloride affects the balance between potassium chloride and potassium malate in guard cells of *Vicia faba* L. Plant Physiol. *62*, 84–87 (1978)

Rawson, H.M., Begg, J.E., Woodward, R.G.: The effect of atmospheric humidity on photosynthesis, transpiration and water use efficiency of leaves of several plant species. Planta *134*, 5–10 (1977)

Salin, M.L., Campbell, W.H., Black, C.C., Jr.: Oxaloacetate as the Hill oxidant in mesophyll cells of plants possessing the C_4-decarboxylic acid cycle of leaf photosynthesis. Proc. Natl. Acad. Sci. USA *70*, 3730–3734 (1973)

Sanchez, S.M: The fine structure of the guard cells of *Helianthus annuus*. Am. J. Bot. *64*, 814–824 (1977)

Sawhney, B.L., Zelitch, I.: Direct determination of potassium ion accumulation in guard cells in relation to stomatal opening in light. Plant Physiol. *44*, 1350–1354 (1969)

Scarth, G.W.: Mechanism of the action of light and other factors on stomatal movement. Plant Physiol. *7*, 481–504 (1932)

Schnabl, H.: Der Einfluß von Aluminiumionen auf den Stärkemetabolismus von *Vica faba*-Epidermen. Z. Pflanzenphysiol. *77*, 167–173 (1976)

Schnabl, H.: Isolation and identification of soluble polysaccharides in epidermal tissue of *Allium cepa*. Planta *135*, 307–311 (1977)

Schnabl, H., Ziegler, H.: The mechanism of stomatal movement in *Allium cepa* L. Planta *136*, 37–43 (1977)

Schulze, E.-D., Lange, O.L., Buschbom, U., Kappen, L., Evenari, M.: Stomatal responses to changes in humidity in plants growing in the desert. Planta *108*, 259–270 (1972)

Schwendener, S.: Über Bau und Mechanik der Spaltöffnungen. Monatsberichte der Königlich Preussischen Akademie der Wissenschaften zu Berlin. Aus dem Jahre 1881. 833–867 (1882)

Schwendener, S.: Die Spaltöffnungen der Gramineen und Cyperaceen. Sitzungsberichte kgl. preuß. Akad. Wiss. Berlin 1889, pp. 65–79

Shaw, M., MacLachlan, G.A.: The physiology of stomata I. Carbon dioxide fixation in guard cells. Can. J. Bot. *32*, 784–794 (1954)

Shoemaker, E.M., Srivastava, L.M.: The mechanics of stomatal opening in corn (*Zea mays* L.) leaves. J. Theor. Biol. *42*, 219–225 (1973)

Singh, A.P., Srivastava, L.M.: The fine structure of pea stomata. Protoplasma *76*, 61–82 (1973)

Squire, G.R., Mansfield, T.A.: Studies of the mechanism of action of fusicoccin, the fungal toxin that induces wilting and its interaction with abscisic acid. Planta *105*, 71–78 (1972a)

Squire, G.R., Mansfield, T.A.: A simple method of isolating stomata on detached epidermis by low pH treatment: Observations of the importance of the subsidiary cells. New Phytol. *71*, 1033–1043 (1972b)

Squire, G.R., Mansfield, T.A.: The action of fusicoccin on stomatal guard cells and subsidiary cells. New Phytol. *73*, 433–440 (1974)
Srivastava, L.M., Singh, A.P.: Stomatal structure in corn leaves. J. Ultrastruct. Res. *39*, 345–363 (1972)
Stålfelt, M.G.: Die Abhängigkeit der Spaltöffnungsreaktionen von der Wasserbilanz. Planta *8*, 287–340 (1929)
Stålfelt, M.G.: Die stomatäre Transpiration und die Physiologie der Spaltöffnungen. In: Handbuch der Pflanzenphysiologie. Ruhland, W. (ed.). Berlin, Göttingen, Heidelberg: Springer, 1956, Vol. III, p. 351–426
Stålfelt, M.G.: The relation between the endogenous and induced elements of the stomatal movements. Physiol. Plant. *18*, 177–184 (1965)
Steinberger, A.-L.: Über Regulation des osmotischen Wertes in den Schließzellen von Luft- und Wasserspalten. Biol. Zentralbl. *42*, 405–419 (1922)
Stevens, R.A., Martin, E.S.: New structure associated with stomatal complex of the fern *Polypodium vulgare*. Nature (London) *265*, 331–334 (1977a)
Stevens, R.A., Martin, E.S.: Ion-absorbent substomatal structures in *Tradescantia pallidus*. Nature (London) *268*, 364–365 (1977b)
Stuart, K.L., Coke, L.B.: The effect of vomifoliol on stomatal aperture. Planta *122*, 307–310 (1975)
Tal, M., Eshel, A., Witztum, A.: Abnormal stomatal behavior and ion imbalance in *Capsicum Scabrous Diminutive*. J. Exp. Bot. *27*, 111–118 (1976)
Thorpe, N., Milthorpe, F.L.: Stomatal metabolism: CO_2 fixation and respiration. Aust. J. Plant Physiol. *4*, 611–621 (1977)
Travis, A.J., Mansfield, T.A.: Studies of malate formation in 'isolated' guard cells. New Phytol. *78*, 541–546 (1977)
Turner, N.C.: Response of adaxial and abaxial stomata to light. New Phytol. *69*, 647–653 (1970)
Turner, N.C.: Action of fusicoccin on the potassium balance of guard cells of *Phaseolus vulgaris*. Am. J. Bot. *60*, 717–725 (1973)
Turner, N.C.: Stomatal behavior and water status of maize, sorghum, and tobacco under field conditions. II. At low soil water potential. Plant Physiol. *53*, 360–365 (1974)
Van Kirk, C.A., Raschke, K.: Presence of chloride reduces malate production in epidermis during stomatal opening. Plant Physiol. *61*, 361–364 (1978a)
Van Kirk, C.A., Raschke, K.: Release of malate from epidermal strips during stomatal closure. Plant Physiol. *61*, 474–475 (1978b)
Vassilyev, A.E., Vassilyeva, G.V.: The ultrastructure of the stomatal apparatus in gymnosperms (with special reference to the stomatal movements). Bot. Zh. (Leningrad) *61*, 449–465 (1976) (In Russian)
Voskresenskaya, N.P., Polyakov, M.A.: Enhancement of CO_2–gas exchange in the leaves of *Convallaria majalis* L., its action spectrum and light saturation. Plant. Sci. Lett. *5*, 333–338 (1975)
Walker, D.A., Zelitch, I.: Some effects of metabolic inhibitions, temperature, and anaerobic conditions on stomatal movement. Plant Physiol. *38*, 390–396 (1963)
Walton, D.C., Galson, E., Harrison, M.A.: The relationship between stomatal resistance and abscisic acid levels in leaves of water-stressed bean plants. Planta *133*, 145–148 (1977)
Wellburn, A.R., Ogunkanmi, A.B., Fenton, R., Mansfield, T.A.: All-*trans*-farnesol: a naturally occurring antitranspirant. Planta *120*, 255–263 (1974)
West, D.W., Gaff, D.F.: The effect of leaf water potential, leaf temperature and light intensity on leaf diffusion resistance and the transpiration of leaves of *Malus silvestris*. Physiol. Plant. *38*, 98–104 (1976)
Willmer, C.M., Dittrich, P.: Carbon dioxide fixation by epidermal and mesophyll tissues of *Tulipa* and *Commelina*. Planta *117*, 123–132 (1974)
Willmer, C.M., Mansfield, T.A.: A critical examination of the use of detached epidermis in studies of stomatal physiology. New Phytol. *68*, 363–375 (1969)
Willmer, C.M., Pallas, J.E., Jr.: A survey of stomatal movements and associated potassium fluxes in the plant kingdom. Can. J. Bot. *51*, 37–42 (1973)
Willmer, C.M., Pallas, J.E., Jr., Black, C.C. Jr.: Carbon dioxide metabolism in leaf epidermal tissue. Plant Physiol. *52*, 448–452 (1973)

Wright, S.T.C., Hiron, R.W.P.: (+)-abscisic acid, the growth inhibitor induced in detached wheat leaves by period of wilting. Nature (London) *224*, 719–720 (1969)
Yamashita, T.: Influences of potassium supply upon various properties and movement of the guard cell. Sieboldia, Acta Biol. *1*, 51–70 (1952)
Zabadal, T.J.: A water potential threshold for the increase of abscisic acid in leaves. Plant Physiol. *53*, 125–227 (1974)
Zeiger, E., Hepler, P.K.: Light and stomatal function: blue light stimulates swelling of guard cell protoplasts. Science *196*, 887–889 (1977)
Zelitch, I., Waggoner, P.E.: Effect of chemical control of stomata on transpiration and photosynthesis. Proc. Natl. Acad. Sci. U.S.A. *48*, 1101–1108 (1962)
Ziegenspeck, H.: Die Micellierung der Turgeszenzmechanismen. Teil I: Die Spaltöffnungen (mit phylogenetischen Ausblicken). Bot. Archiv. *39*, 268–309, 332–372 (1938)

4.2 Leaf Movements and Tendril Curling

R.L. SATTER

1. Introduction

Movements discussed in this chapter fall into three categories: rhythmic leaf movements in nyctinastic plants, more rapid seismonastic movements that occur in a limited number of nyctinastic species, and the thigmonastic or thigmotropic curling of thread-like appendages in certain vines. These movements all depend upon differential turgor changes in key cells in the organ of curvature, which may be the pulvinus of a nyctinastic plant, or the tendril of a vine. Although these organs have highly specialized structures, the regulatory mechanisms controlling curvature are similar to those controlling other aspects of growth and development. Studies of these movements have elucidated some basic scientific truths. Analyses of leaf movements in the early eighteenth century provided the first clue that organisms have internal clocks (DE MAIRAN, 1729), and subsequent studies are providing useful information on light–rhythm interaction in biological timekeeping. Investigation of seismonasty in *Mimosa pudica* and the carnivorous plants has revealed that electrical signals propagate through plants (BURDON-SANDERSON, 1873; BOSE, 1926; HOUWINK, 1935) as well as animals and are an important means of communication among the different cells of the plant body. For a discussion of seismonastic movements in carnivorous plants, see BENTRUP, Chapter 1.3, this volume.

The three types of movements described in this chapter vary considerably in rate, reversibility, and molecular control mechanisms. They range from leisurely light-modulated rhythmic movements which require a circadian period for completion, to rapid movements in *Mimosa*, completed 1–2 s after mechanical stimulation. They include localized responses in nyctinastic plants, where photoreceptors are probably in the same cells that undergo turgor changes, and translocated responses in thigmosensitive plants, where mechanoreceptors and motor cells are spatially separated. Nyctinastic and seismonastic movements in plants with pulvini are completely reversible, while tendril curling involves permanent changes in cell size and structure if the contact stimulus is not removed prematurely. Although none of the control mechanisms is completely understood, transmembrane fluxes of K, Cl, Ca, H, sugars and tannin-like compounds are implicated, as are changes in the titer or activity of auxin, ethylene, and possibly other hormones as well. Thus these movements reveal the diverse ways in which plants regulate cell size and shape.

Abbreviations. R: brief red irradiation, ca. 660 nm; FR: brief far-red irradiation, ca. 730 nm; H, Na, Cl, K, Ca, and Mg are used to designate these elements, even when measurements (e.g., pH) indicate the ionized form.

All of these topics were reviewed in Volume 17 of this series (BAILLAUD, 1959; BÜNNING, 1959a, b, c, d; UMRATH, 1959) and tendril curling was reviewed more recently (JAFFE and GALSTON, 1968b). Material described in these earlier chapters will be treated only briefly.

2. Leaf Movements in Nyctinastic Plants

2.1 Nature and Significance

2.1.1 Definitions

Plants whose leaves or leaflets assume a vertical orientation (close) in darkness are called *nyctinastic* from the Greek *nýktos* = night + *nastós* = closure. Such leaves normally assume a horizontal orientation (open) upon illumination. However leaves in many nyctinastic plants do not remain open or closed for more than ca. 12 h, oscillating between these two states with a circadian period in the absence of environmental pertubations; such movements are called *circadian*. Both light-stimulated and free-running movements are interrelated, since light-dark transitions determine the phase of the oscillations during subsequent free-running conditions. The term *rhythmic* is used in this chapter to describe movements under either of these conditions.

Nyctinastic movements may occur in immature leaves that lack pulvini, if abaxial and adaxial cells elongate during different parts of the diurnal cycle. They also occur in mature leaves and leaflets of certain members of the Leguminoseae and the Oxalidaceae. Movements in these plants are controlled by fully reversible changes in the size and shape of cells in the pulvinus at the base of the leaf or leaflet. Most of the discussion in this chapter relates to pulvinar rather than growth-regulated movements.

2.1.2 Direction of Movement

The five species that have been used most extensively in recent experiments, *Mimosa pudica*, *Albizzia julibrissin*, *Samanea saman*[1], *Trifolium repens* (clover) and *Phaseolus coccineus* (bean) will be designated by their generic names, except for use of *P. coccineus* (also called *P. multiflorus*) to avoid confusion with *P. vulgaris*.

Movements in *Mimosa*, *Albizzia* and *Samanea* are extremely complex. Each has doubly compound leaves that are divided into paired pinnae, in turn subdivided into paired pinnules (Fig. 1). Each leaf is subtended by a primary (1°) pulvinus, each pinna by a secondary (2°) pulvinus, and each pinnule by a tertiary (3°) pulvinus. Daytime movements, called opening, separate the paired pinnules (or pinnae) from each other and from the rachilla (or rachis), while sleep movements, called closure, bring the pinnules (or pinnae) toward each other and the rachilla (or rachis). However, the sleep movements of pinnules in all three plants and of pinnae in *Albizzia* and *Mimosa* occur in an acropetal direction, but those of *Samanea* pinnae occur in a basipetal direction. Although movements in these plants occur in two planes, they are usually monitored in only one. Published *Albizzia* data indicate the angle between the paired pinnules, while *Samanea* data

[1] Although the name of the genus was recently changed to *Pithecolobium*, I use the name *Samanea* in conformity with the literature cited.

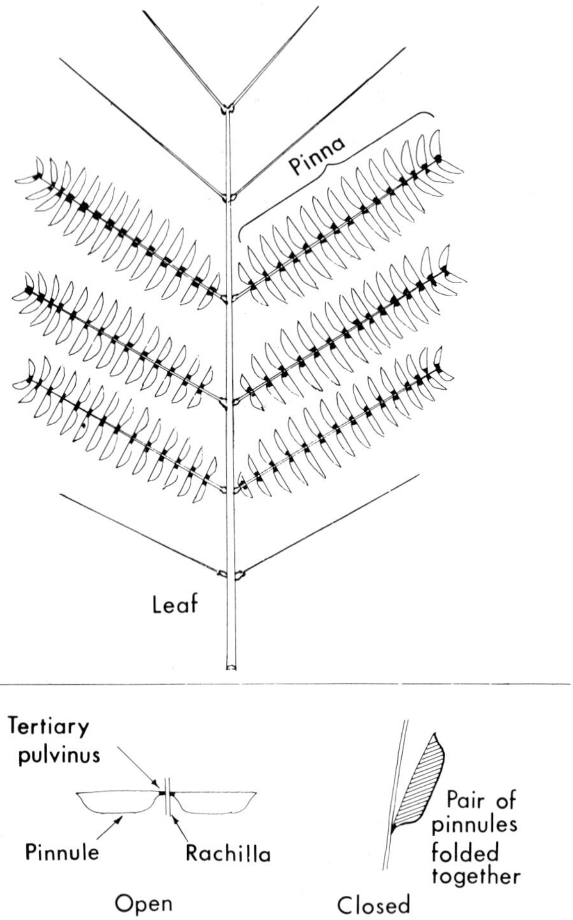

Fig. 1a and b. Leaves of **a** *Albizzia julibrissin* and **b** *Samanea saman*, open and closed. ▭ adaxial; ▨ abaxial. (Adapted from SATTER et al., 1970a and 1974a)

(except for those of SWEET and HILLMAN, 1969) indicate the angle between the rachis and rachilla.

The leaves of *Trifolium* are trifoliate, with each leaflet attached to the petiole by a pulvinus. Movements of the two lateral leaftlets resemble those of *Samanea* pinnae, but the terminal leaflet, used exclusively in recent experiments, moves in a single plane, through an angle of 180° or more. The adaxial surface of the terminal leaflet faces the sky during the day, and faces the ground at night (see Fig. 3 in BÜNNING, 1959d).

The paired, primary, unifoliate leaves of *Phaseolus* contain two pulvini; movement of the 1° pulvinus at the base of the petiole controls the angle between the petiole and the stem, while movement of the 2° pulvinus at the base of the lamina controls the angle between the lamina and the petiole. The angle of motion at the 2° pulvinus exceeds that at the 1° pulvinus and thus has been monitored in most experiments. Data refer to this, unless otherwise indicated. Most of its

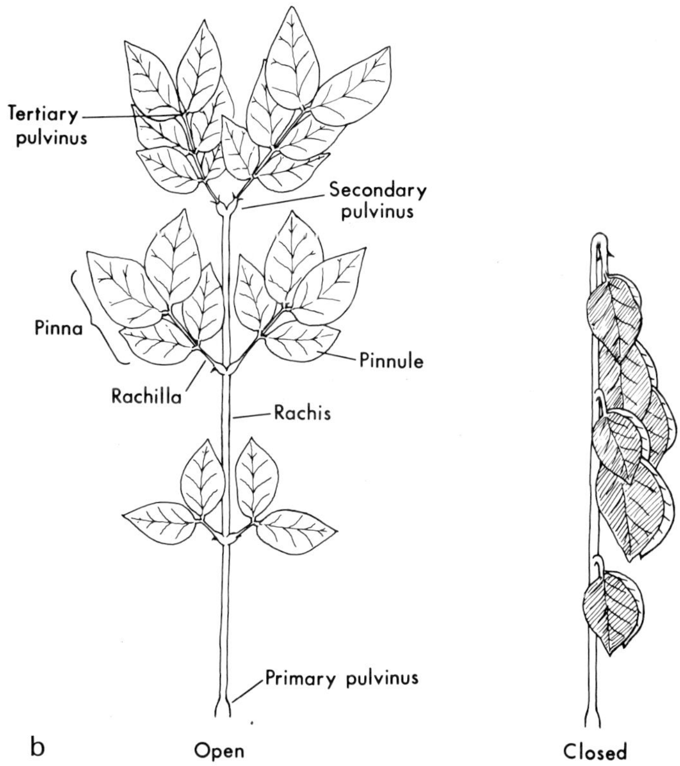

Fig. 1b

movement occurs in a single plane, although high frequency oscillations of the lamina in a lateral direction have been detected (G. GARDNER, personal communication).

2.1.3 Automatic Monitoring Systems

Leaf-movement data have been obtained by manual measurement made at periodic intervals, or by one of several automatic recording techniques involving: a kymograph and leverage system (PALMER and ASPREY, 1958a; BÜNNING and ZIMMER, 1962; BÜNNING and MOSER, 1966a), a strain gauge method (HOSHIZAKI and YOKAYAMA, 1965; ODA and ABE, 1972; AIMI and SHIBASAKI, 1975), time-lapse photography (HOLDSWORTH, 1964; HOSHIZAKI and HAMNER, 1964), a photoelectric system (TIBBITTS et al., 1973; ENGELMANN et al., 1973; APPLEWHITE et al., 1978), or a motor-driven height-sensing unit (HALABAN, 1968). A null method was used by SCOTT and GULLINE (1972), who monitored the force required to prevent movement.

2.1.4 Stimuli that Evoke Movement

Movements regulated by light interacting with a rhythmic oscillator occur each day at each pulvinar junction, in each of the species described above. Under free-running conditions, the movements have a circadian period, with low amplitude higher frequency components of 13.8 h (*P. vulgaris* maintained at 15 °C to 20 °C, HOSHIZAKI and HAMNER, 1969), 4.8 and 8 h (*Trifolium*, SCOTT and GULLINE, 1972), 4–5 h (*Samanea*, SIMON et al., 1976a), 0.5–1 h (*Mimosa*, FONDÉVILLE, 1965), or

0.5–3 h (*Gossypium hirsutum*, MILLER, 1975). Oscillations in the latter species are growth-regulated and dependent upon moisture stress.

Some pulvini are also geotropically sensitive. In *Mimosa*, 3° pulvini are insensitive to gravitational stimulation, 2° pulvini react to gravity during certain parts of the rhythmic cycle, while 1° pulvini are reactive at all times (GAVAUDAN and ROBLIN, 1969). The leaves of a plant that has been inverted 180° reorient within 2 h; if the plant is then reinverted to its original position, the leaves reorient once again. Since triiodobenzoic acid (0.05–0.1 mM), inhibitor of auxin transport and action (CHATTERJEE and LEOPOLD, 1963), interferes with the response (ROBLIN, 1976a), auxin might be an endogenous regulator.

In addition to these leisurely circadian and gravitational movements, rapid or seismonastic movements also occur in *Mimosa* and a few other species, when plants are stimulated mechanically, i.e., by shaking or touch. These rapid movements can also be evoked by completely different stimuli such as electrical or temperature shock, chemical treatment, or high intensity light. Independent of the nature of the stimulus, the rapid movements will be called "seismonastic" in this chapter.

2.1.5 Adaptive Value

The adaptive value of the horizontal daytime position which orients the leaves for maximum absorption of sunlight is obvious, but the function of the sleep movements is less apparent. DARWIN (1881) hypothesized that sleep movements prevent excessive heat loss to the open sky at night. SMITH's (1974) data, obtained with a rosette plant growing at a high altitude, support DARWIN's hypothesis, but nyctinastic movements in soybean have only a minor effect on night leaf temperature, and no measurable effect on frost hardiness (SCHWINTZER, 1971).

According to BÜNNING and MOSER (1969), nyctinastic movements which bring paired leaflets together so that their upper surfaces shade each other (Fig. 1) prevent moonlight from interfering with time measurement in photoperiod-sensitive plants. The intensity of light from a full moon on a cloudless night may reach 0.3 lux at a latitude of 50°, and more than 3 times this value in tropical regions, yet intensities as low as 0.1 lux can influence time measurement in some plants. Sleep movements in soybean, peanut and clover reduce the intensity of moonlight incident on the upper leaf surface by 85% to 95%. Furthermore, nyctinastic movements in 12 varieties of soybean are correlated with photoperiod sensitivity: varieties that require short days for floral initiation have pronounced leaf movements, while day neutral varieties have very weak movements (BÜNNING, 1948).

Experiments with periclinal chimeras composed of the epidermis of one nyctinastic solanaceous species and internal tissue from another, led MAYER et al. (1973) and JUNKER and MAYER (1974) to conclude, likewise, that the epidermis is important for light perception. Their experiments tested the effect of light and temperature on the phase and the effect of light on the frequency of the leaf movement rhythm of the chimeras and their constituent species. They reported that the internal tissue of the leaf acts as motor tissue, while the photoreceptors for phase shifting are localized in the epidermis. The epidermis also affects the light intensity dependence of rhythmic frequency. These reports, together with the finding that the photoreceptor for photoperiodic floral induction is also located in the epidermis of the leaf (SCHWABE, 1968; BÜNNING and MOSER, 1966b; MAYER et al., 1973) support BÜNNING and MOSER's interesting hypothesis, but more data are needed to test its general applicability.

The function of seismonastic movements in *Mimosa* is less debatable. It is generally agreed that these movements discourage predation by making the leaves inconspicuous and exposing thorns on the stems. The secretion of astringent tannin-like compounds through hydathodes on the epidermis of the pulvinus might also ward off potential predators (TORIYAMA, 1954b).

2.2 Structure of the Pulvinus

(see earlier reviews by BAILLAUD (1959) and BÜNNING (1959b, d))

2.2.1 General Features

The pulvinus has a cylindrical shape when leaves are horizontal, with bending facilitated by transverse furrows on the adaxial and abaxial sides of the tissue. It contains a central vascular core with both xylem and phloem surrounded by sclerenchyma, in turn surrounded by 10 to 20 layers of parenchyma (Fig. 2). Cells in the parenchyma, especially those in the inner region, are connected by numerous plasmodesmata, largely confined within pit fields (CHAUDHRI et al., 1977). Cells in the inner cortex are small and rich in cytoplasm, those in the mid cortex are much larger but contain relatively little cytoplasm, while those in the outer cortex are small but show large changes in size and shape during leaf movement. All cortical cells are called motor cells in this chapter, although those in the outer zone are motor cells in the strictest sense, since their shape and volume changes control leaf movement (Fig. 3).

Two types of chloroplasts have been described in the 1° pulvinus of *Mimosa* (FLEURAT-LESSARD and GAVAUDAN, 1973) and the 2° pulvinus of *Samanea* (MORSE and SATTER, in preparation). Numerous round, swollen chloroplasts with very large starch deposits but few grana are localized in the vascular tissue and in one or two surrounding layers of cortical cells; FLEURAT-LESSARD and GAVAUDAN propose that these plastids are statoliths controlling the response to gravitational stimulation. Chloroplasts in the remainder of cortex have a lens-shaped appearance, and contain more grana. In *Samanea*, these plastids contain very large starch granules at the beginning of the dark period, but are virtually devoid of starch at the beginning of the light period. The swollen centrally localized chloroplasts, by contrast, are filled with starch at all times.

Stomata have been detected on the surfaces of pulvini in many species, but they are always closed, and lack the underlying intercellular air spaces that characterize functional stomatal complexes (TORIYAMA, 1954b). In *Mimosa*, numerous hydathodes populate the upper surface of the 1° pulvinus and the "transitional zone" which lies between the pulvinus and the petiole or stem. Some of these form a continuum with intercellular spaces, which are unusually large (TORIYAMA, 1954a). These spaces in *Mimosa* serve an important function, since K, tannins and colloidal substances secreted from shrinking cells during seismonastic movement diffuse through them to the exterior of the tissue (TORIYAMA, 1955b).

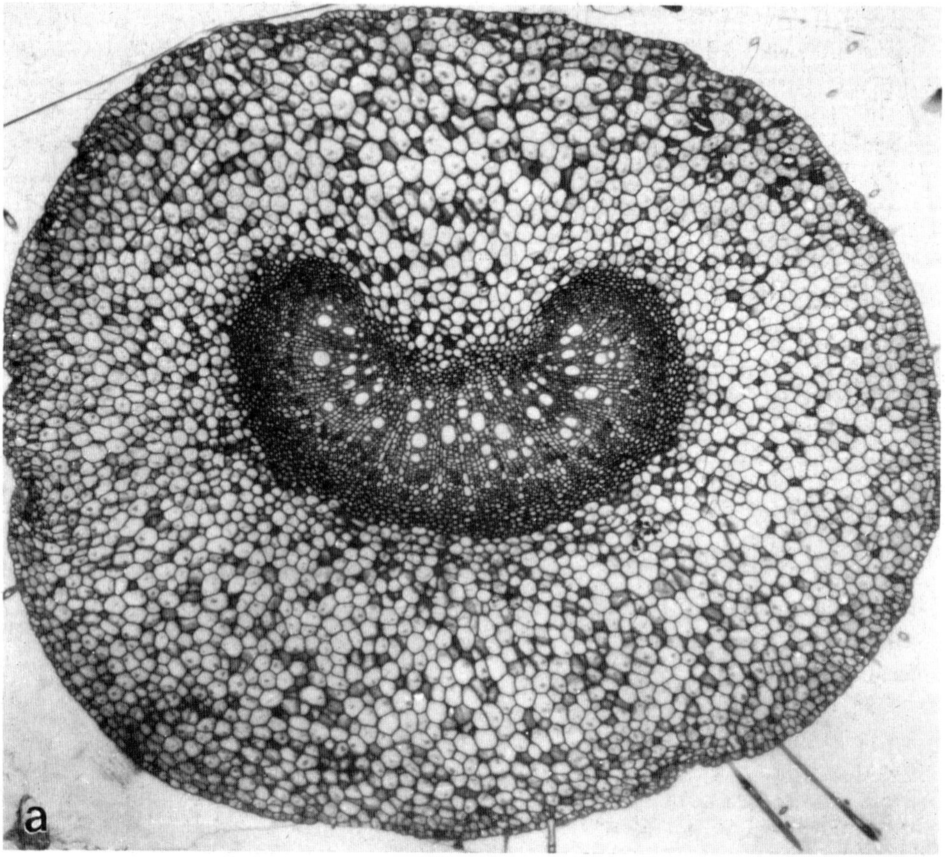

Fig. 2a. Cross section of the *Samanea* 2° pulvinus. The tissue was fixed in 2% formaldehyde and 3% glutaraldehyde, post-fixed in 1% osmium tetroxide and stained with toluidine blue. ×80

2.2.2 Extensor and Flexor Regions

Various authors describing plant movements have given the terms "ventral" and "dorsal" conflicting meanings. This is not surprising, since some dictionaries and textbooks give the correct botanical definition (ESAU, 1953), i.e., the upper or adaxial surface of the leaf is ventral, and the lower or abaxial surface is dorsal, while others use the terminology commonly employed by zoologists, i.e., ventral is lower, while dorsal is upper.

The fact that some leaves move upward at night while others move downward, and that some movements occur in a single plane and others in two planes creates additional confusion, for cells that undergo the largest turgor decrease during closure and increase during opening may be located at mid-abaxial, mid-adaxial, or some intervening region. To facilitate comparison of movements at different pulvini, the terms "extensor" and "flexor" have been used (SATTER et al., 1974a). These regions are defined by function rather than location. They have a different meaning than when used by animal physiologists.

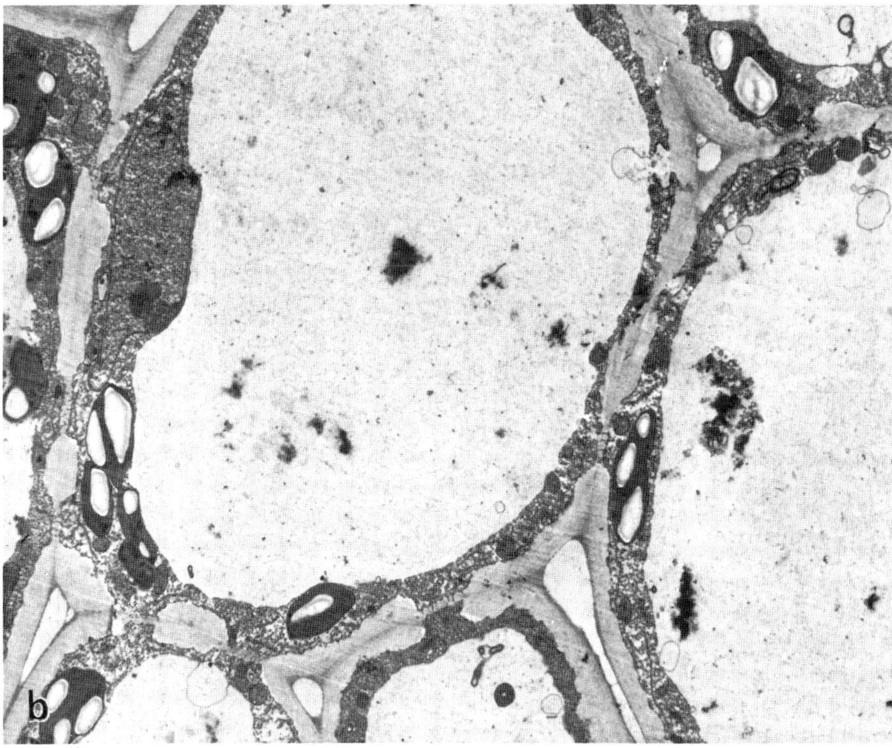

Fig. 2b. Electron micrograph of cells in the inner cortex. Pitfields, each containing an unusually large number of plasmodesmata, are evident in all cell walls. The vacuoles contain a large amount of condensed material, as also reported in pulvini of other species. ×4240. (Courtesy of J. Chaudhri)

Extensor motor cells increase in turgor during leaf opening (sometimes called pulvinar opening) and decrease in turgor during closure, while flexor motor cells change in the opposite manner. The terms extensor and flexor will be used where practical in this review, even though the original investigators may have used the terms upper and lower, ventral and dorsal, or adaxial and abaxial. The terms adaxial and abaxial will, however, be used to describe movement at the 1° pulvinus of *Mimosa*. Its rapid seismonastic movement is basipetal as is its early nocturnal movement, but its later nocturnal movement is acropetal (Vines, 1886, described by Palmer and Asprey, 1958a). Both seismonastic and sleep movements at the 2° and 3° pulvini of *Mimosa* are acropetal.

2.2.3 Motor Cell Structure and Ultrastructure

Cell Walls. In the 1° (Toriyama, 1954a) and 3° (Weintraub, 1952) pulvini of *Mimosa*, cells that lose turgor during seismonastic movement have unusually thin and elastic walls, permitting large, rapid, reversible changes in cell size and shape; by contrast, cells on the opposite side of the pulvinus have thicker walls, consistent with their smaller changes in volume. Weintraub's analysis of wall composition

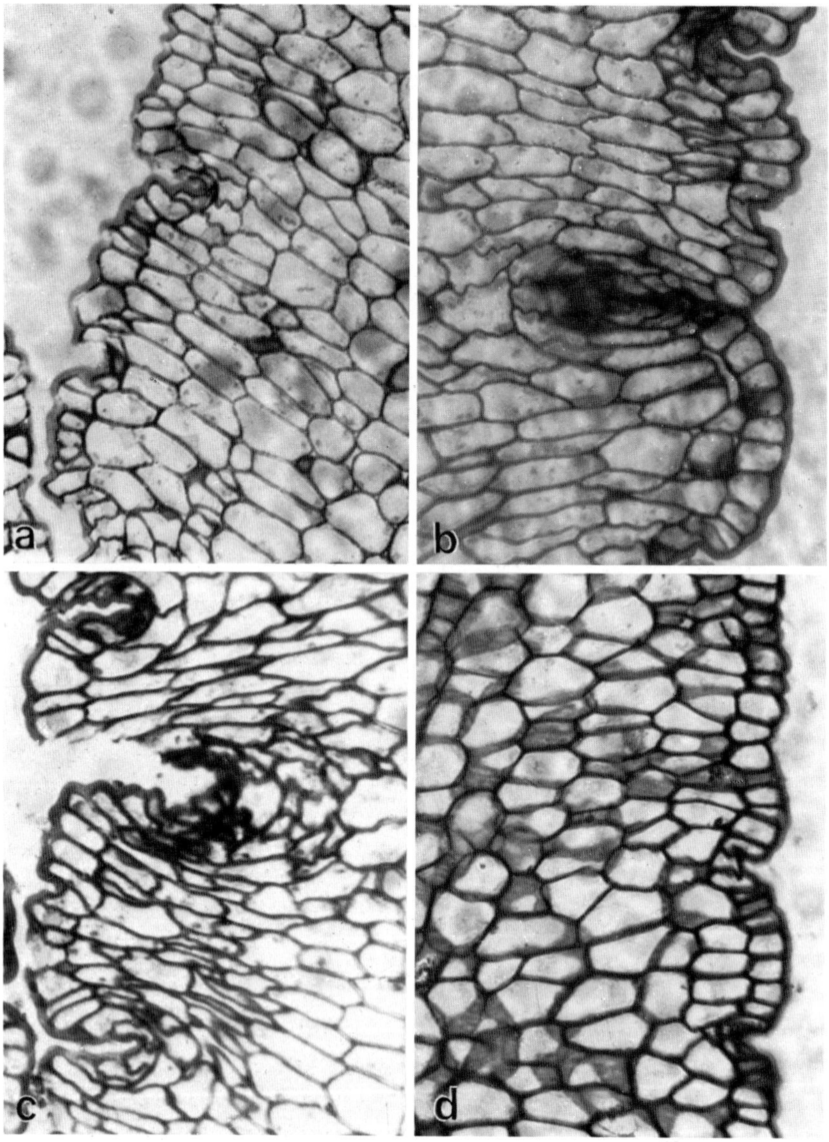

Fig. 3a–d. Abaxial-adaxial longitudinal section of the outer region of the cortex of the 3° pulvinus of *Albizzia*. The numbers refer to the average area of cells in the outer 3–4 layers. **a** Open, adaxial (extensor, 160 μ^2); **b** open, abaxial (flexor, 176 μ^2); **c** closed, adaxial (extensor, 83 μ^2); **d** closed, abaxial (flexor, 225 μ^2). ×420. (From SATTER et al., 1970a)

reveals that the walls of extensor cells in the 3° pulvinus, like those of juvenile cells in other plants, contain a compound, probably protein, which masks the usual cellulose reaction. Furthermore, the middle lamella is composed primarily of pectic acid, which is more pliable than calcium pectate, a more common wall substance. More recently, TORIYAMA and JAFFE (1972) reported that Ca migrates out of the walls of adaxial cells in the 1° pulvinus of *Mimosa* during seismonastic movement, and they suggested that loss of Ca would reduce wall rigidity, thereby facilitating the bending movement.

Leaf movements would not take place if the cell walls were not able to distort as cell turgidity changes, yet this aspect has received relatively little attention. The development of new techniques for analyzing cell wall composition (KEEGSTRA et al., 1973) should stimulate further investigation of this interesting phenomenon.

Vacuoles. Since vacuoles occupy ca. 90% of the volume of most plant cells, changes in cell size are largely due to vacuolar expansion and contraction. WEINTRAUB (1952) witnessed the shrinking of extensor cells through the light microscope, using thick fresh sections of the 3° pulvinus of *Mimosa*. These cells have a large central vacuole that contracts following seismonastic stimulation. Contraction is accompanied by the successive appearance and disappearance of numerous small vacuoles, ca. 1 μm in diameter. Pre-absorbed dye molecules remain sequestered in the central vacuole, which becomes more highly pigmented as it contracts, while the newly formed small vacuoles are colorless. These observations are consistent with the suggestion that the small vacuoles derive their water from the cytoplasm, which in turn derives its water from the central vacuole. Water from the small vacuoles is then secreted to the cellular exterior, but the mechanism remains obscure.

Numerous small vacuoles have also been observed with the electron microscope in motor cells of *Albizzia* (SATTER et al., 1970a), *Mimosa* (CAMPBELL and THOMSON, 1977a) and *Samanea* (MORSE, in preparation). The vacuoles in *Albizzia* and *Mimosa* range in size and number per cell, but are smallest and most numerous in subepidermal cells, which undergo the largest changes in volume during leaflet movement. Spherosome-like bodies are also present in the motor tissue of *Albizzia* and are preferentially localized in the multivacuolate subepidermal cells. These bodies, which are not clearly bounded by a unit membrane although their edges are moderately electron dense, seem to coalesce and give rise to the vacuoles.

It is interesting to note that stomatal guard cells, which also undergo large daily reversible turgor changes, are multivacuolate in some species (THOMSON and DE JOURNETT, 1970; see also RASCHKE, Chap. 4.1, this vol.). Furthermore, the large central vacuole in the tentacle of *Drosera rotundifolia* fragments into numerous small vacuoles as the tentacle curls following a touch stimulus (AKERMANN, 1917). The possibility that large vacuoles in motor cells of nyctinastic plants also fragment when cells shrink and reform when cells swell, is discussed in Section 2.6.2.

Tannins. Tannins, commonly stored in cell vacuoles, are unusually abundant in vacuoles of motor cells of rapidly motile plants such as *Mimosa* and the carnivorous plants *Dionaea* and *Drosera* (TORIYAMA, 1973). Based on observations with the light microscope, TORIYAMA (1953, 1954a, 1955a, 1957, 1958) proposed that tannins in the 1° pulvinus of *Mimosa* are confined within membranous sacs, called tannin vacuoles, that occupy much of the volume of the central vacuole in many but not all motor cells. This view was challenged by DATTA (1959) who confirmed the existence of massive tannin aggregations in the vacuoles but disagreed that they are bound by membranes. Subsequent examinations with the electron microscope support Toriyama's earlier view (TORIYAMA, 1973; TORIYAMA and SATÔ, 1968a, b, c, 1969, 1971) and reveal that tannins are also surrounded by membranes in other species (CHAFE and DURZAN, 1973). Tannin bodies in the 1° pulvinus of *Mimosa* change shape and size following mechanical stimulation, and tannin-like compounds together with K and a colloidal substance, are secreted to intercellular spaces, and from there to hydathodes on the upper epidermis (TORIYAMA, 1955b, 1957).

Tannins have also been reported in vacuoles of motor cells of non-seismonastic nyctinastic plants, although they are less abundant and more highly dispersed in these plants than in *Mimosa* (DATTA, 1959). It is of interest to note that the size and shape of tannin aggregations in both seismonastic and non-seismonastic species change from day to night (DATTA, 1959; TORIYAMA, 1954b, 1973). The size of tannin vacuoles in *Mimosa* is also regulated by phytochrome conversion (SETTY and JAFFE, 1972).

It is difficult to assess the physiological significance of tannins in regulation of leaf movements. TORIYAMA (1973) hypothesizes that they play a fundamental regulatory role in *Mimosa*, complexing Ca when the plant is in its resting state, but releasing it when the plant is stimulated mechanically; such release then triggers a series of processes that lead to vacuolar contraction. However, such a role is difficult to reconcile with TORIYAMA's earlier (1955a) data revealing that tannin deposits are not present in very young pulvini that nevertheless respond to mechanical stimulation. Tannins might function as osmotic agents that aid in turgor regulation (DATTA, 1959; TORIYAMA and KOMADA, 1971) or might release sugars that provide energy for the movements (TORIYAMA, 1954a).

The term "tannin" describes a large heterogenous group of compounds, and techniques for chemical identification of many tannins are available (HASLAM, 1966). Knowledge of the structure of tannins localized in motor cells of nastic plants, and of molecular changes in these molecules during leaf movement, might aid in resolving their possible role in leaf-movement regulation.

Fibrils. A most unusual feature of *Mimosa* motor cell ultrastructure has been reported by TORIYAMA and SATÔ (1968b, 1969, 1971). They describe narrow fibrils in the central vacuole that change appearance after the pulvinus has been stimulated mechanically. These fibrils, which are 20 to 30 nm thick and 200 to 400 nm long before stimulation, are attached to the tonoplast and to the surface of the tannin vacuole (Fig. 4). After stimulation, the tonoplast collapses, while the fibrils fragment and aggregate near the tannin vacuole, which now appears crushed. Noting that the fibrils dissolve in Weber-Edasall solution, a preparation that presumably removes contractile proteins, TORIYAMA and SATÔ proposed that a contractile process occurs, as discussed in Section 2.6.3.

Numerous fibrils oriented parallel to each other and presenting an organized appearance have also been reported in the cytoplasm of motor cells of *Mimosa* (FLEURAT-LESSARD and GAVAUDAN, 1973) and *Albizzia* (SATTER et al., 1970a). Fibrils in *Mimosa* are of particular interest, since the protoplasm pulls away from the cell wall (WATANABE, 1970) and shrinks following mechanical stimulation, aggregating near the crushed tonoplast (TORIYAMA and SATÔ, 1968a, b).

The preceding pages contain numerous references to the work of TORIYAMA and colleagues. His carefully designed experiments and creative use of a variety of staining methods enabled him to discern many unusual features of pulvinar structure, some of which have not yet been fully interpreted (e.g., TORIYAMA, 1962a). As we learn more regarding pulvinar function, his papers will continue to remain important references for finding structural correlates.

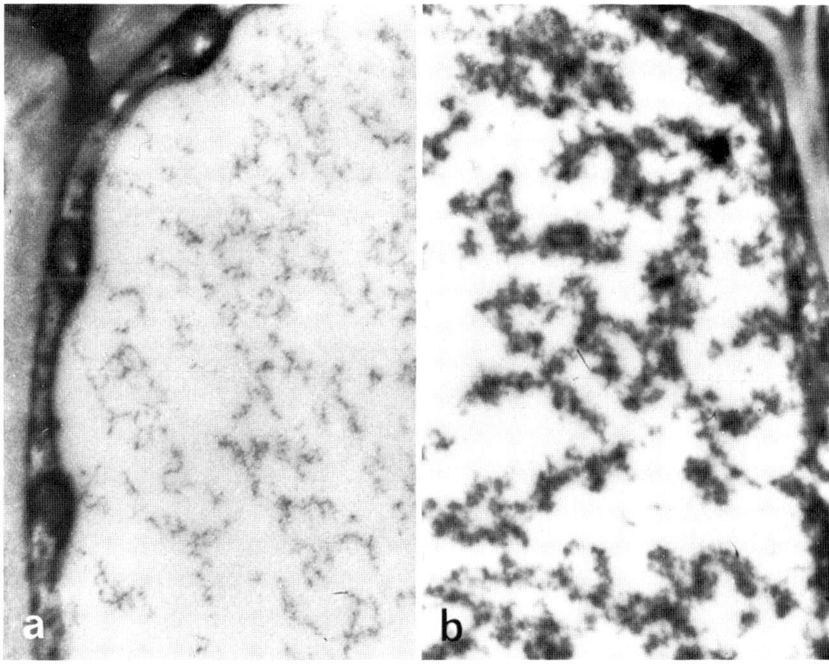

Fig. 4a and b. Fibrils in the central vacuole of a *Mimosa* motor cell. **a** Before stimulation; **b** after stimulation. ×18,000. The tissue was fixed with OsO_4. (From TORIYAMA and SATÔ, 1968c)

2.3 Mechanics of Leaf Movement

2.3.1 Roles of Adaxial and Abaxial Cells

Both seismonastic and rhythmic movements are controlled by changes in the size and/or shape of cells on opposite sides of the pulvinus. To determine whether both upper and lower cells actively expand and contract, or whether cells on one side of the pulvinus respond passively to changes in the volume of cells on the opposite side, selected cells have been removed. Such dissections have very different effects on the two types of movement.

Seismonasty. The petiole of *Mimosa* still moves basipetally in response to seismonastic stimulation and recovers in a normal fashion if adaxial pulvinar cells are removed, but it fails to recover without abaxial cells (data summarized by PFEFFER, 1905), indicating that the latter are necessary for the movements, but the former are not. In fact, the adaxial motor cells might even hinder petiole movement, since these cells contract in response to mechanical stimulation under certain experimental conditions (AIMI, 1960). Water released from abaxial cells moves through intercellular channels to the transitional zones that flank the pulvinus, and from there to hydathodes on the epidermis of the pulvinus (DUTT and GUHA THAKURTA, 1956).

Rhythmic Movements. By contrast, the expansion and contraction of cells on *both* sides of the pulvinus contribute toward regulating rhythmic movements. Extensor cells in *Samanea* continue to swell and shrink after flexor cells have been surgically removed; flexor cells behave similarly after dissection of extensor cells (PALMER and ASPREY, 1958b). Extensor cells, however, characteristically show larger volume changes during the diurnal cycle than do flexor cells. For example, the average fresh weight of cells in the extensor half of the *P. vulgaris* pulvinus is 48% higher at noon than at midnight, while that of cells in the flexor half is only 27% higher at midnight than at noon (KIYOSAWA and TANAKA, 1976). Changes in the volume of the *motile* cells are much larger than these average values, which include the relatively stable inner cortical and vascular cells.

2.3.2 Ion Movement: Methodology and Major Conclusions

During the past two decades, it has become clear that leaf movements are correlated with massive migration of K and Cl within the pulvinus. Evidence has been obtained using histochemical techniques, radioisotopes, flame photometric analysis, and electron microprobe analysis. Each of these methods has provided information not available by the other methods, and thus has made a unique contribution to our understanding of the turgor changes that regulate leaf movement (reviewed by MACROBBIE, 1977).

Histochemical Methods. TORIYAMA (1955b) was the first to demonstrate a relationship between K flux and leaf movement, by use of two histochemical methods: (1) Tissue from the 1° pulvinus of *Mimosa* was fixed in Scott's solution, embedded in paraffin, sectioned and then incinerated; K ash was identified by its blue pigmentation. (2) Tissue was fixed in McCallum's modified solution containing Na-cobaltinitrite, which precipitates K in situ. Both methods reveal that K is concentrated near tannin deposits in the central vacuole of abaxial motor cells prior to mechanical stimulation, but after stimulation it is found in the intercellular spaces outside these cells, and in the transitional zones that flank the pulvinus. Although the cobaltinitrite method has been criticized as permitting K displacement (VAN STEVENINCK 1976), TORIYAMA's conclusions are supported by ALLEN's (1969) experiments with ^{42}K (see Radiochemical Methods below), and by reports (DUTT and GUHA THAKURTA, 1956) that K salts are secreted through hydathodes on the epidermis during seismonastic movement.

Calcium movement in the 1° pulvinus of *Mimosa* was also monitored histochemically (TORIYAMA and JAFFE, 1972). The tissue was fixed with Lillie's neutral buffered formalin and stained with alizarin red sulfate, which complexes with Ca, giving a red pigmentation. These experiments reveal that Ca migrates from the tannin vacuole to the central vacuole following a mechanical stimulus, and a small amount of Ca is also released to the central vacuole from the walls of the upper cells.

Chloride can also be localized histochemically, by treating tissue with $AgNO_3$, which precipitates Cl in situ. Using this method, MAYER (1977) demonstrated circadian changes in the Cl content of the mid-flexor region of the *P. coccineus* pulvinus.

Radiochemical Methods. ALLEN (1969) pre-loaded the 1° pulvinus of *Mimosa* with ^{42}K and then measured efflux of label into the bathing solution following mechanical stimulation. The rate of efflux increased more than 100% following application of the stimulus, providing further evidence that K is secreted from motor cells during seismonastic movement.

Both ^{42}K and ^{45}Ca have been used to study the uptake of elements from the bathing solution into the pulvinus of *Trifolium* (SCOTT and GULLINE, 1975; SCOTT et al., 1977). After leaflets had floated on labeled solution for 2 h, the pulvinus was excised, washed, and separated into extensor and flexor halves prior to counting. Although the rate of ^{45}Ca uptake is enhanced by 40% in white light, it remains constant when light conditions are constant, and thus was used as a standard for determining temporal variations in the rate of ^{42}K uptake. The rate of $^{42}K/^{45}Ca$ uptake varies rhythmically in continuous light, with rhythms in extensor and flexor cells 180° out of phase with each other (Fig. 5). Uptake by cells in each region increases about 3 h prior to the onset of cellular swelling and remains high until about 3 h before shrinkage begins, then decreases and remains low for the remainder of the cycle. These data suggest that K movement through motor cell plasmalemma membranes is rhythmic, although rhythmic uptake of K by cells in the lamina followed by diffusion through the symplast to pulvinar cells, is not excluded. Rhythmic ionic absorption and release from the cell walls is another possibility.

Flame Photometry. The concentrations of K, Na, Ca, and Mg in extracts of tissue from *Albizzia* (SATTER et al., 1970b), *Samanea* (SATTER et al., 1974a), *P. vulgaris* (KIYOSAWA and TANAKA, 1976), *P. coccineus* (MAYER, 1977) and *Trifolium* (SCOTT, et al., 1977) have been measured with the atomic absorption spectrophotometer. K levels are unusually high, particularly in the pulvinus. *Samanea* tissues contain, per g dry wt.: 50 mg K (2° pulvinus), 62 mg K (3° pulvinus), 30 mg K (rachilla) and 25 mg K (lamina); *Albizzia* values are similar. The concentration of K in the pulvinar sap of *Albizzia* and *Samanea* is ca. 0.4–0.5 M. Total K in pulvini of *Albizzia*, *Samanea* and *Phaseolus* does not change during the circadian cycle, indicating that K movements are confined within the pulvinar microstructure. During seismonastic movement in *Mimosa*, by contrast, K migrates from the pulvinus to the petiole and stem (TORIYAMA, 1955b; ALLEN, 1969).

In *P. vulgaris* grown in 12-h light, 12-h dark cycles, K/mg protein in the flexor half of the 2° pulvinus is 38% higher at midnight than at noon, while that in the extensor half is 60% higher at noon than at midnight. However, K/mg fresh wt. in both extensor and flexor halves changes only 8% during the light–dark cycle, indicating that H_2O movement accompanies K movement, so that the concentration of K in the cell sap remains relatively constant. Na, Ca and Mg are much less concentrated than K in all pulvini analyzed, and rhythmic changes in these ions were not detected.

The simple technique of flame photometric analysis has provided useful information on the concentration of cations in the cell sap, and on ion movements between extensor and flexor. However, these data and those obtained in uptake studies obscure the large variations within the extensor and the flexor regions.

X-Ray Analysis with the Electron Microprobe. The electron microprobe has provided the most detailed information available on ion movements in pulvini of nyctinastic plants. The instrument contains an electron beam that is focused on a section of tissue, and excites atoms within the tissue to emit X-rays (ANDERSON, 1967). The X-rays, whose frequency is characteristic of the emitting element, are collected in tuned spectrometers, and converted to scintillations. Qualitative information on the distribution of a single element within a region of the tissue can be obtained by scanning with a narrow beam, and recording the scintillation pattern photographically. To obtain quantitative information on the simultaneous distribution of several elements, a stationary beam of somewhat larger diameter is focused on cells of interest and the number of scintillations emitted by each element during a specified time interval is recorded. The beam is then refocused on other regions of the tissue for equivalent time periods. Maps showing the distribution of elements within the pulvinus can be constructed from these data. This technique has been used to analyze elements in thick (24–40 μm) longitudinal (Fig. 6) and cross-sections of pulvini of *Albizzia* (SATTER and GALSTON, 1971a, b; SATTER et al., 1970b, 1972a, 1974c; SCHREMPF et al., 1976) and *Samanea* (SATTER et al., 1974a, b, 1977); reviewed by GALSTON and SATTER (1972, 1976) and SATTER and GALSTON (1973).

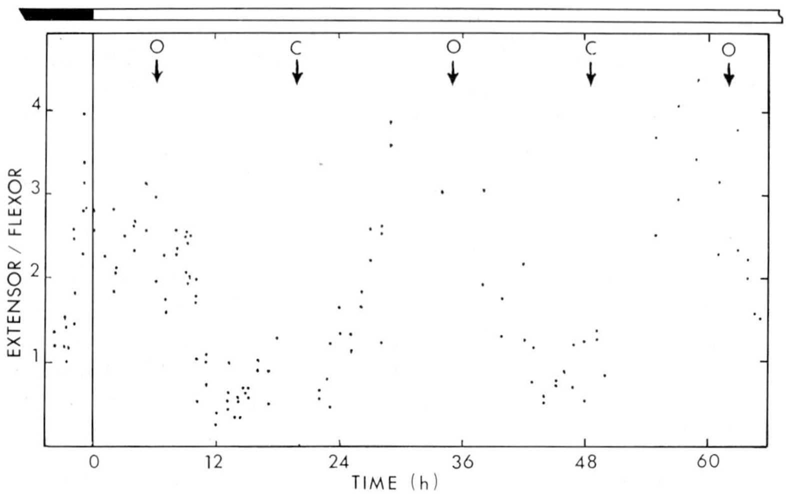

Fig. 5. Potassium: calcium influx ratio in *Trifolium* pulvini during continuous light. *Arrows* indicate the times of mid-opening (*o*) and mid-closure (*c*). (Adapted from SCOTT et al., 1977)

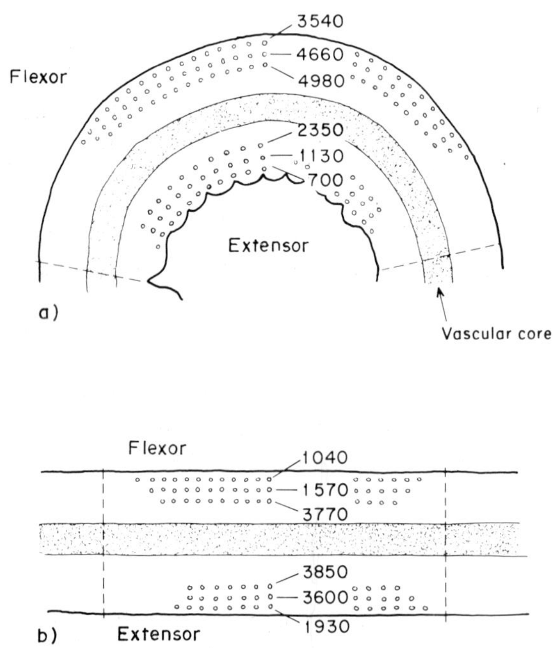

Fig. 6. K distribution in longitudinal sections of a *Samanea* pulvinus that is **a** closed in the dark or **b** open in white light. K was analyzed in the circled regions with an electron microprobe. The numbers represent average number of scintillations/15 s in the outer cortex (30 μm from the epidermis) and mid-cortex (100 and 225 μm from the epidermis). (From SATTER et al., 1974a)

Whether plants are in white light or in darkness, and whether the latter is interrupted by R or by FR, the distribution of K and Cl in the pulvinus of *Samanea* always correlates with leaflet angle; neither Ca nor P exhibits this behavior (see Fig. 1 and Table 1 in SATTER et al., 1977). Both K and Cl, high at mid-flexor when leaflets are closed, migrate toward mid-extensor during opening, and in the reverse direction during closure. Ion fluxes are massive, especially in the extensor, outer cortex, where K varies 15-fold and Cl 21-fold as leaflets oscillate between the open and closed states. It is of interest that the K and Cl content of the inner cortical tissue changes almost as dramatically as does that of the outer cortex, even though the inner cortical cells maintain relatively stable volume during leaflet movement. Both K and Cl are usually higher in the inner than in the outer cortex, but occasionally the situation is reversed, suggesting that ions might shuttle between cells in these two regions.

The content of Cl in the pulvinus is 60% that of K in *Samanea* (SATTER et al., 1977), and slightly lower in *Albizzia* (SCHREMPF et al., 1976). Therefore a large proportion of the K ions must be balanced by anions other than Cl, or by fixed negative charges in the cellular interior or in the walls. Fixed charges might be particularly significant in the flexor region, where a large fraction of the K appears to be immobile.

Although results published to date do not differentiate between intracellular and extracellular ions, new higher-resolution techniques being developed at present should permit such discrimination. Methods which have been used successfully for subcellular localization of elements in other tissues, involve freeze substitution (LÄUCHLI et al., 1970) or ultra-thin sectioning in a cryostat ultramicrotome (APPLETON, 1974), followed by analysis in a scanning or transmission electron microscope equipped with an energy-dispersive X-ray detector.

2.3.3 K, Cl and Motor Cell Turgor

Data summarized above reveal consistent correspondence between the K content and the shrinking and swelling of motor cells in all species examined, by all methods utilized; wherever tested, Cl moves with K, balancing ca. 65–100% of its ionic charge. We might expect this correlation between the ions and water (1) if they move together by a bulk flow mechanism, or (2) if ion flux precedes water movement, which then occurs by osmosis. Evidence that massive changes in K and Cl take place in all cells in the cortex of the *Samanea* pulvinus, although only cells in the outer region undergo large volume changes, argues against the former possibility (SATTER et al., 1977). Thus an osmotic mechanism is the most reasonable hypothesis. Changes in tannins, sugars, carbohydrates and macromolecules in motor cells probably also contribute to osmotic regulation, particularly during seismonastic movement (DATTA, 1959; TORIYAMA and KOMADA, 1971).

2.4 Circadian Movements

Most nyctinastic plants growing in their native habitats begin to open their leaves a few hours before dawn, and to close them a few hours before dusk, although the rate of opening is accelerated and the final angle increased by light, while the rate of closure is accelerated and the final angle decreased by darkness (PALMER and ASPREY, 1958a). In continuous darkness or in light of constant intensity, leaflets alternately open and close with a circadian periodicity. Thus light promotes opening and darkness promotes closure, but a circadian oscillator exerts overriding control. Transitions between light and darkness act as "Zeitgebers" for entraining the oscillations (HOSHIZAKI et al., 1974).

Leaf movements have been monitored in situ, and in explants floated on aqueous solutions (*Albizzia* and *Trifolium*) or supplied with solution via the cut end of the petiole (*Phaseolus*) or rachilla (*Samanea*). Whether pulvini are excised or remain attached to the plant, they oscillate with a circadian period under constant conditions and respond to changes in light intensity or quality, indicating that the clock and the photoreceptors that modulate the movements are localized in the pulvinus (SATTER and GALSTON, 1971b; SATTER et al., 1974a, b; SCOTT and GULLINE, 1972). Selective irradiations of the pulvinus or lamina led to a similar conclusion (KOUKKARI and HILLMAN, 1968; WATANABE and SIBAOKA, 1973).

2.4.1 Light-Promoted Opening and Dark-Induced Closure

Opening and closure movements are influenced by two photosystems: a blue light-absorbing pigment and phytochrome. Prolonged irradiation with blue or far-red light promotes opening in *Mimosa* (BURKHOLDER and PRATT, 1936), *Albizzia* (JAFFE and GALSTON, 1967c), *Arachis* (KADMAN-ZAHAVI, 1972) and *Samanea* (SATTER, unpublished data), with peak activity at 480 nm in *Mimosa* (FONDÉVILLE et al., 1967) and 440 nm in *Albizzia* (EVANS and ALLAWAY, 1972), and a minor peak in the 710–720 nm region in both plants. Although the blue-absorbing pigment has not yet been identified, the flavoprotein that acts as photoreceptor for blue light effects in numerous other plants (BRIGGS, 1976) is a likely candidate. The long wave length peak might be attributable to the phytochrome system. As discussed below (Sect. 2.4.2) a brief FR irradiation that interrupts darkness promotes opening in *Albizzia* (SATTER and GALSTON, 1971b), *Samanea* (SATTER et al., 1974b) and *Arachis* (KADMAN-ZAHAVI, 1972).

Leaflets of *Mimosa* (FONDÉVILLE et al., 1966) *Albizzia* (HILLMAN and KOUKKARI, 1967; JAFFE and GALSTON, 1967c), *Samanea* (SWEET and HILLMAN, 1969; SATTER et al., 1974b) and *Arachis* (KADMAN-ZAHAVI, 1972) close much more rapidly if irradiated briefly with R rather than FR prior to darkness, indicating that P_{fr} promotes closure. However P_{fr} is only effective under certain conditions. Absorption of blue light, for example, overrides the P_{fr} effect, since leaflets remain open in cool white fluorescent light which is rich in blue wavelengths, but also maintains a high P_{fr} level.

Dark-induced closure (commonly called nyctinastic closure) of *Albizzia* leaflets takes 30–90 min at 23° C and requires oxygen and the continuous presence of P_{fr} (SATTER et al., 1970b). At any time during the closure period, conversion of P_{fr} to P_r slows down the reaction, while conversion of P_r to P_{fr} speeds it up.

The effects of temperature alteration are quite complex. Differences between the effects of R and FR pre-irradiation are smaller at 18° C than at 24° C (*Albizzia*) (SATTER et al., 1970b) or 24° C compared to 37° C (*Samanea*) (SATTER et al., 1974b). Nevertheless, the absolute amount of closure of both R and FR pre-irradiated leaflets increases as the temperature is lowered (Fig. 7). Apparently leaflets can close partially by an energy-requiring P_{fr}-dependent process, or they can close more completely by a P_{fr}-insensitive process that is promoted by low temperature. This complex behavior might be related to the fact that either a temperature jump (SCOTT and GULLINE, 1972) or phytochrome conversion (SIMON et al., 1976b) resets the clock regulating leaflet movement.

Fig. 7. The nyctinastic closure of excised *Albizzia* leaflets pre-irradiated with R (O) or FR(X) followed by incubation in the dark at 18° (——) or 24° (---). The experiment was started at h 1.25 of the photoperiod. Leaflets were exposed to R (1.2 $W \cdot m^{-2}$, ca. 660 nm) for 4 min or FR (9 $W \cdot m^{-2}$, ca. 730 nm) for 1.5 min

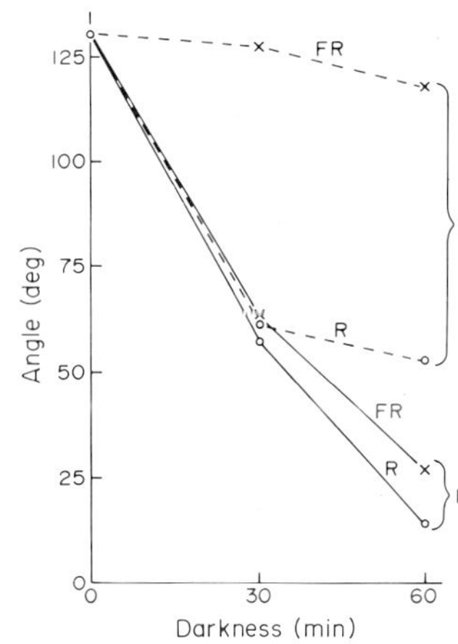

The rate and extent of nyctinastic closure are also influenced by the length of the white light period preceding darkness. *Albizzia* (SATTER and GALSTON, 1971b) and *Samanea* (SWEET and HILLMAN, 1969) leaflets that are darkened soon after they have opened in white light fail to close if pre-irradiated with FR, while those pre-irradiated with R close slowly and incompletely. The longer the white light period preceding darkness, the more rapid and complete the closure, and the less it depends upon P_{fr}. These complex effects are poorly understood, and could be due to some slow light-dependent process. However it is more likely that they are a consequence of interactions between a circadian oscillator and light signals absorbed by phytochrome, and possibly by the blue absorbing pigment. Such interactions are discussed in the next section.

2.4.2 Phytochrome–Clock Interaction During Extended Dark Periods

Albizzia or *Samanea* explants that are darkened at the end of a 16 h photoperiod begin to open their leaves after 4–8 h of darkness. Such opening, controlled by the circadian oscillator, is sensitive to P_{fr}, for if the leaflets are irradiated briefly with R or FR prior to opening, those exposed to FR open to a wider angle than those exposed to R (SATTER and GALSTON, 1971b; SATTER et al., 1973, 1974b). Furthermore, R–FR differences are enhanced by increase in temperature to 33° (*Albizzia*) or 37° (*Samanea*). During circadian opening as during nyctinastic closure, a high P_{fr} level depresses leaflet angle by a temperature-sensitive process.

Nevertheless, if *Albizzia* or *Samanea* explants are exposed briefly to R during each of several successive circadian cycles, the amplitude of the oscillations remains constant; if explants are similarly treated with FR, on the other hand, the amplitude decreases progressively during successive cycles (SIMON et al., 1976b; SATTER et al., 1976). Oscillations of *Arachis* also fade away following FR (KADMAN-ZAHAVI, 1972). In all these cases, P_{fr} is necessary for rhythmic oscillations to persist.

P_{fr} also acts as the effector for rhythmic rephasing and entrainment. Experiments with *P. multiflorus* suggest this possibility, since irradiation with wavelengths between 600 and 700 nm initiates circadian oscillations in leaves of dark grown seedlings, but simultaneous irradiation with wavelengths between 700 and 800 nm

(LÖRCHER, 1958) prevents this effect. However caution must be exercised in interpreting these data, since rhythmic leaf movement in *P. multiflorus* can be rephased either by R absorbed by pigments in the pulvinus, or by FR absorbed by pigments in the lamina (BÜNNING and MOSER, 1966a).

Phytochrome has been interpreted as photoreceptor for rephasing the petal-movement rhythm in *Kalanchoe blossfeldiana* (ZIMMER, 1962; SCHREMPF, 1975), although a 2–4 h red irradiation is required and the effect is not FR-reversible. This interpretation is compatible with data of WILKINS and HARRIS (1976) revealing that phase-shifting effects mediated by P_{fr} need not display typical R, FR reversibility. Brief pulses of R spread over a few-hour time period or prolonged red irradiation rephase a rhythm in CO_2 output in *Bryophyllum*. The phase-shifting effects of the brief pulses are FR-reversible, even though the effect of the long irradiation is not.

Unequivocal evidence that P_{fr} can rephase leaf-movement rhythms comes from experiments with *Samanea*. A single brief (5 min) R irradiation rephases the pulvinar-movement rhythm, and this effect is completely reversible by brief (1.5 min) FR (SIMON et al., 1976b; see Fig. 3 in SWEENEY, Chap. 1.4, this vol.). Furthermore, brief R irradiations repeated at 24 h intervals entrain the rhythm. Since these experiments were conducted with "naked" explants that lack laminar tissue, there is little doubt that phytochrome in the pulvinus is the photoreceptor. The effect of R is based on rephasing of rhythms in K and Cl flux within the pulvinus (SATTER et al., 1977). The light pulse has similar effects on both K and Cl rhythms, and does not alter K:Cl stoichiometry.

It is of interest that the proportion of red to far-red energy of natural daylight decreases rapidly at sunset, and increases rapidly soon after sunrise, implying that the P_{fr} level in plants grown under natural conditions decreases at dusk and increases at dawn (HOLMES and SMITH, 1977). These changes in light quality would be expected to entrain leaf movement rhythms, so that the oscillations remain in phase with environmental cycles.

One might expect the blue-absorbing pigment that promotes opening in *Mimosa, Albizzia* and *Samanea* to act also as photoreceptor for changes in rhythmic phase and frequency, but appropriate experiments have not yet been performed. Blue light delays and red light advances the phase of the leaf-movement rhythm in a *Coleus* cultivar (HALABAN, 1969). However this plant lacks pulvini and phytochrome does not appear to be the photoreceptor for the red light effects. Thus its regulatory systems might be quite different from those in leguminous species with pulvini.

2.4.3 The Membrane Potential

Leaflet movements, whether light-stimulated or free-running, are always accompanied by changes in the membrane potential. The potential was monitored on the surface of *Phaseolus* pulvini with Ag-Ag Cl electrodes (AIMI and SHIBASAKI, 1975) and in motor cells in pulvini of *Samanea* (RACUSEN and SATTER, 1975) and *Trifolium* (SCOTT et al., 1977), using a glass intracellular micro-electrode filled with 3M KCl and a grounded reference electrode in the bathing solution. The potential of flexor cells in *Samanea* begins to depolarize within 1 min after the onset of the usual white light photoperiod, and continues for about 50 min (from

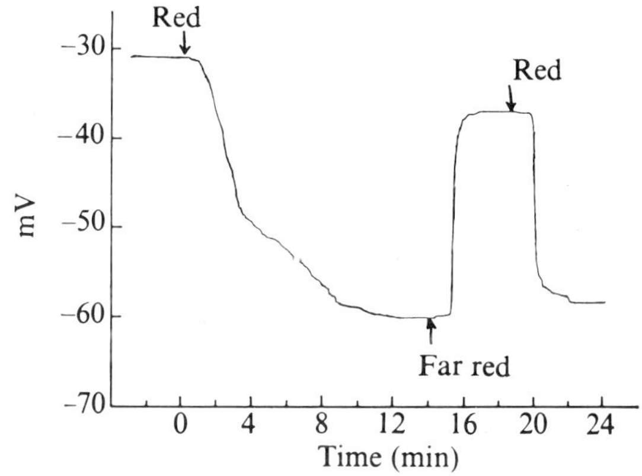

Fig. 8. Phytochrome-controlled changes in the transmembrane potential of flexor motor cells in the *Samanea* pulvinus. Irradiance levels were 350 mW·m^{-2}, ca. 660 nm and 900 mW·m^{-2} ca. 730 nm. (From RACUSEN and SATTER, 1975)

−50 to −20 mV), then slowly reverses to about −30 mV. Blue light is more effective than white (RACUSEN, unpublished data), suggesting that the same photoreceptor influences the potential changes and leaflet opening. If the white light is turned off 2 h later (nyctinastic closure) the potential is regulated by the phytochrome state and is repeatedly photoreversible, with R leading to hyperpolarization and FR to depolarization (Fig. 8). If the first R treatment follows immediately after the white light period, hyperpolarization begins immediately but requires 10 min for completion, whereas R treatments that follow FR reveal a 45 s lag, followed by rapid hyperpolarization, completed within 45 s. The leisurely pace of the first response is presumably due to the blue-light effect, which slowly decays and permits P_{fr} to act. Blue-light effects decay at a similar rate in flavoprotein-mediated systems, as cytochrome b, reduced in the light, slowly oxidizes in the dark (POFF and BUTLER, 1974). The lag when R follows FR might be due in part to the time required for conversion of appreciable P_r to P_{fr}.

In the absence of environmental perturbations (prolonged darkness) the potentials of both extensor and flexor cells oscillate with circadian periods, approximately 8 h out of phase with each other (RACUSEN and SATTER, 1975). Diurnal changes are massive, ranging from −35 mV to 100 mV in the flexor, and −40 mV to −95 mV in the extensor. Increase in external K or Na from 1 to 10 mM leads to ca. 6 mV depolarization in both extensor and flexor, while similar increase in Ca produces a 12 mV depolarization (RACUSEN and GALSTON, 1977). The effects of these ions are independent of the time of application.

Addition of sucrose to the bathing solution produces larger depolarizations during certain parts of the rhythmic cycle, while similar addition at other times is ineffective (Fig. 9). The sucrose-induced depolarization is greater at pH 5.3 than at pH 6.5, and is accompanied by an increase in the pH of the external solution (Fig. 10). The clear implication is that sucrose enhances the uptake of H, and might enter the motor cells with H via a co-transport system, as postulated for H-glucose uptake in *Neurospora crassa* (SLAYMAN and SLAYMAN, 1974) and *Chlorella vulgaris* (KOMOR and TANNER, 1974). A most interesting feature in *Samanea* is that sucrose-mediated H uptake is oscillatory, with rhythms in extensor and flexor cells ca. 4–8 h out of phase with each other. Furthermore, increase in K or Na from 1 to 10 mM prior to the addition of sucrose prevents the sucrose-induced

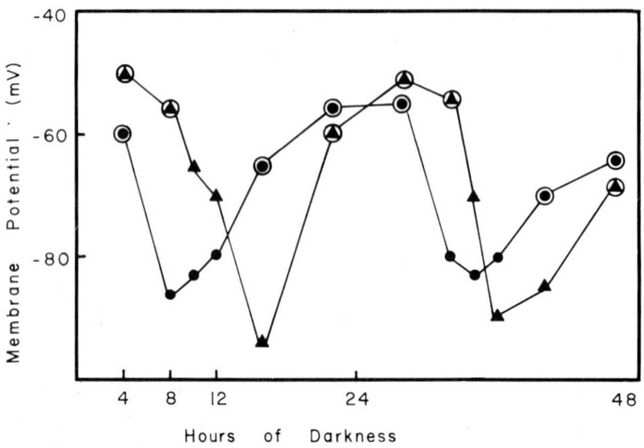

Fig. 9. Membrane potentials of flexor (▲) and extensor (●) cells in the *Samanea* pulvinus were monitored 12 times during a 48-h dark period. The data points enclosed by circles indicate times in which depolarization occurred in response to 10 mM sucrose. The pH was 5.3. (From RACUSEN and GALSTON, 1977)

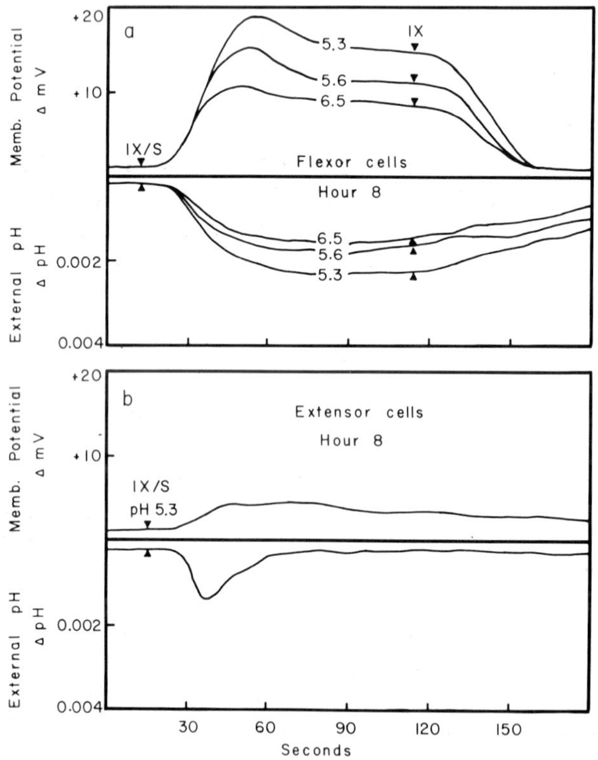

Fig. 10a and b. The effect of 10 mM sucrose on the membrane potential and external pH of flexor and extensor cells in the *Samanea* pulvinus at h 8 of the dark period. Test solutions (1 X) were buffered at pH=5.3, 5.6 or 6.5. 1X/S indicates addition of 10 mM sucrose, which caused: **a** Rapid depolarization of the membrane potential and an increase in the external pH of the flexor cells; **b** little change in potential or external pH of the extensor cells. (From RACUSEN and GALSTON, 1977)

depolarization. If the system also operates in vivo, it could play an important role in regulating the membrane potential, which in turn could act as a driving force regulating the movement of ions through cell membranes.

2.4.4 Sucrose-Light Interactions

Sucrose interacts with light in complex ways to regulate leaf movements in *Albizzia* and *Samanea* (SATTER et al., 1976; SIMON et al., 1976a, b). If supplied during an extended dark period, it acts cooperatively with P_{fr} in maintaining the oscillations, but it also reduces the mesor (average angle), whether the P_{fr} level is high or low. If supplied during light-dark cycles, by contrast, it interferes with rhythmicity and increases the mesor. It is difficult to rationalize all of these responses to sucrose solely in terms of substrate for energy-requiring reactions. Knowledge that sucrose uptake alters H uptake provides an additional possible explanation.

Elucidation of sucrose action in pulvini might aid in interpreting sucrose-light interactions in other systems. Sucrose interferes with floral induction in both long-day (KANDELER, 1968) and short-day (POSNER, 1969) *Lemna* species, even when both are grown with photoperiods that are normally inductive, yet permits floral induction in *Pharbitis nil* (KIMURA, 1963) and *Sinapis alba* (EVANS, 1971) grown with photoperiods that are normally noninductive. Membrane-depolarization in response to organic substances has also been reported in *Impatiens balsamina* (JONES et al., 1975) and might occur in many or all higher plants.

2.4.5 Energetics of Circadian Opening and Closure

The Q_{10} values for circadian opening in the dark in *Albizzia* and *Samanea* lie between 2.5 and 3, while those for circadian closure are below 1.0 in both plants (SATTER et al., 1973, 1974a). As discussed in the preceding section, the Q_{10} for nyctinastic closure in both plants is also less than 1. Furthermore, incubation of *Albizzia* leaflets on sodium azide inhibits the K movements that lead to circadian opening, but promotes the movements that lead to closure. These data suggest that energy-requiring processes, which could involve the active transport of K, Cl, H, or other ions, take place during opening, while diffusive processes predominate during closure (SATTER and GALSTON, 1973). Data acquired subsequently suggest that this hypothesis might need modification.

1. The presence of numerous plasmodesmata between pulvinar cells (CHAUDHRI et al., 1977) raises the possibility that K and Cl move through the symplast rather than (or in conjunction with) the apoplast. This would not invalidate the pump-leak theory, but would place control at the tonoplast rather than (or in addition to) the plasmalemma.

2. Recent analyses with the electron microprobe reveal large gradients of K and Cl in both radial and circumferential directions of the pulvinus (SATTER et al., 1977). The concept of inwardly directed K or Cl pumps in membranes of extensor cells and outwardly directed ion pumps in membranes of flexor cells obviously requires modification, to accord with the heterogenous distribution of K and Cl within both extensor and flexor.

3. The average angle of *Albizzia* (SATTER et al., 1976) or *Samanea* (SIMON et al., 1976a) explants supplied with sucrose during a long dark period is smaller than that of explants supplied with water. If the role of sucrose is that of supplying substrate for the synthesis of ATP, this would suggest that closure is an energy-

requiring process. Knowledge that sucrose affects H uptake and the membrane potential (RACUSEN and GALSTON, 1977) provides an alternative explanation for the sucrose requirement. It is not known whether either of these interpretations is correct.

2.4.6 Control by Chemical Treatments

Compounds that change the phase or frequency of rhythms do so by interacting with the clock or with a mechanism for coupling oscillations to the clock; thus their identification provides valuable information about the nature of the clock. Pulsed incubation in alcohol (BÜNNING and BALTES, 1962), D_2O (BÜNNING and BALTES, 1963), valinomycin (BÜNNING and MOSER, 1972), KCl (BÜNNING and MOSER, 1973), theophylline (MAYER et al., 1975), or caffeine (MAYER and SCHERER, 1975) rephases the leaf movement rhythm in *Phaseolus,* while lithium and D_2O (ENGELMANN et al., 1974) rephase the petal movement rhythm in *Kalanchöe* (see SWEENEY, Chap. 1.4, this vol., for details). All of these compounds presumably alter membrane properties. Experiments in progress in our laboratory reveal that pulsed incubation in cycloheximide (0.01 mM), inhibitor of protein synthesis (APPLEWHITE et al., 1973) and of K flux (SATTER et al., 1974c) in *Albizzia,* also rephases the leaflet movement rhythm. Thus ion transport in pulvinar systems may be linked to membrane protein synthesis (MACROBBIE, 1976), which in turn may be part of the timekeeping mechanism (FELDMAN, 1967).

No consistent picture of hormone involvement can yet be formulated, although such involvement is a distinct possibility. Conversion of P_r to P_{fr} in etiolated plants changes the levels of all the known plant hormones (summarized by WAREING and THOMPSON, 1976). Auxin influences cell-wall elasticity and cell turgor in other systems (ZIMMERMANN et al., 1976) and also promotes H secretion (CLELAND, 1973), which could facilitate K uptake. Furthermore, some effects of ethylene have been attributed to increase in the leakiness of the tonoplast (KENDE and HANSON, 1976).

Indoleacetic acid (10–100 μM) and gibberellic acid 0.1–100 μM) promote premature opening and retard closure of darkened *Mimosa* (WILLIAMS and RAGHAVAN, 1966) and *Albizzia* (MCEVOY and KOUKKARI, 1972; SATTER et al., 1972a) leaflets. Ethylene inhibits leaf movements in *Acacia farnesiana* (mesquite) and *Prosopis juliflora* (huisache), either when supplied in gaseous form or when synthesized by these plants in response to picloram treatment (BAUR and MORGAN, 1969; MORGAN and BAUR, 1970). However abscisic acid is ineffective in *Albizzia* (SATTER et al., 1972a) despite its important role in turgor regulation in stomatal guard cells (JONES and MANSFIELD, 1970).

If any of the hormones participate in regulating leaf movements in vivo, we might expect light-stimulated or rhythmic movements to be accompanied by changes in hormone titer or activity. *Albizzia* pulvini contain acetylcholine, but its titer remains constant during the rhythmic cycle and following phytochrome conversion (SATTER et al., 1972b), in contrast to the rapid phytochrome-controlled changes in the acetylcholine titer of mung bean root tips (JAFFE, 1970a). Although rhythmic variations in diffusible auxin have been reported in pulvini of nyctinastic plants (summarized by BRAUNER and ARSLAN, 1951), these experiments were conducted prior to the discovery of phytochrome, and a dim red light was presumed to be a safelight in some cases. Additional experiments need be conducted under more stringent conditions.

2.5 Seismonastic Leaf Movements

2.5.1 Description of Seismonastic Movements and Recovery

A limited number of nyctinastic plants, including *Mimosa pudica, M. spegazzini, Biophytum sensitivum* and *Cassia fasciculata* can execute rapid seismonastic movements in addition to the slow movements described in the preceding section. Seismonastic movements take place at the 1°, 2°, and 3° pulvini of *Mimosa:* the paired pinnules and pinnae fold together as during sleep movements, while the petiole drops so that it is oriented in an almost vertical direction (see Sect. 2.2.2 for further details). Movement in *Mimosa* begins almost immediately after the plant is shaken, and is completed in 1–2 s (ODA and ABE, 1972). The petiole returns to the nonstimulated position in ca. 8–15 min, although it sometimes overshoots the repose position and undergoes high-frequency oscillations before the angle stabilizes (AIMI, 1963; JONAS, 1976; CHARNLEY et al., 1975; ROBLIN, 1976b; STOECKEL, 1976). It is quite remarkable that a leaf, electrically stimulated to drop every 20 or 30 min during light-dark cycles or continuous light, always returns to the same angle as a nonstimulated control (Fig. 11). The rate of recovery is increased by exposure to white light or, if the plant is darkened, by increase in temperature (STOECKEL, 1976). Recovery depends upon resorption of K and other osmotically effective substances by abaxial motor cells (TORIYAMA, 1955b); thus the promotive effects of light and high temperature are most reasonably interpreted in terms of the need for ATP for ion uptake.

Pinnule pairs and pinnae, in contrast to the petiole, fail to recover after seismonastic closure unless they are illuminated (WATANABE and SIBAOKA, 1973). Since seismonastic movements at both the 2° and 3° pulvini occur in the same direction as sleep movements, and light is required to reverse the latter, it is not possible to determine whether light is necessary per se for recovery from seismonasty.

2.5.2 Localized and Propagated Effects

Mechanical vibration has a direct stimulatory effect on seismonastic movement at all pulvinar junctions in *Mimosa*. Propagation through the petiole is not involved, for the stimulus is directly perceived by receptors in each pulvinus. Seismonastic movements can also be initiated by touch, by electrical, chemical or temperature shock, by wounding, or by rapid increase in light intensity (reviewed by SIBAOKA, 1966, 1969). All parts of the leaf and stem respond to many of these stimuli. The stimulus is transduced to an action potential that propagates through excitable cells in the phloem and protoxylem of the petiole at ca. 2–4 cm/s (SIBAOKA, 1962). Approximately 0.2 s after the petiolar action potential reaches the 1° pulvinus, it initiates a pulvinar action potential (SIBAOKA, 1966) that propagates through excitable cells in the motor tissue. By use of intracellular electrodes, ABE and ODA (1976) determined that cells on both adaxial and abaxial sides of the pulvinus are excitable, although the action potential measured with extracellular electrodes is five times higher on the lower surface than on the upper one (SIBAOKA, 1969).

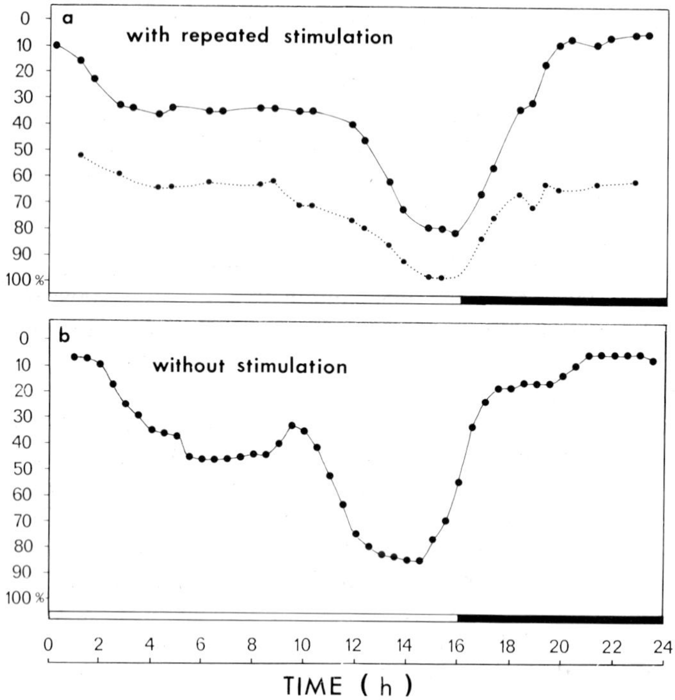

Fig. 11a and b. Movement of the primary pulvinus of *Mimosa* during a light-dark (16 h:8 h) cycle. The amplitude of the movement is measured as percent of possible movement. **a** Seismonastic movement was stimulated at 30–60 min intervals by a voltage pulse (2 volt for 1 s) across the pulvinus. *Dotted line* indicates the angle immediately after stimulation; *solid line* indicates the angle of repose. **b** Unstimulated control. (From STOECKEL, 1976)

Overt movement begins 10–20 ms after the pulvinar action potential is initiated. Further details of stimulus reception, transduction and transmission can be found in SHROPSHIRE, Chapter 1.2, and BENTRUP, Chapter 1.3, both this volume, as well as in earlier reviews by SIBAOKA (1966, 1969) and PICKARD (1973).

An observation described by PFEFFER in 1873 (reported by APPLEWHITE, 1972a), is that a plant habituates to stimuli presented every few seconds by re-opening and remaining open. This adaptive mechanism prevents wind and rain from keeping the leaflets closed. Nevertheless, leaflets habituated to mechanical vibration will close to electrical stimulation, and vice versa (APPLEWHITE, 1972a). Since leaflet movement, irrespective of the stimulus, is due to the shrinkage of extensor cells, in turn a consequence of K efflux, habituation apparently involves events that precede K efflux and are stimulus-specific.

2.5.3 Photostimulation of Seismonastic Movement

High intensity light can substitute for a mechanical stimulus in evoking seismonastic movement at the 1° (DARWIN, 1881; FONDÉVILLE, 1963), 2° (ROBLIN and GAVAUDAN, 1974) and 3° (APPLEWHITE and GARDNER, 1971) pulvini of *Mimosa* and at the 2° pulvini of *Biophytum sensitivum* (ROBLIN and GAVAUDAN, 1974). As discussed above (Sect. 2.5.1) light also promotes recovery from seismonastic movement in *Mimosa* (STOECKEL, 1976); however the irradiance required to stimulate seismonasty is ca. an order of magnitude higher than that required to promote recovery (combined data from FONDÉVILLE et al., 1967, and STOECKEL, 1976). Seismonastic photosensitivity of the 1° pulvinus varies during the diurnal cycle, and is maximal in the middle of the dark period (FONDÉVILLE, 1963). It is reduced by prior subthreshold stimulations (ROBLIN and GAVAUDAN, 1967); the differential between the background and stimulating intensities, as well as the absolute intensity of the stimulating light determine the probability of a response and its latency, which ranges between 30 and 90 s. As with other stimuli that elicit seismonastic movement, the response of an individual leaf or leaflet is an "all or nothing" phenomenon.

The action spectrum for seismonastic petiole drop in *Mimosa* peaks between 400 and 470 nm, with wavelengths longer than 520 nm completely ineffective (FONDÉVILLE et al., 1967). However wavelengths of 470–520 nm and 620–660 nm are most effective in promoting seismonastic closure of pinnule pairs in *Mimosa*, if such light follows a 5–20 min dark period (APPLEWHITE and GARDNER, 1971). Disparity between the spectral responses for petiole drop and pinnule closure makes it difficult to predict what pigment(s) might be involved. Furthermore, it is not clear whether the photoreceptors are within the pulvinus, or in some other tissue, possibly the lamina. If the former, light might have a direct effect on the motor cells, whereas in the latter case, the light stimulus would initiate an action potential that propagates through the petiole. Until this question is resolved, the mode of light action will remain obscure.

2.5.4 Circadian Variations in Seismonastic Movements

Although seismonastic movements can take place during all parts of the rhythmic cycle, sensitivity to stimulation varies rhythmically. If a *Mimosa* plant is shaken with suboptimal intensity so that some leaves respond but others do not, reactivity is greatest toward the end of a 14-h photoperiod, and minimum toward the end of the dark period (GAVAUDAN and ROBLIN, 1971). A fascinating study of seismonasty–rhythm interaction by STOECKEL (1976) reveals that several parameters of the movements at the 1° pulvinus of *Mimosa* are oscillatory. The amplitude of leaf-drop movement oscillates during light-dark cycles (Fig. 11) and under free-running conditions, as does the rate of recovery. Circadian oscillations in the amplitude of leaf drop are not dependent upon oscillations in the angle of repose, since the former rhythm persists after the latter has damped. A rhythm in the rate of recovery suggests that K uptake in *Mimosa*, as in *Trifolium* (Fig. 5) is oscillatory.

When movement is stimulated by a voltage pulse near the apex of the petiole, the duration of the action potential and its speed of transmission through the petiole oscillate with circadian periods (STOECKEL, 1976). When the voltage pulse is applied at the 1° pulvinus, it propagates acropetally through the petiole during certain parts of the diurnal cycle, but fails to propagate in this direction at other times (STOECKEL and PFIRSCH, 1975). This phenomenon, called "inverse transmission", oscillates with a circadian period under free-running conditions and remains synchronized with other rhythms in electrical activity that are described above, and with rhythms in seismonastic leaf drop and recovery (STOECKEL, 1976). All of these rhythms have the same period length in white light (25.5 h) as the sleep-movement rhythm of *Mimosa* pinnules (FONDÉVILLE, 1965), and they are all rephased similarly by the transition from darkness to white light (STOECKEL, 1976). It is apparent that a single circadian oscillator influences both rapid seismonastic movements and the slower sleep movements!

2.6 Analysis of Control Mechanisms

2.6.1 Comparisons Between Rhythmic and Seismonastic Movements

Both seismonastic and rhythmic movements depend upon changes in cellular K, its accompanying anion (mainly Cl during rhythmic movement, but not yet identified during seismonastic movement) and other substances that affect motor cell water potential. Yet turgor regulation during the two types of movement differs in fundamental ways. K and Cl loss from cells on one side of the pulvinus during rhythmic movement is balanced by increase of these ions in cells on the opposite side, but K that effluxes from shrinking cells during seismonastic movement remains in the apoplast until re-absorbed by these same cells during the recovery period (Sects. 2.3.1 and 2.3.2). Rhythmic and seismonastic movements also differ in rate, response to externally supplied reagents, effects of light and the photoreceptors for such effects. Furthermore, plants that can execute seismonastic movements have larger tannin deposits than plants that lack this ability (Sect. 2.2.3). There seems little doubt that seismonastic and rhythmic movements are regulated by different molecular mechanisms.

Nevertheless, both types of movement are influenced by the same oscillator (STOECKEL and PFIRSCH, 1975) and have features in common. Some evidence, largely circumstantial, suggests that Ca might provide a common link. Theophylline (MAYER et al., 1975) and caffeine (MAYER and SCHERER, 1975) rephase the leaf movement rhythm in *Phaseolus* while caffeine inhibits nyctinastic closure in *Mimosa* (APPLEWHITE, 1972b) and aminophylline inhibits opening in white-light and nyctinastic closure in *Albizzia* (SATTER et al., 1974c). These substances alter cyclic nucleotide levels (BUTCHER and SUTHERLAND, 1962) and this might explain their effects on circadian systems (FELDMAN, 1975). However they also interfere with Ca transport and binding (PAUL and GOFF, 1973) in a manner compatible with their effects on leaf-movement rhythms (MAYER et al., 1975). $CaCl_2$ (100 mM) promotes (SATTER et al., 1974c) and EDTA, which preferentially complexes divalent cations, inhibits nyctinastic closure in *Albizzia* (500 mM; MCEVOY and KOUKKARI, 1972); EDTA (1 mM) also inhibits recovery from seismonastic closure in *Mimosa* (CAMPBELL and THOMSON, 1977b). Lanthanum, competitive inhibitor of Ca transport (LEONARD et al., 1975) inhibits seismonastic closure (10 mM; CAMPBELL and THOMSON, 1977b) and transmission of the action potential (30 mM; STOECKEL, 1976) in *Mimosa*, suggesting that Ca might be required both for stimulus transmission and for coupling between the action potential and K efflux. Although Ca movements were not detected in electron microprobe analyses (SATTER et al., 1970b), they could have occurred and been obscured by high background levels. Ca movements might be largely intracellular (Sect. 2.3.2), as reported during seismonastic movement in *Mimosa* (TORIYAMA, and JAFFE, 1972). Effects of Ca on the integrity of membranes in other organisms are well documented (SLAYMAN, 1974).

2.6.2 Circadian Movements

Oscillations in Cell Membranes. An understanding of how rhythmic leaf movements are regulated requires knowledge of how the clock operates, and such knowledge is beyond our grasp. Yet evidence from both animal and plant systems (reviewed by SWEENEY, Chap. 1.4, this vol.) imply that oscillations in membrane ultrastructure, chemical composition and transport properties play a central role in time measurement.

Recent studies by GARDNER (1977) reveal time-dependent changes in the fatty acid composition of phospholipids extracted from extensor and flexor halves of *Phaseolus* pulvini. He reported circadian oscillations in the amount of palmitic and linoleic acids, and in the ratio of total saturated: unsaturated acids during light-dark cycles and prolonged darkness. The tissue homogenates were not fractionated for structural components prior to chemical extraction, so it is not yet known whether some or all of the oscillations occur in plasmalemma or tonoplast lipids. If they do, this could result in circadian oscillations in the fluidity of these membranes, and could play an important role in the timekeeping mechanism, especially aspects related to temperature compensation.

An aspect that has not yet been investigated involves changes that must take place in the tonoplast and plasmalemma, respectively, to accomodate changes in the volume of the vacuoles and protoplasts. Large vacuoles might fragment into smaller ones as cells shrink, thereby reducing the volume: surface ratio of the vacuolar complex, while the reverse process could take place as cells swell. Numerous small vacuoles were detected in motor cells of *Albizzia* by SATTER et al. (1970a) and in *Mimosa* by CAMPBELL and THOMSON (1977a) and WEINTRAUB (1952), yet similar vacuoles have not been described by other investigators studying the ultrastructure of *Mimosa*. Thus the number, size, and distribution of vacuoles might vary during the circadian cycle, as vesiculation in salt glands of *Frankenia* varies with salt status (CAMPBELL and THOMSON, 1976). If the membrane system in the motor cells changes rhythmically, one would expect rhythmic changes in membrane proteins and lipids; total protein synthesis in *Albizzia* is rhythmic (APPLEWHITE et al., 1973) and inhibitors of protein synthesis phase shift the rhythm (Sec. 2.4.6).

Rhythmic changes in membrane transport or permeability could lead to all of the following oscillations: K uptake through the plasmalemma (SCOTT et al., 1977), K and Cl distribution within the pulvinus (SATTER et al., 1977), membrane potential (RACUSEN and SATTER, 1975), sucrose-mediated H transport (RACUSEN and GALSTON, 1977), amplitude of seismonastic leaf drop (STOECKEL, 1976) and rate of recovery. Since the latter two effects are auxin-sensitive (WATANABE, 1971), one might expect them to oscillate if auxin titer or activity is oscillatory.

Rhythms in ion movement are strongly influenced by rhythms in transmembrane potential. Thus an understanding of how the membrane potential rhythms originate is a key toward understanding the molecular basis for turgor regulation in nyctinastic plants. Processes known to affect the potential across plant cell membranes include ATP-driven electrogenic pumps, standing ion gradients, oxidation-reduction systems that separate H^+ and OH^- across membranes (SLAYMAN, 1974), and hydrostatic pressure (ZIMMERMANN and STEUDLE, 1974). Electrogenic transport probably makes the largest quantitative contribution in most cells; therefore the oscillatory sucrose-mediated H transport system described by RACUSEN and GALSTON (1977) would be expected to play a significant role in generating oscillations in potential, if that system functions in vivo. The large hydrostatic pressure gradients that exist within the pulvinus might also be expected to alter membrane electrical parameters. Transmembrane gradients of K and Cl, on the other hand, probably make only a minor contribution, since increase in external KCl from 1 to 10 mM has only a small (6 mV) depolarizing effect, independent of the time of application (RACUSEN and GALSTON, 1977). (Higher KCl concentrations have not yet been tested, but could have a much larger effect.)

Blue-light and phytochrome conversions both have rapid effects on the membrane potential in *Samanea* motor cells (Sect. 2.4.3 and Fig. 8) although the mechanisms have not been defined. If the blue-absorbing pigment is a membrane-localized flavoprotein that uses its excitation energy to reduce cytochrome b, as proposed in other systems (POFF and BUTLER, 1974, 1975; MUÑOZ and BUTLER, 1975), the consequent transmembrane gradients in H^+ and OH^- would be expected to alter the potential. Phytochrome conversions have been shown to affect the surface charge at the plasmalemma (TANADA, 1968; RACUSEN and ETHERTON, 1975), but the molecular details remain obscure.

The Apoplast and the Symplast. The leaf-movement rhythm occurs at the tissue level, yet depends upon rhythms in the numerous cells of the pulvinus. It is clear that individual cells must communicate effectively with each other, in order that their oscillations remain synchronized under free-running conditions. Plasmodesmata are particularly numerous in pulvinar tissue (Fig. 2) and should be important in this regard, for they unite the protoplasts of individual cells into a large continuum.

Electrical coupling through plasmodesmata in etiolated oat coleoptiles is promoted by conversions of P_r to P_{fr} (RACUSEN, 1976); if P_{fr} has a similar function in green tissue, this might explain why it is required for rhythmic persistence (Sect. 2.4.2). Coupling could take place through the endoplasmic reticulum, which according to ROBARDS (1976) is continuous from cell to cell via the desmotubule that lines the plasmodesmatal core.

High concentrations of plasmodesmata are usually associated with high rates of solute flux through the symplast (ROBARDS, 1976). Determination of the number, size and location of plasmodesmata in pulvinar tissue should indicate whether they *could* accomodate the massive K and Cl fluxes that take place during each circadian cycle. If transport is symplastic, gradients in the concentration of K and Cl between adjacent cells might provide a driving force for ion movements (TYREE, 1970). Under these conditions, the transport properties of the tonoplast would be of primary importance in regulating K and Cl movements, and we might expect large rhythmic changes in tonoplast-localized ion pumps and channels. However if the apoplastic route is utilized, ion movements through the plasmalemma should also be oscillatory, as already demonstrated in *Trifolium* (SCOTT et al., 1977).

Histochemical precipitation methods and X-ray analysis techniques for monitoring ion movements at the ultrastuctural level have been developed recently (APPLETON, 1974; CAMPBELL,

1975). Use of these methods, together with ion-specific electrodes, should reveal whether transport is apoplastic, symplastic, or both.

2.6.3 Seismonastic Movements

Action potentials in plants are often characterized by transient efflux of K from excitable cells, as reported in the petiole of *Mimosa* (TORIYAMA, 1962b) and in internodal regions of *Nitella* (SPYROPOULOS et al., 1961) and *Chara* (ODA, 1975). Since motor cells in the 1° pulvinus of *Mimosa* are excitable (ABE and ODA, 1976), it is not surprising that they also lose K during the pulvinar action potential. However, K loss from cells on the abaxial side of the pulvinus is far more massive than usually occurs during transmission of an action potential, and recovery takes longer than usual (ODA, 1975). It is apparent that these cells differ from other excitable cells.

Several investigators have sought parallels between motor cells in *Mimosa* and contractile systems in animals. Actin is probably ubiquitous in living systems and is often found in close association with cell membranes (MOOSEKER and TILNEY, 1975). It has been suggested that an actomyosin-like contractile system might alter the conformation of proteins in the tonoplast membranes of motor cells when *Mimosa* is stimulated, thereby increasing membrane permeability, with consequent loss of K and other osmotic agents (TORIYAMA and JAFFE, 1972; SANBERG, 1976). The following data have been cited to support this view: (1) Protamine sulfate and mersalyl, compounds that interfere respectively with contractile proteins and their linked ATPases, cause *Mimosa* pinnules to close without contact stimulus and inhibit recovery following mechanical stimulation (DRIESSCHE, 1963). (2) ATPase activity has been detected histochemically (TORIYAMA, 1963) and in extracts from *Mimosa* pulvini (LYUBIMOVA et al., 1964, 1969). An ATPase purified from the 1° pulvinus exhibits a viscosity change upon addition of ATP, if (and only if) certain other protein-containing fractions are added to the incubation medium (BISWAS and BOSE, 1972). This is consistent with the concept that the ATPase interacts with another protein in the extract in a contractile process. (3) Analysis of the force-velocity characteristics of movement at the 1° pulvinus gives values for Hill's constants similar to those of animal muscle systems and reconstituted contractile collagen (BALMER and FRANKS, 1975). (4) Electron micrographs (TORIYAMA and SATÔ 1968b, 1969, 1971) reveal that fibrils attached to the tonoplast fragment after stimulation, and the tonoplast collapses (Fig. 4). (5) Ca, which activates contractile proteins in animal systems (EBASHI and ENDO, 1968), migrates from the tannin vacuole to the central vacuole after *Mimosa* has been stimulated mechanically (TORIYAMA and JAFFE, 1972). The latter authors and TORIYAMA (1973) propose that Ca migration stimulates a contractile process in *Mimosa,* which then leads to rapid increase in membrane permeability and K leakage followed by loss of turgor and leaf movement.

The theory is attractive, supported by some evidence and refuted by none, but definitive data are not yet available. The ability of microfilaments to bind the heavy meromyosin portion of myosin is a commonly accepted method for identifying actin (PALEVITZ and HEPLER, 1975), yet experiments testing such binding have not been performed with *Mimosa* motor cells. Thus conclusions are premature at this time.

Other aspects of seismonastic movements are also unresolved. It has been suggested that the small transient vacuoles described by WEINTRAUB (1952), like the contractile vacuoles in lower forms of invertebrates, empty their contents to the cellular exterior through a reverse pinocytotic mechanism; this theory also remains to be tested. Fascinating problems await future investigators!

3. Tendril Curling

3.1 Nature and Significance

Some plants whose stems lack the strengthening tissues to maintain the plant in an upright position support themselves by means of thin appendages called tendrils that grasp onto objects they encounter. This is accomplished by three distinct types of movement. Spontaneous circumnutational movements increase the probability the tendril will contact a supporting structure (TRONCHET, 1958; JAFFE, 1972; JOHNSSON, Chap. 5.3.1, this vol.). When such contact occurs, tendrils of some species form adhesive pads that grow into cracks and crevices in the supporting surface, while others coil tightly around the support. Most tendrils also form "free" coils subjacent to the contact coils, thereby bringing the plant closer to its mechanical support. Fascinating accounts of tendril movements appear in the writings of DARWIN (1893), MACDOUGAL (1896), and PFEFFER (1905). These movements were reviewed by BÜNNING (1959c), TRONCHET (1962) and JAFFE and GALSTON (1968b) and will be treated only briefly here.

Tendrils of some species are thigmotropic (the direction of curling is determined by the surface stimulated) while others are thigmonastic (the direction of curling is morphologically determined). Unfortunately, the term thigmotropic has been used to describe both types of movement. Furthermore, the zoologist's rather than the botanist's definition of dorsal and ventral (see Sect. 2.2.2) has been used in much of the literature on tendril movements, creating additional ambiguity. When thigmonastic tendrils curl, the lower or abaxial surface becomes concave while the upper or adaxial surface becomes convex.

Tendrils of *Eccremocarpus scaber* and *Luffa cylindrica* are thigmonastic; therefore it is not surprising that the distribution of mechanoreceptors in each of these species is asymmetric. Receptor cells are sparse on the adaxial surface, heavily concentrated on the abaxial surface, and most prolific near the tip, the most irritable region (JUNKER and REINHOLD, 1975). The receptors in *L. cylindrica* and in several other members of the Cucurbitaceae are warty protrusions of epidermal cells called tactile blebs, while receptors in *E. scaber* are dome-shaped cells called tactile papillae. Many species lack these specialized cell types, but have a wrinkled or corrugated thigmosensitive epidermal surface (see BENTRUP, Chap. 1.3, this vol. for a discussion of mechanoreception).

The most striking ultrastructural feature of *E. scaber* tendrils is the abundance of plasmodesmata, many of them compound (JUNKER, 1977a). The plasmodesmata, which are often localized in pit fields, appear to unite all protoplasts in the tendril, but are most abundant in the walls between papillae and underlying epidermal cells, and in the anticlinal epidermal cell walls. Thus the epidermal cell layer, including papillae, forms an efficient communication system.

3.2 Physiological Analysis

3.2.1 Kinetics of Movement

Contact coiling begins ca. 0.5–35 min after tactile stimulation, but tendrils of most species straighten if the contact stimulus is removed prematurely (DARWIN, 1893). Coiling of *Pisum sativum* (pea) tendrils can be stimulated by stroking the abaxial surface with a glass rod (JAFFE and GALSTON, 1966a). The magnitude of the response increases with the number and frequency of strokes. During the initial phase, which begins ca. 2 min after stimulation, abaxial cells shrink and adaxial cells elongate, but after 2 h both elongate, the former more slowly than the latter.

Stroking only the adaxial surface of a pea (JAFFE, 1970c) or *Passiflora caerulea* (JAFFE, 1970b) tendril is ineffective, yet if both sides of the tendril are stroked simultaneously, it coils less than if only the abaxial surface were stroked. Furthermore, strokes on the abaxial surface of a pea tendril can have their effectiveness reduced by subsequent stroking of the adaxial surface, if the time interval between the two series of strokes is less than 10 min. Coiling gradually escapes from inhibitability during this interval.

3.2.2 Energetics of Movement

Several types of evidence indicate that ATP is utilized during the coiling of pea tendrils (JAFFE and GALSTON, 1966b, 1967a). The ATPase activity of tissue homogenates is inversely proportional to the amount of coiling prior to homogenization. Respiratory inhibitors decrease coiling of darkened tendrils and exogenous ATP promotes it. Coiling ceases when the temperature is lowered from 25° C to 5° C, but resumes at its initial rate if the temperature is raised to 25° C within 60 min (JAFFE, 1977). Furthermore, coiling is promoted by light, with spectral characteristics suggestive of chlorophyll absorption (JAFFE and GALSTON, 1966b).

Tendrils incubated in the dark for 72 h, and then stroked, remain uncoiled unless illuminated within 2 h after the tactile stimulus (JAFFE, 1977). Thus the sensory function proceeds in the dark, but the motor response requires light. This technique for separating the two processes experimentally should aid in mechanistic analyses.

3.2.3 Analysis of Mechanism

Tendrils of all species tested coil without contact stimulus in a CO_2 atmosphere (REINHOLD, 1967; JUNKER, 1977b) or if floated on IAA (0.01 mM) or if the tip is dipped into a 1 mM IAA solution (REINHOLD, 1967; JAFFE, 1975). Ethylene also promotes coiling in some species (REINHOLD, 1967; BANGERTH, 1974), particularly if applied asymmetrically (JAFFE, 1970d). Coiling starts more rapidly when stimulated by CO_2 rather than by touch (JUNKER, 1977b). The mode of CO_2 action is not known.

IAA-induced coiling starts more slowly than that following contact stimulus, although the final angle is larger. IAA also substitutes for a tactile stimulus in promoting the formation of adhesive pads in tendrils of *Parthenocissus cuspidata* and asymmetric lignification of the xylem in tendrils of *Passiflora caerulea* (REINHOLD et al., 1970). Thus symmetrical application of IAA produces coiling and lignification, both asymmetric responses. Yet the abaxial-adaxial distribution of label in *Passiflora* tendrils pulse-labeled at their tips with ^{14}C-IAA is relatively symmetrical prior to coiling, and remains so during the coiling process (JUNKER, 1976a, 1977b). The logical conclusion is that abaxial and adaxial cells respond differently to auxin.

Auxin again

Auxin might facilitate both the elongation of cells on one side of the tendril and the shrinkage of cells on the opposite side, with the latter effect mediated by ethylene. Auxin (0.5 mM) stimulates ethylene synthesis in cells of *Cucumis sativus* tendrils at a rate that is several times higher in abaxial than in adaxial cells (BANGERTH, 1974; for differential auxin/ethylene interactions see KANG, Chap. 5.3.2, this vol.). Furthermore ethylene (supplied as ethephon) increases the rate of efflux of label from *Cucumis* tendrils labelled with ^{14}C-glucose. Since increase in the efflux of carbon compounds and electrolytes leads to the shrinkage of abaxial cells of pea tendrils during the initial phase of coiling (JAFFE and GALSTON, 1968a) ethylene could participate in regulating this aspect of coiling in vivo.

Numerous correlations between tendril coiling and content of phenolics and flavonoids have been reported (TRONCHET, 1960; JAFFE and GALSTON, 1967b; reviewed by JAFFE and GALSTON, 1968b). Whether these compounds participate in regulating the movements in vivo is not known; if so, effects might be related to alteration of auxin metabolism (JAFFE and GALSTON, 1968b).

Auxin cannot be the primary transmitter of the stimulus that leads to coiling (JUNKER, 1977b). Auxin transport is basipetal, yet stimulation of a single papilla in *E. scaber* tendrils induces coiling both basipetal and acropetal to the stimulus (JUNKER, 1977a). Furthermore, the coiling stimulus is transmitted far more rapidly than auxin (JUNKER, 1976a). UMRATH (1934) reported that tactile stimulation initiates an action potential in tendrils of *Cucumis pepo* but REINHOLD et al. (1970) were unable to detect similar potentials in *Cucumus sativus* or *Passiflora edulis*. JAFFE and GALSTON (1967a) reported that an extract from pea tendrils with ATPase activity decreases in viscosity following addition of ATP, consistent with the possibility that a contractile protein might be involved. It is unfortunate that subsequent authors have treated this speculation as fact, despite lack of further evidence (see Sect. 2.6.3 for discussion of pertinent experiments). Thus the nature of the primary response to touch remains unresolved.

4. Synthesis

Circadian leaf movements are a consequence of large rhythmic changes in K and Cl distribution within the pulvinus, in turn a consequence of rhythmic changes in membrane-localized pumps and/or channels. Several aspects that have received relatively little attention demand close scrutiny. Do K and Cl migrate through the apoplast, symplast, or both as they shuttle between extensor and flexor? If transport is symplastic, how is the rate and polarity of ion movement regulated? Can rhythmic changes in membrane ultrastructure or chemical composition be detected? Does Ca, or do any of the hormones, participate in regulating membrane properties in vivo? Is H-sucrose co-transport part of the endogenous regulatory mechanism?

Seismonastic movements in *Mimosa* are also a consequence of massive flux of K and other solutes, which migrate from the vacuoles of shrinking cells to the apoplast. Solute flux is dependent upon rapid increase in the permeability of the tonoplast and plasmalemma: key questions relate to mechanisms for controlling permeability changes. Are Ca-activated contractile proteins involved? If so, do they open channels in motor cell membranes? Do tannins play a role in the regulatory mechanism?

Recent investigations of tendril curling have focused on the effects of externally supplied auxin and ethylene. It is important to determine whether these compounds regulate coiling in vivo. If so, do they alter membrane properties? Do K, Cl or other osmotically effective substances play an important role in turgor regulation?

It will be necessary to develop new methods for probing membrane structure and ion movements at the ultrastructural level to answer some of these questions. Plants with motile organs should be useful for developing these techniques, since high rates of solute flux occur in specified cells in response to defined stimuli.

Acknowledgments. Some of the ideas expressed in this chapter developed during many years of collaboration with Dr. A.W. Galston. I am grateful for our fruitful partnership. I thank Drs. P.B. Applewhite, R.H. Racusen, M.R. Sussman, and S.E. Williams, graduate student G.T. Geballe, and research assistant M.J. Morse for reading the manuscript and making helpful suggestions. I also appreciate the enthusiastic assistance of N. Baratz, librarian. This work was supported by the National Science Foundation.

Note Added in Proof: A new review of thigmosensitive movements (TRONCHET, 1977) covers many of the topics included in this review.

Three new papers on chemical control of leaf movements have been published since this review went to press. KRIEGER (1978) confirmed earlier reports (Sec. 2.4.6.) that auxin application promotes bending at the 1° pulvinus of *P. vulgaris*, while BONNEMAIN et al. (1978) report that externally supplied abscisic acid causes *Cassia fasciculata* and *Mimosa* leaflets to close in the light (threshold effects at 0.1 μM and 10 μM, respectively), and the toxin fusicoccin (1 μM) induces opening in the dark in both species. Both of these compounds have dramatic effects on stomatal guard cells, abscisic acid causing K^+ loss and stomatal closure and fusicoccin promoting K^+ accumulation and stomatal opening (reviewed by HSIAO, 1976). SCHILDKNECHT et al. (1978) report that substances in extracts of *Mimosa* pulvini, including individual amino acids and mixtures of amino acids, stimulate leaflet movement when applied through the cut end of the rachis.

Ultrastructural features relevant to apoplastic and symplastic transport have also been reported. FLEURAT-LESSARD and BONNEMAIN (1978) describe membrane systems in xylem parenchyma and phloem cells that resemble those in transfer cells in other species, as well as a symplastic continuum extending from the middle of the vascular tissue to the cortex. Plasmodesmata, which are virtually all localized in pit fields, were quantified in the motor tissue of the 2° pulvinus of *Samanea* (MORSE and SATTER, in preparation). They are asymmetrically distributed, being more highly concentrated in the extensor than in the flexor and in the inner cortex than in the outer cortex.

References

Abe, T., Oda, K.: Resting and action potentials of excitable cells in the main pulvinus of *Mimosa pudica*. Plant Cell Physiol. *17*, 1343–1346 (1976)

Aimi, R.: Studies on the mechanism of seismonastic leaf movement in *Mimosa pudica* L. I. Existence of irritability in the upper half of the main pulvinus. Bot. Mag. *73*, 412–416 (1960)

Aimi, R.: Studies on the irritability of the pulvinus of *Mimosa pudica*. Bot. Mag. *76*, 374–380 (1963)

Aimi, R., Shibasaki, S.: Diurnal change in bioelectric potential of *Phaseolus* plant in relation to the leaf movement and light condition. Plant Cell Physiol. *16*, 1157–1162 (1975)

Akermann, A.: Untersuchungen über die Aggregation in den Tentakeln von *Drosera rotundifolia*. Bot. Notiser, 145–192 (1917)

Allen, R.D.: Mechanism of the seimonastic reaction in *Mimosa pudica*. Plant Physiol. *44*, 1101–1107 (1969)

Anderson, C.A.: An introduction to the electron probe microanalyzer and its application to biochemistry. In: Methods of Biochemical Analysis. Glick, D. (ed.). New York-London: Wiley, 1967, Vol. XV, pp. 147–270

Appleton, T.C.: A cryostat approach to ultrathin 'dry' frozen sections for electron microscopy: a morphological and X-ray analytical study. J. Microscopy *100*, 49–74 (1974)

Applewhite, P.B.: Behavioral plasticity in the sensitive plant, *Mimosa*. Behav. Biol. *7*, 47–53 (1972a)

Applewhite, P.B.: Drugs affecting sensitivity to stimuli in the plant *Mimosa* and the protozoan *Spirostomum*. Physiol. Behav. *9*, 869–871 (1972b)

Applewhite, P.B., Gardner, F.T.: Rapid leaf closure of *Mimosa* in response to light. Nature (London) *233*, 279–280 (1971)

Applewhite, P.B., Satter, R.L., Galston, A.W.: Protein synthesis during endogenous rhythmic leaflet movement in *Albizzia*. J. Gen. Physiol. *62*, 707–713 (1973)

Applewhite, P.B., Satter, R.L., Morse, M.J., Galston, A.W.: A simple and automatic leaflet movement monitoring system. Plant Physiol. *62*, 139–140 (1978)

Baillaud, L.: Anatomie physiologique des organes thigmotropiques et thigmonastiques. In: Encyclopedia of Plant Physiology. Ruhland, W. (ed.). Berlin-Heidelberg-New York: Springer, 1959, Vol. XVII, pt 1, pp. 243–253

Balmer, R.T., Franks, J.G.: Contractile characteristics of *Mimosa pudica* L. Plant Physiol. *56*, 464–467 (1975)

Bangerth, F.: Interaktionen von Auxin und Äthylen bei der thigmotropen Bewegung der Ranken von *Cucumis sativus*. Planta *117*, 329–338 (1974)

Baur, J.R., Morgan, P.W.: Effects of picloram and ethylene on leaf movement in huisache and mesquite seedlings. Plant Physiol. *44*, 831–838 (1969)

Biswas, S., Bose, D.M.: An ATPase in sensitive plant *Mimosa pudica*. I. Purification and characterization. Arch. Biochem. Biophys. *148*, 199–207 (1972)

Bonnemain, J.L., Roblin, G., Gaillochet, J., Fleurat-Lessard, P.: Effets de l'acide abscissique et de la fusicoccine sur les réactions motrices des pulvinus du *Cassia fasciculata* Michx. et du *Mimosa pudica* L. C.R. Soc. Biol. Ser. D *286*, 1681–1686 (1978)

Bose, J.C.: The Nervous Mechanism in Plants. London: Longmans, 1926

Brauner, L., Arslan, N.: Experiments on the auxin reactions of the pulvinus of *Phaseolus multiflorus*. Rev. Fac. Sci. Univ. Istanbul *16B*, 257–300 (1951)

Briggs, W.R.: The nature of the blue light photoreceptor in higher plants and fungi. In: Light and Plant Development. Smith, H. (ed.). London-Boston: Butterworths, 1976, pp. 7–18

Bünning, E.: Weitere Versuche über die Beziehung zwischen endogener Tagesrhythmik und Photoperiodismus. Z. Naturforsch. Teil B. *3b*, 457–464 (1948)

Bünning, E.: Allgemeine Gesetze und Phänomene der pflanzlichen Bewegungsphysiologie. In: Encyclopedia of Plant Physiology. Ruhland, W. (ed.). Berlin-Heidelberg-New York: Springer, 1959a. Vol. XVII, Pt. 1, pp. 8–23

Bünning, E.: Die seismonastischen Reaktionen. In: Encyclopedia of Plant Physiology. Ruhland, W. (ed.). Berlin-Heidelberg-New York: Springer, 1959b, Vol. XVII, Pt. 1, pp. 184–242

Bünning, E.: Die thigmonastischen und thigmotropischen Reaktionen. In: Encyclopedia of Plant Physiology. Ruhland, W. (ed.). Berlin-Heidelberg-New York: Springer, 1959c, Vol. XVII, Pt. 1, pp. 254–277
Bünning, E.: Tagesperiodische Bewegungen. In: Encyclopedia of Plant Physiology. Ruhland, W. (ed.). Berlin-Heidelberg-New York: Springer, 1959d, Vol. XVII, Pt. 1, pp. 579–656
Bünning, E., Baltes, J.: Wirkung von Äthylalkohol auf die physiologische Uhr. Naturwissenschaften *49*, 19 (1962)
Bünning, E., Baltes, J.: Zur Wirkung von schwerem Wasser auf die endogene Tagesrhythmik. Naturwissenschaften *50*, 622 (1963)
Bünning, E., Moser, I.: Response-Kurven bei der circadianen Rhythmik von *Phaseolus*. Planta *69*, 101–110 (1966a)
Bünning, E., Moser, I.: Difference in photoperiodic sensitivity of the upper and lower sides of leaves of *Kalanchoe blossfeldiana*. Planta *69*, 296–298 (1966b)
Bünning, E., Moser, I.: Interference of moonlight with the photoperiodic measurement of time by plants, and their adaptive reaction. Proc. Natl. Acad. Sci. USA *62*, 1018–1022 (1969)
Bünning, E., Moser, I.: Influence of valinomycin on circadian leaf movements of *Phaseolus*. Proc. Natl. Acad. Sci. USA *69*, 2732–2733 (1972)
Bünning, E., Moser, I.: Light-induced phase shifts of circadian leaf movements of *Phaseolus*: comparison with the effects of potassium and of ethyl alcohol. Proc. Natl. Acad. Sci. USA *70*, 3387–3389 (1973)
Bünning, E., Zimmer, R.: Zur Deutung der Phasenverschiebungen und "Transients" nach exogener Störung endogener Rhythmen. Planta *59*, 1–14 (1962)
Burdon-Sanderson, J.S.: Note on the electrical phenomena which accompany stimulation of the leaf of *Dionaea muscipula*. Proc. R. Soc. London *21*, 495–496 (1873)
Burkholder, P.R., Pratt, R.: Leaflet movement of *Mimosa pudica* in relationship to the intensity and wave length of the incident radiation. Am. J. Bot. *23*, 212–220 (1936)
Butcher, R.W., Sutherland, E.W.: Adenosine 3′,5′-phosphate in biological materials. J. Biol. Chem. *237*, 1244–1250 (1962)
Campbell, N.A.: Chloride localization in the leaf of *Tamarix*. Protoplasma *83*, 1–14 (1975)
Campbell, N.A., Thomson, W.W.: The ultrastructure of *Frankenia* salt glands. Ann. Bot. *40*, 681–686 (1976)
Campbell, N.A., Thomson, W.W.: Multivacuolate motor cells in *Mimosa pudica* L. Ann. Bot. *41*, 1361–1362 (1977a)
Campbell, N.A., Thomson, W.W.: The effects of lanthanum and ethylenediaminetetraacetate on the leaf movements of *Mimosa*. Plant Physiol. *60*, 635–639 (1977b)
Chafe, S.C., Durzan, D.J.: Tannin inclusions in cell suspension cultures of white spruce. Planta *113*, 251–262 (1973)
Chatterjee, S., Leopold, A.C.: Auxin structure and abscission activity. Plant Physiol. *38*, 268–273 (1963)
Charnley, T., Perrin, R., Porter, D.: Recovery process of the sensitive plant. Nature (London) *257*, 389–390 (1975)
Chaudhri, J., Satter, R.L., Berlyn, G.: Anatomical studies of the *Samanea* pulvinus. Plant Physiol. Ann. Supp. *59*, 85 (1977)
Cleland, R.: Auxin-induced hydrogen ion excretion from *Avena* coleoptiles. Proc. Natl. Acad. Sci. USA *70*, 3092–3093 (1973)
Darwin, C.R.: The Power of Movement in Plants. New York: D. Appleton and Co., 1881. Reprinted New York: De Capo Press, 1966
Darwin, C.R.: The Movements and Habits of Climbing Plants. New York: D. Appleton, 1893. Reprinted New York: AMS Press, 1972
Datta, M.: A new interpretation of the structure of the contractile vacuole in sensitive pulvini. Trans. Bose Res. Inst. *23*, 1–20 (1959)
De Mairan: Observation botanique. Histoire de l'Academie Royale des Sciences Paris 1729 p. 35
Driessche, T.Vanden: Implication of contractile proteins in the leaf movement in *Mimosa pudica* L. Ann. Physiol. Vegetale de l'Université de Bruxelles *8*, 101–112 (1963)
Dutt, B.K., Guha Thakurta, A.: Loss of turgor in the pulvinus of *M. pudica* due to excitatory contraction. Trans. Bose Res. Inst. *21*, 51–59 (1956)

Ebashi, S., Endo, M.: Calcium ions and muscle contraction. Prog. Biophys. Mol. Biol. *18*, 123–183 (1968)
Engelmann, W., Karlsson, H.G., Johnsson, A.: Phase shifts in the *Kalanchoë* petal rhythm, caused by light pulses of different duration. A theoretical and experimental study. Int. J. Chronobiol. *1*, 147–156 (1973)
Engelmann, W., Maurer, A., Mühlback, M., Johnsson, A.: Action of lithium ions and heavy water in slowing circadian rhythms of petal movement in *Kalanchoë*. J. Interdiscip. Cycle Res. *5*, 199–205 (1974)
Esau, K.: Plant Anatomy, 2nd ed. New York-London-Sidney: Wiley, 1953
Evans, L.T.: Flower induction and the florigen concept. Annu. Rev. Plant Physiol. *22*, 365–394 (1971)
Evans, L.T., Allaway, W.G.: Action spectrum for the opening of *Albizzia julibrissin* pinnules, and the role of phytochrome in the closing movements of pinnules and of stomata of *Vicia faba*. Aust. J. Biol. Sci. *25*, 885–893 (1972)
Feldman, J.F.: Lengthening the period of a biological clock in *Euglena* by cycloheximide, an inhibitor of protein synthesis. Proc. Natl. Acad. Sci. USA *57*, 1080–1087 (1967)
Feldman, J.F.: Circadian periodicity in *Neurospora*: Alteration by inhibitors of cyclic AMP phosphodiesterase. Science *190*, 789–790 (1975)
Fleurat-Lessard, P., Bonnemain, J.L.: Structural and ultrastructural characteristics of the vascular apparatus of the sensitive plant (*Mimosa pudica* L.) Protoplasma *94*, 127–143 (1978)
Fleurat-Lessard, P., Gavaudan, P.: Organisation structurale et ultrastructurale de l'organe sensorio-moteur principal en relation avec la physiologie du mouvement chez *Mimosa pudica* L. C.R. Soc. Biol. *167*, 332–337 (1973)
Fondéville, J.C.: La photostimulation chez *Mimosa pudica*. C. R. Soc. Biol. *157*, 1291–1294 (1963)
Fondéville, J.C.: Les rhythmes endogènes chez *Mimosa pudica*. Soc. Savantes de Paris, Nice, *2*, 391–401 (1965)
Fondéville, J.C., Borthwick, H.A., Hendricks, S.B.: Leaflet movements of *Mimosa pudica* L. indicative of phytochrome action. Planta *69*, 357–364 (1966)
Fondéville, J.C., Schneider, M.J., Borthwick, H.A., Hendricks, S.B.: Photocontrol of *Mimosa pudica* L. leaf movement. Planta *75*, 228–238 (1967)
Galston, A.W., Satter, R.L.: A study of the mechanism of phytochrome action. In: Structural and Functional Aspects of Phytochemistry, Recent Advances in Phytochemistry. Runeckles, V.C. Tso, T.C. (eds.). New York: Academic Press, 1972, Vol. V pp. 51–79
Galston, A.W., Satter, R.L.: Light, clocks and ion flux: an analysis of leaf movements. In: Light and Plant Development. Smith, H. (ed.). London-Boston: Butterworths, 1976, pp. 159–184
Gardner, G.: Ph.D. thesis, Yale University, New Haven, 1977
Gavaudan, P., Roblin, G.: Enregistrement et analyse des réactions motrices des feuilles de *Mimosa pudica* L. soumises à diverses orientations dans le champ de gravite. C. R. Soc. Biol. *163*, 743–746 (1969)
Gavaudan, P., Roblin, G.: Étude statistique de la sensibilité et de l'accomodation à l'excitant mécanique chez *Mimosa pudica* L. C. R. Soc. Biol. *165*, 162–166 (1971)
Halaban, R.: The circadian rhythm of leaf movement of *Coleus blumei* × *C. frederici*, a short day plant. I. Under constant light conditions. Plant Physiol. *43*, 1883–1886 (1968)
Halaban, R.: Effects of light quality on the circadian rhythm of leaf movement of a short-day-plant. Plant Physiol. *44*, 973–977 (1969)
Haslam, E.: Chemistry of Vegetable Tannins. London-New York: Academic Press, 1966
Hillman, W.S.: Entrainment of *Lemna* CO_2 output through phytochrome. Plant Physiol. *48*, 770–774 (1971)
Hillman, W.S., Koukkari, W.L.: Phytochrome effects in the nyctinastic leaf movements of *Albizzia julibrissin* and some other legumes. Plant Physiol. *42*, 1413–1418 (1967)
Holdsworth, M.: The leaf movement of soybean, a short day plant. J. Exp. Bot. *15*, 391–398 (1964)
Holmes, M.G., Smith, H.: The function of phytochrome in the natural environment. I. Characterization of daylight for studies in photomorphogenesis and photoperiodism. Photochem. Photobiol. *25*, 533–538 (1977)
Hoshizaki, T., Brest, D.E., Hamner, K.C.: The participation of two rhythms in the leaf

movements of *Xanthium* plants given various light-dark cycles. Plant. Physiol. *53*, 176–179 (1974)
Hoshizaki, T., Hamner, K.C.: Circadian leaf movements: Persistence in bean plants grown in continuous high-intensity light. Science *144*, 1240–1241 (1964)
Hoshizaki, T., Hamner, K.C.: Computer analysis of the leaf movements of Pinto beans. Plant. Physiol. *44*, 1045–1050 (1969)
Hoshizaki, T., Yokayama, K.: Recording leaf movements with a strain gauge. Nature (London) *207*, 880–881 (1965)
Houwink, A.L.: The conduction of excitation in *Mimosa pudica*. Recl. Trav. Bot. Neerl. *32*, 51–91 (1935)
Hsiao, T.C.: Stomatal ion transport. In: Encyclopedia of Plant Physiology, New Series. Lüttge, U., Pitman, M.G. (eds.). Berlin-Heidelberg-New York: Springer, 1976, Vol. II B, pp. 195–221
Jaffe, M.J.: Evidence for the regulation of phytochrome-mediated processes in bean roots by the neurohumor, acetylcholine. Plant Physiol. *46*, 768–777 (1970a)
Jaffe, M.J.: Reversible force transduction in tendrils of *Passiflora coerulea*. Plant Cell Physiol. *11*, 47–53 (1970b)
Jaffe, M.J.: Physiological studies of pea tendrils. VI. The characteristics of sensory perception and transduction. Plant Physiol. *45*, 756–760 (1970c)
Jaffe, M.J.: Physiological studies on pea tendrils. VII. Evaluation of a technique for the asymmetrical application of ethylene. Plant Physiol. *46*, 631–633 (1970d)
Jaffe, M.J.: Physiological studies on pea tendrils. VIII. The relationship of circumnutation to contact coiling. With a description of a laboratory intervalometer using integrated digital circuits. Physiol. Plant. *26*, 73–80 (1972)
Jaffe, M.J.: The role of auxin in the early events of the contact coiling of tendrils. Plant Sci. Lett. *5*, 217–225 (1975)
Jaffe, M.J.: Experimental separation of sensory and motor functions in pea tendrils. Science *195*, 191–192 (1977)
Jaffe, M.J., Galston, A.W.: Physiological studies on pea tendrils. I. Growth and coiling following mechanical stimulation. Plant Physiol. *41*, 1014–1025 (1966a)
Jaffe, M.J., Galston, A.W.: Physiological studies on pea tendrils. II. The role of light and ATP in contact coiling. Plant Physiol. *41*, 1152–1158 (1966b)
Jaffe, M.J., Galston, A.W.: Physiological studies on pea tendrils. III. ATPase activity and contractility associated with coiling. Plant Physiol. *42*, 845–847 (1967a)
Jaffe, M.J., Galston, A.W.: Physiological studies on pea tendrils. IV. Flavonoids and contact coiling. Plant Physiol. *42*, 848–850 (1967b)
Jaffe, M.J., Galston, A.W.: Phytochrome control of rapid nyctinastic movements and membrane permeability in *Albizzia julibrissin*. Planta *77*, 135–141 (1967c)
Jaffe, M.J., Galston, A.W.: Physiological studies on pea tendrils. V. Membrane changes and water movement associated with contact coiling. Plant Physiol. *43*, 537–542 (1968a)
Jaffe, M.J., Galston, A.W.: The Physiology of Tendrils. Annu. Rev. Plant Physiol. *19*, 417–434 (1968b)
Jonas, H.: *Mimosa pudica* L., responses to electrical and mechanical stimuli, cardenolides and light. Z. Pflanzenphysiol. *80*, 395–406 (1976)
Jones, M.G.K., Novacky, A., Dropkin, V.H.: Transmembrane potentials of parenchyma cells and nematode-induced transfer cells. Protoplasma *85*, 15–37 (1975)
Jones, R.J., Mansfield, T.A.: Suppression of stomatal opening in leaves treated with abscisic acid. J. Exp. Bot. *21*, 714–719 (1970)
Junker, G., Mayer, W.: Die Bedeutung der Epidermis für licht- und temperaturinduzierte Phasenverschiebungen circadianer Laubblattbewegungen. Planta *121*, 27–37 (1974)
Junker, S.: Auxin transport in tendril segments of *Passiflora caerulea*. Physiol. Plant. *37*, 258–262 (1976a)
Junker, S.: A scanning electron microscope study on the development of tendrils of *Parthenocissus tricuspidata* Sieb. and Zucc. New Phytol. *77*, 741–746 (1976b)
Junker, S.: Ultrastructure of tactile papillae on tendrils of *Eccremocarpus scaber* R. and P. New Phytol. *78*, 607–610 (1977a)
Junker, S.: Thigmotropic coiling of tendrils of *Passiflora quadrangularis* L. is not caused by lateral redistribution of auxin. Physiol. Plant. *41*, 51–54 (1977b)

Junker, S., Reinhold, L.: A scanning electron microscope survey of the surface of sensitive tendrils. J. Microsc. Biol. Cell *23*, 175–180 (1975)
Kadman-Zahavi, A.: Persistence of P_{fr} and other manifestations of phytochrome in the nyctinastic leaf movements of peanut (*Arachis hypogaea* L.). Israel J. Bot. *21*, 142–149 (1972)
Kandeler, R.: Blühinduktion bei Lemnaceen. Biol. Rundsch. *6*, 49–57 (1968)
Keegstra, K., Talmadge, K.W., Bauer, W.D., Albersheim, P. The structure of plant cell walls. III. A model of the walls of suspension-cultured sycamore cells based on the interconnections of the macromolecular components. Plant Physiol. *51*, 188–197 (1973)
Kende, H., Hanson, A.D.: Relationship between ethylene evolution and senescence in morning-glory flower tissue. Plant Physiol. *57*, 523–527 (1976)
Kimura, K.: Floral induction in *Pharbitis nil* subjected to continuous illumination at relatively low temperatures. II. Effect of some factors in culture medium on floral initiation. Bot. Mag. *76*, 351–358 (1963)
Kiyosawa, K., Tanaka, H.: Change in potassium distribution in a *Phaseolus* pulvinus during circadian movement of the leaf. Plant Cell Physiol. *17*, 289–298 (1976)
Komor, E., Tanner, W.: The hexose-proton symport system of *Chlorella vulgaris*. Eur. J. Biochem. *44*, 219–223 (1974)
Koukkari, W.L., Hillman, W.S.: Pulvini as the photoreceptors in the phytochrome effect on nyctinasty in *Albizzia julibrissin*. Plant Physiol. *43*, 698–704 (1968)
Krieger, K.G.: Early time course and specificity of auxin effects on turgor movement of the bean pulvinus. Planta *140*, 107–109 (1978)
Läuchli, A., Spurr, A.R., Wittkopp, R.W.: Electron probe analysis of freeze-substituted, epoxy resin embedded tissue for ion transport studies in plants. Planta *95*, 341–350 (1970)
Leonard, R.T., Nagahashi, G., Thomson, W.W.: Effect of lanthanum on ion absorption in corn roots. Plant Physiol. *55*, 542–546 (1975)
Lörcher, L.: Die Wirkung verschiedener Lichtqualitäten auf die endogene Tagesrhythmik von *Phaseolus*. Z. Bot. *46*, 207–241 (1958)
Lyubimova, M.N., Demyanovskaya, N.S., Fedorovich, I.B., Itomlenskite, I.V.: Participation of ATP in the motor function of the *Mimosa pudica* leaf. Translated from Biokhimiya *29*, 774–779 (1964)
Lyubimova, M.N., Fain, F. S., Demyanovskaya, N.S.: Ultrasonic isolation of structural protein possessing Mg-ATPase activity from pulvini of *Mimosa pudica*. Translated from Dokl. Acad. Nauk SSSR *185*, 206–209 (1969)
Macdougal, D.T.: The mechanism of curvature of tendrils. Ann. Bot. *10*, 373–402 (1896)
MacRobbie, E.A.C.: Accumulation of ions in plant cell vacuoles. In: Perspectives in Experimental Biology. Sunderland, N. (ed.). New York: Pergamon, 1976, Vol. II, pp. 369–380
MacRobbie, E.A.C.: Functions of ion transport in plant cells and tissues. In: International Review of Biochemistry, Plant Biochemistry II. Northcote, D.H. (ed.). Baltimore: University Park Press, 1977, Vol. 13, pp. 211–247
Mayer, W.: Further indications for the role of the epidermis as the site of light perception in circadian leaf movement. Z. Naturforsch. *28*c, 776 (1973)
Mayer, W.: Kalium und Chloridverteilung im Laminargelenk von *Phaseolus coccineus* L. während der circadianen Blattbewegung im tagesperiodischen Licht-Dunkelwechsel. Z. Pflanzenphysiol. *83*, 127–135 (1977)
Mayer, W., Gruner, R., Strubel, H.: Periodenverlängerung und Phasenverschiebungen der circadianen Rhythmik von *Phaseolus coccineus* L. durch Theophyllin. Planta *125*, 141–148 (1975)
Mayer, W., Moser, I., Bünning, E.: The epidermis as the site of light perception in circadian leaf movement and photoperiod induction. Z. Pflanzenphysiol. *70*, 66–73 (1973)
Mayer, W., Scherer, I.: Phase shifting effect of caffeine in the circadian rhythm of *Phaseolus coccineus* L. Z. Naturforsch *30c*, 855–856 (1975)
McEvoy, R.C., Koukkari, W.L.: Effects of ethylenediaminetetraacetic acid, auxin and gibberellic acid on phytochrome controlled nyctinasty in *Albizzia julibrissin*. Physiol. Plant. *26*, 143–147 (1972)
Miller, C.S.: Short interval leaf movements of cotton. Plant Physiol. *55*, 562–566 (1975)
Mooseker, M.S., Tilney, L.G.: Organization of an actin filament-membrane complex. Filament polarity and membrane attachment in the microvilli of intestinal epithelial cells. J. Cell Biol. *67*, 725–743 (1975)

Morgan, P.W., Baur, J.R.: Involvement of ethylene in picloram-induced leaf movement response. Plant Physiol. *46*, 655–659 (1970)
Muñoz, V., Butler, W.L.: Photoreceptor pigment for blue light in *Neurospora crassa*. Plant Physiol. *55*, 421–426 (1975)
Oda, K.: Recording of the potassium efflux during a single action potential in *Chara corallina*. Plant Cell Physiol. *16*, 525–528 (1975)
Oda, K., Abe, T.: Action potential and rapid movement in the main pulvinus of *Mimosa pudica*. Bot. Mag. *85*, 135–145 (1972)
Palevitz, B.A., Hepler, P.K.: Is P-protein actin-like? Not yet. Planta *125*, 261–271 (1975)
Palmer, J.H., Asprey, G.F.: Studies in the nyctinastic movement of the leaf pinnae of *Samanea saman* (Jacq) Merrill. I. A general description of the effect of light on the nyctinastic rhythm. Planta *51*, 757–769 (1958a)
Palmer, J.H., Asprey, G.F.: Studies in the nyctinastic movement of the leaf pinnae of *Samanea saman* (Jacq) Merrill. II. The behaviour of upper and lower half-pulvini. Planta *51*, 770–785 (1958b)
Paul, D.C., Goff, C.H.W.: Comparative effects of caffeine, its analogues and calcium deficiency on cytokinesis. Exp. Cell Res. *78*, 399–413 (1973)
Pfeffer, W.: The Physiology of Plants. Translated into English by A.J. Ewart. Oxford: Clarendon Press, 1905. Vol. III
Pickard, B.G.: Action potentials in higher plants. Bot. Rev. *39*, 172–201 (1973)
Poff, K.L., Butler, W.L.: Absorbance changes induced by blue light in *Phycomyces blakesleeanus* and *Dictyostelium discoideum*. Nature (London) *248*, 799–801 (1974)
Poff, K.L., Butler, W.L.: Spectral characterization of the photo-reducible b-type cytochrome of *Dictyostelium discoideum*. Plant Physiol. *55*, 427–429 (1975)
Posner, H.B.: Inhibitory effect of carbohydrate on flowering in *Lemna perpusilla*. I. Interaction of sucrose with calcium and phosphate ions. Plant Physiol. *44*, 562–566 (1969)
Racusen, R.H.: Phytochrome control of electrical potentials and intercellular coupling in oat-coleoptile tissue. Planta *132*, 25–29 (1976)
Racusen, R.H., Etherton, B.: Role of membrane-bound, fixed-charge changes in phytochrome-mediated mung bean root tip adherence phenomenon. Plant. Physiol. *55*, 491–495 (1975)
Racusen, R.H., Galston, A.W.: Electrical evidence for rhythmic changes in the cotransport of sucrose and hydrogen ions in *Samanea* pulvini. Planta *135*, 57–62 (1977)
Racusen, R.H., Satter, R.L.: Rhythmic and phytochrome-regulated changes in transmembrane potential in *Samanea* pulvini. Nature (London) *255*, 408–410 (1975)
Reinhold, L.: Induction of coiling in tendrils by auxin and carbon dioxide. Science *158*, 791–793 (1967)
Reinhold, L., Sachs, T., Vislovska, L.: The role of auxin in thigmotropism. In: Plant Growth Substances. Carr, D.J. (ed.). Berlin-Heidelberg-New York: Springer 1970, pp. 731–737
Robards, A.W.: Plasmodesmata in higher plants. In: Intercellular Communication in Plants: Studies on Plasmodesmata. Gunning, B.E.S., Robards, A.W. (eds.). Berlin-Heidelberg-New York: Springer 1976, pp. 15–58
Roblin, G.: Inhibition of the pulvinar geotropic movements in *Mimosa pudica* by triiodobenzoic acid. Naturwissenschaften *63*, 42 (1976a)
Roblin, G.: Experimental analysis of seismonastic movement in the sensitive plant. Nature (London) *261*, 437–438 (1976b)
Roblin, G., Gavaudan, P.: Inhibition par la photostimulation chez *Mimosa pudica* L. par préstimulations lumineuses sous-liminaires. C. R. Soc. Biol. *161*, 666–670 (1967)
Roblin, G., Gavaudan, P.: Caractère différentiel de la photostimulation chez diverses plantes douées de propriétés sensorio-motrices. C.R. Soc. Biol. *168*, 781–787 (1974)
Sanberg, P.R.: "Neural capacity" in *Mimosa*: a review. J. Behav. Biol. *17*, 435–452 (1976)
Satter, R.L., Applewhite, P.B., Chaudhri, J., Galston, A.W.: P_{fr} phytochrome and sucrose requirement for rhythmic leaflet movement in *Albizzia*. Photochem. Photobiol. *23*, 107–112 (1976)
Satter, R.L., Applewhite, P.B., Galston, A.W.: Phytochrome controlled nyctinasty in *Albizzia julibrissin*. V. Evidence against acetylcholine participation. Plant Physiol. *50*, 523–525 (1972b)
Satter, R.L., Applewhite, P.B., Galston, A.W.: Rhythmic potassium flux in *Albizzia*: effect of aminophylline, cations and inhibitors of respiration and protein synthesis. Plant Physiol. *54*, 280–285 (1974c)

Satter, R.L., Applewhite, P.B., Kreis, D.J., Jr., Galston, A.W.: Rhythmic leaflet movement in *Albizzia julibrissin*: effect of electrolytes and temperature alteration. Plant Physiol. *52*, 202–207 (1973)
Satter, R.L., Galston, A.W.: Potassium flux: a common feature of *Albizzia* leaflet movement controlled by phytochrome or endogenous rhythm. Science *174*, 518–520 (1971a)
Satter, R.L., Galston, A.W.: Phytochrome-controlled nyctinasty in *Alibizzia julibrissin*. III. Interaction between an endogenous rhythm and phytochrome in control of potassium flux and leaflet movement. Plant Physiol. *48*, 740–746 (1971b)
Satter, R.L., Galston, A.W.: Leaf movements: rosetta stone of plant behavior? BioScience *23*, 407–416 (1973)
Satter, R.L., Geballe, G.T., Applewhite, P.B., Galston, A.W.: Potassium flux and leaf movement in *Samanea saman*. I. Rhythmic movement. J. Gen. Physiol. *64*, 413–430 (1974a)
Satter, R.L., Geballe, G.T., Galston, A.W.: Potassium flux and leaf movement in *Samanea saman*. II. Phytochrome controlled movement. J. Gen. Physiol. *64*, 431–442 (1974b)
Satter, R.L., Marinoff, P., Galston, A.W.: Phytochrome controlled nyctinasty in *Albizzia julibrissin*. II. Potassium flux as a basis for leaflet movement. Am. J. Bot. *57*, 916–926 (1970b)
Satter, R.L. Marinoff, P., Galston, A.W.: Phytochrome controlled nyctinasty in *Albizzia julibrissin*. IV. Auxin effects on leaflet movement and K flux. Plant Physiol. *50*, 235–241 (1972a)
Satter, R.L., Sabnis, D.D., Galston, A.W.: Phytochrome controlled nyctinasty in *Albizzia julibrissin*. I. Anatomy and fine structure of the pulvinule. Am. J. Bot. *57*, 374–381 (1970a)
Satter, R.L., Schrempf, M., Chaudhri, J., Galston, A.W.: Phytochrome and circadian clocks in *Samanea*: rhythmic redistribution of potassium and chloride within the pulvinus during long dark periods. Plant Physiol. *59*, 231–235 (1977)
Schildknecht, H., Tauscher, B., Pesh-Imam, M., Beltle, W., Kunzelmann, P., Schneider, D.: Aminosäuren als „leaf movement factors". Pflanzenabwehrstoffe X. Über die Biologische Chemie der Mimosaceen II. Naturwissenschaften *65*, 125–129 (1978)
Schrempf, M.: Eigenschaften und Lokalization des Photorezeptors für phasenverschiebendes Störlicht bei der Blütenblattbewegung von *Kalanchoe blossfeldiana*. Ph. D. thesis, Tübingen 1975
Schrempf, M., Satter, R.L. Galston, A.W.: Potassium-linked chloride fluxes during rhythmic leaf movement of *Albizzia julbrissin*. Plant Physiol. *58*, 190–192 (1976)
Schwabe, W.W.: Studies on the role of the leaf epiderm in photoperiodic perception in *Kalanchoe blossfeldiana*. J. Exp. Bot. *19*, 108–113 (1968)
Schwintzer, C.R.: Energy budgets and temperatures of nyctinastic leaves on freezing nights. Plant Physiol. *48*, 203–207 (1971)
Scott, B.I.H., Gulline, H.F.: Natural and forced circadian oscillations in the leaf of *Trifolium repens*. Aust. J. Biol. Sci. *25*, 61–76 (1972)
Scott, B.I.H., Gulline, H.F.: Membrane changes in a circadian system. Nature (London) *254*, 69–70 (1975)
Scott, B.I.H., Gulline, H.F., Robinson, G.R.: Circadian electrochemical changes in the pulvinules of *Trifolium repens*. Aust. J. Plant Physiol. *4*, 193–206 (1977)
Setty, S., Jaffe, M.J.: Phytochrome-controlled rapid contraction and recovery of contractile vacuoles in the motor cells of *Mimosa pudica* as an intracellular correlate of nyctinasty. Planta *108*, 121–131 (1972)
Sibaoka, T.: Excitable cells in *Mimosa*. Science *137*, 226 (1962)
Sibaoka, T.: Action potentials in plant organs. Symp. Soc. Exp. Biol. *20*, 49–73 (1966)
Sibaoka, T.: Physiology of rapid movements in higher plants. Annu. Rev. Plant Physiol. *20*, 165–184 (1969)
Simon, E., Satter, R.L., Galston, A.W.: Circadian rhythmicity in excised *Samanea* pulvini. I. Sucrose-white light interactions. Plant Physiol. *58*, 417–420 (1976a)
Simon, E., Satter, R.L., Galston, A.W.: Circadian rhythmicity in excised *Samanea* pulvini. II. Resetting the clock by phytochrome conversion. Plant Physiol. *58*, 421–425 (1976b)
Slayman, C.L.: Proton pumping and generalized energetics of transport: a review. In: Membrane Transport in Plants. Zimmermann, U., Dainty, J. (eds.). Berlin-Heidelberg-New York: Springer, 1974, pp. 107–119
Slayman, C.L., Slayman, C.W.: Depolarization of the plasma membrane of *Neurospora* during

active transport of glucose: evidence for a proton-dependent cotransport system. Proc. Natl. Acad. Sci. USA *71*, 1935–1939 (1974)

Smith, A.P.: Bud temperature in relation to nyctinastic leaf movement in an Andean giant rosette plant. Biotropica *6*, 263–266 (1974)

Spyropoulos, C.S., Tasaki, I., Hayward, G.: Fractionation of tracer effluxes during action potential. Science *133*, 2064–2065 (1961)

Stoeckel, H.: Ph.D. thesis, L'Université Louis Pasteur de Strasbourg, 1976

Stoeckel, H., Pfirsch, R.: Modification temporaire des relations fonctionelles entre pulvinus primaire et petiole chez feuilles jeunes de *Mimosa pudica* L. Z. Pflanzenphysiol. *76*, 1–13 (1975)

Sweet, H.C., Hillman, W.S.: Phytochrome control of nyctinasty in *Samanea* as modified by oxygen, submergence, and chemicals. Physiol. Plant. *22*, 776–786 (1969)

Tanada, T.: A rapid photoreversible response of barley root tips in the presence of 3-indoleacetic acid. Proc. Natl. Acad. Sci. USA *59*, 376–380 (1968)

Thomson, W.W., De Journett, R.: Studies on the ultrastructure of the guard cells of *Opuntia*. Am. J. Bot. *57*, 309–316 (1970)

Tibbitts, T.W., Nutter, G.D., Hoshizaki, T., Badeau, R.R.: Modified photoelectric device for recording leaf movements. Plant Physiol. *51*, 812–814 (1973)

Toriyama, H.: Observational and experimental studies of sensitive plants. I. The structure of parenchymatous cells of pulvinus. Cytologia *18*, 283–292 (1953)

Toriyama, H.: Observational and experimental studies of sensitive plants. II. On the changes in motor cells of diurnal and nocturnal condition. Cytologia *19*, 29–40 (1954a)

Toriyama, H.: Observational and experimental studies of sensitive plants. III. On the epidermal system of the pulvinus and leaflet. Cytologia *19*, 286–298 (1954b)

Toriyama, H.: Observational and experimental studies of sensitive plants. V. The development of the tannin vacuole in the motor cell of the pulvinus. Bot. Mag. *68*, 203–208 (1955a)

Toriyama, H.: Observational and experimental studies of sensitive plants. VI. The migration of potassium in the primary pulvinus. Cytologia *20*, 367–377 (1955b)

Toriyama, H.: Observational and experimental studies of sensitive plants. VIII. The migration of colloidal substance in the primary pulvinus. Cytologia *22*, 184–192 (1957)

Toriyama, H.: Observational and experimental studies of sensitive plants. IX. On the canaliculated intercellular spaces of the primary pulvinus. Bot. Mag. *71*, 11–15 (1958)

Toriyama, H.: Observational and experimental studies of sensitive plants. XIV. On the changes in a new cellular element of *Mimosa pudica* in diurnal and nocturnal conditions. Cytologia *27*, 276–284 (1962a)

Toriyama, H.,: Observational and experimental studies of sensitive plants. XV. The migration of potassium in the petiole of *Mimosa pudica*. Cytologia *27*, 431–442 (1962b)

Toriyama, H.: Histochemical detection of ATP-ase activity in motor tissue of *Mimosa pudica*. Bot. Mag. *76*, 79–80 (1963)

Toriyama, H.: Tannins and the tannin vacuole in the motor organ of higher plants. Science Reports of Tokyo Woman's Christian Coll. Nos. *29–31*, 368–382 (1973)

Toriyama, H., Jaffe, M.J.: Migration of calcium and its role in the regulation of seismonasty in the motor cell of *Mimosa pudica* L. Plant Physiol. *49*, 72–81 (1972)

Toriyama, H., Komada, Y.: The recovery process of the tannin vacuole in the motor cell of *Mimosa pudica* L. Cytologia *36*, 690–697 (1971)

Toriyama, H., Satô, S.: On the membrane of the tannin vacuole in the *Mimosa* motor cell. Proc. Jpn. Acad. *44*, 528–532 (1968a)

Toriyama, H., Satô, S.: Electron microscope observation of the motor cell of *Mimosa pudica* L. I. A comparison of the motor cell before and after stimulation. Proc. Jpn. Acad. *44*, 702–706 (1968b)

Toriyama, H., Satô, S.: Electron microscope observation of the motor cell of *Mimosa pudica* L. II. On the contents of the central vacuole of the motor cell. Proc. Jpn. Acad. *44*, 949–953 (1968c)

Toriyama, H., Satô, S.: Electron microscope observation of the motor cell of *Mimosa pudica* L. III. The fine structure of the central vacuole in the motor cell. Proc. Jpn. Acad. *45*, 175–179 (1969)

Toriyama, H., Satô, S.: On the contents of the central vacuole in the *Mimosa* motor cell. Cytologia *36*, 359–375 (1971)

Tronchet, A.: Periodicité et nutation révolutive des vrilles. Ann. Sci. Univ. Besancon *12*, 71–80 (1958)
Tronchet, A.: Quelques aspects de le sensibilité des vrilles. Ann. Sci. Univ. Besancon *18*, 3–17 (1962)
Tronchet, A.: La Sensibilité des Plantes. Paris-New York-Barcelone-Milan: Masson, 1977
Tronchet, J.: Étude par chromatographie sur papier des modifications biochimiques réversibles induites par la stimulation de contact dans les vrilles d'*Echinoscystis wrightii*. Ann. Sci. Univ. Besancon *15*, 75–80 (1960)
Tyree, M.T.: The symplast concept. A general theory of symplastic transport according to the thermodynamics of irreversible processes. J. Theor. Biol. *26*, 181–214 (1970)
Umrath, K.: Über die elektrischen Erscheinungen bei thigmischer Reizung der Ranken von *Cucumis pepo*. Planta *23*, 47–50 (1934)
Umrath, K.: Mögliche Mechanismen von Krümmungsbewegungen. In: Encyclopedia of Plant Physiology. Ruhland, W. (ed.). Berlin-Heidelberg-New York: Springer, 1959, Vol. XVII, pt. 1, pp. 111–118
Van Steveninck, R.F.M.: Cytochemical evidence for ion transport through plasmodesmata. In: Intercellular Communication in Plants: Studies on Plasmodesmata. Gunning, B.E.S., Robards, A.W. (eds.). Berlin-Heidelberg-New York: Springer, 1976. pp. 131–148
Wareing, P.F., Thompson, A.G.: Rapid effects of red light on hormone levels. In: Light and Plant Development. Smith, H. (ed.). London-Boston: Butterworths, 1976, pp. 285–294
Watanabe, S.: Electron microscope studies of *Mimosa* pulvinus cells before and after movement. Artes Liberales, Iwate Univ. *6*, 49–51 (1970)
Watanabe, S.: Ouabain and IAA effects on *Mimosa* pulvinus. Artes Liberales, Iwate Univ. *8*, 75–80 (1971)
Watanabe, S., Sibaoka, T.: Site of photo-reception to opening response in *Mimosa* leaflets. Plant Cell Physiol. *14*, 1221–1224 (1973)
Weintraub, M.: Leaf movements in *Mimosa pudica* L. New Phytol. *50*, 357–382 (1952)
Wilkins, M.B., Harris, P.J.C.: Phytochrome and phase setting of endogenous rhythms. In: Light and Plant Development. Smith, H. (ed.). London-Boston: Butterworths, 1976, pp. 399–418
Williams, C.N., Raghavan, V.: Effects of light and growth substances on the diurnal leaf movements of the leaflets of *Mimosa pudica*. J. Exp. Bot. *17*, 742–749 (1966)
Zimmer, R.: Phasenverschiebung und andere Störlichtwirkungen auf die endogen tagesperiodischen Blütenblattbewegungen von *Kalanchoë blossfeldiana*. Planta *58*, 283–300 (1962)
Zimmermann, U., Steudle, E.: The pressure-dependence of the hydraulic conductivity, the membrane resistance and the membrane potential during turgor pressure regulation in *Valonia utricularis*. J. Membr. Biol. *16*, 331–352 (1974)
Zimmermann, U., Steudle, E., Lelkes, P.I.: Turgor pressure regulation in *Valonia utricularis*. Effect of cell wall elasticity and auxin. Plant Physiol. *58*, 608–613 (1976)

Growth movements Tropisms

5. Growth Movements

5.1 Growth Movements Directed by Light

5.1.1 Induction of Polarity

M.H. WEISENSEEL

1. Introduction

The acquisition of polarity is one of the first and fundamental steps in development. It is a very early expression of differentiation and orientation in space. It becomes apparent in the unequal distribution of substances, in unequal cell division and in unequal growth.

The induction of polarity, i.e., the process of establishment of a main morphological axis with two distinct poles, is intimately related to movements.

1. A patterned redistribution of cytoplasmic and surface components underlies and precedes visible polarity in plant cells. In the germinating spores of *Equisetum*, for instance, chloroplasts move into the future thallus region and the nucleus migrates toward the future rhizoid end (v. WETTSTEIN, 1965). In the developing zygotes of *Fucus*, mitochondria, ribosomes, osmiophilic bodies, sulphated polysaccharides, and RNA move toward the prospective rhizoid site (QUATRANO, 1972; QUATRANO and CRAYTON, 1973; NAKAZAWA and TAKAMURA, 1967).

2. Orientation of growth toward optimal conditions is an important consequence of polarization in plants. In *Funaria* spores, for instance, induction of polarity by light causes the photosynthetically active chloronema to grow toward the light and the chloroplast-free rhizoid to grow away from it (v. WETTSTEIN, 1965). Polarization also plays a major role in the orientation of growth in higher plants and their organs. In seedlings, for example, upward growth of the shoot and downward growth of the root is effected by the influence of polarizing environmental factors (light, gravity; see DENNISON, Chap. 5.1.2, VOLKMANN and SIEVERS, Chap. 5.2.2, and WILKINS, Chap. 5.2.3, this vol. for discussion of phototropism and geotropism, respectively).

Most of our knowledge about the induction of polarity has been gained from investigations of individual plant cells. Nearly all plant phyla include species which release apolar gametes, spores or vegetative cells. These cells have a number of advantages for the study of polarization: they are released as individual cells that can be obtained in great numbers, they can be manipulated and polarized by environmental factors, and they can easily be cultured and observed without complications from surrounding tissue. Apolar eggs, gametes, or spores are released into the seawater by marine brown algae (*Fucus, Pelvetia, Cystosira, Hormosira,*

Abbreviations. IAA indole-3-acetic acid, NAA α-NAA or β-NAA, naphthalene-1-acetic acid, or naphthalene-2-acetic acid, AMP adenosine 5′-monophosphate, c-AMP adenosine 3′:5′-cyclic monophosphate, DNP 2,4-dinitrophenol.

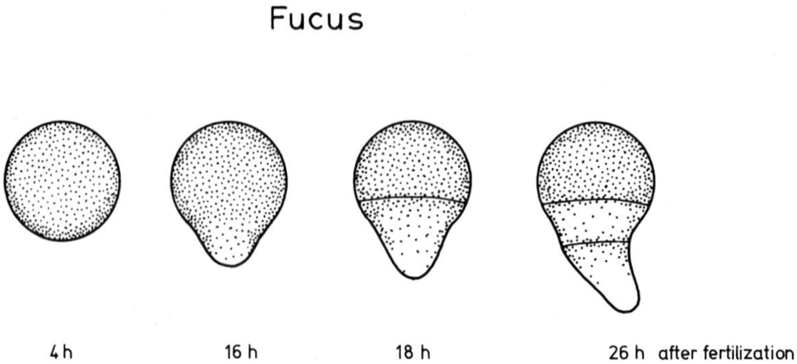

Fig. 1. Early development of *Fucus vesiculosus*. 4 h after fertilization—a spherically symmetrical cell; 16 h after fertilization—a rhizoidal bulge, but no cell wall has formed; 18 h after fertilization—a wall between rhizoid and thallus has formed; 26 h after fertilization—further elongation and division of the rhizoid cell. (Redrawn according to microphotographs; from JAFFE, 1969)

Ascophyllum), by marine green algae (*Ulva, Codium*) and by marine red algae (*Ceramium, Polysiphonia, Chondrus, Bangia*) (JAFFE, 1958a; v. WETTSTEIN, 1965; WEBER, 1969; SAND, 1973). Among the terrestrial plants, apolar spores are known in fungi (*Botrytis*), mosses (*Funaria*), ferns (*Dryopteris, Osmunda, Onoclea*) and horsetails (*Equisetum*) (JAFFE, 1958a, HAUPT, 1962; v. WETTSTEIN, 1965). Also the vegetative cell of a pollen grain is apparently apolar, since external factors can determine the location of outgrowth (e.g., *Vinca,* BEAMS and KING, 1944; MARSH and BEAMS, 1945).

The prototypes of polarization are the zygotes of *Fucus* and *Pelvetia.* Naked eggs of 60–100 μm diameter, which are densely filled with cytoplasm, are released into the sea from the receptacles of the parent plant and sink to the ground. Within the first few hours after fertilization, the spherically symmetrical zygote secretes a wall and a "glue" which fastens it to almost any substratum.

In seawater at 15° C, visible growth in *Fucus* begins at 12–14 h after fertilization (8–10 h in *Pelvetia*), with the appearance of a bulge in the cell surface (Fig. 1). This protuberance will eventually develop into the multicellular rhizoid, the root-like structure that anchors the embryo to the substratum. The emergence of the presumptive rhizoid marks the establishment of the main morphological axis, with a pointed rhizoid pole and a rounded "thallus" pole. Less than a day later, the first cell division occurs, perpendicular to the main axis and closer to the rhizoid pole. Subsequent cell divisions in the "thallus" part of the zygote lead to formation of the thallus, the main body of the plant.

Until shortly before visible growth begins, an axis can be induced by environmental factors. Experiments with *Fucus* zygotes indicate that polarity can really be induced, not just reoriented. When illuminated with plane-polarized light, up to 50% of the zygotes grow two rhizoids in opposite directions (the others grow one rhizoid in either of these two directions) (JAFFE, 1956, 1958b, 1968). Since a preexisting axis cannot rotate in two different directions at once, polarity must have been established de novo. Although a new axis can really be induced, the fucoid zygote already has some weak polarity. This axis is induced by fertilization.

When other inducing factors are absent, the zygote germinates from the site where the sperm had entered (KNAPP 1931. In: v. WETTSTEIN, 1965).

The major problems which will be considered in this review are perception and transduction of polarity-inducing factors (see also BENTRUP, Chap. 1.3, this vol.), and amplification of a labile polarity. The main, or most important inducing factors are probably light, chemical gradients, and electric fields generated by the developing organism. Much has also been learned in recent years about spontaneous cellular events during polarization, by the application of some new cell physiology techniques. The results point to a primary role of the plasma membrane during induction of polarity, and to an electrical mechanism for the process of amplification.

Comprehensive reviews, or reviews stressing a particular vector or mechanism of polarization have been written by HAUPT (1962), v. WETTSTEIN (1965), BLOCH (1965), JAFFE (1968, 1969), and BENTRUP (1971a). More general reviews including the topic of polarity have been written by JAFFE (1958a), HAUPT (1965), STANGE (1965), WARDLAW (1965), GREEN (1969), BOPP (1975), and SEITZ (1975).

2. Light as an Inducing Factor

Unilateral light, illumination of part of the cell, or illumination with plane-polarized light can induce polarity in the zygotes of *Fucus* and *Pelvetia*, as well as in the spores of *Botrytis*, *Funaria*, *Dryopteris*, *Osmunda* and *Equisetum*.

The main morphological axis induced by light is always oriented parallel to the largest gradient of absorption. The direction of this inducing absorption gradient might be different from the direction of the incident light. For instance, spores that are partially illuminated from below, so that half the spore is in the light, and the other half is shaded, germinate at an angle of about 90° to the incident light direction, they germinate either at the shaded pole (*Osmunda*, Fig. 2a) or at the illuminated pole (*Botrytis*, Fig. 2b) (JAFFE and ETZOLD, 1962). In *Fucus* zygotes, strong unilateral light elicits an outgrowth at the rear, self-shaded pole (Fig. 2c), light of moderate intensities (10^{15}–10^{16} quanta cm^{-2}) induces germination at 90–135° away from the light source, the so-called "subequatorial response" (Fig. 2e; JAFFE, 1958b).

The results of experiments with partial illumination and unilateral illumination, together with a consideration of the absorption and refraction properties of the cell content, yield information about the correlation between pattern of absorption and site of growth. The future site of germination is either the site of strongest light absorption (*Botrytis*, Fig. 2d) or the site of least light absorption (*Osmunda*, Fig. 2a; *Fucus*, Fig. 2c and 2e; also *Equisetum* and *Dryopteris*). In the translucent spore of *Botrytis*, the zone of strongest light absorption is a site at the rear part of the cell, onto which the light is focused by the cytoplasm (ETZOLD, 1961). In *Fucus*, the zone of least light absorption is the rear hemisphere, particularly the zone beneath the cell's equator which is bypassed by the light (JAFFE, 1958b).

The inducing absorption gradient for an individual cell must be rather large; JAFFE and ETZOLD (1962, 1965) estimated that across a spore of *Botrytis* or *Os-*

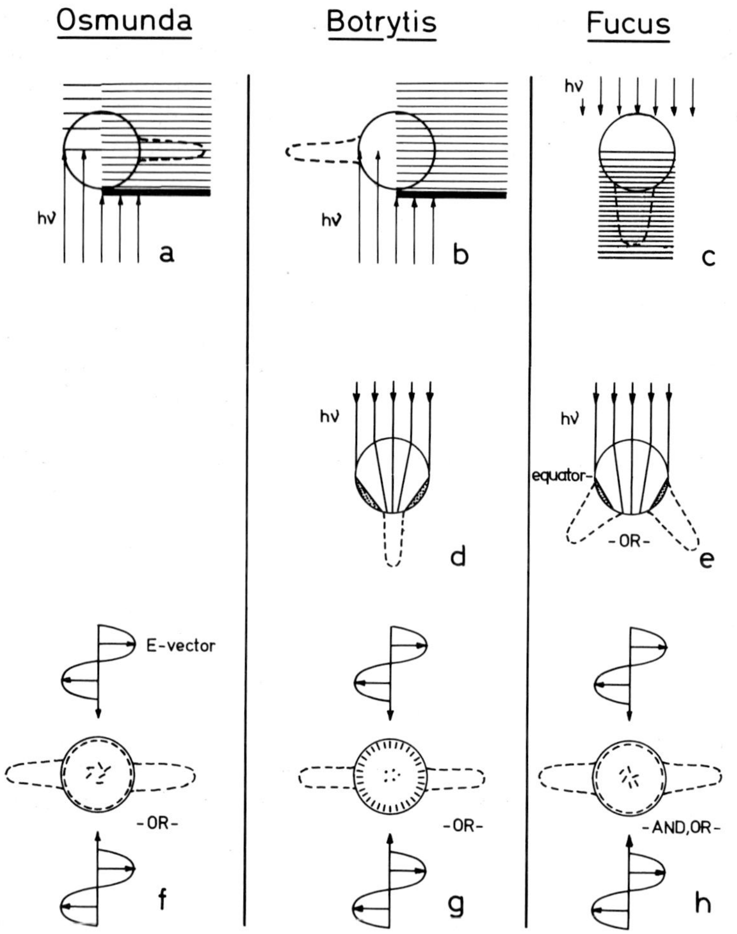

Fig. 2a–h. Light-induced orientation of outgrowth *(dashed line)* of a spore of *Osmunda* **a** and *Botrytis* **b** due to partial illumination and of a zygote of *Fucus* **c** due to strong unilateral light (shaded areas are *hatched*). The pattern of lit and shaded regions *(dotted)* in unilateral light of medium intensity **d**, **e**. The inferred orientation of the photoreceptor molecules from the orientation of the outgrowth in plane-polarized light shining from opposite sides on the cells **f**, **g**, **h**. In **e**, **f**, **g** and **h**, two equally like possibilities for the outgrowth are given, from which one will be realized randomly. (The sizes of spores and zygotes are not drawn to scale and exact shape: the spore of *Osmunda* is spherical and circa 50 μm in diameter, that of *Botrytis* is ovoid and circa 11 μm long and 8 μm wide, the zygote of *Fucus* is spherical and circa 80 μm in diameter)

munda, a light absorption difference must be 60–90% in order to become effective. Generally, light induces a single transcellular axis, as for instance the thallus-rhizoid axis in *Fucus*, and the polarity of that axis is independent of light intensity; the rhizoid always grows at the shaded pole, and the thallus at the illuminated pole. An exception is the spore of *Funaria* (JAFFE and ETZOLD, 1965). In unipolar or bipolar illumination with red light of weak to medium intensity, a chloronema grows from the brightest part of the cell and a rhizoid from the darkest one. In high-intensity red light just the opposite is true. Moreover, the rhizoid-chloronema axis never becomes a transcellular axis; if the direction of the unilateral light is changed between the time of rhizoid growth and of chloronema growth, the chloronema is induced to grow in response to the *second* light direction, whereas the rhizoid responds to the first light direction. In some cases, the chloronema even grows adjacent and parallel to the rhizoid (v. WETTSTEIN, 1965).

In *Fucus* and *Equisetum* there is evidence that polarity-inducing light is absorbed by two different photoreceptors, a low-intensity receptor and a high-intensity receptor (HAUPT, 1957, 1962; BENTRUP, 1963, 1964). This is inferred from the observation that with increasing intensity of unilateral or polarized light, the percentage of orientation in a population increases, decreases, and increases again. BENTRUP (1963) suggests that the receptor which absorbs low intensity light may become photo-inactivated at higher intensities. The action spectra for the low-intensity and the high-intensity induction both show blue light (with a peak at about 450 nm) and near UV light to be maximally effective; yet, according to HAUPT and BENTRUP, small differences between the two action spectra point to a protein-bound carotenoid for the low-intensity receptor, and to a riboflavin for the high-intensity receptor.

In the spores of *Botrytis* and *Osmunda*, maximal activity also occurs in the blue and near-UV region, suggesting that the absorbing pigment is a carotenoid or a flavin (JAFFE and ETZOLD, 1962). In *Botrytis* the receptor molecule has been shown to have a single axis of absorption, i.e., the receptor molecule seems to be uniaxial (L.F. JAFFE, personal communication). This suggests that it is a carotenoid rather than a flavin, since the structure of most carotenoids will support uniaxial absorption, while the planar structure of flavins seems likely to support biaxial absorption. In the spores of *Funaria*, phytochrome seems to mediate induction of chloronema growth by low-intensity red light (JAFFE and ETZOLD, 1965).

In experiments with unilateral light, close observation of the site of outgrowth and a construction of the light path within the cell provide evidence that the photoreceptor molecules must lie at the cell surface (see e.g., *Fucus*, JAFFE, 1958b). Experiments with plane-polarized light furthermore provide evidence that these molecules are dichroic and that their main direction of absorption is oriented either parallel or perpendicular to the nearby cell surface. In *Fucus, Equisetum, Dryopteris* and *Osmunda* (cells which grow from their *darkest* area), outgrowth occurs parallel to the E vector of polarized light. The photoreceptor molecules are thus presumably oriented parallel to the cell surface, so that absorption by the molecules at the site of outgrowth would be minimal (see Fig. 2f, h); (JAFFE, 1956, 1958b; JAFFE and ETZOLD, 1962; HAUPT, 1962). The spores of *Botrytis* also germinate parallel to the E vector of polarized light. However, these cells grow from their *brightest* zone; the dichroic molecules are thus probably oriented perpendicular to the cell surface, allowing for maximal absorption at the site of growth (see Fig. 2g; ETZOLD, 1961).

A more complicated situation seems to exist in *Funaria* spores (JAFFE and ETZOLD, 1965): three photoreceptor systems direct chloronemal primordia. (1) A low intensity system; its photoreceptors seem to be disoriented and probably are dispersed phytochrome molecules. (2) A medium-intensity system; its receptor molecules seem to be tangentially oriented. (3) A high-intensity system; its receptor molecules also seem to be tangentially oriented, but they are probably magnetically excited (unlike all others which are electrically excited).

The existence of a preferential orientation of the photoreceptor molecules suggests that these molecules must be localized in some stable and peripheral cell structure, e.g., in the cell membrane (HAUPT, 1962, 1965). The primary target for inducing light therefore seems to be the cell periphery, and consequently polarization probably starts there.

3. Inducing Chemical Gradients

Self-generated gradients and imposed gradients of organic or inorganic substances can induce polarity in various cells. *Fucus* zygotes germinate toward thallus pieces of various seaweed species, even if the thallus tissue is enclosed in a dialysis membrane. The polarizing substance from thallus tissue is effective up to a distance of 5 mm; it therefore seems to be a small molecule, with a diffusion coefficient of about $10^{-5}\,cm^2 s^{-1}$. The substance is assumed to be either unstable or degradable by cell-surface enzymes, since the diffusion gradient of a stable substance across a cell of 100 μm in diameter at a distance of 5 mm from the source would be negligible and probably ineffective (JAFFE, 1968). Several lines of evidence suggest that the developing zygote itself emits substances that affect rhizoid formation.

1. Individual zygotes of *Fucus* and *Pelvetia* germinate toward a diffusion barrier (at pH 6) or away from it (at pH 8) (LOWRANCE and WHITAKER, 1937, from JAFFE 1968), presumably the region of highest concentration of a diffusible substance.

2. Pairs of zygotes or the zygotes of a group germinate toward each other (at pH 6) or away from each other (at pH 8) (JAFFE and NEUSCHELER, 1969).

3. *Fucus* zygotes exposed to a laminar flow of 0.01–1 $\mu m\, s^{-1}$ (at pH 6.5) germinate downstream, the direction in which a diffusible substance would be transported (BENTRUP and JAFFE, 1968). The spores of *Botrytis* also germinate downstream with flow rates of 0.1–1000 $\mu m\, s^{-1}$ (MÜLLER and JAFFE, 1965).

The stimulating substance produced by the developing *Fucus* zygote has an estimated diffusion coefficient of approximately $10^{-8}\,cm^2\,s^{-1}$, that of the *Botrytis* spore of approximately $10^{-7}\,cm^2\,s^{-1}$. These diffusion coefficents indicate a molecular weight of about 10^7 daltons. In *Botrytis* the halflife of the substance seems to be about 10 s. Each substance is probably part of a natural positive feed-back mechanism which stabilizes a labile cellular axis by amplifying its own release or production at the site of growth.

The chemical nature of these inducing substances, which in *Fucus* was called "rhizin" by JAFFE, is still unknown, but there is experimental evidence that it is *not* identical with any of the following substances: H^+, CO_2, O_2 (JAFFE, 1955), IAA, NAA, AMP or c-AMP (KLEMKE and BENTRUP, 1973).

Polarity in *Fucus* and *Pelvetia* zygotes can also be induced by imposed gradients of H^+, K^+ (BENTRUP et al., 1967), Ca^{2+} (ROBINSON and JAFFE, 1976; ROBINSON, 1977) and the uncoupler DNP (WHITAKER and BERG, 1944). The rhizoid is formed toward the side of higher H^+-concentration (at $pH > 5.5$) and higher DNP concentration. Since it is known that DNP shuttles H^+ through membranes (FINKELSTEIN, 1970), the side with a higher DNP concentration is probably also the side of higher H^+ influx.

The rhizoid of *Fucus* is formed toward the higher K^+-concentration in a gradient of 9 mM/cell to 36 mM/cell, with an average K^+-concentration of over 10 mM (10 mM is the natural K^+-concentration of seawater). In a K^+ gradient of 2 mM/cell to 5 mM/cell, with an average K^+-concentration below 10 mM, the zygotes germinate toward the side of the lower K^+-concentration.

A common feature emerging from the experiments with imposed gradients is that the number of oriented cells of a population is proportional to the magnitude of the gradient evoking orientation. This has clearly been demonstrated in polarization of *Fucus* zygotes with K^+ and H^+ gradients and in the rheotropic responses of *Fucus* and *Botrytis*. These experiments (particularly those with K^+ gradients) suggest that a cell is also sensitive to the average outside concentration of the polarizing substance, not just to its gradient, i.e., it is sensitive to a relative gradient (concentration difference/distance × average concentration).

All of the chemically defined substances which induce polarity (H^+, K^+, Ca^{2+}, DNP) are well-known membrane affectors, causing changes in the properties of the plasma membrane. The establishment of a gradient of such a substance across a cell may cause pole-to-pole differences in membrane permeability and membrane potential, and in internal ion concentrations. The inducing effect of such a difference in permeability and in internal ion concentrations has recently been demonstrated with Ca^{2+} (ROBINSON and JAFFE, 1976; ROBINSON, 1977): In an external Ca^{2+} gradient, *Pelvetia* zygotes germinate toward the side of lower Ca^{2+}. However, influx experiments with radioactive Ca^{2+} show that *more* Ca^{2+} *enters* the zygotes at a *lower* Ca concentration than at a higher one; for instance, at 1.5 mM Ca^{2+} concentration, the influx of Ca^{2+} is seven times greater than at the natural 10 mM of seawater. This indicates that the Ca^{2+} permeability increases with decreasing external Ca^{2+}, and that in a Ca^{2+} gradient the zygotes germinate at the side where the Ca^{2+} permeability is increased and *more* Ca^{2+} enters the cells, i.e., presumably at the side of higher internal Ca^{2+} concentration.

One might ask if each substance which has been found to induce polarity becomes effective via a different mechanism, or if they might have some fundamental mechanism in common. It seems possible that such a common link exists, and it might be a localized increase in internal Ca^{2+} concentration.

4. Induction by Imposed Currents

Application of electrical fields to developing cells is a valuable tool in probing for electrical control mechanisms in polarization and in investigating the role of the plasma membrane.

Applying an electrical field affects a cell in two ways, (1) it drives a current through the cell and (2) it changes the cell's membrane potential. The current driven through the cell is usually rather small, because the membrane resistivity is much higher than the resistivity of the surrounding medium (10^9–10^{11} ohm·cm membrane resistivity versus circa 2 · 10 ohm · cm resistivity of seawater or 10^3–10^4 ohm·cm resistivity of pondwater). Nevertheless, current is flowing through the cell. This current might create an electrical field along the surface and in the cytoplasm which, in turn, could lead to a redistribution of charged subcellular components (see p. 501). The change in membrane potential consists in a hyperpolarization at the side facing the positive electrode (where the current is driven inward through the membrane) and in a depolarization at the opposite end.

The difference in membrane potential (ΔE_m) between two opposite sites on the membrane of a spherical cell, e.g., a *Fucus* zygote, can be described by the equation (NOVAK and BENTRUP, 1973):

$$\Delta E_m = 3 \cdot E \cdot a \cdot \cos\Theta \qquad (1)$$

where E = applied field ($V\,cm^{-1}$), a = cell radius (cm) and Θ = the site's "latitude" (radians).

The zygotes of *Fucus*, *Pelvetia* and *Ulva*, the spores of *Equisetum* and *Funaria*, and the pollen grains of *Vinca* respond to applied fields by orienting their outgrowth parallel to the field. The percentage of polarized cells in a population increases, without any apparent threshold, with increasing magnitude of the field. This is illustrated with *Equisetum* and *Fucus* (BENTRUP 1968a, b), with *Pelvetia* (PENG and JAFFE, 1976) and with *Ulva* (SAND, 1973) in Figure 3.

A rather puzzling result of these experiments is that, in a DC field, the site of growth varies with the species, and even from batch to batch within a species, facing either the positive electrode or the negative one. LUND (1923) reported that the zygotes of *Fucus inflatus* germinate toward the positive electrode. On the other hand, BENTRUP (1968a, b) found that the zygotes of *F. serratus* germinate toward the negative electrode. (*F. serratus* grows toward the positive electrode only at a K^+ concentration above normal and in a field of 5 mV/cell). According to SAND (1973) the zygotes of *Ulva mutabilis* germinate toward the positive pole. In zygotes of *Pelvetia* exposed to fields from 5 h after fertilization until germination, PENG and JAFFE (1976) found that 11 batches germinated toward the positive electrode, 2 batches grew toward the negative electrode and 3 batches showed a mixed response, germinating toward the negative electrode in weak fields, and toward the positive electrode in strong fields. The spores of *Equisetum* germinate toward the positive electrode (BENTRUP, 1968a).

The spores of *Funaria* show a preference for growing a chloronema toward the negative electrode (BENTRUP, 1968a), and the pollen grains of *Vinca* tend to sprout a pollen tube from a pore facing the negative electrode (MARSH and BEAMS, 1945).

The reasons for these differences in response are not clear. To explain the varying responses, PENG and JAFFE (1976) suggest that the outgrowth is always initiated at that site where the most Ca^{2+} enters the cell. Depending on the properties of the plasma membrane of a given cell, this might be the hyperpolarized end, at which the driving force for Ca^{2+} influx is increased, or the depolarized end, at which the Ca^{2+} permeability might have increased. The basis for the proposed intraspecific differences in membrane properties remains to be elucidated.

In comparing electric-field data of different investigators, one cannot rule out the possibility that inconsistencies in reported field effects may have resulted from subtle differences in experimental conditions, e.g., in the ionic composition of the medium, in the time of onset and in the duration of the field, or in the introduction of concentration gradients to which *Fucus* zygotes are to some degree sensitive.

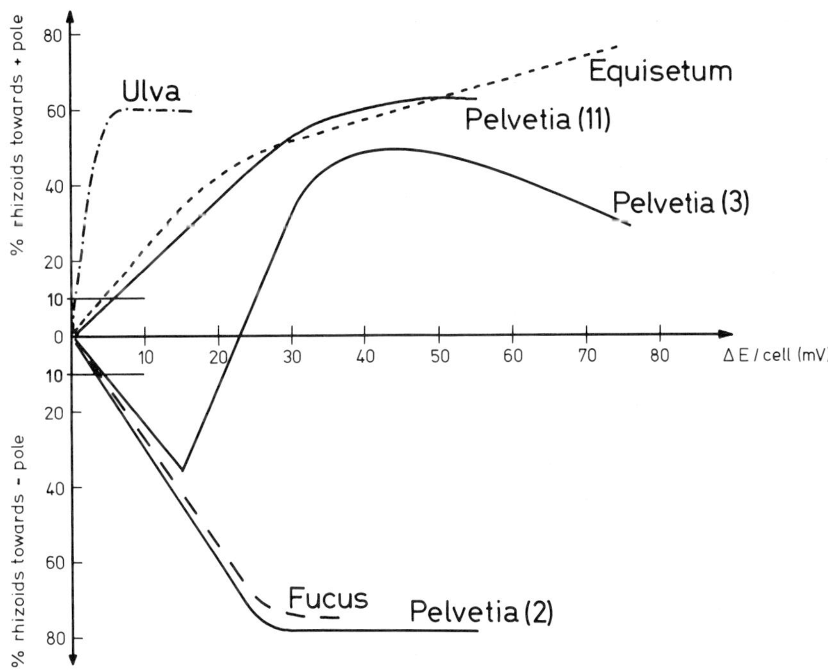

Fig. 3. Polarization of zygotes of *Ulva mutabilis*, *Fucus serratus* and *Pelvetia fastigiata*, and of spores of *Equisetum limosum* by imposed electrical fields. (Data from SAND, 1973; BENTRUP, 1968a; PENG and JAFFE, 1976). *Ordinate:* percent rhizoids toward positive or negative electrode exceeding random distribution. (For details about the calculation of the percentage see e.g., PENG and JAFFE, 1976.) *Abscissa:* field strength expressed as the actual voltage drop across each cell, i.e., applied voltage per cell diameter multiplied by a factor of 1.5 to account for the distortion of the field near the cell (see PENG and JAFFE, 1976). For *Pelvetia* the three different modes of response of 16 batches of zygotes are represented; the number of batches belonging to each mode are shown in brackets. (The diameter of an *Ulva* zygote was estimated as 20 μm in order to adapt SAND's data for Fig. 3)

The fields needed to polarize a population measurably, i.e., by about 10%, are generally quite small: 3–4 mV/cell in *Fucus serratus* (BENTRUP, 1968a), circa 1 mV/cell in *Ulva mutabilis* (SAND, 1973), 6 mV/cell in *Pelvetia* when the response is toward the positive electrode and 3 mV/cell when the response is toward the negative electrode (PENG and JAFFE, 1976), and 4 mV/cell in *Equisetum limosum* (BENTRUP, 1968a).

AC fields can also induce polarity. When *Fucus* zygotes are exposed to AC fields of about 100 mV/cell from 1 to 8 h after fertilization, the site of the germination in about half the population becomes oriented parallel to the field, with the zygotes growing a rhizoid either toward one electrode or toward the other, or, in rare cases, toward both. The most effective frequency is in the range of 1–100 Hz (NOVAK and BENTRUP, 1970, 1973). The polarizing effect of AC fields seems to indicate a change in membrane potential as an essential step in polarization. However, an effect of AC fields on the cytoplasm cannot be excluded, because the effect caused by one half-cycle of the field might not have been reversed by the next half-cycle (PENG and JAFFE, 1976).

The currents driven through cells by fields sufficient to polarize them are comparable in magnitude to the natural currents which polarizing cells drive through themselves (see p. 498/99). However, the ionic composition of the imposed currents is quite uncertain and might be a rather unnatural one. Therefore, caution seems necessary in interpreting the results of experiments with imposed currents.

5. Other Inducing Factors and Inhibiting Factors

Many factors in addition to those discussed above can induce polarity, at least in some systems and under laboratory conditions. A representative listing of factors and observed cases of induction by these factors is given below. The role they play under natural conditions remains to be established.

1. Temperature. In *Fucus* zygotes a temperature gradient of only 0.5°C/cell induces germination at the warmer end of the cell (LOWRANCE, 1937. In: v. WETTSTEIN, 1965).

2. Centrifugation. Zygotes of *Fucus*, *Cystosira* and *Pelvetia* which have been stratified by centrifugation germinate at the centrifugal pole (at pH 8) or at the centripetal pole (at pH 6; WHITAKER, 1940. In: v. WETTSTEIN, 1965).

3. Gravity. Even gravity seems to induce polarity in the spores of *Funaria*: spores which are sown on vertically oriented agar plates germinate from their lower ends (HEITZ, 1941).

4. Deformation. *Fucus* zygotes which have been made oblong by mechanical compression in a capillary tube grow a rhizoid from one end or the other of the long axis (at pH 8) or from the short axis (at pH 6); WHITAKER, 1940. In: v. WETTSTEIN, 1965).

In some cases (e.g., in the absence of an inducing vector from the environment), the site of outgrowth is specified relative to the position of some morphological features of the cell:

5. Haptera position. Spores of *Equisetum* which are kept in darkness germinate from a locus opposite to the site where the haptera are inserted (NAKAZAWA, 1958. In: v. WETTSTEIN, 1965).

6. Site of nucleus. Stalk segments of *Acetabularia* containing a secondary nucleus that dedifferentiates into a primary nucleus grow a rhizoid next to the nucleus, irrespective of the former polarity of the segment (HÄMMERLING, 1955).

Other factors are known to *inhibit* the induction of polarity. Some examples are listed below. At present, no well-founded speculations can be offered as to the mechanism by which these factors might become effective.

1. Glucose. Liverwort spores growing on a medium supplemented with 2% glucose grow rapidly for several months, but fail to show any polarity. When transferred to glucose-free medium, these apolar cells become polar and develop normally (ALLSOPP, 1957. In: JAFFE, 1958a).

2. Auxin. In *Funaria* spores a relatively high concentration of IAA, namely 25 μM, induces equal growth in all directions, leading to giant apolar cells (HEITZ, 1940. In: v. WETTSTEIN, 1965).

3. Colchicine. Spores of the fern *Onoclea* treated with colchicine (which is known to disrupt microtubules) grow but do not polarize (MILLER and STEPHANI, 1971).

However, red light or far-red light reverses the inhibitory effect of colchicine. Evidently, microtubules are not necessary for polarization.

4. Abnormal chromosome number. Spores of the fern *Hypodematium* that develop from parents which have three sets of chromosomes (i.e., 3×41) have an odd number of chromosomes (circa 123/2) and fail to become stably polarized. The plane of the mitotic spindle is different in each consecutive cell division (LOYAL and PAIK, 1971).

6. Receptive Stages

There are only certain receptive stages of development during which a stimulus can be transduced and amplified into a stable polarity. The receptivity for a factor usually increases from zero, peaks and then decreases to zero again during the early stages of development (Table 1). As an exception, in *Fucus* zygotes there seem to be not just one, but two consecutive peaks for light sensitivity, the first one at 12 h and the second one at 15 h after fertilization, each of them of 1 h duration at 17°C (FEUCHT and BENTRUP, 1972).

The development of receptivity appears to involve a series of metabolic events. Exposure to low temperature or treatment with metabolic poisons extends the duration of the receptive stages. In *Fucus* each of the two maxima of light receptivity becomes 4 h long at 7°C (FEUCHT and BENTRUP, 1972). In *Equisetum*, a transfer of spores to 5°C, or poisoning them with KCN at the time of maximal light sensitivity prolongs the maximum for at least 24 h (HAUPT, 1957).

The limiting factor seems to reside in these metabolic transduction steps rather than in the perception of the stimulus. A cell is probably fully sensitive to an inducing stimulus before the peak of the specific receptive phase, but may not yet be able to respond to that stimulus by becoming polarized. In *Equisetum* spores, for example, induction by light shows the same dosage requirement for optimal induction whether the light is presented at 20 min or at 200 min after sowing, but the polarization of a population becomes only 20% at 20 min as compared to 60% at 200 min (HAUPT, 1957).

An induction of polarity is open to counterinduction as long as the polarity axis is not stably fixed. The earlier the original induction takes place, the longer it is open for counterinduction. If a second induction has not enough energy to cancel the first induction, both inductions will become integrated, resulting in an orientation of the polarity axis somewhere in between the two directions of induction.

These results seem to indicate that endogenous processes underlie polarization, and that induction does *not start* these processes. They further indicate that decreasing sensitivity toward a counterinduction coincides with increasing stabilization of polarity.

Within a given type of organism, receptive stages to various polarizing factors occur at different times (Table 1). In some cases the sensitive phases of two different vectors nearly coincide; this is true, for example, for light and for electrical fields in *Equisetum* and in *Ulva*.

Table 1. The receptive stages for various polarizing factors (times in h after fertilization or sowing[a])

Species	Factor	Response	Receptive stage			Time of ca. 50% germination	Reference
			Beginning	Maximum	Low level		
Ulva mutabilis	Light	Rhizoid orientation		40–50		ca. 78	SAND (1973)
	DC field	Rhizoid orientation	12–18	40–50	72–84	ca. 78	SAND (1973)
Codium fragile	Light	Germ tube orientation	4–5	6–7	15–18	10–20	WEBER (1969)
Fucus serratus	Light	Rhizoid orientation	8–9	11–13	15–17	ca. 15	BENTRUP (1963)
	Light	Rhizoid orientation	8–9	12, 15	17–19	ca. 15	FEUCHT and BENTRUP (1972)
	DC field	Rhizoid orientation	1–3	3–6	10–12		BENTRUP (1968a)
Fucus furcatus	Light	Rhizoid orientation	4–5	7–10	17–20	ca. 17	WHITAKER and LOWRANCE (1936) (in JAFFE, 1968)
	Flow of medium	Rhizoid orientation	(0)–2	2	10–12	ca. 17	BENTRUP and JAFFE (1968)
	Centrifugation	Rhizoid orientation	0	0–10	>16	ca. 17	WHITAKER (1940) (in JAFFE, 1968)
Pelvetia fastigiata	Light	Rhizoid orientation	2–3	4–11	>15	ca. 16	JAFFE (1968)
	Light	Rhizoid orientation	1–2	4–11	13–21	ca. 20	MURPHY et al. (1970)
Equisetum variegatum	Light	Rhizoid orientation	0–1	2–4	6–7	ca. 24	HAUPT (1957)
	Light	Rhizoid orientation	0–1	3–6	9–10	ca. 24	FEUCHT and BENTRUP (1972)
	DC field	Rhizoid orientation	0–1	2–3	4–5	ca. 24	BENTRUP (1968a)

[a] The times given in this table may vary to some degree with the particular experimental conditions (e.g., temperature, onset and duration of factor, energy of factor) and with the batch of material.

The evidence discussed above clearly indicates that polarity in a given cell type can be induced by any one of a variety of factors. These various possibilities of induction probably ensure that the cell will become polarized, and will thus be able to develop normally under almost any environmental circumstances. Since some of the factors come to bear on induction of polarity only if other factors are lacking, a hierarchy among factors seems indicated. The differences in the time of peak receptivity for different inducing vectors probably also reflect this hierarchy among vectors, the one with the latest maximum, namely light, being the most important one.

7. Endogenous Events During Early Development

Early in the development of polarizing cells, one observes a variety of remarkable changes in the properties of the cell membrane, probably reflecting a fundamental reassortment of membrane components. Since similar changes occur in quite unrelated cells, namely the fucoid egg and the pollen grain, endogenous membrane changes seem to be a common event during the early phases of polarization.

Significant knowledge about spontaneous events during polarization has been gained through the application of two new techniques. JAFFE and NUCCITELLI (1974) developed a "vibrating electrode" technique that allows the measurement of the strength and the pattern of natural currents which traverse developing cells. The sensor of the measuring system is a metal-filled glass micropipette, bearing a platinum-black ball of about 20 μm diameter at its tip and a platinum-black reference electrode a few mm behind the tip. With the aid of a piezoelectric element, the electrode is vibrated, at a few hundred Hz, with an amplitude of about 30 μm. It records voltage differences in the medium near the cell as they are caused by ion currents. The low input impedance of the platinum tip, and the high frequency with which the electrode is vibrated keep the noise down to almost its theoretical minimum.

The method offers two main advantages over conventional saline-filled microelectrodes: (1) the measurements are done extracellularly, therefore causing little or no disturbance of the system studied, and (2) the low noise allows reliable measurements in the nV range in media of high conductivity; this represents an increase in sensitivity of 100–1000-fold compared to conventional methods.

A new technique for measuring pole-to-pole differences in membrane permeabilities has been developed by ROBINSON and JAFFE (1975) and has been applied successfully to *Fucus* and *Pelvetia* zygotes. Zygotes are set into the closely spaced, funnel-shaped holes of a thin nickel screen which is placed as a diaphragm between two chambers. The egg-filled screen is then illuminated from one side, so that all presumptive thallus ends are exposed to one chamber, and the presumptive rhizoid ends to the other. Asymmetry of influx is determined by slowly passing medium containing radioactive isotope through one chamber and unlabeled medium through the other (to wash away any leaking isotope). After a certain time, the zygotes are removed and the amount of isotope, taken up into the zygotes through that portion of the membrane exposed to the labeled medium, is measured. To measure asymmetries of efflux, the zygotes are preloaded with isotopes, and unlabeled medium is passed through both chambers and collected for assaying.

With the aid of conventional intracellular microelectrodes, a substantial hyperpolarization of the membrane has been found to precede visible growth in *Fucus* and *Pelvetia* zygotes and lily pollen grains, indicating some major changes in the plasma membrane. In *Fucus* and *Pelvetia* eggs, this hyperpolarization starts at the time of fertilization, and brings the membrane potential from between

-20 and -30 mV in the unfertilized egg to between -70 and -80 mV within 7–9 h after fertilization (BENTRUP, 1970; WEISENSEEL and JAFFE, 1972, 1974). In lily pollen grains, a hyperpolarization of similar magnitude, (i.e., circa 50 mV) occurs between sowing and germination (WEISENSEEL, 1977). Comparing the degree of membrane hyperpolarization in *Fucus* and *Pelvetia* with the time course of photosensitivity, it seems quite surprising to find that in both species photosensitivity begins when the membrane potential exceeds -50 mV, and that it reaches its maximum when the membrane potential exceeds -60 mV.

In *Fucus* and *Pelvetia* two factors have been found which seem to be the major causes for the hyperpolarization: (1) the zygotes accumulate KCl, and (2) the cell membrane becomes selectively permeable to K^+.

The accumulation of KCl is evidently mediated by the activation of a membrane pump, which moves Cl^- against its electrochemical gradient into the cell (K^+ seems to follow Cl^- passively) and by the sealing off of Cl^- leaks in the membrane (ROBINSON and JAFFE, 1973). The entry of water, following the accumulation of KCl, leads to development of about 5 bar of turgor pressure (ALLEN et al., 1972). This increase in turgor pressure is a prerequisite for expansion growth but evidently not for polarization. Under high external osmotic pressure, a condition under which the zygotes cannot expand, they nonetheless accumulate sulphated polysaccharides (cell wall precursors) at the future rhizoid pole, just as zygotes under normal conditions do (VREUGHENHIL et al., 1976).

Electrical and tracer-flux measurements indicate that the cell membrane of the unfertilized egg is quite unselectively permeable to K^+, Na^+ and Cl^-. After fertilization, the zygote develops a high degree of K^+ selectivity (BENTRUP, 1971b; WEISENSEEL and JAFFE, 1972). In *Pelvetia,* for example, the K^+ permeability increases from $0.85 \cdot 10^{-7}$ cm s^{-1} to $5 \cdot 10^{-7}$ cm s^{-1} and the Na^+ permeability decreases from $0.23 \cdot 10^{-7}$ cm s^{-1} to $0.06 \cdot 10^{-7}$ cm s^{-1} within a few hours after fertilization (ROBINSON and JAFFE, 1973).

In *Fucus* zygotes there may also be an increase of water permeability at the future site of the rhizoid, since this end plasmolyzes more rapidly than the opposite end (REED and WHITAKER, 1941).

Using the "vibrating electrode", a natural transcellular current has been found in the zygotes of *Pelvetia* and in the pollen grains of *Lilium*. In *Pelvetia,* already at about 1 h after fertilization an electrical current of circa 1 μA cm^{-2} can be measured. Current enters the cell at a certain region of the surface, spreads out through the cell, and exits over a relatively large area of the remaining surface, closing its loop via the surrounding seawater (Fig. 4). Initially the site of current entry seems to shift its position, but by about 5 h after fertilization, the current direction becomes stable. Germination always occurs from the final site of current entry, which, in the case of unilateral light, is the shaded pole (NUCCITELLI and JAFFE, 1976).

Similarly, in lily pollen, a self-generated current starts traversing the cell at about half an hour after sowing the grain into a liquid germination medium. The current density increases from zero to circa 4 μA cm^{-2} within 1–3 h. Here again, germination begins at the site where current enters the cell (WEISENSEEL et al., 1975; Fig. 4).

In regenerating stalk segments of *Acetabularia,* pulses of 100 mV amplitude and 100 s duration appear about 15 h before visible growth begins. These pulses

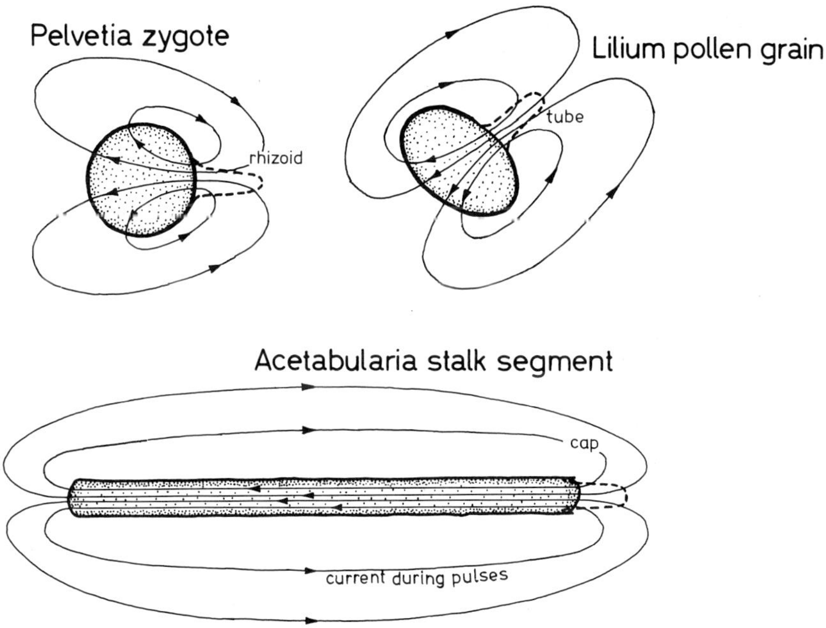

Fig. 4. Schematic diagram of the natural current pattern in a developing *Pelvetia* zygote, a *Lilium* pollen grain, and a regenerating stalk segment of *Acetabularia*, as inferred from external potential measurements before visible growth begins. The site and shape of the later outgrowth is indicated by *dashed lines*. (The different objects are not drawn to scale: *Pelvetia* zygotes are 90–100 μm in diameter, *Lilium* pollen grains are 100–110 μm long and 80–90 μm wide, *Acetabularia* stalk segments are 10–20 mm long and 0.3 mm in diameter)

indicate a current of a few μA cm^{-2} at the future site of growth (NOVAK and BENTRUP, 1972; Fig. 4). Besides this pulse-current, a small steady current traverses the regenerating stalk segment. At first, this steady current enters the regenerating end and later on it leaves it.

In *Pelvetia* zygotes a major component of the early current appears to be Ca^{2+}, which leaks into the future rhizoid region and is presumably pumped out at the opposite end. Experiments with the "nickel screen method" show that at 5–6 h after fertilization a Ca^{2+} current of 2 pA (=0.03 μA cm^{-2}) traverses the cell (ROBINSON and JAFFE, 1975). Later, polarity is also observed in the movements of other ions. Na^+ leaks into the future rhizoid end faster than into the thallus end, while Cl^- is pumped into the thallus end faster than into the rhizoid end. Since the total influx and efflux of Ca^{2+} remains constant during early development, it seems likely that the "leaky" sites and the pumps for this ion are not built into the membrane de novo during polarization, but that they are only redistributed, with the pumps shifted to one end and the leaky sites to the other (JAFFE et al., 1974a).

Electron-micrograph studies provide independent evidence for a redistribution of membrane components. Freeze-fracture replicas of the cell membrane of *Pelvetia* zygotes show isolated intramembrane particles of circa 10 nm in width. At 8 h after fertilization there are about twice as many of these particles at the thallus end as at the rhizoid end. However, these studies also show that new membrane patches are inserted into the cell membrane during polarization. Shortly after unilateral light is introduced, the number of sites where vesicles fuse with the plasma membrane increases at the future rhizoid end and at about 5 h after fertilization, fusion sites are present exclusively at the presumptive rhizoid end (PENG, 1976).

If one assumes that the plasma membrane plays an essential role during polarization of a cell, one can here ask the question: is polarity induced or only reoriented? It seems quite likely that membrane components, as for instance Ca pumps, are not distributed completely symmetrically in the cell membrane, i.e., there may be a certain polarity with respect to their distribution. Therefore, a redistribution of such components, caused by inducing factors, may be regarded as a "reorientation" of an existing polarity. On the other hand, if the localized insertion of new membrane patches into the plasma membrane is effected by inducing factors, this may be considered as "induction" of polarity. It has been shown that unilateral light triggers both events in fucoid zygotes. It therefore seems that the establishment of cell polarity involves both phenomena: "induction" as well as "reorientation".

8. Amplification of a Labile Polarity

The inducing vectors are only triggers. They do not provide the energy for polarization. This fact becomes most obvious in cases, such as that of *Equisetum*, in which a few seconds of stimulation are sufficient to induce polarity (HAUPT, 1957). Inducing vectors do not cause an instantly stable polarization; instead, they induce a feeble and labile axis which needs amplification.

In *Fucus* zygotes, this process of amplification has been shown to be independent of some major cell functions, namely, protein synthesis and development of turgor pressure, since poisoning the zygotes with cycloheximide or application of hypertonic seawater does not prevent polarization (QUATRANO, 1973; VREUGHENHIL et al., 1976). How does a labile polarity become a stable one? A hypothesis of "self-electrophoresis" for amplification and stabilization of polarity was suggested by MATHEWS (1903), see also WENT (1932) and LUND (1947). In recent years this hypothesis has been greatly advanced theoretically and experimentally by JAFFE and coworkers (e.g., JAFFE, 1968, 1969, 1977; JAFFE et al., 1974a, b). The main concept is that an electrical field across a cell, which may be caused by natural transcellular currents, segregates charged entities within the cytoplasm and within the cell membrane. The resulting gradients of cytoplasmic and membrane components may then act back to amplify and stabilize local changes induced by the polarizing factor.

The ability of an electrical field to move charged components is proportional to its magnitude. The magnitude of the electrical field across the cytoplasm depends on the ions carrying the current. When non binding ions like K^+ or Cl^- move, the electrical field (E) is simply proportional to the current density (δ) and the cytoplasmic resistivity (ρ) and is given by Ohm's Law:

$$E = \delta \cdot \rho \ (\mathrm{V\ cm^{-1}}) \qquad (2)$$

If the current consists of ions which are strongly bound to the cytoplasm, e.g., Ca^{2+} and H^+, then the effective resistivity (ρ^*) is much higher than the resistivity (ρ), and the electrical field (E) is given by

$$E = \delta \cdot \rho^* \tag{2a}$$

The resistivity (ρ) is usually in the range of 100–300 ohm·cm, but the effective resistivity (ρ^*) can be 1000 times as large, i.e., 10^5 to $3 \cdot 10^5$ ohm·cm (JAFFE and NUCCITELLI, 1977). This means that a Ca^{2+} current can cause the same electrical field across the cytoplasm as a K^+ current 1000 times its magnitude.

Such a cytoplasmic field might pull negatively charged entities (e.g., certain vesicles and proteins) toward one end, the end where current enters the cell, and positively charged entities toward the other end. If the field acts long enough, this segregation can cause a stable concentration gradient, the magnitude of which is given by the electrophoretic flow (Φ electroph) in one direction and by the flow due to diffusion (Φ diffusion) in the opposite direction [Eq. (3)]. For fairly large particles, an electrical field of only 1 mV could create a concentration gradient of circa 10% in less than 1 h (JAFFE, 1969).

$$\begin{aligned} &\Phi \text{ electroph} = \Phi \text{ diffusion} \\ &-m \cdot C \cdot E = -D \cdot (\Delta C/\Delta x) \\ &\Delta C/(C \cdot \Delta x) = (m/D) \cdot E \cdot 100\ (\%) \end{aligned} \tag{3}$$

ΔC = concentration difference, C = total concentration, Δx = distance, m = electrophoretic mobility, D = diffusion coefficient, E = electrical field.

In developing cells, electrical fields can arise not only across the cytoplasm, but also along the cell surface. Membrane fields may arise when inducing vectors cause a gradient in the membrane potential, or when natural or imposed currents create a field across the cell. Such a field can cause movement of charged and mobile membrane components within the membrane, a phenomenon called "lateral electrophoresis" (JAFFE, 1977). Natural fields seem to be strong enough and to last long enough to cause substantial redistribution of membrane components. For example, POO and ROBINSON (1977) have demonstrated a redistribution for the concanavalin-A receptor in embryonic muscle cells with a field as small as 0.6 mV/cell that lasted for 24 h. Note that similar small fields have been shown to induce polarity in zygotes and spores.

The hypothesis of "self-electrophoresis" is mainly supported by the observations that (1) polarity can be induced by imposed currents and (2) polarizing cells drive currents through themselves. A hypothesis assuming a more direct effect of local differences in membrane potential is suggested by BENTRUP (1968a). According to this concept, the future site of growth is that site where the membrane potential undergoes the greatest displacement from its resting value.

The two hypotheses need not be regarded as mutually exclusive. They seem rather to emphasize only different aspects of a common electrical mechanism which leads to stable polarization. A membrane potential is a large force (i.e., between 10^4 and 10^5 Vcm^{-1}) acting on membrane components; thus a local change in the membrane potential may locally alter activities of membrane enzymes (e.g., pumps) and mobilities of membrane components. On the other hand, a gradient

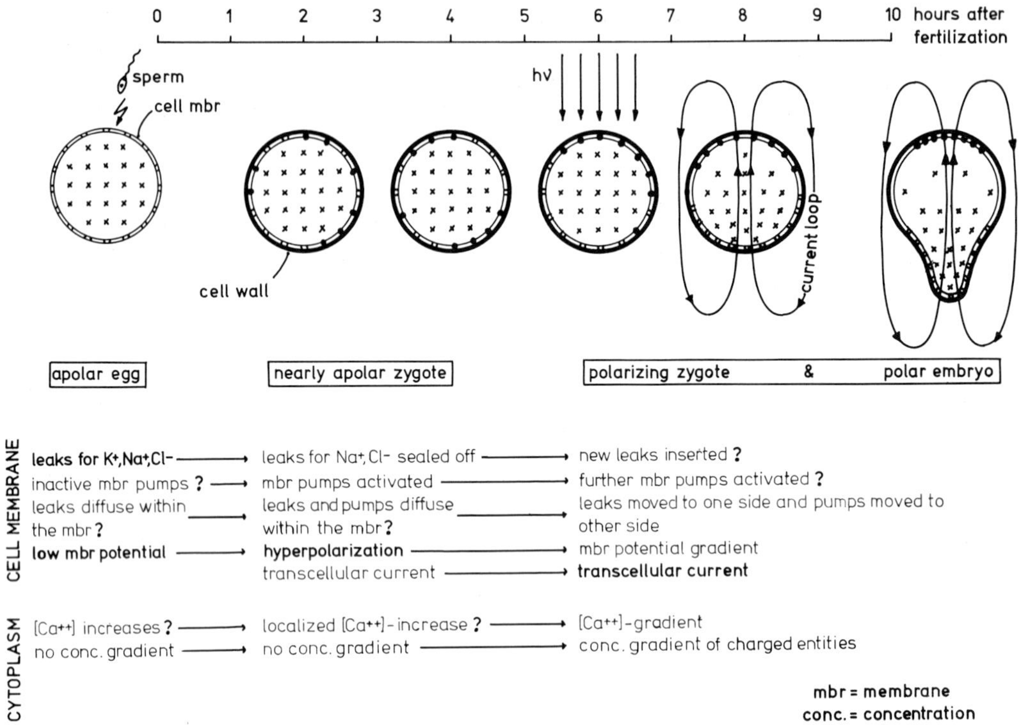

Fig. 5. Hypothetical model of the major events in the membrane and in the cytoplasm during early development of a fucoid egg that becomes polarized by unilateral light. Properties and events that have already been established experimentally are written with *bold letters;* those that seem to need more experimental support are written in *normal letters;* and properties and events that are speculative are labeled with a *question mark.* II, membrane leaky site; •, active membrane pump. (The size of the cell membrane is drawn grossly exaggerated)

of membrane potential drives a current through the cell, which might cause electrophoresis and set up concentration gradients. At present, it seems difficult to decide which of these phenomena is more important for polarization.

Finally, to summarize briefly our present knowledge on induction of polarity, a simple hypothetical scheme is presented, incorporating the major events which have been revealed to take place and those that presumably take place during polarization in a fucoid egg (Fig. 5).

Much progress has been made in the last decade to elucidate the phenomenon of polarization, but our knowledge is still limited. This is particularly true in the fields of vector perception, stability, and the site of polarity. Progress in perception physiology seems likely to advance our understanding of polarization, and to stimulate research. Further progress in our knowledge of polarization seems possible with the study of cases of inhibition, rotation, or loss of polarity. Experiments in which polarization is inhibited by chemical or physical agents of known target might help in elucidating the chain of events that leads to polarity. The study of cases which exhibit a rotation or loss of the morphological axis might give important information about stability and site of polarity. A permanent rotation of polarity occurs, for instance, during normal development of the fern or moss protonema; the beginning of two-dimensional growth, which is induced by blue light, is preceded by a rotation of the axis of the tip cell by 90°. Other cells (e.g., parenchyma cells of the *Begonia* leaf) lose their polarity and become embryonic when isolated from the tissue.

Acknowledgments. I wish to express my gratitude to Professor M.E. Feinleib for her valuable advice during the writing of this review, and to Professor W. Haupt and Professor L.F. Jaffe for critical and helpful comments on the manuscript. The technical help of Mr. A. Dorn is also gratefully acknowledged.

References

Allen, R.D., Jacobsen, L., Joaquin, J., Jaffe, L.F.: Ionic concentrations in developing *Pelvetia* eggs. Dev. Biol. *27*, 538–545 (1972)

Beams, H.W., King, R.L.: The "negative group effect" in the pollen grains of *Vinca rosea*. J. Cell Comp. Physiol. *23*, 39–45 (1944)

Bentrup, F.W.: Vergleichende Untersuchungen zur Polaritätsinduktion durch das Licht an der *Equisetum*-Spore und der *Fucus*-Zygote. Planta *59*, 472–491 (1963)

Bentrup, F.W.: Zur Frage eines Photoinaktivierungs-Effektes bei der Polaritätsinduktion in *Equisetum*sporen und *Fucus*zygoten. Planta *63*, 356–365 (1964)

Bentrup, F.W.: Die Morphogenese pflanzlicher Zellen im elektrischen Feld. Z. Pflanzenphysiol. *59*, 303–339 (1968a)

Bentrup, F.W.: Zur Funktion der Zellmembran bei der Cytomorphogenese. Ber. Dtsch. Bot. Ges. *81*, 311–314 (1968b)

Bentrup, F.W.: Elektrophysiologische Untersuchungen am Ei von *Fucus serratus*: Das Membranpotential. Planta *94*, 319–332 (1970)

Bentrup, F.W.: Räumliche Zelldifferenzierung. Umschau *10*, 335–339 (1971a)

Bentrup, F.W.: Temporal ion permeability pattern of the *Fucus* egg membrane. Proc. Ist Eur. Biophys. Congr. (Baden) *3*, 69–73 (1971b)

Bentrup, F.W., Jaffe, L.F.: Analyzing the "group effect": rheotropic responses of developing *Fucus* eggs. Protoplasma *65*, 25–35 (1968)

Bentrup, F.W., Sandan, T., Jaffe, L.F.: Induction of polarity in *Fucus* eggs by potassium ion gradients. Protoplasma *64*, 254–266 (1967)

Bloch, R.: Polarity and gradients in plants: A survey. In: Encyclopedia of Plant Physiology. Lang, A. (ed.). Berlin-Heidelberg-New York: Springer, 1965, Vol. 15/I, pp. 234–274

Bopp, M.: Entwicklungsphysiologie. Prog. Bot. *37*, 155–176 (1975)

Etzold, H.: Die Wirkungen des linear polarisierten Lichtes auf Pilze und ihre Beziehungen zu den tropistischen Wirkungen des einseitigen Lichtes. Exp. Cell Res. *25*, 229–245 (1961)

Feucht, U., Bentrup, F.W.: Über die photosensible Phase der Polaritätsinduktion bei *Equisetum*sporen und *Fucus*zygoten. Z. Pflanzenphysiol. *66*, 233–242 (1972)

Finkelstein, A.: Weak-acid uncouplers of oxidative phosphorylation. Mechanism of action on thin lipid membranes. Biochim. Biophys. Acta *205*, 1–6 (1970)

Green, P.B.: Cell morphogenesis. Annu. Rev. Plant Physiol. *20*, 365–394 (1969)

Hämmerling, J.: Neuere Versuche über Polarität und Differenzierung bei *Acetabularia*. Biol. Zentralbl. *74*, 545–554 (1955)

Haupt, W.: Die Induktion der Polarität bei der Spore von *Equisetum*. Planta *49*, 61–90 (1957)

Haupt, W.: Die Entstehung der Polarität in pflanzlichen Keimzellen, insbesondere die Induktion durch Licht. Ergeb. Biol. *25*, 1–32 (1962)

Haupt, W.: Perception of environmental stimuli orienting growth and movement in lower plants. Annu. Rev. Plant Physiol. *16*, 267–290 (1965)

Heitz, E.: Die keimende *Funaria*-Spore als physiologisches Versuchsobjekt. Ber. Dtsch. Bot. Ges. *60*, 17–27 (1941)

Jaffe, L.F.: Do *Fucus* eggs interact through a CO_2-pH gradient? Proc. Natl. Acad. Sci. USA *41*, 267–270 (1955)
Jaffe, L.F.: Effect of polarized light on polarity of *Fucus*. Science *123*, 1081–1082 (1956)
Jaffe, L.F.: Morphogenesis in lower plants. Annu. Rev. Plant Physiol. *9*, 359–384 (1958a)
Jaffe, L.F.: Tropistic response of zygotes of the *Fucaceae* to polarized light. Exp. Cell Res. *15*, 282–299 (1958b)
Jaffe, L.F.: Localization in the developing *Fucus* egg and the general role of localizing currents. Adv. Morphogen. *7*, 295–328 (1968)
Jaffe, L.F.: On the centripetal course of development, the *Fucus* egg and self-electrophoresis. Dev. Biol. Suppl. *3*, 83–111 (1969)
Jaffe, L.F.: Electrophoresis along cell membranes. Nature (London) *265*, 600–602 (1977)
Jaffe, L.F., Etzold, H.: Orientation and locus of tropic photoreceptor molecules in spores of *Botrytis* and *Osmunda*. J. Cell Biol. *13*, 13–31 (1962)
Jaffe, L.F., Etzold, H.: Tropic responses of *Funaria* spores to red light. Biophys. J. *5*, 715–742 (1965)
Jaffe, L.F., Neuscheler, W.: On the mutual polarization of nearby pairs of *Fucaceous* eggs. Dev. Biol. *19*, 549–565 (1969)
Jaffe, L.F., Nuccitelli, R.: An ultrasensitive vibrating probe for measuring steady extracellular currents. J. Cell. Biol. *63*, 614–628 (1974)
Jaffe, L.F., Nuccitelli, R.: Electrical controls of development. Annu. Rev. Biophys. Bioeng. *6*, 445–476 (1977)
Jaffe, L.F., Robinson, K.R., Nuccitelli, R.: Local cation entry and selfelectrophoresis as an intracellular localization mechanism. Ann. N.Y. Acad. Sci. *238*, 372–389 (1974a)
Jaffe, L.F., Robinson, K.R., Nuccitelli, R.: Transcellular currents and ion fluxes through developing *Fucoid* eggs. In: Membrane Transport in Plants. Zimmermann, U., Dainty, J. (eds.). Berlin-Heidelberg-New York: Springer, 1974b, pp. 226–233
Klemke, I., Bentrup, F.W.: Zur Frage nach den Signalstoffen beim Gruppeneffekt keimender *Fucus*zygoten. Plant. Sci. Lett. *1*, 315–320 (1973)
Loyal, D.S., Paik, P.: Morphogenetic variations in the gametophyte of a 3x race of a fern—*Hypodematium crenatum*. Curr. Sci. *40*, 384–385 (1971)
Lund, E.J.: Electrical control of organic polarity in the egg of *Fucus*. Bot. Gaz. *76*, 288–301 (1923)
Lund, E.J.: Bioelectric fields and growth. Austin, Texas: University of Texas Press, 1947
Marsh, G., Beams, H.W.: The orientation of pollentubes of *Vinca* in the electric current. J. Cell Comp. Physiol. *25*, 195–204 (1945)
Mathews, A.P.: Electrical polarity in the hydroids. Am. J. Physiol. *8*, 294–299 (1903)
Miller, J.H., Stephani, M.C.: Effects of colchicine and light on cell form in fern gametophytes. Implications for a mechanism of light-induced cell elongation. Physiol. Plant. *24*, 264–271 (1971)
Müller, D., Jaffe, L.F.: A quantitative study of cellular rheotropism. Biophys. J. *5*, 317–335 (1965)
Murphy, T.M., Kahn, A., Lang, A.: Stabilization of the polarity axis in the zygotes of some *Fucaceae*. Planta *90*, 97–108 (1970)
Nakazawa, S., Takamura, K.: An analysis of rhizoid differentiation in *Fucus* eggs. Cytologia *32*, 408–415 (1967)
Novak, B., Bentrup, F.W.: Induction of *Fucus* egg polarity by electric a.c.-fields. Naturwissenschaften *57*, 549 (1970)
Novak, B., Bentrup, F.W.: An electrophysiological study of regeneration in *Acetabularia mediterranea*. Planta *108*, 227–244 (1972)
Novak, B., Bentrup, F.W.: Orientation of *Fucus* egg polarity by electric a.c. and d.c. fields. Biophysik *9*, 253–260 (1973)
Nuccitelli, R., Jaffe, L.F.: Ooplasmic segregation in the *Pelvetia* egg is preceded and predicted by a cell generated electrical current. J. Cell Biol. *70*, 50a (1976)
Peng, H.B.: Polarization of the *Fucoid* egg; galvanotropism and ultrastructure. Ph.D. thesis Purdue Univ. West-Lafayette, Ind. (1976)
Peng, H.B., Jaffe, L.F.: Polarization of *Fucoid* eggs by steady electrical fields. Dev. Biol. *53*, 277–284 (1976)
Poo, M., Robinson, K.R.: Electrophoresis of concanavalin-A receptors along embryonic muscle cell membrane. Nature (London) *265*, 602–605 (1977)

Quatrano, R.S.: An ultrastructural study of the determined site of rhizoid formation in *Fucus* zygotes. Exp. Cell Res. *70*, 1–12 (1972)
Quatrano, R.S.: Separation of processes associated with differentiation of two-celled *Fucus* embryos. Dev. Biol. *30*, 209–214 (1973)
Quatrano, R.S., Crayton, M.A.: Sulfation of fucoidan in *Fucus* embryos. 1. Possible role in localization. Dev. Biol. *30*, 29–41 (1973)
Reed, E.A., Whitaker, D.M.: Polarized plasmolysis of *Fucus* eggs with particular reference to ultraviolet light. J. Cell. Comp. Physiol. *18*, 329–338 (1941)
Robinson, K.R.: Reduced external calcium or sodium stimulated calcium influx in *Pelvetia* eggs. J. Physiol. (submitted) 1977
Robinson, K.R., Jaffe, L.F.: Ion movements in a developing *Fucoid* egg. Dev. Biol. *35*, 349–361 (1973)
Robinson, K.R., Jaffe, L.F.: Polarizing *Fucoid* eggs drive a calcium current through themselves. Science *187*, 70–72 (1975)
Robinson, K.R., Jaffe, L.F.: Calcium gradients and egg polarity. J. Cell Biol. *70*, 37a (1976)
Sand, O.: On orientation of rhizoid outgrowth of *Ulva mutabilis* by applied electric fields. Exp. Cell Res. *76*, 444–446 (1973)
Seitz, K.: Orientation in space: Plants. In: Marine Ecology II 2. Kinne, O. (ed.). New York: Wiley, 1975, pp. 451–497
Stange, L.: Plant cell differentiation. Annu. Rev. Plant Physiol. *16*, 119–140 (1965)
Vreughenhil, D., Dijkstra, M.L., Libbebga, K.R.: Ultrastructure of cell wall of normal and apolar embryos of *Fucus vesiculosus*. Protoplasma *88*, 305–313 (1976)
Wardlaw, C.W.: General physiological problems of embryogenesis in plants. In: Encyclopedia of Plant Physiology. Lang, A. (ed.). Berlin-Heidelberg-New York: Springer, 1965, Vol. 15/I, pp. 424–442
Weber, W.: Morphogenetische und keimungsphysiologische Untersuchungen an einigen Meeresalgen unter besonderer Berücksichtigung der Polarität. Bot. Mar. *12*, 135–178 (1969)
Weisenseel, M.H.: Changes in membrane properties and transcellular ion movements in developing plant cells. In: Regulation of Cell Membrane Activities in Plants. Marré, E., Cifferi, O. (eds.). Amsterdam: Elsevier North-Holland, 1977, pp. 267–273
Weisenseel, M.H., Jaffe, L.F.: Membrane potential and impedance of developing *Fucoid* eggs. Dev. Biol. *27*, 555–574 (1972)
Weisenseel, M.H., Jaffe, L.F.: Return to normal of *Fucus* egg membrane after microelectrode impalement. Exp. Cell Res. *89*, 55–62 (1974)
Weisenseel, M.H., Nuccitelli, R., Jaffe, L.F.: Large electrical currents traverse growing pollen tubes. J. Cell Biol. *66*, 556–567 (1975)
Went, F.W.: Eine botanische Polaritätstheorie. Jahrb. Wiss. Bot. *76*, 528–557 (1932)
Wettstein, D. v.: Die Induktion und experimentelle Beeinflussung der Polarität bei Pflanzen. In: Encyclopedia of Plant Physiology. Lang, A. (ed.). Berlin-Heidelberg-New York: Springer 1965, Vol. 15/I, pp. 275–330
Whitaker, D.M., Berg, W.E.: The development of *Fucus* eggs in concentration gradients. Biol. Bull. *86*, 125–129 (1944)

5.1.2 Phototropism

D.S. DENNISON

1. Introduction

When certain growing plant organs are exposed to asymmetric illumination, they bend towards the source (Fig. 1); such a response is called phototropism. For the purposes of this chapter, it is useful to define this term somewhat more precisely. Phototropism involves the change in growth direction of the organ, caused by the redistribution of growth; in most cases, as curvature develops, the convex side elongates more rapidly than the concave. Phototropism thus includes only those bending movements associated with growth: the synthesis of new cellular material. Phototropism includes movements toward the light source (positive) and away from it (negative). More precisely, we include movements that are oriented only approximately in the source direction, since random and systematic aiming errors may be present.

The definition must be stretched somewhat to include polarotropism, in which the growing organ is confined to a plane perpendicular to the illumination direction, and its orientation within that plane is determined by the angle of the electric vector of plane-polarized light. In this case, the growth direction is indeed oriented by some quality of the illumination, even though that quality is polarization rather than direction.

Phototropism is based on a change in the spatial distribution of growth, but this change may occur in different ways. For example, the shaded side of an optically dense organ may grow faster than the illuminated side, thus causing positive curvature, or, in a cell where the growth is mainly at the tip, the locus of maximum growth may simply shift toward the illuminated side. These two phototropic mechanisms, called bowing and bulging, respectively, can be distinguished by the use of markers on the cell surface. Even though the germination of some spores is oriented by asymmetric illumination, and is quite similar to the bulging phenomenon, this process is excluded from the definition on the rather arbitrary grounds that no growing organ exists prior to the light exposure.

In this chapter, experiments with both higher and lower plants will be considered together when they relate to a common conceptual problem. In this way, it is hoped that some unifying principles will become apparent. We will begin by considering the basic phenomenology of phototropism including the dose–response relation and action spectra. Although difficult to interpret, these observations form a useful factual framework for examining the fundamental problems of phototropism.

Since the publication of the old Handbuch der Pflanzenphysiologie (GALSTON, 1959; BANBURY, 1959), there have been many reviews and other reference works

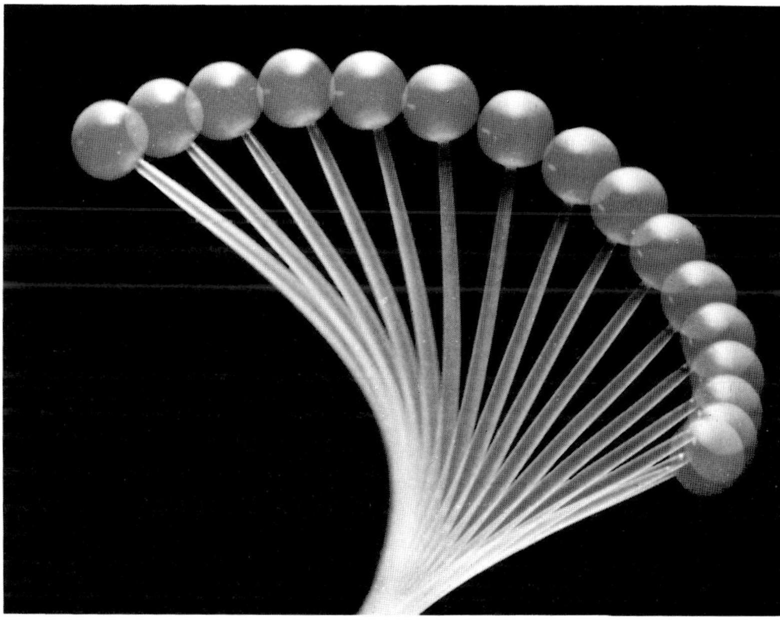

Fig. 1. Phototropic bending of the sporangiophore of *Phycomyces blakesleeanus* towards a blue light source impinging from the left. Two min elapse between successive exposures, which are made using phototropically inert wave lengths. Note that curvature occurs at a constant distance (2 mm) from the sporangium, i.e., at the base of the growth zone

covering the general topic of phototropism, as well as its more specialized aspects. Some of these are presented below in the hope that they may be of use to the reader. Recent textbook treatments of phototropism include those of MOHR (1971) and STILES and COCKING (1969). Reviews of phototropism include those of THIMANN and CURRY (1960, 1961), BRIGGS (1963c, 1964), CURRY (1969), and BALL (1969). STEYER (1965) has surveyed the literature on phototropism in dicotyledonous seedlings. An interesting summary of recent work concerning the blue light receptor has been written by BRIGGS (1976). The topic of auxin relationships in phototropism has been reviewed by GORDON and SHEN-MILLER (1968) and by ZENK (1968), and auxin transport has been very recently reviewed by GOLDSMITH (1977). Cell-wall extensibility has been covered by MORRÉ and EISINGER (1968) and by CLELAND (1971). There have been recent reviews of the photobiology of fungi (CARLILE, 1965, 1970), phototropism in fungi (PAGE, 1968) and in coprophilous zygomycetes (FOSTER, 1977), and the biology of *Phycomyces* (BERGMAN et al., 1969). Perception of stimuli by lower plants has been reviewed by HAUPT (1965).

2. Fundamental Phenomena

To set the stage for consideration of the nature of the photoreceptor molecule and the mechanisms of stimulus transduction, gradient detection, and growth redistribution, we begin by describing the fundamental phenomena of the phototropic response.

2.1 Dose–Response Relations

2.1.1 Experiments on Oats and Corn

We may begin by pointing out that there is no universal agreement as to what constitutes the phototropic response, nor indeed whether "dose" is the appropriate stimulus parameter.

The definition of the response is complicated by uncertainties as to whether the rate of phototropic bending or its final angle is more significant. In either case, the organ's angular position may vary with time in a nonlinear fashion; at what point in this process do we measure the bending rate or the total bend? For geotropic organs irradiated in an initially vertical position, a significant phototropic bending inevitably results in a geotropic stimulation which will begin to interfere with the phototropic response. One solution to this problem is to eliminate geotropism by using a clinostat. However, the more usual practice is to choose a total response duration (usually 100 min in *Avena* or *Zea* coleoptiles) which corresponds to the greatest phototropic excursion before bending reverses.

Comparisons of dose are complicated by differences in intensity and exposure time used by different investigators. If the reciprocity law (Roscoe-Bunsen law of photochemical equivalence) is known to be valid in a given situation, it would be immaterial whether time or intensity is varied, since the final outcome would depend only on the product of the dose rate (intensity) and time, the product being equal to the total energy dose delivered. However, in many cases it is not known whether the reciprocity law is in fact valid.

In addition to these uncertainties, there are complications due to differences in experimental protocols, especially those concerned with light pretreatments, which are apt to affect dose–response relationships.

A typical set of dose–response curves is shown in Figure 2 (ZIMMERMAN and BRIGGS, 1963a, b). The investigators used initially vertical *Avena* coleoptiles pretreated with 2 h of red light, and they exposed the apical 3 mm to three intensities of monochromatic blue light at 436 nm. At each of the intensities the total dose was varied by varying the exposure time from 0.7 s to 37 min, and the curvatures were measured at 100 min after the end of the exposure. Similar curves have been reported by THIMANN and CURRY (1960) and other more recent workers (EVERETT and THIMANN, 1968; MEYER, 1969a, b; BLAAUW and BLAAUW-JANSEN, 1970a, b).

The curve at the highest intensity (C in Fig. 2) shows the characteristic features noted by earlier workers (see BRIGGS, 1964). The initial rising part of the curve and the first peak are termed the first-positive curvature, the subsequent minimum is termed the first-negative curvature, and the second rising part is termed the second-positive curvature. Although not present in the results of Figure 2, it is possible to have another minimum after the second-positive curvature, and another peak, called the third-positive curvature. In correspondence with their

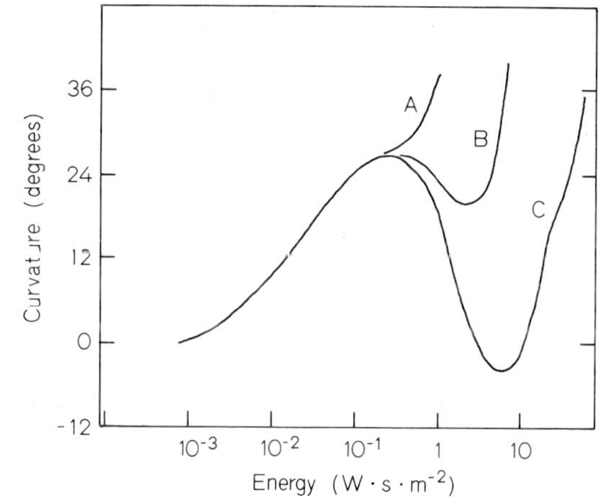

Fig. 2. Dose–response curve for phototropism in *Avena* coleoptiles for monochromatic illumination at 436 nm. Curves *A, B,* and *C* are at intensities of $3.84 \cdot 10^{-4}$, $3.84 \cdot 10^{-3}$ and $3.84 \cdot 10^{-2}$ $W \cdot m^{-2}$, respectively; each curve thus represents a range of exposure times. (Redrawn from ZIMMERMAN and BRIGGS, 1963a)

kinetic model, ZIMMERMAN and BRIGGS (1963b) term the first three regions Systems I, II, and III. A third terminology has been more recently introduced by BLAAUW and BLAAUW-JANSEN (1970a), and will be discussed later on.

While the first-positive region is constant for the three intensities of Figure 2, the other regions are not. As the intensity is lowered, the region of the second-positive curvature (System III) shifts towards lower doses (curves B and A), and in the process the first-negative region (System II) becomes obliterated. For each tenfold reduction in intensity, the dose to produce a given System III curvature, say 36°, is likewise reduced tenfold. In other words this curvature is characteristic of a certain exposure time (in this case 37 min), rather than of a particular dose, and hence, reciprocity does not hold in System III. Thus, at higher doses, the coleoptile response is evidently a superposition of the System I and II responses, which are functions of dose only, and the System III response, which is a function of exposure time. At lower exposure intensities, then, the necessarily long exposure times trigger the System III response along with the System I and II responses, and a composite curve results.

BRIGGS had shown earlier (BRIGGS, 1960) that in white light the reciprocity law begins to fail for *Zea* at doses between 0.40 and 1.260 $W \cdot s \cdot m^{-2}$ (equivalent blue light[1]) and for *Avena* at doses between 0.04 and 0.40 $W \cdot s \cdot m^{-2}$, using exposure times ranging from 2 to 10 min. At doses greater than these, the reciprocity law was shown to be invalid. An exact comparison of dosages is difficult due to the different spectral light qualities used, and is of uncertain significance since in the earlier work (BRIGGS, 1960) the plants were not given a standard red light pretreatment, which, as will be discussed later, affects the phototropic sensitivity and the shape of the dose–response curves.

ZIMMERMAN and BRIGGS proposed an empirical kinetic model (1963b), in which Systems I and II are each explained by a simple two-step reaction characterized by a pair of photosensitive forward rate constants. The solution of the resulting differential equations involves exponential functions of $I \times t$, thus incorporating

---

[1] In this case the intensity units have been converted from meter-candle-seconds to $W \cdot s \cdot m^{-2}$ of phototropically equivalent blue light using the formula, 1 $W \cdot s \cdot m^{-2} = 25000$ mcs (BERGMAN et al., 1969).

the reciprocity law. By appropriate adjustments of the four arbitrary constants (two for each system) the authors succeeded in fitting a theoretical curve to their experimental results in the region of the first-positive and first-negative curvature. In the region of the second-positive curvature, they proposed a very complex scheme, comprising a photoequilibrium step followed by a thermochemical step, and including several additional assumptions, so that at suitably long times the amount of product accumulated becomes a linear function of time, but not intensity. This is in accordance with their finding that, when the theoretical curvatures for Systems I and II are subtracted from the experimental values, the result shows the curvature to be a linear function of exposure time, over the range from 2 to 32 min.

Although not suggested by ZIMMERMAN and BRIGGS, the multiplicity of light-driven steps in their model and the "aberrant" behavior of the second-positive curvature both raise the possibility of multiple photopigments. If this were so, one would then expect that the shape of the dose–response curve might be wavelength-dependent. Some data on this point are provided by THIMANN and CURRY (1961), who compared the dose–response curve at 436 nm (blue) with that at 365 nm (near UV). These wavelengths correspond to rather widely separated peaks in the action spectrum (discussed more fully later), and yet the dose–response curves for the first-positive and first-negative curvatures were found to be essentially identical and similar to those of Figure 2, with the exception that the curve for 365 nm is shifted towards higher doses. This shift amounts to a threefold loss of sensitivity at 365 nm and reflects the action spectrum value at this wavelength relative to 436 nm. The similarity in shapes, however, is evidence against a multiplicity of photopigments for Systems I and II.

Using UV stimuli at 280 nm, CURRY et al. (1956) were able to produce curvatures characterized by a pronounced bend at the base of the coleoptile, near the node. The dose used, 0.6 $W \cdot s \cdot m^{-2}$ delivered in about 10 s, would be considerably below the region of second positive curvature for blue and near UV light. Another remarkable feature of the UV-type base curvature is its dose–response curve. At an intensity of 0.055 $W \cdot m^{-2}$ of 280 nm radiation, curvature increases linearly and monotonically with the logarithm of exposure time and finally reaches a plateau at 60 s exposure; there is no sign of the first-positive or first-negative curvatures typical of the system in blue light. Hence, the dose–response curve at 280 nm is significantly different from those at 365 and 436 nm.

Since the behavior of the second-positive curvature is so different, as reflected in the formulation of the System III kinetic model of ZIMMERMAN and BRIGGS (1963b), this might again argue for a difference in photopigment systems to be found in this case. However, the action spectrum for the second-positive curvature in *Avena* is not different from that for the first-positive curvature (EVERETT and THIMANN, 1968).

These experiments were done using plants pretreated with red light and then exposed to a standard blue stimulus (470 nm) of 5.7 $W \cdot s \cdot m^{-2}$, delivered in 100 s. This dose corresponds to the beginning of the region of the second-positive curvature, and the subsequent exposures at each wavelength result in positive curvature. Since curvature is presumably a function of exposure duration, it would at first glance seem impossible to perform an action spectrum, which of course requires that the photobiological result be a function of the energy delivered to the system. However, the authors found that if the second exposure was held constant at 400 s, the curvature was found to be proportional to intensity over a limited intensity range ($6\cdot10^{-4}$ to $6\cdot10^{-3}$ $W\cdot m^{-2}$). The resulting action spectrum is limited in generality by these restricted experimental conditions; nevertheless it agrees remarkably well with that for the first-positive curvature.

We must therefore conclude that the complexities of the kinetic model of ZIMMERMAN and BRIGGS do not correspond to complexities in the primary photopigment system of the phototropic response. This of course in no way invalidates the model, since its features could well be the result of many other kinds of processes, removed from the primary photoevent.

The phototropic dose–response curve for several species of dicotyledonous plants and for three other monocots have been compared with that for *Avena* (STEYER, 1967). Remarkably, all plants showed the typical pattern: first-positive curvature at low doses and second-positive curvature at high doses. Under the red-pretreatment conditions used (a dim red light), the *Avena* coleoptile showed a negative curvature region between the first- and second-positive regions. None of the other species showed this negative curvature, but instead had an indifferent region of varying extent for different species. Under conditions in which the maximum first-positive *Avena* curvature was about 25°, other species had maximum values of from 15° to 25°, except for *Helianthus annuus*, which had only 4°. Since the red pretreatment was not well controlled, and since the phototropic stimulus was unfiltered tungsten illumination of varying wattage, the species differences found by STEYER cannot yield any definite conclusions about the different systems.

Using *Raphanus sativus* L. seedlings, dark-grown or with a 24-h red light treatment, EVERETT (1974) obtained dose–response curves with no negative or indifferent region; instead, they showed a very broad, plateau-like region of first-positive curvature, which merged smoothly with the rising second-positive peak. Seedlings raised in white light, however, showed no curvature in the first-positive region at all; the response began only after the stimulus energy reached the second-positive region. Since the white light treatment was ended 2 h before the experiment was performed, this effect is probably not similar to the tonic action of blue light (Sect. 2.1.2.), but may instead be due to a more fundamental alteration of the plant's sensitivity to light.

Much has been made of the separation of *Avena* phototropism into the so-called tip and base responses (THIMANN and CURRY, 1960; CURRY, 1969). According to this view, only the tip is sensitive to low dosage stimuli, and only the tip is responsible for the first-positive curvature, which is slowly propagated down the length of the coleoptile. The sensitivity to larger stimuli is not confined to the tip, but is rather distributed throughout the entire coleoptile and is hence called the base response. This base response is thought to be triggered only by large stimuli (in the dosage region of the second-positive curvature), and it is not propagated, but develops at the same time all along the coleoptile. Support for this distinction is largely morphological, in that a coleoptile with a pure tip response is relatively sharply bent near the apex, whereas one with a pure base response is bent more or less uniformly along the coleoptile (THIMANN and CURRY, 1960; BRIGGS, 1964). More recently, BLAAUW and BLAAUW-JANSEN (1970b) have maintained that the distinction between tip and base curvatures is spurious, and arises from uncontrolled red light effects. They found that without a red pretreatment the curvature was uniformly distributed, and that red light shifts the curvature towards the coleoptile tip. Further, they report that stimuli in both the first-positive and the second-positive regions produce the same tip curvature.

Novel experimental and analytical approaches were used by PICKARD et al. (1969) to separate the dose–response relations for the first- and second-positive curvatures in *Avena* coleoptiles pretreated with red light. They first established that, for doses corresponding to the second-positive response, the coleoptile bends for a much longer time if it is rotated on a clinostat following the irradiation. This effect, observed also for first-positive curvature (SHEN-MILLER and GORDON, 1967) is due to the inhibition of geotropism, which would otherwise come into play as the coleoptile begins to deviate from vertical.

In this way, PICKARD et al. were able to work with the maximum obtainable curvatures, reached 135 min after the end of the exposure. They compared the dose–response curves of plants with and without opaque caps shielding the apical 1 mm (Fig. 3). They then multiplied the capped values by 1.65, the correction factor needed to make the two curves coincide at and above their peak intensities (about 1 $W \cdot m^{-2}$; Fig. 3). Finally they obtained the first-positive response by subtracting the capped curve (multiplied by the correction factor) from the uncapped curve. The validity of both the correction and the subtraction procedures depends on the assumption of linearity of response as a function of the amount of tissue irradiated; i.e., the total response is simply the algebraic sum of the individual contributions. The linearity assumption is more likely to be valid when the clinostat technique is used, because no geotropic "clipping" of the response, and hence nonlinearity, could occur. There is no direct quantitative experimental evidence in support of the linearity assumption; however, as we have seen, ZIMMERMAN and BRIGGS (1963b) performed comparable calculations. Underlying this procedure is the reasonable hypothesis that while the sensitive region for the first-positive curvature is restricted to the apical 1 mm, the sensitive region for the second-positive curvature extends along the entire organ, including the apical 1 mm.

The intensity at the peak of the calculated first-positive response is about $5 \cdot 10^{-3}$ $W \cdot m^{-2}$ of blue light, which for a 45-min exposure is an energy value of 13.5 $W \cdot s \cdot m^{-2}$. As comparison with Figure 2 will show, this dosage value is about 50-fold higher than the corresponding value given by ZIMMERMAN and BRIGGS (1963a). This discrepancy was noted by PICKARD et al. (1969), who suggested that it might be explained in part by assuming a reciprocity interval of 10 min, i.e., the reciprocity law fails for exposures longer than 10 min. Thus if only the first 10 min of the 45-min exposure is effective, then the true energy dosage values would be reduced about fivefold. However, this explanation does not hold for the second-positive curvature, since exposures beyond 10 min do in fact contribute to the final result. There is no reason to suppose the first-positive curvature is any more likely to depart from the reciprocity law, and in all likelihood the discrepancy is a result of an error in light-source calibration.

One of the difficulties in understanding the dose–response relations of coleoptiles stems from the fact that even under the simplest set of conditions the measured curvature is a function of two variables: the intensity and the duration of the exposure. Since some experimenters tend to use a constant intensity and vary the exposure time, while others keep the exposure time constant and vary the intensity, it is difficult to compare the two sets of results without invoking the reciprocity law.

Approaching the problem directly. BLAAUW and BLAAUW-JANSEN (1970a) have measured curvature of *Avena* coleoptiles as a function of intensity from 10^{-5} to 12.6 $W \cdot m^{-2}$ and as a function of exposure time from $^{1}/_{100}$ s to 40 min. The

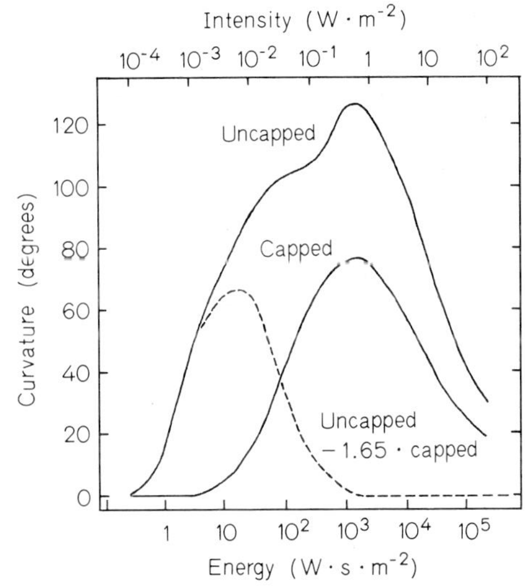

Fig. 3. Dose–response curves for phototropism in *Avena* coleoptiles. Exposures were for 45 min, and in one set of experiments ("capped") the apical 1 mm was shielded by a cap. *Dashed* (---) *curve* is calculated by subtracting from the "capped" values the "uncapped" values, multiplied by a correction factor of 1.65. (Redrawn from PICKARD et al., 1969)

coleoptile was pretreated with red light, and the curvature was measured 100 min after the end of the exposure. The data have the usual appearance when plotted either as a function of light intensity (Fig. 4) or as a function of time (Fig. 5), holding the other factor constant in each case. However, when the curvature is displayed as a function of two variables at once, a three-dimensional surface results (Fig. 6). Although it has a rather unconventional appearance, this surface may be the most meaningful way to comprehend the entire system.

To relate Figures 4 and 5 to the topography of Figure 6, the reader should imagine the latter sectioned by vertical planes running parallel either to the intensity or to the time axis. The latter sections yield response as a function of time and differ markedly when made at different intensities. At the highest intensities ($10^{4.1}$ erg·cm^{-2}·s^{-1}, or 12.6 W·m^{-2}) the curve is quite complex, showing three maxima known classically as first-positive, second-positive, and third-positive curvature, and by BLAAUW and BLAAUW-JANSEN as areas A, B, and C, respectively. As constant-intensity sections are made at progressively lower intensities, areas A and B shift to longer exposure times, as does the negative region between A and B (the classical first-negative response). This shift is in accordance with the reciprocity law; for example, the peak curvature of area A corresponds to a constant energy dosage value (given by the authors as 0.12 W·s·m^{-2}) equal to the product of time and intensity. Area C does not shift along the time axis; rather, its peak and slope increase with decreasing intensities, showing that reciprocity does not hold.

This pattern of changing curve shapes has the intriguing result that at progressively lower intensities area B becomes merged with, and finally obliterated by, the rising flank of area C. The negative region, too, becomes swallowed up and finally vanishes at an intensity of about 10^{-3} W·m^{-2} (1 erg·cm^{-2}·s^{-1}). This strongly implies that low-intensity dose–response curves such as those of Figure 2 (ZIMMERMAN and BRIGGS, 1963a) do not correspond to a classical first-positive, second-positive response, but on the contrary correspond to a classical first-positive,

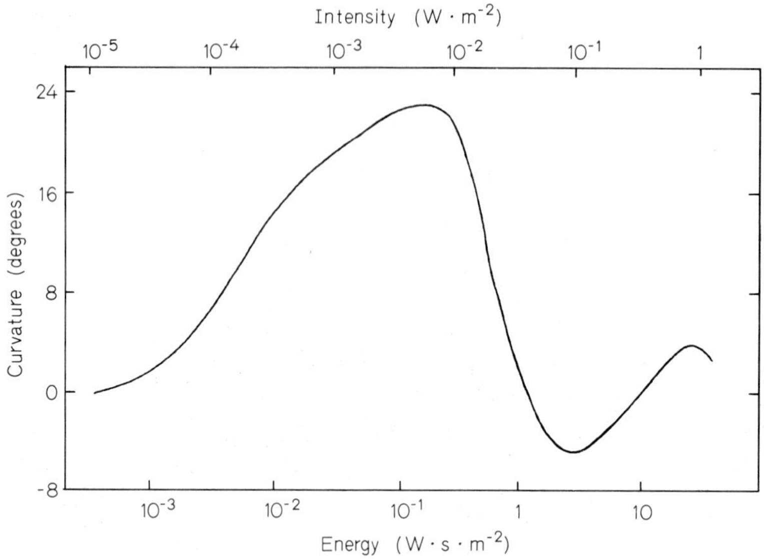

Fig. 4. Dose–response curve for phototropism in *Avena* coleoptiles, using blue light. Curvature was measured as a function of intensity; exposure time was fixed at 30 s. (After BLAAUW and BLAAUW-JANSEN, 1970a)

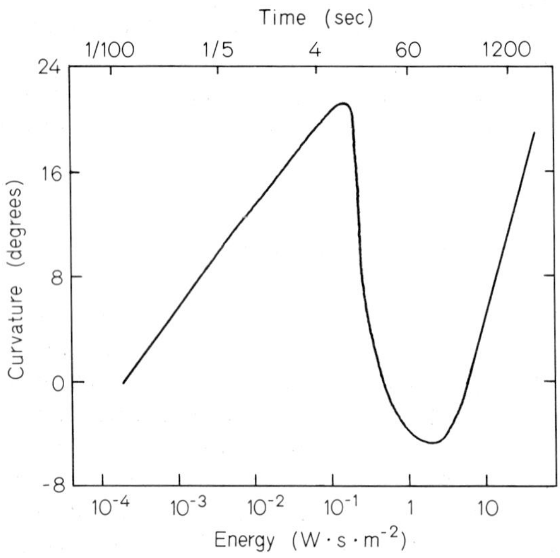

Fig. 5. Dose–response curve for phototropism in *Avena* coleoptiles, using blue light. Curvature was measured as a function of exposure time; light intensity is fixed at $1.86 \cdot 10^{-2}$ $W \cdot m^{-2}$. (After BLAAUW and BLAAUW-JANSEN, 1970a)

third-positive response, the second-positive response (BLAAUW and BLAAUW-JANSEN's area B) having vanished at this intensity.

Dose–response curves at constant exposure times can be visualized as vertical sections running from back to front of the surface in Figure 6. Thus, for example, the curve of Figure 3 (uncapped) is for a constant exposure time of 45 min and should correspond to a section near the extreme right-hand edge of Figure 6,

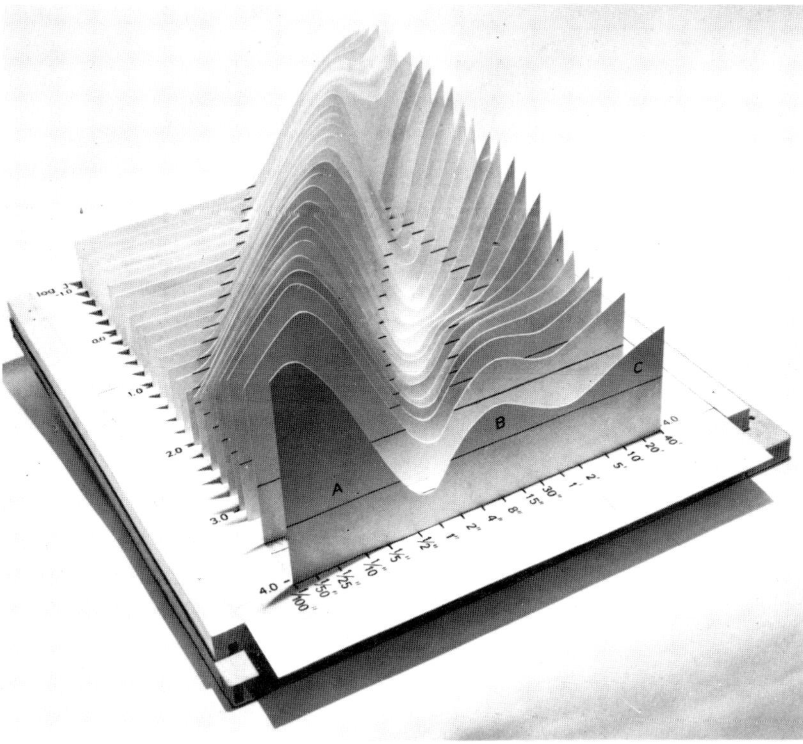

Fig. 6. Phototropic curvature of *Avena* coleoptiles (*ordinate*) as a function of blue light intensity and exposure time. From *back to front* are increasing intensities and from *left to right* are increasing exposure times. A 2-h red pretreatment (intensity 0.4 $W \cdot m^{-2}$) was used. (From BLAAUW and BLAAUW-JANSEN, 1970a)

marking the confluence (at low intensity) of the A ridge and the flank of the C area.

Shorter constant-exposure times may produce curves that do in fact correspond to the classical first-positive, second-positive curvatures, A and B areas respectively. For example imagine the surface of Figure 6 sectioned by a vertical plane at an exposure time of 15 s. This would be very similar to Figure 4, but the second peak would be due to the B area and not the C area as in Figure 5.

The surface representation of dose–response relationships is thus a significant improvement in that one is enabled to synthesize a great variety of experimental results into a single pattern. In particular it allows the disentangling of the second-positive and third-positive responses, and it indicates what conditions of intensity and exposure time are best suited for their individual study.

2.1.2 Tonic Effects of Phototropically Effective Light on the Dose–Response Relation

The complexity of the dose–response relation for oats and corn implies that each energy increment of the phototropic stimulus has two effects. This energy is the trigger for a bending response and in addition it changes the state of the organ so that its sensitivity to a subsequent energy increment is now altered. The latter is a tonic, or nondirectional effect. Delivery of 1 s of blue light at an intensity of 0.038 $W \cdot m^{-2}$ will thus produce positive curvature in an oat coleoptile previously unexposed to blue light. However, if the coleoptile has been exposed to 10 s of this intensity immediately beforehand, the 1-s exposure will now produce a positive curvature of decreased magnitude, indicating that the system is now approaching the region of first-negative curvature.

This tonic effect persists for some time after the first light exposure. BRIGGS (1960) found that corn coleoptiles (not treated with red light) given a white light dose of 1.26 $W \cdot s \cdot m^{-2}$ (equivalent blue light) showed little or no curvature either to this dose or with an additional 0.04 $W \cdot s \cdot m^{-2}$ given within 2 min of the first exposure. This is reasonable, since the first dose is in the region of the first-negative response, which was nearly zero under these conditions. The second dose was simply additive, and had no noticeable effect. However, if a longer dark interval was interposed between the two exposures, positive curvature resulted, and when this interval reached 25 min, the response to the second stimulus was the same as if the first stimulus had not been present. BRIGGS found similar results with oats, and he concluded that the first exposure destroys the light sensitivity of the system, so that it is unable to respond to the second exposure. In the dark this sensitivity returns and is completely restored after 25 min.

MEYER (1969a) showed that in *Avena* the first stimulus can be an effective desensitizer even when given vertically from above. An initial vertical blue light exposure of 6 $W \cdot s \cdot m^{-2}$ in 5 s is followed after varying times by a lateral 5-s exposure of 0.06 $W \cdot s \cdot m^{-2}$. The desensitization produced by the first exposure is complete for 2 min and then begins to fade. After 20 min in the dark the response to the second exposure has returned to normal.

A much smaller first exposure also affects the subsequent response. A vertical exposure of 0.06 $W \cdot s \cdot m^{-2}$ strongly reduces the tropic sensitivity to a subsequent lateral exposure (at the same intensity) for 2 min; however, the sensitivity is fully recovered after 5 min in the dark. Interestingly, if the sequence of exposures is reversed, i.e., if the lateral exposure precedes the vertical one, the curvature is greatly reduced, even if the exposures are separated by a dark period of up to 20 min. This suggests that the later vertical exposure interferes with the asymmetry, or "secondary gradient" set up by the earlier lateral exposure. Unexpectedly, this interference occurs only if there is a dark interval of at least 5 min between the two exposures.

Similar results were found by BLAAUW and BLAAUW-JANSEN (1970b), who worked with red-pretreated *Avena* coleoptiles. They found that the C-type (third-positive) curvature obtained with a 10-min exposure of blue light (intensity 0.0148 $W \cdot m^{-2}$) could also be elicited if as much as the first 9 min of the exposure was given vertically instead of horizontally. Also, the vertical exposure could be shortened to 1 min, as long as at least a 2-min dark period intervened before the subsequent horizontal exposure. This dark period could be as long as 40 min without diminishing the positive (C-type) curvature in response to the second exposure.

Since the horizontal 1-min exposure taken by itself is in the first-negative region, the increased curvature can be interpreted as a re-activation of the first-

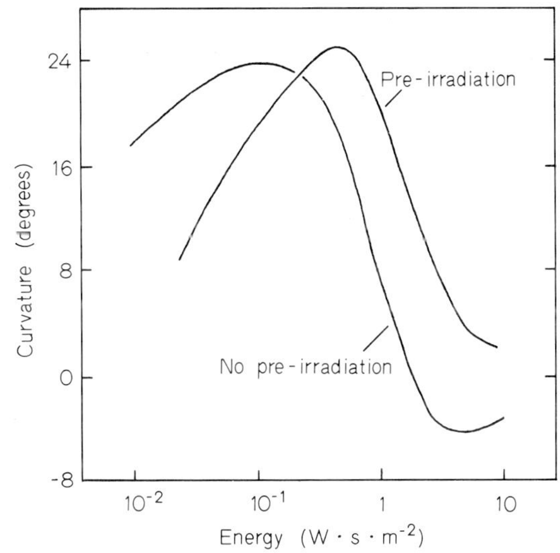

Fig. 7. Dose–response curves for phototropism in *Avena* coleoptiles, showing the effect of a blue light pre-irradiation of 3 $W \cdot s \cdot m^{-2}$. The unilateral stimulus was delivered in a 1-min exposure. (After BLAAUW and BLAAUW-JANSEN, 1970b)

positive response caused by a lowered sensitivity. Evidence for this view is shown in Figure 7, in which the dose–response curve is determined with and without a prior vertical dose of 3 $W \cdot s \cdot m^{-2}$ and 20-min dark period. This blue pretreatment is seen to shift the entire curve towards higher doses. For a horizontal exposure of about 1 $W \cdot s \cdot m^{-2}$, this shift does indeed increase curvature from 4° to about 22°. For smaller horizontal exposures, the effect is quite different, and even the shape of the curve is somewhat altered.

BLAAUW and BLAAUW-JANSEN conclude that the C-type curvature is not actually an independent response at all, but the result of the desensitization of the first-positive response system. They note that, in agreement with MEYER (1969a), this desensitization requires from 5 to 10 min following the exposure to become fully developed, after which it gradually dies out. They visualize this process as the movement of the A-region, or first-positive curvature, towards higher intensities, reaching a maximum shift 5 to 10 min after the first (desensitizing) exposure, and then moving gradually back to its original position.

It must be pointed out that more extensive experiments should be made to document this tonic shift in the dose–response curve more thoroughly, especially at a variety of intensities. These results clearly indicate that the tonic effects of blue light are complex, particularly in regard to time relationships. In this connection it should be recalled that MEYER (1969a) has shown that a tonic effect is also present at lower intensities, and that in this case the time required for sensitivity changes is considerably shorter.

The tonic effect of blue light is especially marked in the phototropism of *Phycomyces* sporangiophores. The response is almost completely determined by its sensitivity (formalized as the "level of adaptation" by DELBRÜCK and REICHARDT, 1956), which in turn depends on the previous blue light illumination and the time elapsed. If, for example, a sporangiophore is exposed to lateral illumination at an intensity significantly below its level of adaptation, there will be no phototropic response at all until the sensitivity has risen by a sufficient amount. The time required for this depends on the difference between the intensity and the level of adaptation. Once phototropism begins, the rate of bending depends only slightly on the intensity of illumination in the range from $3.9 \cdot 10^{-4}$ to $0.4\ \mathrm{W \cdot m^{-2}}$ (REICHARDT and VARJÚ, 1958). Thus, within limits, the sporangiophore adjusts its sensitivity to match the intensity of illumination, and thereafter bends at a constant rate.

2.1.3 Dose–Response Relation with Balancing Stimuli

When a variable phototropic stimulus is balanced against an opposing stimulus of constant intensity, a different kind of dose–response relationship appears. For example, when 45-min stimuli (constant intensity: $0.5\ \mathrm{W \cdot m^{-2}}$) were delivered to opposite sides of *Avena* coleoptiles, PICKARD et al. (1969) found that the curvature (developed after 135 min on a clinostat) was a linear function of the intensity ratio, with a threshold at 0.8.

A linear relationship was also found for *Phycomyces* by CASTLE (1965). His mathematical theory predicts that the bending speed in response to two opposed sources, relative to that for a single source, should be proportional to the difference of the two intensities divided by their sum. This linear relation was confirmed by CASTLE's data.

A very useful form of dose–response relation results when a light stimulus is balanced against a gravity stimulus. When illuminated steadily by a horizontal stimulus, a *Phycomyces* sporangiophore comes to an equilibrium that is a balance between the phototropic stimulus acting horizontally and the (negative) geotropic stimulus acting vertically. The phototropic effect increases from 10^{-9} to 10^{-6} $\mathrm{W \cdot m^{-2}}$, and then the equilibrium angle is rather constant at 20° above horizontal for intensities from 10^{-6} to about $1\ \mathrm{W \cdot m^{-2}}$ (BERGMAN et al., 1969; Fig. 8). At intensities above $1\ \mathrm{W \cdot m^{-2}}$, the phototropic mechanism fails, probably because it saturates and is no longer able to discriminate an intensity gradient across the cell. BERGMAN et al. (1973) have studied carefully the intensity dependence of the equilibrium angle, not only for the wild-type strain, but also for mutants they isolated from spores treated with a chemical mutagen. Figure 8 shows the photo-geo dose–response curve for the wild-type strain. The phototropic mutants show dose–response curves with altered absolute thresholds (curve shifted to higher intensities), a decreased phototropic bending capacity (curve lowered), and various combinations of these. Identification of the molecular basis for the altered phototropism in these mutants would provide important clues as to the mechanism of the photoresponse; however this has not yet been accomplished.

A striking feature of the dose–response curve of Figure 8 is the constancy of its equilibrium angle over such a large range of intensity. This constancy is explained by the sensitivity-adaptation mechanism; since the system adapts completely at all intensities (within limits), the phototropic effectiveness of a single source should be independent of its intensity. It is more difficult to account for the magnitude of the angle, but the value of 20° is evidently a compromise between the rather weak vertical response to gravity and the decreasing phototropic response as the angle of incidence decreases. The bending speed was found by DENNISON (1965) to decrease steadily as the angle between the stimulus beam and the sporangiophore axis decreases from 45° to 14° (at the latter angle, the bending rate

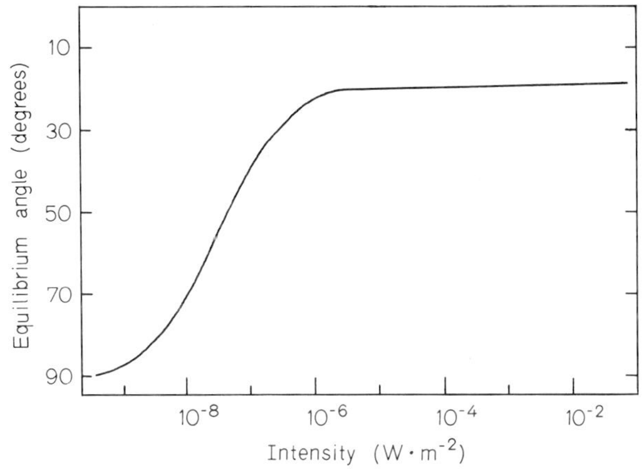

Fig. 8. Phototropic-geotropic equilibrium in *Phycomyces* sporangiophores illuminated by a horizontal light beam. The angle is measured upwards from horizontal. (After BERGMAN et al., 1973)

is essentially zero). Thus a steady angle of 20° provides a very small phototropic stimulus, just sufficient to balance the geotropic stimulus. The reason for the decrease in phototropism is not known in detail, but is undoubtedly linked to the altered internal intensity distribution caused by the altered optical geometry. The rotational theory of *Phycomyces* phototropism (discussed in Sect. 5.1.1.) has been incorporated by MEDINA and CERDÁ-OLMEDO (1977) into a quantitative model which accounts for the constancy of this angle, as well as the shape of various photomutant dose–response curves.

2.2 Effects of Red Light

2.2.1 Effects on Dose–Response Relation

The effect of red light on the dose–response curve of *Avena* coleoptiles was discovered by CURRY (1957, cited in CURRY, 1969), who found that the first-positive region was shifted to higher intensities by a factor of 10. These results were confirmed by ZIMMERMAN and BRIGGS (1963a), who showed that in *Avena* the region of first-positive and first-negative curvature (Systems I and II) was shifted to higher intensities (tenfold shift) by a red light exposure immediately prior to blue light exposure (Fig. 9). They found that the System III region was shifted towards lower intensities (threefold shift) by this same red light pretreatment. Thus the red pretreatment lowers the sensitivity of the first-positive and first-negative response, and raises the sensitivity of the second-positive response (the C-type curvature of BLAAUW and BLAAUW-JANSEN, 1970a).

ZIMMERMAN and BRIGGS propose that the red light exposure increases the amount of System III pigment, which raises the sensitivity of System III. Simultaneously, this increased pigment level is assumed to act as a screen for the pigment of Systems I and II, and thus to reduce their sensitivity. This hypothesis has in its support the increased levels of carotenoids and flavins in *Avena* coleoptiles following red light exposure (ASOMANING and GALSTON, 1961). On the other hand, as discussed earlier, the evidence from action spectra does not support the assumption that System III has a different pigment than do Systems I and II.

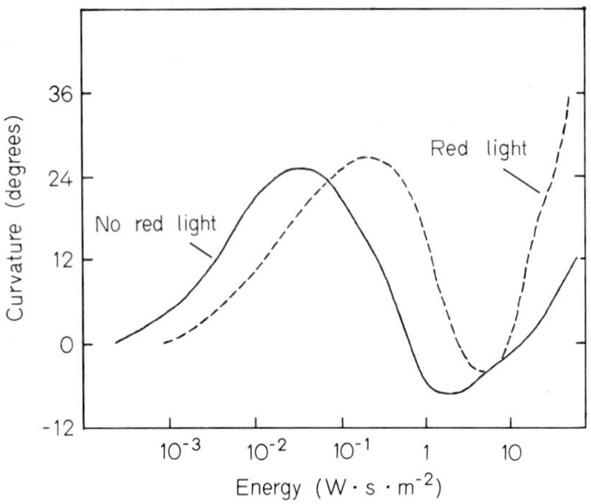

Fig. 9. Dose–response curves for phototropism in *Avena* coleoptiles, showing the effect of a red-light pretreatment. The intensity of 436 nm light was constant at $3.84 \cdot 10^{-2}\ W \cdot m^{-2}$, and exposure times were varied to obtain the energy doses shown. Red light was a 2-h exposure at $3.3\ W \cdot m^{-2}$ from a ruby bulb. (Redrawn from ZIMMERMAN and BRIGGS, 1963a)

BRIGGS (1963b) found that for oats and corn an exposure to red light ($3.3\ W \cdot m^{-2}$) of at least 1 h was necessary to saturate the system. In corn, this effect begins to diminish within 1 h after the end of the red exposure and completely disappears after 3 to 4 h. Subsequently, CHON and BRIGGS (1966) reported that in corn even a 1-s exposure to red light ($1.4\ W \cdot m^{-2}$) was sufficient to produce the maximum sensitivity shift, which was fully developed after 1 h. An action spectrum for this effect shows a single major peak at 660 nm.

BLAAUW and BLAAUW-JANSEN (1964) found that comparable red pretreatments led to a shift of A-area curvature (Fig. 6) toward higher dosages and to a possible shift of C-area curvature toward lower dosages (or to an increase in C-area curvature; the two cannot be distinguished on the basis of the data; BLAAUW and BLAAUW-JANSEN, 1970a). This constitutes further support for the contention that area C is simply the re-appearance of the upper flank of area A in the system desensitized by long blue exposures.

In contrast, a quite different situation exists for area B (BLAAUW and BLAAUW-JANSEN, 1970a). If no red pretreatment is used, area B disappears completely, leaving only a broad indifferent region between the two minima. Also, replacing the 2-h red pretreatment ($0.4\ W \cdot m^{-2}$) with a 2-min red pretreatment followed by 2 h of darkness reduces the B-area curvature from about 5° to about 1.5°.

Using a 4-h blue light stimulus at $0.048\ W \cdot m^{-2}$ (presumably in the second-positive region), KANG and BURG (1974) found that the response of *Pisum sativum* was increased from about 30° to about 50° by a 5-min red pretreatment ($1.25\ W \cdot m^{-2}$) followed by 4 h of darkness. When the dark period was reduced to 2 h, there was scarcely any enhancement. The first-positive curvature was found to be unaffected in magnitude or in sensitivity by red light. Thus in *Pisum* there seems to be no correlation between the red light effects on A- and C-area curvatures, as found in *Avena*.

Two findings by CURRY et al. (1956) are of interest, and yet are difficult to fit into the picture developed thus far. The authors note that the UV-type base curvature in *Avena* is reduced by one-half if a 60-min red light pretreatment is introduced immediately before the UV stimulus. This contradicts the identification

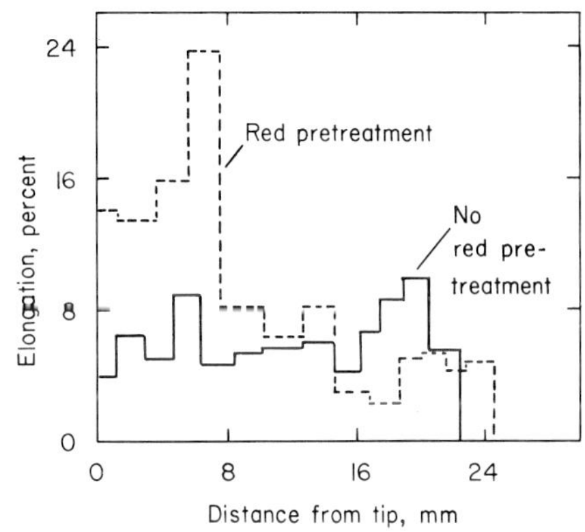

Fig. 10. Distribution of growth rate along an *Avena* coleoptile, showing the effect of a red light pretreatment. The red exposure was 1 h, using filtered red light from a fluorescent lamp at about 12 foot-candles. Growth rates were determined by measuring the separation of lanolin droplets during a 100-min interval. (After CURRY et al., 1956)

of the UV-type base response with the second-positive (C-type) curvature, since the latter is increased by such a red pretreatment. As shown in Figure 10, red light alters the rate of elongation and its distribution along the coleoptile.

The change in elongation distribution may be related to the effect of red light on the localization of curvature (BLAAUW and BLAAUW-JANSEN, 1970b). When blue stimuli in the dosage range of the C-type curvature were used, red pretreatment shifted the locus of maximum curvature towards the coleoptile tip. This is reasonable in view of the increased elongation rate in the same region. Also, as pointed out earlier, the distinction between tip and base curvature may well hinge on poorly controlled red pretreatment protocols.

Although the effects of red light are complex and not well understood, they are highly significant. Clearly, red light causes major and pervasive changes in the pattern of growth of the coleoptile, and in the way this growth is modulated by blue and UV stimuli.

2.2.2 Photoreversibility of the Red Light Effect: the Role of Phytochrome

In spite of the great complexity of the red light effects there exists what would seem to be a relatively simple test for the involvement of phytochrome, namely the photoreversal by far-red illumination (wavelengths greater than 700 nm). Such a test was performed under rather restricted conditions by BRIGGS (1963b). He subjected oat coleoptiles to a white light stimulus (1000 mcs) sufficient to produce a negative curvature of 5° to 10° without a red light pretreatment and a positive curvature of 20° to 25° when four 5-s red exposures were given (spaced at half-hour intervals) during the 2 h before the white stimulus. The addition of a 100-s far-red exposure (intensity 20 $W \cdot m^{-2}$) after each of the red exposures resulted in negative curvatures of about 2°, indicating that a large fraction of the red light effect had been reversed by the far-red exposure.

BRIGGS and CHON (1966) measured the distribution of red light sensitivity (ability to shift the phototropic dose–response curve) along the coleoptile, and found it was fairly well correlated with the distribution of phytochrome (spectrophotometric measure of photoreversibility): both were most heavily concentrated in the apical 3 mm. Dose–response comparison of the two red light effects had quite unexpected results, however (BRIGGS and CHON, 1966). The sensitivity of the physiological effect on phototropism is a factor of 10^3 or 10^4 higher than that of the effect on spectrophotometrically detectable phytochrome. The phototropic effect is saturated at about 0.02 $W \cdot s \cdot m^{-2}$, whereas the measurable phytochrome photoconversion is not detectable until the dose reaches 2 $W \cdot s \cdot m^{-2}$ and saturates at about 2000 $W \cdot s \cdot m^{-2}$.

This enormous difference in sensitivity indicates that if both effects are indeed due to phytochrome, the pigment must be divided into two pools: a large spectrophotometrically detectable one, and a small spectrophotometrically undetectable one that is, however, physiologically active. Further data collected by BRIGGS and CHON (1966), concerning the decay kinetics of P_{FR} and P_R and the shape of the dose-response curves, indicate that if the same phytochrome pigment is responsible for both effects, the pools must be functionally distinct, and the photochemical rate constants for P_R–P_{FR} interconversion must be different within each pool. This "phytochrome paradox" is not confined to effects on phototropism and is more fully discussed in a future volume in this Encyclopedia.

2.3 Phototropic Action Spectra

2.3.1 Blue Light Type

An important step in elucidating a photobiological process is the identification of the primary photoreceptor pigment, and a reasonable first step in such identification is the determination of an action spectrum. Although there are many pitfalls in their interpretation, action spectra are clearly important tools in photopigment identification. We will first describe action spectra of the typical blue light type (in which red is phototropically ineffective), obtained for coleoptiles, *Phycomyces* sporangiophores, and some other phototropic organs.

SHROPSHIRE and WITHROW (1958) obtained an action spectrum in *Avena* by measuring dose-response curves (first-positive region) at each wavelength. Curvatures were then plotted against log dose, and straight lines were fitted. Remarkably, the slopes of these lines were wavelength-dependent, the greatest slope being associated with the 370-nm region. Thus when the action spectrum is given as the quantum effectiveness (the inverse of the quantum requirement for a given curvature) in relation to a fixed response, the shape of the resulting curve depends on the response magnitude chosen. As can be seen in Figure 11, a threshold (zero curvature) response yields a curve with three prominent peaks between 410 nm and 475 nm, and a shoulder from 350 nm to 390 nm. On the other hand, when a response size of 10° is chosen, a peak at about 370 nm becomes more prominent. The authors interpret this variation with response size as a consequence of the screening action of an inert pigment that absorbs strongly near 370 nm. In a later report CURRY and GRUEN (1961) were unable to confirm such a wavelength dependence of the dose-response curve.

An *Avena* action spectrum, with quantum effectiveness related to a constant curvature of 10°, is shown in Figure 12 (THIMANN and CURRY, 1960; CURRY and THIMANN, 1961). The same peaks as in Figure 11 are evident, although the relation between the visible and the near-UV peaks is clearly different.

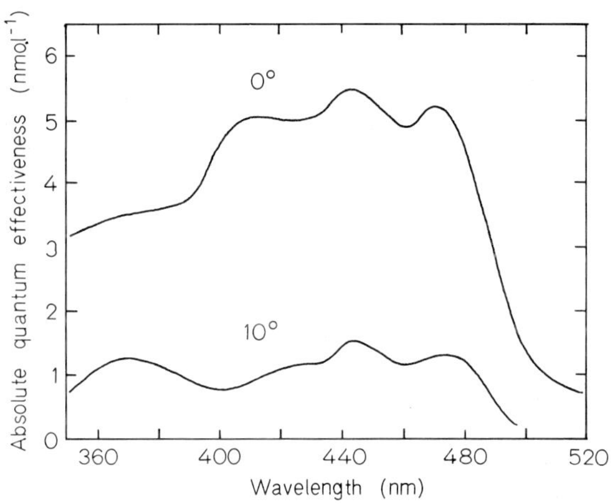

Fig. 11. Action spectra for phototropism in *Avena* coleoptiles, showing quantum effectiveness for curvatures of 0° and 10°. (After SHROPSHIRE and WITHROW, 1958)

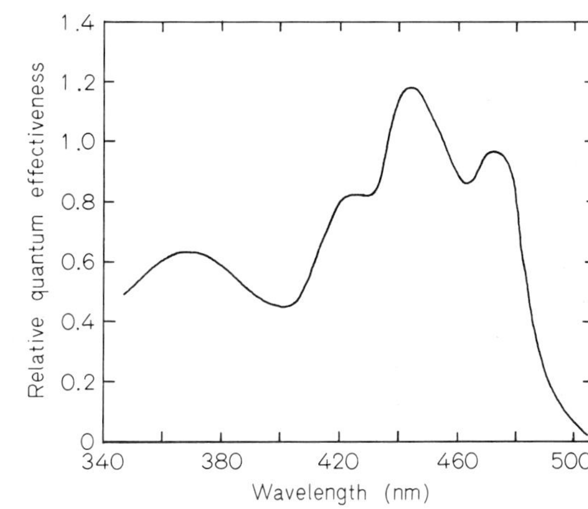

Fig. 12. Action spectrum for phototropism in *Avena* coleoptiles, showing the quantum effectiveness for a curvature of 10° relative to the response at 436 nm. (After THIMANN and CURRY, 1960)

The UV region of an *Avena* action spectrum, for coleoptiles with the apical 2 to 4 mm shielded, is shown in Figure 13 (CURRY et al., 1956). It is difficult to relate this to Figure 12, since the latter is for irradiation of the entire coleoptile. The curvatures of Figure 12 are normalized to the value at 436 nm, whereas those of Figure 13 are normalized to 289 nm; therefore we have no information as to the size of the peak at 290 nm in relation to those in the blue and near UV.

Constant-curvature *Avena* action spectra were obtained by ASOMANING and GALSTON (1961) for responses of 10°, 15°, and 20°. These are very similar to the curves of Figures 11 and 12, but variations in the 370 nm peak with response size could not be seen since no wavelengths below 400 nm were used. These authors also examined the action spectrum of barley coleoptiles, *Hordeum vulgare*. They used exposure times of from 20 to 60 min, which may have placed the system in the second-positive region. Although the spectrum does not reach into the near UV, a steep rise at the short-wavelength end does suggest the existence of the 370 nm peak. More remarkably, and as in Figure 11, the shape of the curve

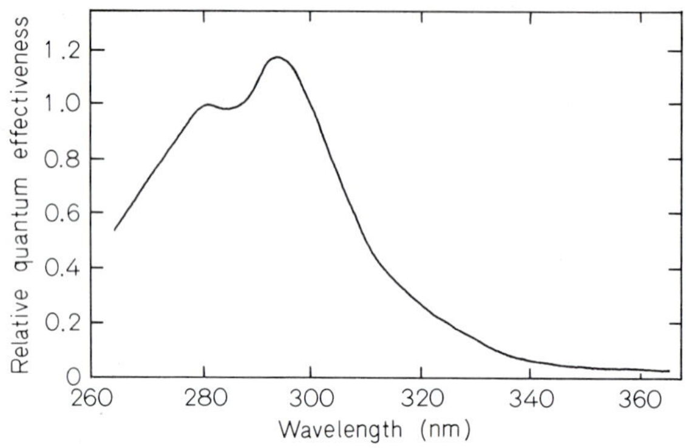

Fig. 13. Action spectrum for phototropism in apically-shielded *Avena* coleoptiles (2–4 mm), showing the quantum effectiveness for a curvature of 8.6° relative to the response at 289 nm. (After CURRY et al., 1956)

depends on the curvature chosen, suggesting a wavelength-dependent dose–response curve, as found by SHROPSHIRE and WITHROW (1958). ASOMANING and GALSTON (1961), however, attributed these results to the rather aberrant barley phototropic system, which, in its relative insensitivity and its reciprocity failure, resembles the second-positive curvature region of *Avena*.

The action spectrum for the second-positive curvature has been obtained using *Avena* coleoptiles that had previously been exposed to 5.7 $W \cdot s \cdot m^{-2}$ of 470 nm radiation (EVERETT and THIMANN, 1968). The resulting curve is similar to Figures 11 and 12, but has a narrower peak in the blue, and less activity below 375 nm. Interestingly, these authors also found that the slope of the dose–response curve increased with decreasing wavelength from 450 nm to 375 nm. This is similar to the result of SHROPSHIRE and WITHROW (1958) and argues for increased screening in the near UV.

Phycomyces action spectra are determined most advantageously with a balance method, since the adaptive mechanism makes it impossible to obtain the usual dose–response curve. One method (CURRY and GRUEN, 1959) balances a standard stimulus against the monochromatic stimulus, each source being located at one end of a 2-m plank. Cultures are placed along the plank, and the point of balance is determined. Action spectra for the visible (Fig. 14), in which the standard stimulus is a broad-band filtered source with peak wavelength at 460 nm, and for the UV (Fig. 15), in which the standard stimulus is a 254-nm germicidal lamp, are very similar to those for *Avena* (Figs. 12 and 13, respectively). However, in contrast to the positive phototropism of coleoptiles in the UV, *Phycomyces* sporangiophores show negative curvatures at these wavelengths, although in the visible they curve positively. As will be discussed later, this reversal of *Phycomyces* phototropism is closely related to the mechanism of the perception of light direction, and is due to the strong UV absorption by the sporangiophore. The *Phycomyces* action spectrum below 250 nm shows a strongly rising region not known to exist in *Avena*. Also, CURRY and GRUEN (1959) note the existence of a limited indifferent region (neither positive nor negative curvature) extending above 310 nm; positive responses were found at 334 nm and above.

DELBRÜCK and SHROPSHIRE also measured a *Phycomyces* action spectrum (1960); they used individual sporangiophores, however, and balanced the stimulus at each wavelength with a standard broad-band source with peak wavelength

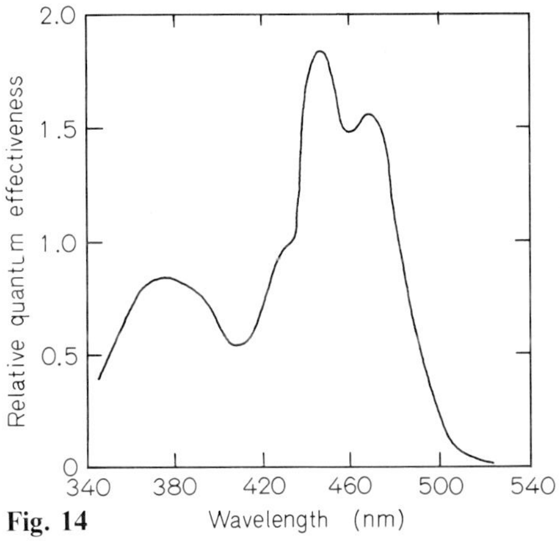

Fig. 14

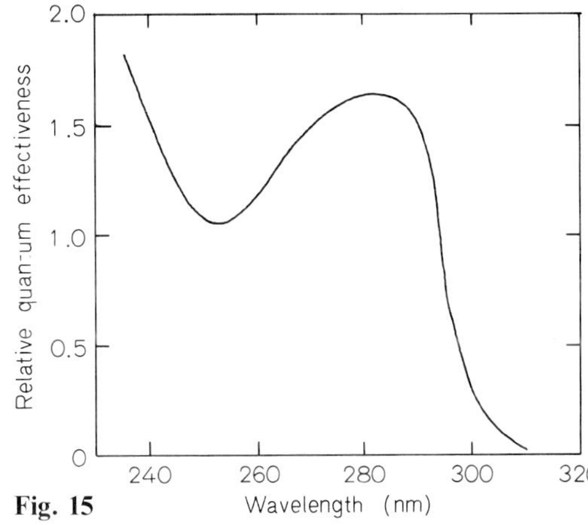

Fig. 15

Fig. 14. Action spectrum for positive phototropism in *Phycomyces* sporangiophores. The determinations were made by balancing the phototropic effectiveness of a standard source at 436 nm against that of the test source at each wavelength. Units on the *ordinate* are reciprocals of the quanta needed for balance at each wavelength, relative to the quanta needed at the wavelength of the standard. (After CURRY and GRUEN, 1959)

Fig. 15. Action spectrum for negative phototropism in *Phycomyces* sporangiophores. The determinations were made by balancing the phototropic effectiveness of a standard source at 254 nm against that of the test source at each wavelength. Otherwise as in Figure 14. (After CURRY and GRUEN, 1959)

at 460 nm. Since a single standard was used in both the UV and visible ranges, the resulting curves can be quantitatively compared. It is particularly noteworthy that the UV peak at 280 nm is about fourfold higher than the main peak at 455 nm. Here also, the reason lies in the great opacity of the cell at 280 nm, leading to an exaggerated intensity gradient across the cell (see Sect. 5.1.2). A growth-response action spectrum (DELBRÜCK and SHROPSHIRE, 1960) is closely similar, except that the magnitude of its 280 nm peak does not exceed that of the 455 nm peak. The light–growth response is positive at all wavelengths.

The phototropic action spectrum of young sporangiophores of *Pilobolus kleinii* was obtained by PAGE and CURRY (1966), using a balance method similar to that used by CURRY and GRUEN (1959) for *Phycomyces*. The result is also similar, although the location of the peak near 360 nm was not precisely determined. The *Pilobolus* action spectrum also mimics that of *Phycomyces* in its reversal for wavelengths below 300 nm. Although no data concerning the UV peak are available, the authors note that the negative phototropism is very pronounced at 280 nm.

A quite different phototropic system is found in the coenocytic alga, *Vaucheria geminata*, investigated in some detail by KATAOKA (1975a, b, 1977a, b). Instead of bending by means of differential cell-wall extension, as in *Phycomyces*, the cells of *Vaucheria* re-orient their growth direction by initiating a new growing point. The cell tip is hemispherical with the point of maximum growth normally located on the center line, at the extreme apex. When illuminated unilaterally, this growth-center migrates from the center line and occupies a position shifted

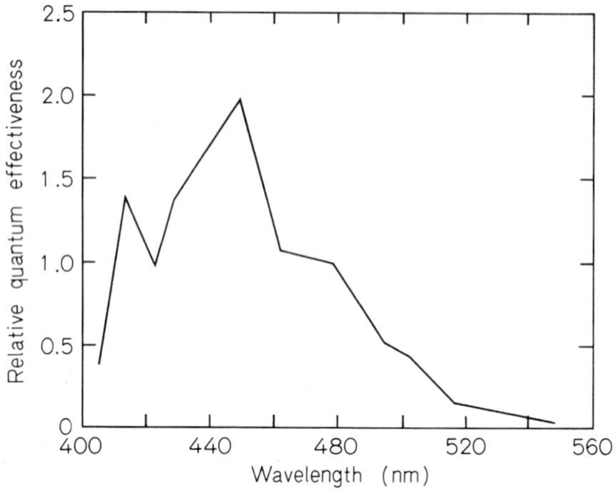

Fig. 16. Action spectrum for phototropism in cells of *Vaucheria geminata*. Determinations were made by balancing the phototropic effectiveness of a standard source at 479 nm against that of the test source at each wavelength. Otherwise as in Figure 14. (After KATAOKA, 1975a)

laterally towards the light source. The cell's growth is now aligned with the new growth-center. One of the earliest and most sensitive indicators of this growth-center shift is the tilt in the axis of the clear hyaline cap at the cell's apex. This tilt is a smooth function of the unilateral stimulus-energy dosage, and KATAOKA has shown (1975a) that the reciprocity law holds over a rather large dosage range. Remarkably, it has recently been found that the *Vaucheria* dose–response curve has features like that of *Avena*, including first-positive, first-negative, and second-positive regions (KATAOKA, 1977b).

The *Vaucheria* action spectrum (Fig. 16) was obtained by a split-field illumination system. The cell was mounted on a horizontal slide and illuminated by two spots of light from below. One flank of the cell was illuminated at a standard wavelength (in this case 479 nm) and the opposite flank at the test wavelength. When the two spots were of unequal effectiveness, the resultant tilt (in the plane of the slide) was measured. Action spectrum values were obtained by adjusting the test spot intensity until the cap tilted neither to one side nor to the other. The action spectrum is not greatly different from the others discussed above, although the long-wavelength cut-off is perhaps more gradual.

It is of some significance, in the light of action spectra for other systems to be presented next, that there is no phototropism at wavelengths above 550 nm in any of the systems discussed so far; thus red light is totally without effect. Similarly, no red effect was found in the polarotropic response of the liverwort, *Sphaerocarpus donnellii* Aust., to be discussed more fully in a later section.

2.3.2 Red Light Type

A strikingly different action spectrum was found for the chloronemal filaments of protonemas of the moss, *Physcomitrium turbinatum* (NEBEL, 1968). These structures grow at the apex, and the initiation of a phototropic bend occurs when the growing point shifts from its normal, axial position to an off-center position, displaced towards the light source. Subsequent growth follows this new direction. The action spectrum was obtained in an aqueous medium using the balance method. Figure 17 shows the results when the standard source is of wavelength 665 nm.

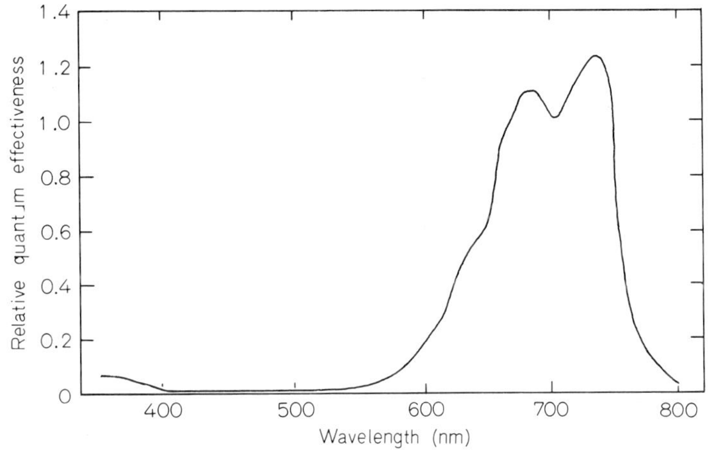

Fig. 17. Action spectrum for phototropism in protonemata of *Physcomitrium turbinatum*. Determinations were made by balancing the phototropic effectiveness of a standard source at 665 nm against that of the test source at each wavelength. Otherwise as in Figure 14. (After NEBEL, 1968)

As can be seen, blue light is almost totally without phototropic effect, although there is a slight suggestion of activity at 365 nm. A second action spectrum, with the standard stimulus at 730 nm, is identical for wavelengths shorter than 680 nm, but shows no decline at 700 nm. This action spectrum could not be extended above 700 nm because the growth action spectrum of these filaments drops to zero between 700 and 730 nm (NEBEL, 1968), thus ruling out phototropic determinations when both test and standard wavelengths are above 700 nm. The author suggests that these complexities indicate that phototropism involves an interaction between two photosystems. The lack of phototropic response to blue light rules out any involvement of chlorophyll, which absorbs strongly in the blue region. The red chlorophyll absorption band, in this case with a peak at about 680 nm, would act as a screen, significantly increasing the intensity gradient across the cell. Therefore the presence of the action spectrum peak at 680 nm may be due to a passive effect of the chlorophyll screen. When a correction is made for such a screening action, the result resembles the absorption spectrum of phytochrome in the P_{FR} form.

Another type of spectral sensitivity is that of the chloronemal filaments of the protonema of the fern *Dryopteris filix-mas* (L.) Schott. ETZOLD (1965) has shown that the filament is positively phototropic in unilateral red light (665 nm) and that the bending mechanism is of the bulging variety, where the growing point shifts to the lighted side of the apical hemisphere. The action spectrum was measured not for phototropism but for polarotropism, a bending response to the polarization of light.

The polarotropic response is observed as follows. Spores are germinated at the bottom of an overturned Petri dish containing nutrient agar, all the while illuminated from above by plane-polarized red light. After 8 or 9 days the chloronemal filaments are observed growing parallel to one another, all of them perpendicular to the electric vector of the red light. Then the polarizing filter is turned through 50°, and the exposure is continued for a further 8 h. Finally the cultures are left in the dark for another 16 h. Examination then shows that the 50° change in polarization plane has caused a 50° change in growth direction of the filaments.

Exposure times shorter than 8 h produced curvatures less than 50°, and in the range from 5° to 30°, the curvature was a linear function of the log dose of the second polarized exposure. An action spectrum for the curvature caused by this second exposure showed two main peaks, one in the red region, at 665 nm, the other in the blue region, at 460 nm. Most remarkably, the blue peak is 100 times the magnitude of the red peak.

Such an extreme difference in phototropic effectiveness between blue and red light led ETZOLD to postulate that in *Dryopteris* there are two entirely different phototropic mechanisms, with correspondingly separate spectral sensitivities. The author states that this two-pigment hypothesis is further buttressed by his observation that the dose–response curves differ markedly in shape as well.

This matter was re-investigated by STEINER (1967a), who found that the *Dryopteris* polarotropic dose–response curve is a function of exposure time, and that this dependence is the same for blue and red light. This conflicts with ETZOLD's observation but does not invalidate his hypothesis of a two-pigment photosystem. Using 240-min exposure times and a standard response of 17.5°, STEINER (1967b) obtained the action spectrum shown in Figure 18. The main result is similar to that of ETZOLD: two peaks with the blue peak about 100-fold higher than the red peak. In addition the fine structure seen in the blue peak is very similar to that of *Phycomyces* (Fig. 14), except that in *Dryopteris* the near-UV peak (375 nm) is more dominant. STEINER (1967b) also determined the polarotropic action spectrum of the liverwort, *Sphaerocarpus donnellii* Aust., and found that it is nearly identical to the *Dryopteris* action spectrum for wavelengths below 500 nm. *Sphaerocarpus* has no sensitivity whatever for wavelenths above 500 nm. The fact that *Sphaerocarpus* seems to lack one of the two pigments of *Dryopteris* is definitely in support of ETZOLD's two-pigment hypothesis.

The involvement of phytochrome in the polarotropic response of *Dryopteris* is suggested by ETZOLD's (1965) observation that under certain conditions, the effect of red light can be partially reversed by far-red light.

If a filament previously aligned to polarized red light is allowed to dark-adapt for $2^1/_2$ h, a 5-min exposure to red light with a 50° change in angle produces a curvature of 8.5°. If the dark period is shortened to 12 min, the curvature is thereby reduced to 2°, indicating that now the system is relatively insensitive to red light, perhaps because much of the phytochrome is shifted to the P_{FR} form. If this 12-min dark period is replaced by a 9-min exposure to unpolarized far red, still retaining the final 5-min red exposure at 50°, the curvature is 4°. Thus it appears that part of the sensitivity to red light has been restored by the unpolarized far-red exposure.

In the filamentous green alga *Mougeotia*, there are two distinct types of bending movements in response to unilateral light (NEUSCHELER-WIRTH, 1970). When filaments are cultured under red light, the terminal cells differentiate into apical cells and rhizoid cells. In the apical cells, which are positively phototropic, the chloroplast occupies the extreme distal end of the cell, so that there is no apical clear region. When illuminated unilaterally in red light, the apical growing center shifts to the illuminated side, and a bend is produced by the bulging mechanism. Also, when illuminated from above by polarized red light, a polarotropic bending

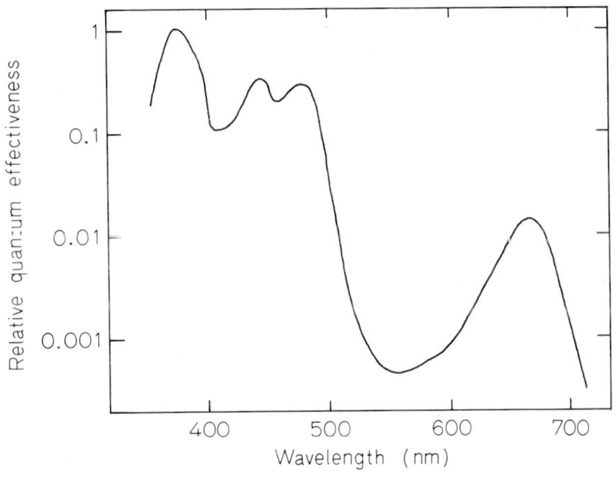

Fig. 18. Action spectrum for polarotropism in *Dryopteris filix-mas* (L.) Schott chloronemal filaments. The units on the ordinate (log scale) are reciprocals of the quanta needed for a curvature of 17.5°, relative to the quanta needed at 376 nm. (After STEINER, 1967b)

reaction ensues, resulting in new growth perpendicular to the electric vector. Filaments grown in blue or white light, which produces no differentiation of the terminal cell, react to unilateral blue (or white) light with a flexing of the entire strand so that it becomes convex facing the light source. This flexure is not due to any reorientation of the growing point, but simply differential growth along the whole filament; the flank receiving the greatest intensity has the greatest elongation rate.

A crude action spectrum for the negative phototropism of the uredospore germ tube of the rust fungus, *Puccinia graminis tritici* was reported by CHANG and CALPOUZOS (1973). Phototropic effectiveness was high at 400 nm, moderate at 500 nm, and essentially zero at 300 nm and 600 nm.

3. The Photoreceptor Pigment for Phototropism

3.1 Identification

It is not a straightforward matter to discover the chemical identity of a receptor pigment responsible for a photobiological process. Knowledge of the action spectrum is clearly a vital first step, but numerous difficulties arise when one attempts to match the action spectrum for a process with the absorption spectrum of a pigment considered a likely candidate. Even comparing the action spectra of different photoprocesses is a rather crude procedure. There are several reasons for these difficulties. First, the data are often of very limited precision, perhaps reflecting the extreme difficulties in obtaining the action spectrum at all. Second, there may be variations in peak wavelengths due to differences in the molecular environment of the pigment molecule, or of the protein portion of a chromoprotein complex. Third, there may be distortions due to screening pigments and photoreversible systems, whose steady state may vary with the nature of the light pretreatment. Fourth, there may be more than one photoreceptor pigment triggering the response.

3.1.1 The Ubiquitous Blue-Sensitive Pigment: Flavin or Carotenoid?

In addition to these general problems, there is a more specific difficulty when one attempts to distinguish between flavin and carotenoid receptor pigments on the basis of phototropic action spectra. This is due to the relative scarcity of action spectra that extend into the UV, or even the near UV, where the absorption spectra of flavins and carotenoids usually have marked differences. It is to be hoped that in the future experimenters will extend their action spectra into the UV, thus making many more interesting comparisons possible.

Evidence from Phototropic Action Spectra. It is thus remarkable that two such diverse systems as *Phycomyces* sporangiophores and *Avena* coleoptiles should have very nearly identical action spectra (see Figs. 12, 13, 14, and 15). This detailed correspondence in both visible and UV regions strongly suggests that both organisms have the same phototropic photopigment. While no other action spectra are sufficiently detailed to be placed in this group with equal confidence, those of *Sphaerocarpus* (Fig. 18, blue portion), *Vaucheria* (Fig. 16), and *Pilobolus* are probably due to the same pigment. This remarkable recurrence suggests that there may be a unity in phototropic mechanism for these, and perhaps a large number of other plant systems, and that such unity is built upon a unity of photopigment. Is this pigment a flavin or a carotenoid?

There is little basis for choice in the visible region, where the absorption spectra of both flavins and carotenoids are generally a good fit with the action spectra. Carotenoids have a characteristic triple-peak structure in the region around 450 nm, and although riboflavin absorption spectra shown in earlier reports (THIMANN and CURRY, 1960) are devoid of fine structure in this region, spectra obtained in ethanol at 77 K by SUN et al. (1972) do indeed have the requisite shape (Fig. 19). More pronounced discrepancies show up in the near UV, however, where the riboflavin absorption peak in ethanol at 77 K is shifted toward shorter wavelengths by about 30 nm, as compared to the action spectrum and to the absorption spectrum in aqueous solution at room temperature (SINSHEIMER, 1955). The carotenoid absorption spectrum typically lacks peaks altogether in the 300 nm–400 nm range. Both pigments generally have another peak near 280 nm, though the magnitudes differ.

Thus one cannot rule out conclusively either flavins or carotenoids on the basis of these spectral comparisons. Further, pigments extracted from the phototropic organs themselves do not fit the action spectrum significantly better.

Evidence from Related Action Spectra. In view of the lack of progress from a straightforward attack on the pigment problem, many investigators have turned to the examination of other related photoeffects, in the reasonable hope that a comparison would produce some enlightenment. Since the phototropic response involves a spatial redistribution of growth, let us first consider action spectra for growth effects other than phototropism.

In *Phycomyces*, action spectra are easily obtained for the light–growth response, the transient change of growth rate as a result of a change in the steady state light intensity. The most straightforward method is to alternate standard and monochromatic test stimuli at 5-min intervals. If the two stimuli are of unequal effectiveness, the growth rate will show a 5-min periodicity whose amplitude is proportional to the magnitude of imbalance. This amplitude can be used as an

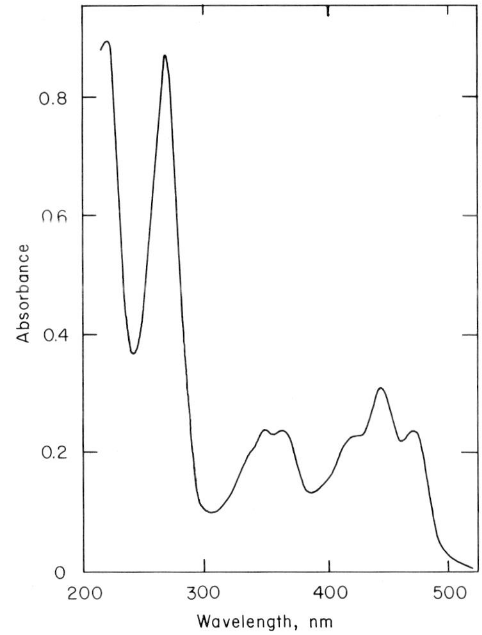

Fig. 19. Absorption spectrum of riboflavin in ethanol at 77 K. (After SUN et al., 1972)

indicator, allowing the intensity of the test stimulus to be adjusted and a null point determined. Using this method, DELBRÜCK and SHROPSHIRE (1960) obtained a growth-response action spectrum that is essentially identical to the phototropic spectrum, except for the height of the UV (280 nm) peak relative to the visible portion of the curve. The screening effect of UV-absorbing substances in the cell causes the magnitude of this peak to be greatly exaggerated in the phototropic action spectrum, relative to its magnitude in the light-growth action spectrum, as explained previously.

This work has been recently extended by DELBRÜCK et al. (1976), who have obtained a highly detailed growth-response action spectrum in the region from 575 nm to 630 nm, using a tunable laser emitting highly monochromatic radiation (blue contamination less than 10^{-10}) at intensities up to $1.8 \cdot 10^5$ W $\cdot$ m^{-2}. The authors discovered a small shoulder at about 600 nm, whose magnitude is about 10^{-9} relative to the main peak at 455 nm. This shoulder is attributed to the "forbidden" triplet excited state of riboflavin, known from spectroscopic data (SUN et al., 1972) to be centered at 600 nm. The detection of such a faint feature in the action spectrum is a remarkable accomplishment and constitutes strong evidence in favor of a flavin photoreceptor.

Sporangiophore induction in *Phycomyces* is partly controlled by illumination. BERGMAN (1972) found that periodic induction occurs when short periods of illumination are delivered at daily intervals to otherwise-darkened cultures, and that the cutoff wavelength for this effect lies between 510 nm and 540 nm. This resemblance to the phototropic action spectrum suggests that the same pigment might be involved in both responses. The existence of mutants defective in both phototropism and light-induced sporangiophore formation is further evidence for the involvement of the same receptor system in both responses. These spectral limits were confirmed by THORNTON (1973), who found that sporangiophore formation could be induced by light of wavelengths from 380 nm to 480 nm, but not at 520 nm or above.

The negative light–growth response in *Avena* coleoptiles has been examined by ELLIOTT and SHEN-MILLER (1976). The experimental technique greatly limited the accuracy of the measurements, but the resulting action spectrum does not differ significantly from that of phototropism, in the region from 360 nm to 680 nm. The authors also present evidence suggesting that the light–growth response becomes positive between 700 nm and 740 nm, a reversal not shown by the phototropic action spectrum. However in view of the method used, it is not clear how much reliance should be placed on this result.

In a variety of other light responses, essentially the same action spectrum as in *Avena* and *Phycomyces* has been found, thus suggesting that the same photopigment is involved.

The light-induced re-orientation of the chloroplasts of the club moss *Selaginella martensii* has an action spectrum very like that of blue-sensitive phototropism, with principal peaks at 475 nm and 350 nm, and a broad minimum centered at 400 nm (MAYER, 1964). Considerable activity is also reported at 254 nm, although no quantitative data are available. Although the data are not sufficiently detailed to allow a positive characterization, they do not conflict with the author's suggestion that a flavin pigment is the primary photoreceptor; additional evidence for that suggestion is given on p. 537.

The action spectrum for phototaxis in *Euglena gracilis* is very similar to that of blue-sensitive phototropism (DIEHN, 1969), and there is strong evidence in favor of a flavin as the photopigment in this system (see HÄDER, Chap. 3.2.1, this vol.). Likewise, similar action spectra have been obtained for the induction of polarity in *Fucus* and *Equisetum* spores (BENTRUP, 1963) and for the induction of fruiting bodies in the basidiomycete *Schizophyllum commune* (PERKINS and GORDON, 1969).

The amoeboid stage of the slime mold, *Dictyostelium discoideum* is positively phototactic, and the cells show two types of light-induced absorbance change (see POFF and WHITAKER, Chap. 3.2.4, this vol.). One of these (Δ absorption at 411 nm) has an action spectrum with peaks at 432 nm and 560 nm, resembling the phototactic action spectrum (POFF et al., 1973). The other absorbance change (Δ absorption at 430 nm) has an action spectrum with a single peak in the visible range at 465 nm, no sensitivity above 520 nm, and does not correspond to any known photoeffect in *Dictyostelium*. However, an absorbance change similar to the latter was found in the intact mycelium of *Phycomyces* (POFF and BUTLER, 1974). It is thought that the absorbance change in *Phycomyces* mycelium may be due to the same photopigment that is responsible for phototropism. An action spectrum for the absorbance change in *Phycomyces* mycelium has recently been obtained (LIPSON and PRESTI, 1977), and it resembles that of *Phycomyces* phototropism, with peaks at 470 nm and 360 nm. POFF and BUTLER (1975) have further characterized the cell-free *Dictyostelium* system and attribute the absorbance change to the reduction of a b-type cytochrome. Light absorption by some accessory photopigment, possibly a flavin, provides the energy for the cytochrome reduction. Whether *Dictyostelium* will prove to be an effective system for the isolation and characterization of the phototropic blue light receptor remains to be seen.

The light-induced absorbance change in the mycelium of *Neurospora crassa* has the spectral features corresponding to the reduction of a b-type cytochrome (MUÑOZ and BUTLER, 1975). The action spectrum between 360 nm and 520 nm matches very closely the *Phycomyces* and *Avena* phototropic action spectrum, even down to the triple-peaked fine structure near 460 nm. Cell-free extracts did not show photoreduction of cytochrome b unless FAD or FMN were added to the extracts. In later work (SCHMIDT and BUTLER, 1976b) an improved culture and isolation technique allowed the isolation of two cell-free extracts: a soluble fraction showing photooxidation of a c-type cytochrome, and a particulate fraction showing photoreduction of a c-type cytochrome and a cytochrome oxidase. In both fractions, no exogenous flavins were necessary. The action spectrum for the photoreduction in the soluble fraction differs from that of MUÑOZ and BUTLER (1975) in the exaggerated height of the 475 nm shoulder. More recently, this absorbance change has been localized in a plasma membrane-rich fraction isolated by centrifugation from *Neurospora* mycelium (BRAIN et al., 1977); a similar finding has been reported in corn coleoptile tissue by JESAITIS et al. (1977). SCHMIDT and BUTLER (1976a) found that an artificial system (i.e., not derived from live tissue) containing riboflavin and cytochrome c similarly showed a light-induced photoreduction; its action spectrum, a good match for riboflavin absorption, lacks the prominent shoulder at 475 nm. These complexities illustrate the spectral perturbation that can result from the various interactions between flavin chromophores and protein molecules, and the dangers of assuming that processes with similar action spectra have similar pigments and vice versa.

The light induction of carotenoid synthesis in *Neurospora* is also due to a blue light receptor, and a recent detailed action spectrum does not fit the flavin pattern at all closely (DE FABO et al., 1976). In particular, the peak at 370 nm is only about one third of the principal one at 450 nm, and the peak at 280 nm is even lower. Also, there is no shoulder between 420 nm and 450 nm, and the peak at 481 nm is higher than the one at 450 nm. Because of these features, the authors favor a carotenoid as the photopigment involved. Thus, *Neurospora* may have two light-sensitive systems, involving two separate blue-absorbing pigments, only one of which is similar to the phototropic pigment.

In the fungus *Fusarium aquaeductum*, however, light-induced carotenoid synthesis was found by RAU (1967) to have an action spectrum very closely resembling that of phototropism in *Avena* and *Phycomyces*. The precision of RAU's curve is sufficient to allow accurate identification of the main peak at 455 nm, shoulders at 475 nm and 430 nm, minimum at 400 nm, and second peak at 375 nm with a magnitude about 70% of the 455 nm peak. Although the data do not extend as far as the 280 nm peak typical of flavins (Fig. 19), carotenoids seem excluded as photoreceptor pigments in this system.

Not only is the blue-sensitive pigment widely distributed in the plant world, it occurs in animals as well. An example of the latter is the effect of a flash of light in causing a phase shift in the emergence of pupae of *Drosophila pseudoobscura*. The highly detailed action spectrum obtained by KLEMM and NINNEMAN (1976) shows a principal peak at 457 nm, shoulders at 435 nm and 473 nm, a minimum at 395 nm, and a second peak at 375 nm, about 25% of the 457 nm peak. This spectrum is attributed by the authors to a flavin photopigment. It is clearly a very good match for the phototropic action spectrum in the visible, somewhat less good in the near UV.

In conclusion, the evidence from action spectra indicates that a remarkably large number of photosystems probably have the same receptor pigment as the phototropic systems of *Avena* and *Phycomyces*, and that this pigment is probably a flavin. Some of the differences between action spectra can undoubtedly be explained by interactions between the flavin chromophore and the different proteins. In comparing these spectra, one is struck by the lamentable scarcity of data at wavelengths below 450 nm; such experiments admittedly pose technical difficulties but might provide uniquely useful information.

Other Evidence. There are several other lines of evidence bearing on the chemical nature of the receptor pigment for blue-sensitive phototropism, and they include the microspectrophotometry of intact tissues, the behavior of photomutants, measured flavin and carotenoid levels in wild-type and mutant strains, and fundamental photochemical arguments.

A direct approach to the identification of the blue-sensitive receptor pigment is the quantitative pigment determination in phototropic organs. This approach has its drawbacks, however, and one of them is that if the receptor pigment is present in very low concentrations, it will be undetectable in the presence of other blue-absorbing, nonphotoreceptor pigments.

One estimate of the minimum concentration of receptor pigment was made by BRIGGS (1964), starting with the assumption that at least $2 \cdot 10^8$ absorbed quanta are required for coleoptile curvature. This leads to a value of $3 \cdot 10^{-16}$ mol of absorbing pigment per coleoptile, (i.e. a concentration of 10^{-9} M). Using *Phycomyces* and somewhat different assumptions, MEISSNER and DELBRÜCK (1968) estimate that at least $3 \cdot 10^9$ receptor molecules are needed per sporangiophore growing zone (i.e., a concentration of $3 \cdot 10^{-7}$ M).

MEISSNER and DELBRÜCK (1968) analyzed the carotenoids of *Phycomyces* strain Cl, an albino mutant with normal phototropism, and were unable to detect any β-carotene. They estimated that the β-carotene content of Cl sporangiophores may be as high as 0.1 μg per g dry weight of sporangiophore (0.05% of the normal amount), equivalent to $1.2 \cdot 10^{10}$ molecules per growing zone, and remain below the limit of detection by the methods used. Recently, PRESTI et al. (1977) have detected no β-carotene in albino *Phycomyces* mutants, doubly blocked for β-carotene synthesis but with normal phototropism, using methods ten times more sensitive than those of MEISSNER and DELBRÜCK. Thus the maximum amount of β-carotene that could be present in the doubly blocked mutants is no more than the absolute minimum amount needed for phototropism, as estimated by MEISSNER and DELBRÜCK. Hence we may conclude that β-carotene is probably not the photoreceptor in *Phycomyces*.

Even if the receptor is a flavin, its absorbance might be undetectable spectrophotometrically in the albino, due to absorbance of β-carotene. Thus it is undoubtedly fruitless to search for the receptor pigment by measuring the absorbance of phototropic organs. In fact, absorbance measurements of albino sporangiophores show only the presence of cytochromes (MEISSNER and DELBRÜCK, 1968), and even these features are obscured by the β-carotene present in normal sporangiophores.

Evidence against β-carotene as the photopigment in *Phycomyces* was reported by ZANKEL et al. (1967), who obtained absorption spectra of intact sporangiophore growing zones. Comparison of their curve with *Phycomyces* action spectra shows that the absorption spectrum, presumably due to β-carotene, is clearly red-shifted by about 10 nm.

WOLKEN (1969) has carried out a microspectrophotometric analysis of *Phycomyces* sporangiophores using scanning spots as small as 1 μm². He found that the large octahedral crystals normally present in the growing zone have an absorption spectrum similar to that of a flavin. THORNTON (1969) concluded that these crystals are composed of an acidic protein. No absorbance data were given for intact crystals, but when dissolved in phosphate buffer at pH 9, the crystalloid material shows no absorbance at wavelengths above 320 nm.

Cell-free extracts of *Phycomyces* sporangiophores were reported by BERNS and VAUGHN (1970) to show fluorescent emission bands at 450 nm and 540 nm, with peak excitation wavelengths of 340 nm and 465 nm, respectively. These emissions could be reversibly reduced by prolonged irradiation at the exciting wavelengths, suggesting that the pigment can be reversibly bleached by light. Extracts of a night-blind mutant (i.e., having a high stimulus threshold for phototropism) did not show this effect. These observations led the authors to suggest that the fluorescing system might be related to the phototropic pigment, and that a flavoprotein is a strong possibility.

Genetics can be a powerful tool in the analysis of biochemical mechanisms, and is beginning to produce interesting results, especially in *Phycomyces*. Two main approaches have been used: the study of phototropism in albino mutants, and the isolation and study of mutant strains with an abnormal phototropic response.

Phycomyces mutants defective in phototropism (*mad* mutants) have been systematically isolated and classified (BERGMAN et al., 1973). The basic method of isolating *mad* mutants is to treat vegetative spores with a chemical mutagen and seed them on plates illuminated only from below. All sporangiophores with normal phototropism grow downwards into the medium, while those in which the phototropic response is weak or missing grow more or less upwards. The authors were then able to classify the *mad* mutants further. Class 1 *mad* mutants retain the ability to bend in response to other unilateral stimuli, such as gravity. Class 1 is further divided into a subclass, 1.1, in which other photoreactions, such as induction of β-carotene synthesis, are also abnormal, and a subclass, 1.2, in which these other photoreactions are normal. Class 2 *mad* mutants do not bend in response to any stimulus, being therefore defective in some "output" stage, governing differential growth on opposite sides of the cell. That Class 2 mutants can still sense light is shown by the fact that they are normal in other photoresponses not related to growth, such as the induction of β-carotene synthesis.

Using the method of sporangiophore grafting to produce heterokaryons, OOTAKI et al. (1974) were able to show that among the 15 *mad* mutants tested, there were five complementation groups, defining five genes. Furthermore, these five genes correspond to the phenotypic classes in the following way. Class 1.1 contains genes *A* and *B*, Class 1.2 contains only gene *C*, and Class 2 contains genes *D* and *E*. More recently, sexual crosses have shown that none of the five genes is linked (ESLAVA et al., 1976).

LIPSON (1975b) analyzed the light–growth response of 7 *mad* mutants and found that the defects correspond to quantitative modifications in various stages of the transduction mechanism, rather than the total absence of photoreceptor pigment. LIPSON also points out that mutants totally without light sensitivity might be unable to survive the conditions of isolation, explaining why no *mad* mutants are totally blind.

Light-induced absorbance changes in the mycelium of four *mad* mutants have been studied by LIPSON and PRESTI (1977). They found that the changes, corresponding to the reduction of a b-type cytochrome, are identical in the mutant and normally responding strains. These results thus do not support the notion that the absorbance change is related to the phototropic pigment. They do not refute this notion either, since it is possible that *mad* mutants have normal photopigment levels. Another problem is that LIPSON and PRESTI calculate a quantum yield of 0.015 for the absorbance change, assuming a flavin receptor pigment, and this is far below the value estimated for the *Phycomyces* growth-response pigment (LIPSON, 1975a).

The phototropic response of albino barley coleoptiles was found to be comparable to that of normal coleoptiles, except for a threefold increase in threshold (ASOMANING and GALSTON, 1961). The total carotenoid level in the tips of the albino coleoptiles was about 40% of that in normal tips, and the flavin level was the same in normal and in albino tips. Thus, these results do not provide evidence for either flavins or carotenoids as the photoreceptor.

A 15-fold reduction in the phototropic sensitivity of sporangiophores of *Pilobolus kleinii* was observed by JACOB (1964) upon addition of diphenylamine to the growth medium. This substance inhibits the production of β-carotene, and the pigmentation level of the sporangiophore was thereby visibly reduced. JACOB attributed this lowered sensitivity to the loss of the internal screening pigment rather than to lowered photopigment content.

LIPSON (1975a) has applied very lengthy theoretical analysis to *Phycomyces* light–growth response data, and has estimated that the product of the photopigment's molar extinction coefficient and its quantum yield for bleaching is equal to 15,000 l/mol cm at 455 nm. This value may be compared to an extinction coefficient of 12,500 l/mol cm for oxidized riboflavin at 446 nm and 40,000 l/mol cm for rhodopsin at 500 nm. Thus LIPSON's value is consistent either with a flavin receptor having a very high quantum yield, essentially unity, or with a carotenoid-based receptor, similar to rhodopsin, with a quantum yield of about 0.4. Since the rhodopsin quantum yield is probably between 0.66 and unity (ABRAHAMSON and WIESENFELD, 1972), the evidence appears to favor flavins.

The addition of water to the ethanol solvent causes greater absorption by some carotenoids (e.g., lycopene and lutein) in the region from 350 to 380 nm, and hence greater resemblance to the phototropic action spectrum (HAGER, 1970). SONG and MOORE (1974) note that this effect is due to dimeric stacking of the molecules and consequent exciton interaction. However, β-carotene does not form such stacked dimers in ethanol–water mixtures, thus apparently excluding such dimers as components of the photoreceptor system. SONG and MOORE further argue that monomeric β-carotene is an unlikely photopigment, since no long-lived singlet states are detectable, and population of long-lived triplet states is too inefficient. Thus on purely photochemical grounds, flavins are more promising than carotenoids as photoreceptors.

Another fundamental approach involves the inhibitory effect of potassium iodide (KI) on flavin-sensitized photochemical reactions. The inhibitory effect observed in purified in vitro systems is thought to occur because of the ability of KI to quench the triplet excited state of riboflavin (SONG and MOORE, 1968). Although it is by no means proven that such a mechanism operates in vivo, inhibition of photoeffects by KI has been observed. In *Euglena gracilis*, MIKOLAJCZYK and DIEHN (1975) have shown that the step-down photophobic response is strongly inhibited by 100 mM KI. The step-up photophobic response, however, is unaffected by the same concentration. The authors conclude that a flavin may be implicated in the former response but not in the latter.

MEYER (1969b) has shown that treatment of *Avena* coleoptiles with 100 mM KI for 2 h before illumination causes a strong reduction in the resulting first-positive phototropic curvature. However, 100 mM potassium nitrate also inhibited the curvature, though to a lesser extent, and even a water treatment caused a slight inhibition. Although it is certainly possible that the inhibition by KI is due to quenching the triplet excited state of a flavin, other effects are obviously occurring as well, and the evidence is not clear as yet. A similar KI-induced inhibition of phototropism has been reported by SCHMIDT et al. (1977) in *Zea* seedlings.

There is additional evidence that KI may act by mechanisms other than by quenching the flavin excited triplet. A light-induced absorbance change in intact wheat (*Triticum aestivum*) coleoptiles has been shown by WIDELL and BJORN (1976) to be about 50% inhibited by a 30-min immersion in 100 mM KI. However, the action spectrum for the absorbance change differs from both flavin and carotenoid absorption spectra and probably indicates that a cytochrome is the photoreceptor in this case.

MAYER (1966) has reported a very clear KI inhibition of the light-induced chloroplast orientation in *Selaginella martensii.* Immersion of the leaves in 100 mM KI reduced the orientation reaction to 23% of its value in water. This inhibition could be reversed after 90 min by flushing the preparation with water. Potassium bromide had essentially the same effect, but potassium chloride and potassium nitrate had no effect. These results, in conjunction with the flavin-like action spectrum for this process (discussed earlier), suggest that these halogen ions exert their effect by their ability to quench the flavin triplet-excited state.

3.1.2 Phytochrome

Although, as we have seen, several phototropic systems have red sensitivity, direct evidence for the involvement of phytochrome is scanty. As discussed earlier, the polarotropic response of *Dryopteris* involves sensitivity to both blue and red light (Fig. 18). There is some degree of resemblance of the *Dryopteris* action spectrum to that of the HIR (high irradiance response) type of phytochrome system, that for example is involved in the light-induced inhibition of hypocotyl growth in *Lactuca sativa* L. (HARTMANN, 1967). Such HIR action spectra include a blue and near-UV portion, very similar to that of *Dryopteris* (Fig. 18), but also a far-red portion consisting of a single sharp peak near 720 nm. This latter feature contrasts strikingly with the broad *Dryopteris* maximum centered at 655 nm. Also, the red and blue maxima are of approximately equal magnitude in the HIR action spectrum, whereas in *Dryopteris* they differ 100-fold. On the other hand, the fact that the red light effect can be partly reversed with far-red light (ETZOLD, 1965) argues for the involvement of phytochrome. Perhaps the safest conclusions at present are that a red, far-red photoreversible pigment, probably phytochrome, is involved in *Dryopteris* polarotropism, that the system is not of the HIR variety, and that the blue sensitivity is due to a different pigment.

As indicated earlier, there is some spectral evidence supporting the involvement of phytochrome in the phototropic responses of *Physcomitrium turbinatum* (NEBEL, 1968). When the action spectrum is taken as a null balance relative to 665 nm and suitably corrected for the screening action of chlorophyll, the result suggests the involvement of phytochrome in the P_{FR} form. However, when the action spectrum is taken relative to 730 nm, no tropic reactions were observed when the test wavelength exceeded 700 nm, thus contradicting the previous interpretation. The fact that the system responds to wavelengths above 700 nm only in the presence of additional light at shorter wavelengths led NEBEL to suggest that the response might be due to phytochrome cycling, i.e., the simultaneous conversion of P_R to P_{FR} and P_{FR} to P_R. However, until direct evidence is available to indicate photoreversibility of the phototropic effect of red or far-red light, the nature of the photoreceptor pigment in *Physcomitrium* remains in question.

3.2 Localization of the Photoreceptor

3.2.1 Within the Multicellular Phototropic Organ

In coleoptiles it is well established that the phototropic sensitivity rises to a maximum in the general region of the tip. Quantitative data concerning the exact distribution of sensitivity are remarkably scanty, perhaps because, as pointed out by CURRY (1969), the technique of stimulating small areas of the coleoptile is hampered by the complications of scattered light. Thus the areas adjacent to an illuminated spot will also be lit, though more dimly, by light scattered by the illuminated tissue itself, and by stray light from the optical apparatus. Using very low light levels and continuous illumination, CURRY re-examined the sensitivity distribution of *Avena* coleoptiles, but could do no more than confirm the classical view. In particular, he found no evidence that the sensitivity is concentrated at the extreme apex. This is confirmed by MEYER (1969b), who illuminated either the apical 350 μm of the coleoptile or the next 350 μm-segment with 5-s blue light stimuli. Clear responses were found in both cases, and curvatures were twice as large in the former case as in the latter. Further, she found that the dose–response curve for the more basal segment was not shifted along the dosage axis, but was merely changed in amplitude, implying that the sensitivity of the more basal section cannot be attributed to scattered light reaching the apex.

CURRY (1969) found that in coleoptiles the number of yellow plastids rises to a maximum value between 0.5 mm and 2 mm below the apex and falls off rapidly below 2 mm. His observation that both plastids and sensitivity are mainly located in the apical 2 mm led CURRY to postulate that the receptor is located in the plastids. Since the dominant plastid pigment is a carotenoid, CURRY's theory seems to require a carotenoid photoreceptor; however, flavins could also be present in small amounts. The postulate that the plastids are the location of the photoreceptor has the attractive feature that the same bodies would act as the primary stimulus source for both phototropism and geotropism, in the latter case acting as statoliths. In fact, THIMANN and CURRY (1961) even suggest that a unilateral light stimulus causes plastids to migrate to the darker side of the cell and there trigger auxin transport across that cell wall. However, experimental support for the plastid migration hypothesis is not well established, and no observations of such plastid migration have been reported in coleoptiles. In addition, according to CURRY (1969), plastid density falls off rapidly in the upper 0.5 mm of the oat coleoptile, and none were found in the apical 50 μm. However, MEYER's finding (1969b, see above) that the highest phototropic sensitivity is found in the uppermost

350 μm is difficult to explain on the basis of CURRY's theory but is consistent with flavin as a photoreceptor.

3.2.2 Within the Unicellular Phototropic Organ

Microbeam stimulus experiments have been employed to map the location of the photoreceptor in *Phycomyces* sporangiophores, using white light (stimulating 0.2 mm segments, COHEN and DELBRÜCK, 1959) and UV (stimulating 0.1 mm segments, DELBRÜCK and VARJÚ, 1961), against a continuous blue background light. In both cases the sensitive zone was found to extend from 0.5 mm to about 2 mm below the sporangium. Since the region of cell-wall elongation extends from 0.1 mm to about 2 mm below the sporangium (COHEN and DELBRÜCK, 1958; ORTEGA et al., 1974), these results suggest that the distribution of sensitivity parallels that of growth everywhere except in the region from 0.1 mm to 0.5 mm. However, we are not able to conclude that the photoreceptor must be confined to the region below 0.5 mm. In the first place, factors other than the photoreceptor may be involved, and secondly, DELBRÜCK and VARJÚ have found that the region above 0.5 mm does have sensitivity to light under certain conditions, namely, when a sustained (not periodic) UV stimulus is used, and the stimulated area is held at a fixed distance from the sporangium. The tropic bend appears, not at the position of the stimulus, but 0.5 mm below the sporangium, and the bend appears later than its usual delay time of 3.3 min. The delay time increases linearly as the stimulus position approaches the sporangium, thus suggesting that when the stimulus is given above 0.5 mm, it causes a latent response which can only be expressed after cell-wall growth has displaced the stimulated region downward to 0.5 mm below the sporangium. This complicated sensitivity map for phototropism has not been satisfactorily explained in terms of cell morphology; however, we can reasonably conclude that no highly concentrated receptor region exists in *Phycomyces* sporangiophores.

An interesting attempt has been made by MEISTRICH et al. (1970) to localize *Phycomyces* photoreceptors with regard to cytoplasm or cell wall. They used a laser beam (488 nm) 270 μm by 3.7 μm in cross-section, with the long dimension parallel to the cell axis. These dimensions are small in relation to the cell diameter (100 μm) and growing zone length (2 mm). The sporangiophore was immersed in oil ($n = 1.35$) to eliminate refraction at the cell surface. The beam was directed at various positions across the cell diameter, and the lateral tropic response was measured for different distances between the beam and the cell axis. As expected, the investigators found no response when the beam passed directly through the cell axis, and they found increasing response magnitudes as the beam traversed the more peripheral portions of the cell. The response was approximately constant when the distance between the beam and the cell axis varied from 40% to 80% of the radius of the cell. Reasoning from simple models of photoreceptor distribution, the authors conclude that this broad maximum is inconsistent with the hypothesis that the photoreceptors are located at or near the cell wall or near the tonoplast membrane of the axial vacuole. If the receptors were near the cell wall, one would expect a rather sharp response maximum when the beam is just inside the cell's surface, contrary to their results. These data thus suggest that the receptors are more or less uniformly distributed throughout the cytoplasm, a situation difficult to reconcile with the polarized-light studies (discussed in the next section) which imply that the receptors are oriented and thus probably fixed to a solid substrate. In criticism of the results of MEISTRICH et al., it may be pointed out that they did not take into account the effect of scattered light. Such scattered light, if strong enough, might obscure the sharp maximum predicted for receptors near the cell wall exposed to grazing illumination.

PAGE and CURRY (1966) investigated the location of phototropic sensitivity of young sporangiophores of the fungus *Pilobolus kleinii* growing in air, using microbeams of white light impinging on the flank of the cell. They found that photosensitivity was confined to the terminal 50 μm of the cell. Illumination of the more basal region, which has a high concentration of a carotenoid pigment, produced little or no phototropic curvature. Thus in these cells the photoreceptor is much more localized than in *Phycomyces*.

An extremely localized receptor is also found in the alga, *Vaucheria geminata* (KATAOKA, 1975a). The cell is immersed in an aqueous medium and illuminated by a split-field arrangement, in which the cell axis is the dividing line between the two half fields. One of the half fields is kept dark, and the other is masked so that the illuminated field does not reach the extreme tip. KATAOKA then studied the lateral bending produced by gradually increasing the fraction of the hemispherical tip illuminated by the half field. He found that bending did not occur until about half of the hemisphere was illuminated; bending reached only 50% of normal when all but the terminal 20% was illuminated. Thus it may be concluded that most of the photosensitivity is concentrated in the extreme tip of this cell, perhaps the apical 10 μm, which represents only about 10% of the cell diameter.

3.3 Orientation of the Photoreceptor: Polarized Light Effects

Flavins, carotenoids, and many other photopigment molecules are dichroic, having preferred electric vector orientations for the absorption of light. If, in addition, the photoreceptors are to some degree oriented parallel to one another in a fixed array, the system should demonstrate an overall sensitivity to the plane of polarization of the stimulating light. Thus, analysis of the effects of polarized light can provide information about the orientation of the receptors.

3.3.1 *Phycomyces*

Transversely polarized light (i.e., the E vector perpendicular to the cell axis; 450 nm) was found by SHROPSHIRE (1959) to have 24% more effectiveness for the light-growth response than longitudinally polarized light. The experimental method was to alternate a bilateral nonpolarized blue illumination of fixed intensity with a unilateral polarized stimulus of variable intensity at 5-min intervals. The sporangiophore was rotated continuously, so that the unilateral stimulus was in effect distributed uniformly around the cell. For each orientation of the plane of polarization, the stimulus intensity was adjusted until a growth-response null was reached, that is, no 5-min periodicity was seen in the growth rate. SHROPSHIRE also made this determination on a sporangiophore immersed in a liquid medium of matching refractive index, and in this case the relative advantage of transversely polarized light was only 6%, not greatly above the error of the method. This reduction led SHROPSHIRE to suggest that the polarized-light effect is caused by Fresnel reflection losses, which in the cylindrical system would be larger for longitudinally polarized light, since these losses would be greatly lowered when the refractive indices of the two media are nearly the same.

However, JAFFE (1960) showed theoretically that for an optically clear cell the differential Fresnel losses at the front surface of the cylinder would be just compensated for by the greater retention of the obliquely refracted rays when they undergo multiple internal reflections. In this view, the 24% increase in effectiveness of transversely polarized light is due to the presence of dichroic-oriented receptors. Further, JAFFE explained qualitatively the effect of increasing the refractive index of the medium by pointing out that the rays striking the cell near its edge are refracted inward very sharply when the cell is in air, but are virtually undeviated when the cell is in the denser medium. Thus in the former case, these peripheral rays strike the cell layer just beneath the wall at an incidence more nearly normal to the wall and would be maximally absorbed by receptor dipoles preferentially oriented in a "barrel-hoop" direction. In the denser medium, however, these peripheral, undeviated rays would strike the hoop-oriented dipoles end-on, and would be less effectively absorbed.

The polarized-light effect in *Phycomyces* is not very strong, and this could either result from only a fraction of the receptor dipoles having the hoop orientation and the rest random, or from an oblique (helical) orientation of all the dipoles, which would then have a minor component along the hoop axis. To distinguish these alternatives, HAUPT and BUCHWALD (1967) illuminated a series of sporangiophores with opposed sources of light, whose planes of polarization were parallel to each other and oblique to the axis of the sporangiophore. Without the action of polarized light there would be an indifferent zone, where both phototropic stimuli balance each other, since the intensity from each source falls off with increasing distance. With the polarizers in place as shown in Figure 20, the assumption of a uniformly oblique (helical) dipole orientation would lead to greatly different absorption on opposite sides of the cell. Hence the rotation of both polarizers through 90° should lead to a large shift in the indifferent zone. The authors found no detectable shift and concluded that only 5% or 10% of the receptor dipoles are hoop-oriented and the rest are random. However, a dipole orientation of 5°, or possibly 10°, from the hoop axis is nevertheless consistent with the sensitivity of this method as estimated by the authors. Even in this case, of course, only a small fraction of receptor dipoles could be so oriented; if a majority were oriented, a much larger difference between the effectiveness of transversely and longitudinally polarized light would have been found.

The existence of ordered, dichroic photoreceptors has been more recently confirmed by JESAITIS (1974). He measured the growth rate changes in alternate 5-min intervals in response to shifting the plane of polarization through 90° every 5 min. As a further refinement, conditions were arranged (Table 1) so that the optical density of the sporangiophore was relatively great (see p. 552) and therefore there were no complications from the stimulus beam passing through the cell and striking the back wall. Under these conditions JESAITIS found that transversely polarized light was more effective than longitudinally polarized light. Further, he determined the orientation of the receptor dipoles. Normally, a comparison

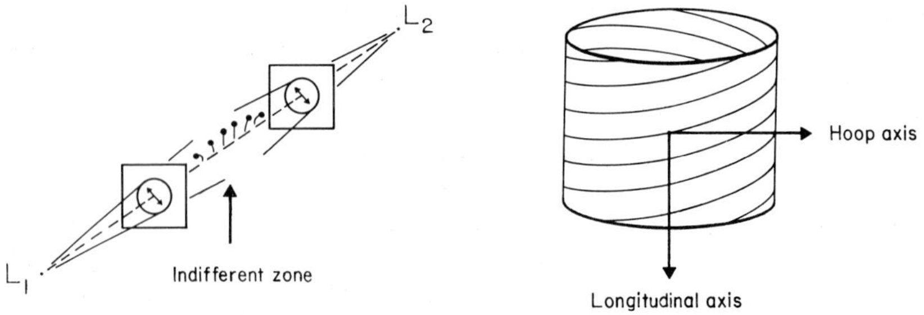

Fig. 20. *Left*: Experimental arrangement for illumination of *Phycomyces* sporangiophores with opposed beams of polarized light. L_1 and L_2 are blue light sources and the *double-pointed arrows* indicate the plane of polarization (E-vector) of each polarizing filter. (After HAUPT and BUCHWALD, 1967.) *Right*: Diagram showing helical orientation of hypothetical receptor dipoles on a cylindrical cell surface. Dipoles are arbitrarily given an orientation of 10° from the hoop axis

Table 1. Polarization angles for maximum response[a]

Strain	Growth medium	Wavelength (nm)	Transmission (%)	Angle of max. response[b] (degrees)	$(\Delta I/I_0)^c_{eq}$ (%)
NRRL 1555 (wild type)	Potato dextrose plus yeast extract	280	1.2	42 ± 4	18 ± 8
C 158 (mutant rich in carotene)	Glucose-asparagine-leucine plus yeast extract	456	5	7 ± 8	14 ± 5

[a] Data from JESAITIS (1974).
[b] Angle measured from hoop axis.
[c] The difference in effectiveness between two stimuli whose polarization (E-vector) angles differ by 90°, expressed as a relative equivalent intensity.

is made of the growth rates resulting from two orientations of the plane of polarization of the stimulus, 90° apart, one longitudinal and one transverse. But now if the initial member of each 90° pair is varied in its orientation with respect to the hoop axis, the orientation for maximum difference in the growth response can be determined. The resulting orientations are given in Table 1, along with values of the equivalent intensity differences. The finding that the angle of maximum response for 456 nm was 7° from the hoop axis, and that for 280 nm was 42° from the hoop axis is no cause for concern: on the contrary, it supports the hypothesis that the receptor molecule is a flavin, since flavins typically have dipole moments for UV and visible excitations that differ in direction by about this amount. After geometrical corrections are made, JESAITIS concludes that the orientation of the blue and UV dipole moments of the receptor molecule are 5.5° and 33.5°, respectively, from the hoop axis. Possible radial dipole moment orientations are rejected by the author as being incompatible with the effect of dense refractive media found by SHROPSHIRE (1959).

As mentioned earlier, the necessity for ordered photoreceptors in *Phycomyces* suggests that they are fixed to a more or less rigid structure, namely the cell wall. Thus these results support a cell-wall location for the phototropic receptor molecule. However, the degree of order is not very high, and could perhaps be explained by particle-bound photoreceptors if the particles are oriented by protoplasmic streaming. As mentioned earlier (p. 539), the results of MEISTRICH et al. (1970) are not consistent with receptors located at or very near the cell wall. Thus at present, the evidence for a dichroic-oriented receptor is very good, but the evidence for its location in the cell is inconclusive.

3.3.2 Polarotropic Spores and Filaments

Although not phototropic in the strict sense of an alteration of growth direction in response to an asymmetric illumination, the effect of polarized light on spore germination can provide some information about photoreceptor orientation and location (for a discussion of this subject, see WEISENSEEL, Chap. 5.1.1, this vol.).

The red light polarotropism of spore germination and chloronema growth of the moss *Funaria hygrometrica* has been studied in some detail by JAFFE and ETZOLD (1965). They found a complex situation, in which the receptor is disoriented at low intensities and tangentially oriented at medium intensities; in the latter case growth is from the areas of maximum absorption. At high intensities, the growth pattern is still different.

ETZOLD (1965) observed that the chloronema of *Dryopteris* grows out on the lit side when partially illuminated with either blue or red light. This, plus the fact that the filaments orient perpendicularly to the plane of polarization of both red and blue illumination, suggests that the photoreceptor dipoles lie in and parallel to the surface of the hemispherical tip of the chloronemal filament. ETZOLD also observed that the polarotropic effect of polarized red light was not reversed by a subsequent exposure to similarly oriented polarized far-red light, but, on the contrary, was enhanced. Further, he found that adaptation to unpolarized red light enhanced the effect of a subsequent polarized far-red illumination. To explain these anomalous results, ETZOLD postulated that only the P_R form of the phytochrome pigment molecule lies with its dipole parallel to the cell-wall surface, and that upon exposure to red light the molecule flips over, so that in the P_{FR} form it is perpendicular to the cell wall. Using this P_R–P_{FR} flip-flop hypothesis, ETZOLD accounted for all his findings, at least qualitatively.

In the protonema of the moss *Physcomitrium* NEBEL (1969) has found that the phototropic effectiveness of red or far-red light is maximal when the plane of polarization of the 45° incident beam is vertical. This corresponds to maximal absorption by receptors in and parallel to the cell surface directly beneath the beam and minimal absorption by receptors on the opposite flank, where these receptors would be perpendicular to the plane of polarization. Thus NEBEL concludes that the receptors for both red and far-red light are near the cell surface and the preferred dipole orientation is parallel to it.

4. Intracellular Transduction Events

The mechanism of phototropism may be visualized as a sequence of events beginning with the absorption of light by the photopigment molecule and ending with a spatial redistribution of growth within the phototropic organ. In multicellular organs some of the later steps are concerned with the production and subsequent action of auxin concentration gradients; the present discussion will exclude this, however, and will be concerned only with events occurring within the initially stimulated cell. What kinds of intracellular transduction event might be detectable? One might hope to discover oxidation-reduction reactions, detected as absorbance changes in appropriate dyes, or changes in metabolite concentrations. Further, transduction events might be manifest in membrane conductance changes, which could be picked up electrically or by tracer fluxes. More indirectly, something might be learned about such mechanisms from the effects of temperature and specific inhibitors.

None of these approaches has yielded much concrete information, however. We are still ignorant of the steps immediately following the excitation of the photoreceptor molecule, which probably involve amplification of the stimulus, and may involve its transmission from one part of the cell to another. Finally, the amplified stimulus, perhaps in a completely altered physical or chemical form, causes a series of unknown events at the cell surface. In the unicellular phototropic organ, these events directly involve the growth of the cell wall and the redistribution of that growth. In multicellular organs, these events must involve the secretion of auxins or auxin precursors. Between photoreception and the final event at the cell surface lie the processes of transduction. The following discussion of transduction events will be restricted to topics that bear mainly on these intermediate processes, insofar as the data can be interpreted.

4.1 Membrane Potential Changes

Coleoptiles develop transverse electric potentials when subjected to lateral illumination (GALSTON, 1959). SCHRANK (1950) also found that if he removed the primary leaf and filled the central cavity with Shive's solution at twice the standard concentration, phototropic curvature was almost completely abolished. He attributed this effect to interference of the electrolyte with the electric potential gradient, since the glycerol solution he used as a control had very little inhibitory effect. BRIDGES and WILKINS (1971) repeated Schrank's experiments, using, in addition to glycerol, solutions of mannitol, polyethylene glycol, and potassium chloride. They found that, when used at the same osmotic potential, the latter three substances inhibited phototropism to very nearly the same extent, even though only one of the three is an electrolyte. This result is clearly inconsistent with the hypothesis that the inhibition is due to the electrical conductivity of the medium introduced into the coleoptile cavity. The authors account for the inhibition on the basis of osmotic effects. BRIDGES and WILKINS confirmed the ineffectiveness of glycerol, and attributed this to its ease of penetration of the cell membrane, and consequent lack of osmotic effectiveness.

We may ask whether the phenomenon of light-induced electric potential is related in any way to the cellular transduction process. The results of JOHNSSON (1965) suggest that it is not. Using a vibrating capacitance electrode that does not contact the plant tissue, he found that for a unilateral stimulus of up to 10 min the electric potential across a corn coleoptile does not appear until 30 min after the stimulus and reaches a peak magnitude 60 min after the stimulus. This slow time course suggests that the potential may be a result of the auxin redistribution rather than its cause. Decapitation of the coleoptile abolished the potential, which could then be restored by adding exogenous IAA.

MOGUS and WOLKEN (1974) have recorded signals 2 mV to 10 mV in amplitude from *Phycomyces* sporangiophores following a pulsed light stimulus. No such signals were recorded by GROVES and GAMOW (1975), however, who found a resting potential of 119 mV (negative inside), but could not detect any change in response to light stimuli. In view of this contradiction, the existence of light-induced electrical potentials in *Phycomyces* remains in doubt.

A light-induced signal of short latency in *Avena* coleoptiles has been briefly reported by BRIGGS (1976). The potential between the tip and the base of the coleoptile changed within 1 s of the beginning of a blue light stimulus. If confirmed, this observation could provide the basis for further investigations of early intracellular transduction events.

4.2 Early Metabolic Changes

One of the first reports of light-induced alterations in metabolism was that of SHROPSHIRE and GETTENS (1966), who found a 30% increase in the ATP extractable from *Phycomyces* sporangiophore growing zones following a 15-s light stimulus. Their method involved plucking the sporangiophores from the mycelium immediately following the stimulus, freezing on dry ice, removing the growing zone, and extracting in boiling water. The ATP levels were determined using luciferin and luciferase from firefly tails. Subsequently these experiments were repeated by SHROPSHIRE and BERGMAN (1968), who then found that the light-induced ATP increase was undetectable in the presence of high variability between replicate determinations. SHROPSHIRE and BERGMAN estimated that the ATP level they found in the steady state (22 pmol/cm growing zone) was only about 1% of the amount of ATP turned over per min. Thus it is likely that small variations in the timing of freezing might have caused the observed variability.

More recently, COHEN (1974a) has reported a 50% decrease in the level of cyclic AMP in whole *Phycomyces* sporangiophores 1 min after the beginning of a sustained light stimulus. His method involved plucking the sporangiophores, freezing in liquid nitrogen, pulverizing, and isolation by anionic exchange chromatography. Interestingly, COHEN found that by 2 min after the stimulus, the cyclic AMP concentration returned to the pre-stimulus level and finally stabilized at about 80% of that level. Also, COHEN found that addition of dibutyryl cyclic AMP to a sporangiophore growing in an inert fluorochemical medium caused a transient 30% decrease in growth rate, resembling the response to a light decrease. Cyclic AMP has several morphological effects on *Phycomyces* mycelium growth when added to the medium (TU and MALHOTRA, 1977). These effects include increased diameter of hyphae, thickened cell wall, and the appearance of particle clusters in the plasma membrane as revealed by freeze-fracture electron microscopy. TU and MALHOTRA attribute these changes to increased synthesis of chitin, the principal component of *Phycomyces* cell wall, thus raising the possibility that cyclic AMP affects chitin synthetase activity. Cyclic AMP has also been implicated in *Vaucheria* phototropism by KATAOKA (1977a), who found that, at a concentration of 10^{-4} M, it greatly reduced phototropic bending without significantly affecting the linear growth rate. KATAOKA proposed that cyclic AMP functions in the cell's "steering mechanism", which can be swamped by an excess of the substance.

4.3 Effects of Low Temperature

The use of low temperature is an attractive experimental approach to the study of transduction processes in phototropism because such a method can clearly separate the primary photoevent, which must be unaffected by temperature, from the subsequent processes, which will in general be affected by temperature, probably to a variable degree. In particular, it should be possible to perform "cold storage" experiments, in which the system is stimulated in the cold, and the response is then observed upon subsequent warming.

Such experiments were performed on *Phycomyces* sporangiophores by PETZUCH and DELBRÜCK (1970). Unfortunately, these cells are not very tolerant to temperature changes. Thus, for example, lowering the temperature from 24°C to 1°C decreased the growth rate from 2.6 mm/h to 0.5 mm/h, but 10 min after rewarming, the growth rate had only returned to 25% of its initial value. To minimize this shortcoming, the authors used unilateral UV stimuli, which produced relatively vigorous tropic responses even at low growth rates. Even so, the response of a cell stimulated immediately after rewarming was slower and peaked later (about 27 instead of 12 min) than that of a cell that had never been cooled. Comparisons were therefore made only between rewarmed cells, in one case stimulated during the cold period and in the other case (the control) stimulated just after rewarming. Under these conditions, it was found that a unilateral stimulus delivered in the cold can be stored for at least 2 h at 2°C and, upon rewarming, give a response similar to that of the control. The nature of the "stored stimulus" is not known; it is presumably some type of chemical intermediate.

Similar experiments were carried out on *Avena* coleoptiles by PICKARD et al. (1969). This system is more tolerant of cold shocks; cooling from 25°C to 2°C reduced the growth rate from 18 μm/min to 0.5 μm/min, but on rewarming the rate returned to 75% of its initial value. A "stored" 45-min unilateral blue light stimulus delivered at 2°C was found to persist undiminished for at least 450 min of additional time in the cold. PICKARD et al. also provide data suggesting that some transduction steps occur even at 2°C: curvature increased linearly with exposure time, but at a much slower rate than at 25°C. At 25°C a 10-min exposure ($0.32\ W \cdot m^{-2}$ blue light) produced a 19° curvature, whereas at 2°C a 30-min exposure at the same intensity was required for the same curvature. Thus some step subsequent to the primary photochemical one continues in the cold, but at a lower rate. Whether this transduction event is intracellular or involves the establishment of a concentration gradient of auxin or auxin precursors across the coleoptile is not known.

5. Light Intensity Gradients and Growth Redistribution

The phototropic response of an organ is usually directed toward or away from a source of light. Excluding the possibility of sensitivity to phase, we must assume that the organ bases its response to the light source on the resulting distribution of light intensity set up within that organ by purely optical mechanisms. Such intensity distributions are often loosely termed intensity gradients, but it should be kept in mind that light may have a very complex spatial intensity distribution within the phototropic organ. Such a distribution may or may not conform to a simple gradient of intensity across the organ.

In this section we will examine gradients within phototropic cells, and the mechanisms by which such gradients may lead to a redistribution of growth within these cells. Two classes of gradients will be considered, those created principally through optical refraction by the cell, and those created principally through transmission losses due to internal scattering or absorption. Finally, we will consider evidence bearing on the mechanism by which light intensity gradients across multicellular organs lead to asymmetries in auxin concentration.

5.1 Unicellular Mechanisms

5.1.1 Focusing Effects

The sporangiophore of *Phycomyces* is undoubtedly the clearest case in which the lens property of a phototropic organ is the principal means of establishing the intracellular intensity gradient. This is nicely shown by the reversal of the direction of phototropism when the cell is immersed in a medium of higher refractive index (e.g., paraffin oil), but SHROPSHIRE (1962) demonstrated it still more elegantly without such immersion. The mechanism of reversal in oil involves the alteration of the normal convergent ray pattern to a divergent ray pattern. In other words, immersion in oil converts a positive lens to a negative one. SHROPSHIRE reasoned that if this change from convergence to divergence is indeed the mechanism for reversal, illumination of a sporangiophore with a diverging ray bundle should also reverse phototropism. Such a divergent bundle was obtained by passing a beam of parallel light first through a cylindrical glass lens of 160 μm diameter, and then through the sporangiophore located 140 μm away. The rays come to a focus just after passing through the glass lens, and thus are divergent again when they strike the sporangiophore. The strong negative tropic response SHROPSHIRE observed under these conditions established conclusively the dioptric basis of the intensity gradient, and hence of the primary phototropic stimulus.

How can the lens property of the cell account for the normal positive phototropic response in air? At first glance the matter seems straightforward, since the convergent rays form a focal band on the far wall of the cylindrical cell (Fig. 21). The part of the cell surface within this band receives considerably higher intensity than is incident on the front wall and, by elongating at a correspondingly greater rate, causes the cell to become concave toward the light source. This mechanism also requires the assumption that the light–growth response is locally autonomous, so that the growth contribution of each element of cell wall is solely

Fig. 21. Focusing of light in a model of a *Phycomyces* sporangiophore viewed in cross-section. Light is in the plane of the paper and enters from above. In the model, the angle between the light direction and the cell axis was 70°, but a similar pattern is found at 90°

determined by the light intensity falling on that element. The UV microbeam experiment of DELBRÜCK and VARJÚ (1961) are not without difficulties of interpretation, but are certainly in accord with the assumption of local autonomy of the light-growth response.

However, it is just this property of local autonomy that makes the above mechanism of phototropism untenable. DELBRÜCK and VARJÚ (1961) showed that the property of local autonomy applies not only to the light-growth response itself, but also to the sensitivity changes (adaptation) of the photo-growth system. When a portion of the growing zone was light-adapted by exposure to the sharp edge of an UV slit stimulus, oriented perpendicular to the cell's long axis, they found that the exposed region still had a lowered sensitivity 10 min later and, more importantly, still had a sharp edge, within the limits of their measurement, about 100 μm in the longitudinal direction. Now, as has long been known (DELBRÜCK and REICHARDT, 1956), the photo-growth system adapts completely to any constant light intensity; thus, for example, a step-up in intensity elicits an increase in growth rate of only a few minutes' duration, and after about 15 min the growth rate has fallen to the pre-step level (for examples, see FOSTER and LIPSON, 1973). Therefore if both the response and the adaptation are locally autonomous, each local element of an illuminated cell should independently arrive at a state of equilibrium with its local intensity, whereupon the growth should be the same everywhere. In consequence, one would expect the phototropic response to die out after 10 to 15 min. Since, on the contrary, *Phycomyces* phototropism is a steady-state phenomenon (DENNISON, 1965) and can continue unabated for many hours, a phototropic mechanism based solely on the intensity gradient and the positive light-growth response is contradicted.

One flaw in the above reasoning is that while the results of DELBRÜCK and VARJÚ (1961) indicate that the light-adapted regions of the cell remain distinct along the longitudinal axis of the cell, they do not rule out the possibility that there is blurring of the adapted areas around the circumference of the cell. In fact BERGMAN et al. (1969) explicitly assume that the local sensitivity level is determined by the average intensity throughout the cell cross-section and is the same at all points around the periphery. Thus, under this assumption, those parts of the cell in the bright focal band would perpetually experience a disequilibrium between the higher intensity and the lower adaptation level, and would therefore show a permanently elevated growth rate, since growth rate is proportional to the ratio of intensity and adaptation level. To settle this point, DENNISON and BOZOF (1973) irradiated sporangiophores unilaterally with UV light (285 nm) and then stimulated with a second unilateral UV stimulus impinging at 90° to the first one. The second stimulus falls upon an inhomogeneous surface; one part of it had previously been exposed, and the other part had not. It was found that the bending response in such circumstances deviates laterally from the path followed when the first UV exposure is made equilaterally. This deviation in response path can be explained if it is assumed that the first exposure causes a strictly local change in sensitivity, which does not spread into the shaded part of the cell. This sensitivity gradient persists and causes deviation of the path of subsequent phototropic bending. This observation contradicts the assumption of uniform sensitivity level averaged over the cell cross-section.

The two properties of local autonomy and complete sensitivity adaptation seem to eliminate the possibility of continued phototropism, because all parts of the cell cross-section are thereby rendered equivalent. To detect and respond to an intensity gradient, some kind of comparison must be made of the conditions on opposite sides of the cell. A novel theory of phototropism proposes that such comparison is made by a physical movement of the photoreceptors from one side to the other. This rotational theory (DENNISON and FOSTER, 1977) proposes that the photoreceptors participate in the rotational motion of the cell wall caused by the spiral pattern of wall growth. Thus, in their journey around the cell circumference, the receptors experience a variety of intensities in sequence. For a cell in air the receptors must pass through a zone of near darkness before striking the bright focal band on the back surface (see Fig. 21). This insures that the receptors will be maximally sensitive when they reach the edge of the bright band; thus they will give a strong growth response. Of course the receptors would also give a negative growth response when they pass out of the bright band, but FOSTER and LIPSON (1973) have found that the magnitude of the negative response to an intensity decrease is considerably smaller than the positive response to an intensity increase; thus the net response would be positive.

What are the basic facts about spiral growth in *Phycomyces?* As most recently shown by ORTEGA et al. (1974), the growing zone of a mature (stage IVb) sporangiophore elongates with a twisting component, so that growth is spiral in the sense of a left-handed screw (i.e., the sporangium rotates clockwise as seen from above). Using an ingenious photographic technique, ORTEGA et al. were able to record the rate of rotation of tin foil markers placed at various positions along the growing zone. The results of one such experiment are shown in Figure 22. The upper 0.7 mm of growing zone typically rotates at about 10°/min (clockwise), and below 1 mm the rate is usually below 3°/min. Thus the region of maximum twist, where the rotation rate varies most rapidly with position along the long axis, is confined to a narrow band between 0.5 and 1.0 mm, in confirmation of the earlier results of COHEN and DELBRÜCK (1958).

In their experimental test of the rotational theory of phototropism, DENNISON and FOSTER (1977) reasoned as follows. If the photoreceptors sense the light gradient by a concerted rotational motion, then by rotating the receptors backwards arti-

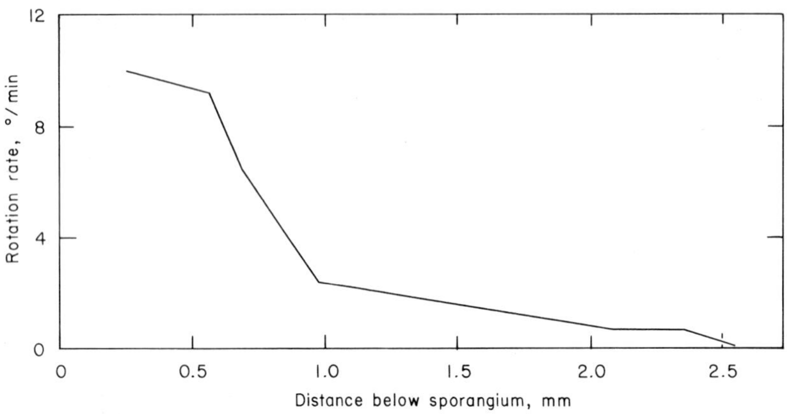

Fig. 22. Cell-wall rotation rate in a *Phycomyces* sporangiophore as a function of position. (After ORTEGA et al., 1974)

ficially it should be possible to bring the net receptor rotation rate, relative to the incident light, to zero, and thus abolish phototropism. The experimental method was to measure the phototropic response of sporangiophores that are rotated on a mechanical turntable at various speeds during the exposure to a unilateral stimulus. The phototropic bending speed in response to a blue (450 nm) stimulus is shown as a function of turntable rotation rate in Figure 23. Clearly phototropism is very sensitive to the speed of turntable rotation. As the turntable speed increases in the counterclockwise sense, the rate of phototropic bending declines sharply, reaching a minimum at a turntable speed of 10°/min. On the other hand, clockwise turntable rotation causes phototropic bending to rise considerably above its normal rate.

DENNISON and FOSTER (1977) concluded from these results that, first, photoreceptor rotation is necessary for phototropism, and, second, the receptors rotate clockwise at 10°/min in the normal cell. The fact that this speed coincides with the rotation rate of the upper portion of the growing zone (Fig. 22) is undoubtedly highly significant. DENNISON and FOSTER measured the sporangium rotation rate of each sporangiophore used in Figure 23 and found a mean value of 9.9°/min (±1.3). From the shape of the curve in Figure 22 it can reasonably be inferred that the top 0.5 mm to 0.7 mm of the cell rotates at nearly the same rate as the sporangium. These results suggest, therefore, that the phototropic receptors are rotating at the same rate as the cell wall in the upper portion of the growing zone. This does not, of course, imply that the receptors must be located at the cell wall, since it is possible that the immediately adjacent underlying layers of cytoplasm may be pulled along by viscous forces at nearly the same rate.

The objection may be raised that while the rate of phototropism is greatly reduced at a turntable rate of 10°/min, it does not vanish completely. One attractive explanation is that not all the receptors are rotating at the same rate, presumably due to their distribution along the growing zone where rotation rates vary. On the other hand, minute-by-minute examination of the phototropic response for 10°/min rotation shows that for long periods (up to 16 min in one case) the rate is indeed zero, thus implying that the cell's rotation varies slightly during the

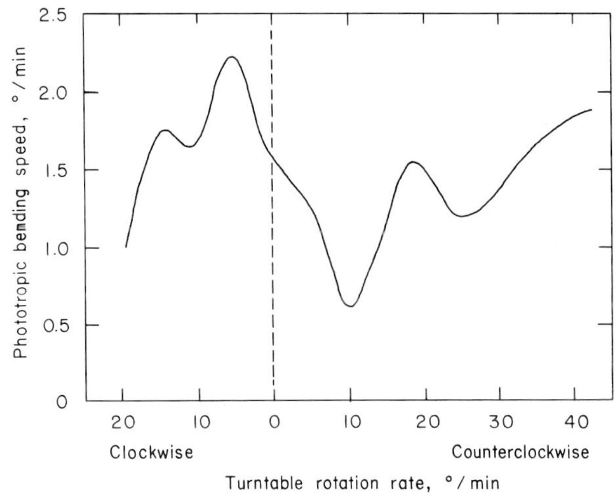

Fig. 23. Phototropic bending speed in *Phycomyces* sporangiophores when steadily rotated on a turntable. (After DENNISON and FOSTER, 1977)

course of each experimental run, making a continuous match of speeds impossible (DENNISON and FOSTER, 1977).

Although this method of gradient detection was discovered in a cell where focusing action is critical, it could easily play a role in any phototropic system where an intracellular light intensity gradient must trigger the response. The rotational mechanism has the inherent advantage of functioning effectively with a system that adjusts its sensitivity within a range of intensities, and this enables phototropism to function well over this intensity range. It will be of considerable interest to know if other phototropic organs also employ the rotational mechanism.

Phototropism in *Avena* coleoptiles is also reversed by immersion in paraffin oil, but it is not clear whether the mechanism is that of a focusing effect. HUMPHRY (1966) concluded that the oil has no effects on the mechanism of redistribution of growth, since geotropism is unaffected by this treatment. MEYER (1969c) found that it is not necessary to illuminate the coleoptile while immersed, but that soaking in oil prior to unilateral illumination in air also produces reversal. She found that this oil treatment must occur at least 1 min before the 5-s light stimulus to cause reversal; a 15-s oil soaking immediately before the stimulus is not sufficient. Penetration of the oil into the intercellular spaces and into the upper 300 μm of the central cavity was observed microscopically. MEYER concluded that the reversal was due to physical changes in the cell wall structure throughout the coleoptile tissue, caused by the penetrating oil. However, it also seems possible that optical effects may play a role at the cellular level, even in coleoptiles.

Direct evidence for a focusing mechanism in *Avena* was provided by SHROPSHIRE (1975), who reported that coleoptiles showed negative phototropism when exposed in air to a diverging beam of light, using a dose in the region of first-positive curvature. A converging beam at the same dose produced positive phototropism.

The situation in *Vaucheria* is not at all clear. Phototropic bending is brought about by a shift in the growing center to the illuminated side, but immersion in paraffin oil did not reverse the direction of bending (KATAOKA, 1975b). Presumably the gradient detection mechanism must involve screening, but no data are available in support of this view.

5.1.2 Screening Effects

In the absence of focusing effects, the light gradient across the cell will be determined by light absorption and scattering due to the cell contents. These two processes are quite distinct, since absorption is due to the presence of pigments and is a much more sensitive function of wavelength than is scattering. Unless otherwise noted, however, we will ignore this distinction and lump the two processes in our consideration of screening.

Although, as has been discussed, *Phycomyces* is a system in which the focusing effect is dominant, there are experimental conditions under which the screening becomes relatively large, and in these cases phototropism reverses sign. For example, the reversal in UV light can be attributed to the great increase in absorbance of the cell at these wavelengths, due to the high concentration of gallic acid in the cell (DELBRÜCK and SHROPSHIRE, 1960). In addition, a mutant strain (C 158), which has a superabundant amount of β-carotene, shows negative phototropism in blue light, but has a normal threshold and light–growth response. Photometry indicates an optical density of 1.19 at 450 nm, about ten times normal (JESAITIS, 1974).

A curious instance of interaction of screening and focusing effects is described in *Pilobolus longipes (*JACOB, 1964). Normally the young (Stage I) sporangiophores are positively phototropic, as are the mature (Stage II) sporangiophores, which have developed the lens-like subsporangial swelling. As shown by microbeam experiments, however, the growth rate is reduced by light in Stage I (curvature toward the illuminated side) and increased by light in Stage II (curvature away from the illuminated side). JACOB also cultured *P. longipes* on medium containing diphenylamine, which inhibits β-carotene synthesis and in this case leads to colorless sporangiophores. These colorless sporangiophores now were strongly negatively phototropic in Stage I but still positively phototropic in Stage II. The explanation appears to be that the screening action of β-carotene in normal Stage I cells causes higher intensity in the front part of the cell, causing it to curve towards the light because of the negative effect of light on the growth rate. In the colorless Stage I cells, however, the lens effect dominates, and the back part of the cell is more brightly lit and therefore inhibited more strongly, thus causing negative phototropism. This explanation is clinched by JACOB's demonstration that immersing the colorless Stage I cells in paraffin oil reverses the reversal, and phototropism is once again positive. In Stage II cells of *P. longipes,* phototropism is caused by enhanced growth at the rear of the cell, due to the lens action of the subsporangial swelling. The fact that variations in β-carotene content have no effect in Stage II indicates that the pigment is located where it does not significantly counteract this lens action.

Different results have been obtained in another species, *Pilobolus kleinii* (JACOB, 1964). Here, in Stage I, no reversal was seen in colorless cells, a result attributed by JACOB to the strong light-scattering action of the lipid droplets present in this species. Thus the screening action would not be dependent on the light absorption of β-carotene and thus would be unaffected by its absence. The microbeam experiments of PAGE and CURRY (1966) on *P. kleinii* indicate that, contrary to JACOB's finding, light enhances the growth of Stage I sporangiophores. PAGE and CURRY maintain that the lens effect, and not screening, is dominant in this case, and that their failure to observe reversal in paraffin oil is due to the abnormally high refractive index of the apex, given by them as 1.57. Thus the relative significance of screening and focusing in *P. kleinii* Stage I sporangiophores is not yet clear.

Screening seems to be the dominant mechanism in *Physcomitrium turbinatum*, where the cell apex contains significant amounts of chlorophyll (NEBEL, 1968). In support of this view is the positive effect of light on the growth rate and the lack of reversal in paraffin oil.

Some interesting results of microbeam illumination of *Avena* coleoptiles have been reported by MEYER (1969b). Curvature was reduced when the magnitude of the light intensity gradient across the apex was reduced by one of two means. One was simply the illumination of the broad side of the organ instead of the narrow side; this reduces the gradient because the light path is now shorter with consequently lower light absorption and scattering. The other means was to deliver half of the stimulus intensity impinging on the narrow side and half simultaneously from above, thus weakening the gradient. Dose–response curves showed a reduction in magnitude but not shift along the dosage axis. In contrast, when MEYER partially illuminated the broad side (Fig. 24) the shape of the dose–response curve was altered, and, unexpectedly, the first-negative response was abolished. She explained the reduction in size of the first-positive curvature by the skewing of the response direction because the two intensity gradients run in mutually perpendicular directions. The loss of negative curvature is more difficult to understand, but might be associated with the lowered intensity on what would normally be the shaded side (facing away from the viewer in Fig. 24A; on the left in Fig. 24B). With the off-center microbeam stimulus (see Fig. 24B), this shaded side may be darker than it is under the control illumination conditions (Fig. 24A), and this may inhibit the negative curvature. Alternatively, the increased gradient may simply extend the dose range of the first-positive curvature to higher doses, so that it now cancels out the negative curvature. Reciprocity problems do not enter, since MEYER kept the stimulus duration constant at 5 s. Although it is not known whether the critical impact of the light gradient is within or between the cells of the apex, the central importance of this gradient in determining the magnitude and sensitivity of the phototropic response is strongly suggested by these results.

Finally, let us consider the question raised by MOHR and PETERS (1960): why are the seedlings of *Sinapis alba* L. not phototropic in red light? They established that there is an intensity gradient in both red and far-red light across the hypocotyl, and that both these radiations can decrease the rate of elongation of this tissue. As an explanation, they propose that the growth inhibition by red and far-red light proceeds by the reduction of the endogenous gibberellin level, and, unlike IAA, no gibberellin gradient can be maintained in the tissue, perhaps due to rapid lateral translocation. Thus light gradients are a necessary, but not sufficient, condition for the redistribution of growth.

5.2 Multicellular Mechanism: Auxin Redistribution

As has been discussed in many previous reviews (e.g., BRIGGS, 1964), phototropism of coleoptiles, and perhaps some other multicellular organs as well, can be described

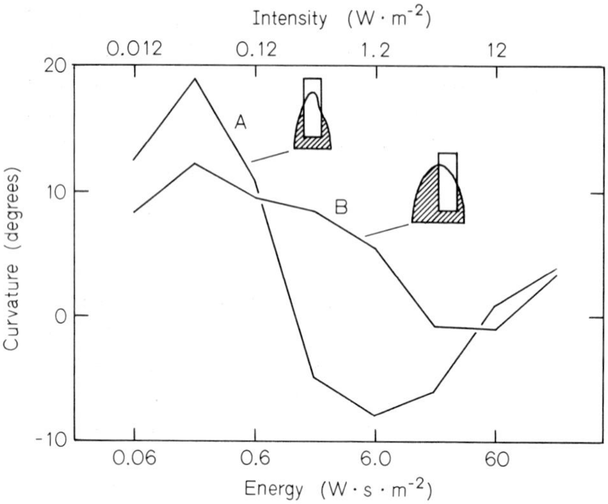

Fig. 24. Dose–response curves for phototropism in *Avena* coleoptiles. Blue light exposures were 5 s long at different intensities. In curve A the microbeam strikes the narrow side of the apex and in curve B the microbeam strikes the broad side of the apex and is off center. In both cases the response is measured in the plane of the larger apical dimension. (Redrawn from MEYER, 1969b)

in terms of the Cholodny-Went theory. This theory holds that the differential growth which underlies phototropic curvature is caused by the establishment of a lateral concentration gradient of a growth hormone, or auxin, across the coleoptile. The light-intensity gradient across the apex leads, in some unspecified way, to the establishment of such an auxin gradient. In this section we will examine some recent experiments bearing on this theory, and showing that the auxin gradient is established by transport mechanisms rather than mechanisms of auxin synthesis or destruction. Finally, we will consider the effects of light on auxin transport.

5.2.1 Phototropism and Lateral Auxin Redistribution

Extensive studies were made by BRIGGS (1963a) on the production of auxin gradients in excised tips of corn coleoptiles. He used white light stimuli and the standard *Avena* curvature test to determine the auxin content of agar blocks placed beneath the basal cut end of the tips. It is important to establish whether light reduces the total level of recoverable auxin, as would be expected if the auxin gradient in phototropism is due to differential photodestruction or to photoinhibition of synthesis. Using doses in the range of either the first-positive or the second-positive curvatures, BRIGGS found no significant differences between the auxin yields of the illuminated tips and those of the dark controls. Using a much higher intensity of white light, only a slight reduction in auxin yield was found for the illuminated tips. Thus, these results indicate that under the usual conditions for phototropism no significant total reduction in auxin yield occurs.

To show that lateral transport does occur, BRIGGS (1963a) used coleoptile tips that had been sectioned longitudinally from the base up to, but not including, the extreme apex (at least 0.5 mm remains intact), and that had thin glass barriers inserted between the halves of the sectioned portions. He found that the shaded half of the tip delivered significantly more auxin into the receiver block, producing an average *Avena* curvature of 30.7°, as compared to 16.2° for the illuminated half. If the tips were completely divided and the halves separated by a thin glass barrier, no auxin differences were noted, thus indicating that some substance must move laterally in the intact part of the apex. This same auxin gradient was observed with doses corresponding to first-positive and second-positive curvatures. A dose in the indifferent range, which produced no curvature in control coleoptiles, produced very nearly equal auxin yields on the lit and shaded sides. These results suggest that phototropic bending is correlated with increased auxin on the convex side and decreased auxin on the concave side.

BRIGGS (1963b) also found that if the intact corn coleoptiles were exposed, immediately before excision, to a 2-h red light exposure sufficient to decrease phototropic sensitivity significantly, the total auxin yield of the tips was reduced by nearly half. On the other hand, the same red light treatment begun immediately after excision of the tips had no effect. Thus, some type of light-induced depression of diffusible auxin does indeed exist in corn coleoptiles (at least for red light), but it seems to function only in the intact organ. BRIGGS (1963b) also examined the auxin gradients induced by unilateral white light and found that the difference in auxin yield between the illuminated and shaded halves was approximately the same, whether or not the red light treatment is given.

In the work just described, the auxin levels are defined in terms of *Avena*-test activity, and it is important to show by direct means that there is a corresponding concentration gradient of a defined growth substance. This has been done by PICKARD and THIMANN (1964), who applied ^{14}C-labeled indoleacetic acid (^{14}C-IAA) to agar blocks and then let these blocks contact the extreme apex of excised 6.5-mm corn coleoptiles the basal 1 mm of which was partitioned by the insertion of razor blades. When a unilateral white light dose corresponding to the first-positive curvature was administered, the radioactivity recovered in the basal agar blocks was in the ratio of 25:75, for the illuminated and shaded sides, respectively. Under similar conditions the radioactivity of the illuminated and dark halves of the coleoptile tissue itself had the ratio of 35:65. Thus, in this system at least, the IAA applied to the apex has been transported laterally, constituting a concentration gradient in the tissue itself, and finally leading, via basipetal transport, to a concentration difference in the agar blocks. Further, since in these experiments the unilateral light stimulus was administered before the ^{14}C-IAA was applied, the light clearly does not act on the applied IAA itself, but rather sets up in the tissue an asymmetry, or polarity, which subsequently moves the IAA laterally. PICKARD and THIMANN also used light exposures corresponding to second-positive curvatures, and in this case the ^{14}C-IAA ratio in the basal agar blocks was about 34:66 (light:dark). When the apical 2 mm was removed, the remaining subapical coleoptile sections were much less effective in setting up a lateral gradient, which was only 46:54 (light:dark). Thus the system that transports IAA laterally is mainly located in the upper 2 mm of the coleoptile, for both the first-positive and the second-positive curvatures. These results are thus consistent with the view that the tip is the locus of sensitivity for both first-positive and second-positive curvatures (see p. 511).

One aspect of the experimental technique of PICKARD and THIMANN may be of significance: when the donor agar block was held rigidly above the growing coleoptile, the tip penetrated a significant distance into the block during the transport period (165 or 180 min). When this happened, a larger amount of ^{14}C-IAA entered the tissue and, unexpectedly, a much smaller lateral IAA gradient resulted in the tissue. Recognizing this problem, PICKARD and THIMANN altered the technique thereafter, so as to minimize this penetration. However, this effect suggests that only the IAA applied very near the extreme apex is transported laterally, and that IAA applied more basally does not enter the lateral transport system.

This effect was also noted by GARDNER et al. (1974), who studied the distribution of ^{3}H-IAA in corn coleoptiles when applied by a glass micropipette (tip diameter 100 μm) that penetrated the coleoptile tip very slightly at an off-axis position. They found that the lateral IAA gradients produced were quite variable when the pipette was applied within 1 mm from the apex, but became more reproducible when the application was restricted to the apical 0.2 mm. This extreme localization of the main lateral IAA transport capacity undoubtedly makes it difficult to obtain consistent data for exogenous auxins.

The technique of asymmetric IAA application was selected by GARDNER et al. (1974) to distinguish lateral transport from unequal longitudinal transport. In some instances they found good evidence for lateral transport: intact corn coleoptiles exposed to blue light producing 11.6° (first-positive) curvature in controls had 31.1% of the radioactivity on the side opposite the point of application when the point of application was illuminated and 15.8% on the opposite side when the point of application was in darkness (Fig. 25). Thus these results correspond to a "net lateral movement" of 15.3% from the illuminated side to the dark side. However, when this experiment was repeated with excised coleoptile tips and agar block receivers, the net lateral movement was only 9.4%. A red light pretreatment introduced further complications. In intact corn coleoptiles, the same blue light exposure produced a slightly smaller curvature (9.0°) when preceded by a 1.5-h red treatment, but the net lateral movement in the tissue was reduced to 3.9%, essentially zero. However, in the case of excised tips and agar blocks the net lateral movement was increased to 19.9%. Similar experiments repeated with *Avena* also gave results difficult to interpret. In general, large values of net lateral movement were accompanied by significant curvatures, but in many cases the reverse was not true, and conditions giving curvatures in excess of 15° did not always produce significant net lateral movement. This may be due to the weakness of the technique, or to the existence of hormones other than IAA capable of causing differential growth.

In excised third internodes of pea seedlings (*Pisum sativum*), KANG and BURG (1974) were able to show that ^{3}H-IAA applied to the apical end of the section in lanolin paste showed a lateral concentration ratio of 35:65 (light:dark) under conditions where a phototropic curvature of 24° developed. Remarkably, a red pretreatment that nearly doubled the curvature did not significantly change the lateral IAA gradient. Thus we again see a lack of correlation between curvature and IAA gradients.

In light-grown sunflower seedlings (*Helianthus annuus* L.), BRUINSMA et al. (1975) found no detectable gradient of IAA, when the concave and convex flanks of phototropically bending plants were extracted and analyzed for IAA by a spectrophotometric method. The authors found that the cut basal end of the hypocotyl secreted considerable auxin into a water or buffer solution, as measured by *Avena*-test activity. However, when spectroscopic IAA determinations were made on these diffusates, only a few percent of this auxin activity could be accounted for. Thus the authors concluded that IAA is not responsible for differential growth in *Helianthus*, and they speculated that this role must be played by some other

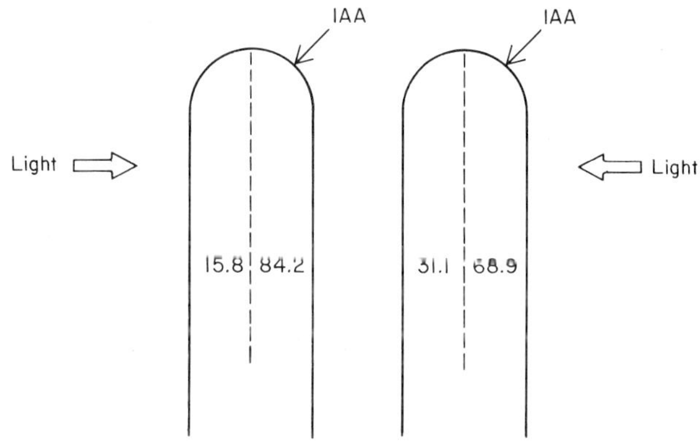

Fig. 25. Lateral transport of radioactive IAA applied asymmetrically to intact *Zea* coleoptile tips. Data are percentages of radioactivity recovered from the two sides of the uppermost 25 mm of the coleoptile. (Data from GARDNER et al., 1974)

hormone. In the same organism, PHILLIPS (1972) found a concentration ratio of 12:88 (light:dark) in the diffusible gibberellins extracted from unilaterally illuminated excised hypocotyls. Hormone activity was measured on chromatographed extracts using a lettuce hypocotyl elongation test. Thus it seems likely that *Helianthus* phototropism involves a hormone distinct from IAA, possibly a gibberellin.

5.2.2 Effect of Light on Polar Auxin Transport

If the transport of auxin through a tissue such as a coleoptile is polarized, it is due to a polarity within the individual cells that make up this tissue. Thus mechanisms by which light-intensity gradients alter such transport are basically mechanisms by which the polarity of each cell is altered. Unfortunately the basic nature of this cellular polarity is unknown, although an attractive recent theory invokes a difference in cell membrane permeability to ionized IAA on opposite ends of the cell (RAVEN, 1975; GOLDSMITH, 1977). Thus if the photoreceptors are located near the cell wall, they would be ideally situated to influence local plasma membrane permeability and hence affect IAA transport. Whether or not this theory proves to be useful, it is of some interest to consider the influence of light on polar auxin transport.

SHEN-MILLER et al. (1969) examined this transport by direct measurement of the distribution of radioactivity in *Avena* coleoptiles to the apices of which ^{14}C-IAA had been applied. They found that the inhibition of longitudinal IAA transport, relative to dark controls, had approximately the same relationship to dose and wavelength as did the phototropic response. In particular they found no inhibition at wavelengths of 500 nm and above.

THORNTON and THIMANN (1967) studied the transport of ^{14}C-IAA applied in agar blocks to the apical ends of 4-mm sections of *Avena* coleoptiles. The transport rate, measured as increased radioactivity in basal receiver blocks, showed a 12% to 17% transient decrease following a 15-min blue light stimulus. For low IAA concentrations in the donor block (0.13 mg/l), this suppression lasted only about 20 min, but at higher IAA concentrations it had a considerably longer duration.

Photoinhibition of polar auxin transport in corn coleoptiles was also found by HAGER and SCHMIDT (1968a). They introduced ^{14}C-IAA to the apex in a block of agar and measured the accumulation of radioactivity in a receiver block at the basal end of the coleoptile section. Light inhibition of auxin transport was observed, even though the illumination was delivered before the ^{14}C-IAA block was added, the transport being measured in darkness. Inhibition was observed with monochromatic illumination at 450 nm, but none was observed at 660 nm. Inhibition was also observed when sections were incubated in a solution of IAA and flavin mononucleotide that had been illuminated (without illumination there was no inhibition). HAGER and SCHMIDT therefore attributed this inhibition to the formation of an IAA oxidation product, 3-methyleneoxindole (3-M), by action of light on a flavin sensitizer. The role of endogenous IAA in the formation of 3-M was shown in experiments with decapitated coleoptile sections that had been incubated in buffer with sugar added for 21 h in darkness. It could be shown that this treatment greatly depletes the tissue of endogenous IAA, and HAGER and SCHMIDT found that in such depleted coleoptile sections the light treatment did not inhibit the subsequent accumulation of ^{14}C-IAA in the receiver blocks. When non-labeled IAA was applied to the apical end of these sections before illumination, the light-induced inhibition of IAA transport was restored. These results indicate that light-induced inhibition of IAA transport occurs only in the presence of an adequate concentration of IAA, either endogenous or exogenous.

Further, it was reported that 3-M is formed from ^{14}C-IAA in intact illuminated coleoptiles, and that the application of 3-M inhibits IAA transport. In a subsequent paper (HAGER and SCHMIDT, 1968b), the authors present evidence that 3-M enhances the uptake of IAA by coleoptile sections but inhibits its release. In view of this, they suggest that 3-M acts by inhibition of the enzymatic secretion of IAA from the cell membrane.

Contradictory results were recently obtained in *Avena* by MENSCHICK et al. (1977), who were unable to detect any significant change in the 3-M content of coleoptiles due to illumination. Also, the rate of IAA decarboxylation, as measured by the output of $^{14}CO_2$ from ^{14}C-IAA-treated coleoptiles, was found to be unaffected by illumination, indicating that the breakdown of IAA to 3-M is not light-sensitive in *Avena*.

Although the results with corn support a very plausible mechanism whereby a light gradient could result in polar auxin transport, the results with *Avena* do not bear out the contention that the inhibition is due to the formation of 3-M. Clearly, further experimentation is needed to settle this question.

6. Control of Cell-Wall Extension

As discussed in the previous section, light gradients across multicellular phototropic organs lead to lateral gradients in auxin concentration, and hence to differential growth. It seems logical, therefore, to complete this discussion of phototropism with a detailed consideration of the mechanisms by which auxins influence the growth of plant cells. However, in spite of this and other good reasons for doing so, it is felt that the inclusion of such a discussion in the present chapter would be inappropriate. In the first place, there is a vast literature on the general subject of plant growth hormones, and, in fact, this subject is treated at great length in Volumes 9, 10, 11 in this Series. Secondly, it can be argued that although phototropism is fundamentally a problem of growth, it is principally concerned with the light-induced spatial redistribution of growth, not the growth itself. In other words, the important questions in phototropism deal with the nature of the photoreceptor, and the mechanism of the establishment of differential concentrations of substances that influence growth; but, once such differential concentrations have been established, the remaining steps constitute problems in growth regulation, and are hardly peculiar to phototropism. Thus, in place of a thorough review of the field, we present here a few general remarks on studies of the mechanical properties of cell walls.

The growth of plant cells involves the extension of the cell wall under the influence of turgor pressure. One of the most straightforward experimental approaches would seem to be the investigation of the mechanical properties of plant cell walls and the modification of these properties by auxin. However, as CLELAND (1971) points out, the mechanical properties of plant cell walls are neither purely elastic nor purely viscous, but viscoelastic. In such a system, an applied stress gives rise to a rapid extension followed by a slower component, or creep. If the stress is removed, the relaxation usually accounts for only part of the extension and such an extension can be considered partly elastic (reversible) and partly plastic (irreversible). Whether the extension is plastic, elastic, or a combination, depends in some cases on the previous history; if the stress is applied and removed repeatedly, the system may become mechanically conditioned (strain-hardened) and hence purely elastic, as long as the conditioning stress is not exceeded. Regardless of these complexities, the action of auxin seems to involve a "loosening", or increase in extensibility, of the cell wall. The mechanisms of such loosening and the role played by auxin are, of course, the principal unresolved questions. CLELAND (1971) suggests that some unidentified loosening factor is released from the cell, perhaps actively transported across the plasma membrane, and acts by the cleaving and reforming of cross-linking bonds in the molecular framework of the cell wall. Thus, auxin may act by influencing the release of this loosening factor, perhaps by affecting the transport proteins of the plasma membrane.

No auxin is known to be involved in the regulation of cell-wall extension in *Phycomyces*, and in this case one would expect that the wall extensibility might be rather directly affected by the light stimulus. Such a light-induced change in extensibility has been found by ORTEGA and GAMOW (1976; see also ORTEGA et al., 1975), who used sporangiophores that had been strain-hardened by the repeated application of a 240 mg load, thus insuring that the extensibility is purely elastic (except for a minor component due to growth). After a 4-min "rest" period (in red light), the extensibility was determined. The extensibility invariably increased as a result of the rest period, and the amount of the increase was enhanced by a 2-min blue light stimulus delivered at the beginning of the rest period. When measured at various times after the stimulus, the extensibility increased rapidly to a maximum 4 min after the stimulus, and then decreased more gradually, a time course remarkably similar to that of the light–growth response. Thus it appears likely that the differential growth in the phototropic response of *Phycomyces* is due to the differential light-induced softening of the chitin cell wall. A chitinase (COHEN, 1974b) and a chitin synthetase (JAN, 1974) have been isolated from sporangiophores and characterized. Partially purified chitin synthetase is not directly influenced by light, but, in a crude mycelial homogenate, exposure to bright blue light for 2 h caused a 30% increase in synthetase activity. We may thus conjecture that the action of light on the photoreceptor leads to the softening of the cell wall by the production of activators or inhibitors of the enzymes involved in the synthesis and destruction of chitin.

7. Concluding Comments

Upon reaching the end of this chapter, the reader may well agree that, while a few generalizations are possible, phototropism shows a distressingly high degree of diversity among the various systems. Although many systems have similar action spectra, and in some cases a flavin is clearly suggested, there is considerable dissent. Neither the mode of bending nor the mechanism of establishing an intensity gradient is the same for all systems. Further, it is likely that, as new phototropic mechanisms are worked out, fresh novelties will appear.

It seems unfortunate that so many systems have been investigated, but in no case has a satisfactory level of understanding been achieved. For example, the identity of the photoreceptor remains unknown in every case. The coleoptile, a classical research system, is convenient in many ways for the observation of phototropic behavior, but it is simply not suitable for the application of modern molecular knowledge and methods. Clearly, the significant frontiers of phototropism are at the subcellular level, where problems of enzyme activation, membrane transport, protein synthesis, and so forth, should be attacked. However, a coleoptile is too complex a tissue for such investigation, nor is it suitable for the application of the powerful methods of molecular genetics.

Significant progress in this field may well depend on finding a phototropic system which can be exploited using modern biochemical and genetic approaches. An attractive candidate is a unicellular organ, which would be devoid of the complications of growth hormones. Although *Phycomyces* is perhaps not an ideal system for several reasons, it does nevertheless possess several advantages, and it has the additional merit that it is by far the most extensively studied of all the unicellular phototropic systems. It seems quite possible, therefore, that a concerted and multidisciplinary study of this one system might be the most effective way to advance our understanding of the mechanism of phototropism.

References

Abrahamson, E.W., Wiesenfeld, J.R.: The structure, spectra, and reactivity of visual pigments. In: Handbook of Sensory Physiology. Dartnall, H.J.A. (ed.). Berlin-Heidelberg-New York: Springer 1972, Vol. VII, pt. 1, pp. 69–121

Asomaning, E.J.A., Galston, A.W.: Comparative study of phototropic response and pigment content in oat and barley coleoptiles. Plant Physiol. *36*, 453–464 (1961)

Ball, N.G.: Tropic, nastic and tactic responses. In: Plant Physiology, A Treatise. Steward, F.C. (ed.). New York: Academic Press 1969 vol. Va (Analysis of Growth: Behavior of Plants and their Organs), pp. 119–228

Banbury, G.H.: Phototropism of lower plants. In: Encyclopedia of Plant Physiology. Ruhland, W. (ed.). Berlin-Heidelberg-New York: Springer 1959, vol. 17/1. pp. 530–578

Bentrup, F.W.: Vergleichende Untersuchungen zur Polaritätsinduktion durch das Licht an der *Equisetum*-Spore und der *Fucus*-Zygote. Planta *59*, 472–491 (1963)

Bergman, K.: Blue-light control of sporangiophore initiation in *Phycomyces*. Planta *107*, 53–67 (1972)
Bergman, K., Burke, P.V., Cerdá-Olmedo, E., David, C.N., Delbrück, M., Foster, K.W., Goodell, E.W., Heisenberg, M., Meissner, G., Zalokar, M., Dennison, D.S., Shropshire, Jr., W.: *Phycomyces*. Bacteriol. Rev. *33*, 99–157 (1969)
Bergman, K., Eslava, A.P., Cerdá-Olmedo, E.: Mutants of *Phycomyces* with abnormal phototropism. Mol. Gen. Genet. *123*, 1–16 (1973)
Berns, D.S., Vaughn, J.R.: Studies on the photopigment system in *Phycomyces*. Biochem. Biophys. Res. Commun. *39*, 1094–1103 (1970)
Blaauw, O.H., Blaauw-Jansen, G.: The influence of red light on the phototropism of *Avena* coleoptiles. Acta Bot. Neerl. *13*, 541–552 (1964)
Blaauw, O.H., Blaauw-Jansen, G.: The phototropic responses of *Avena* coleoptiles. Acta Bot. Neerl. *19*, 755–763 (1970a)
Blaauw, O.H., Blaauw-Jansen, G.: Third positive (c-type) phototropism in the *Avena* coleoptile. Acta Bot. Neerl. *19*, 764–776 (1970b)
Brain, R.D., Freeberg, J.A., Weiss, C.V., Briggs, W.R.: Blue light-induced absorbance changes in membrane fractions from corn and *Neurospora*. Plant Physiol. *59*, 948–952 (1977)
Bridges, I.G., Wilkins, M.B.: Effects of electrolyte and nonelectrolyte solutions on the tropic responses of *Avena* coleoptiles. J. Exp. Bot. *22*, 208–212 (1971)
Briggs, W.R.: Light dosage and phototropic responses of corn and oat coleoptiles. Plant Physiol. *35*, 961–962 (1960)
Briggs, W.R.: Mediation of phototropic responses of corn coleoptiles by lateral transport of auxin. Plant Physiol. *38*, 237–247 (1963a)
Briggs, W.R.: Red light, auxin relationships and the phototropic responses of corn and oat coleoptiles. Am. J. Bot. *50*, 196–207 (1963b)
Briggs, W.R.: The phototropic responses of higher plants. Annu. Rev. Plant Physiol. *14*, 311–352 (1963c)
Briggs, W.R.: Phototropism in higher plants. In: Photophysiology. Giese, A.C. (ed.) New York: Academic Press 1964, pp. 223–271
Briggs, W.R.: The nature of the blue light photoreceptor in higher plants and fungi. In: Light and Development. Smith, H. (ed). London: Butterworths 1976, pp. 7–18 (proceedings of the 22nd Nottingham Easter School in Agricultural Sciences)
Briggs, W.R., Chon, H.P.: The physiological versus the spectrophotometric status of phytochrome in corn coleoptiles. Plant Physiol. *41*, 1159–1166 (1966)
Bruinsma, J., Karssen, C.M., Benschop, M., Van Dort, J.B.: Hormonal regulation of phototropism in the light-grown sunflower seedling, *Helianthus annuus* L.: Immobility of endogenous indoleacetic acid and inhibition of hypocotyl growth by illuminated cotyledons. J. Exp. Bot. *26*, 411–418 (1975)
Carlile, M.J.: The photobiology of fungi. Annu. Rev. Plant Physiol. *16*, 175–202 (1965)
Carlile, M.J.: The photoresponses of fungi. In: The Photobiology of Microorganisms. Halldal, P. (ed.). London: Wiley-Interscience 1970, pp. 309–344
Castle, E.S.: Differential growth and phototropic bending in *Phycomyces*. J. Gen. Physiol. *48*, 409–423 (1965)
Chang, H., Calpouzos, L.: Phototropism of the uredospore germ tubes of *Puccinia graminis tritici*. Bot. Bull. Acad. Sin. *14*, 35–40 (1973)
Chon, H.P., Briggs, W.R.: Effect of red light on the phototropic sensitivity of corn coleoptiles. Plant Physiol. *41*, 1715–1724 (1966)
Cleland, R.: Cell wall extension. Annu. Rev. Plant Physiol. *22*, 197–222 (1971)
Cohen, R., Delbrück, M.: Distribution of stretch and twist along the growing zone of the sporangiophore of *Phycomyces* and the distribution of response to a periodic illumination program. J. Cell. Comp. Physiol. *52*, 361–388 (1958)
Cohen, R., Delbrück, M.: Photoreactions in *Phycomyces*. Growth and tropic responses to the stimulation of narrow test areas. J. Gen. Physiol. *42*, 677–695 (1959)
Cohen, R.J.: Cyclic AMP levels in *Phycomyces* during a response to light. Nature (London) *251*, 144–146 (1974a)
Cohen, R.J.: Some properties of chitinase from *Phycomyces blakesleeanus*. Life Sci. *15*, 289–300 (1974b)
Curry, G.M.: Studies on the Spectral Sensitivity of Phototropism. Ph.D. Thesis, Harvard University 1957

Curry, G.M.: Phototropism. In: The Physiology of Plant Growth and Development. Wilkins, M.B. (ed.). New York: McGraw-Hill 1969, pp. 241–273

Curry, G.M., Gruen, H.E.: Action spectra for the positive and negative phototropism of *Phycomyces* sporangiophores. Proc. Natl. Acad. Sci. U.S.A. *45*, 797–804 (1959)

Curry, G.M., Gruen, H.E.: Dose-response relationships at different wavelengths in phototropism of *Avena*. In: Progress in Photobiology. Christensen, B.C., Buchmann, B. (eds.). New York: Elsevier Publishing Co. 1961, pp. 155–157

Curry, G.M., Thimann, K.V., Ray, P.M.: The base curvature of *Avena* seedlings to the ultraviolet. Physiol. Plant. *9*, 429–440 (1956)

Curry, G.M., Thimann, K.V.: Phototropism: The nature of the photoreceptor in higher and lower plants. In: Progress in Photobiology. Christensen, B.C., Buchmann, B. (eds.). New York: Elsevier Publishing Co. 1961, pp. 127–134

Defabo, E.C., Harding, R.W., Shropshire, W.: Action spectrum between 260 and 800 nanometers for photoinduction of carotenoid biosynthesis in *Neurospora crassa*. Plant Physiol. *57*, 440–445 (1976)

Delbrück, M., Katzir, A., Presti, D.: Responses of *Phycomyces* indicating optical excitation of the lowest triplet state of riboflavin. Proc. Natl. Acad. Sci. U.S.A. *73*, 1969–1973 (1976)

Delbrück, M., Reichardt, W.: System analysis for the light growth reaction in *Phycomyces*. In: Cellular Mechanisms in Differentiation and Growth. Rudnick, D. (ed.). Princeton, N.J.: Princeton University Press 1956, pp. 3–44

Delbrück, M., Shropshire, Jr., W.: Action and transmission spectra of *Phycomyces*. Plant Physiol. *35*, 194–204 (1960)

Delbrück, M., Varjú, D.: Photoreactions in *Phycomyces*. Responses to the stimulation of narrow test areas with ultraviolet light. J. Gen. Physiol. *44*, 1177–1188 (1961)

Dennison, D.S.: Steady-state phototropism in *Phycomyces*. J. Gen. Physiol. *48*, 393–408 (1965)

Dennison, D.S., Bozof, R.P.: Phototropism and local adaptation in *Phycomyces* sporangiophores. J. Gen. Physiol. *62*, 157–168 (1973)

Dennison, D.S., Foster, K.W.: Intracellular rotation and the phototropic response of *Phycomyces*. Biophys. J. *18*, 103–123 (1977)

Diehn, B.: Action spectra of the phototactic responses in *Euglena*. Biochem. Biophys. Acta *177*, 136–143 (1969)

Elliott, W.M., Shen-Miller, J.: Similarity in dose response, action spectra and red light responses between phototropism and photoinhibition of growth. Photochem. Photobiol. *23*, 195–199 (1976)

Eslava, A.P., Alvarez, M.I., Lipson, E.D., Presti, D., Kong, K.: Recombination between mutants of *Phycomyces* with abnormal phototropism. Mol. Gen. Genet. *147*, 235–241 (1976)

Etzold, H.: Der Polarotropismus und Phototropismus der Chloronemen von *Dryopteris filix-mas* (L.) Schott. Planta *64*, 254–280 (1965)

Everett, M.: Dose-response curves for radish seedling phototropism. Plant Physiol. *54*, 222–225 (1974)

Everett, M., Thimann, K.V.: Second positive phototropism in the *Avena* coleoptile. Plant Physiol. *43*, 1786–1792 (1968)

Foster, K.W.: Phototropism of coprophilous zygomycetes. Annu. Rev. Biophys. Bioeng. *6*, 419–443 (1977)

Foster, K.W., Lipson, E.D.: The light growth response of *Phycomyces*. J. Gen. Physiol. *62*, 590–617 (1973)

Galston, A.W., Phototropism of stems, roots and coleoptiles. In: Encyclopedia of Plant Physiology. Ruhland, W. (ed.). Berlin-Heidelberg-New York: Springer 1959, Vol. 17/1, pp. 492–529

Gardner, G., Shaw, S., Wilkins, M.B.: IAA transport during the phototropic responses of intact *Zea* and *Avena* coleoptiles. Planta *121*, 237–251 (1974)

Goldsmith, M.H.M.: The polar transport of auxin. Annu. Rev. Plant Physiol. *28*, 439–478 (1977)

Gordon, S.A., Shen-Miller, J.: Auxin relations in phototropism of the coleoptile: a re-examination. In: Biochemistry and Physiology of Plant Growth Substances. Wightman, F., Setterfield, G. (eds.). Ottawa: The Runge Press 1968 (Proceedings of the 6th International Conference on Plant Growth Substances, July 24–29, 1967), pp. 1097–1108

Groves, P.M., Gamow, R.I.: Intracellular recordings from *Phycomyces*. Plant Physiol. *55*, 946–947 (1975)
Hager, A.: Ausbildung von Maxima im Absorptionsspektrum von Carotinoiden im Bereich um 370 nm; Folgen für die Interpretation bestimmter Wirkungsspektren. Planta *91*, 38–53 (1970)
Hager, A., Schmidt, R.: Auxintransport und Phototropismus. I. Die lichtbedingte Bildung eines Hemmstoffes für den Transport von Wuchsstoffen in Koleoptilen. Planta *83*, 347–371 (1968a)
Hager, A., Schmidt, R.: Auxintransport und Phototropismus. II. Der Hemmechanismus des aus IES gebildeten Photooxidationsproduktes 3-Methylenoxindol beim Transport von Wuchsstoffen. Planta *83*, 372–386 (1968b)
Hartmann, K.M.: Ein Wirkungsspektrum der Photomorphogenese unter Hochenergiebedingungen und seine Interpretation auf der Basis der Phytochroms (Hypokotylwachstumshemmung bei *Lactuca sativa* L.). Z. Naturforsch. *22b*, 1172–1175 (1967)
Haupt, W.: Perception of environmental stimuli orienting growth and movement in lower plants. Annu. Rev. Plant Physiol. *16*, 267–290 (1965)
Haupt, W., Buchwald, M.: Die Orientierung der Photoreceptor-Moleküle im Sporangienträger von *Phycomyces*. Z. Pflanzenphysiol. *56*, 20–26 (1967)
Humphry, V.R.: The effects of paraffin oil on phototropic and geotropic responses in *Avena* coleoptiles. Ann. Bot. *30*, 39–45 (1966)
Jacob, F.: Über die Funktion eines Karotin-Lichtschirmes bei dem Phototropismus von Sporangienträgern chromosporer *Pilobolus-Arten*. Flora *155*, 209–222 (1964)
Jaffe, L.F.: The effect of polarized light on the growth of a transparent cell. A theoretical analysis. J. Gen. Physiol. *43*, 897–911 (1960)
Jaffe, L., Etzold, H.: Tropic responses of *Funaria* spores to red light. Biophys. J. *5*, 715–742 (1965)
Jan, Y.N.: Properties and cellular localization of chitin synthetase in *Phycomyces blakesleeanus*. J. Biol. Chem. *249*, 1973–1979 (1974)
Jesaitis, A.J.: Linear dichroism and orientation of the *Phycomyces* photopigment. J. Gen. Physiol. *63*, 1–21 (1974)
Jesaitis, A.J., Heners, P.R., Hertel, R., Briggs, W.R.: Characterization of a membrane fraction containing a *b*-type cytochrome. Plant Physiol. *59*, 941–947 (1977)
Johnsson, A.: Photoinduced lateral potentials in *Zea mays*. Physiol. Plant. *18*, 574–576 (1965)
Kang, B.G., Burg, S.P.: Red light enhancement of the phototropic response of etiolated pea stems. Plant Physiol. *53*, 445–448 (1974)
Kataoka, H.: Phototropism in *Vaucheria geminata* I. The action spectrum. Plant Cell Physiol. *16*, 427–437 (1975a)
Kataoka, H.: Phototropism in *Vaucheria geminata* II. The mechanism of bending and branching. Plant Cell Physiol. *16*, 439–448 (1975b)
Kataoka, H.: Phototropic sensitivity in *Vaucheria geminata* regulated by 3′, 5′-cyclic AMP. Plant Cell Physiol. *18*, 431–440 (1977a)
Kataoka, H.: Second positive- and negative phototropism in *Vaucheria geminata*. Plant Cell Physiol. *18*, 473–476 (1977b)
Klemm, E., Ninnemann, H.: Detailed action spectrum for the delay shift in pupae emergence of *Drosophila pseudoobscura*. Photochem. Photobiol. *24*, 369–371 (1976)
Lipson, E.D.: White noise analysis of *Phycomyces* light growth response system. II. Extended intensity ranges. Biophys. J. *15*, 1013–1031 (1975a)
Lipson, E.D.: White noise analysis of *Phycomyces* light growth response system. III. Photomutants. Biophys. J. *15*, 1033–1045 (1975b)
Lipson, E.D., Presti, D.: Light-induced absorbance changes in *Phycomyces* photomutants. Photochem. Photobiol. *25*, 203–208 (1977)
Mayer, F.: Lichtorientierte Chloroplasten-Verlagerungen bei *Selaginella martensii*. Z. Bot. *52*, 346–381 (1964)
Mayer, F.: Lichtinduzierte Chloroplasten-Verlagerungen bei *Selaginella martensii*. Z. Pflanzenphysiol. *55*, 65–70 (1966)
Medina, J.R., Cerdá-Olmedo, E.: A quantitative model of *Phycomyces* phototropism. J. Theor. Biol. *69*, 709–719 (1977)
Meissner, G., Delbrück, M.: Carotenes and retinal in *Phycomyces* mutants. Plant Physiol. *43*, 1279–1283 (1968)

Meistrich, M.L., Fork, R.L., Matricon, J.: Phototropism in *Phycomyces* as investigated by focused laser radiation. Science *169*, 370–371 (1970)
Menschick, R., Hild, V., Hager, A.: Decarboxylierung von Indolylessigsäure im Zusammenhang mit dem Phototropismus in *Avena*-Koleoptilen. Planta *133*, 223–228 (1977)
Meyer, A.M.: Versuche zur Trennung von 1. positiver und negativer Krümmung der *Avena*koleoptile. Z. Pflanzenphysiol. *60*, 135–146 (1969a)
Meyer, A.M.: Versuche zur 1. positiven und zur negativen phototropischen Krümmung der *Avena*koleoptile: I. Lichtperception und Absorptionsgradient. Z. Pflanzenphysiol. *60*, 418–433 (1969b)
Meyer, A.M.: Versuche zur 1. positiven und zur negativen phototropischen Krümmung der *Avena*koleoptile: II. Die Inversion durch Paraffinöl. Z. Pflanzenphysiol. *61*, 129–134 (1969c)
Mikolajczyk, E., Diehn, B.: The effect of potassium iodide on phototropic responses in *Euglena*: evidence for two photoreceptor pigments. Photochem. Photobiol. *22*, 269–271 (1975)
Mogus, M.A., Wolken, J.J.: *Phycomyces*: Electrical response to light stimuli. Plant Physiol. *53*, 512–513 (1974)
Mohr, H.: Lehrbuch der Pflanzenphysiologie. Berlin-Heidelberg-New York: Springer: 1969, pp. 226–242
Mohr, H., Peters, E.: Der Phototropismus und das lichtabhängige Längenwachstum der Hypokotyls von *Sinapsis alba*. L. Planta *55*, 637–646 (1960)
Morré, D.J., Eisinger, W.R.: Cell wall extensibility: its control by auxin and relationship to cell elongation. In: Biochemistry and Physiology of Plant Growth Substances. Wightman, F., Setterfield, G. (eds.). Ottawa: The Runge Press 1968 (Proceedings of the 6th International Conference on Plant Growth Substances, July 24–29, 1967), pp. 625–645
Muñoz, V., Butler, W.L.: Photoreceptor pigment for blue light in *Neurospora crassa*. Plant Physiol. *55*, 421–426 (1975)
Nebel, B.J.: Action spectra for photogrowth and phototropism in protonemata of the moss *Physcomitrium turbinatum*. Planta *81*, 287–302 (1968)
Nebel, B.J.: Responses of moss protonemata to red and far-red polarized light: evidence for disc-shaped phytochrome photoreceptors. Planta *87*, 170–179 (1969)
Neuscheler-Wirch, H.: Photomorphogenese und Phototropismus bei *Mougeotia*. Z. Pflanzenphysiol. *63*, 238–260 (1970)
Ootaki, T., Fischer, E.P., Lockhart, P.: Complementation between mutants of *Phycomyces* with abnormal phototropism. Mol. Gen. Genet. *131*, 233–246 (1974)
Ortega, J.K.E., Gamow, R.I.: An increase in mechanical extensibility during the period of light-stimulated growth. Plant Physiol. *57*, 456–457 (1976)
Ortega, J.K.E., Gamow, R.I., Ahlquist, C.N.: *Phycomyces*: A change in mechanical properties after a light stimulus. Plant Physiol. *55*, 333–337 (1975)
Ortega, J.K.E., Harris, J.F., Gamow, R.I.: The analysis of spiral growth in *Phycomyces* using a novel optical method. Plant Physiol. *53*, 485–490 (1974)
Page, R.M.: Phototropism in fungi. In: Photophysiology. Giese, A.C. (ed.). New York: Academic Press 1968, vol. III, pp. 65–90
Page, R.M., Curry, G.M.: Studies on phototropism of young sporangiophores of *Pilobolus kleinii*. Photochem. Photobiol. *5*, 31–40 (1966)
Perkins, J.H., Gordon, S.A.: Morphogenesis in *Schizophyllum commune*. II. Effects of monochromatic light. Plant Physiol. *44*, 1712–1716 (1969)
Petzuch, M., Delbrück, M.: Effects of cold periods on the stimulus-response system of *Phycomyces*. J. Gen. Physiol. *56*, 297–308 (1970)
Phillips, I.D.J.: Diffusible gibberellins and phototropism in *Helianthus annuus*. Planta *106*, 363–367 (1972)
Pickard, B.G., Dutson, K., Harrison, V., Donegan, E.: Second positive phototropic response patterns of the oat coleoptile. Planta *88*, 1–33 (1969)
Pickard, B.G., Thimann, K.V.: Transport and distribution of auxin during tropistic response. II. The lateral migration of auxin in phototropism of coleoptiles. Plant Physiol. *39*, 341–350 (1964)
Poff, K.L., Butler, W.L.: Absorption changes induced by blue light in *Phycomyces blakesleeanus* and *Dictyostelium discoideum*. Nature (London) *248*, 799–801 (1974)
Poff, K.L., Butler, W.L.: Spectral characterization of the photoreducible b-type cytochrome of *Dictyostelium discoideum*. Plant Physiol. *55*, 427–429 (1975)

Poff, K.L., Butler, W.L., Loomis, W.F., Jr.: Light-induced absorbance changes associated with phototaxis in *Dictyostelium*. Proc. Natl. Acad. Sci. U.S.A. *70* 813–816 (1973)
Presti, D., Hsu, W.-J., Delbrück, M.: Phototropism in *Phycomyces* mutants lacking β-carotene. Photochem. Photobiol. *26*, 403–405 (1977)
Rau, W.: Untersuchungen über die Lichtabhängige Carotinoidsynthese. I. Das Wirkungsspektrum von *Fusarium aquaeductuum*. Planta *72*, 14–28 (1967)
Raven, J.A.: Transport of indoleacetic acid in plant cells in relation to pH and electrical potential gradients, and its significance for polar IAA transport. New Phytol. *74*, 163–172 (1975)
Reichardt, W., Varjú, D.: Eine Inversionsphase der phototropischen Reaktion (Experimente an dem Pilz *Phycomyces blakesleeanus*). Z. Phys. Chem. Neue Folge *15*, 297–320 (1958)
Schmidt, W., Butler, W.L.: Flavin-mediated photoreactions in artificial systems: a possible model for the blue-light photoreceptor pigment in living systems. Photochem. Photobiol. *24*, 71–75 (1976a)
Schmidt, W., Butler, W.L.: Light-induced absorbance changes in cell-free extracts of *Neurospora crassa*. Photochem. Photobiol. *24*, 77–80 (1976b)
Schmidt, W., Hart, J., Filner, P., Poff, K.L.: Specific inhibition of phototropism in corn seedlings. Plant Physiol. *60*, 736–738 (1977)
Schrank, A.R.: Inhibition of curvature responses by shunting the inherent electrical field. Plant Physiol. *25*, 583–593 (1950)
Shen-Miller, J., Gordon, S.A.: Gravitational compensation and the phototropic response of oat coleoptiles. Plant Physiol. *42*, 352–360 (1967)
Shen-Miller, J., Cooper, P., Gordon, S.A.: Phototropism and photoinhibition of basipolar transport of auxin in oat coleoptiles. Plant Physiol. *44*, 491–496 (1969)
Shropshire, W., Jr.: Growth responses of *Phycomyces* to polarized light stimuli. Science *130*, 336 (1959)
Shropshire, W., Jr.: The lens effect and phototropism of *Phycomyces*. J. Gen. Physiol. *45*, 949–958 (1962)
Shropshire, W., Jr.: Phototropism. In: Progress in Photobiology. Proceedings of the VI International Congress on Photobiology. Schenk, G.O. (ed.). Berlin-Heidelberg-New York: Springer 1975, pp. 1–5
Shropshire, W., Jr., Bergman, K.: Light induced concentration changes of ATP from *Phycomyces* sporangiophores: a reexamination. Plant Physiol. *43*, 1317–1318 (1968)
Shropshire, W., Jr., Gettens, R.H.: Light induced concentration changes of adenosine-triphosphate in *Phycomyces* sporangiophores. Plant Physiol. *41*, 203–207 (1966)
Shropshire, W., Jr., Withrow, R.B.: Action spectrum of phototropic tip-curvature of *Avena*. Plant Physiol. *33*, 360–365 (1958)
Sinsheimer, R.L.: Ultraviolet absorption spectra. In: Radiation Biology. Hollaender, A. (ed.). New York: McGraw-Hill 1955, vol. II, pp. 165–201
Song, P-.S., Moore, T.A.: Mechanism of the photodephosphorylation of menadiol diphosphate. A model for bioquantum conversion. J. Am. Chem. Soc. *90*, 6507–6514 (1968)
Song, P-.S., Moore, T.A.: On the photoreceptor pigment for phototropism and phototaxis: is a carotenoid the most likely candidate? Photochem. Photobiol. *19*, 435–441 (1974)
Steiner, A.M.: Dose-response curves for polarotropism in germlings of a fern and a liverwort. Naturwissenschaften *54*, 497 (1967a)
Steiner, A.M.: Action spectra for polarotropism in germlings of a fern and a liverwort. Naturwissenschaften *54*, 497–498 (1967b)
Steyer, B.: Der Phototropismus dicotyler Keimpflanzen (Sammelreferat) Wiss. Z. Univ. Rostock, Math-Naturwiss. Reihe, *5/6*, 493–502 (1965)
Steyer, B.: Die Dosis-Wirkungsrelationen bei geotroper und phototroper Reizung: Vergleich von Mono- mit Dicotyledonen. Planta *77*, 277–286 (1967)
Stiles, W., Cocking, E.C.: An Introduction to the Principles of Plant Physiology, 3rd ed. London: Methuen and Co., Ltd., 1969, pp. 498–524
Sun, M., Moore, T.A., Song, P-.S.: Molecular luminescence studies of flavins. I. The excited states of flavins. J. Am. Chem. Soc. *94*, 1730–1740 (1972)
Thimann, K.V., Curry, G.M.: Phototropism and phototaxis. In: Comparative Biochemistry: A Comparative Treatise. Florkin, M., Mason, H.S. (eds.). New York: Academic Press 1960, vol. I, Sources of Free Energy, pp. 243–309
Thimann, K.V., Curry, G.M.: Phototropism. In: Light and Life. McElroy, W.D., Glass, B. (eds.). Baltimore: The Johns Hopkins Press 1961, pp. 646–670

Thornton, R.M.: Crystalloids of *Phycomyces* sporangiophores: nature and photosensitive accumulation. Plant Physiol. *44*, 861–865 (1969)
Thornton, R.M.: New photoresponses of *Phycomyces*. Plant Physiol. *51*, 570–576 (1973)
Thornton, R.M., Thimann, K.V.: Transient effects of light on auxin transport in the *Avena* coleoptile. Plant Physiol. *42*, 247–257 (1967)
Tu, J.C., Malhotra, S.K.: The significance of cAMP induced alterations in the cellular structure of *Phycomyces*. Can. J. Microbiol. *23*, 378–388 (1977)
Widell, S., Bjorn, L.O.: Light-induced absorption changes in etiolated coleoptiles. Physiol. Plant. *36*, 305–309 (1976)
Wolken, J.J.: Microspectrophotometry and the photoreceptor of *Phycomyces* I. J. Cell Biol. *43*, 354–360 (1969)
Zankel, K.L., Burke, P.V., Delbrück, M.: Absorption and screening in *Phycomyces*. J. Gen. Physiol. *50*, 1893–1906 (1967)
Zenk, M.H.: The action of light on the metabolism of auxin in relation to phototropism. In: Biochemistry and Physiology of Plant Growth Substances. Wightman, F., Setterfield, G. (eds.). Ottawa: The Runge Press 1968 (Proceedings of the 6th International Conference on Plant Growth Substances, July 24–29, 1967), pp. 1109–1128
Zimmerman, B.K., Briggs, W.R.: Phototropic dosage-response curves for oat coleoptiles. Plant Physiol. *38*, 248–253 (1963a)
Zimmerman, B.K., Briggs, W.R.: A kinetic model for phototropic responses of oat coleoptiles. Plant Physiol. *38*, 253–261 (1963b)

5.2 Growth Movements Directed by Gravity

5.2.1 Gravitropism[1] in Single Cells

A. SIEVERS and D. VOLKMANN

1. The Gravitropic Reaction Chain of the *Chara* Rhizoid

The *Chara* rhizoid is a cylindrical cell with an ellipsoid-shaped tip (Fig. 1). Its diameter is 30 μm, and it can reach a length of several cm. The rhizoid grows in a strictly positive orthogravitropic direction, and at room temperature reaches a growth rate of 180 μm/h. The cell grows only at its tip, with no further growth of the cylindrical section (HEJNOWICZ et al., 1977). The arrangement of the cell

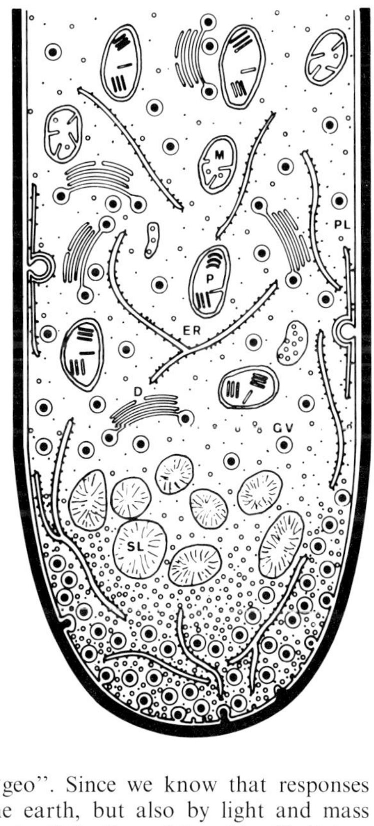

Fig. 1. Distribution of cytoplasmatic components in a vertically growing *Chara* rhizoid, showing polar organization of the cell tip. The dictyosomes producing Golgi vesicles are only located basal to the statoliths. Golgi vesicles are incorporated into the plasmalemma apical to the statoliths. *D*, dictyosome; *ER*, endoplasmic reticulum; *GV*, Golgi vesicle; *M*, mitochondrion; *P*, plastid; *PL*, plasmalemma; *SL*, statolith

---

[1] The prefix "gravi" replaces the hitherto customary "geo". Since we know that responses in organisms are induced not only by the sun, and the earth, but also by light and mass acceleration, we should not only use the term "phototropism" (in lieu of the older "heliotropism") but also "gravitropism". This revised terminology will be of special importance when experiments in space are carried out.

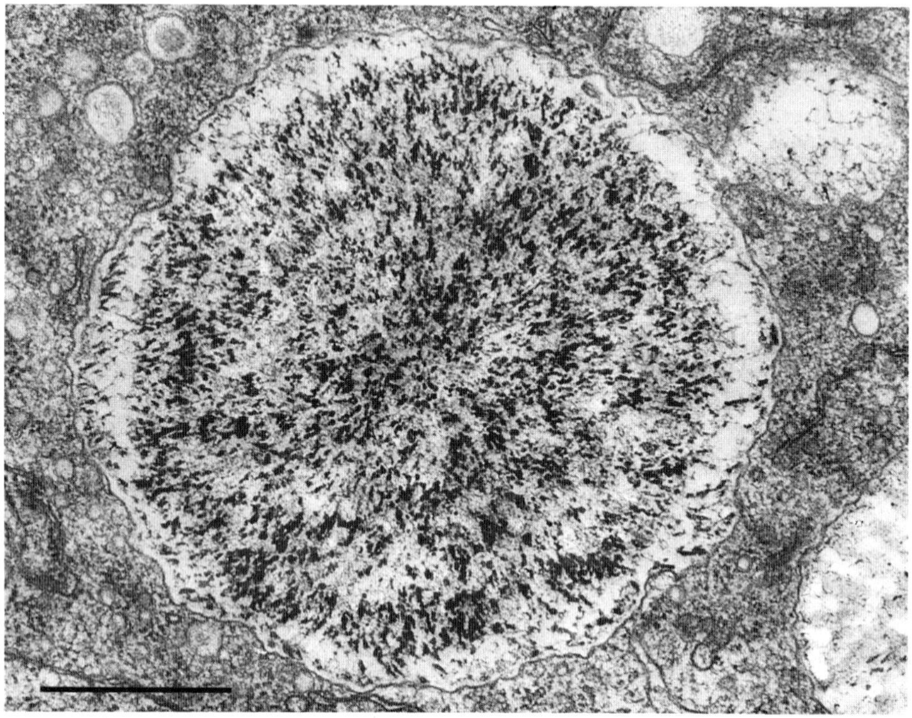

Fig. 2. Statolith of the *Chara* rhizoid: a membranebound compartment filled with electron dense crystallites of $BaSO_4$. Fixation: OsO_4. Scale bar: 0.5 μm

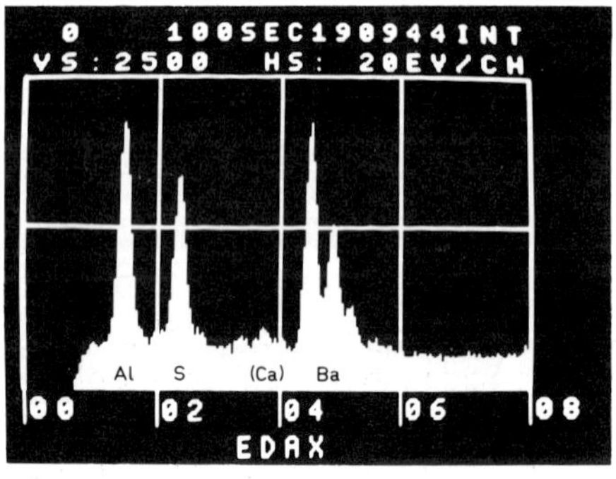

Fig. 3. Energy dispersive X-ray spectrum of a single statolith (*Chara* rhizoid fixed in glutaraldehyde, embedded in araldite, section 1 μm thick, coated with aluminium). The only elements present are sulphur and barium (SCHRÖTER et al., 1975)

organelles in the rhizoid is radially symmetrical and exhibits a clear longitudinal polarity (Fig. 1). Distributed over the cross section of the cell are 50–60 statoliths, 10–20 μm basal to the rhizoid tip. A statolith is a membrane-limited compartment approximately 2 μm in diameter (Fig. 2). It is probably formed from the endoplasmic reticulum (SCHRÖTER et al., 1973). By electron microprobe analysis (Fig. 3), atomic absorption spectrophotometry and selected area electron diffraction, it was proved that the statolith compartments are filled with crystallites of $BaSO_4$ (SCHRÖTER et al., 1975; STEUDLE et al., 1978).

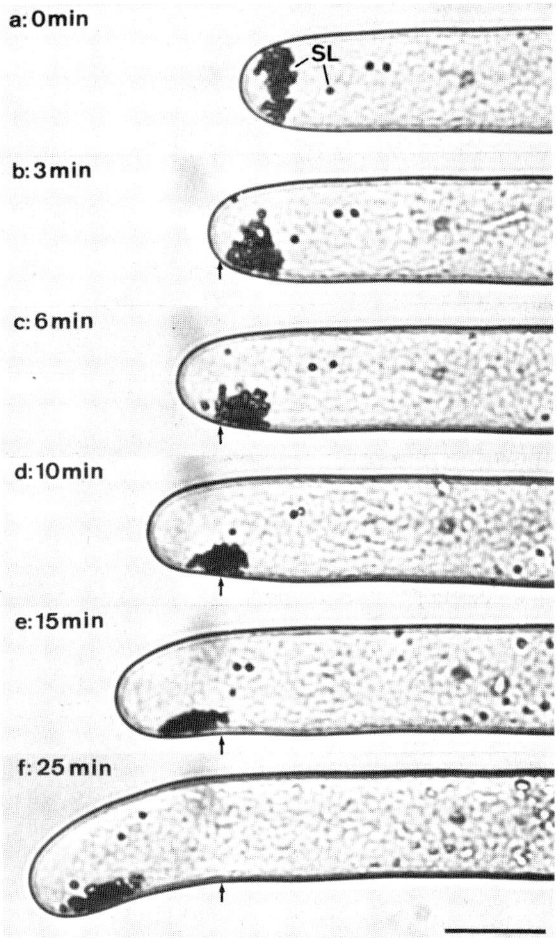

Fig. 4a–f. A series of time-lapse photographs illustrating gravitropic response of a *Chara* rhizoid after deviation from the vertical by 90°. Note sedimentation of the grouped statoliths, followed by downward bending of the rhizoid tip. The *arrow* indicates the same point on the cell wall in each micrograph. Note the reversible local inhibition of cell wall growth where the statoliths sedimented first (indentation in *d*, *arrow*). *SL* statoliths. Scale bar: 30 µm (SIEVERS and SCHRÖTER, 1971)

The polarity of the rhizoid becomes especially visible in the distribution of the Golgi apparatus elements. Dictyosomes are located basal to the statoliths, while the contents of the Golgi vesicles are incorporated only in the apical cell wall region, where they supply substances for cell wall growth. Since the statoliths block the centre of the cell, the Golgi vesicles must be transported to the cell tip in the narrow peripheral region between statoliths and plasmalemma (SIEVERS, 1967b).

When a rhizoid is turned from the vertical into the horizontal position, the statoliths sediment within a few min onto the physically lower sub-apical flank (Fig. 4). Initially the cell continues to grow straight. After a latent period of 10 min however, the tip of the cell shows an asymmetrical lateral outline. After 15 min, a clear curvature can be observed. After 2–3 h the graviresponse is completed, and the statoliths are redistributed normally (SIEVERS and SCHRÖTER, 1971). If the statoliths are translocated from their subapical site into the base by centrifuga-

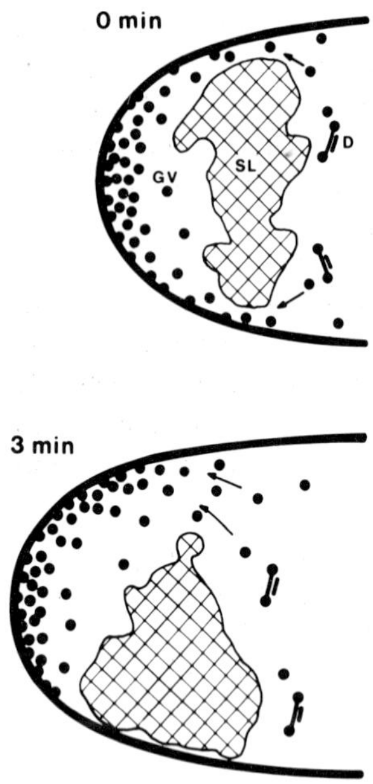

Fig. 5. Scheme of the displacement of statoliths and Golgi vesicles in the apex of the *Chara* rhizoid at the beginning of graviresponse. *D*, dictyosome; *GV*, Golgi vesicle; *SL*, complex of statoliths (SIEVERS and SCHRÖTER, 1971)

tion (6 min, 400–450 **g**; BUDER, 1961), graviresponse does not take place and growth remains unchanged. Statoliths first reappear in the tip of the cell several hours later, and the ability to perform a graviresponse returns simultaneously. This confirms that the statoliths bring about the perception of gravity.

Together with the sedimentation of the statoliths a further disturbance of the radial symmetry of the cell organelles occurs in the sub-apical cross section of the cell: more Golgi vesicles and increased vesicle incorporation into the plasmalemma were observed in the physically upper cell flank than in the lower one (SIEVERS, 1967a and c; Fig. 5). This asymmetry leads to differential growth of the sub-apical flanks and thus results in gravitropic curvature. After a damped oscillation of the growth direction, symmetrical cell growth is reestablished by the renewed symmetrical distribution of the statoliths (SIEVERS and SCHRÖTER, 1971). In the *Chara* rhizoid, therefore, statolith distribution and cell wall growth marked by Golgi vesicle incorporation, are coupled by a very simple feed-back principle. Still unknown, however, are the mechanisms for maintaining statolith position at a fixed distance from the tip and statolith inhibition of Golgi vesicle exocytosis.

Similarly, in *Micrasterias*, the growth of the cell wall is locally inhibited if Golgi vesicles are prevented by centrifugation from being incorporated into the plasmalemma (KIERMAYER and DORDEL, 1976). In this case slime vesicles displace the Golgi vesicles that are necessary for cell wall growth.

It has been proved that statoliths inhibit local cell wall growth when they are sedimented, so that they lie near the plasmalemma. The slight indentation of the cell wall at the underside of the rhizoid in Figure 4d (arrow) is a clear indication of a local inhibition of cell wall growth by the presence of the statoliths. The inhibition occurs exactly where the statoliths are sedimented first. Later it disappears (Fig. 4f). By placing resin spheres on a rhizoid tip, the growth of both flanks can be measured exactly (HEJNOWICZ et al., 1977). After horizontal stimulation the extreme tip and the physically upper, sub-apical flank continue growing, while the growth of the lower flank is immediately inhibited by the sedimented statoliths. If the statoliths happen to be displaced somewhat basipetally–as is often observed–the cell wall area that is now free starts growing again. This growth stops at the latest when the statoliths are relocated to the apical area in the course of gravitropic curvature (SIEVERS et al., 1979). If statoliths, in a vertically growing rhizoid, are forced to sediment to the apex of the cell where maximal cell growth occurs, longitudinal growth is inhibited irreversibly (SIEVERS and SCHRÖTER, 1971).

The dependence of gravitropic curvature of the rhizoid on the rate of centrifugal acceleration (FRIEDRICH and HERTEL, 1973) can readily be explained by the high sensitivity of the growing cell wall area toward statolith sedimentation. Increasing the pressure of the statoliths by 5 or 10 **g** may increase the reversible local inhibition of growth and result in a stronger response.

2. The Graviresponse of the *Phycomyces* Sporangiophore

In the growing negatively orthogravitropic sporangiophore of *Phycomyces* two different responses to mass acceleration have been demonstrated by centrifugation (DENNISON, 1961). One response is a transient but rapid bending, which is caused by the mechanical deformation of the cell, resulting from the action of centrifugal acceleration on the sporangium. The long-term response is a slow but steady bending, which continues until the direction of sporangiophore growth is opposite to the vector of acceleration. This response is caused by an intracellular effect of acceleration. The statoliths in this system have not yet been identified.

References

Buder, J.: Der Geotropismus der Characeenrhizoide. Ber. Dtsch. Bot. Ges. *74*, (14)–(23) (1961)

Dennison, D.S.: Tropic responses of *Phycomyces* sporangiophores to gravitational and centrifugal stimuli. J. Gen. Physiol. *45*, 23–38 (1961)

Friedrich, U., Hertel, R.: Abhängigkeit der geotropischen Krümmung der *Chara*-Rhizoide von der Zentrifugalbeschleunigung. Z. Pflanzenphysiol. *70*, 173–184 (1973)

Hejnowicz, Z., Heinemann, B., Sievers, A.: Tip growth: patterns of growth rate and stress in the *Chara* rhizoid. Z. Pflanzenphysiol. *81*, 409–424 (1977)

Kiermayer, O., Dordel, S.: Elektronenmikroskopische Untersuchungen zum Problem der Cytomorphogenese von *Micrasterias denticulata* Bréb. II. Einfluß von Vitalzentrifugierung auf Formbildung und Feinstruktur. Protoplasma *87*, 179–190 (1976)

Schröter, K., Läuchli, A., Sievers, A.: Mikroanalytische Identifikation von Bariumsulfat-Kristallen in den Statolithen der Rhizoide von *Chara fragilis*, Desv. Planta *122*, 213–225 (1975)

Schröter, K., Rodriguez-Garcia, M.I., Sievers, A.: Die Rolle des endoplasmatischen Retikulums bei der Genese der *Chara*-Statolithen. Protoplasma *76*, 435–442 (1973)

Sievers, A.: Zum Wirkungsmechanismus der Statolithen in der pflanzlichen Zelle. Naturwissenschaften *54*, 252–253 (1967a)

Sievers, A.: Elektronenmikroskopische Untersuchungen zur geotropischen Reaktion. II. Die polare Organisation des normal wachsenden Rhizoids von *Chara foetida*. Protoplasma *64*, 225–253 (1967b)

Sievers, A.: Elektronenmikroskopische Untersuchungen zur geotropischen Reaktion. III. Die transversale Polarisierung der Rhizoidspitze von *Chara foetida* nach 5 bis 10 Minuten Horizontallage. Z. Pflanzenphysiol. *57*, 462–473 (1967c)

Sievers, A., Heinemann, B., Rodriguez-Garcia, M.I.: Nachweis des subapikalen differentiellen Flankenwachstums im *Chara*-Rhizoid während der Graviresponse. Z. Pflanzenphysiol. *91*, 435–442 (1979)

Sievers, A., Schröter, K.: Versuch einer Kausalanalyse der geotropischen Reaktionskette im *Chara*-Rhizoid. Planta *96*, 339–353 (1971)

Steudle, E., Läuchli, A., Sievers, A.: X-ray microanalysis of barium and calcium in plant material: significance for the analysis of statoliths. Z. Naturforsch. *33c*, 444–446 (1978)

5.2.2 Graviperception[1] in Multicellular Organs

D. VOLKMANN and A. SIEVERS

1. Introduction

The reaction chain of gravity-dependent orientation of plant organs can be divided, as is the case with other reaction chains, into perception, transduction, and response (see also BENTRUP, Chap. 1.3, and SHROPSHIRE, Chap. 1.2, this vol.). Numerous reports have been published on this subject since 1962 (AUDUS, 1962, 1969, 1975; WILKINS, 1966, 1976, 1977; BALL, 1969; HOSHIZAKI, 1973; JUNIPER, 1976). The gravitropic orientation of multicellular systems ultimately results—as is the case with unicellular systems—from a growth gradient over the cross section of the responsive organs even though preliminary results (IVERSEN, 1973) support the view that alterations of turgor ought to precede a growth gradient. However, in a multicellular organism, unlike a single cell, integration among cells has to take place. This integration involves perception in specialized cells (statocytes), and also response, in which numerous cells participate. The present article is limited to results which are either concerned directly with perception, or are of importance to this latter. Transduction and response are dealt with in a separate chapter (see WILKINS, Chap. 5.2.3, this vol.).

The main problem of perception is the transformation of a physical signal into a physiological one. In contrast to other stimuli such as light, chemicals, or contact, gravitational force acts on all cells in the same way. Nevertheless, as with other stimuli, a gradient is required at the site of perception, if a directed response is to occur. This gradient must be transduced as a differential spatial signal to the responsive cells, thus resulting in the growth gradient. For analyzing stimulus perception, the following points have to be clarified: the sensitivity of the system, the site of perception, the cell structures taking part in perception, and finally the mechanism of perception.

2. The Sensitivity of Graviperceiving Organs

The sensitivity of an organ or of its stimulus-perceiving system is usually assessed by the minimum stimulation (threshold) necessary to elicit a response. The quantity of stimulation or dose (d) is a function of time (t) and mass acceleration ($\mathbf{a}$) with its components direction or stimulus angle (r) and absolute value (a):

$$d = F(t, r, a).$$

[1] See A. SIEVERS and D. VOLKMANN, Chap. 5.2.1, this vol.

The thresholds of the variables *t*, *r* and *a* (as described in the following sections) demonstrate a high sensitivity of plant organs to gravity, although some methodological problems still have to be considered.

2.1 Threshold Time (t)

The shortest time sufficient to elicit a visible response (*presentation time*) has been measured as 12 s in roots of *Lepidium* (IVERSEN and LARSEN, 1973), 18 s in *Vicia* roots (GRIFFITHS, 1963, after AUDUS, 1975), and less than 30 s in coleoptiles of *Avena* (JOHNSSON, 1971), using a constant angle r=90° and a constant acceleration of **a**=982 cm s^{-2}=1 **g**. The *perception time* is, in contrast, the minimum time of stimulation necessary for the perception of a stimulus even if it does not elicit a visible response. It can be investigated using intermittent stimulation either on clinostats (FITTING, 1905a, b), or by alternating vertical and horizontal exposures (PICKARD, 1973). In *Vicia, Helianthus, Phaseolus,* and *Avena,* stimuli of 0.5 s could be summated to produce one stimulus that induced response (see p. 576). According to PICKARD (1973), accelerations of 0.3 **g** during the change of position can be neglected.

2.2 Threshold Angle (r)

FITTING (1905a, b) found a sensitivity of directional change of 0,5°–10° for *Vicia, Helianthus,* and *Phaseolus*. The optimum angle of stimulation is about 90° deviation from the vertical orientation, according to the sine rule. As a consequence, the differential sensitivity (i.e., increment in sensitivity over increment in angle of stimulation) is highest at small angles (2°–5°) and lowest at large angles (85°–90°).

Deviations from the sine rule with an optimum angle of stimulation above 90° were always obtained with longer-lasting stimulations considerably above the threshold (LARSEN, 1962b, 1969; IVERSEN and LARSEN, 1971; PICKARD, 1973; PERBAL, 1974, 1977; PERBAL and PERBAL, 1976). With such stimulations perception, transduction, and response overlap each other. Thus, it is not possible to draw conclusions on perception alone.

2.3 Threshold Value (a)

The threshold of the absolute value was investigated under two different experimental conditions. (1) In order to obtain unstimulated plants, gravity was compensated for by rotation of the plant about its horizontal axis at 1–2 rpm. Defined accelerations were then achieved by additional faster rotation around a vertical axis (SHEN-MILLER et al., 1968). (2) SOBICK and SIEVERS (1979) worked with a fast rotating clinostat after MULLER (1958). Advantage is taken of the fact that, during rapid rotation (55–120 rpm) about a horizontal axis neither acceleration due to gravity nor to centrifugation occurs right in that axis. Thus, sedimentable particles are distributed statistically, whether in an ideal liquid or in a plant cell (SIEVERS and VOLKMANN, 1977b). In principle, the results confirm the value already found by

CZAPEK of $\mathbf{a} \approx 10^{-3}$ **g**. *Avena* coleoptiles, in which $10^{-3}\,\mathbf{g} < \mathbf{a} < 10^{-2}\,\mathbf{g}$ (SHEN-MILLER et al., 1968) were reported to be less sensitive than roots, in which $10^{-4}\,\mathbf{g} < \mathbf{a} < 10^{-3}\,\mathbf{g}$ (SHEN-MILLER et al., 1968, *Avena;* SOBICK and SIEVERS, 1979, *Lepidium*). However, the experiments were always carried out under continuous stimulation and therefore these values can hardly be considered as absolute threshold doses.

2.4 Stimulation Quantity (d)

The value for the quantity of stimulation has routinely been examined with the validity of the reciprocity rule

$\mathbf{a} \cdot t = d$ $\mathbf{a}$ = mass acceleration in **g**
t = time in s
d = dose in $\mathbf{g} \cdot \mathrm{s}$ (reciprocity constant)

(BÜNNING, 1959; ANKER, 1962; LARSEN, 1962b). Working with *Avena* coleoptiles, JOHNSSON (1965) and SHEN-MILLER (1970) confirmed the validity of the reciprocity rule over a wide range from 0.09 **g** to 12.5 **g**. The dose for the stimulation quantity has been reported by JOHNSSON (1965) to be 240 **g**·s (22.5° C) and by SHEN-MILLER (1970) to be 120 **g**·s (27° C). Different experimental temperatures explain the differences between the values found (ANKER, 1962; JOHNSSON, 1965). These values were obtained by extrapolation to zero response from the response curve where the curvature degree was plotted against the stimulation time. However, JOHNSSON (1971) showed that the response curve is biphasic: the derivative of it changes near zero response. This results in a threshold stimulation time of less than half a minute with 1 **g** (see p. 574). Therefore, the true threshold value for the quantity of stimulation must be much lower than the extrapolated values of 240 **g**·s and 120 **g**·s, respectively.

2.5 Criticism of Methods

Experiments on clinostats. In all investigations using threshold values, the stimulation must be removed some time after its beginning. The gravitational field of the earth, by contrast, cannot be extinguished but can only be compensated for with the aid of clinostats. Thereby the unilateral stimulus is replaced by multilateral stimuli, the effect of which is extinguished with regard to the visible response. The plant, therefore, is under nondirected constant stimulation. This causes physiological and structural changes.

Avena roots respond to this constant stimulation with increased growth and with increased respiration (DEDOLPH et al., 1965, 1967). In *Lepidium* roots that have been clinostated, response to gravitropic stimulation starts earlier, but the growth rate is not changed (HENSEL and SIEVERS, 1976). Coleoptiles react after clinostating with an increased gravitropic sensitivity: the graviresponse is three times larger than that in the control, whereas the growth rate remains equal (DEDOLPH et al., 1965). The coleoptiles also show a stronger phototropic response after horizontal rotation than do control plants, that rotated in the vertical position (Shen-Miller and Gordon, 1967). Tomato plants produce twice as much ethylene during the first 2 h of treatment on the horizontal clinostat (Leather et al., 1972).

Remarkable structural alterations in gravity-perceiving cells have been found by Hensel and Sievers (1976) in *Lepidium* roots (Fig. 8f). The polar arrangement of cell structures, a characteristic feature of these cells (see p. 583), is lost if the plants, after cultivation in the vertical orientation, are rotated horizontally for 2 h at 2 rpm. The ER cisternae, normally located at the distal cell pole, are distributed at random. The starch content of the amyloplasts decreases. The number of lysosomes and the size of the lipid droplets increase. By contrast, in *Avena* coleoptiles the starch content of the amyloplasts is increased (Hinchman and Gordon, 1974); this might be an explanation for the increased sensitivity of the coleoptiles mentioned above. If the plants are rotated on a slow or fast rotating clinostat right from the beginning of seed-soaking, the perceiving cells show no alterations in polarity (Sievers et al. 1976; cf. also Fig. 8c).

Parameters of Measurement. The visible response is the measurable parameter for all investigations on threshold values, but this is an effect that can be registered only a certain time after the beginning of perception and transduction. The response is influenced by numerous unknown factors (Fitting, 1905a, b) and does not allow direct conclusions about perception (Bünning, 1959; Anker, 1962). Johnsson (1965, 1971), therefore, does not limit his investigations to the visible response, but also studies the potential differences between the opposite flanks, measurable prior to response (see Wilkins, Chap. 5.2.3, this vol.).

Perception Time. In measurements of the perception time (see p. 574) there is the additional problem of stimulus summation (Bünning, 1959; Larsen, 1962a, b). Summation is generally possible under certain conditions. To make a statement about the perception time, both the intervals between the single stimuli and the stimuli themselves should be chosen such that after stimulus summation, the same response rate occurs as after the presentation time with a continuous stimulus. This condition is largely satisfied with Fitting's (1905a, b) and Pickard's (1973) experiments.

Cytological Facts. In model systems (Audus, 1964; Perbal, 1974) and experiments on clinostats (Johnsson, 1965, 1971), generalized assumptions are frequently made; e.g., about cell shape or the position of cell structures, that do not correspond to cytological facts (Sievers and Volkmann, 1972). This can lead to wrong conclusions.

2.6 Conclusions

Although the variety of experimental methods, plants, and organs makes it difficult to evaluate the results quantitatively, it is certain that plant organs possess an extremely high sensitivity towards mass acceleration. The cell structures participating in the perception of the stimulus (see p. 577ff.) have very little space and time in which to transform the physical into the physiological signal (Sievers and Volkmann, 1977a, b).

3. Site of Perception

Investigations on the site of perception (LARSEN, 1962b) normally entail the search for sedimentable cell particles, grouped under the general term "statoliths". The most commonly found candidates for the statolith role are amyloplasts; occasionally crystals are found. A clear statement on the site of perception is possible only if the phases following perception are not influenced by the experimental conditions. The evidence for this is an unchanged growth rate or even better an undisturbed phototropic reaction under the conditions of the experiment (WILKINS, 1966).

3.1 Primary Root

JUNIPER et al. (1966) removed the root cap (calyptra) of *Zea* with such great care that the growth of the roots was not affected. The de-capped roots failed to show any gravitropic response, clearly demonstrating that the primary root perceives the gravitational stimulus in the calyptra. Resection experiments with *Pisum* (KONINGS, 1968; ZINKE, 1968), *Hordeum* (CERČEK, 1970), *Lens* (PILET, 1971, 1972), and *Triticum* (BARLOW, 1974), with corresponding results, allow a generalization of this important finding. The response, in contrast, is localized in the subapical elongation zone (LARSEN, 1962b; PILET et al., 1969; SHEN-MILLER et al., 1978). This implies transduction of the signal via many cells (see WILKINS, Chap. 5.2.3, this vol.).

3.2 Coleoptile, Hypocotyl, Shoot, and Grass Node

The site of perception is not yet known for all negatively gravitropic organs (ANKER, 1962). The presentation time of *Avena* coleoptiles, for example, is not influenced after removal of the apical 5 mm, and the gravity-induced potential difference in *Zea* (see WILKINS, Chap. 5.2.3, this vol.) remains constant even after removal of the apical 20 mm (JOHNSSON et al., 1971).

If the peripheral cell layers of *Helianthus* hypocotyls are removed, the peeled hypocotyls are no longer capable of responding, whereas the growth rate is still unchanged (FIRN and DIGBY, 1977). The site of perception is, therefore, assumed to be in the peripheral cell layers. Gravitropic response in the *Helianthus* hypocotyls appears simultaneously in all regions along the length of the organ (FIRN et al., 1978; KÖHLER, 1978). Thus, there is no evidence for the longitudinal signal transduction which had frequently been assumed to occur. This observation supports the suggestion that, in sunflower hypocotyls, the sites of graviperception and graviresponse may be intimately associated, as has been found in the grass node (OSBORNE and WRIGHT, 1977; WRIGHT and OSBORNE, 1977).

4. Cell Structures Participating in Perception

4.1 Perception Without the Participation of Statoliths

For more than 30 years the theory of the geoelectric effect has been considered as an alternative to the statolith theory (BRAUNER, 1962). However, the geoelectric effect is not part of perception. The potential difference measurable directly upon stimulation is a purely physical phenomenon, appearing under anoxia and even in dead tissue (GRAHM and HERTZ, 1962; BRAUNER et al., 1964). The potential difference between the opposite flanks measurable 15 min after the beginning of stimulation has been attributed to the difference in auxin content (GRAHM and HERTZ, 1964; WILKINS and WOODCOCK, 1965). Therefore, potential difference as a result of gravistimulation is part of the transductory or motor phase rather than of perception (see WILKINS, Chap. 5.2.3, this vol.).

AUDUS (1962) and WESTING (1971) discuss the pressure of the cytoplasm on the plasmalemma as a trigger for perception. However, the pressure in the range of 1 dyne cm^{-2} probably is too low to influence the cell membrane mechanically.

KUTSCHERA-MITTER (1971 a, b) assumed the root to react by a mechanism involving differential supply of water molecules in the gravitational field. However, this hypothesis does not need to be discussed any further, as graviresponses can also be induced under water (RUFELT, 1962; HABERKORN and SIEVERS, 1977).

Thus, there is no serious evidence for graviperception without statoliths, with the exception of the physical deformation of the *Phycomyces* sporangiophore that induces a rapid curvature (see SIEVERS and VOLKMANN, Chap. 5.2.1, this vol.).

4.2 Perception with the Participation of Statoliths

4.2.1 Theoretical Considerations on Sedimentation

Only intracellular particles can be responsible for triggering perception. Whether or not mass acceleration can result in a stratification of these particles according to their density, is determined after EINSTEIN by the BOLTZMANN distribution

$$n/n_o = e^{-V\Delta\rho \mathbf{a} h/kT},$$

whereby n/n_o gives the relationship of the number of particles in the distribution related to height h and o.

V = particle volume,
$\Delta\rho$ = difference in density between particles and medium,
a = mass acceleration,
k = BOLTZMANN constant,
T = absolute temperature.

Under cellular conditions: $1\ \mu m < h < 10\ \mu m$; $0.1\ g \cdot cm^{-3} < \Delta\rho < 1.7\ g \cdot cm^{-3}$, and for mass acceleration: $\mathbf{a} = 982\ cm\ s^{-2}$ (=1 **g**), POLLARD (1971) and MARKL (1977) calculated that the radius of spherical particles has to be at least 0.1 μm in order to obtain stratification. The conclusion is that microscopically visible cell organelles such as nuclei, plastids, mitochondria, dictyosomes, and even nucleoli and starch grains in amyloplasts can sediment under 1 **g** condition, but ribosomes cannot. In addition, as the time factor plays a decisive part in the perception of the gravitational stimulus (see p. 574), the speed of sedimentation is of particular importance. It is determined after STOKE's law for spherical particles in a NEWTONIAN liquid that

$$4/3\ \pi r^3 \Delta\rho \mathbf{a} = 6\pi\eta r v.$$

i.e., r = particle radius,
η = plasma viscosity,
$\Delta\rho$ = difference in density between particle and medium,
a = mass acceleration,
v = speed of sedimentation.

In his calculations, AUDUS (1962) chose 20 centipoise for the plasma viscosity (after HEILBRONN, 1914). Accordingly, at **a** = 1 **g** amyloplasts (density 1.5 g · cm^{-3}) need 36 s to cover a distance that corresponds to their diameter (for more realistic data see p. 593). Mitochondria (density: 1.2 g · cm^{-3}) would need 6 min for such a distance. This seems to exclude mitochondria as statoliths, at least in those cases where high sensitivity has been found, i.e., a presentation time in the range of seconds.

4.2.2 Cytological Prerequisites for Sedimentation

However, such theoretical calculations can only be rough estimates, because cytological parameters were not considered. A lower viscosity value, e.g., of 3 instead of 20 centipoise, corresponding to that of the blood at 37° C, would increase the sedimentation speed almost sevenfold. Apart from this, ideal conditions such as spherical shape of the particles and NEWTONIAN properties of the medium are taken as a basis in such calculations. Both conditions are necessary for the application of STOKE's law. Inside the cell these conditions are practically nonexistent. Therefore, the sedimentation of cell particles will be influenced, in addition, by the following factors:
Structure of the ground cytoplasm,
cytoplasmic streaming,
compartmentalization and electrostatic forces.

Structure of the Ground Cytoplasm; Cytoplasm Viscosity. Unlike the above mentioned condition described by STOKE's law, the ground cytoplasm does not behave like a NEWTONIAN liquid. It possesses properties such as elasticity and compressibility, and its viscosity does not correspond to that of an ideal liquid. This delays the sedimentation of cell particles. Numerous earlier experiments clearly show that there must be filamentous structures in the ground cytoplasm influencing its physical qualities, mainly its viscosity (FREY-WYSSLING, 1955). In the cases of *Amoeba proteus* (POLLARD and ITO, 1970) and *Physarum polycephalum* (ISENBERG and WOHLFARTH-BOTTERMANN, 1976) it is known that the plasma viscosity is influenced by the transformation from globular to filamentous actin units and vice versa (see also WAGNER, Chap. 2.2, this vol.).

Cytoplasmic Streaming. Active cytoplasmic streaming also delays or prevents the separation of particles. However, this streaming has not been observed to date in graviperceiving cells. Considering the distinct polar arrangement of cell structures, cytoplasmic streaming cannot be expected to occur in cells of root caps (see p. 583). It may exist, however, in strongly vacuolated statocytes of negatively gravitropic organs.

Compartmentalization – Electrostatic Forces. While the structure of the ground cytoplasm mainly influences the sedimentation *speed* of particles, compartmentalization could even determine the *pathway* of sedimentable particles around obstacles such as vacuoles, nuclei, or ER membranes. All root-cap cells investigated so far, which are assumed to perceive gravity, show small vacuoles (SIEVERS and VOLKMANN, 1977a). As these easily collapse after fixation with $KMnO_4$, they are difficult to identify (however, see JUNIPER, 1976). They probably can influence the sedimentation pathway of particles only slightly. However, the large central vacuoles of the negatively gravitropic organs—coleoptiles and grass nodes (OSBORNE and WRIGHT, 1977; WRIGHT and OSBORNE, 1977),—do not permit sedimentation along the shortest pathway; rather, sedimentation is restricted to the thin cytoplasmic layer near the cell wall. Furthermore, in the statocytes of the root cap, sedimentation is stopped by the endoplasmic reticulum. This process may be influenced by electrostatic forces between the membranes of the two organelles (see pp. 585, 594).

4.2.3 Amyloplasts as Statoliths

Since the formulation of the statolith theory, the possibility that sedimentable amyloplasts ("statolith starch") function in gravity perception has been under consideration (WILKINS, 1966; WESTING, 1971; SIEVERS and VOLKMANN, 1972; SHEN-MILLER and HINCHMAN, 1974; AUDUS, 1975; JUNIPER, 1976). Convincing evidence for this theory has recently emerged from experiments involving: (1) mutants, (2) hormonal degradation of starch in the statoliths, and (3) removal of statolith starch-containing tissues. The recent work has in part resolved the contradictory conclusions of earlier destarching experiments.

Amyloplasts of Mutants. Mutants have proved to be especially valuable for investigations of stimulus perception. This is true not only for phototropism (DENNISON, Chap. 5.1.2, this vol.) but also for gravitropism. The maize strains "wild type" and "amylomaize" differ in the size of their amyloplasts (HERTEL et al., 1969; FILNER et al., 1970). The magnitudes of all investigated effects of gravity: percentage and speed of sedimentation, lateral auxin transport, and degree of curvature of coleoptiles are strictly correlated with the size of the amyloplasts. "Wild type", which possesses larger amyloplasts than the "amylomaize" mutant, always shows larger gravity effects. These results are particularly convincing, since measurements were made on effects that appear prior to the visible response as well as on the curvature itself. Furthermore, it is possible to increase lateral auxin transport by increased mass acceleration (10 **g**) in both strains of maize. Finally, the response system seems to be identical in both strains, as concluded from identical phototropic reaction (FILNER et al., 1970).

Destarching of Amyloplasts in Intact Organs. The plant hormones gibberellin and kinetin are involved in the degradation of starch (JONES, 1973). Destarching experiments with coleoptiles and roots have been carried out using these hormones (PICKARD and THIMANN, 1966; IVERSEN, 1969, 1974; PILET and NOUGARÈDE, 1971; NOUGARÈDE and PILET, 1971; PILET et al., 1971; NOUGARÈDE et al., 1973). At first sight the results are contradictory. *Avena* coleoptiles that are reported not to contain any starch after hormone treatment curve in response to continuous stimulation after 7 to 10 h. The rate of curvature relative to the growth rate is almost identical for hormone-treated coleoptiles and for controls (PICKARD and THIMANN, 1966). IVERSEN (1974) found the opposite result, using the same organism and experimental arrangement. If the coleoptiles are incubated with hormones at 30° C, as in the work of PICKARD and THIMANN, some starch is left in the plastids.

After 24 h of continuous stimulation these coleoptiles show a clear response which, however, is considerably slower than that of the controls. At an incubation temperature of 34° C, the coleoptiles are completely deficient in starch and, correspondingly, show no gravity response.

Similarly, IVERSEN (1969, 1974) treated roots of *Lepidium* and *Trifolium* with gibberellin and kinetin. At an incubation temperature of 35° C, the roots were free of starch after 29 h. If such roots were placed in a horizontal position for 30 min and then rotated for 3 h on a horizontal clinostat they did not react gravitropically, in contrast to the controls. The growth rates were 0.48 $mm \cdot h^{-1}$ for the treated roots and 0.64 $mm \cdot h^{-1}$ for the controls. After exposure to light the roots regenerated starch after 20–24 h. Simultaneously, response to gravity reappeared. Thus, the only serious argument against the statolith theory (PICKARD and THIMANN, 1966) has lost its significance.

Removal of Starch-Containing Organs. After removal of the root cap primary roots of *Zea* and *Triticum* do not react gravitropically at first. The capability for graviresponse, however, returns in both plants before a new root cap has regenerated. After the removal of the root cap, amyloplasts develop from proplastids in the cells of the root apex which obviously could take over the statolith function (BARLOW, 1974; BARLOW and GRUNDWAG, 1974).

4.2.4 Non-Starch-Containing Particles

After electron microscopy came into use, special attention was given to mitochondria and to dictyosomes by SHEN-MILLER (1972); SHEN-MILLER and MILLER (1972a, b); SHEN-MILLER and HINCHMAN (1974). These authors found a differential distribution of both organelles in the avascular tip cells of *Avena* coleoptiles after gravitropic stimulation. The differential distribution of dictyosomes is said to start after 6 min, that of mitochondria somewhat later. Clearly, neither organelle sediments because of its density (mitochondria: 1.2 $g \cdot cm^{-3}$, AUDUS, 1962; dictyosomes: 1.125 $g \cdot cm^{-3}$, MORRÉ et al., 1974). Instead, they can only be shifted passively due to sedimentation of the denser amyloplasts.

Dictyosomes in the cells of the root tip of *Pisum* do not show a clear sedimentation even at 20.000 **g** after 1 h at 6° C (BOUCK, 1963a, b). Neither in the *Chara* rhizoid (SIEVERS, 1971) nor in the statocytes of *Lepidium* (SIEVERS and VOLKMANN, 1972) can a sedimentation of dictyosomes be observed at 1 **g**. Thirty min after the onset of stimulation, a differential activation of the dictyosomes has been observed to start (SHEN-MILLER and MILLER, 1972b), reportedly caused by "differentially sensitive regions of the plasmalemma" (cf. Sect. 5.1). This is, of course, too slow to be a perception process, since the presentation time for graviresponse is in the range of a few seconds.

4.2.5 Ultrastructure of Statocytes

A common feature of gravity-perceiving cells (=statocytes) is the presence of sedimentable amyloplasts (approx. 3 μm in diameter; approx. 25 per statocyte). Statocytes of negatively gravitropic organs (grass nodes: OSBORNE and WRIGHT, 1977; WRIGHT and OSBORNE, 1977) possess a large central vacuole just like other differentiated plant cells (Fig. 1). The amyloplasts sediment in the thin cytoplasmic

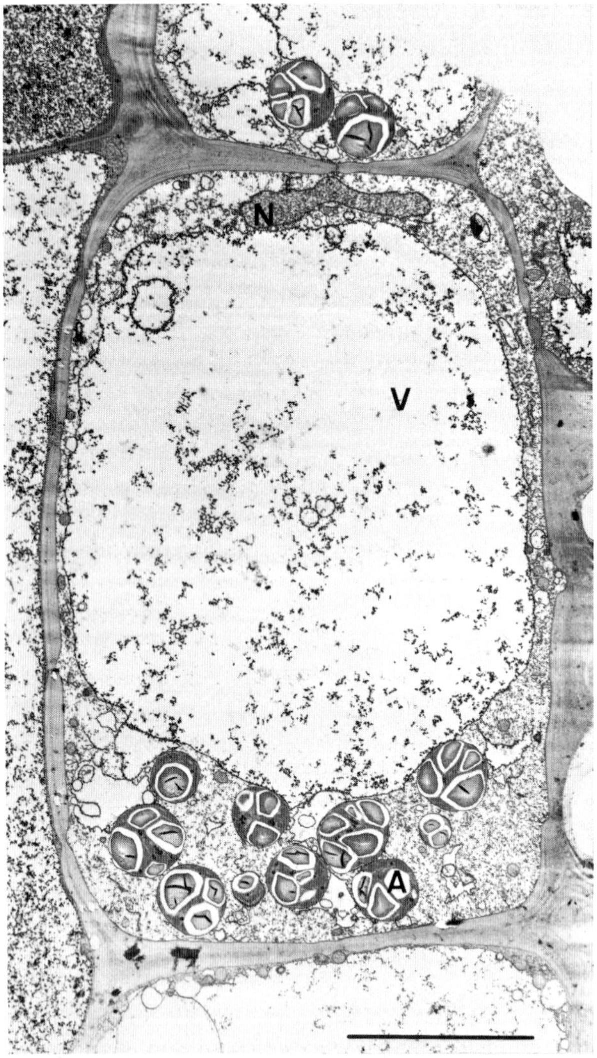

Fig. 1. Statocyte of a grass node showing amyloplasts at the base and the nucleus at the apex of the cell. Note the large vacuole and the paucity of other organelles. *A*, amyloplast; *N*, nucleus; *V*, vacuole. Fixation: glutaraldehyde + OsO_4. Scale bar: 10 µm. Photograph by Dr. MARY PARKER. Courtesy of OSBORNE and WRIGHT (1977)

layer, since the vacuole blocks the shortest pathway of sedimentation. Therefore, a random distribution of the amyloplasts in the statocyte—such as achieved by rotation on the clinostat—is not possible. In the regions where the amyloplasts are located, the cytoplasmic layer is thicker than elsewhere in the cell. This is also true in statocytes that have had their orientation changed, e.g., from vertical to horizontal (see Figs. 10–13 in OSBORNE and WRIGHT, 1977). Whether the ground cytoplasm itself sediments, or whether it is shifted by the sedimenting amyloplasts remains open. The plasmalemma would seem to be the most likely substratum for the sedimenting amyloplasts, although contact between the two cannot be seen in the electron microscopic photographs available so far. No specific cell structures have been observed in this area.

In the root cap, however, the statocytes have distinctive ultrastructural features. Various plants of different systematic groups have been investigated so far:

Apiaceae:	
Daucus carota L.,	VOLKMANN, 1974
Brassicaceae:	
Lepidium sativum L.,	IVERSEN, 1969
	SIEVERS and VOLKMANN, 1972, 1977a, b
	SIEVERS and SCHMITZ, 1973
	IVERSEN, 1974
	VOLKMANN, 1974, 1976
	VOLKMANN and SIEVERS, 1975
	SIEVERS et al., 1976
	SCHMITZ, 1977
Raphanus sativus L.,	SCHMITZ, 1977
Caryophyllaceae:	
Vaccaria pyramidata Med.,	SCHMITZ, 1977
Convolvulaceae:	
Convolvulus arvensis L.,	PHILLIPS and TORREY, 1974
Fabaceae:	
Genista tinctoria L.,	VOLKMANN, 1974
Lens culinaris Med.,	PERBAL, 1971, 1977
	VOLKMANN, 1974
Pisum sativum L.,	JUNIPER, 1972
	GRUNDWAG (after JUNIPER and FRENCH, 1973)
Vicia faba L.,	AUDUS, 1962,
	GRIFFITHS and AUDUS, 1964
Liliaceae:	
Allium cepa L.,	VOLKMANN, 1974
Pinaceae:	
Picea abies L.,	HESTNES and IVERSEN, 1978
Poaceae:	
Zea mays L.,	JUNIPER and FRENCH, 1970 and 1973

The statocytes of nearly all these roots are characterized by their polar arrangement of cell organelles. This can be seen clearly in *Lepidium*, the root cap of which is built up of a precise geometrical array of parabolically shaped cell layers (Fig. 2). The statenchyma (i.e., tissue consisting of statocytes) makes up four of these layers. The cells of the statenchyma and those of the peripheral secretory tissue of the root cap originate from the meristem in a distal direction. In a meristematic cell the nucleus is situated in the center (Fig. 3a). It is surrounded by the other cell organelles which do not show a specific spatial orientation or arrangement.

During early germination of the seed the embryonic radicle is at first nongravitropic, but soon develops into a gravitropic root. Simultaneously there arises a striking polar arrangement of cell organelles within the statocytes (Figs. 3b, 8a). This gradient does not depend on the density of the organelles (BOUCK, 1963a). Both the development of the root and the special development of its statocytes are inhibited by abscisic acid (Fig. 8b; VOLKMANN, 1976).

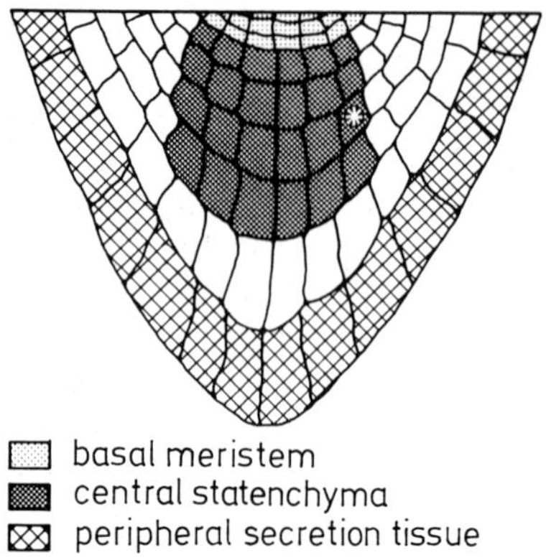

Fig. 2. Diagram of a longitudinal section through the root cap of *Lepidium*. Note the number and geometry of the cell layers and the topography of the tissues. Periclines run parallel and anticlines perpendicular to the outline and to the parabolic layers. The *asterisk* denotes the cell shown in Figure 4

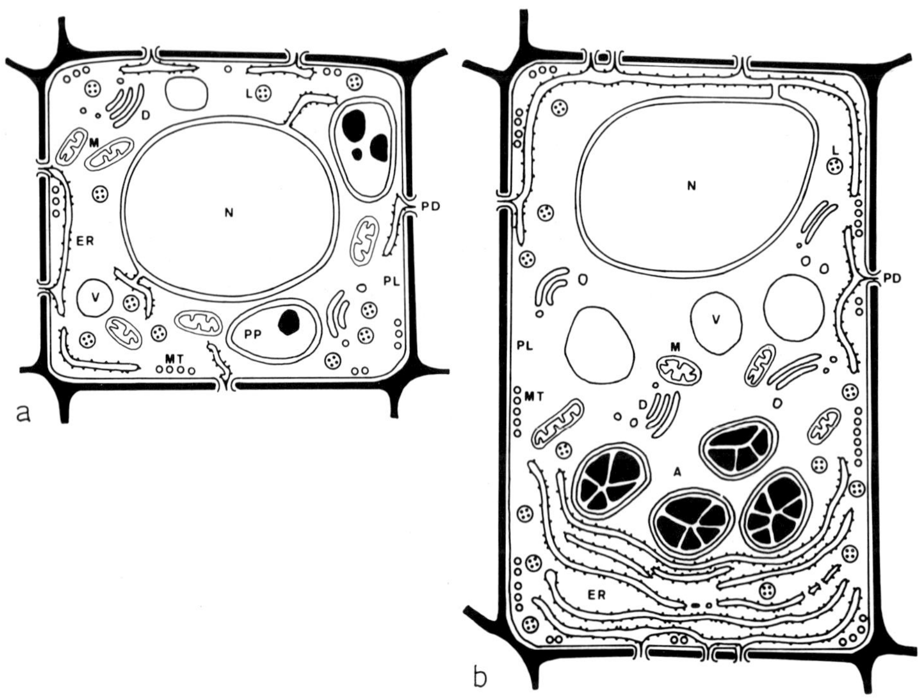

Fig. 3a and b. Diagrams of a meristematic cell (**a**) and of a central statocyte (**b**) of the root cap of *Lepidium*. The meristematic cell is isodiametrical. Its nucleus is situated in the center; all the other organelles are distributed randomly, including the proplastids. The statocyte is elongated in the direction of the root axis, and it is characterized by a polar arrangement of organelles. The nucleus represents the proximal cell pole, the ER complex the distal one. The amyloplasts are sedimented on the ER complex. *A*, amyloplast; *D*, dictyosome; *ER*, endoplasmic reticulum; *L*, lipid droplet; *M*, mitochondrion; *MT*, microtubule; *N*, nucleus; *PD*, plasmodesma; *PP*, proplastid; *V*, vacuole

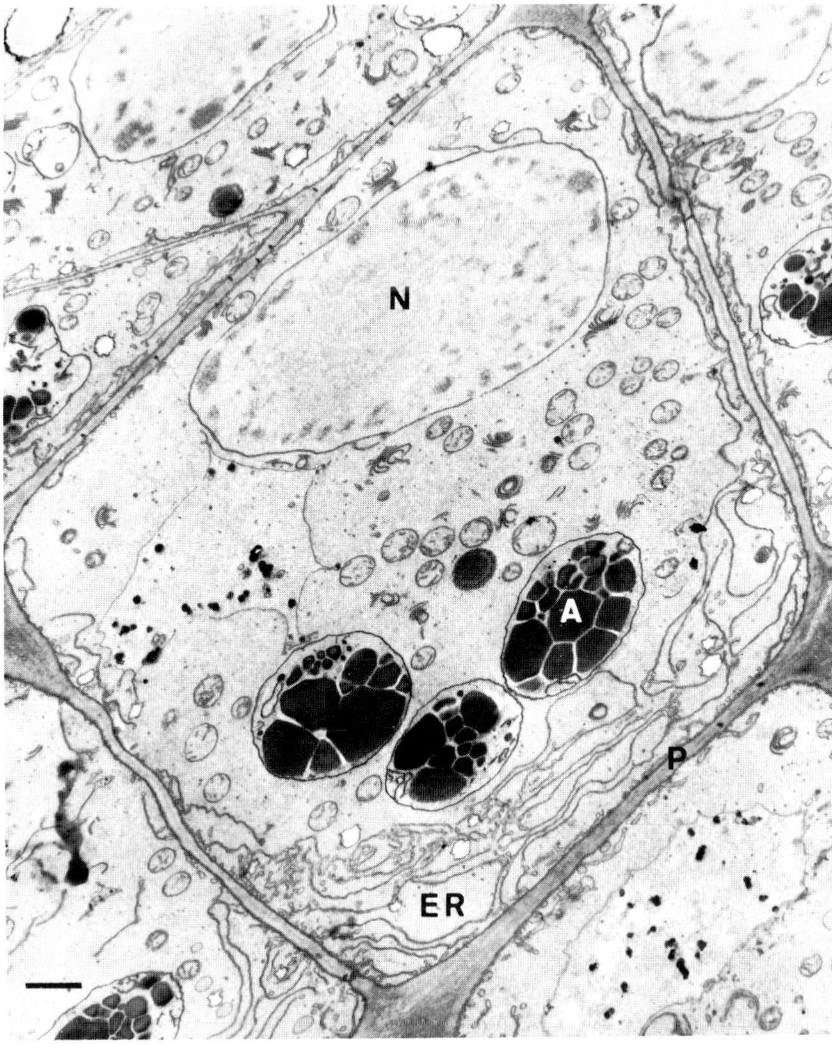

Fig. 4. Lateral statocyte (see Fig. 2) of a *Lepidium* root. The periclines and the ER complex form an angle of approximately 55° to the root axis. *A*, amyloplast; *ER*, ER complex; *N*, nucleus; *P*, periclinal cell wall. Fixation: $KMnO_4$. Scale bar: 1 μm

Near the distal cell pole, a multiple layer of cisternae of the rough endoplasmic reticulum (ER complex) is formed, onto which the amyloplasts sediment (Figs. 3, 4; SIEVERS and VOLKMANN, 1972). The nucleus is located near the proximal cell wall. Other cell organelles such as small vacuoles, mitochondria and dictyosomes are distributed at random as in the meristematic cell. Microtubules and lipid droplets are located near the plasmalemma. Further ultrastructural components of the ground cytoplasm such as microfilaments have not yet been found.

There are more plasmodesmata in the periclinal than in the anticlinal cell walls (in *Zea* 10:1, JUNIPER and BARLOW, 1969). As each plasmodesma contains an ER tubule, not only are all the root cells integrated into a symplast, but all the ER complexes of the statocytes are connected to each other. Therefore, the whole ER, including that of the statenchyma, represents in effect an intracellular and intercellular compartment. It may function as a communicating system that is independent of cell borders.

The lateral cells of the statenchyma are of special significance to graviperception (Fig. 4). Due to the parabolic shape of the cell layers, the periclinal wall of each lateral cell forms an angle of approximately 55° with the root axis. Since the distal ER complex lies along the periclinal cell wall, its orientation is also tilted by the same angle (see Fig. 5).

5. Mechanisms of Graviperception

As has been shown in the preceding sections, the statolith function of amyloplasts is now well established. This raises the question of the primary mechanism of statolith action within the cell. Since cytological and physiological data are more complete for roots than for negatively gravitropic organs, roots will be treated in detail in the following sections. The following alternative assumptions can be made about the primary action of the statoliths upon gravitropic stimulation by horizontal exposure:

1. The *new distribution* of the statoliths *after* gravity-induced translocation is the important factor.

2. The *sliding* of the statoliths along cytoplasmic structures *during* translocation is the intracellular stimulus.

3. Whether or not true translocation occurs, the critical factor is the change in amount and direction of *pressure* exerted on the *original* site of sedimentation, i.e., on a sensitive structure.

There are experimental results in favor of each of these models, but no final proof is available. If, however, the models are judged according to their efficiency in explaining sensitivity, signal transduction, and regulation, the third model is the most compelling.

5.1 New Distribution of Statoliths as a Result of Gravistimulation

In 1902, HABERLANDT proposed that, after an organ is moved out of the vertical position, the amyloplasts reach a sensitive region where they trigger graviperception. This idea was taken up again in a speculative article by AUDUS (1971). Accordingly, the plasmalemma of the anticlinal cell walls in roots and of the periclinal cell walls in shoots, respectively, should be differently sensitive towards the pressure of the sedimenting amyloplasts.

Several models are based on the existence of such a sensitive plasmalemma, they all deal with negatively gravitropic organs, viz. coleoptiles (HAGER, 1971; SHEN-MILLER and HINCHMAN, 1974) and grass nodes (OSBORNE and WRIGHT, 1977;

WRIGHT and OSBORNE, 1977). However, the assumption of a sensitive region (AUDUS, 1971) is not proved.

A variation of this hypothesis is proposed for roots, it postulates the existence of a "plasmodesmatal statolith valve system" (JUNIPER, 1976, 1977). This suggestion is based on the large number of plasmodesmata in the statocytes (JUNIPER and BARLOW, 1969). Wherever the statoliths are sedimented, the plasmodesmata are closed like valves by the pressure of the statoliths on the ER cisternae. Thus, the symplastic intercellular transport of some substance is blocked locally. Accordingly, in vertically oriented roots, transport in the axial direction is blocked, whereas in horizontally oriented roots the transverse transport is blocked. Stimulating as this hypothesis is, it does not agree with the actual directions of hormone transport in the root cap (WILKINS, Chap. 5.2.3, this vol.).

5.2 Sliding of the Statoliths Along Sensitive Structures

If the *pathway* of translocation of amyloplasts were the decisive factor, graviperception would take place *during* rather than *after* the translocation. This concept of *dynamic* graviperception goes back to inversion experiments by ZIMMERMANN and v. UBISCH (LARSEN, 1962b) and to experiments by LARSEN (1965, 1969, 1971), IVERSEN et al. (1968), IVERSEN and LARSEN (1971), PERBAL (1977).

IVERSEN and LARSEN (1973) and IVERSEN (1974) treated *Lepidium* roots right from germination according to the following chart. The sequence of treatments for each set of roots proceeds from left to right.

After the experiment the length of the translocation pathway of the amyloplasts was determined. Also, the degree of curvature of the roots was measured after rotation on a horizontal clinostat at 2 rpm. Those roots in which the amyloplasts were able to slide along the entire length of the anticlinal walls showed the largest graviresponse. Therefore, the authors concluded that the translocation pathway of the amyloplasts is an essential component of graviperception.

Although there is no doubt about the effect of pre-inversion on graviperception, the following considerations have to be borne in mind:

1. The responses being compared result from different numbers of stimuli. Here the problem of stimulus summation arises (Sect. 2.5).

2. The growth rate of roots in the vertical and inverted position was not examined. In 1927, ZIMMERMANN reported that roots of *Lepidium* that had been

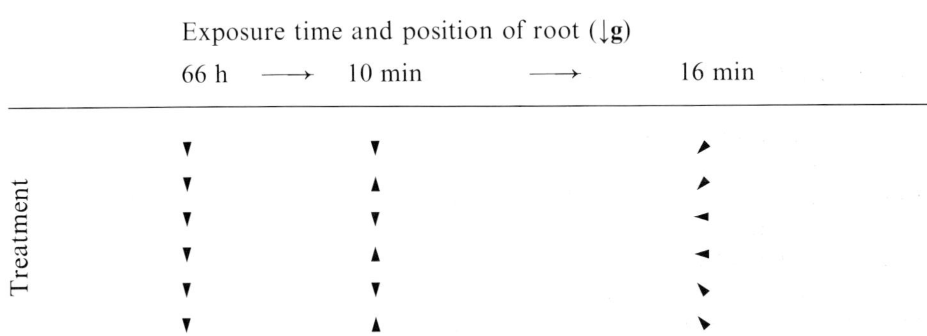

inverted for approximately 10 min showed a growth promotion for 40 min; this is exactly how long the pre-inversion effect lasts.

5.3 Pressure of Statoliths on a Sensitive Structure

The polarity described in Section 4.2.5 can be found without exception in all the statocytes of *Lepidium*. We chose cress roots for our research as they have both a regular structure and a favorable graviresponse. Because of the sedimentation of the amyloplasts onto the distal ER complex, and because of the distinct geometry of the statenchyma, it has been assumed that the pressure exerted by the amyloplasts on the ER complex (or the contact between these structures), is the basis of graviperception (SIEVERS and VOLKMANN, 1972). Therefore, the ER complexes are considered to be the sensitive substratum. While the root is oriented vertically, the pressure on the ER complex is equal in statocytes on either side of the root axis. Because of this, a symmetrical signal could develop in all the interconnected ER complexes, to regulate the normal growth of the root. Any deviation from the vertical orientation, especially a horizontal exposure, causes an inequality in pressure, which becomes very obvious in symmetrically located pairs of lateral statocytes. Figure 5 demonstrates this inequality by means of arrows of differing lengths. This could be the basis for an asymmetrical signal, resulting in gravitropic curvature.

The ER complex remains localized at the distal cell pole during the subsequent response period of approximately 2 h. Thus, the polarity of *Lepidium* statocytes described previously is relatively stable (SIEVERS and VOLKMANN, 1972; VOLKMANN, 1974). In statocytes of other plants, however, where polar organization is less strict, displacements of the ER have been described (GRIFFITHS and AUDUS, 1964; JUNIPER and FRENCH, 1973). In *Lepidium*, the amyloplasts sediment within a few minutes. Later, however, when the root curves, the amyloplasts continuously sediment back onto the distal ER complexes. In this way the disequilibrium between amyloplasts and ER complex caused by gravitropic stimulation is returned to the original equilibrium. It is important to note, however, that the mechanism proposed here would work even if no measurable translocation of the statoliths occurred (see below, and Fig. 5a, b).

Assuming that the ER complex functions as a gravisensor, what is the nature of the interactions between the amyloplasts and the ER complex? Electron microscope investigations of vertically growing roots have already demonstrated that the membranes of the ER complex are affected by the sedimented amyloplasts. An amyloplast often presses its shape—or even that of its starch grains—so deeply into the ER complex that a print is formed (Fig. 6a, b). Directly underneath the amyloplasts the ER cisternae are compressed and the distances between successive elements of the stack is reduced to a minimum. In the neighboring regions, by contrast, the cisternae are farther apart.

The locally limited pressure effect of individual amyloplasts on the membranes of the ER complex can be even more clearly observed by comparing the preservation of the membranes after the use of different fixatives. On using the fixative $KMnO_4$, which diffuses quickly, large cisternae are visible in the ER complex. The cisternae are located parallel to, and frequently cover, the whole distal cell wall and the

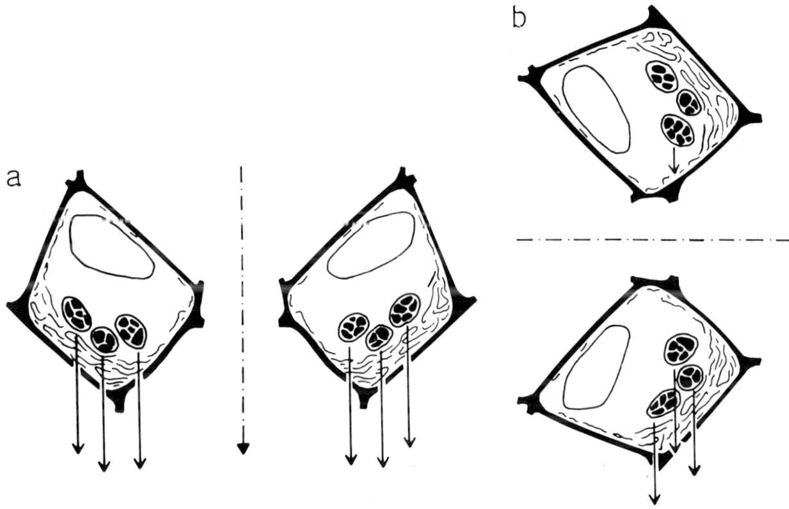

Fig. 5a and b. Pairs of symmetrically located lateral statocytes (diagram according to Fig. 4). The *dashed arrows* point to the root tip. **a** In the case of vertical orientation of the root the pressure exerted by the amyloplasts on the ER is equal in statocytes on either side of the root axis (*equal arrows*). **b** Horizontal exposure causes an inequality (*unequal arrows*)

distal part of the side walls. However, after fixation with glutaraldehyde-OsO_4, which penetrates and kills tissues more slowly (WOHLFARTH-BOTTERMANN and KOMNICK, 1966; SITTE, 1977), the ER cisternae are best preserved only in those areas where the amyloplasts are sedimented furthest (SIEVERS and VOLKMANN, 1977b). In neighboring regions, the cisternae often disintegrate into vesicles as an artifact of the fixation procedure. It can be concluded that the slower fixation allows more damage to occur before the cells are totally fixed. However, the presence of amyloplasts pressed into the membranes protects those regions from such damage.

When roots are centrifuged with the force directed distally; e.g., 20 min at 100 **g**, the amyloplasts displace the distal ER complexes. The cisternae are then located at the anticlines (Figs. 7, 8e). After a further 20 min at only 1 **g**, however, the ER complex has returned to its original site, beneath the amyloplasts and parallel to the distal pericline. These observations prove that the amyloplasts exert pressure on the ER complex. Furthermore, the results confirm that the polarity of the statocytes is relatively stable.

When roots are tilted from the vertical orientation by 30°, the amyloplasts do not only sediment in the direction of gravity but they are also shifted basally, i.e., in the formerly upward direction, and this effect is even more pronounced at 60° (HABERKORN, 1974; SIEVERS and VOLKMANN, 1977a, b). This upward movement can only be explained by the fact that the normal bed of amyloplasts—the ER complex—is elastic to a certain extent. After total inversion of the root, the ER complexes that are freed from the pressure of the amyloplasts are loosened; the cisternae are not stacked as closely, and do not run as strictly parallel to

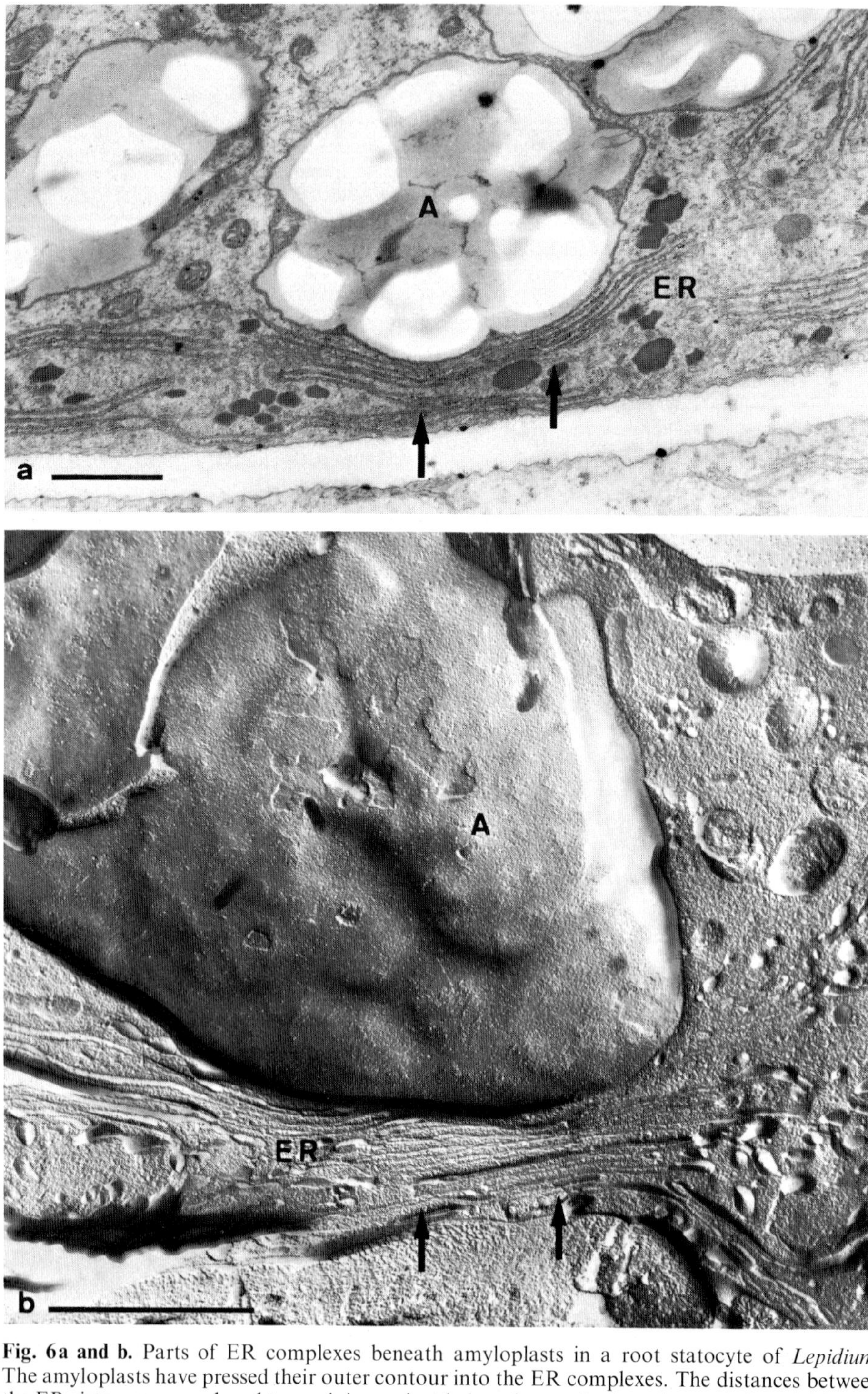

Fig. 6a and b. Parts of ER complexes beneath amyloplasts in a root statocyte of *Lepidium*. The amyloplasts have pressed their outer contour into the ER complexes. The distances between the ER cisternae are reduced to a minimum just below the amyloplasts (*arrows*). In neighboring regions, they are further apart. *A*, amyloplast; *ER*, ER complex. Fixation: **a** Glutaraldehyde-OsO_4, **b** frozen in melting nitrogen without previous chemical fixation but after cryoprotection for some seconds. Scale bars: 1 μm

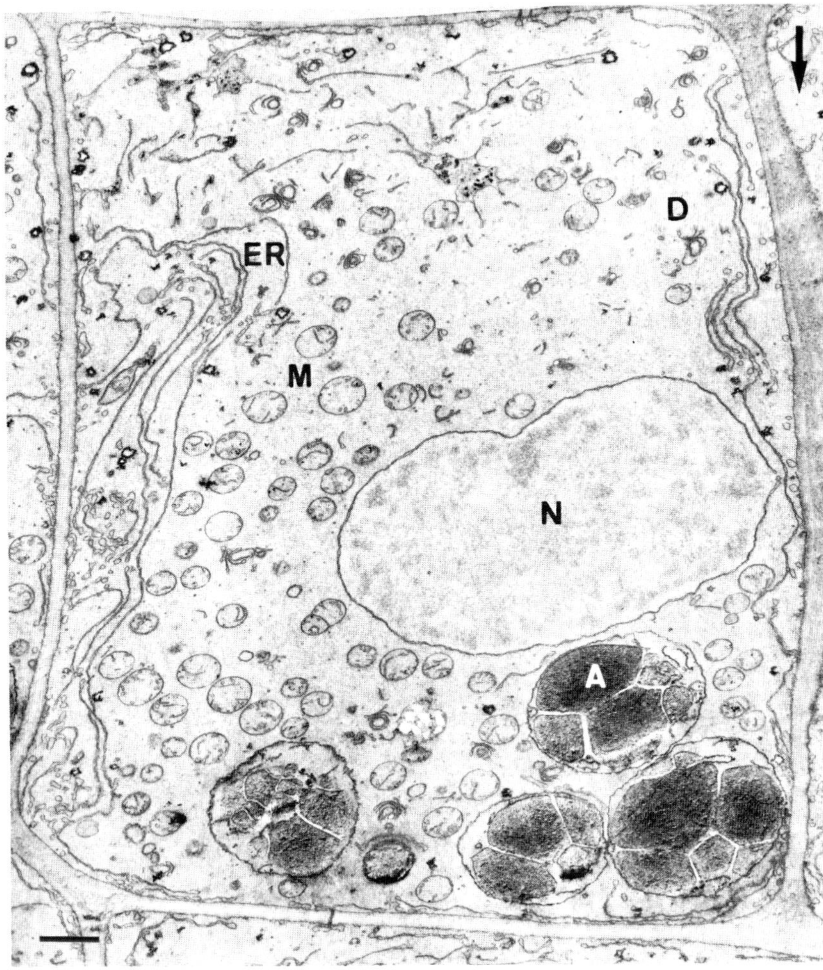

Fig. 7. Central statocyte of a *Lepidium* root. The root has been centrifuged for 20 min at 100 g, with the force acting in the distal direction. The distal ER complex is displaced to the anticlines by the pressure of the amyloplasts. Only one ER cisterna is still located near the distal pericline. The nucleus has sedimented. Dictyosomes and mitochondria are still distributed almost randomly. *A*, amyloplast; *D*, dictyosome; *ER*, ER complex near the anticline; *M*, mitochondrion; *N*, nucleus; *arrow*, direction of centrifugation. Fixation: $KMnO_4$. Scale bar: 1 μm. Photograph by LIANE CASPERS

the distal cell wall as they do in the normal position (VOLKMANN and SIEVERS, 1975).

The question arises whether the arrangement of the ER complex at the distal cell pole of vertically growing roots arises endogenously, as has been shown in plagiotropic lateral roots (SIEVERS and SCHMITZ, 1973; SCHMITZ, 1977), or if it is induced by gravity. In order to exclude the gravity vector as the inducer for the distal location of ER complex, cress seeds were germinated on a horizontal clinostat at 2 and 55 rpm for 24 h (SIEVERS et al., 1976). The distal ER complex in the statocytes develops normally, although the gravity vector has never influenced

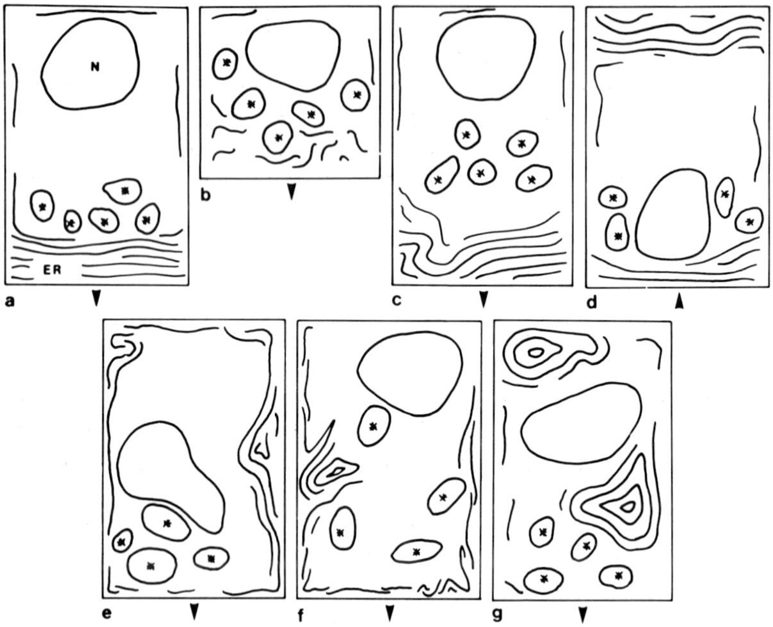

Fig. 8a–g. Schematic drawings of central root statocytes of *Lepidium* demonstrating the arrangement of ER cisternae (*lines*), amyloplasts (*) and nucleus (*N*) after different treatment of the roots at room temperature. ▼ direction of the root tip. For further explanation see pp. 576 (8c, f); 583 (8a, b); 589 (8e); 592 (8c, d, g)

the root in the longitudinal direction (Fig. 8c). The ER complex is just as loosened as is the case after inversion. Therefore, the polarity of the statocytes is caused endogenously.

Graviresponse is prevented mechanically if roots are forced to grow for several hours in capillaries in the inverted position (VOLKMANN and SIEVERS, 1975). After 2 h the statocytes of the inverted roots already start to form a second ER complex near the proximal cell wall beneath the amyloplasts, which are sedimented within the first few minutes of inversion. After 18 h the new proximal ER complex contains the same number of cisternae as the distal one (Fig. 8d). It is thus possible to induce the localization of an ER complex by the direction of gravity in addition to the endogenously determined localization. It can be assumed that the roots which had been inverted for several hours can only regain their orientation in space by completing the system of amyloplasts and ER complex.

An electric field can, under certain conditions, induce a stronger response in the plant than can the gravitational field (BRAUNER and BÜNNING, 1930). Cress roots were placed in a normal vertical position in a horizontal static electric field at 2000–3000 V cm^{-1}. After 6–14 h a distinct electrotropic response was observed. Pronounced changes appear only in the statocytes (Fig. 8g). There is a disruption of cell polarity, similar to that observed after rotating a root on a clinostat for several hours (Fig. 8f; SIEVERS and VOLKMANN, 1977a, b; BEHRENS, 1977). The distal ER complex has disappeared; instead, large clumps of ER and single cisternae are randomly distributed in the cell. The total surface area of the ER is doubled. The nucleus no longer maintains the normal position at the proximal cell pole and the number of dictyosomes is greatly reduced. The amyloplasts are sedimented. If the current is switched off, the roots regain their gravitropic orientation. At the same time the usual polarity of the statocytes is restored. These observations again indicate the high sensitivity of statocytes toward exogenous influences. Furthermore, these results show the strong correlation of structural polarity and graviperception that seems to be a characteristic feature of the gravisensory organ of the root. Above all, however, they demonstrate again the relatively stable cell polarity.

5.4 Critical Evaluation of the Hypotheses

In the preceding sections, the hypotheses for statolith action have been explained assuming the special case of a horizontal stimulation. In such an extreme case, amyloplasts are, of course, translocated from their natural substratum. As shown in Section 2.2, however, the sensitivity of plants is much higher than such experiments might suggest: they already perceive the gravity stimulus at a deviation of $< 10°$ from the vertical orientation. In these cases one cannot expect a *remarkable* translocation of the amyloplasts, especially if the geometry of the root cap of *Lepidium* is considered. Furthermore, it must be noted that within a presentation time of 12 s an amyloplast can traverse a maximum distance of 0.2 μm at the sedimentation speed of approximately 1 $\mu m\ min^{-1}$ (IVERSEN et al., 1968; IVERSEN and LARSEN, 1973). This is 1/15 of its diameter and 1/40 of the cell width (for theoretical data see pp. 579, 580). However, taking the perception time as a basis of this calculation (0.5 s, FITTING, 1905a, b; PICKARD, 1973), an amyloplast can theoretically be displaced only by 1/360 of its diameter (corresponding to 1/960 of the cell width) during the process of perception. Therefore, translocation of amyloplasts cannot be the main factor in graviperception. It may be concluded that amyloplasts can trigger graviperception only by changing the pressure on their presumably sensitive substratum (NĚMEC, 1901; BRAUNER, 1962; SIEVERS and VOLKMANN, 1972). This augments the arguments already given against the other two hypotheses, viz. new distribution and sliding.

The idea that differential pressure at endomembranes induces graviperception is supported by a series of effects on membranes under defined pressure. In *Valonia* it has been shown by pressure probe that pressure can influence the thickness of membranes (COSTER and ZIMMERMANN, 1976) and that it is thereby possible to modify active transport processes (STEUDLE and ZIMMERMANN, 1974). Pressure changes of $2 \cdot 10^6$ $dyne \cdot cm^{-2}$ induce action potentials (ZIMMERMANN and BECKERS, 1978). Enzyme activities can also be influenced by pressure. For example, the ATPase of tomatoes needs a pressure as high as $5 \cdot 10^6$ $dyne \cdot cm^{-2}$ (KUIPER, 1971). On the other hand, isolated and spread acetylcholinesterase is activated at ≈ 10 $dyne \cdot cm^{-2}$ only (SKOU, 1959). This latter is comparable to the pressure of 2–4 $dyne \cdot cm^{-2}$ exerted by amyloplasts (AUDUS, 1962; VOLKMANN, 1974); but even the much higher pressures in the examples mentioned before may still have some relevance if one considers the conditions of the experiments. Here, the pressure is exerted on the stretched plasmalemma of the whole cell, against the underlying cell wall. The amyloplasts, however, press on locally limited parts of endomembranes which do not possess a support similar to that of the cell wall. In vivo they are weakly stressed, which is proved by the disintegration into vesicles in the experiments reported above (Sect. 5.3). Such "weak membranes" (POLLARD, 1971) should be especially sensitive to pressure differences. Our observations indicate that the stress (=force per unit area) exerted on the ER is limited not only spatially, but also in time: sedimented amyloplasts do not rest immobile on their substratum. Instead, they show an irregular motion, similar to BROWNIAN movement.

However, as long as no direct proof exists for an effect of amyloplast pressure on endomembranes, other ideas should not be completely excluded. For instance, interactions between the two membranes may result from electrostatic forces, providing a possible explanation for the effect of electrical field on statocytes. As storage organelles, amyloplasts could secrete metabolites of carbohydrate metabolism. Finally, analogous to the *Chara* rhizoid (see SIEVERS and VOLKMANN, Chap. 5.2.1, this vol.) the amyloplasts could function as a transport barrier for vesicles or molecules needed by the ER membranes (AUDUS, 1971; VOLKMANN, 1974).

6. Concluding Remarks

The high sensitivity of plants toward mass acceleration forces us to assume that sedimentable amyloplasts induce graviperception in their substratum. In the root, where the spatial separation between site of perception and site of response demands transduction via many cells, the substratum consists of cisternae of the endoplasmic reticulum. This intra- and intercellulary communicating compartment seems to be a suitable candidate for both graviperception and transduction.

Statocytes of negatively gravitropic organs are lacking in such a distinct endomembrane system. A transduction via many cells also seems to be missing, at least in the direction of the organ axis. The plasmalemma seems to be the substratum of the amyloplasts in these organs. It is possible that there is a direct connection between the pressure effect of the amyloplasts on the plasmalemma and the response induction, maybe via the secretion of auxins.

The gravisensitivity of the plant seems to lie in the sensitivity of biomembranes to pressure. This also appears to be the case in the statocyst of a coelenterate that does not possess cilia and nerve cells (CAMPBELL, 1972). The pressures exerted are extraordinarily small. The weight of all sedimentable amyloplasts in a *Lepidium* root only amounts to 50 ng. In larger root caps (*Zea, Vicia, Pisum*) the total weight of the amyloplasts is likely to be 10 times higher. In spite of this, it is realistic to assume that this total weight is significant in the process of stimulus perception. A comparison with thigmoreceptors clarifies the situation if one considers that PFEFFER's cotton thread that induces a response in tendrils, weighs only 250 ng. Moreover, the cotton thread touches the epidermis outer wall, which is still far away from the sensitive plasmalemma, whereas amyloplasts press directly on biomembranes.

References

Anker, L.: Ortho-geotropism in shoots and coleoptiles. In: Encyclopedia of Plant Physiology. Ruhland, W. (ed.). Berlin-Göttingen-Heidelberg: Springer, 1962, Vol. XVII, pt. 2, pp. 103–152

Audus, L.J.: The mechanism of the perception of gravity by plants. In: Biological Receptor Mechanisms. Beament, J.W.L. (ed.). Soc. Exp. Biol. Symp. XVI, Cambridge: University Press, 1962, pp. 197–226
Audus, L.J.: Geotropism and the modified sine rule; an interpretation based on the amyloplast statolith theory. Physiol. Plant. *17*, 737–745 (1964)
Audus, L.J.: Geotropism. In: Physiology of Plant Growth and Development. Wilkins, M.B. (ed.). London: McGraw-Hill, 1969, pp. 205–242
Audus, L.J.: Linkage between detection and the mechanisms establishing differential growth factor concentrations. In: Gravity and the Organism. Gordon, S.A., Cohen, M.J. (eds.). Chicago-London: University of Chicago Press, 1971, pp. 137–150
Audus, L.J.: Geotropism in roots. In: The Development and Function of Roots. Torrey, J.G., Clarkson, D.T. (eds.). London: Academic Press, 1975, pp. 327–363
Ball, N.G.: Tropic, nastic, and tactic responses. In: Plant Physiology—a Treatise. Steward, F.C. (ed.). Vol. VA: Analysis of Growth: Behavior of Plants and Their Organs. New York-London: Academic Press, 1969, pp. 119–228
Barlow, P.W.: Recovery of geotropism after removal of the root cap. J. Exp. Bot. *25*, 1137–1146 (1974)
Barlow, P.W., Grundwag, M.: The development of amyloplasts in cells of the quiescent centre of *Zea* roots in response to removal of the root cap. Z. Pflanzenphysiol. *73*, 56–64 (1974)
Behrens, H.M.: Untersuchungen an Gartenkressewurzeln (*Lepidium sativum* L.) im statischen elektrischen Feld. Diplomarbeit. Bot. Inst. Univ. Bonn, 1977
Bouck, G.B.: An examination of the effects of ultracentrifugation on the organelles in living root tip cells. Am. J. Bot. *50*, 1046–1054 (1963a)
Bouck, G.B.: Stratification and subsequent behavior of plant cell organelles. J. Cell Biol. *18*, 441–457 (1963b)
Brauner, L.: Primäreffekte der Schwerkraft bei der geotropischen Reaktion. In: Encyclopedia of Plant Physiology. Ruhland, W. (ed.). Berlin-Göttingen-Heidelberg: Springer, 1962, Vol. XVII, pt. 2, pp. 74–102
Brauner, L., Bünning, E.: Geoelektrischer Effekt und Elektrotropismus. Ber. Dtsch. Bot. Ges. *48*, 470–476 (1930)
Brauner, L., v. Dellingshausen, M., Böck, A.: Neue Versuche zum geoelektrischen Effekt. Planta *62*, 195–220 (1964)
Bünning, E.: Allgemeine Gesetze und Phänomene der pflanzlichen Bewegungsphysiologie. In: Encyclopedia of Plant Physiology. Ruhland, W. (ed.). Berlin-Göttingen-Heidelberg: Springer, 1959, Vol. XVII, pt. 1, pp. 8–23
Campbell, R.D.: Statocyst lacking cilia in the coelenterate *Corymorpha palma*. Nature (London) *238*, 49–51 (1972)
Cerček, L.: Effect of X-ray irradiation on regeneration and geotropic function of barley root caps. Int. J. Radiat. Biol. *17*, 187–194 (1970)
Coster, H.G.L., Zimmermann, U.: Transduction of turgor pressure by cell membrane compression. Z. Naturforsch. *31*c, 461–463 (1976)
Dedolph, R.R., Naqvi, S.M., Gordon, S.A.: Effect of gravity compensation on the geotropic sensitivity of *Avena* seedlings. Plant Physiol. *40*, 961–965 (1965)
Dedolph, R.R., Oemick, D.A., Wilson, B.R., Smith, G.R.: Causal basis of gravity stimulus nullification by clinostat rotation. Plant Physiol. *42*, 1373–1383 (1967)
Filner, B., Hertel, R., Steele, C., Fan, V.: Some aspects of geotropism in coleoptiles. Planta *94*, 333–354 (1970)
Firn, R., Digby, J.: The role of the peripheral cell layers in the geotropic curvature of sunflower hypocotyls: a new model of shoot geotropism. Aust. J. Plant Physiol. *4*, 337–347 (1977)
Firn, R.D., Digby, J., Riley, H.: Shoot geotropic curvature—the location, magnitude and kinetics of the gravity-induced differential growth in horizontal sunflower hypocotyls. Ann. Bot. *42*, 465–468 (1978)
Fitting, H.: Untersuchungen über den geotropischen Reizvorgang. Teil I. Die geotropische Empfindlichkeit der Pflanzen. Jahrb. Wiss. Bot. *41*, 221–330 (1905a)
Fitting, H.: Untersuchungen über den geotropischen Reizvorgang. Teil II. Weitere Erfolge mit der intermittierenden Reizung. Jahrb. Wiss. Bot. *41*, 331–398 (1905b)

Frey-Wyssling, A.: Die submikroskopische Struktur des Cytoplasmas. Protoplasmatologia II, A 2. Wien: Springer, 1955.
Grahm, L., Hertz, C.H.: Measurements of the geoelectric effect in coleoptiles by a new technique. Physiol. Plant. *15*, 96–114 (1962)
Grahm, L., Hertz, C.H.: Measurements of the geoelectric effect in coleoptiles. Physiol. Plant. *17*, 186–201 (1964)
Griffiths, H.J., Audus, L.J.: Organelle distribution in the statocyte cells of the root-tip of *Vicia faba* in relation to geotropic stimulation. New Phytol. *63*, 319–333 (1964)
Haberkorn, H.-R.: Movement of amyloplasts and reaction of the ER-complex in statocytes. Int. Assoc. Plant Physiol. 1st Meeting Würzburg, Sept. 23–29, 1974. Abstracts.
Haberkorn, H.-R., Sievers, A.: Response to gravity of roots growing in water. Naturwissenschaften *64*, 639–640 (1977)
Haberlandt, G.: Ueber die Statolithenfunction der Stärkekörner. Ber. Dtsch. Bot. Ges. *20*, 189–195 (1902)
Hager, A.: Das differenzielle Wachstum bei photo- und geotropischen Krümmungen. Ber. Dtsch. Bot. Ges. *84*, 331–350 (1971)
Heilbronn, A.: Zustand des Plasmas und Reizbarkeit. Ein Beitrag zur Physiologie der lebenden Substanz. Jahrb. Wiss. Bot. *54*, 357–390 (1914)
Hensel, W., Sievers, A.: Influence of the slowly rotating clinostat on the polarity of statocytes and on the gravitropic reaction of roots. In: Collected abstracts of the paper-demonstrations. 9th Int. Conf. Plant Growth Substances. Pilet, P.-E. (ed.). Lausanne 1976, pp. 139–140
Hertel, R., de la Fuente, R.K., Leopold, A.C.: Geotropism and the lateral transport of auxin in the corn mutant amylomaize. Planta *88*, 204–214 (1969)
Hestnes, A., Iversen, T.-H.: Movement of cell organelles and the geotropic curvature in roots of Norway spruce (*Picea abies*). Physiol. Plant. *42*, 406–414 (1978)
Hinchman, R.R., Gordon, S.A.: Amyloplast size and number in gravity-compensated oat seedlings. Plant Physiol. *53*, 398–401 (1974)
Hoshizaki, T.: Influence of gravitational forces on plants. Environ. Biol. Med. *2*, 47–79 (1973)
Isenberg, G., Wohlfarth-Bottermann, K.E.: Transformation of cytoplasmic actin. Importance for the organization of the contractile gel reticulum and the contraction—relaxation cycle of cytoplasmic actomyosin. Cell Tissue Res. *173*, 495–528 (1976)
Iversen, T.-H.: Elimination of geotropic responsiveness in roots of cress (*Lepidium sativum*) by removal of statolith starch. Physiol. Plant. *22*, 1251–1262 (1969)
Iversen, T.-H.: Geotropic curvatures in roots of cress (*Lepidium sativum*). Physiol. Plant. *28*, 332–340 (1973)
Iversen, T.-H.: The roles of statoliths, auxin transport, and auxin metabolism in root geotropism. K. norske Vidensk. Selsk. Mus. Miscellanea *15*, 1–216 (1974)
Iversen, T.-H., Larsen, P.: The starch satolith hypothesis and the optimum angle of geotropic stimulation. Physiol. Plant. *25*, 23–27 (1971)
Iversen, T.-H., Larsen, P.: Movement of amyloplasts in the statocytes of geotropically stimulated roots. The pre-inversion effect. Physiol. Plant. *28*, 172–181 (1973)
Iversen, T.-H., Pedersen, K., Larsen, P.: Movement of amyloplasts in the root cap cells of geotropically sensitive roots. Physiol. Plant. *21*, 811–819 (1968)
Johnsson, A.: Investigations of the reciprocity rule by means of geotropic and geoelectric measurements. Physiol. Plant. *18*, 945–967 (1965)
Johnsson, A.: Investigations of the geotropic curvature of the *Avena* coleoptile. I. The geotropic response curve. Physiol. Plant. *25*, 35–42 (1971)
Johnsson, A., Rengman, K., Grahm, L.: Investigations of the geotropic curvature of the *Avena* coleoptile. II. The presentation time as a function of distance from apex. Physiol. Plant. *25*, 43–47 (1971)
Jones, R.L.: Gibberellins: their physiological role. Annu. Rev. Plant Physiol. *24*, 571–598 (1973)
Juniper, B.: Mechanisms of perception and patterns of organisation in root caps. In: The Dynamics of Meristem Cell Populations. Miller, M.W., Kuehnert, C.C. (eds.). New York: Plenum Press, 1972, pp. 119–131
Juniper, B.E.: Geotropism. Annu. Rev. Plant Physiol. *27*, 385–406 (1976)

Juniper, B.E.: The perception of gravity by a plant. Proc. R. Soc. Lond. B. *199*, 537–550 (1977)
Juniper, B.E., Barlow, P.W.: The distribution of plasmodesmata in the root tip of maize. Planta *89*, 352–360 (1969)
Juniper, B.E., French, A.: The fine structure of the cells that perceive gravity in the root tip of maize. Planta *95*, 314–329 (1970)
Juniper, B.E., French, A.: The distribution and redistribution of endoplasmic reticulum (ER) in geoperceptive cells. Planta *109*, 211–224 (1973)
Juniper, B.E., Groves, S., Landau-Schachar, B., Audus, L.J.: Root cap and the perception of gravity. Nature (London) *209*, 93–94 (1966)
Köhler, D.: "Reizleitung" beim negativen Geotropismus des Sprosses? Z. Pflanzenphysiol. *86*, 269–272 (1978)
Konings, H.: The significance of the root cap for geotropism. Acta Bot. Neerl. *17*, 203–211 (1968)
Kuiper, P.J.C.: Potato ATPase: sensitivity to hydrostatic pressure of a cold labile form. Biochim. Biophys. Acta *250*, 443–445 (1971)
Kutschera-Mitter, L.: Das positiv geotrope Wachstum des Keimblattes von *Phoenix dactylifera* als Beweis einer neuen Erklärung des Geotropismus. Österr. Bot. Z. *119*, 154–168 (1971a)
Kutschera-Mitter, L.: Über das geotrope Wachstum der Wurzel. Beitr. Biol. Pflanz. *47*, 371–436 (1971b)
Larsen, P.: Geotropism. An introduction. In: Encyclopedia of Plant Physiology. Ruhland, W. (ed.). Berlin-Göttingen-Heidelberg: Springer, 1962a, Vol. XVII, pt. 2, pp. 34–73
Larsen, P.: Orthogeotropism in roots. In: Encyclopedia of Plant Physiology. Ruhland, W. (ed.). Berlin-Göttingen-Heidelberg: Springer, 1962b, Vol. XVII, pt. 2, pp. 153–199
Larsen, P.: Geotropic responses in roots as influenced by their orientation before and after stimulation. Physiol. Plant. *18*, 747–765 (1965)
Larsen, P.: The optimum angle of geotropic stimulation and its relation to the starch statolith hypothesis. Physiol. Plant. *22*, 469–488 (1969)
Larsen, P.: The susception of gravity by higher plants. In: Gravity and the Organism. Gordon, S.A., Cohen, M.J. (eds.). Chicago-London: University of Chicago Press, 1971, pp. 73–87
Leather, G.R., Forrence, L.E., Abeles, F.B.: Increased ethylene production during clinostat experiments may cause leaf epinasty. Plant Physiol. *49*, 183–186 (1972)
Markl, H.: Geo-Biophysik: Schwerefeld, Magnetfeld und Organismen. In: Biophysik. Hoppe, W., Lohmann, W., Markl, H., Ziegler, H. (eds.). Berlin-Heidelberg-New York: Springer, 1977, pp. 609–619
Morré, D.J., Lembi, C.A., van der Woude, W.J.: Golgi-Apparat und verwandte Zellbestandteile. In: Biochemische Cytologie der Pflanzenzelle. Jacobi, G., (ed.). Stuttgart: Georg Thieme, 1974, pp. 147–172
Muller, H.J.: Approximation to a gravity-free situation for the human organism achievable at moderate expense. Science *128*, 772 (1958)
Němec, B.: Ueber die Wahrnehmung des Schwerkraftreizes bei den Pflanzen. Jahrb. Wiss. Bot. *36*, 80–178 (1901)
Nougarède, A., Pilet, P.-E.: Action de l'acide gibbérellique (GA_3) sur l'ultrastructure des amyloplastes du statenchyme de la racine du *Lens culinaris* L. C.R. Acad. Sci. Paris *273*, Série D, 348–351 (1971)
Nougarède, A., Simonin, P.-A., Pilet, P.-E.: Activité α-amylasique et comportement des plastes amylifères du coléoptile du *Triticum durum* traité par la kinétine ou par l'acide gibbérellique. C.R. Acad. Sci. Paris *276*, Série D, 3299–3302 (1973)
Osborne, D.J., Wright, M.: Gravity-induced cell elongation. Proc. R. Soc. Lond. B. *199*, 551–564 (1977)
Perbal, G.: Action de la pesanteur sur la répartition des organites dans les cellules axiales centrales (statenchyme) de la coiffe du *Lens culinaris* L. C.R. Acad. Sci. Paris *273*, Série D, 789–792 (1971)
Perbal, G.: L'action des statolithes dans la réponse géotropique des racines de *Lens culinaris*. Planta *116*, 153–171 (1974)
Perbal, G.: L'action des statolithes dans la perception géotropique des racines du *Lens culinaris* L. C.R. Acad. Sci. Paris *284*, Série D, 2347–2350 (1977)

Perbal, G., Perbal, P.: La perception géotropique dans la coiffe des racines de lentille. Physiol. Plant. *37*, 42–48 (1976)
Phillips, H.L., Jr., Torrey, J.G.: The ultrastructure of the root cap in cultured roots of *Convolvulus arvensis* L. Am. J. Bot. *61*, 879–887 (1974)
Pickard, B.G.: Geotropic response patterns of the *Avena* coleoptile. I. Dependence on angle and duration of stimulation. Can. J. Bot. *51*, 1003–1021 (1973)
Pickard, B.G., Thimann, K.V.: Geotropic response of wheat coleoptiles in absence of amyloplast starch. J. Gen. Physiol. *49*, 1065–1086 (1966)
Pilet, P.-E.: Rôle de l'apex radiculaire dans la croissance, le géotropisme et le transport des auxines. Ber. Schweiz. Bot. Ges. *81*, 52–65 (1971)
Pilet, P.-E.: Géoperception et géoréaction racinaires. Physiol. Vég. *10*, 347–367 (1972)
Pilet, P.-E.: Abscisic acid as a root growth inhibitor: physiological analyses. Planta *122*, 299–302 (1975)
Pilet, P.-E., Greppin, H., Bonzon, M.: Action de l'acide gibbérellique sur la densité des amyloplastes racinaires. C.R. Acad. Sci. Paris *272*, Série D, 1760–1763 (1971)
Pilet, P.-E., Nougarède, A.: Action de l'acide gibbérellique sur la croissance et le géotropisme radiculaires. C.R. Acad. Sci. Paris *272*, Série D, 418–421 (1971)
Pilet, P.-E., Nougarède, A., Perbal, G.: Modifications cytochimiques et infrastructurales des cellules de la zone de courbure au cours de la réaction géotropique de la racine du *Lens culinaris* L. C.R. Acad. Sci. Paris *268*, Série D, 2056–2059 (1969)
Pollard, E.C.: Physical determinants of receptor mechanisms. In: Gravity and the Organism. Gordon, S.A., Cohen, M.J. (eds.). Chicago-London: University of Chicago Press, 1971, pp. 25–34
Pollard, T.D., Ito, S.: Cytoplasmic filaments of *Amoeba proteus*. I. The role of filaments in consistency changes and movement. J. Cell Biol. *46*, 267–289 (1970)
Rufelt, H.: Plagiogeotropism in roots. In: Encyclopedia of Plant Physiology. Ruhland, W. (ed.). Berlin-Göttingen-Heidelberg: Springer, 1962, Vol XVII, pt. 2, pp. 322–343
Schmitz, U.: Ultrastruktur und Geoperception in der Kalyptra von Seitenwurzeln. Dissertationes Botanicae Vol. 37. Vaduz: J. Cramer, 1977
Shen-Miller, J.: Reciprocity in the activation of geotropism in oat coleoptiles grown on clinostats. Planta *92*, 152–163 (1970)
Shen-Miller, J.: The Golgi apparatus and geotropism. In: Hormone Regulation of Plant Growth and Development. Kaldewey, H., Vardar, Y. (eds.). Weinheim: Verlag Chemie, 1972, pp. 365–376
Shen-Miller, J., Gordon, S.A.: Gravitational compensation and the phototropic response of oat coleoptiles. Plant Physiol. *42*, 352–360 (1967)
Shen-Miller, J., Hinchman, R.R.: Gravity sensing in plants: a critique of the statolith theory. BioScience *24*, 643–651 (1974)
Shen-Miller, J., Hinchman, R.R., Gordon, S.A.: Thresholds for georesponse to acceleration in gravity-compensated *Avena* seedlings. Plant Physiol. *43*, 338–344 (1968)
Shen-Miller, J., McNitt, R.E., Wojciechowski, M.: Regions of differential cell elongation and mitosis, and root meristem morphology in different tissues of geotropically stimulated maize root apices. Plant Physiol. *61*, 7–12 (1978)
Shen-Miller, J., Miller, C.: Intracellular distribution of mitochondria after geotropic stimulation of the oat coleoptile. Plant Physiol. *50*, 51–54 (1972a)
Shen-Miller, J., Miller, C.: Distribution and activation of the Golgi apparatus in geotropism. Plant Physiol. *49*, 634–639 (1972b)
Sievers, A.: Gravity receptors in lower plants. In: Gravity and the Organism. Gordon, S.A., Cohen, M.J. (eds.). Chicago-London: University of Chicago Press, 1971, pp. 51–63
Sievers, A., Schmitz, U.: Dorsiventralität der Statocyten in plagiogeotropen Seitenwurzeln von *Lepidium sativum* L. Planta *114*, 373–378 (1973)
Sievers, A., Volkmann, D.: Verursacht differentieller Druck der Amyloplasten auf ein komplexes Endomembransystem die Geoperzeption in Wurzeln? Planta *102*, 160–172 (1972)
Sievers, A., Volkmann, D.: Ultrastructure of gravity-perceiving cells in plant roots. Proc. R. Soc. Lond. B. *199*, 525–536 (1977a)
Sievers, A., Volkmann, D.: Ultrastructural aspects of georeceptors in roots. In: Plant Growth Regulation. 9th Int. Conf. Plant Growth Substances. Pilet, P.-E. (ed.). Berlin-Heidelberg-New York: Springer, 1977b, pp. 208–217

Sievers, A., Volkmann, D., Hensel, W., Sobick, V., Briegleb, W.: Cell polarity in root statocytes in spite of simulated weightlessness. Naturwissenschaften *63*, 343 (1976)
Sitte, P.: Functional organization of biomembranes. In: Lipids and Lipid Polymers in Higher Plants. Tevini, M., Lichtenthaler, H.K. (eds.). Berlin-Heidelberg-New York: Springer, 1977, pp. 1–28
Skou, J.C.: Studies on the influence of the degree of unfolding and the orientation of the side chains on the activity of a surface-spread enzyme. Biochim. Biophys. Acta *31*, 1–10 (1959)
Sobick, V., Sievers, A.: Reponses of roots to simulated weightlessness on the fast-rotating clinostat. In: Cospar, Life Sciences and Space Research Vol. XII. Holmquist, R. (ed.). Oxford: Pergamon Press, in press, 1979
Steudle, E., Zimmermann, U.: Turgor pressure regulation in algal cells: pressure-dependence of electrical parameters of the membrane in large pressure ranges. In: Membrane Transport in Plants. Zimmermann, U., Dainty, J. (eds.). Berlin-Heidelberg-New York: Springer, 1974, pp. 72–78
Volkmann, D.: Amyloplasten und Endomembranen: Das Geoperzeptionssystem der Primärwurzel. Protoplasma *79*, 159–183 (1974)
Volkmann, D.: In cress roots (*Lepidium sativum* L.), abscisic acid prevents the development of the central cap cells into statocytes. Planta *130*, 89–91 (1976)
Volkmann, D., Sievers, A.: Wirkung der Inversion auf die Anordnung des Endoplasmatischen Reticulums und die Polarität von Statocyten in Wurzeln von *Lepidium sativum*. Planta *127*, 11–19 (1975)
Westing, A.H.: A case against statoliths. In: Gravity and the Organism. Gordon, S.A., Cohen, M.J. (eds.). Chicago-London: University of Chicago Press, 1971, pp. 97–101
Wilkins, M.B.: Geotropism. Annu. Rev. Plant Physiol. *17*, 379–408 (1966)
Wilkins, M.B.: Gravity-sensing guidance systems in plants. Sci. Prog., Oxford *63*, 187–217 (1976)
Wilkins, M.B.: Gravity and light-sensing guidance systems in primary roots. In: Integration of Activity in the Higher Plant. Jennings, D.H. (ed.). Soc. Exp. Biol. Symp. XXXI, Cambridge: University Press, 1977, pp. 275–335
Wilkins, M.B., Woodcock, A.E.R.: Origin of the geoelectric effect in plants. Nature (London) *208*, 990–992 (1965)
Wohlfarth-Bottermann, K.E., Komnick, H.: Die Gefahren der Glutaraldehyd-Fixation. J. Microsc. *5*, 441–452 (1966)
Wright, M., Osborne, D.J.: Gravity-regulation of cell elongation in nodes of the grass *Echinochloa colonum*. Biochem. Physiol. Pflanz. *171*, 479–492 (1977)
Zimmermann, U., Beckers, F.: Generation of action potentials in *Chara corallina* by turgor pressure changes. Planta *138*, 173–179 (1978)
Zimmermann, W.: Beiträge zur Kenntnis der Georeaktionen. I. Geotonische Längskraftwirkungen auf orthotrope Hauptwurzeln. Jahrb. Wiss. Bot. *66*, 631–677 (1927)
Zinke, H.: Versuche zur Analyse der positiven und der negativen geotropischen Reaktionen von Keimwurzeln. Planta *82*, 50–72 (1968)

Additional Remarks

In gravitropism of higher plants, the most important progress has been made in analyzing the perception of gravity on the one hand, and in elucidating the hormonal connections between perception and response on the other. Accordingly, the two main chapters cover these processes. There are, however, additional investigations which deal with gravitropism, but which cannot be assigned unambiguously either to perception or to the hormonal relationships. Omitting these investigations does not mean that they are unimportant; they may well open new fields of research in gravitropism.

First, if a gravitropic response of corn coleoptiles is observed carefully from the very beginning of stimulation, an early phase of response is found where the coleoptiles bend downward for a few minutes, i.e., the normal (and expected) negative gravitropism is preceded by a short positive gravitropism (cf. FILNER et al., 1970; HILD and HERTEL, 1972). This effect, which may be related to the link connecting graviperception and auxin redistribution, will be discussed to some detail in KANG, Chap. 5.3.2, this vol.

Second, there are many investigations dealing with light effects on gravitropism. These investigations are largely aimed at collecting data on phytochrome action; but they are also interesting from the point of view of gravitropism. It is, however, not yet known whether these light effects modulate the perception process, one of the transductory steps, or the response proper. Only a few examples will be mentioned.

In some mutants of *Zea mays*, light is an absolute requirement for the gravitropic response of roots (e.g., SCOTT and WILKINS, 1969; PILET, 1975; see also WILKINS, Chap. 5.2.3, this vol.), and the same has been found to be true of a few other plants (TEPFER and BONNETT, 1972).

In oat coleoptiles, gravitropic response is increased or decreased by prior or simultaneous irradiation, depending on the conditions of the experiment (e.g., WILKINS and GOLDSMITH, 1964; WILKINS, 1965). Some of these are red light effects, mediated via phytochrome (e.g., WILKINS and GOLDSMITH, 1964; WILKINS, 1965), while others can be produced either by red or by blue light (SCOTT and WILKINS, 1969).

Light may selectively affect specific phases of the total response: in corn coleoptiles, the brief early positive bending is reduced by pretreatment with blue light; the subsequent negative bending is much less influenced by blue (HILD, 1977). After a preirradiation with red light (far-red reversible), the positive phase may even disappear completely and be replaced by a phase of early negative bending (HILD and HERTEL, 1972).

Additional References

Filner, B., Hertel, R., Steele, C., Fan, V.: Some aspects of geotropism in coleoptiles. Planta *94*, 333–354 (1970)

Hild, V.: Wirkung von Vorbestrahlung mit Rot- oder Blaulicht auf die geotropische Empfindlichkeit von Maiskoleoptilen. Planta *133*, 309–314 (1977)

Hild, V., Hertel, R.: Initial phases of gravity-induced lateral auxin transport and geotropic curvature in corn coleoptiles. Planta *108*, 245–258 (1972)

Pilet, P.-E.: Effects of light on the georeaction and growth inhibitor content of roots. Physiol. Plant. *33*, 94–97 (1975)

Scott, T.K., Wilkins, M.B.: Auxin transport in roots. IV. Effects of light on IAA movement and geotropic responsiveness in *Zea* roots. Planta *87*, 249–258 (1969)

Tepfer, D.A., Bonnett, H.T.: The role of phytochrome in the geotropic behavior of roots of *Convolvulus arvensis*. Planta *106*, 311–324 (1972)

Wilkins, M.B.: Red light and the geotropic response of the *Avena* coleoptile. Plant Physiol. *40*, 24–34 (1965)

Wilkins, M.B., Goldsmith, M.H.M.: The effects of red, far-red, and blue light on the geotropic response of coleoptiles of *Zea mays*. J. Exp. Bot. *15*, 600–615 (1964)

5.2.3 Growth-Control Mechanisms in Gravitropism

M.B. WILKINS

1. Introduction

The gravitropic curvature developed by primary shoots and roots involves the fine control of the rate of cell extension on the opposite sides of the organ. During the 15 years which have elapsed since the appearance of the previous volume (XVII/2) on plant movement in this series, there have been many investigations of the way in which this fine and localised control of cell growth is achieved, and very significant advances have been made in our understanding of the processes involved. In addition, attention has been given to the development of gravitropic curvature in mature shoots of cereals and grasses which have completed their normal extension growth. This response differs from that in primary roots and shoots in that specialised areas of the shoot, the leaf sheath bases, retain the capacity to grow if stimulated, and it is the *initiation* of growth only on the lowermost side when such shoots are placed in the horizontal position that has been investigated in some detail.

The advances which have been made have, for the most part, been due, firstly, to the employment of physicochemical techniques to identify unequivocally the hormones present in the responding organs, and, secondly, to the execution of relatively simple experiments which have revealed the distribution and movement of growth regulators *within the tissues of the organs* rather than in agar blocks in contact with one or other of the cut surfaces of an organ segment. The amounts of growth hormones which *accumulate* in agar blocks at the basal end of segments cannot be used to assess the amount present in the tissues with which the blocks are in contact for obvious reasons (see p. 609), and this kind of experimental procedure has now largely been abandoned. Some discussion of growth regulators in agar receiver blocks will, nevertheless, be included in this review simply because they are the only data available at the present time.

A number of reviews have been published in the past few years on gravitropic response mechanisms in plants. Those by AUDUS (1962) and WILKINS (1966) cover the earlier work, while more recent articles have been written by AUDUS (1975), TORREY (1976), JUNIPER (1976), WILKINS (1975, 1976a, 1976b, 1977, 1978) and PILET (1977). Most of the more recent studies of the gravitropic response mechanism have been made on the primary roots and coleoptiles of *Zea mays* seedlings, although some attention has been given to dicotyledonous organs such as the shoots of *Helianthus annuus* and the roots of *Phaseolus coccineus*. In order to avoid confusion, the studies on different organs will be reviewed separately. Space precludes an exhaustive and encyclopaedic review of all studies of gravitropic response mechanisms in the past 15 years. This article obviously reflects a personal view and assessment of the investigations that have been published. In the author's opinion there is now a considerable amount of evidence to indicate the kind of mechanism involved in the fine control of cell extension in both coleoptiles and primary roots, at least of *Zea mays* seedlings. If the reader is not convinced by this evidence, and is stimulated to begin experimental investigation in this field, then this review will have been successful, for there is no doubt that many aspects of gravitropism require further detailed investigation, and provide potentially fruitful areas of research for the years ahead.

The principal hypothesis to explain the gravitropic responses of shoots and roots is that proposed independently by CHOLODNY and WENT in 1926. This stated that in vertical organs auxin synthesised at the apex was transported basipetally in a *symmetrical* manner into the subapical zone of cell extension, and as a result the organs grew straight; in a horizontal organ, however, the auxin underwent a downward lateral displacement into the lowermost half of the organ, thus becoming *asymmetrically* distributed. The increased amount of auxin in the lowermost half of a horizontal shoot increased the rate of cell extension, giving rise to upward curvature, whereas the increased amount of auxin in the lowermost half of a horizontal root was held to inhibit cell extension, thereby giving rise to downward curvature.

The validity of the Cholodny-Went hypothesis rests on two criteria: first, the identity of the growth-regulating hormone(s) at the apex of the organ must be established unequivocally by the rigorous physicochemical methods of modern analytical chemistry, and second, the compound(s) must be shown to undergo downward lateral transport and accumulate in the lowermost half of the horizontal organ. Studies which have critically examined these criteria will now be reviewed separately for different organs.

2. Shoots

2.1 Coleoptiles

The growth-rate measurements of NAVEZ and ROBINSON (1933) supported the idea that a growth-regulating hormone was redistributed within a horizontal cereal coleoptile, since the growth rate of the upper half was found to decrease, and that of the lower half to increase, by approximately equal amounts. DOLK (1930) was the first to attempt to determine whether or not a difference was established in the auxin content of the upper and lower halves of a horizontal coleoptile. He used detached coleoptile apices, and subapical coleoptile segments supplied with an apical donor block containing a natural growth-promoting diffusate from coleoptile apices. In each case two agar receiver blocks separated by an impermeable barrier were placed at the basal end of the tissue as illustrated in Figure 1A and B. The lowermost receiver blocks of the horizontal assemblies were found to have a greater capacity to promote cell extension in the *Avena* curvature test than the upper blocks.

Great caution needs to be exercised in interpreting these results for two reasons. Firstly, it is the *net growth-promoting* activity of the receiver blocks which is mea-

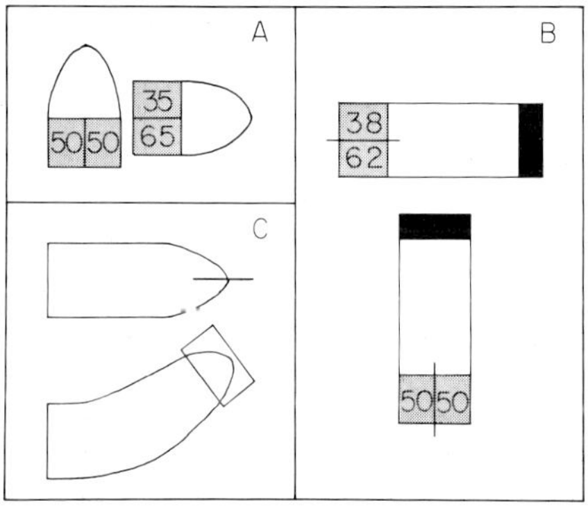

Fig. 1A–C. Diagrams to illustrate the classical procedures used to study the validity of the Cholodny-Went hypothesis. Two receiver blocks (*shaded*) of agar, separated by an impermeable barrier, were placed in contact with the basal end of a detached coleoptile apex (**A**) or a subapical coleoptile segment 10–15 mm in length (**B**). In the latter case, growth-promoting diffusates from a coleoptile apex were applied in an agar donor block (*black*) at the apical end of the segment. The net growth-promoting effect (%) of the diffusate in the receiver blocks is shown by the numbers in the shaded areas. In later experiments the arrangement shown in (**B**) has involved the application of radioactively labelled IAA in the donor block (*black*) in place of the natural tip diffusate employed in earlier studies. **C** The technique of BRAUNER and APPEL (1960) in which an impermeable barrier is inserted into the apex of horizontal coleoptiles so as to prevent (upper diagram) or allow (lower diagram) the lateral movement of growth-regulating compounds from the upper to the lower half of the organ or *vice versa*. The curvature developed after these treatments is indicated by the diagrams. For further discussion see text

sured with the *Avena* curvature test, and this may or may not reflect the amount of auxin they contain, since we now know that a number of growth-promoting and growth-inhibiting compounds are present in plant tissues. DOLK's (1930) results could thus be attributed to the establishment of a difference in the amounts of an inhibitory compound in the two blocks just as readily as to the classic explanation based on a growth-promoting substance (auxin). Secondly, the *only* fact that either of the two experimental systems shown in Figure 1A and B can establish is that a difference in net growth-promoting activity of the two blocks occurs; they cannot provide evidence to show *by what mechanism* this difference arises. The asymmetric distribution might well have arisen from the downward lateral transport of auxin, the explanation usually offered for this type of result, but it could equally well have arisen from the upward transport of an inhibitor. It is, however, quite unnecessary to attribute this result to the lateral transport of a growth-regulating molecule, since the difference in the net growth-promoting activity of the two receiver blocks could have arisen from different rates of synthesis, release, longitudinal transport, metabolism, immobilisation, or secretion of growth hormone by the upper and lower halves of the horizontal organ, or, in addition, by differential uptake of auxin from the donor block in Figure 1B. It is also important to realise that the net growth-promoting activity *accumulates* in the receiver blocks with time and may not reflect the amount of activity in the adjacent tissues with which they are in contact. Obviously it is only the distribution of growth regulators *in the tissues* of the responding organ that is of importance in establishing the validity of the Cholodny-Went hypothesis.

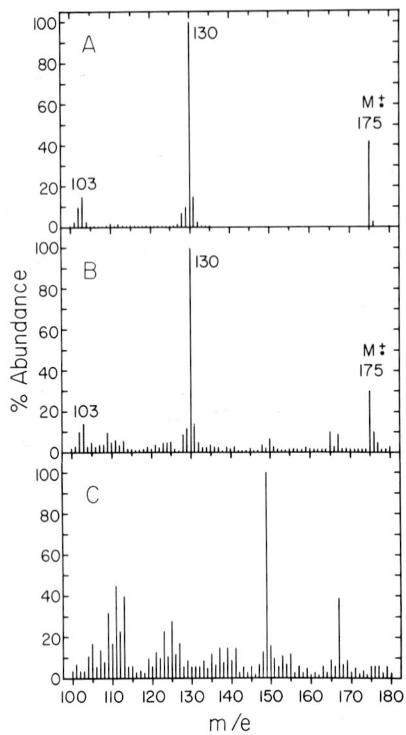

Fig. 2. Mass spectra of (**A**) authentic IAA at a sample concentration of 50 µg in 0.05 cm^{-3} of methanol and a probe temperature 280° C; (**B**) the violet fluorescent spot from the IAA R_f obtained after the chromatographic purification of the diffusate from 15,000 *Zea* coleoptile tips, probe temperature 270° C; and (**C**) the same zone of a chromatogram as used in **B**, except that the extracts were made of blank agar blocks; probe temperature 220° C. (From GREENWOOD et al., 1972)

The experiment of BRAUNER and APPEL (1960), shown in Fig. 1C, provided evidence that the development of gravitropic curvature depended upon the lateral movement of a growth regulator in the horizontal coleoptile. However, the virtual abolition of response in the presence of the horizontal barrier cannot be attributed to the abolition of downward movement of a promoter any more than to the upward movement of an inhibitor.

Work proceeded on auxin physiology for more than 40 years without the chemical identity of the molecule being firmly established. The rather unsatisfactory evidence for the identification of auxin as indol-3yl-acetic acid in the storage tissue (endosperm) of *Zea mays* fruits (HAAGEN-SMIT et al, 1942; BERGER and AVERY, 1944; HAAGEN-SMIT et al., 1946) is really irrelevant to the question of whether IAA is the critical growth-regulating molecule present in the apices and extension zones of growing and developing plant organs where it is held to exercise such an important regulating role.

The identity of the auxin present in coleoptile tips of *Zea mays* was recently established using high-resolution mass spectroscopy (GREENWOOD et al., 1972). Diffusates from 15,000 coleoptile apices were purified chromatographically, and the eluate of the Rf zone at which IAA would be expected to occur was injected directly into a mass spectrometer. The fractionation pattern of the molecule was identical to that of authentic IAA (Fig. 2). The high-resolution molecular-ion mass determination on the sample gave a value of 175,0633, while the calculated, theoretical molecular-ion mass of authentic IAA was 175,0628. Similar evidence was ob-

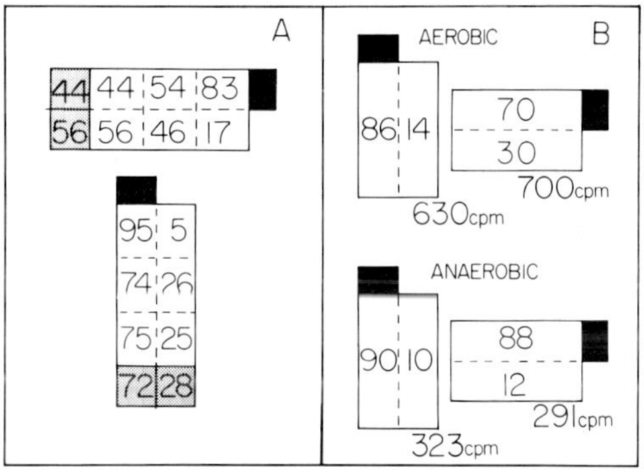

Fig. 3A and B. Diagram illustrating lateral transport of IAA towards the lower side of a horizontal subapical segment of *Zea mays* coleoptile (**A**) and its dependence upon aerobic metabolism (**B**). Agar donor blocks containing [1-^{14}C]IAA are shown in *black*, and agar receiver blocks are *shaded*. *Numbers* refer to the percentage distribution of IAA across each of the three portions of the 15-mm segment and the receiver blocks (**A**) or between the two halves of the 10-mm segment (**B**). Total cpm in the segments is indicated in (**B**) for both aerobic and anaerobic conditions. (Data in **A** from GOLDSMITH and WILKINS, 1964, and in **B** from WILKINS and WHYTE, 1968)

tained when coleoptile tips were extracted in cold methanol. The auxin in *Zea* coleoptile apices is therefore IAA, and since IAA regulates the rate of extension of the cells in the growing zone of *Zea* coleoptiles at low concentrations, its redistribution laterally in a horizontal organ could obviously give rise to differential growth and hence curvature.

The downward lateral transport of IAA in horizontal shoot tissue was first demonstrated unequivocally by GOLDSMITH and WILKINS (1964). Donor blocks were supplied *asymmetrically* to the apical ends of *Zea* coleoptile segments as shown in Figure 3A and B, and the relative distribution of total radioactivity was determined between the two halves of segments that were orientated either vertically or horizontally. The critical point of this experiment is that radioactive IAA found in the non-donated half of the coleoptile *must* have come from the half in contact with the donor block. If the proportion of the total radioactivity found in the non-donated half of the segment differs when the segment is in different orientations with respect to gravity, then a change in the lateral movement of IAA is established with certainty. Approximately twice as much IAA moves to the lower half of a horizontal segment supplied with an upper donor as moves across in a vertical segment (Fig. 3A, B). This polarised lateral transport has subsequently been shown (Fig. 3B) to be dependent upon metabolic energy (WILKINS and WHYTE, 1968).

Results obtained with excised sections of plant organs may not, of course, reflect the behaviour to be found in the intact organ, and so SHAW et al. (1973) ascertained whether or not lateral transport of IAA occurred in intact coleoptiles of *Zea mays* following gravitropic stimulation. Such an investigation was made possible by the development of a micro-application technique for supplying tritiated IAA (5-^{3}H-IAA) of very high specific activity to intact organs at a predetermined point (SHAW and WILKINS, 1974). Fine pipettes, identical to the glass micro-electrodes used in electrophysiology, but having a larger end diameter of 50–100 μm allow the precise application of a pulse of IAA to a tissue by acting as a point source for diffusion. The tip of the pipette just penetrates the cuticle of the coleoptile, and is allowed to remain in position for a few seconds before being withdrawn. Radioactive IAA which diffuses from the pipette can then be traced in the coleoptile tissue as a function of both time and orientation in the gravitational field. The very small amount of tissue damage at the point of penetration of the pipette, and the minute amount of IAA supplied, appeared to have no effect on the coleoptile.

Using this technique, SHAW et al. (1973) demonstrated a strongly polarized, downward, lateral transport of IAA in coleoptiles of intact *Zea mays* seedlings and a less marked, but highly significant downward lateral movement of IAA in coleoptiles of intact *Avena sativa* seedlings (Fig. 4). Chromatographic analysis has shown the distribution of radioactivity in the tissue to reflect the distribution of IAA.

The Cholodny-Went hypothesis can therefore be regarded as valid for *Zea* and *Avena* coleoptiles to the extent that apically synthesised IAA is laterally transported downwards within the horizontal coleoptile and the asymmetric distribution of this growth-promoting hormone in the tissue (GILLESPIE and THIMANN, 1963; GOLDSMITH and WILKINS, 1964) will lead to differential growth and upward curvature. Despite its validity, however, the Cholodny-Went hypothesis may represent only a part of the overall response mechanism even in *Zea* coleoptiles. There are two reasons for believing this to be so. Firstly, another phenomenon, namely an increase in the basipetal longitudinal transport of IAA in the lower half of a horizontal coleoptile, as compared with that in the upper half, is also induced by stimulation of the gravity-sensing mechanism, and will undoubtedly contribute to the asymmetric distribution of IAA in the growing zone of the organ (NAQVI and GORDON, 1966; CANE and WILKINS, 1969). Secondly, there is some evidence of growth hormones other than IAA becoming asymmetrically distributed when the gravity-sensing mechanism is stimulated. RAILTON and PHILLIPS (1973) found endogenous gibberellin-like activity to diffuse from the basal end of a detached 4-mm apex of a *Zea* coleoptile into an agar receiver block over a period of 20 h. Horizontal apices yielded five times more activity than vertical ones. When horizontal apices were placed with the upper and lower halves of their basal cut surfaces in contact with two separate receiver blocks, the lower block was found to have more activity than the upper one in the ratio of approximately 80:20. The increased total yield of gibberellin-like activity from horizontal coleoptile apices suggests a gravity-induced increase in the amount and release of active gibberellins in the tissues, though a decrease in the amount of an inhibiting complex could also account for the observations. The latter possibility is suggested by the occurrence of significant growth-inhibiting activity in the upper receiver block of a horizontal coleoptile apex, and not in the lower block. Exactly how the asymmetry in net gibberellin-like activity is established is not known; downward lateral transport of gibberellins, upward lateral transport of inhibitors, and a differential synthesis, release and basipetal transport of gibberellins, or of inhibiting compounds, by the upper and lower halves of the horizontal coleoptile are obvious possibilities.

There is no evidence of a downward lateral transport of externally applied gibberellin (^{3}H-GA_3) in horizontal *Zea* coleoptiles even after 24 h (WILKINS and NASH, 1974; WEBSTER and WILKINS, 1974), although an upward movement of

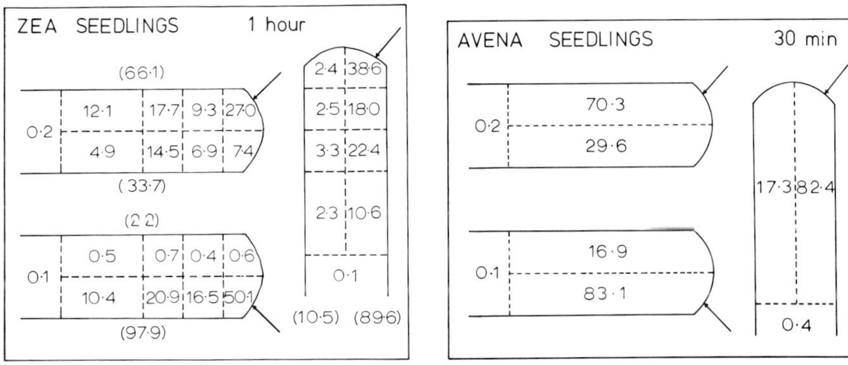

Fig. 4. Diagram illustrating lateral transport of IAA in coleoptiles of intact seedlings of *Zea mays* (*left*) and *Avena sativa* (*right*) after asymmetric point-source micro-application of [^{3}H]IAA at the point shown by the *arrows*. For *Zea* the percentage distribution of IAA is shown in the several portions of the tissue, and the total percentage of IAA in each half of the organ by *numbers in brackets* at the side of each horizontal coleoptile and at the base of the vertical one. For *Avena*, percentage distribution of IAA is shown in the two halves of the tissue. (From SHAW et al., 1973)

radioactivity has been detected using both ^{3}H-GA$_3$ (WEBSTER and WILKINS, 1974) and ^{14}C-GA$_3$ (BRIDGES and WILKINS, unpublished) but not with ^{3}H-GA$_1$ (WEBSTER and WILKINS, 1974). In these studies, little or no basipetal movement of radioactivity was observed in the coleoptile tissue, and none into agar-receiving blocks. It thus seems unlikely that the gibberellin-like activity in the receiver blocks assayed by RAILTON and PHILLIPS (1973) could be attributed to GA$_1$ or GA$_3$. PHILLIPS and HARTUNG (1974) and HARTUNG and PHILLIPS (1974) have also reported only small amounts of radioactivity in receiver blocks which had been in contact with root and shoot segments of *Phaseolus coccineus* supplied at their opposite ends with radioactive gibberellins.

Before further progress can be made, it is essential that the naturally occurring gibberellins in the tissues be identified, and the possible occurrence of asymmetries in their natural distribution ascertained following gravitropic stimulation. Only then, using the appropriate radioactively labelled gibberellins, will it be possible to establish whether or not lateral transport plays a role in the establishment of their asymmetric distribution.

2.2 Dicotyledonous Shoots

Most investigations of the gravitropic response mechanism in dicotyledonous shoots have been made with *Helianthus annuus* using segments of the hypocotyl, segments of the internode above the insertion of the cotyledons or detached apices of green shoots with expanded leaves. Some studies have used green shoots, others have used dark-grown shoots. The tissue being used, and whether it is green or etiolated, is of substantial significance because different mechanisms may be involved in different regions of a shoot, and the amounts and metabolism of growth hormones, particularly the gibberellins (BOWN et al., 1975), are not identical in green and etiolated shoot tissue. The young etiolated shoot just emerging from the seed is the structure which normally executes the gravitropic response beneath the soil in agricultural crops, though for a natural seed falling on the surface, it is likely that exposure to light will take place even before the shoot is long enough to have responded to the gravitational stimulus. Studies of both green and etiolated shoots are therefore equally justified in relation to the environment in which the response naturally occurs.

Several years ago, my colleague I.G. BRIDGES and I briefly studied the movement of IAA in gravitropically stimulated, etiolated hypocotyl segments of *Phaseolus vulgaris* and *Helianthus annuus* and found a small but significant downward lateral movement of radioactivity to occur in horizontal segments supplied asymmetrically with either ^{14}C-IAA or ^{3}H-IAA. The lateral movement was very much less than that found in *Zea* coleoptiles. Although IAA is present in the apical bud of *Phaseolus vulgaris* seedlings (WHITE et al., 1975) and is transported basipetally, the magnitude of the lateral transport of IAA observed in the hypocotyl segments seemed scarcely enough, in itself, to establish an adequate asymmetry in distribution to give rise to the differential growth underlying the gravitropic response. It must be recognized, however, that lateral transport in a segment is probably much less than in the intact organ (GOLDSMITH and WILKINS, 1964; SHAW et al., 1973).

In horizontal, green, first-internode segments of *Helianthus* no asymmetric distribution or lateral transport of either ^{14}C-IAA or ^{14}C-GA_1 could be detected by PHILLIPS and HARTUNG (1976). This pattern of behaviour for IAA thus differs from that found in etiolated hypocotyl segments by BRIDGES and myself.

Removal of the apical bud strongly suppressed internode elongation in seedlings of *Helianthus annuus* and *Phaseolus multiflorus*. Application of GA_3 to the cut shoot restored elongation to that observed in the control, but IAA was without effect (PHILLIPS and HARTUNG, 1976). Simultaneous application of GA_3 and IAA did not lead to a restoration of the normal growth rate; the IAA obviously antagonised the effectiveness of the GA_3. Since the young leaves of the apical bud are the sites of synthesis of gibberellins which appear to regulate the cell elongation underlying internodal growth (JONES and PHILLIPS, 1966; PHILLIPS, 1971), the induction of differential growth during the gravitropic response of a green dicotyledonous shoot could be the result of an asymmetry in the gibberellin distribution being established.

It is not known whether an asymmetrical distribution of naturally occurring IAA develops in horizontal internodes of intact shoots; certainly no lateral transport of externally applied IAA occurs in the internode segments. Even if an asymmetry in IAA distribution were to occur, however, it would be unlikely to cause differential growth since this hormone does not appear to have the capacity to promote cell extension in the green *Helianthus* internode (PHILLIPS and HARTUNG, 1976). Moreover, if an asymmetric distribution of IAA in favour of the lower half of a horizontal shoot of *Phaseolus* or *Helianthus* were to occur, it might, because of its antagonistic effect (PHILLIPS and HARTUNG, 1976), abolish the effectiveness of an asymmetry in the distribution of gibberellins in inducing differential growth and hence upward curvature.

PHILLIPS (1972) has shown that in excised apical buds of light-grown *Helianthus annuus* seedlings, an asymmetrical distribution of gibberellin-like activity develops in basally applied receiver blocks following gravitropic stimulation. The ratio of growth activity between the lower and upper blocks was, on average, about 10:1. Furthermore, the yield of GA_3-equivalents, on a per bud basis, was approximately three times as great in horizontally orientated buds as in vertical ones.

For a number of reasons it is difficult to assess whether or not this asymmetry in gibberellin-like activity is causally related to the gravitropic response of the *Helianthus* shoot. Firstly, the response begins to develop within 1 h and PHILLIPS' data were obtained with buds that had been left for 20 h in a particular orientation with respect to gravity. Hormone distribution data are obviously required after a period of 1–2 h. Secondly, the apparent hormone content of agar receiver blocks must be viewed with caution, since they do not necessarily reflect the amounts of hormone in the adjacent tissues with which they are in contact. Differences in longitudinal transport and secretion of gibberellins, and of other compounds that may antagonise their activity, by the upper and lower halves of the horizontal shoot will in time lead to differences in the *accumulation* of activity in the two blocks without there necessarily being a difference in the amount of hormone per cell in the two halves of the tissue. Thirdly, more than one molecular species is almost certainly present in the receiver blocks. Chromatographic separation of the gibberellins in the blocks revealed activity over the whole of the chromatogram in some cases, and over 7 or 8 of the Rf zones in virtually all of the published results. Furthermore, the total amount of gibberellin-like activity in the blocks could, using a bioassay system, be appreciably affected by acidic inhibiting compounds (the β-inhibitor complex) which might come through the partitioning procedures, irrespective of the Rf zone to which these inhibitors might run. If more β-inhibitor compounds were present in the upper receiver block than in the lower one, then the smaller net gibberellin-like activity in the upper block, and especially its virtual absence from Rf 0.6–1.0, evident quite often in PHILLIPS' (1972) data, might give rise to differences in total gibberellin-like activity between the two receiver blocks. Such an occurrence would give rise only to a difference in gibberellin-like activity in a bioassay system, and not to a real difference in gibberellin content of the blocks.

Notwithstanding these points, if it is accepted that a real redistribution of gibberellins occurs within the tissues, then it has to be established by what mechanism the redistribution is achieved. The threefold increase in the amount of gibberellin-like activity in the receiver blocks of horizontal buds suggests a gravity-stimulated increase in synthesis or release of gibberellins, or an increase in the longitudinal basipetal transport and secretion of gibberellins into the receiver blocks. One or all of these changes might be occurring to a greater extent in the lower half than in the upper half of the horizontal bud, and could lead to the asymmetrical distribution of gibberellin-like activity. Another possibility is that there is an overall decrease or an asymmetric redistribution of inhibiting substances, with little or no change in the level or distribution of gibberellins. Downward lateral transport of gibberellin into the lower half of the horizontal shoot might, of course, take place. Although PHILLIPS and HARTUNG (1976) could find no evidence of this in green internode segments of *Helianthus annuus* using labelled GA_1, it may, in fact, occur only in the most apical part of the shoot or in the bud. It has not been established that GA_1 occurs naturally in *Helianthus*; but even if it does, an absence of its lateral movement does not preclude the possibility that lateral movement of other naturally occurring gibberellins may take place on gravitropic stimulation. Activity broadly spread over the chromatogram (PHILLIPS, 1972) must indicate the presence of a number of gibberellins.

The findings of PHILLIPS (1972) and PHILLIPS and HARTUNG (1976) have opened up a most interesting field of enquiry and raise many important questions. What is now required is an unequivocal identification of the individual gibberellins present in the tissues of the apical buds and subapical parts of the shoot, using physicochemical analytical techniques, and an assessment of whether an asymmetry in the distribution of any of the naturally occurring gibberellins occurs within the tissues of the shoot following gravitropic stimulation.

Whether or not a gradient in naturally occurring IAA develops between the upper and lower halves of the intact, green, horizontal internode of *Helianthus* shoots also needs to be resolved by analytical procedures, since the absence of lateral transport reported by PHILLIPS and HARTUNG (1976) does not rule out the development of an asymmetrical distribution of IAA following gravitropic stimulation.

The lateral transport of IAA in dark-grown, horizontal *Helianthus* hypocotyl segments observed by BRIDGES and myself is compatible with an earlier report that a gradient of ^{14}C was established between the upper and lower halves of segments of hypocotyls of green seedlings supplied symmetrically with ^{14}C-IAA (GILLESPIE and THIMANN, 1963). Furthermore, ABROL and AUDUS (1973a, b) found that ^{14}C-2,4-D underwent lateral displacement towards the lower side of segments of light-grown *Helianthus* hypocotyls. Thus, lateral movement of radioactivity from ^{14}C-IAA, ^{3}H-IAA and ^{14}C-2,4-D in *hypocotyl* segments of *Helianthus* has been observed, and contrasts with the lack of a detectable lateral transport in segments of internodes above the insertion of the cotyledons (PHILLIPS and HARTUNG, 1976). These findings suggest that auxin may be involved in the gravitropic response of the hypocotyl, but not in that of the internodes, in which, on present evidence, the gibberellins may play a significant role. The gravitropic responses of dicotyledonous shoots are in need of further study (WILKINS, 1977).

Several recent investigations have focussed attention on the importance of the epidermis in regulating the rate of extension of the two halves of a gravitropically stimulated shoot, and the control of epidermal cell extension by endogenous hormones. YAMAMOTO et al. (1970) and TAMIMOTO and MASUDA (1971) found auxin-induced extension of pea stem sections to be controlled by regulation of the growth rate of the epidermal cells. Auxin was shown to modify the extensibility of the epidermal cell walls. IWAMI and MASUDA (1973) reported that a strip of filter paper saturated with a buffer solution at pH=3.0 placed on the upper side of a horizontal, light-grown hypocotyl of *Cucumis sativus* eliminated the negative gravitropic response by promoting the growth of the cells with which it was in contact. Similar unilateral treatment of a vertical hypocotyl caused curvature and induced cell-wall loosening of the epidermis.

IWAMI and MASUDA (1974) studied in further detail the gravitropic response in *Cucumis* hypocotyls especially in relation to epidermal cell extension and its control by auxin. Application of auxin to the apical end of intact seedlings promoted elongation of the hypocotyl especially in the uppermost 5-mm portion (Zone I) but also in the next two 5-mm portions (Zones II and III) beneath. In intact, horizontal seedlings gravitropic curvature began within 30 min, commencing in Zone I and moving basipetally. Removal of the apex and the cotyledons did not appreciably affect the gravitropic response, but removal of Zone I of the hypocotyl substantially reduced the response, which could be restored by the application of IAA. The extensibility of the walls of the lower epidermis cells in horizontal hypocotyls was increased significantly after 15 to 30 min of gravistimulation, that is before the onset of gravitropic curvatures. No change in the extensibility of the upper epidermal cells was observed. IWAMI and MASUDA (1974) conclude that Zone I or the region immediately above Zone I is the source of IAA necessary for the gravitropic response, that it undergoes downward lateral transport on gravitropic stimulation, and that the increased amount of IAA in the lower epidermal cells increases proton secretion; this then gives rise to an increase in cell-wall extensibility, and hence growth.

Although most interest has been centred on the asymmetric distribution of growth-regulating molecules in gravitropically stimulated shoots, there have been several studies of the establishment of asymmetry in other cellular components. In particular, the distribution of the cations K^+ and Ca^{2+} have been investigated in gravitropically stimulated hypocotyls of *Helianthus annuus*. BODE (1959, 1960) found a higher proportion of K^+ in the ash of the lower half of the hypocotyl than in that of the upper half, and there was an indication that Ca^{2+} showed the reverse trend. The distribution of Ca^{2+} was also investigated by ARSLAN-CERIN (1966). The seedlings were supplied with $^{45}CaCl_2$ and then placed horizontally for 1, 2, or 3 h. More $^{45}Ca^{2+}$ was found in the upper half of the hypocotyl than in the lower half, especially after 3 h. No such asymmetry could be detected when gravitropic stimulation took place under anoxic conditions, suggesting that metabolic energy was involved in the establishment of the asymmetry. Confirmation of these findings has recently been published by GOSWAMI and AUDUS (1976), who investigated the kinetics of the redistribution. The maximum difference in Ca^{2+} level of about 20% in favour of the upper half of the hypocotyl was attained after about 1 h in the horizontal position, by which time gravitropic curvature was just beginning. An asymmetry in K^+ was also observed in favour of the lower half of the horizontal hypocotyl; it increased linearly with time, having reached significance after 1 h. An asymmetry in ^{32}P appeared to develop in favour of the lower half and was similar in its kinetics to that of K^+ ions. Further study of the $^{45}Ca^{2+}$ asymmetry showed that it did not develop at 4° C or when the hypocotyls were treated with N-l-naphthylphthalamic acid, which also abolishes the gravitropic response.

The role of these ion asymmetries and the mechanism by which they arise are not yet understood. It appears that they could be a consequence of the asymmetry in IAA which arises in gravitropically stimulated hypocotyls. GOSWAMI and AUDUS (1976) established artificial IAA gradients in vertical *Helianthus* hypocotyls following the unilateral application procedure of WILKINS and WOODCOCK (1965); they found more $^{45}Ca^{2+}$ in the non-donated concave side, and more $^{42}K^+$ and ^{32}P in the IAA-rich, donated side of the hypocotyl, as would have been anticipated on the basis of a downward lateral transport of IAA in horizontal hypocotyls. These ionic asymmetries do not appear to arise as a consequence of the differential expansion of cells on the two sides of the gravitropically responding organ, but rather to be caused by the redistribution of growth hormones, particularly IAA. The phenomenon of ion redistribution may thus fall into the category of the "geoelectric" effect which is a consequence of the downward redistribution of IAA in horizontal *Zea* coleoptiles and *Helianthus* hypocotyls. This redistribution of IAA probably gives rise to differences in proton transport across the plasma membranes of cells in the upper and lower halves of the organ, thus generating an electro-potential difference between its upper and lower external surfaces (GRAHM, 1964; GRAHM and HERTZ, 1962, 1964; WOODCOCK and WILKINS, 1969a, 1969b, 1970, 1971). The asymmetric distribution of IAA may induce the asymmetry in Ca^{2+}, K^{+} and ^{32}P distribution by initiating a lateral redistribution of these ions, or differences in their longitudinal movement along the seedling axis. Whether or not the redistribution of these ions has a role in regulating differential cell extension on the two sides of a gravitropically responding hypocotyl remains to be established.

Although all the data discussed hitherto have implied that differential growth in the gravitropic response of a shoot is attributable to the establishment of an asymmetric distribution of growth hormones, FIRN and DIGBY (1976) have suggested that gravity may act directly on the cells of the epidermis, increasing their rate of growth when they are orientated with their outermost side directed downwards. This situation appears to be not unlike the one in grass nodes (BRIDGES and WILKINS, 1973a, b; OSBORNE and WRIGHT, 1977) although in the latter organs the effect is not limited to the epidermal cells. This idea now needs experimental evidence to establish, first, how the epidermal cells detect their orientation in the gravitational field, since they appear to lack sedimentable particles, and second, the precise changes that occur in the extensibility of their walls in different orientations with respect to gravity when they are *completely* isolated from other tissues of the stem.

2.3 Grass Nodes

An important gravitropic response of considerable economic significance is found in the mature shoots of cereals and grasses. The stem within the leaf sheath base is quite supple, and depends for support upon the surrounding sheath. The leaf-sheath base in wheat, for example, retains the potential for growth which does not occur if the organ is maintained in the vertical position (ARSLAN and BENNET-CLARK, 1960). When placed in the horizontal position, however, cells on the lower side begin to grow, thus giving rise to an upward gravitropic curvature. The response is therefore quite different mechanistically from those to be found in primary roots and shoots in that it does not involve modification of the growth rate of already extending cells, but rather the *initiation* of cell extension only in that part of the leaf-sheath base in which the outer epidermis is directed downwards. The implication of this observation is that it is only when the moveable starch grains come to lie against the outermost walls of the cells that the processes resulting in growth are induced. There must therefore be a difference in the relative sensitivities of the different internal faces of the cells to contact with the amyloplasts (see also VOLKMANN and SIEVERS, Chap. 5.2.2, this vol.).

BRIDGES and WILKINS (1973a, b, 1974) made a detailed investigation of the gravitropic response mechanism in the leaf sheath base of wheat. The gravitropic response in these structures does not involve the redistribution of hormonal or other molecules, since growth can be initiated in shoot preparations which have been bisected longitudinally providing they are placed so that the outer epidermis is directed downwards. Furthermore, growth can be induced, or not, in a small piece of excised leaf sheath base according to its precise orientation in the gravitational field. The growth occurs only as long as the cells are held in the growth-inducing orientation with respect to gravity, and stops abruptly when they are replaced in the vertical position.

Of the known major hormones, only IAA was able to initiate growth in vertical leaf sheath base tissue (BRIDGES and WILKINS, 1973a). However, its presence in an active form in normal vertical tissue must be minimal in view of the absence of growth in stems so orientated, though the de novo synthesis, release, or activation of the molecule might occur upon gravity stimulation, especially since the latent period between the onset of gravitropic stimulation and the first detectable curvature is 2.3 h. It is unlikely that a pool of a growth regulator is built up in this time because of the rapid cessation of growth on restoration of the tissue to the vertical position, although a mechanism may exist for the rapid breakdown of IAA on restoration of the tissue to the vertical position. However, no evidence for the presence of IAA in wheat leaf sheath base tissue was obtained using either biological assay or mass-spectrometric methods. Other evidence which tends to argue against the involvement of IAA in the gravitropic response of the leaf sheath bases of wheat is the finding that the morphactin CFM (the methyl ester of 2-chloro-9-hydroxyfluorene-9-carboxylic acid) does not abolish IAA-induced growth, whereas it does eliminate the gravity-induced growth in the tissue.

Growth can be initiated in the leaf sheath base by lowering the pH to 3, and terminated by raising the pH again to 5 or 7. The induction of growth by gravity stimulation and by lowering the pH to 3 are similar in that both responses involve only an increase in cell length, are rapidly terminated on removing the stimulus, have a similar Q_{10}, and are dependent upon cell turgor. They differ, however, in that the acid response is very rapid, occurring about 1.5 min after transfer to a buffer solution at pH 3, and is unaffected by anoxia. The acid-induced growth can be started and stopped repeatedly on sequential transfer to buffers at acid and neutral or alkaline pH (Fig. 5). The maximum growth induced by low pH in non-gravity-stimulated segments of leaf sheath base tissue cannot be further increased by gravity stimulation.

Whether a proton secretion mechanism is involved in the induction of growth that underlies the gravitropic response of the leaf sheath base tissue is not known at present, but it is an interesting possibility in view of the apparent localisation of the induction of growth and the apparent absence of an involvement of hormones. Gravity-induced changes in the secretion of protons from the cytoplasm of cells orientated with their outermost walls directed downwards could undoubtedly initiate the growth necessary for the gravitropic response but this would imply that gravity had some direct control of metabolic processes in the cells which result in the production or secretion of protons from the cytoplasm through the plasmalemma and into the cell wall.

One of the most exciting observations made with the leaf sheath base tissue is that gravity stimulation does have a large effect on at least one enzyme system (BRIDGES and WILKINS, 1974), and this renders more plausible the hypothesis outlined above. ARSLAN and BENNET-CLARK (1960) found a substantial increase in reducing sugars in the lowermost, that is the growing, half of a horizontally placed

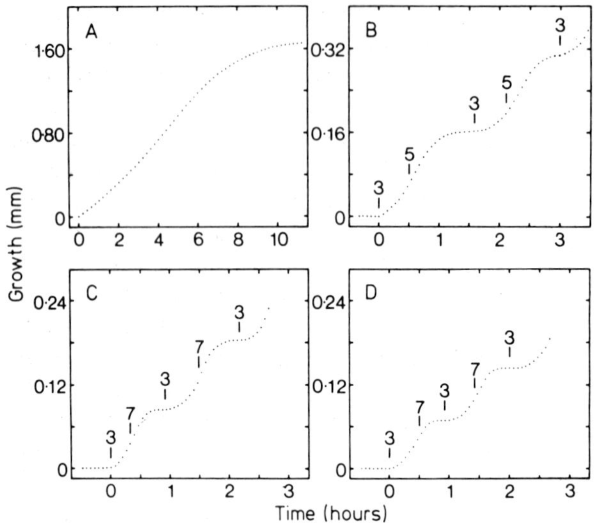

Fig. 5 A–D. Time courses for the development of acid-induced growth in wheat leaf-sheath base tissue at pH 3 (**A**) and for the effects of alternation between buffers of pH 3 and pH 5 (**B**) or pH 3 and pH 7 (**C**, **D**). Growth is initiated using either citrate:phosphate (**A**, **B**, **C**) or glycine:HCl (**D**) buffers and terminated using citrate:phosphate (**B**, **C**) or phosphate:phosphate (**D**) buffers. The times of the pH changes are denoted by *vertical bars.* (Data from BRIDGES and WILKINS, 1973b)

leaf sheath base. Since the relative increases in reducing sugars and in the fresh weight of the tissue were similar, they thought that the increase in cellular sugar concentration might be responsible for the induction of growth. The studies of BRIDGES and WILKINS (1974) confirm the occurrence of a marked increase in reducing sugars in the lower half of the leaf sheath base, but these increases, which involve virtually identical increases in the amounts of both fructose and glucose, are much greater than the increase in fresh weight. The growth is thus unlikely to arise from an increased turgor pressure on the cell wall, a view which is confirmed by the observation that curvature, and hence growth, stops immediately when a leaf sheath base is returned to the vertical position, despite the fact that the large increase in the level of reducing sugars continues to persist for at least 3 h after growth has ceased.

The reducing-sugar content of the lower half of the leaf sheath base in wheat increases to 300%–400% of that of the upper half. The increase involves the hydrolysis of sucrose and will develop even in very mature nodes that have lost the capacity for growth initiation, and hence for developing a gravitropic response. The very marked increase in the capacity of leaf sheath base cells, orientated in a particular way with respect to gravity, to hydrolyse sucrose, draws attention to the most important fact that there may be a direct linkage between the mechanisms controlling the activity, synthesis, or release of an enzyme and the gravity-sensing system in this organ. The concept of a direct control of the level of enzyme activity in a tissue by a gravity-perception mechanism is an exciting possibility which does not appear to have received much attention.

Two further investigations have recently given attention to the gravitropic responses of grass and cereal nodes, and have, for the most part, confirmed the results of ARSLAN and BENNET-CLARK (1960) and BRIDGES and WILKINS (1973a, b, 1974). DAYANANDAN et al. (1976) found that initiation of growth occurs only on the lower side of horizontal leaf sheath bases in *Avena sativa, Triticum sativum, Hordeum vulgare*, and *Secale cereale*. In vertical organs growth was initiated by the application of IAA, 2,4-D and 1-naphthalene acetic acid, but not by gibberellic acid and kinetin, and 2,3,5-TIBA interfered with the IAA-induced and gravity-induced growth response. OSBORNE and WRIGHT (1977) used leaf sheath bases of the grass *Echinochloa colonum*. Some growth (20% in 24 h) was found to occur in *excised* vertical nodes, but not in those in which more than 3 mm of the leaf sheath and stem were retained above the excised organ or in the vertical intact shoot. Significant growth took place only on the lower sides of the horizontal leaf sheath base, and in small longitudinally excised segments only when the outer epidermis is directed downwards. Growth could be initiated in the node tissue by application of IAA, and as found by BRIDGES and WILKINS (1973a), gravity-induced growth could not be further increased by the application of IAA. Estimation of the amount of IAA in the separated upper and lower halves of a horizontally placed node gave values of 5–10 ng/g fresh mass and up to 30 ng/g fresh mass respectively. Although the detailed data are not published in the paper by OSBORNE and WRIGHT (1977), the amount of IAA in the lower half is stated to have increased by at least 40% within 30 min after the tissue is placed in the horizontal position. In this respect the results for *Echinochloa* differ from those in *Triticum* where no IAA was detected (BRIDGES and WILKINS, 1973a). Nevertheless, the observation of a gravity-induced increase in the synthesis or release of a growth hormone such as IAA lays further emphasis on the possible occurrence of a direct control of enzymic activity by the gravity-sensing mechanism as suggested by BRIDGES and WILKINS (1974).

Several aspects of the gravitropic response of grass nodes are of interest in a wider context; it offers a fresh opportunity of studying growth-control mechanisms in plant cells, and provides a tissue in which the regulation of enzyme activity by a gravity-sensing system could be subjected to further investigation.

3. Roots

The growth-regulating mechanism underlying the gravitropic response of roots is less well understood than that in coleoptiles. Indeed, scarcely anything is really known about the mechanisms controlling the growth of roots in their natural environment. The effects of hormones and synthetic growth regulators have, for the most part, been studied by their application to the external surface of excised root segments or cultured roots. It is hard to assess what such studies reveal about the natural mechanisms regulating growth and development, because roots have the capacity to accumulate and metabolise hormones, especially by the tissues of the cortex (GREENWOOD et al., 1973). In their natural environment roots derive neither their major organic nutrients nor their hormones from the exterior; they will, after all, grow perfectly well in humid air. These compounds or their precursors are supplied by other parts of the plant through the sophisticated longitudinal transport systems in the stelar core.

The very existence of growth hormones in roots has been a matter of speculation for a long time, but it is now certain that IAA, cytokinins, and gibberellins are present, even though their precise physiological functions are not known. The transport of IAA is highly polarised towards the root tip, and takes place only in the tissues of the stele (SCOTT and WILKINS, 1968; WILKINS and SCOTT, 1968; BOWEN et al., 1972; SHAW and WILKINS, 1974). Substances which inhibit growth are also present, and at least one of these has its origin in the root cap. Of particular interest in the context of this review is the evidence which shows that a growth regulator from the root cap is involved in bringing about the downward gravitropic response of the primary root.

The observation that removal of the cap abolished the gravitropic response in primary roots of *Zea mays* (JUNIPER et al., 1966) initiated the most recent phase of work on the gravitropic response mechanism in roots. Since the cap is spatially separated from the zone of cell extension in which curvature develops, there must be a communication mechanism between the former and latter regions of the organ. The most likely possibility is that a chemical regulator of growth arises in the root cap, and is transported basipetally to the growing zone. The easiest and most profitable way to investigate whether the apex of an organ is the source of growth regulators, and whether their net effect is a promotion or an inhibition of growth is to remove one half of the apex. By removing one half of the cap of the primary root of *Zea mays* seedlings GIBBONS and WILKINS (1970) showed that the root always developed a large curvature towards the side upon which the remaining half-cap was located, regardless of the orientation of the root with respect to gravity. The cap must therefore be the source of a net growth-inhibiting influence, and for simplicity and ease of description this will be referred to as a growth inhibitor. Clearly more than one substance may be involved, but the simplest hypothesis to explain all the published observations requires no more than one growth-inhibiting compound to be present in the cap.

Detailed studies of the transport of the root cap hormone in primary roots of both *Zea* and *Pisum* have been made by WILKINS et al. (1972) and SHAW and WILKINS (1973), using experimental treatments such as those shown in Figure 6. While removal of one half of the cap always resulted in curvature of the root towards the remaining half-cap (Fig. 6A), excision of one half of the extreme apex of a root from which the whole cap had been removed did not give rise to curvature (Fig. 6B). Thus the root apex, as distinct from the cap, is not the source of any growth regulators, either of a promoting or inhibiting nature. The insertion of small pieces of mica and metal foil into the side of the root cap parallel to the long axis of the root did not induce curvature. It is unlikely, therefore, that surgical trauma is responsible for the curvatures induced by the removal of one half of the cap, a view strongly supported by the fact that replacing one half of the cap immediately after its removal prevents the development of curvature (PILET, 1973).

The insertion of mechanical barriers to longitudinal transport of growth regulators on one side of a root has shown that only growth compounds moving basipetally from the cap to the growing zone are involved in the gravitropic response of the primary root. Barriers inserted on the apical side of the extension zone induce large curvatures in roots with intact caps (Fig. 6C), and none in those from which the caps have been completely removed (Fig. 6D). On the other hand,

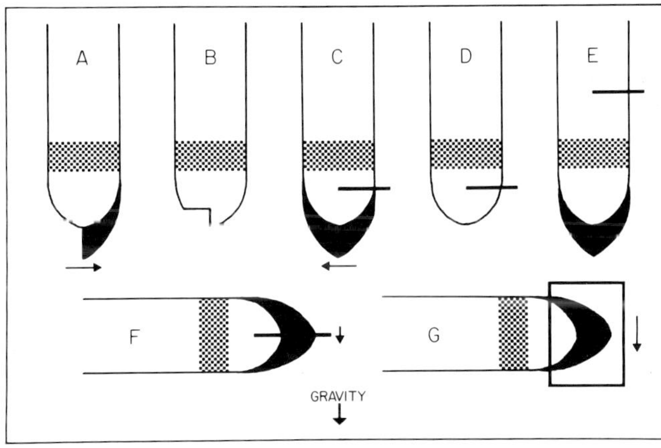

Fig. 6A–G. Diagrammatic representation of the various treatments to which *Zea mays* roots were subjected by SHAW and WILKINS (1973), and to which reference is made in the text. The root cap is shown in *black* and the growing zone of the root is *shaded*. Barriers inserted into the sides of vertical roots are shown in *black* (**C**, **D**, **E**) and the orientation of the apical barriers in the horizontal roots is *horizontal* in (**F**) and *vertical* in (**G**). The direction of any curvature which developed as the result of the treatments is shown by the *black arrows* at the apex of the roots. *Small arrow* in (**F**) indicates a very much smaller curvature than that which was developed in (**G**)

insertion of similar barriers into the side of the root behind the growing zone (Fig. 6E) have no effect on the curvature developed by gravitropically stimulated roots with an intact cap, and do not induce curvature either in roots which have had their caps completely removed, or in vertical roots with intact caps.

If an inhibitor arising in the cap is responsible for the induction of gravitropic curvature in horizontally placed roots, then it is clear that either an asymmetry has to be established in its distribution between the lower and upper halves of the root, or some change in the sensitivity of the cells in the two halves of the root to the inhibitor must develop. There are many ways in which such an asymmetry in inhibitor distribution could arise in a horizontal root. Different rates of synthesis or release of the substance could occur in the upper and lower halves of the root cap, different rates of basipetal transport to the extension zone could take place in the upper and lower halves of the root, or there could be differences in the rate of breakdown or immobilisation of the inhibitor in the two halves of the root. In addition, there could occur a downward lateral transport of inhibitor into the lowermost half of the horizontal root. Experiments by SHAW and WILKINS (1973) have shown that a downward lateral transport of inhibitor from the upper to the lower half does take place in horizontal roots, and that this lateral transport of inhibitor is involved in the gravitropic response mechanism. It must be emphasised, however, that the occurrence of lateral transport of a growth inhibitor in the gravitropic response of both *Pisum* and *Zea* roots does not eliminate the involvement, in addition, of any or all of the other suggested mechanisms which could lead to the asymmetry in inhibitor distribution.

The lateral transport of inhibitor in gravitropically stimulated *Zea* roots has been established by comparing the curvatures developed by roots which have

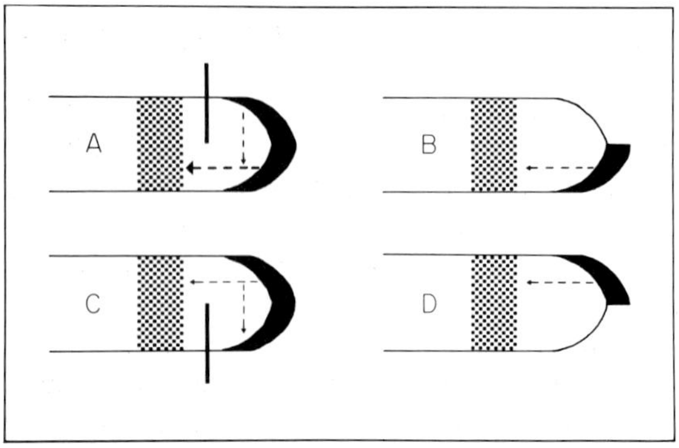

Fig. 7 A–D. Diagram to explain the differences in curvature developed by roots of *Zea mays* into which a barrier had been inserted or from which one half of the cap had been removed. Root caps are shown in *black* and *broken arrows* indicate the movement of inhibitor. The lateral movement of inhibitor takes place largely in the cap, but for clarity is shown occurring in the root apex. The growing zone of the root is *shaded*. Root **A** bends down to a greater extent than root **B**, while root **C** bends up to a smaller extent than root **D**. (After SHAW and WILKINS, 1973)

had either one half of their caps removed, or their caps left intact and a barrier to longitudinal transport of inhibitor inserted into either the upper or the lower half of the root between the cap and the extension zone. These treatments are illustrated in Figure 7. Root A developed a larger downward curvature than root B. The only explanation for this difference is that more inhibitor reaches the lower half of the growing zone in A than in B, and the only place whence this extra inhibitor can come is the upper half cap. If this deduction is correct and lateral transport does occur, then the opposite result would be expected in the cases of roots C and D; C must bend upwards less than D because lateral transport of the inhibitor will occur in C and reduce the amount of inhibitor moving basipetally into the upper half of the extension zone. The laterally transported inhibitor will accumulate on the apical side of the barrier and be unable to reach the sensitive cells in the lower half of the growing zone. In root D there is no lower half-cap into which the inhibitor can be laterally transported, and more will therefore be transported basipetally into the upper half of the extending zone than is the case in root C. The experimental results are in accord with this prediction. Root C was found to bend less than root D (SHAW and WILKINS, 1973). The validity of the deduction is amenable to further experimental test since, if it is correct, insertion of barriers to downward movement of growth substances in the root cap *must* decrease the gravitropic response. The experimental treatments are shown in Figure 6F and G, and it was found that the horizontal barrier (Fig. 6F) reduced gravitropic curvature to about 10°, or 30% of that achieved by roots having a vertical barrier (Fig. 6G). These results support the view that lateral transport of an inhibitor does take place, and show the process to be an integral part of the gravitropic response system (SHAW and WILKINS, 1973).

In summary, therefore, at least one inhibitor of growth arises in the root cap and undergoes downward lateral movement in horizontal roots to give rise, at least in part, to the positive gravitropic response. Whether or not other mechanisms contribute to the differential growth underlying the gravitropic response has yet to be resolved.

The chemical identity of the inhibitory compounds in the root cap obviously needs to be established. According to the Cholodny-Went hypothesis, the inhibitor involved in the gravitropic response of roots was auxin (IAA) but this now seems to be very unlikely. The presence of IAA in roots has only recently been unequivocally established with mass spectrometry (BRIDGES et al., 1973; ELLIOTT and GREENWOOD, 1974). In *Zea* roots, BRIDGES et al. (1973) found that IAA was virtually confined to the stele, with very little occurring in the cortex. Some IAA was present in the root apex. Auxin is transported in a strongly polarised manner from the base towards the apex in roots (SCOTT and WILKINS, 1968) and the polar transport is confined to the tissues of the stele in which IAA naturally occurs (BOWEN et al., 1972; SHAW and WILKINS, 1974). Recently RIVIER and PILET (1974) demonstrated the presence of IAA in the root cap of *Zea mays* by mass spectrometry. Broadly speaking, the polarity of IAA *transport* is in the wrong direction for this hormone to be involved in the gravitropic response of roots, and SHAW and WILKINS (unpublished), using their micro-application technique (cf. p. 10), were unable to detect a downward lateral transport of IAA in gravitropically stimulated *Zea mays* roots. Several other studies of the growth hormones present in root caps have failed to reveal the presence of growth-active compounds at the Rf of IAA when the zones of the chromatograms have been examined with bioassay procedures. KUNDU and AUDUS (1974) found only one inhibitor of root growth to be present in root caps of *Zea*. It occurred at an Rf closely similar to that of abscisic acid in the one solvent system employed, and it was certainly not IAA. H. WILKINS and WAIN (1974) also examined the inhibitors in caps of the primary roots of *Zea mays* seedlings, using both a physicochemical method and a wheat root bioassay procedure. After three successive purifications of the zone at the Rf of abscisic acid (ABA) in different chromatographic systems, optical rotatory dispersion curves showed ABA to be the only growth-active substance present in the final chromatogram. Abscisic acid and two other, as yet unidentified, inhibitors were present in the caps of roots that had been exposed to white light, but not in those which had been kept in darkness. They found no evidence for the presence of IAA. RIVIER et al. (1977) have recently unequivocally demonstrated with combined gas chromatography–mass spectrometry the presence of ABA in root caps of *Zea mays* seedlings. The cap and apex contained 36.1 μg kg^{-1} and 66.5 μg kg^{-1} respectively. At present, there is but one report of IAA occurring in the root cap of *Zea* (RIVIER and PILET, 1974), while there are a number of studies in which this compound was not detected. Further study of the presence of IAA in root caps is needed to clarify these conflicting reports.

There is no doubt, however, about the presence of ABA in root caps of *Zea mays*, and the question which immediately arises is whether ABA is the critical growth inhibitor responsible for bringing about gravitropic curvature. There is now a substantial body of evidence to suggest that this is so, and it may be summarised as follows:

1. At concentrations between 10^{-8} M and 10^{-4} M, AB inhibits root cell elongation when applied to detipped roots of *Zea* in agar blocks, and the roots develop a substantial curvature when it is supplied asymmetrically even at 10^{-8} M. The asymmetric application of IAA did not induce curvature (PILET, 1975).

2. In some varieties of *Zea mays*, the primary roots do not have a positive gravitropic response in darkness, but become responsive after exposure to light (SCOTT and WILKINS, 1969). Unpublished data of GIBBONS and WILKINS show that the responsiveness to gravity induced by exposure to light declines with time in darkness and is lost after about 8–10 h. It can be reinduced, however, by another exposure to light. These results suggest that light makes available a critical component in the gravitropic response mechanism which is absent in dark-grown roots; the simplest possibility is that a specific amount of a growth inhibitor is synthesised or released in the cap following irradiation, and then transported basipetally and ultimately metabolised after having exerted its effect in the growing zone of the root.

3. Light inhibits the rate of elongation of *Zea* roots. By separating the cap from the root, irradiating each independently, and then replacing the cap on the root, the site of perception of the light stimulus was shown to be confined solely to the cap (H. WILKINS and WAIN, 1974). Moreover, when orientated in the horizontal position a light-treated cap placed on a dark-grown root apex gave a gravitropic curvature similar to that of an intact illuminated root, whereas a cap kept in darkness did not give rise to any gravitropic curvature when placed on a light-treated root (H. WILKINS and WAIN, 1975a). Light therefore appears to lead to the production or release of an inhibitor of growth in the root cap, and without this compound no gravitropic response can develop.

4. Transfer of a cap from dark-grown roots of the Anjou variety of *Zea mays*, which is gravitropically responsive in the dark, to a decapped, dark-grown root of the Kelvedon 33 variety, which is light-requiring, leads to the development of a gravitropic response even in darkness (PILET, 1976).

5. H. WILKINS and WAIN (1974) have clearly demonstrated that in darkness ABA is not present in the root caps of the light-requiring variety of *Zea* LG11. After exposure of the caps to light, however, ABA is present. The evidence cited above suggests that the absence of gravitropic responsiveness in a dark-grown *Zea* root is due to the absence of ABA in the cap. If this is so, then a *dark-grown*, horizontal root should develop a gravitropic response when supplied in the dark with exogenous ABA. When roots were kept wholly in darkness, those which were intact developed a positive gravitropic response on immersion in a 10^{-9} M solution of ABA, whereas those which had been decapped did not (H. WILKINS and WAIN, 1975b).

The evidence summarised above strongly supports the view that abscisic acid is the growth inhibitor which is present in the root cap, and upon which the gravitropic response of the *Zea mays* root depends. The response appears to involve the lateral transport of ABA in the root cap and the subsequent basipetal transport if the asymmetrically distributed ABA into the zone of cell extension of the root. The difference in the amounts of ABA reaching the upper and lower halves of the growing zone of the horizontal root leads to unequal cell extension, and the appearance of downward curvature.

Despite the weight of evidence supporting this thesis and its inherent attractiveness over the older notion involving supra-optimal "concentrations" of IAA, for which, incidentally, there is no evidence of any substance, two further pieces of information are required before its validity can be fully accepted. Both involve considerable difficulty from the technical point of view. It must be established first, that a gradient in naturally occurring ABA develops between the upper and lower halves of a horizontally placed *Zea mays* root, and second, that a downward lateral transport of authentic, radioactively-labelled ABA takes place in horizontal *Zea* root caps. Both phenomena must be shown to occur before, or simultaneously with, the onset of the gravitropic curvature.

The possible occurrence of a lateral transport of ABA in gravitropically stimulated roots of *Phaseolus coccineus* seedlings was investigated recently by HARTUNG (1976). Unilateral microapplication of ^{14}C-ABA to the upper or lower side of the caps of horizontal roots revealed a significant *upward* lateral movement of ^{14}C in the tissue after 1 h and 3 h, but no difference was detectable after 6 h. Although HARTUNG's results are clearly contrary to the thesis being advanced in this review, there are several points which need to the considered carefully in their evaluation. Firstly, these experiments were carried out on roots of *Phaseolus coccineus* rather than *Zea mays*, and it is just possible that in dicotyledonous roots the mechanism may be different from that found in cereal roots. Secondly, the ^{14}C-ABA was of very low specific activity, 45 μCi mg^{-1}, and thus had to be prepared at a high concentration (1 mg ml^{-1}) for use with the micropipette method. With such a high concentration of a physiologically active molecule, the amount in the tissue after a 20-s application might well attain toxic levels. Thirdly, and of the greatest importance, the point of application was stated to be 1–1.5 mm behind the root tip. Since the length of the root cap is probably not more than about 2.0 mm, the application of ABA was apparently made to its most basal region. It might have been better to apply ABA nearer the apical extremity of the cap so that it would have passed through the cap tissues in which the gravity-sensing mechanism is located. HARTUNG (1976) has clearly attempted one of the critical experiments to test the current hypothesis of root gravitropism, but the technical difficulties arising from the low specific activity of the ABA, and the point of application of the inhibitor, make full assessment of his findings somewhat difficult. A repetition of these experiments on roots of *Zea mays* with ABA of higher specific activity, application nearer the apical extremity of the cap, and analysis of the radioactivity in the tissue would be of the utmost interest.

The production or release of abscisic acid in the root cap, its lateral transport towards the lower half of the organ upon gravity stimulation, and its basipetal movement to the growing zone of the root therefore appear to be major features of the gravitropic response mechanism in *Zea* roots. The *principle* of this mechanism is obviously that suggested in the Cholodny-Went hypothesis but, of course, the original hypothesis involved IAA rather than ABA, and this now appears to be incorrect. However, the extent to which other processes are also involved in the development of gravitropic curvature is totally unknown at the present time. The possible occurrence of differences (a) in the longitudinal transport of inhibitor in the upper and lower halves of the horizontal root, (b) in the production, release or activation of the inhibitor in the upper and lower halves of the horizontal root cap and (c) in the sensitivity of the target cells in the upper and lower halves of the growing zone of the root must be investigated.

What then is the role of IAA in roots of *Zea mays*? The acropetal flux of IAA in the stele may have an important part to play in several aspects of the growth and development of the root system. It may be necessary for the occurrence of cell extension and cell division in the root, and also for the initiation of lateral roots. What the author of the present review now seriously questions is that IAA is associated with the *fine control* of the rate of cell extension in a gravitropically stimulated root; the fine control that leads to a slightly lower growth rate on the lower side of the root as compared to the upper side and hence to the downward gravitropic response.

Two recent studies have, however, examined the possibility that basipetally moving IAA might have a role in the gravitropic response of roots in addition to that played by ABA. PERNET and PILET (1976) applied 5-^{3}H-IAA in an agar donor block to the apex of the cap of a detached 10-mm root of *Zea mays*, and noted the movement of radioactivity into the root after 1 and 3 h. The basipetal movement of radioactivity was very slow, scarcely any having reached the growing zone after 1 h, and very little after 3 h. Most of the radioactivity appeared to accumulate in the basal region of the cap itself. PERNET and PILET (1976) conclude that the rate of movement of radioactivity from the cap to the root apex was too slow for IAA to be involved in the gravitropic response. Unfortunately it was not established whether the radioactivity in the root was still associated with the IAA molecule. This is highly unlikely since roots of *Zea mays* have a very high capacity to metabolise IAA applied externally (GREENWOOD et al., 1973).

DAVIES et al. (1976) loaded anion exchange resin beads with ^{3}H-IAA and then applied a bead to a specific area of the surface of the root cap or the root of *Zea mays* seedlings. The distribution of IAA along the root was monitored after 4 h. From such a continuous point source of IAA, radioactivity was found to move basipetally through the root cap, into the growing zone of the root and to a distance of 8 mm behind the apex. The profile of radioactivity distribution was not identical with the exponential decrease expected on the basis of physical diffusion, and the authors conclude that a basipetal transport of IAA was taking place. It would have been of interest to establish whether a similar profile occurred under anaerobic conditions, and whether a pulse of IAA moved basipetally, since this would have indicated whether diffusion, an active transport, or both types of movement were involved. DAVIES et al. (1976) did not establish that the radioactivity in the root was confined to the IAA molecule. However, when a bead loaded with a higher level of unlabelled IAA was applied asymmetrically to the cap of a vertical root a significant curvature developed. This finding does not mean, however, that the physiological effect must be associated directly with the applied compound (IAA) or one of its metabolically produced derivatives. The asymmetrically applied IAA could, for example, have induced the release of an endogenous inhibitory molecule on that side of the cap, and this alone, or together with the applied compound or its metabolic product, could have been transported basipetally and given rise to the curvature.

Finally, there remains the question of whether any of the growth hormones other than IAA are involved in root gravitropism. The report by EL-ANTABLY and LARSEN (1974) that more gibberellin-like activity can be extracted from the upper half than from the lower half of horizontal roots of *Vicia faba* raises the possibility that an asymmetry in gibberellin distribution is established as the result of gravitropic stimulation. Such a possibility is also evident in data of WEBSTER and WILKINS (1974) which show a net upward movement of radioactivity supplied asymmetrically in the GA_3 molecule in horizontal roots of *Zea mays* seedlings.

While the significance of an upward movement of a gibberellin which, at low concentrations, promotes cell elongation at least in roots of *Vicia faba* (EL-ANTABLY and LARSEN, 1974) can be appreciated, much more extensive experimental evidence will be required before it can be established whether gibberellins have a role in root gravitropism.

4. Conclusions

Progress has been made over the past decade in elucidating the mechanisms of the gravitropic responses in shoots and roots, but much remains to be done since the mechanisms are almost certainly more complex than those described in this review. One aspect of the mechanisms about which nothing is known is the way in which stimulation of the gravity-sensing system initiates the lateral transport of growth regulators. The many unsolved problems present a formidable intellectual and technological challenge even in orthotropic organs, while the mechanisms involved in plagiogeotropic and diageotropic organs are virtually unexplored. The continued application of the most rigorous methodology of analytical chemistry (HILLMAN, 1978) and of increasingly sophisticated physiological techniques is of the utmost importance if the advance in our understanding of the gravitropic response systems is to be maintained. Many of the studies that are now required are either tremendously tedious or very labour-intensive, involving the manipulation of very large amounts of carefully dissected tissue, but it is in such studies that future progress lies.

References

Abrol, B.K., Audus, L.J.: The lateral transport of 2,4-dichlorophenoxyacetic acid in horizontal hypocotyl segments of *Helianthus annuus*. J. Exp. Bot. *24*, 1209–1223 (1973a)

Abrol, B.K., Audus, L.J.: The effects of N-1-naphthylphthalamic acid and (2-chloroethyl)-phosphonic acid on the gravity-induced lateral transport of 2,4-dichlorophenoxyacetic acid. J. Exp. Bot. *24*, 1224–1230 (1973b)

Arslan, N., Bennet-Clark, T.A.: Geotropic behaviour of grass nodes. J. Exp. Bot. *11*, 1–12 (1960)

Arslan-Cerin, N.: The redistribution of radioactivity in geotropically stimulated hypocotyls of *Helianthus annuus* pretreated with radioactive calcium. J. Exp. Bot. *17*, 236–240 (1966)

Audus, L.J.: The mechanism of the perception of gravity by plants. Symp. Soc. Exp. Biol. *16*, 197–226 (1962)

Audus, L.J.: Geotropism in roots. In: The Development and Function of Roots. Torrey, J., Clarkson, D.T. (eds.). London: Academic Press, 1975, pp. 327–363

Berger, J., Avery, G.S.: Isolation of an auxin precursor and an auxin (indoleacetic acid) from maize. Am. J. Bot. *31*, 199–203 (1944)

Bode, H.R.: Über den Einfluß des Heteroauxins auf die Kationenzusammensetzung der Blattasche der Tomate. Planta *53*, 212–218 (1959)

Bode, H.R.: Über den Einfluß der Geoinduktion auf die Kationenverteilung im Hypokotyl von *Helianthus annuus*. Planta *54*, 15–33 (1960)

Bowen, M.R., Wilkins, M.B., Cane, A.R., McCorquodale, I.: Auxin transport in roots. VIII. The distribution of radioactivity in the tissues of *Zea* root segments. Planta *105*, 273–292 (1972)

Bown, A.W., Reeve, D.R., Crozier, A.: The effect of light on gibberellin metabolism and growth of *Phaseolus coccineus* seedlings. Planta *126*, 83–91 (1975)

Brauner, L., Appel, E.: Zum Problem der Wuchsstoffquerverschiebung bei der geotropischen Induktion. Planta *55*, 226–234 (1960)
Bridges, I.G., Hillman, J.R., Wilkins, M.B.: Identification and localisation of auxin in primary roots of *Zea mays* by mass spectroscopy. Planta *115*, 189–192 (1973)
Bridges, I.G., Wilkins, M.B.: Growth initiation in the geotropic response of the wheat node. Planta *112*, 191–200 (1973a)
Bridges, I.G., Wilkins, M.B.: Acid-induced growth and the geotropic response of the wheat node. Planta *114*, 331–339 (1973b)
Bridges, I.G., Wilkins, M.B.: The role of reducing sugars in the geotropic response of the wheat node. Planta *117*, 243–250 (1974)
Cane, A.R., Wilkins, M.B.: Independence of lateral and differential longitudinal movement of indoleacetic acid in geotropically stimulated coleoptiles of *Zea mays*. Plant Physiol. *44*, 1481–1487 (1969)
Cholodny, N.: Beiträge zur Analyse der geotropischen Reaktion. Jahrb. Wiss. Bot. *65*, 447–459 (1926)
Davies, P.J., Doro, J.A., Tarbox, A.W.: The movement and physiological effect of indoleacetic acid following point applications to root tips of *Zea mays*. Physiol. Plant. *36*, 333–337 (1976)
Dayanandan, P., Hebard, F.V., Kaufman, P.B.: Cell elongation in the grass pulvinus in response to geotropic stimulation and auxin application. Planta *131*, 245–252 (1976)
Dolk, H.E.: Geotropie en groestof. Dissertation, Utrecht (1930)
El-Antably, H.M.M., Larsen, P.: Redistribution of endogenous gibberellins in geotropically stimulated roots. Nature (London) *250*, 76–77 (1974)
Elliott, M.C., Greenwood, M.S.: Indol-3yl-acetic acid in roots of *Zea mays*. Phytochemistry *13*, 239–241 (1974)
Firn, R.D., Digby, J.: Shoot geotropism: new interpretations of old observations. What's New Plant Physiol. *8* (1976)
Gibbons, G.S.B., Wilkins, M.B.: Growth inhibitor production by root caps in relation to geotropic responses. Nature (London) *226*, 558–559 (1970)
Gillespie, B., Thimann, K.V.: Transport and distribution of auxin during tropistic response. I. The lateral migration of auxin in geotropism. Plant Physiol. *38*, 214–225 (1963)
Goldsmith, M.H.M., Wilkins, M.B.: Movement of auxin in coleoptiles of *Zea mays* L. during geotropic stimulation. Plant Physiol. *39*, 151–162 (1964)
Goswami, K.K.A., Audus, L.J.: Distribution of calcium, potassium and phosphorus in *Helianthus annuus* hypocotyls and *Zea mays* coleoptiles in relation to tropic stimuli and curvatures. Ann. Bot. *40*, 49–64 (1976)
Grahm, L.: Measurement of geoelectric and auxin induced potentials in coleoptiles using a refined vibrating electrode technique. Physiol. Plant. *17*, 231–261 (1964)
Grahm, L., Hertz, C.R.: Measurement of geoelectric effect in coleoptiles by a new technique. Physiol. Plant. *15*, 96–114 (1962)
Grahm, L., Hertz, C.R.: Measurement of the geoelectric effect in coleoptiles. Physiol. Plant. *17*, 186–201 (1964)
Greenwood, M.S., Hillman, J.R., Shaw, S., Wilkins, M.B.: Localization and identification of auxin in roots of *Zea mays*. Planta *109*, 369–374 (1973)
Greenwood, M.S., Shaw, S., Hillman, J.R., Richie, A., Wilkins, M.B.: Identification of auxin from *Zea* coleoptile tips by mass spectrometry. Planta *108*, 179–183 (1972)
Haagen-Smit, J.A., Dandliker, W.B., Wittwer, S.H., Murmeek, A.E.: Isolation of 3-indole acetic acid from immature corn kernels. Am. J. Bot. *33*, 118–120 (1946)
Haagen-Smit, A.J., Leech, W.D., Bergren, W.R.: The estimation, isolation and identification of auxins in plant materials. Am. J. Bot. *29*, 500–506 (1942)
Hartung, W.: Der basipetale [2-^{14}C] Abscisinsäuretransport in Wurzeln intakter Bohnenkeimlinge und seine Bedeutung für den Wurzelgeotropismus. Planta *128*, 59–62 (1976)
Hartung, W., Phillips, I.D.J.: Basipetally polarised transport of [^{3}H] gibberellin A_1 and [^{14}C] gibberellin A_3 and acropetal polarity of [^{14}C] indole-3-acetic acid transport in stelar tissues of *Phaseolus coccineus* roots. Planta *118*, 311–322 (1974)
Hillman, J.R. (Ed.): Isolation of plant growth substances. Cambridge University Press. London, New York, Melbourne (1978)
Iwami, S., Masuda, Y.: Hydrogen ion-induced curvature in cucumber hypocotyls. Plant Cell Physiol. *14*, 757–762 (1973)

Iwami, S., Masuda, Y.: Geotropic response of cucumber hypocotyls. Plant Cell Physiol. *15*, 121–129 (1974)
Jones, R.L., Phillips, I.D.J.: Organs of gibberellin synthesis in light grown sunflower plants. Plant Physiol. *41*, 1381–1386 (1966)
Juniper, B.J.: Geotropism. Annu. Rev. Plant Physiol. *27*, 385–406 (1976)
Juniper, B.E., Groves, S., Landau-Schacher, B., Audus, L.J.: Root cap and the perception of gravity. Nature (London) *209*, 93–94 (1966)
Kundu, K.K., Audus, L.J.: Root growth inhibitors from root cap and root meristem of *Zea mays*. J. Exp. Bot. *25*, 479–489 (1974)
Naqvi, S.M., Gordon, S.A.: Auxin transport in *Zea mays* L. coleoptiles. I. Influence of gravity on the transport of indoleacetic acid-2-^{14}C. Plant Physiol. *41*, 1113–1118 (1966)
Navez, A.E., Robinson, T.W.: Geotropic curvature of *Avena* coleoptiles. J. Gen. Physiol. *16*, 133–145 (1933)
Osborne, D.J., Wright, M.: Gravity induced cell elongation. Proc. R. Soc. Lond. B. *199*, 551–564 (1977)
Pernet, J.J., Pilet, P-E.: Indoleacetic acid movement in the root cap. Planta *128*, 183–184 (1976)
Phillips, I.D.J.: Maintenance of gibberellin biosynthesis in excised sunflower apical buds by exogenous sucrose and mevalonate. Planta *96*, 27–34 (1971)
Phillips, I.D.J.: Endogenous gibberellin transport and biosynthesis in relation to geotropic induction of excised sunflower shoot tips. Planta *105*, 234–244 (1972)
Phillips, I.D.J., Hartung, W.: Basipetal and acropetal transport of [3,4-^{3}H] gibberellin A_1 in short and long segments of *Phaseolus coccineus* second internode. Planta *116*, 109–121 (1974)
Phillips, I.D.J., Hartung, W.: Longitudinal and lateral transport of [3,4-^{3}H] gibberellin A_1 and 3-indolyl (acetic acid-2-^{14}C) in upright and geotropically responding green internode segments from *Helianthus annuus*. New Phytol. *76*, 1–9 (1976)
Pilet, P-E.: Growth inhibitor from the root cap of *Zea mays*. Planta *111*, 275–278 (1973)
Pilet, P-E.: Abscisic acid as a root growth inhibitor: physiological analyses. Planta *122*, 299–302 (1975)
Pilet, P-E.: The light effect on the growth inhibitors produced by the root cap. Planta *130*, 245–249 (1976)
Pilet, P-E.: Root georeaction: gravity effect on hormone balance. In: Life Sciences Research in Space. Burke, W.R., Guyenne, T.D. (eds.). Noordwijk, The Netherlands: E.S.A. Scientific and Technical Publications Branch, 1977, pp. 223–227
Railton, I.D., Phillips, I.D.J.: Gibberellins and geotropism in *Zea mays* coleoptiles. Planta *109*, 121–126 (1973)
Rivier, L., Milon, H., Pilet, P-E.: Gas chromatography—Mass spectrometric determinations of abscisic acid levels in the cap and the apex of maize roots. Planta *134*, 23–27 (1977)
Rivier, L., Pilet, P-E.: Indolyl-3-acetic acid in cap and apex of maize roots: identification and quantification by mass fragmentography. Planta *120*, 107–112 (1974)
Scott, T.K., Wilkins, M.B.: Auxin transport in roots. II. Polar flux of IAA in *Zea* roots. Planta *83*, 323–334 (1968)
Scott, T.K., Wilkins, M.B.: Auxin transport in roots. IV. Effects of light on IAA movement and geotropic responsiveness in *Zea* roots. Planta *87*, 249–258 (1969)
Shaw, S., Gardner, G., Wilkins, M.B.: The lateral transport of IAA in intact coleoptiles of *Avena sativa* L. and *Zea mays* L. during geotropic stimulation. Planta *115*, 97–111 (1973)
Shaw, S., Wilkins, M.B.: The source and lateral transport of growth inhibitors in geotropically stimulated roots of *Zea mays* and *Pisum sativum*. Planta *109*, 11–26 (1973)
Shaw, S., Wilkins, M.B.: Auxin transport in roots. X. Relative movement of radioactivity from IAA in the stele and cortex of *Zea* root segments. J. Exp. Bot. *25*, 199–207 (1974)
Tamimoto, E., Masuda, Y.: Role of the epidermis in auxin-induced elongation of light grown pea stem segments. Plant Cell Physiol. *12*, 663–673 (1971)
Torrey, J.G.: Root hormones and plant growth. Annu. Rev. Plant Physiol. *27*, 435–459 (1976)
Webster, J.H., Wilkins, M.B.: Lateral movement of radioactivity from [^{14}C] gibberellic acid (GA_3) in roots and coleoptiles of *Zea mays* L. seedlings during geotropic stimulation. Planta *121*, 303–308 (1974)

Went, F.W.: On growth accelerating substances in the coleoptile of *Avena sativa*. Proc. K. Akad. Wet. Amsterdam *30*, 10–19 (1926)

White, J.C., Medlow, G.C., Hillman, J.R., Wilkins, M.B.: Correlative inhibition of lateral bud growth in *Phaseolus vulgaris* L. Isolation of indoleacetic acid from the inhibitory region. J. Exp. Bot. *26*, 419–424 (1975)

Wilkins, H., Wain, R.L.: The root cap and control of root elongation in *Zea mays* L. seedlings exposed to white light. Planta *121*, 1–8 (1974)

Wilkins, H., Wain, R.L.: The role of the root cap in the response of the primary roots of *Zea mays* L. seedlings to white light and gravity. Planta *123*, 217–222 (1975a)

Wilkins, H., Wain, R.L.: Abscisic acid and the response of the roots of *Zea mays* seedlings to gravity. Planta *126*, 19–23 (1975b)

Wilkins, M.B.: Geotropism. Annu. Rev. Plant Physiol. *17*, 379–408 (1966)

Wilkins, M.B.: The role of the root cap in root geotropism. Curr. Adv. Plant Sci. *13*, 317–328 (1975)

Wilkins, M.B.: Identification of auxin in growing plant organs and its role in geotropism. In: Perspectives in Experimental Biology, Vol. 2, Sutherland, N. (ed). Oxford-New York: Pergamon 1976a pp. 77–78

Wilkins, M.B.: Gravity-sensing guidance systems in plants. Science Progress, Oxford *63*, 187–217 (1976b)

Wilkins, M.B.: Gravity and light-sensing guidance systems in primary roots and shoots. In: Integration of Activity in the Higher Plant. Jennings, D.H. (ed.). Proc. Symp. Soc. Exp. Biol. *31*, 1977 pp. 275–335

Wilkins, M.B.: Gravity-sensing guidance systems in roots and shoots. Bot. Mag. Tokyo. Special Issue No. 1. 255–277 (1978)

Wilkins, M.B., Gibbons, G.S.B., Shaw, S.: The source and transport of growth regulators responsible for the geotropic response of *Zea mays* roots. In: Plant Growth Regulators 1970. Carr, D.J. (ed.). Proc 7th Int. Conf. Plant Growth Substance, Canberra, 1970. Heidelberg: Springer, 1972, pp. 717–724

Wilkins, M.B., Nash, L.J.: Movement of radioactivity from [^{3}H]GA_3 in geotropically stimulated coleoptiles of *Zea mays*. Planta *115*, 245–251 (1974)

Wilkins, M.B., Scott, T.K.: Auxin transport in roots. III. Dependence of the polar flux of IAA in *Zea* roots upon metabolism. Planta *83*, 335–346 (1968)

Wilkins, M.B., Whyte, P.: Relationship between metabolism and the lateral transport of IAA in corn coleoptiles. Plant Physiol. *43*, 1435–1442 (1968)

Wilkins, M.B., Woodcock, A.E.R.: Origin of the geo-electric effect in plants. Nature (London) *208*, 990–992 (1965)

Woodcock, A.E.R., Wilkins, M.B.: The geoelectric effect in plant shoots. I. The characteristics of the effect. J. Exp. Bot. *20*, 156–169 (1969a)

Woodcock, A.E.R., Wilkins, M.B.: The geoelectric effect in plant shoots. II. Sensitivity of concentration chain electrodes to reorientation. J. Exp. Bot. *20*, 687–697 (1969b)

Woodcock, A.E.R., Wilkins, M.B.: The geoelectric effect in plant shoots. III. Dependence upon auxin concentration gradients and aerobic metabolism. J. Exp. Bot. *21*, 985–986 (1970)

Woodcock, A.E.R., Wilkins, M.B.: The geoelectric effect in plant shoots. IV. Inter-relationship between growth, auxin concentration and electrical potentials in *Zea* coleoptiles. J. Exp. Bot. *22*, 512–525 (1971)

Yamamoto, R., Shinozaki, K., Masuda, Y.: Stress relaxation properties of plant cell walls with special reference to auxin action. Plant Cell Physiol. *11*, 947–956 (1970)

5.3 Growth Movements Not Directed Primarily by External Stimuli

5.3.1 Circumnutation

A. JOHNSSON

1. Introduction

The term circumnutation is used to denote a rotary movement of plant organs in space: the helical growth movement of tips of several seedlings can be put forward as an example, as can leaf movements and tendril movements. Leaf movements will be dealt with in Chapter 4.2 of this volume and will accordingly not be discussed here. Circumnutations often occur around a reference direction; a particularly simple example is circumnutation around the plumbline.

The period of circumnutation is very variable. Leaflets can move with a typical period of some minutes, while circumnutations of many shoots have a period of some hours. The movements can be very regular in time and space, but this is not always the case. They can for example, start as plane movements across the reference direction, and then change into more elliptical movements around the reference direction. Variation in period often occurs, even in a single plant. Some movements are restricted to a plane and do not change into rotary movements. In such cases the term circumnutation is, strictly speaking, a misnomer and could be replaced by e.g., pendulum movement.

The present contribution concentrates on some aspects of circumnutation. It does not cover all experimental results and all the literature in the field. The reader should consult previous excellent surveys by BAILLAUD (1962a, b), which also summarize much work done by French researchers who have contributed abundantly to observations of circumnutation. Relevant review articles have also appeared on the relation between circumnutation and gravity (JOHNSSON and HEATHCOTE, 1973) and on the physiology of tendrils (JAFFE and GALSTON, 1968a).

Instead of aiming at completeness, this chapter tries to give a more formal outline of some mechanisms, or models, which might lie behind circumnutation. Recent experiments will also be summarized and discussed, with the proposed mechanisms as a background.

2. Examples of Circumnutation

Some examples of circumnutation will be given for illustrative purposes; the reader should consult BAILLAUD (1962a, b) for further details and examples.

Circumnutation of a pea tendril, seen from the side and from above, is depicted in Figure 1a (from JAFFE, 1972). Figure 1b shows a projection onto a horizontal

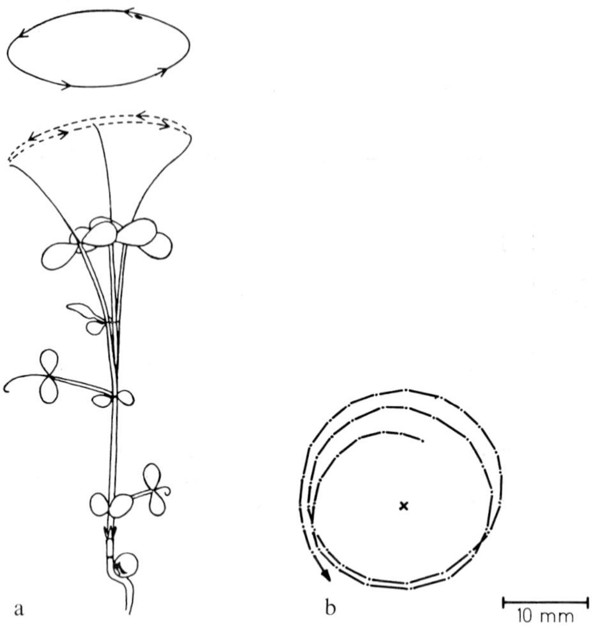

Fig. 1. a Schematic representation of tendril circumnutation (fifth node of a pea plant; from JAFFE, 1972). **b** Circumnutation of *Dioscora batatas* as seen from above. (From BAILLAUD, 1962b.) *Points* denote position every 5 min, the *cross* indicates the fixed base of the moving organ. The movement is regular and has a period of about 80 min

plane of tip movements of a shoot rotating in space. The shoot is kept fixed well below its moving tip, the point of attachment is denoted x (from BAILLAUD, 1962b).

Figure 2 illustrates growth and movements of a barley coleoptile. The movements are seen from the side and consist of more or less sinusoidal oscillations across the plumbline. The period is about 3h (from ARNAL, 1953b).

Figure 3 shows a side projection of circumnutations of a pea root growing in nutrient solution (from SPURNÝ, 1968c).

Finally Figure 4 illustrates that in bean hypocotyls the moving upper part of the plant generates a wave of curvature which moves basipetally during the circumnutation (after HEATHCOTE, 1965, from JOHNSSON and HEATHCOTE, 1973).

3. Oscillatory Phenomena

Circumnutations arise as a result of different extension rates on different sides of the organ. To produce circumnutations, the extension rate must vary rhythmically: hence our interest in oscillatory processes in plants.

It is pertinent to look at some oscillatory systems from a more general point of view. The aim is to emphasize that many relevant results have emerged during

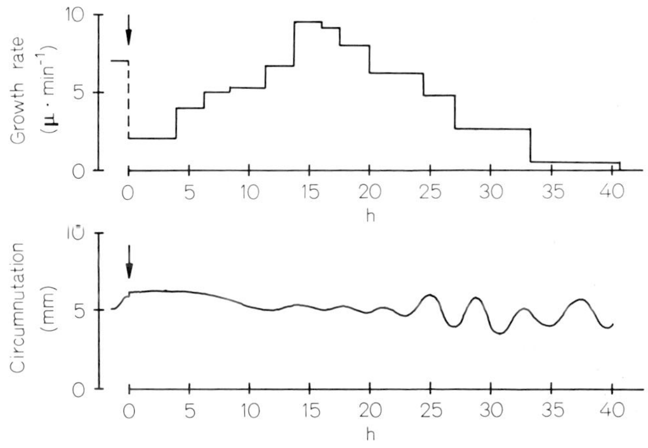

Fig. 2. Growth rate and magnitude of circumnutation (i.e., linear deviation of the tip as seen in Fig. 1a) of barley coleoptile. (After ARNAL, 1953b.) *Arrows* indicate time of decapitation of the coleoptile During the subsequent regeneration of the apex growth rate varies and circumnutation appears

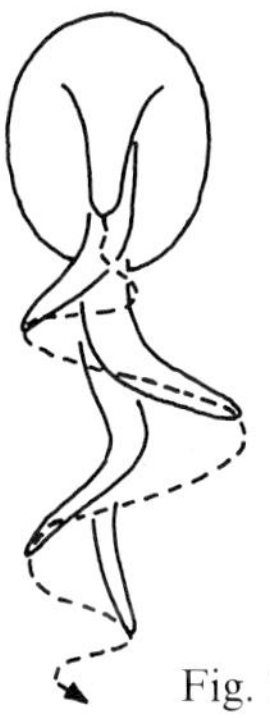
Fig. 3

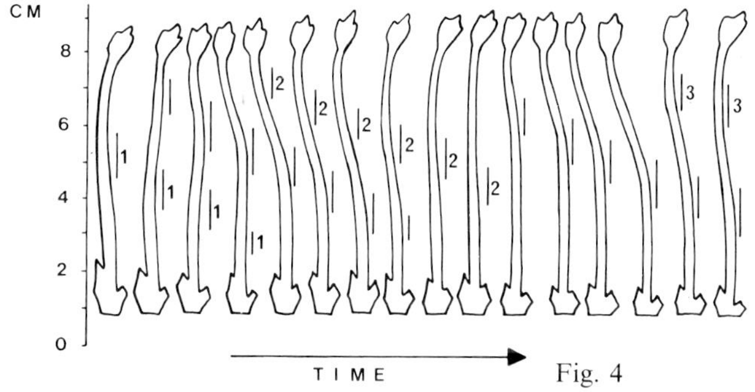

Fig. 4

Fig. 3. Front view of pea root. (From SPURNÝ, 1968c.) The path of the root tip is indicated by the *broken line*

Fig. 4. Silhouettes of the first internode of *Phaseolus multiflorus* at 15-min intervals. (From JOHNSSON and HEATHCOTE, 1973.) *Vertical bars* trace the course of waves passing down the internode

recent studies of oscillatory processes, both on a basic, molecular level, and on a "system" level.

3.1 Molecular Level

As is well known, many biochemical reactions show oscillatory features (see e.g., HESS and BOITEAUX, 1971). The glycolytic reactions can be put forward as an example of a molecular reaction chain showing self-sustained oscillations in sub-

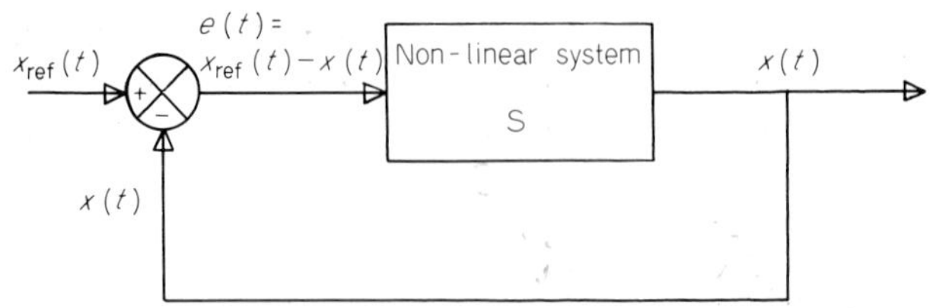

Fig. 5. Simple feedback system

stance concentration (for references see, e.g., CHANCE et al., 1973). Both in vivo (yeast cells) and in vitro the glycolytic pathway can oscillate with a typical period of some minutes. Details of the dynamics of the system will not be mentioned; it suffices here to point out that individual cells often show oscillations in fundamental biochemical processes.

Inorganic chemical reactions can also show sustained oscillations as exemplified by the well-known Belousov-Zhabotinskii reaction (TYSON, 1976). This reaction shows not only temporal, but also spatial variations: in a bulk solution, "waves" of different reaction phases can be detected. Such reaction patterns are of course of interest in connection with circumnutation.

3.2 "System" Level

Oscillations are frequently encountered in control systems of plants. As an introductory example, to be elaborated later, it is useful to consider the following fairly general picture of a control system (Fig. 5).

The reader can, for example, associate the variable $x(t)$ with a position in a balance system. The reference position is $x_{ref}(t)$. If $x(t)$ deviates from $x_{ref}(t)$, the error signal $e(t)$, i.e., $x_{ref}(t) - x(t)$, will be nonzero. The input signal to the system S under consideration is provided by $e(t)$. When $e(t)$ is nonzero, the system changes its output signal $x(t)$ according to the system characteristics. The output signal $x(t)$ will then acquire a new value. This value is fed back to the input, compared again with $x_{ref}(t)$ etc. We have a feedback system which acts as a control system to keep $e(t)$ as small as possible or equivalently, $x(t)$ as close to $x_{ref}(t)$ as possible.

If, specifically, $x_{ref}(t)$ is a constant, one often denotes the control system a regulatory system (e.g., ELGERD, 1967).

It is important to realize that oscillations can only arise when two conditions are fulfilled. The first is that the loop gain should exceed a certain minimum level: i.e., changes in the output level caused by changes in the $e(t)$ level should not be too low. The second requirement is that the phase change between the $e(t)$ and the $x(t)$ signals, caused by the system, should be of a suitable magnitude. A special case is encountered when the system contains a simple time lag: an oscillatory tendency in the input signal $e(t)$ can then be amplified via the feedback if $x(t)$ is delayed by the system so as to coincide in phase with the oscillatory input $e(t)$.

The content of the block S consists of several lumped sub-systems. Although a detailed, ideally a molecular, description of the sub-systems would be of great interest, many features of the total system can be understood without having this detailed knowledge. This makes control-system analysis a powerful tool.

We will later discuss a model of circumnutations based on a regulatory system of the type mentioned above.

3.3 General Properties of Nonlinear Oscillations

It is important to realize that self-sustained stable biological oscillations occur only in nonlinear systems. Mechanisms which should account for circumnutations must therefore be nonlinear, and nonlinear functions or elements are also included in the block *S* of Figure 5. Only to a certain extent can the theory for linear systems be used in the study of nonlinear ones.

Self-sustained oscillations can change their phase after a suitable perturbation. Such a phase change is called a delay or an advance, depending on whether the perturbed oscillation has been delayed or advanced with respect to an unperturbed control.

Oscillators – let us say in individual cells – can influence each other by being coupled in different ways (for example through diffusion of substances from cell to cell). A group of such coupled, nonlinear oscillators can show an overall rhythm with new features, unpredictable from the characteristics of the individual oscillator. The frequency of the overall rhythm can be reduced, spontaneous frequency changes can appear after perturbations etc. The mathematics of coupled, nonlinear oscillators is now attracting considerable interest and has important bearings for many biological fields (for references see PAVLIDIS, 1973).

4. Suggested Mechanisms for Circumnutation

Different mechanisms have been proposed to explain the origin of circumnutation. Considering the variety of observed movements, it is by no means unnatural to assume that different mechanisms are also implicated. However, it is useful to group the mechanisms suggested so far into three main groups, Sect. 4.1 to 4.3 below.

4.1 Movements Attributed to Autonomous Processes

Autonomous processes mean processes which are self-produced (i.e., internal, endogenous). The term has been, and is in many cases, used as an equivalent to "unknown processes" in the plant organ itself. Some circumnutation hypotheses of this type will be mentioned below. Since they are fairly vaguely outlined in the literature, I have found it constructive to try to extend or specify the mechanisms, although the proposals certainly are open to criticism and need elaboration and critical tests.

a) A transverse section of a circumnutating plant organ is symbolized in Figure 6. Volume changes, elastic or plastic (i.e., connected with growth), can occur

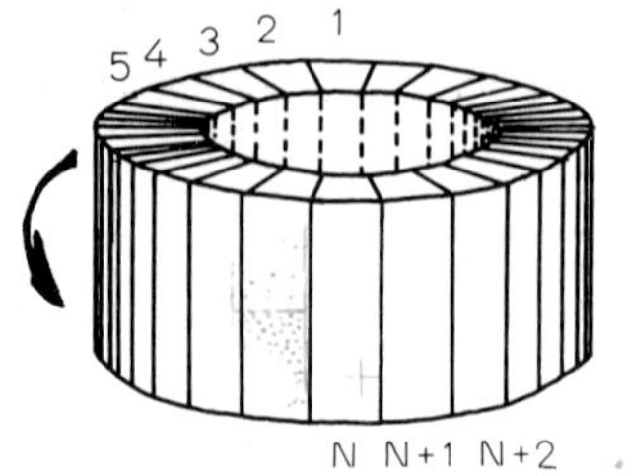

Fig. 6. Ring pattern of cells, representing a transverse section of a circumnutating organ. *Arrow* shows an assumed direction of increased extension

in the cells. It is evident that if the volume changes are synchronized over the whole section, no bending of the plant organ will occur. However, if the cells show oscillatory volume changes, and if neighbor cells are fairly closely phased, one can imagine that a wave of cell extension can propagate around the section. At a certain time maximum extension is shown by the cells 1, 2 and 3, for example, while the cells N, N+1, and N+2 on the other side of the organ do not extend. After a certain time, cells 4, 5 and 6 extend maximally. At a later time, cells N, N+1, and N+2 will show maximum extension, while cells 1, 2 and 3 no longer extend. Such a process of cell extension will lead to bending reactions, and a bending wave, illustrated by the arrow, propagates around the center. In this way circumnutations will occur.

The crucial point is, of course, to explain the oscillatory volume changes. This has not been elaborated in the literature but, as we have seen, oscillatory processes in individual cells are known (Sect. 3.1) and oscillatory volume changes can thus be visualized. They could, for example, be mediated via rhythmic ion-concentration changes in the cells (much in the same way as rhythmic volume changes of guard cells can be produced, cf. for example JOHNSSON, 1976).

HEATHCOTE and ASTON (1970) used the mechanism above to explain circumnutations. The oscillatory processes in the cells were assumed to be regular changes between "expansive" and "consolidative" phases of growth. Not all cells of the section were thought to expand simultaneously, but neighbor cells were. Extensive strains between neighbor cells were said to be avoided in this way (see review by JOHNSSON and HEATHCOTE, 1973).

One can imagine a system of coupled, nonlinear cell oscillators (Sect. 3.3) across the section of Figure 6. The mathematics of such a system is relevant for studies of circumnutation. We should note that the system can show a frequency change: if the coupling between the cells alters, the overall rhythm can switch to double the frequency (PAVLIDIS, 1973).

b) The ring pattern of cells in Figure 6 has also been drawn to recall the classic paper by TURING on morphogenesis (TURING, 1952, cf. ROSEN, 1970). TURING studied chemical substances (he called them morphogens) which diffused and reacted throughout a ring of cells. He showed that such substances, diffusing from cell to cell, could very well develop *stationary* concentration waves on the ring. As an example the concentration in cell 1 could oscillate rhythmically out of phase with the concentration in cell N. Somewhere in between these cells, say in cell N/2, the concentration remained constant. More complicated stationary waves were likewise possible. *Traveling* concentration waves along the ring could also appear, i.e., the peaks, nodes, and troughs moved around the ring.

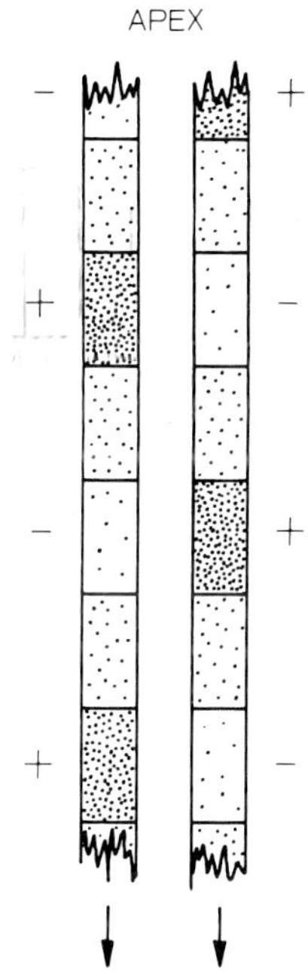

Fig. 7. Diagrammatic representation of pulses of growth hormone, traveling basipetally in the two opposite sides of a longitudinal section of a circumnutating plant organ. *Intensity of stippling* indicates concentration, maxima denoted by + and minima by −. The pulse trains on the two sides are out of phase, but travel with the same rate toward the base (cf. Fig. 4)

We could, tentatively, look at growth-hormone concentrations instead of morphogen concentrations. Concentration gradients would then produce growth movements in a plane or, when traveling waves were present, rotary movements of the plant organ.

One should note that the production of waves in this case is a "collective" phenomenon: if we studied an isolated, individual cell, no oscillations would be found. Oscillations arise in the cell system through the cell-to-cell coupling. In Section 4.1a we assumed that every cell contained an individual self-sustained oscillator. This is evidently not a prerequisite to explain a wave of extension traveling around the plant organ, as the TURING mechanism shows.

c) An autonomous mechanism was proposed by ARNAL (1953a, 1954) to explain the circumnutation of barley coleoptiles. Apically produced auxin, or a precursor to auxin, was assumed to be distributed sequentially to different sides of the coleoptile in order to give rise to plant movements across the plumbline (see Fig. 2). It was not suggested how the transport to the different sides of the coleoptile was controlled by the apex. Experiments have disproved the concept that the apex is the site of such an oscillator (see Sect. 5).

However, if we exclude the role of the apex on an experimental basis, an alternative growth hormone-based mechanism can be sketched. Assume a substance – auxin or not – to be transported basipetally in the plant organ and, through a sequence of reactions, to influence its own transport rate. A concentration gradient of the substance, either along or across the plant stem, can in this way affect its own magnitude. We have then a feedback system and, as mentioned in Section 3, oscillations can arise under certain conditions.

In Figure 7 such a case is depicted schematically. The transport system has caused concentration pulses at the two opposite sides of an organ, the plus sign indicating higher, the minus sign lower concentration. The two pulse trains are out of phase, but travel at the same rate basipetally. The different phase can be due, for example, to an initial transverse transport caused by a tropic stimulation. If one applied growth hormone symmetrically at the apical side, the pulses would of course be in phase instead and reach the basipetal side at the same time.

These oscillatory changes of growth-hormone concentration in the tissue could of course lead to rhythmic extension of cells and, consequently, to rhythmic plant movements. Oscillatory transport of auxin has, in fact, been observed both for longitudinal and for transverse auxin transport in coleoptiles (cf. HERTEL and FLORY, 1968). Elaboration of the hypothesis might therefore be relevant in connection with circumnutation studies.

d) As a final autonomous mechanism to be contemplated here, we should pay attention to the *Phycomyces* work by DENNISON and ROTH (1967). For the sporangiophore it was shown that mechanical compression of the cell caused a subsequent interval of faster growth, while mechanical extension caused a subsequent slower growth. Thus, if we dare to generalize to the case of multicellular organs: extension of one side of a curved organ could initiate a subsequent decrease in the growth rate of that side. A nonextension of the other side could initiate a subsequent increased growth of that side. As a consequence, the curved organ would curve in the opposite direction; then the process starts over again, etc. As a final result, overshoot bendings, and thus oscillations, would occur.

Such overshoot movements, which can be encompassed in the general framework of a feedback mechanism like the one in Figure 5, could arise in a simple way due to geotropic reactions. This should be dealt with in the next section, since geotropically induced processes are usually not regarded as autonomous ones.

4.2 Movements Attributed to Geotropic "Overcompensation"

The circumnutations of several geotropicallysensitive seedlings have an amplitude of more than 10° with respect to the plumbline—in sunflower hypocotyls, often 20°. Any hypothesis for circumnutation in such plants must take geotropic reactions into consideration. The same is of course, not true for species where the amplitude of the circumnutation is very small, or where the plant organ is geotropically fairly insensitive. For a discussion of geotropism (gravitropism) see VOLKMANN and SIEVERS (Chap. 5.2.2, this vol.) and WILKINS (Chap. 5.2.3, this vol.).

Circumnutations around the plumbline have been explained on the basis of geotropic overshoot movements in several species; e.g., sunflower seedlings, *Carthamus* seedlings, roots of *Pisum* etc. (Some relevant references in this context are GRADMANN, 1921a, b, 1922, 1927, 1928; KARVÉ and SALANKI, 1964a, b, c; KARVÉ and JINGAJINNI, 1965; KARVÉ and SAWANT, 1966; SPURNÝ, 1966, 1968a, b, c, 1974a; ISRAELSSON and JOHNSSON, 1967; JOHNSSON, 1971, 1974; JOHNSSON and ISRAELSSON, 1968, 1969; ANDERSEN and JOHNSSON, 1972a, b; ANDERSEN, 1976). A mathematical formulation of this hypothesis has been put forward and studied in a series of papers (cited works by ANDERSEN, ISRAELSSON and JOHNSSON). Only a summary of this formulation will be given here.

The mechanism is easily explained in the case of movements in a plane across the plumbline. A deviation from the plumbline, the normal growth direction, causes geotropic movements of the plant. It bends after a while toward the plumbline. The geotropic regulatory system reacts in such a way that the plant ultimately overshoots the plumbline and receives a new geotropic stimulation, but now in the opposite direction. Oscillations in a plane across the plumbline can therefore be built up.

In Figure 5, the reference value $x_{ref}(t)$ is thus the fixed plumbline direction. and the variable $x(t)$ equals the angle of the plant with respect to the plumbline. Denoting this angle $\alpha(t)$ and putting $x_{ref}(t)=0$ by definition, the authors cited arrived at a quantitative description in the following manner: the bending rate was assumed to be proportional to accumulated geotropic stimulations, received by the hypocotyl studied. However, these stimulations cannot cause a reaction until the geotropic reaction time has passed. Therefore the system has a time lag which makes overshoots of the plumbline possible.

In order to quantify the accumulation of geotropic stimulations the authors introduced a weighting function $f(t')$. It determines to what extent a geotropic stimulation acquired at time t' adds to the bending rate at time t.

The geotropic movement will therefore in this simplified case be described by

$$\frac{d\alpha(t)}{dt} = -\text{const} \cdot \int_0^\infty f(t') \cdot \sin \alpha(t-t')\, dt'$$

(the sin of the bending angle is introduced since the transverse (sin-) component of the gravity vector is of interest in the geotropic stimulation).

The corresponding block diagram is seen in Figure 8. The block labeled nonlinear functions contains information about geotropic stimulations and hormone actions in the plant. Block f contains the weighting function assumed to have an exponentially decaying appearance. The last block contains an integrator in order to achieve the position of the plant, $\alpha(t)$, from the known bending rate explicitly given in the formula.

With suitable parameter values, the system is unstable, and even minute deviations from the plumbline direction will lead to oscillations which grow in amplitude until limited by the nonlinearities. The model has also been elaborated for the case of movements *around* the plumbline (ANDERSEN, 1976). This extension is not a trivial one since oscillatory movements in two perpendicular planes cannot be directly added for such a nonlinear system.

In the geotropically based explanation of circumnutation, the differential growth is caused by growth-hormone gradients across the stem. For circumnutations around the plumbline, a wave of cell extension therefore travels around the stem, as in the autonomous models, but the wave is in this case caused by the geotropic reactions.

4.3 "Intermediate Mechanisms." A Comparison Between the Models

One can easily imagine combinations of the two types of mechanism outlined above (as emphasized by JOHNSSON, 1977). Self-sustained oscillations in the growth-hormone gradient across a stem could be amplified by the successive geotropic stimulations implicated in the geotropic overshoot model. Sustained circumnutations of large amplitude would then be the result. Likewise, damped oscillations in the growth-hormone system can be amplified by geotropic stimulation. As a consequence of this amplification, sustained oscillations can arise.

How would one distinguish between the endogenous and the exogenous components of these intermediate mechanisms? How could one discriminate among the main groups of models of circumnutation? Johnsson and Heathcote tried to give an answer in their review (1973): one should study circumnutation under satellite conditions where no mass acceleration is present (with reference to the satellite system).

Hypotheses based on autonomous processes would predict that if mass acceleration were absent, as under favorable conditions in a satellite, circumnutation would continue without many changes. Gravity could indirectly influence energy supplies, growth processes etc, but these effects are likely to play a minor role in a fairly short-term experiment under "gravity-free" conditions. It is obvious that if circumnutations occur due, for example, to compression/extension-initiated mechanisms on both sides of an organ (Sect. 4.1), they should continue under satellite conditions. The geotropic overshoot model however, would predict a cessation of circumnutations, since no geotropic reactions could go on without mass acceleration.

Finally, if circumnutations are due to "intermediate mechanisms", one would expect the movements to damp out (or at least show a very reduced amplitude) when mass acceleration is absent.

5. Some Experimental Studies of Circumnutation

Experimental studies during recent years have concentrated on effects of gravity on circumnutation and on the participation of growth hormones and chemical substances in circumnutation reactions. A summary of some pertinent experiments will be given below.

5.1 Occurrence of Circumnutation

The presence of circumnutations seems to be correlated with growth, circumnutations being, in general, vigorous during the growth period of plants. Careful studies on this point have been made, for example, on twining shoots of *Periploca graeca* (Melin, 1975a, b). The shoots must evidently reach a threshold value of the growth rate (about 12.5 mm/day) before circumnutation appears. Arnal (1953b, 1959), Joerrens (1959) and Anker (1972) removed the extreme tip of coleoptiles; circumnutation then stopped, but reappeared when the physiological apex was restored or IAA was applied and growth resumed.

Removal of parts of the geotropically most sensitive regions in *Helianthus* leads to a reduction in amplitude and frequency of circumnutation. Similar results have been found for *Phaseolus* (Heathcote, 1968).

Taken together these facts point to a parallelism between growth rate and presence of circumnutation. However, growth rate can be high while circumnutation disappears and vice versa, as shown by the example in Figure 2. In *Pisum* roots the large circumnutations halt completely when the extreme tip is removed (SPURNÝ, 1968b, c). Growth itself is not stopped by this decapitation, and one has therefore another example of decoupling between growth and presence of circumnutation. It therefore seems unwise to presume, in the general case, a causal relationship between growth rate and presence of circumnutation. Interestingly enough, however, there seems to be even an influence of circumnutation on the growth rate in some species; if circumnutation is mechanically prevented, the growth rate decreases (CLAIRE, 1974; MELIN, 1975a).

In *Triticum* and *Phaseolus*, the point of maximum curvature of the stem has been shown to move basipetally during circumnutation (ARNAL, 1953a, HEATHCOTE, 1965). A wave of bending thus travels downward as exemplified in Figure 4, the rate being comparable with that of basipetal auxin transport.

Circumnutation occurs both clockwise and counter-clockwise, dependent on the species (examples are given by BAILLAUD, 1962a, b). Many species show rotation in both directions with a higher preference for one of them and spontaneous change of direction of rotation has been reported (BAILLAUD, 1962a). If the movements are studied as viewed from the side, changes of rotation direction could be observed as phase changes of the oscillations (cf. HEATHCOTE, 1968). Geotropic stimulations given so as to reverse the direction of rotation could actually do so in plants such as *Helianthus* and *Phaseolus* (GRADMANN, 1921b; ANDERSEN, 1976). Such transitions from one direction of rotation to the opposite often occur via plane movements across the plumbline (circumnutation can also be restricted to such plane movements, e.g., in a certain percentage of sunflower hypocotyls, cf. p. 627).

5.2 Period of Circumnutation

The basic period is strongly dependent on the temperature in the temperature interval 10–30 °C, and decreases with temperature. At about 30 °C there is a minimum in the period, and above 30 °C it may even increase again. Further details are given by BAILLAUD (1962a, b) and JOHNSSON and HEATHCOTE (1973).

Attention should be drawn to plant species which can circumnutate with different periods. ARNAL (1953b) states that normally the period of barley coleoptiles is roughly 150 min, but that in a certain percentage or under special experimental conditions, it can be half of that, i.e., 70–80 min. HEATHCOTE (1968) reports the same observation for bean plants. In *Phaseolus aureus* some plants show double the normal period (HEATHCOTE, 1977).

Two different frequencies can also be superimposed. In *Sicyos* tendrils the movements in one plane may occur five times as fast as in the plane perpendicular to it (GRADMANN, 1922). A similar observation seems to have been made for *Passiflora* (QUANTIN, 1952). HEATHCOTE (1966) discovered small-amplitude, high-frequency oscillations in bean seedlings. They were superimposed on the normal elliptical circumnutations, and the period of these "micronutations" was about

15–30 min (dependent on the temperature). They were seven to nine times as rapid as the basic circumnutations.

ROBLIN (1975) found that in *Mimosa*, the circumnutating pulvinus leaves had a circumnutation period which increased with the amplitude. Some experiments indicate that the period can depend on the direction of rotation (sunflower seedlings, ANDERSEN, 1976; *Phaseolus*, GRADMANN, 1928).

In *Passiflora* the circumnutation period of tendrils and shoots decreases if the plant stem is partially immobilized (BOYER, 1973). Generally, the period is dependent on the length of the free-moving part (as can be studied when the organ is fixed at different points.)

In several species the period of the circumnutation is roughly four times as long as the so-called geotropic reaction time (i.e., the delay in the geotropic response of the plant). This ratio was found experimentally to be fairly independent of temperature in several cases (see JOHNSSON and HEATHCOTE, 1973). In *Phaseolus aureus* the ratio seems to be lower than 4 at higher temperatures (HEATHCOTE, 1977).

5.3 Interactions Between Circumnutation and Geotropic Stimulation

Since the different models in Section 4 ascribe totally different roles to gravity and the geotropic reactions, much attention has been paid to the effects of geotropic stimulation on circumnutation. Figure 8 shows how geotropic stimuli can be introduced in the geotropic overshoot model for circumnutation. α_{stim} represents the stimulus applied.

Two main ideas can be traced in the geotropic experiments published.

a) The geotropic stimulations are given as pulses or continuously. The aim is to study, for example, possible phase shifts of circumnutation.

A *single* geotropic pulse causes phase shifts of some types of circumnutation, the magnitude of the delays and advances being dependent on the phase at which the pulse is given (KARVÉ and SALANKI, 1964b; JOHNSSON and ISRAELSSON, 1969).

Circumnutation in *Helianthus* seedlings has been studied while repetitive geotropic pulses were administered. The circumnutations locked to the pulses in a certain frequency range. The amplitude of the forced circumnutations, the phase between pulses and movement and, finally, the locking range were studied (ANDERSEN and JOHNSSON, 1972).

If seedlings are left in a horizontal position, they bend geotropically upward, but not always in a smooth, continuous fashion. At least in some species (*Phaseolus* and *Zea mays*), the upward movement has an oscillatory character (HEATHCOTE and ASTON, 1970, SHEN-MILLER et al., 1970). This might be interpreted as an upward geotropic movement superimposed on an autonomous circumnutation (HEATHCOTE and ASTON, 1970).

b) The geotropic stimuli are applied in such a way that $[\alpha_{\text{stim}}(t)+\alpha(t)]$ remains effectively zero, i.e., one tries to keep the error signal in Figure 8 equal to zero. If the error signal is constant and zero, the output $\alpha(t)$ should be constant.

There are at least three different ways of reducing the magnitude of the error signal.

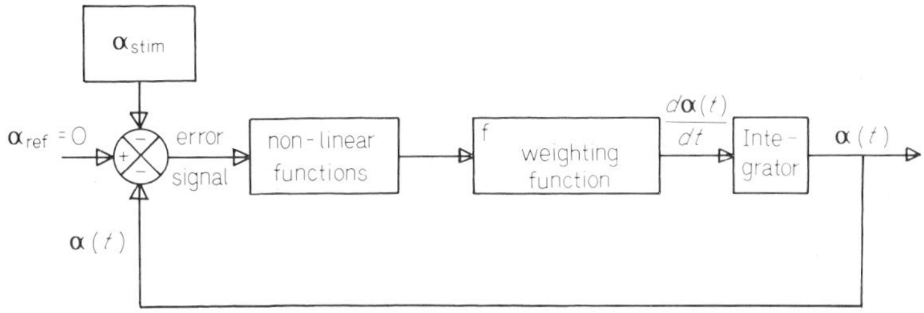

Fig. 8. Geotropic feedback system, used to describe circumnutations on the basis of geotropic movements

1. One can tilt the whole plant in such a way that the geotropically most sensitive region will always be in the vertical position $[\alpha_{\text{stim}}(t)+\alpha(t)=0]$. Such "compensation" experiments lead to drastically reduced circumnutation amplitude in sunflower seedlings (JOHNSSON, unpublished). However, all parts of the curved stem cannot be kept vertical at the same time, and some parts of it will therefore be continuously stimulated during such an experiment.

2. A seedling can be held in the vertical by a force applied to its tip. The force necessary should eventually equal zero, since the error signal and accordingly $\alpha(t)$ will be zero (cf. Fig. 8). A drawback in this type of approach is, for example, that a restriction of movement could by itself retard growth (see Sect. 5.1).

3. The plant can continuously be given large geotropic stimuli in opposite directions. This is usually achieved on a clinostat. An additional stimulation angle of 90° is then given to the plants in the case of a clinostat with a horizontal axis and the plants mounted horizontally. A plant on such a clinostat will receive stimuli of angles fairly close to 90° to the plant axis during the rotation: $90°+\alpha(t)$ at a certain time and $90°-\alpha(t)$ half a turn later. If these two stimulations cause approximately the same geotropic reactions (but in opposite directions) the output signal will be constant, exactly as if the error signal were zero. This case is depicted in Figure 9. Curve a describes a hypothetical case in which geotropic stimuli at angles of $90°+\alpha(t)$ and $90°-\alpha(t)$ cause the same reactions. Due to the clinostat rotation they are given in opposite directions and, as a net result, will cancel each other out. This means that the effective error signal will be zero.

If, however, the reaction follows curve b, the geotropic stimulations in position $90°+\alpha(t)$ and $90°-\alpha(t)$ do *not* produce the same reactions. As a consequence they do not compensate on a clinostat, the average error signal is different from zero and circumnutation might continue, even on a horizontal clinostat (JOHNSSON, 1974). The action of the clinostat is therefore very much dependent on the actual shape of the nonlinearities in the geotropic system.

For plants following curve type b in Figure 9, another clinostat inclination might possibly be found, for which equal but opposite geotropic reactions are produced during clinostat action, leading to a standstill of the circumnutations. Such a clinostat angle, α_{clin}, might be the one indicated in Figure 9, curve b.

Circumnutations of sunflower seedlings halt or reduce their amplitude considerably on a clinostat with horizontal axis (JOHNSSON, 1974; BROWN and CHAPMAN, 1976). Other circumnutations, mainly those of tendrils, do not stop under the same conditions GRADMANN, 1921, etc.; BAILLAUD, 1959, 1962a). When the inclina-

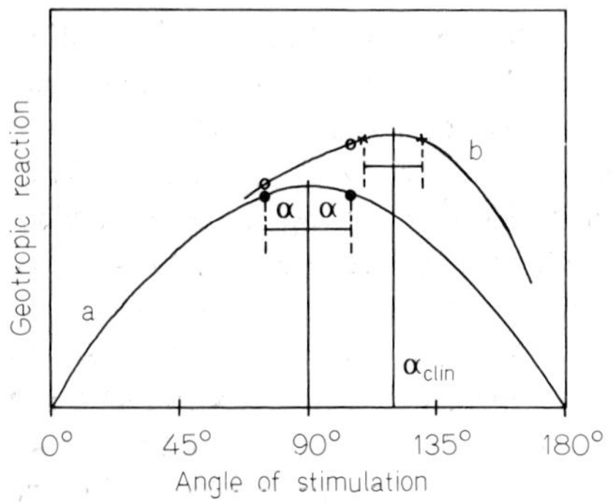

Fig. 9. Geotropic bending (schematic) of plants stimulated at different angles with respect to the plant axis. If the reaction follows curve a, responses to stimulations at $(90+\alpha)$ and $(90-\alpha)$ degrees are equal (●). If the reaction follows curve b they differ (○). For plants which follow curve b one can imagine an angle α_{clin} which lies symmetrically in such a way that deviations on both sides of it cause the same reactions (x)

tion of the clinostat axis is between 0° and 80° from the vertical, sunflower circumnutations continue with reduced amplitude and period (BROWN and CHAPMAN, 1976).

5.4 Circumnutation Under Different Mass Accelerations

BROWN and CHAPMAN (1977) have studied circumnutation in sunflower hypocotyls on a centrifuge. The amplitude increased, but the period of the movement changed only slightly in the stimulus-region investigated (1–10 g). Above 18 g, the plants were unable to stand the increased weight.

No decisive experiments on circumnutation are as yet reported under free-fall conditions, as in a satellite.

5.5 Light and Circumnutation

Light interferes with circumnutation. ARNAL (1954), GALSTON et al. (1964), and others found an increased circumnutation amplitude in light-treated seedlings. Etiolated plants showed small or no circumnutation (e.g., SPURNÝ, 1974b). Data for light influence on circumnutation are, however, rare and scattered (GESSNER and HAMMER, 1958; GRADMANN, 1922; KARVÉ and SALANKI, 1964c; HEATHCOTE, 1968). This field needs more intensive investigation, especially with regard to possible perturbation of circumnutation by light pulses.

5.6 Growth Hormone Studies. Chemically Oriented Investigations

All models mentioned above implicate growth hormones as prerequisites for circumnutation. The detailed roles played by growth-promoting substances, however, are not known. The existence of basipetal IAA transport has been demonstrated in stems of many circumnutating plants, as well as in tendrils of *Passiflora caerulea* (JUNKER, 1976). Oscillatory transport of auxin has been found and is relevant in this context. Oscillations are reported either in auxin transport itself, or in functions likely to be governed by auxin gradients (NEWMAN, 1963; JOHNSSON,

1967; SCOTT, 1967; HERTEL and FLORY, 1968; SHEN-MILLER et al., 1970; SHEN-MILLER, 1973). The period involved is often of the order of 30 min, but even shorter periods can be found (about 15 min, SHEN-MILLER et al., 1970).

JOERRENS (1959) and ANKER (1972) worked with decapitated coleoptiles. After addition of IAA to the cut surfaces, circumnutations could again be recorded. The experiments thus indicate that the apex provides the growth hormone necessary for the occurrence of circumnutations, but does not function as an oscillator.

The action of gibberellic acids on circumnutation is not clear. Growth and circumnutation amplitude are often stimulated simultaneously (e.g., BAILLAUD and MONNIER, 1960; TRONCHET, A., 1961; TRONCHET et al., 1961a; TRONCHET et al., 1961b, TRONCHET et al., 1960; DEQUEUDRE et al., 1966). However, in *Triticum* circumnutations are diminished by gibberellic acid (ARNAL, 1958) and, in *Ipomoea*, growth is augmented, but there is no change in the circumnutation period (CLAIRE, 1970). The action of gibberellic acids is therefore somewhat contradictory and, as has been suggested, e.g., by GALSTON et al. (1964), any such action on circumnutation might be indirect via effects on growth.

Important studies of tendril movements after physical stimulation have been pursued by JAFFE and GALSTON (e.g., 1966a, b, 1967a, b, 1968a, b) who investigated, for example, the action of growth hormones, flavonoids and ATPase activity on the contact coiling process. (According to MELIN, 1974, flavonoids are not directly involved in the generation of circumnutation; cf. TRONCHET, J. 1960, 1961.) Contact coiling and circumnutation seem, at least in *Pisum*, to be connected via a negative feedback mechanism: when the circumnutations are rapid, the rate of contact coiling is slow and vice versa (JAFFE, 1972). This feedback can be exerted via a limited ATP pool, providing energy for both processes. It seems worthwhile to pursue these interesting studies of coiling mechanisms as they relate to circumnutation.

The amino acid proline is present in higher concentration in twining shoots than in creeping shoots of *Periploca graeca*, which might indicate its participation in the basic circumnutation process (MELIN, 1975b). Several substances have been tested in efforts to isolate factors responsible for circumnutation (e.g., CLAIRE, 1974).

I think it is fair to conclude that although much effort has gone into investigations of the basic biochemical events behind circumnutation we still lack good biochemical models. This is at least partly due to our general ignorance about growth processes, to which circumnutation is evidently closely connected.

6. Experiments and Models

The experimental results outlined in the previous section do not allow decisive discrimination among the models mentioned in Section 4. Many researchers agree that it is doubtful that a single mechanism is responsible for the many different types of circumnutation present in the many different species studied. It would take us too long to discuss the applicability of the mechanisms to the individual species investigated. Instead, only some comments on the models will be given in the light of the experimental findings.

1. An autonomous mechanism based on oscillators in the individual cells could explain why different periods appear in different individuals of the same species, and why several periods appear in a single plant. The existence of circumnutation with half the usual period can, for example, be explained by this mechanism (see p. 637). Micronutations could reflect the period of the individual oscillators which, when coupled, also produce the much longer period normally encountered in circumnutation.

The autonomous mechanism with self-sustained oscillators in the individual cells must, of course, be elaborated in order to account for geotropic interference with circumnutation. One might assume for instance, that the growth hormone acts on the individual oscillator or on the coupling mechanism between the cells.

Should circumnutation continue on a clinostat according to this model? No definite predictions can be made. If the system is fairly insensitive towards geotropic stimuli, circumnutation would probably continue on the clinostat. On the other hand, one could imagine that geotropic stimulation during clinostat treatment changes the growth hormone pattern in the plant drastically. As a result, the individual oscillators or the coupling between the cells might be changed and the circumnutations might disappear.

A disadvantage of this type of model is its vagueness. Neither the nature of the individual oscillator nor that of the intercellular coupling is specified. It is difficult to study testable consequences of and predictions from this type of model.

2. Autonomous mechanisms based on the idea that oscillations can arise in growth-hormone concentrations and gradients (without self-sustained oscillations being present in isolated, individual cells) were outlined on p. 632. Such mechanisms are attractive from many points of view. Rhythmic transport of growth hormones with periods of the right magnitude has been reported (HERTEL and FLORY, 1968). Micronutations might be explained by the transverse, short periodic growth reactions detected in *Zea mays*. Experiments with decapitated coleoptiles indicate that the presence of IAA is necessary for the occurrence of circumnutation in several species. Furthermore, such mechanisms could be in agreement with the outcome of geotropic experiments, since the latter involve changes in growth-hormone patterns.

Factors influencing the auxin gradient should also affect circumnutation. In geotropically sensitive species, one would therefore expect circumnutation to be affected by clinostat treatments. Light could exert influence on circumnutation via auxin processes.

At the moment it is not clear if such mechanisms can explain the different periods sometimes encountered in circumnutating organs.

3. The geotropic overshoot model is attractive, in part, because of the simple concepts involved. This model emphasizes the fact that the circumnutations of some plants are of such a magnitude that large geotropic reactions must occur during the movements.

Arguments in favor of this mechanism have come mainly from experiments on seedlings growing around the plumbline direction. In the case of sunflower, this model fits, — even quantitatively, — all circumnutation experiments published to date. It also predicts that the period of the movements depends on the direction of rotation; this might also be the case experimentally (ANDERSEN, 1976). Evidence of circumnutation in *Pisum* roots supports this mechanism, since removal of the geotropically-sensitive root parts abolishes the circumnutatons without stopping growth.

The model predicts that circumnutation should be diminished on a horizontal clinostat, but does not necessarily require that it should disappear (JOHNSSON, 1974, see also p. 639). The model predicts in its basic form that the geotropic reaction time should be about one fourth of the circumnutation period. This is also experimentally found in many species over a range of temperatures, but as stated on p. 638, interesting exceptions occur and require further studies of this "rule."

Some experimental results difficult to reconcile with this overcompensation model certainly exist. A doubling or halving of circumnutation frequency cannot be accounted for without changes in the time-determining parameters of the model (above all the so-called weighting function). Micronutations in *Phaseolus* have not been explained; on the other hand, the *Sicyos* movements mentioned in Sect. 5.2 can be accounted for.

The oscillatory upward bending during continuous geotropic stimulation found in *Phaseolus, Zea mays,* etc. (see p. 638) is an argument against the model.

4. A comment on the intermediate mechanisms, discussed in Section 4.3 should also be given. Many experiments which can be explained by the autonomous mechanisms outlined in Section 4.1 could also be explained on the basis of a combination of oscillatory autonomous processes together with geotropic reactions. Quantitative predictions on the behavior of the circumnutations can, of course, not be specified as long as the oscillatory autonomous processes participating in the generation of the circumnutations are unknown. This type of model is fairly "vague", and is therefore difficult to disprove.

5. It is evident that many features of circumnutation are in agreement with all the mechanisms and models outlined: the general linkage between growth and presence of circumnutation, the basipetal movement of the point of greatest curvature on a stem etc., are compatible with all mechanisms discussed. Growth-hormone studies and biochemical work in the field can also be encompassed within the different models. As has been mentioned, careful experiments under free-fall conditions, as in a satellite, could remove some of the ambiguities concerning the validity of the mechanisms (see p. 636). Fortunately such an experiment is now at the planning stage: Spacelab I should carry an experiment in which sunflower circumnutations are to be studied (BROWN, personal communication).

7. Final Remark

I have pointed out some recent theoretical and experimental results of interest for future understanding of circumnutation. Other authors would certainly have emphasized other achievements or aspects. However, it is my firm belief that development and criticism of as detailed models as possible could help us to achieve rapidly a better understanding of circumnutations and of growth processes in general. As is well known, "models are caricatures of Nature". They distort the truth but, at the same time, they help us to see.

Acknowledgement. I would like to thank my friend David Heathcote for the many stimulating discussions we have had on the many stimulating problems in this field.

References

Andersen, H.: A mathematical model for circumnutations. Report 2/1976. Dept. Elec. Measurements, Lund Inst. Technol. (1976)

Andersen, H., Johnsson, A.: Entrainment of geotropic oscillations in hypocotyls of *Helianthus annuus*—an experimental and theoretical investigation. I and II. Physiol. Plant. *26*, 44–61 (1972)

Anker, L.: The circumnutation of the *Avena coleoptile,* its autonomous nature and its interference with the geotropic reaction. Acta Bot. Neerl. *21*, 71–76 (1972)

Arnal, C.: Recherches sur la nutation des coléoptiles I. Nutation et croissance. Ann. Univ. Sarav. Naturwiss. Sci. *2*, 92–105 (1953a)

Arnal, C.: Recherches sur la nutation des coléoptiles II. Nutation et régeneration physiologique. Ann. Univ. Sarav. Naturwiss. Sci. *2*, 186–203 (1953b)

Arnal, C.: Recherches sur la nutation des coléoptiles III. Nutation et étiolement. Ann. Univ. Sarav. Naturwiss. *3*, 85–98 (1954)

Arnal, C.: Action de la gibberelline sur la croissance et la nutation de coléoptile de blé. Ann. Sci. Univ. Besançon 2nd Ser. Bot. *12*, 59–61 (1958)

Arnal, C.: Determinism auxinique de la nutation du coléoptile de blé. 8th Congr. Soc. Sav. 461–466 (1959)

Baillaud, L.: Recherches sur les mouvements spontanés des plantes grimpantes. Thèse Sci. Univ. Besançon, no 5 1957: Ann. Sci. Univ. Besançon, ser. 2. Bot. 11 (1959)

Baillaud, L.: Mouvements autonomes des tiges, vrilles et autres organes à l'exception des organes volubiles et des feuilles. Handbuch der Pflanzenphysiologie. Berlin-Heidelberg-New York: Springer, 1962a, Vol 17/2, pp. 562–634

Baillaud, L.: Les mouvements d'exploration et d'enroulement des plantes volubiles. Handbuch der Pflanzenphysiologie. Berlin-Heidelberg-New York, 1962b, Vol 17/2, pp. 637–715

Baillaud, L., Monnier, Y.: La circumnutation de la tige d'un *Phaseolus* rendu volubile par l'acide gibbérellique. C.R. Acad. Sci. (Paris) *250*, 4032–4034 (1960)

Boyer, N.: Mouvement périodique de circumnutation et croissance des vrilles. Ann. Sci. Nat. Bot. (Paris) *XIII*, 417–475 (1972)

Boyer, N.: Corrélations de croissance dans les plantes à vrilles. Ann. Sci. Nat. Bot. (Paris) 14, 301–366 (1973)

Brown, A.H., Chapman, D.K.: Nutations of Sunflower Seedlings on Tilted Clinostats. Report. Biol. Dept. Univ. Pennsylvania, 1976

Brown, A.H., Chapman, D.K.: Effects of increased gravity force on nutation of sunflower hypocotyls. Plant Physiol. *59*, 636–640 (1977)

Chance, B., Kendall Pye, E., Ghosh, A.K., Hess, B.: Biological and Biochemical Oscillators. New York-London, Academic Press, 1973

Claire, A.: Recherches sur la mouvement révolutif, la croissance et la torsion homodrome des tiges volubiles chez l'*Ipomoea purpurea.* Thèse Doct. Etat, Univ. Clermont-Ferrand, 204p (1970)

Claire, A.: Relation réciproque entre la croissance et la mouvement révolutif des tiges volubiles chez l'*Ipomoea purpurea*. Physiol. Vég. *12*, 327–373 (1974)

Dennison, D.S., Roth, C.C.: *Phycomyces* sporangiophores: fungal stretch receptors. Science *156*, 1386–1388 (1967)

Dequeudre, G., Courtot, Y., Baillaud, L.: Mouvements révolutifs et mouvements d'enroulement liés à des facteurs activants la croissance chez une espèce subvolubile (*Vincetoxicum officinale*). Z. Pflanzenphysiol. *54*, 254–269 (1966)

Elgerd, O.: Control Systems Theory. New York: McGraw-Hill, 1967

Galston, A.W., Tuttle, A.A., Penny, P.J.: A kinetic study of growth movements and photomorphogenesis in etiolated pea seedlings. Am. J. Bot. *51*, 853–858 (1964)

Gessner, F., Hammer, L.: Lichtbedingte Wachstumsschwingungen bei *Helianthus annuus*. Österr. Bot. Z. *105*, 529–549 (1958)

Gradmann, H.: Die Bewegungen der Windepflanzen. Z. Bot. *13*, 337–393 (1921a)

Gradmann, H.: Die Überkrümmungsbewegungen der Ranken. Jahrb. Wiss. Bot. *60*, 411–457 (1921b)

Gradmann, H.: Die Fünfphasenbewegungen der Ranken. Jahrb. Wiss. Bot. *61*, 169–204 (1922)
Gradmann, H.: Die Überkrümmungsbewegungen etiolierter Keimpflanzen. Jahrb. Wiss. Bot. *66*, 339–357 (1927)
Gradmann, H.: Die Lateralwirkung bei der Windepflanze. Jahrb. Wiss. Bot. *68*, 46–78 (1928)
Heathcote, D.G.: Generation of nutational movements. Nature (London) *208*, 909–910 (1965)
Heathcote, D.G.: A new type of rhythmic plant movement: micronutation. J. Exp. Bot. *17*, 690–695 (1966)
Heathcote, D.G.: Studies on the rhythmic movement of plant organs. Ph. D. thesis. Univ. Wales (1968)
Heathcote, D.G.: Circumnutation and gravity. Eur. Symp. Life Sci. Res. in Space, Cologne/Porz, Germany 1977. Proc. 195–201 (1977)
Heathcote, D.G., Aston, T.J.: The physiology of plant nutation. I. Nutation and geotropic response. J. Exp. Bot. *21*, 997–1002 (1970)
Hertel, R., Flory, R.: Auxin movement in corn coleoptiles. Planta *82*, 123–144 (1968)
Hess, B., Boiteaux, A.: Oscillatory phenomena in biochemistry. Annu. Rev. Biochem. *40*, 237–258 (1971)
Israelsson, D., Johnsson, A.: A theory for circumnutations in Helianthus annuus. Physiol. Plant. *20*, 957–976 (1967)
Jaffe, M.J.: Physiological studies on Pea tendrils. VIII. The relationship of circumnutation to contact coiling. Physiol. Plant. *26*, 73–80 (1972)
Jaffe, M.J., Galston, A.W.: Physiological studies on Pea tendrils. I. Growth and coiling following mechanical stimulation. Plant Physiol. *41*, 1014–1025 (1966a)
Jaffe, M.J., Galston, A.W.: Physiological studies on Pea tendrils. II. The role of light and ATP in contact coiling. Plant Physiol. *41*, 1152–1158 (1966b)
Jaffe, M.J., Galston, A.W.: Physiological studies on Pea tendrils. III. ATPase activity and contractility associated with coiling. Plant Physiol. *42*, 845–847 (1967a)
Jaffe, M.J., Galston, A.W.: Physiological Studies on Pea tendrils. IV. Flavonoids and contact coiling. Plant Physiol. *42*, 848–850 (1967b)
Jaffe, M.J., Galston, A.W.: The physiology of tendrils. Annu. Rev. Plant Physiol. *91*, 417–434 (1968a)
Jaffe, M.J., Galston, A.W.: Physiological studies of Pea tendrils. V. Membrane changes and water movement associated with contact coiling. Plant Physiol. *43*, 537–542 (1968b)
Joerrens, G.: Nutationsbewegungen bei *Triticum*-Koleoptilen. Z. Bot. *47*, 403–420 (1959)
Johnsson, A.: Relationships between photo-induced and gravity-induced electrical potentials in *Zea mays*. Physiol. Plant. *20*, 562–579 (1967)
Johnsson, A.: Geotropic responses in *Helianthus* and their dependence on the auxin ratio – with a refined mathematical description of the course of geotropic movements. Physiol. Plant. *24*, 419–425 (1971)
Johnsson, A.: Gravitational stimulations inhibit oscillatory growth movements. Z. Naturforsch. *29c*, 717–724 (1974)
Johnsson, A.: Oscillatory water regulation in plants. Cambridge. England: Bull. Inst. Math. Applications *12*, 22–26 (1976)
Johnsson, A.: Circumnutations in space? Proc. R. Soc. London B199, 505–512 (1977)
Johnsson, A., Heathcote, D.: Experimental evidence and models on circumnutations. Z. Pflanzenphysiol. *70*, 371–405 (1973)
Johnsson, A., Israelsson, D.: Application of a theory for circumnutations to geotropic movements. Physiol. Plant. *21*, 282–291 (1968)
Johnsson, A., Israelsson, D.: Phase-shift in geotropical oscillations – a theoretical and experimental study. Physiol. Plant. *22*, 1226–1237 (1969)
Junker, S.: Auxin transport in tendril segments of *Passiflora caerulea*. Physiol. Plant. *37*, 258–262 (1976)
Karvé, A.D., Jingajinni, S.A.: An entrained oscillation as a model of a two-oscillator system. Z. Pflanzenphysiol. *53*, 363–365 (1965)
Karvé, A.D., Salanki, A.S.: A feedback oscillation of relatively high frequency in hypocotyls of *Carthamus tinctorius* L. Z. Bot. *52*, 98–102 (1964a)
Karvé, A.D., Salanki, A.S.: The phenomenon of phase shift in geotropically induced feedback oscillations in *Carthamus tinctorius* L. seedlings. Z. Bot. *52*, 113–117 (1964b)
Karvé, A.D., Salanki, A.S.: Effect of alcohol and light upon geotropically induced oscillations. Z. Bot. *52*, 186–192 (1964c)

Karvé, A.D., Sawant, D.S.: The phenomenon of damping in geotropically induced feedback oscillations. Naturwissenschaften *53*, 160–161 (1966)

Melin, D.: Etude des flavonols dans quelques espèces à rameux volubiles. Rôle dans la circumnutation? Ann. Sci. Univ. Besançon *15*, 139–143 (1974)

Melin, D.: Croissance et mouvement révolutif des tiges de *Periploca graeca* L. Z. Pflanzenphysiol. *76*, 384–399 (1975a)

Melin, D.: Acides aminés libres, port et activité révolutive des rameaux du *Periploca graeca* L. Planta *125*, 105–114 (1975b)

Newman, I.A.: Electrical potentials and auxin translocation in *Avena*. Aust. J. Biol. Sci. *16*, 629–646 (1963)

Pavlidis, Th.: Biological Oscillators: Their Mathematical Analysis. New York: Academic Press, 1973

Quantin, A.: Contribution à l'étude de la nutation des vrilles de Passifloracées. Ann. Sci. Univ. Besançon 6/7, Bot. No 1, 11–15, 1952 (as cited in Baillaud, 1962a)

Roblin, G.: Chronobiologie et proprietes sensorio-motrices de la sensitive (*Mimosa pudica* L.). Trav. Stat. Biol. Beau-site, Poitiers (1975)

Rosen, R.: Dynamical System Theory in Biology. I. Stability Theory and its Applications. New York: Wiley-Interscience, 1970

Scott, I.H.: Electric fields in plants. Annu. Rev. Plant Physiol. *18*, 409–418 (1967)

Shen-Miller, J.: Rhythmic differences in the basipetal movement of indoleacetic acid between separated upper and lower halves of geotropically stimulated corn coleoptiles. Plant Physiol. *52*, 166–170 (1973)

Shen-Miller, J., Noack, N.G., Baker, J.E.: Kinetics and periodicity of auxin transport, geotropic curvature and growth. Arg. Nat. Rep. ANL-7770 (1970)

Spurný, M.: Spiral feedback oscillations of growing hypocotyl with radicle in *Pisum sativum* L. Biol. Plant *8*, 381–392 (1966)

Spurný, M.: Spiral oscillations of the growing radicle in *Pisum sativum* L. Naturwissenschaften *55*, 46–47 (1968a)

Spurný, M.: Evidence of a feedback control system in the growing hypocotyl with radicle as related to the effect of spiral oscillations and geoperception by the root tip in *Pisum sativum* L. Folia Prírodovedcké Fak. Univ. J.E. Purkyne v. Brne Biol. *18*, 91–97 (1968b)

Spurný, M.: Effect of root tip amputation on spiral oscillations of the growing hypocotyl with radicle of the Pea (*Pisum sativum* L.). Biol. Plant. *10*, 98–111 (1968c)

Spurný, M.: Spiral oscillations resulting from a feedback control system in the growing pea root (*Pisum sativum* L.). 2nd Int. Symp. Ecol. Physiol. Root Growth. Berlin: Akademie Verlag 1974a

Spurný, M.: Interactions of photo- and geotropism with periodical oscillations of growing pea root (*Pisum sativum* L.). Biol. Plant. *16*, 43–49 (1974b)

Tronchet, A.: Les effets du traitment gibbérellique sur des plants de *Lepidium sativum, Mercurialis annua* et *Ocimum basilicum*. Bull. Soc. Hist. Nat. Doubs. *63*, 85–86 (1961)

Tronchet, A., Tronchet, J., Bressot, J.-M., Marchand, G., Blanc, B.: Sur les mouvements révolutifs des tiges traitées par l'acide gibbérellique. Ann. Sci. Univ. Besançon. 2e Ser. Bot. *17*, 3–26 (1961a)

Tronchet, A., Tronchet, J., Marchal, J.: Transformation de feuilles en vrille par traitement gibbérellique (*Lactua saligna* var. annuelle). C.R. Acad. Sci. (Paris) *252*, 2927–2929 (1961b)

Tronchet, A., Tronchet, J., Perney, J.: Modifications de croissance et mouvements révolutifs de plantes de *Zinnia* traité par l'acide gibbérellique. 85th Congr. Soc. Sav. Chambery, Annecy, 419–425 (1960)

Tronchet, J.: Sur la présence d'hétérosides des flavonols kempferol et quercétine dans des vrilles et tiges volubiles. Étude par chromatographie sur papier. Bull. Soc. Fr. Physiol. Vég. *6*, 2, 99 (1960)

Tronchet, J.: Sur la présence de flavonoides dans les tissus corticaux des vrilles et des tiges volubiles et leur intervention possible dans la circumnutation. 86th Cong. Soc. Sav. Montpellier 493–495 (1961)

Turing, A.M.: The chemical basis of morphogenesis. Phil. Trans. R. Soc. London Ser B *237*, 37–72 (1952)

Tyson, J.J.: The Belousov-Zhabotinskii Reaction. Lecture Notes in Biomathematics 10. Berlin-Heidelberg-New York, Springer, 1976

Tropisms.

5.3.2 Epinasty

B.G. KANG

1. Introduction

Epinasty and hyponasty are curvatures resulting from differential growth. Epinasty normally results from more rapid growth on the adaxial (inner) side of an organ relative to the abaxial (outer) side[1]. Conversely, hyponasty results from a relatively faster growth on the abaxial side of an organ.

The "typical" example of epinasty is the downward curvature of a morphologically dorsi-ventral organ, such as a leaf and certain floral structures, but movements of this nature also occur in radially symmetrical and/or orthotropic organs such as the stem and the root (see PALMER in Vol. 11 of this series). In these latter cases, induction of physiological dorsi-ventrality is a prerequisite for epinastic movements (see KALDEWEY, 1962).

Epinastic movements are thought to be primarily autonomous, i.e., not induced or oriented by specific external factors. This separates epinasty from all tropisms, which are oriented with respect to external stimuli, and from other nastic movements as, e.g., seismonasty or nyctinasty. These latter movements can easily be confused in appearance with epinasty; but they are of a different nature – in addition to their being controlled primarily by external stimuli – in that they may involve changes in the turgor of particular cells in a pulvinus (see SATTER, Chap. 4.2, this vol.). However, external factors are of course not completely without effect on epinastic responses, which at least can be influenced or modified by factors such as gravity, light and temperature. PALMER (in Vol. 11 of this series) reviews the effects of external factors, as well as the induction of epinasty by such factors as plant pathogens and edaphic conditions.

In contrast to tropic movements, where auxin has been thought to be involved as the major regulatory hormone, at least until recently (PHILLIPS, 1972; RAILTON and PHILLIPS, 1973), epinastic curvatures were often found to be controlled by ethylene (e.g., NELJUBOW, 1901; HARVEY, 1915).

Early observations of epinasty were made as a result of the effect of illuminating gas and cigarette smoke (WÄCHTER, 1905); HARVEY (1915) first demonstrated that the active component was ethylene. Subsequently, extensive studies were carried out by investigators at the Boyce Thompson Institute on the epinasty of petioles and lateral shoots in relation to actions of growth substances – primarily ethylene and related gases, and various auxins (e.g., CROCKER et al., 1932, 1935; ZIMMERMAN and WILCOXON, 1935).

Results of experiments involving both auxin and ethylene suggested the existence of hormonal interaction rather than the action of only one hormone.

[1] Often, but not always, adaxial is equivalent to dorsal and abaxial is equivalent to ventral; (see SATTER, Chap. 4.2, this vol.).

Unlike the extensive reviews on the subject in the previous series of this publication (CHAMPAGNAT, 1961; KALDEWEY, 1962), this chapter is not intended to cover all aspects of epinasty. Instead, this review deals with three selected areas of the general subject; namely, leaf petiole epinasty, leaf unrolling of grass seedlings, and epinastic formation and hyponastic opening of plumular hooks of dicotyledonous seedlings. Particularly the role of phytohormones in these responses will be described at the present stage of knowledge.

2. Leaf Petiole Epinasty

2.1 Characteristics

When young leaves emerge from the bud, as has been described for *Helianthus* leaf epinasty (PALMER and PHILLIPS, 1963), their axes appear to lie almost parallel to the shoot axis, but as they grow and their petioles elongate, they bend downward to form an angle with the shoot axis until a proper orientation is achieved for photosynthesis and for other activities. During early stages of leaf development, curvature develops in the lamina along the midrib due to more rapid cellular growth on the adaxial (upper) side of the leaf as compared with the abaxial (lower) side. This curvature appears to move basipetally along the petiole as the leaf matures. At maturation, the leaf will reach a final orientation, and no further petiole bending is observed under natural conditions. Nevertheless, there are several experimental conditions under which even these leaves again start epinastic bending. Most of the paragraphs of this following section will deal mainly with this type of epinasty.

The various regions and the various ages of the leaf have varying sensitivities to epinasty-inducing agents, resulting in different kinetics for the process and making leaf epinasty a complex phenomenon (PALMER, 1972). For instance, in young, actively growing leaves, ethylene induces leaf epinasty along the entire petiole, whereas in older leaves, where growth has essentially ceased, epinasty is observed only at the extreme base of the petiole, close to the shoot axis (DOUBT, 1917), and in addition, this bending is much slower than in young leaves (PALMER, 1972).

2.2 Regulatory Mechanisms

2.2.1 Gravity Factor

In an early review by RAWITSCHER (1932), the problem was discussed as to whether leaf orientation under natural conditions is a geotropic or an epinastic response. Plants were rotated on the horizontal axis of a clinostat, where gravitational stimulation is often believed to be nullified. These plants exhibit epinastic curvature in those older leaves which under natural conditions had already reached their final orientation (see above). Epinasties under these conditions could be interpreted as being either a manifestation of the intrinsic dorsiventrality of an organ, which normally would be counteracted by gravity, and/or due to an internal adaptation which might persist even after the stimulus (i.e., gravity) is removed (see Sect. 2.2.3). The former alternative is consistent with the concept of epinasty as a movement primarily controlled by internal factors. In addition, however, epinasty can be influenced by geotropic stimulation. CROCKER et al. (1932) pointed out that inverting plants to an upside down position modified the epinastic movement of the petioles.

The use of a clinostat as a tool to compensate for or nullify stimulation through gravity, by providing essentially equilateral stimulation around the plant axis, has received some criticism, since various conditions of rotation impose different physical states upon the cells (DEDOLPH and DIPERT, 1971; see VOLKMANN and SIEVERS, Chap. 5.2.2, this vol., for a critical discussion of the use of a clinostat in studies on plant movement).

A valid way to attain true conditions of weightlessness became available for biological experiments through the Biosatellite II mission in 1967 (SAUNDERS, 1971). Along with other biological specimens, pepper (*Capsicum annuum*) plants were put into orbit (10^{-5} *g* for a 45-h period), and changes in the liminal angle of the petioles in the spacecraft were compared with earth-based, clinostat-simulated controls. Both groups of plants underwent epinastic movement (JOHNSON and TIBBITTS, 1968). THIMANN (1968) concluded from this experiment that "weightlessness can be closely imitated by rotation on the horizontal clinostat". However, there seem to be some quantitative differences in the response of the orbited and the clinostated plants (JOHNSON and TIBBITTS, 1968; BROWN et al., 1974). Obvious differences in physical conditions were the presence of rotational force in the clinostat, but not in the spacecraft (JOHNSON and TIBBITTS, 1968), and the presence of convection in the clinostat, but not in space (BROWN et al., 1974).

2.2.2 Role of Phytohormones

There is ample evidence that ethylene is the principal substance causing epinasty. Agents that induce ethylene production in plant tissues are also known to induce epinasty; these include, most notably, auxins (JAMES and WAIN, 1969; HARPER and WAIN, 1971). Also effective are the herbicides ethrel (GARCIA-TORRES and GOMEZ-CAMPO, 1972), picloram (BAUR and MORGAN, 1969), dicamba (STACEWICZ-SAPUNCAKIS et al., 1973), and mechanical or physical stimulation such as shaking (MITCHELL et al., 1975) and flooding (see PALMER in Vol. 11 of this series).

Those substances that counteract ethylene, on the other hand, are found to inhibit epinasty. These include CO_2 (LEATHER et al., 1972), $AgNO_3$ (BEYER, 1976a), and benzothiadiazole (JACKSON and CAMPBELL, 1976; PARUPS, 1973).

It must be noted, however, that auxin has an action on epinasty in addition to inducing ethylene production. For instance, it can be seen from the work of SOEKARJO (1965b) that the IAA concentration curve for the epinasty of isolated *Coleus* petioles (Fig. 1) has a small maximum in epinasty followed by a minimum (still at low concentrations), and a second maximum at high concentrations. The apparent dual effect of auxin on epinasty may best be interpreted as a direct action of the auxin on cell expansion in the upper side of the petiole at low concentrations, plus an auxin action to induce ethylene production at higher concentrations.

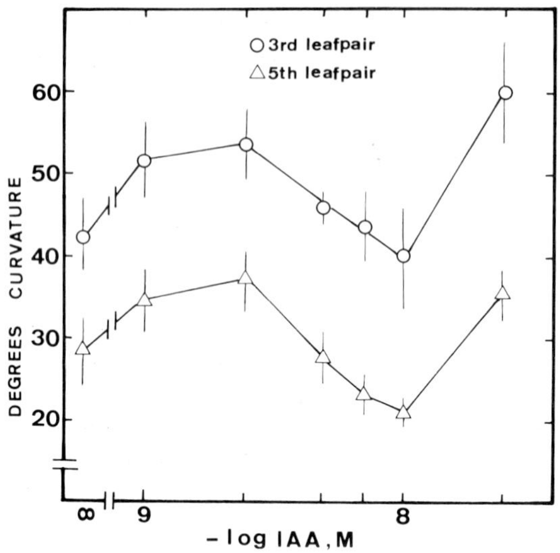

Fig. 1. The epinastic curvature of isolated petioles of the third and the fifth leaf pair of *Coleus*, after 24 h in solutions of IAA. (From SOEKARJO, 1965b)

Removal of the leaf blade results in decreased epinastic curvature of petioles of clinostated plants, which can be reversed by IAA application (LEIKE and GUTTENBERG, 1961; LYON, 1963a, b). These findings indicate that auxin is required for epinastic responses to occur, and that the lamina functions as a source of endogenous auxin that is transported basipetally through the petiole. How then is asymmetric growth of the petiole brought about?

VARDAR (1953) suggested that the two opposite flanks of the petiole have different sensitivities or responsiveness to auxin, but experimental evidence for this hypothesis is still lacking. An obvious alternative possibility is asymmetric distribution of auxin (FREYTAG, 1930; FISCHNICH, 1935). This possibility could be strengthened by taking ethylene action into account. Ethylene has been known to have strong influence on auxin transport (see ABELES, 1972). In *Vicia faba* exhibiting horizontal growth under the influence of ethylene (VAN DER LAAN, 1934), and in leafy branches of *Coleus* undergoing epinastic curvature (LYON, 1963a), auxin bioassay revealed more auxin in the adaxial than in the abaxial side. When labeled auxin is applied to the distal cut end of a petiole or branch explant undergoing ethylene-induced epinasty, the upper side shows higher radioactivity than the lower side (LYON, 1970; KANG and BURG, 1974). The same is true for epinasty induced by clinostat rotation (LYON, 1963b, 1965).

It appears from these findings that epinasty represents a manifestation of auxin-ethylene cooperation in which auxin induces ethylene production, and ethylene, in turn, acts on the auxin transport to cause asymmetric auxin distribution toward the adaxial (upper) side of a lateral organ. The question remains by what factor this ethylene-induced auxin transport is directed toward the adaxial side. There is some evidence that ethylene always moves auxin against the gravitational vector. This suggests gravity as an important factor in epinasty, as will be explained in more detail on p. 652ff. In favor of this view also is the fact that this ethylene effect is apparently not restricted to lateral organs such as petioles and branches, but is also found to be operative in orthotropic organs such as *Vicia* and *Pisum* stems under geotropic conditions (VAN DER LAAN, 1934; KANG and BURG, 1974).

On the other hand, PALMER (1976) was not able to find ethylene-induced auxin redistribution in a petiole after he had applied labeled auxin to the distal region of a *Coleus* leaf blade.

Surprisingly, in these experiments higher radioactivity was found in the lower half than in the upper half of the petiole, both in the presence and in the absence of applied ethylene. However, this evidently cannot be regarded as auxin transport in a true sense; the unusually high rate of auxin movement suggests that auxin had been translocated through the vascular tissue (see GOLDSMITH et al., 1974).

STEWART and FREEBAIRN (1969) suggest that auxin is necessary for petiolar epinasty only because it induces ethylene production. IAA-treated tomato petioles fail to exhibit epinastic curvature when given a heat treatment (40° C for 4 h), known to inhibit ethylene synthesis. In the presence of added ethylene, however, epinasty occurs even with the same heat treatment. PALMER (1972) found that ethylene induced rapid cell growth on the upper side of debladed petioles which were otherwise nongrowing, and thereby caused epinasty. He claimed that this phenomenon could not be accounted for in terms of auxin asymmetry. He concluded that ethylene-induced epinasty could be attributed to a direct stimulatory effect of the gas on growth of cells in the upper side of the petioles (PALMER, 1976).

Several other instances have been reported in which ethylene stimulates cell elongation rather than inhibiting it (e.g., KU et al., 1970; SUGE, 1971; MUSGRAVE et al., 1972). However, it is not clear whether ethylene directly promotes cell growth, or whether the action is mediated through another endogenous hormone.

In fact, auxin has been assumed to be involved in at least some of the examples of ethylene-induced growth promotion (IMASEKI and PJON, 1970; MUSGRAVE and WALTERS, 1973). One possibility suggested by MITCHENER (1938) is that ethylene increases the sensitivity of the tissue to auxin. This could then explain PALMER'S (1972) data that ethylene promotes the growth of the upper side of debladed petioles without added auxin (i.e., enhanced responsiveness to endogenous auxin).

OSBORNE (1976) classifies plant cells into two types, those where growth is promoted by auxin and inhibited by ethylene, and certain "specialized" cells responding to auxin and ethylene in just the reverse way; an example of the latter is found in the abscission zone of bean plants (WRIGHT and OSBORNE, 1974). OSBORNE (1974) developed the idea that in the first type, the shape of a cell is determined by a balance between the two hormones. Since ethylene can diffuse freely, at least across a cylindrical organ, its concentration is assumed to be the same on both sides. Growth distribution might be controlled by a combined effect of ethylene and differential auxin concentration, as follows. Auxin accumulates on one side of an organ (e.g., as a result of geotropic stimulation) and promotes growth on that side. On the opposite side, meanwhile, where auxin has been depleted by lateral migration, auxin-induced ethylene inhibits growth. According to that assumption, the presence of auxin "buffers" the cells against ethylene action (see OSBORNE, 1976). During the epinastic process, growth inhibition of cells in the lower side of the petioles does in fact occur (PALMER, 1972).

Gibberellic acid alleviates or slightly reverses epinastic curvature of the petioles, i.e., it induces hyponasty (PALMER, 1964; SOEKARJO, 1965a). It seems possible, in view of the opposite effects of GA and ethylene on auxin transport (PALMER and HALSALL, 1969), that the actions of GA and ethylene in epinasty are mediated through a common regulatory system. The cytokinin benzyladenine also inhibits epinastic curvature under certain conditions, probably indirectly through an interaction with other hormones (RAILTON and REID, 1973). Although gibberellins and cytokinins may participate in the regulation of epinasty under certain abnormal circumstances, it has not been clearly established that they normally do so (see PALMER in Vol. 11 of this series).

Thus, we are left mainly with auxin and ethylene as the most important factors in epinasty. To summarize the possible interactions of these two substances: auxin can induce ethylene synthesis; ethylene can influence auxin transport; ethylene and auxin can act antagonistically on cell growth.

2.2.3 Interaction of Gravity with Hormones

Induction of leaf epinasty by rotation on a clinostat has been interpreted in various ways, implicating both auxin and ethylene.

In several species of potted plants rotated at 1 and 6 rph on a horizontal clinostat, LYON (1963b) observed development of leaf epinasty and also detected asymmetric movement of labeled auxin into the upper side of the petioles. He suggested that there exist in the petioles both a gravity-directed auxin movement toward the abaxial side and an upward auxin movement of unknown origin toward the adaxial side. According to LYON (1963a, b, 1970) epinasty results when the former component of the transport system is blocked either by ethylene or by the lack of stimulation by gravity, i.e., on a clinostat. For a discussion of geotropism (gravitropism) see VOLKMANN and SIEVERS (Chap. 5.2.2, this vol.) and WILKINS (Chap. 5.2.3, this vol.).

However, LEATHER et al. (1972) attributed clinostat-induced leaf epinasty to increased ethylene production resulting from clinostat treatment itself. They based this interpretation on their findings that epinasty induced on a clinostat was alleviated by CO_2, a competitive inhibitor of ethylene action. PALMER (1973) reached the same conclusion from studies in which he observed a transitory petiole epinasty on a continuously rotated clinostat. His reasoning was based primarily on the observation that a short exposure of the plants to ethylene simulated the clinostat effect. However, he did not measure the ethylene produced in plants rotated on a clinostat. Hence, the physiological basis for the effects of reduced g-forces is not well understood.

A model has been proposed to explain epinasty caused by weightlessness and by ethylene as well (KANG and BURG, 1974). This model will be explained, in this section, in two steps.

For the *first step,* the idea was derived from the system of light–growth response in *Phycomyces* (DELBRÜCK and REICHARDT, 1956; BERGMAN et al., 1969). This system describes a relationship between the magnitude of the external stimulus (I) and a time-dependent internal development of a counteracting adaptation (A). The effective stimulus (i) on which the reaction is dependent is given by the quotient $i = I/A$, where $i = 1$ means effective stimulation at equilibrium conditions.

This concept has been adapted to geotropism in shoots or coleoptiles (HILD and HERTEL, 1972). It has been suggested that lateral auxin transport, which is one important step in the geotropic reaction chain, is caused by asymmetric auxin secretion from cells in the sensitive region, and that this in turn is due to asymmetric activity of auxin permeases at the membranes of these cells. This activity of the auxin permeases is thought to be controlled by the pressure of the intracellular statoliths. In terms of the above-mentioned model, the stimulus I is the pressure of the statoliths to the membrane (which, of course, is proportional to the gravity vector).

The auxin permease is taken, then, as the next transducer. If this system can undergo adaptation, the following results can be predicted: after a sufficiently long continuous exposition to gravity, A should approach and eventually reach I, and hence i should equal unity. In this case, a symmetry in effective stimulation is reached, and there is no difference in the activity of the permease on opposite sides: no lateral auxin transport occurs.

According to this model, constant geostimulation is counteracted by adaptation, and in lateral organs, this adaptation level might even be intrinsic. Indeed, from experimental results it has been concluded that plagiotropic roots are insensitive to constant gravistimulation, and react only to changes in the direction of the gravity vector (BENNET-CLARK, et al., 1959; BURSTRÖM, 1971).

When, in such a system, the geostimulation is nullified, either on a clinostat or even in weightlessness, the pressure of the statoliths on what had been the lower cell membrane decreases, $I < A$, and hence the auxin permease on that side becomes less active. As a consequence, an inverse asymmetry of activity is established, and auxin will move upwards. The adaptation level or the difference in sensitivity may persist nearly indefinitely (i.e., at least for a couple of days) in the absence of stimulation; this is seen in the pepper plants of the Biosatellite II experiments (JOHNSON and TIBBITTS, 1968). In other cases, the adaptation may slowly disappear upon removal of stimulation; this results in a transient epinasty as found in clinostated sunflower plants (PALMER, 1973).

In order to explain the effect of ethylene on auxin transport (and therefore also on epinasty), two further assumptions have to be made in a *second step*. We assume that ethylene inhibits the adaptation process (dA/dt), and we make use of the concept of overstimulation as proposed by HILD and HERTEL (1972). For many physiological systems, the dose–response curve increases to an optimum. As stimulus dose is increased still further, response decreases, and may even reverse direction (from "positive" to "negative"). This is schematically shown in Figure 2 for the presumed activity of auxin permease as a function of the effective geotropic stimulus, i. According to this dose–response curve, very strong effective stimulation should lead to inactivation of the auxin permease at the lower cell membrane, and hence to auxin transport in the "wrong" direction, i.e., against gravity. This is regularly observed in coleoptiles within the first minutes of geotropic stimulation, i.e., before adaptation has occurred (HILD and HERTEL, 1972). The same effect must be expected to continue, then, if adaptation is inhibited by ethylene.

With these additional assumptions, the model explains the observed ethylene-induced auxin movement into the upper sides – against gravity – for geotropically reacting *Pisum* stems (thus resulting in downward bending) and growing *Vicia* stems (thus forcing them to bend horizontally). It also explains the observed upward auxin movement in lateral petioles and branches (thus enlarging the "opening angle" in adult organs, as mentioned earlier, Sect. 2.1).

Obviously, this effect of ethylene on auxin transport is not uniform in all plants. Existing differences can, however, be accounted for by the model. CROCKER et al. (1932) observed ethylene-induced epinasty in *Pisum sativum* but failed to

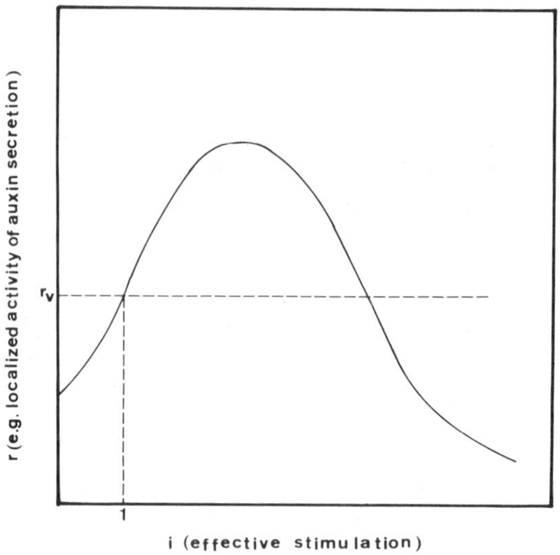

Fig. 2. Hypothetical response r as a function of i, r_v = steady-state activity at the level adapted to statolith pressure, compensating for the activity at the opposite flank of an unstimulated cell. (From HILD and HERTEL, 1972)

do so in *Phaseolus vulgaris*. Accordingly, the action of ethylene in reversing gravity-directed lateral auxin transport should be operative in peas but not in beans. This has been verified directly with pea epicotyls and bean hypotyls (KANG and BURG, 1974). It should be investigated whether, in beans, the inhibition of adaptation by ethylene is reduced, or whether there is no overstimulation by supraoptimal stimulation.

Finally, the model can account for the observation (CROCKER et al., 1932) that ethylene fails to induce epinasty in plants held upside down. Under these conditions, the gravity factor tends to move the auxin to the adaxial side, and hence to bend the petiole physically upwards. Morphologically, this would mean an epinasty. This gravity action is, however, counteracted by ethylene by the mechanism explained above.

3. Unrolling of Grass Leaves

3.1 Characteristics

Tightly rolled young leaves emerging from the coleoptile of grass seedlings normally undergo unrolling processes upon exposure to light. BURSTRÖM (1942) described this phenomenon as growth of the mesophyll cells due to increased cell-wall extensibility. Since unrolling results from preferential growth of the adaxial side relative to the abaxial side of the leaf, this is a case of leaf epinasty. This kind of leaf epinasty seems to be of a nature similar to that of auxin-induced hyponasty in immature primary leaf blades of *Phaseolus* seedlings (LIPPINCOTT and LIPPINCOTT, 1971; HAYES and LIPPINCOTT, 1976). These epinastic and hyponastic movements may be distinguishable, at least superficially, from those of petioles and plumular hooks which are cylindrical in shape.

3.2 Regulatory Mechanisms

3.2.1 Response to Light

VIRGIN (1962) was the first to present an action spectrum for the light-induced unrolling of etiolated wheat-leaf segments, showing a peak at about 660 nm. He demonstrated that the effect could be reversed by far-red irradiation, indicating phytochrome as the responsible photoreceptor.

Studies on the light-induced unrolling of barley leaf segments (DEUTCH and DEUTCH, 1974) indicate that dose–response curves for the phytochrome-induced unrolling at fixed light intensities for varying irradiation times resemble those for the phototropism of the *Avena* coleoptile (see DENNISON, Chap. 5.1.2, this vol.). This kind of dose–response curve is found in only a few phytochrome-mediated cases (BLAAUW et al., 1968; RAVEN and SPRUTT, 1972; SEITZ, 1974). A generally accepted explanation is still lacking.

Like many other phytochrome-regulated phenomena, barley leaf unrolling is also induced by blue light (450, 475 nm; VIRGIN, 1962), and this induction can also be reversed by far-red light (DEUTCH and DEUTCH, 1975). The relative effectiveness of the red and the blue regions depends on the levels of incident energy. However, when simultaneous (mixed) irradiation with red and blue light is used, the blue light appears to inhibit the red-light-induced leaf unrolling. Thus, we are concerned with rather complicated interactions of different wavelength regions.

In etiolated wheat-leaf segments, WAGNÉ (1964) noticed that some unrolling occurred along the entire segment, even when only a part of the segment had been irradiated. Moreover, a nonirradiated part still showed the phytochrome effect, even if it had been isolated from the irradiated part immediately after irradiation (WAGNÉ, 1965). These observations could implicate involvement of a rapidly transmitted factor. Thus, this phenomenon seems to be similar to the interorgan correlations which will be described in more detail in Section 4.2.3.

3.2.2 Role of Phytohormones

The nature of the cellular growth in the adaxial mesophyll tissue during leaf unrolling is probably unique as far as its response to auxin is concerned. Auxin at concentrations as low as 10^{-7} M inhibits leaf unrolling (BURSTRÖM, 1942), but this auxin inhibition of cell expansion cannot be explained by auxin-induced ethylene production, since ethylene itself has no significant effect on the unrolling process (KANG, 1971; LOVEYS and WAREING, 1971). Despite the marked effect of auxin, the regulatory role of this hormone on leaf unrolling under physiological conditions has not yet been confirmed (MENHENETT, 1972).

In contrast, the role of gibberellins (GA) in leaf unrolling is much better understood. Experimental evidence indicates that the photocontrol of leaf unrolling is mediated, at least in part, by photo-induced changes in the endogenous GA level. Application of GA to leaf segments results in enhancement of light-induced leaf unrolling and can substitute in the dark for the red light action (BEEVERS et al., 1970; POULSON and BEEVERS, 1970b). Both AMO-1618 and CCC inhibit light-induced unrolling, suggesting that gibberellin biosynthesis is required for the unrolling process (LOVEYS and WAREING, 1971). Furthermore, it has been reported that red illumination causes a rapid and transient increase in the gibberellin level in etiolated barley (REID et al., 1968) and wheat (BEEVERS et al., 1970) leaves. Interestingly, a short pretreatment of wheat leaf segments with GA, which simulates photoinduction of unrolling, is found to be more effective than a continuous treatment with GA throughout the incubation period, and as effective as treatment with red light (LOVEYS and WAREING, 1971). Phytochrome (P_{fr}) may induce synthesis of gibberellin in the plastid and/or it may act on the plastid membrane to release gibberellin from a bound form (COOKE and SAUNDERS, 1975a, b; COOKE et al., 1975).

Abscisic acid (ABA) strongly inhibits both light-induced and GA-induced leaf unrolling (BEEVERS et al., 1970). POULSON and BEEVERS (1970b) found that preincubation of a barley leaf segment with buffer (1 mM sodium acetate buffer at pH 5.5) hastened unrolling during subsequent photoinduction, and they advanced the speculation that a natural inhibitor, possibly ABA, had been leached out of the leaf during preincubation. Although it has been found that ABA reduces the endogenous GA level in wheat leaves, this action alone cannot fully account for the inhibitory effects of ABA on the unrolling process (BEEVERS et al., 1970). Furthermore MENHENETT and CARR (1973) isolated an inhibitor from etiolated barley leaves, and established that its level was indeed decreased by pre-illumination, but they found evidence that the inhibitor was not ABA.

Kinetin, like GA, also stimulates leaf unrolling both in the light and in the dark. This effect has been thought to result from increases in the endogenous gibberellin level through action of kinetin (LOVEYS and WAREING, 1971).

3.2.3 Metabolic Changes Associated with Unrolling

Leaf unrolling represents only one visible manifestation of the complex set of metabolic events initiated by phytochrome and/or hormones. One thus needs to distinguish metabolic changes that may lead to leaf unrolling from those that occur in parallel with, or even in consequence of, leaf unrolling. One approach to this problem is to study the kinetics of light-induced leaf unrolling. Time-course analyses indicate an apparent lag for the photo-response of between 4 and 12 h, depending on the experimental system (PRICE et al., 1965; POULSON and BEEVERS, 1970; KANG, 1971; LOVEYS and WAREING, 1971). Any events that may occur after this period, therefore, cannot be among the causes of leaf unrolling. Another possible approach is to study metabolic events that still occur under conditions where the final response of the effector system is inhibited. For example, KANG (1971) noticed light-induced increases in protein synthesis in etiolated wheat leaf segments in which unrolling is osmotically inhibited by mannitol. These results indicate that the light effect on protein synthesis is not a consequence of light-induced growth promotion.

In etiolated corn leaves, treatment with photomorphogenically active light causes decreases in the level of both starch and soluble sugars (KLEIN et al., 1963; PRICE et al., 1964, 1965). Based primarily on their time-course data, KLEIN and PRICE speculated that a causal relationship of these light-induced metabolic changes to leaf unrolling exists. Experiments in which labeled glucose is fed to the corn leaves (MITRAKOS et al., 1965, 1967) indicate that such decreases can be attributed to increased synthesis of matrix polysaccharides in the cell wall, a phenomenon also known for auxin-promoted growth of the *Avena* coleoptile

(RAY and BAKER, 1965). Red irradiation of corn leaves also increases the rate of phosphate turnover (STEINER et al., 1968), but whether this increased phosphorylating activity is a cause or a result of the changes in carbohydrate metabolism is unknown.

GA-promoted and ABA-inhibited leaf unrolling is also correlated with GA-increased and ABA-decreased activities in RNA polymerase, although the hormonal action in this case may not be a direct one (POULSON and BEEVERS, 1970c). Furthermore, a correlation has been established between leaf unrolling and increase in polysomal formation in response to light; however, reports on the hormonal (GA and ABA) effects on polysomal changes are conflicting (PEARSON and WAREING, 1970; POULSON and BEEVERS, 1970c). The resulting increase in capacity for RNA synthesis in illuminated leaf segments could bring about increased protein synthesis, which may be responsible for leaf unrolling (POULSON and BEEVERS, 1970a).

More important are several observations that suggest that leaf unrolling requires synthesis of *specific* RNA species, possibly mRNA, and thus synthesis of *specific* proteins. POULSON and BEEVERS (1970a) reported that actinomycin D inhibited unrolling but not greening, whereas 5-fluorouracil, which inhibits the bulk of RNA synthesis, did not inhibit unrolling. Furthermore, leaf unrolling seems to require protein synthesis on cytoplasmic ribosomes, whereas greening is associated with synthesis of plastid protein: cycloheximide inhibits unrolling without affecting synthesis of plastid proteins (KANG, 1971), whereas chloramphenicol inhibits greening but not unrolling (POULSON and BEEVERS, 1970a). Therefore, leaf unrolling and greening seem to be metabolically separable phenomena at the level of RNA and protein synthesis.

4. Plumular Hook Development

4.1 Characteristics

The young seedling of many dicotyledons is characterized by a hooked portion on the shoot axis immediately subjacent to the apical terminus. The apex points downward as the young shoot emerges through the soil during germination, and is thus mechanically protected. The formation of the seedling hook results from asymmetric growth of the developing shoot axis. The capacity for asymmetric growth is probably an intrinsic characteristic of the tissue (RUBINSTEIN, 1972), although in one case a hook may be formed in response to gravitational stimulation (KARVÉ, 1964). This capacity seems to be established in the early stages of embryonic development, since embryos excised from immature seeds have been shown to grow into shoots with hooks under appropriate culture conditions (SKENE, 1969).

More attention has been given to the opening of preformed hooks than to hook formation. The smaller cortical cells on the inner (concave) side of the hook elbow elongate, while the cells on the outer (convex) side remain unchanged. As a result, the cells on both sides become equal in size, and the hook straightens (KLEIN, 1959; MOHR and HAUG, 1962; KANG and RAY, 1969a; GEE and VINCE-PRUE, 1976). If intact seedlings are grown under conditions where opening is not induced, this kind of differential growth may nevertheless proceed, but hook opening is continuously counteracted by formation of hook due to a greater elongation on the outer, apical portion of the elbow (RUBINSTEIN, 1971a). The net result is that the hook appears to move up the hypocotyl as it grows and eventually becomes part of the epicotyl (KLEIN et al., 1957).

4.2 Regulatory Mechanisms

4.2.1 Response to Light

Light is the most important external factor in the regulation of seedling hook configuration. In nature, plumular hooks generally straighten out upon exposure to light when emerging from the soil. An action spectrum for light-induced hook opening in bean seedlings (*Phaseolus vulgaris*) clearly indicates that phytochrome is the responsible photoreceptor (WITHROW et al., 1957). Spectrophotometric measurements of the phytochrome of bean seedlings in vivo indicate that hook opening is closely correlated with the amount of this photoreversible pigment; for instance, phytochrome is found to be most abundant in the physiologically responsive region of the hypocotyl, the hook elbow. Furthermore, the dose–response curves for photoinduction of hook opening by red light at 660 nm as well as those for photoinhibition by far red at 730 nm correlate well with phytochrome absorbance changes (KLEIN et al., 1967). The induction of bean hook opening by blue light (WITHROW et al., 1957) is also known to be mediated by the phytochrome system. The implication that the blue light action entails riboflavin-mediated photoinactivation of IAA, especially in the presence of riboflavin phosphate, (GALSTON, 1959; REINERT, 1959) has been ruled out by SHROPSHIRE et al. (1964). They observed enhanced hook opening by the fluorescence spectrum of a solution of riboflavin phosphate (FMN) which was excited by blue light; this effect is probably due to the long-wavelength tail of the fluorescence spectrum absorbed by phytochrome.

The so-called "high intensity reaction" (HIR) is involved in the photocontrol of hook opening in lettuce seedlings (MOHR and NOBLE, 1960), as well as in the interesting related phenomenon of twining in dodder seedlings (*Cuscuta*) (LANE and KASPERBAUER, 1965; KUJAWSKI and TRUSCOTT, 1974). Dodder plants are heterotrophic, and the seedlings have no cotyledons (TRUSCOTT, 1966), but they develop seedling hooks similar, at least superficially, to the hypocotyl hook of bean and the epicotyl hook of pea seedlings (GALSTON et al., 1964). As in a typical phytochrome-mediated response, a single saturating red irradiance can fully induce a complete opening response, and this red light effect can be just as fully reversed by subsequent far-red illumination (LANE and KASPERBAUER, 1965). However, the rate of opening is also increased by continuous irradiation with blue light, an effect which might be due to the involvement of the HIR system (KUJAWSKI and TRUSCOTT, 1974) in this photoresponse.

Response of seedling hooks to red light seems to vary, not only quantitatively, but also qualitatively among plant species. Plumular hooks of etiolated lettuce (*Lactuca sativa*) seedlings respond to red light by further tightening their configuration (JANES et al., 1976), a response exactly opposite to that observed in other dicot seedlings, and reopening of the hook occurs under continuous blue or far-red irradiation, i.e. under HIR conditions (MOHR and NOBLE, 1960).

4.2.2 Role of Phytohormones

In the regulation of hook development, the action of light seems to be causally related to hormonal levels in the tissue, but the particular hormones and mechanisms involved depend on the type of plant and on its physiological state.

Hook opening in isolated bean hypocotyls is inhibited by auxin even at low concentrations (KLEIN et al., 1956) and at higher concentration, hook opening can even be reversed, i.e., the hook is reformed (KLEIN et al., 1957). The presence of an IAA-like substance in the bean hypocotyl has been demonstrated in the *Avena* curvature test (KANG and RAY, 1969a) and by fluorometric assay (RUBINSTEIN, 1971b). Since hook opening can be induced, to some extent, in darkness by auxin antagonists such as naphthylmethyl sulfide acetate (KLEIN, 1965) and p-chlorophenoxyisobutyric acid (KANG and RAY, 1969a), one might postulate that hook opening is caused by decreases in the endogenous auxin level. However, direct measurements of diffusible auxin from the hook tissue (KANG and RAY, 1969a) fail to support this possibility. On the other hand, RUBINSTEIN (1971b) suggests that a small increase in the rate of auxin transport along the inner side of the hook elbow due to red illumination may result in lowering of the auxin level in that part of the hook, and may lead ultimately to hook opening.

According to KANG and RAY (1969b), auxin inhibition of hook opening, due to growth inhibition of cells on the inner side of the hook elbow, and auxin induction of hook formation, apparently due to growth promotion of cells on the apical outer side, are both mediated by auxin-induced ethylene production. Accordingly, externally applied ethylene inhibits hook opening, and can even induce the hook configuration in newly emerging shoots of light-grown pea plants (BURG and BURG, 1968). Several lines of evidence suggest that ethylene functions as a natural agent for seedling hook formation (KANG and BURG, 1972). First, the internal concentration of endogenous ethylene in the hook tissue is sufficiently high to account for the hook configuration (KANG and RAY, 1969b). Secondly, red-light-induced hook opening can be attributed to inhibition of ethylene production by red light; i.e., the rates of both endogenous and auxin-induced ethylene production in the illuminated tissue have been found to be lower than those in the dark control (KANG et al., 1967).

The major question pertaining to leaf petiole epinasty (see Sect. 2.2.2) may also be raised with regard to plumular hook formation: does ethylene itself induce asymmetric growth, and, if so, how? Unlike branch and petiole epinasty, asymmetric distribution of auxin or other growth factor(s) has not been clearly demonstrated in the hook elbow, although it remains a possibility. Alternatively, ethylene itself may selectively promote or inhibit growth of various cells, depending on a differential sensitivity of those cells to ethylene. The unusual positive growth response toward ethylene occurs not only in a specialized cell type, but also at a particular physiological age during the life of that cell (HALE et al., 1970). In bean and *Xanthium* hypocotyls, GOTO and ESASHI (1974a, b) showed that cells at various locations along the hypocotyl axis (i.e., at different physiological ages) exhibit differential responses to ethylene—either growth promotion or inhibition.

Although auxin is generally known to prevent hook opening, IAA can slightly stimulate the hook opening process of isolated cotton hypocotyls (POWELL and MORGAN, 1970) and intact mung bean seedlings (GEE, 1977), but these auxin effects may not be significant, either because the auxin concentration applied is so high, or because the magnitude of the auxin response is so small.

GA also has varying effects on seedling hook configuration. Hook opening is promoted by GA in bean (KLEIN, 1965), pea (NAKAMURA et al., 1966), and cotton (POWELL and MORGAN, 1970) seedlings. In the bean hypocotyl, however, gibberellin may not play a regulatory role, but may simply be required for the growth involved in the opening processes (KANG and RAY, 1969a). By contrast, GA can strongly induce hook closure of lettuce seedlings, especially in the presence of ethylene. This observation indicates a possible interaction between a growth-promoting factor on one side of the hypocotyl (GA) and a growth-inhibiting factor on the other side (ethylene), bringing about development of a marked curvature (POOVAIAH and LEOPOLD, 1974).

The most pronounced action of a cytokinin on hook configuration is observed in mung-bean seedlings (GEE, 1977). Basally applied kinetin, especially in the presence of GA, can maintain the closed hook posture of the hypocotyl explant, and can block the inductive effect of red irradiation. Thus a combination of the two hormones coming from the root seems to constitute an endogenous regulatory factor in these seedlings.

Although ABA prevents light-induced opening of the bean hypocotyl hook, it fails to induce hook closure even at high concentrations (KANG and RAY, 1969a), and its role in the natural control of the opening process is unknown. It may well be that ABA simply inhibits the cellular growth necessary for the light-induced opening process.

A variety of other substances have been reported to have effects on seedling hook configuration. The effects of these substances can be explained mainly by their actions on the endogenous hormone systems.

Triiodobenzoic acid inhibits hook opening of isolated hypocotyls, presumably by inhibiting transport of auxin from the hook to the basal part of the hypocotyl (RUBINSTEIN, 1971b). By contrast, a morphactin applied to the whole seedling of *Cardamine chenopodifolia* can induce hook opening (LORENZEN, 1968), possibly by inhibiting auxin transport into the hook. Coumarin (MORGAN and POWELL, 1970) and malformin (CURTIS, 1969; CURTIS and JOHN, 1975), both of which are known to induce ethylene production in plant tissues, inhibit hook opening of bean hypocotyls. On the other hand, application of agents that inhibit ethylene biosynthesis, such as acetylcholine (PARUPS, 1976) cobaltous ion (KANG and RAY, 1969a; GROVER and PURVES, 1976; LAU and YANG, 1976), and some antibiotics (KANG and RAY, 1969c) cause hook opening. This anti-ethylene effect can also be produced by CO_2 (KANG and RAY, 1969b), hypobaric treatment (KANG and BURG, 1972), benzothiadiazole (PARUPS, 1973), and silver ion (BEYER, 1976b). Some inorganic ions at high concentrations change the effectiveness of hormones (probably through their interactions with proteins), and strongly influence the hook curvature response of lettuce seedlings (POOVAIAH and LEOPOLD, 1974).

4.2.3 Interorgan Correlation

In the bean hypocotyl hook, the presence of the cotyledons, regarded as a source of endogenous auxin, significantly retards light-induced opening (KLEIN et al., 1956), and moreover, is actually required for hook formation (RUBINSTEIN, 1972).

This is consistent with the finding that high auxin concentrations induce hook closure of isolated (decapitated) hypocotyls.

It has been well known since the work of KLEIN (1965) that the presence of the tissue basal to the hook elbow, the so-called shank, accelerates the opening process. The presence of the root is also known to influence the opening behavior of cotton (POWELL and MORGAN, 1970) and mung-bean (GEE and VINCE-PRUE, 1976) hypocotyl hooks. It appears that the control of the hook configuration represents an interaction among organs, with the apical parts supplying some regulatory substance(s) (e.g., auxin), and the basal ones providing others (e.g., gibberellins and/or cytokinins).

In cotton seedlings, where hook opening can be promoted slightly by auxin, the presence of the cotyledons indeed stimulates the opening process (POWELL and MORGAN, 1970). However, the auxin promotion of opening is relatively small compared to the cotyledonary effect; thus, the simple explanation of the cotyledon serving solely as an auxin source is inadequate. One must realize that the system controlling the hook configuration involves interactions of more than one hormonal factor and that more than one part of the seedling may participate in controlling the levels of these endogenous substances.

Most interesting, however, are some recent findings implying interorgan correlations which cannot be explained by the action of phytohormones. According to CAUBERGS and DE GREEF (1975), when only the primary leaves of a bean seedling are irradiated, the hook subsequently appears to undergo opening. Moreover, removal of the irradiated leaves immediately after illumination does not seem to prevent or otherwise change the opening effect exerted on the hook elbow. The investigators postulate the existence of a system of rapid stimulus transmission from the leaves to the hook, presumably nonhormonal in nature. Similar physiological processes may be involved in the rapid transmission of phytochrome action from cotyledons to hypocotyls (BLACK and SHUTTLEWORTH, 1974) and from hooks to cotyledons (OELZE-KAROW and MOHR, 1973, 1974). Changes in the photoequilibria of phytochrome are thought to be associated with interorgan correlation (MARCHAL-GABRIAC et al., 1974). The nature of these communication systems is not clearly defined at present.

5. Concluding Remarks

Epinasty is a general term for various kinds of dorsiventrality curvature of plant organs, and may occur in almost all parts of the plant. Various kinds of epinastic (and hyponastic) movement have different underlying mechanisms, but all have a strong autonomous component. Most epinastic responses are also affected or even controlled by external physical stimuli (e.g., light or gravity), which interact with the internal chemical (i.e., hormonal) control system. Petiole and branch epinasty, as well as formation and opening of seedling hooks, clearly require ethylene and auxin action as major components of the hormonal mechanism, although other hormones may also participate as controlling agents. Of the three types of epinasty (and hyponasty) discussed in this chapter, leaf unrolling of grass seedlings, a response of monocotyledons, has a unique hormonal feature in that GA and possibly ABA play a major role in its regulation.

There are different ideas as to how asymmetric growth is brought about by auxin-ethylene interaction, especially in petioles. One view is that ethylene, which is produced either by endogenous or by added auxin, directly stimulates the growth of certain cells selectively, e.g., those on the upper side of the lateral organ. The hypothesis of selective cell stimulation by ethylene assumes a preexisting asymmetry in the sensitivity to ethylene of cells on opposite sides of the organ. Another idea is essentially an extension of the Cholodny-Went theory: the auxin induces ethylene production, but the ethylene, in turn, modifies the auxin transport with regard to gravitational stimulation. The result is asymmetric auxin distribution toward the upper side, and thus asymmetric growth. This relation represents a feedback control. The action of auxin is different in these two hypotheses. Either it is by itself the factor responsible for the differences in growth, or it is the basal requirement for growing upon which another factor acts differentially. This latter case would reflect WENT's (1928) "ohne Wuchsstoff kein Wachstum"; i.e., whatever the final response, it cannot occur unless growth is enabled by auxin.

Taking into account the complexity of these possible interactions, epinasty also seems to be an excellent example of what THIMANN (1974) has called "hormonal multiplicity".

References

Abeles, F.B.: Ethylene in Plant Biology. New York: Academic Press, 1973

Baur, J.R., Morgan, P.W.: Effects of picloram and ethylene on leaf movement in huisache and mesquite seedlings. Plant Physiol. *44*, 831–838 (1969)

Beevers, L., Loveys, B., Pearson, J.A., Wareing, P.F.: Phytochrome and hormonal control of expansion and greening of etiolated wheat leaves. Planta *90*, 284–294 (1970)

Bennet-Clark, T.A., Younis, A.F., Esnault, R.: Geotropic behaviour of roots. J. Exp. Bot. *10*, 69–86 (1959)

Bergman, K., Burke, P.V., Cerda-Olmedo, E., David, C.N., Delbrück, M., Foster, K.W., Goodell, E.W., Heisenberg, M., Meissner, G., Zalokar, M., Dennison, D.S., Shropshire, Jr, W.: *Phycomyces*. Bacteriol. Rev. *33*, 99–157 (1969)

Beyer, Jr. E.: Silver ion: A potent antiethylene agent in cucumber and tomato. Hortic. Sci. *11*, 195–196 (1976a)

Beyer, Jr. E.: A potent inhibitor of ethylene action in plants. Plant Physiol. *58*, 269–271 (1976b)

Blaauw, O.H., Blaauw-Jansen, G., Van Leeuwen, W.J.: An irreversible red-light-induced growth response in *Avena*. Planta *82*, 87–104 (1968)

Black, M., Shuttleworth, J.E.: The role of cotyledones in the photocontrol of hypocotyl extension in *Cucumis sativus* L. Planta *117*, 57–66 (1974)

Brown, A.H., Chapman, D.K., Liu, S.W.W.: A comparison of leaf epinasty induced by weightlessness or by clinostat rotation. Bioscience *24*, 518–520 (1974)

Burg, S.P., Burg, E.A.: Auxin stimulated ethylene formation: Its relationship to auxin inhibited growth, root geotropism and other plant processes. In: Biochemistry and Physiology of Plant Growth Substances. Wightman, F., Setterfield, G. (eds.). Ottawa: Runge Press, 1968, *pp*. 1275–1294

Burström, H.: Über Entfaltung und Einrollen eines mesophilen Grasblattes. Bot. Notis. 1942, 351–361 (1942)

Burström, H.G.: Orientation and geotropic reaction of seminal roots of wheat. Physiol. Plant. *25*, 283–293 (1971)

Caubergs, R., De Greef, J.A.: Studies on hook-opening in *Phaseolus vulgalis* L. by selective R/FR pretreatments of embryonic axis and primary leaves. Photochem. Photobiol. *22*, 139–144 (1975)
Champagnat, P.: Dominance apicale, tropismes, epinastie. In: Handbuch der Pflanzenphysiol. Ruhland, W. (ed.). Berlin-Heidelberg-New York: Springer 1961, Vol. XIV, pp. 872–908
Cooke, R.J., Saunders, P.F.: Phytochrome mediated changes in extractable gibberellin activity in a cell-free system from etiolated wheat leaves. Planta *123*, 299–302 (1975a)
Cooke, R.J., Saunders, P.F.: Photocontrol of gibberellin levels as related to the unrolling of etiolated wheat leaves. Planta *126*, 151–160 (1975b)
Cooke, R.J., Saunders, P.F., Kendrick, R.E.: Red light induced production of gibberellin-like substances in homogenates of etiolated wheat leaves and in suspensions of intact etioplasts. Planta *124*, 319–328 (1975)
Crocker, W., Hitchcock, A.E., Zimmerman, P.W.: Similarities in the effects of ethylene and the plant auxins. Contrib. Boyce Thompson Inst. *7*, 231–248 (1935)
Crocker, W., Zimmerman, P.W., Hitchcock, A.E.: Ethylene-induced epinasty of leaves and the relation of gravity to it. Contrib. Boyce Thompson Inst. *4*, 177–218 (1932)
Curtis, R.W.: Stimulation of ethylene or ethane production by malformin. Plant Cell Physiol. (Tokyo) *10*, 909–916 (1969)
Curtis, R.W., John, W.W.: Effect of malformin on phytochrome- and ethrel-mediated responses. Plant Cell Physiol. (Tokyo) *16*, 719–728 (1975)
Dedolph, R.R., Dipert, M.H.: The physical basis of gravity stimulus nullification by clinostat rotation. Plant Physiol. *47*, 756–764 (1971)
Delbrück, M., Reichardt, W.: System analysis for the light growth reaction in *Phycomyces*. In: Cellular Mechanisms in Differentiation and Growth. Rudnick, D. (ed.). Princeton: Univ. Press 1956, pp. 3–44
Deutch, B.: Barley leaf unfolding under mixtures of red and blue light. Physiol. Plant *38*, 57–60 (1976)
Deutch, B., Deutch, B.I.: A kinetic theory of first order cyclical processes-Phytochrome controlled red light induced cereal leaf unfolding compared with theory. Physiol. Plant *32*, 273–281 (1974)
Deutch, B., Deutch, B.I.: Blue light induction of barley leaf unfolding. A Phytochrome reaction? Physiol. Plant *35*, 322–327 (1975)
Doubt, S.L.: The response of plants to illuminating gas. Bot. Gaz. *63*, 209–224 (1917)
Fischnich, O.: Über den Einfluß von β-Indolylessigsäure auf die Blattbewegungen und die Adventivwurzelbildung von *Coleus*. Planta *24*, 552–583 (1935)
Freytag, H.: Untersuchungen über die Plagioytropie der Blätter von *Tropaeolum maius*. Planta *12*, 267–292 (1930)
Galston, A.W.: Phototropism of stems, roots and coleoptiles. In: Handbuch der Pflanzenphysiol. Ruhland, W. (ed.). Berlin-Heidelberg-New York: Springer 1959, Vol. XVII, pp. 492–529
Galston, A.W., Tuttle, A.A., Penny, P.J.: A kinetic study of growth movements and photomorphogenesis in etiolated pea seedling. Am. J. Bot. *51*, 853–858 (1964)
Garcia-Torres, I., Gomez-Campo, C.: Increased tuberization in potatoes by ethrel (2-chloroethyl-phosphonic acid). Potato Res. *15*, 76–80 (1972)
Gee, H.M.: The control of hypocotyl hook angle in *Phaseolus mungo* L. The role of growth substances. J. Exp. Bot. *28*, 367–376 (1977)
Gee, H.M., Vince-Prue, D.: Control of the hypocotyl hook angle in *Phaseolus mungo* L.: The role of parts of the seedling. J. Exp. Bot. *27*, 314–323 (1976)
Goldsmith, M.H., Cataldo, D.A., Karn, J.: The rapid nonpolar transport of auxin in the phloem of intact *Coleus* plants. Planta *116*, 301–307 (1974)
Goto, N., Esashi, Y.: Stimulation by ethylene of axis and hypocotyl growth in bean and cocklebur seedlings. Physiol. Plant. *31*, 204–210 (1974a)
Goto, N., Esashi, Y.: Regulation of hypocotyl growth by ethylene. In: Plant Growth Substances *1973*. Tokyo: Hirokawa, 1974b, pp. 853–863
Grover, S., Purves, W.K.: Cobalt and plant development. Interactions with ethylene in hypocotyl growth. Plant Physiol. *57*, 886–889 (1976)
Hale, C.R., Coombe, B.G., Hawker, J.S.: Effects of ethylene and 2-chloroethane phosphonic acid on the ripening of grapes. Plant Physiol. *45*, 620–623 (1970)

Harper, D.B., Wain, R.L.: Studies on plant growth-regulating substances. XXXI. The metabolism of certain 3,6-disubstituted phenols within plant tissues. Ann. Appl. Biol. *67*, 395–408 (1971)
Harvey, E.M.: Some effects of ethylene on the metabolism of plants. Bot. Gaz. *60*, 193–214 (1915)
Hayes, A.B., Lippincott, J.A.: Growth and gravitational response in the development of leaf blade hyponasty. Am. J. Bot. *63*, 383–387 (1976)
Hild, V., Hertel, R.: Initial phases of gravity-induced lateral auxin transport and geotropic curvature in corn coleoptiles. Planta *108*, 245–258 (1972)
Imaseki, H., Pjon, C.-J.: The effect of ethylene on auxin-induced growth of excised rice coleoptile segments. Plant Cell Physiol. (Tokyo) *11*, 827–829 (1970)
Jackson, M.B., Campbell, D.J.: Waterlogging and petiole epinasty in tomato: The role of ethylene and low oxygen. New Phytol. *76*, 21–29 (1976)
James, C.S., Wain, R.L.: Studies on plant growth-regulating substances. XXVIII. Halogen-substituted benzoic acids. Ann. Appl. Biol. *63*, 205–210 (1969)
Janes, H.W., Loercher, L., Frenkel, C.: Effects of red light and ethylene on growth of etiolated lettuce seedlings. Plant Physiol. *57*, 420–423 (1976)
Johnson, S.P., Tibbitts, T.W.: The liminal angle of a plageotropic organ under weightlessness. Bio. Sci. *18*, 655–661 (1968)
Kaldeway, H.: Plagio- und Diageotropismus der Sproses und Blätter, einschließlich Epinastie, Hyponastie, Entfaltungsbewegungen. In: Handbuch der Pflanzenphysiol. Berlin-Heidelberg-New York: Springer, 1962, Vol. XVII, pt. 2, pp. 200–321
Kang, B.G.: Phytochrome-controlled leaf unrolling and protein synthesis, Plant Physiol. *47*, 352–356 (1971)
Kang, B.G., Burg, S.P.: Ethylene as a natural agent inducing plumular hook formation in pea seedlings. Planta *104*, 275–281 (1972)
Kang, B.G., Burg, S.P.: Ethylene action on lateral auxin transport in tropic responses, leaf epinasty, and horizontal nutation. In: Plant Growth Substances 1973, Tokyo: Hirokawa 1974, pp. 1090–1094
Kang, B.G., Ray, P.M.: Role of growth regulators in the bean hypotyl hook opening response. Planta *87*, 193–205 (1969a)
Kang, B.G., Ray, P.M.: Ethylene and carbon dioxide as mediators in the response of the bean hypocotyl hook to light and auxins. Planta *87*, 206–216 (1969b)
Kang, B.G., Ray, P.M.: Effects of inhibitors of RNA and protein synthesis on bean hypocotyl hook opening and their implications regarding phytochrome action. Planta *87*, 217–226 (1969c)
Kang, B.G., Yocum, C.S., Burg, S.P., Ray, P.M.: Ethylene and carbon dioxide; mediation of hypocotyl hook-opening response. Science *156*, 958–959 (1967)
Karvé, A.D.: Upon the formation and opening of plumular hooks in seedlings of *Carthamus tinctorius* L. Naturwissenschaften *51*, 441–442 (1964)
Klein, R.M.: Photomorphogenesis of the bean plumular hook. Physiol. Plant. *18*, 1026–1033 (1965)
Klein, W.H.: Interaction of growth factors with photoprocesses in seedling growth. In: Photoperiodism and Related Phenomena in Plants and Animals. Withrow, R.B. (ed.). Washington: AAAS 1959, *55*, pp. 207–215
Klein, W.H., Edwards, J.L., Shropshire, W., Jr.: Spectrophotometric measurements of phytochrome *in vivo* and their correlation with photomorphogenic responses of *Phaseolus*. Plant Physiol. *42*, 264–270 (1967)
Klein, W.H., Price, L., Milrakos, K.: Light stimulated starch degradation in plants and leaf morphogenesis. Photochem. Photobiol. *2*, 233–240 (1963)
Klein, W.H., Withrow, R.B., Elstad, V.B.: Response of the hypocotyl hook of bean seedlings to radiant energy and other factors. Plant Physiol. *31*, 289–294 (1956)
Klein, W.H., Withrow, R.B., Elstad, V.B., Price, L.: Photocontrol of growth and pigment synthesis in the bean seedling as related to irradiance and wavelength. Am. J. Bot. *44*, 15–19 (1957)
Ku, H.S., Suge, H., Rappaport, L., Pratt, H.K.: Stimulation of rice coleoptile growth by ethylene. Planta *90*, 333–339 (1970)
Kujawski, R.F., Truscott, F.H.: Photocontrol of hook opening in *Cuscuta gronovii* Wild. Plant Physiol. *53*, 610–614 (1974)

Lane, H.C., Kasperbauer, M.J.: Photomorphogenic responses of dodder seedlings. Plant Physiol. *40*, 109–116 (1965)
Lau, O-L., Yang, S.F.: Inhibition of ethylene production by cobaltous ion. Plant Physiol. *58*, 114–117 (1976)
Leather, G.R., Forrence, L.E., Abeles, F.B.: Increased ethylene production during clinostat experiments may cause leaf epinasty. Plant Physiol. *49*, 183–186 (1972)
Leike, H., von Guttenberg, H.: Über die Bedeutung des Auxins für die epinastische Bewegung plagiotroper Seitensprosse. Naturwissenschaften *48*, 604–605 (1961)
Leike, H., von Guttenberg, H.: Die Rolle des Auxins bei den epinastischen Krümmung plagiotroper, Seitensprosse von *Coleus blumei* Benth. Planta *58*, 453–470 (1962)
Lippincott, B.B., Lippincott, J.A.: Auxin-induced hyponasty of the leaf blade of *Phaseolus vulgaris*. Am. J. Bot. *58*, 817–826 (1971)
Lorenzen, H.: Öffnung der Hypocotylhaken von *Cardamine chenopodifolia* Pers. durch Morphactin im Dunkeln. Naturwissenschaften *9*, 451–452 (1968)
Loveys, B.R., Wareing, P.F.: The hormonal control of wheat leaf unrolling. Planta *98*, 117–127 (1971)
Lyon, C.J.: Auxin factor in branch epinasty. Plant Physiol. *38*, 145–152 (1963a)
Lyon, C.J.: Auxin transport in leaf epinasty. Plant Physiol. *38*, 567–574 (1963b)
Lyon, C.J.: Action of gravity on basipetal transport of auxin. Plant Physiol. *40*, 953–961 (1965)
Lyon, C.J.: Ethylene inhibition of auxin transport by gravity in leaves. Plant Physiol. *45*, 644–646 (1970)
Marchal-Gabriac, P., Veyron, P., Rollin, P., Boisard, J.: Mesure des Photoequilibres du Phytochrome dans les platules de courge. Plant Sci. Lett. *3*, 191–198 (1974)
Menhenett, R.: Hormonal and phytochrome-induced unrolling of barley leaves. Ph.D. Thesis, Australian National University, 1972
Menhenett, R., Carr, D.J.: Growth inhibitors from etiolated leaves of barley (*Hordeum vulgare* L.). Aust. J. Biol. Sci. *26*, 527–537 (1973)
Mitchell, C.A., Severson, C.J., Wott, J.A., Hammer, P.A.: Seismomorphogenic regulation of plant growth. J. Am. Soc. Hortic. Sci. *100*, 161–165 (1975)
Mitchener, H.D.: The action of ethylene on plant growth. Am. J. Bot. *25*, 711–720 (1938)
Mitrakos, K., Klein, W.H., Price, L.: Soluble sugar changes in etiolated corn leaf tissue as influenced by red light treatment. Planta *66*, 207–215 (1965)
Mitrakos, K., Price, L., Klein, W.H., Steiner, A.: Red-light effect on tracer distribution in etiolated leaf tissue. Planta *76*, 190–196 (1967)
Mohr, H., Haug, A.: Die histologischen Vorgänge während der lichtabhängigen Schließung und Öffnung des Plumulahakens bei den Keimlingen von *Lactuca sativa* L. Planta *59*, 151–164 (1962)
Mohr, H., Noble, A.: Die Steuerung der Schließung und Öffnung des Plumulahakens bei Keimlingen von *Lactuca sativa* durch sichtbare Strahlung. Planta *55*, 327–342 (1960)
Morgan, P.W., Powell, R.D.: Involvement of ethylene in responses of etiolated bean hypocotyl to coumarin. Plant Physiol. *45*, 553–557 (1970)
Musgrave, A., Jackson, M.B., Ling, E.: *Callitriche* stem elongation is controlled by ethylene and gibberellin. Nature (New Biol.) *238*, 93–96 (1972)
Musgrave, A., Walters, J.: Ethylene-stimulated growth and auxin transport in *Ranunculus sceleratus* petioles. New Phytol. *72*, 783–789 (1973)
Nakamura, T., Yamada, T., Takahashi, N.: Effect of gibberellic acid on the growth of the plumular hook section of etiolated pea seedling. Bot. Mag. (Tokyo) *70*, 404–413 (1966)
Neljubow, R.: Über die horizontale Nutation der Stengel von *Pisum sativum* und einiger anderer Pflanzen. Bot. Centr. *10*, 128–139 (1901)
Oelze-Karow, H., Mohr, H.: Quantitative correlation between spectrophotometric phytochrome assay and physiological response. Photochem. Photobiol. *18*, 319–330 (1973)
Oelze-Karow, H., Mohr, H.: Interorgan correlation in a phytochrome-mediated response in the mustard seedling. Photochem. Photobiol. *20*, 127–141 (1974)
Osborne, D.J.: Auxin, ethylene and the growth of cells. In: Mechanisms of Regulation of Plant Growth. (eds. R.L. Bieleski et al.). Bulletin *12*, pp. 645–654. Wellington: The Royal Society of New Zealand 1974

Osborne, D.J.: Control of cell shape and cell size by the dual regulation of auxin and ethylene. In: Perspectives in Experimental Biology. Sunderland, N. (ed.). Oxford-New York: Pergamon Press, 1976, pp. 89–102

Palmer, J.H.: Comparative study of the effects of applied indoleacetic acid and horizontal orientation of the primary shoots, upon internode extension and petiole orientation in *Helianthus annuus* and the modifying influence of gibberellic acid. Planta *61*, 283–297 (1964)

Palmer, J.H.: Roles of ethylene and indol-3yl-acetic acid in petiole epinasty in *Helianthus annuus* and the modifying influence of gibberellic acid. J. Exp. Bot. *23*, 733–743 (1972)

Palmer, J.H.: Ethylene as a cause of transient petiole epinasty in *Helianthus annuus* during clinostat experiments. Physiol. Plant. *28*, 188–193 (1973)

Palmer, J.H.: Failure of ethylene to change the distribution of indoleacetic acid in the petiole of *Coleus blumei* × *frederici* during epinasty. Plant Physiol. *58*, 513–515 (1976)

Palmer, J.H., Halsall, D.M.: Effect of transverse gravity stimulation, gibberellin and indoleacetic acid upon polar transport of IAA-C^{14} in the stem of *Helianthus annuus*. Physiol. Plant. *22*, 59–67 (1969)

Palmer, J.H., Phillips, I.D.J.: The effect of the terminal bud, indoleacetic acid, and nitrogen supply on the growth and orientation of the petiole of *Helianthus annuus*. Physiol. Plant. *16*, 575–584 (1963)

Parups, E.V.: Control of ethylene-induced responses in plants by a substituted benzothiadiazole. Physiol. Plant. *29*, 365–370 (1973)

Parups, E.V.: Acetylcholine and synthesis of ethylene in etiolate bean tissues. Physiol. Plant. *36*, 154–156 (1976)

Pearson, J.A., Wareing, P.F.: Polysomal changes in developing wheat leaves. Planta *93*, 309–313 (1970)

Phillips, I.D.J.: Diffusible gibberellins and phototropism in *Helianthus annuus*. Planta *106*, 363–367 (1972)

Poovaiah, B.W., Leopold, A.C.: Hormone-solute interactions in the lettuce hypocotyl hook. Plant Physiol. *54*, 289–293 (1974)

Poulson, R., Beevers, L.: Nucleic acid metabolism during greening and unrolling of barley leaf segments. Plant Physiol. *46*, 315–319 (1970a)

Poulson, R., Beevers, L.: Effects of light and growth regulators on leaf unrolling in barley. Plant Physiol. *46*, 509–514 (1970b)

Poulson, R.L., Beevers, L.: Effects of growth regulators on ribonucleic acid metabolism of barley leaf segments. Plant Physiol. *46*, 782–785 (1970c)

Powell, R.D., Morgan, P.W.: Factors involved in the opening of the hypocotyl hook of cotton and beans. Plant Physiol. *45*, 548–552 (1970)

Price, L., Mitrakos, K., Klein, W.H.: Photomorphogenesis and carbohydrate changes in etiolated leaf tissue. Q. Rev. Biol. *39*, 11–18 (1964)

Price, L., Mitrakos, K., Klein, W.H.: Some kinetical aspects of light-induced carbohydrate utilization in etiolated leaf tissue. Physiol. Plant. *18*, 540–549 (1965)

Railton, I.D., Phillips, I.D.J.: Gibberellins and geotropism in *Zea mays* coleoptiles. Planta *109*, 121–126 (1973)

Railton, I.D., Reid, D.M.: Effects of benzyladenine on the growth of waterlogged tomato plants. Planta *111*, 261–266 (1973)

Raven, C.W., Sprutt, C.J.P.: Induction of rapid chlorophyll accumulation in dark grown seedling. I. Action spectrum for pea. Acta Bot. Neerl. *21*, 219–230 (1972)

Rawitscher, F.: Der Geotropismus der Pflanzen. Jena: Fischer, 1932

Ray, P.M., Baker, D.B.: The effect of auxin on synthesis of oat coleoptile cell wall constituents. Plant Physiol. *40*, 353–360 (1965)

Reid, D.M., Clements, J.B., Carr, D.J.: Red light induction of gibberellin synthesis in leaves. Nature (London) *217*, 580–582 (1968)

Reinert, J.: Phototropism and phototaxis. Annu. Rev. Plant Physiol. *10*, 441–458 (1959)

Rubinstein, B.: The role of various regions of the bean hypocotyl on red light-induced hook opening. Plant Physiol. *48*, 183–186 (1971a)

Rubinstein, B.: Auxin and red light in the control of hypocotyl hook opening in beans. Plant Physiol. *48*, 187–192 (1971b)

Rubinstein, B.: Characteristics of hook formation by bean seedlings. Plant Physiol. *49*, 640–643 (1972)

Saunders, J.F. (ed.): The experiments of Biosatellite II. Doc. No. NASA SP-204. Washington, D.C.: Scientific and Technical Information Office. National Aeronautics and Space Administration 1971

Seitz, K.: Irradiance dependent far-red effects upon red induced germination of lettuce seeds. Z. Pflanzenphysiol. *71*, 49–56 (1974)

Shropshire, W. Jr., Klein, W.H., Edwards, J.L.: Photomorphogenesis induced by flavin-mononucleotide fluorescence. Physiol. Plant. *17*, 676–682 (1964)

Skene, K.G.M.: Stimulation of germination of immature bean embryos by gibberellic acid. Planta *87*, 188–192 (1969)

Soekarjo, R.: The effect of caffeic acid on the epinastic curvature of isolated petioles of *Coleus* sp. in comparison with the effect of gibberellic acid and indoleacetic acid. Acta Bot. Neerl. *14*, 266–277 (1965a)

Soekarjo, R.: On the formation of adventitious roots in cuttings of *Coleus* in relation to the effect of indoleacetic acid on the epinastic curvature of isolated petioles. Acta Bot. Neerl. *14*, 373–399 (1965b)

Stacewicz-Sapuncakis, M., Marsh, H.V. Jr., Vengris, J., Jennings, P.H., Robinson, T.: Participation of ethylene in common purslane responses to dicamba. Plant Physiol. *52*, 466–471 (1973)

Steiner, A., Price, L., Mitrakos, K., Klein, W.H.: Red light effects on uptake of ^{14}C and ^{32}P into etiolated corn leaf tissue during photomorphogenic leaf opening. Physiol. Plant. *21*, 895–901 (1968)

Stewart, E.R., Freebairn, H.T.: Ethylene, seed germination and epinasty. Plant Physiol. *44*, 955–958 (1969)

Suge, H.: Stimulation of oat and rice mesocotyl growth by ethylene. Plant Cell Physiol. (Tokyo) *12*, 831–837 (1971)

Thimann, K.V.: Biosatellite II experiments: Preliminary results. Proc. Natl. Acad. Sci. USA *60*, 347–361 (1968)

Thimann, K.V.: Opening lecture. In: Plant Growth Substances 1973. Tokyo: Hirokawa 1974, pp. 3–6

Truscott, F.H.: Some aspects of morphogenesis in *Cuscuta gronovii* Am. J. Bot. *53*, 739–750 (1966)

van der Lann, P.A.: Der Einfluß von Aethylen auf die Wuchsstoffbildung bei *Avena* und *Vicia*. Recl. Trav. Bot. Neer. *31*, 691–742 (1934)

Vardar, Y.: A study of the auxin factor in epinastic and hyponastic movements. Rev. Fac. Univ. Istanbul *18*, 317–352 (1953)

Virgin, H.I.: Light-induced unfolding of the grass leaf. Physiol. Plant. *15*, 380–389 (1962)

Wächter, W.: Chemonastische Bewegungen der Blätter von *Callisia repens*. Ber. Dtsch. Bot. Ges. *23*, 379–382 (1905)

Wagné, C.: The distribution of the light effect in partly irradiated grass leaves. Physiol. Plant. *17*, 751–756 (1964)

Wagné, C.: The distribution of the light effect from irradiated to non-irradiated parts of grass leaves. Physiol. Plant. *18*, 1001–1006 (1965)

Went, F.W.: Wuchsstoff und Wachstum. Recl. Trav. Bot. Neerl. *25*, 1–114 (1928)

Withrow, R.B., Klein, W.H., Elstad, V.B.: Action spectra of photomorphogenic induction and its photoinactivation. Plant Physiol. *32*, 453–462 (1957)

Wright, M., Osborne, D.J.: Abscission in *Phaseolus vulgaris*. The positional differentiation and ethylene-induced expansion growth of specialized cells. Planta *120*, 163–170 (1974)

Zimmerman, P.W., Wilcoxon, F.: Several chemical growth substances which cause initiation of roots and other responses in plants. Contrib. Boyce Thompson Inst. *7*, 209–229 (1935)

Author Index

Page numbers in *italics* refer to the references

Subject Index

Encyclopedia of Plant Physiology

New Series

Editors:
A. Pirson, M.H. Zimmermann

Volume 1
Transport in Plants I
Phloem Transport
Editors: M.H. Zimmermann, J.A. Milburn
With contributions by numerous experts.
1975. 93 figures. XIX, 535 pages
ISBN 3-540-07314-0
Contents: Structural Considerations in Phloem Transport. – Nature of Substances in Phloem. – Phloem Transport: Assessment of Evidence. – Possible Mechanisms of Phloem Transport. – Phloem Loading: Storage and Circulation. – Author Index. Subject Index.

Volume 2 (in 2 Parts)
Transport in Plants II
Part A: Cells
Editors: U. Lüttge, M.G. Pitman
With a foreword by R.N. Robertson
With contributions by numerous experts.
1976. 97 figures, 64 tables. XVI, 419 pages
ISBN 3-540-07452-X
Contents: Theoretical and Biophysical Approaches. – Particular Cell Systems. – Regulation, Metabolism and Transport.

Part B: Tissues and Organs
Editors: U. Lüttge, M.G. Pitman
With contributions by numerous experts.
1976. 129 figures, 45 tables. XII, 475 pages
ISBN 3-540-07453-8
Contents: Pathways of Transport in Tissues. – Particular Tissue Systems. – Control and Regulation of Transport in Tissues and Integration in Whole Plants.

Volume 3
Transport in Plants III
Intracellular Interactions and Transport Processes
Editors: C.R. Stocking, U. Heber
With contributions by numerous experts.
1976. 123 figures. XXII, 517 pages
ISBN 3 540 07818 5
Contents: Membrane Structure. – Intracellular Interactions. – Intracellular Transport in Relation to Energy Conservation. – Theory of Membrane Transport.

Volume 4
Physiological Plant Pathology
Editors: R. Heitefuss, P.H. Williams
With contributions by numerous experts.
1976. 92 figures. XX, 890 pages
ISBN 3-540-07557-7
Contents: General. – Spore Germination and Its Regulation. – Cytology and Physiology of Penetration and Establishment. – Forces by Which the Pathogen Attacks the Host Plant. – Physiology of Host Response to Infection. – Modification of the Host Response. – Predisposition. – Biothropic Parasites in Culture. – Genetics of Host-Parasite Interactions. – Author Index. – Index of Microorganisms and Nematodes. – Subject Index. – List of Symbols and Abbreviations.

Volume 5
Photosynthesis I
Photosynthetic Electron Transport and Photophosphorylation
Editors: A. Trebst, M. Avron
With contributions by numerous experts.
1977. 128 figures. XXIV, 730 pages
ISBN 3-540-07962-9
Contents: Introduction. – History. – Electron Transport. – Energy Conservation. – Structure and Function. – Algal and Bacterial Photosynthesis.

Volume 6
Photosynthesis II
Regulation of Photosynthetic Carbon Metabolism and Related Processes
Editors: M. Gibbs, E. Latzko
With contributions by numerous experts
Scheduled to come out in December 1978

Springer-Verlag
Berlin Heidelberg New York

Greenhouse Management

J.J. Hanan, W.D. Holley, K.L. Goldsberry

1978. 283 figures, 150 tables. XIV, 530 pages
(Advanced Series in Agricultural Sciences, Volume 5)
ISBN 3-540-08478-9
Distribution rights for India: Allied Publishers Private Ltd., New Delhi

This is the first comprehensive American treatment of greenhouse production since Post's book on Florist Crop Production and Marketing, published in 1950. The authors develop principles of crop behaviour in greenhouses on basic fundamentals of plant physiology, illustrating the practical application of these principles with numerous examples. For the first time in a greenhouse management text, attention is given in detail to the effects of climate, disease and insect control, pollution and the use of herbicides, personnel management, growth regulation, greenhouse engineering and all phases of marketing. The treatment of each major phase provides a "bench-mark", or reference point, for the state-of-the-art. It is intended as a text upon which serious practitioners of greenhouse production may build for the future.

Contents:
Introduction. – Light. – Greenhouse Construction. – Temperature. – Water. – Soils and Soil Mixtures. – Nutrition. – Carbon Dioxide and Pollution. – Insect and Disease Control. – Chemical Growth Regulation. – Business Management. – Marketing. – Appendices: Conversion Tables. Symbolism. Definitions.

Springer-Verlag
Berlin
Heidelberg
New York